CHESLEY'S HYPERTENSIVE DISORDERS IN PREGNANCY

FOURTH EDITION

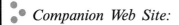

CHESLEY'S HYPERTENSIVE DISORDERS IN PREGNANCY

FOURTH EDITION

Edited by

ROBERT N. TAYLOR, MD, PHD
Professor and Vice Chair for Research,
Department of Obstetrics and Gynecology,
Wake Forest School of Medicine, Winston-Salem, North Carolina, USA

JAMES M. ROBERTS, MD
Senior Scientist, Magee Women's Research Institute,
Professor, Obstetrics, Gynecology,
Reproductive Sciences and Epidemiology,
University of Pittsburgh, Pittsburgh, Pennsylvania, USA

F. GARY CUNNINGHAM, MD
Beatrice and Miguel Elias Distinguished Chair in
Obstetrics and Gynecology, University of Texas,
Southwestern Medical Center, Dallas, Texas, USA

MARSHALL D. LINDHEIMER, MD
Professor Emeritus, Department of Obstetrics and
Gynecology and Medicine, and the Committe on
Clinical Pharmacology, The University of Chicago,
Chicago, Illinois, USA

ELSEVIER

AMSTERDAM • BOSTON • HEIDELBERG • LONDON • NEW YORK • OXFORD
PARIS • SAN DIEGO • SAN FRANCISCO • SINGAPORE • SYDNEY • TOKYO
Academic Press is an imprint of Elsevier

Academic Press is an imprint of Elsevier
32 Jamestown Road, London NW1 7BY, UK
525 B Street, Suite 1800, San Diego, CA 92101-4495, USA
225 Wyman Street, Waltham, MA 02451, USA
The Boulevard, Langford Lane, Kidlington, Oxford OX5 1GB, UK

Fourth edition

Cover Figure Legend: Left and middle panels show syncytial knots derived from a preeclamptic placenta, transduced with adenovirus carrying enhanced green fluorescent protein (left panel) and stained for PKH26, a yellow dye that stably integrates into plasma membranes (middle panel). The right panel shows a syncytial knot from a normal placenta stained with MitoTracker® Red, to identify mitochondria. All sections were stained with DAPI (4',6'-diamidino-2-phenylindole) a blue dye that binds strongly to A-T rich DNA, and identifies the syncytialized nuclei. Courtesy of Augustine Rajakumar, PhD and S. Ananth Karumanchi, MD, Harvard Medical School.

British Library Cataloguing-in-Publication Data
A catalogue record for this book is available from the British Library

Library of Congress Cataloging-in-Publication Data
A catalog record for this book is available from the Library of Congress

ISBN: 978-0-12-407866-6

For information on all Academic Press publications
visit our website at www.store.elsevier.com

Typeset by MPS Limited, Chennai, India
www.adi-mps.com

15 16 17 18 19 10 9 8 7 6 5 4 3 2 1

Contents

List of Contributors

EDGARDO J. ABALOS, MD, Vice Director, Centro Rosarino de Estudios Perinatales, Rosario, Santa Fe, Argentina

JAMES M. ALEXANDER, MD, Professor of Obstetrics & Gynecology, Southwestern Medical Center at Dallas, Texas, USA

PHYLLIS AUGUST, MD, MPH, Ralph A. Baer MD Professor of Research in Medicine, Professor of Obstetrics & Gynecology, Medicine, and Public Health, Weill Cornell Medical College, New York, New York, USA; Theresa Lang Director of Lang Center for Research and Education, New York Hospital Queens, New York, USA

MARILYN J. CIPOLLA, PhD, Professor, Department of Neurological Sciences, Obstetrics, Gynecology and Reproductive Sciences; Pharmacology, University of Vermont, Vermont, USA

AGUSTIN CONDE-AGUDELO, MD, MPH, Perinatology Research Branch, Intramural Division, NICHD/NIH/DHHS, Bethesda, Maryland and Detroit, Michigan, USA

KIRK P. CONRAD, MD, Professor, Departments of Physiology and Functional Genomics and Obstetrics and Gynecology, University of Florida, College of Medicine, Gainesville, Florida, USA

F. GARY CUNNINGHAM, MD, Beatrice and Miguel Elias Distinguished Chair in Obstetrics and Gynecology, University of Texas Southwestern Medical Center at Dallas, Texas, USA

SANDRA T. DAVIDGE, PhD, Professor, Departments of OB/GYN and Physiology, Canada Research Chair in Women's Cardiovascular Health, University of Alberta, Edmonton, Alberta, Canada

RALF DECHEND, MD, Department of Cardiology, Franz-Volhard-Clinic, Max Delbruck Center for Molecular Medicine, Berlin, Germany

CHRISTIANNE J.M. DE GROOT, MD, PhD, Professor, Department of Obstetrics & Gynecology, Vrije Universiteit (VU) Medical Center, Amsterdam, The Netherlands

SUSAN J. FISHER, PhD, The Ely and Edythe Broad Center of Regeneration Medicine and Stem Cell Research, Center for Reproductive Sciences, Department of Obstetrics, Gynecology and Reproductive Sciences, Division of Maternal Fetal Medicine, and Department of Anatomy, University of California at San Francisco, San Francisico, California, USA

ERIC M. GEORGE, PhD, Assistant Professor of Physiology and Biochemistry, University of Mississippi Medical Canter, Jackson, Mississippi, USA

JOEY P. GRANGER, PhD, Billy S. Guyton Distinguished Professor, Professor of Physiology and Medicine, Director, Center for Excellence in Cardiovascular-Renal Research, Dean, School of Graduate Studies in the Health Sciences, University of Mississippi Medical Canter, Jackson, Mississippi, USA

JUDITH U. HIBBARD, MD, Professor and Vice Chair of Obstetrics, Department of Obstetrics & Gynecology, Medical College of Wisconsin, Milwaukee, Wisconsin, USA

CARL A. HUBEL, PhD, Associate Professor, Magee-Womens Research Institute; Department of Obstetrics and Gynecology and Reproductive Sciences, University of Pittsburgh School of Medicine; Department of Environmental and Occupational Health, Pittsburgh, Pennsylvania, USA

ARUN JEYABALAN, MD, MS, Department of Obstetrics and Gynecology and Reproductive Sciences, University of Pittsburgh School of Medicine, Magee-Womens Research Institute, Clinical and Translational Research Institute, University of Pittsburgh, Pennsylvania, USA

S. ANANTH KARUMANCHI, MD, Associate Professor of Medicine, Obstetrics & Gynecology Beth Israel Deaconess Medical Center and Harvard Medical School, Boston, Massachusetts, USA

LOUISE C. KENNY, MBChB, PhD, Senior Lecturer & Consultant-Obstetrician & Gynaecologist, Cork University Maternity Hospital, Cork, Ireland

BABBETTE LAMARCA, PhD, Associate Professor, Director, Research Division, University of Mississippi Medical Canter, Jackson, Mississippi, USA

MARSHALL D. LINDHEIMER, MD, Professor Emeritus, Department of Obstetrics and Gynecology and Medicine, and the Committee on Clinical Pharmacology and Pharmacogenetics, The University of Chicago, Chicago, Illinois, USA

KEITH R. MCCRAE, MD, Hematologic Oncology and Blood Disorders, Taussig Cancer Center, Department of Cellular and Molecular Medicine, Cleveland Clinic Lerner Research Institute, Cleveland, Ohio, USA

MICHAEL T. MCMASTER, PhD, The Ely and Edythe Broad Center of Regeneration Medicine and Stem Cell Research and Department of Cell and Tissue Biology, University of California at San Francisco, San Francisco, California, USA

ROBERTA B. NESS, MD, MPH, Dean and M. David Low Chair in Public Health, The University of Texas School of Public Health, Houston, Texas, USA

SAROSH RANA, MD, Assistant Professor, Obstetrics, Gynecology and Reproductive Biology, Beth Israel Deaconess Medical Center/ Harvard Medical School, Boston, Massachusetts, USA

CHRISTOPHER W.G. REDMAN, MB, BChr, Professor of Obstetric Medicine, Nuffield Department of Obstetrics and Gynecology, John Radcliffe Hospital, Oxford, London, United Kingdom

JANET W. RICH-EDWARDS, ScD, MPH, Associate Professor, Department of Medicine, Connors Center for Women's Health and Gender Biology, Brigham and Women's Hospital; Associate Professor, Department of Epidemiology, Harvard School of Public Health, Boston, Massachusetts, USA

JAMES M. ROBERTS, MD, Senior Scientist, Magee-Women's Research Institute, Professor of Obstetrics, Gynecology, Reproductive Sciences and Epidemiology and Clinical and Traslational Research University of Pittsburgh, Pittsburgh, Pennsylvania, USA

ROBERTO ROMERO, MD, D. Hon. C, Chief, Perinatology Research Branch Program Director of Obstetrics and Perinatology, Intramural Division NICHD, NIH, DHHS, Professor, Department of Obstetrics and Gynecology, University of Michigan, Ann Arbor, Michigan, Professor, Department of Epidemiology and Biostatistics, Michigan State University, Michigan, USA

IAN L. SARGENT, BSc, PhD, Professor of Reproductive Science, Nuffield Department of Obstetrics and Gynecology, University of Oxford, John Radcliffe Hospital, Oxford, United Kingdom

SANJEEV G. SHROFF, PhD, Distinguished Professor of & Gerald E. McGinnis Chair in Bioengineering and Professor of Medicine, Department of Bioengineering, Swanson School of Engineering, University of Pittsburgh, Pittsburgh, Pennsylvania, USA

BAHA M. SIBAI, MD, Professor, Department of Obstetrics and Gynecology College of Medicine, University of Cincinnati, Cincinnati, Ohio, USA

ANNE CATHRINE STAFF, MD, PhD, Professor II, Head of the Research Center of Obstetrics and Gynecology, Oslo University Hospital, Oslo, Norway

ISAAC E. STILLMAN, MD, Associate Professor of Pathology, Beth Israel Deaconess Medical Center and Harvard Medical School, Boston, Massachusetts, USA

ROBERT N. TAYLOR, MD, PhD, Professor and Vice Chair for Research, Department of Obstetrics and Gynecology, Wake Forest School of Medicine, Winston-Salem, North Carolina, USA

JASON G. UMANS, MD, PhD, MedStar Health Research Institute, the Departments of Medicine and of Obstetrics and Gynecology, Georgetown University; Georgetown-Howard Universities Center for Clinical and Translational Science, Washington, DC, USA

KENNETH WARD, MD, President and CEO, Juneau Biosciences, LLC, Salt Lake City, Utah, USA

VIRGINIA D. WINN, MD, PhD, Department of Obstetrics and Gynecology, University of Colorado School of Medicine, Aurora, Colorado, USA

GERDA G. ZEEMAN, MD, PhD, Department of Obstetrics and Gynecology, Division of Obstetrics and Prenatal Medicine, Erasmus Medical Center, Rotterdam, The Netherlands

Preface

PREFACE TO THE SECOND EDITION

In 1996, Gary Cunningham, senior author of *Williams' Obstetrics*, approached me with the idea of organizing a second edition of *Chesley's Hypertensive Disorders in Pregnancy*. This first edition, though published in 1978, was still a major reference text for investigators in the field, often kept locked in drawers for safe-keeping, as this out-of-print classic was difficult to locate. Next we contacted Dr. James Roberts, Director of the Magee-Womens' Research Institute, whose investigative team is among the leading contributors to the literature on preeclampsia this decade. Of interest, their conference room, dedicated by Leon Chesley, and bearing his name, contains his framed photograph positioned to keep a watchful eye on the quality of their deliberations. With three enthusiastic editors on board the second edition was born.

The 1999 edition is a multi-authored text, reflecting the enormous progress in the field since 1978 (and the editors' recognition that there are few Leon Chesleys among us today as well!). To stay in the spirit of the initial edition, your editors developed the following strategy. Each chapter would be first-authored by an acknowledged expert in the topic assigned, and would also be co-authored by one of the editors. This was meant to insure the timeliness of the material, increase cohesiveness, minimize redundancy, and above all, to achieve the goals set for the text. We believe that the strategy worked, though it may have been taxing to the co-author editor, especially when the topic deviated several degrees from his own area of expertise.

This edition is aimed both at investigators and practicing physicians. Thus, one will find chapters designed primarily as scholarly assessments of a given research area, and others with didactic advice for the care of pregnant women with hyper tension. Of interest is the considerable variability in focus from chapter to chapter, some authors reviewing the literature, primarily as those cited have presented it, others combining this with commentary, at times almost philosophical (in Chapter 4, an editors' comment was inserted to tell the reader why!). Some authors tried to mimic the encyclopedic approach (with all encompassing tables) in Chesley's first edition, others preferred a more concise approach, (but with extensive bibliography). Your editors decided to leave these approaches intact, the co-editor comparing the chapter's goal and the initial Chesley text in the introductory comments. Finally, Chesley's Chapter 2 in the first edition, "History," is reprinted in its entirety as part of our own first chapter. This historical review is a must read, and we doubt if it will be equaled for quite some time.

We consider Leon Chesley the father of modern research in the hypertensive disorders of pregnancy (some might say of modern research in obstetrics as well). He is now over 90 years of age, having weathered two strokes with mind and sense of humor intact. We visited him several times during the genesis of this edition, receiving his blessing on each occasion. We are proud to honor him with this edition.

COMMENTS ADDED TO THE PREFACE BY EDITORS IN 2008

Leon Chesley died shortly after publication of the second edition, and your editors missed those periodic exhilarating visits to Leon that highlighted preparation of that publication. *Thus this edition is both a tribute and memorial to him* and we worked hard to attain the perfection he always insisted on, and hope that, wherever he is, he is proud of his disciples.

The third edition is patterned after the second but there are changes. In the previous edition normal and abnormal aspects of organ systems were usually separated into distinct chapters, while in this edition each of these topics, such as the cardiovascular system, or kidney, is combined into a single chapter. This led to another small change. In the second edition each chapter, with the exception of the introductory first, was first authored by but one expert coupled to only one of the three editors as co-author, a style designed to ensure continuity and decrease excessive repetition. In the current edition the complexity of several chapters required "two experts" (now able to bully their editor co-author more effectively!). Finally, there are three new chapters, each devoted to recent and exciting observations in areas unknown or given short shrift in the preceding edition. These topics are angiogenic factors, agonistic antibodies to the angtiotensin II type 1 receptor and immunology and inflammation. Finally, the lead editor (MDL), having celebrated his diamond birthday in 2007 and golden wedding anniversary in 2008, will retire from this chore following this third edition of *Chesley's Hypertensive Disorders in Pregnancy*. The field is progressing quickly and developing new giants and the editors assure you that the next edition will meet the standards that Leon set with his first, and we strove to repeat in these last two editions.

Marshall D. Lindheimer, MD

PREFACE TO THE FOURTH EDITION

The third edition of *Chesley's Hypertensive Disorders in Pregnancy*, as the two prior ones, was widely acclaimed as an essential scholastic resource and enthusiastically endorsed by clinicians and scientists, alike. In the modern digital era, "downloads" are the currency of readership. Since its publication in 2009, the third edition of Chesley's has been one of the most highly downloaded and viewed biomedical references on Science Direct. Hypertension during pregnancy remains amongst the three leading killers of pregnant women and their unborn children; yet surveys show that half of physicians fail to elicit a history of pregnancy hypertension during routine encounters, despite its association with remote cardiovascular disease, and research support remains dismal compared with equally mortal diseases in terms of DALYs (Disability Adjusted Lost Years).

We have optimistic expectations for this fourth edition of the textbook, which witnesses the passing of the torch from lead editor Marshall Lindheimer to me (RNT). With the prior edition, Prof. Lindheimer stated his intention to retire as senior editor, having achieved a number of personal and professional milestones. Fortunately for his colleagues and audience, Marshall generously supported the current effort in his critical role as "editor emeritus" and spiritual guru of the present edition. It is an honor for me to accept the responsibility to assure that future generations of physicians, physiologists, educators and scholars of preeclampsia will continue to be galvanized by the information and ideas contained in these pages and challenged by the knowledge gaps that remain.

The overall format of the fourth edition follows that of its immediate precursor. We have continued the tradition of presenting Leon Chesley's masterfully written "History" chapter in its entirety within Chapter 1. Internationally renowned experts in the field, including several new authors, have been invited to collaborate with the editors to present state-of-the-art chapters designed to summarize and illuminate the horizons of knowledge in each academic domain. Fundamental aspects of the principles that underpin preeclampsia pathogenesis are updated in the first half of the text, with an emphasis on new clinical and conceptual insights and their therapeutic applications in the later chapters.

Lest one forgets the remarkable contributions of our predecessors, we proffer the distinguished photograph below of Leon Chesley, whom all four editors had the inspiring opportunity to claim as a mentor. I was struck by his curious necktie choice, with its kangaroo motif. While making that specific sartorial selection for his pose on the frontispiece, Dr. Chesley might have been wondering: "Had humans evolved from marsupials, with their transient embryonic attachment to a simple yolk-sac placenta, would our species have ever developed preeclampsia?" You may have to wait for the fifth edition to find out.

Robert N. Taylor

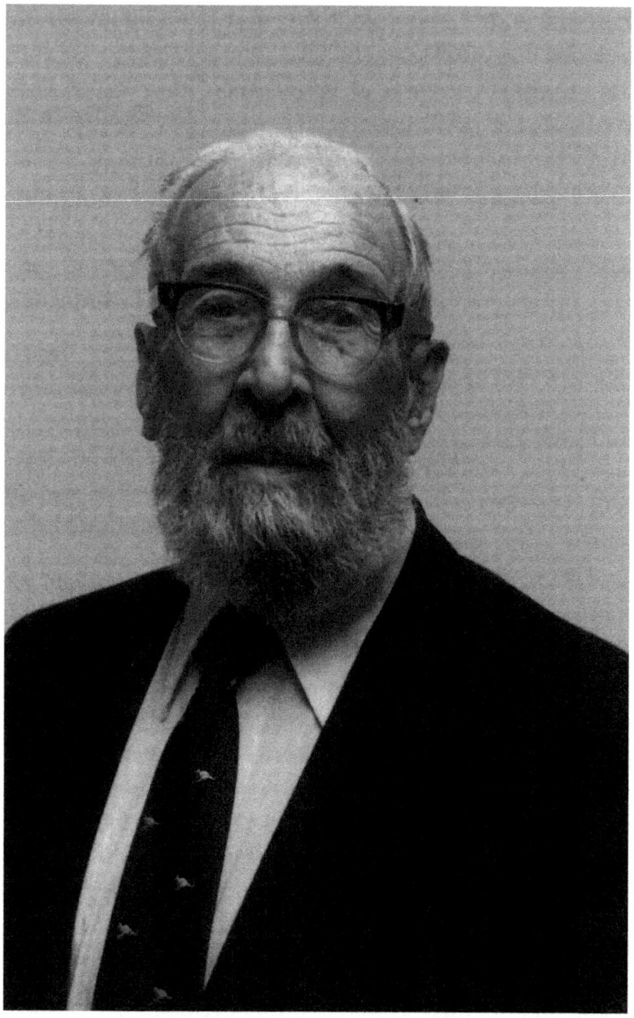

Leon C. Chesley, PhD (1908–2000)

CHAPTER 1

Introduction, History, Controversies, and Definitions

MARSHALL D. LINDHEIMER, ROBERT N. TAYLOR, JAMES M. ROBERTS, F. GARY CUNNINGHAM AND LEON CHESLEY†

Leon Chesley's *Hypertensive Disorders in Pregnancy* was initially published in 1978.[1] Then, as now, hypertension complicating pregnancy was a major cause of fetal and maternal morbidity and death, particularly in less developed nations. Most of this morbidity was and remains associated with preeclampsia, a disorder with devastating effects in many organ systems, high blood pressure being but one aspect of the disease. The first edition was single authored, written entirely by Dr. Chesley, a PhD in physiology, who originally found employment as a chemist at the Newark New Jersey's Margaret Hague Maternity Hospital, during the Great Depression of the 1930s. Curious why certain tests were being performed on convulsing pregnant women, he went to the wards, observed, and was stimulated to study that enigmatic disorder preeclampsia, the result being signal contributions published from the late 1930s through the early 1980s. His contributions included major observations in such diverse areas as epidemiology, remote prognosis, vascular and renal pathophysiology, and treatment, all focusing on hypertension in pregnancy. A compendium of his achievements is but one aspect of the initial edition of this text, which for the next two decades was a leading resource for clinicians and investigators who wished to learn more about high blood pressure in pregnant women.

In 1978 a text devoted to the hypertensive disorders in pregnancy could be single authored, due in part to the energy, intellect, and other attributes of Leon Chesley, but also because research in this important area of reproductive medicine was still sporadic and unfocused, and progress regrettably slow. Leon almost singly energized the field, and the editors of this text are among many of those for whom he served as a role model, nurturing three of us in early and mid-career. The initial edition, and other signal events during the 1970s (summarized further later in this chapter as "EDITORS' UPDATE"), spurred rapid progress in many areas including prevention trials, observations regarding pathogenesis, and management considerations. Thus, just like the second and third, the fourth edition again aims to be a leading reference text, multi-authored by leaders in the field. Again our stated goal of the previous editions, that this text will do scholarly justice to Dr. Chesley's 1978 *tour de force*, remains the major and obvious goal for the fourth edition.

The remainder of this chapter is as follows: We reproduce Dr. Chesley's original chapter entitled "History" in its entirety. Unable to improve on it, we add an EDITORS' UPDATE, and then conclude by republishing Dr. Chesley's 1975 workshop banquet address "False Steps in the Study of Preeclampsia."[2] That meeting led to the formation of the International Society for the Study of Hypertension in Pregnancy, and Dr. Chesley's message about how to study preeclampsia remains valid today.

HISTORY (FIG. 1.1)

Several German authors, such as von Siebold, Knapp, Kossmann, Fasbender, Fischer, and Bernhart, have written on the history of eclampsia, but all too often they did not document their sources and made errors that live on in second-, third-, and *n*th-hand reviews.[3–8]

Bernhart wrote that eclampsia was mentioned in the ancient Egyptian, Chinese, Indian, and Greek medical literature.[8] One of the oldest sources that he cited, without specific reference, was the Kahun (Petrie) papyrus dating from about 2200 BC. His source is likely to have been Menascha.[9] Griffith had translated Prescription No. 33, on the third page of the papyrus, as: "To prevent (the uterus) of a woman from itching (?) auit pound — upon her jaws the day of birth. It cures itching of the womb excellent truly millions of times."[10] Menascha cited Griffith's paper but rendered the translation (in German) as: "To prevent a woman from biting her tongue auit pound — upon her jaws the day of birth. It is a cure of biting excellent truly millions of times."[9] He suggested that the untranslated word "*auit*"

† Deceased.

Chesley's Hypertensive Disorders in Pregnancy.
ISBN: 978-0-12-407866-6

DOI: http://dx.doi.org/10.1016/B978-0-12-407866-6.00001-8

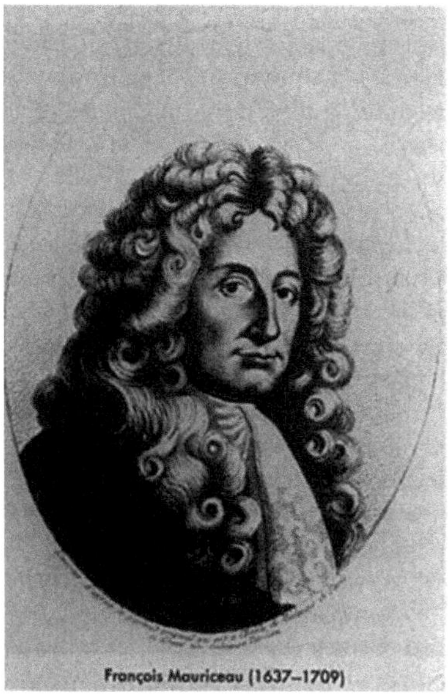

FIGURE 1.1 This portrait of Francois Mauriceau inaugurated the "History" chapter in the first edition of this text.

means "small wooden stick." In a later book on the Kahun papyrus, Griffith changed his translation to: "To prevent a woman from biting (her tongue?) beans, pound — upon her jaws the day of birth."[10,11] Curiously, Menascha did not cite Griffith's later translation and he included the word "*auit*" from the first version. Possibly the ancient scribe had eclampsia in mind, but that interpretation is tenuous at best.[12]

Bernhart also wrote, again without references, that both the Indian Atharva-Veda and the Sushruta, of old but unknown dates, mention eclampsia. He said that the Atharva-Veda described an amulet to be worn in late pregnancy for warding off convulsions during childbirth.[8] There are several references to pregnancy in the Atharva-Veda (translated by Whitney).[12] One is a description of a protective amulet to be put on in the 8th month of gestation (Bk. VIII, 6, pp. 493–498), but there is not the remotest indication of any specific disorder such as convulsions. The ceremonial verses are clearly directed toward protecting the woman's genital organs against demons and rapists, who are characterized by such epithets as "after-snuffling," "fore-feeling," and "much licking" (to name the milder ones).

There are two possible references to eclampsia in the Sushruta (English translation edited by Bhishagratna).[13] In Volume II, Chapter 8, p. 58: "A child, moving in the womb of a dead mother, who had just expired (from convulsions etc.)" should be delivered by cesarean section. The parenthetic "from convulsions etc." was supplied by the editor and comparison with the Latin translation (Hessler)[13] indicates that it probably was not in the original text. In Chapter 1, p. 11 of Volume II: "An attack of Apatànkah due to excessive hemorrhage, or following closely upon an abortion or miscarriage at pregnancy (difficult labor) or which is incidental to an external blow or injury (traumatic) should be regarded as incurable." Again the parenthetic words are editorial explanations and the "Apatànkah" (convulsions) might well be those associated with severe hemorrhage. By comparison with the Latin translation, the English version seems to have been embellished, for the Latin version specifies only abortion and hemorrhage. An editorial note (pp. 58–60, Vol. II) asserts that the ancient Indians delivered living eclamptic women by cesarean section, but the editor provided no documentation whatever.

Bernhart's reference to the old Chinese literature was to Wang Dui Me, whose work was translated into German by Lo.[14] The work, originally published in 1832 AD, was thought to be free of any influence of Western medicine but even it if were, there is no indication that it recorded only ancient observations. In several respects it seems to have been contemporary; the author described what Lo translated as "Eklampsie" and wrote: "I use recipe No. 232 ...."

Several of the German authors cite Hippocrates as commenting on the susceptibility of pregnant women to convulsions and on their prognosis. None of the quotations appears in *The Genuine Works of Hippocrates* as translated by Adams, or in any of the half-dozen other translations that I have seen. Some of the quotations can be found in other Greek sources.[15] Earlier translators, for instance, had attributed the *Coacae Praenotiones* to Hippocrates, but most scholars agree that it was written before Hippocrates's time. One such quotation, appearing in several German papers is: "In pregnancy, drowsiness and headache accompanied by heaviness and convulsions, is generally bad." It comes from the *Coacae Praenotiones* (*Coan Prognosis*), XXXI, No. 507. The Greeks of that time recognized preeclampsia, for in the *Coan Prognosis*, XXXI, No. 523, we find: "In pregnancy, the onset of drowsy headaches with heaviness is bad; such cases are perhaps liable to some sort of fits at the same time" (translated by Chadwick and Mann).[16] Hippocrates (4th century Be), in his Aphorisms (Sec. VI, No. 30), wrote: "It proves fatal to a woman in a state of pregnancy, if she be seized with any of the acute diseases." Galen, in the 2nd century AD commented that epilepsy, apoplexy, convulsions, and tetanus are especially lethal (Vol. 17, pt. II, p. 820, Kühn [ed]).[17] It may be significant that Galen specified convulsive disorders and perhaps he had in mind what we now call eclampsia, which was not to be differentiated from epilepsy for another 1600 years.

Celsus, in the first century AD, mentioned often fatal convulsions in association with the extraction of dead fetuses (Bk. VII, Chapter 29, translated by Lee).[18] In the same connection, Aetios, in the 6th century AD, wrote: "Those who are seriously ill are oppressed by a stuporous condition ...," "Some are subject to convulsions ...," and "The pulse is strong and swollen" (translated by Ricci).[19]

There is a possible reference to eclampsia in Rösslin's *Der Swangern Frawen und Hebammen Rosengarten*, a book that was the standard text of midwifery in Europe and England for almost two centuries.[20] In discussing the maternal prognosis in difficult labor with fetal death, Rösslin listed among the ominous signs unconsciousness and convulsions (Bk. I, Chapter 9, p. 67). The book was largely based upon the older classics, and the relevant section is reminiscent of Celsus, Aetios, and, especially, Paul of Aegina (translated by Adams).[21] The book was translated into English from a Latin version of what probably was the second edition and appeared in 1540 as *The Byrth of Mankinde*. Raynalde revised and amplified the second edition of 1545, and the text was little altered thereafter. Ballantyne's quotation of the relevant paragraph in Book II, Chapter 9, from the edition of 1560 is virtually identical with that published 53 years later (Raynalde), except for the variable and carefree spelling of the times.[22,23]

Gaebelkhouern (variously, Gabelchoverus, Gabelkover) distinguished four sorts of epilepsy in relation to the seats of their causes, which he placed in the head, the stomach, the uterus, and chilled extremities.[24] He further specified that only the pregnant uterus causes convulsions, particularly if it carries a malformed fetus. "The mothers feel a biting and gnawing in the uterus and diaphragm that leads them to think that something is gnawing on their hearts (epigastric pain?)." The description of that symptom is usually credited to Chaussier, 1824, 228 years later).[25]

Although eclampsia is dramatic, it is not astonishing that there are so few references to it in the older writings, which covered the whole field of medicine. Eclampsia had not been differentiated from epilepsy, and obstetrics was largely in the hands of midwives. Even some relatively modern textbooks of obstetrics have barely noticed eclampsia, and those of Burton and Exton made no mention whatever of convulsions.[26,27] In the first edition of Mauriceau's book, the only comment on convulsions relates to those associated with severe hemorrhage, of which his sister died.[28] The literature of eclampsia, for practical purposes, began in France because it was there that male physicians first took up the practice of obstetrics on a significant scale. Viardel, Portal, Peu, and de la Motte each published notable books in the late 17th and early 18th centuries.

In later editions of his book, Mauriceau devoted more and more attention to what we now call eclampsia. Hugh Chamberlen published purported translations of Mauriceau's later editions, but they seem to have been impostures and really were reissues of the translation of the first edition.[29] Such fraud befits a family that kept so important an invention as the forceps secret through three generations for personal profit, and befits the man who sold the secret. In the edition of 1694, and possibly earlier, Mauriceau set forth several aphorisms dealing with the subject. Among them were (No. 228): The mortal danger to mother and fetus is greater when the mother does not recover consciousness between convulsions.[30] (No. 229): Primigravidas are at far greater risk of convulsions than are multiparas. (No. 230): Convulsions during pregnancy are more dangerous than those beginning after delivery. (No. 231): Convulsions are more dangerous if the fetus is dead than if it is alive. He attributed the convulsions to an excess of heated blood rising from the uterus and stimulating the nervous system and thought that irritation of the cervix would aggravate the situation. He also believed that if the fetus were dead, malignant vapors arising from its decomposition might cause convulsions. His assigning convulsions to such specific causes carries the implication that he had distinguished eclampsia from epilepsy.

Kossmann wrote that in 1760, before he had bought (gekauft hatte) his title "de Sauvages," Bossier first introduced the word *eclampsia*.[5] He said that de Sauvages was a typical Frenchman in that he took it badly whenever his title was omitted, that he had mistaken the meaning of the Greek word from which he derived eclampsia, that none of the supporting references that he cited was correct, and that we owe the word to de Sauvages's slovenly scholarship.

Kossmann was in error. De Sauvages published under that name at least as early as 1739, and there is no indication in the *Biographisches Lexikon* (Hirsch) that he had not been born as de Sauvages.[31] He did acquire the title "de Lacroix" later. De Sauvages differentiated eclampsia from epilepsy in his *Pathologia Methodica*, the three editions of which were forerunners of his *Nosologia* that Kossmann cited. He indicated that epilepsy is chronic and that the fits recur over long periods of time; all convulsions of acute causation de Sauvages called *eclampsia*, spelled with one c in the first and second editions and with two in later publications.[32,33] He attributed the source of the words to Hippocrates, in the sense of *Epilepsia puerilis*, which Kossmann considered to be erroneous. In later editions, he cited de Gorris's *Definitionem Medicarum*, Hippocrates, and the *Coan Prognosis*, in none of which the word occurs, according to Kossmann.

Part of the discrepancy is explained by the questionable authorship of many writings that have been attributed to Hippocrates.[15,16,34] Most scholars do not accept the sixth book of *Epidemics* as being his, but in Section I, No.5, the word does appear and has been translated as "epilepsy," both before and after de Sauvages's time. Galen (Vol. 17, pt. I, p. 824, Kühn [ed]: 1829) translated εχλαμψιεζ as "fulgores" (lightning, shining, brilliance) but after four half-pages of discussion as to its significance, concluded that here it means epilepsy.[17] Nearly a century after de Sauvages, Grimm translated the word as "Fallsucht" (epilepsy).[34] The word does not appear in the edition of de Gorris's definitions that I have seen, but it may be in others.[35] Perhaps de Sauvages cited the wrong dictionary, for he is vindicated by another one. Castelli, in his *Lexicon Medicum*, defined *eclampsis* as brightness, lightning, effulgence, or shining forth, as in a flashing glance ("splendorum, fulgorum, effulgescentium et emicationem, qualis ex oculis aliquando prodeunt").[36] He cited several writings attributed to Hippocrates in which the word was used metaphorically to mean the shining vital flame in puberty and the vigorous years of life ("emicatione flamme vitalis in pubertate et aetatis vigora"). Under *Effulgescentia* he wrote "vide eclampsis." In an earlier edition (1651), *eclampsis* did not appear, but *effulgescentia* had several definitions, the first of which is a disorder characteristic of boys, the most familiar being epilepsy ("quas Graeci εχλαμψιαζ vocant Hipp. praesertim significant morbum puerorum proprium, aut certe perquam familiarissimum, id est, Epilepsium").[37] Castelli, who followed Galen's discussion just mentioned, wrote that to some the word denoted the temperamental change to warmth, or the effulgent vital flame of youth and early manhood. Others considered the interpretation to be the bodily development and perfection during early adulthood.

Blancard (variously Blancardo, Blankaard) in his *Lexicon Medicum*, defined *eclampsis* as "effulgio" and wrote that some authors had called the circulation eclampsis because they thought that a flashing principle in the heart ("luminoso principe in corde") impelled the blood.[38] The word disappeared from his later editions.

In the third edition (1759) of *Pathologia Methodica*, de Sauvages listed several species of eclampsia in relation to such acute causes as severe hemorrhage, various sources of pain, vermicular infestations, and other such factors as had been noted by Hippocrates.[33] One species was *Eclampsia parturientium* and de Sauvages indicated that Mauriceau had described the disease.

Vogel, Cullen, and Sagar, in their classifications of diseases, adopted de Sauvages's *Ecclampsia parturientium*, but dropped one of the two *c*'s.[39–41] Interestingly, the taxonomists defined both *Convulsio gravidarum* and *Eclampsia parturientium* (or *parturientes*) as different genera and without cross-reference between the two.

Gutsch, a student of J. C. Gehler in Leipsig, may have been the first German obstetrician to take up the word, and for a generation the German use of it seems to have been confined to that center.[42] Kossmann wrote that the word reappeared in France in 1844, but Ryan said that it was generally used there in his time.[5,43] That is confirmed by the listing of publications in the *Index-Catalogue of the Library of the Surgeon General's Office* (1890), where the word *eclampsia* appears in the titles of 31 books or monographs from six European countries before 1845; there were many from France.

Ryan recognized the specificity of what he called dystocia convulsiva: He gave as synonyms "labor with convulsions," "convulsio apoplectica," "apoplexia hysterica," "apoplexia lactusa," "apoplexia sympathetica," and "eclampsia."[43] When consciousness returned between fits, Ryan called them epileptiform; when coma or stertor supervened, he called them apoplectic or eclamptic convulsions. He wrote that the convulsions may occur during the last 3 months of pregnancy, in labor, or after delivery and that the prognosis is unfavorable "as a third of those afflicted are destroyed." Postpartum eclampsia is less dangerous, he said.

Bossier de Sauvages's use of the word *eclampsia* as a generic term for convulsions having an acute cause persisted for more than 200 years.[32] *Stedman's Medical Dictionary* (1957) defined eclampsia as "convulsions of an epileptoid character" and listed several varieties. Puerperal eclampsia was defined as "convulsions of uremic or other origin, occurring in the latter part of pregnancy or during labor"; there was no mention of the puerperium. The 20th edition, in 1961, discarded all but the obstetric definition; "Coma and convulsions that may develop during or immediately after pregnancy, related to proteinuria, edema, and hypertension." Puerperal eclampsia was described as following delivery, which is technically correct, but a misleading guide to interpretation of much of the literature of the 19th century.

SIGNS

Edema

Fasbender wrote that Demanet was the first to relate convulsions with edema.[6,44] An anonymous author included Demanet's entire paper in a review that is more accessible than the original.[45] All six of Demanet's eclamptic patients were edematous, and he suggested that anasarca be added to the three recognized causes of convulsions, i.e., depletion, repletion, and labor pains. Most eclamptic women have such marked edema that it could not have escaped notice before 1797 and, in fact, several writers had commented on it before then. Mauriceau remarked on the severe edema of one of his patients (Observation No. 90), but he usually did not describe the women other than to specify age and parity.[30] De la Motte considered edema to be benign unless associated with convulsions.[46] Smellie presented his cases as exemplifying methods of delivery and said nothing about the appearance of the patients.[47] Van Swieten, in his commentary on Boerhaave's Aphorism No. 1302, specified edema as one of the indications for phlebotomy in women threatened with convulsions.[48]

Proteinuria

Rayer found protein in the urines of three pregnant edematous women.[49] From his descriptions, it seems probable that the first one had preeclampsia and the other two had Bright's disease.

Lever is generally credited with the discovery of proteinuria in eclampsia.[50] He was stimulated to look for it by the clinical resemblance between eclampsia and Bright's disease, and he found it in nine of ten convulsive women. The description of the postmortem findings in the one woman who did not have proteinuria is suggestive of meningitis and perhaps her convulsions were not of eclamptic origin. Because of the rapid abatement of the proteinuria after delivery, Lever concluded that eclampsia is different from Bright's disease, although others of his era were not so astute. Lever attributed the proteinuria to renal congestion caused by compression of the renal veins by the bulky uterus. He speculated that such compression might be absent in the "upwards of 50" normal women in labor whose urines he found to be normal, unless "symptoms have presented themselves, which are readily recognized as precursors of puerperal fits."

Simpson should share credit with Lever, for in the same year he wrote: "(I) had publically taught for the last two sessions, viz., that patients attacked with puerperal convulsions had almost invariably albuminous urine, and some accompanying, or rather, preceding dropsical complications."[51] Unfortunately, one of his fatal cases of eclampsia did have chronic nephritis and he found granular kidneys at autopsy, which led him to believe that eclampsia was a manifestation of nephritis.

Hypertension

Old-time clinicians surmised the presence of eclamptic hypertension from the hard, bounding pulse, but confirmation was long delayed for want of methods for measuring the blood pressure. Sphygmographic tracings were interpreted as showing arterial hypertension, but no absolute values could be specified. Mahomed reported that such tracings indicated the presence of hypertension in nearly all pregnant women, and he concluded that "Puerperal convulsions and albuminuria were accounted for by the predisposing condition of high tension in the arterial system existing during pregnancy."[52,53] The sphygmographic features pointing to hypertension were: (1) the increased external pressure required to obtain optimal tracings, (2) a well-marked percussion wave separated from the tidal wave, (3) a small dicrotic wave, and (4) a prolonged tidal wave. We now know that the hemodynamic changes of normal pregnancy do not include hypertension, but the increased cardiac output changes the character of the pulse. The ancient Chinese recognized the altered pulse perhaps as long as 4500 years ago; in the *Yellow Emperor's Classic of Internal Medicine* we find: "When the motion of her pulse is great she is with child" (translation by Veith).[54]

Ballantyne, from sphygmograms made in two eclamptic and one severely preeclamptic women, concluded that arterial blood pressure is considerably increased.[55] One of the patients died 10 hours after delivery, and the tracings suggested that "after the completion of labor there is a great tendency to complete collapse (of the arterial pressure) and that unless checked will go on till death closes the scene." His description of terminal hypotension is descriptive of many cases of fatal eclampsia, although he generalized too broadly. Galabin wrote: "From sphygmographic tracings taken during the eclamptic state, I have found that the pulse is ... one of abnormally high tension, like that in Bright's."[56] In discussing the management of eclampsia, he wrote: "The first treatment should be to give an active purgative. This lowers arterial pressure ...."

Despite the efforts of earlier investigators, indirect methods for the measurement of arterial blood pressure did not become available until 1875. The instruments of Marey, Potain, von Basch, and others led to overestimates of the blood pressure but did give relative values. Thus, Lebedeff and Porochjakow, using von Basch's sphygmomanometer, found that the blood pressure is higher during labor than in the early puerperium.[57] Vinay, using Potain's device, observed that the blood pressure was increased in pregnant women with proteinuria (180–200 mm Hg as compared with the normal of up to about 160, by his method).[58] The discovery of eclamptic hypertension is widely credited to Vaquez and Nobecourt, who remarked that they had confirmed Vinay's observations published in his textbook 3 years earlier.[59] Vinay, however, said nothing about the blood pressure in eclampsia and regarded his hypertensive albuminuric patients as having Bright's disease.[58] Wiessner reported that the blood pressure fluctuates widely during eclampsia.[60]

Cook and Briggs used an improved model of Riva Rocci's sphygmomanometer that has not been greatly changed to this day.[61] They observed that normal pregnancy has little effect on the blood pressure until the onset of labor, when it increases with uterine contractions. Women with proteinuria were found to have hypertension, and the authors wrote that the detection of increased blood pressure in a pregnant woman should "excite the apprehension of eclampsia." They observed that proteinuria was usually associated with hypertension and thought that the blood pressure was the better guide to prognosis.

The differentiation of preeclampsia–eclampsia from renal disease and essential hypertension was long delayed, and although we now recognize that they are separate entities, the correct diagnosis is often difficult. Although Lever looked for proteinuria in eclamptic women because of their clinical resemblance to patients with glomerulonephritis, he concluded that the diseases are different because eclamptic proteinuria cleared rapidly after delivery.[50] Others of that era, however, cited his discovery of proteinuria as evidence for the identity of the diseases. Frerichs, in his textbook, wrote that eclampsia represents uremic convulsions and the concept persisted for half a century.[62] Autopsies of women dying of eclampsia often uncovered no renal abnormalities detectable by methods then available, but that objection was countered by Spiegelberg,[63] for example. He wrote, in italics, "*True eclampsia depends upon uremic poisoning in consequence of deficient renal excretion.*" He attributed the deficiency to chronic nephritis aggravated by pregnancy or to disease of the renal arteries secondary to vasospasm. He suggested, as had others before him, that the renal vasospasm arose reflexively from stimulation of the uterine nerves, a hypothesis revived in modern times by Sophian.[64]

The *Zeitgeist* was reflected in the 1881 issue of the *Index-Catalogue of the Library of the Surgeon General's Office*. Under "Bright's disease" it specified "see, also, –Puerperal *convulsions*."

Toward the end of the 19th century the development of cellular pathology and of improved histologic methods led to the detection of a characteristic hepatic lesion and the recognition of eclampsia as an entity, distinct from Bright's disease (Jürgens; Schmorl).[65,66] The differentiation of the nonfatal, nonconvulsive hypertensive disorders remained confused for many years. The terms "nephritic toxemia," "Schwangerschaftsniere," and "Nephropathie" persisted through the 1930s and the term "low reserve kidney" was introduced as late as 1926.

The recognition of primary or essential hypertension is relatively recent, but its relevance to pregnancy was not appreciated for many years after it had been accepted as an entity. Allbutt observed that middle-aged and older men and especially women often develop hypertension and that the increase in blood pressure is not accompanied by any other evidence of renal disease.[67] He referred to the condition as "senile plethora" or "hyperpiesis"; later it was termed "essential hypertension" by Frank or "hypertensive cardiovascular disease" by Janeway.[68,69] The appellation "senile" had a lingering effect, and obstetricians thought that women of childbearing age were not old enough to have developed essential hypertension.

Herrick and coworkers recognized essential hypertension as an important and frequent component of the hypertensive disorders in pregnancy.[70–72] They showed that what the obstetricians called chronic nephritis in and following pregnancy was more often essential hypertension. Herrick wrote: "Viewed largely, then, the toxemias of pregnancy are probably not toxemias. Rather they are evidences of underlying tendencies to disease."[70] He thought that about a quarter of the cases have renal disease, either frank or brought to light by pregnancy. The rest, he thought, have frank or latent essential hypertension. In some papers, he seemed not to have decided whether eclampsia and severe preeclampsia caused vascular disease or were manifestations of it that were revealed and peculiarly colored by pregnancy. In one of his last papers on the subject (Herrick and Tillman), he wrote: "When these are fully delineated it is our opinion that we shall find nephritis concerned in but a small fraction of the toxemias; that the larger number, including the eclampsias, the preeclampsias, and the variously designated milder types of late toxemia ... will be found to have unit characteristics based upon cardiovascular disease with hypertension."[72]

Fishberg, in the fourth edition of his book *Hypertension and Nephritis*, denied the specificity of preeclampsia–eclampsia, which he regarded as manifestations of essential hypertension.[73] Although he retreated from that view in the following edition (1954), he continued to regard eclampsia as "a typical variety of hypertensive encephalopathy."[74]

Dieckmann, in his book *The Toxemias of Pregnancy*, said that about half of the women with hypertensive disorders in pregnancy have either nephritis or essential hypertension, but that primary renal disease accounted for not more than 2%.[75] That opinion, in which he both followed and led, gained wide acceptance. Herrick's estimate of the prevalence of chronic renal disease, however, seems to have been closer to the truth. Several studies of renal biopsies have indicated that 10 to 12% of women in whom preeclampsia is diagnosed clinically have the lesions of primary renal disease, usually chronic glomerulonephritis.

HYPOTHESES AND RATIONAL MANAGEMENT

Zuspan and Ward wrote that in the treatment of the eclamptic patient, "she has been blistered, bled, purged, packed, lavaged, irrigated, punctured, starved, sedated, anesthetized, paralyzed, tranquilized, rendered hypotensive, drowned, been given diuretics, had mammectomy, been dehydrated, forcibly delivered, and neglected."[76] Many procedures could be added to the list. Aside from the great variety of medications, surgical approaches have included ureteral catheterization, implantation of the ureters in the colon, renal decapsulation, drainage of spinal fluid, cisternal puncture, trepanation, ventral suspension of the uterus, postpartum curettage, oophorectomy, and so on. I do not rehearse the list in any spirit of levity, for it is important to remember that each of the treatments was rational in the light of some hypothesis as to the cause or nature of eclampsia. That is more than we can say for our present management, which is purely empiric, perhaps too often symptomatic, and in some respects based upon imitative magic.

Eclampsia was not differentiated from epilepsy until 1739, and the distinction was not generally accepted for another century. Merriman discussed dystocia convulsiva and wrote: "The cases alone deserving the appellation of puerperal convulsions, which have fallen under my observation, have borne a very exact resemblance to the epilepsy."[77] Ryan, in his *Manual of Midwifery* (1831), recognized eclampsia as an entity, but 25 years later his countryman Churchill, in his *Theory and Practice of Midwifery* (1856), classified gestational convulsions as hysteric, epileptic, and apoplectic.[43,78] By the time the differentiation from epilepsy was generally accepted, eclampsia had been confused with uremic Bright's disease, and the proliferation of hypotheses as to its cause was delayed until late in the 19th century.

Hippocrates, in his Aphorisms, Section VI, No. 39, wrote: "Convulsions take place either from repletion or depletion" (translated by Adams).[15] Hippocrates referred to convulsions generally, as did Galen, who iterated his view (Vol. 18, pt. I, p. 61, Kühn [ed]: 1829).[17] Accordingly, obstetricians divided on the question of which factor accounted for convulsions in childbirth. Mauriceau recommended bloodletting, except in the convulsions associated with severe hemorrhage.[30] According to Gutsch, who did not give a reference, van Swieten wrote that depletion was the cause and attributed the convulsions to collapse of the cerebral blood vessels.[42] Gutsch was completely wrong, but virtually every history of eclampsia has perpetuated his error. Van Swieten, in commenting on Boerhaave's Aphorism No. 1322, was in agreement with the concept that the sudden reduction in intraabdominal pressure at delivery might lead to a pooling of blood diverted from the brain and thus account for weakness and syncope.[48] Van Swieten went on to say that if the uterus did not contract, "then lying-in women run with blood, and, by the sudden inanition of the (cerebral) vessels, die in convulsions; pretty nearly in the same manner that the strongest animals, when their arteries are cut open by the butcher, their blood being entirely exhausted, are seized with violent convulsions before they die" (English translation of 1776).[48] Clearly, van Swieten was not referring to eclamptic convulsions. In his comments on Aphorisms Nos. 1010, 1295, and 1302, van Swieten indicated unequivocally that cerebral congestion is the cause of what we now call eclamptic convulsions. He attributed the cerebral repletion to compression of the abdominal organs by the large uterus, to blockage of the aorta by the uterus, and to the violent expulsive efforts at delivery, all of which diverted blood to the brain. Accordingly, he wrote: "no one can doubt but the letting of blood must prove of the greatest service, especially if these symptoms (including edema) happen near the time of delivery; for then by the violent efforts of labour, the blood may be forcibly thrown into the vessels of the encephalon, and all its functions thereby suppressed; or even a fatal apoplexy may ensue from a rupture of the vessels; convulsions too may often follow" (comment on Aphorism No. 1302).

In addition to the factors specified by van Swieten as leading to repletion, other writers had suggested reflex effects arising from stimulation of the uterine nerves and suppression of the menses during pregnancy. The opposite hypothesis, that the convulsions were caused by depletion or cerebral anemia, had its proponents and still lingers on in terms of cerebral vasospasm and edema.

Old-time physicians and barber-surgeons resorted to bloodletting in the treatment of many disorders and they noted the extraordinary tolerance of pregnant women for hemorrhage. By the end of the 18th century the "plethora of pregnancy" was a widely accepted concept that seems to have tipped the scales in favor of repletion and cerebral congestion as the cause of gestational convulsions. Phlebotomy and purgation, which were the sheet anchors in the management of eclampsia one and two centuries ago, probably were of late origin. Section V of Hippocrates's Aphorisms specified contraindications to those measures. No. 29: "Women in state of pregnancy may be purged, if there be any urgent necessity, from the fourth to seventh month, but less so in the latter case. In the first and last periods it must be avoided." No. 31: "If a woman with child be bled, she will have an abortion, and this will be the more likely to happen, the larger the foetus" (translated by Adams, 1849). Galen agreed (Vol. 17, pt. II, pp. 652, 821, Kühn [ed]: 1829).[17]

Although Celsus, in the first century AD, disputed the adverse effect of bleeding, the doctrine persisted (Bk. II, Chapter 10, translated by Lee).[18] In the 6th century, Aetios reiterated the deleterious effect of phlebotomy; he cited Hippocrates when he recommended bleeding as a means of inducing abortion (Chapter 18, translated by Ricci).[19] Avicenna, in the 11th century, advised against both bleeding and purgation during pregnancy (translated by Krueger).[79] Maimonides, in the 12th century, seems to have contradicted himself.[80] His 12th Treatise, Aphorism 5 is: "The conditions and complications that mitigate [sic] against bloodletting, although signs of filling may be apparent, are as follows: Convulsive disorders ..." But Aphorism 22 says: "Venesection is an utmost necessity at the very onset: (in) patients suffering from ... convulsions ..." (translated by Rosner and Muntner).

The prime object of phlebotomy was to decrease cerebral congestion and to that end, some physicians preferred bleeding from the temporal artery or jugular vein. Leeches and cups were applied to the scalp, neck, and even face to draw blood away from the brain. Blisters and sinapisms were placed in various areas for the same purpose, and the scalp was shaved for the closer application of cold packs. Sometimes the physician recognized that repeated phlebotomies had so weakened the woman that another would be hazardous, so that rather than subject her to another general hemorrhage, he placed leeches or cups on the head for the local diversion of blood from the brain. Ryan, who attributed eclampsia to cerebral congestion, wrote: "In these kingdoms, copious depletion with camphor mixture, ether, etc, are chiefly employed," along with repeated bleeding.[43] Ether, which was popular in France, was given in mixtures by mouth or subcutaneously.

Those who believed that the convulsions were caused by irritation of the uterus also used sinapisms, blistering, and the like as counterirritants, and they bled patients from veins in the feet, which they believed to be a revulsive measure.

Later, when a circulating toxin was postulated as the cause of eclampsia, phlebotomy was retained as a rational treatment because it directly removed the noxious substance. To the same end, all of the emunctories were stimulated and the use of diuretic, purgative, emetic, and sudorific drugs became popular. High colonic irrigation and gastric lavage were used for the same purpose. Tincture or extract of jaborandi was used to induce intense sweating and when its most active alkaloid was identified as pilocarpine, that drug came into use. It was tried for a time in Edinburgh, but abandoned when it was found to have doubled the maternal mortality from 30 or 35% to 67%; the women drowned in their own secretions (Hirst).[81] A parallel situation exists today in the common use of potent diuretic drugs, which probably do no real good and are dangerous, though not so dramatically as in the case of pilocarpine.

The concept of eclampsia as a toxemia is more than a century old. The earliest reference that I have found is by William Tyler Smith, who wrote: "It deserves to be borne in mind, that the depurgatory functions ought, in order to preserve health, to be increased during gestation, as the *debris* of the foetal, as well as the maternal system, have to be eliminated by the organs of the mother.[82] Besides these forms of toxaemia, the state of the blood which obtains during fevers, or during the excitement of the first secretion of milk, may excite the convulsive disorder." He used the word *toxaemia* so casually as to suggest that it might have been a current concept. Murphy, in his *Lectures on the Principles and Practice of Midwifery*, wrote 13 years later: "Predisposing causes of convulsions are hyperaemia, anaemia, and toxaemia" and " *The direct proximate cause* of convulsions is impure blood" (his italics).[83] Mahomed, explaining the hypertension that he thought to be present in all pregnant women, wrote: "The blood of the mother is overcharged with effete material, for she has to discharge the excrementitious matters of the fetus by her own excretory organs; her blood is therefore in a measure poisoned ..." and "Thus puerperal eclampsia and albuminuria were accounted for ...."[52,53] The later controversy as to whether Fehling or van der Hoeven had priority in suggesting fetal waste products as the cause of eclampsia was obviously an exercise in futility.[84,85]

Actually, Mauriceau had attributed convulsions in many cases to decompositional products of the dead fetus.[30]

In the symposium on eclampsia, held in Giessen in 1901, the almost unanimous opinion was that the disease is caused by a toxin, but there was no agreement as to its source. The uremic hypothesis had not yet succumbed, and some writers held out renal insufficiency, either intrinsic or secondary to uterine compression of the renal veins or ureters, as the cause. As previously mentioned, a variant explanation was renal vascular spasm arising as a reflex from nervous stimulation in the uterus. Other hypotheses included fetal catabolic products, bacterial toxins, autointoxication by noxious substances absorbed from the gut, toxins from the placenta released directly or by lytic antibodies against it. Several French investigators, of whom Delore was probably the first, suggested bacterial infection.[86] Gerdes attributed eclampsia to a bacillus that was later identified as *Proteus vulgaris* and Favre alleged the same role for *Micrococcus eclampsia*.[87,88] That hypothesis was quickly overthrown, but the idea that bacterial toxins released in focal infections had a role was advocated for another half century. Proponents of autointoxication as the cause of eclampsia pointed to the predominance of the hepatic lesion in the periportal areas of the hepatic lobules, which receive most of the blood draining the gut.

Rosenstein, adopting Traube's explanation of uremic convulsions, suggested that the proteinuria depleted the plasma proteins and that the combination of watery blood and hypertension led to cerebral edema, convulsions, and coma.[89] Munk tested Traube's hypothesis by ligating the ureters and jugular veins of dogs and injecting water through the carotid artery, thereby evoking convulsions and coma that he regarded as uremia.[90] A modern clinical counterpart is the water intoxication produced in an occasional patient by the injudicious and prolonged infusion of oxytocin in large volumes of dextrose solution.

An enormous amount of work was expended in trying to identify the toxin. Frerichs, who equated eclampsia with uremic convulsions, postulated an enzyme that converted urea to ammonium carbonate.[62] Some thought that a precursor of urea, carbamic acid, was the toxin. Other substances "identified" as the toxin included creatine, creatinine, xanthine, acetone, lactic acid, urobilin, leucomaines akin to ptomaines, globulins, and water.

Then, as now, new hypotheses were introduced with supporting observations or experiments that either could not be confirmed or were interpreted differently by other investigators. Many examples could be cited. The French school developed the concept that pregnant women excreted less than normal amounts of endogenous toxins, which therefore accumulated in the bloodstream. In support of their hypothesis, they reported that the urine of eclamptic women was less toxic and the serum more toxic than the same fluids from normal pregnant and nonpregnant subjects. Volhard and Schumacher reviewed their work critically, repeated their experiments, and demolished their conclusions.[91,92] Dixon and Taylor reported pressor activity in alcoholic extracts of placenta, but Rosenheim showed that bacterial contamination accounted for the effect.[93,94] More recently, the many reports of antidiuretic activity in blood, urine, and cerebrospinal fluid of women with preeclampsia–eclampsia have been called into question for the same reason (Krieger, Butler, and Kilvington).[95] The earlier reports of the lethal effect of placental extracts, press juices, and autolysates given intravenously or their production of proteinuria when injected into the abdominal cavity were largely explained by Lichtenstein.[96] He found (1) that extracts of other organs are equally toxic, (2) that the particulate matter in the preparations blocked the pulmonary capillaries, (3) that the lethal effect and intravascular coagulation could be duplicated by the injection of inorganic particles suspended in saline solution, (4) that the proteinuria could be duplicated by the injection of other foreign proteins, and (5) that filtered extracts were innocuous. Schneider identified the "toxin" in placental extracts free of particles as thromboplastin, which causes intravascular coagulation.[97]

PROPHYLAXIS

Mauriceau recommended two or three phlebotomies during the course of pregnancy as prophylaxis against eclampsia, but he disparaged a colleague who had bled one woman 48 times and another 90 times.[30]

Dietary taboos originated in the superstitions of antiquity and have persisted, with modifications, to the present day. Meat, especially red meat, has had a bad name and has been forbidden or restricted in the dietary treatment of many disorders, including preeclampsia–eclampsia. Thus Miquel, in discussing prophylaxis against convulsions in pregnancy, recommended a farinaceous vegetable diet in the form of a slop or, preferably, one of milk products together with avoidance of spices.[98] De Wees attributed to overeating the one case of eclampsia that he mentioned in his *Treatise on the Diseases of Females*. Prenatal care was unusual in the first half of the 19th century, but some physicians did see private patients before labor.[99] Johns wrote that every physician should see his obstetric patients at intervals during the latter months of pregnancy. He described edema of the hands and face, headache, giddiness, ringing in the ears, loss of vision, pain in the stomach, and a flushed face as denoting an increased risk of convulsions.[100] He wrote that the risk was converted to certainty if (1) the women were pregnant for the first time or had had convulsions in a previous pregnancy; (2) if the head of the child presented, or (3) if the women were of full and plethoric habit. (In passing, it was widely believed at that time that convulsions occurred only in association with vertex presentations.) To prevent convulsions, Johns advocated a diet of fruits, vegetables, and milk, as well as laxatives or purgatives, diuretics, moderate exercise and plenty of fresh air, phlebotomy, and, if the signs and symptoms were marked, emetics. Meigs boasted that although he had seen a good many cases of eclampsia, he was very sure that he had prevented a far greater number.[101]

Sinclair and Johnston wrote that admittances of women to the Dublin Lying-in Hospital, in all cases except dire emergencies, were by arrangement made before the end of pregnancy. Each woman was given a ticket to be signed by a priest or by a respectable citizen, which she then took to the Dispensary for the countersignature of a physician.[102] The physician checked on her signs and symptoms, and if she had edema, headaches, dizziness, or proteinuria, she was either admitted to the hospital or seen regularly in the Dispensary. She was purged freely and repeatedly, kept in bed, and allowed nothing but the mildest and lightest nutriment. The authors stated: "Very often have convulsions been most certainly warded off altogether." When convulsions had not been prevented, they thought that the severity of the disease had been decreased by their treatment.

After Lever's discovery of proteinuria in eclamptic and preeclamptic women, more and more physicians recommended periodic urinalyses in the latter months of pregnancy.[50] When proteinuria was found, they prescribed dietary restrictions along with laxative and diuretic agents and, often, phlebotomy. The diet usually was limited to fruits, vegetables, and milk and was low in protein. Low protein diets were advocated for an ever-increasing number of reasons. They were thought to be more easily digestible and to minimize gastric irritation; nervous stimuli from the uterus and gastrointestinal tract were long thought to cause cerebral repletion (or depletion) and thus to trigger convulsions. Another objective was to lessen the "plethora of pregnancy." When eclampsia came to be regarded as uremic Bright's disease, the diets seemed rational because they reduced the load of nitrogenous catabolites and supposed toxins to be excreted by the kidney. Still later, the Dublin school, especially, argued that incompletely digested fragments of the protein molecule were absorbed from the intestine and had a toxic effect. The bodily defenses against the so-called split proteins were normally adequate but during pregnancy the fetus and placenta represented an additional source of such noxious substances. When the combined invasions overwhelmed the defenses, toxemia and eclampsia resulted. Another hypothesis was that the amino acids from the digested proteins were decarboxylated but not deaminated, with the production of toxic amines.

One of the circumstantial evidences for the efficacy of low protein diets was the observation that eclampsia was predominantly a disease of middle- and upperclass women. That widely held opinion may have been influenced by the fact that many physicians who published had private practices, but Fitzgibbon, who saw all classes in Dublin's Rotunda, wrote: "Toxaemia is unquestionably a disease of the well-to-do classes of society."[103] Ruiz-Contreras, of Barcelona, reported that the incidences of proteinuria and eclampsia were far greater in his private patients than in the charity patients he managed in the clinic.[104] When the nutrition improved and the dietary intake of protein increased among the masses, the incidences of proteinuria and eclampsia rose.

During World War I the incidence of eclampsia decreased significantly in Germany and rose again after the Armistice. Germans are reputed to eat heavily, and the nutritionists reasoned that in good times they eat too much. During the war years they might have eaten less protein and benefited by a relative immunity to eclampsia. An editorial (1917) in the *Journal of the American Medical Association* stated: "The conclusion seems inevitable that restrictions of fat and meat tend to ward off eclampsia."[105] That interpretation had an effect that persists to the present day. As late as 1945, Stander, in the ninth edition of *Williams' Obstetrics*, advocated the dietary restriction of protein in the treatment of hypertensive disorders in pregnancy.[106] Many obstetricians prescribed such diets for all of their patients as prophylaxis against preeclampsia. The current practice of nearly all American obstetricians in limiting weight gain during pregnancy to 15 or 20 lbs stems from the same source.

CLASSIFICATION OF THE HYPERTENSIVE DISORDERS IN PREGNANCY

Classifications are of relatively recent origin. Women with prodromal signs of gestational or puerperal convulsions were designated as having threatening or imminent eclampsia, but a specific name of the condition was long delayed. During much of the latter half of the 19th century, eclampsia was thought to be uremic Bright's disease and women with proteinuria and edema were thought to have nephritis, although a few authors simply called it "albuminuria," a term that persisted for many years in England.

Leyden described "the kidney of pregnancy" (Schwangerschaftsniere) and, in 1886, pointed out that the renal changes in eclampsia are similar (Chesley referred to his own Chapter 4, [our ref. 1]; his other citations are our refs [107,108]). He reviewed the meager literature, citing several authors who had suggested that prolonged duration of the kidney of pregnancy sometimes leads to chronic nephritis and wrote that he had seen several such cases. Probably the most vigorous proponent of that view was Schroeder, who thought that what we call preeclampsia was acute nephritis.[109] He wrote that prompt termination of pregnancy would prevent the progression to chronic nephritis, whereas delay would favor it. Needless to say, the kidneys of survivors were not examined during pregnancy and the differential diagnosis was highly uncertain.

As a result of Leyden's work, nonconvulsive hypertensive disorders in pregnancy were called "kidney of pregnancy" or "nephropathy," and later, "nephritic toxemia." Some of those terms still persist. In the United States, just after 1900, the most common designation was "the toxemia of pregnancy." Webster, in his *Textbook of Obstetrics*, referred to the "pre-eclamptic state" and Bar introduced the word "eclampsisme" as meaning eclampsia without convulsions or threatening eclampsia.[110,111]

Once the concept of circulating toxins gained acceptance at the turn of the century, many disorders of obscure origin came to be classified as toxemias of pregnancy. Included were such diverse conditions as hyperemesis, acute yellow atrophy of the liver, ptyalism, gingivitis, pruritus, herpes, severe dermatitides, neuritis, psychosis, chorea, anemia, abruptio placentae, and all forms of hypertension. We see a comparable situation today, when so many physicians attribute almost any disorder of unknown origin, or even malaise, to "a virus."

The unitarians thought that a single toxin might be responsible for the array of effects and suggested that hyperemesis protected a woman from eclampsia because she vomited out much of the noxious substance. Williams, in the third edition of his *Obstetrics*, wrote that the unitarians were impeding progress and that investigators should look for toxins specific for each disorder.[112] His classification of the "toxemias of pregnancy" was: *pernicious vomiting, acute yellow atrophy of the liver, nephritic toxemia, preeclamptic toxemia, eclampsia, presumable toxemia* (under which he included most of the diverse array just cited).

The many classifications proposed before 1940 were essentially variations on Williams' theme, although there was a progressive disappearance of the presumable toxemias. Some writers differentiated hepatic eclampsia from renal eclampsia.

In 1940, the American Committee on Maternal Welfare (Bell et al.)[113] proposed the following classification:

Group A. Diseases not peculiar to pregnancy
- I. Hypertensive disease (hypertensive cardiovascular disease) – benign, mild, severe, or malignant
- II. Renal diseases
  - a. Chronic vascular nephritis (nephrosclerosis)
  - b. Glomerulonephritis, acute or chronic
  - c. Nephrosis, acute or chronic
  - d. Other forms of renal disease

Group B. Disease dependent on or peculiar to pregnancy
- I. Preeclampsia, mild or severe
- II. Eclampsia
  - a. Convulsive
  - b. Nonconvulsive (coma with findings at necropsy typical of eclampsia)

Group C. Vomiting of pregnancy

Group D. Unclassified toxemia, in which the above categories cannot be separated for want of information.

Acute yellow atrophy was dropped, but vomiting was retained "because of precedent." Mild hypertensive disease was defined by the absence of marked vascular changes and blood pressures below 160/100; no lower limit of blood pressure was specified. Mild preeclampsia was defined by the appearance after the 24th week of blood pressures of 140 to 160 systolic and 90 to 100 diastolic, proteinuria of less than 6 g/L, and slight or no edema.

In 1952, another subcommittee of the American Committee on Maternal Welfare (Eastman et al.) revised the classification.[114] They dropped vomiting as unrelated to the hypertensive disorders and deleted renal diseases in the mistaken belief that they are easily differentiable from essential hypertension and preeclampsia. Two new categories were added: preeclampsia or eclampsia superimposed upon chronic hypertension, and recurrent toxemias. Clumsy diction inadvertently permitted the diagnosis of preeclampsia on the basis of any one of the three cardinal signs, even persistent edema or rapid weight gain alone.

Obstetricians in Aberdeen, and in some other areas, follow Nelson's definition of preeclampsia.[115] Edema is ignored and the diagnosis is made if the diastolic pressure rises to 90 mm Hg or more after the 25th week and is found on at least 2 days. In the absence of proteinuria, the disorder is called mild; if proteinuria appears, it is called severe preeclampsia. The mildly preeclamptic group, thus classified, must include many women with latent hypertension brought to light by pregnancy, as well as chronic hypertensive women whose blood pressures have abated during midgestation. The severe group would include many cases that would be called mild in other classifications.

Various schemes of classification have been published within the past few years in *Gynaecologia*. They have been compared by Rippmann, who has compiled a list of more than 60 names in English and more than 40 in German that have been applied to the hypertensive disorders in pregnancy.[116] *(Dr Chesley's original chapter ends here.)*

EDITORS' UPDATE

Chesley's original introductory chapter, reproduced above, had been composed before 1976, while at about the same time, and with Leon's support and blessing, Drs. Lindheimer and Zuspan, at the University of Chicago, were organizing a National Institutes of Health supported International Workshop focusing on the Hypertensive Disorders of Pregnancy. The primary goal of the meeting, held 25–27 September 1975, as stated in the preface of its proceedings,[117] was that: "*Hypertension, especially preeclampsia, is a major complication of pregnancy causing significant morbidity and mortality in both fetus and mother. Nevertheless, research on the hypertensive complications of pregnancy has been sporadic, and scientists studying this and related fields have rarely communicated.* " The primary goal, further stated, was to "*stimulate investigative efforts in the field and to establish avenues of communication between clinical and laboratory scientists in various disciplines*."

The workshop, composed of ~60 invitees with diverse research and clinical disciplines, shared views, data, argued profusely at times, and parted energized with new perspectives. The meeting achieved its immediate goals, communication between disciplines and stimulation of further research, and was also the impetus to the establishment of a new organization, the International Society for the Study of Hypertension in Pregnancy (ISSHP) that has held biennial meetings starting in 1978. More important, the workshop's long-term goals, implied as well in Leon's sole authored first edition, have been realized, namely, there are now more and more investigators focusing attention on a critical but neglected area of maternal and fetal health. Thus, as noted in this chapter's introduction, the 1980s and 1990s were characterized by an awakening of the need for focused research, producing at first new and important findings that affected both clinical and research approaches, followed by a logarithmic growth in investigative breakthroughs that have characterized the first decade of this millennium. Physicians now recognize preeclampsia as a systemic disease, more than just hypertension and proteinuria,[118] and investigators keep uncovering a panoply of pathophysiological changes associated with preeclampsia. In addition we are now witnessing initial probes into specific therapies based on findings that relate to uncovered mechanisms. (e.g., ref. 119).

One should note that this renewed emphasis on reversing the morbidity and mortality associated with hypertension and pregnancy has not gone by unnoticed. National and international agencies including the World Health Organization, the National Institutes of Health in the United States (e.g., their Heart Lung and Blood, as well as Child and Health Development Institutes), and the Royal College of Obstetrics and Gynaecology (London), among them, have all either convened study groups that summarized needs, or formulated policy statements underscoring the status of knowledge in the field, or the enormous underfunding that exists in this important area of reproductive medicine that must be addressed. Still it is fair to say that as of 2014 research support especially in regard to preeclampsia remains near the bottom in terms of Disability Adjusted Lost Years (DALYs), despite the fact this disease remains a leading cause of maternal and fetal death.

The chapters that follow include, among other topics, attention to new findings regarding normal and abnormal placentation, roles of circulating molecules that include altered hormone levels and proteins and/or other substances of placental origin, autocoids, as well as the physiology and pathophysiology of the cardiovascular system (especially the endothelium). New, also, is a rapidly expanding interest in the contribution of the maternal constitution, both genetic and environmental, and a better understanding of the normal and abnormal cerebral circulation. Also highlighted is exciting new work relating to immunology, epigenetics, the systemic inflammatory and antioxidant states, as well as potential mechanisms behind the occurrence of eclampsia. A topic of special interest, virtually unstudied until 2003, is the exciting role of antiangiogenic factors in the production of preeclampsia phenotypes, and its emerging role in triage of women with early preeclampsia. Finally, we note that Leon Chesley's landmark study[120] of the remote prognosis of eclampsia in which he personally investigated each individual's outcome for as long as 40 years after their the index pregnancy has spurred newer and larger epidemiological surveys that have established preeclampsia as a risk marker for future cardiac and metabolic disease. We now know women who have manifested *de novo* hypertension in pregnancy, including preeclampsia, require closer scrutiny throughout life and interventions such as early life-style changes may alter their remote outcomes. All of the accomplishments alluded to above and more will be amply cited in the chapters that follow.

This optimistic introduction, however, must be tempered a bit, as unsolved problems and controversies remain. For example, the exact etiologies of both preeclampsia and eclampsia are still unclear (and such clarification is a key step if definitive preventative and therapeutic strategies are to be achieved). The remainder of this Editors' Update focuses on two of the controversies, measurement of blood pressure during gestation, this controversy now resolved, and progress toward agreement on classification schemas.

Measurement of Blood Pressure

Until the 1990s, literature regarding measurement of blood pressure in pregnancy was quite confusing. There was no unanimity regarding the preferable posture for evaluating the subject (i.e., lateral recumbency versus quiet sitting), and most important which Korotkoff sound, K4 (muffling) or K5 (disappearance of sounds), correctly measured diastolic pressure in pregnant women.[121] The debate over posture related to observations suggesting blood pressure increases in some women destined to develop preeclampsia when they change from a lateral to a supine posture (once called the "roll over" phenomenon),[122] many noting that the lowest blood pressures were obtained when pregnant women were positioned on their sides. It is now appreciated lower values with the woman in lateral recumbency merely reflect the difference in hydrostatic pressure when the cuff is positioned substantially above the left ventricle.[123] In this respect, the "roll over" test once utilized to predict preeclampsia has been discredited, though one group has claimed that the differences in blood pressure when measurements in lateral and supine recumbency are compared are more than just a postural phenomenon.[124]

The controversy concerning whether K4 or K5 represented diastolic pressure stemmed from a belief that the hyperdynamic circulation of pregnant women *frequently* led to large blood pressure gaps between the appearance of each sound, with the latter often approaching zero. For example in the early 1990s, groups such as the World Health Organization and the British Hypertension Society defined diastolic blood pressure in pregnant women as K4, while the National High Blood Pressure Education Program (NHBPEP) recommended K5.[125–127] One can imagine the confusion this causes in reviewing publications, many of whose authors neglected to state which sound was actually used. The newly formed ISSHP became one forum for this debate and of course new research ensued. We now know that large differences between the two sounds are very infrequent, that K5 more closely approximates true (intraarterial) diastolic levels, and that K4 is often unreliable. It also appears that even where and when K4 was designated the diastolic level the majority of health providers continued to measure K5, the putative descriptor of diastolic blood pressure in nonpregnant populations worldwide.[128–130] Current recommendations are now universal; that is, blood pressure should be measured in pregnant women in a manner similar to that in nonpregnant populations: the patient should be seated and rest for about 5 minutes, and the cuff over the arm should be at the level of the heart. There is also periodic interest in the role of home (including 24h) monitoring and as mercury-based blood pressure monitors are being phased out the literature to validate various automatic measuring apparatus for use in pregnant women is growing.

Concerning classification: in 1969 Rippman stated that a major detriment to understanding hypertension in pregnancy was the multiple conflicting terminologies, a fact underscored in Chesley's original chapter as well.[1,116] In this respect perusing the literature, at the end of the millennium one encountered terms such as toxemia, pregnancy-induced hypertension, pregnancy-associated hypertension, preeclampsia, and preeclamptic toxemia, EPH gestosis, etc. This plethora of terminology can be confusing, especially when the same term, e.g., pregnancy-induced hypertension (PIH), is defined differently by various authors. (In this text we adhere to the classification schema in the 2013 American College of Obstetrics and Gynecology Task Force report as outlined below, which is actually quite similar to the NHBPEP schema cited in the previous two editions).

Further complicating classification were questions regarding the definition of the blood pressure level that defined hypertension, including whether to consider absolute blood pressures or incremental changes. Another dilemma was whether or not it was necessary to include proteinuria in defining the pregnancy-specific disorder preeclampsia.[127,131–136] As of 2014, however, virtually all guidelines include criteria where preeclampsia can be diagnosed without proteinuria, as discussed below.

The primary goal of a successful classification schema is to differentiate a pregnancy-specific disorder associated with increased fetal and maternal risk from other more "benign" preexisting or gestational forms of hypertension. Why so many and at times discrepant classifications? First, signs and symptoms selected to identify preeclampsia were occasionally chosen for convenience rather than relevance to the pathophysiology or outcome. Second, the values used to discriminate normal from abnormal were selected more or less arbitrarily, and remain to be validated. These include cut-off levels for systolic (≥140mm Hg) and diastolic (≥90mm Hg) blood pressure (probably too high) as well as criteria for determining abnormal proteinuria (especially qualitative vs. quantitative criteria). Perhaps the most important confounder is the purpose for which the classification was developed, that is whether sensitivity or specificity was the most important criterion of the classification scheme being developed. For clinical care sensitivity was important, while for research specificity needed to prevail.

Regarding classification schema using absolute levels vs. those that permitted incremental rises in blood pressure one notes that a pathophysiologic approach would support incremental rises because the predominant pathological change is the appearance in late pregnancy of vasoconstriction. However, an argument against incremental blood pressure is that it is too nonspecific, and will lead to substantial over-labeling of normal gravidas as gestational hypertensives and/or preeclamptics. The problem of specificity and sensitivity becomes even more complicated when proteinuria is omitted as a diagnostic requirement in the classification of preeclampsia. For instance, the National High Blood Pressure Education Program (NHBPEP) Working Group's updated report[136] and the 2002 American College of Obstetricians and Gynecologists (ACOG) practice bulletin[137] classified preeclampsia by but two criteria, hypertension and proteinuria. Contrarily in the older literature (reviewed in[1]) one often finds *de novo* hypertension alone in late pregnancy defined as *mild* preeclampsia, while a 2001 position paper published by the Australasian Society for the Study of Hypertension in Pregnancy[138] recommended that when faced with *de novo* hypertension without proteinuria after mid pregnancy the presence of developing renal insufficiency, as well as a variety of neurologic, hematologic, or hepatic system signs or laboratory abnormalities, define preeclampsia. The 2013 report of ACOG's multidisciplinary task force (see below) has incorporated some of these criteria and gives definitions that permit diagnosing non-proteinuric preeclampsia.

One important conclusion from the debate on the place of proteinuria when defining the disorder is as follows: for clinical care alone we believe that it is appropriate to over-diagnose preeclampsia, as such a policy ensures proper surveillance. For research, however, classification schema must be stricter.[138] Leon Chesley continually underscored how our understanding of preeclampsia has been both compromised and impeded by the use of broad clinical criteria, such as some described above for research purposes.[139] He continually advocated rigorous diagnostic criteria which included *de novo* hypertension, proteinuria, and hyperuricemia in nulliparous women who have a normal medical history or preferably with all of the abnormalities returning to normal postpartum.[140]

We note here that investigation of factors other than just blood pressure measurements and laboratory studies may be required, if future research is to increase our understanding of preeclampsia. One area needing further exploration is consideration that subtypes of preeclampsia may exist, as through 2014 certain subtypes have rarely been investigated separately. In this respect a disorder that we currently diagnose clinically by such nonspecific findings as gestational hypertension and proteinuria (and hyperuricemia) may likely include more than one condition.

The philosophy espoused above is more than just semantics. How complete would be our understanding of diabetes if Types 1 and 2 diabetes were considered but one disease? Currently we have no pathogenic reasons to divide preeclampsia into subsets, but epidemiological information clearly indicates that early-onset preeclampsia, for instance, identifies subjects in whom the disease recurs more frequently, and who have increased cardiovascular morbidity later in life when compared to those whose disease was at term. Years ago Chesley (see below) urged the study of only primiparous subjects. Perhaps that is too restrictive but it is clearly wise and in keeping with the spirit of his recommendation that primiparous and multiparous preeclampsia and early- and late-onset preeclampsia be at least looked at separately in studies of the disorder.

Finally, in the second edition we reviewed the 1990 recommendations of the NHBPEP Working Group,[127] in which three of this text's editors participated, and its 2000 update was incorporated into edition 3.[136] As no further NHBPEP working groups in pregnancy had been convened in the first decade of this millennium (we all know how federal research funds wax and wane!), ACOG's president, James Martin, 2010–12, convened a multidisciplinary task force (two of the editors participated; one, JMR, its chair), whose aim amongst others was to update the NHBPEP report, its recommendations summarized below. However, this chapter's first author has a "lament."

The 1990 NHBPEP report came about during a period when bedlam and acrimony reigned concerning preeclampsia management. There were a myriad of different national guidelines and in the Untied States, at least, internists and obstetricians might meet at the bedside and view each other as species of baboons. Aware of this, Claude Lenfant, then head of the NIH's National Heart Blood and Lung Institute, convened a working group comprising equal numbers of internists and obstetricians (and even a woman!) that under the auspices of NHBPEP produced guidelines endorsed by 43 national organizations and several international advisors. It is fair to say that for about two decades the report became a standard endorsed and mimicked by many other national societies including ACOG, and was also the focus of guidelines in other nations. However, this is a history chapter, and to paraphrase an old cliché, history has repeated itself. The first 12 years of this millennium have seen numerous new guidelines, most without the care and endorsement that went into NHBPEP reports, some by organizations with under a thousand members, and of course there are many disagreements (for example refs. 141–149). The ACOG hypertension task force that produced the guideline recommendations we summarize below sought a multidisciplinary group, similar to the NHBPEP working group report, and included a representative from NICHD, but it is still a national guideline, and without the endorsements that marked the ascendancy of previous NHBPEP publications. Alas those cycles in history. Here we end history (this Editors' Update) and summarize the ACOG document's definitions and classification schema, while noting that newer subdivisions or reclassifications of preeclampsia (such as early and late, by severity or markers) are already being debated.[150]

Classifications and Definitions: The American College of Obstetricians and Gynecologists Hypertension in Pregnancy Hypertension in Pregnancy Task Force (2013 report)

The ACOG Task Force document[148] that based it recommendations in terms of evidence recommended by the Grade system[151] maintained the NHBPEP's Working Group's clinical classification schema's four categories: chronic hypertension, preeclampsia–eclampsia, preeclampsia superimposed upon chronic hypertension, and transient hypertension.

Preeclampsia–Eclampsia

The diagnosis of preeclampsia now includes in addition to the *de novo* appearance of both hypertension and proteinuria after mid pregnancy the appearance of high blood pressure in association with thrombocytopenia, impaired liver function, renal insufficiency, pulmonary edema, cerebral or visual disturbances when these latter findings are new in onset. *We wish to underscore here that, while previously only suggested, the ACOG task force now stressed that a diagnosis of preeclampsia can be made in the absence of protieinuria.* Also, though the onset is defined as appearing after mid-pregnancy, it rarely occurs earlier, usually with trophoblastic diseases such as hydatidiform mole. Hypertension in pregnancy is defined as a blood pressure ≥140mm Hg systolic or ≥90mm Hg (K5) diastolic, and proteinuria is defined as the urinary excretion ≥0.3g in a 24-hour collection or a protein to creatinine ratio in a single voided specimen ≥0.3. It was agreed that when quantitative evaluations were unavailable qualitative testing could be used, noting that a 24-hour measurement ≥0.3g frequently did not correlate with 30mg/dL (1 + dipstick) or greater in a random urine determination, suggesting its use be screening followed by quantitative confirmation.

Also as emphasized in previous editions, neither edema with hypertension in the absence of proteinuria, nor increases of ≥ 15mm Hg or ≥ 30mm Hg in diastolic and systolic pressures respectively to values below 140/90mm Hg are to be considered diagnostic criteria. The reasons these older criteria had been abandoned related to the substantial presence of edema in normal gestation and a failure to note increases in adverse events when blood pressures remain below 140/90mm Hg.[152] However, your editors remain uncomfortable here, advising, as did the 2000 NHBPEP working group report, that it is their collective opinion *that women who have a rise of 30mm Hg systolic or 15mm Hg diastolic blood pressure warrant close observation, especially if proteinuria and hyperuricemia (uric acid greater than or equal to 6mg/dL) are also present.*

Preeclampsia occurs as a spectrum and its severity can increase rapidly. The task force thus replaced "mild" and "severe" as either preeclampsia without or with severe features as the previous term "mild" appears to have produced a false sense of security in some caregivers that led to less than optimal care. The management of preeclampsia is detailed in Chapter 20, including those most severe signs that suggest imminent delivery, but we note here that the magnitude of the proteinuria, which can be altered by circumstances independent of disease progression,[153] is no longer included as a sign of severity.

Eclampsia is the occurrence in women with preeclampsia of seizures that cannot be attributed to other causes. This complication is discussed further in Chapter 20.

Chronic Hypertension

Defined as hypertension known or detected prior to conception or diagnosed before the 20th gestational week.

Preeclampsia Superimposed upon Chronic Hypertension

Preeclampsia can occur in women with preexistent hypertension of both the essential and secondary variety, the incidence exceeding 20%.[154] Distinguishing superimposed preeclampsia from worsening chronic hypertension may be difficult, and again it is prudent to over-diagnose this complication.

Gestational Hypertension

This describes women who develop *de novo* hypertension without proteinuria. A substantial number of these women later develop proteinuria and are reclassified as preeclamptics. (Not discussed in the ACOG guidelines, but noted in the NHBEP document we discussed in the previous edition is that failure of blood pressure to normalize after pregnancy means the woman was a chronic hypertensive, the diagnosis missed when pressures were normalized by the physiologic fall in blood pressure during the initial trimester. Your editors suggest reclassification when this happens.)

Finally, the spectrum of how preeclampsia manifests clinically is detailed in Chapter 2 and management of all hypertensive complications of pregnancy in Chapter 20. In adition the Executive Summary of the ACOG Hypertension task report is printed in the appendix.

DENOUEMENT

This introductory chapter would be incomplete without a commentary by Leon Chesley made two decades ago at the banquet of the 1975 Workshop on Hypertension in Pregnancy[2] and reprinted below. Bear it in mind as you read the ensuing chapters.

False Steps in the Study of Preeclampsia

A century ago, an American humorist, Josh Billings, made a good generalization that applies to our views of preeclampsia–eclampsia when he said: "The trouble with people isn't that they don't know, but that so much of what they know ain't so." In talking about false starts in the study of eclampsia, I am acutely aware that much of what we believe today will be added to the list.

An observation by Mauriceau,[30] made in the 17th century, has stood the test of time: preeclampsia–eclampsia is predominantly a disease of primigravidas. It does occur in multiparas, but when it does, there is usually some predisposing factor such as multiple gestation, diabetes, or preexisting primary or secondary hypertension. Hinselmann's analysis of 6498 cases from the literature showed that 74% of cases occurred in primigravidas, although they contributed only a quarter to a third of all pregnancies.[155] He calculated that primigravidas are six times as likely to develop eclampsia as are multiparas.

McCartney has made an outstanding contribution in his electron microscopic studies of renal biopsies.[156] He carefully selected 152 multiparas with known chronic hypertension, all of whom fulfilled the criteria of the American Committee on Maternal Welfare for the diagnosis of superimposed preeclampsia. Renal biopsies of only five of the women showed the pure "preeclamptic lesion" of glomerular capillary endotheliosis, while in another 16 the lesion was superimposed upon renal arteriolar sclerosis. If the definitive diagnosis were based upon anatomic findings, the clinical diagnosis would be erroneous in 86% of the cases.

Admittedly, it has not been proven that every woman with preeclampsia has the typical lesion, and the uncertainties of clinical diagnosis are such that a one-to-one ratio is not likely to be established. Nearly half of the multiparas had lesions of renal disease, usually nephrosclerosis, and it seems that preeclampsia was grossly overdiagnosed clinically.

There are diagnostic difficulties in primigravidas as well. McCartney selected 62 women who were apparently normal in mid pregnancy and who developed hypertension, proteinuria, and edema in the third trimester. The clinical diagnosis was preeclampsia. Only 71%, however, had the renal lesion of preeclampsia. That finding might mean merely that the lesion is not always present in preeclampsia, except that 26% had lesions of chronic renal disease and 13 of these 16 women had chronic glomerulonephritis that had been silent during much of pregnancy. Three had no renal lesion at all, and they may have had latent essential hypertension revealed by pregnancy.

The finding of lesions of chronic renal disease in women who were apparently normal in mid pregnancy is consistent with certain older clinical observations. Theobald found the prevalence of proteinuria to be lower in mid pregnancy than in nonpregnant women of the same ages as the pregnant subjects.[157] In a later paper, he suggested that both proteinuria and hypertension present before conception often disappear during mid pregnancy and reappear in the third trimester.[158] Thus, apparently normal women may have a seemingly acute onset of hypertension and proteinuria in the third trimester, and the erroneous diagnosis of preeclampsia is likely to be made. Reid and Teel, in studying women with known hypertension, observed that a significant proportion of them had normal blood pressures during the second trimester; characteristically, the pressure rose again early in the third trimester.[159]

Neither Theobald nor Reid and Teel specified the frequency of such changes, and when I first read their papers I naively dismissed their observations as curiosities of no statistical significance in the differential diagnosis of the hypertensive disorders in pregnancy.[160] In 1946, however, John Annitto and I analyzed 301 pregnancies of women with known hypertension and found that in 39% the blood pressures were significantly decreased until the third trimester.[161] Many had normal pressures during mid pregnancy. One extreme case was that of a woman whom we saw during five pregnancies and many times between pregnancies. She was always severely hypertensive while not pregnant, and always normotensive during pregnancy. Near term, she might have a pressure such as 140/84 mm Hg in one pregnancy and 130/90 mm Hg in another. In short, we had her in the hospital at bedrest and under sedation for 8 days at about 18 months after her 4th pregnancy. Her lowest diastolic pressure was 150 mm Hg, and it ranged up to 180, with systolic pressures always greater than 250 mm Hg. Nearly a year later, she registered in the antepartum clinic in the 4th month of pregnancy; her blood pressure was 110/60 mm Hg, and she remained normotensive throughout pregnancy, as she had in her earlier gestations. She deviated from the usual pattern in not becoming hypertensive early in the 3rd trimester.

In brief, hypertension of apparently acute onset seldom justifies the diagnosis of preeclampsia in a multipara, who is much more likely to have frank essential hypertension or renal disease that abated during mid pregnancy, or latent essential hypertension that is brought to light and sometimes peculiarly altered by gestation. I have developed this theme at some length because now that eclampsia has become rare, most studies of the disease are made in women bearing the diagnosis of preeclampsia. Any study of preeclampsia that includes multiparas will lead to erroneous conclusions and I have done more than my share to confuse the field. The Pennsylvania Dutch have a saying that describes my situation: "Too soon Oldt, too late Schmart."

The relation between preeclampsia–eclampsia and later chronic hypertension has been a source of contention and confusion for many years. In 1940, our group at the Margaret Hague Maternity Hospital reexamined most of the women who had had hypertensive disorders in their pregnancies during 1935 and 1936. The prevalence of hypertension among the 319 thought to have had preeclampsia was about 36%, and there seemed to be an almost linear relation between the duration of the supposed preeclampsia and the frequency of hypertension at follow-up. A similar relation had been described by Harris,[162] Gibberd,[163] Peckham[164] and others, and we erroneously accepted their conclusion that prolonged preeclampsia had caused chronic hypertension.

When Annitto and I completed our study of pregnancies in women with documented essential hypertension, I looked again at the data in the follow-up study of 1940. Multiparas constituted about half of the patients who had satisfied the criteria for the diagnosis of preeclampsia as laid down by the American Committee on Maternal Welfare. Many of them had had hypertension in earlier pregnancies, and they undoubtedly had latent or frank hypertensive disease, rather than preeclampsia.

The woman with chronic hypertension whose blood pressure falls to normal levels in mid pregnancy and rises again early in the 3rd trimester is the patient who is likely to be carried for several weeks with supposed preeclampsia.

Termination of the pregnancy when the pressure rises would jeopardize the fetus, and so the physician temporizes. The clinical picture does not worsen as weeks go by, because all the patient has is hypertension. When she is found to have hypertension at follow-up, she has it because she had it before pregnancy and *not* because she was carried for several weeks with preeclampsia.

We dropped our follow-up studies of women with nonconvulsive hypertensive disorders and retracted our conclusion that prolonged preeclampsia causes chronic hypertension. In an effort to ascertain the relation, if any, between acute gestational hypertension and chronic hypertension, we have reexamined at intervals of 6 or 7 years nearly all of the 270 women who survived eclampsia at the Margaret Hague Maternity Hospital from 1931 through 1951. Only one of these patients has had no follow-up, and in 1974 we traced all but three. The diagnosis of eclampsia is not always correct, but it is far more reliable than that of preeclampsia. Moreover, 75% of the patients were primigravidas, and the multiparas have been analyzed separately. The greatly different prognoses for primiparous and multiparous eclamptic women are instructive.

In brief, the prevalence of chronic hypertension among women who had eclampsia as primiparas is not different from that in unselected women of the same ages. Eclampsia, then, is not a sign of latent hypertension, as some internists had believed, and eclampsia, whatever the duration of the acute hypertensive phase, does not cause chronic hypertension. That conclusion has been drawn by Theobald,[165] Browne,[166] Adams and MacGillivray[167] and others in the British Isles and by Dieckmann,[75] Tillman,[168] Bryans[169] and others in the United States. Tillman's study is notable, for he accumulated a large group of women whose blood pressures were known before, during, and after pregnancies. He found that whether the pregnancies were normotensive or hypertensive they had no effect on blood pressure at follow-up.

Women who had eclampsia as multiparas have had a remarkably different prognosis. Nearly three times the expected number have died, and 80% of the remote deaths have resulted from the lethal consequences of hypertensive disease. The prevalence of chronic hypertension among the survivors is increased over that found in several series of unselected women matched for age. The high prevalence of hypertensive disease among women having eclampsia as multiparas is explained by their having had hypertension before pregnancy, which predisposed them to eclampsia.

Clearly, women who have eclampsia as multiparas are different from those who have eclampsia as primiparas, and the two groups must be analyzed separately. The inclusion of multiparas in studies of preeclampsia or eclampsia leads to erroneous conclusions, and, as a result, I now regard many of my own earlier publications as virtually worthless.

I should mention one paper that has received wide attention. Epstein,[170] an internist then at Yale, reexamined 48 women at about 15 years after hypertensive pregnancies thought to be preeclamptic. He found a higher prevalence of hypertension among them than in control women selected because their pregnancies had been normotensive, and he concluded that prolonged preeclampsia causes hypertension. One of his tables shows that 24 of the 48 women were multiparas at the time the pregnancies were studied. It seems probable that few of the women really had preeclampsia, and history has repeated itself. Moreover, the selection as controls of women who had escaped gestational hypertension constituted a strong bias, for it excluded many who would develop hypertension as they grew older. Many future hypertensive women manifest the diathesis by gestational hypertension, and such women have apparent but not "true" preeclampsia.

Many of the hypotheses about the cause of eclampsia cannot stand in light of the predominant occurrence of the disease in nulliparas. When I came into the field 40 years ago, the standard management of preeclampsia was restriction of dietary protein, and many obstetricians put all of their patients on low protein diets as prophylaxis against preeclampsia. It was a time-honored custom, reinforced by the 200-year-old obstetric belief that eclampsia is a disease of upper-class women and by the observation that the incidence of eclampsia decreased in some parts of Germany during World War 1. Germans are reputed to eat heavily and, according to the nutritionists, too much. During the war they might have eaten less, and the nutritionists seized on protein as the source of the eclamptic toxin because protein had long been anathema. An editorial in the *Journal of the American Medical Association* said "The conclusion seems inevitable that restriction of fat and meat tends to ward off eclampsia."[105] Ruiz-Contreras wrote that the incidence of proteinuria and eclampsia was far higher among his private patients than in the women whose pregnancies he managed in the charity wards.[104] He observed a rise in the incidence of proteinuria and eclampsia in ward patients when their nutrition improved, as the intake of dietary protein increased among the masses.

The restriction of protein seemed rational a century ago, when eclampsia was thought to be a form of uremia. When that hypothesis was disproved, others took its place. One proposed incomplete digestion of proteins with the production of toxic "split products." That led to the Dublin or Rotunda method of management, in which gastric lavage, high colonic irrigation, and purgation were aimed at washing out all toxins within reach. In addition, an initial phlebotomy removed some of the circulating toxin; a second bleeding 2 hours later was aimed at getting toxins that had been mobilized into the bloodstream in the interim. Another hypothesis was that the amino acids resulting from the digestion of protein were decarboxylated but not deaminated, thus resulting in toxic amines.

The decreased incidence of eclampsia in some parts of Germany during the war was explained in various ways. Some thought that increased physical activity in pregnant working women was the answer. Others postulated toxins or antigens in semen as the cause of eclampsia; with the men away at the front, the women supposedly were not realizing their wonted (and perhaps wanted) chronic exposure to poisoning. The hypothesis relating eclampsia to dietary protein prevailed, with effects that persist to the present day.

Germany suffered no serious shortage of food during the war; however, the incidence of eclampsia began to decline toward the end of the first year of the war. The severe shortages occurred during the first year following the armistice, while the Allies continued the blockade, and the incidence of eclampsia then rose. The converse of the German experience occurred during the siege of Madrid, about 30 years later. The incidence of eclampsia rose strikingly, only to subside in the following year.

Lehmann[171] obtained the statistics for all of Baden, from the Landsamt in Karlsruhe, and found that the proportionate representation of primiparous deliveries had decreased in proportion to the reduction in incidence of eclampsia during the war. Some months after the armistice, the proportion of primiparous deliveries rose dramatically, with a parallel increase in the incidence of eclampsia. During the war, many of the men remaining home were fathers of large families, and their wives continued to have pregnancies as multiparas. After the young men came home there were, in due course, parallel surges in primiparous deliveries and eclampsia. Hinselmann,[142] however, did note a decrease in the incidence of eclampsia in primigravidas, but Lehmann's study explains most of the fluctuation.

Diaz del Castillo found that in Madrid there had been a surge in primiparous deliveries in proportion to the high incidence of eclampsia.[172] During the siege, the boys had been cooped up with the nubile nulliparas, and in due time there was a wave of primiparous deliveries.

Fifty years ago eclampsia was common in New York City, and it had a seasonal incidence, with the peak in March and April. The followers of Hippocrates attributed it to the variable, unsettled weather and the nutritionists to the lack of fresh vegetables and fruit. In Denmark in the 1920s, there were two seasonal peaks, 6 months apart. Many young women worked as domestic servants under contracts that expired the same day throughout the nation. Lehmann wrote that at the expiration of the contracts, many of the women quit and married.[171] The peak incidence of primiparous deliveries and of eclampsia occurred together, 9 to 10 months later. To return to New York, March and April fall at just the right interval after June weddings.

It has been known for more than two centuries that the incidence of eclampsia is higher in city women than in their country cousins. The moralistic nutritionists have explained the phenomenon by saying that the effete city woman lounges around her apartment drinking cocktails, smoking cigarettes, and eating snacks of junk food, for which she pays the penalty of an increased susceptibility to eclampsia. The country wife, supposedly, is out in the fresh air and sunshine, does physical labor such as shoveling hay, and eats a wholesome diet. Again, Lehmann found that the incidence of eclampsia follows the incidence of primiparous deliveries.[171] In those days, the average family size in the cities was smaller than in the countryside, where the farmers begat their future farm hands. The proportion of primigravidous deliveries accounted for nearly all of the difference, and admittance of country women to hospitals in the city accounted for the rest.

The nutritionists are now taking a different approach and attributing eclampsia to a dietary deficiency of protein, or of this or that vitamin or one or another of various minerals. As one example, Siddall,[173] pointed out that the highest incidence of eclampsia in the nation was in the southeastern block of states, the region in which almost all cases of pellagra occurred. He concluded that deficiency of thiamin is the cause of eclampsia. H. L. Mencken once assessed the cultural level of each of the then 48 states, using criteria such as the prevalence of literacy, high school and college education, libraries, musical organizations, etc. The states came out in nearly the same order by each criterion, and the final evaluation put Massachusetts, Connecticut, and New York at the top, with Mississippi, Arkansas, and Alabama at the bottom. Characteristically, Mencken put another column into the table showing the prevalence of membership in churches. There was a good inverse correlation – the higher the membership in churches, the lower the cultural level. About 25 years ago, some joker put in a third column showing eclamptic deaths, and from the correlation concluded that the cause of eclampsia must be church membership. This conclusion seems as valid as the one drawn by the nutritionists.

The difficulty with the nutritional hypothesis as to the cause of eclampsia is that the disease is predominantly one of the first viable pregnancy. Impoverished and malnourished women are the very ones who have pregnancies in rapid succession. The fetal drain thus imposed upon whatever poor reserves they begin with must deplete them progressively. If malnutrition were the cause of eclampsia, its frequency should rise with parity. It does not.

We have alluded to the high incidence of eclamptic deaths in the southeastern states, which brings up the question of the reliability of statistics and their interpretation. We know nothing of the incidence of preeclampsia anywhere but in our own hospitals, and even there, that knowledge depends upon who makes the diagnoses. In studying the familial factor in eclampsia, I have seen the delivery charts of daughters, daughters-in-law, and granddaughters of eclamptic women in well over 100 different hospitals. Considering only first pregnancies, where the risk of preeclampsia is high, no blood pressure was recorded in 23 charts, and there was but a single recording in 27, 7 of which were from 140/90 to 178/110 mm Hg. Twenty-four percent of the charts showed no urinalysis, and 31% carried no notation about the presence or absence of edema. In nine charts, there was no recording of blood pressure, no urinalysis, and nothing about edema. In such hospitals, the incidence of diagnosed preeclampsia must be low. Moreover, many charts bearing the diagnosis of normal pregnancy showed clear evidence of mild or severe preeclampsia and even, in one case, of eclampsia – with nine convulsions, gross proteinuria, anasarca, and acute hypertension in the range of 150–174/100–120 mm Hg; this patient was on the "danger list" for 24 hours. Is that a normal pregnancy?

Even if hospital statistics were accurate, they would not reflect the incidence of preeclampsia in a given community. In many regions, including the southeastern states, many women have no prenatal care and are delivered at home unless some serious complication occurs. I (Leon Chesley) used to visit Frederick Zuspan when he was at the Medical College of Georgia, and I once saw seven eclamptic patients on the wards; not one was from Augusta. They had been brought in from as much as 200 miles away because of convulsions.

The other source of geographic statistics is deaths from eclampsia. These statistics are not any better. What they reflect chiefly is the quality, availability, and utilization of prenatal care, for when preeclampsia is detected early, a progression to severe preeclampsia to eclampsia to death usually can be prevented.

In 1831, Ryan wrote that women particularly susceptible to eclampsia are "those who are in labor for the first time and who are illicitly in this condition."[43] The nutritionists have suggested a poor diet in unmarried pregnant women as the explanation for their predisposition to eclampsia. The psychosomaticists have suggested mental stress. Lehmann confirmed the difference in Danish women, but when he analyzed by parity he found identical incidences in married and unmarried primigravidas.[171] The incidence in multiparas was much lower but the same in married and unmarried women. Except among slum-dwelling relief clients, most illegitimate pregnancies in the past have occurred in young nulliparas. The current wave of sexual liberation may change that, although access to abortion is an ameliorative factor.

As previously mentioned, obstetricians believed for two centuries that eclampsia was a disease of upper-class women. In 1768, Denman wrote that frail, delicate, educated, and highly intelligent city women who cultivated music were at particular risk.[174] Perhaps the belief arose because obstetricians who served upper-class women wrote papers, whereas midwives, who served the lower classes, seldom published. It was not until about 1925 that Kosmak[175] suggested that eclampsia might be equally common in the impoverished and the wealthy, and today the widespread belief is that it is far more common in the poor than in private patients. Reverting for a moment to the undiagnosed case of eclampsia, it was the patient's private physician who wrote on the diagnosis sheet, "Normal pregnancy, normal labor, normal delivery, normal puerperium."

In Aberdeen, Scotland, the medical school has access to data for nearly every delivery in the city, and it appears from Nelson's analysis that there is little difference in the incidence of preeclampsia among the five social classes.[115] A slight increase was noted in social class 3, and further analysis has shown an increase in class 3-C, which includes such women as the wives of streetcar conductors.

Another popular belief is that black American women are more susceptible to preeclampsia–eclampsia than are white women, and the belief has been linked to social class as well as to race. Mengert[176] reviewed his wide experience in three different major centers and concluded that there is no racial difference; I agree. In the data from the Margaret Hague Maternity Hospital, which I mentioned above, 8% of the eclamptic women were black, and 8% of all deliveries were of black women. At the King's County Hospital, where black women account for 78% of obstetric patients, the incidence of preeclampsia is identical in black women and in white women. The prevalence of chronic hypertension among black women is nearly three times that of white patients, which accords with several epidemiologic studies in nonpregnant populations. Erroneous differential diagnosis has been responsible for the idea that black women are more susceptible to preeclampsia.

Old-time obstetricians believed that certain bodily builds predispose to eclampsia, but they could never agree as to what the characteristics were. Denman said frail, delicate women, but others thought that it was strong, plethoric subjects.[174] Systematic studies were long delayed, and, when they were made in the late 1920s, they unfortunately were conducted with patients thought to have preeclampsia and included large proportions of multiparas. Another flurry of papers appeared in the early 1960s, and they are not better. The conclusion was that short, squat, obese women are more susceptible than long, lean, lank ones. For the past 40 years I (Leon Chesley) have been seeing formerly eclamptic women in follow-up studies, and there are few obese women among them. In several indices of bodily build, the data for these women fall on normal distribution curves for unselected women. The bodily build alleged to predispose to eclampsia is that often associated with essential hypertension, and the inclusion of multiparas in studies of preeclampsia again seems to have led to an erroneous conclusion.

I have alluded to the German experience during World War 1. The nutritionists' misinterpretation of the role of dietary protein is now behind us, but for half a century virtually all American obstetricians have tried to limit the weight gained in pregnancy to 20, 15, or even 12 lbs as prophylaxis against preeclampsia. That practice stems from the same source as the restriction of dietary protein. In 1923, Carl Henry Davis was the first to advocate the measure.[177] He wrote that the gain should be limited to 20 lbs and that overweight women should gain less or should even lose weight. "The marked decrease of eclampsia in Central Europe during the period of war rationing led us to realize that eclamptics usually are women who have gained weight excessively during pregnancy."

In the same year, Calvin R. Hannah estimated the "reproductive weight" as 12 lbs, which he derived by adding together the average weights of fetus, placenta, and amniotic fluid.[178] He wrote: "Patients whose weight is above standard and who continue to gain over the reproductive weight of 12 lbs, manifest preeclamptic symptoms."[1]

A flood of papers appeared describing the gain in weight during normal and hypertensive pregnancies, usually with no separation of primigravidas from multiparas. Each author ascertained the average gain in his normal subjects, which was 24 lbs when all of the data were pooled. In a statistical maneuver that defies comprehension, each author then set his average as the upper limit of normal, thereby denying the normality of half of his own normal patients. From the pooled data of many papers, it appears that two-thirds of normal women had gained between 13 and 35 lbs, with one-sixth having gained more than 35 lbs, despite frequent attempts to limit the gain.

Most authors found the combined incidence and prevalence of hypertensive disorders to increase somewhat with greater gains in weight. When one evaluates the published data, however, it becomes evident that 90% of women who gained more than 30 lbs did not develop any sort of hypertension. Conversely, 90% of women with what was called preeclampsia had gained less than 30 lbs, and 60% had gained less than 22 pounds. In relation to preeclampsia, the total gain during pregnancy probably means nothing, unless a large component of the gain is fluid, which is detected better by high rates of gain in the latter half of gestation.

The original rationale for the limitation of weight gained in pregnancy was fallacious, and there is no good evidence that it prevents preeclampsia. It may even be harmful. Several recent studies have shown that the average birth weights of the infants are lower when maternal gain is restricted. The Netherlands, the Scandinavian countries, and some other countries have lower rates of perinatal mortality than does the United States, and birth weights tend to be higher in the countries with superior salvage of infants. Some extremists suggest that our general practice of limiting maternal gain has increased the rate of prematurity and the birth of smaller weaklings at all stages of pregnancy.

For many years, obstetricians have been frightened by the appearance of edema in their patients, especially if it involves the hands and face, for they regard it as an early sign of preeclampsia. British studies have shown, however, that such edema usually is physiologic and occurs in a high percentage of pregnancies. Moreover, in the absence of proteinuric preeclampsia, infants born to women with edema of the hands or face weigh more at birth than do infants born to nonedematous women. It is not that the infants share the edema, for they do not lose weight excessively in the crib.

Tentatively, it appears that there may be two sorts of gestational edema – one is physiologic and good; the other may be the edema of preeclampsia. I do not know how to differentiate between them. The prevalence of edema is higher in preeclampsia than in normal pregnancy, and the edema is often more severe in preeclampsia.

Edema of the hands and face has long been accepted as an early sign of preeclampsia, but that may have been another of our false steps. The belief stems from retrospective studies of the charts of women with preeclampsia–eclampsia. One goes back in the record looking for premonitory signs, and in perhaps half the cases one finds that the patient's wedding ring was tight at, say, the 30th week of gestation. Aha! Edema of the fingers pointed to oncoming preeclampsia. But how many such studies have included reviews of the records of patients with normal pregnancies? None that I know of. The edema at 30 weeks gestation may have been physiologic and have had nothing to do with preeclampsia. Certainly Robertson,[179] in his prospective study of gestational edema, could find no relation between generalized edema and later preeclampsia.

Double-blind studies have shown that prevention or dissipation of gestational edema with saluretic agents has no effect upon the incidence of clinical preeclampsia. The control or prevention of a sign does not strike at the basic disorder and, in this case, we may not even be dealing with a sign. I consider the use of diuretics as another of our false steps and can enumerate eight contraindications to their use in pregnant women, especially in preeclampsia.

The use of diuretics brings pilocarpine to mind. A century ago, the induction of sweating was an integral part of the management of eclampsia. Tincture of jaborandi was widely used for this purpose, and when pilocarpine was identified as the active principle, the purified drug was used. The rather poor results were rationalized in this manner: the intense sweating had distilled off water, leaving the eclamptic toxin behind in higher concentration. The drug was given a controlled trial in Edinburgh and abandoned when the usual case mortality doubled from 30 or 35% to 67%.

Since the space allotted to me is limited, I shall close with a few remarks about the management of eclampsia. Zuspan and Ward, in discussing it, wrote, "The eclamptic patient has certainly tested the ingenuity of physicians throughout the centuries as she has been blistered, bled, purged, packed, lavaged, irrigated, punctured, starved, sedated, anesthetized, paralyzed, tranquilized, rendered hypotensive, drowned, been given diuretics, had mammectomy, been dehydrated, been forcibly delivered, and neglected."[76] The list could be extended considerably. Surgical procedures have included ventral suspension of the uterus, drainage of spinal fluid, cisternal puncture, trepanation, ureteral catheterization, implantation of the ureters in the colon, renal decapsulation, oophorectomy, and postpartum curettage. The medical treatments have been legion.

We should remember that each of the treatments was rational in the light of some hypothesis as to the cause or nature of eclampsia. That is more than I can say for

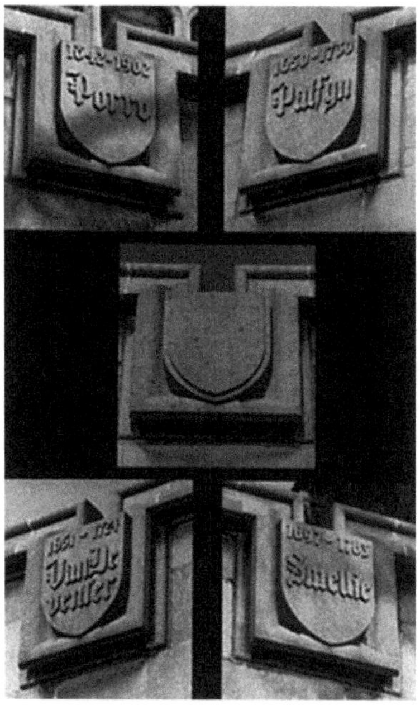

FIGURE 1.2 Placards honoring famous physicians who made major contributions to the field of obstetrics and gynecology adorn Chicago Lying-in Hospital. The empty plaque is reserved for the individual who discovers the cause and/or cure of preeclampsia (perhaps a contributor to this text!).

present day management, which is empiric, too often symptomatic, and in some respects, based upon imitative magic (Fig. 1.2).

References

1. Chesley LC. *Hypertensive Disorders in Pregnancy*. New York: *Appleton Century Crofts*; 1978:628.
2. Chesley LC. False steps in the study of preeclampsia. In: Lindheimer MD, Katz AI, Zuspan FP, eds. *Hypertension in Pregnancy*. New York: John Wiley & Sons; 1976:1–10.
3. von Siebold ECJ. *Versuch einer Geschichte der Geburtshülfe*, Vols I, II. Berlin: Enslin; 1839, 1845.
4. Knapp L. Bieträge zur Geschichte der Eklampsie. *Mtschr Geburtsh Gynaekol*. 1901;141:65–109.
5. Kossmann R. Zur Geschichte des Wortes "Eclampsie". *Mtschr Geburtsch Gynaekol*. 1901;14:288–290.
6. Fasbender H. *Geschichte der Geburtshülfe*. Jena: Fisher; 1906:777–804.
7. Fischer I. Geschichte der Gynäkologie Halban J. Sietz L, editors. *Biologie und Pathologie des Wiebes*, Vol. 1. Berlin: Urban und Schwatzenberg; 1924.
8. Bernhart F. Geschichte, Wesen und Behandlung der Eklampsie. *Wien Klin Wchschr*. 1939;52(1003–1009):1036–1043.
9. Menascha I. Die Geburtshilfe bei den alten Ägyptern. *Arch Gynaekol*. 1927;131:425–461.
10. Griffith FL. *The Petrie Papyri. Hieratic Papyri from Kahun and Gurob*, Vol. I. London: Quaritch; 1898:11.
11. Griffith FL. A medical papyrus from Egypt. *Br Med J*. 1893;1:1172–1174.
12. Whitney WD (Trans.). *Atharva-Veda Såmhitā*. Cambridge, MA: Harvard University; 1905, vols 7, 8, Harvard Oriental Series, Bk VIII, Sec 6, vol. 8:493–498.
13. Suśrutas. Áyurvédas. *Id est Medicinae Systema a Venerabili d'Hanvantare Demonstratum a Susruta Discipulo Compositum*, Bk II, Sec VIII. Translated and annotated by F Hessler, Erlange Enke; 1853:188.
14. Wang DM *Schou Schen Hsiau* [Lo JH, Trans. (into German)]. Abhandlung Med Facultät Sun Yatsen Universität; 1930;2:19–126.
15. Hippocrates. *The Genuine Works of Hippocrates* [Adams F, Trans.]. London: The Sydenham Society; 1849, Vol. 2: 715, 743, 758, 766.
16. Hippocrates. *The Medical Works of Hippocrates* [Chadwick J, Mann WN, Trans.]. Oxford: Blackwell Scientific; 1950.
17. Galeni C.Kühn DCG, editor. *Opera Omnia*, Vol. 17B. Leipzig: Car Cnoblochii; 1829:821. pt II:652.
18. Celsus AC. *On Medicine* [Lee A, Trans.]. London: Cox; 1831: 99, 347.
19. Aetios of Amida. *The Gynecology and Obstetrics of the VIth century* AD [Ricci JV, Trans.]. from the Latin edition of Cornarius, 1542. Philadelphia: Blakeston; 1950: 27, 31, 32.
20. Rösslin E. *Der Swangern Frawen und Hebammê Rossegarte . (Facsimile, Eucharius Rösslin's "Rosengarten" gendruckt im Jahre 1513. Beigleit-Text von Gustav Klein.)*. Munich: Kuhn; 1910:67.
21. Paulus Aegineta. *The Seven Books* [Adams F, Trans.]. London: Sydenham Society; 1844, Vol. I:4, 5; Vol. II:387.
22. Ballantyne JW. The Byrth of Mankinde. *J Obstet Gynaecol Br Emp*. 1906;10:297–325. 1907;12:175–104, 255–274.
23. Raynalde T. *The Byrth of Mankinde, Otherwise Named the Woman's Booke*. London: Adams; 1613.
24. Gabelchoverus *Artzneybuch, darninnen vast für alle de menschlichen Leibs, anlingen und Gebrechen, ausserlesene und bewehrte Artzneyen usw*. Tübingen: Gruppenbach; 1596.
25. Chaussier F. *Considérations sur les convulsions qui attaquent les femmes enceintes*. 2nd ed. Paris: Compére Juene; 1824.
26. Burton J. *An Essay Towards a Complete New System of Midwifery, Theoretical and Practical*. London: Hodges.
27. Exton B. *A New and General System of Midwifery*. 3rd ed. London: Owen (no date given, first edition published, 1751).
28. Mauriceau F. *Des Maladies des Femmes Grosses et Accouchées avec la Bonne et Veritable Méthode, etc*. Paris: Cercle du Livre précieux; 1668.
29. Mauriceau F. *The Accomplisht Midwife, Treating of the Diseases of Women with Child, and in Childbed, etc* [Chamberlen H, Trans.]. London: Darby; 1673.
30. Mauriceau F. *Traité des Maladies des Femmes Grosses, et celles Qui Sont Accouchées, Einseignant la Bonne et Veritable Méthode pour Bien Aider, etc*. Paris: d'Houry; 1694. bk II, [Chapter 28].
31. Hirsch A. *Biographische Lexikon der Hervorragender Aerzte*. Wien und Liepzig: Urban und Schwatzenberg; 1887.
32. de Sauvages F. *Pathologia methodical, seu de Cognoscendis morbis*. Monspelli: Martel; 1739:120.
33. de Sauvages FBS. Pathologia methodical, seu de Cognoscendis morbis. 3rd ed. Lieden: Fratum de Tournes; 1759:286.
34. Hippocrates. *Hippokrates Werke aus dem Griechischen übersetz und mit Erläuterungen*, trans into German by JFC Grimm. Glogau: Prausnitz; 1838.

35. de Gorris J. *Definitionem Medicarum, Libre XXIIII, Literis Graecis Distincti*. Francoforti: Wecheli; 1578.
36. Castelli B. *Castellus Renovatus: Hoc Est, Lexicon Medicum, Quondam à Barth*. Norembergae: JD Tauberi; 1662:484.
37. Castelli B. *Lexicon Medicum Graeco-Latinum*. Roterodami: Leers; 1651.
38. Blancardo S. *Lexicon Medicum Graeco-Latinum, in Quo Termini Totius Artis Medicae, Secundum Neotermicorum Placita Definiunter Vel Circumscribunter Graeca Item Vocabula ex Originibus Suis Deducunter Antehac*. Jena: Literis Müllerianis; 1683.
39. Vogel RA. *Definiones Generum Morbum*. Wittwe: Göttingham A: Vandenhöks; 1964.
40. Cullen G. *Synopsis Nosologiae Methodicae. In Usum Studiosorum. Part IV; Genera Morborum*. Edinburgh: Kincaid and Creech; 1771.
41. Sagar JBM. *Systema Morborum Symptomaticum*. Vienensis: Kraus; 1776:437.
42. Gutsch JG. *De Eclampsia Parturientium, Morbo Gravi Quidem Neque Adeo Funesta, Sectio Prior Pathologica*. Inaugural dissertation. Leipzig; 1776.
43. Ryan M. *Manual of Midwifery, or Compendium of Gynaecology and Paidonosology, etc*. 3rd ed. London: Renshaw and Rush; 1831.
44. Demanet G. Observations sur une cause particulière de convulsions, qui arrivent aux femmes durant la grossesse ou pendant l' accouchement. *Actes Soc Méd Chir Pharmacal Bruxelles an VI*. 1797;1(pt 2):21–28. Cited by Anonymous: *Arch Gén Méd* 1855;6 (5th series):464–472.
45. Anonymous Revue critique. Note pour servir à l'histoire de l'anasarque des femmes enceintes et de l'eclampsie puerpérale. *Arch Gén Méd*. 1855 6 (5th series):464–472.
46. de la Motte GM. *Traité Complet des Accouchemens Naturela, non Naturels, et contre Nature. Expliqués dans un Grand Nombre d' Observations et de Réflexions sur l' Arc d'Accoucher*. Paris: Gosse; 1726:307–318.
47. Smellie W. 3rd ed. *Theory and practice of Midwifery*, 176 London: Wilson and Durham; 1756:257.
48. van Swieten GLB. *Commentaria in Hermanni. Boerhaave Aphorismos de Cognescendis et Curandis Morbis*. 2nd ed. Lugduni: Batavorum, Verbeck: 1745 (translated from the Latin, Edinburgh: Eliot; 1776).
49. Rayer P. *Traité des Maladies des Reins et des Altérations de la sécretion urinaire, etudiees en elles-memes et dans leurs rapports avec les maladies des uretères, de la vessie, de la prostate, etc*, Vol. II. Paris: Bailière; 1839–1841:399–407.
50. Lever JCW. Cases of puerperal convulsions, with remarks. *Guy's Hosp Reports*; 1843;2nd series:495–517.
51. Simpson JY. Contributions to the pathology and treatment of diseases of the uterus. *London Edinburgh Monthly J Med Sci*. 1843;3:1009–1027.
52. Mahomed FA. The etiology of Bright's disease and the prealbuminuric stage. *Med Chir Trans London*. 1874;39:197–228.
53. Mahomed FA. The etiology of Bright's disease and the prealbuminuric stage. *Br Med J*. 1874;1:585–586.
54. Veith I, trans. *Yellow Emperior's Classic of Internal Medicine*. Baltimore: Williams & Wilkins; 1949:141, 147, 172.
55. Ballantyne JW. Sphygmographic tracings in puerperal eclampsia. *Edinburgh Med J*. 1885;30:1007–1020.
56. Galabin AI. *A Manual of Midwifery*. Philadelphia: Blakistin; 1886:276–280.
57. Lebedeff A. Porochjakow: Basch's sphygmomanometer und der Blutdruck wahrend der Geburt und des Wochenbettes im Zusammenhange mit Puls, Temperatur und Respiration. *Centralbl Gynaekol*. 1884;8:1–6.
58. Vinay C. *Traité de Maladies de la Grossesse et des Suites de Couches*. Paris: Bailliere; 1894:386.
59. Vaquez N. De la pression artérielle dans l'eclampsie puerperale. *Bull Mem Soc Méd Hôp Paris*. 1897;14:117–119.
60. Wiesnner Über Blutdruckmessungen während der Menstruation and Schwangerschaft. *Centralbl Gynaekol*. 1899;23:1335.
61. Cook HW, Briggs JB. Clinical observations on blood pressure. *Johns Hopkins Hosp Rep*. 1903;11:451–534.
62. Frerichs FT. *Die Bright'sche Nierenkrankheit und deren Behandlung*. Braunchweig: Friedwich Vieweg und Sohn; 1851:211–220.
63. Spiegelberg O. The pathology and treatment of puerperal eclampsia. *Trans Am Gynecol Soc*. 1878;2:161–174.
64. Sophian J. *Toxaemias of Pregnancy*. London: Butterworth; 1953.
65. Jürgens. Berliner medicinische Gesellschaft, Sitzung vom 7 Juli 1886, Discussion. *Berl Klin Wchschr*. 1886;23:519–520.
66. Schmorl G. *Pathologisch-anatomische Untersuchungen über Puerperal-Eklampsie*. Leipzig: FCW Vogel; 1893.
67. Allbutt C. Senile plethora or high arterial pressure in elderly persons *Trans Hunterian Soc*. London: Headly Brothers; 1896:38–57.
68. Frank E. Bestehen Beziehungen zwischen Chromaffinem system und der chronischer Hypertonie des Menschen? *Deutsch Arch Klin Med*. 1911;103:397–412.
69. Janeway TC. A clinical study of hypertensive cardiovascular disease. *Arch Intern Med*. 1913;81:749–756.
70. Herrick WW. The toxemias of pregnancy and their end results from the viewpoint of internal medicine. *Illinois Med J*. 1932;62:210–220.
71. Herrick WW, Tillman AJB. Toxemia of pregnancy (its relation to cardiovascular and renal disease; clinical and necropsy findings with a long follow-up). *Arch Int Med*. 1935;55:643–664.
72. Herrick WW, Tillman AJB. The mild toxemias of late pregnancy: their relation to cardio-vascular and renal disease. *Am J Obstet Gynecol*. 1936;31:832–844.
73. Fishberg AM. *Hypertension and Nephritis*. 4th ed. Philadelphia: Lea & Febiger; 1938:746.
74. Fishberg AM. *Hypertension and Nephritis*. 5th ed. Philadelphia: Lea & Febiger; 1954.
75. Dieckmann WJ. *The Toxemias of Pregnancy*. 2nd ed. St. Louis: Mosby; 1952.
76. Zuspan FP, Ward MC. Treatment of eclampsia. *South Med J*. 1964;57:954–959.
77. Merriman S. *A Synopsis of the Various Kinds of Difficult Parturition: With Practical Remarks on the Management of Labours*. 3rd ed. London: Callow; 1820:132–133.
78. Churchill F. *On the Theory and Practice of Midwifery: A New American, from the Last Improved (2d) Dublin Edition*. Philadelphia: Blanchard and Lea; 1856:445–446.
79. Avicenna. *Poem on Medicine*, Krueger HC, Trans Springfield, IL: Thomas; 1963:60.

80. Maimonides M. *The Medical Aphorisms of Moses Maimonides* [Rosner F, Muntner S, Trans.]. New York: Yeshiva University Press; 1970, Vol. I: 234–239.
81. Hirst BC. *A textbook of obstetrics*. Philadelphia: Saunders; 1909:637.
82. Smith WT. *Parturition and the Principles and Practice of Obstetrics*. Philadelphia: Lea and Blanchard; 1849:281–345.
83. Murphy EW. *Lectures on the Principles and Practice of Midwifery*. London: Walton and Maberl; 1862:497.
84. Fehling Die Pathogenese und Behandlung der Eklampsie in Lichte der heutigen Anschauung. *Münch Med Wchschr*. 1899;46:714–715.
85. van der Hoeven PCT. *Die ätiologie der eklampsie*, inaugural dissertation. Leiden; 1896.
86. Delore L'eclampsie reconnâitrait une origine bactérienne. *Lyon Med*. 1884;47:186–187.
87. Gerdes E. Zur ätiologie der Puerperaleklampsie. *Centralbl Gynaekol*. 1892;16:379–384.
88. Favre A. Über Puerpenaleklampsie. *Virchows Arch [Pathol Anat]*. 1891;124:177–216.
89. Rosenstein S. Über Eclampsie. *Mtschr Geburtsk Frauenk*. 1864;23:413–426.
90. Munk P. Über Urämie. *Berl Klin Wchschr*. 1864;1:111–113.
91. Volhard F. Experimentelle und kritische Studien zur Pathogenase der Eklampsie. *Mtschr Geburtsh Gynaekol*. 1897;5:411–437.
92. Schumacher H. Experimentelle Beiträge zur Eklampsiefrage. *Beitr Geburtsch Gynaekol*. 1901;5:257–309.
93. Dixon WE, Taylor FE. Physiological action of the placenta. *Lancet*. 1907;2:1158–1159.
94. Rosenheim O. The pressor principles of placental extracts. *J Physiol*. 1909;38:337–342.
95. Krieger VI, Butler HM, Kilvington TB. Antidiuretic substance in the urine during pregnancy and its frequent association with bacterial growth. *J Obstet Gynaecol Br Emp*. 1951;58:5–17.
96. Lichtenstein F. Kritische und experimentelle Studien zur Toxicologie der Placenta, zugleich ein Beitrag gegen die placentare Theorie der Eklampsieätiologie. *Arch Gynaekol*. 1908;86:434–504.
97. Schneider CL. The active principle of placental toxin: Thromboplastin; its inactivator in blood: antithromboplastin. *Am J Physiol*. 1947;149:123–129.
98. Miquel A. *Traité des Convulsions chez les Femmes Enceintes en Travail et en Couche*. Paris: Gazette de Sante; 1824.
99. De Wees WP. *A Treatise on the Diseases of Females*. 2nd ed. Philadelphia: Carey, Lea and Carey; 1828:156–157.
100. Johns R. Observations on puerperal convulsions. *Dublin J Med Sci*. 1843;24:101–115.
101. Meigs CD. *Females and Their Diseases*. Philadelphia: Lea and Blanchard; 1848:632.
102. Sinclair EB, Johnson G. Practical Midwifery: Comprizing an Account of 13,748 Deliveries Which Occurred in the Dublin Lying-in Hospital During a Period of Seven Years Commencing November, 1847. London: Churchill; 1858.
103. Fitzgibbon G. The relationship of eclampsia to the other toxaemias of pregnancy. *J Obstet Gynaecol Br Emp*. 1922;29:402–415.
104. Ruiz-Contreras JM. Bemerkugen über den Einfluss der Lebensmittal auf die Einstetehung der Eklampsie und Albuminerre. *Zentralbl Gynaekol*. 1992;46:764.
105. Editorial Eclampsia rare on war diet in Germany. *JAMA*. 1917;68:732.
106. Stander HJ. *Textbook of Obstetrics*. New York: Appleton; 1945. 599 (n.b., actually *Williams Obstetrics*, 9th ed.).
107. Leyden E. Klinische Untersuchungen über Morbus Brightii. *Z Klin Med*. 1881;2:133–191.
108. Leyden E. Über Hydrops and Aalbuminurie der Schwangeren. *Z Klin Med*. 1886;11:26–49.
109. Schroeder. Discussion. Gesellschaft für Geburtschülfe und Gynaekologie in Berlin, Sitzung von 14 Mai, 1878. *Berliner Klin Wchschr*. 1878;15:599.
110. Webster JC. *A Text-Book of Obstetrics*. Philadelphia: Saunders; 1903:375–376.
111. Bar T. Éclampsisme, éclampsie sans attaques. *J Sages-Femmes*. 1908;36:153.
112. Williams JW. *Obstetrics*. 3rd ed. New York: Appleton; 1912.
113. Bell ET, Deckmann W, Eastman NJ. Classification of the toxemias of pregnancy. *Mother*. 1940;1:13–17.
114. Eastman NJ, Bell ET, Dieckmann WI, et al. *Definition and Classification of Toxemias Brought Up to-Date*. Chicago: American Committee on maternal Welfare; 1952.
115. Nelson TR. A clinical study of pre-eclampsia. Pts I and II. *J Obstet Gynaecol Br Emp*. 1955;62:48–66.
116. Rippmann ET. Prä-eklampsie oder Schwangerschaftsspätgestose? *Gynaecologia*. 1969;167:478–490.
117. Lindheimer M, Katz AI, Zuspan FP. *Hypertension in Pregnancy*. New York: John Wiley & Sons; 1976. 443.
118. Roberts JM, Redman CWG. Pre-eclampsia: more than pregnancy induced hypertension. *Lancet*. 1993;341:1447–1451.
119. Thadhani R, Kisner T, Hagmann H, et al. Pilot study of extracorporeal removal of soluble fms-like tyrosine kinase 1 in preeclampsia. *Circulation*. 2011;124:940–950.
120. Chesley LC, Annito JE, Cosgrove RA. The remote prognosis of eclamptic women: sixth periodic report. *Am J Obstet Gynecol*. 1976;124:446–459.
121. Johenning AR, Barron WM. Indirect pressure measurement in pregnancy: Korotkoff phase 4 versus phase 5. *Am J Obstet Gynecol*. 1992;167:577–580.
122. Gant NF, Chang S, Worley RI, et al. A clinical test useful for predicting the development of acute hypertension in pregnancy. *Am J Obstet Gynecol*. 1974;1290:1–7.
123. Van Dongen PWI, Eskes TAKB, Martin CB, et al. Postural blood pressure differences in pregnancy: a prospective study of blood pressure differences between supine and left lateral position as measured by ultrasound. *Am J Obstet Gynecol*. 1980;138:1–5.
124. Hallak M, Bottoms SF, Knudson K, et al. Determining blood pressure in pregnancy: positional hydrostatic effects. *J Reprod Med*. 1997;42:333–336.
125. Petrie JC, Obrien ET, Littler WA, et al. Recommendations on blood pressure measurement. British hypertension society. *Br J Med*. 1986;293:611–615.
126. *The hypertensive disorders of pregnancy*. Geneva: World Health Organization; WHO Tech Bull 1987:758.
127. Report on high blood pressure in pregnancy. *Am J Obstet Gynecol* 1990;163:1689–1712.

128. Blank DG, Helseth G, Pickering TG, et al. How should diastolic pressure be defined during pregnancy? *Hypertension*. 1994;24:234–240.
129. Shennan A, Gupta M, Halligan A, et al. Lack of reproducibility of Korotkoff phase 4 as measured by mercury sphygmomanometry. *Lancet*. 1996;346:139–142.
130. deSwiet M, Shennan A. Blood pressure measurement in pregnancy. *Br J Obstet Gynecol*. 1996;102:862–863.
131. Redman CW, Jeffries M. Revised definition of pre-eclampsia. *Lancet*. 1988;i:892–898.
132. Davey DA, MacGillivray I. The classification and definition of the hypertensive disorders of pregnancy. *Am J Obstet Gynecol*. 1989;158:892–898.
133. Australian Society of Hypertension in Pregnancy: Consensus statement management of hypertension in pregnancy: executive summary. *Med J Australia* 1993;158:700–702.
134. ACOG Committee on Technical Bulletins Hypertension in pregnancy. *ACOG Tech Bull* 1996;219:1–8.
135. Brown MA, Buddie ML. What's in a name? Problems with the classification of hypertension in pregnancy. *J Hypertens*. 1997;15:1049–1054.
136. Report of the national high blood pressure education program, working group on high blood pressure in Pregnancy. *Am J Obstet Gynecol*. 2000;183:S1–S22.
137. ACOG practice bulletin. Diagnosis and management of preeclampsia and eclampsia. *Obstet Gynecol*. 2002;99:159–167.
138. Brown MA, Lindheimer MD, deSwiet M, Van Assche A, Moutquin JM. The classification and diagnosis of the hypertensive disorders of pregnancy: Statement of the International Society for the Study of Hypertension in Pregnancy (editorial). *Hypertens Preg*. 2001;20:IX–XIV.
139. Chesley LC. Mild preeclampsia: potentially lethal for women and for the advancement of knowledge. *Clin Exp Hypertens*. 1989;B8:3–12.
140. Chesley LC. Diagnosis of preeclampsia. *Obstet Gynecol*. 1985;65:423–425.
141. Von Dadelszen P, Magee LA, Roberts JM. Subclassification of preeclampsia. *Hypertens Preg*. 2003;22:143–148.
142. Brown M, Lindheimer MD, De Swiet M, et al. The classification and diagnosis of the hypertensive disorders of pregnancy: Statement from the International Society for the Study of Hypertension in Pregnancy. *Hypertens Pregnancy*. 2001;20:IX–XIV.
143. Milne F, Redman C, Walker J, et al. The pre-eclampsia community guidelines (PRECOG): how to screen and detect preeclampsia in the community. *BMJ*. 2005;330:376–380. (see also PRECOGII 339:b3129).
144. Lowe S, Brown MA, Decker G, et al. Society of obstetric medicine of Australia and New Zealand guidelines for the management of hypertensive disorders of pregnancy 2008. *Aus NZ J Obstet Gynaecol*. 2009;49:242–246.
145. MaGee LA, Helewa MZ, Moutquin JM, et al. Diagnosis, evaluation, and management of the hypertensive disorders of pregnancy. Hypertension guidelines' committee; strategic training initiative in research in the reproductive health science scholars (STRRHS). *J Obstet Gynaecol Can*. 2008;suppl 3:S1–S48.
146. Steegers EA, Von Dadelzen P, Pinjenburg R. Pre-eclampsia (seminar). *Lancet*. 2010;376:631–641.
147. Lindheimer MD, Taler SJ, Cunningham FG. Hypertension in pregnancy. *J Am Soc Hypertens*. 2010;4:68–78. (invited position paper).
148. Hypertension in pregnancy. NICE clinical guidelines 107. Guidance: nice.uk/cg107.
149. WHO recommendations for prevention and treatment of preeclampsia and eclampsia.
150. Report of the American College of Obstetricians and Gynecologists Task Force on Hypertension in Pregnancy. Excecutive summary. *Obstetric Gynecol* 2013; 122:1122–1231.
151. The GRADE working group Grading quality of evidence and strength of recommendations. *BMJ*. 2004;328:1490–1494. printed abridged version (see also see < http://www.gradeworkinggroup.org/index.htm) > .
152. Levine RJ, Ewell MG, Hauth JC. Should the definition of preeclampsia include a rise in diastolic blood pressure of > / = 15 mm Hg to a level < 90 mm Hg in association with proteinuria? *Am J Obstet Gynecol*. 2000;183:787–792.
153. Lindheimer MD, Kanter D. Interpreting abnormal proteinuria in pregnancy: the need for a more pathophysiological approach. *Obstet Gynecol*. 2010;115:365–375.
154. Caritis S, Sibai B, Hauth J, et al. Low-dose aspirin to prevent preeclampsia in women at high risk. *N Engl J Med*. 1998;338:701–705.
155. Hinselmann H. *Die Elklamsie*. Bonn: Cohen; 1924.
156. McCartney CP. Pathological anatomy of acute hypertension of pregnancy. *Circulation*. 1964;30:37–42.
157. Theobald GW. The incidence of albumin and sugar in the urine of normal women. *Lancet*. 1931;2:1380–1383.
158. Theobald GW. Further observations on the relation of pregnancy to hypertension and chronic nephritis. *J Obstet Gynaecol Br Emp*. 1936;43:1037–1052.
159. Reid DE, Teel HM. Nonconvulsive pregnancy toxemias; their relationship to chronic vascular and renal disease. *Am J Obstet Gynecol*. 1939;37:886–896.
160. Chesley LC, Somers WH, Gorenberg HR, et al. An analysis of some factors associated with posttoxemic hypertension. *Am J Obstet Gynecol*. 1941;41:751–764.
161. Chesley LC, Annitto JE. Pregnancy in the patient with hypertensive disease. *Am J Obstet Gynecol*. 1947;53:372–381.
162. Harris JW. The after-effects of the late toxemias of pregnancy. *Bull Johns Hopkins Hosp*. 1924;35:103–107.
163. Gibberd GF. Albuminuria complicating pregnancy. *Lancet*. 1931;2:520–525.
164. Peckham CH, Stout MI. A study of the late effects of the toxemias of pregnancy (excluding vomiting and eclampsia). *Bull Johns Hopkins Hosp*. 1931;49:225–245.
165. Theobald GW. The relationship of albuminuria of pregnancy to chronic nephritis. *Lancet*. 1933;1:626–630.
166. Browne FJ, Sheumack DR. Chronic hypertension following pre-eclamptic toxaemia: the influence of familial hypertension on its causation. *J Obstet Gynaecol Br Emp*. 1956;63:677–679.
167. Adams EA, MacGillivray I. Long-term effect of pre-eclampsia on blood pressure. *Lancet*. 1961;2:1373–1375.
168. Tillman AJB. Long-range incidence and clinical relationship of toxemias of pregnancy to hypertensive vascular disease. *Circulation*. 1964;30:76–79.

169. Bryans CI. The remote prognosis in toxemia of pregnancy. *Clin Obstet Gynecol.* 1966;9:973–990.
170. Epstein FH. Late vascular effects of toxemia of pregnancy. *N Engl J Med.* 1964;271:391–395.
171. Lehmann K. *Eklampsien i Danmark i Aarene 1918–1927.* Copenhagen: Busck; 1933.
172. Díaz del Castillo FO. Influencia de la alimentacion sobre la eclampsia (la eclampsia en Madrid durant la guerra). *Rev Clin Esp.* 1942;6:166–170.
173. Siddall AC. Vitamin B1 deficiency as an etiologic factor in pregnancy toxemias. *Am J Obstet Gynecol.* 1940;39:818–821.
174. Dennman T. *Essays on the Puerperal Fever, and on Puerperal Convulsions.* London: Walter; 1758.
175. Kosmak GW. *The Toxemias of Pregnancy.* New York: Appleton; 1924.
176. Mengert WF. Racial contrasts in obstetrics and gynecology. *J Natl Med Assoc.* 1966;58:413–415.
177. Davis CH. Weight in pregnancy; its value as a routine test. *Am J Obstet Gynecol.* 1923;6:575–583.
178. Hannah CR. Weight during pregnancy. *Texas State J Med.* 1923;19:224–226.
179. Robertson EG. The natural history of oedema during pregnancy. *J Obstet Gynaecol Br Commonw.* 1971;78:520–529.

CHAPTER 2

The Clinical Spectrum of Preeclampsia

F. GARY CUNNINGHAM, JAMES M. ROBERTS AND ROBERT N. TAYLOR

Editors' comment: *The fourth edition of Chesley's Hypertensive Disorders of Pregnancy is a revision of the previous two multi-authored editions. The fact that four editors and an additional 34 co-authors contributed to the current edition is a testament to Leon Chesley's remarkable single-authored accomplishment published in 1978.*[1] *This chapter describes the vast clinical spectrum of preeclampsia with the goal of setting the stage for further description by other authors through the eyes of their specific expertise. Emphasized will be knowledge accrued during the 35 years following Chesley's epic first edition – keeping apace of the information that seems to have increased exponentially in the past three decades. To be sure, the many advances that are chronicled in the chapters that follow stress giant strides in basic research leading to new insights regarding the pathophysiology of preeclampsia, some made with cutting-edge tools in areas such as genetics, immunology, and molecular biology. But before this we thought it important to describe the spectrum of the clinical disease on which the book is focused. Thus, this chapter outlines the vast array of clinical observations that have prompted evidence-based management schemes to improve maternal and perinatal outcomes of the millions of women whose pregnancies are affected by preeclampsia worldwide each year.*

Recall that Leon Chesley was a scientist – a PhD who did not practice medicine – but the goals of his studies were always designed with the object to improve pregnancy outcomes as a common denominator. To this end, he espoused the application of scientific observations in the clinic, the hospital, the labor and delivery unit, and importantly, to long-term follow-up of women found to be hypertensive during pregnancy. What follows is a legacy of that keen scientific application of new knowledge to improve the practice of clinical medicine that Dr. Chesley so well communicated in words and deeds.

INTRODUCTION

Preeclampsia is much more than hypertension and proteinuria complicating pregnancy – it is a *syndrome* affecting virtually every organ system. An appreciation of this has been a principal reason for improved clinical management over the past two to three decades.[2] This does not mean, however, that systemic effects were unknown until recently. In the previous chapter we reprinted Chesley's original and elegant text describing the history of eclampsia from ancient Chinese, Egyptian, Greek, and Indian literature, as well as its management – and mismanagement – through the 20th century. While the anatomical effects of the disorder on brain, kidney, and liver were well described, the focus was placed clinically on convulsions until the end of the 19th century when blood pressure measurement became possible. Also of interest, preeclampsia was termed "toxemia" – it is still utilized by some – because a prevailing thought was that it was caused by circulating poisons. The latter is an obvious reflection in the "purging" era of therapy. Though the term "toxemia" was subsequently downplayed when the designation "preeclampsia" gained preeminence, the following chapters attest that we have returned to an age that focuses on and speculates that circulating proteins such as cytokines, antiangiogenic factors, antioxidants, and other putative compounds activate the endothelium to cause preeclampsia phenotypes. Indeed, some authors now compare the systemic inflammatory state of preeclampsia to that described in the sepsis syndrome.[3]

The concept that preeclampsia is a protean syndrome is important. Like other syndromes, in individual patients, some organ systems are predominantly affected more than others. For example, in some women the primary pathological manifestation may be asymptomatic, albeit dangerously elevated blood pressures of 230/120 mm Hg or higher, while other women may have an eclamptic convulsion with only minimally elevated blood pressure, and with the majority having a clinical picture somewhere in between these. Of cardinal importance is recognition and differentiation of preeclampsia – whether pure or superimposed – from other causes of hypertension because this is the disorder most likely to cause serious maternal and fetal complications. Precise diagnosis, however, may not be possible, in which case, and as underscored throughout the report of the National High Blood Pressure Education Program's

Chesley's Hypertensive Disorders in Pregnancy.
ISBN: 978-0-12-407866-6

DOI: http://dx.doi.org/10.1016/B978-0-12-407866-6.00002-X

Working Group,[4] and more recently the Task Force of the American College of Obstetricians and Gynecologists on Hypertension in Pregnancy,[5] it is best to manage such women as if they had preeclampsia.

CLINICAL MANIFESTATIONS OF PREECLAMPSIA SYNDROME

Although the cause of the preeclampsia syndrome remains unknown, evidence for its manifestation begins early in pregnancy with covert pathophysiological changes that gain momentum across gestation and eventually become clinically apparent. Unless delivery supervenes, these changes ultimately result in multiorgan involvement with a clinical spectrum ranging from *forme fruste* manifestations to one of cataclysmic pathophysiological deterioration that is life-threatening for both mother and fetus. As discussed, these are thought to be a consequence of endothelial dysfunction, vasospasm, and ischemia. Whilst the myriad of maternal consequences of the preeclampsia syndrome are usually described in terms of individual organ systems, they frequently are multiple and they overlap clinically.

This chapter is a capsule of the clinical spectrum of this disorder, and some of the subsequent chapters address individual organ system involvement in greater detail. Preeclampsia, characterized by hypertension, proteinuria, edema, and overt or subclinical coagulation and liver involvement, occurs more commonly in nulliparas, usually after 20 weeks gestation, and becomes more common as term approaches.[6,7] The earlier the onset, the less likely it is to be "pure" preeclampsia, and the higher the probability the disease is superimposed on underlying essential hypertension or a renal disorder. It also has become customary to categorize the disease as "severe," with the remainder being either "mild" or "nonsevere." Severe preeclampsia is usually defined by diastolic and systolic levels of >110 and >160 mm Hg, respectively, the appearance of nephrotic-range proteinuria, sudden oliguria, or neurological symptoms such as headache or hyperreflexia, as well as by laboratory tests demonstrating thrombocytopenia and/or hepatocellular disruption. Because a woman with seemingly mild disease – for example, a teenage girl with a blood pressure of 140/90 mm Hg and minimal proteinuria – can suddenly convulse, designations such as mild and severe could be misleading. In fact, *de novo* hypertension alone in a third-trimester nulliparous woman is sufficient reason to manage as preeclampsia.

Another manifestation of the syndrome is that of early preeclampsia, i.e., that with an onset <34 weeks, and this variant is associated with greater morbidity than late-onset preeclampsia.[8] The possibility that the heterogeneous expression of preeclampsia might indicate subtypes that could have different pathogeneses directing different preventive strategies is important and should be kept in mind as results of clinical trials are evaluated. Nonetheless, at our current stage of knowledge modifying therapy based on such concepts is inappropriate as all preeclampsia is potentially explosive.

The eclamptic convulsion – certainly the most dramatic – and one of the most life-threatening complications of preeclampsia, was once associated with a maternal mortality of 30%.[1] Improved and aggressive obstetrical management has decreased but not eliminated convulsions. Eclampsia is usually preceded by various premonitory signs including headache, visual disturbances, severe epigastric pain, sensation of constriction of the thorax, apprehension, excitability, hyperreflexia, and hemoconcentration. That said, convulsions can still develop suddenly and without warning in a seemingly stable patient with only minimal elevations of blood pressure. It is the capricious nature of this disorder that underlies the need for early hospitalization of women with suspected preeclampsia.

One aspect of the preeclampsia syndrome that is repeatedly discussed throughout this book is that of the *HELLP syndrome*. This is an acronym for *H*emolysis, *E*levated *L*iver enzymes, and *L*ow *P*latelets – and constitutes an emergency requirement for termination of the pregnancy. The syndrome is characterized by hemolysis with marked evidence of both liver and coagulation abnormalities that include serum transaminase and lactic dehydrogenase increasing to >1000 U/L, platelet counts decreasing to <100,000/μL, with schizocytes seen on the blood smear. Although usually fulminant in nature, the HELLP syndrome can be atypical.[9] A deceptively benign form that has an initially mild clinical presentation, with the woman presenting with borderline thrombocytopenia, and perhaps slightly abnormal serum transaminase levels, normal or minimally elevated blood pressure, and little or no renal dysfunction can manifest. But these seemingly mild complications may rapidly become life-threatening when within 24–48 hours there is progression to a more fulminant form.

Finally, there is another uncommon manifestation of atypical preeclampsia that is termed "late postpartum eclampsia," defined as hypertension and convulsions that develop 48 hours to several weeks after delivery.[9,10] Some of these women have typical manifestations of preeclampsia and/or eclampsia prodroma. They often present at emergency rooms, where personnel are unfamiliar with gestation-related problems.[11]

Cardiovascular System

Hemodynamic changes normally induced by pregnancy, as well as the severe disturbances of cardiovascular function with preeclampsia or eclampsia, are discussed in detail in Chapter 14. To summarize, cardiovascular effects can be divided into those that impact myocardial and ventricular functions.

I. Myocardial Function

Serial echocardiographic studies have documented that in preeclampsia there is evidence for ventricular remodeling which is accompanied by diastolic dysfunction in 40% of women.[12] In some of these women, functional differences persisted at 16 months.[13] Ventricular remodeling was judged to be an adaptive response to maintain normal contractility with the increased afterload of preeclampsia. In the otherwise healthy pregnant woman, these changes are usually clinically inconsequential, but if there is underlying concentric hypertrophy, diastolic dysfunction can develop with congestive heart failure and pulmonary edema.

II. Ventricular Function

Despite the relatively high frequency of diastolic dysfunction with preeclampsia, in the majority of women ventricular function is appropriate. Changes are centered on increased cardiac afterload caused by hypertension. At the same time, cardiac filling or preload, whilst substantively affected by normal pregnancy-induced hypervolemia, may be diminished to absent due to preeclampsia, or it may be increased by intravenous crystalloid or oncotic solutions. Importantly, both normally pregnant women, as well as those with severe preeclampsia, have normal ventricular function as shown in Fig. 2.1 and as plotted on the Braunwald ventricular function graph. In both of these groups of women, cardiac output is appropriate for left-sided filling pressures.

Data from studies of preeclamptic women obtained by invasive hemodynamic assessment are confounded because of the heterogeneity of populations and interventions that also may significantly alter these measurements. Ventricular function studies of preeclamptic women from a number of investigations discussed in Chapter 14 are plotted on the graph shown in Fig. 2.2. Although cardiac function was hyperdynamic in all women, filling pressures were dependent on intravenous fluid infusions. Specifically, aggressive hydration resulted in hyperdynamic ventricular function in most women. As noted, pulmonary edema may develop despite normal ventricular function because of endothelial–epithelial leak that is compounded by decreased oncotic pressure from low serum albumin concentrations.[20] Similar values have been reported using noninvasive whole-body impedance cardiography.[21]

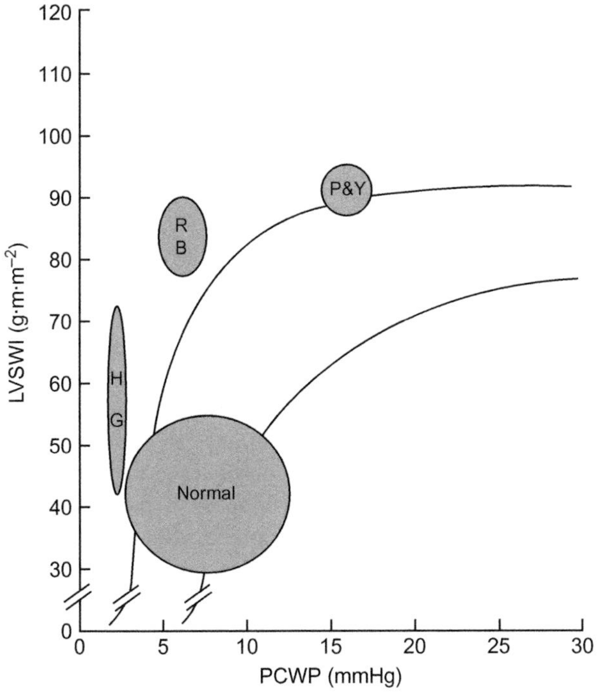

FIGURE 2.1 Ventricular function in normal pregnancy and in severe preeclampsia–eclampsia. Left-ventricular stroke work index (LVSWI) and pulmonary capillary wedge pressure (PCWP) are plotted on a Braunwald ventricular function curve. The normal values reported by Clark et al.[14] fall approximately within the larger circle. The other data points represent mean values obtained in each of five studies and the letters adjacent to the data points represent each investigator: P&Y = Phelan and Yurth;[15] R = Rafferty and Berkowitz;[16] B = Benedetti et al.;[17] H = Hankins et al.;[18] G = Groenendijk et al.[19]

Loss of Pregnancy Hypervolemia

It has been known for over 100 years that *hemoconcentration* is a hallmark of eclampsia. Zeeman et al.[22]

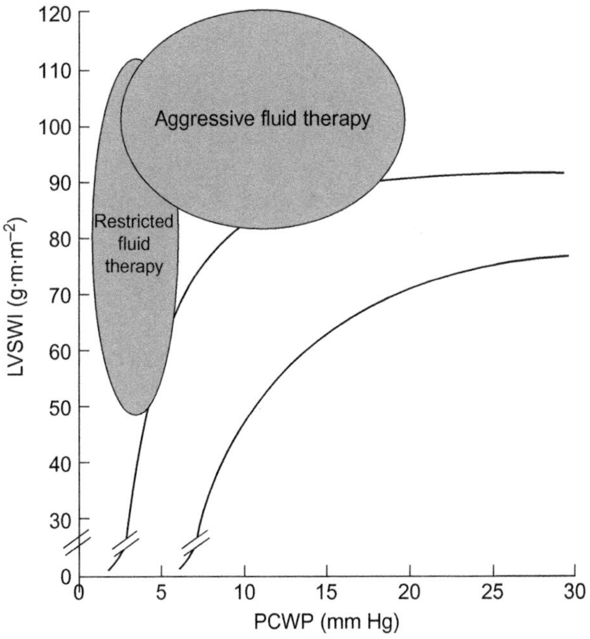

FIGURE 2.2 Ventricular function in women with severe preeclampsia–eclampsia plotted on a Braunwald graph. In almost all of these, there is hyperdynamic function evidenced by elevated left-ventricular stroke work index (LVSWI). The pulmonary capillary wedge pressures (PCWP) as a group are higher in those managed with aggressive fluid administration to expand their intravascular volume compared with restricted fluid therapy. Eight women in those managed with aggressive fluids developed clinical pulmonary edema; despite that all but one had normal to hyperdynamic ventricular function.

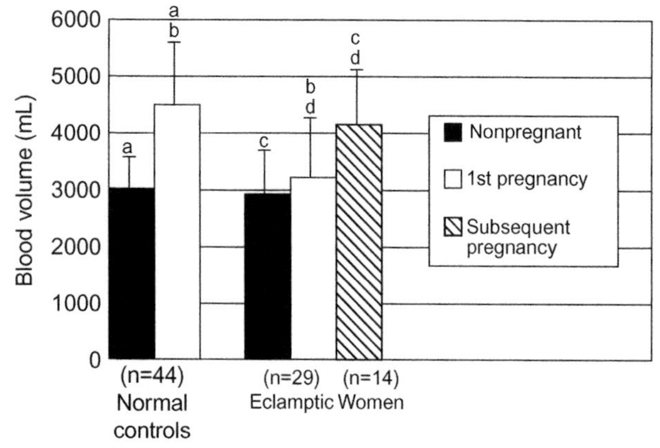

FIGURE 2.3 Bar graph comparing nonpregnant mean blood volumes with those obtained at the time of delivery in a group of women with normal pregnancy, eclampsia in their first pregnancy, and subsequent normal pregnancy in some of the women who had eclampsia. Extensions above bars represent one standard deviation. Comparisons between values with identical lowercase letters, *viz.*, a-a, b-b, c-c, d-d, are significant ($p < .001$).[22]

expanded the previous observations of Pritchard et al.[23] that eclamptic women have severe curtailment of normally expected pregnancy-induced hypervolemia (Fig. 2.3). Mechanisms for this are discussed in detail in Chapter 14. Hemoconcentration is of immense clinical significance. Women of average size should have a blood volume of nearly 4500 mL during the last several weeks of a normal pregnancy, compared with about 3000 mL when they are not pregnant. With eclampsia, however, much or all of the anticipated excess 1500 mL is not available. Clinically, this is of vital importance even when there is average blood loss with cesarean delivery or episiotomy that frequently exceeds 1000 mL In addition, postpartum hemorrhage is more common in women with severe preeclampsia because of factors such as placental abruption, labor induction, and increased risk of operative vaginal or cesarean delivery.

Blood and Coagulation

Hematological abnormalities develop in some women with preeclampsia. Indeed, thrombocytopenia, which was described almost a century ago,[24] can be so severe as to be life-threatening. In addition, the levels of some plasma clotting factors may be decreased, as discussed in detail in Chapter 17. In some women with severe disease, erythrocytes may display bizarre shapes and undergo rapid hemolysis. The clinical significance of thrombocytopenia, in addition to any impairment in coagulation, is that it reflects the severity of the pathological process. In general, the lower the platelet count is, the greater the maternal and fetal morbidity and mortality.[25] In 1954, Pritchard and colleagues[23] called attention to thrombocytopenia accompanied by intravascular hemolysis in women with eclampsia. As discussed in Chapter 17, in his first edition of this text,[1] Chesley summarized the available data and concluded that elevated serum hepatic transaminase levels were common in severe forms of preeclampsia and in eclampsia. To call attention to the seriousness of these, Weinstein[26,27] later referred to this combination of events as the *HELLP syndrome*, which was described above. This moniker is now used worldwide and we use it throughout this book.

There are also subtle changes consistent with intravascular coagulation, and less often erythrocyte destruction, that commonly accompany preeclampsia and especially eclampsia (see Chapter 17). In the first edition of this book,[1] Chesley had already concluded that there was little evidence that these abnormalities are clinically significant. This view was further supported by the results of a large study reported by Pritchard et al.,[28] and by the findings of Barron et al.[29] The latter investigators suggested that routine laboratory assessment of coagulation other than platelet count, including prothrombin time, activated partial thromboplastin time, and plasma fibrinogen level, was unnecessary in the management of pregnancy-associated hypertensive disorders. One exception is perhaps when moderate to severe thrombocytopenia – platelet count <100,000/μL – is identified.[30] Others indications are events that coexist and predispose to consumptive coagulopathy. Examples include placental abruption with rapid defibrination, or extensive hepatic necrosis with preeclampsia, or acute fatty liver of pregnancy.

The Kidney

During normal pregnancy, renal blood flow and glomerular filtration rate are increased appreciably. With the preeclampsia syndrome, a number of usually reversible anatomical and pathophysiological changes may occur and are discussed in detail in Chapter 16. Of clinical importance, renal perfusion and glomerular filtration are reduced. These changes are usually minimal, and a rise in serum creatinine levels to values above those of normal nonpregnant women is infrequent and usually the consequence of very severe preeclampsia or its superimposition on a previous unrecognized kidney disorder. For example, if severe intrarenal vasospasm is profound, thus plasma creatinine may be elevated several times the nonpregnant normal value – up to 2 to 3 mg/dL.[23,31] As discussed in Chapter 16, plasma uric acid concentration is typically elevated out of proportion to the reduction in glomerular filtration rate that accompanies preeclampsia.[32]

In most women, urine sodium concentration is often elevated. Levels of urine osmolality, urine:plasma creatinine ratio, and fractional excretion of sodium may suggest that a prerenal mechanism is involved. Clinically, this is important because in severe preeclampsia oliguria may develop despite normal ventricular filling pressures such as shown in Fig. 2.1.[33] And as shown in Fig. 2.2, crystalloid infusion increases left ventricular filling pressure, and even though oliguria improves temporarily, this is with a danger of clinically apparent pulmonary edema. *Intensive intravenous fluid therapy is not indicated for most women with oliguria.*

Proteinuria

At least in the recent past, when protein excretion was detected together with *de novo* hypertension in late pregnancy, then this combination satisfied the "official" diagnosis of preeclampsia. But with the 2013 Task Force classification, facets other than demonstration of proteinuria are used to make the diagnosis of preeclampsia. These are discussed in Chapter 1 and shown in Table 2.1. This major change is because proteinuria is not invariably present. First, proteinuria may be a late phenomenon, and some women with preeclampsia may be delivered – or suffer eclamptic convulsions – before it appears. In one example, Sibai[34] noted 10–15% of women with HELLP syndrome do not have proteinuria at discovery. Second, the optimal way of establishing either abnormal levels of urine protein or albumin in pregnant women remains to be defined. Dipstick qualitative determinations depend on the concentration of the urine tested and are notorious for false-positive and -negative results. While a quantitative value of >300 mg/24 h – or its interpolated equivalent in shorter collections – is the standard "consensus" cut-off, this is not irrefutably established. Indeed, Tun et al.[35] proposed a 12-hour collection with a 165-mg upper limit of normal.

Because timed urine specimens can be quite difficult for pregnant women to collect, this has led to recent suggestions that determination of the protein or albumin:creatinine ratio would supplant 24-hour collections.[36] In a systematic review, Papanna et al.[37] concluded that random urine protein:creatinine ratios that are below 130–150 mg/g indicate that the likelihood of proteinuria exceeding 300 mg/day is very low. Midrange ratios, *viz.*, 300 mg/g have poor sensitivity and specificity and they recommend that a 24-hour specimen be collected for accuracy. Finally, there are several methods used to measure proteinuria, and none detects all of the various proteins normally excreted in urine. In reality, albeit impractical, each hospital laboratory would need to establish its own norms. Finally, there is now available technology that permits the measurement of urinary albumin:creatinine ratios within an outpatient setting.[36]

Also relegated to the past by the 2013 Task Force was the notion that worsening proteinuria indicated worsening preeclampsia. In the new schema, "heavy," "nephrotic range," or worsening proteinuria are not necessarily signs of severe disease with ominous indications. Thus, proteinuria has a binary function – either positive or negative, present or absent, or "yes" or "no" – when it is new-onset and is quantified to be 300 mg/day or more in a woman with *de novo* hypertension. The magnitude of the proteinuria alone as risk marker, in the absence of other ominous signs, has been questioned for decades.[38]

Normally proteins pass the glomerular barrier poorly and filtration of albumin is far greater than that of the other larger globulins. In addition, there are tubular proteins that also enter the urine. The quantity of all urinary proteins is probably below 100 mg/day in most nonpregnant women; however, empirically the upper limit of normal is considered to be >300 mg/day during normal gestation. When the glomerular barrier to proteins is disrupted as it is in many glomerular diseases, preeclampsia included, most of the increased protein in the urine is albumin.

In addition to barrier disruption – the "leak" – the amount of protein in the urine will be determined by the amount of albumin brought to the glomerulus per unit of time, *viz.*, plasma albumin × renal plasma flow. This is related to the plasma albumin concentration, with less excretion with lower plasma levels for a given leak. Also in the equation is production and catabolism of albumin, all affected by both liver and renal involvement. Thus when considering normal renal physiology and urinary protein excretion, it seems almost inevitable that the magnitude of changes in proteinuria would not make an efficient prognosticator of severity. That said, and given that all confounding variables are not always identifiable, it is prudent to view moderate to heavy proteinuria as both diagnostic of preeclampsia and therefore not to be ignored.

Recovery of Renal Function

In most women, there is minimal renal dysfunction and recovery is the rule.[39] There are, however, some women with very severe forms of preeclampsia who are at risk of varying degrees of acute renal failure. Also designated as *acute kidney injury*,[40] such degrees of renal failure are usually due to *acute tubular necrosis*, characterized by oliguria (rarely anuria) and rapidly developing azotemia. Depending on the severity of renal shutdown, the serum creatinine level may increase rapidly, often as much as 1–2 mg/dL daily.

Most often, acute kidney injury is seen in neglected cases, or it is induced in the preeclamptic woman by hypovolemic shock associated with obstetrical hemorrhage, for which adequate blood replacement is not given. Drakeley et al.[41] described 72 women with preeclampsia and renal failure – half had HELLP syndrome and a third had a placental abruption. Looked at another way, Haddad et al.[42] reported that 5% of 183 women with HELLP syndrome developed acute renal failure. Half of these also had a placental abruption, and most had postpartum hemorrhage. Frangieh et al.[43] reported an incidence of 3.8% with eclampsia. In the past, irreversible *renal cortical necrosis* went on to develop in many of these women, but with moderate resuscitative algorithms this is rare today, and generally seen most often in undeveloped areas of the world.

The Liver

From a clinical standpoint, hepatic involvement usually testifies to the severity of the preeclampsia syndrome. That

TABLE 2.1 Some Distinguishing Clinical and Laboratory Findings in Differential Diagnosis of Severe Preeclampsia Syndrome

		Laboratory Findings						
Disorder	**Clinical Findings**	**Hemolysis and Anemia**	**Bilirubin Elevation**	**Thrombocytopenia**	**Creatinine Elevation**	**AST/ALT Elevation**	**Intravascular Coagulation**	**Hypoglycemia**
Preeclampsia, eclampsia, HELLP syndrome	Variable HTN, HA, proteinuria, convulsions, epigastric pain	Common, usually mild	Usually mild	Mild to moderate	Variable, mild to moderate	Variable, mild to moderate	Rare, mild	No
Fatty liver	N&V, mild HTN, impending liver failure	Common, mild to moderate	Common, moderate	Moderate to severe	Common, moderate to severe	Common, moderate	Common, severe (fibrinogen destruction *and* hypoproduction)	Common, moderate to severe
Thrombotic microangiopathy[a]	Marked hemolysis, thrombocytopenia, CNS findings, occasional HTN, renal involvement	Common, moderate to severe	Variable, mild to moderate	Common, moderate to severe	Variable, mild to moderate	Unusual, mild	Unusual, mild	No
Exacerbation SLE[b]	HTN, proteinuria, lupus flare symptoms, thrombocytopenia	Occasional, mild to moderate	Unusual	Variable, mild to severe	Common, mild to moderate	Unusual, mild	Unusual, mild	No

[a]Includes thrombotic thrombocytopenia purpura (TTP) and hemolytic uremic syndrome (HUS).
[b]SLE = systemic erythematosus. May also have antiphospholipid antibody (APA) syndrome.

said, hepatic hemorrhage and cellular necrosis are seldom extensive enough to be clinically relevant. Differentiating characteristics from other obstetrical hepatic disorders are shown in Table 2.1. As discussed in Chapter 17, in 1856 Virchow[44] described characteristic lesions commonly found with fatal cases of eclampsia to be regions of periportal hemorrhage in the liver periphery. In their autopsy studies, Sheehan and Lynch[45] reported that some hepatic infarction accompanied hemorrhage in almost half of women with eclampsia. Reports of elevated serum hepatic transaminase levels began to surface in the late 1950s, and over the next 20 years it became appreciated that these changes were also seen with severe preeclampsia and they usually paralleled severity of disease. As discussed, such involvement was termed *HELLP syndrome* by Weinstein.[26,27]

From a pragmatic point, liver involvement with preeclampsia may be clinically significant in the following circumstances:

1. Symptomatic involvement, usually manifest by moderate to severe right-upper to midepigastric pain and tenderness, is seen with severe disease. In many cases such women also have elevations in serum hepatic aminotransferase levels – aspartate transferase (AST) or alanine transferase (ALT). In some cases, however, the amount of hepatic tissue involved with infarction may be surprisingly extensive, yet still clinically insignificant. In our experiences, infarction may be made worse by hypotension from obstetrical hemorrhage. Shown in Fig. 2.4 is a CT scan from a woman with severe preeclampsia and massive obstetrical hemorrhage. While there was extensive hepatic infarction visible on the CT scan, there were no clinical symptoms of liver dysfunction.[47]
2. Asymptomatic elevations of serum hepatic transaminase levels – AST and ALT – are considered markers for severe preeclampsia. Values seldom exceed 500 U/L or so, but can be over 2000 U/L in some women. In general, they inversely follow platelet levels, and they normalize within 3 days following delivery.
3. Hepatic hemorrhage from areas of cellular necrosis and infarction can be identified using CT- or MR-imaging as shown in Fig. 2.4. Hepatic hemorrhage may form an intrahepatic hematoma or a subcapsular hematoma (see Fig. 17.3). These hematomas – especially the subcapsular variety – may rupture with hemorrhage into the peritoneal cavity. It is likely, however, that hemorrhages without rupture are more common than clinically suspected, especially in women with HELLP syndrome.[48,49] Although once considered a surgical condition, most currently favor observation and conservative management of hematomas unless extensive hemorrhage is ongoing.[50] That said, in some cases, prompt surgical intervention may be life-saving. Rinehart[51] and Vigil De Gracia[52] and their colleagues reviewed over 300 cases and reported that maternal mortality was about 25%. The majority of these hematomas were in women with HELLP syndrome and most were ruptured. In a few cases liver transplant has proved life-saving.[48,49]
4. Acute fatty liver of pregnancy is sometimes confused with preeclampsia and HELLP syndrome.[53,54] It too has an onset in late pregnancy, and often there is accompanying hypertension, elevated serum transaminase levels, and thrombocytopenia. It is thought to arise from any of several gene mutations of lipid β-oxidation enzymes that cause accumulation of intracellular fat vesicles causing varying degrees of liver failure. Unlike in women with preeclampsia there is frequently persistent nausea and vomiting, jaundice, and systemic symptoms of 1- to 2-weeks duration. The accompanying coagulopathy from liver failure may be profound, and hemorrhage with operative delivery can be fatal. In most cases, liver function usually improves 2–3 days following delivery.

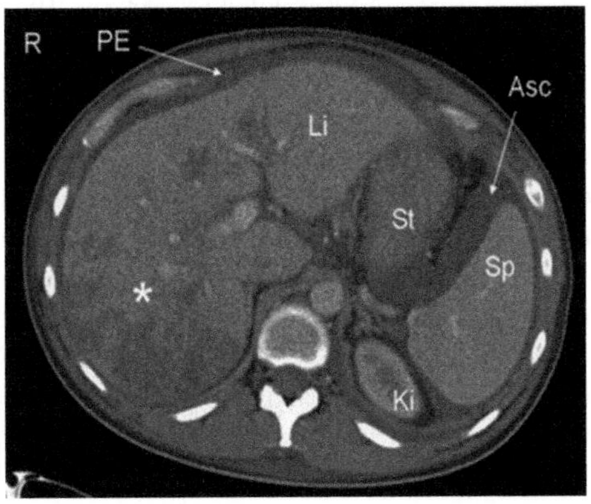

FIGURE 2.4 Abdominal CT scan showing subcapsular liver hematoma with areas of decreased attenuation consistent with hepatic infarction. (From van Lonkhuijzen,[47] with permission.)

HELLP Syndrome

As discussed on page 28 and in Chapter 17, thrombocytopenia is reflective of the severity of the preeclampsia syndrome. As hepatic involvement characterized by elevated serum transaminase levels also became well known as a marker for severity, these two began to be used in combination. As discussed, this constellation of findings is now commonly termed HELLP syndrome.[26,27] Because there is no strict definition, its incidence varies according to the investigator. In one large study, it was identified in almost 20% of women with severe preeclampsia or eclampsia.[55] In earlier reports, severe morbidity and mortality were described. From a multicenter study, Haddad et al.[42] described 183 women with HELLP syndrome. Adverse outcomes developed in almost 40% of cases, and two women died. Complications included eclampsia (6%), placental abruption (10%), acute renal failure (5%), pulmonary edema (10%), and subcapsular liver hematoma (1.6%). Citing autopsy data, Isler et al.[56] identified associated conditions that included stroke, coagulopathy, acute respiratory distress syndrome, renal failure, and sepsis. Subsequent studies continue to chronicle substantively increased maternal and perinatal morbidity and mortality.[57–59]

Some clinicians have advocated corticosteroid therapy for amelioration of HELLP syndrome.[60,61] In a Cochrane Database review, Matchaba and Moodley[62] concluded that there was insufficient evidence that adjunctive steroid use is beneficial. In a later review, Vidaeff and Yeomans[63] reached similar conclusions. At least two recent randomized trials have shown no benefits to corticosteroid therapy in women.[64,65] These results are discussed in more detail in Chapters 17 and 20.

The Brain

Headaches and visual symptoms are common with severe forms of preeclampsia, and associated convulsions define eclampsia. Neuroanatomical and pathophysiological aspects of the brain with preeclampsia syndrome are detailed in Chapter 13. The earliest anatomical descriptions of brain involvement came from autopsy specimens, but imaging and Doppler studies have added new insights into cerebrovascular involvement. Chapter 13 also discusses the clinical utility of CT scans and MRI, including numerous advances since the previous edition.

Neurological manifestations of the preeclampsia syndrome are several. Each signifies severe involvement and requires immediate attention:

1. Headache and scotomata are thought to arise from cerebrovascular hyperperfusion, which has a predilection for the occipital lobes. According to Sibai,[66] 50–75% of women have headaches and 20–30% have visual changes preceding eclamptic convulsions. The headaches may be mild to severe, intermittent to constant, and, in our experiences, they usually improve after magnesium sulfate infusion is initiated.
2. Convulsions are diagnostic for eclampsia and their proposed pathogenesis is discussed in Chapter 13.
3. Blindness is rare with preeclampsia alone, but it complicates eclamptic convulsions in 10–15% of women.[67] Blindness has been reported to develop up to a week or more following delivery.[68]
4. Generalized cerebral edema with mental status changes can vary from confusion to coma.

Blindness

As indicated, visual disturbances are common with severe preeclampsia and especially eclampsia. Scotomata are the most common symptoms and these usually abate with magnesium sulfate therapy and/or lowered blood pressure. Visual defects, including blindness, are less common, and usually reversible except in rare cases. They may arise in three potential areas: the visual cortex of the occipital lobe, the lateral geniculate nuclei, and retinal lesions that include ischemia, infarction, and detachments.

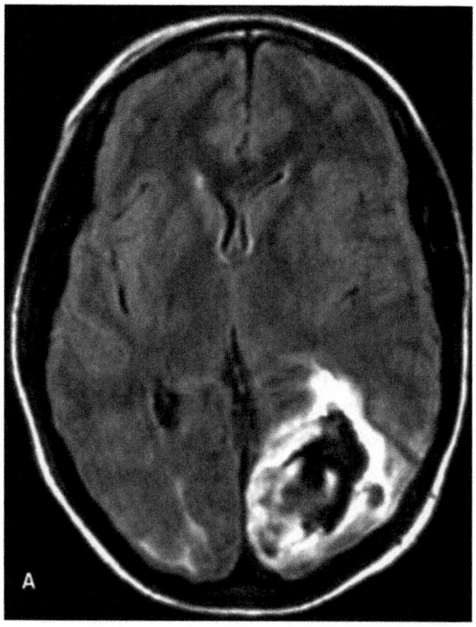

FIGURE 2.5 Cranial magnetic resonance image done 3 days postpartum in a woman with eclampsia and HELLP syndrome. Neurovisual defects persisted at 1 year, causing job disability.[69]

Occipital blindness is also called *amaurosis* – from the Greek "dimming" – and affected women usually have evidence of extensive occipital lobe vasogenic edema on CT scanning and MR imaging. Of 15 such women cared for at Parkland Hospital, blindness lasted from 4 hours to 8 days, but it resolved completely in all cases.[67] Rarely, cerebral infarctions may result in total or partial visual defects (Fig. 2.5). Retinal lesions occasionally complicate the preeclampsia syndrome. These are caused by either retinal ischemia or infarction and are also termed *Purtscher retinopathy* (Fig. 2.6). Moseman and Shelton[71] described a woman with permanent blindness due to a combination of infarctions in the lateral geniculate nucleus bilaterally as well as retinal infarction. In many cases visual acuity improves,[70] but if caused by retinal artery occlusion, the prognosis is worse.[72] Conditions associated with retinal neovascularization may be more vulnerable to the development of permanent visual defects.[73] In women studied remotely from eclampsia, preliminary observations indicate possible interference with higher-order visual function.[74]

Retinal detachment may also cause altered vision, although it is usually unilateral and seldom causes total visual loss. Occasionally it coexists with cortical edema and visual defects. Asymptomatic serous detachment apparently is quite common.[75] Detachment is obvious by examination.[76] Surgical management is seldom indicated, the prognosis generally is good, and vision usually returns to normal within a week.

Cerebral Edema

Clinical manifestations suggesting more widespread cerebral edema are worrisome. During a 13-year period,

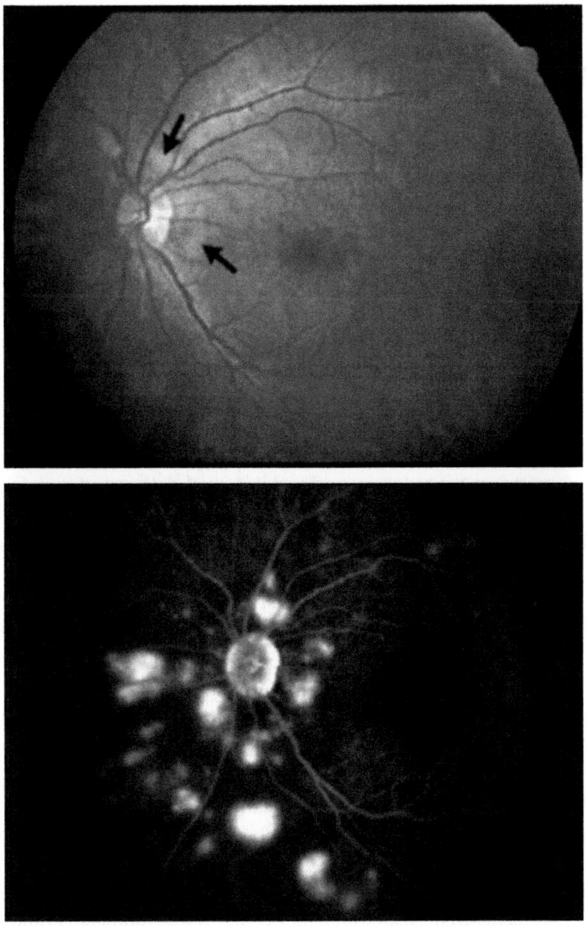
(A) (B)

FIGURE 2.6 Purtscher retinopathy caused by choroidal ischemia in preeclampsia syndrome. (A) Ophthalmoscopy shows scattered yellowish, opaque lesions of the retina (arrows). (B) The late phase of fluorescein angiography shows areas of intense hyperfluorescence representing pooling of extravasated dye.[70] (This figure is reproduced in color in the color plate section.)

10 of 175 women with eclampsia at Parkland Hospital were diagnosed with symptomatic cerebral edema.[77] Symptoms ranged from lethargy, confusion, and blurred vision to obtundation and coma. In some cases, symptoms waxed and waned. Mental status changes generally correlated with the degree of involvement seen with CT and MRI studies. Three women with generalized edema were comatose and had imaging findings of impending transtentorial herniation, and one of these died from herniation.

Clinically, these women are very susceptible to sudden and severe blood pressure elevations, which acutely worsen the widespread vasogenic edema. Thus, careful blood pressure control is essential. Consideration is given for treatment with mannitol or dexamethasone.

Uteroplacental Perfusion

Defects in placentation and of uteroplacental perfusion germane to the preeclampsia syndrome are discussed in Chapter 5. Abnormal trophoblastic invasion and defective placentation are believed to be etiological in at least some cases of preeclampsia.[78–80] And of immense clinical importance, compromised uteroplacental perfusion from vasospasm is almost certainly a major culprit in the genesis of increased perinatal morbidity and mortality associated with the disease. Thus, measurement of uterine and thus placental blood flow would likely be informative. Attempts to assess human maternal and placental blood flow have been hampered by several obstacles, including inaccessibility of the placenta, the complexity of its venous effluent, and the need for radioisotopes or invasive techniques that are unsuitable for human research.

Measurement of uterine artery blood flow velocity has been used to estimate resistance to uteroplacental blood flow. Vascular resistance is estimated by comparing arterial systolic and diastolic velocity waveforms. By the time of completion of placentation, impedance to uterine artery blood flow is markedly decreased, but with abnormal placentation, abnormally high resistance persists.[81,82] Earlier studies were done to assess this by using Doppler velocimetry to determine the systolic:diastolic ratio of the uterine and umbilical arteries in preeclamptic pregnancies. The results were interpreted as showing that in some cases, but certainly not all, there was increased resistance.[83,84]

Another Doppler waveform – *uterine artery "notching"* – has been reported to be associated with increased risk of preeclampsia.[85] In the Maternal-Fetal Medicine Units Network study reported by Myatt et al.,[86] however, there was a low predictive value of notching except for early-onset disease.

Matijevic and Johnson[87] used color pulsed Doppler velocimetry to measure resistance in *uterine spiral arteries*. Impedance was higher in peripheral than in central vessels. This has been termed a "ring-like" distribution. Mean resistance was higher in all women with preeclampsia compared with normotensive controls. Ong et al.[88] used magnetic resonance scanning and intravoxel incoherent techniques to assess placental perfusion ex vivo in myometrial arteries removed from women with preeclampsia or fetal growth restriction. They confirmed that in both conditions myometrial arteries exhibited endothelium-dependent vasodilatory response. Moreover, other pregnancy conditions such as fetal growth restriction are also associated with increased resistance.[89]

Placental vascularity has been assessed with 3-D power Doppler histograms.[90] Compared with normal pregnancy, the *placental vascularity index* was decreased in women with any form of pregnancy hypertensive disorder – 15.2 versus 11.1%, respectively.

Thus, evidence for compromised uteroplacental circulation is found in only a minority of women who go on to develop preeclampsia. Indeed, when preeclampsia develops during the third trimester, only a quarter of women have abnormal uterine artery velocimetry, and with severe disease perhaps a third do so.[91] In one study of 50 women with HELLP syndrome,[92] only a third of fetuses had abnormal uterine artery waveforms. Importantly, the extent of abnormal waveforms correlates with severity of fetal involvement.[81,85]

DIFFERENTIAL DIAGNOSIS

Given the fact that preeclampsia syndrome is a multisystem disorder, one might anticipate that the tendency to confuse it clinically with other pregnancy-related entities – or even nonrelated pregnancy diseases – would be increased. One would also anticipate considerable overlap in both clinical presentation and laboratory abnormalities. In fact, preeclampsia has been referred to as "Another Great Imitator."[93] Sibai[53] compared preeclampsia and a number of pregnancy-related and unrelated disorders that included acute fatty liver of pregnancy, the thrombotic microangiopathies, and exacerbation of systemic lupus erythematosus (SLE). The clinical homology of these disorders with preeclampsia syndrome was summarized in Table 2.1. For example, increased SLE activity in the second half of pregnancy is frequently indistinguishable from new-onset severe preeclampsia.[94] In those cases in which differentiation cannot be made, delivery is indicated so that severe preeclampsia–eclampsia, if present, does not go untreated.

References

1. Chesley LC. *Hypertensive Disorders in Pregnancy*. New York: Appleton-Century-Crofts; 1978: 628.
2. Roberts JM, Redman CWG. Pre-eclampsia: more than pregnancy-induced hypertension. *Lancet*. 1993;341:1447–1451.
3. Redman CWG, Sargent IL. Preeclampsia and the systemic inflammatory response. *Semin Nephrol*. 2004;24:565–570.
4. National high blood pressure education program working group on high blood pressure in pregnancy. *Am J Obstet Gynecol*. 2000;183:S1–S22.
5. American College of Obstetricians and Gynecologists. Task Force on Hypertension in Pregnancy. November 2013.
6. Cunningham FG, Leveno KJ, Bloom SL, Hauth J, Gilstrap LC, Wenstrom KD, eds. *Hypertensive disorders in pregnancy Williams Obstetrics*. New York: McGraw-Hill; 2005. 761–808.
7. Lindheimer MD, Conrad KP, Karumanchi SA. Renal physiology and disease in pregnancy. In: Alpern RJ, Hebert SC, eds. *Seldin and Geibasch's The Kidney*. New York: Academic Press; 2007:2339–2398.
8. Ingrens HU, Irgens HU, Reisaeter L, Irgens LM, Lie RT. Long-term mortality of mothers and fathers after pre-eclampsia: population based cohort study. *BMJ*. 2001;323:1213–1217.
9. Sibai BM, Stella CL. Diagnosis and management of atypical preeclampsia-eclampsia. *Am J Obstet Gynecol*. 2009;200(5):481. e1–e7.
10. Brown CE, Cunningham FG, Pritchard JA. Convulsions in hypertensive, proteinuric primiparas more than 24 hours after delivery. Eclampsia or some other cause. *J Reprod*. 1987;32:499–503.
11. Hirshfeld-Cytron J, Lam C, Karumanchi SA, Lindheimer M. Late postpartum eclampsia: examples and review. *Obstet Gynecol Surv*. 2006;61:471–480.
12. Melchiorre K, Thilaganathan B. Maternal cardiac function in preeclampsia. *Curr Opin Obstet Gynecol*. 2011;23(6):440–447.
13. Evans CS, Gooch L, Flotta D, Lykins D, et al. Cardiovascular system during the postpartum state in women with a history of preeclampsia. *Hypertension*. 2011;58(1):57–62.
14. Clark SL, Cotton DB, Wesley L. Central hemodynamic assessment of normal term pregnancy. *Am J Obstet Gynecol*. 1989;161:1439–1442.
15. Phelan JP, Yurth DA. Severe eclampsia. I. peripartum hemodynamic observations. *Am J Obstet Gynecol*. 1982;144:17–22.
16. Rafferty TD, Berkowitz RL. Hemodynamics in patients with severe toxemia during labor and delivery. *Am J Obstet Gynecol*. 1980;138:263–270.
17. Benedetti TJ, Cotton DB, Read JC, Miller FC. Hemodynamic observations in severe preeclampsia with a flow-directed pulmonary artery catheter. *Am J Obstet Gynecol*. 1980;136:465–470.
18. Hankins GDV, Wendel GW, Cunningham FG, Leveno KL. Longitudinal evaluation of hemodynamic changes in eclampsia. *Am J Obstet Gynecol*. 1984;150:506–512.
19. Groenedijk R, Trimbros JBM, Wallenburg HCS. Hemodynamic measurements in preeclampsia: preliminary observations. *Am J Obstet Gynecol*. 1984;150:232–236.
20. American college of obstetricians and gynecologists: diagnosis and management of preeclampsia and eclampsia. *ACOG Practice Bulletin No. 33*. 2002;99:159–167.
21. Tihtonen K, Kööbi T, Yli-Hankala A, Huhtala H, Uotila J. Maternal haemodynamics in pre-eclampsia compared with normal pregnancy during caesarean delivery. *BJOG*. 2006;113:657–663.
22. Zeeman GG, Cunningham FG, Pritchard JA. The magnitude of hemoconcentration with eclampsia. *Hypertens Preg*. 2009;28:127–137.
23. Pritchard JA, Cunningham FG, Pritchard SA. The Parkland memorial hospital protocol for treatment of eclampsia: evaluation of 245 cases. *Am J Obstet Gynecol*. 1984;148:951.
24. Stahnke E. Über das Verhatten der Blutplättchen bei Eklampie. *Zentralbe fur Gynäk*. 1922;46:391–394.
25. Leduc L, Wheeler JM, Kirshon B. Coagulation profile in severe preeclampsia. *Obstet Gynecol*. 1992;79:14–18.
26. Weinstein L. Syndrome of hemolysis, elevated liver enzymes and low platelet count: a severe consequence of hypertension in pregnancy. *Am J Obstet Gynecol*. 1982;142:159.
27. Weinstein L. Preeclampsia-eclampsia with hemolysis, elevated liver enzymes, and thrombocytopenia. *Obstet Gynecol*. 1985;66:657.
28. Pritchard JA, Cunningham FG, Mason RA. Coagulation changes in eclampsia: their frequency and pathogensis. *Am J Obstet Gynecol*. 1976;124:855–864.
29. Barron WM, Heckerling P, Hibbard JU. Reducing unnecessary coagulation testing in hypertensive disorders of pregnancy. *Obstet Gynecol*. 1999;94:364.
30. Royal College of Obstetricians and Gynaecologists. The management of severe pre-eclampsia. *RCOG Guideline*. 2006;10A:1–11.

31. Cornelis T, Adutayo A, Keunen J, et al. The kidney in normal pregnancy and preeclampsia. *Semin Nephrol.* 2011;31:4.
32. Chesley LC, Williams LO. Renal glomerular and tubular function in relation to the hyperuricemia of preeclampsia and eclampsia. *Am J Obstet Gynecol.* 1945;50:367.
33. Lee W, Gonik B, Cotton DB. Urinary diagnostic indices in preeclampsia-associated oliguria: correlation with invasive hemodynamic monitoring. *Am J Obstet Gynecol.* 1987;156:100.
34. Sibai BM. Diagnosis, controversies, and management of the syndrome of hemolysis, elevated liver enzymes, and low platelet count. *Obstet Gynecol.* 2004;103:981–991.
35. Tun C, Quiñones JN, Kurt A, et al. Comparison of 12-hour urine protein and protein:creatinine ratio with 24-hour urine protein for the diagnosis of preeclampsia. *Am J Obstet Gynecol.* 2012;207:223e1–228e.
36. Kyle PM, Fielder JN, Pullar B, Horwood LJ, Moore MP. Comparison of methods to identify significant proteinuria in pregnancy in the outpatient setting. *BJOG.* 2008;115:523–527.
37. Papanna R, Mann LK, Kouides RW, Glantz JC. Protein/creatinine ratio in preeclampsia: a systematic review. *Obstet Gynecol.* 2008;112:135–144.
38. Airoldi J, Weinstein L. Clinical significance of proteinuria in pregnancy. *Obstet Gynecol Surv.* 2007;62:117–124.
39. Spaan JJ, Ekhart T, Spaanderman ME, Peeters L. Renal function after preeclampsia: a longitudinal pilot study. *Nephron Clin Pract.* 2012;120(3):c156–c161.
40. American society of nephrology renal research report. *Am J Soc Nephrol.* 2005;16:886–903.
41. Drakeley AJ, LeRoux PA, Anthony J, Penny J. Acute renal failure complicating severe preeclampsia requiring admission to an obstetric intensive care unit. *Am J Obstet Gynecol.* 2002;186:253–256.
42. Haddad B, Barton JR, Livingston JC. Risk factors for adverse maternal outcomes among women with HELLP (hemolysis, elevated liver enzymes, and low platelet count) syndrome. *Am J Obstet Gynecol.* 2000;183:444.
43. Frangieh SA, Friedman SA, Audibert F. Maternal outcome in women with eclampsia. *Am J Obstet Gynecol.* 1996;174:453.
44. Virchow R. *Gesammelte Abhandlungen zur Wissenschaftlichen Medicin.* Frankfurt a M: Meidinger Sohn; 1856:778.
45. Sheehan HL, Lynch JB. Cerebral lesions *Pathology of Toxaemia of Pregnancy.* Edinburgh: Churchill Livingstone; 1973.
46. Stella CL, Malik KM, Sibai BM. HELLP syndrome: an atypical presentation. *Am J Obstet Gynecol.* 2008;198:e6–e8.
47. Van Lonkhuijzen LR, Westeriaan HE, Oonk MH. *Am J Obstet Gynecol.* 2007;197:215. **e1-e2**.
48. Hunter SK, Martin M, Benda JA. Liver transplant after massive spontaneous hepatic rupture in pregnancy complicated by preeclampsia. *Obstet Gynecol.* 1995;85:819.
49. Wicke C, Pereira PL, Neeser E. Subcapsular liver hematoma in HELLP syndrome: evaluation of diagnostic and therapeutic options-a unicenter study. *Am J Obstet Gynecol.* 2004;190:106.
50. Barton JR, Sibai BM. HELLP and the liver diseases of preeclampsia. *Clin Liver Dis.* 1999;3:31–48.
51. Rinehart BK, Terrone DA, Magann EF. Preeclampsia-associated hepatic hemorrhage and rupture: mode of management related to maternal and perinatal outcome. *Obstet Gynecol Surv.* 1999;54:3.
52. Vigil-De Gracia P, Ortega-Paz LA. Pre-eclampsia/eclampsia and hepatic rupture. *Int J Gynaecol Obstet.* 2012;118(3):186–189.
53. Sibai BM. Imitators of severe preeclampsia. *Obstet Gynecol.* 2007;109:956–966.
54. Nelson DB, Yost NP, Cunningham FG. Acute fatty liver of pregnancy: clinical outcomes and expected duration of recovery. *Am J Obstet Gynecol.* 2013;209:456.
55. Sibai BM, Ramadan MK, Usta I. Maternal morbidity and mortality in 442 pregnancies with hemolysis, elevated liver enzymes, and low platelets (HELLP syndrome). *Am J Obstet Gynecol.* 1993;169:1000.
56. Isler CM, Rinehart BK, Terrone DA. Maternal mortality associated with HELLP (hemolysis, elevated liver enzymes, and low platelets) syndrome. *Am J Obstet Gynecol.* 1999;181:924.
57. Kozic JR, Benton SJ, Hutcheon JA, Payne BA, Magee LA, von Dadelszen P. Preeclampsia Integrated estimate of risk study group. Abnormal liver function tests as predictors of adverse maternal outcomes in women with preeclampsia. *J Obstet Gynaecol Can.* 2011;33(10):995–1004.
58. Martin Jr JN, Brewer JM, Wallace K, et al. Hellp syndrome and composite major maternal morbidity: importance of Mississippi classification system. *J Matern Fetal Neonatal Med.* 2013;26(12):1201–1206.
59. Keiser SD, Owens MY, Parrish MR, et al. HELLP syndrome with and without eclampsia. *Am J Perinatol.* 2011;28(3): 187–194.
60. Isler CM, Barrilleaux PS, Magann EF. A prospective, randomized trial comparing the efficacy of dexamethasone and betamethasone for the treatment of antepartum HELLP hemolysis, elevated liver enzymes, and low platelet count syndrome. *Am J Obstet Gynecol.* 2001;184:1332.
61. Martin JN, Thigpen BD, Rose CH. Maternal benefit of high-dose intravenous corticosteroid therapy for HELLP syndrome. *Am J Obstet Gynecol.* 2003;189:830.
62. Matchaba P, Moodley J. Corticosteroids for HELLP syndrome in pregnancy. *Cochrane Database.* 2004.
63. Vidaeff AC, Yeomans ER. Corticosteroids for the syndrome of hemolysis, elevated liver enzymes, and low platelets (HELLP): what evidence? *Minerva Ginecol.* 2007;59:183–190.
64. Fonseca JE, Méndez F, Cataño C, Arias F. Dexamethasone treatment does not improve the outcome of women with HELLP syndrome: a double-blind, placebo-controlled, randomized clinical trial. *Am J Obstet Gynecol.* 2005;193:1591–1598.
65. Katz L, de Amorim MMR, Figueroa JN, Pinto e Silva JL. Postpartum dexamethasone for women with hemolysis, elevated liver enzymes, and low platelets (HELLP) syndrome: a double-blind, placebo-controlled, randomized clinical trial. *Am J Obstet Gynecol.* 2008;198:283. **e1 283.e8**.
66. Sibai BM. Diagnosis, prevention, and management of eclampsia. *Obstet Gynecol.* 2005;105:402–410.
67. Cunningham FG, Fernandez CO, Hernandez C. Blindness associated with preeclampsia and eclampsia. *Am J Obstet Gynecol.* 1995;172:1291.
68. Chambers KA, Cain TW. Postpartum blindness: two cases. *Ann Emerg Med.* 2004;43:243.

69. Murphy MAM, Ayazifar M. Permanent visual deficits secondary to the HELLP syndrome. *J Neurophthalmol.* 2005;25:122–127.
70. Lam DS, Chan W. Images in clinical medicine. Choroidal ischemia in preeclampsia. *New Engl J Med.* 2001;344:739.
71. Moseman CP, Shelton S. Permanent blindness as a complication of pregnancy induced hypertension. *Obstet Gynecol.* 2002;100:943–945.
72. Blodi BA, Johnson MW, Gass JD, Fine SL, Joffe LM. Purtscher's-like retinopathy after childbirth. *Ophthalmology.* 1990;97:1654–1659.
73. Ghaem-Maghami S, Cook H, Bird A, Williams D. High myopia and pre-eclampsia: a blinding combination. *BJOG.* 2006;113:608–609.
74. Wiegman MJ, de Groot JC, Jansonius NM, et al. Long-term visual functioning after eclampsia. *Obstet Gynecol.* 2012;119(5):959–966.
75. Saito Y, Tano Y. Retinal pigment epithelial lesions associated with choroidal ischemia in preeclampsia. *Retina.* 1998;18:103–108.
76. Younis MT, McKibbin M, Wright A. Bilateral exudative retinal detachment causing blindness in severe preeclampsia. *J Obstet Gynaecol.* 2007;27:847–848.
77. Cunningham FG, Twickler D. Cerebral edema complicating eclampsia. *Am J Obstet Gynecol.* 2000;182:94.
78. Karumanchi SA, Maynard SE, Stillman IE. Preeclampsia: a renal perspective. *Kidney Int.* 2005;67:2101–2113.
79. Karumanchi SA, Lindheimer MD. Advances in the understanding of eclampsia. *Curr Hypertens Rep.* 2008;10:305–312.
80. Brosens I, Pijnenborg R, Vercruysse L, Romero R. The "Great Obstetrical Syndromes" are associated with disorders of deep placentation. *Am J Obstet Gynecol.* 2011;204:193–201.
81. Ghidini A, Locatelli A. Monitoring of fetal well-being: role of uterine artery Doppler. *Semin Perinatol.* 2008;32:258–262.
82. Everett TR, Mahendru AA, McEniery CM. Raised uterine artery impedance is associated with increased maternal arterial stiffness in the late second trimester. *Placenta.* 2012;33(7):572–577.
83. Fleischer A, Schulman H, Farmakides G. Uterine artery Doppler velocimetry in pregnant women with hypertension. *Am J Obstet Gynecol.* 1986;154:806–813.
84. Trudinger BJ, Cook CM. Doppler umbilical and uterine flow waveforms in severe pregnancy hypertension. *Br J Obstet Gynaecol.* 1990;97:142–148.
85. Groom KM, North RA, Stone PR, et al. Patterns of change in uterine artery Doppler studies between 20 and 24 weeks of gestation and pregnancy outcomes. *Obstet Gynecol.* 2009;113:332–338.
86. Myatt L, Clifton RG, Roberts JM, et al. The utility of uterine artery Doppler velocimetry in prediction of preeclampsia in a low-risk population. *Obstet Gynecol.* 2012;120(4):815–822.
87. Matijevic R, Johnston T. In vivo assessment of failed trophoblastic invasion of the spiral arteries in pre-eclampsia. *Br J Obstet Gynaecol.* 1999;106:78.
88. Ong SS, Moore RJ, Warren AY. Myometrial and placental artery reactivity alone cannot explain reduced placental perfusion in pre-eclampsia and intrauterine growth restriction. *BJOG.* 2003;110:909–915.
89. Urban G, Vergani P, Ghindini A, et al. State of the art: non-invasive ultrasound assessment of the uteroplacental circulation. *Semin Perinatol.* 2007;31:232–239.
90. Pimenta E, Ruano R, Francisco R, et al. 3D-power Doppler quantification of placental vascularity in pregnancy complicated with hypertensive disorders. Abstract No. 652. *Am J Obstet Gynecol.* 2013;208(1 Suppl):S264.
91. Li H, Gudnason H, Olofsson P, Dubiel M, Gudmundsson S. Increased uterine artery vascular impedance is related to adverse outcome of pregnancy but is present in only one-third of late third-trimester pre-eclampsia women. *Ultrasound Obstet Gynecol.* 2005;25:459–463.
92. Bush KD, O'Brien JM, Barton JR. The utility of umbilical artery Doppler investigation in women with HELLP (hemolysis, elevated liver enzymes, and low platelet count) syndrome. *Am J Obstet Gynecol.* 2001;184:1087–1089.
93. Goodlin RC. Severe pre-eclampsia: another great imitator. *Am J Obstet Gynecol.* 1976;125:747–753.
94. Petri M. The Hopkins Lupus Pregnancy center: ten key issues in management. *Rheum Dis Clin North Am.* 2007;33:227–235.
95. Napolitano R, Melchiorre K, Arcangeli T, et al. Screening for pre-eclampsia by using changes in uterine artery Doppler indices with advancing gestation. *Prenat Diagn.* 2012;32(2):180–184.

CHAPTER 3

Epidemiology of Pregnancy-Related Hypertension

JANET W. RICH-EDWARDS, ROBERTA B. NESS AND JAMES M. ROBERTS

Editors' comment: *The epidemiology literature relating to hypertension in pregnancy grew substantially since the third edition, much of it focusing on preeclampsia's relationship to subsequent cardiovascular and renal disease. There is now little doubt that formerly preeclamptic women have significantly more remote disease, especially if their preeclampsia occurred preterm, when compared to women with normotensive deliveries, but whether preeclampsia is a "marker" of already at risk populations, or that preeclampsia itself is causal, remains to be determined. Either way there is a need for balanced programs that focus on prevention, forestalling, and early intervention for this at-risk population. Finally, we welcome a new first author, Janet Rich-Edwards an epidemiologist and co-director of the Reproductive, Perinatal, and Pediatric Track at the Harvard University School of Public Health. Her research focus includes effects of the perinatal environment on future cardiovascular disease in mothers and their offspring.*

INTRODUCTION

Hypertensive disorders specific to pregnancy include gestational hypertension (formerly known as transient hypertension of pregnancy), preeclampsia, and eclampsia. These syndromes, marked by elevations in blood pressure that return to normal after delivery, are distinguished clinically by the presence (preeclampsia and eclampsia) or absence (gestational hypertension) of proteinuria or systemic findings. Eclampsia is a more life-threatening syndrome, characterized by seizures and thought to represent the progression of preeclampsia. Preeclampsia arising in the context of chronic hypertension preceding pregnancy is known as superimposed preeclampsia; in this case, blood pressures do not return to normal after pregnancy. As many women's blood pressures are unknown before pregnancy, when hypertension fails to resolve after pregnancy, superimposed preeclampsia is assumed *post hoc*.

Here, we will review the epidemiologic data that characterize the frequency of occurrence, risk factors and predictors for, and the natural history and later life implications of hypertensive disorders in pregnancy. It is the purpose of this chapter to review these epidemiologic data, critically assess the methods used in conducting previous studies, and suggest new research areas, focusing on preeclampsia. We will review evidence suggesting that preeclampsia, in itself, is a syndrome composed of at least two underlying pathophysiologies, usually interactive.[1] One pathophysiological pathway may be the result of reduced placental perfusion and is the component of preeclampsia unique to pregnancy. The other may be related to preexisting maternal pathology, in particular, an often subclinical maternal predisposition to cardiovascular risk. We have proposed in the past that these causes converge to predispose to preeclampsia, with its hallmark placental lesions and deficit of angiogenesis, by the common mechanism of endothelial dysfunction. The combination of both maternal and placental factors is particularly damaging. As such, we will discuss the natural heterogeneity among the risk factors and clinical predictors as well as among the outcomes for mother and infant within the spectrum of preeclampsia.

DEFINITIONS OF THE HYPERTENSIVE DISORDERS OF PREGNANCY

There are differences in reports of the frequency with which preeclampsia and gestational hypertension occur, due in part to disparities in the many definitions used to classify the hypertensive disorders of pregnancy. In recent years, there has been increasing convergence in definitions of preeclampsia, including those of the American College of Obstetrics and Gynecology (ACOG) Task Force on Hypertension in Pregnancy and the International Society for the Study of Hypertension in Pregnancy. Both bodies require the new onset of systolic blood pressure of ≥140 mm Hg or a diastolic blood pressure of ≥90 mm Hg as the foundation for both gestational hypertension and preeclampsia.[2,3] Both definitions recognize the combination of hypertension and proteinuria as preeclampsia, but both bodies now recognize new definitions of

Chesley's Hypertensive Disorders in Pregnancy.
ISBN: 978-0-12-407866-6

DOI: http://dx.doi.org/10.1016/B978-0-12-407866-6.00003-1

preeclampsia that do not require proteinuria. In recognition of the syndromic nature of preeclampsia other organ involvement may substitute for proteinuria and can be used to complete the diagnosis of preeclampsia. They recommended documentation of the elevated blood pressure and proteinuria on at least two occasions. For research purposes, both bodies require documentation of normotension before 20 weeks gestation and after 12 weeks postpartum; in 2000, the International Society for the Study of Hypertension in Pregnancy (ISSHP) recommended researchers increase the specificity of preeclampsia diagnosis by requiring proteinuria.[4] Other, less well-accepted definitions, often including clinical symptoms as diagnostic criteria, have also been used. Thus, the frequency of hypertensive disorders has been variously estimated based on different definitions that vary across geography and time. It is likely that the changes in diagnostic criteria by the ACOG and ISSHP will increase the frequency of preeclampsia.

Most, but not all, studies of hypertension in pregnancy exclude women with preexisting hypertension, so that few estimates of the prevalence of preeclampsia include superimposed preeclampsia. Recently, studies have begun to consider superimposed preeclampsia in its own right, due to its apparent increase with rising obesity and chronic hypertension in young women.

Recently, the ISSHP issued consensus definitions for severe and early-onset preeclampsia.[5] 'Severe' preeclampsia required systolic blood pressure >160mm Hg or diastolic blood pressure >110mm Hg, with evidence of proteinuria (spot urine >30mg/mmol creatinine). Degree of proteinuria was not deemed a criterion for severity of preeclampsia. The ACOG definition of severe preeclampsia has similar blood pressure criteria, but departs from the ISSHP definition by not requiring proteinuria, and instead requiring any of the following: thrombocytopenia, impaired liver function, renal insufficiency, pulmonary edema, or new onset cerebral of visual disturbances.[2] The ISSHP has defined early onset preeclampsia as that arising before 34 weeks gestation. "Preterm preeclampsia" was defined as that occurring between 34 + 1 to <37 + 0 gestation, and "term" preeclampsia as that occurring after 37 weeks. It should be noted that most large studies use the gestational age at delivery, rather than the gestational age at diagnosis of preeclampsia, to distinguish "preterm" from "term" preeclampsia, as the latter is not usually available and is subject to the frequency of prenatal visits.

Finally, changes in the international classification of hypertensive pregnancy have made it difficult to compare preeclampsia prevalence across time and place; for example, the ICD-10 system combines mild preeclampsia and gestational hypertension, while the ICD-9 system did not, resulting in a seeming drop in the prevalence of preeclampsia when statistics are compared across old and new versions. With these caveats in mind, we review below the literature on the prevalence of the hypertensive disorders of pregnancy. Illustrative studies, organized by case ascertainment method, are presented in Table 3.1.

PREVALENCE OF HYPERTENSIVE DISORDERS OF PREGNANCY

Eclampsia

The most reliable estimate of the prevalence of eclampsia is probably that reported from a national survey in the United Kingdom by Douglas and Redman.[10] All obstetricians and all hospitals with an obstetrics unit were asked to participate in an active surveillance program in 1992. Each presumptive case was reviewed by a single obstetrician and was defined as eclamptic if there were seizures in the setting of hypertension, proteinuria, and either thrombocytopenia or an increased plasma aspartate transaminase concentration. The prevalence of eclampsia was estimated at 0.049% pregnancies. Most seizures occurred despite prenatal care (70%) and even after admission to the hospital (77%).

Table 3.1 shows a broad range of estimates of eclampsia incidence, ranging from 0.02% to 0.1% when ascertained in birth and other statistical registries and from 0.02% to 0.6% in studies with medical record review. The prevalence of eclampsia has declined rapidly during the 20th century in high-income countries, from prevalences of 0.3% or more before 1930 to prevalences of 0.03% or less in the past decade.[21] Chesley showed a marked reduction at Margaret Hague Maternity Hospital in Jersey City from 1931 to 1951 (Table 3.2). In recent decades, the risk of eclampsia dropped in the USA from an average annual rate of 0.10% in 1987–1995 to 0.08% of deliveries from 1996–2004, even as rates of preeclampsia have risen.[6] As pointed out several times by Leon Chesley,[22] the reduction in eclampsia is largely related to improved medical care rather than a changing natural history of preeclampsia. The prevalence of eclampsia and eclampsia-associated case-fatality in many lower-income countries remains high.[21] A recent systematic review estimated rates as high as 2.7% in Africa and as low as 0.1% in Europe.[20]

Preeclampsia

Because of the decline in eclampsia in the developed world, much of the recent epidemiologic research has focused on preeclampsia. Table 3.1 shows estimates derived from registries, ranging from 1% to 4%, and from medical record review, ranging somewhat higher, at 1% to 7%. In the USA, where reporting of preeclampsia is not mandatory, Wallis, Saftlas and colleagues[6,22] estimated the prevalence of hypertensive pregnancies from 1979 through 2004 using a nationally representative sample of hospital discharge records. Women with discharge diagnoses of gestational hypertension (ICD-9 642.3), preeclampsia (ICD-9 642.4 or 643.5) or eclampsia (ICD-9 643.6) were included. The authors estimated that preeclampsia complicated 2.9% of pregnancies in 2003–2004, a 25% increase since 1987.[6] Similar prevalence of preeclampsia was estimated from hospital discharge data in the Danish National

TABLE 3.1 Prevalence of Hypertensive Disorders of Pregnancy Since 1980 in Illustrative Studies, by Case Ascertainment Method

Study	Population	Size (ICD)	P(HDP)	P (GH)	P (PE)	P (ECL)
Hypertensive disorders ascertained by general population registry or administrative database						
Wallis, 2008[6]	US National Hospital Discharge Survey, 1987–2004	~200,000 births per year (ICD-9)		2.1% 3.0% in 2004	2.7% 3.2% in 2004	0.09%
Goldenberg, 2011	Iceland, USA, UK, Singapore since 1980					0.02–0.1%
Roberts, 2011[7]	Alberta, Canada	256,137 (ICD-10)	6.0%		1.4%	
	New South Wales, Australia	732,288 (ICD-10)	8.8%		3.3%	
	Western Australia	149,624 (ICD-10)	9.1%		2.9%	
	Denmark	645,993 (ICD-10)	3.6%		2.7%	
	Norway	456,353 (ICD-10)	5.8%		4.0%	
	Scotland	531,622 (ICD-10)	5.9%		2.2%	
	Sweden	913,779 (ICD-10)	3.9%		2.9%	
	Massachusetts, USA	762,723 (ICD-9)	7.0%		3.3%	
Gaio, 2001[8]	Brazilian Multicentre Cohort: 5 state capitals	4,892 hospital discharge records	7.5%	0.7%	2.3%	0.16%
Roberts CL, 2005[9]	New South Wales, 2000–2002	250,173	9.8%	4.3%	4.2%	0.06%
Hypertensive disorders ascertained by medical record review or study protocol (in placebo or standard care arms, where relevant)						
Douglas, 1994[10]	UK, 1992. Case review.	All UK births				0.05%
Knight, 2007[11]	UK, Feb 2005 – Feb 2006. Case review.	All UK births				0.03%
Souza, 2013[12]	WHO hospital survey in 29 countries in Asia, Africa, and Latin America	314,623			2.5%	0.3%
Conde-Agudelo, 2000[13]	Perinatal information system, Latin America & Caribbean (ICD-10)	878,680			4.8%	0.2%
Sibai, 1993[14]	Aspirin trial: USA, healthy nulliparas	1,565		5.9%	6.3%	
Villar, 2001[15]	WHO Antenatal Care trial: Argentina, Cuba, Saudi Arabia, Thailand	11,121		5.0%	1.3%	0.08%
Lumbiganon, 2007[16]	22 hospitals, Mexico City	18,288			5.5%	0.6%
	18 hospitals, Thailand	17,525			1.9%	0.3%
Chalumeau, 2002[17]	6 West African countries, 1994–1996	20,326	8.0%			0.02%
Rotchell, 1998[18]	Barbados Aspirin Study: Maternity Hospital	1,822		7.3%	4.6%	
Roberts, 2010[19]	Antioxidant trial: USA, healthy nullipara	4,993		19.9%	6.7%	0.1%
Hypertensive disorders defined by mixed methods						
Abalos, 2013[20]	Africa (AFRO)	93,613 (57% Nigerian)			4.0%	2.7%
	Americas (AMRO)	36,693,594 (99% U.S.)			2.3%	1.1%
	Eastern Mediterranean (EMRO)	148,909			1.2%	0.5%
	European (EURO)	1,093,782			3.8%	0.1%
	Southeast Asian (SEARO)	203,159			2.7%	1.3%
	Western Pacific (WPRO)	361,402 (69% Australian)			4.2%	0.1%

ICD = International classification of diseases; P = percentage; HDP = hypertensive disorders of pregnancy; GH = gestational hypertension; PE = preeclampsia; ECL = eclampsia.

TABLE 3.2 Incidence of Eclampsia in Clinic Patients at the Margaret Hague Maternity Hospital

	1931–1934	1935–1939	1940–1945	1946–1951	Totals
Registrations	12,604	17,407	12,022	12,208	54,241
Cases of eclampsia	51	41	11	4	107
% incidence	0.40	0.23	0.09	0.03	0.20

Birth Cohort of over 100,000 women recruited early in pregnancy, where the overall prevalence of preeclampsia was 3.0%, including 4.2% of nulliparous women and 1.3% of parous women.[24]

The recent survey of multiple datasets published by Abalos (summarized in Table 3.1) reveals variability in regional preeclampsia rates, but does not suggest a pattern of especially high preeclampsia rates in lower-income regions.[20] Observational studies and randomized clinical trials that follow women prospectively with the express intent of documenting hypertensive disorders of pregnancy with standardized protocols often report modestly higher prevalences of these disorders than are captured retrospectively by hospital discharge or birth registry databases. In a somewhat more selected population in the United States, the incidence of preeclampsia was estimated among control women enrolled in the NICHD Maternal Fetal Medicine Network for Clinical Trials (MFMU Network) trial of low-dose aspirin to prevent preeclampsia terminating in 1993.[14] Included in the study were nulliparous women presenting to a series of academic medical centers for prenatal care. Women with a history of chronic hypertension, diabetes mellitus, renal disease, and other medical illnesses as well as women with a baseline blood pressure above 135/85 mm Hg were excluded. Among the 1500 women in the placebo group who were followed throughout pregnancy, 94 (6.3%) developed preeclampsia as defined by hypertension (systolic blood pressure of ≥140 mm Hg or a diastolic blood pressure of ≥90 mm Hg) plus proteinuria (either ≥300 mg/24 h or 2+ or more by dipstick on two or more occasions 4 hours apart). In a more recent 2010 study of more than 4500 control pregnancies using the same patient inclusion and exclusion and diagnostic criteria, the preeclampsia rate was remarkably similar, 6.7%.[19] Other randomized clinical trials in the United States and Europe have demonstrated similar incidence rates for preeclampsia.[25–27]

Superimposed preeclampsia, the development or worsening features of preeclampsia among women with chronic hypertension, has often been excluded from the study of hypertensive pregnancies. Estimates of the prevalence of superimposed preeclampsia are hard to come by, although many commentators have observed that it may be increasing as the prevalence of overweight and chronic hypertension rises among women of reproductive age. Among women with preexisting hypertension, preeclampsia will develop in 10–25%, as compared to a general population rate of 3–7%.[2–20,22–30] The more severe and longstanding the hypertension prior to pregnancy, the greater is the risk of preeclampsia.[31–33] In data from the MFMU Network,[33] women with hypertension for at least 4 years duration had a remarkably high rate of preeclampsia: 31%.

Gestational hypertension has been especially difficult to characterize, as it is notoriously under-reported in registry databases. Estimated prevalences range from 1% to 4% from registries, but are considerably higher in studies that review medical records, on the order of 5% to 7% (Table 3.1). Wallis et al.'s analysis of the U.S. National Hospital Discharge survey estimated that gestational hypertension complicated 3.0% of pregnancies in 2003–2004, a prevalence that had increased by 184% since 1987.[6] Probably the best estimates come from trials. The MFMU Network aspirin trial in 1993 estimated that 5.9% of control women developed gestational hypertension (hypertension as defined above without proteinuria) but the percentage was much higher, 19.9%, in a 2010 MFMU Network study[19] and in another 1993 NICHD trial of healthy US nulliparous women (17.3%).[14,34] Rates in most other studies are in agreement with the aspirin trial.[26,27]

Discussion of Differential Frequency Estimates

Hospital-based incidence estimates for preeclampsia systematically differ from national estimates. There are probably five main reasons for this discrepancy. First, many studies, such as that by Saftlas et al., rely on discharge diagnoses.[23] Ales and Charlson[35] performed a validation study of medical records at the New York Hospital and found that 25% of ICD-9 codes incorrectly diagnosed preeclampsia and that 53% of true cases were missed by ICD-9 coding. Similarly, Eskenazi et al.[36] found that 47% of 263 women who received a discharge diagnosis of severe preeclampsia or eclampsia did not meet a rigorous set of criteria. Klemmensen et al. reported 2.9% prevalence of PE recorded in the Danish National Patient Registry, compared with 2.7% by medical record review and 3.4% by maternal recall.[37] However, 26% of cases in the registry proved not to have preeclampsia by record review, and 31% of preeclampsia cases identified by record review were not recorded as such by the registry. The registry severely underestimated gestational hypertension compared with medical record review. Such under-reporting of milder forms of pregnancy hypertension is widely reported.[37,38] Second, the MFMU Network studies included only nulliparous women, a group known to be at five-fold or greater risk of developing preeclampsia as compared to parous women.[19,39] Third, women electing to enroll in a randomized clinical trial to prevent preeclampsia may also be a group with characteristics that would suggest a tendency to developing hypertension in pregnancy. However, the strict definition of preeclampsia used in the MFMU Network trials should have resulted in a decreased incidence estimate. Fourth, women seeking prenatal care at academic medical centers are a selected group who are probably at higher risk of developing pregnancy complications than would be reflected in a national sample. Fifth, data from the British Commonwealth in years past considered *de novo* hypertension without proteinuria as mild preeclampsia, and in these studies the diagnosis is substantially contaminated with women with transient hypertension. Nonetheless, a reasonable estimated range for the rate of preeclampsia in developed countries is 3–7%.

RISK FACTORS FOR PREECLAMPSIA

Risk factors consistently shown to be associated with an increased rate of preeclampsia (Table 3.3) include elevated prepregnancy or early pregnancy blood pressure, prepregnancy adiposity, age (Fig. 3.1), family history of preeclampsia or of cardiovascular disease, African-American ethnicity (Fig 3.2), preexisting medical conditions such as hypertension or diabetes, obstetric characteristics such as multiple gestation and hydrops fetalis, nulliparity, and history of a previous preeclamptic pregnancy (Table 3.4).

Most of these factors can be understood in relation to a maternal predisposition to cardiovascular disease. The last factors, obstetric characteristics and nulliparity, may represent the placental or uniquely pregnancy-related component of preeclampsia. For each of these factors, we will relate the data suggesting its association with preeclampsia. We will also evaluate each factor's ability to accurately predict the development of preeclampsia.

TABLE 3.3 Risk Factors for Preeclampsia and a Subjective Rating of the Weight of Supporting Evidence

Risk factor [ref.]	Rating of Evidence
Older age[23,40,41]	++
High blood pressure in 2nd/early 3rd trimester[22,31,39,42–47]	++
Prepregnancy hypertension[31,48]	+++
Prepregnancy diabetes[22,42,49,50]	+++
Caloric excess[22,51,52]	+
Elevated body mass index[24,36,53–56]	+++
Weight gain during pregnancy[57–59]	+
Family history[60–63]	+++
Nulliparity[64,65]	+++
New paternity[66–68]	++
Lack of previous abortion[36,42,65,69–71]	++
Barrier contraception[70,71]	++
Excessive placental size[41,72–74]	+++
Lack of smoking[71,75–78]	+++
Specific dietary factors[79–81]	+
African American[36,42]	+

Cardiovascular Risk Factors

Cardiovascular risk factors related to preeclampsia include: preexisting hypertension,[29] elevated prepregnancy or early pregnancy blood pressure,[42,90] family history of heart disease,[63] diabetes mellitus,[91] and in this category we also place African-American ethnicity,[82] adiposity,[36,92] lack of physical activity,[93] and older age.[40] Biochemical markers associated with cardiovascular risk, including elevated glucose/insulin, lipids, coagulation factors, inflammatory

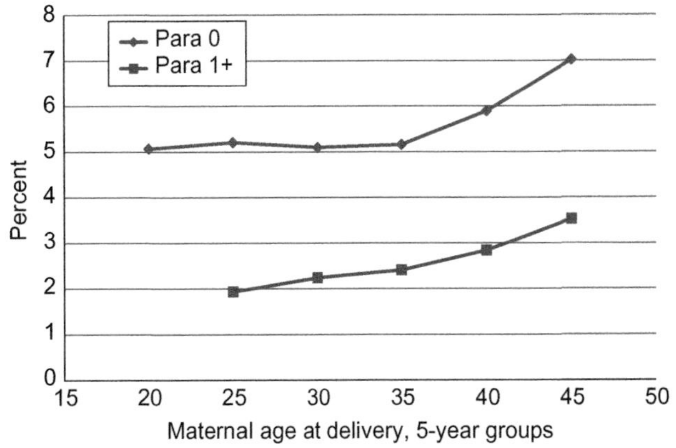

FIGURE 3.1 Prevalence of preeclampsia by age and parity, Norway, 1999–2012. (K. Klunsoyr, personal communication.) (This figure is reproduced in color in the color plate section.)

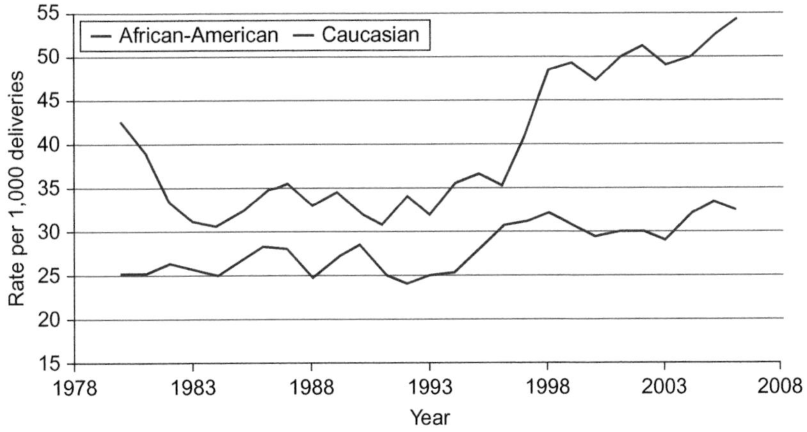

FIGURE 3.2 Cases of preeclampsia per 1000 births by maternal race, U.S. National Hospital Discharge Survey, 1979 to 2006.[82] Reprinted from Breathett K, Muhlestein D, Foraker R, Gulati M. The incidence of pre-eclampsia remains higher in African-American women compared to Caucasian women: trends from the National Hospital Discharge Survey 1979–2006. *Circulation.* 2013;127:AP192, with permission. (This figure is reproduced in color in the color plate section.)

TABLE 3.4 Prevalence of Hypertensive Disorders of Pregnancy in Illustrative Studies, by Parity, Multiple Pregnancy, and Recurrence

Study	Population	Size (ICD)	P (GH)	P (PE)	P (ECL)
By parity					
Wu, 2009[83]	Denmark, 1978–2004	1.6 million		6.3% nulliparous 1.9% parous	0.10% nulliparous 0.03% parous
Hernandez-Diaz, 2009[84]	Sweden, 1987–2004	763,795 (ICD-9 & 10)		4.1% nulliparous 1.7% parous	
By singleton versus multiple pregnancies					
Ros, 1998[85]	Sweden, 1987–1993	10,666 nulliparas (ICD-9)	4.4% single 6.4% multi	6.4% singleton 18.0% multiple pregnancy	
Douglas, 1994[10]	UK, 1992	All UK births			0.05% single 0.28% multi
Prevalence of preeclampsia in second pregnancies following first pregnancies that were preeclamptic					
			P (PE)		
Mostello, 2008[86]	Missouri, USA, 1989–1997	103,860	14.7% Risk falling with increased length of 1st pregnancy: 12.9–38.6%		
Lykke, 2009[87]	Denmark	22,198	14.1–37.9% Risk falling with increased length of 1st pregnancy		
Hernandez-Diaz, 2009[84]	Sweden, 1987–2004	763,795 (ICD-9 & 10)	14.7%		
Klungsoyr, 2012[88]	Norway, 1967–2006	2,416,501	15.2%		
Hnat, 2002[89]	USA, 1991–1995	598	17.9%		

ICD = International classification of diseases; P = percentage; GH = gestational hypertension; PE = preeclampsia; ECL = eclampsia.

mediators, homocysteine, uric acid, and angiogenic and endothelial cell activation markers, also differentiate women developing preeclampsia from women with unaffected pregnancies. Many of these cardiovascular risk factors predate preeclamptic pregnancies, albeit at subclinical levels.[90] Even in healthy nulliparous women, higher blood pressures prior to 27 weeks gestation are associated with an elevated risk of developing preeclampsia; the higher the blood pressure, the higher the risk of preeclampsia.[42] Prepregnancy diabetes also increases the risk for developing preeclampsia.[42] The more severe the diabetes prior to pregnancy, the greater is the risk of developing preeclampsia.[91] Among women without preexisting diabetes, hyperglycemia and insulin resistance elevate the risk of developing preeclampsia.[94,95] Women with polycystic ovarian syndrome (PCOS) have an excess of circulating androgens, are insulin resistant and have an increased frequency of preeclampsia.[94] It is posited that increased insulin resistance explains the increased risk; however, in women without PCOS, preeclampsia is also associated with increased circulating androgens.[96]

Body Mass Index

Elevations in body mass index, a marker of adiposity, have repeatedly been associated with risk of preeclampsia.[24,36,53,97–99] Measured either at the first prenatal visit or prior to pregnancy, the magnitude of this effect ranges from 2.5 to 5-fold. This increase is in mild,[54] severe,[24,55] and early-onset[24] preeclampsia, and is present in both black and white women.[55] Even a modest increase in adiposity can increase preeclampsia risk. Women who fail to return to their prepregnant weight between pregnancies are at increased risk of preeclampsia in their second pregnancy compared to women who return to prepregnancy weight.[15] Among morbidly obese women, those who obtained gastric bypass surgery before pregnancy had a 2.5% risk of preeclampsia or eclampsia, compared with a 14.5% risk among those who had their surgery after pregnancy; similar reductions were observed for gestational hypertension and superimposed preeclampsia/eclampsia.[100] As with cardiovascular disease, it is likely that visceral obesity is most relevant. This is supported by the observation that waist circumference in excess of 80 cm, a more specific marker of biologically active abdominal fat, was associated with a 2.7-fold elevation in the development of preeclampsia in one recent study.[92]

Physical Activity

Paralleling the known benefits of physical activity in reducing the risk of cardiovascular disease and type 2 diabetes mellitus, limited evidence suggests that physical activity may protect against preeclampsia. A recent review found that the role of exercise in reducing preeclampsia was supported in six observational studies, but not in prospective studies. Nonetheless, the weight of opinion favors a beneficial effect.[101] The mechanisms are not established, but it has been shown that exercise can reduce weight gain during pregnancy.[102,103] In a study of acute and chronic exercise in relationship to angiogenic activity, acute exercise in pregnant women was associated with an increase in the pro-angiogenic placental growth factor, and a reduction in the antiangiogenic factors, soluble endoglin and soluble fms-like tyrosine kinase.[104]

Diet

Diet may also affect preeclampsia risk in a similar fashion to its effect on coronary artery disease risk. Women with low concentrations of erythrocyte membrane omega-3 (marine oil) fatty acids are reportedly more likely to have had preeclamptic pregnancies than women with high erythrocyte omega-3 fatty acids.[105] Dietary intake studies largely from Scandinavia support an observation from older and smaller studies that increased sugar intake, largely from soft drinks, is associated with increased risk of preeclampsia.[79] Interestingly, in a very large study the uncorrected effect of added sugar but not sugars in food was associated with an increased risk of preeclampsia. However, after adjustment for maternal age at delivery, education, prepregnant BMI, height, smoking, leisure exercise in the first pregnancy trimester, total energy intake and dietary fiber, only the effects of sweetened drinks persisted. A unique feature of such drinks is sweetening with high-fructose corn syrup, which has been suggested to have unique metabolic features increasing the risk of obesity and perhaps cardiovascular disease.[106] In a study of dietary patterns from Norway, women with high intake of fresh and cooked vegetables, olive oil, fruits and berries and poultry had a reduced incidence of preeclampsia while women with an intake of predominantly processed food (including sweetened drinks) had an increased risk.[107] Studies of micronutrients also suggest a relationship of increased risk with deficiencies of certain micronutrients. These deficits unfortunately have not yet resulted in effective therapy with supplements. Women who ingested less than 85 mg of vitamin C daily (below the recommended daily allowance) were at two-fold increased risk of developing preeclampsia,[81] while women at high risk of preeclampsia with the highest quartile of vitamin C concentration at 18 weeks gestation had a 40% lower risk of preeclampsia.[108] Nonetheless, there is no evidence that in unselected high- or low-risk populations the administration of large doses of vitamins C and E reduces the frequency of preeclampsia.[109,110] One of the older relationships is that of low calcium intake and preeclampsia;[111] however, replacement studies with up to 2 grams of calcium have had only a small effect.[112] Other micronutrients may also be relevant. A self-reported history of ingestion of multivitamins prior to gestation was associated with a 71% reduction in preeclampsia.[113] Interestingly, this reduction was not present in obese women.

Vitamin D

In recent years there has been extensive attention paid to the relationship of circulating concentrations of vitamin D to adverse pregnancy outcomes, including preeclampsia. Based on data from Pittsburgh, 25-hydroxyvitamin D (25(OH)D) was deficient (<37.5 nmol/L) in 5.0 % of white women and insufficient (37.5–80 nmol/L) in 42.1%.[114] The frequency of inadequate circulating concentrations of 25(OH)D was even greater in black women, with 29.2% at deficient concentrations and 54.1% insufficient concentrations, perhaps reflecting the lower conversion of vitamin D from 7-dehydrocholesterol in dark skin or lower intake of vitamin D-rich or enriched foods or supplements. Lower plasma vitamin D concentrations have been posited as a potential explanation for the increase in adverse pregnancy outcomes in African-Americans.[115] In keeping with the importance of ultraviolet light exposure, in northern countries with less ultraviolet exposure the rates of deficiency are even higher. In white individuals the rate of deficiency was 47%.[116] A recent meta-analysis found that in 31 studies of the relationship of vitamin D with preeclampsia the rate of preeclampsia was 80% higher (1.79, 1.25 to 2.58) in individuals with vitamin D insufficiency.[117] There are no large published randomized controlled trials of vitamin D to prevent preeclampsia. However, in a large Norwegian study it was possible to estimate vitamin D intake from diet and supplements and as vitamin D intake increased from 3 μg to 15–20 μg/ day there was a 23% reduction in the frequency of preeclampsia.[107] Nonetheless, increasing the dose of vitamin D by supplementation remains controversial, with two major scientific organizations presenting conflicting recommendations.[118]

Smoking

The one traditional coronary artery disease risk factor associated with a reduction, rather than an elevation, in risk of preeclampsia is cigarette smoking.[21,36] The reason for this inverse association with preeclampsia is unclear but may reflect an idiosyncratic effect of smoking on angiogenesis,[119,120] a survival bias wherein the most affected fetuses are aborted and thus do not develop preeclampsia (M. Williams, personal communication), or an antiinflammatory effect of pregnancy.[121] There is also evidence that the effect of smoking to reduce preeclampsia is not present in women beyond the age of 30.[122] This could suggest that there is an early pharmacological effect to prevent preeclampsia, which is overcome by "damage" with more prolonged smoking. Although the identity of this pharmacological agent is not established, it does not appear to be nicotine. Unlike smoking, the intake of nicotine in snuff does not reduce preeclampsia.[123]

Family Patterns

The best estimates of heritability and non-shared environmental effect on preeclampsia are an 0.54 index for heritability and a 0.46 index for non-shared environment, suggesting equivalent contributions of genes and environment in determining preeclampsia risk; corresponding estimates for gestational hypertension are 0.24 for heritability and 0.76 for non-shared environment.[124] In the Swedish Twin Registry that examined female twin pairs in which at least one twin had a pregnancy complicated by preeclampsia, there were 16 of 61 monozygotic pairs in which both twins were affected (concordance rate 0.25); and 4 of 63 dizygotic pairs in which both sisters were affected (concordance rate 0.06), These results suggest an important role of maternal genes in determining preeclampsia, and a lesser role in determining gestational hypertension.[124] These estimates rely on linkage of mothers and their sisters, and do not comment upon the contribution of paternal (and therefore fetal) genes.

That fathers and fetuses help drive the risk of preeclampsia is suggested by the observation that men who father preeclamptic pregnancies with one woman are at 80% higher likelihood of fathering a preeclamptic pregnancy with another woman.[125] Furthermore, there appears to be a "mother-in-law effect" in which a son born of a preeclamptic pregnancy has an elevated chance of fathering a preeclamptic pregnancy.[126] In fact, women born of a preeclamptic pregnancy have a 3-fold higher and men born of preeclamptic pregnancy have a 2-fold higher risk of triggering severe preeclampsia in their own (or their partner's) pregnancy.[126]

These patterns imply an interacting role for both maternal and fetal genes in preeclampsia. Further complexity is added by the possibility that the impact of parentally imprinted genes controlling trophoblast growth and fetal development depends on whether the genes are of maternal or paternal origin.[127]

FIRST BIRTH AND OTHER PLACENTAL FACTORS

Nulliparity is a particularly strong risk factor for the development of preeclampsia (Fig. 3.1, Table 3.3, Table 3.4). Large studies from Denmark and Sweden indicate that nulliparous women have preeclampsia risks on the order of 4% to 6%, in contrast to parous women, whose risk falls to roughly 2% (Table 3.4).[83,84] MacGillivray noted in 1958 in a well-characterized population in Scotland that preeclampsia occurred in 5.6% of nulliparas and in only 0.3% of secundi-paras.[39] Other authors confirmed the observation that nulliparas were anywhere from five to ten times more likely to experience preeclampsia than parous women. Unadjusted data on the relationship between age and preeclampsia have frequently shown younger women to be at an increased risk. However, nulliparity may be the underlying factor driving this apparent association as younger women are more likely to be nulliparous. In fact, after accounting for parity, preeclampsia risk increases with maternal age (Fig. 3.1). There is evidence that the risk associated with advanced maternal age may be narrowing over time.[88]

The relationship between nulliparity and preeclampsia suggests something about the uniqueness of a first placentation vs. a later baby's placental implantation. Redman suggested that in later pregnancies there is the development of protective immunologic mechanisms against paternal antigens.[64] Other epidemiologic data support this suggestion. For example, studies have shown that the relative protection afforded by further pregnancies was reduced or eliminated with new paternity, that previous induced abortion was protective,[36,42,65,69] that women using barrier methods of contraception were at increased risk,[70,71] and that risk was reduced with increased duration of sexual activity antedating pregnancy.[66] All of these risk factors would reduce maternal recognition of paternal antigens prior to pregnancy and all of the protective factors would enhance maternal recognition of paternal antigens. Other studies have reported that change in sexual partner elevated the risk of preeclampsia in a second pregnancy to a level almost as high as for first pregnancies.[128] However, subsequent cohort studies demonstrated that a longer interval between pregnancies might account for the apparent partner change effect.[129] Further supporting the antigenic theory are observations that teenage pregnancies, out-of-wedlock pregnancies, and donor sperm insemination[130] each represent reproductive events wherein the female is likely to be relatively naïve to sperm antigens and each has been shown to increase the risk of preeclampsia. These observations must be viewed with caution since age, educational level, carriage of sexually transmitted infections, and access to healthcare may all confound the relationship between shorter-duration partnerships and preeclampsia.

Other observations also support the idea that novel exposure to paternal antigens may elevate preeclampsia risk. Women undergoing intrauterine insemination with washed donor sperm have a higher risk of preeclampsia than women undergoing intrauterine insemination with washed partner (autologous) sperm, implicating exposure to unfamiliar sperm. A particularly interesting comparison of *in vitro* fertilization or intracytoplasmic sperm injections (ICSI) with ICSI involving surgically obtained sperm showed a two-fold elevation in preeclampsia risk with the latter, suggesting that exposure to sperm, in the absence of semen, might increase preeclampsia risk.[131]

Even if implantation is normal, relative placental perfusion may be reduced because of excessive placental size. Obstetric complications associated with increased placental size such as twin pregnancies,[125] hydatidiform moles,[72] and hydrops fetalis[73] all markedly increase the risk of developing preeclampsia. Zhang et al. pooled the results from six studies that compared both twin and singleton gestations and found that pregnant women with twin pregnancies were three times more likely to develop hypertension in pregnancy than women with singletons[41] (Table 3.4). Women with these hypertensive twin pregnancies do not appear to have heightened risk of recurrence in subsequent singleton pregnancies.

CLINICAL PREDICTORS

Beyond these pre-pregnancy risk factors, there have been a number of studies evaluating the predictive value for preeclampsia of several readily available biochemical tests performed during pregnancy (see Chapter 11). Many of these tests may reflect the pathophysiological perturbations that link the maternal and placental contributions to preeclampsia. Other tests to predict the appearance of preeclampsia are markers related to inflammation, oxidative stress, lipid metabolism, vitamin status, angiogenesis, etc. These areas and the markers tested are discussed in Chapters 6, 7, 8, 9 and 17.

NATURAL HISTORY

Maternal Morbidity Immediately Related to Preeclampsia

Preeclampsia can be a major contributor to maternal morbidity and mortality as an immediate consequence of the progression to eclampsia. Douglas and Redman[10] estimated that among eclamptic women in the United Kingdom in 1992, 1.8% died and 35% had at least one major complication. The most common major complications encountered included a requirement for assisted ventilation (23%), renal failure (6%), disseminated intravascular coagulation (9%), hemolysis, elevated liver enzymes, and low platelets (HELLP syndrome) (7%), and pulmonary edema (5%). The findings from investigators in the United States[132,133] confirmed the very low mortality rates among eclamptic women. Despite the very real morbidity that still accompanies eclampsia, modern-day maternal outcomes for this condition are far better than those of the past. In the early 1900s, before the advent of magnesium sulfate therapy, mortality associated with eclampsia was estimated to be between 10 and 15%.[134] Chesley pointed out[22] that these sobering statistics from the past were primarily the result of overaggressive medical management, rather than the result of marked advances in pathophysiological understanding.

Recurrence of Preeclampsia in Subsequent Pregnancies

The woman who has had preeclampsia has increased risk of preeclampsia in subsequent pregnancies; recurrence risk, overall, is roughly 15% to 18% (Table 3.4). However, the risk of preeclampsia in a pregnancy following preterm preeclampsia may range as high as 38%,[86,135] and is inversely related to the length of the first preeclamptic pregnancy (Table 3.4).[32,89,136–138] Recurrence is slightly more common in black women.[86]

Reduced Risk of Later-Life Breast Cancer

Since the early 1990s, studies have largely reported that women with a history of preeclampsia have a reduced risk of breast cancer.[139–141] In a 2013 review by Kim et al., 10 out of 13 qualifying studies in their meta-analysis reported breast cancer hazard ratios from 0.24 to 0.87 (overall HR: 0.86) for women with a history of preeclampsia as compared to woman with no such history.[141] Two of the studies reporting the opposite trend come from the same Jerusalem Perinatal Cohort,[142,143] a population where it has been suggested that the increased prevalence of the BRCA mutation obscures the relative risk of breast cancer associated with non-genetic factors.[144] Studies have been inconclusive as to why preeclampsia seems to have this protective effect.[145–153] Initial research suggested that the protective effect of preeclampsia on breast cancer risk came from decreased levels of circulating estrogens in preeclampsia,[148] though this was not supported by later studies.[153] However, androgen levels have been found to be significantly higher in preeclamptic pregnancies,[141,151,153] possibly suggesting a protective androgenic effect on breast cancer risk. Consistent with this theory, women with preeclamptic pregnancies which resulted in a son showed a stronger breast cancer risk reduction than those women who delivered a daughter.[146,150] Blood samples from a Norwegian cohort did not show a correlation between pregnancy levels of three angiogenic factors and later diagnosis of breast cancer, regardless of preeclampsia status,[147] though placental weight was shown to be positively correlated with risk of breast cancer in later life.[149] Kim and others[141,154] have proposed that AFP (alpha-fetoprotein), shown in rodent fetuses to prevent the crossing of estrogen across the placenta by binding to it, plays some role in the relationship between breast cancer and preeclampsia. AFP has been shown in some studies to inhibit the response of estrogen-sensitive cells to estrogen by binding the hormone, which may decrease mammary gland estrogen exposure and breast cancer risk.[141] Some studies have found that women born from preeclamptic pregnancies have a 50–60% lower risk of developing breast cancer compared to women born from normotensive pregnancies,[155,156] while other studies have not found any relationship between maternal preeclampsia and breast cancer.[145] Although there are intriguing patterns between intrauterine factors, preeclampsia, and breast cancer, the mechanisms behind the interactions are still to be elucidated.

Increased Risk of Later Life Cardiovascular Disease

Preeclampsia foretells at least a two-fold higher rate of later heart disease.[2,157,158] This was established by the classic follow-up studies of eclampsia by Leon Chesley (Fig. 3.3).

More recently, studies from the USA, Iceland, the UK, Norway, Sweden, Jerusalem. Scotland and Taiwan have

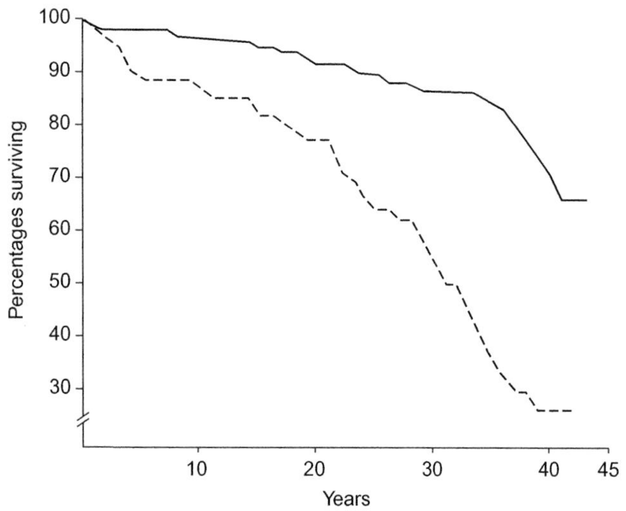

FIGURE 3.3 Survival following eclampsia. Solid line represents women who had eclampsia in the first pregnancy; broken line represents those who had eclampsia as multiparas. Note the poorer survival in the latter group, their deaths being primarily cardiovascular in origin. (Reprinted with permission.[22])

extended Chesley's conclusions.[158] These more recent studies, substantially larger than those of Chesley, have consistently reported that death from premature coronary artery disease was about twice as common among women experiencing prior eclampsia or preeclampsia than among women without these conditions.[159–162] More limited data suggest a similar excess risk of cardiovascular disease among women with a history of gestational hypertension.[135,163] Among women with prior preeclampsia, the greatest jeopardy for developing coronary artery disease later in life is borne by women with recurrent preeclampsia or early-onset preeclampsia.[32,164–167] Compared with normotensive term pregnancies, women delivering preterm preeclamptic pregnancies have relative risks of future cardiovascular disease ranging from 2.5 to 9.5.[135,161,168,169] The challenge now facing care providers is how to use this information in terms of assessing women with a history of a preeclamptic pregnancy.[170]

Physiological Mechanisms Linking Preeclampsia to Maternal CVD Risk

Pathways that link preeclampsia to later-life maternal CVD are just beginning to be explored by investigation of CVD risk factors before, during and after pregnancies complicated by preeclampsia.[158] These CVD risk factors include lipid metabolism, vascular dysfunction, metabolic risk, inflammation, coagulation, and angiogenesis.

As reviewed recently,[158] several CVD risk factors have been associated with preeclampsia. These include cardiometabolic risk factors measured before pregnancy, such as dyslipidemia,[90,171] increased blood pressure,[90] overweight, and family history of diabetes.[172,173] During pregnancy, women with preeclampsia are more likely than women without preeclampsia to exhibit relative hypertriglyceridemia and hypercholesterolemia,[174–177] gestational diabetes,[178] and an activated, perhaps dysfunctional, fibrinolytic cascade.[179–181] Pregnancies complicated by preeclampsia may have increased resistance in the uterine arteries,[182,183] and impaired endothelial response.[184,185] Finally, after pregnancy, women with a history of preeclampsia continue to exhibit elevated CVD risk factors, including impaired endothelial function,[186] and higher risks of impaired glucose tolerance, insulin resistance, and a three-fold risk of developing type 2 diabetes.[135,187–192] Women with a history of preeclampsia have higher levels of total cholesterol, LDL cholesterol, and triglycerides, although differences are not always statistically significant.[187–190,191–193] Lower HDL cholesterol concentrations among women with a history of preeclampsia have been reported by some,[96,188,189,192] but not all,[187,190,193] studies. After pregnancy, plasma C-reactive protein is elevated among women with prior eclampsia and indicated preterm births, suggesting that systemic low-grade inflammation may link some adverse pregnancy outcomes and later CVD.[194,196] Finally, women with a history of preeclampsia have significantly increased serum concentrations of antiangiogenic sFlt-1 more than one year postpartum.[191]

Thus, cardiovascular risk factors are elevated before, during, and after preeclamptic pregnancies, and risk of both short-term and long-term cardiovascular events is at least double that of women with no history of preeclampsia. These data suggest that preeclampsia serves as a "stress test" of underlying, subclinical cardiovascular disease.[157,158] While these observations indicate that the associations of preeclampsia with cardiovascular disease risk are correlational, ongoing experimental models are testing the extent to which induced preeclampsia may actually increase maternal cardiovascular risk through endothelial and organ injury.[197] This idea is suggested obliquely by observations that sisters of women with a history of preeclampsia do not share their elevated risk of end-stage renal disease (sororal risk of cardiovascular disease has not yet been examined).[198] Whether or not preeclampsia causes cardiovascular disease, a history of preeclampsia may be useful to identify a group of women with increased risk of later-life cardiovascular disease. The increased risk of preeclampsia for later-life cardiovascular disease has been acknowledged by the American Heart Association, who recommend including pregnancy history in the assessment of cardiovascular health in women and that a history of preeclampsia be included as a major risk factor in assessing if a woman is "at risk" for CVD.[199] This offers a unique opportunity for early identification of a large group of women whose hypertensive pregnancy history signals an exceptionally high risk of developing CVD. Ongoing studies are testing the extent to which preeclampsia might be productively added to CVD risk scoring systems, such as the Framingham Risk Score, to better predict CVD in women. More work is needed to determine the extent to which standard CVD prevention protocols are effective in women with a history of preeclampsia, or whether specific prevention and treatment therapies can be developed to reduce the risk of future CVD in women with a history of preeclampsia.[170]

The key to such a prevention goal is the ability of primary care clinicians to determine a history of preeclampsia in their patients. At present in the United States, it is unusual for a women's antenatal record to be accessible to the primary care clinician responsible for her cardiovascular preventive care. Even electronic medical record systems rarely link antenatal records to primary care records, and most primary care clinicians rely on maternal recall to identify history of pregnancy complications, if they query such history at all. Several studies have investigated maternal recall of preeclampsia. While the amount of information that is communicated by clinicians to mothers with preeclampsia has presumably increased in recent decades, studies of maternal recall suggest that the hypertensive disorders are poorly recalled by mothers. A systematic review of maternal recall of hypertensive disorders of pregnancy concluded that, while specificity of maternal recall of preeclampsia was >95%, sensitivity – the ability of mothers to recall preeclampsia when they had experienced it – was lower, on the order of 75% to 85% in cohort studies that validated maternal recall against records.[200] Severe disease and preterm disease were more likely to be recalled accurately, however, and these forms of preeclampsia are the most likely to predict future CVD. The maternal recall validation studies are affected by several factors, including the likelihood that doctors clearly informed women that they had preeclampsia, the nature of the recall questions asked of mothers (preeclampsia/toxemia versus particular diagnostic criteria, such as proteinuria), and the nature and stringency of criteria applied to the medical record to define preeclampsia. Newer validation studies conducted in the context of current medical practice may yield different results.

IMPACT ON CHILDREN

Perinatal Outcomes

Seventy percent of infants born to preeclamptic women are of normal birth weight for their gestational age.[201,202] Indeed, conditions associated with large infants, such as obesity[22,36,203] and gestational diabetes,[204] increase the risk of preeclampsia. Nevertheless, perinatal mortality is higher for infants of preeclamptic women (Table 3.5).[206] These demises are the result of intrauterine death from placental insufficiency and abruptio placentae, or the result of prematurity from delivery needed to control the disease.[207]

Early/severe/recurrent preeclampsia and adverse neonatal outcomes comprise the morbidities that make preeclampsia a major cause of maternal morbidity and neonatal mortality in many countries, including the United States.[208] Early onset of preeclampsia heralds not only poor later maternal health but also poor perinatal outcomes in the index

TABLE 3.5 Rate Ratios for Fetal Death by Diastolic Pressure and Proteinuria Combinations[205]

	Diastolic Blood Pressure (mm Hg)						
Proteinuria	**<65**	**65–74**	**75–84**	**85–94**	**95–104**	**105+**	**Total**
None	2.5[a]	1.5	1.0[b]	1.4	3.1[a]	3.3[a]	1.0[c]
Trace	2.2	1.3	1.2	1.5	2.8[a]	4.5[a]	1.1
1+	1.0	0.9	1.0	3.8[a]	4.3[a]	10.1[a]	1.4
2+	0.0	5.3[a]	3.1[a]	0.0	9.0[a]	11.1[a]	2.7[a]
3+	0.0	6.7[a]	0.0	3.6	18.6[a]	20.2[a]	5.1[a]
4+	0.0	0.0	0.0	0.0	23.1[a]	17.9[a]	6.6[a]
Total	2.0	1.3	1.0[d]	1.5	3.7	6.1	

[a]Statistically significant, $p < 0.01$.
[b]Referent rate 0.62%.
[c]Referent rate 0.86%.
[d]Referent rate 0.68%.

pregnancy.[29,48,136,209–214] Beyond this, early onset of preeclampsia in the index pregnancy is related to perinatal morbidity in subsequent pregnancies. Lopez-Llera and Horta examined 110 women in whom the onset of eclampsia was divided into gestational ages of <35 weeks, 35–37 weeks, and ≥38 weeks.[210] With earlier onset in the index pregnancy, recurrent preeclampsia and small-for-gestational age infants (SGA) in subsequent pregnancies were more common (42%, 15%, and 10% of infants were SGA in the three groups). Similarly, Sibai et al. showed that earlier-onset disease in the first pregnancy was associated with more frequent perinatal death in subsequent pregnancies.[209] Finally, neonatal morbidity is substantially higher in women with previous preeclampsia as compared to nulliparous preeclampsia.[89]

The interaction of underlying maternal disease and preeclampsia results in particularly high perinatal morbidity and mortality. Hypertension detected prior to 20 weeks gestation without superimposed preeclampsia has been shown to increase the rate of perinatal mortality in most, but not all studies,[48,214] whereas chronic hypertension with superimposed preeclampsia has been associated with high perinatal morbidity and mortality rates in all reports. Page and Christianson prospectively studied 10,074 white and 2880 African-American women divided into four groups: with and without chronic hypertension and with and without superimposed preeclampsia.[47] In comparison to women with normal blood pressures without proteinuria, perinatal death rates were 2–3-fold higher among chronic hypertensives and 6–10-fold higher among chronic hypertensives with superimposed preeclampsia. Rates of stillbirth and SGA followed similar patterns to those noted for perinatal death. These results have been replicated in studies of chronic hypertensives and control pregnant women.[213,215] Rey and Couturier in Montreal enrolled 298 women who had a well-documented diagnosis of hypertension prior to pregnancy or who had hypertension before 20 weeks gestation and again after 6 weeks postdelivery.[29] Perinatal deaths were substantially higher among chronic hypertensives with superimposed preeclampsia (101/1000) and modestly higher among chronic hypertensives without preeclampsia (29/1000) than among controls (12/1000). In a study of the MFMU Network, preexisting hypertension with superimposed preeclampsia resulted in elevated rates of preterm delivery at both <37 weeks (56%) and <35 weeks (36%).[93] Forty percent of babies among hypertensive, preeclamptic women were admitted to the neonatal intensive care unit and 8% had perinatal death.[33]

A similar relationship exists between diabetes that antedates pregnancy and preeclampsia. Women with diabetes and superimposed preeclampsia are more likely to experience a perinatal death than are women with diabetes alone. Perinatal mortality was substantially higher in preeclamptic diabetic patients (60/1000) than normotensive diabetic patients (3/1000).[49] This was despite a lack of difference between preeclamptic and normotensive diabetic patients for age, parity, glucose control, blood urea nitrogen, and serum creatinine. In another MFMU Network study, women with pre-gestational diabetes mellitus and superimposed early proteinuria as compared to women with only gestational diabetes had high rates of preterm delivery (29% vs. 13%) and SGA infants (14% vs. 3%) but lower rates of large-for-gestational-age infants (14% vs. 40%).[91]

Thus, underlying maternal diabetes and hypertension adversely affect perinatal outcomes but preeclampsia superimposed on these maternal conditions has even greater adverse perinatal consequences. Furthermore, biochemical risk markers for coronary artery disease, such as uric acid, predict neonatal morbidity and mortality among women with preeclampsia and are elevated early in pregnancy in preeclamptic women.[216–220]

Remote Impact on Cardiovascular Function in Offspring of Preeclamptic Pregnancies

A meta-analysis by Davis and colleagues concluded that children and young adults born of preeclamptic pregnancies have, on average, a 0.6 kg/m^2 higher BMI, 2.5 mm Hg higher systolic and 1.4 mm Hg higher diastolic blood pressures.[221] These findings are consistent with the two-fold higher risk of stroke reported for offspring of severely preeclamptic pregnancies followed in Finland; smaller elevations in coronary heart disease risk were not statistically significant.[222] That this risk may be pursuant to *in utero* exposure to preeclampsia, rather than simply the function of shared genetic factors, is suggested by the finding that offspring of preeclamptic pregnancies, but not their siblings from non-preeclamptic pregnancies, have evidence of vascular dysfunction.[223]

CRITIQUE OF STUDIES

Methodological issues that cloud the interpretation of epidemiologic results include misclassification of diseased vs. nondiseased individuals, biased patient selection, differential loss to follow-up, lack of control for potential confounders, and difficulties in selecting controls. Classification of women into those with preeclampsia, those with gestational hypertension, and controls has been inconsistent and difficult. Some of the differences in findings between studies may relate to misclassification. For this reason it is appealing to rely on reports about women with more severe disease, such as eclampsia. However, because more severe disease may be more strongly associated with underlying maternal disease, such reports may result in associations that are not generalizable to more mild hypertensive disease in pregnancy. The reports reviewed generally did not indicate the proportion of subjects studied as a function of the underlying population affected. Thus, patients with early-onset, severe, complicated, or confusing disease, perhaps indicative of a maternal etiology, may have been more likely to be included in these studies.

CONCLUSION

Preeclampsia and gestational hypertension continue to be relatively common. Although preeclampsia appears to be much less commonly complicated by eclampsia than it was in the past, the frequency of preeclampsia may be increasing, particularly preeclampsia complicated by underlying maternal morbidities such as obesity, diabetes, and chronic hypertension.[6]

Risk factors for the occurrence of preeclampsia include those associated with maternal cardiovascular risk as well as those that are unique to pregnancy and may represent poor placentation. The natural history of preeclampsia also suggests a heterogeneity in long-term maternal effects as well as in perinatal outcomes depending upon the presence or absence of maternal predisposition to and severity of cardiovascular disease risk. Nonetheless, preeclampsia in all presentations is associated with later-life cardiovascular disease and should be used to identify women at risk, as recommended by the American Heart Association.[199] Placental and cardiovascular markers may provide prediction tools for diagnosing preeclampsia.

Acknowledgment

We would like to acknowledge the assistance of Audrey Carr in the preparation of this chapter.

References

1. Ness RB, Roberts JM. Heterogeneous causes constituting the single syndrome of preeclampsia: a hypothesis and its implications. *Am J Obstet Gynecol.* 1996;175(5):1365–1370.
2. Tranquilli AL, Dekker G, Magee L, Roberts JM, Sibai BM, Steyn W, Zeeman GG, Brown MA. "The classification, diagnosis and management of the hypertensive disorders of pregnancy: A revised statement from the ISSHP. *Pregnancy Hypertension: An International Journal of Women's Cardiovascular Health.* 2014;4(2):97–104.

3. Brown MA, Davis G, Magee L, Roberts J, et al. The classification, diagnosis and management of the hypertensive disorders of pregnancy: a revised statement from the ISSHP. *Pregnancy Hypertens*. 2014;4(2):97–104.
4. Brown MA, Lindheimer MD, de Swiet M, Assche AV, Moutquin J-M. The classification and diagnosis of the hypertensive disorders of pregnancy: statement from the international society for the study of hypertension in pregnancy (ISSHP). *Hypertens Pregnancy*. 2001;20(1):ix–xiv.
5. Tranquilli AL, Brown MA, Zeeman GG, Dekker G, Sibai BM. The definition of severe and early-onset preeclampsia. Statements from the International Society for the Study of Hypertension in Pregnancy (ISSHP). *Pregnancy Hypertens*. 2012;3(1):44–47.
6. Wallis AB, Saftlas AF, Hsia J, Atrash HK. Secular trends in the rates of preeclampsia, eclampsia, and gestational hypertension, United States, 1987–2004. *Am J Hypertens*. 2008; 21(5):521–526.
7. Roberts CL, Ford JB, Algert CS, et al. Population-based trends in pregnancy hypertension and pre-eclampsia: an international comparative study. *BMJ Open*. 2011;1:1.
8. Gaio DS, Schmidt MI, Duncan BB, Nucci LB, Matos MC, Branchtein L. Hypertensive disorders in pregnancy: frequency and associated factors in a cohort of Brazilian women. *Hypertens Pregnancy*. 2001;20(3):269–281.
9. Roberts CL, Algert CS, Morris JM, Ford JB, Henderson-Smart DJ. Hypertensive disorders in pregnancy: a population-based study. *Med J Aust*. 2005;182(7):332–335.
10. Douglas KA, Redman C. Eclampsia in the United Kingdom. *BMJ*. 1994;309(6966):1395–1400.
11. Knight M. Eclampsia in the United Kingdom 2005. *BJOG*. 2007;114(9):1072–1078.
12. Souza JP, Gülmezoglu AM, Vogel J, et al. Moving beyond essential interventions for reduction of maternal mortality (the WHO Multicountry Survey on Maternal and Newborn Health): a cross-sectional study. *Lancet*. 2013;381(9879):1747–1755.
13. Conde-Agudelo A, Belizán JM. Risk factors for pre-eclampsia in a large cohort of Latin American and Caribbean women. *BJOG*. 2000;107(1):75–83.
14. Sibai BM, Caritis SN, Thom E, et al. Prevention of preeclampsia with low-dose aspirin in healthy, nulliparous pregnant women. *New Engl J Med*. 1993;329(17):1213–1218.
15. Villamor E, Cnattingius S. Interpregnancy weight change and risk of adverse pregnancy outcomes: a population-based study. *Lancet*. 2006;368(9542):1164–1170.
16. Lumbiganon P, Gülmezoglu AM, Piaggio G, Langer A, Grimshaw J. Magnesium sulfate is not used for pre-eclampsia and eclampsia in Mexico and Thailand as much as it should be. *B World Health Organ*. 2007;85(10):763–767.
17. Chalumeau M, Salanave B, Bouvier-Colle MH, Bernis Ld, Prua A, Breart G. Risk factors for perinatal mortality in West Africa: a population-based study of 20 326 pregnancies. *Acta Paediatr*. 2000;89(9):1115–1121.
18. Rotchell Y, Cruickshank J, Gay MP, et al. Barbados Low Dose Aspirin Study in Pregnancy (BLASP): a randomised trial for the prevention of pre-eclampsia and its complications. *BJOG*. 1998;105(3):286–292.
19. Roberts JM, Myatt L, Spong CY, et al. Vitamins C and E to prevent complications of pregnancy-associated hypertension. *New Engl J Med*. 2010;362(14):1282–1291.
20. Abalos E, Cuesta C, Grosso AL, Chou D, Say L. Global and regional estimates of preeclampsia and eclampsia: a systematic review. *Eur J Obstet Gynecol Reprod Biol*. 2013
21. Goldenberg RL, McClure EM, MacGuire ER, Kamath BD, Jobe AH. Lessons for low-income regions following the reduction in hypertension-related maternal mortality in high-income countries. *Int J Gynecol Obstet*. 2011; 113(2):91–95.
22. Chesley LC. *Hypertensive Disorders in Pregnancy*. New York: Appleton-Century-Crofts; 1978.
23. Saftlas AF, Olson DR, Franks AL, Atrash HK, Pokras R. Epidemiology of preeclampsia and eclampsia in the United States, 1979–1986. *Am J Obstet Gynecol*. 1990;163(2):460–465.
24. Catov JM, Ness RB, Kip KE, Olsen J. Risk of early or severe preeclampsia related to pre-existing conditions. *Int J Epidemiol*. 2007;36(2):412–419.
25. Kyle P, Buckley D, Kissane J, De Swiet M, Redman C. The angiotensin sensitivity test and low-dose aspirin are ineffective methods to predict and prevent hypertensive disorders in nulliparous pregnancy. *Am J Obstet Gynecol*. 1995; 173(3):865–872.
26. Carroll G, Duley L, Belizán JM, Villar J. Calcium supplementation during pregnancy: a systematic review of randomised controlled trials. *BJOG*. 1994;101(9):753–758.
27. Hauth JC, Goldenberg RL, Parker Jr CR, et al. Low-dose aspirin therapy to prevent preeclampsia. *Am J Obstet Gynecol*. 1993;168(4):1083–1093.
28. Khan KS, Wojdyla D, Say L, Gülmezoglu AM, Van Look PF. WHO analysis of causes of maternal death: a systematic review. *Lancet*. 2006;367(9516):1066–1074.
29. Rey E, Couturier A. The prognosis of pregnancy in women with chronic hypertension. *Am J Obstet Gynecol*. 1994; 171(2):410–416.
30. Roberts J, Perloff D. Hypertension and the obstetrician-gynecologist. *Am J Obstet Gynecol*. 1977;127(3):316–325.
31. Moutquin J, Rainville C, Giroux L, et al. A prospective study of blood pressure in pregnancy: prediction of preeclampsia. *Am J Obstet Gynecol*. 1985;151(2):191–196.
32. Sibai BM, El-Nazer A, Gonzalez-Ruiz A. Severe preeclampsia-eclampsia in young primigravid women: subsequent pregnancy outcome and remote prognosis. *Am J Obstet Gynecol*. 1986;155(5):1011–1016.
33. Sibai BM, Lindheimer M, Hauth J, et al. Risk factors for preeclampsia, abruptio placentae, and adverse neonatal outcomes among women with chronic hypertension. *New Engl J Med*. 1998;339(10):667–671.
34. Levine RJ, Hauth JC, Curet LB, et al. Trial of calcium to prevent preeclampsia. *New Engl J Med*. 1997;337(2):69–77.
35. Ales KL, Charlson ME. Epidemiology of preeclampsia and eclampsia. *Am J Obstet Gynecol*. 1991;165(1):238.
36. Eskenazi B, Fenster L, Sidney S. A multivariate analysis of risk factors for preeclampsia. *JAMA*. 1991;266(2):237–241.
37. Klemmensen ÅK, Olsen SF, Østerdal ML, Tabor A. Validity of preeclampsia-related diagnoses recorded in a national hospital registry and in a postpartum interview of the women. *Am J Epidemiol*. 2007;166(2):117–124.
38. Roberts CL, Bell JC, Ford JB, Hadfield RM, Algert CS, Morris JM. The accuracy of reporting of the hypertensive disorders of pregnancy in population health data. *Hypertens Pregnancy*. 2008;27(3):285–297.

39. MacGillivray I. Some observations on the incidence of pre-eclampsia. *BJOG*. 1958;65(4):536–539.
40. Hansen J. Older maternal age and pregnancy outcome: a review of the literature. *Obstet Gynecol Surv*. 1986;41(11):726.
41. Zhang J, Zeisler J, Hatch MC, Berkowitz G. Epidemiology of pregnancy-induced hypertension. *Epidemiol Rev*. 1996; 19(2):218–232.
42. Sibai BM, Gordon T, Thom E, et al. Risk factors for pre-eclampsia in healthy nulliparous women: a prospective multicenter study. *Am J Obstet Gynecol*. 1995;172(2):642–648.
43. Chambers JC, Fusi L, Malik IS, Haskard DO, De Swiet M, Kooner JS. Association of maternal endothelial dysfunction with preeclampsia. *JAMA*. 2001;285(12):1607–1612.
44. Tillman A. The effect of normal and toxemic pregnancy on blood pressure. *Am J Obstet Gynecol*. 1955;70(3):589.
45. Fallis N, Langford H. Relation of second trimester blood pressure to toxemia of pregnancy in the primigravid patient. *Am J Obstet Gynecol*. 1963;87:123.
46. Villar LMA, Sibai BM. Clinical significance of elevated mean arterial blood pressure in second trimester and threshold increase in systolic or diastolic blood pressure during third trimester. *Am J Obstet Gynecol*. 1989;160(2):419–423.
47. Page E, Christianson R. Influence of blood pressure changes with and without proteinuria upon outcome of pregnancy. *Am J Obstet Gynecol*. 1976;126(7):821.
48. Mabie WC, Pernoll ML, Biswas MK. Chronic hypertension in pregnancy. *Obstet Gynecol*. 1986;67(2):197–205.
49. Garner PR, D'Alton ME, Dudley DK, Huard P, Hardie M. Preeclampsia in diabetic pregnancies. *Am J Obstet Gynecol*. 1990;163(2):505–508.
50. Siddiqi T, Rosenn B, Mimouni F, Khoury J, Miodovnik M. Hypertension during pregnancy in insulin-dependent diabetic women. *Obstet Gynecol*. 1991;77(4):514–519.
51. MacGillivray I. *Pre-eclampsia: The Hypertensive Disease of Pregnancy*. London: WB Saunders Company; 1983.
52. Smith CA. The effect of wartime starvation in Holland upon pregnancy and its product. *Am J Obstet Gynecol*. 1947; 53(4):599.
53. Stone J, Lockwood C, Berkowitz G, Alvarez M, Lapinski R, Berkowitz R. Risk factors for severe preeclampsia. *Obstet Gynecol*. 1994;83(3):357–361.
54. Bodnar LM, Ness RB, Markovic N, Roberts JM. The risk of preeclampsia rises with increasing prepregnancy body mass index. *Ann Epidemiol*. 2005;15(7):475–482.
55. Bodnar LM, Catov JM, Klebanoff MA, Ness RB, Roberts JM. Prepregnancy body mass index and the occurrence of severe hypertensive disorders of pregnancy. *Epidemiology*. 2007; 18(2):234–239.
56. Sibai B, Ewell M, Levine R, et al. Risk factors associated with preeclampsia in healthy nulliparous women. *Am J Obstet Gynecol*. 1997;177(5):1003–1010.
57. Nelson T. Clinical study of preeclampsia part II. *BJOG*. 1955;62(1):55–66.
58. Thomson A, Billewicz W. Clinical significance of weight trends during pregnancy. *BMJ*. 1957;1(5013):243.
59. Thomson A, Hytten F, Billewicz W. The epidemiology of oedema during pregnancy. *BJOG*. 1967;74(1):1–10.
60. Adams EM, Finlayson A. Familial aspects of pre-eclampsia and hypertension in pregnancy. *Obstet Gynecol Surv*. 1962; 17(3):350–351.
61. Chesley L, Cooper D. Genetics of hypertension in pregnancy: possible single gene control of pre-eclampsia and eclampsia in the descendants of eclamptic women. *BJOG*. 1986;93(9):898–908.
62. Roberts J, Redman C. Pre-eclampsia: more than pregnancy-induced hypertension. *Lancet*. 1993;341(8858):1447–1451.
63. Ness RB, Markovic N, Bass D, Harger G, Roberts JM. Family history of hypertension, heart disease, and stroke among women who develop hypertension in pregnancy. *Obstet Gynecol*. 2003;102(6):1366–1371.
64. Redman C. Immunology of preeclampsia. *Semin Perinatol*. 1991;15(3):257–262.
65. Seidman DS, Ever-Hadani P, Stevenson DK, Gale R. The effect of abortion on the incidence of pre-eclampsia. *Eur J Obstet Gynecol Reprod Biol*. 1989;33(2):109–114.
66. Robillard P-Y, Périanin J, Janky E, Miri E, Hulsey T, Papiernik E. Association of pregnancy-induced hypertension with duration of sexual cohabitation before conception. *Lancet*. 1994;344(8928):973–975.
67. Feeney J, Scott J. Pre-eclampsia and changed paternity. *Eur J Obstet Gynecol Reprod Biol*. 1980;11(1):35–38.
68. Chng P. Occurrence of pre-eclampsia in pregnancies to three husbands. *BJOG*. 1982;89(10):862–863.
69. Strickland D, Guzick D, Cox K, Gant N, Rosenfeld C. The relationship between abortion in the first pregnancy and development of pregnancy-induced hypertension in the subsequent pregnancy. *Am J Obstet Gynecol*. 1986;154(1):146–148.
70. Marti JJ, Herrmann U. Immunogestosis: a new etiologic concept of essential EPH gestosis, with special consideration of the primigravid patient. *Am J Obstet Gynecol*. 1977; 128:489–493.
71. Klonoff-Cohen HS, Savitz DA, Cefalo RC, McCann MF. An epidemiologic study of contraception and preeclampsia. *Obstet Gynecol Surv*. 1991;46(6):349–350.
72. Page EW. The relation between hydatid moles, relative ischemia of the gravid uterus, and the placental origin of eclampsia. *Am J Obstet Gynecol*. 1939;37:291–293.
73. Scott JS. Pregnancy toxaemia associated with hydrops fetalis, hydatidiform mole and hydramnios. *BJOG*. 1958; 65(5):689–701.
74. Bulfin M, Lawler P. Problems associated with toxemia in twin pregnancies. *Am J Obstet Gynecol*. 1957;73(1):37.
75. Underwood P, Kesler K, O'Lane J, Callagan D. Parental smoking empirically related to pregnancy outcome. *Obstet Gynecol*. 1967;29(1):1–8.
76. Russell CS, Taylor R, Law C. Smoking in pregnancy, maternal blood pressure, pregnancy outcome, baby weight and growth, and other related factors. A prospective study. *Brit J Prevent Soc Med*. 1968;22(3):119–126.
77. Marcoux S, Brisson J, Fabia J. The effect of cigarette smoking on the risk of preeclampsia and gestational hypertension. *Am J Epidemiol*. 1989;130(5):950–957.
78. Zabriskie J. Effect of cigarette smoking during pregnancy: study of 2000 cases. *Obstet Gynecol*. 1963;21(4):405–411.
79. Clausen T, Slott M, Solvoll K, Drevon CA, Vollset SE, Henriksen T. High intake of energy, sucrose, and polyunsaturated fatty acids is associated with increased risk of pre-eclampsia. *Am J Obstet Gynecol*. 2001;185(2):451–458.

80. Roberts JM, Balk JL, Bodnar LM, Belizán JM, Bergel E, Martinez A. Nutrient involvement in preeclampsia. *J Nutrit.* 2003;133(5):1684S–1692S.
81. Zhang C, Williams M, King I, et al. Vitamin C and the risk of preeclampsia-results from dietary questionnaire and plasma assay. *Epidemiology.* 2002;13(4):409–416.
82. Breathett K, Muhlestein D, Foraker R, Gulati M. The incidence of pre-eclampsia remains higher in African-American women compared to Caucasian women: trends from the National Hospital Discharge Survey 1979–2006. *Circulation.* 2013;127:AP192.
83. Wu CS, Nohr EA, Bech BH, Vestergaard M, Catov JM, Olsen J. Health of children born to mothers who had preeclampsia: a population-based cohort study. *Am J Obstet Gynecol.* 2009;201(3):269e1–269e10.
84. Hernández-Díaz S, Toh S, Cnattingius S. Risk of preeclampsia in first and subsequent pregnancies: prospective cohort study. *BMJ.* 2009:338.
85. Ros HS, Lichtenstein P, Lipworth L, Cnattingius S. Genetic effects on the liability of developing pre-eclampsia and gestational hypertension. *Am J Med Genet.* 2000;91(4): 256–260.
86. Mostello D, Kallogjeri D, Tungsiripat R, Leet T. Recurrence of preeclampsia: effects of gestational age at delivery of the first pregnancy, body mass index, paternity, and interval between births. *Am J Obstet Gynecol.* 2008;199(1):55e1–55e7.
87. Lykke JA, Paidas MJ, Langhoff-Roos J. Recurring complications in second pregnancy. *Obstet Gynecol.* 2009; 113(6):1217–1224.
88. Klungsøyr K, Morken NH, Irgens L, Vollset SE, Skjærven R. Secular trends in the epidemiology of pre-eclampsia throughout 40 years in Norway: prevalence, risk factors and perinatal survival. *Paediatr Perinat Epidemiol.* 2012;26(3):190–198.
89. Hnat MD, Sibai BM, Caritis S, et al. Perinatal outcome in women with recurrent preeclampsia compared with women who develop preeclampsia as nulliparas. *Am J Obstet Gynecol.* 2002;186(3):422–426.
90. Magnussen EB, Vatten LJ, Lund-Nilsen TI, Salvesen KÅ, Smith GD, Romundstad PR. Prepregnancy cardiovascular risk factors as predictors of pre-eclampsia: population based cohort study. *BMJ.* 2007;335(7627):978.
91. Sibai BM, Caritis S, Hauth J, et al. Risks of preeclampsia and adverse neonatal outcomes among women with pregestational diabetes mellitus. *Am J Obstet Gynecol.* 2000; 182(2):364–369.
92. Sattar N, Clark P, Holmes A, Lean ME, Walker I, Greer IA. Antenatal waist circumference and hypertension risk. *Obstet Gynecol.* 2001;97(2):268–271.
93. Sorensen TK, Williams MA, Lee I-M, Dashow EE, Thompson ML, Luthy DA. Recreational physical activity during pregnancy and risk of preeclampsia. *Hypertension.* 2003;41(6):1273–1280.
94. Khan K, Hashmi F, Rizvi J. Are non-diabetic women with abnormal glucose screening test at increased risk of pre-eclampsia, macrosomia and caesarian birth? *JPMA.* 1995;45(7):176.
95. Joffe GM, Esterlitz JR, Levine RJ, et al. The relationship between abnormal glucose tolerance and hypertensive disorders of pregnancy in healthy nulliparous women. *Am J Obstet Gynecol.* 1998;179(4):1032–1037.
96. Carlsen SM, Romundstad P, Jacobsen G. Early second-trimester maternal hyperandrogenemia and subsequent preeclampsia: a prospective study. *Acta Obstet Gynecol Scand.* 2005;84(2):117–121.
97. Bianco AT, Smilen SW, Davis Y, Lopez S, Lapinski R, Lockwood CJ. Pregnancy outcome and weight gain recommendations for the morbidly obese woman. *Obstet Gynecol.* 1998;91(1):60–64.
98. Thadhani R, Stampfer MJ, Chasan-Taber L, Willett WC, Curhan GC. A prospective study of pregravid oral contraceptive use and risk of hypertensive disorders of pregnancy. *Contraception.* 1999;60(3):145–150.
99. Ogunyemi D, Hullett S, Leeper J, Risk A. Prepregnancy body mass index, weight gain during pregnancy, and perinatal outcome in a rural black population. *J Matern Fetal Neonat Med.* 1998;7(4):190–193.
100. Bennett WL, Gilson MM, Jamshidi R, et al. Impact of bariatric surgery on hypertensive disorders in pregnancy: retrospective analysis of insurance claims data. *BMJ.* 2010:340.
101. Kasawara KT, Nascimento SL, Costa ML, Surita FG, Silva E, Pinto JL. Exercise and physical activity in the prevention of pre-eclampsia: systematic review. *Acta Obstet Gynecol Scand.* 2012;91(10):1147–1157.
102. Jiang H, Qian X, Li M, et al. Can physical activity reduce excessive gestational weight gain? Findings from a Chinese urban pregnant women cohort study. *Int J Behav Nutr Phys Act.* 2012;9:12.
103. Shirazian T, Monteith S, Friedman F, Rebarber A. Lifestyle modification program decreases pregnancy weight gain in obese women. *Am J Perinatol.* 2010;27(05):411–414.
104. Weissgerber TL, Davies GA, Roberts JM. Modification of angiogenic factors by regular and acute exercise during pregnancy. *J Appl Physiol.* 2010;108(5):1217–1223.
105. Williams MA, Zingheim RW, King IB, Zebelman AM. Omega-3 fatty acids in maternal erythrocytes and risk of preeclampsia. *Epidemiology.* 1995:232–237.
106. Heinig M, Johnson RJ. Role of uric acid in hypertension, renal disease, and metabolic syndrome. *Clev Clin J Med.* 2006;73(12):1059–1064.
107. Meltzer HM, Brantsæter AL, Nilsen RM, Magnus P, Alexander J, Haugen M. Effect of dietary factors in pregnancy on risk of pregnancy complications: results from the Norwegian Mother and Child Cohort Study. *Am J Clin Nutr.* 2011;94(6 suppl):1970S–1974S.
108. Poston L, Briley A, Seed P, Kelly F, Shennan A. Vitamin C and vitamin E in pregnant women at risk for pre-eclampsia (VIP trial): randomised placebo-controlled trial. *Lancet.* 2006;367(9517):1145–1154.
109. Basaran A, Basaran M, Topatan B. Combined vitamin C and E supplementation for the prevention of preeclampsia: a systematic review and meta-analysis. *Obstet Gynecol Surv.* 2010;65(10):653–667.
110. Conde-Agudelo A, Romero R, Kusanovic JP, Hassan SS. Supplementation with vitamins C and E during pregnancy for the prevention of preeclampsia and other adverse maternal and perinatal outcomes: a systematic review and metaanalysis. *Am J Obstet Gynecol.* 2011;204(6):503e1–503e12.
111. Villar J, Belizan JM, Fischer PJ. Epidemiologic observations on the relationship between calcium intake and eclampsia. *Int J Gynecol Obstet.* 1983;21(4):271–278.

112. Hofmeyr GJ, Lawrie TA, Atallah ÁN, Duley L. Calcium supplementation during pregnancy for preventing hypertensive disorders and related problems. *Cochrane Database Syst Rev*. 2010:8.
113. Bodnar LM, Tang G, Ness RB, Harger G, Roberts JM. Periconceptional multivitamin use reduces the risk of preeclampsia. *Am J Epidemiol*. 2006;164(5):470–477.
114. Bodnar LM, Simhan HN, Powers RW, Frank MP, Cooperstein E, Roberts JM. High prevalence of vitamin D insufficiency in black and white pregnant women residing in the northern United States and their neonates. *J Nutrit*. 2007; 137(2):447–452.
115. Bodnar LM, Simhan HN. Vitamin D may be a link to black-white disparities in adverse birth outcomes. *Obstet Gynecol Surv*. 2010;65(4):273.
116. Hyppönen E, Power C. Hypovitaminosis D in British adults at age 45 y: nationwide cohort study of dietary and lifestyle predictors. *Am J Clin Nutr*. 2007;85(3):860–868.
117. Aghajafari F, Nagulesapillai T, Ronksley PE, Tough SC, O'Beirne M, Rabi DM. Association between maternal serum 25-hydroxyvitamin D level and pregnancy and neonatal outcomes: systematic review and meta-analysis of observational studies. *BMJ*. 2013:346.
118. Rosen CJ, Abrams SA, Aloia JF, et al. IOM committee members respond to Endocrine Society vitamin D guideline. *J Clin Endocr Metab*. 2012;97(4):1146–1152.
119. Lain KY, Powers RW, Krohn MA, Ness RB, Crombleholme WR, Roberts JM. Urinary cotinine concentration confirms the reduced risk of preeclampsia with tobacco exposure. *Am J Obstet Gynecol*. 1999;181(5):1192–1196.
120. Lain KY, Wilson JW, Crombleholme WR, Ness RB, Roberts JM. Smoking during pregnancy is associated with alterations in markers of endothelial function. *Am J Obstet Gynecol*. 2003;189(4):1196–1201.
121. Ness RB, Zhang J, Bass D, Klebanoff MA. Interactions between smoking and weight in pregnancies complicated by preeclampsia and small-for-gestational-age birth. *Am J Epidemiol*. 2008;168(4):427–433.
122. Engel SM, Janevic TM, Stein CR, Savitz DA. Maternal smoking, preeclampsia, and infant health outcomes in New York City, 1995–2003. *Am J Epidemiol*. 2009;169(1):33–40.
123. Wikström A-K, Stephansson O, Cnattingius S. Tobacco use during pregnancy and preeclampsia risk effects of cigarette smoking and snuff. *Hypertension*. 2010;55(5):1254–1259.
124. Salonen Ros H, Lichtenstein P, Lipworth L, Cnattingius S. Genetic effects on the liability of developing pre-eclampsia and gestational hypertension. *Am J Med Genet*. 2000; 91(4):256–260.
125. Lie RT, Rasmussen S, Brunborg H, Gjessing HK, Lie-Nielsen E, Irgens LM. Fetal and maternal contributions to risk of pre-eclampsia: population based study. *BMJ*. 1998; 316(7141):1343.
126. Skjærven R, Vatten LJ, Wilcox AJ, Rønning T, Irgens LM, Lie RT. Recurrence of pre-eclampsia across generations: exploring fetal and maternal genetic components in a population based cohort. *BMJ*. 2005;331(7521):877.
127. Williams PJ, Broughton Pipkin F. The genetics of pre-eclampsia and other hypertensive disorders of pregnancy. *Best Pract Res Cl Ob*. 2011;25(4):405–417.
128. Li D-K, Wi S. Changing paternity and the risk of preeclampsia/eclampsia in the subsequent pregnancy. *Am J Epidemiol*. 2000;151(1):57–62.
129. Skjærven R, Wilcox AJ, Lie RT. The interval between pregnancies and the risk of preeclampsia. *New Engl J Med*. 2002;346(1):33–38.
130. Robillard P-Y, Dekker GA, Hulsey TC. Revisiting the epidemiological standard of preeclampsia: primigravidity or primipaternity? *Eur J Obstet Gynecol Reprod Biol*. 1999;84(1):37–41.
131. Wang JX, Knottnerus A-M, Schuit G, Norman RJ, Chan A, Dekker GA. Surgically obtained sperm, and risk of gestational hypertension and pre-eclampsia. *Lancet*. 2002; 359(9307):673–674.
132. Siba B, Lipshitz J, Anderson G, Dilts Jr P. Reassessment of intravenous $MgSO_4$ therapy in preeclampsia-eclampsia. *Obstet Gynecol*. 1981;57(2):199–202.
133. Pritchard JA, Pritchard S. Standardized treatment of 154 consecutive cases of eclampsia. *Am J Obstet Gynecol*. 1975; 123(5):543–552.
134. Roberts J. *Pregnancy-Related Hypertension. Maternal Fetal Medicine*. 4th ed. Philadelphia: WB Saunders; 1998: 833–872.
135. Lykke JA, Langhoff-Roos J, Sibai BM, Funai EF, Triche EW, Paidas MJ. Hypertensive pregnancy disorders and subsequent cardiovascular morbidity and type 2 diabetes mellitus in the mother. *Hypertension*. 2009;53(6):944–951.
136. Sibai BM, Mercer B, Sarinoglu C. Severe preeclampsia in the second trimester: recurrence risk and long-term prognosis. *Am J Obstet Gynecol*. 1991;165(5):1408–1412.
137. Caritis S, Sibai B, Hauth J, et al. Predictors of pre-eclampsia in women at high risk. *Am J Obstet Gynecol*. 1998;179(4):946–951.
138. Rasmussen S, Irgens LM, Albrechtsen S, Dalaker K. Predicting preeclampsia in the second pregnancy from low birth weight in the first pregnancy. *Obstet Gynecol*. 2000;96(5, Part 1):696–700.
139. Lagiou P. Intrauterine factors and breast cancer risk. *Lancet Oncol*. 2007;8(12):1047–1048.
140. Miller W. Breast cancer: Hormonal factors and risk of breast cancer. *Lancet*. 1993;341(8836):25–26.
141. Kim JS, Kang EJ, Woo OH, et al. The relationship between preeclampsia, pregnancy-induced hypertension and maternal risk of breast cancer: a meta-analysis. *Acta Oncol*. 2012;0:1–6.
142. Paltiel O, Friedlander Y, Tiram E, Barchana M, Xue X, Harlap S. Cancer after pre-eclampsia: follow up of the Jerusalem perinatal study cohort. *BMJ*. 2004;328(7445):919.
143. Calderon-Margalit R, Friedlander Y, Yanetz R, et al. Preeclampsia and subsequent risk of cancer: update from the Jerusalem Perinatal Study. *Am J Obstet Gynecol*. 2009; 200(1):63e1–63e5.
144. King M-C, Marks JH, Mandell JB. Breast and ovarian cancer risks due to inherited mutations in BRCA1 and BRCA2. *Science*. 2003;302(5645):643–646.
145. Troisi R, Doody DR, Mueller BA. A linked-registry study of gestational factors and subsequent breast cancer risk in the mother. *Cancer Epidem Biomar*. 2013;22(5):835–847.
146. Troisi R, Potischman N, Hoover RN. Exploring the underlying hormonal mechanisms of prenatal risk factors for breast cancer: a review and commentary. *Cancer Epidem Biomar*. 2007;16(9):1700–1712.

147. Vatten LJ, Romundstad PR, Jenum PA, Eskild A. Angiogenic balance in pregnancy and subsequent breast cancer risk and survival: a population study. *Cancer Epidem Biomar.* 2009; 18(7):2074–2078.
148. Innes KE, Byers TE. Preeclampsia and breast cancer risk. *Epidemiology.* 1999;10(6):722–732.
149. Cnattingius S, Torrång A, Ekbom A, Granath F, Petersson G, Lambe M. Pregnancy characteristics and maternal risk of breast cancer. *JAMA.* 2005;294(19):2474–2480.
150. Vatten L, Forman M, Nilsen T, Barrett J, Romundstad P. The negative association between pre-eclampsia and breast cancer risk may depend on the offspring's gender. *Brit J Cancer.* 2007;96(9):1436–1438.
151. Acromite MT, Mantzoros CS, Leach RE, Hurwitz J, Dorey LG. Androgens in preeclampsia. *Am J Obstet Gynecol.* 1999;180(1):60–63.
152. Troisi R, Innes K, Roberts J, Hoover R. Preeclampsia and maternal breast cancer risk by offspring gender: do elevated androgen concentrations play a role? *Brit J Cancer.* 2007; 97(5):688–690.
153. Troisi R, Potischman N, Roberts JM, et al. Maternal serum oestrogen and androgen concentrations in preeclamptic and uncomplicated pregnancies. *Int J Epidemiol.* 2003;32(3): 455–460.
154. Waller DK, Lustig LS, Cunningham GC, Feuchtbaum LB, Hook EB. The association between maternal serum alpha-fetoprotein and preterm birth, small for gestational age infants, preeclampsia, and placental complications. *Obstet Gynecol.* 1996;88(5):816–822.
155. Nilsen T, Vatten L. Fetal growth, pre-eclampsia and adult breast cancer risk. *Tidsskr Norske Lægefor Tidsskr Prakt Med.* 2002;122(26):2525.
156. Ekbom A, Adami H-O, Hsieh C-c, Lipworth L, Trichopoulos D. Intrauterine environment and breast cancer risk in women: a population-based study. *J Natl Cancer Inst.* 1997;89(1):71–76.
157. Sattar N, Greer IA. Pregnancy complications and maternal cardiovascular risk: opportunities for intervention and screening? *BMJ.* 2002;325(7356):157.
158. Rich-Edwards JW, Fraser A, Lawlor DA, Catov JM. Pregnancy characteristics and women's future cardiovascular health: an underused opportunity to improve women's health? *Epidemiol Rev.* 2013 [Epub ahead of print].
159. Sjónsdóttir L, Arngrimsson R, Geirsson RT, Slgvaldason H, Slgfússon N. Death rates from ischemic heart disease in women with a history of hypertension in pregnancy. *Acta Obstet Gynecol Scand.* 1995;74(10):772–776.
160. Hannaford P, Ferry S, Hirsch S. Cardiovascular sequelae of toxaemia of pregnancy. *Heart.* 1997;77(2):154–158.
161. Smith GN, Pell JP, Walsh D. Pregnancy complications and maternal risk of ischaemic heart disease: a retrospective cohort study of 129 290 births. *Lancet.* 2001;357(9273):2002–2006.
162. Irgens HU, Reisæter L, Irgens LM, Lie RT. Long term mortality of mothers and fathers after pre-eclampsia: population based cohort study. *BMJ.* 2001;323(7323):1213.
163. Kestenbaum B, Seliger SL, Easterling TR, et al. Cardiovascular and thromboembolic events following hypertensive pregnancy. *Am J Kidney Dis.* 2003;42(5):982–989.
164. Chesley S, Annitto J, Cosgrove R. The remote prognosis of eclamptic women. Sixth periodic report. *Am J Obstet Gynecol.* 1976;124(5):446.
165. Funai EF, Friedlander Y, Paltiel O, et al. Long-term mortality after preeclampsia. *Epidemiology.* 2005;16(2): 206–215.
166. Bermax S. Observations in the toxemic clinic, Boston Lying-in Hospital, 1923–1930. *New Engl J Med.* 1930;203: 361–363.
167. Adams E, MacGillivray I. Long-term effect of pre-eclampsia on blood-pressure. *Lancet.* 1961;278(7217):1373–1375.
168. Mongraw-Chaffin ML, Cirillo PM, Cohn BA. Preeclampsia and cardiovascular disease death prospective evidence from the Child Health and Development Studies Cohort. *Hypertension.* 2010;56(1):166–171.
169. Skjaerven R, Wilcox AJ, Klungsøyr K, et al. Cardiovascular mortality after pre-eclampsia in one child mothers: prospective, population based cohort study. *BMJ.* 2012:345.
170. Roberts JM, Catov JM. Pregnancy is a screening test for later life cardiovascular disease: now what? Research recommendations. *Women Health Iss.* 2012;22(2):e8–e123.
171. Magnussen EB, Vatten LJ, Myklestad K, Salvesen KÅ, Romundstad PR. Cardiovascular risk factors prior to conception and the length of pregnancy: population-based cohort study. *Am J Obstet Gynecol.* 2011;204(6):526e1–526e8.
172. Qiu C, Williams MA, Leisenring WM, et al. Family history of hypertension and type 2 diabetes in relation to preeclampsia risk. *Hypertension.* 2003;41(3):408–413.
173. O'Brien TE, Ray JG, Chan W-S. Maternal body mass index and the risk of preeclampsia: a systematic overview. *Epidemiology.* 2003;14(3):368–374.
174. Clausen T, Djurovic S, Henriksen T. Dyslipidemia in early second trimester is mainly a feature of women with early onset pre-eclampsia. *BJOG.* 2001;108(10):1081–1087.
175. Hubel CA, McLaughlin MK, Evans RW, Hauth BA, Sims CJ, Roberts JM. Fasting serum triglycerides, free fatty acids, and malondialdehyde are increased in preeclampsia, are positively correlated, and decrease within 48 hours post partum. *Am J Obstet Gynecol.* 1996;174(3):975–982.
176. Hubel CA, Lyall F, Weissfeld L, Gandley RE, Roberts JM. Small low-density lipoproteins and vascular cell adhesion molecule-1 are increased in association with hyperlipidemia in preeclampsia. *Metabolism.* 1998;47(10):1281–1288.
177. Sattar N, Bendomir A, Berry C, Shepherd J, Greer IA, Packard CJ. Lipoprotein subfraction concentrations in preeclampsia: pathogenic parallels to atherosclerosis. *Obstet Gynecol.* 1997; 89(3):403–408.
178. Ostlund I, Haglund B, Hanson U. Gestational diabetes and preeclampsia. *Eur J Obstet Gynecol Reprod Biol.* 2004;113(1):12–16.
179. Heilmann L, Rath W, Pollow K. Hemostatic abnormalities in patients with severe preeclampsia. *Clin Appl Thromb-Hem.* 2007;13(3):285–291.
180. Catov J, Bodnar L, Hackney D, Roberts J, Simhan S. Activation of the fibrinolytic cascade early in pregnancy among women with spontaneous preterm birth. *Obstet Gynecol.* 2008; 112(5):1116–1122.
181. Hackney DN, Catov JM, Simhan HN. Low concentrations of thrombin-inhibitor complexes and the risk of preterm delivery. *Am J Obstet Gynecol.* 2010;203(2):184e1–184e6.
182. Ducey J, Schulman H, Farmakides G, et al. A classification of hypertension in pregnancy based on Doppler velocimetry. *Am J Obstet Gynecol.* 1987 Sep;157(3):680–685.

183. Campbell S, Griffin DR, Pearce JM, et al. New Doppler technique for assessing uteroplacental blood flow. *Lancet.* 1983;321(8326):675–677.
184. Savvidou MD, Hingorani AD, Tsikas D, Frölich JC, Vallance P, Nicolaides KH. Endothelial dysfunction and raised plasma concentrations of asymmetric dimethylarginine in pregnant women who subsequently develop pre-eclampsia. *Lancet.* 2003;361(9368):1511–1517.
185. Savvidou MD, Kaihura C, Anderson JM, Nicolaides KH. Maternal Arterial Stiffness in women who subsequently develop Pre-Eclampsia. *PLoS ONE.* 2011;6(5):e18703.
186. Agatisa PK, Ness RB, Roberts JM, Costantino JP, Kuller LH, McLaughlin MK. Impairment of endothelial function in women with a history of preeclampsia: an indicator of cardiovascular risk. *Am J Physiol-Heart C.* 2004;286(4):H1389–H1393.
187. Manten GT, Sikkema MJ, Voorbij HA, Visser GH, Bruinse HW, Franx A. Risk factors for cardiovascular disease in women with a history of pregnancy complicated by preeclampsia or intrauterine growth restriction. *Hypertens Pregnancy.* 2007;26(1):39–50.
188. Smith GN, Walker MC, Liu A, et al. A history of preeclampsia identifies women who have underlying cardiovascular risk factors. *Am J Obstet Gynecol.* 2009;200(1):58e1–58e8.
189. Magnussen EB, Vatten LJ, Smith GD, Romundstad PR. Hypertensive disorders in pregnancy and subsequently measured cardiovascular risk factors. *Obstet Gynecol.* 2009 Nov;114(5):961–970.
190. Laivuori H, Tikkanen MJ, Ylikorkala O. Hyperinsulinemia 17 years after preeclamptic first pregnancy. *J Clin Endocrinol Metab.* 1996 Aug;81(8):2908–2911.
191. Wolf M, Hubel CA, Lam C, et al. Preeclampsia and future cardiovascular disease: potential role of altered angiogenesis and insulin resistance. *J Clin Endocrinol Metab.* 2004;89(12):6239–6243.
192. Fraser A, Nelson SM, Macdonald-Wallis C, et al. Associations of pregnancy complications with calculated cardiovascular disease risk and cardiovascular risk factors in middle age: a clinical perspective. *Circulation.* 2012;125(11):1367–1380.
193. Sattar N, Ramsay J, Crawford L, Cheyne H, Greer IA. Classic and novel risk factor parameters in women with a history of preeclampsia. *Hypertension.* 2003;42(1):39–42.
194. Hubel CA, Powers RW, Snaedal S, et al. C-reactive protein is elevated 30 years after eclamptic pregnancy. *Hypertension.* 2008;51(6):1499–1505.
195. Romundstad PR, Magnussen EB, Smith GD, Vatten LJ. Hypertension in pregnancy and later cardiovascular risk: common antecedents? *Circulation.* 2010;122(6):579–584.
196. Hastie CE, Smith GC, Mackay DF, Pell JP. Association between preterm delivery and subsequent C-reactive protein: a retrospective cohort study. *Am J Obstet Gynecol.* 2011; 205(6):556e1–4.
197. Rich-Edwards JW, McElrath TF, Karumanchi SA, Seely EW. Breathing life into the lifecourse approach: pregnancy history and cardiovascular disease in women. *Hypertension.* 2010;56(3):331–334.
198. Vikse BE, Irgens LM, Leivestad T, Skjærven R, Iversen BM. Preeclampsia and the risk of end-stage renal disease. *New Engl J Med.* 2008;359(8):800–809.
199. Mosca L, Benjamin E, Berra K. Effectiveness-based guidelines for the prevention of cardiovascular disease in women: 2011 update: a guideline from the American Heart Association. *J Am Coll Cardiol.* 2012;59(18):1404–1423.
200. Stuart JJ, Bairey Merz CN, Berga SL, et al. Maternal recall of hypertensive disorders in pregnancy: a systematic review. *J Womens Health.* 2013;22(1):37–47.
201. Pietrantoni M, O'brien W. The current impact of the hypertensive disorders of pregnancy. *Clin Exp Hypertension.* 1994;16(4):479–492.
202. Eskenazi B, Fenster L, Sidney S, Elkin EP. Fetal growth retardation in infants of multiparous and nulliparous women with preeclampsia. *Am J Obstet Gynecol.* 1993;169(5):1112–1118.
203. Dekker G, Sibai BM. Early detection of preeclampsia. *Am J Obstet Gynecol.* 1991;165(1):160–172.
204. Suhonen L, Teramo K. Hypertension and pre-eclampsia in women with gestational glucose intolerance. *Acta Obstet Gynecol Scand.* 1993;72(4):269–272.
205. Friedman E, Neff R. Pregnancy outcomes as related to hypertension, edema, and proteinuria. In: Lindheimer M, Katz A, Zuspan F, eds. *Hypertension in Pregnancy.* New York: John Wiley & Sons; 1976:13–22.
206. Plouin P, Chatellier G, Breart G, Blot P, Ioan A, Azoulay M. Frequency and perinatal consequences of hypertensive disease of pregnancy. *Adv Nephrol Necker Hosp.* 1985;15:57–69.
207. Naeye R, Friedman E. Causes of perinatal death associated with gestational hypertension and proteinuria. *Am J Obstet Gynecol.* 1979;133(1):8.
208. What we have learned about preeclampsia Sibai B.M. Caritis S, Hauth J, editors. *Semin Perinatol*, 27; 2003239–246.
209. Sibai BM, Sarinoglu C, Mercer BM. Eclampsia: VII. Pregnancy outcome after eclampsia and long-term prognosis. *Am J Obstet Gynecol.* 1992;166(6):1757–1763.
210. Lopez-Llera M, Hernández HJ. Pregnancy after eclampsia. *Am J Obstet Gynecol.* 1974;119(2):193.
211. Long P, Abell D, Beischer N. Fetal growth retardation and preeclampsia. *BJOG.* 1980;87(1):13–18.
212. Moore M, Redman C. Case-control study of severe pre-eclampsia of early onset. *BMJ (Clinical research ed).* 1983;287(6392):580.
213. Sibai BM, Spinnato JA, Watson DL, Hill GA, Anderson GD. Pregnancy outcome in 303 cases with severe preeclampsia. *Obstet Gynecol.* 1984;64(3):319–325.
214. Ferrazzani S, Caruso A, De Carolis S, Martino IV, Mancuso S. Proteinuria and outcome of 444 pregnancies complicated by hypertension. *Am J Obstet Gynecol.* 1990;162(2):366–371.
215. Sibia B, Abdella T, Anderson G. Pregnancy outcome in 211 patients with mild chronic hypertension. *Obstet Gynecol.* 1983;61(5):571–576.
216. Redman C, Beilin L, Bonnar J, Wilkinson R. Plasma-urate measurements in predicting fetal death in hypertensive pregnancy. *Lancet.* 1976;307(7974):1370–1373.
217. Wakwe V, Abudu O. Estimation of plasma uric acid in pregnancy induced hypertension (PIH). Is the test still relevant? *Afr J Med Med Sci.* 1999;28(3-4):155.
218. Ferrazzani S, De Carolis S, Pomini F, Testa AC, Mastromarino C, Caruso A. The duration of hypertension in the puerperium of preeclamptic women: relationship with renal impairment and week of delivery. *Am J Obstet Gynecol.* 1994;171(2):506–512.

219. Powers RW, Bodnar LM, Ness RB, et al. Uric acid concentrations in early pregnancy among preeclamptic women with gestational hyperuricemia at delivery. *Am J Obstet Gynecol.* 2006;194(1):160e1–160e8.
220. Roberts JM, Bodnar LM, Lain KY, et al. Uric acid is as important as proteinuria in identifying fetal risk in women with gestational hypertension. *Hypertension.* 2005;46(6):1263–1269.
221. Davis EF, Lazdam M, Lewandowski AJ, et al. Cardiovascular risk factors in children and young adults born to preeclamptic pregnancies: a systematic review. *Pediatrics.* 2012;129(6):e1552–e1561.
222. Kajantie E, Eriksson JG, Osmond C, Thornburg K, Barker DJ. Pre-eclampsia is associated with increased risk of stroke in the adult offspring the helsinki birth cohort study. *Stroke.* 2009;40(4):1176–1180.
223. Jayet P-Y, Rimoldi SF, Stuber T, et al. Pulmonary and systemic vascular dysfunction in young offspring of mothers with preeclampsia. *Circulation.* 2010;122(5):488–494.

CHAPTER 4

Genetic Factors in the Etiology of Preeclampsia/Eclampsia

KENNETH WARD AND ROBERT N. TAYLOR

Editors' comment: *Recognition that preeclampsia might have a genetic predeliction dates back to late 1800s reports of familial clustering of eclampsia. But in his 1941 treatise on "The Toxemias of Pregnancy" Dieckmann does not mention genetic or familial propensity. Chesley's first-edition text summarized early investigations in mother–daughter pairs, and he himself reported that among daughters of women with eclampsia, 26% had preeclampsia in their first pregnancies. By contrast, the daughters-in-law control group had only an 8% rate of first-pregnancy preeclampsia. This comprehensively updated chapter will "fast-forward" the reader 35 years into the post-genomic era. Rendering and analysis of increasingly precise molecular platforms and bioinformatic tools to manage megadata have galvanized the field of preeclampsia genetics, which promises to revolutionize how we think about the complexity of the etiologies, management strategies and prevention of this syndrome.*

Preeclampsia is a complex familial disorder and likely involves multiple genes in multiple biological pathways. Various types of genetic studies have been employed to solve the preeclampsia puzzle, including family reports, twin studies, segregation analyses, linkage analyses, association studies, and next-generation sequencing studies. The association studies comprise most of the literature on the genetics of preeclampsia, but because they typically study only a few pre-selected genes at a time and utilize fewer than 500 human subjects, they are insufficient for advancing the research. Genome-wide studies are the future of preeclampsia genetic research. Multicenter efforts are needed to define clinical and pathologic subsets, and high-dimensional systems biology needs to be employed so that these studies can account for preeclampsia's heterogeneity. New directions in research could lead to a greater understanding of preeclampsia's primary pathophysiology and the development of genetic screening and diagnostic tests and appropriate treatments and therapies.

DEDICATION

Current interest in the genetics of preeclampsia can be traced back to the signal study by Leon Chesley, who single-handedly followed the remote course of 267 women who survived eclampsia during the years 1931–1951. He postulated that in the absence of a renal biopsy, a convulsion – especially in a primiparous hypertensive patient – was the most convincing clinical evidence that diagnosis of preeclampsia was correct. These patients were interviewed and reexamined periodically, the data from some participants spanning more than 40 years after their eclamptic convulsion. In the course of his evaluations, Dr. Chesley noted the increased occurrence of preeclampsia–eclampsia within families.[1] His observations and his reviews of other data that suggested familial factors are involved in the etiology of preeclampsia caught the fancy of genetic investigators. With the advent of molecular genetics and the still more recent mapping of the human genome, this field is advancing rapidly. This chapter, dedicated to the pioneering studies of Leon Chesley, summarizes research into the genetics of preeclampsia through 2013. A glossary of genetics terminology is provided for readers who are less familiar with genetic concepts.

INTRODUCTION

Preeclampsia and eclampsia* are familial, as genetic research on these conditions over the last century has shown. Due to monumental improvements in neonatal care and decreased mortality rates of newborns over the last 50 years, generational trends can now be observed: eclamptic or

* Throughout the rest of this chapter, preeclampsia will be the word used when the discussion could refer to either preeclampsia or eclampsia. Distinctions will be made between the two conditions as necessary.

Chesley's Hypertensive Disorders in Pregnancy.
ISBN: 978-0-12-407866-6

DOI: http://dx.doi.org/10.1016/B978-0-12-407866-6.00004-3

preeclamptic mothers, aunts, and grandmothers have had female descendants who show an increased risk of preeclampsia over the general population. Preeclampsia tends to cluster in families; a heritability study using a Utah genealogy database determined the coefficient of kinship for preeclampsia cases to be more than 30 standard deviations higher than for controls[2] (and unpublished data). The recurrence risk for preeclampsia in the daughters of either eclamptic or preeclamptic mothers is in the 20–40% range. For sisters it is in the 11–37% range. Much lower rates are seen in relatives by marriage, such as daughters in-law and mothers in-law.[3] African-American mothers at all socioeconomic levels experience a higher rate of preeclampsia than the general population in the United States, suggesting that ethnicity, rather than socioeconomic status, has a greater impact on incidence of preeclampsia.[4] And finally, twin studies estimate that approximately 22% to 47% of preeclampsia risk is heritable, as opposed to environmentally influenced.[3,5–10]

While the ultimate causes of preeclampsia remain unknown, it is perhaps obvious that genes should play a role. As discussed in other chapters of this book, preeclampsia occurs when placental ischemia or inflammation causes various mediators to be released directly into the maternal circulation. The maternal endothelium and arterioles respond initially in an adaptive manner, but ultimately cause profound dysfunction of various major organs. Any number of disorders with a genetic component might interfere with maternal vascular responses, affect trophoblast function, or increase the placental mass, causing fetal demands to outpace the supply. Every ligand, every receptor, every amplification cascade, every aspect of the programmed responses that orchestrate the pathophysiological response is under the control of either the mother's or the fetus's genes.

Geneticists finally have the technology and genomic data to find practical clinical solutions for the genetic testing, prevention, and treatment of preeclampsia. The completion of the Human Genome Project has enabled researchers to investigate the genetic causes of diseases both rare and common, and new technologies and statistical models are able to detect increasingly subtle variations within the human genome, gene effects on physiology, and their interactions with environmental factors.

Studies conducted over the last decade suggest that the genetic contributions to preeclampsia are likely to be complex, involving non-Mendelian transmission and numerous variants, gene–gene interactions, and environmental variables. The genes involved would not directly cause preeclampsia but rather would lower a woman's biological threshold at which she would develop the condition.

Indeed, as noted throughout this chapter, preeclampsia is best understood as a multifactorial, polygenic condition. In a Mendelian disease (e.g., cystic fibrosis), an allelic variant or mutation is directly involved in causing the disease, and penetrance is complete or nearly complete for the relatively few persons in the population with that disease genotype. By contrast, complex diseases are the result of numerous common variants, usually at multiple loci, which contribute to varying degrees to a person's susceptibility to that disease.[11] Because of environmental and other variables, genotype alone may not result in phenotypic manifestation of the disease, but it will increase disease susceptibility and risk.[12,13] These polygenic multifactorial conditions are familial, usually affecting multiple generations, but their inheritance does not follow Mendelian ratios. Polygenic disease occurs under diverse environmental conditions, and the genes underlying the disease can show highly variable expression usually resulting in a continuum of phenotypes with patients above some threshold classified as having the disease.

Genetic background plays a critical role in polygenic inheritance since several factors must collaborate to cause a bodily function to go awry, and only after these factors reach some critical point is the phenotypic effect seen. Thus recurrence risk of a polygenic multifactorial disease is higher within populations with a high incidence of the disorder. In disorders with a relatively high heritability, the recurrence risk of the disorder approximates the square root of the population incidence. Marked variation in the incidence and expression may occur in different ethnic groups. Within families, the greater the number of family members who have already been affected with a multifactorial condition, the more likely it is that the genetic background is favorable for expression of this condition. Consanguinity also increases the risk because of the greater likelihood of deleterious genes being shared. The severity of the disorder often correlates with the recurrence risk.

In the case of preeclampsia, the relative importance of each environmental risk factor would vary among women, depending on their genotype. One woman may have a genotype that requires an accumulation of several environmental variables, such as substance abuse and multiple gestation, to exceed this threshold. Another woman may have no identifiable environmental risk factors but may have prepregnancy diabetes, which may result in an extremely low, easily exceeded preeclampsia threshold.

Since the diagnostic criteria of preeclampsia force arbitrary thresholds on the continuous distributions of blood pressure and proteinuria, it is likely that no single etiology or genetic marker will account for all cases of preeclampsia. Given the clinical heterogeneity of the condition, it is all the more important to first correctly classify preeclampsia cases into subtypes, based on apparent etiology. This concept is stressed in Chapters 2 and 8 in the current edition of this text.

Table 4.1 summarizes many of the genetic terms you will find in this chapter.

TABLE 4.1 Glossary of Genetics Terminology

Term	Definition
Allele	The DNA sequence on a single chromatid at a particular genetic locus; at each locus, a single allele is inherited separately from each parent.
Base pair (bp)	Two nucleotides on opposite complementary DNA or RNA strands that are connected via hydrogen bonds. The human genome is composed of approximately 3 billion base pairs. Another unit frequently used is the *kilobase (kb)*, which denotes 1000 base pairs.
Candidate gene	A gene researched and suspected of being involved in a particular trait or disease. Frequently identified in gene expression or genetic association studies.
Coefficient of kinship	The probability that a gene at a given locus, picked at random from each of two individuals, will be identical due to familial relationship.
Epigenesis	Environment-induced variations in the expression or operation of a functional gene even though the underlying genomic code is stable and unchanging.
Epistasis	Interaction between or among non-allelic genes in which one combination of such genes has a dominant effect over other combinations (for instance when one gene suppresses the expression of another)
Exon	Only about 1% of our genome is the "genes" which are "translated" into the proteins. The roughly 22,000 human genes are divided into 180,000 functional segments referred to as exons.
Exome	The portion of the genome (the 180,000 exons) that codes for proteins. Currently, over two-thirds of the DNA errors known to cause genetic disorders occur in the exome.
Family clustering	The repeated occurrence of a phenotype in a given family. Often detected in genetic family studies conducted for this purpose.
Fine-mapping	The determination of the sequence of nucleotides and their relative distances from one another in a specific region or locus of the genome.
Founder effect	The loss of genetic variation that occurs when a relatively small number of individuals establish a new colony distinct from their original, larger population. As a result of the loss of genetic variation, the new population may be genetically and phenotypically different from the parent population but have relatively low genetic variation within itself. It may show increased sensitivity to genetic drift and an increase in inbreeding.
Gene expression	The process by which gene-coded information is converted into structural proteins and enzymes. Expressed genes are transcribed into mRNA, which in turn is translated into protein, or they are transcribed into RNA that does not translate into protein (e.g., transfer, noncoding and ribosomal RNAs).
Genetic association	The hypothesis that a given candidate gene or SNP causes or is otherwise related to a particular genetic condition, based on the differences in single-locus alleles or genotype frequency between a case group (with the genetic condition) and a control group (without the genetic condition).
Genetic drift	The evolutionary process of change in allele frequencies that occurs entirely from chance, from one generation to the next.
Genomic	Of or relating to the entire genome, which in humans comprises 23 chromosome pairs, or 3.2 billion DNA base pairs.
Genotype	The specific allele makeup of an individual, usually referring to one or more particular alleles being studied.
GWAS	Genome-wide association study – a discovery approach, widely used in recent genetics research, in which genetic variations (genotypes) across the genomes of many cases and controls are assayed in order to find genetic variants associated with a disease or trait (phenotypes).
Haplotype	1. Two or more alleles or SNPs at distinct loci on one chromosome that are transmitted together. 2. A set of SNPs on a single chromatid that is statistically associated.
Heritability	The proportion of phenotypic variation in a population that is attributable to genetic variation among individuals.
Heterogeneity	The variability of phenotypes despite identical genetic information.
Heterozygous	When, of the two possible alleles at a given locus on homologous chromosomes, one of each allele is present.
Human Genome Project	A 13-year research project completed in 2003 and coordinated by the US Department of Energy and the National Institutes of Health with contributions from numerous industrial nations. It identified all genes in the human genome (approximately 23,000) and determined the sequence of the 3 billion nucleotide base pairs that make up human DNA. It also developed infrastructure for the storage of these data and their future analysis, which is expected to continue in the private sector for many years.
Immunogenetic	1. The genetic basis of susceptibility to immune response and disease. 2. The relationship between immunity to disease and genetic makeup.
Linkage	Occurs when particular genetic loci on a chromosome, or alleles, are inherited jointly due to their close proximity to one another. It is measured by the percentage recombination between loci (unlinked genes showing 50% recombination). A linkage analysis studies this phenomenon in the context of candidate genes and markers.
Linkage disequilibrium (LD)	The occurrence of some alleles at two or more loci (like linkage, though not necessarily on the same chromosome) more or less often than would be expected, presumably because the combination confers some selective (evolutionary) advantage or disadvantage, respectively. Non-random associations between polymorphisms at different loci are measured by the degree of LD and can be demonstrated with haplotype analysis.

(*Continued*)

TABLE 4.1 (Continued)

Term	Definition
Locus (plural is loci)	The position on a chromosome of a gene or other marker; also, the DNA at that position. The meaning of locus is sometimes restricted to regions of DNA that are expressed (see *Gene expression*).
Logarithm of the odds (LOD) score	A measure of the likelihood of two loci being within a measurable distance of each other.
Marker	An identifiable physical location on the genome, the inheritance of which can be monitored. Markers can be expressed regions of DNA (genes), a restriction enzyme cutting site, or some segment of DNA with no known coding function but with a distinguishable inheritance pattern. Markers must be linked with a clear-cut phenotype; they are used as a point of reference when mapping new variants.
Multifactorial	A disease or condition influenced in its expression by many factors, both genetic and environmental.
Mutation	Any heritable, permanent change in DNA sequence; also, the process by which genes undergo a structural change.
Next-generation sequencing	DNA sequencing is the process of determining the order of the four nucleotide bases – adenine, guanine, cytosine, and thymine – in a strand of DNA. Next generation sequencing uses high-throughput, automated methods to sequence hundreds of thousands of sequences in parallel so that an entire human genome can be sequenced in a day.
Nucleotide	One of the monomeric units from which DNA or RNA polymers are constructed; it consists of a purine or pyrimidine base, a pentose sugar, and a phosphoric acid group.
Pedigree	A diagram of the relationships within a given family with symbols to represent people and lines to represent genetic relationships. Often used to determine the mode of inheritance (dominant, recessive, etc.) of genetic diseases.
Penetrance	The extent to which individuals who carry the gene for a particular genetic condition express that gene as an expected phenotype. Penetrance may be described as complete, incomplete, low, or high, or may be quantitatively expressed as a percentage.
Polygenic disorder	A genetic disorder resulting from the combined action of alleles of more than one gene (e.g., heart disease, diabetes, and some cancers). Although such disorders are inherited, they depend on the simultaneous presence of several alleles; thus the hereditary patterns are usually more complex than those of single-gene disorders. Also referred to as *complex disorders*.
Polymorphism	Genetic differences in the DNA sequence that naturally occur among individuals. A genetic variation that occurs in more than 1% of a population would be considered a useful polymorphism for genetic linkage analysis.
Population admixture	The unwitting inclusion of members of a genetic population different from the genetic population being studied, having been selected to increase genetic homogeneity. It has the potential to give false-positive results in studies of genes underlying complex traits or to mask, change, or even reverse true genetic effects.
Predisposition	An increased likelihood of, or an advanced tendency toward, a specific medical condition.
Promoter	A site on DNA to which RNA polymerase will bind and initiate transcription; classically 5′ to its coding region.
Quantitative trait loci (QTL)	Stretches of DNA that are closely linked to the gene(s) that underlie the trait in question, though they are not necessarily genes themselves. Can be molecularly identified to help map regions of the genome that contain genes involved in specifying a quantitative trait.
Race/Ethnicity	Indicators of evolutionary ancestral geographic origin (by continent, in the broadest terms) that can be detected via certain markers on a person's genome. In population genetics, these distinctions are useful in increasing the homogeneity of a population for detection of differences between diseased and control subjects.
Recessive	A gene that is phenotypically manifest in the homozygous state but is masked in the presence of a dominant allele.
Recombination	The process by which offspring derive a specific combination of genes different from that of either parent.
Recurrence risk	The chance that a genetic disease present in a family will recur in that family and affect another person (or persons).
Segregation analysis	A statistical test to determine the pattern of genetic inheritance for a trait or genetic condition (e.g., Mendelian, dominant autosomal, epistatic, polygenic, age-dependent).
Single nucleotide polymorphism (SNP)	A common (occurring in more than 1% of the population), single base substitution in the DNA sequence that may or may not cause a difference in gene expression (called functional or nonfunctional, respectively). Functional SNPs include: (1) SNPs in coding regions of genes resulting in amino acid substitutions (non-synonymous SNPs), which may in turn alter protein sequence, structure, function, or interaction; enzyme stability; catalytic activity; and/or substrate specificity; (2) SNPs in the non-coding, regulatory regions of genes that affect the genes' transcription, translation, regulation, or mRNA stability (synonymous SNPs); (3) SNPs in genes that are duplicated, resulting in higher product levels; (4) SNPs in genes that are completely or partially deleted, resulting in no product; (5) SNPs that cause splice site variants that result in truncated or alternatively spliced protein products. Also look up *promoter*, *enhancer*, *repressor*, *regulator*, *intron*, *exon*, *codon*, *deletion*, *duplication*, *insertion*, *inversion*, *translocation*, and *copy number variant (CNV)*.
Single-gene disorder	Hereditary disorder caused by a mutant allele of a single gene (e.g., cystic fibrosis, myotonic dystrophy, sickle cell disease).

(*Continued*)

TABLE 4.1 (Continued)

Term	Definition
Subtype	One of two or more genetic pathways to the same disease phenotype. A disease may be defined by clinical characteristics, but once its genetic contributions are discovered, that same disease may be separated into subtypes based on which genes contribute to which type. This classification may or may not lead to refinements in the definition of the phenotype and/or clinical characteristics.
Wellcome Trust Case Control Consortium	A collaboration of 24 human geneticists who analyzed over 19,000 DNA samples from patients suffering from different complex diseases to identify common genetic variations for each condition. Conditions included tuberculosis, coronary heart disease, type 1 diabetes, type 2 diabetes, rheumatoid arthritis, Crohn's disease, bipolar disorder, and hypertension. Two thousand patients were recruited for each disease, and 3000 were recruited as controls. The research was conducted at a number of institutes throughout the UK.

BIOLOGICAL PATHWAYS OF PREECLAMPSIA

Although preeclampsia may be a fairly homogeneous entity when it is defined by glomerular endotheliosis – the characteristic histopathologic feature of the condition – we are forced to depend on other criteria for probing the genetic aspects of the condition, as the diagnostic signs are nonspecific. The diagnostic blood pressure and proteinuria criteria in common use are, in fact, arbitrary cutoffs along a continuous distribution of values. Even if perfect diagnostic criteria existed, there is a great deal of debate about what the proper phenotype is for study, whether it is proteinuric hypertension, gestational hypertension, a placental phenotype such as reduced trophoblast invasion, a renal phenotype such as glomerular endotheliosis, or other phenotypes. Up to one-third of infants born of preeclamptic pregnancies are affected by intrauterine growth restriction, and this subset may have different variants involved.[10] Some have argued that early-onset disease needs to be considered a different subtype.[14] Co-morbidities related to diabetes, thrombophilia, renal disease, auto-immunity, et cetera might allow more optimal classification. The non-genetic research of preeclampsia suggests that its essential pathophysiology comprises oxidative stress and endothelial decompensation originating from and/or contributing to poor placental perfusion, which cumulatively lead to the observable symptoms of the disease.

Preeclampsia is a difficult disorder to study using genetic methodologies. A major lesson of modern genetics is that syndromes defined on the basis of clustering of clinical symptoms often reveal marked heterogeneity until they are better understood at a molecular level. In this respect, the boundaries around preeclampsia, gestational hypertension, and HELLP syndrome are likely to be redrawn once genetic determinants can be examined directly.

Research in the last decade has suggested that preeclampsia, despite its appearance only in pregnancy, has a systemic pathophysiology involving distinct yet diverse biological pathways in both the fetus and the mother.[10,15] Any of these pathways – including immunologic, inflammatory, hypoxic, and thrombophilic pathways – can push inadequate placentation or pregnancy-induced hypertension over the threshold into preeclampsia, and all of these pathways are influenced by the biological products of genetic expression and genotype. As in all genetic studies of pregnancy disorders, not just one but two genotypes must be considered: the genotype of the fetus as well as the mother.

Any genetic hypothesis of preeclampsia must explain the first pregnancy effect. It is widely known that most women will not have preeclampsia with future pregnancies unless another condition exists (e.g., underlying renal disease, twins, diabetes). This has suggested an immunogenetic mechanism to many investigators, in the form of desensitization or tolerance to paternal antigens in subsequent gestations. An increased risk for first pregnancies with new partners (i.e., primipaternity with new paternal antigens presented in the placenta) has also been noted.[16] Limited evidence exists that couples who use condoms for contraception, who accept an ovum donation, or who have a shorter length of cohabitation prior to conception have an increased risk of preeclampsia; couples that practice oral sex and women who have had multiple blood transfusions have a lower risk. Other explanations for the first pregnancy phenomenon must also be considered, however: certain enzymes in pregnancy are permanently induced and never go back to baseline levels after delivery.[17,18] Similarly, permanent changes occur in maternal blood volume and in the vascular architecture of the uterus after a term gestation, so that a greater volume expansion is generally achieved with subsequent pregnancies at later parities.[19]

While increased compared to normotensive deliveries, remote essential hypertension in first pregnancy preeclamptic women is surprisingly low and may hold a clue as to why preeclamptic patients do not have a particularly high rate of developing essential hypertension. If, for example, the *AGT* T235 polymorphism increases the risk of both preeclampsia and essential hypertension through a blood volume mechanism, sustained changes in baseline blood volumes that occur after delivery might explain both the first-pregnancy effect and possible reduction in an affected woman's lifelong risk of essential hypertension.

Fetal/Placental Components of Preeclampsia

There have been numerous reports of circumstantial evidence of a paternal/fetal genetic effect.[16,20,21] Astin et al. reported a man who lost two consecutive wives to eclampsia,[22] and had a severely preeclamptic third wife. The observation that preeclampsia is extremely common in molar pregnancies (in which all the fetal chromosomes are derived from the father) is considered further evidence for the role of paternal genes. Triploidy, although unusual in advanced gestations, frequently presents with preeclampsia.[23] The increase in paternal genetic material associated with the triploid diandric placenta may support the role of paternal genes in the development of preeclampsia. Hydatidiform moles, which have two sets of paternal chromosomes, commonly cause a preeclampsia-like illness.[24]

In the segregation analyses described above, Cooper found an increased rate of preeclampsia if the proband's own mother was eclamptic with their pregnancy.[25,26] Others have investigated a small increase in the incidence of preeclampsia in the daughters in-law of women who had pregnancy-induced hypertension.[27–29]

Males presumably do not manifest a phenotype when they carry susceptibility genes, which may be much greater in number than the existing literature indicates, considering how few studies have been done on paternal carriers. Considering that for most of human history, eclampsia was a common but frequently fatal disease for mother or child or both, a large percentage of modern cases must be due to new mutations without a direct familial pattern.[30]

Immunogenetic Factors (see also Chapter 8)

The suppression of the maternal immune response to the fetus/placenta in normal and abnormal pregnancies has been frequently and thoroughly studied.[31–52] The immunological aspects of preeclampsia get to the heart of this query because symptoms and other epidemiological evidence of the mother's abnormal immune response to paternal antigens – and the fetoplacental allograft derived from those antigens – are curiously prominent in many cases of preeclampsia:

- Preeclampsia is more common during a first conception.
- Preeclampsia may be more common after a mother has switched partners.
- Oral or long-term semen exposure decreases the risk of preeclampsia.
- Preeclampsia is more common in women who use barrier contraception.
- The frequency of preeclampsia in the cases of donated gametes is very high.
- HIV-related T cell immune deficiency is associated with a low rate of preeclampsia.[53,54]

On the molecular level, immune mediators are closely involved in many aspects of pregnancy from implantation and placentation to labor. A normal pregnancy is accompanied by a pregnancy-specific, immunomodulated inflammatory response to the antigenic stimulus presented by the fetal-placental semiallograft.[55] The largest surface area of contact between maternal immunocompetent T cells and the fetus is at the level of the villous trophoblasts. These cells originate in the embryo and lack expression of major histocompatibility complex (MHC) class I and class II antigens. The extravillous trophoblasts (EVT) only express human leukocyte antigens (HLA) C (weakly), Ib, G, F, and E, rather than the strong transplantation antigens HLA-A, -B, -D, -Ia and -II. Of these, only HLA-C is signaling paternal (foreign) alloantigens.[15,54,56–58] There is new evidence that maternal immune cells cross the placenta, colonize fetal lymph nodes and remain to tolerize fetal T regulatory cells until early adulthood.[59] Other inflammatory factors in preeclampsia are an abnormal immunological maternal response, comprising a change in the role of monocytes and natural killer (NK) cells for the release of circulating cytokines and an activation of proinflammatory angiotensin II subtype 1 (AT1) receptors. The significance of these receptors is discussed further in Chapter 15 by Dechend, LaMarca and Taylor. Activated neutrophils, monocytes, and NK cells initiate inflammation, which in turn induces endothelial dysfunction, if activated T cells support inadequate tolerance during pregnancy.[54,60]

In preeclampsia, genetic or non-genetic factors such as hypoxia or oxidative stress can induce necrosis or aponecrosis of trophoblasts.[58] Macrophages or dendritic cells that phagocytose these trophoblasts produce type 1 cytokines such as tumor necrosis factor alpha (TNFα), interleukin (IL) 12, and IFN-γ that augment inflammation,[61,62] such that the cytokine profile in preeclampsia, in contrast to that of normal pregnancy, is one of type 1 proinflammatory cytokine dominance, while the production of type 2 immunomodulatory cytokines is suppressed. This is supported by studies reporting elevated plasma levels of TNFα and IL-1β in preeclampsia.[63] Furthermore, if type 2 regulatory cytokines are reduced system-wide regulatory T cell function can be inhibited.[54,60]

Redman and Sargent[53] argue that the immunological interaction between the mother and fetus might be mediated predominantly by NK cells instead of T cells (see their chapter – Chapter 8). For example, the invading trophoblasts predominantly encounter maternal decidual lymphocytes that are NK cells with an unusual phenotype. Notably, the NK cells express receptors (such as killer immunoglobulin-like receptors, or KIRs) that recognize the exact combination of HLAs associated with invasive cytotrophoblasts, particularly polymorphic HLA-C. These NK cells express a unique array of KIRs for binding the combinations of HLAs expressed by intermingling cytotrophoblast and NK cells that mediate immune recognition.[53]

Numerous haplotypes, differing in gene content and allele combinations, are associated with the multigene KIRs. Some haplotypes inhibit NK-cell function (cytokine production in these cells), whereas others are stimulatory, depending on both the KIR phenotype of the NK cells and the HLA-C phenotype of the stimulating cells. KIR gene haplotypes can be divided into two functional groups: the simpler A group codes mainly for inhibitory KIR, and the more complex B group codes for receptors that stimulate NK cells. Preeclampsia is much more prevalent in women homozygous for the inhibitory KIR haplotypes (AA) than in women homozygous for the stimulator KIR BB. The effect is strongest if the fetus is homozygous for the HLA-C2 haplotype.[53]

Essentially, normal placentation is more likely and preeclampsia is less likely when trophoblasts strongly stimulate uterine (maternal) NK cells. This interaction between trophoblasts and NK cells serves as an essential component of immunity suppression/stimulation issues in normal and abnormal pregnancies, specifically in preeclampsia.[56–58,60,64–67]

There is evidence for increased release of syncytiotrophoblast microvesicles and other cellular "debris" into the maternal plasma that also influence immune stimulation, cytokine production and vascular remodeling.[68,69] The ability of the mother to mount an adequate response to these released pro-oxidant molecules may be of key importance for oxygen supply throughout pregnancy. This view is endorsed by the increased risk of preeclampsia in women with preexisting medical conditions that frequently lead to oxidative stress, including chronic hypertension, diabetes, and renal disease. Furthermore, long-term follow-up studies have shown that women who were unaffected by these conditions prior to conception are nevertheless more likely to develop them later in life following an episode of preeclampsia, as reviewed in Chapter 3. It is now well known that chronic hypertension and diabetes have a genetic component, which raises the possibility that preeclampsia shares susceptibility genes with these conditions, particularly related to oxidative stress.[10]

TYPES OF GENETIC STUDIES CONDUCTED

Family Reports

The first hints that preeclampsia is a genetic disease came from reports of familial clustering. Numerous, early case series were summarized by Chesley.[70] Elliott was the first to report familial incidence of eclampsia in 1873. He reported a woman who died of eclampsia during her fifth pregnancy. Three of her four daughters subsequently died of eclampsia as well. One of the first systematic studies of the genetics of preeclampsia was presented in 1960 by Humphries, who studied mother–daughter pairs delivering at the Johns Hopkins Hospital.[71] Preeclampsia occurred in 28% of the daughters of women who had toxemia in pregnancy, compared with 13% of the comparison group. As noted above, Chesley's own remarkable study was well under way in the 1960s, when he reported data on the pregnancies of the daughters, daughters in-law, and sisters of the eclamptic probands, in whom he observed a relative risk of eight-fold for eclampsia in the daughters.[70] Remarkably persistent in this effort, Dr. Chesley was able to find information on 96% of all the daughters, greatly reducing the possibility of ascertainment bias. At the time of his death, he was preparing to publish additional data on the granddaughters and granddaughters-in-law of the original cohort.

Another large body of information comes from the Aberdeen Maternity Hospital in Scotland. Research teams headed by Adams, Cooper, and Sutherland studied this population in 1961, 1979, and 1981, respectively.[72–74] These studies were unique because of consistent diagnostic criteria and classification and careful recording of births through several decades. Arngrimsson studied the Icelandic population.[27] Because of the population's small size and interest in genealogy, as well as the concentration of maternity records at only one hospital, relatively complete information was available on the relatives of the index cases with preeclampsia. Daughters had a prevalence of preeclampsia or eclampsia of 23%, whereas those syndromes occurred in just 10% of the daughters-in-law.

Alexander notes that both male and female children of preeclamptic mothers are at an increased risk of having or fathering a preeclamptic pregnancy in turn, providing evidence of fetal contribution to preeclampsia susceptibility. However, according to pedigrees and heritability studies, this susceptibility remains greater in female than in male offspring, suggesting that transmission from mother to daughter is a critical component.[58]

Table 4.2 summarizes the family reports discussed in this section.

Twin Studies

Once there are indications that a disorder is genetic, twin studies can be used to measure the heritability of the condition (the proportion of preeclampsia risk that is attributable to genetics, as opposed to environmental factors). Several studies wholly or partially dedicated to this question regarding preeclampsia have been conducted.[3,5,6,9,75] Using twin concordance to estimate heritability is never a straightforward method because of the mechanisms and associations underlying the monozygotic twinning process.[76] Indeed, Hall postulated that the development of a discordant cell line (by any mechanism including chromosomal, single gene mutation, mitochondrial mutation, uniparental disomy, somatic crossing over, X-inactivation, imprinting, etc.) early in embryonic development may in fact be an underlying cause of human monozygotic (MZ) twinning;

TABLE 4.2 Noteworthy Family Clustering Studies of Preeclampsia

Author(s)[ref]	Year	Focus of Study
Humphries[68]	1960	Mother–daughter pairs.
Adams and Finlayson[69]	1961	Disease of preeclampsia, sisters of preeclamptic women.
Chesley et al.[67]	1968	Pregnancies of daughters and granddaughters of eclamptics compared to daughters and granddaughters "in-law."
Cooper and Liston[70]	1979	"Severe" preeclampsia.
Sutherland et al.[71]	1981	Increased preeclampsia in mothers and daughters of preeclamptic women.
Armgrimsson et al.[26]	1990	Increased rate of preeclampsia in mothers and daughters of preeclamptic women.
Esplin et al.[2]	2001	Utah database: both men and women who were born to a mother with preeclampsia were significantly more likely to have a child who was the product of a pregnancy complicated by preeclampsia.
Cnattingius et al.[30]	2004	Swedish registries. Heritability estimated at 0.55, maternal genes contribute more than fetal genes.
Alexander[56]	2007	Mother–fetus pairs, including male transmission.

thus artificially inflating the number of discordant MZ twin pairs, leading to biased estimates of heritability.[77]

Thornton and Macdonald[5] studied a cohort of female twins in the UK to estimate the maternal genetic contribution to preeclampsia and non-proteinuric hypertension. However, their study ran into complications when many of the twin pairs' self-reported preeclampsia diagnoses could not be confirmed with medical records. Based on self-reported diagnoses only, preeclampsia had a heritability of 0.221, whereas non-proteinuric hypertension heritability was 0.198. Using hospital records and various models to predict the heritability of preeclampsia based on the heritability of non-proteinuric hypertension (0.375), the study concluded that neither preeclampsia nor non-proteinuric hypertension was as heritable as previously believed, and certainly non-Mendelian in transmission.

O'Shaughnessy et al.[6] conducted a study in the UK on monozygotic twin concordance for preeclampsia, using four pairs of monozygotic twin mothers and one monozygotic triplet mother. The study indicated that concordant monozygotic siblings were no more likely to develop preeclampsia than discordant ones. Finally, an Australian study[3] examined a large cohort of twin pairs to determine the maternal versus fetal genetic causes of preeclampsia by evaluating concordance among several degrees of relatives: between monozygotic and dizygotic female co-twins, between female partners of male monozygotic and dizygotic twin pairs, and between female twins and partners of their male co-twins in dizygotic, opposite-sex pairs. The study determined preeclampsia to have a very low genetic recurrence risk; the maternal genetic contribution was also much lower than expected.

The accumulated evidence on twin studies of preeclampsia, including those conducted by Thornton and Onwude,[7] Lachmeijer et al.,[8] and Salonen Ros et al.,[9] suggests that penetrance in preeclampsia is generally less than 50%, and that the accumulated confidence interval is quite wide (95% CI, 0–0.71).[10] This may suggest a greater diversity of inheritance models and modes across the spectrum of women who exhibit the preeclampsia phenotype, encompassing Mendelian, single-gene dominance through polygenic, multifactorial inheritance.

Segregation Analyses

Segregation analyses attempt to fit the recurrence risk data from family studies into a genetic model and are useful, if not for proving inheritance via a particular model, for eliminating alternative models. Several segregation analyses of preeclampsia have been published, with varying conclusions.

In the aggregate, pre-2000 segregation analyses consistently suggested a relatively common allele acting as a "major gene" conferring susceptibility to preeclampsia. The marked increase in the incidence of preeclampsia in blood relatives but not in relatives by marriage implies that maternal genes are more important than fetal genes. Alternatively, this inheritance pattern supports the hypothesis of transmission of preeclampsia from mother to fetus by a recessive gene. More recent analyses suggest a multifactorial, polygenic inheritance with strong epigenetic contributions.[78] Some examples of methylated CpG islands in introns or promoters of differentially expressed genes have been proposed, e.g., the *STOX1* transcription factor alleles (see below), but these remain controversial.[79]

One alternative not adequately addressed in these models is whether a very high new mutation rate (as would be expected for a common but deadly condition) exists.[80] Chesley has reviewed the recorded history of eclampsia, and it is clear that mortality from eclampsia was high until the last few generations. Given its high lethality, it is unclear how a preeclampsia gene would become so common in the population. With the exception of one report describing a gorilla pedigree with preeclampsia, we could find no evidence that preeclampsia occurs spontaneously in our recent primate ancestors.[81] (However, surgical induction of uteroplacental ischemia in baboons and rhesus can result in a preeclampsia-like syndrome.)[82,83] Usually the only way a lethal gene will stay common in the population is if there are frequent new mutations or if the gene is positively selected on some other basis. Both of these are possibilities have been suggested in preeclampsia.

New studies are determining which preeclampsia-causing alleles are common in the population, and alleles at which mutations are frequent by tracking their occurrence through populations of women with and without preeclampsia.[75,78,84,85] In one of the most significant new sequence analyses, van Dijk et al.[84] narrowed a minimal critical region at 10q22 linked with preeclampsia in women of Dutch ancestry to 444 kb. The *STOX1* gene in this region contains five different missense mutations, identical between affected sisters, cosegregating with the preeclamptic phenotype and following matrilineal inheritance. The predominant Y153H variation of this gene is highly mutagenic by conservation criteria but subject to incomplete penetrance. Given the maternal effect, a causal relation of this gene with preeclampsia can exist only if the mutations are maternally derived and transmitted to the children born from the affected pregnancies. Substitution of the conserved amino acid (either tyrosine or phenylalanine) from the Y153H mutation may lead to disease.

Oudejans et al.[78] provide a concise review of segregation studies conducted as well as models for the transcription mechanisms at the various sites throughout the genome. Regarding the van Dijk analysis, Oudejans's study group proposes that a second variation, also on 10q22, is needed to explain the full phenotype in Dutch females. Oudejans's study group also noted that all of the susceptibility loci with significant linkage to preeclampsia detected previously – 2p12 (Iceland), 2p25 (Finland), and 9p13 (Finland), in addition to the 10q22 region – display evidence of epigenetic effects, which could contribute to the persistence of preeclampsia despite its historically high mortality. To date, this work has not been confirmed in other populations.

Table 4.3 summarizes the segregation analyses discussed in this section.

TABLE 4.3 Noteworthy Segregation Analyses of Preeclampsia

Author(s)[ref]	Year	Study Conclusions
Rana et al.[75]	2003	Familial forms of focal segmental glomerulosclerosis (FSGS) determined to be primarily autosomal recessive; preeclampsia developed in successive pregnancies.
Van Dijk et al.[74]	2005	Narrowed a minimal critical region at 10q22 linked with preeclampsia in women of Dutch ancestry.
Laivuori[72]	2007	Polygenic; also notes maternally inherited missense mutations in the *STOX1* gene of the fetus.
Oudejans et al.[76]	2007	Polygenic, with segregation of preeclampsia into early-onset, placental and late-onset. Notes contribution of epigenetics.
Berends et al.[86]	2008	Cosegregation of preeclampsia with intrauterine growth restriction. High rate of consanguinity.

Linkage Analyses

Linkage studies use a "positional" approach to gene identification using regions of the genome that segregate with the disease of interest, in contrast to physiologic hypotheses. Linkage studies require an accurate diagnosis of the disease under study and precise histories of family relationships among the study participants. Furthermore, linkage analysis of pedigrees requires that the appropriate model be used in the LOD (logarithm of the odds) score calculation. Markers are tested in family studies to find any violations of Mendel's second law, which states that independent traits segregate independently. Whenever two independent traits are closely located on the same chromosome, Mendel's second law is violated. These aberrations from independent segregation can be used to map the chromosomal location of a disease gene by comparing these markers with the other chromosomes.

With technological advances in the last decade, genome-wide linkage studies have become possible and preferable to analysis of pre-selected regions on pre-selected chromosomes for detecting susceptibility loci for complex diseases. After a genome-wide scan highlights certain regions and rules out others, the fine mapping and sequencing of the suspect regions can zero in on candidate single nucleotide polymorphisms (SNPs) and genes and detect parent-of-origin effects. Below is a summary of the genome-wide linkage studies and the genomic regions they have investigated. Other studies have been conducted based on biological hypotheses regarding functional genes in predetermined regions of the genome, or in an effort to fine-map areas first detected with genome-wide studies.

- Harrison et al. published a genome-wide linkage study of 15 Australian family pedigrees in 1997.[87] They found a 2.8-cM candidate region between D4S450 and D4S610 on 4q with a barely significant, maximum multipoint LOD score of 2.9. Because of uncertainties concerning inheritance and diagnosis, four different inheritance models were used to carry out LOD score analysis.
- Arngrimsson et al.[88] published their results of a genome-wide linkage study of 124 Icelandic family pedigrees in 1999. They found a maternal susceptibility locus on 2p13 with a significant LOD score of 4.70. Their data supported a primarily dominant inheritance model.

- Moses et al.[89] published their results of a medium-density genome-wide scan of 34 Australian/NZ families in 2000. They found loci on chromosome 2 at an LOD score of 2.58 (tentatively supporting Arngrimsson's group[78]) and at 11q23–24 (LOD score 2.02). The model used was multipoint nonparametric.
- Lachmeijer et al.[90] published their results of a genome-wide scan of 67 Dutch families in 2001. The highest LOD score of 1.99 was determined on the long arm of chromosome 12, associated primarily with HELLP-afflicted families. When HELLP families were eliminated, the remaining 38 families were evaluated and returned LOD scores of 2.38 on chromosome 10q and of 2.41 on chromosome 22q. No chromosome 2 loci were detected.
- Laivuori et al.[91] published results of a genome-wide scan of 15 Finnish families in 2003. Two loci were detected: 2p25, with a nonparametric linkage (NPL) score of 3.77, and 9p13, with an NPL score of 3.74. A third locus of slightly weaker score was found at 4q32 (NPL 3.13).
- In 2006, Kalmyrzaev et al.[92] published results of a genome-wide scan on a single Kyrgyz family, so selected for the notable early onset of preeclampsia and the geographic isolation of the Kyrgyz population, which was expected to reduce heterogeneity of the condition. Nonparametric analysis detected a region at 2q23–q37, and with 2-point parametric analysis an LOD score of 2.67 was obtained at 2q24.3.
- Moses et al. published results of a genome scan of 34 Australian and New Zealand families in 2006,[93] in which they detected a significant locus at 2q with an LOD score of 3.43.
- In 2007, Johnson et al.[94] reanalyzed a previous genome-wide scan of 34 Australian/New Zealand families[89,93] with a more refined and powerful variance components model represented by quantitative trait loci (QTL). Doing this analysis returned two novel QTLs at 5q and 13q, with LOD scores of 3.12 and 3.10, respectively.

Table 4.4 summarizes the linkage analyses discussed in this section.

Our poor understanding of the underlying heterogeneity of preeclampsia has hampered all of the linkage studies conducted to date. Because the diagnostic criteria for preeclampsia are nonspecific, patients who represent "phenocopies" of the preeclampsia patients with strong genetic factors contributing are likely to be included in the pedigrees. Nonparametric approaches may be preferable given our incomplete knowledge of the segregation patterns expected.

Association Studies

Geneticists have always had at their disposal two main approaches to mapping and identifying genetic associations in complex diseases: linkage (a positional approach), and association (a functional approach). The preponderance of and preference for association studies thus far could be partially attributed to the ease, logistically and financially, of recruiting large numbers of unrelated cases and controls, as opposed to recruiting multiple subjects from extended family pedigrees necessary for linkage studies.

Genetic association studies begin with a hypothesis regarding the participation, or function, of a specific gene or limited number of genes in a disease phenotype. These candidate genes are typically identified when physiological information suggests that their protein products influence the biochemistry of the studied condition. An association study may then be instigated to attempt to establish a statistical association between this hypothesized gene (or a polymorphism in or close to the relevant causal change in the candidate gene) and the disease in question. DNA samples are collected from cases and controls, the samples from each population are genotyped for the functional gene variant of interest, and chi-square comparisons are tested.

Candidate gene studies require the prioritization of specific genes – and only genes – for investigation into links with identifiable biological functions that are consistent with the pathogenesis of the studied disease. For years, candidate gene association studies have been the method *de rigueur* for matching genes with diseases. Biological pathways are frequently well known long before the genetic and proteomic contributions are discerned, and not unreasonably,

TABLE 4.4 Noteworthy Genome-wide Linkage Analyses of Preeclampsia

Author(s)[ref]	Year	Inheritance Model	Results
Harrison et al.[77]	1997	Four different models	Linkage with 4q
Arngrimsson et al.[78]	1999	Dominant	Linkage with 2p13
Moses et al.[79]	2000	Multipoint nonparametric	Linkage on chromosome 2 and at 11q23–24
Lachmeijer et al.[80]	2001	Various	Linkage on chromosome 12, 10q, and 22q
Laivuori et al.[81]	2003	Various	Linkage with 2p25 and 9p13
Kalmyrzaev et al.[82]	2006	Two-point parametric	Linkage on chromosome 2
Moses et al.[83]	2006	Various	Quantitative trait locus on 2q22
Johnson et al.[84]	2007	Variance components	Linkage with 5q and 13q
Majander et al.[95]	2013	Fetal phenotype	Fetal susceptibility locus on chromosome 18

researchers typically break the biological pathways down into their molecular components and attempt to match them to genetic polymorphisms of known function.

It is possible that the demonstrable successes since the late 1980s in utilizing this approach to identify Mendelian, single-locus genetic diseases have set a precedent for immediate gratification that the study of complex diseases cannot provide: either the gene variant in question affects disease risk directly, or it does not; either it is a marker for a nearby gene that affects risk, or it is not.[12,13] In complex diseases, a series of well-hypothesized candidate gene studies can quickly turn into a fishing expedition, particularly when one or even all of the known candidate genes do not explain all phenotypes, as in preeclampsia. The numerous other genetic and environmental factors that can contribute to a person's susceptibility to a complex disease like preeclampsia, including other functional SNPs that might indirectly but very crucially impact downstream genetic coding, can distort the results of any candidate gene study.

This distortion is demonstrated in the persistent, biological hypothesis-driven selection of candidate genes for study (see Table 4.5) and the inconclusiveness of the collective data. This compilation of studies regarding preeclampsia yields no candidate genes with even a modest population attributable risk. We note the same trends as in genetic association studies of complex diseases in pregnancy conducted by Genc and Schantz-Dunn.[13]

- Primarily hypothesis-based case-control association studies.
- Primarily small sample sizes of fewer than 500 individuals, cases and controls combined.
- Primarily focused on individual polymorphisms rather than haplotypes.
- Focused inordinately on single SNPs heretofore shown to be of dubious association, such as *MTHFR*.
- Typically maternal or fetal genotype analyzed, rarely both, and even more rarely combined.
- Association often studied without correction for population admixture or environmental factors.
- Seemingly similar, independent studies often yield conflicting results regarding association of the polymorphism with preeclampsia, and few studies are independently replicated, due to type I error due to small sample size, population stratification, variable linkage disequilibrium between the studied gene variant and the true causal variant, or population-specific gene–environment interactions.[105]

If performed carefully, positive data can point to a factor as being either a predisposing or causative gene or a marker linked to such a gene (linkage disequilibrium). However, association studies cannot prove biologic causation. What has slowly become clear is that the genes that contribute to preeclampsia vary from case to case and that each contributor may have moderate or even low levels of association. Very large studies with detailed clinical phenotyping will be required.

Many genetic association studies of preeclampsia have been published to date. Below is a review of the research of some of the most likely biological pathways involved in preeclampsia and the most promising candidate genes hypothesized to affect these pathways.

Summary of Candidate Genes most Frequently Studied

The approach to the study of preeclampsia candidate genes has been scattershot in one respect, and pointedly selective in another. More than 100 genes have been studied for association, but researchers have collectively fixated on a handful of genes based on biological hypotheses of their contributions to preeclampsia (Table 4.5).

Methylenetetrahydrofolate reductase (*MTHFR*) at 1p36.3. The 677T variant of the methylenetetrahydrofolate reductase gene (*MTHFR*) has been identified as a risk factor for vascular disease. The *MTHFR* gene is on the short arm

TABLE 4.5 Summary of Research on the Seven Genes most Frequently Studied for Association with Preeclampsia. See Several Recent Reviews[96,97]

Chromosome Location	Gene	Primary Polymorphism Studied	Recent Review	Presumed Biological Association with Preeclampsia
1p36.3	*MTHFR*	C677T	Wang, 2013[98]	Vascular diseases
1q23	*F5*	Leiden (rs6205)	Buurma, 2013[96]	Thrombophilia
1q42–q43	*AGT*	M235T	Lin, 2012[99]	Blood pressure regulation, uterine vascular remodeling
7q36	*NOS3*	Glu298Asp	Dai, 2013[100]	Vascular endothelial function
11p11–q12	*F2*	G20210A	Buurma, 2013[96]	Blood coagulation
17q23	*ACE*	I/D at intron 16	Zhong, 2012[101] Chen, 2012[102] Zhu, 2012[103] Buurma, 2013[96]	Blood pressure regulation
3q24	*AGTRI*	rs5186	Zhao, 2012[104]	Vascular biology

of chromosome 1 at location 36.3. As a major gene associated with homocysteine levels and clinical thrombosis, it is the gene most frequently studied for association with preeclampsia: more than 50 studies were published between 1997 and 2013. All investigated the C677T polymorphism causing decreased enzymatic activity. Most were case-control studies with population sizes varying from 28 to 2000 individuals, including a variety of geographic regions and populations (e.g., Finnish, Norwegian, German/Croatian, Italian, Dutch, Irish, East Anglian, Indonesian, Mexican, Brazilian, sub-Saharan African, East Asian). Several studies finding a negative association for the *MTHFR* polymorphism alone detected an association with preeclampsia when *MTHFR* was analyzed in conjunction with other polymorphisms such as Factor II and Factor V Leiden.[106,107] Several meta-analyses suggest that the *MTHFR* C677T polymorphism was associated with risk of PE in overall, Caucasian, and East Asia populations.[98,108,109] None of the studies have included haplotype analysis.

Catechol-*O*-methyltransferase (*COMT*). In a murine model created by the global deletion of both *COMT* alleles, Kanasaki et al. described signs consistent with a preeclampsia-like phenotype that was reversed following administration of the catechol-estrogen derivative, 2-methoxyestradiol (2-ME). The authors discounted the mild elevation of monoamines, which also are important metabolites of the COMT pathway and attribute the effect to 2-ME.[110] Following that report, some studies have suggested that variant alleles of *COMT* are associated with preeclampsia, but findings have been mixed.[111]

***Fms*-like tyrosine kinase 1 (*Flt-1*).** Tagging SNPs of relevant angiogenic pathway genes were studied independently in African-American and Caucasian cases and controls.[112] Among white women, a SNP in the *Flt-1* gene was associated with an odds ratio of 2.12 (95% CI, 1.07–4.19; P=0.03).

Factor V Leiden (*F5*) at 1q23. The Leiden polymorphism of *F5* has been the focus of over 40 preeclampsia association studies between 1996 and 2013, several of which indicated a significant, positive association. Three of the studies utilized more than 500 subjects, cases and controls combined. Approximately half reported the ethnicity of cases and controls. One study included haplotype analysis. *F5* (Leiden) is a missense mutation that prevents normal degradation of activated factor V.[113,114] Because *F5* Leiden molecules retain their procoagulant activity, patients with this polymorphism are predisposed to clotting disorders, including thromboembolic complications that occur in pregnancy.

How can a gene responsible for thrombophilia lead to preeclampsia? The Utah group hypothesizes that low-pressure intervillous blood flow in the presence of a maternal hypercoaguable state and trophoblast dysfunction may trigger excessive fibrin deposition in the placenta. In turn, trophoblast and placental infarction occur, frequent features of preeclamptic pathophysiology. Indeed, when Dizon-Townson et al. studied a series of placentas with large infarcts (involving greater than 10% of the placental mass), there was a greater than ten-fold increase in the *F5* Leiden carrier rate. The odds ratio with 95% confidence intervals calculated were 37.24 (10.98< OR <130.43).[115]

Angiotensinogen (*AGT*) at 1q42–q43. Over 50 studies have considered whether a common *AGT* variant is associated with preeclampsia. As molecular variants of *AGT* appear to predispose to essential hypertension, several arguments can be advanced to postulate a similar implication in preeclampsia: (1) plasma angiotensinogen is elevated in estrogenic states, (2) the relative hypovolemia of preeclampsia suggests an abnormality in the control of extracellular fluid volume, and (3) the renin–angiotensin–aldosterone axis is frequently perturbed in preeclamptic patients. By three sets of observations – genetic linkage, allelic associations, and differences in plasma angiotensinogen concentrations among *AGT* genotypes – Jeunemaitre et al. demonstrated involvement of the T235 variant of the *AGT* gene in essential hypertension.[116] It is uncertain whether the T235 variant directly mediates a predisposition to hypertension, or whether an unidentified risk factor is associated with the T235 haplotype.

Human leukocyte antigen (*HLA*) at 6p21.3. The human leukocyte antigen locus has been the focus of over 20 preeclampsia association studies between 1997 and 2013, several of which indicated a significant, positive association. The majority of studies (nine) genotyped mother–child pairs or parent–child trios, which may be expected considering that *HLA* is associated with immunity and, hence, the suppression of the maternal immune response to the fetus/placenta during pregnancy. The majority of studies reported the ethnicity of cases and controls. No studies included haplotype analysis.

Endothelial nitric oxide synthase (*NOS3*) at 7q36. The endothelial nitric oxide synthase gene has been the focus of at least 25 preeclampsia association studies between 1999 and 2013, most investigating the Glu298Asp polymorphism. Several studies utilized more than 500 human subjects, cases and controls combined, one of which was a meta-analysis. Two studies included haplotype analysis. The evidence of association is weak.

Prothrombin, Factor II (*F2*) at 11p11–q12. The G20210A polymorphism of *F2* has been the focus of more than 20 preeclampsia association studies between 1999 and 2013, two of which indicated a positive association, though not significant. Notably, this gene was studied by itself only once; all other studies investigated it as one of multiple "thrombophilia" genes. A meta-analysis investigating thrombophila genes found that the odds ratio for the *F2* 20210 polymorphism in all preeclampsia was 1.37 (95% CI 0.72–2.57) and 1.98 (.94–4.17) for severe preeclampsia.[109]

Angiotensin converting enzyme (*ACE*) at 17q23. The insertion/deletion polymorphism in intron 16 of *ACE* has

TABLE 4.6 Examples of Recent Genome-wide Association Studies of Preeclampsia

First Author, Year	Initial Sample Size	Replication Sample Size	Platform [SNPs passing QC]	Findings
Ward, 2011[117]	500 cases 1000 controls European ancestry	None	Illumina 620,000	Five SNPs with *p*-trend values reaching genome-wide significance (nearest genes are *SCAPER, NEDD4L, NMNAT2, PTH2R*)
Johnson, 2012[118]	538 cases, 540 controls, European ancestry	1,894 cases, 3,022 controls European ancestry	Illumina 648,175	Probable risk locus for preeclampsia on 2q14, near the inhibin, beta B gene
Zhao, 2013[119]	133 cases, 2917 controls mixed ancestry	177 cases, 116 controls Caucasian	Illumina, varies by subset	Suggested candidate copy number variant regions

been the focus of over 30 preeclampsia association studies and several meta-analyses between 1999 and 2013. Larger studies find that the DD genotype is associated with preeclampsia in Asian and Caucasian patients.

Genome-Wide Association Studies (GWAS)

In a genome-wide association study (GWAS), hundreds of thousands to millions of common genetic variants (typically SNPs) are tested in cases and controls to see whether any variant is associated with a disease or trait. Since the SNPs tested have known genomic locations, finding a disease association with a SNP can "mark" a chromosomal region as influencing the risk of the disease being studied.

A GWAS can only show association; it cannot prove causation or pathophysiology. GWAS studies have the advantage of investigating the entire genome rather than focusing on one gene or one pathway at a time. Thus, GWAS is said to be non-candidate-driven, hypothesis-free, "discovery science." Since the first successful GWAS in 2005, thousands of GWAS have been completed, providing many important insights into complex diseases.

Only a few GWAS studies have been performed for preeclampsia to date (see Table 4.6). GWAS are expensive to conduct and there are practical difficulties in defining cases and controls without introducing population stratification that can lead to false-positive results. GWAS have an important limitation since they depend on the assumption that common genetic variation plays a large role in explaining the heritable variation of common disease. Thus, GWAS will only find older ancestral alleles contributing to a disease, and not the newer mutations which are likely to be very important in any disease associated with both maternal mortality and reproductive loss. Replication across multiple studies is often seen as the key to confirming GWAS results, but insisting on replication *across* ancestral groups leads to a smaller number of (much older) confirmed markers that tend to have minor effects. Replication within an ancestral group is more important than replication across groups. Insisting on broad replication and Bonferroni (or similar) corrections for multiple testing has resulted in false-negative findings. Many GWAS have additional limitations because they exclude SNPs with a minor allele frequency of <5%.

Functional SNPs that increase disease risk are most often found in important protein coding domains and most often change protein expression levels or alter their structure.[120] In the case of a complex phenotype like preeclampsia, considering the many different roles that many functional SNPs (known and unknown) might fulfill, and considering the possible combinations of SNPs that work in concert to effect a phenotypic change, it is extremely difficult to assign a definitive pathogenic role to any single, common functional SNP. Non-functional SNPs are more common than functional SNPs and are dispersed throughout the genome. These SNPs can be used as genetic markers for nearby disease-causing events via association analyses, linkage disequilibrium studies in founder populations, and classic linkage analyses, but their full potential is realized in the identification of haplotype blocks in high-density, genome-wide association studies. In SNP-based studies of complex diseases, the importance of each SNP, functional or non-functional, coding or non-coding, is correlated to its occurrence in the population and its level of expression in disease-specific cases and controls. A SNP whose role is no greater than acting as a susceptibility marker of another co-inherited SNP yields far more precise information than a functional SNP whose quantifiable link to phenotype is variable. The genome-wide association study is essentially a random search, but it may ultimately be the ideal way to thoroughly identify the genetic risk factors involved in risk of preeclampsia.[12]

To date, the software used for GWAS studies has limited power to detect the polygenic effects of epistasis (the interaction between non-allelic genes). For instance, the epistatic effects of neighboring loci within a haplotype appear as part of the genetic variance in pedigree studies but as unmeasured gene-gene interactions in GWAS. As David Haig points out, pedigree-based heritability measures the phenotypic effects of much larger chunks of chromosome

TABLE 4.7 Next-generation Sequencing Studies

First Author, Year	Initial Sample Size	Replication Sample Size	Target	Findings
Johnson, 2012[123]	7 cases, 11 controls from two Australian families	Unknown	Exome	Three damaging missense SNPs (*LYSMD3*, *GPR98* and *LIG4* genes)
Ding, 2012[124]	30 cases, 38 controls African-American	162 cases, 171 controls African-American	Mitochondria	Excess rate of mitochondrial variants in women with preeclampsia

than GWAS-based heritability.[121] If one considers two variants inherited as a haplotype having an effect on a disease only in certain haplotype combinations, it is possible that neither variant would show association in a GWAS, but the haplotype which is associated with the disease phenotype would be reliably transmitted from parents to offspring and contribute to estimates of pedigree- based heritability.

Next-Generation Sequencing

The rapidly decreasing price of complete genome sequencing now provides a realistic alternative or complement to array-based GWAS. Next-generation sequencing technologies have lowered the cost of whole-genome sequencing (WGS) by more than 2000-fold in recent years. WGS provides a higher-resolution look at the genome (examining 6.4 billion features rather than the one million features typically considered in a GWAS). With WGS, the status of both chromosomes may be delineated at virtually every nucleotide position and all candidate genes may be considered and interrogated simultaneously. At present, there are some regions that are difficult to sequence and extensive bioinformatics analysis is required to distinguish true DNA variants from noise. Usually, WGS is performed to at least an average 30-fold coverage (90 billion bases of sequence for a 3 billion base haploid genome) with data quality score aiming for a base call accuracy of greater than 99%. Often analytical specificity (the accuracy of genotypes at positions with variants) is sacrificed in favor of analytical sensitivity (the proportion of true sequence variants identified).

Because the cost and difficulty of WGS analysis are still relatively high, many studies are focusing on the "exome" (coding regions for all genes) or the "Mendelianome" (~0.25% of the genome composed of the coding regions of 3000 known disease-causing genes).[122] These focused sequencing approaches require enrichment of the desired genomic regions before sequencing the sample. None of the target enrichment methods currently available are perfect; all target enrichment methods create some bias.

One obvious benefit of WGS is that new mutations can be detected along with ancestral mutations. Haplotypes can be examined directly and prediction algorithms can be used to suggest the most likely disease-associated variants. Family-based study designs will be critical for optimal interpretation and to probe epistatic interactions within a chromosomal region. Table 4.7 summarizes the next-generation sequencing studies reported to date.

Johnson et al. were the first to report on exome sequencing in preeclampsia. They sequenced the exomes of 18 women (7 preeclamptics, 11 controls) from two Australian families whose previous linkage studies implicated loci on 5q and 13q.[123] In one family, they identified two rare missense SNPs (rs62375061 within the *LYSMD3* gene, predicted to "possibly" damage the protein, and rs111033530 within the *GPR98* gene, predicted to "probably" damage the protein). In the second family they identified a more common missense SNP (rs1805388, with a 17% population prevalence) residing within the *LIG4* gene and predicted to be highly deleterious, which segregated in the preeclamptic women but not in the unaffected women. Neither of these variants line up precisely with the regions implicated in preeclampsia linkage studies, but a recent GWAS reported a significant association with diastolic blood pressure and the *GPR98* gene.

More recently, Ding et al. used next-generation sequencing of placental mitochondrial DNA in 30 African-American preeclamptic women and 38 controls.[124] They found 88% of preeclamptic women and 53% of controls carried at least one nonsynonymous substitution ($P = 0.005$). These results were not replicated in a sample of African-American preeclamptic women ($N = 162$) and controls ($N = 171$) from Detroit.

A GENOMICS APPROACH TO PREECLAMPSIA

Two main questions to be answered by genomics are: (1) when, where, and how are particular genes expressed during preeclampsia; and (2) how do quantitative patterns of expression differ between cases and controls? Study of the entire human genome, rather than individual genes or candidate regions that have been narrowed too quickly, is bound to yield information regarding the participation of novel genes, non-coding areas, and pathways in gene regulation and protein expression that have been previously unsuspected and unstudied.[105]

The SNP, rather than the gene, is the currency of large-scale, high-throughput studies of the human genome, which are the key to complex diseases like preeclampsia. The particular biological impacts of many SNPs are still poorly known. The candidate gene association studies listed in Table 4.5 analyzed one or two particular SNPs at particular loci in given genes (e.g., *F5* Leiden and *MTHFR* C677T) and are, by definition, limited to functional SNPs that directly affect promoters, enhancers, repressors, regulators, introns, exons, or codons and influence the subject's susceptibility to or pathophysiology of a particular disease.[125]

On a more macrogenomic level, researchers are not only detecting participation of SNPs in various complex disease phenotypes, but they are detecting structural variation, as well as new patterns and sequence phenomena, throughout the genome. For their purposes, Genc and Schantz-Dunn[13] defined structural variants as genomic alterations that involve segments of DNA larger than 1 kb, including deletions, duplications, insertions, inversions and translocations, and copy-number variants (CNVs). Ultimately, in whole-genome mapping, functional and non-functional (marker) CNPs need to be considered and coded like SNPs in terms of their biological or informational value regarding the complex disease.

- In 2006, Zintzaras et al.[126] reported a unique, heterogeneity-based meta-analysis on genome searches (HEGESMA) for identifying susceptibility loci for preeclampsia. HEGESMA synthesizes the available genome data and identifies genetic regions that rank highly in terms of linkage statistics across all studies Weighted and unweighted analyses were conducted, as well as heterogeneity testing. Four genomic regions were determined to be formally significant using both unweighted and weighted methods for general preeclampsia, and five regions were determined to be formally significant in both unweighted and weighted analysis for severe preeclampsia. The associated p-values were modest (rank p between 0.06 and 0.05). The primary contribution of this study was to demonstrate how quantification of heterogeneity is an important metric in meta-analysis as well as in genome-wide studies in general.
- The study reported by Goddard et al. in 2007[127] is a small-scale model for future studies based on genome-wide association searches. This study evaluated 775 SNPs (as well as haplotypes) in 190 candidate genes selected for potential roles in obstetrical complications. High-throughput genotyping was accomplished using the MassARRAY TM System. Sample sizes were much larger than previous association studies examining only a few SNPs: in this case, 394 cases and their 324 offspring were compared with 602 control women and their 631 offspring. All subjects were from one hospital in Chile. This study not only detected SNPs in both mothers and infants (with p-values ranging from <0.001 to 0.014), it also employed a genetic model for maternal–fetal interaction. But none of the statistically significant results remained significant after adjusting for multiple testing; this model would have needed a p-value less than 6.7×10^{-5}. This power calculation illustrates the ease with which type I error can plague even a large association study.

The genome-wide association strategy has been used most effectively to identify susceptibility loci for the development of disease. Serrano[15] provides a timeline of the investigations of chromosome 2 that started with linkage analyses using families with preeclampsia from Australia, New Zealand, Iceland, and Finland.[88,89,91,93,94] The mapped locus in the population from Australia and New Zealand was named *PREG1*.[89] A follow-up fine-mapping study of the *PREG1* locus detected two significant linkage peaks: 2p11 and 2p23.[128] Two genes located in those regions with plausible biological connection to preeclampsia were chosen for the SNP analysis: the *TACR1* gene, which codes for a vasoactive neuropeptide found in elevated concentrations in preeclamptic pregnancies, and the *TCF7L1* gene, which produces a transcription factor that functions as a mediator in the *Wnt*-β catenin signaling pathway and contributes to the synthesis of the extracellular matrix components associated with decidualization. Two SNPs showed linkage in the fine-mapping analysis, one for each gene, with significant LOD scores.

The above studies, and others like them, are a step in the right direction. Nonetheless, they are all notable for the very low number of individuals tested and the low number of SNPs evaluated. Truly high-throughput genetic screening involving many cohorts and many genes simultaneously, as demonstrated in the Wellcome Trust Case Control Consortium,[129] is needed to establish any recognizable pathways in preeclampsia because large numbers of diverse, candidate SNPs are likely to contribute to the various etiologies of preeclampsia.[125]

It may be helpful to consider the lessons from other complex diseases for which gene discovery has reached a more advanced stage in setting expectations for preeclampsia gene research. Breast cancer is a polygenic multifactorial disease that occurs with a similar incidence, but is less familial and less heritable than preeclampsia. Even though only 5% to 10% of all breast cancers are "heritable," critically important disease genes have been discovered using linkage analysis in high-risk families. For instance, women with mutations in their *BRCA1* or *BRCA2* gene have markedly increased risks of breast and ovarian cancer. Disease genes rarely play a role in all affected patients; despite their importance, *BRCA1* and *2* are mutated in <10% of women with breast cancer. Genetic findings will usually vary among geographic or ethnic groups: particular mutations in *BRCA1/2* that are very common in people of Ashkenazi Jewish heritage are extremely rare in other ancestral groups. The existing gene discovery methodologies each lead to different insights; while linkage analysis can find high-impact genes, GWAS finds more common ancestral alleles with smaller effects and next-generation sequencing can uncover rare and new mutations. At least 25 genes have been associated with breast cancer through GWAS and more recently another 50 have been implicated using next-generation sequencing approaches. Discovery of these disease genes has not identified a simple "cause" of breast cancer, but it has advanced counseling, diagnosis, classification, treatment, and understanding of those tumors. We anticipate similar developments in preeclampsia.

ESSENTIAL VARIABLES TO CONSIDER

Numerous variables contribute to complex genetic diseases like preeclampsia, and any study that attempts to definitively determine genetic contributions to these diseases must corral and manage all variables in a manner that can be reproduced in independent studies. As yet, there are no universal methods to manage these variables, partially because complex diseases – especially those in which the genetic contribution affects susceptibility but not contraction of the disease – have unique combinations of variables affecting the phenotype. Selection of analytic tools must therefore be tailored to the disease of interest, and interpretation of results must be conservative.

Interacting Genomes

One variable that must be considered as unique to complex disorders of pregnancy is the interaction of at least two genomes: maternal and fetal. Researchers have long debated the tension between mother and fetus during gestation, and whether pregnancy represents cooperation, interaction, and/or competition between the two organisms.[130]

It is likely that both the maternal genotype and the fetal genotype are involved in producing the preeclampsia phenotype. However, it is also possible that the different components of the disease (e.g., oxidative stress, immunology) contribute to varying degrees to the phenotype, depending on the particular interaction of the individual mother and individual fetus. Looking at the candidate gene association studies of preeclampsia, the maternal genotype has been studied far more than the fetal genotype or both genotypes. When studies have considered both maternal and fetal genotypes, only a few genetic loci were examined.

The degree to which the maternal and fetal genotypes in a pair are interrelated during gestation, let alone how they might create preeclampsia between them, has been largely beyond the scope of studies conducted to date, though it is perhaps this information that will contribute the most solid foundation upon which to build theories regarding deviations from the gestational norm.

According to Romero et al.,[131] the maternal and fetal genotypes interact with one another via the admixture of cells and their gene products at the placental–decidual interface. This interaction may lead to either reproductive success or pregnancy-specific diseases that affect the fetus or the mother or both. Several researchers[12,120] have developed means of analyzing the compatibility of maternal genotypes with fetal genotypes.

Subgroups

Stringently defined phenotypic and etiologic subgroups are fundamental means of controlling the interpretation and replication of studies on the genetic contributions to a complex disease like preeclampsia. As suggested in Chapter 1, more precise nosology of preeclampsia may benefit from subclassification, based possibly on etiology or genetic fingerprint. For example, association studies conducted thus far suggest that there are notable differences in the frequency of preeclampsia among ethnic groups. Dividing preeclampsia cases into subgroups increases each subgroup's trait homogeneity, which should, in turn, increase the sensitivity of studies aiming to detect differences in genetic epidemiology studies of preeclampsia.

Perhaps the first and most obvious subgroup classification would be ethnicity, as determined by genetic markers. From there, other broad subtypes of cases may be identified (e.g., women with prepregnancy hypertension or diabetes; primigravid women). Alternating classification with evaluation of genotypes and haplotypes should gradually refine subgroups until consistent results are achieved.

Adams and Eschenbach[105] and Esplin and Varner[132] recommend precise definitions of the etiologies and pathogeneses of pregnancy complications, particularly in light of genomics and proteomics as research tools into novel genes and proteins. Another rationale for precise definitions is replication of results in subsequent studies to indicate that genetic heterogeneity has been satisfactorily overcome. To achieve this, Menon et al.[12] recommend meta-analyses of past and present studies according to carefully delineated subgroups. Esplin and Varner[132] also urge future studies to note experimental conditions and pharmacological exposures, and to report differentially regulated genes using standardized accession numbers (e.g., GenBank).

Other researchers have offered alternative means of refining subgroups. Bo and Jonassen[133] discuss the various methods of establishing subgroups of cases using feature subgroup selection methods on gene pairs. Pennell et al.[120] suggest that an alternative to subgroup analyses would be covariate-based analyses that can incorporate quantitative traits directly.

Genomic Ethnicity

Frequencies of genetic variants and haplotypes differ across the world due to evolutionary forces such as genetic drift, founder effect, and selection. Anthropological and genetic evidence suggests a world population founder effect in a small African subgroup, ca. 60,000–125,000 years ago, that migrated in one or more waves out of northern Africa to populate the rest of the world. Researchers can trace human beings to approximate geographic origins, and thus into subgroups, based on the similarity of certain sections of the genome. These general subgroups/geographic origins are Africans, Caucasians, Pacific Islanders, East Asians, and Native Americans.[134,135] Ethnic and racial admixture must be considered as it will either complicate or enhance the analysis depending on the methodology used. A racially admixed individual may need to be considered African for certain loci, Asian for others, and Caucasian for the remaining loci.

As different ethnic groups carry alleles at different frequencies, the need to consider ethnicity as a variable increases as the genetic responsibility for a complex disease is spread more thinly among more and more polymorphisms throughout the genome. The risks of population stratification similarly increase, so it is crucial to match the proportions of ethnicity in both cases and controls,[105] and to consider carefully the impact of ethnicity on genetic results.

Population Size

One self-evident factor that has yet to be implemented consistently in large-scale analyses is sufficient population sizes for case and control groups, regardless of study type or structure method. Insufficient population size is a major reason for the lack of reproducibility in past studies and is inextricably linked to issues of subgroup delineation for complex diseases. It is admirable to start a study of a complex disease like preeclampsia with 1000 cases and 1000 controls, but if subgroups are delineated, each should be sufficiently populated to achieve reproducible results. To detect small or weak genetic effects important to each subgroup, one must minimize the occurrence of false-positive/false-negative associations.[12,105] This may mean adding cases and/or controls to a particular subgroup during the study to achieve sufficient power.

In the case of preeclampsia subgroups, allelic and genotypic variation within the population may be lost or overlooked if the population size is insufficient to discern distribution. In simpler, Mendelian disease phenotypes, power depends mainly on relative risk and frequency of risk alleles in the population, estimated based on SNPs used and/or the phenotype of interest. However, for complex genetic models like preeclampsia, explicit power calculations are difficult or impossible to derive. It may be possible to calculate power empirically using simulated models of preeclampsia subgroups to provide an estimate of the sample size needed, but these calculations may not apply to all subgroups.[12]

Selection of appropriate controls, who accurately represent the base population from which the cases arise, cannot be underscored enough. A cohort study design, with cases and controls arising from a single group, offers the benefit of being predicated on prospective collection of subjects from a defined population. By contrast, selection of random controls in straight case-control genetic association studies may skew interpretation of results. Careful description of inclusion and exclusion criteria and corrections for bias should be incorporated into each study design.[120] Most of the extant association studies on preeclampsia have fewer than 500 cases and controls combined, and do not distinguish among subgroups.

Gene–Gene Interactions

Gene–gene interaction, or epistasis, has been widely accepted as an important genetic contributor to the manifestation of complex diseases like preeclampsia. It occurs when specific alleles and/or products of two or more genes affect one another and, in turn, the individual's phenotype. Genetic association studies that investigate one or a few individual genes of known function, but whose results cannot be replicated, may be suffering from a failure to recognize the underlying architecture of gene–gene interactions, either among the genes studied or involving other unnamed, unstudied genes. Genetic studies of complex diseases that ignore epistasis are likely to reveal only part of this genetic architecture.

Such interactions undoubtedly play a part in the hallmark phenotype of preeclampsia, considering the discretion of some pathways and of the genes implicated in a given pathway. Examples of epistasis can be seen in the group of genes associated with the thrombophilic pathway, as studied by Dalmaz et al.[106] and Tempfer et al.[136] In each study, individual alleles failed to achieve significant association with preeclampsia but collectively, with the other thrombophilia genes studied, demonstrated significant levels of interaction. These variants are unlikely to be the functional alleles and are more likely to be markers (through LD) of the true functional genes.

After GWAS for preeclampsia are replicated, a major next step is to systematically determine which of the genes represent true causative biochemical mechanisms in the maternal and fetal tissues involved and which ones represent secondary responses downstream of the path way(s).[10,89–91,93,94,126,128,137–143] The epistasis phenomenon has been theorized since the early 20th century, but the means and methods for detecting and determining its effect in clinical studies have only recently become available. These include various multi-analytic approaches; the genotype–pedigree disequilibrium test and multifactor-dimensionality reduction methods; classification and regression trees, multivariate adaptive regression spline, focused interaction testing framework, neural networks; a two-stage, Backward Genotype-Trait Association algorithm; and models for linking epistasis with pathophysiology have been proposed.[120,144–146]

Epigenetics

Epigenetic phenomena are variations in the expression or operation of a functional gene even though the underlying genomic code is stable and unchanging.[147] In humans, epigenetic phenomena are most readily observed in discordant pairs of monozygotic twins and explain why one twin is larger than the other at birth, or why identical twins diverge in appearance as they age.[148] Epigenesis is frequently caused by environmental factors, but, particularly relevant to discussions of pregnancy, it can also be caused by cellular differentiation and division, as occurs throughout a pregnancy, well after the fetal genome is established.

Endothelial dysfunction is a reported epigenetic consequence in infants who were born small, in normal and abnormal pregnancies. Therefore, women who were themselves born small, possibly as a result of preeclampsia,[149] may have an increased risk not only of chronic hypertension and diabetes, but also of preeclampsia and hypertensive disorders of pregnancy, as well as reduced placental perfusion. This type of dysfunction is associated with all maternal vascular beds and organs, not just the placenta, suggesting that preexisting maternal endothelial dysfunction could serve to exacerbate preeclampsia or other complications of pregnancy.[58] It is possible that women born small in normotensive pregnancies are one mechanism by which preeclampsia has persisted from generation to generation, despite its historical lethality. Numerous researchers have hypothesized a heavy epigenetic effect in preeclampsia.[78,84,150–153]

Gene–Environment Interactions

Gene–environment interactions are situations in which environmental factors affect different individuals differently, depending upon genotype, and in which genetic factors have a differential effect, depending upon attributes of the environment. The possible number of gene–environment interactions involved in a complex disease should daunt researchers more than the genomics of that disease. The genome, at least (with 3 billion nucleotides), is finite!

Though it is impossible to detect all miniscule fluctuations in the response of an organism's genome to its environment, it is important to remember that (1) researchers only need to detect and quantify gene–environment interactions rigorously enough to achieve reproducible results, and (2) investigators should identify gene–environment interactions influencing susceptibility to preeclampsia.

Practically speaking, one way to determine which environmental effects to study is to evaluate data from the control subjects themselves. Similar to specific gene–gene interactions being associated with particular biochemical pathways, as represented in subgroups of the case population, gene–environment interactions will be associated with particular subgroups that share a common geographical location (continuous), behavior or history (discrete; including sequences of life events and response to stress), physical/medical condition independent of pregnancy (including preexisting conditions such as diabetes), diet, or toxin exposure.

Because environmental factors do not appear to interact precisely in a closed system to the degree that genes do, there can be more leeway in how environmental factors are analyzed, quantified, and interpreted. Methods proposed by Ottman,[154] Yang et al.,[155] and Humphries et al.,[156] as well as many of the same analytical methods for detecting gene–gene interactions, have great potential for isolating, detecting, and analyzing gene–environment interactions.

Confounding Variables

Confounding factors may hide actual associations or falsely create an apparent association between study variables where no real association between them exists. If confounding factors are not measured and controlled experimentally, bias may result in the conclusion of the study.

Any of the variables mentioned previously has the potential to be a confounding factor. Medical conditions that are independent of preeclampsia but may co-occur with or have symptoms similar to preeclampsia also can be confounding factors (e.g., diabetes, chronic hypertension, intrauterine fetal growth restriction).

Because we can never be certain that observational data are not hiding a confounding variable, it is never safe to conclude that a regression model demonstrates a causal relationship with 100% certainty, no matter how strong the association. In a technical report, Pearl[157] attempts to define and non-parametrically formulate analytical principles, incidental and stable, and asserts that confounding cannot be completely eliminated without a commitment to stable unbiasedness. Pearl's discussion could easily apply to subject selection criteria for genetic studies of complex diseases, as well as genomic phenomena (e.g., certain classes of repeats, quasi-linkage, alternative splice forms, transcripts from related gene sequences, expressed pseudogenes) that occur amidst targeted genetic and genomic phenomena but are ignored or minimized because they are outside the scope of the study.

More relevant to the discussion of preeclampsia, Kist et al.[158] utilize meta-analysis of case-control studies between 1966 and 2006 to explore the influences of confounding factors – including ethnicity, severity of illness and method of testing – in thrombophilic adverse pregnancy outcome. Using the Mantel–Haenszel method, they sorted and weighted data depending on heterogeneity and which confounders were in play. The results of the meta-analysis demonstrate differential impact of these confounders and emphasize the importance of uniform patient populations and study design.

HIGH-DIMENSIONAL BIOLOGY

Use of the high-dimensional biology (HDB) paradigm to reexamine some fundamental assumptions about preeclampsia will continue to increase as its utility is proven.[159] HDB refers to the study of an organ, tissue, or organism in health and disease by simultaneously looking at the genetic variants (DNA sequence variation), mRNA transcription, peptides and proteins, and metabolites of the system. This method of study is suited to the investigation of evolutionary complexity and change and utilizes tools, methods, and premises that have not been tractable with conventional biochemical approaches. HDB is a systems biology paradigm that serves as an alternative/complement to the reductionism upon which so much of modern-day biology is based.

An example of the reductionist approach is illustrated by the effective diagnostic use of a single hormone, human chorionic gonadotrophin, to confirm an early pregnancy. This singular analyte cannot, however, describe the complexity of pregnancy disorders such as preeclampsia, involving numerous biological systems and genotype–environment–phenotype interactions.[131]

HDB studies can accommodate large populations (1000s of cases and controls), accurate characterization of multiple phenotypes, DNA quality, population stratification, and a large number of samples for replication.[131]

A PREDICTIVE GENETIC TEST

A predictive genetic test for preeclampsia has the potential to provide a sound basis for treatment and prevention. However, each individual's genome has redundant control mechanisms that influence the regulation of biological functions. Because of this redundancy, it is likely that detection of combinations of genes and/or proteins, in conjunction with obstetric family histories, will be required to achieve the goal of a truly sensitive, specific, and useful screening test for preeclampsia.[132,159] This strategy appears promising using biomarkers discovered through proteomics.[160]

In the quest for ideal predictive genetic tests for preeclampsia, accurately placing cases into appropriate subgroups will be the key to producing consistent and precise results to identify predisposition at the biochemical or molecular level. The process of refining such a predictive genetic test will likely involve the incorporation of environmental factors and amassing and replicating data from a range of populations.

PHARMACOGENOMICS

The field of pharmacogenomics holds great promise for the treatment and prevention of preeclampsia for at-risk women.[161] Pharmacogenomics is a rapidly evolving discipline that studies how interacting systems of genes determine drug responses. It has been recognized for many decades that individual differences in response to pharmacologic treatment, exhibited as drug toxicity or a lack of therapeutic effect, are due in part to genetic differences in the metabolites of those drugs.[13] The difference between traditional pharmacologic research and pharmacogenomics parallels the difference between candidate gene studies and whole-genome association studies. Traditional pharmacologic studies attempt to identify biochemical agonist or antagonist targets and are limited by the need to know the target in question (i.e., hypothesis-based treatment). A pharmacogenomic test would consist of an agnostic approach to determine an individual's variations based on whole genome or SNP maps, haplotype markers, alterations in gene expression, and/or inactivation that may be correlated with pharmacological function and therapeutic response. Such a test would be able to identify and characterize an individual's unique drug efficacy and/or toxicity profile and could, in turn, guide clinicians to optimize effectiveness and minimize risks of medical treatments for individual women.[132] As patient-specific pharmacogenomic profiles are developed, it will be possible to develop patient-specific treatment regimens and institute patient-specific preventive measures.[132]

The fruits of this research will be particularly helpful for potentially preeclamptic women who enter pregnancy with high blood pressure, diabetes, obesity, or other predisposing conditions. Necessary medications for these conditions could easily be added to the pharmacogenomic model for an individual woman, to determine unforeseen risks or side effects that might result in an adverse pregnancy outcome.

THE FUTURE OF PREECLAMPSIA GENETIC RESEARCH

The fields of genomics and proteomics continue to evolve rapidly to meet the needs of studies of the complex diseases to which they are applied, including preeclampsia. These technologies will identify and associate etiological and serologic protein patterns, use them as the basis for a better understanding of the pathophysiology of preeclampsia and its variants whether causative or associative, discover novel target molecules and diagnostic biochemical markers, and ultimately aid in formulating more effective and timelier interventions to prevent preeclampsia.

CONCLUSIONS

Preeclampsia is a familial disorder and one or more major predisposing genes are likely to be characterized in the next several years. Multi-center efforts are needed to define clinical and pathologic subsets. Genome-wide analyses and

HDB technologies hold the greatest promise for solving the problem of preeclampsia. The search for preeclampsia genes could lead to the development of blood tests for susceptibility to preeclampsia, a greater understanding of its primary pathogenesis, and the development of rational treatments or preventive strategies. Clearly, the potential results warrant continued investment into the expanding scope of current genetics research.

References

1. Chesley LC. Hypertension in pregnancy: definitions, familial factor, and remote prognosis. *Kidney Int.* 1980;18(2):234–240.
2. Esplin MS, Fausett MB, Fraser A, et al. Paternal and maternal components of the predisposition to preeclampsia. *New Engl J Med.* 2001;344(12):867–872.
3. Treloar SA, Cooper DW, Brennecke SP, Grehan MM, Martin NG. An Australian twin study of the genetic basis of preeclampsia and eclampsia. *Am J Obstet Gynecol.* 2001;184(3):374–381.
4. Shen JJ, Tymkow C, MacMullen N. Disparities in maternal outcomes among four ethnic populations. *Ethn Dis.* 2005;15(3):492–497.
5. Thornton JG, Macdonald AM. Twin mothers, pregnancy hypertension and pre-eclampsia. *Brit J Obstet Gynecol.* 1999;106(6):570–575.
6. O'Shaughnessy KM, Ferraro F, Fu B, Downing S, Morris NH. Identification of monozygotic twins that are concordant for preeclampsia. *Am J Obstet Gynecol.* 2000;182(5):1156–1157.
7. Thornton JG, Onwude JL. Pre-eclampsia: discordance among identical twins. *BMJ.* 1991;303(6812):1241–1242.
8. Lachmeijer AM, Aarnoudse JG, ten Kate LP, Pals G, Dekker GA. Concordance for pre-eclampsia in monozygous twins. *Brit J Obstet Gynecol.* 1998;105(12):1315–1317.
9. Salonen Ros H, Lichtenstein P, Lipworth L, Cnattingius S. Genetic effects on the liability of developing pre-eclampsia and gestational hypertension. *Am J Med Genet.* 2000;91(4):256–260.
10. Chappell S, Morgan L. Searching for genetic clues to the causes of pre-eclampsia. *Clin Sci (Lond).* 2006;110(4):443–458.
11. Giarratano G. Genetic influences on preterm birth. *MCN Am J Matern Child Nurs.* 2006;31(3):169–175. quiz 76–77.
12. Menon R, Fortunato SJ, Thorsen P, Williams S. Genetic associations in preterm birth: a primer of marker selection, study design, and data analysis. *J Soc Gynecol Invest.* 2006;13(8):531–541.
13. Genc MR, Schantz-Dunn J. The role of gene-environment interaction in predicting adverse pregnancy outcome. *Best Pract Res Cl Ob.* 2007;21(3):491–504.
14. Abildgaard U, Heimdal K. Pathogenesis of the syndrome of hemolysis, elevated liver enzymes, and low platelet count (HELLP): a review. *Eur J Obstet Gynec Reprod Biol.* 2013;166(2):117–123.
15. Serrano NC. Immunology and genetic of preeclampsia. *Clin Dev Immunol.* 2006;13(2–4):197–201.
16. Dekker G, Robillard PY, Roberts C. The etiology of pre-eclampsia: the role of the father. *J Reprod Immunol.* 2011;89(2):126–132.
17. Brown MA. The physiology of pre-eclampsia. *Clin Exp Pharmacol Physiol.* 1995;22(11):781–791.
18. Lunell NO, Nylund LE, Lewander R, Sarby B. Uteroplacental blood flow in pre-eclampsia measurements with indium-113m and a computer-linked gamma camera. *Clin Exp Hypertens B Hypertens Pregnancy.* 1982;1(1):105–117.
19. Wallenburg HCS. Hemodynamics in hypertensive pregnancy. In: Rubin PC, ed. *Hypertension in Pregnancy.* Amsterdam: Elsevier; 1980:66–101.
20. Need JA. Pre-eclampsia in pregnancies by different fathers: immunological studies. *Br Med J.* 1975;1(5957):548–549.
21. Feeney JG, Scott JS. Pre-eclampsia and changed paternity. *Eur J Obstet Gynec Reprod Biol.* 1980;11(1):35–38.
22. Astin M, Scott JR, Worley RJ. Pre-eclampsia/eclampsia: a fatal father factor. *Lancet.* 1981;2(8245):533.
23. Boyd PA, Lindenbaum RH, Redman C. Pre-eclampsia and trisomy 13: a possible association. *Lancet.* 1987;2(8556):425–427.
24. Broekhuizen FF, Elejalde R, Hamilton PR. Early-onset preeclampsia, triploidy and fetal hydrops. *J Reprod Med.* 1983;28(3):223–226.
25. Cooper DW, Hill JA, Chesley LC, Bryans CI. Genetic control of susceptibility to eclampsia and miscarriage. *Brit J Obstet Gynecol.* 1988;95(7):644–653.
26. Rasmussen S, Irgens LM. Pregnancy-induced hypertension in women who were born small. *Hypertension.* 2007;49(4):806–812.
27. Arngrimsson R, Bjornsson S, Geirsson RT, Bjornsson H, Walker JJ, Snaedal G. Genetic and familial predisposition to eclampsia and pre-eclampsia in a defined population. *Brit J Obstet Gynecol.* 1990;97(9):762–769.
28. Skjaerven R, Vatten LJ, Wilcox AJ, Ronning T, Irgens LM, Lie RT. Recurrence of pre-eclampsia across generations: exploring fetal and maternal genetic components in a population based cohort. *BMJ.* 2005;331(7521):877.
29. Dekker GA, Robillard PY. Preeclampsia: a couple's disease with maternal and fetal manifestations. *Curr Pharmaceut Des.* 2005;11(6):699–710.
30. Cnattingius S, Reilly M, Pawitan Y, Lichtenstein P. Maternal and fetal genetic factors account for most of familial aggregation of preeclampsia: a population-based Swedish cohort study. *Am J Med Genet Part A.* 2004;130A(4):365–371.
31. Blaschitz A, Hutter H, Dohr G. HLA Class I protein expression in the human placenta. *Early Pregnancy.* 2001;5(1):67–69.
32. Gaunt G, Ramin K. Immunological tolerance of the human fetus. *Am J Perinatol.* 2001;18(6):299–312.
33. Snyder SK, Wessner DH, Wessells JL, et al. Pregnancy-specific glycoproteins function as immunomodulators by inducing secretion of IL-10, IL-6 and TGF-beta1 by human monocytes. *Am J Reprod Immunol.* 2001;45(4):205–216.
34. Froen JF, Moyland RA, Saugstad OD, Stray-Pedersen B. Maternal health in sudden intrauterine unexplained death: do urinary tract infections protect the fetus? *Obstet Gynecol.* 2002;100(5 Pt 1):909–915.
35. Petroff MG, Chen L, Phillips TA, Hunt JS. B7 family molecules: novel immunomodulators at the maternal-fetal interface. *Placenta.* 2002;23(suppl A):S95–101.
36. Pijnenborg R. Implantation and immunology: maternal inflammatory and immune cellular responses to

implantation and trophoblast invasion. *Reprod Biomed Online*. 2002;4(suppl 3):14–17.
37. Power LL, Popplewell EJ, Holloway JA, Diaper ND, Warner JO, Jones CA. Immunoregulatory molecules during pregnancy and at birth. *J Reprod Immunol*. 2002;56(1-2):19–28.
38. Piccinni MP. Role of immune cells in pregnancy. *Autoimmunity*. 2003;36(1):1–4.
39. Bulla R, Fischetti F, Bossi F, Tedesco F. Feto-maternal immune interaction at the placental level. *Lupus*. 2004;13(9):625–629.
40. Juretic K, Strbo N, Crncic TB, Laskarin G, Rukavina D. An insight into the dendritic cells at the maternal-fetal interface. *Am J Reprod Immunol*. 2004;52(6):350–355.
41. Matthiesen L, Berg G, Ernerudh J, Ekerfelt C, Jonsson Y, Sharma S. Immunology of preeclampsia. *Chem Immunol Allergy*. 2005;89:49–61.
42. Piccinni MP. T cells in pregnancy. *Chem Immunol Allergy*. 2005;89:3–9.
43. Wicherek L, Klimek M, Dutsch-Wicherek M. The level of maternal immune tolerance and fetal maturity. *Neuroendocrinol Lett*. 2005;26(5):561–566.
44. Hauguel-de Mouzon S, Guerre-Millo M. The placenta cytokine network and inflammatory signals. *Placenta*. 2006;27(8):794–798.
45. Romero R, Gotsch F, Pineles B, Kusanovic JP. Inflammation in pregnancy: its roles in reproductive physiology, obstetrical complications, and fetal injury. *Nutr Rev*. 2007;65(12 Pt 2):S194–S202.
46. Rusterholz C, Hahn S, Holzgreve W. Role of placentally produced inflammatory and regulatory cytokines in pregnancy and the etiology of preeclampsia. *Semin Immunopathol*. 2007;29(2):151–162.
47. Vigano P, Cintorino M, Schatz F, Lockwood CJ, Arcuri F. The role of macrophage migration inhibitory factor in maintaining the immune privilege at the fetal-maternal interface. *Semin Immunopathol*. 2007;29(2):135–150.
48. Koga K, Mor G. Expression and function of toll-like receptors at the maternal-fetal interface. *Reprod Sci*. 2008;15(3):231–242.
49. Kumar V, Medhi B. Emerging role of uterine natural killer cells in establishing pregnancy. *Iran J Immunol*. 2008;5(2):71–81.
50. Mor G. Inflammation and pregnancy: the role of toll-like receptors in trophoblast-immune interaction. *Ann NY Acad Sci*. 2008;1127:121–128.
51. Riley JK, Yokoyama WM. NK cell tolerance and the maternal-fetal interface. *Am J Reprod Immunol*. 2008;59(5):371–387.
52. van Rijn BB, Franx A, Steegers EA, et al. Maternal TLR4 and NOD2 gene variants, pro-inflammatory phenotype and susceptibility to early-onset preeclampsia and HELLP syndrome. *PloS One*. 2008;3(4):e1865.
53. Redman CW, Sargent IL. Latest advances in understanding preeclampsia. *Science*. 2005;308(5728):1592–1594.
54. Saito S, Shiozaki A, Nakashima A, Sakai M, Sasaki Y. The role of the immune system in preeclampsia. *Mol Aspects Med*. 2007;28(2):192–209.
55. Robillard PY, Dekker G, Chaouat G, Hulsey TC, Saftlas A. Epidemiological studies on primipaternity and immunology in preeclampsia--a statement after twelve years of workshops. *J Reprod Immunol*. 2011;89(2):104–117.
56. Alexander BT. Placental insufficiency leads to development of hypertension in growth-restricted offspring. *Hypertension*. 2003;41(3):457–462.
57. Alexander BT. Fetal programming of hypertension. *Am J Physiol Regul Integr Comp Physiol*. 2006;290(1):R1–R10.
58. Alexander BT. Prenatal influences and endothelial dysfunction: a link between reduced placental perfusion and preeclampsia. *Hypertension*. 2007;49(4):775–776.
59. Mold JE, Michaelsson J, Burt TD, et al. Maternal alloantigens promote the development of tolerogenic fetal regulatory T cells in utero. *Science*. 2008;322(5907):1562–1565.
60. Schiessl B. Inflammatory response in preeclampsia. *Mol Aspects Med*. 2007;28(2):210–219.
61. Huppertz B, Kingdom J, Caniggia I, et al. Hypoxia favours necrotic versus apoptotic shedding of placental syncytiotrophoblast into the maternal circulation. *Placenta*. 2003;24(2-3):181–190.
62. Abrahams VM, Kim YM, Straszewski SL, Romero R, Mor G. Macrophages and apoptotic cell clearance during pregnancy. *Am J Reprod Immunol*. 2004;51(4):275–282.
63. Heiskanen J, Romppanen EL, Hiltunen M, et al. Polymorphism in the tumor necrosis factor-alpha gene in women with preeclampsia. *J Assist Reprod Genet*. 2002;19(5):220–223.
64. Keelan JA, Mitchell MD. Placental cytokines and preeclampsia. *Front Biosci*. 2007;12:2706–2727.
65. Makris A, Xu B, Yu B, Thornton C, Hennessy A. Placental deficiency of interleukin-10 (IL-10) in preeclampsia and its relationship to an IL10 promoter polymorphism. *Placenta*. 2006;27(4-5):445–451.
66. Kamali-Sarvestani E, Kiany S, Gharesi-Fard B, Robati M. Association study of IL-10 and IFN-gamma gene polymorphisms in Iranian women with preeclampsia. *J Reprod Immunol*. 2006;72(1-2):118–126.
67. Daher S, Sass N, Oliveira LG, Mattar R. Cytokine genotyping in preeclampsia. *Am J Reprod Immunol*. 2006;55(2):130–135.
68. Knight M, Redman CW, Linton EA, Sargent IL. Shedding of syncytiotrophoblast microvilli into the maternal circulation in pre-eclamptic pregnancies. *Brit J Obstet Gynecol*. 1998;105(6):632–640.
69. Redman CW, Tannetta DS, Dragovic RA, et al. Review: does size matter? Placental debris and the pathophysiology of preeclampsia. *Placenta*. 2012;33(suppl):S48–54.
70. Chesley LC, Annitto JE, Cosgrove RA. The familial factor in toxemia of pregnancy. *Obstet Gynecol*. 1968;32(3):303–311.
71. Humphries J. Occurrence of hypertensive toxemia of pregnancy in mother-daughter pairs. *Bull Johns Hopkins Hosp*. 1960;107:271–277.
72. Adams EM, Finlayson A. Familial aspects of preeclampsia and hypertension in pregnancy. *Lancet*. 1961;2(7217):1375–1378.
73. Cooper DW, Liston WA. Genetic control of severe preeclampsia. *J Med Genet*. 1979;16(6):409–416.
74. Sutherland A, Cooper DW, Howie PW, Liston WA, MacGillivray I. The indicence of severe pre-eclampsia amongst mothers and mothers-in-law of pre-eclamptics and controls. *Brit J Obstet Gynecol*. 1981;88(8):785–791.
75. Laivuori H. Genetic aspects of preeclampsia. *Front Biosci*. 2007;12:2372–2382.

76. Machin GA. Some causes of genotypic and phenotypic discordance in monozygotic twin pairs. *Am J Med Genet.* 1996;61(3):216–228.
77. Hall JG. Twins and twinning. *Am J Med Genet.* 1996;61(3):202–204.
78. Oudejans CB, van Dijk M, Oosterkamp M, Lachmeijer A, Blankenstein MA. Genetics of preeclampsia: paradigm shifts. *Hum Genet.* 2007;120(5):607–612.
79. George EM, Bidwell GL. STOX1: a new player in preeclampsia? *Hypertension.* 2013;61(3):561–563.
80. Brown EA, Ruvolo M, Sabeti PC. Many ways to die, one way to arrive: how selection acts through pregnancy. *Trends Gen.* 2013;29:585–592.
81. Thornton JG, Onwude JL. Convulsions in pregnancy in related gorillas. *Am J Obstet Gynecol.* 1992;167(1):240–241.
82. Hennessy A, Phippard AF, Harewood WF, Painter DM, Kirwan PD, Horvath JS. Histomorphometry of the renal lesions in baboons with placental ischaemia. *Clin Exp Hypertens B.* 1993.
83. Combs CA, Katz MA, Kitzmiller JL, Brescia RJ. Experimental preeclampsia produced by chronic constriction of the lower aorta: validation with longitudinal blood pressure measurements in conscious rhesus monkeys. *Am J Obstet Gynecol.* 1993;169(1):215–223.
84. van Dijk M, Mulders J, Poutsma A, et al. Maternal segregation of the Dutch preeclampsia locus at 10q22 with a new member of the winged helix gene family. *Nat Genet.* 2005;37(5):514–519.
85. Rana K, Isbel N, Buzza M, et al. Clinical, histopathologic, and genetic studies in nine families with focal segmental glomerulosclerosis. *Am J Kidney Dis.* 2003;41(6):1170–1178.
86. Berends AL, Steegers EA, Isaacs A, et al. Familial aggregation of preeclampsia and intrauterine growth restriction in a genetically isolated population in The Netherlands. *Eur J Hum Genet.* 2008;16(12):1437–1442.
87. Harrison GA, Humphrey KE, Jones N, et al. A genome-wide linkage study of preeclampsia/eclampsia reveals evidence for a candidate region on 4q. *Am J Hum Genet.* 1997;60(5):1158–1167.
88. Arngrimsson R, Siguroardottir S, Frigge ML, et al. A genome-wide scan reveals a maternal susceptibility locus for pre-eclampsia on chromosome 2p13. *Hum Mol Genet.* 1999;8(9):1799–1805.
89. Moses EK, Lade JA, Guo G, et al. A genome scan in families from Australia and New Zealand confirms the presence of a maternal susceptibility locus for pre-eclampsia, on chromosome 2. *Am J Hum Genet.* 2000;67(6):1581–1585.
90. Lachmeijer AM, Arngrimsson R, Bastiaans EJ, et al. A genome-wide scan for preeclampsia in the Netherlands. *Eur J Hum Genet.* 2001;9(10):758–764.
91. Laivuori H, Lahermo P, Ollikainen V, et al. Susceptibility loci for preeclampsia on chromosomes 2p25 and 9p13 in Finnish families. *Am J Hum Genet.* 2003;72(1):168–177.
92. Kalmyrzaev B, Aldashev A, Khalmatov M, et al. Genome-wide scan for premature hypertension supports linkage to chromosome 2 in a large Kyrgyz family. *Hypertension.* 2006;48(5):908–913.
93. Moses EK, Fitzpatrick E, Freed KA, et al. Objective prioritization of positional candidate genes at a quantitative trait locus for pre-eclampsia on 2q22. *Mol Hum Reprod.* 2006;12(8):505–512.
94. Johnson MP, Fitzpatrick E, Dyer TD, et al. Identification of two novel quantitative trait loci for pre-eclampsia susceptibility on chromosomes 5q and 13q using a variance components-based linkage approach. *Mol Hum Reprod.* 2007;13(1):61–67.
95. Majander KK, Villa PM, Kivinen K, Kere J, Laivuori H. A follow-up linkage study of Finnish pre-eclampsia families identifies a new fetal susceptibility locus on chromosome 18. *Eur J Hum Genet.* 2013;21(9):1024–1026.
96. Buurma AJ, Turner RJ, Driessen JH, et al. Genetic variants in pre-eclampsia: a meta-analysis. *Hum Reprod Update.* 2013;19(3):289–303.
97. Staines-Urias E, Paez MC, Doyle P, et al. Genetic association studies in pre-eclampsia: systematic meta-analyses and field synopsis. *Int J Epidemiol.* 2012;41(6):1764–1775.
98. Wang XM, Wu HY, Qiu XJ. Methylenetetrahydrofolate reductase (MTHFR) gene C677T polymorphism and risk of preeclampsia: an updated meta-analysis based on 51 studies. *Arch Med Res.* 2013;44(3):159–168.
99. Lin R, Lei Y, Yuan Z, Ju H, Li D. Angiotensinogen gene M235T and T174M polymorphisms and susceptibility of pre-eclampsia: a meta-analysis. *Ann Hum Genet.* 2012;76(5):377–386.
100. Dai B, Liu T, Zhang B, Zhang X, Wang Z. The polymorphism for endothelial nitric oxide synthase gene, the level of nitric oxide and the risk for pre-eclampsia: a meta-analysis. *Gene.* 2013;519(1):187–193.
101. Zhong WG, Wang Y, Zhu H, Zhao X. Meta analysis of angiotensin-converting enzyme I/D polymorphism as a risk factor for preeclampsia in Chinese women. *Genet Mol Res.* 2012;11(3):2268–2276.
102. Chen Z, Xu F, Wei Y, Liu F, Qi H. Angiotensin converting enzyme insertion/deletion polymorphism and risk of pregnancy hypertensive disorders: a meta-analysis. *JRAAS.* 2012;13(1):184–195.
103. Zhu M, Zhang J, Nie S, Yan W. Associations of ACE I/D, AGT M235T gene polymorphisms with pregnancy induced hypertension in Chinese population: a meta-analysis. *J Assist Reprod Genet.* 2012;29(9):921–932.
104. Zhao L, Dewan AT, Bracken MB. Association of maternal AGTR1 polymorphisms and preeclampsia: a systematic review and meta-analysis. *J Matern-Fetal Neonat Med.* 2012;25(12):2676–2680.
105. Adams KM, Eschenbach DA. The genetic contribution towards preterm delivery. *Semin Fetal Neonat Med.* 2004;9(6):445–452.
106. Dalmaz CA, Santos KG, Botton MR, Tedoldi CL, Roisenberg I. Relationship between polymorphisms in thrombophilic genes and preeclampsia in a Brazilian population. *Blood Cell Mol Dis.* 2006;37(2):107–110.
107. Vefring H, Lie RT, ØDegård R, Mansoor MA, Nilsen ST. Maternal and fetal variants of genetic thrombophilias and the risk of preeclampsia. *Epidemiology.* 2004;15(3):317–322.
108. Zusterzeel PL, Visser W, Blom HJ, Peters WH, Heil SG, Steegers EA. Methylenetetrahydrofolate reductase polymorphisms in preeclampsia and the HELLP syndrome. *Hypertens Pregnancy.* 2000;19(3):299–307.

109. Lin J, August P. Genetic thrombophilias and preeclampsia: a meta-analysis. *Obstet Gynecol.* 2005;105(1):182–192.
110. Kanasaki K, Palmsten K, Sugimoto H, et al. Deficiency in catechol-O-methyltransferase and 2-methoxyoestradiol is associated with pre-eclampsia. *Nature.* 2008;453(7198):1117–1121.
111. Lim JH, Kim SY, Kim do J, et al. Genetic polymorphism of catechol-O-methyltransferase and cytochrome P450c17alpha in preeclampsia. *Pharmacogenet Genom.* 2010;20(10):605–610.
112. Srinivas SK, Morrison AC, Andrela CM, Elovitz MA. Allelic variations in angiogenic pathway genes are associated with preeclampsia. *Am J Obstet Gynecol.* 2010;202(5):445e1–445e11.
113. Bertina RM, Koeleman BP, Koster T, et al. Mutation in blood coagulation factor V associated with resistance to activated protein C. *Nature.* 1994;369(6475):64–67.
114. Dahlback B. Inherited resistance to activated protein C, a major cause of venous thrombosis, is due to a mutation in the factor V gene. *Haemostasis.* 1994;24(2):139–151.
115. Dizon-Townson DS, Meline L, Nelson LM, Varner M, Ward K. Fetal carriers of the factor V Leiden mutation are prone to miscarriage and placental infarction. *Am J Obstet Gynecol.* 1997;177(2):402–405.
116. Jeunemaitre X, Soubrier F, Kotelevtsev YV, et al. Molecular basis of human hypertension: role of angiotensinogen. *Cell.* 1992;71(1):169–180.
117. Ward K, Chettier R, Ward J, Nelson L. Genome-wide association study of preeclampsia points to several novel genes. *Am J Obstet Gynecol.* 2011;204(1):s285.
118. Johnson MP, Brennecke SP, East CE, et al. Genome-wide association scan identifies a risk locus for preeclampsia on 2q14, near the inhibin, beta B gene. *PloS One.* 2012;7(3):e33666.
119. Zhao L, Bracken MB, Dewan AT. Genome-wide association study of pre-eclampsia detects novel maternal single nucleotide polymorphisms and copy-number variants in subsets of the Hyperglycemia and Adverse Pregnancy Outcome (HAPO) study cohort. *Ann Hum Genet.* 2013.
120. Pennell CE, Jacobsson B, Williams SM, et al. Genetic epidemiologic studies of preterm birth: guidelines for research. *Am J Obstet Gynecol.* 2007;196(2):107–118.
121. Haig D. Does heritability hide in epistasis between linked SNPs? *Eur J Hum Genet.* 2011;19(2):123.
122. Isakov O, Perrone M, Shomron N. Exome sequencing analysis: a guide to disease variant detection. *Methods Mol Biol.* 2013;1038:137–158.
123. Johnson M, Løset M, Brennecke S, et al. OS049. Exome sequencing identifies likely functional variantsinfluencing preeclampsia and CVD risk. *Pregnancy Hypertens.* 2012;2(3):203–204.
124. Ding D, Scott NM, Thompson EE, et al. Increased protein-coding mutations in the mitochondrial genome of African American women with preeclampsia. *Reprod Sci.* 2012;19(12):1343–1351.
125. Orsi NM, Gopichandran N, Simpson NA. Genetics of preterm labour. *Best Pract Res Cl Ob.* 2007;21(5):757–772.
126. Zintzaras E, Kitsios G, Harrison GA, et al. Heterogeneity-based genome search meta-analysis for preeclampsia. *Hum Genet.* 2006;120(3):360–370.
127. Goddard KA, Tromp G, Romero R, et al. Candidate-gene association study of mothers with pre-eclampsia, and their infants, analyzing 775 SNPs in 190 genes. *Hum Hered.* 2007;63(1):1–16.
128. Fitzpatrick E, Goring HH, Liu H, et al. Fine mapping and SNP analysis of positional candidates at the preeclampsia susceptibility locus (PREG1) on chromosome 2. *Hum Biol.* 2004;76(6):849–862.
129. Wellcome Trust Case Control Consortium. Genome-wide association study of 14,000 cases of seven common diseases and 3,000 shared controls. *Nature.* 2007;447(7145):661–678.
130. Haig D. Altercation of generations: genetic conflicts of pregnancy. *Am J Reprod Immunol.* 1996;35(3):226–232.
131. Romero R, Espinoza J, Gotsch F, et al. The use of high-dimensional biology (genomics, transcriptomics, proteomics, and metabolomics) to understand the preterm parturition syndrome. *BJOG.* 2006;113(suppl 3):118–135.
132. Esplin MS, Varner MW. Genetic factors in preterm birth--the future. *BJOG.* 2005;112(suppl 1):97–102.
133. Bo T, Jonassen I. New feature subset selection procedures for classification of expression profiles. *Genome Biol.* 2002;3(4) RESEARCH0017.
134. Lamont RF. Looking to the future. *BJOG.* 2003;110(suppl 20):131–135.
135. Conrad DF, Jakobsson M, Coop G, et al. A worldwide survey of haplotype variation and linkage disequilibrium in the human genome. *Nat Genet.* 2006;38(11):1251–1260.
136. Tempfer CB, Jirecek S, Riener EK, et al. Polymorphisms of thrombophilic and vasoactive genes and severe preeclampsia: a pilot study. *J Soc Gynecol Invest.* 2004;11(4):227–231.
137. Roberts JM, Cooper DW. Pathogenesis and genetics of pre-eclampsia. *Lancet.* 2001;357(9249):53–56.
138. Lachmeijer AM, Dekker GA, Pals G, Aarnoudse JG, ten Kate LP, Arngrimsson R. Searching for preeclampsia genes: the current position. *Eur J Obstet Gynec Reprod Biol.* 2002;105(2):94–113.
139. Bernard N, Giguere Y. Genetics of preeclampsia: what are the challenges? *JOGC.* 2003;25(7):578–585.
140. Laasanen J, Hiltunen M, Romppanen EL, Punnonen K, Mannermaa A, Heinonen S. Microsatellite marker association at chromosome region 2p13 in Finnish patients with preeclampsia and obstetric cholestasis suggests a common risk locus. *Eur J Hum Genet.* 2003;11(3):232–236.
141. Austgulen R. Recent knowledge on mechanisms underlying development of pre-eclampsia. *Tidsskr Norske Lægefor Tidsskr Prakt Med.* 2004;124(1):21–24.
142. Descamps OS, Bruniaux M, Guilmot PF, Tonglet R, Heller FR. Lipoprotein metabolism of pregnant women is associated with both their genetic polymorphisms and those of their newborn children. *J Lipid Res.* 2005;46(11):2405–2414.
143. Rogers MS, D'Amato RJ. The effect of genetic diversity on angiogenesis. *Exp Cell Res.* 2006;312(5):561–574.
144. Musani SK, Shriner D, Liu N, et al. Detection of gene x gene interactions in genome-wide association studies of human population data. *Hum Hered.* 2007;63(2):67–84.
145. Kotti S, Bickeboller H, Clerget-Darpoux F. Strategy for detecting susceptibility genes with weak or no marginal effect. *Hum Hered.* 2007;63(2):85–92.

146. Manolio TA, Collins FS. Genes, environment, health, and disease: facing up to complexity. *Hum Hered.* 2007;63(2):63–66.
147. Calicchio R, Doridot L, Miralles F, Mehats C, Vaiman D. DNA methylation, an epigenetic mode of gene expression regulation in reproductive science. *Curr Pharmaceut Des.* 2013.
148. Fraga MF, Ballestar E, Paz MF, et al. Epigenetic differences arise during the lifetime of monozygotic twins. *Proc Natl Acad Sci USA.* 2005;102(30):10604–10609.
149. Davis EF, Newton L, Lewandowski AJ, et al. Pre-eclampsia and offspring cardiovascular health: mechanistic insights from experimental studies. *Clin Sci (Lond).* 2012;123(2):53–72.
150. Chelbi ST, Mondon F, Jammes H, et al. Expressional and epigenetic alterations of placental serine protease inhibitors: SERPINA3 is a potential marker of preeclampsia. *Hypertension.* 2007;49(1):76–83.
151. Arngrimsson R. Epigenetics of hypertension in pregnancy. *Nat Genet.* 2005;37(5):460–461.
152. Chelbi ST, Vaiman D. Genetic and epigenetic factors contribute to the onset of preeclampsia. *Mol Cell Endocrinol.* 2008;282(1-2):120–129.
153. Choudhury M, Friedman JE. Epigenetics and microRNAs in preeclampsia. *Clin Exp Hypertens.* 2012;34(5):334–341.
154. Ottman R. Analysis of genetically complex epilepsies. *Epilepsia.* 2005;46(suppl 10):7–14.
155. Yang HC, Pan CC, Lin CY, Fann CS. PDA: Pooled DNA analyzer. *BMC Bioinformat.* 2006;7:233.
156. Humphries SE, Donati MB. Analysis of gene-environment interaction in coronary artery disease. *Ital Heart J.* 2002;3(1):3–5.
157. Pearl J. Why there is no statistical test for confounding, why many think there is, and why they are almost right. UCLA Computer Science Department, Technical Report (R-256) 1998.
158. Kist WJ, Janssen NG, Kalk JJ, Hague WM, Dekker GA, de Vries JI. Thrombophilias and adverse pregnancy outcome - A confounded problem! *Thromb Haemost.* 2008;99(1):77–85.
159. Founds SA, Shi H, Conley YP, Jeyabalan A, Roberts JM, Lyons-Weiler J. Variations in discovery-based preeclampsia candidate genes. *Clin Translat Sci.* 2012;5(4):333–339.
160. Myers JE, Tuytten R, Thomas G, et al. Integrated proteomics pipeline yields novel biomarkers for predicting preeclampsia. *Hypertension.* 2013;61(6):1281–1288.
161. Williams PJ, Morgan L. The role of genetics in pre-eclampsia and potential pharmacogenomic interventions. *Pharmacogenom Personal Med.* 2012;5:37–51.

CHAPTER 5

The Placenta in Normal Pregnancy and Preeclampsia

SUSAN J. FISHER, MICHAEL MCMASTER AND JAMES M. ROBERTS

Editors' comment: *Without the placenta we would not have preeclampsia (or human civilization either). Thus the recent explosion of research regarding this organ and preeclampsia is not surprising, and now includes work not only on the origin of the disease but research that may lead to prevention and treatment. Given the logarithmic increase in reportable observations your authors thought it best to divide Chapter 5 into two parts. The first and main chapter extends the overview provided in the previous edition, describing placental development and function during normal gestation and in women developing preeclampsia. In the new edition, this is followed by an Appendix focusing on exciting new molecular biology contributions from the authors in the area entitled "Trophoblast gene expression in normal pregnancy and preeclampsia". The reader should be aware that this is really an entrée into the placental story. Gems appearing elsewhere, including Chapters 6 (devoted to pro- and antiangiogenic proteins), 8 (immunology) and 9 (endothelial dysfunction), also deal with the roles of placental "debris" and other factors in this disorder.*

INTRODUCTION

This chapter focuses on the unique process by which the human placenta normally forms, and how changes can lead to serious pregnancy complications such as preeclampsia. Special emphasis will be placed on work from the principal author's laboratory that led to the discovery that the subset of placental cells, termed cytotrophoblasts, that invade the uterus and form vascular connections with the resident maternal vessels undergo a novel transformation from an epithelial cell of ectodermal origin to a vascular-like cell with a myriad of endothelial-like properties.[1]

THE MICROANATOMY OF NORMAL HUMAN PLACENTATION

The human placenta's unique anatomy (Figs. 5.1 and 5.2) is due in large part to differentiation of its ectodermally derived progenitors, termed cytotrophoblasts.[2] How these cells differentiate determines whether chorionic villi, the placenta's functional units, float in maternal blood or anchor the conceptus to the uterine wall. In floating villi, cytotrophoblasts differentiate by fusing to form multinucleate syncytiotrophoblasts whose primary function – transport – is ideally suited to their location at the villus surface. In anchoring villi, cytotrophoblasts also fuse, but many remain as single cells that detach from their basement membrane and aggregate to form cell columns. Cytotrophoblasts at the distal ends of these columns attach to, then deeply invade the uterus (interstitial invasion) and its arterioles (endovascular invasion). As a result of endovascular invasion, the cells replace the endothelial and muscular linings of uterine arterioles, a process that initiates maternal blood flow to the placenta and greatly enlarges the vessel diameter. Paradoxically, the cells invade only the superficial portions of uterine venules.

Cells that participate in endovascular invasion have two types of interactions with maternal arterioles. In the first, large aggregates of these fetal cells are found primarily inside the vessel lumen. These aggregates can either lie adjacent to the apical surface of the resident endothelium or replace it such that they appear directly attached to the vessel wall. In the second type of interaction, cytotrophoblasts are found within the vessel wall rather than in the lumen. In this position, they colonize the smooth-muscle layer of the vessel and lie subjacent to the endothelium. These different types of interactions may be progressive stages in a single process, or indicative of different strategies by which cytotrophoblasts accomplish endovascular invasion. In either case, the stage in which fetal cytotrophoblasts cohabit with maternal endothelium in the spiral arterioles is transient. By late second trimester these vessels are lined exclusively by cytotrophoblasts, and endothelial cells are no longer visible in either the endometrial or the superficial portions of their myometrial segments.

THE MICROANATOMY OF ABNORMAL HUMAN PLACENTATION IN PREECLAMPSIA

Preeclampsia is a disease that adversely affects 7–10% of first pregnancies in the United States.[3] The mother shows

Chesley's Hypertensive Disorders in Pregnancy.
ISBN: 978-0-12-407866-6

DOI: http://dx.doi.org/10.1016/B978-0-12-407866-6.00005-5

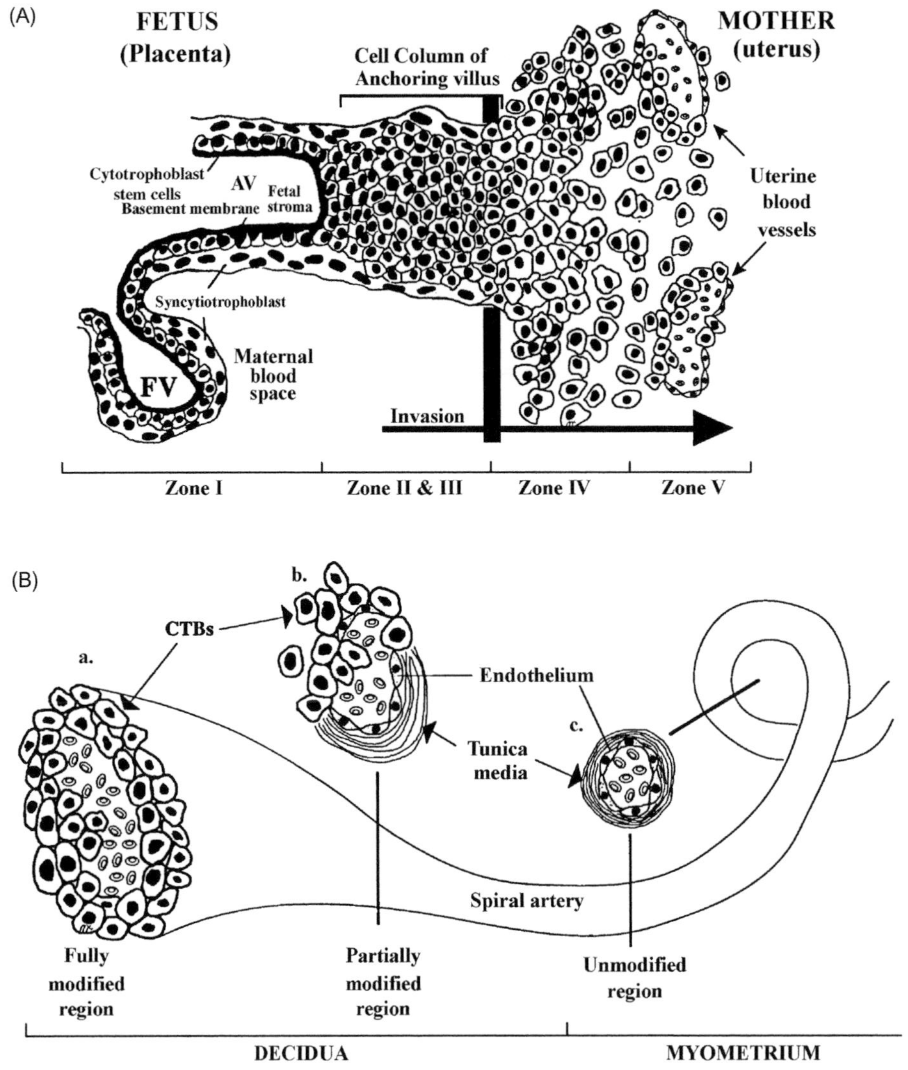

FIGURE 5.1 (A) Diagram of a longitudinal section of an anchoring chorionic villus (AV) at the fetal–maternal interface at about 10 weeks of gestational age. The anchoring villus (AV) functions as a bridge between the fetal and maternal compartments, whereas floating villi (FV) are suspended in the intervillous space and are bathed by maternal blood. Cytotrophoblasts (CTB) in AV (Zone I) form cell columns (Zones II and III). CTB then invade the uterine interstitium (decidua and first third of the myometrium: Zone IV) and maternal vasculature (Zone V), thereby anchoring the fetus to the mother and accessing the maternal circulation. Zone designations mark areas in which CTB have distinct patterns of stage-specific antigen expression. (B) Diagram of a uterine (spiral) artery in which endovascular invasion is in progress (10–18 weeks of gestation). Endometrial and then myometrial segments of spiral arteries are modified progressively. (a) In fully modified regions the vessel diameter is large. CTB are present in the lumen and occupy the entire surface of the vessel wall. A discrete muscular layer (tunica media) is not evident. (b) Partially modified vessel segments. CTB and maternal endothelium occupy discrete regions of the vessel wall. In areas of intersection, CTB appear to lie deep to the endothelium and in contact with the vessel wall. (c) Unmodified vessel segments in the myometrium. Vessel segments in the superficial third of the myometrium will become modified when endovascular invasion reaches its fullest extent (by 22 weeks), while deeper segments of the same artery will retain their normal structure.

signs and symptoms that suggest widespread alterations in endothelial function (e.g., high blood pressure, proteinuria, and edema[4]). In some cases fetal growth slows, which leads to intrauterine growth retardation. The severity of the disease varies greatly (see Chapter 2). In its mildest form the signs/symptoms appear near term and resolve after birth, with no lasting effects on either the mother or the child. In its severest form the signs/symptoms often occur in the second or early third trimesters. If they cannot be controlled, the only option is delivery, with consequent iatrogenic fetal prematurity. Owing to the latter form of the disease, preeclampsia and hypertensive diseases of pregnancy are leading causes of maternal death and contribute significantly to premature deliveries in the United States.[4]

Although the cause of preeclampsia is unknown, the accumulated evidence strongly implicates the placenta.[5]

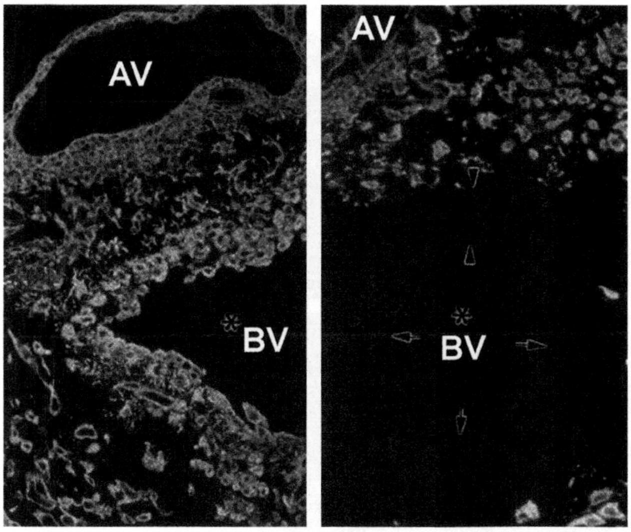

FIGURE 5.2 In normal pregnancy (left) fetal cytotrophoblasts (stained with anti-cytokeratin) from the anchoring villi (AV) of the placenta invade the maternal uterine blood vessels (BV). In preeclampsia (right) fetal cells fail to penetrate the uterine vasculature (arrows and arrowheads).

Anatomic examination shows that the area of the placenta most affected by this syndrome is the fetal–maternal interface (Fig. 5.2). Cytotrophoblast invasion of the uterus is shallow, and endovascular invasion does not proceed beyond the terminal portions of the spiral arterioles.[6] The effect of preeclampsia on endovascular invasion is particularly evident when interactions between fetal cytotrophoblasts and maternal endothelial cells are studied in detail.[7,8] Serial sections through placental bed biopsies show that few of the spiral arterioles contain cytotrophoblasts. Instead, most cytotrophoblasts remain at some distance from these vessels. Where endovascular cytotrophoblasts are detected, their invasion is limited to the portion of the vessel that spans the superficial decidua. Even when the cytotrophoblasts gain access to the lumina, they usually fail to form tight aggregates among themselves, or to spread out on the vessel wall, as is observed for cytotrophoblasts in control samples matched for gestational age. Instead they tend to remain as individual rounded cells, suggesting that they are poorly anchored to the vessel wall. Thus, cytotrophoblasts in preeclampsia not only have a limited capacity for endovascular invasion but also display an altered morphology in their interactions with maternal arterioles.

Because of these alterations in endovascular invasion, the maternal vessels of preeclamptic patients do not undergo the complete spectrum of physiologic changes that normally occur (e.g., loss of their endothelial lining and musculoelastic tissue); the mean external diameter of the myometrial vessels is less than half that of similar vessels from uncomplicated pregnancies.[9–11] In addition, not as many vessels show evidence of cytotrophoblast invasion.[9,10,12] Thus, the architecture of these vessels precludes an adequate response to gestation-related fetal demands for increased blood flow.

THE ROAD TO PREECLAMPSIA

Several reviews summarize the current state of knowledge regarding preeclampsia. This interest reflects the clinical importance of this condition, which continues to be a primary driver of obstetric care in the developed world.[13–15] This is due, in large part, to the sometimes rapid onset of preeclampsia – hence the name "preeclampsia" (derived from the Greek εχλαμψιεζ, sudden flash or development). A somewhat clearer picture of the pathogenesis of preeclampsia has begun to emerge. A two-stage model has been proposed in which the initiating event, poor placentation, is thought to occur early on.[16] This concept is supported by several studies that document the association between reduced blood flow to the placenta before 20 weeks of gestation, as determined by color Doppler ultrasound evaluation of spiral arterial blood flow, and a greatly increased risk of developing preeclampsia[17,18] (Fig. 5.1).

The second stage of preeclampsia is the maternal response to abnormal placentation. Systemic endothelial dysfunction is thought to be an important common denominator.[4,16,19] For example, serum from pregnancies complicated by preeclampsia induces activation of umbilical endothelial cells *in vitro*,[20] experimental evidence in support of this theory. Interestingly, increased pressor sensitivity and abnormal-flow-induced vasodilatation can be demonstrated before the clinical signs appear,[21–23] as can sensitivity to angiotensin II.[24] Much of the current data point to an imbalance in circulating factors with relevant biological activities. For example, molecules that are associated with endothelial dysfunction, such as fibronectin, factor VIII antigen, and thrombomodulin, are found at higher levels in the blood of women with preeclampsia.[25–27] The concentration of *S*-nitrosoalbumin, a major nitric oxide reservoir, also increases[28] as does endothelin.[29] Concomitantly, the levels of endothelial-derived vasodilators, such as prostacyclin, decrease.[30] Activators of peroxisome proliferator-activated receptor gamma, which functions in metabolic and immune responses, show the same pattern, with abnormally low levels evident weeks and in some cases months before the clinical diagnosis of preeclampsia is made.[31] Maternal blood levels of the soluble form of VEGFR-1 (sFlt-1) rise as do those of the TGFβ receptor, endoglin, with a concomitant failure of placental growth factor levels to reach normal values (see Chapter 6). These changes have been attributed by one group of investigators[32] to a deficiency in catechol-*O*-methyltransferase and 2-methoxyestradiol.[32] Recent work in animal models has also implicated agonistic autoantibodies to the angiotensin receptor in the cascade of events that lead to preeclampsia.[33]

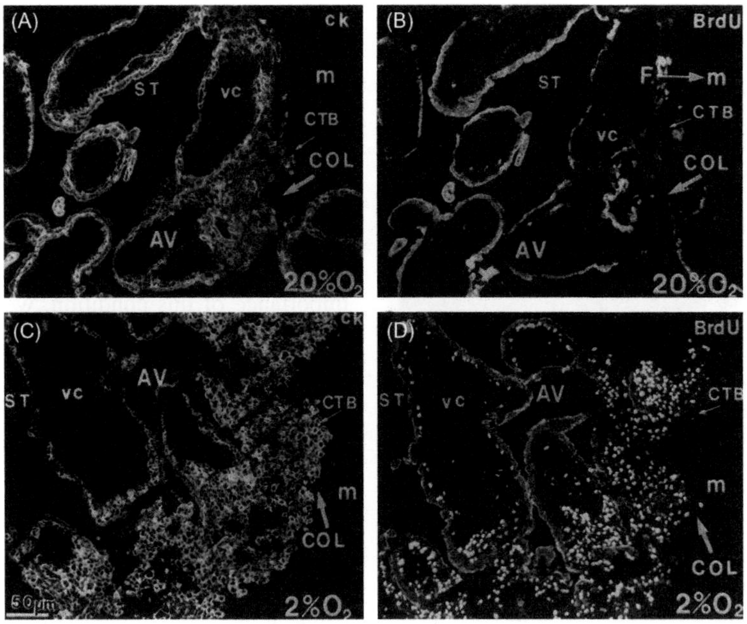

FIGURE 5.3 Low oxygen (2% O_2) stimulates cytotrophoblast BrdU incorporation *in vitro*. Anchoring villi (AV) from 6–8-week placentas were cultured on Matrigel (m) for 72 hours in either 20% O_2 (A, B) or 2% O_2 (C, D). By the end of the culture period, fetal cytotrophoblasts migrated into the Matrigel (F→m). To assess cell proliferation, BrdU was added to the medium. Tissue sections of the villi were stained with anti-cytokeratin (ck; A, C), which recognizes syncytiotrophoblasts (ST) and cytotrophoblasts (CTB) but not cells in the villus core (vc); and with anti-BrdU (B, D), which detects cells in S phase. Villus explants maintained in 2% O_2 (C) formed much more prominent columns (COL) with a larger proportion of CTB nuclei that incorporated BrdU (D) than explants cultured in 20% O_2 (A, B). (Reprinted with permission.[36] Copyright 1997 American Association for the Advancement of Science.)

OXYGEN TENSION REGULATES HUMAN CYTOTROPHOBLAST PROLIFERATION AND DIFFERENTIATION *IN VITRO*

Information about the morphological and molecular aspects of cytotrophoblast invasion in normal pregnancy and in preeclampsia has been used to formulate hypotheses about the regulatory factors involved. Specifically, in normal pregnancy cytotrophoblasts invade large-bore arterioles, where they are in contact with well-oxygenated maternal blood. But in preeclampsia, invasive cytotrophoblasts are relatively hypoxic. Another important consideration is that placental blood flow changes dramatically during early pregnancy. During much of the first trimester there is little endovascular invasion, so maternal blood flow to the placenta is at a minimum. The oxygen pressures of the intervillous space (that is, at the uterine surface) and within the endometrium are estimated to be approximately 18 mm Hg and 40 mm Hg, respectively, at 8–10 weeks of gestation.[34] Afterwards, endovascular invasion proceeds rapidly; cytotrophoblasts are in direct contact with blood from maternal spiral arterioles, which could have a mean oxygen pressure as high as 90–100 mm Hg. Thus, as cytotrophoblasts invade the uterus during the first half of pregnancy, they encounter a steep, positive oxygen tension gradient. These observations, together with the results of our own experiments conducted on isolated cytotrophoblasts,[35] suggested that oxygen tension might regulate cytotrophoblast proliferation and differentiation along the invasive pathway.[36]

First, we used immunolocalization techniques to study the relationship between cytotrophoblast proliferation and differentiation *in situ*. Cytotrophoblasts in columns (i.e., cells in the initial stages of differentiation) reacted with an antibody against the Ki67 antigen,[36] which is indicative of DNA synthesis.[37] Distal to this region, anti-Ki67 staining abruptly stopped, and the cytotrophoblasts intricately modulated their expression of stage-specific antigens, including integrin cell adhesion molecules,[38] matrix metalloproteinase-9,[39] HLA-G (a cytotrophoblast class Ib major histocompatibility complex molecule[40,41]), and human placental lactogen.[42] These results suggested that during differentiation along the invasive pathway, cytotrophoblasts first undergo mitosis, then exit the cell cycle and modulate their expression of stage-specific antigens.

As an *in vitro* model system for testing this hypothesis, we used organ cultures of anchoring villi explanted from early gestation (6–8-week) placentas onto an extracellular matrix substrate. Some of the anchoring villi were cultured for 72 hours in a standard tissue culture incubator (20% O_2 or 98 mm Hg). Figure 5.3A shows a section of one such control villus that was stained with an antibody that recognizes cytokeratin to demonstrate syncytiotrophoblasts and cytotrophoblasts. The attached cell columns were clearly visible. To assess the cells' ability to synthesize DNA, the villi were incubated with bromodeoxyuridine (BrdU). Incorporation was detected in the cytoplasm, but not the nuclei, of syncytiotrophoblasts. Few or none of the cells in columns incorporated BrdU (Fig. 5.3B). Other anchoring villi were maintained in a hypoxic atmosphere (2% O_2 or 14 mm Hg). After 72 hours, cytokeratin staining showed prominent cell columns (Fig. 5.3C), and the nuclei of many of the cytotrophoblasts in these columns incorporated BrdU (Fig. 5.3D). Because cytotrophoblasts were the only cells that entered S phase, we also compared the ability of anchoring villus explants cultured under standard and hypoxic conditions to incorporate [^{3}H]thymidine. Villus explants cultured under hypoxic conditions (2% O_2) incorporated 3.3±1.2 times more [^{3}H]thymidine than villi cultured under standard conditions (20% O_2). In contrast, [^{3}H]thymidine incorporation by explants cultured in a 6% O_2 atmosphere (40 mm Hg) was no different than in control villi. Taken together, these results suggest that a hypoxic environment, comparable to that encountered by early gestation cytotrophoblasts in the intervillous space, stimulates the cells to enter S phase.

Cytokeratin staining also showed that the cell columns associated with anchoring villi cultured under hypoxic conditions were larger than cell columns of control villi cultured under standard conditions (compare Figs. 5.3A and 5.3C). To quantify this, we made serial sections of villus explants maintained in either 20% or 2% O_2, and then counted the number of cells in columns. Under hypoxic culture conditions, the columns contained triple the number of cells present in columns maintained in 20% oxygen. These results indicate that hypoxia stimulates cytotrophoblasts in cell columns to proliferate.[35,36]

The next series of studies were based on the hypothesis that the hypoxia-induced changes in the cells' proliferative capacity would be reflected by changes in their expression of proteins that regulate passage through the cell cycle (Fig. 5.4). With regard to the G2 to M transition, we were particularly interested in their cyclin B expression, since threshold levels of this protein are required for cells to enter mitosis.[43] Immunoblotting of cell extracts showed that after 3 days in culture, anchoring villi maintained in 2% O_2 contained 3.1 times more cyclin B than did villi maintained in 20% O_2 (Fig. 5.4A). Immunolocalization experiments confirmed that cyclin B was primarily expressed by cytotrophoblasts (not shown). Since p21$^{WF1/CIP1}$ abundance has been correlated with cell cycle arrest,[44] we also examined the effects of oxygen tension on cytotrophoblast expression of this protein. Very little p21$^{WF1/CIP1}$ expression was detected in cell extracts of anchoring villi maintained for 72 hours in 2% O_2, but expression increased 3.8-fold in anchoring villi maintained for the same time period in 20% O_2 (Fig. 5.4B). Immunolocalization experiments confirmed that p21$^{WF1/CIP1}$ was primarily expressed by cytotrophoblasts. These results, replicated in five separate experiments, confirm that culturing anchoring villi in 20% O_2 induces cytotrophoblasts in the attached cell columns to undergo cell cycle arrest, whereas culturing them in 2% O_2 induces them to enter mitosis.[35,36]

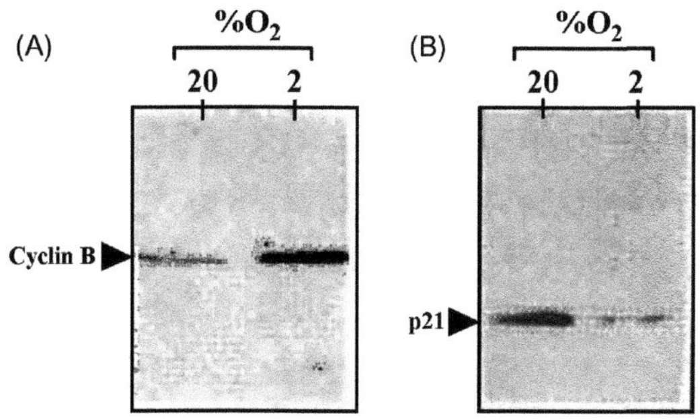

FIGURE 5.4 Hypoxia induces changes in cytotrophoblast expression of proteins that regulate progression through the cell cycle. (A) Villus explants cultured for 72 hours in 2% O_2 contained 3.1 times more cyclin B than did villi maintained under standard culture conditions (20% O_2) for the same length of time. (B) Expression of p21 increased 3.8-fold in anchoring villi cultured for 72 hours in 20% O_2 as compared to 2% O_2. (Reprinted with permission.[36] Copyright 1997 American Association for the Advancement of Science.)

Changes in proliferative capacity are often accompanied by concomitant changes in differentiation. Accordingly, we investigated the effects of hypoxia on the ability of cytotrophoblasts to differentiate along the invasive pathway (Fig. 5.5). Under standard tissue culture conditions, cytotrophoblasts migrated from the cell columns and modulated their expression of stage-specific antigens, as they do during uterine invasion *in vivo*.[38] For example, they began to express integrin α1, a laminin-collagen receptor that is required for invasiveness *in vitro*.[45] Both differentiated cytotrophoblasts and villus stromal cells expressed this antigen (Fig. 5.5B). When cultured under hypoxic conditions, cytotrophoblasts failed to stain for integrin α1, but stromal cells continued to express this molecule, suggesting the observed effects were cell-type specific (Fig. 5.5E). Hypoxia also reduced cytotrophoblast staining for human placental lactogen, another antigen that is expressed once the cells differentiate. However, lowering the O_2 tension did not change cytotrophoblast expression of other stage-specific antigens, such as HLA-G (Figs. 5.5C and 5.5F) and integrins α5β1 and αVβ3 (not shown). These results suggest that hypoxia produces selective deficits in the ability of cytotrophoblasts to differentiate along the invasive pathway.

The effects of oxygen tension on the proliferative capacity of cytotrophoblasts could help explain some of the interesting features of normal placental development. Before cytotrophoblast invasion of maternal vessels establishes the uteroplacental circulation (≤10 weeks), the conceptus is in a relatively hypoxic environment. During this period, placental mass increases much more rapidly than that of the embryo proper. Histological sections of early-stage pregnant human uteri show bi-laminar embryos surrounded

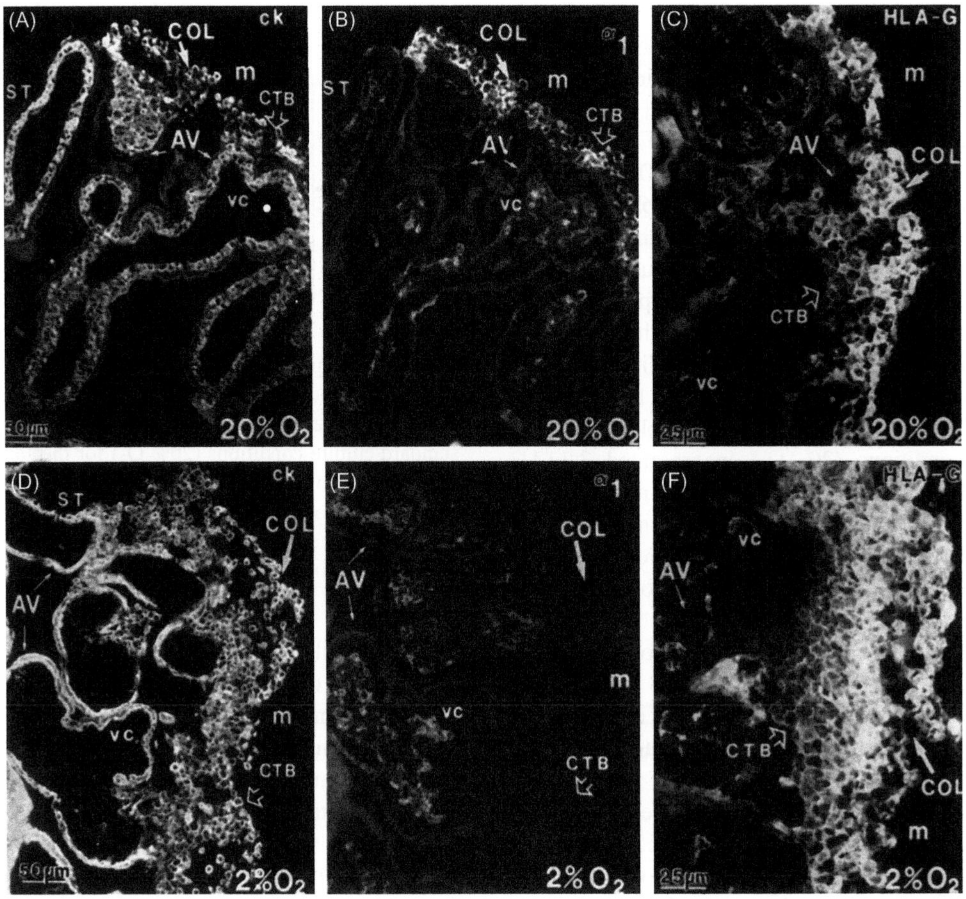

FIGURE 5.5 Some aspects of cytotrophoblast differentiation/invasion are arrested in hypoxia. Anchoring villi (AV) from 6–8-week placentas were cultured on Matrigel (m) for 72 hours in either 20% O_2 (A–C) or 2% O_2 (D–F). Tissue sections of the villi were stained with anti-cytokeratin (ck; A, D), anti-integrin α1 (B, E) or anti-HLA-G (C, F). Cytotrophoblasts (CTB) that composed the cell columns (COL) of villus explants that were cultured in 20% O_2 upregulated both integrin α1 (B) and HLA-G expression (C). In contrast, cytotrophoblasts in anchoring villus columns maintained in 2% O_2 failed to express integrin α1, although constituents of the villus core continued to express this adhesion molecule (E). But not all aspects of differentiation were impaired; the cells upregulated HLA-G expression normally (F). ST, syncytiotrophoblast; vc, villus core. (Reprinted with permission.[36] Copyright 1997 American Association for the Advancement of Science.)

by thousands of trophoblast cells.[46] The fact that hypoxia stimulates cytotrophoblasts, but not most other cells,[47] to undergo mitosis could help account for the discrepancy in size between the embryo and the placenta, which continues well into the second trimester of pregnancy.[48] Although this phenomenon is poorly understood at a mechanistic level, we have recently shown (see below) that cytotrophoblasts within the uterine wall mimic a vascular adhesion molecule phenotype.[1] In other tissues hypoxia induces vascular endothelial growth factor production, which stimulates endothelial cell proliferation.[49] This raises the possibility that similar regulatory pathways operate during placental development.

The effects of oxygen tension on cytotrophoblast differentiation/invasion could also have important implications. Relatively high oxygen tension promotes cytotrophoblast differentiation and could help explain why these cells extensively invade the arterial rather than the venous side of the uterine circulation. Conversely, if cytotrophoblasts do not gain access to an adequate supply of maternal arterial blood, their ability to differentiate into fully invasive cells may be impaired. We suggest that the latter scenario could be a contributing factor to pregnancy-associated diseases, such as preeclampsia, that are associated with abnormally shallow cytotrophoblast invasion and faulty differentiation, as evidenced by their inability to upregulate integrin α1 expression.[7] These results also prompted us to consider the possibility that the profound effects of oxygen on invasive cytotrophoblasts might be indicative of their ability to assume a vascular-like phenotype.

DURING NORMAL PREGNANCY, INVASIVE CYTOTROPHOBLASTS MODULATE THEIR ADHESION MOLECULE REPERTOIRE TO MIMIC THAT OF VASCULAR CELLS

We tested the hypotheses that invasive cytotrophoblasts mimic broadly the adhesion phenotype of the endothelial cells they replace, and that these changes in adhesion phenotype have the net effect of enhancing cytotrophoblast motility and invasiveness.[1] To test these hypotheses, we first stained tissue sections of the fetal–maternal interface for specific integrins, cadherins, and immunoglobulin family adhesion receptors that are characteristic of endothelial cells and leukocytes. Subsequent experiments tested the functional consequences for cytotrophoblast adhesion and invasion of expressing the particular adhesion receptors that were upregulated during cytotrophoblast differentiation.

First, we examined the distribution patterns of αV integrin family members. These molecules are of particular interest because of their regulated expression on endothelial cells during angiogenesis and their upregulation on some types of metastatic tumor cells.[50,51] αV family members displayed unique and highly specific spatial staining patterns on cytotrophoblasts in anchoring villi and the placental bed. An antibody specific for the αVβ5 complex stained the cytotrophoblast monolayer in chorionic villi. Staining was uniform over the entire cell surface. The syncytiotrophoblast layer, and cytotrophoblasts in cell columns and the placental bed, did not stain for αVβ5. In contrast, anti-αVβ6 stained only those chorionic villus cytotrophoblasts that were at sites of column formation. The cytotrophoblast layer still in contact with basement membrane stained brightly, while the first layer of the cell column showed reduced staining. The rest of the cytotrophoblasts in chorionic villi, cytotrophoblasts in more distal regions of cell columns, and cytotrophoblasts within placental bed and vasculature did not stain for αVβ6, documenting a specific association of this integrin with initiation of column formation. In yet a different pattern, staining for anti-αVβ3 was weak or not detected on villus cytotrophoblasts or on cytotrophoblasts in the initial layers of cell columns. However, strong staining was detected on cytotrophoblasts within the uterine wall and vasculature. Thus, individual members of the αV family, like those of the β1 family,[38] are spatially regulated during cytotrophoblast differentiation. Of particular relevance is the observation that αVβ3 integrin, whose expression on endothelial cells is stimulated by angiogenic factors, is prominent on cytotrophoblasts that have invaded the uterine wall and maternal vasculature.

Since blocking αVβ3 function suppresses endothelial migration during angiogenesis, we determined whether perturbing its interactions also affects cytotrophoblast invasion *in vitro*. Freshly isolated first-trimester cytotrophoblasts were plated for 48 hours on Matrigel-coated Transwell filters in the presence of control mouse IgG or the complex-specific anti-αVβ3 IgG, LM609. Cytotrophoblast invasion was evaluated by counting cells and cellular processes that had invaded the Matrigel barrier and extended through the holes in the Transwell filters. LM609 reduced cytotrophoblast invasion by more than 75% in this assay, indicating that this receptor, like the α1β1 integrin,[45] contributes significantly to the invasive phenotype of cytotrophoblasts.

Next, we examined cadherin switching during cytotrophoblast differentiation *in vivo* (Fig. 5.6). The cytotrophoblast epithelial monolayer stained strongly for the ubiquitous epithelial cadherin, E-cadherin, in a polarized pattern (Fig. 5.6A). Staining was strong on the surfaces of cytotrophoblasts in contact with one another and with the overlying syncytiotrophoblast layer, and was absent at the basal surface of cytotrophoblasts in contact with basement membrane. In cell columns, E-cadherin staining intensity was reduced on cytotrophoblasts near the uterine wall and on cytotrophoblasts within the decidua. This reduction in staining was particularly pronounced in second-trimester tissue. At this stage, E-cadherin staining was also very weak or undetectable on cytotrophoblasts that had colonized maternal blood vessels and on cytotrophoblasts in the surrounding myometrium. All locations of reduced E-cadherin staining were areas in which invasion is active during the first half of gestation. Interestingly, the staining intensity of E-cadherin was strong on cytotrophoblasts in all locations in term placentas, at which time cytotrophoblast invasive activity is poor. Taken together, these data are consistent with the idea that cytotrophoblasts transiently reduce E-cadherin function at times and places of their greatest invasive activity.

Cadherin switching occurs frequently during embryonic development when significant morphogenetic events take place. We therefore stained sections of first- and second-trimester placental tissue with antibodies to other classical cadherins. These tissues did not react with antibodies against P-cadherin, but did stain with three different monoclonal antibodies that recognize the endothelial cadherin, VE-cadherin (Fig. 5.6B). In chorionic villi, antibody to VE cadherin did not stain villus cytotrophoblasts, although it stained the endothelium of fetal blood vessels within the villus stroma. In contrast, anti-VE-cadherin stained cytotrophoblasts in cell columns and in the decidua, the very areas in which E-cadherin staining was reduced. VE-cadherin staining was stronger in these areas in second-trimester tissues. In maternal vessels that had not yet been modified by cytotrophoblasts, anti-VE-cadherin stained the endothelial layer strongly. Following endovascular invasion, cytotrophoblasts lining maternal blood vessels also stained strongly for VE-cadherin (Fig. 5.6D). Thus, cytotrophoblasts that invade the uterine wall and vasculature express a cadherin characteristic of endothelial cells.[1]

Next, we used function-perturbing anti-cadherin antibodies, in conjunction with the Matrigel invasion assay, to assess the functional consequences of cadherin modulation for cytotrophoblast invasiveness. We plated isolated second-trimester cytotrophoblasts for 48 hours on Matrigel-coated filters in the presence of control IgG or function-perturbing antibodies against VE-cadherin or E-cadherin. By 48 hours, significant invasion was evident in control cytotrophoblasts. In cultures treated with anti-E-cadherin, cytotrophoblast invasiveness increased more than 3-fold, suggesting that

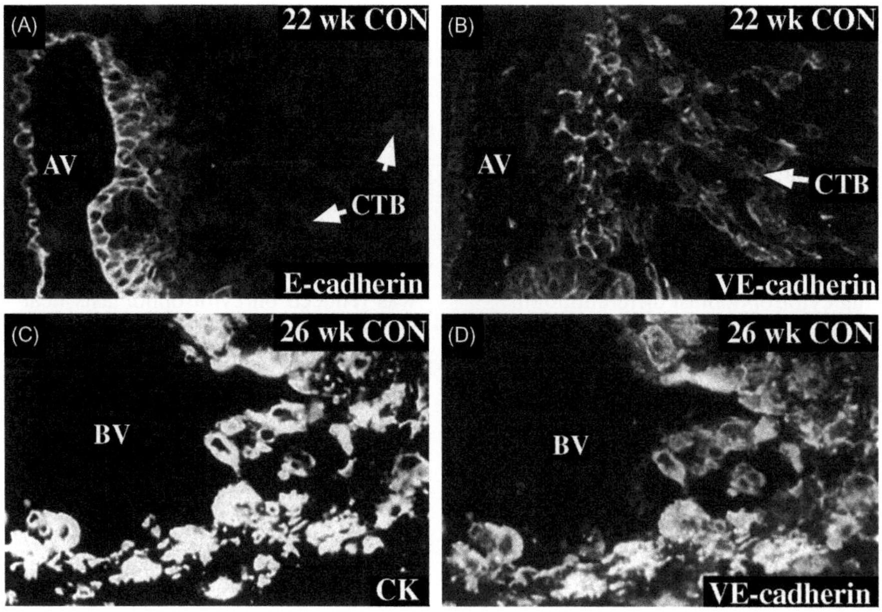

FIGURE 5.6 E-cadherin staining is reduced and VE-cadherin staining is upregulated in normal, differentiating second-trimester CTB. Sections of second-trimester placental bed tissue were stained with antibody against E-cadherin (A), VE-cadherin (B and D) or cytokeratin (CK, 7D3: C). (A) E-cadherin staining was strong on anchoring villus (AV) CTB and on CTB in the proximal portion of cell columns (Zone II). Staining was sharply reduced on CTB in the distal column (Zone III) and in the uterine interstitium (Zone IV). Staining was not detected on CTB within maternal vessels (Zone V). (B) VE-cadherin was not detected on CTB in AV (although fetal blood vessels in villus stromal core are stained). VE-cadherin was detected on column CTB (B) and on interstitial and endovascular CTB (D). VE-cadherin is also detected on maternal endothelium in vessels that have not been modified by CTB (For description of zones, see Fig. 5.1).

E-cadherin normally has a restraining effect on invasiveness. In contrast, antibody against VE-cadherin reduced the invasion of cytotrophoblasts to about 60% of control. This suggests that the presence of VE cadherin normally facilitates cytotrophoblast invasion. Taken together, these functional data suggest that as they differentiate, the cells modulate their cadherin repertoire to one that contributes to their increased invasiveness.

Our data presented thus far indicate that, as they differentiate, cytotrophoblasts downregulate adhesion receptors highly characteristic of epithelial cells (integrin $\alpha6\beta4$[38] and E-cadherin) and upregulate analogous receptors that are expressed on endothelial cells (integrins $\alpha1\beta1$[38] and $\alpha V\beta3$, and VE-cadherin). These observations support our hypothesis that normal cytotrophoblasts undergo a comprehensive switch in phenotype so as to resemble the endothelial cells they replace during endovascular invasion.

We hypothesize that this unusual phenomenon plays an important role in the process whereby these cells form vascular connections with the uterine vessels. Ultimately these connections are so extensive that the spiral arterioles become hybrid structures in which fetal cytotrophoblasts replace the maternal endothelium and much of the highly muscular tunica media. As a result, the diameter of the spiral arterioles increases dramatically, allowing blood flow to the placenta to keep pace with fetal growth. Circumstantial evidence suggests that several of the adhesion molecules whose expression we studied could play an important role in forming these novel vascular connections. In the mouse, for example, targeted disruption of either vascular cell adhesion molecule (VCAM)-1 or $\alpha4$ expression results in failure of chorioallantoic fusion. It is very interesting to find that cytotrophoblasts are the only cells, other than the endothelium, that express VE-cadherin. In addition, VE-cadherin and platelet-endothelial cell adhesion molecule (PECAM)-1 are the first adhesion receptors expressed by differentiating endothelial cells during early development. $\alpha V\beta3$ expression is upregulated on endothelial cells during angiogenesis by soluble factors that regulate this process. Thus, adhesion receptors that are upregulated as normal cytotrophoblasts differentiate/invade play vital roles in differentiation and expansion of the vasculature.

IN PREECLAMPSIA, INVASIVE CYTOTROPHOBLASTS FAIL TO SWITCH THEIR ADHESION MOLECULE REPERTOIRE TO MIMIC THAT OF VASCULAR CELLS

The next hypothesis tested was that preeclampsia impairs the ability of cytotrophoblasts to express the adhesion molecules that are normally modulated during the unique epithelial-to-vascular transformation that occurs in normal pregnancy.[8] First, we compared cytotrophoblast expression

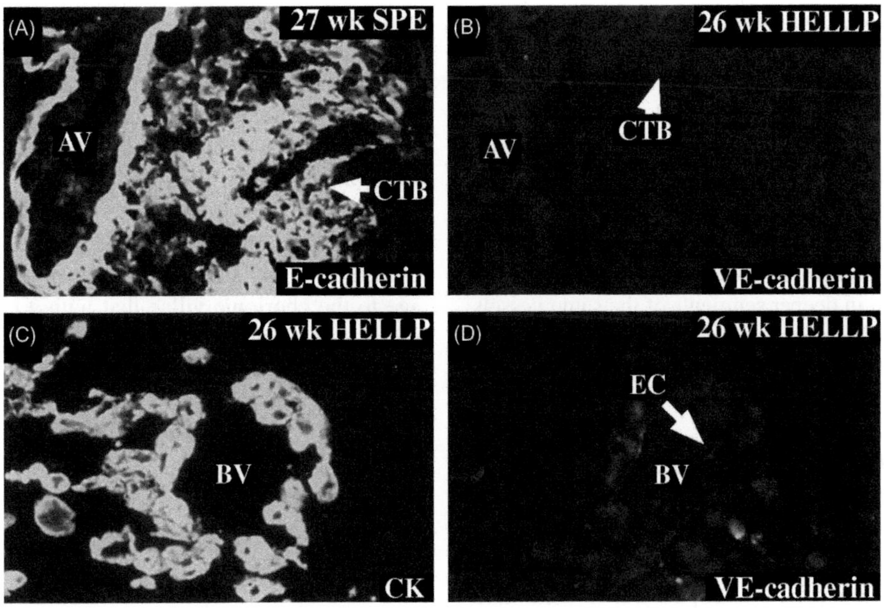

FIGURE 5.7 In preeclampsia, E-cadherin staining is retained on placental bed CTB and VE-cadherin staining is not detected. Sections of 27-week severe preeclamptic (SPE: A) and 26-week HELLP tissue (B, C, D) were stained with antibody against E-cadherin (A), VE-cadherin (B and D) or cytokeratin (CK, 7D3: C). (A) E-cadherin staining was strong on CTB in almost all locations. CTB also appeared to be in large aggregates. VE-cadherin was not detected on CTB in cell columns (B) or near blood vessels (D). But the endothelial cells (EC) that line the vessel did stain (arrows).

of three members of the αV family (αVβ5, αVβ6, and αVβ3) in placental bed biopsies obtained from control and preeclamptic patients that were matched for gestational age. Preeclampsia changed cytotrophoblast expression of all three αV-family members. When samples were matched for gestational age, fewer preeclamptic cytotrophoblast stem cells stained with an antibody that recognized integrin β5. In contrast, staining for β6 was much brighter in preeclamptic tissue and extended beyond the column to include cytotrophoblasts within the superficial decidua. Of greatest interest, staining for β3 was weak on cytotrophoblasts in all locations; cytotrophoblasts in the uterine wall of preeclamptic patients failed to show strong staining for β3, as did cytotrophoblasts that penetrated the spiral arterioles. Thus, in preeclampsia, differentiating/invading cytotrophoblasts retain expression of αVβ6, which is transiently expressed in remodeling epithelium, and fail to upregulate αVβ3, which is characteristic of angiogenic endothelium. Therefore, as was the case for integrin α1,[7] our analyses of the expression of αV-family members suggest that in preeclampsia, cytotrophoblasts start to differentiate along the invasive pathway but cannot complete this process.[8]

Preeclampsia also had a striking effect on cytotrophoblast cadherin expression (Fig. 5.7). In contrast to control samples, cytotrophoblasts in both the villi and decidua showed strong reactivity with anti-E-cadherin (Fig. 5.7A), and staining remained strong even on cytotrophoblasts that had penetrated the superficial portions of uterine arterioles (data not shown). Interestingly, in preeclampsia cytotrophoblasts within the uterine wall tended to exist as large aggregates, rather than as smaller clusters and single cells, as is the case in normal pregnancy. This observation is in accord with the likelihood that E-cadherin mediates strong intercellular adhesion between cytotrophoblasts, as it does in all other normal epithelia examined.

Strikingly, no VE-cadherin staining was detected on cytotrophoblasts in any location in placental bed specimens obtained from preeclamptic patients; neither cytotrophoblasts in the cell columns (Fig. 5.7B) nor the few cells that were found in association with vessels in the superficial decidua expressed VE-cadherin (Fig. 5.7D). However, staining for this adhesion molecule was detected on maternal endothelium in the unmodified uterine vessels in preeclamptic placental bed biopsy specimens. Thus, cadherin modulation by cytotrophoblasts in preeclampsia was defective, as shown by the persistence of strong E-cadherin staining and the absence of VE-cadherin staining on cytotrophoblasts in columns and in the superficial decidua.

The results summarized above raise the interesting possibility that the failure of preeclamptic cytotrophoblasts to express vascular-type adhesion molecules, as normal cytotrophoblasts do, impairs their ability to form connections with the uterine vessels. This failure ultimately limits the supply of maternal blood to the placenta and fetus, an effect thought to be closely linked to the pathophysiology of the disease. We also hypothesize that the failure of preeclamptic cytotrophoblasts to make a transition to a vascular cell adhesion phenotype might be part of a broader-spectrum defect in which the cells fail to function properly as endothelium. Such a failure would no doubt have important effects on the maintenance of vascular integrity at the maternal–fetal interface. Clearly, in preeclampsia undifferentiated cytotrophoblasts that fail to mimic the adhesion phenotype of endothelial cells are present in the termini of maternal spiral arterioles. Whether or not the observed defects are related to their propensity to undergo apoptosis is not yet known[52] (reviewed in[53]). Whether their presence also affects the phenotype of maternal endothelium in deeper segments of the same vessels and/or is linked to the maternal endothelial pathology that is a hallmark of this disease remains to be investigated.

THE PATHOLOGICAL CONSEQUENCES OF ABNORMAL CYTOTROPHOBLAST INVASION AND FAILED SPIRAL ARTERY REMODELING

The major effect of abnormal cytotrophoblast invasion and failed remodeling of the maternal spiral arteries is abnormal perfusion of the intervillus space. Before conception the spiral arteries, the terminal branches of the uterine arteries, are typical small muscular vessels, and are richly innervated.[54] During pregnancy there is modification of much of the uterine artery, the most prominent being that of the spiral branches that perfuse the placenta. with a 4 fold dilatation of the terminal portion of the arteries.[55] However, there is also remodeling of the vascular wall throughout the spiral artery's length, extending as far as the inner third of the myometrium.[56] There is a striking loss of smooth muscle and elastic tissue, rendering the artery unresponsive to neural or humoral signals.[56,57] The depth of the remodeling is also important. In normal pregnancy the loss of muscle extends beyond the decidua into the inner third of the myometrium. In the non-pregnant state the spiral artery at the junction of the uterine mucosa ("endometrium" before conception, decidua in pregnancy) acts as a "functional sphincter" during menses.[58] Constriction of this portion of the vessel serves to terminate bleeding after endometrial shedding. During normal pregnancy the depth of remodeling eliminates this functional sphincter and its constrictor responsiveness.

In preeclampsia the net result of failed remodeling is failure of terminal spiral artery dilatation to occur. The depth of remodeling is also compromised and does not extend beyond the junction of decidua and myometrium, leaving the "functional sphincter" intact, resulting in major consequences. This failure of terminal dilatation is the pathology that has received most attention, proposed, for example, to dramatically reduce intervillus perfusion (according to Poiseuille's law, flow increases with the fourth power of the radius). However, as has been elegantly pointed out by Graham Burton[54] and colleagues this vascular dilatation is essentially confined to the terminal portion of the vessel and thus has minimal effect on perfusion. The major impact actually is upon the velocity of blood flow as it leaves the spiral artery. The increase in cardiac output and redistribution of blood that characterizes normal pregnancy would result in an enormous increase in the velocity of blood exiting the spiral artery (2–3 meters/second). The terminal dilatation of the spiral artery can be predicted to reduce blood flow velocity to 10 centimeters/second. The dramatically increased blood velocity with the failed modeling and consequent non-dilated terminal arteries would be predicted to lead to damage to the chorionic villae that with a hemochorial placenta are in direct contact with maternal blood. The accelerated velocity of blood also reduces time for extraction of blood and nutrients from the intervillus blood.[54]

Another aspect of the failed remodeling should also have important consequences. The maintenance of smooth muscle in the spiral arteries results in vessels, which, unlike the situation in normal pregnancy, remains responsive to external signals. The reduced depth of remodeling of the vessels in preeclampsia that does not extend beyond the decidua could be particularly relevant since this results in the maintenance of the "functional sphincter" at the junction of the decidua and myometrium in the spiral artery.

The consequence of the maintenance of responsiveness in the unremodeled spiral arteries is an increased risk of intermittent reduction of blood flow to the intervillus space. This intermittent reduction would set the stage for a hypoxia reperfusion scenario with subsequent oxidative stress.[59] Oxidative stress occurs when the production of reactive oxygen species exceeds the local capacity for buffering by labile antioxidants or antioxidant enzymes.[60] Once the balance is tipped in favor of oxidative stress there is an explosive feed-forward generation of free radicals that damage proteins, lipids, and DNA.[60,61] In a setting of high oxygen demand (as is present in the placenta) and intermittent flow reduction there is inadequate oxygen available for ATP function and ATP is degraded eventually to uric acid and either NADPH or superoxide. The enzyme that forms uric acid as the terminal step in nucleotide metabolism, xanthine oxidase/dehydrogenase, is a bifunctional enzyme forming either NADPH or the reactive oxygen species, superoxide and H_2O_2.[62] With hypoxia, the preferential end products are superoxide and H_2O_2.[63] Further free radicals are generated by NADPH oxidase that is activated by inflammatory activators either from injury secondary to superoxide and H_2O_2 or as a response to the augmented inflammation characteristic of preeclampsia.[63] The superoxide generated also affects the function of nitric oxide synthase, the enzyme responsible for the formation of nitric oxide. The enzyme becomes uncoupled and no longer makes NO (which is an antioxidant) but rather more superoxide.[64] To compound the problem reactive oxygen species downregulate one of the major antioxidant enzymes, superoxide dismutase.[65] These concepts also are reviewed in Chapter 9.

Closely linked with oxidative stress is endoplasmic reticulum stress.[66] Reduced oxygen delivery can result in both oxidative and endoplasmic reticulum stress.[61] Inflammatory mediators can induce both and both can produce these signaling molecules. Endoplasmic reticulum stress is a cellular mechanism to reduce protein synthesis in settings in which nutrient and oxygen delivery is not sufficient to fully process proteins. This results in a characteristic response, the "unfolded protein response" (UPR). The UPR turns off protein synthesis and could account for the small placenta associated with growth restriction.[67] Apoptotic cell death results with profound stress.[68] UPR also induces the formation of reactive oxygen species. Oxidative stress and endoplasmic reticulum stress not only affect local placental function but also participate in signal generation leading to the systemic features of preeclampsia. Free radicals modify lipid structure and the altered structure and apoptosis increase the shedding of trophoblast fragments. The shedding is augmented by the increased velocity of intervillus blood flow through the unremodeled spiral arteries. Redman and Sargent have championed the idea that such fragments have the capacity to activate immune cells[69] and perhaps directly injure endothelial cells (Chapter 8).[70] Because of the oxidized lipids in these particles they have the potential to transfer oxidative stress systemically. In addition, inflammatory cells passing through the intervillus space, which is replete with free radicals, could also be activated. Cytokines produced in the placenta also enter the systemic circulation. Furthermore, with tissue damage xanthine oxidase/dehydrogenase can be released into the circulation where it can target endothelial cells.[71]

NOVEL UNBIASED APPROACHES FOR ADDRESSING THE COMPLEXITIES OF THE PREECLAMPSIA SYNDROME

An important outstanding question has been the extent to which the defects in cytotrophoblast differentiation/invasion that are observed in preeclampsia, particularly the severe forms that occur early in gestation, are a unique feature of this pregnancy complication rather than a default pathway that is activated in response to other pathological conditions. This point has been particularly difficult to address because tissue samples of the maternal–fetal interface from uncomplicated pregnancies are not available after 24 weeks of gestation, when elective terminations are no longer permitted in many countries. Thus, it is impossible to obtain true control samples for studies of the severe forms of PE, which usually occur in the 24- to 32-week interval. The importance of this problem was recently emphasized by the results of a microarray study in which we compared changes in gene expression at the maternal–fetal interface over five time intervals between 14 to 24 and 37 to 40 weeks of gestation. The results showed surprisingly few differences before 24 weeks and hundreds of changes by term, evidence that gestational age is an important variable during this time period.[72]

Accordingly, we tested the hypothesis that the constellation of morphological and molecular defects that are associated with preeclampsia are unique to this condition.[33] Specifically, we compared the histology of the maternal–fetal interface and cytotrophoblast expression of stage-specific antigens in preeclampsia and in preterm labor, with or without inflammation. In the absence of inflammation, biopsies obtained after preterm labor were near normal at histological and molecular levels. In accord with previously published data, preeclampsia had severe negative effects on the endpoints analyzed; biopsies obtained after preterm labor with inflammation had an intermediate phenotype. Thus, our results suggest that the maternal–fetal interface from cases of preterm labor without inflammation can be used for comparative purposes, e.g., as age-matched controls, in studies of the effects of preeclampsia on cells in this region.

As a further proof of principle, we conducted a global analysis of gene expression at the maternal–fetal interface in preeclampsia ($n = 12$; 24–36 weeks) vs. samples from women who delivered due to preterm labor with no evidence of infection (PTL; $n = 11$; 24–36 weeks).[72] Using the HG-U133A&B Affymetrix GeneChip platform and statistical significance set at log odds-ratio of $B > 0$, 55 genes were differentially expressed in preeclampsia. They encoded proteins previously associated with preeclampsia (such as VEGFR-1, leptin, CRH, and inhibin) and novel molecules, e.g., sialic acid binding immunoglobulin-like lectin 6 (Siglec-6), a potential leptin receptor, and pappalysin 2 (PAPP-A2), a proteinase that cleaves insulin-like growth factor binding proteins. We used quantitative-PCR to validate the expression patterns of a subset of the up- or downregulated genes. At the protein level, we confirmed preeclampsia-related changes in the expression of Siglec-6 and PAPP-A2, which localized to invasive cytotrophoblasts and syncytiotrophoblasts. Notably, Siglec-6 placental expression is uniquely human, as is spontaneous preeclampsia. The functional significance of these observations may provide new insights into the pathogenesis of preeclampsia, and assaying the circulating levels of these proteins could have clinical utility for predicting and/or diagnosing this syndrome. These new molecular concepts are developed further in the Appendix on trophoblast gene expression that follows this chapter.

SUMMARY AND FUTURE DIRECTIONS

We now understand a great deal about cytotrophoblast defects in the placentas of patients whose pregnancies are complicated by preeclampsia. In a landmark study published over 35 years ago,[9] Brosens, Robertson, and Dixon first described the abnormally shallow cytotrophoblast invasion that is observed in preeclampsia and a substantial proportion of pregnancies complicated by intrauterine growth retardation. These investigators considered the lack of invasion of the spiral arterioles to be particularly significant. Building on this foundation, our recent studies have shown that cytotrophoblast invasion of the uterus is actually a unique differentiation pathway in which the fetal cells adopt certain attributes of the maternal endothelium they normally replace. In preeclampsia, this differentiation process goes awry.

Currently, we are very interested in using these findings as a point of departure for studies of the disease process from its inception to the appearance of the maternal signs. With regard to its inception, understanding the nature of the phenotypic alterations that are characteristic of cytotrophoblasts in preeclampsia offers us the exciting opportunity to test hypotheses about the causes. From a reductionist viewpoint, preeclampsia can be considered as a two-component system in which the two parts – the placenta and the mother – fail to connect properly. In theory, this failure could be due to either component. For example, it is inevitable that cytotrophoblast differentiation must sometimes go awry. The high frequency of spontaneous abortions that are the results of chromosomal abnormalities is a graphic illustration of the consequences of catastrophic failure of cytotrophoblast differentiation. But the observation that confined placental mosaicism can be associated with IUGR[73] is especially relevant to the studies described in this chapter. Conversely, there is interesting evidence that in certain cases the maternal environment may not permit normal trophoblast invasion. For example, patients with preexisting medical conditions, such as lupus erythematosus and diabetes mellitus, or with increased maternal weight,[74] are prone to developing pregnancy complications, including preeclampsia.[75] Finally, the mother's genotype may also play a role; expression of an angiotensinogen genetic variant has been associated with the predisposition to develop preeclampsia.[76]

An equally interesting area of study is how the faulty link between the placenta and the uterus leads to the fetal and maternal signs of the disease. It is logical that a reduction in maternal blood flow to the placenta could result in fetal IUGR, a fact that has been confirmed in several animal models.[77] But how this scenario also leads to the maternal signs is much less clear. Since the latter signs rapidly resolve once the placenta is removed, most investigators believe that this organ is the source of factors that drive the maternal disease process. Another important consideration is that the local placental abnormalities eventually translate into maternal systemic defects. Thus, it is likely that the causative agents are probably widely distributed in the maternal circulation. But their identities are not yet known. Candidates include macromolecular entities such as fragments of the syncytiotrophoblast microvillous membrane that are shed from the surface of floating villi and can damage endothelial cells.[21] Molecular candidates include the products of hypoxic trophoblasts, whose *in vitro* vascular effects mimic *in vivo* blood vessel alteration in patients with preeclampsia[78] and vasculogenic molecules of placental origin (see Chapters 6 and 8).

In the end, the utility of the observations discussed in this chapter will rest in our ability to use this newfound knowledge to make improvements in the clinical care offered to pregnant women. In this regard, the time is right to mount a systematic attack for discovering potential biomarkers of preeclampsia that circulate in maternal blood, before the signs appear and/or at the time of diagnosis. We think that all the requisite elements for these studies are in place. On the biology side, we now have a good understanding of the underlying defects in placentation that are thought to eventually lead to the full-blown manifestations of this condition. On the technology side, many new platforms for biomarker discovery are being developed, including those that employ powerful mass spectrometry-based approaches.[79] Although current efforts are focusing on prediction and/or diagnosis, these studies might also reveal therapeutic targets. Thus, we are entering an exciting period when years of basic science will move from a research setting to clinical laboratories, a prospect that has the potential to transform obstetrical care from a 20th-century enterprise to a 21st-century venture.

References

1. Zhou Y, Fisher SJ, Janatpour M, et al. Human cytotrophoblasts adopt a vascular phenotype as they differentiate. A strategy for successful endovascular invasion? *J Clin Invest.* 1997;99:2139–2151.
2. Cross JC, Werb Z, Fisher SJ. Implantation and the placenta: key pieces of the development puzzle. *Science.* 1994;266:1508–1518.
3. Roberts JM, Taylor RN, Friedman SA, Goldfien A. New developments in pre-eclampsia. *Fetal Med Rev.* 1990;2:125–141.
4. Roberts JM, Taylor RN, Musci TJ, Rodgers GM, Hubel CA, McLaughlin MK. Preeclampsia: an endothelial cell disorder. *Am J Obstet Gynecol.* 1989;161:1200–1204.
5. Redman CW. Current topic: pre-eclampsia and the placenta. *Placenta.* 1991;12:301–308.
6. Robertson WB, Brosens I, Dixon HG. The pathological response of the vessels of the placental bed to hypertensive pregnancy. *J Pathol Bacteriol.* 1967;93:581–592.
7. Zhou Y, Damsky CH, Chiu K, Roberts JM, Fisher SJ. Preeclampsia is associated with abnormal expression of adhesion molecules by invasive cytotrophoblasts. *J Clin Invest.* 1993;91:950–960.
8. Zhou Y, Damsky CH, Fisher SJ. Preeclampsia is associated with failure of human cytotrophoblasts to mimic a vascular adhesion phenotype. One cause of defective endovascular invasion in this syndrome? *J Clin Invest.* 1997;99:2152–2164.
9. Brosens IA, Robertson WB, Dixon HG. The role of the spiral arteries in the pathogenesis of preeclampsia. *Obstet Gynecol Annu.* 1972;1:177–191.
10. Gerretsen G, Huisjes HJ, Elema JD. Morphological changes of the spiral arteries in the placental bed in relation to preeclampsia and fetal growth retardation. *Br J Obstet Gynaecol.* 1981;88:876–881.
11. Moodley J, Ramsaroop R. Placental bed morphology in black women with eclampsia. *S Afr Med J.* 1989;75:376–378.

12. Khong TY, De Wolf F, Robertson WB, Brosens I. Inadequate maternal vascular response to placentation in pregnancies complicated by pre-eclampsia and by small-for-gestational age infants. *Br J Obstet Gynaecol*. 1986;93:1049–1059.
13. Lain KY, Roberts JM. Contemporary concepts of the pathogenesis and management of preeclampsia. *JAMA*. 2002;287:3183–3186.
14. Lam C, Lim KH, Karumanchi SA. Circulating angiogenic factors in the pathogenesis and prediction of preeclampsia. *Hypertension*. 2005;46:1077–1085.
15. Redman CW, Sargent IL. Latest advances in understanding preeclampsia. *Science*. 2005;308:1592–1594.
16. Roberts JM, Cooper DW. Pathogenesis and genetics of pre-eclampsia. *Lancet*. 2001;357:53–56.
17. Gomez O, Martinez JM, Figueras F. Uterine artery Doppler at 11–14 weeks of gestation to screen for hypertensive disorders and associated complications in an unselected population. *Ultrasound Obstet Gynecol*. 2005;26:490–494.
18. Hershkovitz R, de Swiet M, Kingdom J. Mid-trimester placentation assessment in high-risk pregnancies using maternal serum screening and uterine artery Doppler. *Hypertens Pregnancy*. 2005;24:273–280.
19. Roberts JM. Endothelial dysfunction in preeclampsia. *Semin Reprod Endocrinol*. 1998;16:5–15.
20. Rodgers GM, Taylor RN, Roberts JM. Preeclampsia is associated with a serum factor cytotoxic to human endothelium. *Am J Obstet Gynecol*. 1988;159:908–914.
21. Cockell AP, Learmont JG, Smarason AK, Redman CW, Sargent IL, Poston L. Human placental syncytiotrophoblast microvillous membranes impair maternal vascular endothelial function. *Br J Obstet Gynaecol*. 1997;104:235–240.
22. McCarthy AL, Woolfson RG, Raju SK, Poston L. Abnormal endothelial cell function of resistance arteries from women with preeclampsia. *Am J Obstet Gynecol*. 1993;168:1323–1330.
23. Savvidou MD, Hingorani AD, Tsikas D, Frolich JC, Vallance P, Nicolaides KH. Endothelial dysfunction and raised plasma concentrations of asymmetric dimethylarginine in pregnant women who subsequently develop pre-eclampsia. *Lancet*. 2003;361:1511–1517.
24. Gant NF, Daley GL, Chand S, Whalley PJ, MacDonald PC. A study of angiotensin II pressor response throughout primigravid pregnancy. *J Clin Invest*. 1973;52:2682–2689.
25. Friedman SA, Schiff E, Emeis JJ, Dekker GA, Sibai BM. Biochemical corroboration of endothelial involvement in severe preeclampsia. *Am J Obstet Gynecol*. 1995;172:202–203.
26. Hsu CD, Iriye B, Johnson TR, Witter FR, Hong SF, Chan DW. Elevated circulating thrombomodulin in severe preeclampsia. *Am J Obstet Gynecol*. 1993;169:148–149.
27. Taylor RN, Crombleholme WR, Friedman SA, Jones LA, Casal DC, Roberts JM. High plasma cellular fibronectin levels correlate with biochemical and clinical features of preeclampsia but cannot be attributed to hypertension alone. *Am J Obstet Gynecol*. 1991;165:895–901.
28. Tyurin VA, Liu SX, Tyurina YY. Elevated levels of S-nitrosoalbumin in preeclampsia plasma. *Circ Res*. 2001;88:1210–1215.
29. Clark BA, Halvorson L, Sachs B, Epstein FH. Plasma endothelin levels in preeclampsia: elevation and correlation with uric acid levels and renal impairment. *Am J Obstet Gynecol*. 1992;166:962–968.
30. Mills JL, DerSimonian R, Raymond E. Prostacyclin and thromboxane changes predating clinical onset of preeclampsia: a multicenter prospective study. *JAMA*. 1999;282: 356–362.
31. Waite LL, Louie RE, Taylor RN. Circulating activators of peroxisome proliferator-activated receptors are reduced in preeclamptic pregnancy. *J Clin Endocrinol Metab*. 2005;90:620–626.
32. Kanasaki K, Palmsten K, Sugimoto H. Deficiency in catechol-O-methyltransferase and 2-methoxyoestradiol is associated with pre-eclampsia. *Nature*. 2008;453:1117–1121.
33. Zhou CC, Zhang Y, Irani RA. Angiotensin receptor agonistic autoantibodies induce pre-eclampsia in pregnant mice. *Nat Med*. 2008;14:855–862.
34. Rodesch F, Simon P, Donner C, Jauniaux E. Oxygen measurements in endometrial and trophoblastic tissues during early pregnancy. *Obstet Gynecol*. 1992;80:283–285.
35. Genbacev O, Joslin R, Damsky CH, Polliotti BM, Fisher SJ. Hypoxia alters early gestation human cytotrophoblast differentiation/invasion in vitro and models the placental defects that occur in preeclampsia. *J Clin Invest*. 1996;97:540–550.
36. Genbacev O, Zhou Y, Ludlow JW, Fisher SJ. Regulation of human placental development by oxygen tension. *Science*. 1997;277:1669–1672.
37. Schwarting R. Little missed markers and Ki-67. *Lab Invest*. 1993;68:597–599.
38. Damsky CH, Fitzgerald ML, Fisher SJ. Distribution patterns of extracellular matrix components and adhesion receptors are intricately modulated during first trimester cytotrophoblast differentiation along the invasive pathway, in vivo. *J Clin Invest*. 1992;89:210–222.
39. Librach CL, Werb Z, Fitzgerald ML. 92-kD type IV collagenase mediates invasion of human cytotrophoblasts. *J Cell Biol*. 1991;113:437–449.
40. McMaster MT, Librach CL, Zhou Y. Human placental HLA-G expression is restricted to differentiated cytotrophoblasts. *J Immunol*. 1995;154:3771–3778.
41. McMaster M, Zhou Y, Shorter S, et al. HLA-G isoforms produced by placental cytotrophoblasts and found in amniotic fluid are due to unusual glycosylation. *J Immunol*. 1998;160:5922–5928.
42. Kurman RJ, Young RH, Norris HJ, Main CS, Lawrence WD, Scully RE. Immunocytochemical localization of placental lactogen and chorionic gonadotropin in the normal placenta and trophoblastic tumors, with emphasis on intermediate trophoblast and the placental site trophoblastic tumor. *Int J Gynecol Pathol*. 1984;3:101–121.
43. King RW, Jackson PK, Kirschner MW. Mitosis in transition. *Cell*. 1994;79:563–571.
44. Gartel AL, Serfas MS, Tyner AL. p21 – negative regulator of the cell cycle. *Proc Soc Exp Biol Med*. 1996;213:138–149.
45. Damsky CH, Librach C, Lim KH. Integrin switching regulates normal trophoblast invasion. *Development*. 1994;120:3657–3666.
46. Hertig AT. On the eleven-day pre-villous human ovum with special reference to the variations in its implantation site. *Anat Rec*. 1942;82:420.

47. Graeber TG, Osmanian C, Jacks T, et al. Hypoxia-mediated selection of cells with diminished apoptotic potential in solid tumours. *Nature*. 1996;379:88–91.
48. Boyd JD, Hamilton WJ. Development and structure of the human placenta from the end of the 3rd month of gestation. *J Obstet Gynaecol Br Commonw*. 1967;74:161–226.
49. Stone J, Itin A, Alon T, et al. Development of retinal vasculature is mediated by hypoxia-induced vascular endothelial growth factor (VEGF) expression by neuroglia. *J Neurosci*. 1995;15:4738–4747.
50. Hayashi K, Madri JA, Yurchenco PD. Endothelial cells interact with the core protein of basement membrane perlecan through beta 1 and beta 3 integrins: an adhesion modulated by glycosaminoglycan. *J Cell Biol*. 1992;119:945–959.
51. Hynes RO, George EL, Georges EN, Guan JL, Rayburn H, Yang JT. Toward a genetic analysis of cell-matrix adhesion. *Cold Spring Harb Symp Quant Biol*. 1992;57:249–258.
52. DiFederico E, Genbacev O, Fisher SJ. Preeclampsia is associated with widespread apoptosis of placental cytotrophoblasts within the uterine wall. *Am J Pathol*. 1999;155:293–301.
53. Heazell AE, Crocker IP. Live and let die—regulation of villous trophoblast apoptosis in normal and abnormal pregnancies. *Placenta*. 2008;29:772–783.
54. Burton GJ, Woods AW, Jauniaux E, Kingdom JCP. Rheological and physiological consequences of conversion of the maternal spiral arteries for uteroplacental blood flow during human pregnancy. *Placenta*. 2009;30(6):473–482.
55. Brosens I. A study of the spiral arteries of the decidua basalis in normotensive and hypertensive pregnancies. *J Obstet Gynaecol Br Commonw*. 1964;71:222–230.
56. Pijnenborg R, Vercruysse L, Hanssens A. The uterine spiral arteries in human pregnancy: facts and controversies. *Placenta*. 2006;27(9–10):939–958.
57. Lyall F. Priming and remodelling of human placental bed spiral arteries during pregnancy—a review. *Placenta*. 2005;26:S31–S36.
58. Brosens JJ, Pijnenborg R, Brosens IA. The myometrial junctional zone spiral arteries in normal and abnormal pregnancies: a review of the literature. *Am J Obstet Gynecol*. 2002;187(5):1416–1423.
59. Poston L, Raijmakers MT. Trophoblast oxidative stress, antioxidants and pregnancy outcome--a review. *Placenta*. 2004;25(suppl A):S72–S78.
60. Schulz E, Gori T, Munzel T. Oxidative stress and endothelial dysfunction in hypertension. *Hypertension Res Clin Exp*. 2011;34(6):665–673.
61. Burton GJ, Jauniaux E. Oxidative stress. *Best Pract Res Cl Ob*. 2011;25(3):287–299.
62. Kelley EE, Khoo NKH, Hundley NJ, Malik UZ, Freeman BA, Tarpey MM. Hydrogen peroxide is the major oxidant product of xanthine oxidase. *Free Radic Biol Med*. 2010;48(4):493–498.
63. Al Ghouleh I, Khoo NKH, Knaus UG, et al. Oxidases and peroxidases in cardiovascular and lung disease: new concepts in reactive oxygen species signaling. *Free Radic Biol Med*. 2011;51(7):1271–1288.
64. Cai H, Harrison DG. Endothelial dysfunction in cardiovascular diseases: the role of oxidant stress. *Circ Res*. 2000;87(10):840–844.
65. Jackson RM, Parish G, Ho YS. Effects of hypoxia on expression of superoxide dismutases in cultured ATII cells and lung fibroblasts. *Am J Physiol*. 1996;271(6 Pt 1):L955–L962.
66. HUBEL CA, Oxidative stress and preeclampsia. *Fetal and Maternal Medicine Review*. 1997;9:73–101.
67. Burton GJ, Yung HW, Cindrova-Davies T, Charnock-Jones DS. Placental endoplasmic reticulum stress and oxidative stress in the pathophysiology of unexplained intrauterine growth restriction and early onset preeclampsia. *Placenta*. 2009;30:S43–S48.
68. Wang S, Kaufman RJ. The impact of the unfolded protein response on human disease. *J Cell Biol*. 2012;197(7):857–867.
69. Redman CWG, Sargent IL. Placental stress and pre-eclampsia: a revised view. *Placenta*. 2009;30:S38–S42.
70. Knight M, Redman CW, Linton EA, Sargent IL. Shedding of syncytiotrophoblast microvilli into the maternal circulation in pre-eclamptic pregnancies. *Br J Obstet Gynaecol*. 1998;105(6):632–640.
71. Kelley EE, Hock T, Khoo NK, et al. Moderate hypoxia induces xanthine oxidoreductase activity in arterial endothelial cells [see comment]. *Free Radic Biol Med*. 2006;40(6):952–959.
72. Winn VD, Gormley M, Paquet AC. Severe preeclampsia-related changes in gene expression at the maternal-fetal interface include Siglec-6 and Pappalysin-2. *Endocrinology*. 2009;150(1):452–62.
73. Kalousek DK, Vekemans M. Confined placental mosaicism. *J Med Genet*. 1996;33:529–533.
74. Cnattingius S, Bergstrom R, Lipworth L, Kramer MS. Prepregnancy weight and the risk of adverse pregnancy outcomes. *New Engl J Med*. 1998;338:147–152.
75. Ness RB, Roberts JM. Heterogeneous causes constituting the single syndrome of preeclampsia: a hypothesis and its implications. *Am J Obstet Gynecol*. 1996;175:1365–1370.
76. Ward K, Hata A, Jeunemaitre X. A molecular variant of angiotensinogen associated with preeclampsia. *Nat Genet*. 1993;4:59–61.
77. Combs CA, Katz MA, Kitzmiller JL, Brescia RJ. Experimental preeclampsia produced by chronic constriction of the lower aorta: validation with longitudinal blood pressure measurements in conscious rhesus monkeys. *Am J Obstet Gynecol*. 1993;169:215–223.
78. Gratton RJ, Gandley RE, Genbacev O, McCarthy JF, Fisher SJ, McLaughlin MK. Conditioned medium from hypoxic cytotrophoblasts alters arterial function. *Am J Obstet Gynecol*. 2001;184:984–990.
79. Jaffe JD, Keshishian H, Chang B, Addona TA, Gillette MA, Carr SA. Accurate inclusion mass screening: a bridge from unbiased discovery to targeted assay development for biomarker verification. *Mol Cell Proteomics*. 2008;7:1952–1962.

APPENDIX

Trophoblast Gene Expression in Normal Pregnancy and Preeclampsia

MICHAEL T. MCMASTER, VIRGINIA D. WINN AND SUSAN J. FISHER

INTRODUCTION

In the first part of Chapter 5, immediately preceding this new section on gene expression, we have described how human villous cytotrophoblasts (vCTBs) of the placenta differentiate, establishing the maternal–fetal interface.[1] In floating villi, vCTBs fuse into syncytiotrophoblasts (STBs) with functions that include transport and hormone production (Fig. 5.8A). In anchoring villi, vCTBs acquire tumor-like properties that enable invasion of the decidua and the adjacent third of the myometrium (interstitial invasion). They also breach uterine spiral arterioles, transiently replacing much of the maternal endothelial lining and intercalating within the muscular walls (endovascular invasion).[3] Consequently, high-resistance spiral arterioles are transformed into low-resistance, high-capacitance vessels that divert uterine blood flow to the floating villi. In contrast, CTBs breach and line only the termini of veins. At a molecular level, CTB invasion is accompanied by dramatic phenotypic changes in which these ectodermal derivatives mimic many aspects of the vascular cell surface, e.g., an adhesion molecule repertoire that includes VE-cadherin,[4] Ephs/ephrins that confer an arterial identity[5] and Notch family members that play important roles in vessel functions.[6] In parallel, the cells modulate the expression of a wide range of angiogenic/vasculogenic molecules including VEGF family members.[7]

Given this organ's complexity, the explosive nature by which it develops and its many critical functions, it is not surprising that some of the most clinically significant pregnancy complications are associated with placental anomalies. Preeclampsia (PE), a syndrome that adversely affects the mother (by altering vascular function) and the fetus (by intrauterine growth restriction; IUGR), is a prime example. A two-stage theory of PE pathogenesis has been proposed – abnormal placentation followed by maternal responses that eventually lead to the clinical presentation.[8] In support of this concept, color Doppler ultrasound suggests that, in PE, deficient endovascular invasion precedes the signs.[9] This finding, which is generally associated with the severe forms of this syndrome that occur during the late second trimester/early third trimester period, is consistent with the major placental pathologies, which include shallow uterine invasion (Fig. 5.8B).[10] Anchoring villi, in particular invasive CTBs, are most consistently impacted. Interstitial invasion is often shallow and endovascular invasion is incomplete with fewer spiral arterioles modified in toto.[10–12] At a molecular level, many aspects of the unusual phenotypic switch to a vascular-type cell that normally accompanies CTB invasion fail in PE[6,7,13] with attendant increases in apoptosis.[14]

What is known about the molecular underpinnings of the two stages of PE? A great deal of attention has been focused on the second-stage factors that play a role in the maternal signs. Among other molecules, sFlt,[15] endoglin[16] (discussed in Chapter 6) and adrenomedullin[17] have been implicated. In contrast, the mechanisms that precipitate the first stage – defective CTB differentiation/invasion – are largely unknown. We reasoned that understanding the normal patterns of gene expression at the maternal–fetal interface as well as PE-associated changes would give us important clues about the causes of this syndrome. The studies presented here, as well as the work of other groups who have investigated genes that are dysregulated in placentas from pregnancies complicated by PE,[18–23] have led to a deeper understanding of the causes and pathophysiology of the syndrome.

GESTATION DEPENDENT CHANGES IN GENE EXPRESSION AT THE MATERNAL–FETAL INTERFACE

We have carried out several transcriptional microarray studies to define the transcriptome at the maternal fetal interface and how it is impacted by PE.[2,24,25] In the first set of experiments we collected basal plate biopsy specimens were from 36 placentas (14–40 weeks) from women who had normal pregnancies. RNA was isolated, processed

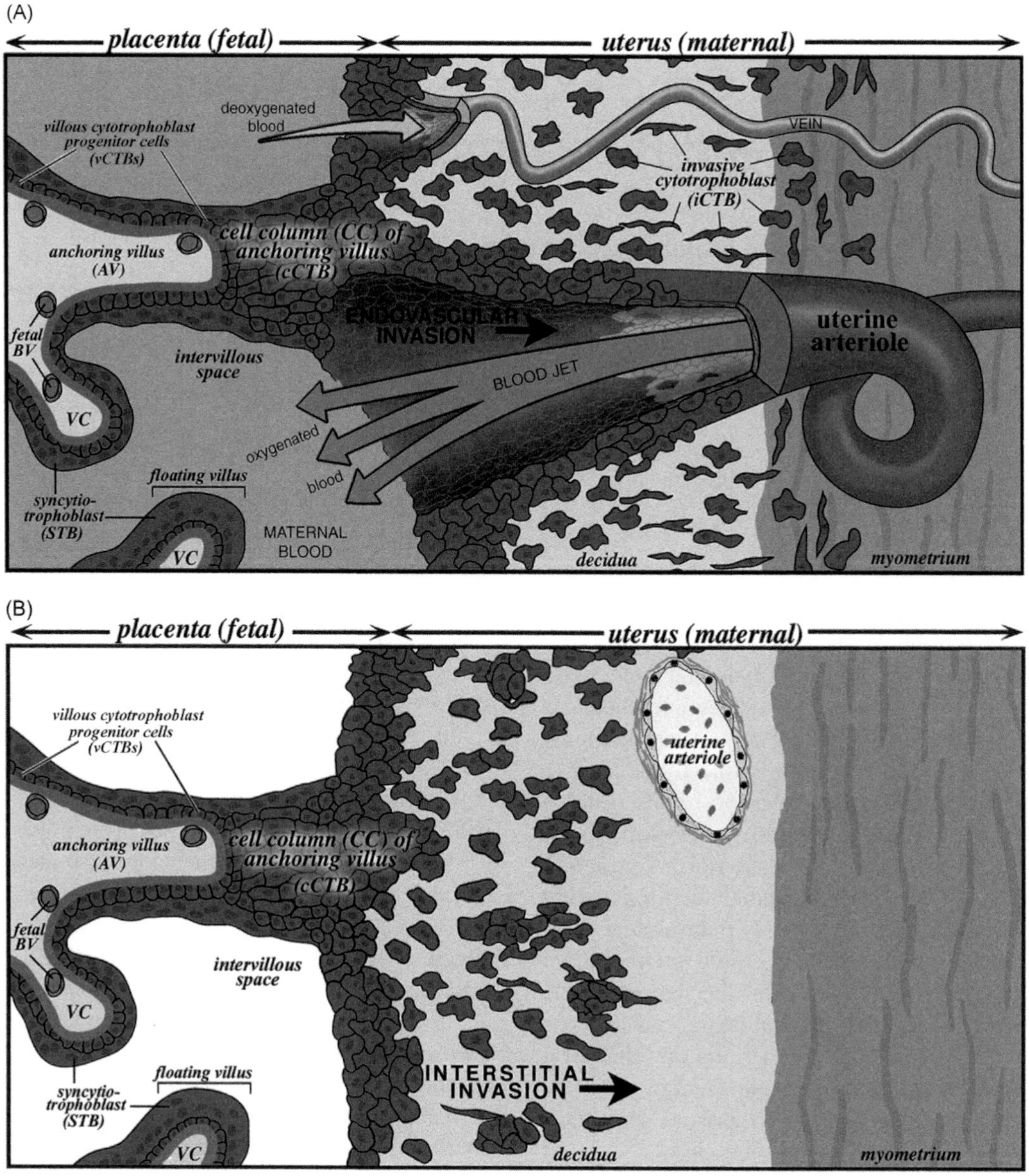

FIGURE 5.8 Diagram of the cellular organization of the human maternal–fetal interface in normal pregnancy and in preeclampsia. (A) Villous cytotrophoblasts (vCTBs) progenitors, the specialized (fetal) epithelial cells of the placenta, differentiate and invade the uterine wall (interstitial invasion; iCTBs), where they also breach maternal blood vessels (endovascular invasion). The basic structural units of the placenta are the chorionic villi, composed of a stromal villous core (VC) with fetal blood vessels, surrounded by a basement membrane and overlain by vCTBs. During differentiation, these cells detach from the basement membrane and adopt one of two fates. They either fuse to form the multinuclear syncytiotrophoblasts (STBs) that cover floating villi or join a column of cytotrophoblasts (cCTBs) at the tips of anchoring villi (AV). The syncytial covering of floating villi mediates the nutrient, gas and waste exchange between fetal and maternal blood. The anchoring villi, through the attachment of cCTBs, establish physical connections between the fetus and the mother. iCTBs penetrate the uterine wall through the first third of the myometrium. A subset of these cells home to uterine spiral arterioles and remodel these vessels by replacing the endothelial lining and intercalating within the muscular walls. To a lesser extent, they also remodel uterine veins. (B) In PE, the interstitial and the endovascular components of CTB invasion are restricted. As a result, interstitial invasion is shallow and many uterine arterioles retain their original structures. Reprinted with permission from *The Journal of Clinical Investigation*.[2]

and hybridized to HG-U133A and HG-U133B Affymetrix GeneChips. Surprisingly, the expression of very few genes was modulated during the 14- to 24-week interval. In contrast, hundreds of genes, including those already known to be regulated over gestation, were modulated between mid-pregnancy (14–24 weeks) and term. These data allowed us to identify molecules that play potentially important roles in the formation of the maternal–fetal interface during the second trimester or in preparation of this area for parturition.

Our analysis revealed a total of 418 genes/expressed sequence tags that were differentially regulated between term and mid-gestation.[24] Based on gene ontogeny (GO) annotations, the differentially expressed genes were involved in a variety of biological processes. At least one-sixth were expressed sequence tags or hypothetical proteins and thus lacked annotations. Of the known differentially expressed genes, 17 were related to lipid metabolism, 10 were involved with formation or regulation of the extracellular matrix (ECM), 21 were immune effectors or modulators, 24 were transcription factors, and 6 had angiogenic/vasculogenic functions.

Ingenuity Pathway Analysis (IPA) software was used to further evaluate the participation of the differentially expressed genes in metabolic and signaling pathways. Analysis of genes with at least a two-fold change highlighted two metabolic pathways – folate biosynthesis and *N*-glycan degradation involving mannose-containing structures. With regard to signaling, eleven of the differentially expressed genes mapped to the Wnt-β-catenin pathway. We also used the IPA software to map networks of the differentially expressed genes. The largest network contained genes that were involved in cell motility, cell-to-cell signaling/interaction and tissue development.

We were also interested to find that genes encoding molecules that are involved in immune defense are highly regulated. For example, defensin alpha 1 was upregulated about three-fold at the RNA level at term as compared with the second trimester. Production of this antimicrobial peptide is constitutive in some cells (e.g., neutrophils) and induced in others (e.g., monocytes and CD8 T lymphocytes) in response to proinflammatory mediators.[26] The presence of defensins in human term placental tissue has been previously reported.[27] Increased expression at term could occur in preparation for labor and placental separation, which increases the risk of infection. In contrast, the expression of another antimicrobial molecule, granulysin, which localizes to the cytolytic granules of T cells, natural killer (NK) cells[28] and certain dendritic cells,[29] is downregulated at term. We speculate that the decreased granulysin expression we observed parallels the decrease in T cell and NK cell numbers at the maternal–fetal interface at term.[30] The downregulation of Ly96 expression, another NK-cell-specific molecule, provides further support for this concept. Although the mechanisms that lead to the eventual disappearance of decidual leukocytes from the maternal–fetal interface are not known, the observed concurrent decrease in expression of chemotactic molecules, such as chemokine-like factor superfamily 6 (CKLFSF6) and secreted phosphoprotein 1 (SPP1), could be a related phenomenon.

A trophoblast-derived noncoding RNA (TncRNA) was one of the most interesting of the highly upregulated differentially expressed genes in the immune function category. This transcript, which directly suppresses MHC class II expression by interacting with the MHC IITA-PIII transactivator, likely accounts for the lack of trophoblast MHC class II expression.[31] As such, this molecule could play an important role in promoting maternal immunotolerance of the hemi-allogeneic fetus. Why TncRNA expression increases at term is unclear, but this phenomenon could be related to the continuing need to suppress MHC class II expression in trophoblasts, particularly as they are shed into maternal blood at the time of delivery. In this regard, it is interesting to note that expression of carcinoembryonic antigen-related cell adhesion molecule 1 (CEACAM1), which plays a role in regulating decidual immune responses, is also upregulated at term.[32]

Additionally, our group has also been interested in the functions of the myriad angiogenic factors that are produced at the maternal–fetal interface.[33,34] In broad terms, we know that these molecules have at least three targets – the intrinsic placental vasculature, the maternal vasculature, and the CTB subpopulation that executes an unusual epithelial-to-endothelial transition as the cells invade the uterine wall and remodel the maternal vasculature in this region. Thus, we anticipated that molecules involved in vasculogenesis/angiogenesis would be upregulated during the active phases of placentation, i.e., in the second trimester rather than at term, which is what we found. Consistent with our previously published work, the downregulated genes included angiopoietin (ANGPT)-2.[35]

As with every microarray analysis, we made a number of interesting observations that warrant additional follow-up. For example, the cluster analysis showed a striking codownregulation at term of ANGPT-2 and microcephalin (MCPH1). Interestingly, ANGPT-2 and MCPH1 genes are transcribed from opposite strands of the same region (chromosome 8p23.1). Their tight coexpression suggests that transcription from this area could be silenced at term, perhaps by local chromatin modifications or the recruitment of inhibitory protein complexes to the same promoter element. It will be interesting to determine whether the pattern of coexpression of ANGPT-2 and MCPH1 occurs in other tissues or is specific to our data set. It is known that MCPH1 controls brain size in humans by regulating the proliferative and, hence, differentiative capacity of neuroblasts, ultimately exerting its effects through cell cycle regulators.[36,37] Furthermore, during human evolution there is evidence that strong genetic selection has been exerted on MCPH1.[38] While the most obvious consequence is brain size, another interesting possibility is that placental form and function have been affected as well.

In summary, we found that gene expression patterns in the basal plate region change dramatically between the second trimester and term. Thus, it is important to control for this variable when studying the effect of pregnancy complications that occur during this timeframe. For our purposes, understanding the normal development and formation of the maternal–fetal interface is an important first step toward understanding PE-related changes.

GENE EXPRESSION AT THE MATERNAL–FETAL INTERFACE IMPACTED BY PREECLAMPSIA

As in our studies of gestation-related change in gene expression at the maternal–fetal interface, we used an unbiased approach to analyze basal plate biopsies from pregnancies afflicted by PE. We focused on preeclampsia that presented in the preterm period (24–36 weeks) as this disease is thought to be more severe and also associated with the greatest morbidity. In this regard, we exploited our observation that preterm labor (PTL) without signs of inflammation is associated with normal CTB differentiation/invasion.[39] Thus, basal plate specimens from these patients served as gestational age-matched controls.

The microarray analysis revealed 55 differentially expressed genes, of which the majority were not previously known to be dysregulated in PE.[25] This list includes molecules that were previously reported to be present at higher than normal levels in maternal serum, chorionic villi and/or cord blood in pregnancies complicated by PE, a finding that gives added confidence to the novel genes that we identified as similarly regulated. However, even for these previously reported molecules, in most cases this was the first description of their increased expression in the basal plate region of the placenta.

For example, our data demonstrated increased leptin expression in the basal plate of PE placentas as compared to control tissue. Numerous investigators have reported a PE-associated increase in circulating levels of leptin,[40–48] and a leptin gene polymorphism has been linked to an increased risk of developing this pregnancy complication.[49] However, a clear picture of how an increase in leptin expression is linked to the pathophysiology of PE has yet to emerge. Interestingly, although the classic leptin receptors were not differentially expressed, we observed elevated levels of the mRNA that encodes Siglec-6, a transmembrane protein that also binds leptin. These findings suggest that this molecule may play an important role as a placental leptin receptor, and that increased Siglec-6 levels could contribute to the pathogenesis of PE. While the cloning strategy for Siglec-6 was based on its ability to interact with leptin, the other Siglec family members bind sialic acid-containing glycans. Siglec-6 has binding specificity for the sialyl-Tn epitope (Siaα 2-6 Gal-NAcα 1-O-R, where R is a serine or threonine). Published data suggest that, in the placenta, leptin is a Siglec-6 ligand, but the endogenous binding partners have yet to be identified.[50] Additionally, Siglec-6 expression has other interesting features. For example, in humans, it is restricted to the placenta and B-lymphocytes. In other species, including non-human primates, placental cells lack Siglec-6 expression, which B cells retain.[51] The fact that Siglec-6 is expressed only in human placentas and not in non-human primate placentas[52] is intriguing, as PE is thought to be a uniquely human disease; spontaneous PE has not been reported in other animals, even non-human primates.[53]

A PE-associated increase in the expression of pappalysin (PAPP-A2) was another novel observation that emerged from our work. PAPP-A2, which has 46% sequence identity with PAPP-A, is a metalloproteinase that cleaves insulin-like growth factor (IGF) binding protein-5 (IGFBP-5).[54] In a fibroblast model, an increase in IGFBP-5 proteolysis attenuates IGF-I stimulatory effects on cell migration.[37] If CTBs respond in an analogous manner, then the observed PE-associated increase in PAPP-A2 levels could inhibit CTB invasion by mechanisms that include an increase in IGFBP-5 proteolysis. These two interesting novel observations, enhanced Siglec-6 and PAPP-A2 expression, were validated at both the RNA and the protein level. In toto, these results suggest fundamental alterations in important biological processes including pathways that are regulated by leptin and IGF signals.

COMBINED ANALYSIS OF GENE EXPRESSION PROFILES AT THE HUMAN MATERNAL–FETAL INTERFACE

To develop a more global view of the maternal–fetal interface we combined the gestational age and preeclampsia microarray datasets. Intensity measures from the Affymetrix U133 A and B chips for each basal plate biopsy were stored as CEL files and probe level normalization performed using robust multichip averaging. The two classes, preeclampsia and "other" (second trimester, PTL and term), were compared using the linear analysis of microarrays (limma; Bioconductor) and Comparative Marker Selection (GenePattern). Genes considered significant had a Bonferonni-corrected $p < 0.01$ and an absolute fold change of ≥ 1.5. This analysis resulted in the identification of 33 genes as increased and 89 genes as decreased in PE. A heat map of the top up- and downregulated genes is presented (Fig. 5.9; complete heat map available in Winn et al., 2011[51]).

When considering this combined analysis there are a few clinical features to keep in mind. First, the second trimester and term samples were from non-labored deliveries while all the PTL and the majority of the PE subjects were from labored deliveries. Interestingly, labor had relatively little impact. Also the clinical outcome for the second-trimester samples was not known. Given the ~5% incidence of PE, it is likely that 1 or 2 samples included in the analysis were from pregnancies that would have been impacted by PE. It is tempting to consider the few samples that clearly have different expression patterns that correspond to PE-specific changes (e.g., leptin, FSTL3, coagulation factor 5 and semaphoring 6D) as being in this category.

Several interesting patterns emerge from this analysis. Some genes were upregulated in PE and term samples as compared to second trimester and PTL. These included

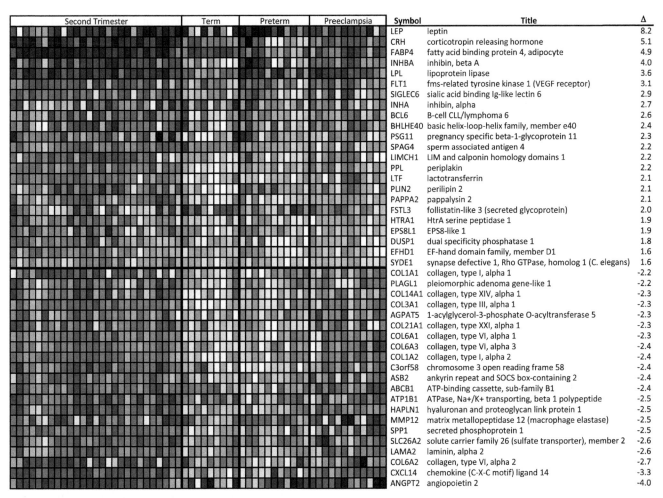

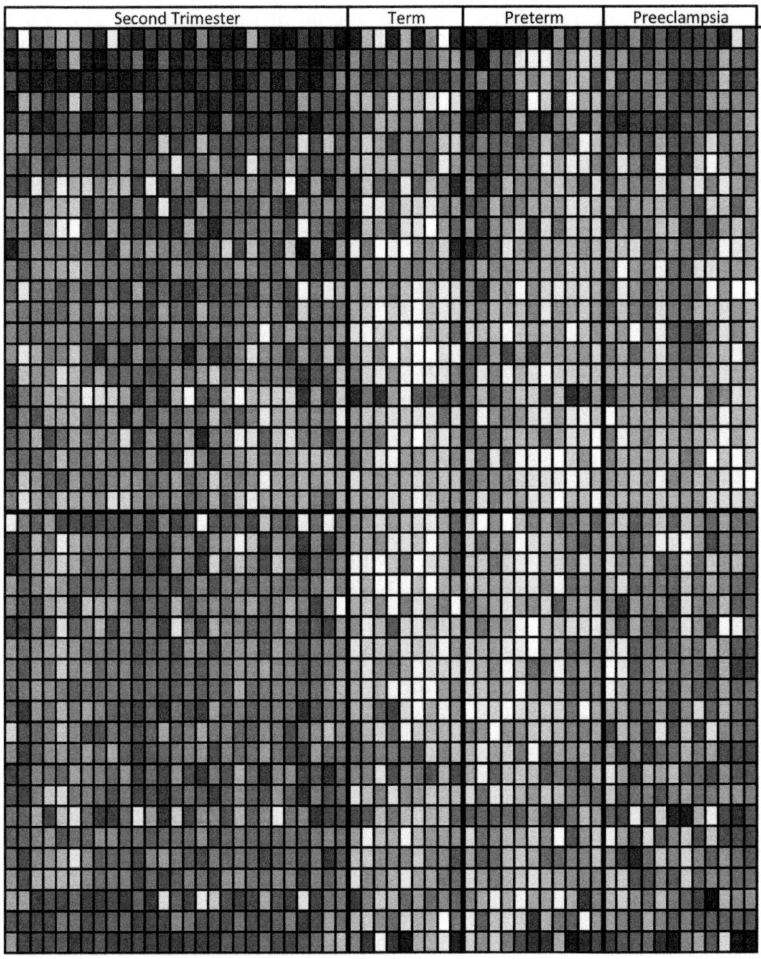

Symbol	Title	Δ
LEP	leptin	8.2
CRH	corticotropin releasing hormone	5.1
FABP4	fatty acid binding protein 4, adipocyte	4.9
INHBA	inhibin, beta A	4.0
LPL	lipoprotein lipase	3.6
FLT1	fms-related tyrosine kinase 1 (VEGF receptor)	3.1
SIGLEC6	sialic acid binding Ig-like lectin 6	2.9
INHA	inhibin, alpha	2.7
BCL6	B-cell CLL/lymphoma 6	2.6
BHLHE40	basic helix-loop-helix family, member e40	2.4
PSG11	pregnancy specific beta-1-glycoprotein 11	2.3
SPAG4	sperm associated antigen 4	2.2
LIMCH1	LIM and calponin homology domains 1	2.2
PPL	periplakin	2.2
LTF	lactotransferrin	2.1
PLIN2	perilipin 2	2.1
PAPPA2	pappalysin 2	2.1
FSTL3	follistatin-like 3 (secreted glycoprotein)	2.0
HTRA1	HtrA serine peptidase 1	1.9
EPS8L1	EPS8-like 1	1.9
DUSP1	dual specificity phosphatase 1	1.8
EFHD1	EF-hand domain family, member D1	1.6
SYDE1	synapse defective 1, Rho GTPase, homolog 1 (C. elegans)	1.6
COL1A1	collagen, type I, alpha 1	-2.2
PLAGL1	pleiomorphic adenoma gene-like 1	-2.2
COL14A1	collagen, type XIV, alpha 1	-2.3
COL3A1	collagen, type III, alpha 1	-2.3
AGPAT5	1-acylglycerol-3-phosphate O-acyltransferase 5	-2.3
COL21A1	collagen, type XXI, alpha 1	-2.3
COL6A1	collagen, type VI, alpha 1	-2.3
COL6A3	collagen, type VI, alpha 3	-2.4
COL1A2	collagen, type I, alpha 2	-2.4
C3orf58	chromosome 3 open reading frame 58	-2.4
ASB2	ankyrin repeat and SOCS box-containing 2	-2.4
ABCB1	ATP-binding cassette, sub-family B1	-2.4
ATP1B1	ATPase, Na+/K+ transporting, beta 1 polypeptide	-2.5
HAPLN1	hyaluronan and proteoglycan link protein 1	-2.5
MMP12	matrix metallopeptidase 12 (macrophage elastase)	-2.5
SPP1	secreted phosphoprotein 1	-2.5
SLC26A2	solute carrier family 26 (sulfate transporter), member 2	-2.6
LAMA2	laminin, alpha 2	-2.6
COL6A2	collagen, type VI, alpha 2	-2.7
CXCL14	chemokine (C-X-C motif) ligand 14	-3.3
ANGPT2	angiopoietin 2	-4.0

FIGURE 5.9 Heat map of the most highly upregulated and downregulated differentially expressed genes in basal plates of PE placentas as compared to the second-trimester, term and preterm labor samples. The normalized log intensity values for the differentially expressed probe sets were centered to the median value of each probe set and colored on a range of −2.5 to +2.5. Red denotes upregulated and blue denotes downregulated expression levels as compared with the median value. Columns contain data from a single basal plate specimen, and rows correspond to a single probe set. Samples within each category are arranged from left to right, ordered by increasing gestational age. Rows are ranked by fold change. Reprinted with permission from *Pregnancy Hypertension*.[51] (This figure is reproduced in color in the color plate section.)

corticotropin releasing hormone (CRH), fatty acid binding protein 4 (FABP4) and lipoprotein lipase (LPL). Interestingly, these are also the genes that tightly clustered in the gestational age dataset discussed above, suggesting co-regulation. Conversely, matrix metallopeptidase 12 (MMP12) and several of the collagens are examples of genes that were downregulated in both PE and term samples.

This analysis provides a glimpse at what may be the molecular profile of premature placental aging in PE. However, it is also clear that this is not the only process involved in PE, as some genes are distinctly dysregulated by this disease process as compared to all other samples regardless of gestational age or condition. Examples of these PE-specific genes are leptin, Flt-1 (the parent molecule of sFlt-1), PAPPA2 and inhibin A (INHA), which are upregulated in PE, and laminin alpha 2 (LAMA2), secreted phosphoprotein 1 (SPP1) and ankyrin repeat and SOCS box-containing 2 (ASB2), which are downregulated in PE. Interestingly, although there is a strong correlation between elevated leptin and sFlt-1 with PE, these molecules are clearly not elevated in several of the subjects despite a clear clinical diagnosis of PE upon review of the clinical data. This is additional evidence that PE is a syndrome and may be the end result of divergent pathogenesis pathways. In fact, we observed substantial variability in gene expression for both the PTL and PE groups as compared to the ostensibly normal samples, a reflection of the heterogeneity of the underlying pathologies. Our data suggest that molecular signatures could be used to better classify these disease entities.

A third group of genes were distinctly expressed in the second-trimester samples. These include ANGPT2, chemokine (C-X-C motif) ligand 14 (CXCL14) and hyaluronan and proteoglycan link protein 1 (HAPLN1). These may be genes with critical activities during the active process of

CTB interstitial and endovascular invasion, which forms the maternal–fetal interface.

Further analyses using IPA showed that the most differentially expressed genes mapped to the PPAR (as was seen with the gestational age dataset) and the neuronal guidance pathways. The latter result was particularly interesting given the importance of neuronal guidance molecules, EPH and ephrins, in CTB invasion and vascular remodeling.[5] These differences may be related at the molecular level to the impaired invasion observed in the basal plate regions of PE placentas.

THE EFFECTS OF PE ON GENE EXPRESSION PROFILES IN PURIFIED CYTOTROPHOBLASTS

Ex Vivo Normalization of sPE CTB Gene Expression

We then turned our attention to transcriptomic analyses of the specific population of invasive CTBs that are implicated in the pathogenesis of PE. We isolated CTBs from the placentas of women who were diagnosed with different forms of severe PE (sPE) ± IUGR including superimposed hypertension and HELLP syndrome (*h*emolysis, *e*levated *l*iver enzymes; *l*ow *p*latelet count). The cells were cultured for 48 h, enabling differentiation/invasion. After various times in culture, we isolated RNA and performed global transcriptional profiling to explore mRNA changes that underlie CTB defects in sPE. Villous CTBs were also isolated from preterm labor patients with no signs of infection (nPTL), which served as gestation-matched controls. Our previous work showed that CTB invasion is essentially normal in the latter group[39] and we confirmed this finding for the samples in this study. However, we cannot eliminate the possibility of other placental pathologies relative to normal controls, which are not possible to collect at the relevant gestational ages.

Clinical information for the subjects is shown in Table 5.1. The newborns did not differ by birth weight or gestational age at delivery. The sPE and nPTL patients had comparable BMIs and ages, but women with sPE had higher systolic/diastolic blood pressures and proteinuria. To better understand the CTB phenotype in the context of sPE variants, we included patients diagnosed with the most clinically significant forms of this condition that necessitated preterm delivery:[55,56] women with sPE ± IUGR, superimposed hypertension or HELLP syndrome ± IUGR. We profiled the gene expression patterns of the case and control groups before plating (0 h) and at 12, 24 and 48 h after culture using the Affymetrix HG-U133Plus 2.0 GeneChip platform. With LIMMA, we identified numerous genes that were differentially expressed – a common CTB fingerprint – in nearly all the sPE samples at one or more time points ($\geq$2-fold; $p \leq 0.05$; Fig. 5.10). Surprisingly, after 48 h in culture most were expressed at control levels. Unsupervised hierarchical clustering showed that the gene expression patterns of sPE and nPTL samples, which segregated into their respective groups at 0 h, merged at 48 h (data not shown).

The initially upregulated molecules included factors previously associated with PE (e.g., growth hormone 2, corticotropin releasing hormone, inhibin A, KISS-1, ADAM-12),[25,57–59] a transcriptional regulator (HOPX) and an angiogenic factor (SEMA3B). In addition to growth hormone 2, other placenta-specific products, including PLAC1 and 4 and seven pregnancy-specific glycoprotein (PSG) family members, were also upregulated, as was an enzyme involved in fat metabolism (oleoyl-ACP hydrolase). Many fewer genes were downregulated. A subset of the results was confirmed by qRT-PCR.[2] The fact that a common set of dysregulated genes was associated with a broad spectrum of the maternal signs suggested that a complex interplay between abnormal placentation and patient-specific factors ultimately determined the clinical features. The finding that most sPE-related aberrations in CTB gene expression normalized when the cells were cultured for two days supported the theory that an unfavorable *in vivo* environment contributed to placental defects in this syndrome. Next, we asked whether any of the dysregulated genes were autocrine regulators of the CTB phenotype that is the hallmark of PE.

Upregulated Trophoblast Expression of SEMA3B in sPE

In PE, TB expression of angiogenic factors is dysregulated. This phenomenon plays a central role in restricting CTB

TABLE 5.1 Maternal and Infant Characteristics

	PE (n = 5)	nPTL (n = 5)	*P* Value
Maternal age, yr	29.0 (5.9)*	25.6 (6.3)	0.40
BMI, kg/m^2	27.4 (4.2)	26.0 (5.0)	0.68
Systolic blood pressure, mm Hg	148 (11)	112 (7)	<0.001
Diastolic blood pressure, mm Hg	88 (5)	65 (13)	<0.01
Proteinuria, designation	+1 to +3	0	NA
Gestational age at delivery, week	31.2 (2.5)	29.5 (3.7)	0.45
Birth weight, g	1365 (528)	1572 (694)	0.61

*Mean ± SD, two-tailed Student's *t*-test.

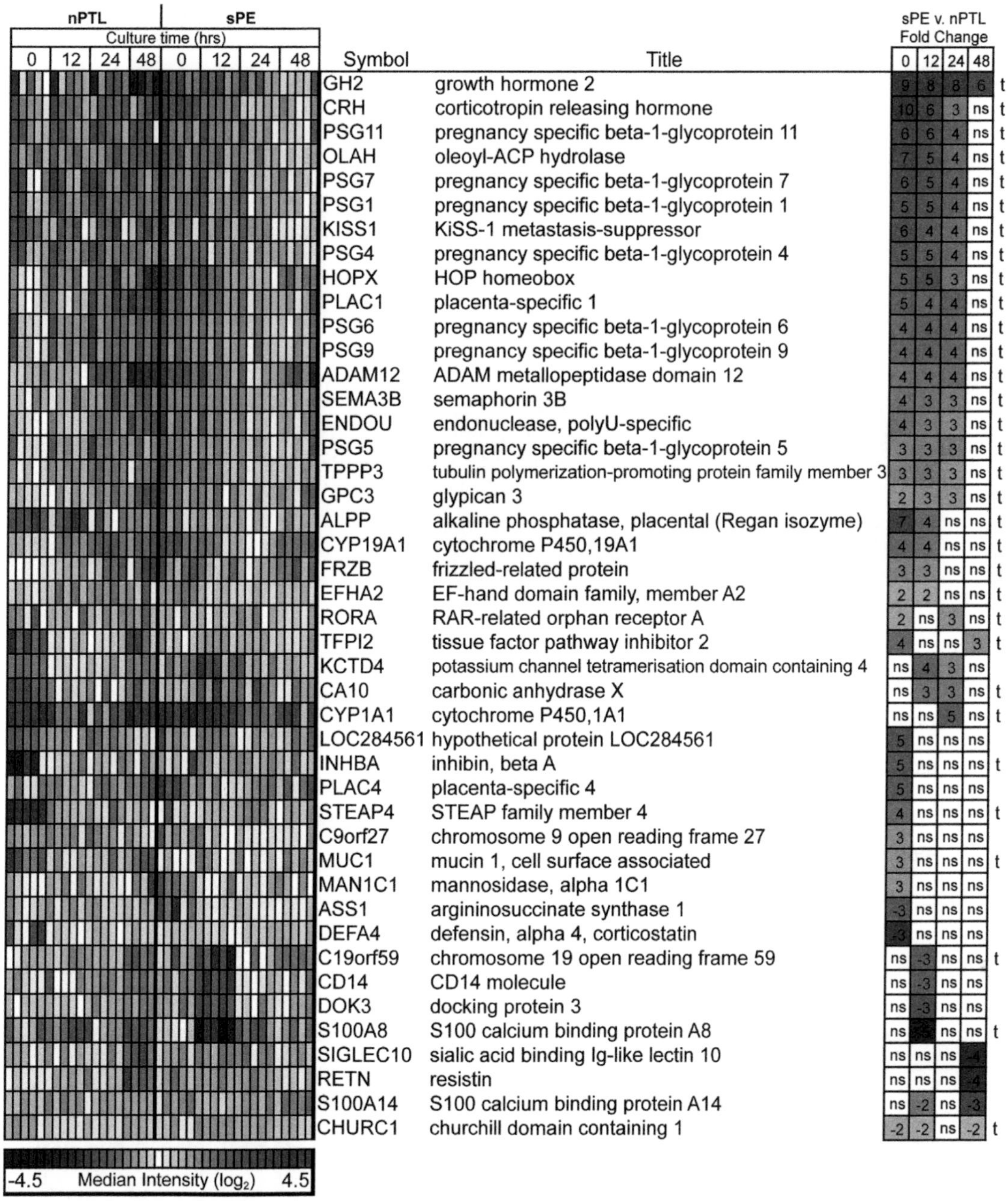

FIGURE 5.10 Severe preeclampsia-associated aberrations in cytotrophoblast gene expression returned to control values after 48 h of culture. RNA was analyzed immediately after the cells were isolated (0 h) and after 12, 24 and 48 h in culture. The relative gene expression levels for cytotrophoblasts (CTBs) isolated from placentas of patients who delivered due to preterm labor with no sign of infection (nPTL; $n = 5$) or a severe form of preeclampsia (sPE; $n = 5$) are shown as a heat map, ranging from high (red) to low (blue). The sPE CTBs were from the following cases (tiled from left to right): (1) hemolysis, elevated liver enzymes and low platelets (HELLP) syndrome and intrauterine growth restriction (IUGR); (2) sPE; (3) sPE and IUGR; (4) superimposed sPE; and (5) HELLP syndrome. One sample of nPTL CTBs collected at 48 h was omitted for technical reasons. The fold changes for each time point (sPE vs. nPTL) are shown on the right. ns, no significant difference (LIMMA); t, no significant difference in expression (sPE vs. nPTL) by 48 h (maSigPro). Reprinted with permission from *The Journal of Clinical Investigation.*[2] (This figure is reproduced in color in the color plate section.)

invasion[7] and in the etiology of the maternal signs including elevated blood pressure and proteinuria.[15,16] In this context, we addressed the functions of SEMA3B. SEMA3 family members play important roles in neuronal wiring,[60] and SEMA3B is an angiogenesis inhibitor and tumor suppressor. As a first step, we profiled SEMA3B mRNA expression in a variety of human cells and organs. Placenta gave the strongest signal (Fig. 5.11A). Northern blot (NB) analyses of mRNA from control (1st, 2nd, and 3rd trimester) and experimental placentas from sPE patients showed that the abundance of SEMA3B mRNA increased as a function of gestational age and was highest in sPE samples

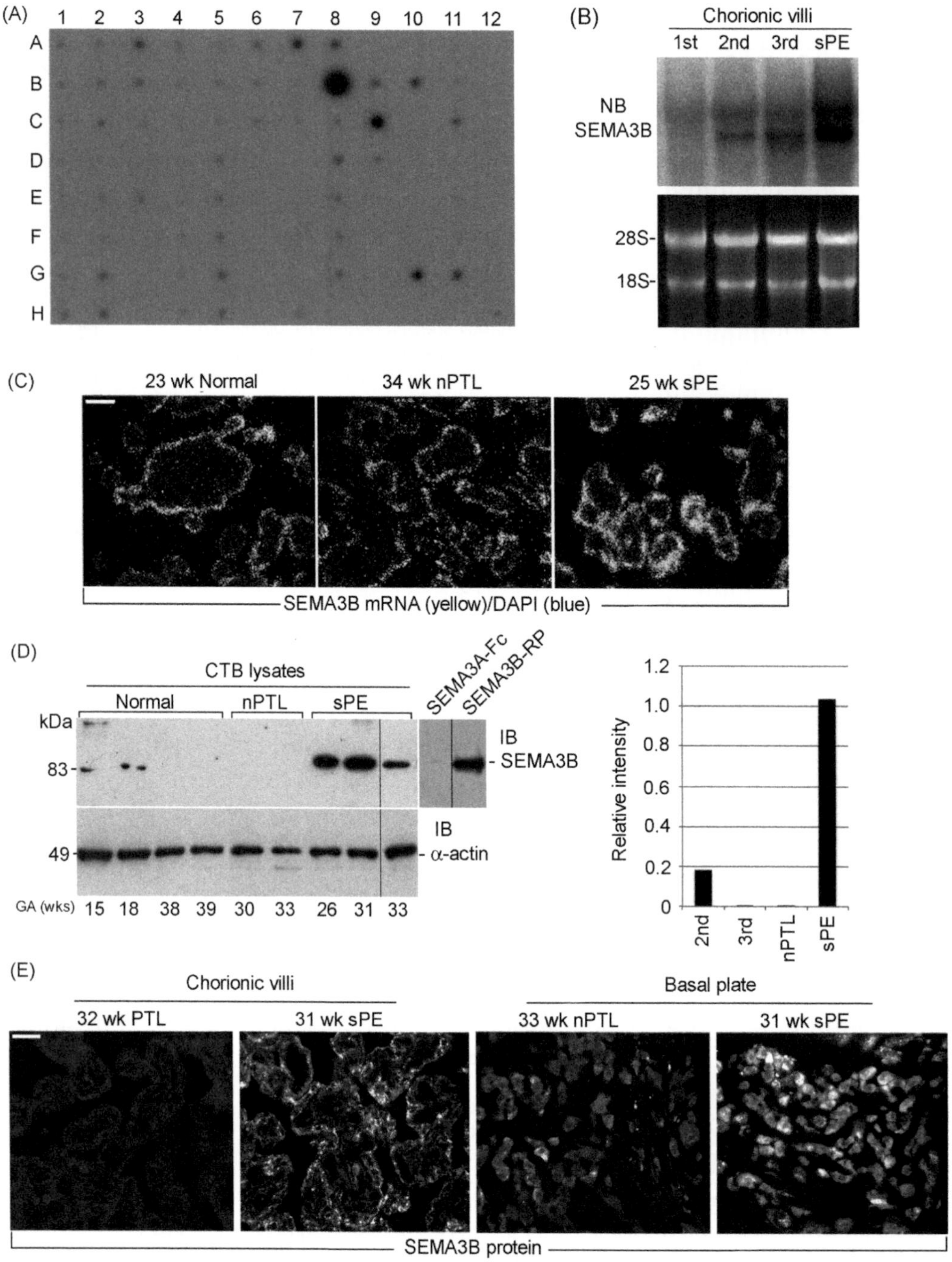

FIGURE 5.11 SEMA3B expression was high in the placenta and upregulated in severe preeclampsia. (A) Binding of a ^{32}P-SEMA3B probe to a multiple tissue expression array revealed high placental expression (coordinate B8). (B) Northern hybridization of polyA^{+} RNA extracted from chorionic villi and pooled from three placentas showed that SEMA3B expression increased over gestation and was highest in sPE (n = 3 replicates). (C) *In situ* hybridization (three placentas/group) confirmed enhanced SEMA3B mRNA expression in the syncytiotrophoblast layer of the chorionic villi in sPE (25 wk) as compared to normal pregnancy (23 wk) and nPTL (34 wk). (D, left panel) Immunoblotting of CTB lysates (15 µg/lane) showed that SEMA3B protein expression was low to undetectable in control cells from normal placentas (15–39 wks). In all cases, expression was higher in sPE (26–33 wks) as compared to nPTL (30, 33 wks). A protein of the expected M_r was detected in COS-1 cells transfected with SEMA3B, but not in the SEMA3A-Fc lane. Vertical lines denote noncontiguous lanes from the same gel. (D, right panel) The relative intensity of the bands was quantified by densitometry. The values for each sample type were averaged and expressed relative to the α-actin loading controls. The entire experiment was repeated twice. (E) Staining tissue sections with anti-SEMA3B showed a sPE-associated upregulation of immunoreactivity associated with the trophoblast components of chorionic villi and among extravillous CTBs within the basal plate (n = 5/group). Trophoblasts were identified by staining adjacent tissue sections with anti-cytokeratin-8/18 (data not shown). Scale bars (C, E): 100 µm. NB, northern blot; GA, gestational age; IB, immunoblot; RP, recombinant protein. Reprinted with permission from *The Journal of Clinical Investigation*.[2] (This figure is reproduced in color in the color plate section.)

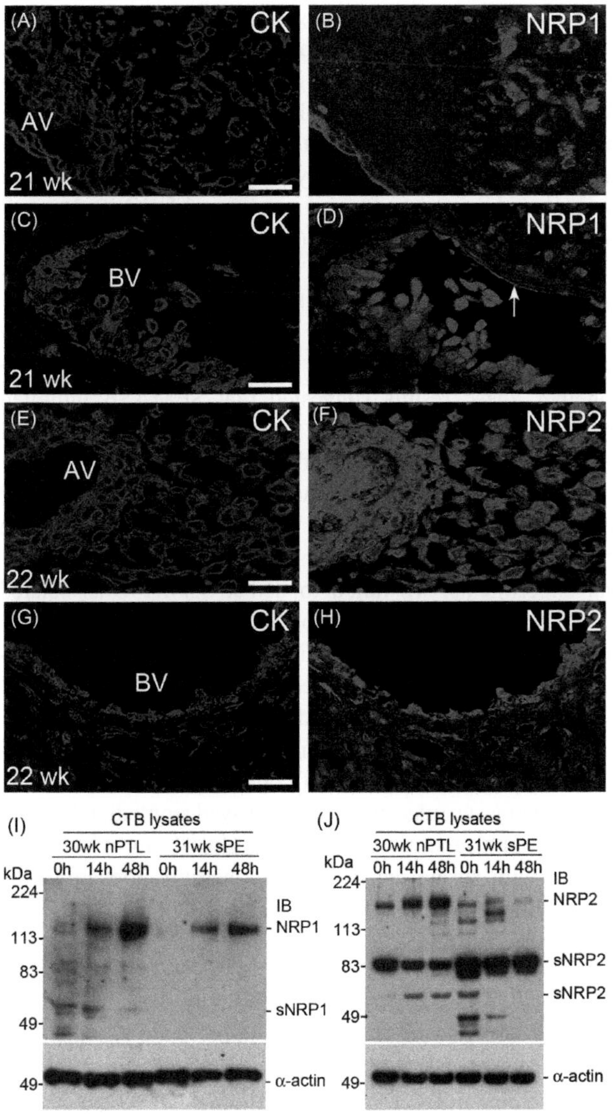

FIGURE 5.12 NRP-1 and -2 (protein) expression at the maternal–fetal interface in normal pregnancy and in sPE. Tissue sections were double-stained with anti-cytokeratin (CK)-7, which reacts with all trophoblast (TB) subpopulations, and anti-NRP-1 or -2. (A, B) NRP-1 expression was detected in association with villous TBs. Within the uterine wall, immunoreactivity associated with invasive CTBs was upregulated as the cells moved from the surface to the deeper regions. (C, D) Endovascular cytotrophoblasts (CTBs) that lined a maternal blood vessel (BV) also stained. (E-H) Anti-NRP-2 reacted with TB and non-TB cells in anchoring villi (AV) as well as interstitial and endovascular CTBs. Essentially the same staining patterns, but with weaker intensity, were observed in sPE (data not shown). CTBs were isolated from the placentas of control nPTL cases and from the placentas of women who experienced sPE. (I, J) Over 48h in culture, NRP-1 expression was upregulated in both instances, but to a lesser degree in sPE. (K, L) Control nPTL CTBs also upregulated NRP2. Expression of this receptor was reduced in sPE and the soluble form was more abundant. (A–L) The data shown are representative of the analysis of a minimum of three samples from different placentas. Scale bars, 100 μm. IB, immunoblot. Reprinted with permission from *The Journal of Clinical Investigation*.[2] (This figure is reproduced in color in the color plate section.)

(Fig. 5.11B). The two bands likely reflect alternative splicing. *In situ* hybridization of placental chorionic villi demonstrated that SEMA3B mRNA expression, which was limited to TBs, was lower in normal second-trimester and nPTL samples as compared to sPE chorionic villi (Fig. 5.11C). Immunoblot (IB) analyses of CTB lysates showed that expression of SEMA3B was either undetectable or low during the second and third trimesters of normal pregnancy and in cells isolated from nPTL placentas (Fig. 5.11D). In contrast, higher levels of SEMA3B were detected in sPE CTBs immediately after isolation. For these experiments, SEMA3A-Fc served as a negative control and recombinant SEMA3B protein (RP) served as a positive control for antibody specificity. Immunolocalization showed that trophoblasts of chorionic villi from control nPTL placentas had much lower anti-SEMA3B immunoreactivity as compared to samples of similar gestational ages from sPE placentas (Fig. 5.11E, left panels). Within the basal plate, extravillous CTBs in the setting of sPE also exhibited stronger staining for SEMA3B as compared to the nPTL samples (Fig. 5.11E, right panels).

Neuropilin Expression and SEMA3B Actions

Next, we assessed the expression of the SEMA3B receptors, neuropilin (NRP)-1 and NRP-2,[60] in tissue sections of the maternal–fetal interface. Co-staining with anti-cytokeratin-8/18 (CK) identified trophoblasts (Fig. 5.12A, C, E and G). Immunolocalization analyses of normal second-trimester samples showed that NRP-1, which was expressed by villous TBs, was upregulated as the CTBs invaded the uterine wall (Fig. 5.12B). Strong staining was also detected in association with endovascular CTBs and the endothelial lining of uterine vessels (Fig. 5.12D; arrow). NRP-2 immunoreactivity was associated with villous TBs and invasive CTBs as well as the villous stroma (Fig. 5.12F). In the uterine wall, NRP-2 expression was strongly upregulated on endovascular CTBs (Fig. 5.12H) and a subset of endothelial cells (data not shown).

Immunoblot analyses of CTB lysates from control nPTL placentas showed upregulation of NRP-1 expression over 48 h of culture, which was blunted in sPE (Fig. 5.12I). As to NRP-2, control nPTL CTBs also upregulated this receptor and soluble forms were detected (Fig. 5.12J). In sPE, NRP2 was expressed at reduced levels and the relative abundance of the major soluble form of this receptor increased as compared to the control nPTL CTBs. Together these data suggested that placenta-derived SEMA3B could have autocrine effects on CTBs and paracrine actions on uterine endothelial cells.

We tested this hypothesis in the context of VEGF actions using our *in vitro* model of CTB invasion. Previously we showed that CTBs produce large amounts of VEGF and its autocrine actions include promoting invasion and inhibiting apoptosis.[7] Blocking VEGF signals (anti-VEGF-A) or the addition of recombinant SEMA3B protein reduced invasion by ~60% as compared to cells that were cultured with a control CD6-Fc protein (Fig. 5.13A). In contrast, removal of both ligands (anti-VEGF plus NRP-1-Fc or NRP-2-Fc) restored invasion to control levels. With regard to apoptosis, the removal of VEGF-A or the addition of SEMA3B doubled the rate and the absence of both ligands returned levels to below control values (Fig. 5.13B). Taken together, these results suggested that SEMA3B opposed the actions of VEGF to restrain CTB invasion by promoting apoptosis of these cells.

As to vascular effects, a monolayer of uterine microvascular endothelial cells (UtMVEC) was disrupted with a scratch and the effects of SEMA3B, in terms of migration, were tracked by video microscopy. The results of a typical experiment are shown in Fig. 5.13C. As with many ECs, the addition of VEGF strongly promoted directed UtMVEC migration; exogenous SEMA3B decreased levels to ~50% of control values with a loss of directionality. Figure 5.13D summarizes the results of three experiments. Under the same conditions, the opposite effects were observed on apoptosis; VEGF was protective and SEMA3B was a strong inducer (Fig. 5.13E). These findings suggested that SEMA3B is primarily antiangiogenic, as was previously proposed. To test this theory, we employed the chick chorioallantoic membrane (CAM) assay in which filter paper discs delivered VEGF, SEMA3B or CD6-Fc (Fig. 5.13F, top row). Removal of the discs showed that SEMA3B significantly inhibited angiogenesis as compared to the positive control VEGF or CD6-Fc (Fig. 5.13F, bottom row). Together these results suggested that the autocrine effects of enhanced SEMA3B expression recapitulated significant aspects of the CTB phenotype in PE, with paracrine actions including impaired UtMVEC functions.

Exogenous SEMA3B Alters CTB Signaling, Phenocopying sPE Effects

Next, we investigated the CTB signaling pathways that were involved. First, we asked whether SEMA3B opposed VEGF signaling by inhibiting the activation of PI3K as measured by the production of phosphatidylinositol 3,4,5-triphosphate (PIP_3). The addition of SEMA3B to first- or second-trimester CTBs reduced PIP_3 concentrations to levels that were comparable to the effects of the PI3K inhibitor, wortmannin (FA). DMSO, the vehicle, had no effect. Addition of VEGF increased PIP_3 production 2.5-fold over control levels. These results suggested SEMA3B as a negative regulator of PI3K.[61]

Then we sought to explain the mechanisms involved. A previous study demonstrated that VEGF-mediated VEGFR-2 phosphorylation creates a docking site for the p85 subunit of PI3K.[62] However, preliminary experiments showed that SEMA3B did not interfere with VEGFR-2 phosphorylation (data not shown). Thus, we studied the interactions between the regulatory subunits of PI3K. Uterine microvascular endothelial cells (UtMEVC) were cultured in medium containing SEMA3B and VEGF or in the absence of one or the other factor. Cell lysates were immunoprecipitated with an antibody that specifically recognized the p85 regulatory subunit of PI3K. The pull downs were immunoblotted with anti-VEGFR-2, anti-NRP-2 and anti-p110α PI3K (Fig. 5.14B); NRP-1 was not expressed (data not shown). VEGFR-2 and NRP-2 levels remained constant under all the test conditions. The addition of SEMA3B resulted in the dissociation of p85 and p110α, which was rescued by the addition of VEGF. Taken together, these results suggested that SEMA3B inhibited PI3K activity (Fig. 5.14A) by preventing the association of p85 and p110α, to our knowledge a novel mechanism.

Downstream of PI3K activation, Akt is phosphorylated at Thr308 and/or Ser473.[63,64] Thus, we were interested in the effects of SEMA3B on this process (Fig. 5.14C). In initial experiments, we failed to detect SEMA3B-associated changes in phosphorylation of Thr308. Thus, we focused on modification of Ser473. First, COS-1 cells were transfected with either an empty vector or SEMA3B. No Ser473 phosphorylation was observed in the latter case. Next we evaluated Akt phosphorylation as a function of CTB differentiation in culture. Lysates of cells isolated from first- and second-trimester placentas were assayed immediately upon isolation (0 h) and after 12 h in culture, during which time a band with strong anti-p-Ser473 reactivity appeared. The time course was rapid. After initial CTB adhesion (1 h), the addition of wortmannin for 30 min downregulated Ser473 phosphorylation as did SEMA3B; VEGF had the opposite effect. In each case, the results were compared with the total amount of Akt, which was determined by stripping the blots and reprobing with an antibody that recognized all forms of this molecule. Together, these data suggested that, in CTBs, SEMA3B strongly downregulated Akt signaling.

Akt inactivates GSK3α and β by phosphorylating Ser21 and Ser9,[65,66] respectively. In COS-1 cells, expressing SEMA3B abolished phosphorylation of GSK3α and β (Fig. 5.14D). In first and second trimester CTBs, GSK3β phosphorylation on Ser9 increased during 12 h of culture whereas GSK3α phosphorylation on Ser21 was variable (Fig. 5.14D). LiCl, a GSK3 inhibitor, increased phosphorylation of Ser9 (data not shown), with wortmannin having the opposite effect (Fig. 5.14D). Consistent with the Akt results, the addition of exogenous SEMA3B decreased phosphorylation of GSK3β, which was increased by the addition of VEGF. Immunoblot analysis with an antibody that recognized the GSK3 protein backbone showed that levels did not change under any of the experimental conditions. Since GSK3, often a negative regulator, intersects several critical signaling pathways,[66,67] it is likely that overexpression of SEMA3B has important consequences.

Based on our analysis of SEMA3B/VEGF effects on PI3K/Akt and GSK3β signaling, we predicted that this

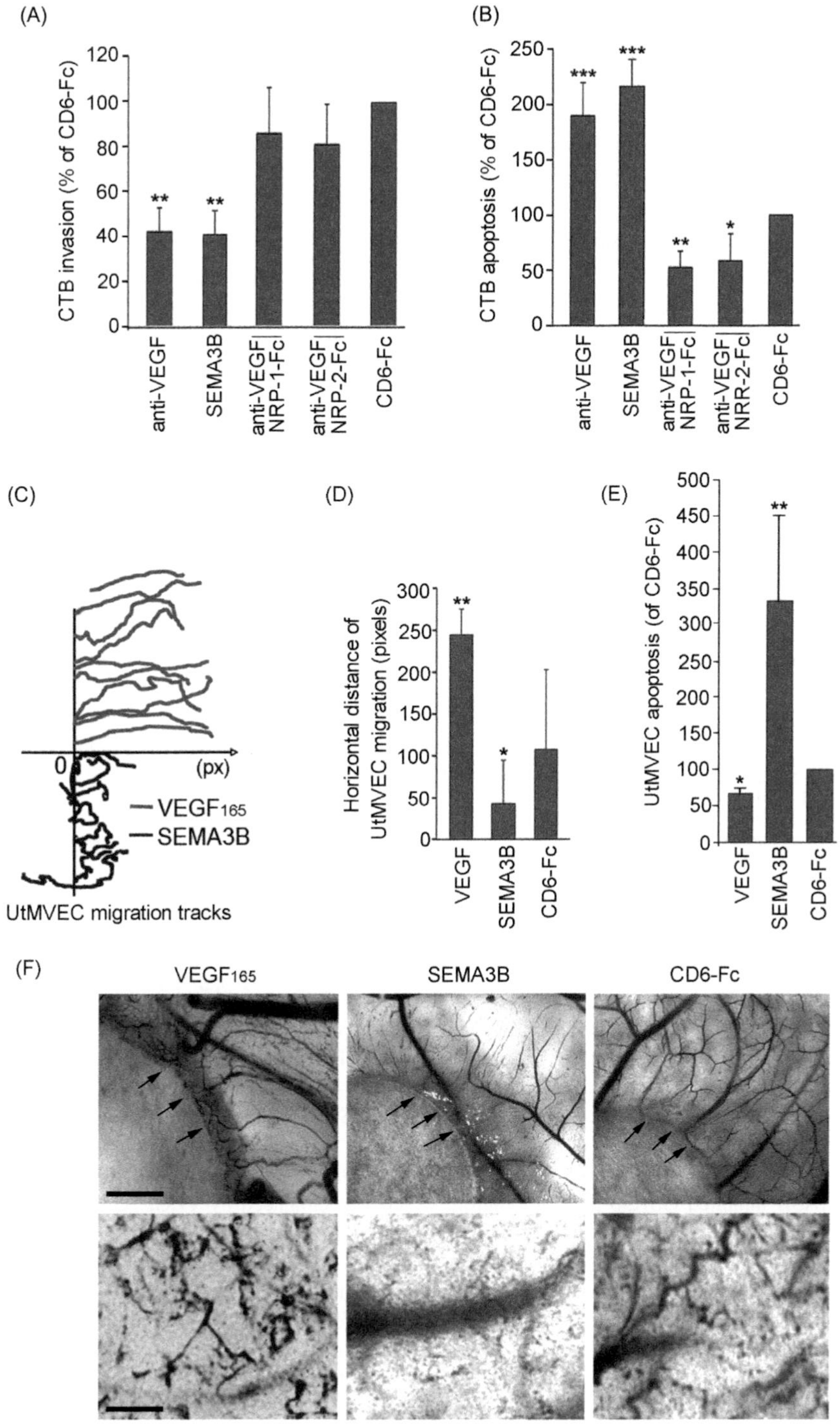

FIGURE 5.13 Exogenous SEMA3B mimicked the effects of sPE on CTBs and endothelial cells, and inhibited angiogenesis. (A) The addition of anti-VEGF or SEMA3B protein significantly inhibited cytotrophoblast (CTB) invasion as compared to the addition of a control protein, CD6-Fc. The removal of both ligands (anti-VEGF/NRP1-Fc, anti-VEGF/NRP2-Fc) restored invasion to control levels. (B) The variables tested in panel A had the opposite effects on CTB apoptosis, suggesting that increased programmed cell death contributed to decreased invasion. (C) Exogenous VEGF stimulated the migration of uterine microvascular endothelial cells (UtMVECs), which was inhibited by SEMA3B. (D) The results in panel C were quantified relative to the addition of CD6-Fc. (E) In UtMVECs, VEGF promoted survival and SEMA3B increased apoptosis relative to control levels. (F) In the chick chorioallantoic membrane (CAM) angiogenesis assay, VEGF promoted angiogenesis by ~3-fold and SEMA3B inhibited this process ~5-fold relative to the effects of CD6-Fc. Top row: arrows mark the edge of the filter paper used to apply the protein. Scale bar: 200 μm. Bottom row: the area of the CAM beneath the filter paper. Scale bar: 100 μm. A–D, $n = 6$ replicates; E, F, $n = 3$ replicates. A,B,D,E, mean ± SEM, two-tailed Student's t-test. $*p < 0.05$, $**p < 0.01$, $***p < 0.001$. Reprinted with permission from *The Journal of Clinical Investigation*.[2] (This figure is reproduced in color in the color plate section.)

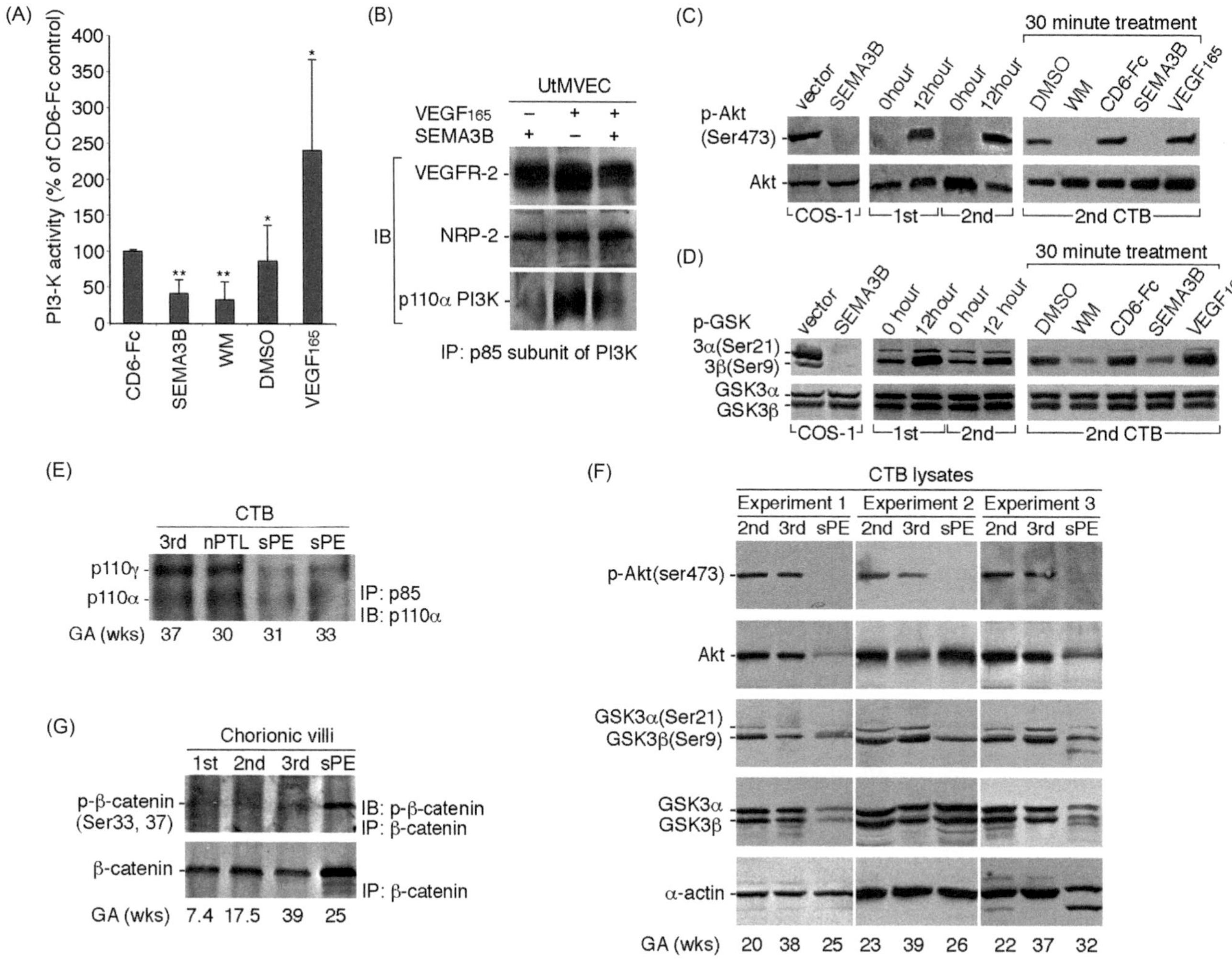

FIGURE 5.14 SEMA3B inhibited PI3K/Akt and GSK3β signaling in CTBs and the same effects were observed in sPE. (A) SEMA3B and wortmanin (WM) inhibited PI3K activity, which was stimulated by VEGF. DMSO, vehicle control. (B) The addition of SEMA3B to uterine microvascular endothelial cells (UtMVEC) resulted in the dissociation of the p85 and the p100α subunits of PI3K, which was rescued by the addition of VEGF. (C) In COS-1 cells, SEMA3B inhibited Akt Ser473 phosphorylation (activation), which increased during CTB differentiation/invasion (0–12 h). The addition of SEMA3B inhibited Akt phosphorylation, which was enhanced by exogenous VEGF. (D) In COS-1 cells, SEMA3B inhibited GSK3β Ser9 phosphorylation (inactivation), which increased during CTB differentiation/invasion (0–12 h). Exogenous SEMA3B inhibited GSK3β phosphorylation, which was enhanced by VEGF. GSK3α Ser21 phosophorylation was variable. (E) In CTBs, sPE correlated with dissociation of the p85 and p110α (and γ) subunits of PI3K relative to control cells isolated from normal third-trimester placentas. (F) In freshly isolated CTBs, sPE was associated with decreased phosphorylation of Akt Ser473 and GSK3β Ser9. α-Actin, loading control. (G) In chorionic villi, sPE was associated with phosphorylation (inactivation) of β-catenin. (A–D) The same results were obtained in three separate experiments that utilized different preparations of cells. (F, G) The results shown are representative of analyses of a total of six CTB isolates from different placentas of women diagnosed with sPE. A, mean ± SEM, two-tailed Student's *t*-test. $*p < 0.05$, $**p < 0.01$. GA, gestational age; IB, immunoblot; IP, immunoprecipitation. Reprinted with permission from *The Journal of Clinical Investigation*.[2]

pathway would be dysregulated in sPE. CTBs from control placentas throughout gestation and from those of patients who were diagnosed with sPE were analyzed immediately after isolation. As with UtMVEC (Fig. 5.14B), an IP/IB strategy showed a significant disassociation of the p85 and p110α subunits of PI3K in sPE (Fig. 5.14E). In this case, the expression of P110γ was also detected and relative expression was reduced in sPE. The results of the Akt and GSK3 analyses were interpreted using the expression of α-actin as a control for protein loading (Fig. 5.14F). The two bands observed in the 32 wk sPE sample were attributed to proteolysis, which is sometimes observed in these samples. Two patterns were seen in sPE. In one (26 wk), no differences were detected at the protein level, but phosphorylation was markedly decreased. The other was characterized by downregulation at both levels (25 and 32 wk). In

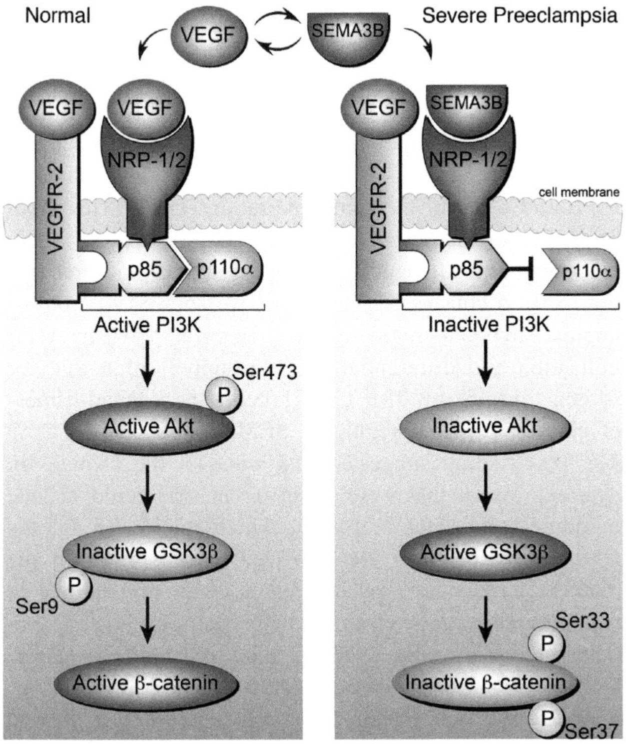

FIGURE 5.15 Model of SEMA3B effects on CTBs in sPE vs. normal pregnancy. Reprinted with permission from *The Journal of Clinical Investigation.*[2] (This figure is reproduced in color in the color plate section.)

the case of Akt, p-Ser473 was either undetectable or nearly absent in the sPE samples (Fig. 5.14F). Likewise, p-Ser9 of GSK3β was lower in abundance. Finally, we reasoned that an increase in GSK3 activity would have important effects on pathways that we know are critical to CTB invasion. In chorionic villi from sPE placentas, we observed a large increase in the phosphorylated form of β-catenin, which leads to ubiquitination and proteosomal degradation of this molecule (Fig. 5.14G). Given that activation of β-catenin is associated with tumorigenesis,[68] inhibiting this pathway could restrict CTB invasion perhaps by altering cell adhesion or Wnt signaling.[69]

Based on these data, we propose a model that integrates SEMA3B and VEGF functions in normal pregnancy and in sPE (Fig. 5.15). We found that SEMA3B competed with VEGF binding to neuropilins. High SEMA3B levels led to the dissociation of the p85 and p110α subunits of PI3K, a novel mechanism. The downstream consequences included inactivation of AKT and activation of GSK3, which led to apoptosis and degradation of β-catenin. Together, these data suggested an autocrine mechanism by which elevated SEMA3B levels contributed to the sPE-associated phenotype of invasive CTBs in terms of the signaling pathways we analyzed.

In summary, global gene expression profiling immediately after isolation of CTBs revealed a common set of upregulated mRNAs despite the different diagnoses. Surprisingly, they were downregulated to control levels over 48 h. The differentially expressed genes included molecules previously identified as dysregulated in PE and many that were not known to be involved in this pregnancy complication. We hypothesized that their autocrine actions could contribute to the CTB phenotypic alterations that are the hallmark of PE. We proved this theory using a novel molecule that was identified in this study, SEMA3B, and the VEGF signaling pathways it engages. Together, these results suggested that the CTB defects in sPE, which can be attributed to the *in vivo* environment, are reversible, evidence that therapeutic interventions may be possible. We also showed that a common signature of misexpressed genes that spans all the sPE subtypes we studied results in the diverse maternal signs that define these variants. We propose that, in severe PE, the *in vivo* environment dysregulates CTB gene expression, the autocrine actions of the upregulated molecules, including SEMA3B, impair differentiation/invasion/signaling and patient-specific factors determine the signs.

DISCUSSION

The causes of PE remain poorly understood and are under intense investigation. We focus on the placental component of this pregnancy complication, thought to be the first link in the chain of events leading to the syndrome. As such, we reasoned that the findings might give us a better understanding of the instigating factors. We targeted for analysis the CTB subpopulation that invades the uterine wall, as defects in this process, particularly remodeling the arterial side of the uterine vasculature, are the hallmark of this condition. As anticipated from our study design, global transcriptional profiling revealed the gene expression patterns of CTBs from the placentas of affected patients. The molecules encoded by the dysregulated genes – the sPE signature – included factors that have been associated with this syndrome, others that have not been studied in relationship to the placental component of sPE and novel participants. Additionally, a significant number were placenta-specific gene products that are unique to humans, an interesting observation given the fact that PE is confined to our species.

We reasoned that the differentially expressed genes might play autocrine roles in CTB functions that go awry in PE. Given our discovery that these cells undergo a vascular-like transformation as they invade the uterine wall,[4] we focused our mechanistic analyses on SEMA3B, which has been proposed as an angiogenesis inhibitor. The results suggested that this molecule is a major driver of the CTB aberrations in sPE. The endpoints we analyzed included CTB invasion, which was inhibited, and apoptosis, which increased. We confirmed the antiangiogenic properties of this molecule and described the signaling pathways it engages, a combination of novel and known mechanisms that are also dysregulated in sPE. These data extended the concept that antiangiogenic factors contribute to the maternal signs of PE[70] by showing that they also play an important role in the observed CTB phenotypic alterations that underlie the etiology of this syndrome. We previously proposed this idea in relationship to dysregulated CTB production of VEGF family members in sPE.[7] In support of this concept, altering maternal levels of adrenomedulin also affects placentation.[17]

Other CTB genes that were upregulated in sPE included growth hormone 2, a placental isoform, which increases the invasiveness of primary CTBs *in vitro*.[71] In contrast, KISS-1 inhibits trophoblast (and tumor cell) invasion.[72] The product of the HOP homeobox gene (HOPX), which marks intestinal epithelial stem cells,[73] interacts with HDAC2 to enable GATA4 deacetylation, thereby inhibiting proliferation of embryonic cardiomyocytes.[74] We described human trophoblast progenitor expression of GATA4,[75] making it possible that this same relationship regulates placental growth. In mice, deletion of *hopx* expands the trophoblast giant cell population with a commensurate reduction in spongiotrophoblasts.[76] Experiments in progress confirmed a sPE-associated upregulation of HOPX expression in CTBs (data not shown). Frizzle-related proteins, a component of the sPE gene signature, have actions that are often context dependent. They usually act as Wnt inhibitors, which could also negatively regulate β-catenin, a downstream target of SEMA3B signaling.[77] Placenta-specific-1 is unique to this organ.[78] Its function(s) are as yet unknown, but it is overexpressed in several pregnancy complications.[79] CTB production of numerous isoforms of pregnancy-specific beta-1 glycoprotein (PSG-1) were also upregulated. Although their functions are not well understood, they appear to have vasculogenic[80,81] and immune properties.[82,83] Glypican 3, proposed as a receptor for some PSG family members,[84] was also upregulated. ADAM-12 is involved in STB formation[85] and promotes CTB invasion.[86] Thus, as a group, the functions of many of these molecules were relevant to the impact of sPE on the placenta. Furthermore, many of them are expressed by both CTBs and STBs. Since placental hypoperfusion is associated with increased numbers of giant cells,[87] the dysregulated molecules could bias the cells toward fusion *in vivo*.

Additionally, the microarray analyses produced several surprising results. For one, nearly all of the CTB genes that were dysregulated in sPE reverted to control levels over 48 h of culture. This unexpected finding was not due to apoptosis because we failed to observe upregulation of genes involved in this process.[2] This is in contrast to our previous report that sPE is associated with many TUNEL-positive invasive CTBs *in situ*.[14] Thus, we concluded that the *in vivo* environment rather than intrinsic CTB defects was involved in the observed gene dysregulation, which resolved *in vitro* and progressed to apoptosis *in situ*. Given the importance of stromal factors in influencing the behavior of epithelial cells,[88] it may be that signals from chorionic villous fibroblasts or decidual cells impede invasion in PE. As to the possible involvement of other maternal factors, metabolic syndrome,[89] vascular disease[90] and/or advanced maternal age[91] increase PE risk. As to fetal factors, Trisomy 21 is associated with an increased incidence of PE and some of the same aberrations in CTB invasion/differentiation are observed in these pregnancies.[92]

Other findings suggest that maternal and placental factors work in concert. For example, we showed that physiological hypoxia regulates the balance between CTB replication and differentiation.[93] In support of a role for oxygen tension, by 48 h COS-1 cells and second-trimester chorionic villi upregulated SEMA3B in 2% vs. 20% O_2.[2] This finding suggested that some of the changes in gene expression that were observed in sPE could be due to reduced placental perfusion, which might explain the *in vitro* "rescue" that we observed. However, given the number of hypoxia-responsive elements in the genome, it was impossible to determine whether genes with this promoter sequence were overrepresented in the sPE signature. Additionally, querying the Ingenuity Knowledge Database failed to identify any pathways that linked the dysregulated molecules, suggesting a complex etiology that will be interesting to unravel. For example, aberrations in the dialogue between trophoblasts and decidua that is initiated at implantation and required for normal CTB invasion could begin the process that culminates in PE.[94]

Another surprising finding was that CTBs from the severe forms of PE have common gene signatures. Since we were assaying the same CTB preparation at four 12-h intervals, which yielded highly statistically significant data, we had the opportunity to compare sPE with other variants of the syndrome. For this purpose we included superimposed hypertension (one case) and HELLP (two cases). We also added IUGR as a variable. Despite the different maternal and fetal manifestations, we found essentially the same pattern of CTB gene dysregulation in all cases. We took this as preliminary evidence that the various forms of sPE diverged at the level of maternal (and fetal) responses, a theory that will need further validation given the low number of samples that we analyzed from the sPE variants. Nevertheless, our findings are possible evidence of individual differences in the mechanisms that lead to the signs. Although it is not clear what these are, they could include the same risk factors (discussed above) that predispose women to develop this pregnancy complication. However, why PE is more common in nulliparous women remains enigmatic.[95] One explanation could be that CTB remodeling of the uterine stroma and/or spiral arterioles becomes progressively easier with each subsequent pregnancy.

Our data demonstrating that sPE-associated aberrations in CTB gene expression are reversible bolster the rationale for developing therapeutic interventions and biomarkers for gauging their utility. One strategy is to target the CTB population that we studied. A small molecule inhibitor that blocks binding of SEMA3A to NRP-1 has been described,[96] raising the possibility that a similar compound could be developed for disrupting SEMA3B actions in sPE. Of note, a portion of the molecules that we discovered are dysregulated in this pregnancy complication, e.g., the particular combination of PSGs whose expression was upregulated, could be useful biomarkers for predicting women with elevated risk and/or assessing therapeutic efficacy. Another strategy would be to target maternal responses. However, our data also suggest that potential therapeutics may need to be tailored to specific patient populations according to the signs, which are different despite a common set of CTB defects. Finally, if faulty decidual signals play a role in restraining CTB invasion, they could also be therapeutic targets.

In summary, our data suggest new concepts or reinforce proposed mechanisms that suggest a possible theory regarding the pathophysiology of PE. First, *in vivo* signals drive CTB gene dysregulation. Cells that were cultured from affected placentas normalized their gene expression patterns over 48 h. These data also suggested the second concept – CTB phenotypic alterations in sPE are reversible and recovery is possible. The third concept is that the autocrine actions of the dysregulated molecules contribute to the CTB functional defects that are the hallmarks of this syndrome – shallow invasion and apoptosis that are associated with deficits in particular signaling pathways. In this context, we identified previously unknown molecules with relevant actions, for example, SEMA3B, an angiogenesis inhibitor. The fourth concept is that unified CTB defects manifest as diverse signs in different patients. Thus, our data suggested that individual differences in maternal responses are driving the clinical presentation of the various forms of sPE. In this regard, we also noted disparate fetal responses in terms of growth effects. Together, our findings provide an important rationale and framework for pursuing treatments, a research area that is usually relegated to the back burner because of questions regarding feasibility, which our data support.

References

1. Maltepe E, Bakardjiev AI, Fisher SJ. The placenta: transcriptional, epigenetic, and physiological integration during development. *J Clin Invest.* 2010;120(4):1016–1025.
2. Zhou Y, Gormley MJ, Hunkapiller NM, et al. Reversal of gene dysregulation in cultured cytotrophoblasts reveals possible causes of preeclampsia. *J Clin Invest.* 2013;123(7):2862–2872.
3. Pijnenborg R, Vercruysse L, Hanssens M. The uterine spiral arteries in human pregnancy: facts and controversies. *Placenta.* 2006;27(9–10):939–958.
4. Zhou Y, Fisher SJ, Janatpour M, et al. Human cytotrophoblasts adopt a vascular phenotype as they differentiate. A strategy for successful endovascular invasion? *J Clin Invest.* 1997;99(9):2139–2151.
5. Red-Horse K, Kapidzic M, Zhou Y, Feng KT, Singh H, Fisher SJ. EPHB4 regulates chemokine-evoked trophoblast responses: a mechanism for incorporating the human placenta into the maternal circulation. *Development.* 2005;132(18):4097–4106.
6. Hunkapiller NM, Gasperowicz M, Kapidzic M, et al. A role for Notch signaling in trophoblast endovascular invasion and in the pathogenesis of pre-eclampsia. *Development.* 2011;138(14):2987–2998.
7. Zhou Y, McMaster M, Woo K, et al. Vascular endothelial growth factor ligands and receptors that regulate human cytotrophoblast survival are dysregulated in severe preeclampsia and hemolysis, elevated liver enzymes, and low platelets syndrome. *Am J Pathol.* 2002;160(4):1405–1423.
8. Redman CW, Sargent IL. Latest advances in understanding preeclampsia. *Science.* 2005;308(5728):1592–1594.
9. Gebb J, Dar P. Colour Doppler ultrasound of spiral artery blood flow in the prediction of pre-eclampsia and intrauterine growth restriction. *Best Pract Res Clin Obstet Gynaecol.* 2011;25(3):355–366.
10. Naicker T, Khedun SM, Moodley J, Pijnenborg R. Quantitative analysis of trophoblast invasion in preeclampsia. *Acta Obstet Gynecol Scand.* 2003;82(8):722–729.
11. Brosens IA, Robertson WB, Dixon HG. The role of the spiral arteries in the pathogenesis of preeclampsia. *Obstet Gynecol Annu.* 1972;1:177–191.
12. Zhou Y, Damsky CH, Chiu K, Roberts JM, Fisher SJ. Preeclampsia is associated with abnormal expression of adhesion molecules by invasive cytotrophoblasts. *J Clin Invest.* 1993;91(3):950–960.
13. Zhou Y, Damsky CH, Fisher SJ. Preeclampsia is associated with failure of human cytotrophoblasts to mimic a vascular adhesion phenotype. One cause of defective endovascular invasion in this syndrome? *J Clin Invest.* 1997;99(9):2152–2164.
14. DiFederico E, Genbacev O, Fisher SJ. Preeclampsia is associated with widespread apoptosis of placental cytotrophoblasts within the uterine wall. *Am J Pathol.* 1999;155(1):293–301.
15. Maynard SE, Min JY, Merchan J, et al. Excess placental soluble fms-like tyrosine kinase 1 (sFlt1) may contribute to endothelial dysfunction, hypertension, and proteinuria in preeclampsia. *J Clin Invest.* 2003;111(5):649–658.
16. Venkatesha S, Toporsian M, Lam C, et al. Soluble endoglin contributes to the pathogenesis of preeclampsia. *Nat Med.* 2006;12(6):642–649.
17. Li M, Yee D, Magnuson TR, Smithies O, Caron KM. Reduced maternal expression of adrenomedullin disrupts fertility, placentation, and fetal growth in mice. *J Clin Invest.* 2006;116(10):2653–2662.
18. Vaiman D, Calicchio R, Miralles F. Landscape of transcriptional deregulations in the preeclamptic placenta. *PLoS One.* 2013;8(6):e65498.
19. Sitras V, Paulssen RH, Gronaas H, et al. Differential placental gene expression in severe preeclampsia. *Placenta.* 2009;30(5):424–433.
20. Meng T, Chen H, Sun M, Wang H, Zhao G, Wang X. Identification of differential gene expression profiles in placentas from preeclamptic pregnancies versus normal pregnancies by DNA microarrays. *Omics J Integrat Biol.* 2012;16(6):301–311.
21. Tsai S, Hardison NE, James AH, et al. Transcriptional profiling of human placentas from pregnancies complicated by preeclampsia reveals disregulation of sialic acid acetylesterase and immune signalling pathways. *Placenta.* 2011;32(2):175–182.

22. Nishizawa H, Pryor-Koishi K, Kato T, Kowa H, Kurahashi H, Udagawa Y. Microarray analysis of differentially expressed fetal genes in placental tissue derived from early and late onset severe pre-eclampsia. *Placenta*. 2007;28(5–6):487–497.
23. Nishizawa H, Ota S, Suzuki M, et al. Comparative gene expression profiling of placentas from patients with severe pre-eclampsia and unexplained fetal growth restriction. *Reprod Biol Endocrinol*. 2011;9:107.
24. Winn VD, Haimov-Kochman R, Paquet AC, et al. Gene expression profiling of the human maternal-fetal interface reveals dramatic changes between midgestation and term. *Endocrinology*. 2007;148(3):1059–1079.
25. Winn VD, Gormley M, Paquet AC, et al. Severe preeclampsia-related changes in gene expression at the maternal-fetal interface include sialic acid-binding immunoglobulin-like lectin-6 and pappalysin-2. *Endocrinology*. 2009;150(1):452–462.
26. Oppenheim JJ, Biragyn A, Kwak LW, Yang D. Roles of antimicrobial peptides such as defensins in innate and adaptive immunity. *Ann Rheum Dis*. 2003;62(suppl 2):ii17–ii21.
27. Svinarich DM, Gomez R, Romero R. Detection of human defensins in the placenta. *Am J Reprod Immunol*. 1997;38(4):252–255.
28. Krensky AM. Granulysin: a novel antimicrobial peptide of cytolytic T lymphocytes and natural killer cells. *Biochem Pharmacol*. 2000;59(4):317–320.
29. Raychaudhuri SP, Jiang WY, Raychaudhuri SK, Krensky AM. Lesional T cells and dermal dendrocytes in psoriasis plaque express increased levels of granulysin. *J Am Acad Dermatol*. 2004;51(6):1006–1008.
30. King A, Loke YW, Chaouat G. NK cells and reproduction. *Immunol Today*. 1997;18(2):64–66.
31. Geirsson A, Paliwal I, Lynch RJ, Bothwell AL, Hammond GL. Class II transactivator promoter activity is suppressed through regulation by a trophoblast noncoding RNA. *Transplantation*. 2003;76(2):387–394.
32. Markel G, Wolf D, Hanna J, et al. Pivotal role of CEACAM1 protein in the inhibition of activated decidual lymphocyte functions. *J Clin Invest*. 2002;110(7):943–953.
33. Levine RJ, Karumanchi SA. Circulating angiogenic factors in preeclampsia. *Clin Obstet Gynecol*. 2005;48(2):372–386.
34. Red-Horse K, Zhou Y, Genbacev O, et al. Trophoblast differentiation during embryo implantation and formation of the maternal-fetal interface. *J Clin Invest*. 2004;114(6):744–754.
35. Zhou Y, Bellingard V, Feng KT, McMaster M, Fisher SJ. Human cytotrophoblasts promote endothelial survival and vascular remodeling through secretion of Ang2, PlGF, and VEGF-C. *Dev Biol*. 2003;263(1):114–125.
36. Wyatt SM, Kraus FT, Roh CR, Elchalal U, Nelson DM, Sadovsky Y. The correlation between sampling site and gene expression in the term human placenta. *Placenta*. 2005;26(5):372–379.
37. Xu Q, Yan B, Li S, Duan C. Fibronectin binds insulin-like growth factor-binding protein 5 and abolishes Its ligand-dependent action on cell migration. *J Biol Chem*. 2004;279(6):4269–4277.
38. Evans PD, Gilbert SL, Mekel-Bobrov N, et al. Microcephalin, a gene regulating brain size, continues to evolve adaptively in humans. *Science*. 2005;309(5741):1717–1720.
39. Zhou Y, Bianco K, Huang L, et al. Comparative analysis of maternal-fetal interface in preeclampsia and preterm labor. *Cell Tissue Res*. 2007;329(3):559–569.
40. Laivuori H, Kaaja R, Koistinen H, et al. Leptin during and after preeclamptic or normal pregnancy: its relation to serum insulin and insulin sensitivity. *Metab Clin Exp*. 2000;49(2):259–263.
41. Laivuori H, Gallaher MJ, Collura L, et al. Relationships between maternal plasma leptin, placental leptin mRNA and protein in normal pregnancy, pre-eclampsia and intrauterine growth restriction without pre-eclampsia. *Mol Hum Reprod*. 2006;12(9):551–556.
42. Li RH, Poon SC, Yu MY, Wong YF. Expression of placental leptin and leptin receptors in preeclampsia. *Int J Gynecol Pathol*. 2004;23(4):378–385.
43. Lu D, Yang X, Wu Y, Wang H, Huang H, Dong M. Serum adiponectin, leptin and soluble leptin receptor in pre-eclampsia. *Int J Gynaecol Obstet*. 2006;95(2):121–126.
44. McCarthy JF, Misra DN, Roberts JM. Maternal plasma leptin is increased in preeclampsia and positively correlates with fetal cord concentration. *Am J Obstet Gynecol*. 1999;180(3 Pt 1):731–736.
45. Mise H, Sagawa N, Matsumoto T, et al. Augmented placental production of leptin in preeclampsia: possible involvement of placental hypoxia. *J Clin Endocrinol Metab*. 1998;83(9):3225–3229.
46. Ramsay JE, Ferrell WR, Crawford L, Wallace AM, Greer IA, Sattar N. Divergent metabolic and vascular phenotypes in preeclampsia and intrauterine growth restriction: relevance of adiposity. *J Hypertens*. 2004;22(11):2177–2183.
47. Tommaselli GA, Pighetti M, Nasti A, et al. Serum leptin levels and uterine Doppler flow velocimetry at 20 weeks' gestation as markers for the development of pre-eclampsia. *Gynecol Endocrinol*. 2004;19(3):160–165.
48. Vitoratos N, Chrystodoulacos G, Kouskouni E, Salamalekis E, Creatsas G. Alterations of maternal and fetal leptin concentrations in hypertensive disorders of pregnancy. *Eur J Obstet Gynecol Reprod Biol*. 2001;96(1):59–62.
49. Muy-Rivera M, Ning Y, Frederic IO, Vadachkoria S, Luthy DA, Williams MA. Leptin, soluble leptin receptor and leptin gene polymorphism in relation to preeclampsia risk. *Physiol Res*. 2005;54(2):167–174.
50. Patel N, Brinkman-Van der Linden EC, Altmann SW, et al. OB-BP1/Siglec-6. a leptin- and sialic acid-binding protein of the immunoglobulin superfamily. *J Biol Chem*. 1999;274(32):22729–22738.
51. Winn VD, Gormley M, Fisher SJ. The impact of preeclampsia on gene expression at the maternal-fetal interface. *Pregnancy Hypertens*. 2011;1(1):100–108.
52. Brinkman-Van der Linden EC, Hurtado-Ziola N, Hayakawa T, et al. Human-specific expression of Siglec-6 in the placenta. *Glycobiology*. 2007;17(9):922–931.
53. Chez RA. Nonhuman primate models of toxemia of pregnancy. *Perspect Nephrol Hypertens*. 1976;5:421–424.
54. Overgaard MT, Boldt HB, Laursen LS, Sottrup-Jensen L, Conover CA, Oxvig C. Pregnancy-associated plasma protein-A2 (PAPP-A2), a novel insulin-like growth factor-binding protein-5 proteinase. *J Biol Chem*. 2001;276(24):21849–21853.
55. ACOG Committee on Obstetric Practice. ACOG practice bulletin. Diagnosis and management of preeclampsia and eclampsia. Number 33, January 2002. American College of Obstetricians and Gynecologists. *Int J Gynaecol Obstet*. 2002;77(1):67–75.

56. Haram K, Svendsen E, Abildgaard U. The HELLP syndrome: clinical issues and management. A review. *BMC Pregnancy Childbirth*. 2009;9:8.
57. Mittal P, Espinoza J, Hassan S, et al. Placental growth hormone is increased in the maternal and fetal serum of patients with preeclampsia. *J Matern Fetal Neonatal Med*. 2007;20(9):651–659.
58. Zhang H, Long Q, Ling L, Gao A, Li H, Lin Q. Elevated expression of KiSS-1 in placenta of preeclampsia and its effect on trophoblast. *Reprod Biol*. 2011;11(2):99–115.
59. Gack S, Marme A, Marme F, et al. Preeclampsia: increased expression of soluble ADAM 12. *J Mol Med (Berl)*. 2005;83(11):887–896.
60. Kolodkin AL, Tessier-Lavigne M. Mechanisms and molecules of neuronal wiring: a primer. *Cold Spring Harb Perspect Biol*. 2011;3:6.
61. Castro-Rivera E, Ran S, Brekken RA, Minna JD. Semaphorin 3B inhibits the phosphatidylinositol 3-kinase/Akt pathway through neuropilin-1 in lung and breast cancer cells. *Cancer Res*. 2008;68(20):8295–8303.
62. Dayanir V, Meyer RD, Lashkari K, Rahimi N. Identification of tyrosine residues in vascular endothelial growth factor receptor-2/FLK-1 involved in activation of phosphatidylinositol 3-kinase and cell proliferation. *J Biol Chem*. 2001;276(21):17686–17692.
63. Datta SR, Dudek H, Tao X, et al. Akt phosphorylation of BAD couples survival signals to the cell-intrinsic death machinery. *Cell*. 1997;91(2):231–241.
64. Kandel ES, Hay N. The regulation and activities of the multifunctional serine/threonine kinase Akt/PKB. *Exp Cell Res*. 1999;253(1):210–229.
65. Cross DA, Alessi DR, Cohen P, Andjelkovich M, Hemmings BA. Inhibition of glycogen synthase kinase-3 by insulin mediated by protein kinase B. *Nature*. 1995;378(6559): 785–789.
66. Shaw M, Cohen P, Alessi DR. Further evidence that the inhibition of glycogen synthase kinase-3beta by IGF-1 is mediated by PDK1/PKB-induced phosphorylation of Ser-9 and not by dephosphorylation of Tyr-216. *FEBS Lett*. 1997;416(3): 307–311.
67. Doble BW, Woodgett JR. GSK-3: tricks of the trade for a multi-tasking kinase. *J Cell Sci*. 2003;116(Pt 7):1175–1186.
68. Zucman-Rossi J, Benhamouche S, Godard C, et al. Differential effects of inactivated Axin1 and activated beta-catenin mutations in human hepatocellular carcinomas. *Oncogene*. 2007;26(5):774–780.
69. Daugherty RL, Gottardi CJ. Phospho-regulation of Beta-catenin adhesion and signaling functions. *Physiology (Bethesda)*. 2007;22:303–309.
70. Powe CE, Levine RJ, Karumanchi SA. Preeclampsia, a disease of the maternal endothelium: the role of antiangiogenic factors and implications for later cardiovascular disease. *Circulation*. 2011;123(24):2856–2869.
71. Lacroix MC, Guibourdenche J, Fournier T, et al. Stimulation of human trophoblast invasion by placental growth hormone. *Endocrinology*. 2005;146(5):2434–2444.
72. Hiden U, Bilban M, Knofler M, Desoye G. Kisspeptins and the placenta: regulation of trophoblast invasion. *Rev Endocr Metab Disord*. 2007;8(1):31–39.
73. Takeda N, Jain R, LeBoeuf MR, Wang Q, Lu MM, Epstein JA. Interconversion between intestinal stem cell populations in distinct niches. *Science*. 2011;334(6061):1420–1424.
74. Trivedi CM, Zhu W, Wang Q, et al. Hopx and Hdac2 interact to modulate Gata4 acetylation and embryonic cardiac myocyte proliferation. *Dev Cell*. 2010;19(3):450–459.
75. Genbacev O, Donne M, Kapidzic M, et al. Establishment of human trophoblast progenitor cell lines from the chorion. *Stem Cells*. 2011;29(9):1427–1436.
76. Asanoma K, Kato H, Yamaguchi S, et al. HOP/NECC1, a novel regulator of mouse trophoblast differentiation. *J Biol Chem*. 2007;282(33):24065–24074.
77. Mii Y, Taira M. Secreted Wnt "inhibitors" are not just inhibitors: regulation of extracellular Wnt by secreted Frizzled-related proteins. *Dev Growth Differ*. 2011;53(8):911–923.
78. Rawn SM, Cross JC. The evolution, regulation, and function of placenta-specific genes. *Annu Rev Cell Dev Biol*. 2008;24:159–181.
79. Fant M, Farina A, Nagaraja R, Schlessinger D. PLAC1 (Placenta-specific 1): a novel, X-linked gene with roles in reproductive and cancer biology. *Prenat Diagn*. 2010;30(6): 497–502.
80. Ha CT, Wu JA, Irmak S, et al. Human pregnancy specific beta-1-glycoprotein 1 (PSG1) has a potential role in placental vascular morphogenesis. *Biol Reprod*. 2010;83(1):27–35.
81. Lisboa FA, Warren J, Sulkowski G, et al. Pregnancy-specific glycoprotein 1 induces endothelial tubulogenesis through interaction with cell surface proteoglycans. *J Biol Chem*. 2011;286(9):7577–7586.
82. Motran CC, Diaz FL, Gruppi A, Slavin D, Chatton B, Bocco JL. Human pregnancy-specific glycoprotein 1a (PSG1a) induces alternative activation in human and mouse monocytes and suppresses the accessory cell-dependent T cell proliferation. *J Leukoc Biol*. 2002;72(3):512–521.
83. Snyder SK, Wessner DH, Wessells JL, et al. Pregnancy-specific glycoproteins function as immunomodulators by inducing secretion of IL-10, IL-6 and TGF-beta1 by human monocytes. *Am J Reprod Immunol*. 2001;45(4):205–216.
84. Sulkowski GN, Warren J, Ha CT, Dveksler GS. Characterization of receptors for murine pregnancy specific glycoproteins 17 and 23. *Placenta*. 2011;32(8):603–610.
85. Huppertz B, Bartz C, Kokozidou M. Trophoblast fusion: fusogenic proteins, syncytins and ADAMs, and other prerequisites for syncytial fusion. *Micron*. 2006;37(6):509–517.
86. Beristain AG, Zhu H, Leung PC. Regulated expression of ADAMTS-12 in human trophoblastic cells: a role for ADAMTS-12 in epithelial cell invasion? *PLoS One*. 2011;6(4):e18473.
87. Redline RW, Boyd T, Campbell V, et al. Maternal vascular underperfusion: nosology and reproducibility of placental reaction patterns. *Pediatr Dev Pathol*. 2004;7(3):237–249.
88. Glaire MA, El-Omar EM, Wang TC, Worthley DL. The mesenchyme in malignancy: a partner in the initiation, progression and dissemination of cancer. *Pharmacol Ther*. 2012;136(2):131–141.
89. Hubel CA, Roberts JM. Metabolic syndrome and preeclampsia. In: Lindheimer MD, Roberts JM, Cunningham FG, eds. *Chesley's Hypertensive Disorders in Pregnancy*. Amsterdam: Elsevier; 2009:105–128.

90. Harville EW, Viikari JS, Raitakari OT. Preconception cardiovascular risk factors and pregnancy outcome. *Epidemiology*. 2011;22(5):724–730.
91. Chibber R. Child-bearing beyond age 50: pregnancy outcome in 59 cases "a concern?" *Arch Gynecol Obstet*. 2005;271(3):189–194.
92. Wright A, Zhou Y, Weier JF, et al. Trisomy 21 is associated with variable defects in cytotrophoblast differentiation along the invasive pathway. *Am J Med Genet A*. 2004;130A(4):354–364.
93. Genbacev O, Zhou Y, Ludlow JW, Fisher SJ. Regulation of human placental development by oxygen tension. *Science*. 1997;277(5332):1669–1672.
94. Dey SK. How we are born. *J Clin Invest*. 2010;120(4):952–955.
95. George EM, Granger JP. Mechanisms and potential therapies for preeclampsia. *Curr Hypertens Rep*. 2011;13(4):269–275.
96. Kikuchi K, Kishino A, Konishi O, et al. In vitro and in vivo characterization of a novel semaphorin 3A inhibitor, SM-216289 or xanthofulvin. *J Biol Chem*. 2003;278(44):42985–42991.

CHAPTER 6

Angiogenesis and Preeclampsia

S. ANANTH KARUMANCHI, SAROSH RANA AND ROBERT N. TAYLOR

Editors' comment: *When Chesley wrote his original monograph he, like most of the scientific community, was unfamiliar with the concept of angiogenesis. Even in the multi-authored second edition, in which the endothelial cell hypothesis of preeclampsia was first proposed, the subject "angiogenesis" was not elaborated. This millennium has witnessed a very different story. Starting with preliminary observations of the roles of angiogenic proteins in normal and abnormal placentation*[1–4] *a signal paper by Maynard et al., published in the* Journal of Clinical Investigation *in 2003*[5] *and now cited widely, research into angiogenesis and preeclampsia took off with rocket speed. This publication, followed by an explosion of related research into pro- and antiangiogenic factors, opened the field, which now suggests a role for these proteins in (1) predicting and diagnosing the disorder, (2) in mediating the signs and symptoms of preeclampsia, and (3) as potential therapeutic or even preventive use in the disease. In the current edition, the authors update exciting advances in our understanding of the role of angiogenic factors in the pathogenesis of preeclampsia and expand their use as biomarkers of disease complications.*

INTRODUCTION

Vascular development occurs through angiogenesis and vasculogenesis.[6,7] Angiogenesis is the process of neovascular sprouting or branching from pre-existing blood vessels, while vasculogenesis is the process of blood vessel generation *de novo* from angioblast precursor cells.

The human placenta undergoes extensive angiogenesis and vasculogenesis throughout development.[8] Additionally, the developing placenta undergoes a process of vascular mimicry (also referred to as pseudo-vasculogenesis) as cytotrophoblasts convert from an epithelial to an endothelial phenotype[9] (see Chapter 5). When placental vascular development is deranged, the success of the pregnancy is jeopardized and serious complications such as preeclampsia and fetal growth restriction can occur. This chapter will discuss placental vascular development during health and in disease, with an emphasis on the role of placental antiangiogenic factors in the pathogenesis of the maternal syndrome of preeclampsia. Also discussed is our view that preeclampsia, at least the form of the disorder that is associated with adverse maternal/fetal outcomes, is a specific entity whose phenotypes relate to angiogenic imbalance, and that measurements of these factors help identify the severe form of the disease, and suggest useful management strategies for these patients.

PLACENTAL VASCULAR DEVELOPMENT IN HEALTH

The placenta is a highly vascular organ, containing both embryonic and maternal blood vessels.[10] Thus to understand the role of angiogenesis in successful placentation, a summary of fundamental steps necessary for this critical developmental milestone follows. These include: (1) trophoblast invasion of the superficial maternal decidua; (2) vascularization of the placental bed to establish and maintain fetoplacental nutrient delivery and waste disposal; and (3) subsequent remodeling of the maternal spiral arteries by the trophoblast, enabling robust uteroplacental perfusion. This latter function is critical as placental trophoblasts must establish a circulation that permits adequate maternal-fetal exchange. Both vasculogenesis and angiogenesis are involved in these processes.

Placental Vasculogenesis

During embryonic development, the blastocyst differentiates into two cell populations: an outer polarized trophoectoderm and the non-polarized inner cell mass. The trophoectoderm gives rise to extra-embryonic membranes, whereas the inner cell mass is destined to form the embryo proper.

Extra-embryonic mesenchymal cells give rise to cores that penetrate into the center of the cytotrophoblast columns. These mesenchymal cells differentiate into endothelial cells, forming the first capillaries of the placental

Chesley's Hypertensive Disorders in Pregnancy.
ISBN: 978-0-12-407866-6

DOI: http://dx.doi.org/10.1016/B978-0-12-407866-6.00006-7

vasculature. In humans, placental vasculogenesis is evident by approximately 21–22 days post-conception.[8] At this stage, cords of hemangiogenic cells are present and some demonstrate primitive lumen formation. These cords further develop so that by approximately day 32 post-conception, most villi show the presence of capillary structures.[8] The highly proliferative trophoblasts ultimately invade the entire endometrium, the outer one-third of the myometrium, and the maternal circulation. Hypoxia is an important driving force for trophoblast proliferation.[11]

As trophoblasts invade the uterine vasculature, they are exposed to increasing concentrations of oxygen, at which time they exit the cell cycle and differentiate. Further development leads to the penetration of cytotrophoblastic cones into the syncytiotrophoblastic mass and the development of lacunae which eventually become the intervillous space. Continuing growth and differentiation of the trophoblasts leads to branching of the trophoblast villi and the shaping of a placental labyrinth or intervillous space, where the fetal/maternal exchange of oxygen and nutrients occurs. Trophophoblasts invade maternal tissue with a variable depth of invasion among species, the deepest known being in humans.

Maternal Vascular Remodeling

The formation of adequate maternal–placental circulation requires remodeling of maternal blood vessels (namely, the spiral arteries). In humans, during the mid-late first trimester, the trophoblasts invade deeply through the endometrium and into the superficial part of the myometrium, completely remodeling the proximal ends of the maternal spiral arteries. Through the open endings of the maternal vessels that are created by trophoblast invasion, maternal blood is released into the intervillous space, flows around the trophoblast villi, and is drained by spiral veins (see Fig. 6.1).

When cytotrophoblasts invade maternal spiral arteries, they replace the luminal endothelial cells, a process common in all species with hemochorial placentation. During this process, the endovascular cytotrophoblasts convert from an epithelial to endothelial phenotype, a process referred to as pseudovasculogenesis or vascular mimicry (Fig. 6.1; see also Chapter 5).[9] Thus cytotrophoblast stem cells lose epithelial markers, such as E-cadherin and $\alpha_6\beta_3$ integrin, and gain endothelial markers, such as vascular endothelial-cadherin (VE-cadherin) and $\alpha_v\beta_3$ integrin.[9,13]

In addition to replacing maternal endothelial cells, cytotrophoblasts also remodel the highly muscular tunica media of spiral arteries, a process that is dependent on the enzyme membrane metalloproteinase 9 (MMP-9).[14,15] This transforms the maternal high resistant vessels into larger low-resistance capacitance vessels.[16] Uterine blood flow during pregnancy increases more than 20-fold, and the functional consequence of spiral artery remodeling maximizes the capacity of the maternal–placental circulation by providing sufficient blood supply for placenta and fetus at low blood pressure.[17] Of note, spiral arteries from both the implantation and non-implantation regions display these physiological changes.

The mechanisms underlying these changes remain unknown.[18] Placental oxygen tension has been suggested to be one of the major regulators of cytotrophoblast migration and differentiation.[11] It has further been hypothesized that decidual natural killer (NK) cells and/or activated macrophages play a role in this vascular remodeling.[19]

Fetal Circulation and Placental Villous Angiogenesis

The fetal circulation enters the placenta via the umbilical vessels. Inside the placenta, fetal vessels branch successively into units within the cotyledons and then into capillary loops within the chorionic villi.[10] From post-conception day 32 until the end of the first trimester, the endothelial tube segments formed by vasculogenesis in the placental villi are transformed into primitive capillary networks by the balanced interaction of two parallel mechanisms; (a) elongation of pre-existing tubes by non-branching angiogenesis, and (b) ramification of these tubes by lateral sprouting (sprouting angiogenesis). A third process, termed intussusceptive microvascular growth, rarely contributes. In the third month of pregnancy, some of the centrally located endothelial tubes of immature intermediate villi achieve large diameters of 100 μm and more. Within a few weeks, they establish thin media- and adventitia-like structures by concentric fibrosis in the surrounding stroma and by differentiation of precursor pericytes and smooth muscle cells expressing α- and γ-smooth muscle actins in addition to vimentin and desmin. This is followed quickly by the expression of smooth muscle myosin.[8,20] These vessels are forerunners of the villous arteries and veins and are developmentally regulated through the platelet-derived growth factor (PDGF) pathway.[21]

After post-conception week 24 and continuing through term, patterns of villous vascular growth switch from the prevailing branching angiogenesis to a prevalence of non-branching angiogenesis. Analysis of proliferation markers at this stage reveals a relative reduction of trophoblast proliferation and an increase in endothelial proliferation along the entire length of these villous structures, resulting in non-sprouting angiogenesis by proliferative elongation. The final length of these peripheral capillary loops exceeds 4000 μm and they grow at a rate which exceeds that of the villi themselves, resulting in coiling of the capillaries.[22,23] The looping capillaries bulge towards, and obtrude into, the trophoblastic surface and thereby contribute to formation of the terminal villi. Each of the latter is supplied by one or two capillary coils and is covered by an extremely thin (<2 μm) layer of trophoblasts that contributes to the so-called vasculosyncytial membranes. These are the

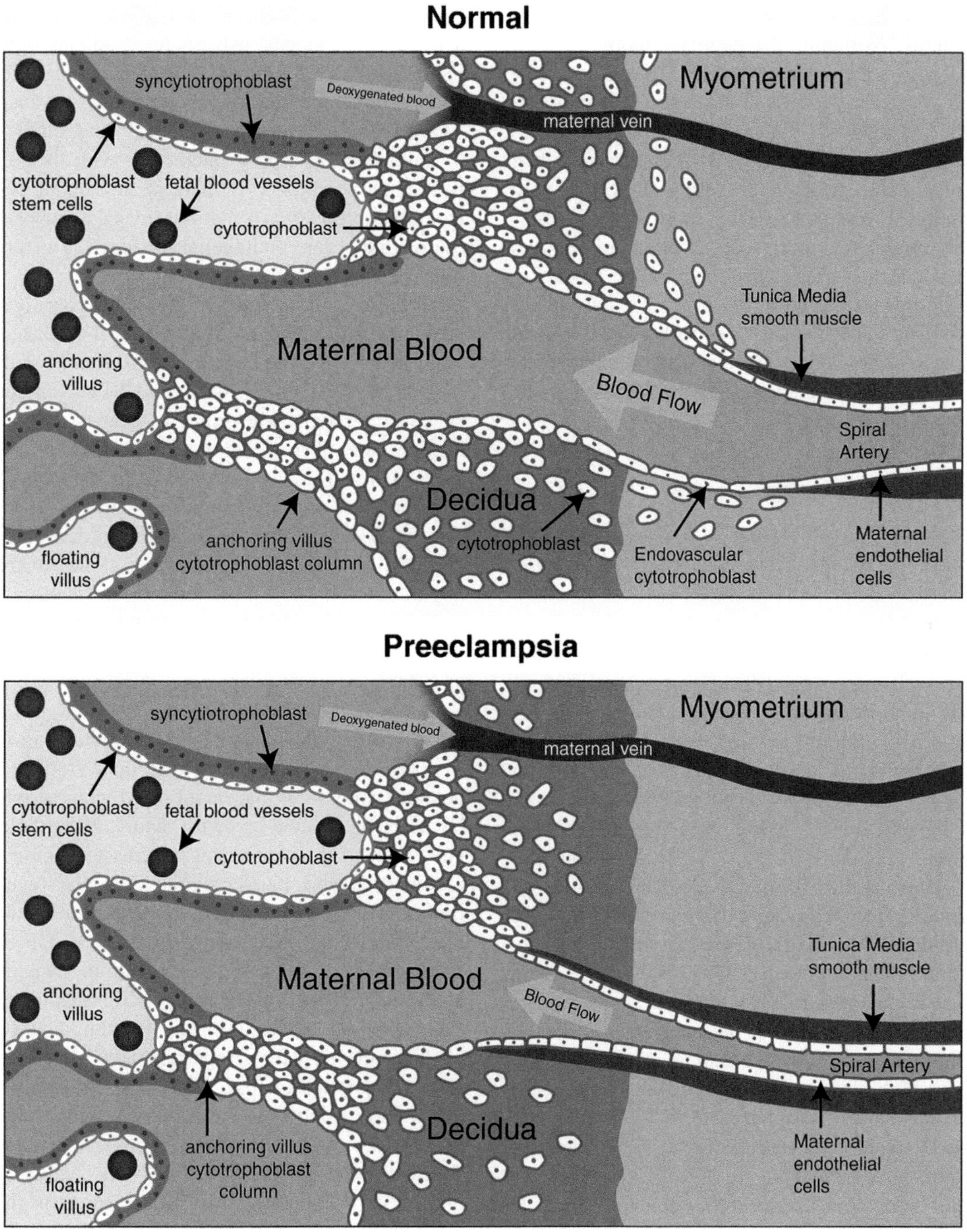

FIGURE 6.1 A schematic of placental vascular remodeling in health (upper panel) and in disease – preeclampsia (lower panel). Exchange of oxygen, nutrients, and waste products between the fetus and the mother depends on adequate placental perfusion by maternal spiral arteries. Blood from the intervillous space is returned to the mother's circulation via spiral maternal veins noted above. In normal placental development, cytotrophoblasts of fetal origin invade the maternal spiral arteries, transforming them from small-caliber resistance vessels to high-caliber capacitance vessels capable of providing adequate placental perfusion to sustain the growing fetus. During the process of vascular invasion, the cytotrophoblasts undergo a transformation from an epithelial to an endothelial phenotype, a process referred to as "pseudovasculogenesis" (upper panel). In preeclampsia, cytotrophoblasts fail to adopt an invasive endothelial phenotype. Instead, invasion of the spiral arteries is shallow and they remain small-caliber, resistance vessels (lower panel). This is thought to lead to placental ischemia and secretion of antiangiogenic factors. Figure reproduced with permission from Lam et al.[12] (This figure is reproduced in color in the color plate section.)

principal sites of diffusional exchange of gases between mother and fetus. Normally, the capillary loops of 5–10 such terminal villi are connected to each other in series by the slender, elongated capillaries of the central mature intermediate villus.

The fetal vessels (chorionic vessels) from the individual cotyledons of the placenta unite at the placental surface to form the umbilical vessels that then traverse the umbilical cord. The umbilical cord consists of one vein and two arteries. The connective tissue surrounding these vessels in the umbilical cord is referred to as Wharton's jelly. Most umbilical cords are twisted at birth, probably related to fetal activity *in utero*. The umbilical vein carries oxygenated blood from the placenta to the fetus and the umbilical arteries carry deoxygenated blood back to the placenta.

Angiogenic Factors and Placentation

Placental vascularization involves a complex interaction of several regulatory factors.[24] The list of pro- and antiangiogenic molecules involved in placental vascular development has expanded exponentially over the past decade and is reviewed in detail elsewhere.[25,26] The families of vascular endothelial growth factor (VEGF) and angiopoietin (Ang) gene products are those most extensively studied.[27]

VEGF-A was initially defined, characterized, and purified for its ability to induce vascular leak and permeability, as well as for its ability to promote vascular endothelial cell proliferation.[28,29] Members of this family include VEGF-A, VEGF-B, VEGF-C, VEGF-D and placental growth factor (PlGF). VEGF-A is an endothelial-specific mitogen and survival factor that exists in four major isoforms: VEGF121, VEGF165, VEGF189, and VEGF206. Inactivation of a single VEGF allele results in embryonic lethality in heterozygous embryos at days 11–12. Additionally, significant defects in placental vasculature are observed, implicating VEGF in placental vascular development.[30] The high-affinity receptor tyrosine kinases for VEGF-A include Flt-1 (also referred to as VEGFR-1) and Kinase-insert Domain containing Receptor (KDR, human)/Flk-1 (murine), also known as VEGFR-2.[31] KDR mediates the major growth and permeability actions of VEGF, whereas Flt-1 may have a negative role, either by acting as a decoy receptor or by suppressing signaling through KDR. Placental growth factor (PlGF), the first VEGF relative identified, was found to be abundantly expressed in the placenta. It acts by binding to Flt-1 but not KDR.[32]

PIGF can potentiate the angiogenic activity of VEGF. Although reproduction and embryonic angiogenesis defects were not observed in mice with an isolated PIGF –/– genotype,[33–35] the cross-sectional area of the midpregnancy placentas was decreased by 40% in PlGF null mice.[36] VEGF-C and VEGF-D, based on their ability to bind the lymphatic-specific Flt-3 receptor (also known as VEGFR-3), appear to be important for lymphatic development.[37,38]

Alternative splicing of Flt-1 results in the production of a truncated, endogenously secreted antiangiogenic protein referred to as sFlt-1, which lacks the cytoplasmic and transmembrane domain but retains the ligand-binding domain.[39] Thus, sFlt-1 can antagonize VEGF and PlGF by binding to them and preventing interaction with their endogenous full-length receptors.[3,39] There is substantial evidence that increased production of sFlt-1 plays a major pathogenic role in the severe endothelial dysfunction of preeclampsia (discussed below).[40]

The Tie-2 receptor binds a family of four ligands termed angiopoietins (Ang-1 to Ang-4). Ang-1-mediated activation of Tie-2 promotes endothelial survival and capillary sprouting.[41–43] The effects of Ang-1 and VEGF on sprouting are synergistic,[41] reflecting different receptor and signal transduction pathways. In addition to its direct effects on endothelial cells, Tie-2 activation has been reported to induce maturation of adjacent smooth muscle and pericyte precursors via paracrine action of endothelial-derived factors such as PDGF-B.[44] Ang-1 has also been shown to inhibit capillary permeability,[45,46] preventing plasma leakage in response to both VEGF and histamine. In the presence of abundant VEGF, Ang-2 is thought to destabilize vascular networks and facilitate sprouting, e.g., during tumor growth.[47,48] Conversely, with low ambient VEGF levels, Ang-2 may cause vessel regression, e.g., in corpus luteum involution.[49]

There are few *in vivo* functional studies in humans demonstrating how placentation is regulated on a molecular basis by vascular growth factors. Nevertheless what is known about the expression patterns of angiogenic factors provides insights into potential molecular regulatory processes. VEGF and its receptors have been localized in the placenta of both primates and humans. VEGF is expressed in maternal decidual cells[50–52] and in invading cytotrophoblasts.[51,53] As for the two VEGF receptors, only Flt-1 expression is localized in the invading extravillous trophoblast during early pregnancy.[50,51,54] Such observations raise the possibility that VEGF-Flt-1 interactions contribute to early trophoblast invasion. Later in gestation, VEGF is localized to cytotrophoblast cells,[50,53,55] and Flt-1 to extravillous trophoblasts,[50,53,55,56] all within the trophoblast tree or columns. This pattern suggests that in addition to regulating trophoblast invasion, the VEGF-Flt-1 axis plays an important role in the coordination of trophoblast differentiation and migration. Interestingly, more recently sFlt-1 has been primarily localized to the syncytiotrophoblast,[57] suggesting that its primary role may be secretion into the maternal bloodstream to regulate systemic vascular homeostasis.

VEGF-A induces cytotrophoblast invasion *in vitro*, an effect blocked by exogenous sFlt-1.[55] Although homozygous knock-out studies for both Flt-1 and Flk-1 in mice show defective fetal and placental vasculogenesis and angiogenesis,[58,59] we are unaware of definitive *in vivo* evidence of impaired trophoblast invasion. This may be because the normal invasion of trophoblasts is relatively shallow even at baseline in rodents,[60] contrasting with human placentas, where invasion is robust. *In vitro* studies using trophoblast cultures suggest that PlGF and VEGF-C (two related growth factors) are also expressed by the invading cytotrophoblasts and may contribute to cytotrophoblast invasion and differentiation via Flt-1 and Flt-3 signaling.[61]

The development of the villous tree occurs contemporaneously with the formation of the fetoplacental vascular system. The latter process involves differentiation and proliferation of fetal endothelial cells, tubule formation and vessel stabilization. The KDR receptor is exclusively expressed on endothelial or mesenchymal cells from which endothelial cells differentiate,[50,53,54,56,62] whereas VEGF expression is localized to the trophoblast.[57] This suggests that endothelial cell differentiation, migration and proliferation, essential steps for building the primary vascular network, are mediated by VEGF/KDR ligand–receptor signaling in a paracrine system.

The angiopoietins are also expressed during the early placentation period in marmosets,[63] indicating their involvement in the regulation of trophoblast growth. In this species Ang-1 is highly expressed in the syncytiotrophoblast, whereas its receptor, Tie-2, is located in the cytotrophoblast.[50,53] Similar observations have been made in the human placenta,[53,64] where Ang-1 has been shown to stimulate trophoblast growth and migration *in vitro*,[64] and Ang-1 gene expression increases as gestation progresses.[50] Thus, Ang-1 expression appears to trigger the in-growth of cytotrophoblast cones into the syncytiotrophoblast, whereas its relatively higher expression later in pregnancy may be required for branching of the villi and shaping the intervillous space.

Fetal capillaries within the placental villi are thin-walled in order to allow oxygen diffusion, while the chorion vessels are stabilized by a thick wall of pericytes, smooth muscle cells or both that ensure their task of collecting and draining fetal placental blood. Tie-2 has been shown to be expressed at high levels in the endothelium of chorion vessels, and at low levels in the fetal capillaries of the villi.[50] These results suggest that the Ang-1/Tie-2 ligand–receptor pair acts on the fetal vasculature, especially on chorion vessels to induce stability. In humans, Ang-1 is secreted into the media of stem villus vessels at term,[64] which is also consistent with its reported paracrine role in vessel maturation and stabilization.

Maternal vessels must be remodeled to attain an effective uteroplacental circulation. In humans, the trophoblast invasion process is so deep that the proximal parts of maternal spiral arteries become completely digested and the intervillous space is filled by their open endings. In addition, with Ang-2 produced by the cytotrophoblast and Tie-2 expressed in the maternal endothelium,[59,64,65] a paracrine mechanism for maternal vascular remodeling is established.[66]

In non-human primates, highest expression of Ang-2 mRNA is detected during early gestation, when maternal vascular remodeling takes place. Ang-2 is a plausible candidate for induction of maternal vascular transformation because it destabilizes the vasculature.[49] This destabilization process is a local event, which may be driven in part by Ang-2/Tie-2. In humans, although the remodeling process of maternal vessels may be supported by Ang-2/Tie-2, it is primarily driven by aggressive trophoblast invasion. Finally, TGF-β1 and 3, two angiogenic morphogens made in the maternal decidua and the syncitiotrophoblast, also affect cytotrophoblast invasion and differentiation through endoglin signaling.[67,68]

In summary, substantial evidence supports a critical role for angiogenic growth factors in hemochorial placentation of both humans and other primates. In our estimation, the most comprehensive studies investigating VEGF and angiopoietins, and their receptors across gestation, has been undertaken in marmoset monkeys, with results indicating a tight spatial and temporal regulation of placentation and angiogenesis.[50,63] These findings led to the hypothesis that VEGF/Flt-1 and Ang1/Tie-2 pairs are critically involved in trophoblast differentiation and invasion; VEGF/KDR and Ang-1/Tie-2 may trigger fetoplacental vascular development, whereas Ang-2/Tie-2 may support the remodeling processes of the maternal vasculature.[50]

The characterization of antiangiogenic proteins in the placenta, in contrast to pro-angiogenic proteins, has received less attention. One important antiangiogenic protein that has been well studied is sFlt-1, a potent antiangiogenic molecule and produced in abundant quantities within the placenta.[2] Its production, predominantly from cytotrophoblasts, increases as pregnancy progresses.[3,39] Other antiangiogenic proteins expressed in the placenta include thrombospondin-1, endostatin, and truncated fragments of prolactin, but their roles during normal placentation are unclear. More recently, soluble endoglin, a novel antiangiogenic protein made by the syncytiotrophoblast that acts by inhibiting TGF-β signaling, was reported to play an important role in the pathogenesis of the maternal syndrome of preeclampsia.[69] However, the precise role of soluble endoglin during normal placentation remains unknown.

In general, the pro-angiogenic proteins are most highly expressed early in pregnancy and probably account for both placental angiogenesis and the increase in placental mass that accompanies fetal development.[24] Toward term, antiangiogenic factors are increasingly expressed, possibly in preparation for delivery.[24] In addition to the gestational age-dependent distribution of these proteins, there are also regional differences in their expression. For example, term human placental extracts derived from the decidual plate express stronger antiangiogenic activity compared with chorionic villus extracts.[70]

Natural Killer Cells and Placental Vascular Development

Natural killer (NK) cells are considered important mediators of innate immunity, and those present at the maternal–fetal interface have been recently noted to play key roles during normal placental vascular remodeling. During the first trimester of human pregnancy, uterine NK (uNK) cells are a major cell population at the maternal–fetal interface, accounting for 70% of the local lymphocytes.[71] In contrast, peripheral blood NK cells (pNK) account for only up to 15% of circulating lymphocytes. The abundance of NK cells in the decidua has prompted the idea that these cells might play a role in pregnancy support and maintenance. Initial evidence came from studies involving NK-cell-deficient mice, which manifested defective decidual vessel remodeling.[71] Several candidate molecules of NK cell origin, such as gamma interferon, have been proposed to account for the NK-cell-mediated vascular remodeling; however, definitive evidence of this mechanism in humans is still lacking. Recent genetic studies suggest that susceptibility to preeclampsia may be influenced by polymorphic HLA-C ligands and killer cell receptors (KIR) present on NK cells. Preeclampsia is much more prevalent in women homozygous for the inhibitory KIR haplotypes (AA) than in women homozygous for the stimulator KIR BB. The effect is strongest if the fetus is homozygous for the HLA-C2 haplotype.[72] Based on these observations, it is believed that in normal pregnancies uNK cell activation, through interaction with HLA-C on extravilous trophoblasts, promotes placental development and maternal decidual spiral artery modification by extra-villous cytotrophoblasts. Insufficient uNK cell activation might therefore blunt this process, resulting in incomplete decidual artery remodeling, thereby increasing the risk of pregnancy complications.[72,73] Recent evidence showing secretion of trophoblast migration promoting factors and angiogenic factors by human uNK cells upon activation supports this hypothesis.[74] Interestingly, consistent with this hypothesis, uNK cell interactions with trophoblasts were found to be impaired in humans at risk of subsequent preeclampsia.[75]

ANGIOGENIC IMBALANCE IN PREECLAMPSIA

During the first two trimesters, as detailed above, the spiral arteries undergo extensive remodeling (see Fig. 6.1). Teleologically this is believed to reduce maternal blood flow resistance and to increase uteroplacental perfusion. Abnormal spiral artery remodeling is a key pathological feature of both preeclampsia and intrauterine growth retardation. Other pregnancy complications such as placental abruption and intrauterine fetal death (IUFD) may also be due to defective placental vascular development and coagulation abnormalities, but the mechanisms mediating the latter complications are less well understood. The hypothesis that defective trophoblastic invasion with resulting uteroplacental hypoperfusion leads to preeclampsia is supported by both animal and human studies.[76,77] Placentas from women with severe preeclampsia frequently show evidence of infarction, and histological examination often reveals narrower maternal vessels, and a distinctive lesion know as acute atherosis (see Chapter 8).[78] None of these lesions are absolutely specific for preeclampsia, and they may be absent in up to half of placentas of women diagnosed clinically as preeclampsia.

Doppler ultrasound estimation of uteroplacental blood flow is usually diminished while uterine vascular resistance is often increased in preeclamptic women.[79] Placental ischemia induced by mechanical constriction of the uterine arteries or the aorta produces hypertension, proteinuria and, variably, glomerular endotheliosis in several animal species[76,79,80] (see Chapter 10). However, placental ischemia alone may not be sufficient to produce preeclampsia, as it is detected in many instances of intrauterine growth restriction in women without preeclampsia. Thus, though uteroplacental ischemia may be an important trigger of preeclampsia, it may be absent in some cases, or alternatively the maternal response to placental ischemia is variable.

Soluble Antiangiogenic Factors in Preeclampsia

The role of maternal endothelial dysfunction in producing the clinical manifestations of preeclampsia has been studied extensively since the 1980s.[81,82] Roberts et al. proposed that endothelial cell injury caused increased sensitivity to pressor agents, vasoconstriction, and activation of the coagulation cascade that served as the basis of preeclampsia.[83] Evidence of endothelial involvement in the kidneys was noted as early as the 1920s and by the 1960s, through the lens of electron microscopy, "glomerular capillary endotheliosis" became considered the characteristic renal lesion of preeclampsia. The lesion, described in detail in Chapters 15 and 16, is characterized by occlusion of the glomerular capillaries by swollen endothelial cells.[84]

Blood from preeclamptic patients demonstrates markers of endothelial injury, including increased levels of von Willebrand Factor,[85–87] cellular fibronectin,[88–90] and thrombomodulin.[91,92] Circulating prostacyclins (normally produced in healthy endothelial cells) are decreased in preeclamptic patients.[93] Blood vessels from preeclamptic patients reveal decreased endothelial-mediated vasodilator ability,[94,95] whereas plasma endothelin-1 levels are increased.[96] In summary, extensive data from multiple studies support the notion that the maternal serum in preeclampsia has soluble factors that mediate endothelial dysfunction. Moreover, serum or plasma from preeclampsia patients alters endothelial cell phenotype *in vitro*, including altered expression of vascular cell adhesion molecules, nitric oxide, and changes in prostaglandin balance.[97,98] Interestingly, preeclamptic patients have also been noted to have decreased skin capillary density compared with healthy pregnant patients, a finding suggesting generalized angiogenesis may be defective in preeclampsia.[99] Finally, women with a prior history of preeclampsia have an increased risk of remote ischemic heart disease, stroke, and hypertension.[100] The association between preeclampsia and subsequent development of cardiovascular disease also points to the existence of systemic endothelial dysfunction in these individuals.

Soluble Fms-Like Tyrosine Kinase I (sFLT-1 or sVEGFRI)

In early 2000, studies were designed whose aim was to identify the soluble factors that mediate maternal endothelial dysfunction in preeclampsia. Gene expression profiling of placental tissue from women with and without preeclampsia, using microarray chips, revealed that messenger RNA for sFlt-1 was dramatically upregulated in preeclamptic placentas.[5] sFlt-1 is a secreted protein, derived from a splice variant of the VEGF (vascular endothelial growth factor) receptor Flt-1 mRNA that lacks the transmembrane and cytoplasmic domain of the membrane-bound receptor.[39] Circulating in the blood, it acts as a potent antagonist to VEGF and PlGF, factors made by normal placenta. VEGF is also constitutively expressed by glomerular podocytes as well as by other supporting cells and is thought to maintain fenestrations within the glomerular and hepatic sinusoidal endothelia.[101] Systemic levels of sFlt-1 in patients with preeclampsia are greatly increased prior to delivery and decrease to baseline 48–72 h after delivery. Using the endothelial tube formation assay, an established *in vitro* model of angiogenesis, the authors observed that serum from patients with preeclampsia inhibited endothelial tube formation, but by 48 h postpartum this antiangiogenic effect had disappeared from the serum, suggesting that the inhibition of angiogenesis was caused by a circulating factor released from the placenta. When sFlt-1 was added to serum from normotensive pregnant subjects (at concentrations mimicking levels measured in subjects with preeclampsia), endothelial tube formation was blocked, recreating the effects noted with serum from preeclamptic patients. These effects could be reversed by adding exogenous VEGF and PlGF.[5] These findings were also confirmed using supernatants from cultured preeclamptic villous explants.[102] Hence, it is believed that an excess of circulating sFlt-1 levels may lead to an antiangiogenic state, causing endothelial dysfunction, and the clinical syndrome of preeclampsia.

Contemporaneous studies evaluating the expression of placental growth factor (PlGF) concentrations in the plasma of normal and preeclamptic pregnancies indicated that this also is an attractive biomarker. High sFlt-1 levels correlate inversely with decreased free PlGF levels during preeclampsia.[5,103–107] In one report, uterine vein sFlt-1 concentrations were 4- to 5- fold higher than peripheral venous levels, suggesting that the predominant source of maternal sFlt-1 was of placental origin.[108] More recently, several new isoforms of sFlt-1 with varying mRNA lengths have been described to be upregulated in preeclamptic placentas.[109,110] In particular, a novel isoform of sFlt-1, referred to as sFlt-1-14, was recently described, expressed only in humans and not in rodents.[109] Interestingly, this sFlt-1-14 has been noted to be the predominant VEGF inhibitor circulating in preeclamptic circulation.[109] Furthermore, the source of sFlt-1-14 was located to the syncitial knots, suggesting that degeneration and/or aging of the placenta may be one mechanism for the upregulated sFlt-1 in preeclampsia. Work from numerous groups (summarized in Chapter 8) suggests that syncytial knots in the preeclamptic tissue are shed systemically in circulation and that circulating syncytial microparticles carry sFlt-1 to distal target organs.[57,111–113]

Exogenous gene transfer of sFlt-1 into pregnant rats using an adenoviral vector produced hypertension, proteinuria, and glomerular endotheliosis, the renal lesion characteristic of preeclampsia.[5] The glomerular lesion in these experimental animals, consisting of severe glomerular endothelial swelling (with loss of endothelial fenestrae) and generally preserved foot processes in the setting of heavy proteinuria, is striking in its resemblance to the renal histological findings in human preeclampsia[5] (see Fig. 6.2). Some of these findings have now been reproduced in mice administered adenovirus expressing sFlt-1 or recombinant sFlt-1 protein.[115,116]

The effects described above were also observed with sFlt-1 gene transfer in non-pregnant animals, suggesting that the effects of sFlt-1 on the maternal vasculature were direct and not dependent on the presence of a placenta. Furthermore, when a soluble form of VEGF receptor-2 (sFlk-1) (which does not antagonize PlGF) was given exogenously it did not induce a preeclamptic phenotype in pregnant rats, suggesting that antagonism of both VEGF and PlGF is necessary to induce the maternal syndrome. Finally, exogenous administration of VEGF-121, a soluble, circulating isoform of VEGF that lacks the heparin binding domain, rescued the preeclamptic phenotype in animal models of preeclampsia without adverse effects to the fetus.[117] Similarly recombinant PlGF has also been shown to ameliorate preeclamptic signs and symptoms in the sFlt-1 overexpression mouse model of preeclampsia.[118] Hence, it has been concluded that excess production of sFlt-1 by the placentas of preeclamptic women might be responsible for the hypertension and proteinuria by VEGF and PlGF signaling and that reduction of sFlt-1 may be one strategy to combat preeclampsia. It is still unknown whether the excess sFlt-1 made in preeclampsia is a primary phenomenon or secondary to a pathophysiological trigger such as placental ischemia or agonistic autoantibodies to the angiotensin 1 receptor (reviewed in Chapter 15) or impaired hemoxygenase expresion.[119]

Several risk factors for preeclampsia can also be correlated with increased sFlt-1 levels. These include multigestational pregnancies, high-altitude pregnancies, trisomy 13, and nulliparity.[57,120–123] Furthermore, the low level of circulating sFlt-1 typically found in smokers may explain

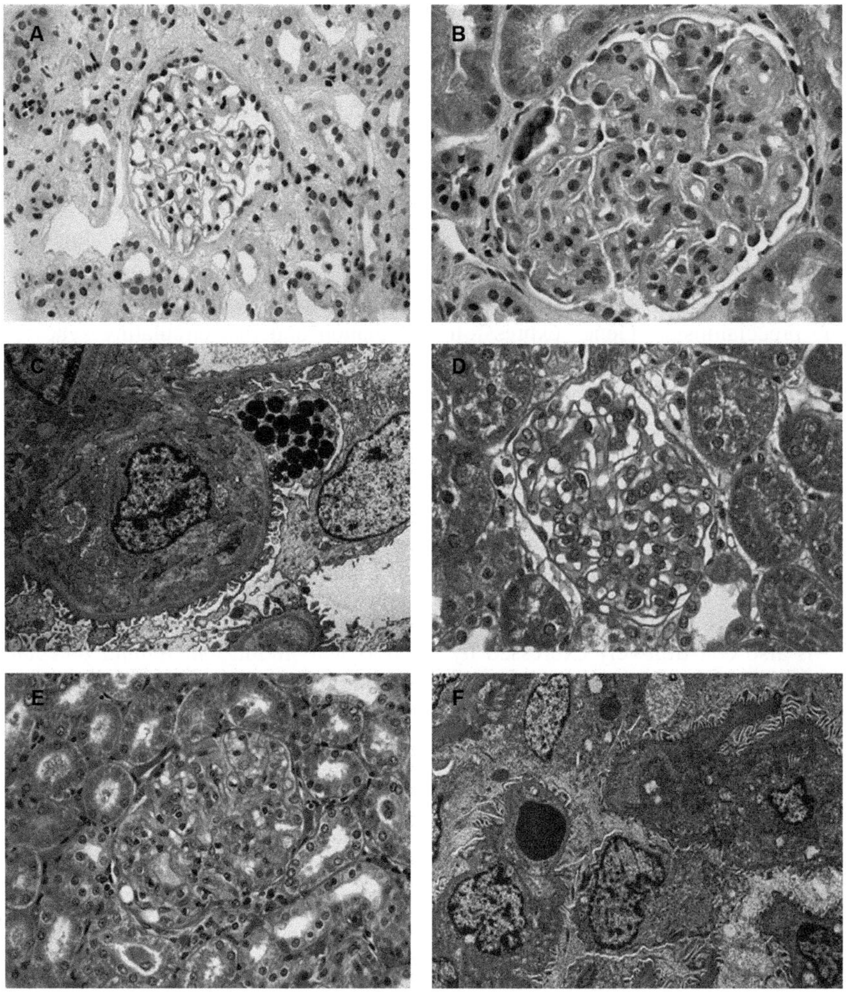

FIGURE 6.2 Glomerular endotheliosis. (A) Normal human glomerulus. (B) Human preeclamptic glomerulus of 33-yr-old woman with a twin gestation and severe preeclampsia at 26 weeks gestation. The urine protein/creatinine ratio was 26 at the time of biopsy. (C) Electron microscopy of a glomerulus from the same patient. Note the occlusion of the capillary lumens by the swollen cytoplasm of endocapillary cells. Podocyte cytoplasm shows protein resorption droplets but relatively intact foot processes. Original magnification 1500×. (D) Control rat glomerulus: note normal cellularity and open capillary loops. (E) sFlt-1 treated rat: note similar occlusion of the capillary lumens by swollen endothelial cells with minimal increase in cellularity. (F) Electron microscopy of a sFlt-1 treated rat: note similar occlusion of capillary loops by swollen endocapillary cell cytoplasm accompanied by the relative preservation of podocyte foot processes. Original magnification 2500×. All light photomicrographs are of H&E sections taken at the identical original magnification of 40×. Figure reproduced with permission from Karumanchi et al.[114] (This figure is reproduced in color in the color plate section.)

the surprising decreased incidence of preeclampsia in this group.[124,125]

Antagonism of VEGF and PlGF may have a pathogenic role in the hypertension and proteinuria noted in preeclampsia. VEGF induces nitric oxide and vasodilatory prostacyclins in endothelial cells, suggesting a role in decreasing vascular tone and blood pressure.[126,127] Exogenous VEGF has been noted to accelerate renal recovery in rat models of glomerulonephritis and experimental thrombotic microangiopathy.[128,129] More recently, exogenous VEGF was shown to ameliorate cyclosporine-related hypertension, endothelial dysfunction, and nephropathy.[130] The tissues targeted in preeclampsia (such as the glomerulus or the hepatic sinusoids) have fenestrated endothelia, and it has been shown that VEGF induces endothelial fenestrae *in vitro*.[131] Additionally reduction by 50% of VEGF production in +/− transgenic mouse glomerulus leads not only to glomerular endotheliosis but also to loss of glomerular endothelial fenestrae.[132] Interestingly, several antiangiogenic compounds, including anti-VEGF antibodies used to treat malignancy-related angiogenesis, have also been associated with hypertension, proteinuria and reversible posterior leukoencephalopathy.[133–135] Collectively, these data suggest that VEGF is important not only in blood pressure regulation, but also in maintaining the integrity of the glomerular filtration barrier. Thus antagonism of VEGF signaling, such as with excess sFlt-1, might lead to endothelial dysfunction, proteinuria, and hypertension. Finally, the excess sFlt-1 production in preeclampsia is consistent with the evolutionary

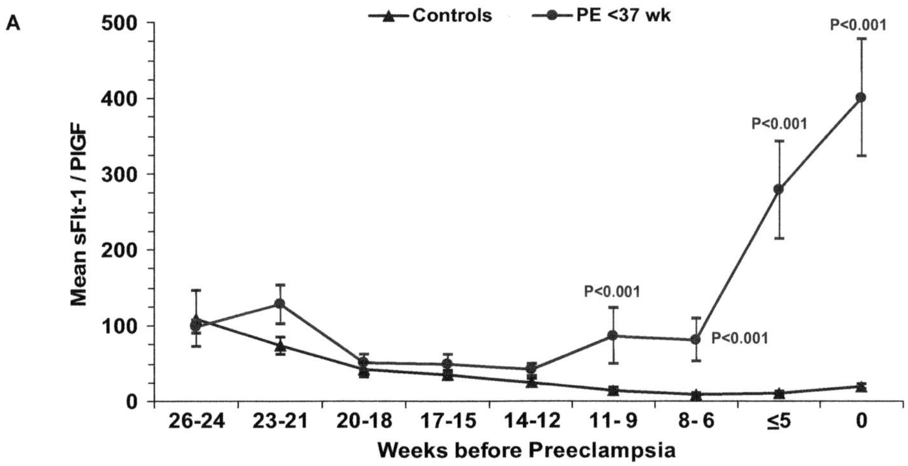

Specimen pairs	4	16	19	16	9	25	25	21	40
Specimen gestational age (day)									
Controls	86	98	118	123	122	171	196	198	241
PE < 37 wk	82	96	118	123	122	170	196	198	240

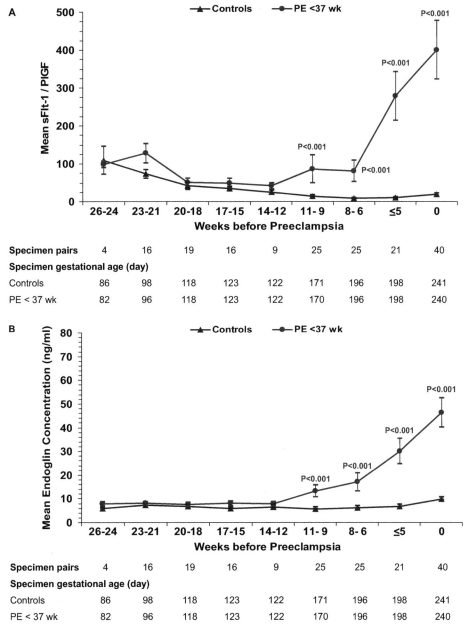

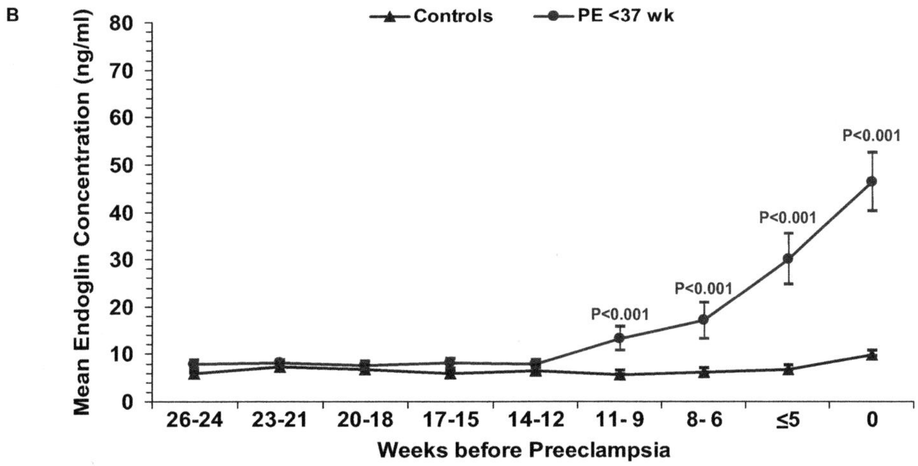

Specimen pairs	4	16	19	16	9	25	25	21	40
Specimen gestational age (day)									
Controls	86	98	118	123	122	171	196	198	241
PE < 37 wk	82	96	118	123	122	170	196	198	240

FIGURE 6.3 Mean levels of sFlt-1/PlGF and soluble endoglin (sEng) and by weeks before the onset of preeclampsia. (A) This panel shows the mean concentrations of sFlt-1/PlGF according to the number of weeks before the onset of preterm preeclampsia (PE < 37 weeks) and the mean concentrations in normotensive controls with appropriate- or large-for-gestational-age infants. Control specimens were matched within 1 week of gestational age to specimens from women who later developed preterm preeclampsia. (B) This panel shows the mean levels of sEng in case and control specimens shown in panel A according to the number of weeks before the onset of preterm preeclampsia (PE < 37 weeks). Figures reproduced with permission from Levine et al. and Hagmann et al.[125,139] (This figure is reproduced in color in the color plate section.)

explanation for preeclampsia pathogenesis that has been hypothesized by Haig,[136] namely to improve fetal nutrient delivery by increasing maternal peripheral vascular resistance.

Soluble Endoglin

A large body of human and animal studies suggest that another placental antiangiogenic protein, soluble endoglin (sEng), may also contribute to the pathogenesis of preeclampsia.[69,125,137,138] Soluble Eng was also elevated in patients with preeclampsia during clinical disease and prior to onset of symptoms[69,125] (see Fig. 6.3).

Endoglin (Eng) is an angiogenic receptor expressed mainly on the surface of endothelial cells and placental syncytiotrophoblast.[140–142] Eng acts as a co-receptor for transforming growth factor-β (TGF-β, a potent pro-angiogenic molecule) signaling in endothelial cells. We recently

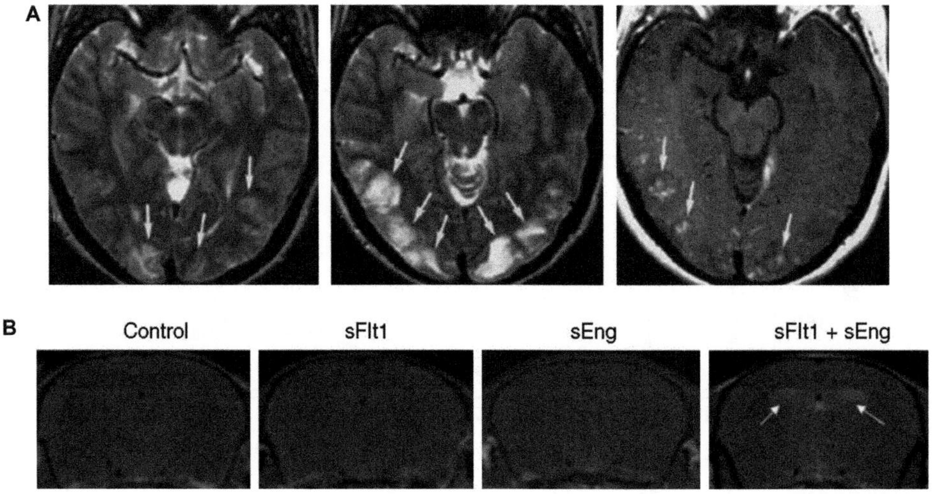

FIGURE 6.4 Cerebral edema in eclamptic subjects and in animal models of preeclampsia/eclampsia. (A) Serial MR images obtained in the brain of a patient with eclampsia that developed 2 days after delivery. Panel on the left (admission MRI scan) and middle panel (obtained at the time of maximal signs) demonstrate cerebral edema in the posterior cerebral cortex. Panel on the right demonstrates MR images from the same subject obtained after IV gadolinium contrast demonstrating disruption of blood–brain barrier. Figures reproduced with permission from Schwartz et al.[144] (B) MR images of brain from mice overexpressing sFlt-1 or sEng or both. Animal exposed to both sFlt-1 and sEng demonstrates edema in the posterior cerebral cortex. Figures reproduced with permission from Maharaj et al.[143]

observed that the Eng mRNA, like that of sFlt-1, is upregulated in the placentas of preeclamptic women. It was further noted that the soluble endoglin (sEng) is released in excess quantities into the circulation of preeclamptic patients. Soluble endoglin acts as an antiangiogenic factor by inhibiting TGF-β signaling in the vasculature. sEng amplifies the vascular damage mediated by sFlt-1 in pregnant rats, inducing a more severe preeclampsia-like syndrome with features of the HELLP syndrome.[69] Furthermore, overexpression of sFlt-1 and sEng in rodents was also found to induce focal vasospasm, hypertension and increased vascular permeability associated with brain edema,[143] producing images reminiscent of reversible posterior leukoencephalopathy associated with human eclampsia (reviewed in Chapter 13) (see Fig. 6.4). The contributions of sEng and sFlt-1 to the pathogenesis of the maternal syndrome of preeclampsia are at least in part, related to their inhibition of VEGF, PlGF and TGF-β stimulation of eNOS activation and vasomotor effects (see Fig. 6.5). Another candidate molecule that may be central to the pathogenesis of the pro-coagulant state and the thrombocytopenia noted in preeclampsia is prostacyclin (PGI_2). Both VEGF and TGFβ1 have been shown to stimulate the production of the anti-thrombotic prostacyclin-PGI_2.[127,146,147] Clinical studies demonstrating decreased endothelial PGI_2 production, even before onset of clinical preeclampsia, support this hypothesis.[93] In summary, maternal endothelial dysfunction caused by placenta-derived soluble factors such as sFlt-1 and sEng is emerging as the final common pathway that mediates the clinical syndrome of preeclampsia. Future studies are needed to clarify the molecular nature of the circulating sEng protein and its downstream signaling pathways that mediate endothelial dysfunction.

Other Circulating Antiangiogenic Factors

Endostatin (a circulating fragment of collagen XVIII and an endogenous inhibitor of angiogenesis) has also been reported to be modestly increased in patients with preeclampsia.[148] More recently, antiangiogenic urinary prolactin fragments referred to as vasoinhibins were also found to be increased in preeclamptic women.[149,150] However, it is unclear at the present time whether these alterations are directly related to the pathogenesis of preeclampsia or are secondary to increased endothelial cell turnover. In exciting studies, it was recently reported that semaphorin 3B, a novel trophoblastic secreted antiangiogenic protein, was upregulated in preeclamptic placentas and that semaphorin 3B inhibits trophoblast migration and invasion by downregulating VEGF signaling.[151] To further define the role of this novel pathway in human preeclampsia, more studies are needed to evaluate whether semaphorin 3B alterations antedate clinical signs and symptoms of preeclampsia and whether semaphorin 3B synergizes with circulating sFlt-1 to induce placental and systemic vascular dysfunction.

Changes in circulating endothelial progenitor cells (EPCs) have also been hypothesized to explain the endothelial dysfunction of preeclampsia.[152] EPCs enhance angiogenesis, promote vascular repair, improve endothelial vasodilator function, and inhibit progression of vascular injury. Limited evidence has been published showing that preeclampsia is characterized by decreased circulating EPCs.[153,154] However, it is unknown whether the abnormalities in circulating EPCs noted in preeclampsia are primary or secondary to other soluble factors such as sFlt-1.

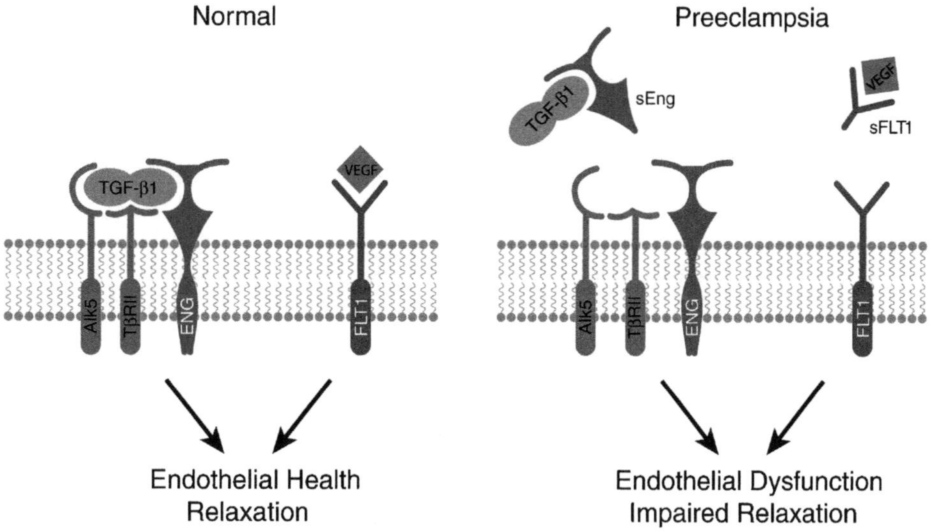

FIGURE 6.5 sFlt-1 and sEng causes endothelial dysfunction by antagonizing VEGF and TGF-β signaling. There is mounting evidence that VEGF and TGF-β are required to maintain endothelial health in several tissues including the kidney and perhaps the placenta. During normal pregnancy, vascular homeostasis is maintained by physiological levels of VEGF and TGF-β signaling in the vasculature. In preeclampsia, excess placental secretion of sFlt-1 and sEng (two endogenous circulating antiangiogenic proteins) inhibits VEGF and TGF-β1 signaling respectively in the vasculature. This results in endothelial cell dysfunction, including decreased prostacyclin, nitric oxide production and release of procoagulant proteins. Figure reproduced with permission from Karumanchi and Epstein.[145] (This figure is reproduced in color in the color plate section.)

Upstream Pathways and Mechanisms of Preeclampsia

The studies described above are compelling and support the hypothesis that the maternal syndrome of preeclampsia, particularly in its more severe forms, is "an antiangiogenic state".[155] Placental hypoxia has been suggested as the cause of excess antiangiogenic factors, but the precise mechanisms of excess sFlt-1 and sEng production by this organ are not known. Various pathways including deficient heme-oxygenase (HO) expression, angiotensin receptor 1 autoantibodies, oxidative stress, inflammation, altered natural killer cell signaling and, more recently, deficient catechol-*O*-methyl transferase have been proposed to play key roles in inducing placental disease.[156–160] Of these factors, alterations in the hemoxygenase system have emerged as an important upstream regulator of placental health.[161] In cell and organ culture studies, HO1 and its downstream metabolite carbon monoxide act as vascular protective factors by inhibiting the production of sFlt-1 and sEng.[156] Women with preeclampsia exhale less carbon monoxide than women with normal pregnancies and decreased HO1 has been noted in the placentas as early as the first trimester of pregnancy.[162] These findings may also explain why cigarette smoking is paradoxically associated with a protective effect in preeclampsia[163] More recently, animal studies suggest that dysregulated transthyretin or low levels of the enzyme cystathionine gamma lyase (which regulates hydrogen sulfide production) may contribute to preeclampsia by disrupting placental angiogenesis.[164,165]

There are other questions to be resolved. Although sFlt-1 levels are increased in most patients with preeclampsia, they are not universally elevated in every diagnosed case. This is especially true when the disease is mild.[124] Moreover, the relationship of sFlt-1 to some of the known risk factors for preeclampsia (such as obesity, and preexisting hypertension) is unclear. One hypothesis proposes that there is a threshold for sFlt-1 to cause disease. This theory evolves from the fact that sFlt-1 levels increase across gestation in all pregnant women. Thus it is theorized that if circulating sFlt-1 concentrations remain below a certain level, pregnancy proceeds normally, but when values exceed the cutoff level preeclampsia develops. In this respect, risk factors effectively lower the threshold, rendering women more sensitive to sFlt-1 and therefore resulting in preeclampsia despite levels that match those of normal pregnancy.

Still another reason for the appearance of preeclampsia phenotypes without increased sFlt-1 levels may be related to the fact that the disease may be misclassified in some patients.[166] This is not entirely surprising as renal biopsy studies by Fisher et al. in 1981 had already suggested that ~15% of such diagnoses were incorrect in nulliparous women, and >40% of multiparas were misdiagnosed.[84] Misclassification of preeclampsia diagnosis can be due to the presence of preconception chronic hypertension and/or obesity with minimal glomerulosclerosis, insufficient to produce proteinuria. A paucity of of major adverse maternal or fetal outcomes in preeclamptics characterized by normal circulating angiogenic profiles supports this latter hypothesis.[167]

If antiangiogenic factor production is an important cause of preeclampsia, there might be at least two kinds of predisposing factors. One could involve overproduction of sFlt-1 and sEng and occur in instances such as multiple gestation, hydatiform mole, trisomy 13, and, possibly, nulliparity.[121–123,168] Another set of predisposing factors would include disorders such as chronic hypertension or thrombophilia that sensitize the maternal vascular endothelium to the antiangiogenic effects of sFlt-1 and sEng. We do not yet know whether diabetes, hypertension and preexisting renal disease predispose to preeclampsia by increasing the production of sFlt-1 and sEng or by sensitizing the vascular endothelium to their presence, or, as discussed above, are likely to be misdiagnosed because of lack of specific criteria for the diagnosis of superimposed preeclampsia.

Hypoxia is known to increase the production of sFlt-1 by placental trophoblasts[169] and to inhibit their production of PlGF,[170] so that placental ischemia might trigger the preeclamptic syndrome. There is strong evidence for placental ischemia in many patients with preeclampsia but not in all. Whereas growth restriction, a corollary of placental ischemia, frequently accompanies preeclampsia, as many as one-third of the neonates of preeclamptic women are large for gestational age.[171] Placental infarction unaccompanied by preeclampsia is a common finding in mothers with sickle cell anemia and those with fetuses who have intrauterine growth retardation. Placental overproduction of sFlt-1, whatever its cause, might impair angiogenesis locally and result in placental ischemia, thereby initiating a vicious circle leading to even more sFlt-1 production.

We propose that three factors conspire, in variable degree, to produce the clinical syndrome of preeclampsia: (1) a disturbance in the balance of circulating factors controlling angiogenesis/antiangiogenesis, attributable to placental overproduction of sFlt-1 and underproduction of PlGF, (2) increased vascular endothelial sensitivity to such factors,[172] and (3) placental ischemia exaggerating the imbalance in angiogenic factors by further inducing the production of antiangiogenic factors. Human pregnancy is initially characterized by rapid angiogenesis, localized to the placenta, followed, as pregnancy concludes, by a slow regression of blood vessel growth. It would therefore not be surprising if derangements of this remarkably complex process might occur and lead to the systemic manifestations of preeclampsia.

Angiogenesis and the Remote Consequences of Preeclampsia

Considerable evidence suggests that preeclampsia is a marker of remote cardiac and vascular diseases,[173,174] but whether preeclampsia per se is a cause of future hypertension and heart disease or whether this represents the sequestration of women in the preeclamptic population to be predestined to these disorders has yet to be determined. In this respect, one small study published in 1964 suggested that the tendency to late hypertension is the result of preeclampsia, rather than a heritable tendency, demonstrated by the absence of hypertension in the siblings of preeclamptic patients.[175] It has also been suggested that subtle renal injury, such as that induced by preeclampsia, can eventually lead to the development of chronic hypertension in animal models.[176] Recently sFlt-1 has been suggested as pathogenic in peripartum cardiomyopathy (PPCM), a rare complication of pregnancy that is often associated with preeclampsia.[177] It is also tempting to speculate that the long-term cardiovascular complications noted in some patients who had preeclampsia may be due to a chronic antiangiogenic state resulting from polymorphisms in genes such as sFlt-1. Alternatively, persistence of autoantibodies against the AT1 receptor could be responsible for the long-term cardiovascular sequlae (see Chapter 15).

Additionally, patients with preeclampsia are said to have a decreased long-term incidence of malignancy,[178] a provocative observation disputed by some,[179] that may suggest that a long-term antiangiogenic state associated with preeclampsia may provide a protective milieu.

Role of Angiogenic Biomarkers in Preeclampsia

Over the past decade, numerous cross-sectional and longitudinal studies have shown that high sFlt-1 and low PlGF are present during preeclampsia and prior to clinical disease (see Fig. 6.3).[139]

The recent availability of automated platforms has allowed researchers to validate these biomarkers in several cohort studies.[180–182] A number of studies have demonstrated the ability of sFlt-1 and PlGF to identify women with preeclampsia using newly developed automated assays with excellent sensitivities and specificities for preterm preeclampsia,[181,183–186] More recently, several reports have suggested that angiogenic factors measured in the last trimester can predict development of severe preeclampsia-related adverse outcomes with very high sensitivity and specificity. These algorithms appear to distinguish patients whose course will be relatively benign from those likely to experience severe complications.[187–190] Measurement of angiogenic factors appears to be useful, especially in distinguishing preterm gravidas who can be managed conservatively from those with substantial risk of developing adverse outcomes and requiring close surveillance and a low threshold for delivery. In addition, circulating angiogenic factors may have the specificity to differentiate preeclampsia from other diseases that mimic preeclampsia such as chronic hypertension, gestational hypertension, and other chronic kidney diseases.[181,185,191–193]

Therapeutic Strategies for Preeclampsia

Studies of angiogenic pathways are helping devise specific therapies for preeclampsia. In a pilot study limited to three severe early preeclamptics (24–32 week gestation) Thadhani et al. depleted sFlt-1 30% by dextran sulfate apheresis and prolonged pregnancy by 2–4 weeks.[194] If confirmed this approach could lead to targeted therapy for a specific group of patients with premature disease and an abnormal angiogenic profile. More recently statin therapy that promotes angiogenesis was shown to prevent or ameliorate disease in an animal model of preeclampsia.[195,196] Pilot human trials to test the safety and efficacy of statins in severe preeclampsia are ongoing.[197] Relaxin, a naturally occurring vasodilator of pregnancy and which induces local VEGF and endothelial progenitor cells, is also being investigated for its therapeutic potential in preeclampsia.[198] Finally, dietary choline supplementation, shown to reduce placental sFlt-1 expression, has been suggested as a strategy to improve placental angiogenesis.[199] The future for specific therapies that reduce sFlt-1 production, antagonize its actions or enhance PlGF or VEGF levels is therefore promising.

Anti-Angiogenic Versus Normal Angiogenic Forms of Preeclampsia

As of 2014, there have been discussions described in several reviews concerning whether preeclampsia is "a syndrome of multiple disorders or multiple subtypes" or a "single homogenous disease".[166,200,201] Characterization of circulating angiogenic factors in clinical studies of preeclampsia has provided some answers to this debate.[167]

Based on epidemiological and experimental studies, we believe that alterations in angiogenic factors are both necessary and sufficient for the development of preeclampsia phenotypes, and their related complications.[166] When preeclamptic patients with anti-angiogenic (defined by high sFlt-1/PlGF ratio $\geq$ 85) and normal angiogenic plasma profiles (defined by sFlt-1/PlGF ratio $<$ 85) were studied, the serious adverse outcomes traditionally ascribed to preeclampsia were only found in subjects with abnormalities in angiogenic factors.[167] These data suggested that patients presenting signs and symptoms of preeclampsia but with relatively normal angiogenic profiles may have been misclassified.[166] This was not surprising as patients who presented with non-angiogenic forms of preeclampsia were obese or had other chronic conditions such as diabetes or chronic hypertension and therefore were less likely to be reliably diagnosed based on the traditional but nonspecific criteria of blood pressure and proteinuria. While there may be different causes that stimulate the placenta to produce an excess of antiangiogenic factors, the latter alone may be sufficient to produce preeclamptic phenotypes. Thus, severe complications of preeclampsia are directly correlated with more dramatic alterations in circulating angiogenic factors.

In contrast, it has been argued by others that preeclampsia is a heterogenous disease with multiple subtypes. Despite decades of research, no single test has been able to predict preeclampsia with the appropriate likelihood ratios required for an adequate clinical test.[201–203] Because various forms exist (e.g., placental or maternal [discussed in Chapter 8]), the search for a single marker of preeclampsia may be futile.[200] Unfortunately, it is very difficult to obtain the evidence necessary to clarify this debate; in developed nations women classified as preeclamptic are not permitted to develop adverse outcomes unless they present very prematurely, and aggressive delivery of patients with suspected preeclampsia is the common practice.

Of considerable interest relative to the two views discussed above is that increasing sFlt-1 and sEng levels in experimental animals produces the histological lesions described in kidney, and liver tissue of a woman dying of either preeclampsia or eclampsia reported by Sheehan and Lynch in 1973.[69,204] Taken together with human epidemiological studies, these data support the hypothesis that increased circulating antiangiogenic factors are necessary to induce the most severe forms of preeclampsia.[166]

Prospective clinical trials are still needed to evaluate whether subjects with a normal angiogenic profile can be expectantly managed and delivered at 37 weeks or beyond. These tests should be easier to perform now that precise automated angiogenic factor assays that produce rapid results are available.

In summary, measuring plasma circulating angiogenic factors may provide a new approach to identify what we consider the "true or multi-systemic preeclampsia". It may also improve classification and lead to better research on causality, prediction, and management of both immediate and remote outcomes.

PERSPECTIVES

Preeclampsia is a systemic disorder characterized by widespread endothelial dysfunction. Substantial experimental animal data and a large body of human data support the hypothesis that an excess of placental antiangiogenic factors, such as sFlt-1 and sEng, and a paucity of proangiogenic factors like PlGF, may be responsible for the clinical syndrome of preeclampsia. Prospective longitudinal studies have established that levels of angiogenic factors are altered in women with the diagnosis of preeclampsia. Recent studies have shown the association of these factors with preeclampsia-related adverse outcomes especially when measured in the third trimester of pregnancy. Larger prospective studies are now needed to test whether measurement of angiogenic factors in clinical practice will help predict adverse outcomes and improve maternal and fetal health. Finally, targeted therapy based on these factors, which could allow clinicians to safely postpone delivery, needs to be developed to ultimately improve neonatal outcomes. Given the global incidence of this condition, there is tremendous potential to improve both the suffering and cost associated with this disease. Recent work has begun to unravel the mechanisms behind preeclampsia. It is our hope that current insights will lead, in turn, to effective preventive strategies.

References

1. Banks RE, Forbes MA, Searles J, et al. Evidence for the existence of a novel pregnancy-associated soluble variant of the vascular endothelial growth factor receptor, Flt-1. *Mol Hum Reprod*. 1998;4(4):377–386.
2. Clark DE, Smith SK, He Y, et al. A vascular endothelial growth factor antagonist is produced by the human placenta and released into the maternal circulation. *Biol Reprod*. 1998;59(6):1540–1548.
3. He Y, Smith SK, Day KA, Clark DE, Licence DR, Charnock-Jones DS. Alternative splicing of vascular endothelial growth factor (VEGF)-R1 (FLT-1) pre-mRNA is important for the regulation of VEGF activity. *Mol Endocrinol*. 1999;13(4):537–545.
4. Ahmed A, Dunk C, Ahmad S, Khaliq A. Regulation of placental vascular endothelial growth factor (VEGF) and placenta growth factor (PIGF) and soluble Flt-1 by oxygen—a review. *Placenta*. 2000;21(suppl A):S16–S24.
5. Maynard SE, Min JY, Merchan J, et al. Excess placental soluble fms-like tyrosine kinase 1 (sFlt1) may contribute to endothelial dysfunction, hypertension, and proteinuria in preeclampsia. *J Clin Invest*. 2003;111(5):649–658.
6. Risau W, Flamme I. Vasculogenesis. *Annu Rev Cell Dev Biol*. 1995;11:73–91.
7. Risau W. Mechanisms of angiogenesis. *Nature*. 1997; 386(6626):671–674.
8. Demir R, Kaufmann P, Castellucci M, Erbengi T, Kotowski A. Fetal vasculogenesis and angiogenesis in human placental villi. *Acta Anat (Basel)*. 1989;136(3):190–203.
9. Zhou Y, Fisher SJ, Janatpour M, et al. Human cytotrophoblasts adopt a vascular phenotype as they differentiate. A strategy for successful endovascular invasion? *J Clin Invest*. 1997;99(9):2139–2151.
10. Cerdeira AS, Karumanchi SA. Angiogenic factors in preeclampsia and related disorders. *Cold Spring Harb Perspect Med*. 2012;2:11.
11. Genbacev O, Zhou Y, Ludlow JW, Fisher SJ. Regulation of human placental development by oxygen tension. *Science*. 1997;277(5332):1669–1672.
12. Lam C, Lim KH, Karumanchi SA. Circulating angiogenic factors in the pathogenesis and prediction of preeclampsia. *Hypertension*. 2005;46(5):1077–1085.
13. Jaffe R. First trimester utero-placental circulation: maternal-fetal interaction. *J Perinat Med*. 1998;26(3):168–174.
14. Roth I, Fisher SJ. IL-10 is an autocrine inhibitor of human placental cytotrophoblast MMP-9 production and invasion. *Dev Biol*. 1999;205(1):194–204.
15. Lim KH, Zhou Y, Janatpour M, et al. Human cytotrophoblast differentiation/invasion is abnormal in pre- eclampsia. *Am J Pathol*. 1997;151(6):1809–1818.
16. Brosens IA, Robertson WB, Dixon HG. The role of the spiral arteries in the pathogenesis of preeclampsia. *Obstet Gynecol Annu*. 1972;1:177–191.
17. Khong TY, De Wolf F, Robertson WB, Brosens I. Inadequate maternal vascular response to placentation in pregnancies complicated by pre-eclampsia and by small-for-gestational age infants. *Br J Obstet Gynaecol*. 1986;93(10):1049–1059.
18. Craven CM, Morgan T, Ward K. Decidual spiral artery remodelling begins before cellular interaction with cytotrophoblasts. *Placenta*. 1998;19(4):241–252.
19. Croy BA, He H, Esadeg S, et al. Uterine natural killer cells: insights into their cellular and molecular biology from mouse modelling. *Reproduction*. 2003;126(2):149–160.
20. Kohnen G, Kertschanska S, Demir R, Kaufmann P. Placental villous stroma as a model system for myofibroblast differentiation. *Histochem Cell Biol*. 1996;105(6):415–429.
21. Taylor RN, Williams LT. Developmental expression of platelet-derived growth factor and its receptor in the human placenta. *Mol Endocrinol*. 1988;2(7):627–632.
22. Kaufmann P, Bruns U, Leiser R, Luckhardt M, Winterhager E. The fetal vascularisation of term human placental villi. II. Intermediate and terminal villi. *Anat Embryol (Berl)*. 1985;173(2):203–214.
23. Kaufmann P, Luckhardt M, Schweikhart G, Cantle SJ. Cross-sectional features and three-dimensional structure of human placental villi. *Placenta*. 1987;8(3):235–247.
24. Bdolah Y, Sukhatme VP, Karumanchi SA. Angiogenic imbalance in the pathophysiology of preeclampsia: newer insights. *Semin Nephrol*. 2004;24(6):548–556.
25. Zygmunt M, Herr F, Munstedt K, Lang U, Liang OD. Angiogenesis and vasculogenesis in pregnancy. *Eur J Obstet Gynecol Reprod Biol*. 2003;110(suppl 1):S10–S18.
26. Wulff C, Weigand M, Kreienberg R, Fraser HM. Angiogenesis during primate placentation in health and disease. *Reproduction*. 2003;126(5):569–577.
27. Demir R, Seval Y, Huppertz B. Vasculogenesis and angiogenesis in the early human placenta. *Acta Histochem*. 2007;109(4):257–265.
28. Dvorak HF. Vascular permeability factor/vascular endothelial growth factor: a critical cytokine in tumor angiogenesis and a potential target for diagnosis and therapy. *J Clin Oncol*. 2002;20(21):4368–4380.
29. Ferrara N, Gerber HP. The role of vascular endothelial growth factor in angiogenesis. *Acta Haematol*. 2001;106(4):148–156.
30. Gerber HP, Condorelli F, Park J, Ferrara N. Differential transcriptional regulation of the two vascular endothelial growth factor receptor genes. Flt-1, but not Flk-1/KDR, is up-regulated by hypoxia. *J Biol Chem*. 1997;272(38):23659–23667.
31. Shibuya M. Structure and function of VEGF/VEGF-receptor system involved in angiogenesis. *Cell Struct Funct*. 2001;26(1):25–35.
32. Persico MG, Vincenti V, DiPalma T. Structure, expression and receptor-binding properties of placenta growth factor (PIGF). *Curr Top Microbiol Immunol*. 1999;237:31–40.
33. Carmeliet P, Moons L, Luttun A, et al. Synergism between vascular endothelial growth factor and placental growth factor contributes to angiogenesis and plasma extravasation in pathological conditions. *Nat Med*. 2001;7(5):575–583.
34. Charnock-Jones DS, Burton GJ. Placental vascular morphogenesis. *Bailliere Best Pract Res*. 2000;14(6):953–968.
35. Luttun A, Brusselmans K, Fukao H, et al. Loss of placental growth factor protects mice against vascular permeability in pathological conditions. *Biochem Biophys Res Commun*. 2002;295(2):428–434.
36. Tayade C, Hilchie D, He H, et al. Genetic deletion of placenta growth factor in mice alters uterine NK cells. *J Immunol*. 2007;178(7):4267–4275.
37. Jussila L, Alitalo K. Vascular growth factors and lymphangiogenesis. *Physiol Rev*. 2002;82(3):673–700.

38. Olofsson B, Jeltsch M, Eriksson U, Alitalo K. Current biology of VEGF-B and VEGF-C. *Curr Opin Biotechnol.* 1999;10(6):528–535.
39. Kendall RL, Thomas KA. Inhibition of vascular endothelial cell growth factor activity by an endogenously encoded soluble receptor. *Proc Natl Acad Sci U S A.* 1993;90(22):10705–10709.
40. Powe CE, Levine RJ, Karumanchi SA. Preeclampsia, a disease of the maternal endothelium: the role of antiangiogenic factors and implications for later cardiovascular disease. *Circulation.* 2011;123(24):2856–2869.
41. Papapetropoulos A, Garcia-Cardena G, Dengler TJ, Maisonpierre PC, Yancopoulos GD, Sessa WC. Direct actions of angiopoietin-1 on human endothelium: evidence for network stabilization, cell survival, and interaction with other angiogenic growth factors. *Lab Invest.* 1999;79(2):213–223.
42. Koblizek TI, Weiss C, Yancopoulos GD, Deutsch U, Risau W. Angiopoietin-1 induces sprouting angiogenesis in vitro. *Curr Biol.* 1998;8(9):529–532.
43. Davis S, Aldrich TH, Jones PF, et al. Isolation of angiopoietin-1, a ligand for the TIE2 receptor, by secretion-trap expression cloning. *Cell.* 1996;87(7):1161–1169.
44. Carmeliet P. Mechanisms of angiogenesis and arteriogenesis. *Nat Med.* 2000;6(4):389–395.
45. Thurston G, Rudge JS, Ioffe E, et al. Angiopoietin-1 protects the adult vasculature against plasma leakage. *Nat Med.* 2000;6(4):460–463.
46. Thurston G, Suri C, Smith K, et al. Leakage-resistant blood vessels in mice transgenically overexpressing angiopoietin-1. *Science.* 1999;286(5449):2511–2514.
47. Stratmann A, Acker T, Burger AM, Amann K, Risau W, Plate KH. Differential inhibition of tumor angiogenesis by tie2 and vascular endothelial growth factor receptor-2 dominant-negative receptor mutants. *Int J Cancer.* 2001;91(3):273–282.
48. Tanaka S, Mori M, Sakamoto Y, Makuuchi M, Sugimachi K, Wands JR. Biologic significance of angiopoietin-2 expression in human hepatocellular carcinoma. *J Clin Invest.* 1999;103(3):341–345.
49. Maisonpierre PC, Suri C, Jones PF, et al. Angiopoietin-2, a natural antagonist for Tie2 that disrupts in vivo angiogenesis. *Science.* 1997;277(5322):55–60.
50. Wulff C, Wilson H, Dickson SE, Wiegand SJ, Fraser HM. Hemochorial placentation in the primate: expression of vascular endothelial growth factor, angiopoietins, and their receptors throughout pregnancy. *Biol Reprod.* 2002;66(3):802–812.
51. Charnock-Jones DS, Sharkey AM, Boocock CA, et al. Vascular endothelial growth factor receptor localization and activation in human trophoblast and choriocarcinoma cells. *Biol Reprod.* 1994;51(3):524–530.
52. Cooper JC, Sharkey AM, McLaren J, Charnock-Jones DS, Smith SK. Localization of vascular endothelial growth factor and its receptor, flt, in human placenta and decidua by immunohistochemistry. *J Reprod Fertil.* 1995;105(2):205–213.
53. Geva E, Ginzinger DG, Zaloudek CJ, Moore DH, Byrne A, Jaffe RB. Human placental vascular development: vasculogenic and angiogenic (branching and nonbranching) transformation is regulated by vascular endothelial growth factor-A, angiopoietin-1, and angiopoietin-2. *J Clin Endocrinol Metab.* 2002;87(9):4213–4224.
54. Clark DE, Smith SK, Sharkey AM, Charnock-Jones DS. Localization of VEGF and expression of its receptors flt and KDR in human placenta throughout pregnancy. *Hum Reprod.* 1996;11(5):1090–1098.
55. Zhou Y, McMaster M, Woo K, et al. Vascular endothelial growth factor ligands and receptors that regulate human cytotrophoblast survival are dysregulated in severe preeclampsia and hemolysis, elevated liver enzymes, and low platelets syndrome. *Am J Pathol.* 2002;160(4):1405–1423.
56. Hildebrandt VA, Babischkin JS, Koos RD, Pepe GJ, Albrecht ED. Developmental regulation of vascular endothelial growth/permeability factor messenger ribonucleic acid levels in and vascularization of the villous placenta during baboon pregnancy. *Endocrinology.* 2001;142(5):2050–2057.
57. Nevo O, Soleymanlou N, Wu Y, et al. Increased expression of sFlt-1 in in vivo and in vitro models of human placental hypoxia is mediated by HIF-1. *Am J Physiol Regul Integr Comp Physiol.* 2006;291(4):R1085–R1093.
58. Shalaby F, Rossant J, Yamaguchi TP, et al. Failure of blood-island formation and vasculogenesis in Flk-1-deficient mice. *Nature.* 1995;376(6535):62–66.
59. Fong G, Rassant J, Gertenstein MMB. Role of Flt-1 receptor tyrosine kinase in regulation of assembly of vascular endothelium. *Nature.* 1995;376:66–67.
60. Ain R, Canham LN, Soares MJ. Gestation stage-dependent intrauterine trophoblast cell invasion in the rat and mouse: novel endocrine phenotype and regulation. *Dev Biol.* 2003;260(1):176–190.
61. Zhou Y, Bellingard V, Feng KT, McMaster M, Fisher SJ. Human cytotrophoblasts promote endothelial survival and vascular remodeling through secretion of Ang2, PlGF, and VEGF-C. *Dev Biol.* 2003;263(1):114–125.
62. Helske S, Vuorela P, Carpen O, Hornig C, Weich H, Halmesmaki E. Expression of vascular endothelial growth factor receptors 1, 2 and 3 in placentas from normal and complicated pregnancies. *Mol Hum Reprod.* 2001;7(2):205–210.
63. Rowe AJ, Wulff C, Fraser HM. Localization of mRNA for vascular endothelial growth factor (VEGF), angiopoietins and their receptors during the peri-implantation period and early pregnancy in marmosets (Callithrix jacchus). *Reproduction.* 2003;126(2):227–238.
64. Dunk C, Shams M, Nijjar S, et al. Angiopoietin-1 and angiopoietin-2 activate trophoblast Tie-2 to promote growth and migration during placental development. *Am J Pathol.* 2000;156(6):2185–2199.
65. Zhang EG, Smith SK, Baker PN, Charnock-Jones DS. The regulation and localization of angiopoietin-1, -2, and their receptor Tie2 in normal and pathologic human placentae. *Mol Med.* 2001;7(9):624–635.
66. Goldman-Wohl DS, Ariel I, Greenfield C, Lavy Y, Yagel S. Tie-2 and angiopoietin-2 expression at the fetal-maternal interface: a receptor ligand model for vascular remodelling. *Mol Hum Reprod.* 2000;6(1):81–87.
67. Caniggia I, Taylor CV, Ritchie JW, Lye SJ, Letarte M. Endoglin regulates trophoblast differentiation along the invasive pathway in human placental villous explants. *Endocrinology.* 1997;138(11):4977–4988.
68. Caniggia I, Grisaru-Gravnosky S, Kuliszewsky M, Post M, Lye SJ. Inhibition of TGF-beta 3 restores the invasive

capability of extravillous trophoblasts in preeclamptic pregnancies. *J Clin Invest.* 1999;103(12):1641–1650.

69. Venkatesha S, Toporsian M, Lam C, et al. Soluble endoglin contributes to the pathogenesis of preeclampsia. *Nat Med.* 2006;12(6):642–649.
70. Stallmach T, Duc C, van Praag E, et al. Feto-maternal interface of human placenta inhibits angiogenesis in the chick chorioallantoic membrane (CAM) assay. *Angiogenesis.* 2001;4(1):79–84.
71. Moffett-King A. Natural killer cells and pregnancy. *Nat Rev Immunol.* 2002;2(9):656–663.
72. Hiby SE, Walker JJ, O'Shaughnessy KM, et al. Combinations of maternal KIR and fetal HLA-C genes influence the risk of preeclampsia and reproductive success. *J Exp Med.* 2004;200(8):957–965.
73. Parham P. NK cells and trophoblasts: partners in pregnancy. *J Exp Med.* 2004;200(8):951–955.
74. Hanna J, Goldman-Wohl D, Hamani Y, et al. Decidual NK cells regulate key developmental processes at the human fetal-maternal interface. *Nat Med.* 2006;12(9):1065–1074.
75. Wallace AE, Host AJ, Whitley GS, Cartwright JE. Decidual natural killer cell interactions with trophoblasts are impaired in pregnancies at increased risk of preeclampsia. *Am J Pathol.* 2013;183(6):1853–1861.
76. Makris A, Thornton C, Thompson J, et al. Uteroplacental ischemia results in proteinuric hypertension and elevated sFLT-1. *Kidney Int.* 2007;71(10):977–984.
77. North RA, Ferrier C, Long D, Townend K, Kincaid-Smith P. Uterine artery Doppler flow velocity waveforms in the second trimester for the prediction of preeclampsia and fetal growth retardation. *Obstet Gynecol.* 1994;83(3):378–386.
78. Moldenhauer JS, Stanek J, Warshak C, Khoury J, Sibai B. The frequency and severity of placental findings in women with preeclampsia are gestational age dependent. *Am J Obstet Gynecol.* 2003;189(4):1173–1177.
79. Harrington K, Cooper D, Lees C, Hecher K, Campbell S. Doppler ultrasound of the uterine arteries: the importance of bilateral notching in the prediction of pre-eclampsia, placental abruption or delivery of a small-for-gestational-age baby. *Ultrasound Obstet Gynecol.* 1996;7(3):182–188.
80. Gilbert JS, Babcock SA, Granger JP. Hypertension produced by reduced uterine perfusion in pregnant rats is associated with increased soluble fms-like tyrosine kinase-1 expression. *Hypertension.* 2007;50(6):1142–1147.
81. Roberts JM, Taylor RN, Musci TJ, Rodgers GM, Hubel CA, McLaughlin MK. Preeclampsia: an endothelial cell disorder. *Am J Obstet Gynecol.* 1989;161(5):1200–1204.
82. Ferris TF. Pregnancy, preeclampsia, and the endothelial cell. *N Engl J Med.* 1991;325(20):1439–1440.
83. Roberts JM. Endothelial dysfunction in preeclampsia. *Semin Reprod Endocrinol.* 1998;16(1):5–15.
84. Fisher KA, Luger A, Spargo BH, Lindheimer MD. Hypertension in pregnancy: clinical-pathological correlations and remote prognosis. *Medicine (Baltimore).* 1981;60(4):267–276.
85. Thorp Jr. JM, White II GC, Moake JL, Bowes Jr. WA. von Willebrand factor multimeric levels and patterns in patients with severe preeclampsia. *Obstet Gynecol.* 1990;75(2):163–167.
86. Redman S, Sargent IL. Preeclampsia: an excessive maternal inflammatory response to pregnancy. *Am J Obstet Gynecol.* 1999;180(2):499–506.
87. Calvin S, Corrigan J, Weinstein L, Jeter M. Factor VIII: von Willebrand factor patterns in the plasma of patients with preeclampsia. *Am J Perinatol.* 1988;5(1):29–32.
88. Lockwood CJ, Peters JH. Increased plasma levels of ED1+ cellular fibronectin precede the clinical signs of preeclampsia. *Am J Obstet Gynecol.* 1990;162(2):358–362.
89. Friedman SA, de Groot CJ, Taylor RN, Golditch BD, Roberts JM. Plasma cellular fibronectin as a measure of endothelial involvement in preeclampsia and intrauterine growth retardation. *Am J Obstet Gynecol.* 1994;170(3):838–841.
90. Taylor RN, Crombleholme WR, Friedman SA, Jones LA, Casal DC, Roberts JM. High plasma cellular fibronectin levels correlate with biochemical and clinical features of preeclampsia but cannot be attributed to hypertension alone. *Am J Obstet Gynecol.* 1991;165(4 Pt 1):895–901.
91. Minakami H, Takahashi T, Izumi A, Tamada T. Increased levels of plasma thrombomodulin in preeclampsia. *Gynecol Obstet Invest.* 1993;36(4):208–210.
92. Boffa MC, Valsecchi L, Fausto A, et al. Predictive value of plasma thrombomodulin in preeclampsia and gestational hypertension. *Thromb Haemost.* 1998;79(6):1092–1095.
93. Mills JL, DerSimonian R, Raymond E, et al. Prostacyclin and thromboxane changes predating clinical onset of preeclampsia: a multicenter prospective study. *JAMA.* 1999;282(4):356–362.
94. Ashworth JR, Warren AY, Baker PN, Johnson IR. A comparison of endothelium-dependent relaxation in omental and myometrial resistance arteries in pregnant and nonpregnant women. *Am J Obstet Gynecol.* 1996;175(5):1307–1312.
95. Pascoal IF, Lindheimer MD, Nalbantian-Brandt C, Umans JG. Preeclampsia selectively impairs endothelium-dependent relaxation and leads to oscillatory activity in small omental arteries. *J Clin Invest.* 1998;101(2):464–470.
96. Taylor RN, Varma M, Teng NN, Roberts JM. Women with preeclampsia have higher plasma endothelin levels than women with normal pregnancies. *J Clin Endocrinol Metab.* 1990;71(6):1675–1677.
97. Roberts JM, Edep ME, Goldfien A, Taylor RN. Sera from preeclamptic women specifically activate human umbilical vein endothelial cells in vitro: morphological and biochemical evidence. *Am J Reprod Immunol.* 1992;27(3-4):101–108.
98. Roberts JM. Preeclampsia: what we know and what we do not know. *Semin Perinatol.* 2000;24(1):24–28.
99. Hasan KM, Manyonda IT, Ng FS, Singer DR, Antonios TF. Skin capillary density changes in normal pregnancy and preeclampsia. *J Hypertens.* 2002;20(12):2439–2443.
100. Jonsdottir LS, Arngrimsson R, Geirsson RT, Sigvaldason H, Sigfusson N. Death rates from ischemic heart disease in women with a history of hypertension in pregnancy. *Acta Obstet Gynecol Scand.* 1995;74(10):772–776.
101. Risau W. Development and differentiation of endothelium. *Kidney Int Suppl.* 1998;67:S3–S6.
102. Ahmad S, Ahmed A. Elevated placental soluble vascular endothelial growth factor receptor-1 inhibits angiogenesis in preeclampsia. *Circ Res.* 2004;95(9):884–891.

103. Levine RJ, Maynard SE, Qian C, et al. Circulating angiogenic factors and the risk of preeclampsia. *N Engl J Med.* 2004;350(7):672–683.
104. Polliotti BM, Fry AG, Saller DN, Mooney RA, Cox C, Miller RK. Second-trimester maternal serum placental growth factor and vascular endothelial growth factor for predicting severe, early-onset preeclampsia. *Obstet Gynecol.* 2003;101(6):1266–1274.
105. Tsatsaris V, Goffin F, Munaut C, et al. Overexpression of the soluble vascular endothelial growth factor receptor in preeclamptic patients: pathophysiological consequences. *J Clin Endocrinol Metab.* 2003;88(11):5555–5563.
106. Chaiworapongsa T, Romero R, Espinoza J, et al. Evidence supporting a role for blockade of the vascular endothelial growth factor system in the pathophysiology of preeclampsia. Young Investigator Award. *Am J Obstet Gynecol.* 2004;190(6):1541–1547. [discussion 7–50].
107. Taylor RN, Grimwood J, Taylor RS, McMaster MT, Fisher SJ, North RA. Longitudinal serum concentrations of placental growth factor: evidence for abnormal placental angiogenesis in pathologic pregnancies. *Am J Obstet Gynecol.* 2003;188(1):177–182.
108. Bujold E, Romero R, Chaiworapongsa T, et al. Evidence supporting that the excess of the sVEGFR-1 concentration in maternal plasma in preeclampsia has a uterine origin. *J Matern Fetal Neonatal Med.* 2005;18(1):9–16.
109. Sela S, Itin A, Natanson-Yaron S, et al. A novel human-specific soluble vascular endothelial growth factor receptor 1: cell-type-specific splicing and implications to vascular endothelial growth factor homeostasis and preeclampsia. *Circ Res.* 2008;102(12):1566–1574.
110. Thomas CP, Andrews JI, Liu KZ. Intronic polyadenylation signal sequences and alternate splicing generate human soluble Flt1 variants and regulate the abundance of soluble Flt1 in the placenta. *Faseb J.* 2007;21(14):3885–3895.
111. Tannetta DS, Dragovic RA, Gardiner C, Redman CW, Sargent IL. Characterisation of syncytiotrophoblast vesicles in normal pregnancy and pre-eclampsia: expression of Flt-1 and endoglin. *PLoS One.* 2013;8(2):e56754.
112. Rajakumar A, Cerdeira AS, Rana S, et al. Transcriptionally active syncytial aggregates in the maternal circulation may contribute to circulating soluble fms-like tyrosine kinase 1 in preeclampsia. *Hypertension.* 2012;59(2):256–264.
113. Buurma AJ, Penning ME, Prins F, et al. Preeclampsia is associated with the presence of transcriptionally active placental fragments in the maternal lung. *Hypertension.* 2013;62:608–613.
114. Karumanchi SA, Maynard SE, Stillman IE, Epstein FH, Sukhatme VP. Preeclampsia: a renal perspective. *Kidney Int.* 2005;67(6):2101–2113.
115. Lu F, Longo M, Tamayo E, et al. The effect of over-expression of sFlt-1 on blood pressure and the occurrence of other manifestations of preeclampsia in unrestrained conscious pregnant mice. *Am J Obstet Gynecol.* 2007;196(4):396.e1–396.e7. [discussion e7].
116. Sugimoto H, Hamano Y, Charytan D, et al. Neutralization of circulating vascular endothelial growth factor (VEGF) by anti-VEGF antibodies and soluble VEGF receptor 1 (sFlt-1) induces proteinuria. *J Biol Chem.* 2003;278(15):12605–12608.
117. Li Z, Zhang Y, Ying MaJ, et al. Recombinant vascular endothelial growth factor 121 attenuates hypertension and improves kidney damage in a rat model of preeclampsia. *Hypertension.* 2007;50(4):686–692.
118. Suzuki H, Ohkuchi A, Matsubara S, et al. Effect of recombinant placental growth factor 2 on hypertension induced by full-length mouse soluble fms-like tyrosine kinase 1 adenoviral vector in pregnant mice. *Hypertension.* 2009;54(5):1129–1135.
119. Karumanchi SA, Lindheimer MD. Preeclampsia pathogenesis: "triple a rating"-autoantibodies and antiangiogenic factors. *Hypertension.* 2008;51(4):991–992.
120. Bdolah Y, Lam C, Rajakumar A, et al. Preeclampsia in twin pregnancies: hypoxia or bigger placental mass? (Abstract). *J Soc Gynecol Investig.* 2006;13(2):A670.
121. Bdolah Y, Palomaki GE, Yaron Y, et al. Circulating angiogenic proteins in trisomy 13. *Am J Obstet Gynecol.* 2006;194(1):239–245.
122. Maynard SE, Moore Simas TA, Solitro MJ, et al. Circulating angiogenic factors in singleton vs multiple-gestation pregnancies. *Am J Obstet Gynecol.* 2008;198(2):200.e1–200.e7.
123. Wolf M, Shah A, Lam C, et al. Circulating levels of the antiangiogenic marker sFLT-1 are increased in first versus second pregnancies. *Am J Obstet Gynecol.* 2005;193(1):16–22.
124. Powers RW, Roberts JM, Cooper KM, et al. Maternal serum soluble fms-like tyrosine kinase 1 concentrations are not increased in early pregnancy and decrease more slowly postpartum in women who develop preeclampsia. *Am J Obstet Gynecol.* 2005;193(1):185–191.
125. Levine RJ, Lam C, Qian C, et al. Soluble endoglin and other circulating antiangiogenic factors in preeclampsia. *N Engl J Med.* 2006;355(10):992–1005.
126. Morbidelli L, Chang CH, Douglas JG, Granger HJ, Ledda F, Ziche M. Nitric oxide mediates mitogenic effect of VEGF on coronary venular endothelium. *Am J Physiol.* 1996;270(1 Pt 2):H411–H415.
127. He H, Venema VJ, Gu X, Venema RC, Marrero MB, Caldwell RB. Vascular endothelial growth factor signals endothelial cell production of nitric oxide and prostacyclin through flk-1/KDR activation of c-Src. *J Biol Chem.* 1999;274(35):25130–25135.
128. Kim YG, Suga SI, Kang DH, et al. Vascular endothelial growth factor accelerates renal recovery in experimental thrombotic microangiopathy. *Kidney Int.* 2000;58(6):2390–2399.
129. Masuda Y, Shimizu A, Mori T, et al. Vascular endothelial growth factor enhances glomerular capillary repair and accelerates resolution of experimentally induced glomerulonephritis. *Am J Pathol.* 2001;159(2):599–608.
130. Kang DH, Kim YG, Andoh TF, et al. Post-cyclosporine-mediated hypertension and nephropathy: amelioration by vascular endothelial growth factor. *Am J Physiol Renal Physiol.* 2001;280(4):F727–F736.
131. Esser S, Wolburg K, Wolburg H, Breier G, Kurzchalia T, Risau W. Vascular endothelial growth factor induces endothelial fenestrations in vitro. *J Cell Biol.* 1998;140(4):947–959.
132. Eremina V, Sood M, Haigh J, et al. Glomerular-specific alterations of VEGF-A expression lead to distinct congenital and acquired renal diseases. *J Clin Invest.* 2003;111(5):707–716.

133. Eremina V, Jefferson JA, Kowalewska J, et al. VEGF inhibition and renal thrombotic microangiopathy. *N Engl J Med.* 2008;358(11):1129–1136.
134. Patel TV, Morgan JA, Demetri GD, et al. A preeclampsia-like syndrome characterized by reversible hypertension and proteinuria induced by the multitargeted kinase inhibitors sunitinib and sorafenib. *J Natl Cancer Inst.* 2008;100(4):282–284.
135. Vaughn C, Zhang L, Schiff D. Reversible posterior leukoencephalopathy syndrome in cancer. *Curr Oncol Rep.* 2008;10(1):86–91.
136. Haig D. Genetic conflicts in human pregnancy. *Q Rev Biol.* 1993;68(4):495–532.
137. Noori M, Donald AE, Angelakopoulou A, Hingorani AD, Williams DJ. Prospective study of placental angiogenic factors and maternal vascular function before and after preeclampsia and gestational hypertension. *Circulation.* 2010;122(5):478–487.
138. Romero R, Nien JK, Espinoza J, et al. A longitudinal study of angiogenic (placental growth factor) and anti-angiogenic (soluble endoglin and soluble vascular endothelial growth factor receptor-1) factors in normal pregnancy and patients destined to develop preeclampsia and deliver a small for gestational age neonate. *J Matern Fetal Neonatal Med.* 2008;21(1):9–23.
139. Hagmann H, Thadhani R, Benzing T, Karumanchi SA, Stepan H. The promise of angiogenic markers for the early diagnosis and prediction of preeclampsia. *Clin Chem.* 2012;58(5):837–845.
140. Cheifetz S, Bellon T, Cales C, et al. Endoglin is a component of the transforming growth factor-beta receptor system in human endothelial cells. *J Biol Chem.* 1992;267(27):19027–19030.
141. Gougos A, St Jacques S, Greaves A, et al. Identification of distinct epitopes of endoglin, an RGD-containing glycoprotein of endothelial cells, leukemic cells, and syncytiotrophoblasts. *Int Immunol.* 1992;4(1):83–92.
142. St-Jacques S, Forte M, Lye SJ, Letarte M. Localization of endoglin, a transforming growth factor-beta binding protein, and of CD44 and integrins in placenta during the first trimester of pregnancy. *Biol Reprod.* 1994;51(3):405–413.
143. Maharaj AS, Walshe TE, Saint-Geniez M, et al. VEGF and TGF-beta are required for the maintenance of the choroid plexus and ependyma. *J Exp Med.* 2008;205(2):491–501.
144. Schwartz RB, Feske SK, Polak JF, et al. Preeclampsia-eclampsia: clinical and neuroradiographic correlates and insights into the pathogenesis of hypertensive encephalopathy. *Radiology.* 2000;217(2):371–376.
145. Karumanchi SA, Epstein FH. Placental ischemia and soluble fms-like tyrosine kinase 1: cause or consequence of preeclampsia? *Kidney Int.* 2007;71(10):959–961.
146. Tatsumi M, Kishi Y, Miyata T, Numano F. Transforming growth factor-beta(1) restores antiplatelet function of endothelial cells exposed to anoxia-reoxygenation injury. *Thromb Res.* 2000;98(5):451–459.
147. Ristimaki A, Ylikorkala O, Viinikka L. Effect of growth factors on human vascular endothelial cell prostacyclin production. *Arteriosclerosis.* 1990;10(4):653–657.
148. Hirtenlehner K, Pollheimer J, Lichtenberger C, et al. Elevated serum concentrations of the angiogenesis inhibitor endostatin in preeclamptic women. *J Soc Gynecol Investig.* 2003;10(7):412–417.
149. Gonzalez C, Parra A, Ramirez-Peredo J, et al. Elevated vasoinhibins may contribute to endothelial cell dysfunction and low birth weight in preeclampsia. *Lab Invest.* 2007;87(10):1009–1017.
150. Leanos-Miranda A, Marquez-Acosta J, Cardenas-Mondragon GM, et al. Urinary prolactin as a reliable marker for preeclampsia, its severity, and the occurrence of adverse pregnancy outcomes. *J Clin Endocrinol Metab.* 2008;93(7):2492–2499.
151. Zhou Y, Gormley MJ, Hunkapiller NM, et al. Reversal of gene dysregulation in cultured cytotrophoblasts reveals possible causes of preeclampsia. *J Clin Invest.* 2013;123(7):2862–2872.
152. Gammill HS, Lin C, Hubel CA. Endothelial progenitor cells and preeclampsia. *Front Biosci.* 2007;12:2383–2394.
153. Matsubara K, Abe E, Matsubara Y, Kameda K, Ito M. Circulating endothelial progenitor cells during normal pregnancy and pre-eclampsia. *Am J Reprod Immunol.* 2006;56(2):79–85.
154. Sugawara J, Mitsui-Saito M, Hayashi C, et al. Decrease and senescence of endothelial progenitor cells in patients with preeclampsia. *J Clin Endocrinol Metab.* 2005;90(9):5329–5332.
155. Romero R, Chaiworapongsa T. Preeclampsia: a link between trophoblast dysregulation and an antiangiogenic state. *J Clin Invest.* 2013;123(7):2775–2777.
156. Cudmore M, Ahmad S, Al-Ani B, et al. Negative regulation of soluble Flt-1 and soluble endoglin release by heme oxygenase-1. *Circulation.* 2007;115(13):1789–1797.
157. Hubel CA. Oxidative stress in the pathogenesis of preeclampsia. *Proc Soc Exp Biol Med.* 1999;222(3):222–235.
158. Kanasaki K, Palmsten K, Sugimoto H, et al. Deficiency in catechol-O-methyltransferase and 2-methoxyoestradiol is associated with pre-eclampsia. *Nature.* 2008;453(7198):1117–1121.
159. Redman CW, Sargent IL. Latest advances in understanding preeclampsia. *Science.* 2005;308(5728):1592–1594.
160. Zhou CC, Zhang Y, Irani RA, et al. Angiotensin receptor agonistic autoantibodies induce pre-eclampsia in pregnant mice. *Nat Med.* 2008;14(8):855–862.
161. Ahmed A, Cudmore MJ. Can the biology of VEGF and haem oxygenases help solve pre-eclampsia? *Biochem Soc Trans.* 2009;37(Pt 6):1237–1242.
162. Ahmed A. New insights into the etiology of preeclampsia: identification of key elusive factors for the vascular complications. *Thromb Res.* 2011;127(suppl 3):S72–S75.
163. Karumanchi SA, Levine RJ. How does smoking reduce the risk of preeclampsia? *Hypertension.* 2010;55(5):1100–1101.
164. Wang K, Ahmad S, Cai M, et al. Dysregulation of hydrogen sulfide producing enzyme cystathionine gamma-lyase contributes to maternal hypertension and placental abnormalities in preeclampsia. *Circulation.* 2013;127(25):2514–2522.
165. Kalkunte SS, Neubeck S, Norris WE, et al. Transthyretin is dysregulated in preeclampsia, and its native form prevents the onset of disease in a preclinical mouse model. *Am J Pathol.* 2013;183(5):1425–1436.
166. Rana S, Karumanchi SA, Lindheimer MD. Angiogenic factors in diagnosis, management, and research in preeclampsia. *Hypertension.* 2014;63:198–202.

167. Rana S, Schnettler WT, Powe C, et al. Clinical characterization and outcomes of preeclampsia with normal angiogenic profile. *Hypertens Pregnancy*. 2013;32(2):189–201.
168. Bdolah Y, Lam C, Rajakumar A, et al. Twin pregnancy and the risk of preeclampsia: bigger placenta or relative ischemia? *Am J Obstet Gynecol*. 2008;198(4):428.e1–428.e6.
169. Nagamatsu T, Fujii T, Kusumi M, et al. Cytotrophoblasts up-regulate soluble fms-like tyrosine kinase-1 expression under reduced oxygen: an implication for the placental vascular development and the pathophysiology of preeclampsia. *Endocrinology*. 2004;145(11):4838–4845.
170. Debieve F, Depoix C, Gruson D, Hubinont C. Reversible effects of oxygen partial pressure on genes associated with placental angiogenesis and differentiation in primary-term cytotrophoblast cell culture. *Mol Reprod Dev*. 2013;80(9):774–784.
171. Xiong X, Demianczuk NN, Buekens P, Saunders LD. Association of preeclampsia with high birth weight for age. *Am J Obstet Gynecol*. 2000;183(1):148–155.
172. Thadhani R, Ecker JL, Mutter WP, et al. Insulin resistance and alterations in angiogenesis: additive insults that may lead to preeclampsia. *Hypertension*. 2004;43(5):988–992.
173. Funai EF, Friedlander Y, Paltiel O, et al. Long-term mortality after preeclampsia. *Epidemiology*. 2005;16(2):206–215.
174. Irgens HU, Reisaeter L, Irgens LM, Lie RT. Long term mortality of mothers and fathers after pre-eclampsia: population based cohort study. *BMJ*. 2001;323(7323):1213–1217.
175. Epstein FH. Late vascular effects of toxemia of pregnancy. *N Engl J Med*. 1964;271:391–395.
176. Johnson RJ, Herrera-Acosta J, Schreiner GF, Rodriguez-Iturbe B. Subtle acquired renal injury as a mechanism of salt-sensitive hypertension. *N Engl J Med*. 2002;346(12):913–923.
177. Patten IS, Rana S, Shahul S, et al. Cardiac angiogenic imbalance leads to peripartum cardiomyopathy. *Nature*. 2012;485(7398):333–338.
178. Vatten LJ, Romundstad PR, Trichopoulos D, Skjaerven R. Pre-eclampsia in pregnancy and subsequent risk for breast cancer. *Br J Cancer*. 2002;87(9):971–973.
179. Mogren I, Stenlund H, Hogberg U. Long-term impact of reproductive factors on the risk of cervical, endometrial, ovarian and breast cancer. *Acta Oncol*. 2001;40(7):849–854.
180. Benton SJ, Hu Y, Xie F, et al. Angiogenic factors as diagnostic tests for preeclampsia: a performance comparison between two commercial immunoassays. *Am J Obstet Gynecol*. 2011;205:469.e1–469.e8.
181. Sunderji S, Gaziano E, Wothe D, et al. Automated assays for sVEGF R1 and PlGF as an aid in the diagnosis of preterm preeclampsia: a prospective clinical study. *Am J Obstet Gynecol*. 2010;202(1):40.e1–40.e7.
182. Verlohren S, Galindo A, Schlembach D, et al. An automated method for the determination of the sFlt-1/PIGF ratio in the assessment of preeclampsia. *Am J Obstet Gynecol*. 2010;202(2):161.e1–161.e11.
183. Ohkuchi A, Hirashima C, Suzuki H, et al. Evaluation of a new and automated electrochemiluminescence immunoassay for plasma sFlt-1 and PlGF levels in women with preeclampsia. *Hypertens Res*. 2010;33(5):422–427.
184. Schiettecatte J, Russcher H, Anckaert E, et al. Multicenter evaluation of the first automated Elecsys sFlt-1 and PlGF assays in normal pregnancies and preeclampsia. *Clin Biochem*. 2010;43(9):768–770.
185. Verlohren S, Herraiz I, Lapaire O, et al. The sFlt-1/PlGF ratio in different types of hypertensive pregnancy disorders and its prognostic potential in preeclamptic patients. *Am J Obstet Gynecol*. 2012;206(1):58.e1–58.e8.
186. Wothe D, Gaziano E, Sunderji S, et al. Measurement of sVEGF R1 and PlGF in serum: comparing prototype assays from Beckman Coulter, Inc. to R&D Systems microplate assays. *Hypertens Pregnancy*. 2011;30(1):18–27.
187. Chaiworapongsa T, Romero R, Korzeniewski SJ, et al. Plasma concentrations of angiogenic/anti-angiogenic factors have prognostic value in women presenting with suspected preeclampsia to the obstetrical triage area: a prospective study. *J Matern Fetal Neonatal Med*. 2014;27:132–144.
188. Chaiworapongsa T, Romero R, Savasan ZA, et al. Maternal plasma concentrations of angiogenic/anti-angiogenic factors are of prognostic value in patients presenting to the obstetrical triage area with the suspicion of preeclampsia. *J Matern Fetal Neonatal Med*. 2011;24(10):1187–1207.
189. Chappell LC, Duckworth S, Seed PT, et al. Diagnostic accuracy of placental growth factor in women with suspected preeclampsia: a prospective multicenter study. *Circulation*. 2013;128(19):2121–2131.
190. Rana S, Powe CE, Salahuddin S, et al. Angiogenic factors and the risk of adverse outcomes in women with suspected preeclampsia. *Circulation*. 2012;125(7):911–919.
191. Perni U, Sison C, Sharma V, et al. Angiogenic factors in superimposed preeclampsia: a longitudinal study of women with chronic hypertension during pregnancy. *Hypertension*. 2012;59:740–746.
192. Verdonk K, Visser W, Russcher H, Danser AH, Steegers EA, van den Meiracker AH. Differential diagnosis of preeclampsia: remember the soluble fms-like tyrosine kinase 1/placental growth factor ratio. *Hypertension*. 2012;60(4):884–890.
193. Rolfo A, Attini R, Nuzzo AM, et al. Chronic kidney disease may be differentially diagnosed from preeclampsia by serum biomarkers. *Kidney Int*. 2013;83(1):177–181.
194. Thadhani R, Kisner T, Hagmann H, et al. Pilot study of extracorporeal removal of soluble fms-like tyrosine kinase 1 in preeclampsia. *Circulation*. 2011;124(8):940–950.
195. Kumasawa K, Ikawa M, Kidoya H, et al. Pravastatin induces placental growth factor (PGF) and ameliorates preeclampsia in a mouse model. *Proc Natl Acad Sci U S A*. 2011;108(4):1451–1455.
196. Bauer AJ, Banek CT, Needham K, et al. Pravastatin attenuates hypertension, oxidative stress, and angiogenic imbalance in rat model of placental ischemia-induced hypertension. *Hypertension*. 2013;61(5):1103–1110.
197. Costantine MM, Cleary K. Pravastatin for the prevention of preeclampsia in high-risk pregnant women. *Obstet Gynecol*. 2013;121(2 Pt 1):349–353.
198. Conrad KP. Maternal vasodilation in pregnancy: the emerging role of relaxin. *Am J Physiol Regul Integr Comp Physiol*. 2011;301(2):R267–R275.
199. Jiang X, Bar HY, Yan J, et al. A higher maternal choline intake among third-trimester pregnant women lowers placental and circulating concentrations of the antiangiogenic factor fms-like tyrosine kinase-1 (sFLT1). *Faseb J*. 2013;27(3):1245–1253.

200. Staff AC, Benton SJ, von Dadelszen P, et al. Redefining preeclampsia using placenta-derived biomarkers. *Hypertension.* 2013;61(5):932–942.
201. Roberts JM, Bell MJ. If we know so much about preeclampsia, why haven't we cured the disease? *J Reprod Immunol.* 2013;99(1-2):1–9.
202. Powers RW, Jeyabalan A, Clifton RG, et al. Soluble fms-Like tyrosine kinase 1 (sFlt1), endoglin and placental growth factor (PlGF) in preeclampsia among high risk pregnancies. *PLoS One.* 2010;5(10):e13263.
203. Myatt L, Clifton RG, Roberts JM, et al. First-trimester prediction of preeclampsia in nulliparous women at low risk. *Obstet Gynecol.* 2012;119(6):1234–1242.
204. Sheehan HL, Lynch JP. *Pathology of Toxaemia of Pregnancy.* Baltimore: Williams and Wilkins; 1973.

CHAPTER 7

Metabolic Syndrome and Preeclampsia

ARUN JEYABALAN, CARL A. HUBEL AND JAMES M. ROBERTS

Editors' comment: *In the early part of the previous century the association between hypertension, increased blood sugar levels and obesity had already been commented on, while during the mid 1900s French medical texts mentioned "goutte diasthesique," the association of "hypertension, hyperglycemie, hyperuricemie, and hypercholesterolemie." Years later this became syndrome X, metamorphosing quickly into "metabolic syndrome" (with the latter's multi-defined criteria). Chesley was aware of such associations too, and in his landmark study of the remote prognosis of eclamptics sought and demonstrated increased hyperglycemia in this population. Since then preeclampsia has been associated with increased insulin resistance, shown to have higher incidence in diabetic and obese women, and subject to scores of observations regarding dyslipidemia. The obesity epidemic, with half of women of reproductive age in the United States obese or overweight, has resulted in this disorder being the leading attributable risk for preeclampsia. The metabolic changes associated with obesity are likely to be related to this increased risk and are reviewed in this chapter.*

INTRODUCTION

In normal pregnancy, cytotrophoblast cells from the placenta invade the maternal spiral arteries, causing them to lose their smooth muscle and enabling the expansion of vascular capacity necessary to support fetal growth. In some, but not all, cases of preeclampsia there is insufficient remodeling of the spiral arteries, resulting in reduced uteroplacental perfusion. Compromised uteroplacental perfusion is thought to lead to the release of signals (many unidentified), from the placenta into the maternal circulation, that target the maternal vascular endothelium. In women who develop preeclampsia, the effects include widespread evidence of inflammation and endothelial cell dysfunction.[1,2] These changes are thought to reflect a disturbance of the delicate balance/conflict between mechanisms intended to preserve fetal growth and those that protect the health of the mother.[3]

A salient feature of preeclampsia, however, is its heterogeneity. Indeed, there are inconsistencies with the notion that the poorly perfused placenta is the sole origin of factors altering endothelial function and causing preeclampsia. Failure of physiologic transformation of the spiral arteries may also be observed in patients with (1) fetal growth restriction without preeclampsia, (2) preterm labor and intact membranes,[4] (3) preterm premature rupture of membranes,[5] and occasionally (4) clinically uncomplicated pregnancies.[6] Conversely, the maternal preeclampsia syndrome may occur with minimal evidence of failed spiral artery transformation. Maternal constitutional susceptibility is probably a determining factor. Gestational diabetes[7] and prepregnancy obesity[8] are often associated with larger babies yet predispose to preeclampsia. Preeclampsia superimposed on chronic hypertension or diabetes mellitus confers a higher risk of poor perinatal outcome compared to preeclampsia alone.[9] Furthermore, not only is there an excess of cardiovascular disease in women with a history of preeclampsia, but preeclampsia and cardiovascular disease share obesity, insulin resistance, diabetes, and inflammation as risk factors.[9] These data are consistent with maternal factors as important determinants of preeclampsia.

The metabolic syndrome (MetS), also known as syndrome X, or the insulin resistance syndrome, or Reaven syndrome, comprises a constellation of metabolic factors and physical conditions associated with increased risk of development of type 2 diabetes and cardiovascular disease. The chapter first overviews some of the ways in which the MetS is thought to promote cardiovascular disease. It then uses this as a platform to focus on evidence that the components of the MetS, particularly chronic insulin resistance, inflammation and dyslipidemia, contribute to maternal endothelial cell dysfunction and the preeclampsia syndrome. The chapter then summarizes evidence suggesting that the same underlying MetS-related disturbances that contribute to preeclampsia also increase a woman's risk of developing cardiovascular disease in later life. Perhaps diverging from conventional thinking, the review then extends this hypothesis to suggest that the MetS actually contributes to poor placental development, with maternal and placental processes then cooperating in a vicious circle of inflammation and endothelial dysfunction. The intended "take-home" message is that obesity and other factors related to the

Chesley's Hypertensive Disorders in Pregnancy.
ISBN: 978-0-12-407866-6

DOI: http://dx.doi.org/10.1016/B978-0-12-407866-6.00007-9

metabolic syndrome construct should be a focus for clinical and research strategies for prevention of preeclampsia and related pregnancy complications, and their cardiovascular sequelae.

METABOLIC SYNDROME

The clustering of metabolic and physiological abnormalities was first described in 1923 by Eskil Kylin as a syndrome consisting of hypertension, hyperglycemia, obesity, and hyperuricemia.[10] Gerald Reaven, in his 1988 Banting Lecture, described "syndrome X" as the clustering, more often than predicted by chance, of resistance to insulin-stimulated glucose uptake, hyperinsulinemia, hyperglycemia, elevated very low-density lipoprotein (VLDL) triglycerides, reduced high-density lipoprotein (HDL) cholesterol, and hypertension.[11] Criteria for the MetS have evolved over the past decade (Table 7.1) based on the prevailing views of its pathogenesis.

In 1998, the World Health Organization Task Force on Diabetes identified insulin resistance as the dominant cause of the MetS.[12] By this criterion, clinical indicators of insulin resistance were required for diagnosis (Table 7.1). With growing evidence of a paramount role for obesity, however, the latter has assumed a more important position among diagnostic criteria. Indeed, obesity and physical inactivity are currently considered to be the driving forces behind the MetS, modulated by susceptibility factors including adipose tissue disorders, genetic factors, race, aging, and endocrine disorders.[13] Insulin resistance is often secondary to obesity but can have genetic components as well.[14] These concepts led to the National Cholesterol Education Program (NCEP) criteria for the MetS in which the need for demonstration of insulin resistance was supplanted by increased waist circumference (abdominal obesity).[12]

The majority of epidemiological studies have used NCEP criteria, defining MetS as the presence of three or more of the following five risk factors: abdominal obesity, high triglycerides, low HDL cholesterol, hypertension, and elevated fasting glucose (Table 7.1). The more recent International Diabetes Federation definition requires central obesity plus any two of the other four 2004 NCEP factors (Table 7.1).

TABLE 7.1 Metabolic Components and their Positive cut-off Points in the Definition of the Metabolic Syndrome by different Criteria

Components	WHO 1998	NCEP 2001	NCEP 2004	IDF 2005
Insulin resistance	Glucose uptake below lowest quartile under hyperinsulinemic and euglycemic condition			
Hyperglycemia				
Fasting plasma glucose (mg/dL)	≥110	≥110	≥100	≥100
Impaired glucose tolerance	Included			Recommended
Previously diagnosed diabetes	Included	Included	Included	Included
Central obesity				
Waist-to-hip ratio and/or	>0.85 (>0.9)			
Body mass index (kg/m^2)	>30			
Waist circumference (cm)		>88 (> 102)	>88 (> 102)	≥80 (≥ 94) for Europids, ethnic specific
Elevated blood pressure				
Blood pressure (mm Hg)	≥140/90	≥130/85	≥130/85	≥130/85
Treatment for hypertension	Not mentioned	Included	Included	Included
Dyslipidemia				
Triglycerides (mg/dL)	≥150	≥150	≥150	≥150 or on triglyceride Rx
HDL-C (mg/dL)	<39 (<34)	<50 (<40)	<50 (<40)	<50 (<40) or on HDL-C Rx
Treatment for dyslipidemia	Not mentioned	Not mentioned	Not mentioned	Included
Urinary albumin excretion or	≥20 μg/min			
Albumin/creatinine ratio	≥30 mg/g			

Cut-off points in parentheses are for men. WHO, World Health Organization; NCEP, National Cholesterol Education Program Adult Treatment Expert Panel III; IDF, International Diabetes Federation. The WHO definition requires the presence of insulin resistance and/or hyperglycemia plus any two of the three other factors. The NCEP 2001 and NCEP 2004 (revised version) definition requires any three of the five factors. The IDF definition requires the presence of central obesity plus any two of the other four factors. Raised triglycerides and reduced HDL are counted as two individual factors in the NCEP and the IDF definitions but as one factor in the WHO definition. HDL-C, high-density lipoprotein cholesterol.

(Adapted from Qiao et al. (2007) Metabolic syndrome and cardiovascular disease. *Ann Clin Biochem* **44**, 232–263.)

METABOLIC SYNDROME AND CARDIOVASCULAR DISEASE

This section of the chapter is intended to illustrate several of the key pathways by which the MetS is thought to promote cardiovascular disease in the general population. Comprehensive reviews are available on the subject of MetS and cardiovascular disease.[15–23] Contemporary reviews are also available that specifically deal with the question of how prenatal and early postnatal environment/exposures increase later-life susceptibility to obesity, MetS, cardiovascular disease, and type 2 diabetes.[24–27]

The MetS is becoming increasingly prevalent worldwide. Using the 2004 NCEP glucose threshold of 100 mg/dL (Table 7.1), each of the major components – obesity, elevated triglycerides, low HDL-cholesterol, and elevated fasting plasma glucose – occurs in approximately one-third of the United States population.[13] The MetS often coexists with other medical conditions, including fatty liver, obstructive sleep apnea, gout, depression, and polycystic ovarian syndrome. People with the MetS are at roughly twice the risk of atherosclerotic cardiovascular disease compared to those without. It also raises the risk of type 2 diabetes by about 5-fold.[13] Large-scale clinical trials such as the Women's Health Study and the Framingham Offspring Study have confirmed the contribution of the MetS to cardiovascular disease in women.[28,29] Data suggest that the presence of the MetS is more predictive of future cardiovascular disease in women than in men.[29]

Although individuals with the MetS are at increased risk of cardiovascular disease, and although the MetS risk factors are found in combination more often than by chance, it is unclear whether the MetS construct is clinically more useful as a marker of risk above and beyond its individual components (i.e., the whole is not greater than the sum of its parts).[30–33] By extension, rather than focusing just on the MetS construct as a whole, it will be important to evaluate the connection of preeclampsia with individual MetS-related factors.

Obesity

Obesity is usually defined as a body mass index (BMI) $\geq 30\,kg/m^2$, and overweight as BMI of 25.0 to 29.9, with BMI calculated as weight in kilograms divided by the square of the height in meters. BMI as a measure of obesity has limitations because it does not distinguish between fat and lean tissue (nearly half of all National Basketball Association players would be inaccurately classified as overweight or obese on the basis of BMI) or fat distribution. Regardless of whether adiposity is gauged by BMI, waist circumference, bioimpedance analysis of fat mass, or dual energy X-ray absorptiometry scans, it is clear that obesity is a major driving force for the increased prevalence of cardiovascular disease world-wide.[34] This includes the developing world, where rates of obesity have tripled in regions that have been adopting a Western lifestyle.[34] Nations of the Middle East, Pacific Islands, Southeast Asia, and China are facing this threat. Some developing countries experience the paradox of families in which the adults are overweight but the children are underweight, the latter potentially attributable to intrauterine growth restriction which may predispose to later-life obesity via acquisition of the "thrifty phenotype."[34] In the United States, medical costs and lost productivity attributable to excess adiposity totaled about $117 billion in the year 2000.[35] At least one-third of all pregnant women in the United States are obese; data released in 2013 by the Centers for Disease Control and Prevention show that 57% of African-American women, 44% of Hispanic women, and 33% of white women are obese.[8,36]

Prospective cohort studies of non-pregnant populations suggest that obesity precedes the development of insulin resistance.[37,38] The increased risk of adverse health outcomes among obese individuals is significantly influenced by MetS-related pathologies, including inflammation, oxidative stress, insulin resistance, dyslipidemia, and endothelial dysfunction,[18] all of which are significantly more prevalent in women with preeclampsia compared to women with uncomplicated pregnancy.[39] However, approximately 29% of obese men and 35% of obese women (obesity gauged by BMI) in the United States are "metabolically healthy" as defined by having either none or not more than one of the following MetS abnormalities: elevated blood pressure, fasting triglycerides, fasting glucose, C-reactive protein, homeostasis model assessment of insulin resistance (HOMA) value, or reduced HDL cholesterol level.[40] Independent characteristics enriched in this "uncomplicated obese" subgroup include non-Hispanic black race, smaller waist circumference and higher level of physical activity. Therefore, an obese individuals' cardiovascular disease risk depends jointly on adiposity and metabolic profile.

Adipose tissue is pleiotropic and functions not just as a lipid storage depot but as an active endocrine organ that secretes an array of bioactive molecules called adipokines.[19] Several adipokines are listed in Table 7.2 according to their apparent role in the pathogenesis of endothelial dysfunction and atherosclerosis, and their association with preeclampsia. It is interesting that the majority of these mediators seem to be produced more actively in adipose tissue from obese individuals; that is, the production is not greater simply because of more tissue but also because of predominant location of the tissue (visceral versus subcutaneous) and/or intrinsic differences in the adipocyte with obesity.[50] Visceral and subcutaneous fat are quite different metabolically. Materials produced by visceral fat are drained directly to the liver where they can upregulate the hepatic production of acute-phase reactants and inflammatory cytokines. This is reflected in the increased circulating concentrations of C-reactive protein, plasminogen activator inhibitor 1 (PAI-1) and inflammatory cytokines in

TABLE 7.2 Adipokines and Cardiovascular Disease

	Endothelial Cell Dysfunction	Atherosclerosis	Preeclampsia
Leptin	Increased NO-induced vasorelaxation	Stimulation of thrombus formation	Elevated in plasma from 20 weeks gestation in women who later develop preeclampsia; rises markedly from 32 weeks as preeclampsia develops[41]
Adiponectin	Decreased expression of adhesion molecules Stimulation of NO-induced relaxation	Inhibition of transformation of macrophages to foam cells Stabilization of atherosclerotic plaques (TIMP-1)	Paradoxically upregulated in preeclampsia;[41] more recent reports variable;[42] correlates negatively with fasting insulin values; possible feedback mechanism to counter insulin insensitivity
IL-6	Stimulation of CRP synthesis in the liver (weak evidence that CRP decreases NO production in endothelial cells)	Stimulation of monocyte/ macrophages infiltration of atherosclerotic plaque Destabilization of atherosclerotic plaque (matrix metalloproteinases)	Many, although not all, studies show higher third-trimester plasma levels of IL-6 and TNF-α in women with preeclampsia; first-trimester IL-6 concentrations may be elevated in women who later develop preeclampsia[43,44]
TNF-α	Increased expression of adhesion molecules	Stimulation of oxidized LDL uptake by macrophages (scavenger receptors)	The source of excess circulating TNF-α in preeclampsia is apparently from macrophages of adipose tissue as placental mRNA is not increased[43–45]
	Increased expression of MCP-1 and M-CSF	Stimulation of proinflammatory cytokines release by monocytes/ macrophages Destabilization of atherosclerotic plaques (matrix metalloproteinases)	
Resistin	—	Promotes fat deposition and has proinflammatory properties	Higher first-trimester resistin in women who developed preeclampsia[46]
FABP4	—	Correlated with carotid intimal thickening, insulin resistance and later-life cardiovascular diseae	Higher circulating FABP4 in first and second trimesters in women who developed preeclampsia[47]
PAI-1	—	Inhibition of fibrin clot breakdown	PAI-1 is principally of placental vascular endothelial origin and is raised in the maternal circulation of women with preeclampsia[48]
Angiotensin II	Vasoconstriction	Stimulation of intimal infiltration by monocytes	Preeclamptic patients manifest exaggerated pressor responses to Ang II, although the circulating Ang II concentrations are lower compared to control pregnancies[48]
	Increased expression of adhesion molecules, MCP-1 and M-CSF Stimulation of NADPH oxidase (reactive oxygen species production)	Stimulation of migration and proliferation of smooth muscle cells	Evidence for the existence and importance of a local, utroplacental renin-angiotensin system[48] Agonistic autoantibodies to the angiotensin II type I receptor are implicated in preeclampsia[49]

Data on endothelial cell dysfunction and atherosclerosis are by permission from Table 7.3 of Chudek and Wiecek (2006) Adipose tissue, inflammation and endothelial dysfunction. *Pharmacological Reports* **58**, 81–88.

individuals with visceral obesity.[50] Also, visceral fat *in vitro* produces more C-reactive protein and inflammatory cytokines than subcutaneous fat.[51] The importance of adipose tissue versus the placenta as the source of elevated circulating leptin, adiponectin, interleukin-6 (IL-6), tumor necrosis factor (TNF)-α, and plasminogen activator inhibitor (PAI-1) during preeclampsia is, however, unclear.

It is clear that visceral fat is a much better predictor of cardiovascular disease than BMI or percent body fat.[51] The central distribution of body fat (increased ratio of waist to hip circumference) is associated with a higher risk of morbidity and mortality than a more peripheral distribution of body fat. Although the information for preeclampsia is limited, there is an increased risk of preeclampsia determined by early pregnancy waist circumference, a surrogate for visceral fat.[52,53]

Free Fatty Acids and TNF-α

Visceral adipose tissue of obese/insulin resistant individuals releases excess amounts of non-esterified ("free") fatty acids (FFA) and the inflammatory cytokine TNF-α into the circulation. Insulin inhibits adipocyte hormone-sensitive lipase, decreasing adipocyte triglyceride hydrolysis and thus limiting concentrations of FFA and glycerol in the circulation. Insulin resistance thus increases hormone-sensitive lipase activity and adipocyte lipolysis (Fig. 7.1). The resulting hepatic overload with FFAs increases hepatic triglyceride production and export into the circulation in the form of VLDL; it also leads to hepatic triglyceride deposition, predisposing to fatty liver. Overproduction of VLDL is a fundamental feature of the MetS that initiates a sequence of further lipoprotein changes, namely higher concentrations of atherogenic remnant lipoprotein particles and smaller-sized (more oxidizable) low-density lipoprotein (LDL) particles, and lower concentrations of HDL cholesterol (Fig. 7.1).[54] The MetS is also associated with higher plasma concentrations of oxidatively modified LDL, which, in turn, are associated with increased risk of myocardial infarction even after adjustment for LDL cholesterol and other established risk factors.[17] These lipid abnormalities are also features of preeclampsia.[39]

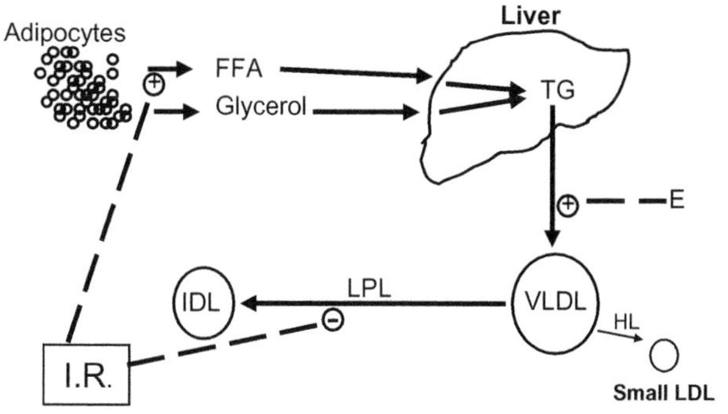

FIGURE 7.1 Interactions of lipoprotein metabolism with insulin resistance and estrogen during pregnancy. The normal physiologic response to pregnancy includes excursions into insulin resistance, dyslipidemia, and systemic inflammation. Preeclampsia (and obesity in pregnancy) is often an extreme of the pregnancy continuum with respect to insulin resistance, dyslipidemia and inflammatory cell activation. Key: (+), activation; (−), inhibition; I.R., insulin resistance; E, estrogen; TG, triglyceride; FFA, free fatty acid; LPL, lipoprotein lipase; VLDL, very low-density lipoprotein; IDL, intermediate-density lipoprotein.

FFA and TNF-α can reduce endothelium-dependent vasodilatation, increase blood pressure, and further impair insulin sensitivity via mechanisms that are not well understood. Production of reactive oxygen species by FFA and TNF-α is probably heavily involved.[55] Free fatty acids increase the production of superoxide anion radical (O_2^-) and hydrogen peroxide in several cell types by stimulating the enzyme NADPH-oxidase.[55] In states of adiposity, processing of excess intracellular free fatty acid by the mitochondria causes mitochondrial uncoupling and overproduction of reactive oxygen species.[56] Reactive oxygen species can oxidatively destroy the endothelium-dependent vasodilator nitric oxide (NO), thus reducing vascular relaxation responsiveness. Raised concentrations of FFA can also blunt insulin-induced phosphatidylinositol (PI3) kinase activation in endothelial cells, thus reducing the bioavailability of nitric oxide by inhibiting endothelial nitric oxide synthase (eNOS) (Fig. 7.2).[19,57] FFA may also increase synthesis of the vasoconstrictor endothelin-1 (ET-1) (Fig. 7.2).

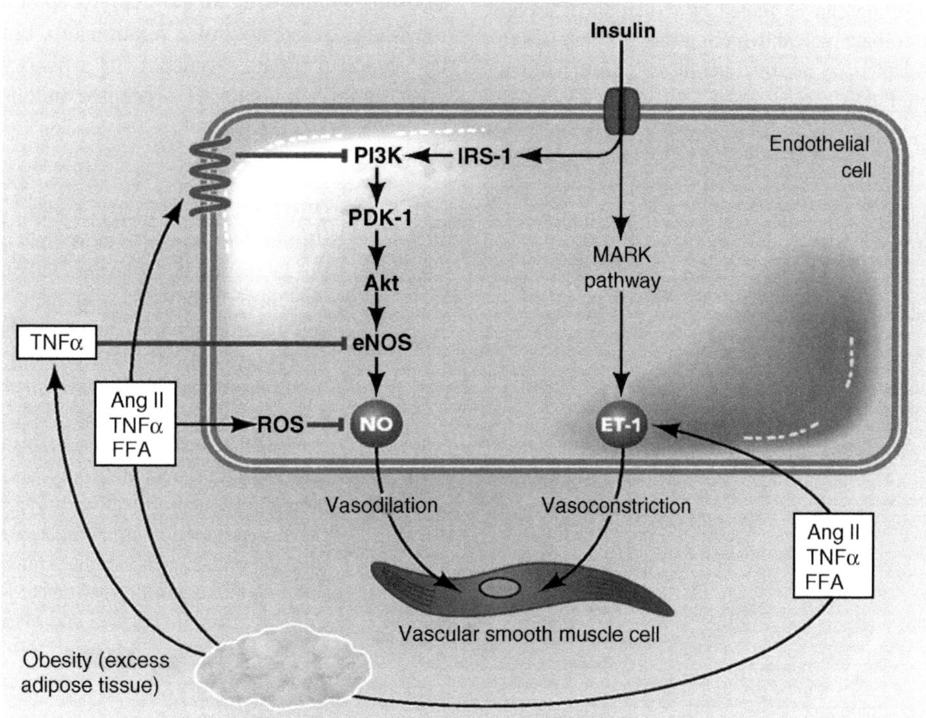

FIGURE 7.2 Mechanisms of insulin-mediated nitric oxide (NO) and endothelin 1 (ET-1) production leading to vasodilatation and vasoconstriction, respectively. Angiotensin II (Ang II), tumor necrosis factor-α (TNFα), and free fatty acids (FFA) inhibit the phosphatidylinositol kinase (PI3K) pathway and stimulate the mitogen-activated protein kinase (MAPK) pathway. IRS-1, insulin receptor substrate 1; PDK-1, phosphoinositide-dependent kinase 1; Akt, protein kinase B; eNOS, endothelial nitric oxide synthase. Reproduced, by permission, from Jonk et al. 2007.[19]

FFAs circulate in the bloodstream bound to albumin. Serum albumin functions as an antioxidant by sequestering copper and by scavenging hydroxyl radicals, the latter by virtue of albumin's reduced cysteine residue (Cys34).[58] As illustrated in Fig. 7.3, and as will be discussed in more detail in the context of preeclampsia, oversaturation of albumin by free fatty acids results in a conformational change in albumin such that the copper bound to albumin becomes redox active; albumin is thereby converted from an antioxidant to a prooxidant.[59–62] The excess ROS produced by these various mechanisms may adversely affect cellular metabolism and subvert endothelium-dependent relaxation by multiple means, including destroying the vasodilator nitric oxide before it can reach its intended targets (Fig. 7.2).

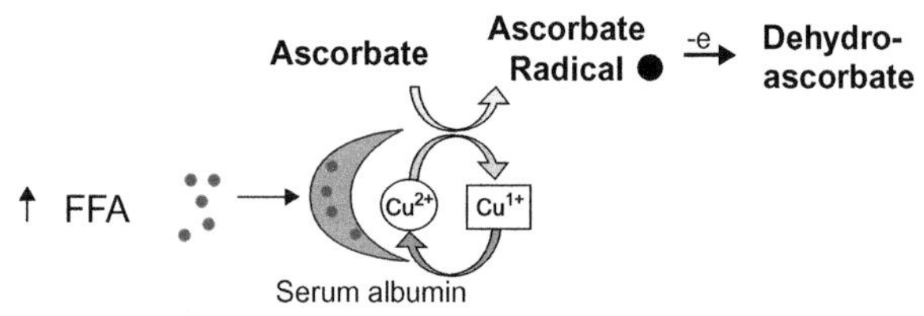

FIGURE 7.3 A mechanism of fatty acid-induced oxidative stress. Excess binding of fatty acids (FFA) to human serum albumin results in a conformational shift in the albumin molecule accompanied by an alteration in copper (Cu)-albumin interactions, with the appearance of "loosely bound" Cu capable of redox-cycling. This redox-cycling can be fueled by oxidation of ascorbate (vitamin C) to ascorbate radical, which further decomposes to dehydroascorbate. (Adapted, by personal permission, from a drawing by Valerian Kagan, PhD.)

TNF-α inhibits insulin action in adipocytes, possibly through inhibition of insulin receptor substrate 1 (one primary substrate of the insulin receptor) by c-Jun N-terminal kinases (JNK) (Fig. 7.4). In animal models of pre-diabetic metabolic syndrome, endothelial cell expression of TNF-α is increased in association with endothelial dysfunction; blockade of TNF-α restored endothelium-mediated vasodilatation.[63] TNF-α, like free fatty acids, may mediate endothelial dysfunction partly by lowering bioavailable nitric oxide, secondary to both stimulation of ROS and inhibition of eNOS (Fig. 7.2).

Renin-Angiotensin System

Human adipose tissue expresses all components of the renin-angiotensin system. The expression of angiotensinogen in omental adipose tissue is higher in obese compared to lean subjects, and plasma angiotensinogen levels are positively correlated with body mass index (BMI).[16,64] Besides being potentially important in systemic vasoconstriction[65] angiotensin II may induce oxidative stress by activation of NADPH-oxidase.[66–69] Several additional angiotensin II signaling pathways are thought to contribute to the chronic subacute inflammatory state characteristic of obesity.[16,18,70]

Both angiotensin II and circulating autoantibodies that activate the angiotensin type I (AT1) receptor have been implicated in the etiology and pathogenesis of preeclampsia.[49,71–74] Information is lacking, however, about any relationship of these agonistic AT1-receptor autoantibodies to the MetS. Suggestive of a relationship, however, the minority of non-pregnant women with a history of preeclampsia who retain agonistic AT1-receptor autoantibodies in their

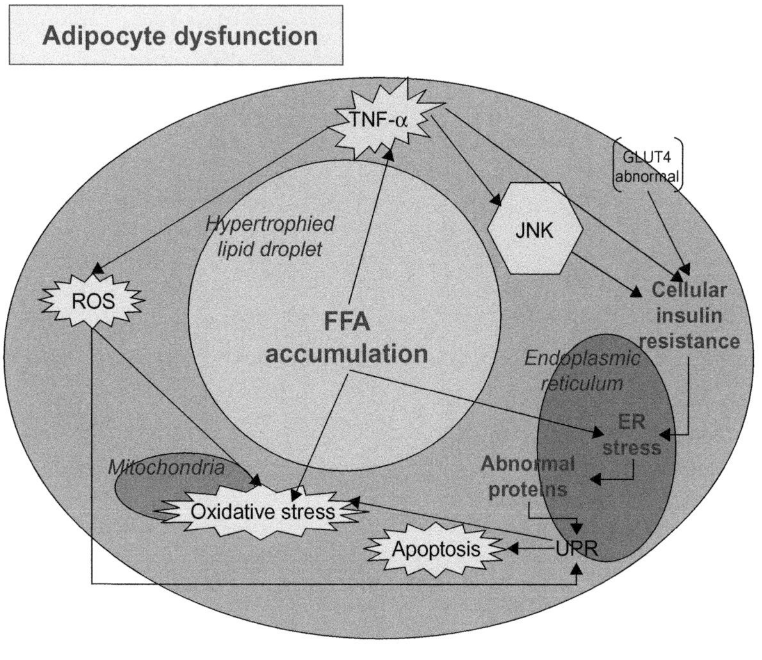

FIGURE 7.4 Aspects of adipocyte dysfunction due to nutrient excess. Excess lipid accumulation leads to increased endoplasmic reticulum (ER) activity, which ultimately can overwhelm the capacity of the ER to properly fold nascent proteins. The unfolded protein response (UPR) can compensate for this situation to some extent. However, if the process proceeds unchecked, apoptosis may result. ER stress can lead to oxidative stress in the mitochondrion, as does the presence of excess free fatty acids (FFA). Oxidative stress produces reactive oxygen species (ROS). TNF-α production is stimulated by FFAs, which in turn act on c-Jun N-terminal kinases (JNK) to contribute to cellular insulin resistance. Reproduced, by permission, from de Ferranti and Mozaffarian (2008).[23]

circulation 18 months postpartum also have higher homeostasis model assessment of insulin resistance (HOMA) indices on average than women with prior preeclampsia who do not have evidence of these autoantibodies.[75]

Adipocyte Hypertrophy and Endoplasmic Reticulum Stress

Calories consumed in excess of calories expended (a fundamental cause of obesity) leads to increased storage of the surplus energy in the form of adipocyte intracellular lipid droplets (Fig. 7.4). Whether adipocyte hypertrophy or adipocyte hyperplasia, or both, occur in response to excess postprandial lipids and glucose depends upon the type of adipose tissue. Subcutaneous fat deposition occurs early in development of obesity, with visceral deposition of fat occurring only after subcutaneous capacity has been reached.[23,76] Macrophages are observed more frequently, along with increased expression of the proinflammatory cytokine monocyte chemoattractant protein-1 (MCP-1), in human omental compared to subcutaneous fat, correlating with waist circumference.[23,77] Such data support the notion that visceral fat has more adverse health consequences than subcutaneous fat.

Hypertrophy of adipocyte lipid droplets may underlie the endoplasmic reticulum stress and mitochondrial abnormalities that are central to the adverse effects of obesity.[23] One manifestation of endoplasmic reticulum stress is the "unfolded protein response" in which abnormally folded proteins aggregate in the cytosol and interfere with normal cellular functions (Fig. 7.4). The unfolded protein response can further increase cellular insulin resistance, with further increases in lipids and endoplasmic reticulum stress, generating a vicious circle of worsening insulin resistance and apoptosis (programmed cell death) (Fig. 7.4). It has been suggested that endoplasmic reticulum stress and the unfolded protein response are involved in the placental, endothelial, and oxidative stress associated with preeclampsia and intrauterine fetal growth restriction (IUGR)[78] and data support this.[79,80] However, the relationship of endoplasmic reticulum stress in preeclampsia to maternal obesity and insulin resistance has not, to our knowledge, been explored.

PREGNANCY-INDUCED METABOLIC CHANGES

Carbohydrate and Lipid Metabolism

Profound changes in carbohydrate and lipid metabolism occur with normal pregnancy, and these changes are important for fetal development.[8,81–84] This section summarizes some of these changes in order to provide a framework for information on disturbed insulin and lipid metabolism in preeclampsia.

Pregnancy results in remarkable maternal metabolic adjustments that optimize nutrient availability for the growing fetal-placental unit. Two metabolic stages of pregnancy can be distinguished, the first roughly corresponding to the first two-thirds of pregnancy when fetal growth is limited and differentiation and organogenesis are dominant, and the second occurring during the last third of pregnancy when fetal and placental growth accelerate (Table 7.3). Maternal metabolism is predominantly anabolic during the first two trimesters, with storage of a greater proportion of nutrients, as evidenced by accumulation of maternal fat mass. At this stage, maternal glucose tolerance is normal or slightly increased compared to non-pregnant women, and insulin sensitivity in peripheral tissues and hepatic glucose production is normal.[85] By late pregnancy, however, the mother's metabolism switches to a more catabolic state to provide the fuels needed by the fetus.[8] Because glucose is the preferred fuel of the fetus, a state of modest insulin resistance normally develops during the last half of pregnancy to augment maternal plasma glucose, lipid and amino acid concentrations for diffusion across the placenta.[8]

One result of the mobilization of maternal adipose lipid depots during the second half of pregnancy is a striking increase in circulating FFA concentrations. Human placental

TABLE 7.3 Summary of Metabolic Changes in Pregnancy and Preeclampsia

	Normal Pregnancy		Preeclampsia	
	Early	**Late**	**Early**	**Late**
Maternal Metabolism	Anabolic	Catabolic	Anabolic	Catabolic
Glucose tolerance	~	↓	~↓	↓↓
Insulin sensitivity	~	↓	~↓	↓↓
Free fatty acids	↑	↑↑	↑↑	↑↑↑
Triglycerides	↑	↑↑	↑↑	↑↑↑
Cholesterol	~	↑	~↑	↑

~: Similar compared to non-pregnant controls.
↑,↑↑,↑↑↑: elevated (relative degree increasing by number of arrows) compared to non-pregnancy controls;
↓,↓↓: lower compared to non-pregnancy controls.
(Adapted by permission from von Versen-Hoeynck and Powers (2007) Maternal-fetal metabolism in normal pregnancy and preeclampsia. *Frontiers in Bioscience* **12**, 2457–2470.)

lactogen, which reaches very high levels during late gestation, appears to contribute to the increase in FFAs by its direct lipolytic action on adipose tissue.[86] Suggesting a feed-forward process, the increasing levels of plasma FFAs may contribute to the gestational insulin resistance.[87] TNF-α and leptin probably also foster gestational insulin resistance.[88,89]

Much of the glycerol and FFA released from fat are taken up by the liver and re-esterified for synthesis of very low-density lipoprotein (VLDL) triglycerides (Fig. 7.1). Higher estrogen concentrations and decreased hepatocyte beta-oxidation also lead to increased hepatocyte VLDL production. By term, plasma triglycerides increase by 50–300% over non-pregnancy levels, at which time higher triglycerides are found not only in VLDL but also in intermediate-density lipoprotein (IDL), low-density lipoprotein (LDL) and high-density lipoprotein (HDL).[84,90–92] Lipoprotein lipase, tethered to the luminal side of capillaries and arteries of extrahepatic tissues, hydrolyzes lipoprotein triglycerides to produce FFA and monoacylglycerol (Fig. 7.1). These lipolysis products are predominantly taken up by tissue locally to meet the needs of that tissue. This promotes triglyceride clearance from the circulation. However, adipose tissue lipoprotein lipase activity decreases substantially during the last third of pregnancy due to insulin resistance and other hormonal influences. The result is a decrease in the rate of removal of triglyceride-rich lipoproteins from the circulation.[93] Therefore, both increased production and decreased removal contribute to the dramatic late gestational increase in postprandial and fasting triglycerides.[93]

Both circulating FFAs and glycerol, but not triglycerides, cross the membranes of the placental syncytiotrophoblast to the fetal circulation.[94,95] However, triglyceride-rich lipoprotein particles are an important source of essential fatty acids for the fetus. A variant of lipoprotein lipase in the trophoblast microvillous membrane appears to be relatively unresponsive to insulin; this enzyme makes FFAs available by hydrolyzing lipoprotein triglycerides.[96,97] An unusually low activity of placental lipoprotein lipase was observed in preterm pregnancies complicated by intrauterine growth restriction (IUGR).[96,97] This suggests that syncytiotrophoblast lipoprotein lipase activity contributes to FFA delivery to the fetus and, as a consequence, fetal growth. Membrane-bound and cytosolic fatty acid binding proteins are important in determining the direction and net flux of fatty acids toward the fetus.[98] After uptake by trophoblast cells, fatty acids are re-esterified to provide a fat reservoir. After intracellular hydrolysis releases fatty acids to fetal plasma, they bind α-fetoprotein and are transported to the fetal liver.[83]

A direct relationship between maternal triglycerides and birthweight has been found in humans. Obesity is strongly correlated with elevated triglyceride and VLDL in pregnancy, both in the mother and in the macrosomic newborn.[8,97] Severe correction of maternal hypertriglyceridemia (as with hypolipidemic drugs), however, has negative effects on fetal growth and development.[83]

Maternal cholesterol levels increase during pregnancy, with a roughly 50% rise over prepregnant levels by term.[99] Maternal cholesterol is important for the fetus during early pregnancy but its importance lessens by late pregnancy owing to the ability of fetal tissues to synthesize cholesterol.[83] HDL cholesterol increases by week 12 of pregnancy in response to estrogen and remains elevated for the rest of pregnancy. Reversal of the physiologic hyperlipidemia of pregnancy begins within hours of delivery and is essentially complete by 6 to 10 weeks postpartum.[91]

Maternal Weight Gain in Pregnancy

Maternal weight gain during pregnancy is anticipated and attributable to the fetus, placenta, amniotic fluid, as well as the physiologic uterine hypertrophy, increase in blood and fluid volume, breast enlargement and fat stores. *Excessive* weight gain is primarily related to the increase in maternal adiposity. Prepregnancy obesity is linked to a host of perinatal problems related to adverse outcome (gestational diabetes, cesarean delivery, infant macrosomia, preterm birth, neural tube and other birth defects, antepartum stillbirth, preeclampsia, eclampsia).[8,100,101] The risk of perinatal death more than doubles with maternal obesity.[102] Recognizing that a greater percentage of women are entering pregnancy overweight or obese and that many are gaining too much weight during pregnancy, the Institute of Medicine convened a task force to re-evaluate the 1990 guidelines for weight gain in pregnancy. Numerous studies published between 1990 and the 2009 guidelines also demonstrated that prepregnancy BMI above normal values (overweight and obese) as well as excessive gestational weight gain during pregnancy are associated with insulin resistance, gestational diabetes, preeclampsia, cesarean section, large for gestational age infants and breast-feeding problems.[103–107] An important consideration for the 2009 guidelines (Table 7.4) was not only the welfare of the infant, but also the health of the mother given the substantial research that prepregnancy weight and gestational weight gain impact pregnancy outcomes as well as the short- and long-term health of the mother.[108] Major differences in the 2009 guidelines compared to those published in 1990 are that the weight gain recommendations are based on World Health Organization BMI categories and that a relatively narrow range of weight gain is recommended for obese women during pregnancy. In addition, optimizing BMI prior to conception is emphasized (Table 7.4).[108]

Additional evidence published since the release of the new guidelines has confirmed the association of excessive gestational weight gain with adverse clinical outcomes including hypertensive disorders[109–103] and gestational diabetes.[112] In an assessment of pregnancy outcomes in nulliparous women with weight gain above or below the 2009 IOM guidelines, excess weight gain occurred in 73% of women and was associated with an increased risk of hypertensive disorders, cesarean delivery, and large for gestational

TABLE 7.4 2009 Institute of Medicine Weight Gain Recommendations for Pregnancy[108]

Prepregnancy BMI	BMI (kilogram/meters^2)	Total Weight Gain Range (pounds)	Rates of Weight Gain in the Second or Third Trimester (mean range in pounds/week)
Underweight	<18.5	28–40	1 (1–1.3)
Normal weight	18.5–24.9	25–35	1 (0.8–1)
Overweight	25.0–29.9	15–25	0.6 (0.5–0.7)
Obese (includes all classes)	≥30.0	11–20	0.5 (0.4–0.6)

age neonates.[109] There were no consistent associations with insufficient weight gain and adverse outcomes in that study. Gestational weight gains below, within, or above the IOM guidelines were all associated with significant postpartum weight retention, but this retention was substantially higher 15 years after delivery in women with gestational weight gain above the recommendations.[113] Of importance is that weight gain during pregnancy is associated with a higher accrual of abdominal or central adiposity over time. McClure and colleagues studied women 4 to 12 years after delivery and also found that excessive gestational weight gain was associated with abdominal adiposity.[114] Fraser and colleagues recently reported associations between excessive gestational weight gain and waist circumference and risk of central adiposity in mothers 16 years after pregnancy.[115] Visceral adiposity is more metabolically active and has been linked to a more adverse cardiometabolic profile.

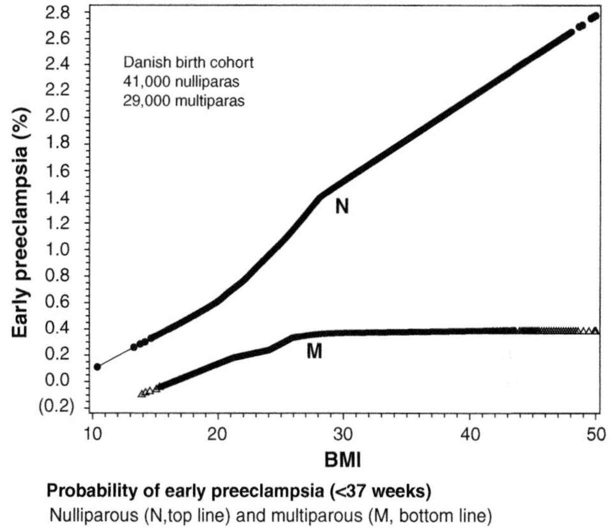

FIGURE 7.5 Unadjusted probability of early preeclampsia (<37 weeks) according to body mass index (BMI) among nulliparous (top line, circles) and multiparous (bottom line, triangles) women. Reproduced, by permission, from Catov et al. (2007).[123]

PREECLAMPSIA AND METABOLIC SYNDROME

Obesity and Insulin Resistance

As previously noted, a strong relationship exists between prepregnancy obesity and preeclampsia.[116–121] Obesity and overweight contribute to the risk of both preterm preeclampsia and severe preeclampsia, a finding with potentially profound public health implications. Analysis of the National Hospital Discharge Survey public-use data set shows an increase in the United States of almost 25% in the rate of preeclampsia during the 18-year period ending in 2004.[122] Much of this rise might reflect the growing "epidemic" of obesity. Stone et al. compared 70 women with severe preeclampsia to 18,000+ normotensive controls, all without history of prepregnancy hypertension, and observed that severe obesity and a history of preeclampsia were the only outstanding risk factors for development of severe preeclampsia.[121] Eskanazi et al., using similar standard criteria for severe cases and controls, noted that, regardless of parity, women with severe preeclampsia were more likely to have had a high prepregnancy BMI (adjusted OR 2.7; 95% CI 1.2 to 6.2).[120]

Analysis of the Danish birth cohort (41,000 nulliparas and 29,000 multiparas) revealed that, whether primiparous or multiparous, women with obesity are overrepresented in cases of both mild preeclampsia and severe or early-onset preeclampsia (Fig. 7.5).[123] Data from the USA Collaborative Perinatal Project (a prospective, cohort study of 19,053 black and 19,135 white women delivering from 1958 to 1964) similarly support a clear relationship between prepregnancy BMI and the risk of severe and mild preeclampsia in both white and black races.[124] The results of both studies show a dose-response relationship between prepregnancy BMI, even in the nonobese BMI range, and risk of both mild and severe (including preterm) preeclampsia.

Women with abnormally low BMI in the first trimester are significantly less likely to develop preeclampsia or gestational hypertension than women with normal BMI.[125] However, excessive gestational weight gain during pregnancy is associated with an increased risk of gestational hypertension and preeclampsia regardless of prepregnancy BMI, with the risk being higher with higher BMI.[109,111] There must be some reservation about the relationship between gestational weight gain and preeclampsia because of the increase in fluid retention characteristic of preeclampsia that would increase weight independent of adipose accumulation. The results of a large, prospective population-based study showed that weight gain during the inter-pregnancy interval is strongly associated with risk of major maternal and perinatal complications, including preeclampsia, independent of whether women are overweight or lean.[126] These striking findings suggest that adipose accrual, even below the obese range, progressively increases the risk of developing preeclampsia. One would assume that central adiposity is the culprit.

Most women with preeclampsia are of normal weight but as a group these same women gain significantly more weight over subsequent decades postpartum than women who did not experience preeclampsia, suggesting that preeclampsia unmasks a latent predisposition to obesity (obesogenic phenotype).[127] If this is the case, the population attributable risk (PAR)[128] of preeclampsia as a result of adiposity, estimated in the Pittsburgh population as 20 to 30% for primiparous women (unpublished data), might be the "tip of the iceberg."

Gestational insulin resistance is accentuated in women with preeclampsia[129] and this difference can be demonstrated in pregnant women weeks before clinically evident preeclampsia,[130–132] and in postpartum women years after a preeclamptic pregnancy.[133–137] This suggests that insulin resistance is an important underlying risk factor for preeclampsia, as it is for cardiovascular disease.

By analogy to the relationship of cardiovascular disease or Type 2 diabetes to obesity and insulin resistance,[37,38] it is unlikely that insulin resistance is the only metabolic change in obesity responsible for the increased risk of preeclampsia. Both lean and obese individuals who are insulin-resistant are at increased risk of cardiovascular disease and Type 2 diabetes. In obese and lean individuals with equivalent insulin resistance, however, Type 2 diabetes still develops more frequently in the obese, indicating that the effect of obesity to increase morbidity is due to more than insulin resistance.[138] Using animal models investigators have shown that adipose accrual reduces vasorelaxation to endothelium-dependent vasodilators before the development of (and thus independently from) insulin resistance, but in association with increased tissue oxidative stress and oxidative destruction of the vasodilator nitric oxide, the latter evidenced by tyrosine nitration in vascular tissue.[37] Therefore, there may be effects of adiposity on the endothelium that are independent of insulin sensitivity. There are data that have led to the hypothesis that factors more prevalent in obese but also relevant to nonobese populations – inflammation, thrombosis, abnormalities of the renin-angiotensin system, abnormalities of angiogenesis, poor nutrition – predispose the gravida to develop preeclampsia.[2,9,139–144]

Obesity promotes endothelial dysfunction in the setting of pregnancy. Obese gravid women have poorer endothelial function on average compared to lean pregnant women; endothelial-dependent vasodilatation was lower at each trimester (51%, 41% and 39%, respectively) in obese women (without previous history of cardiovascular or metabolic disease) than lean women.[145]

Inflammation

An important advance in obesity research came with the understanding that obesity is characterized by chronic, low-grade inflammation.[18] The term "metaflammation" was coined in 2006 to describe this metabolically triggered low-grade inflammation orchestrated by metabolic cells in response to excess nutrients and energy.[146] It is increasingly evident that inflammation contributes to the development of insulin resistance, dislipidemia, oxidative stress, and cardiovascular problems associated with obesity. Obesity is associated with both local adipose inflammation and systemic inflammation. Obese individuals on average manifest increased plasma concentrations of several inflammatory markers and signaling cytokines, including IL-6, TNF-α, and C-reactive protein.[38] Peripheral blood mononuclear cells from obese patients are in a proinflammatory state.[147]

Intriguingly, macrophages and adipocytes share a distant evolutionary origin in the fat body present in insects.[15] Obesity with enlarged fat cells is associated with increased numbers of macrophages in the adipose tissue surrounding individual adipocytes.[15,148] Adipocyte progenitor cells (preadipocytes), especially proliferating preadipocytes, have phagocytic capacity under appropriate stimulation and thus appear to be macrophage-like cells.[15,149,150] Like adipocytes, macrophages are extraordinarily adept at taking up and storing lipids, and share genes and fat metabolism proteins including fatty acid transporters (i.e., fatty acid binding protein 4 (FABP4, also known as A-FABP or aP2)) and peroxisome proliferator-activated receptor-gamma (PPAR-γ), the latter a key nuclear receptor protein that upregulates genes that mediate fatty acid uptake and trapping by adipocytes.[15]

Chronic inflammation is implicated in the mechanisms underlying the adiposity-preeclampsia relationship. Obese or diabetic women are more likely than nonobese women to enter pregnancy in a subclinical inflammatory state.[8] Prepregnancy obesity in the human results in placental over-accumulation of macrophages and increased expression of the proinflammatory cytokines IL-1, TNF-α, and IL-6 in the placenta.[151] The strong linear relation between prepregnancy BMI and preeclampsia risk is partly accounted for by inflammation (using C-reactive protein as a marker) and dyslipidemia (triglycerides).[152] Analyses of correlative data, however, reveal that inflammation in preeclampsia is not always linked to the MetS or insulin resistance.[153]

There is evidence of enhanced neutrophil activation during preeclampsia including: (1) greater basal[154] and fMLP-induced superoxide production,[155,156] (2) increased CD11b expression basally in neutrophils from women with preeclampsia compared to normal pregnancies,[154] and (3) increased neutrophil elastase.[157] Information is lacking, however, on the impact of obesity on these abnormalities.

The transit of neutrophils and monocytes through the placental intervillous space does not normally result in their activation, but trans-placental neutrophil and monocyte activation does appear to occur during preeclampsia.[158] Trophoblast debris shed into the maternal circulation, and maternal neutrophils activated during transit through the placenta, may induce a widespread systemic vascular inflammatory response.[158–161] Furthermore, the systemic vasculature of women with preeclampsia evidences increased infiltration of neutrophils compared to normal non-pregnant women or women with uncomplicated pregnancy (matched for BMI).[162]

Inflammatory Cytokines

C-reactive protein (CRP), TNF (tumor necrosis factor)-α and IL (interleukin)-6 are all increased in the circulation of women with preeclampsia.[43–45,152]

CRP, an acute-phase reactant and inflammatory mediator produced by the liver as well as adipose tissue, is elevated in obese individuals and associated with cardiovascular morbidity. CRP is also elevated in early pregnancy in women who later develop preeclampsia.[116] Studies from our group indicated that CRP has a stronger association with preeclampsia among obese women than triglycerides and together inflammation and triglycerides could account for approximately one-third of the effect of BMI on preeclampsia risk.[152]

TNF-α is produced in adipose tissue both as a locally acting agent by adipocytes as well as a circulating cytokine by infiltrating macrophages of adipose tissue.[163] TNF-α is associated with insulin resistance, oxidative stress, and endothelial dysfunction. Concentrations are higher in obese individuals.[116] TNF-α is increased in preeclampsia;[43,44] the likely source is adipose tissue as placental mRNA is not increased.[164] Although appealing as a mechanism by which obesity might increase the risk of preeclampsia, plasma TNF-α concentrations were not observed to be higher in obese pregnant women than in nonobese pregnant women.[145,165]

IL-6, another potent inflammatory mediator, is associated with obesity, insulin resistance, vascular damage, and later-life cardiovascular disease.[166] IL-6 produced by adipose tissue is estimated to account for 30% of circulating IL-6.[163] Circulating concentrations are higher in obese individuals[166] as well as in women with preeclampsia,[44] indicating a potential relationship.

Adipokines

Leptin and adiponectin, two well-recognized cytokines produced by adipose tissue, affect metabolism and have been linked to cardiovascular disease.[167,168] There are a growing number of adipokines that may have implications for metabolic syndrome; a few that may have relevance to preeclampsia are briefly discussed below.

Leptin is secreted by adipocytes in positive relation to adipocyte size and obesity.[15] It is important for appetite regulation and energy expenditure. The increased concentrations of leptin in obesity are associated with leptin resistance.[15] Leptin may contribute to obesity-associated hypertension through stimulation of central sympathetic pathways.[15] Leptin has structural similarity to proinflammatory cytokines IL-6 and IL-12 and has cytokine-like functions to activate endothelial cells and monocytes.[50,169] Leptin is produced by the placenta with message concentration approaching that present in adipose tissue.[170,171] Maternal plasma leptin concentrations rise gradually to 32 weeks and thereafter decline slightly in women with uncomplicated pregnancy followed by a drastic decline postpartum.[41,172,173] Leptin's functional significance during pregnancy may be related to nutrient availability and delivery to the fetus, including amino acids, by upregulating placental System A amino acid transport activity.[173–175] Leptin concentrations are also positively correlated with fetal weight and insulin resistance.[176,177] Leptin concentrations in women destined to develop preeclampsia are reportedly higher at 20 weeks gestation and rise markedly from 32 weeks as the manifestations of preeclampsia develop.[41] The increased leptin in the maternal circulation is thought to be of placental origin.[48,171,172] Nonetheless, even in late pregnancy when leptin is highest, its concentration still correlates with maternal BMI.[178,179]

Adiponectin is also produced in adipose tissue and has insulin-sensitizing effects. It increases free fatty acid oxidation and is associated with reduced serum free fatty acids, triglyceride, and glucose concentration. It also has anti-inflammatory effects by inhibiting TNF-α-induced monocyte adhesion and adhesion molecule expression.[163] Circulating adiponectin concentrations are lower in obese individuals and inversely correlated with cardiovascular risk. During pregnancy, adiponectin levels reach their lowest concentrations in the third trimester when maternal insulin resistance is greatest[180,181] and these low concentrations are accentuated further with gestational diabetes.[182,183] Concentrations are inversely related to BMI in the third trimester.[184] Our current understanding is that adiponectin is not produced by the placenta.[181,185] Despite the recognized association of insulin resistance and preeclampsia, the precise relationship between adiponectin and preeclampsia remains unclear. Early studies reported higher adiponectin concentrations in women with preeclampsia.[186–188] It has since been noted that there are several forms of circulating adiponectin; the high molecular weight oligomers are considered the most active in inducing insulin sensitivity. As of this writing, there is no consensus on the precise relationship of adiponectin concentrations and preeclampsia, as some studies report lower concentrations[42] and other higher concentrations[189,190] with preeclampsia.

Resistin, also referred to as adipose tissue-specific secretory factor, is secreted primarily by adipocytes and mononuclear cells.[191,192] Resistin impairs glucose intake by adipocytes, increases glucose concentrations, and promotes insulin resistance. It has also been shown to promote fat deposition and has proinflammatory properties.[193,194] Although not consistent across all studies, serum concentrations are reported to be higher in obese compared with lean individuals[195] and implicated in insulin resistance and cardiovascular disease.[196,197] In pregnancy, circulating resistin is present during the first trimester and progressively increases with gestational age.[198,199] Reports are conflicting with regard to maternal serum resistin with established preeclampsia.[200] Recently, Nanda and colleagues evaluated maternal serum resistin concentrations in the first trimester in normal and pathologic pregnancies and reported higher median resistin concentrations in women who later developed preeclampsia compared to controls but no correlation with maternal weight.[46] The authors suggest that resistin may have a pathogenic role in the development of preeclampsia; clearly, further investigation is needed to understand the significance of this relationship.

Fatty acid binding protein 4 (FABP4), also known as adipocyte fatty acid binding protein or aP2, is a member of the intracellular fatty acid binding protein family expressed in adipose tissue, macrophages, trophoblasts and endothelium. Fatty acid binding proteins are linked to inflammatory and metabolic pathways; specifically, FABP4 is secreted from adipocytes into the circulation and is positively correlated with BMI, waist circumference, and insulin resistance.[201,202] Higher baseline levels independently predicted the risk of developing metabolic syndrome over a 5 year period and were positively correlated with carotid intima media thickness in a Chinese population.[202,203] Recent studies indicate a role in preeclampsia. Fasshauer and colleagues demonstrated higher maternal serum FABP4 concentrations in women with preeclampsia compared to normotensive controls.[204] Scifres et al. compared maternal FABP4 concentrations in the first and second trimesters in women who developed preeclampsia to matched, normotensive pregnant controls.[47] Maternal serum FABP4 levels were elevated before the clinical onset of preeclampsia and this increase occurred independently of maternal BMI. There was a modest univariate correlation between first-trimester BMI and maternal FABP4 concentrations (correlation coefficient 0.7, $p < 0.01$). FABP4 is highly upregulated in the placental basal plate with advancing uncomplicated gestation, and further upregulated in preeclampsia as compared to preterm labor with no evidence of inflammation.[205] Placental FABP4 is localized to invasive cytotrophoblasts.[206] Expression of FABP4 is upregulated by hypoxia in primary human trophoblasts, suggesting that upregulation of this protein supports fat accumulation in the hypoxic placenta.[207] Further investigation is needed to elucidate the precise role of FABP4 in preeclampsia.

A number of other novel adipokines such as visfatin, retinol binding protein 4 and vaspin have been identified; however, further study is needed to understand their physiologic role in normal and pathologic pregnancies.[200]

Dyslipidemia

The dyslipidemia that occurs with preeclampsia closely resembles the dyslipidemia of the MetS[39,90,99,129,140] and, in many ways, it represents an accentuation of normal pregnancy changes.[39] Heightened gestational insulin resistance, abnormally increased concentrations of TNF-α (an inhibitor of lipoprotein lipase), and increased human placental lactogen are thought to contribute to this dyslipidemia.[86,99,129,208]

The high prevalence of metabolic complications among obese pregnant women suggests that obesity adversely affects the usual metabolic adjustments in fuel metabolism that accompany pregnancy.[8] Hyperlipidemia is exaggerated in obese pregnant women, in the form of higher FFA, lower HDL cholesterol, and higher triglycerides packaged into VLDL, LDL, and HDL.[8] Meyers and colleagues reported that obese women commence pregnancy with higher plasma triglycerides, reach the same maximum, and then return to higher post-natal levels than normal weight women.[184] Importantly, the atherogenic LDL subfraction measured in the third trimester was two-fold higher in obese women compared to normal weight controls. Recent work confirms prior findings of higher baseline cholesterol, LDL, and triglycerides, but did not demonstrate an effect of excessive gestational weight gain on the rate of change in lipid profiles in either normal or overweight/obese pregnant women.[209] More work is needed, however, to delineate the impact of prepregnancy obesity and gestational weight gain on the metabolic milieu of pregnancy.

Mean plasma triglyceride and free fatty acid concentrations increase about two-fold on average in women with preeclampsia relative to women with uncomplicated pregnancy (Fig. 7.6).[211–213] About one-third of women with preeclampsia develop plasma triglyceride values above 400 mg/dL,[212] greater than the 90th percentile measured in randomly selected women at 36 weeks gestation.[84] This difference reflects a pronounced increase in triglyceride-rich lipoproteins (especially $VLDL_1$).[214] The hypertriglyceridemia of preeclampsia is accompanied by a decrease in cardioprotective HDL cholesterol relative to normal pregancy.[214,215] Not

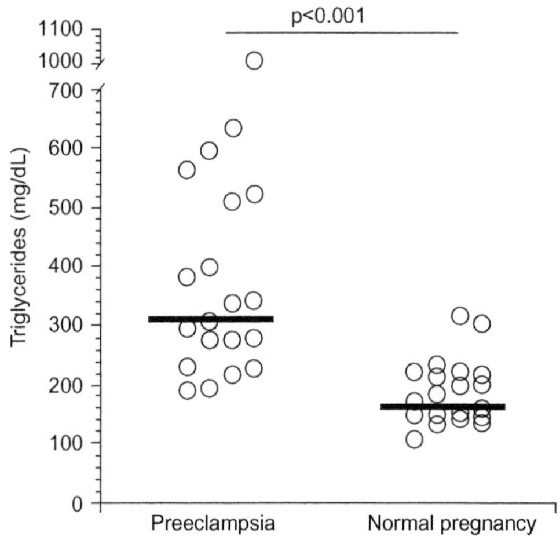

FIGURE 7.6 Scatterplot of serum triglyceride concentrations during late pregnancy. Each symbol corresponds to a different individual. Thick horizontal bars correspond to median values for each group. $p < 0.001$, preeclampsia vs. normal pregnant by Mann-Whitney U test.[210]

all women with preeclampsia become dyslipidemic, and some women with gestational dyslipidemia have normal outcome, fitting with a multi-factorial syndrome.[9]

Consistent with a pathogenic role, dyslipidemia becomes evident during the first and second trimester, antedating the clinical detection of preeclampsia.[48,211,216–218] Higher mean concentrations of the major FFAs (oleic, linoleic, palmitic) are found as early as 16 to 20 weeks gestation in women who later develop preeclampsia compared to women who do not. [211,217,218]

Fasting triglycerides are elevated at earliest reported measurement (10 weeks of gestation) and the difference from controls increases with advancing gestation.[218,219] Early hypertriglyceridemia is mainly associated with increased risk of early-onset preeclampsia (preeclampsia developing before 36 weeks of gestation).[220] HDL-cholesterol is reduced at earliest measurement (20 weeks gestation) and then throughout gestation in women who later develop the disease.[48] Decreased HDL may result from increased triglycerides since the two are metabolically linked.[221,222]

Total cholesterol and LDL cholesterol levels are not typically elevated during preeclampsia compared to normal pregnancy,[210] nor are they generally a component of the MetS. Nevertheless, women with raised total cholesterol during the first trimester are at increased risk of preeclampsia.[131,223] Furthermore, high levels of LDL-cholesterol, or non-HDL-cholesterol, or triglycerides occurring 4 to 5 years before first pregnancy are each associated with significantly increased odds of developing preeclampsia.[224] This is consistent with the hypothesis that pre-existing dyslipidemia, perhaps beyond the umbrella of MetS, contributes to development of preeclampsia.

Free Fatty Acids and Vascular Dysfunction in Preeclampsia

When sequestered by albumin or ceruloplasmin, copper is usually redox-inactive (incapable of producing reactive oxygen species). On average, however, plasma samples from women with preeclampsia display an elevated endogenous redox-cycling activity of copper that is inhibited by copper chelators.[61] This activity may result from increased free fatty acids in the circulation (Fig. 7.3). Circulating FFAs are complexed with albumin. The molar ratio of plasma FFA to albumin is two- to three-fold greater in preeclampsia (up to 2.5:1) than normal pregnancy (0.8:1 to 1:1).[59,213,225] The excess binding of FFA to albumin in preeclampsia is accompanied by conformational change in albumin such that the copper bound to albumin becomes redox-active.[61] The reactive oxygen species generated after excessive FFA binding to Cu-albumin are capable of destroying nitric oxide.[61,226] Vitamin C actually accelerates this free radical redox cycling process, fueling the generation of reactive oxygen species (Fig. 7.3).[62]

When endothelial cells are exposed to preeclampsia plasma or 2.5:1 molar ratio FFA:human serum albumin complex, they take up the FFA in proportion to extracellular concentration; this results in a dramatic increase in cytoplasmic lipid droplets (also called cytoplasmic lipid inclusions).[213] The accumulation of cytoplasmic lipid droplets is associated with decreased secretion of the vasodilator prostaglandin PGI_2 (prostacyclin) and a reduction in the nitric oxide-induced cGMP level.[225] Lipid droplets have made recent headlines as dynamic, complex structures that are major contributors not only to lipid homeostasis but also to diverse cellular functions.[227,228] Beyond adipocytes and macrophages, all cell types observed to date have the ability to generate cytoplasmic lipid droplets in response to elevated FFA, and to metabolize and disperse these droplets when conditions are reversed.[228] The role of lipid droplets in the vascular pathology of obesity and preeclampsia should be fertile ground for research.

Triglycerides and Vascular Dysfunction in Preeclampsia

Both VLDL and atherogenic triglyceride-rich lipoprotein remnant particles are increased in women with preeclampsia.[229] Triglyceride-rich lipoproteins have prothrombotic activity.[230] Human VLDL is toxic to human umbilical vein endothelial cells in culture and this toxicity is prevented by the addition of normal human serum albumin, but not fatty acid-laden human serum albumin, to the culture medium.[59]

Several prospective studies have shown that a preponderance of small-diameter LDL is associated with the MetS and increased risk of coronary heart disease.[231–234] Measured by nondenaturing polyacrylamide gradient gel electrophoresis, the rise in plasma triglyceride during normal pregnancy is accompanied by a progressive shift from predominantly large-diameter LDL to predominantly intermediate and small-sized LDL, with reversal by 6 weeks postpartum.[222,235] Three studies have reported that the average diameter of LDL particles is further decreased in preeclampsia relative to normal pregnancy.[210,236,237] LDL isolated from plasma of women with preeclampsia is more susceptible to oxidation upon exposure to exogenous free radicals than normal pregnancy LDL.[238] This is consistent with a sub-type of LDL capable of impairing vascular function.

Angiogenic Factors

The balance of antiangiogenic factors (soluble Flt-1 (sFlt-1) and soluble endoglin) and proangiogenic factors (placental growth factor (PlGF) and vascular endothelial growth factor (VEGF)), are altered in preeclampsia compared with normal pregnancy even weeks prior to the onset of the clinical condition.[239,240] PlGF concentrations are lower in preeclamptic women, probably due to the higher circulating concentration of sFlt, which binds and inactivates PlGF and VEGF.[241] Soluble endoglin serves as an antagonist to the angiogenic factor TGF-β and is increased with and prior to preeclampsia[242] (see Chapter 6 for a detailed discussion). In the non-pregnant population, obesity is associated with increased circulating angiogenic factors including VEGF, which may represent excess from adipose tissue production, particularly visceral fat.[243] Due to the very high circulating concentrations of sFlt-1 in pregnancy, free VEGF is essentially absent while free PlGF is measurable. PlGF concentrations are significantly lower in both overweight and obese pregnant women compared to normal-weight controls, and this difference is maintained in both women who develop preeclampsia and those who do not.[244] Others have reported that PlGF, sFlt-1, and soluble endoglin are all inversely correlated with BMI.[245] In contrast, Faupel-Badger and colleagues found higher sFlt-1 and sFlt-1/PlGF ratio in women with higher BMI, suggestive of an antiangiogenic milieu.[246] Although inconsistent, the angiogenic milieu associated with obesity may have implications in the development of preeclampsia.

Uric Acid and ADMA

Uric Acid

Uric acid, the end-product of purine metabolism in humans, is an independent risk marker (and may be an important risk factor) for cardiovascular disease, particularly in women.[21,247] Serum uric acid concentrations are frequently elevated in obese patients with the MetS, and concentrations increase with the number of components of MetS. The hyperuricemia of MetS has been related to increased uric acid production or decreased renal uric acid excretion, the latter mediated by enhanced proximal tubular sodium reabsorption and hyperinsulinemia.[248,249]

Hyperuricemia in rats (by pharmacologic inhibition of the enzyme uricase) results in hypertension and renal injury, mediated by stimulation of the renin-angiotensin system and inhibition of nitric oxide synthase.[250,251] Hyperuricemia may also activate the immune response; uric acid release from damaged cells during infection enhances macrophage activation.[252] Uric acid released from damaged cells following an acute ischemia-reperfusion event in mice results in mobilization of endothelial progenitor cells into the circulation; conversely, chronic elevation of uric acid in mice blunts ischemia-induced endothelial progenitor cell mobilization and could thereby limit endothelial repair.[253]

Most women with preeeclampsia are hyperuricemic, typically defined as greater than one standard deviation above the norm for gestational age.[254] Average uric acid values are higher as early as 10 weeks of gestation in groups of women who later develop preeclampsia compared to women whose pregnancies remain uncomplicated.[255] Women with preeclampsia and hyperuricemia have a more severe form of preeclampsia with an increase in indicated preterm births and smaller infants for gestational age.[254] Adjusting for differences in glomerular filtration by serum creatinine accounts for part, but not all, of the increase in serum uric acid among women with preeclampsia, suggesting that hyperuricemia is not entirely due to decreased glomerular filtration, and changes in tubular urate handling and/or uric acid production may play a role.[255] (See also Chapter 16.) More than a decade ago it was suggested that hyperuricemia be "revisited" as a candidate player in the pathogenesis of preeclampsia;[256] this concept may be gaining momentum.[257] The elevated uric acid that many women with the MetS bring into pregnancy might contribute significantly to gestational hyperuricemia, and it seems biologically plausible that gestational hyperuricemia would adversely affect placental and maternal vascular function.

Asymmetric Dimethylarginine

Asymmetric dimethylarginine (ADMA) is an endogenous inhibitor of nitric oxide synthase (NOS). Plasma ADMA concentrations are elevated in association with various components of the MetS, including obesity.[38,258–260] The mechanism that mediates this association, however, is unknown. There is evidence, however, that dimethylarginine dimethylaminohydrolase, the enzyme that metabolizes ADMA to citrulline and methylamines, is sensitive to inactivation by reactive oxygen species.[261]

ADMA may be a major cause of impairment of the nitric oxide-mediated response in several disease states by its ability to compete with L-arginine, the substrate for NOS. ADMA may also "uncouple" endothelial NOS such that molecular oxygen becomes the substrate for electron transfer rather than the guanidino nitrogen of L-arginine.[262] Under these conditions, endothelial NOS generates superoxide anions which in turn can oxidize and thus destroy nitric oxide, and thereby induce additional endothelial dysfunction. There is substantial evidence that ADMA regulates vascular resistance in humans.[259,260,262–265] ADMA induces endothelial dysfunction and impaired angiogenesis *in vivo* and *in vitro*, and is closely correlated with coronary disease and acute coronary events.[266–268] ADMA has been coined an "über marker," as increased ADMA appears to reflect the summative effects of various risk factors (age, hypertension, obesity, insulin resistance, hypertriglyceridemia, hypercholesterolemia, hyperhomocysteinemia) on endothelial health.[262]

Plasma ADMA concentrations are greater in women with preeclampsia compared to uncomplicated pregnancy, and interestingly this difference manifests before clinical evidence of preeclampsia.[269,270] Our recent findings indicate that plasma ADMA concentrations are greater in obese pregnant women compared to lean pregnant women, and ADMA is elevated during the first half of pregnancy more often in obese than lean women who later develop preeclampsia (unpublished data).

Lifestyle Factors

In addition to metabolic and inflammatory factors, lifestyle factors such as diet, sleep disorders, and physical activity are associated with obesity and cardiovascular disease. Many of these factors are also implicated in preeclampsia, thus raising the possibility of a mechanistic link through which obesity may increase the risk of preeclampsia. A detailed discussion is beyond the scope of this chapter, but is reviewed extensively in Chapter 3.[271]

Later-Life Cardiovascular Risk (also see Chapter 3)

Clinical risk factors common to both coronary artery disease and preeclampsia include adiposity,[120] hypertension,[272] diabetes mellitus,[273] systemic lupus erythematosus,[274] polycystic ovarian syndrome,[275] lack of physical activity,[276] and family history of heart disease.[277] Although the hypertension and proteinuria that define preeclampsia typically resolve within a few days after delivery, women with a history of preeclampsia are at significantly increased risk of developing hypertension and coronary artery disease in later years.[278–281] A study of 37,000 women who delivered in Jerusalem showed that women with a history of preeclampsia have a two-fold increased risk of death (mainly due to cardiovascular disease) compared to women with a history of normal pregnancy.[280] The increase in cardiovascular disease-related mortality becomes evident about 20 years after delivery in women with a history of non-recurrent preeclampsia (women whose other pregnancies were without preeclampsia), substantially later than women with a history of recurrent preeclampsia. These findings do not mean that every preeclampsia survivor is destined to develop heart disease, but, rather, that a history of preeclampsia may identify a population at increased risk. On the other hand, women with a history of normal pregnancy may have lower blood pressures on average, and may be less prone to develop cardiovascular disease, than the general (at-large) female population.[282–284]

Endothelial dysfunction is consistently apparent years after preeclamptic compared to normotensive first pregnancy.[285–287] In a study of non-pregnant women 1 year postpartum, by Agatisa et al., the response to mental stress-induced forearm blood flow after venous occlusion was influenced both by pregnancy history and by presence or absence of obesity. Nonobese women with prior uncomplicated pregnancy had twice the percent increase in forearm blood flow exhibited by obese women with prior preeclampsia; intermediate between these two groups, nonobese women with prior preeclampsia had only slightly poorer vasodilatory function than obese women with prior uncomplicated pregnancy.[286]

A study was performed comparing the remote cardiovascular outcomes of postmenopausal Icelandic women who had had eclampsia (the convulsive stage of preeclampsia) to women who had experienced only uncomplicated pregnancies. Icelandic women in both arms were matched for date of birth, age at pregnancy, and parity. The groups did not differ by average BMI but nevertheless the previous-eclampsia group showed significant elevations in plasma apolipoprotein B (an index of total atherogenic lipoprotein particles) and a reduction in the predominant diameter of LDL particles.[137] The combination of high apolipoprotein B and small-sized LDL commonly accompanies the MetS in the general population, and it confers an especially high cardiovascular risk.[233] The Icelandic postmenopausal women with a history of eclampsia were also distinguished from their controls by three-fold elevations in median C-reactive protein, an inflammation marker linked to the MetS.[288] The Icelandic women with prior eclampsia clustered into two evenly populated subgroups, those with high-risk C-reactive protein values and those without (Fig. 7.7). Although these eclampsia

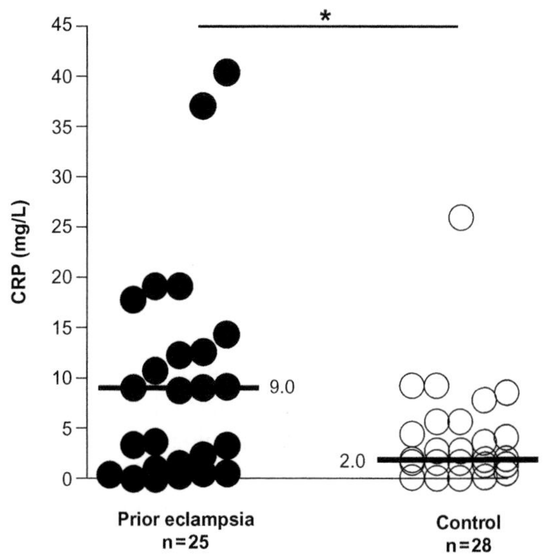

FIGURE 7.7 Scatterplot of serum C-reactive protein (CRP) concentrations in Icelandic postmenopausal women with a history of eclampsia (prior eclampsia) or history of uncomplicated pregnancy (prior NL pregnancy). Each symbol corresponds to a different individual. Thick horizontal bars with adjacent numbers indicate median values for each group. This difference remained significant ($p < 0.04$) after adjustment for current body mass index, smoking, hormone replacement, and age. The difference also remained significant after removal of outliers ($p < 0.05$). From this scatterplot of CRP values, the women with a history of eclampsia could be segregated into two evenly sized subgroups on the basis of very high risk (median 12.8 mg/L, range 9.0 to 40.6 mg/L, $n = 13$) and lower risk (median 0.8 mg/L, range 0.05 to 3.8, $n = 12$) CRP values. The eclampsia-high CRP and eclampsia-low CRP subgroups did not differ by body mass index. However, the eclampsia-high CRP subgroup manifested several adverse lipid and insulin differences compared to controls with a history of normotensive pregnancy – higher fasting insulin and HOMA values, lower HDL cholesterol and higher apolipoprotein B. In contrast, the eclampsia-low CRP subgroup did not show these differences from controls (except for marginally increased apolipoprotein B ($p = 0.055$)). Systolic blood pressure was significantly elevated in the eclampsia-high CRP subgroup compared to controls (mean mm Hg (SD): 148.1 (16.5) versus 130.0 (14.7); $p < 0.05$) but not so in the eclampsia-low CRP subgroup (136.1 (13.9)). Diastolic blood pressures did not differ between the three groups. The percentage of women currently using antihypertensive medications was substantially higher in the eclampsia-high CRP subgroup (62% (8/13)) than the eclampsia-low CRP subgroup (17% (2/12)) ($p < 0.05$). The number of smokers (current or former) was fairly evenly distributed between the eclampsia-high CRP (three current, two former) and eclampsia-low CRP (two current, four former) subgroups.[288]

subgroups did not differ by BMI, the prior-eclampsia subgroup with high-risk C-reactive protein values evidenced significant increases in fasting insulin, HOMA values, systolic blood pressures, and decreased HDL cholesterol compared to controls, whereas the prior-eclampsia/low C-reactive protein subgroup did not.[288] This suggests a close link between dyslipidemia and inflammation in women with a history of eclampsia.

Higher fasting insulin and HOMA values, higher triglycerides, blood pressures, plasma isoprostanes (a marker of oxidative stress), and altered expression of angiogenesis-related proteins have been noted in women with prior preeclampsia compared to women with prior normotensive pregnancies.[75,133–136,289,290] Although the literature on postpartum women with prior preeclampsia is reasonably consistent in showing an association with insulin resistance, it has been more variable in regard to dyslipidemia or inflammation.[284,291,292]

Data from the Medical Birth Registry of Norway reveal that a history of preeclampsia portends an increased risk of later end-stage renal disease, and that having a low-birth-weight or preterm infant increased this relative risk of future chronic kidney disease.[293,294] Antiangiogenic factors, obesity, latent hypertension, insulin resistance and latent endothelial dysfunction were cited as factors possibly contributing to both preeclampsia and end-stage renal disease. However, the authors acknowledge a limitation of current postpartum studies in not distinguishing between possible mechanisms linking preeclampsia and later cardiovascular and renal disease – in this case including (1) underlying risk factors predisposing to both preeclampsia and chronic kidney disease, (2) preeclampsia permanently exacerbating previously unrecognized kidney disease that antedated pregnancy, (3) preeclampsia itself causing permanent renal injury, (4) uncomplicated pregnancy selecting for, or resulting in, women with an unusually low prevalence of chronic kidney disease, or (5) some combination of the above.

The prevailing notion is that women at risk of cardiovascular disease are also at risk of preeclampsia. In other words, preeclampsia may be a harbinger, rather than a direct cause, of future vascular disease.[119] In any case, examining postpartum women with a history of preeclampsia offers a unique approach to *in vivo* investigation of both the maternal contributions to preeclampsia (in the absence of the placenta) and early mechanisms of cardiovascular disease.[119,290]

METABOLIC SYNDROME: A CAUSE OF PLACENTAL DYSFUNCTION?

The results of a study using a metabolic syndrome score (0, 1 or 2 or more criteria), based upon the three NCEP criteria that could be approximated with clinical data (obesity, hypertension prior to 20 weeks gestation, and elevated fasting glucose/diabetes mellitus), showed that increasing score was associated independently with developing preeclampsia, particularly severe disease.[295] It is unlikely that the metabolic syndrome is a late complication of preeclampsia; rather it is more likely a chronic condition that would become increasingly evident in many of these women even without pregnancy.[296] Although insulin sensitivity normally declines during the last half of pregnancy, it is noteworthy in this context that obese women do not show this decline; they begin pregnancy more insulin-resistant and they stay more insulin-resistant (Fig. 7.8).[297]

Preeclampsia is not the only abnormal placentation-related pregnancy disorder associated with an increased risk of the development of cardiovascular disease later in life. Women with a history of recurrent spontaneous abortions are reportedly at increased risk of cerebrovascular disease later in life.[298] Irgens et al. studied a very large cohort of women who delivered preterm (before 37 weeks), and whose pregnancies were not complicated by preeclampsia. Such women then on average 13 years post the index pregnancy had a three-fold increased risk of cardiovascular death compared to matched women who delivered at term, the results shown to be independent of lifestyle or socioeconomic factors.[279]

Pre-existing MetS or endothelial dysfunction may contribute to both placentation defects and later cardiovascular disease. This hypothesis is consistent with prepregnancy, pregnancy and post-partum data. High cholesterol or triglycerides at less than or equal to 15 weeks gestation were associated with a 2.8-fold (1.0–7.9) and 2.0-fold (1.0–3.9) increased risk of preterm birth <34 weeks and ≥34 – <37 weeks, respectively.[299] Overweight women who delivered before 34 weeks evidenced particularly elevated early pregnancy concentrations of cholesterol and low-density

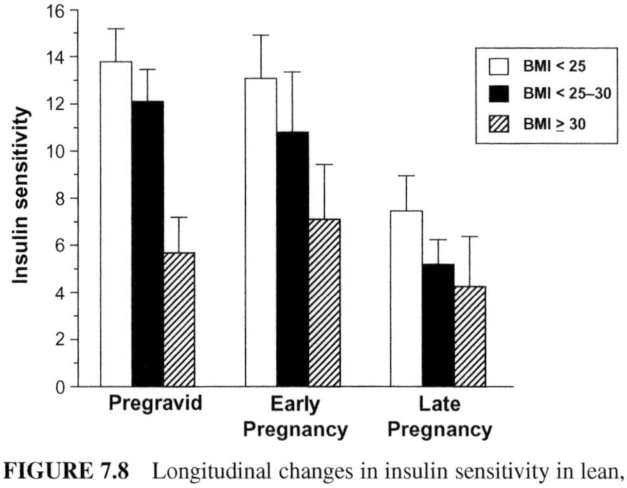

FIGURE 7.8 Longitudinal changes in insulin sensitivity in lean, overweight and obese women, before conception (pregravid) and in early (12–14 weeks) and late (34–36 weeks) gestation, using the hyperinsulinemic–euglycaemic clamp. The obese subjects were significantly less insulin-sensitive, or more insulin-resistant, than the lean women ($p = 0.0001$) and overweight women ($p = 0.004$), particularly pregravid and in early gestation. Reproduced, by permission from Catalano and Ehrenberg (2006).[297]

lipoprotein.[299] As previously mentioned, both prepregnancy and early gestational dyslipidemia are associated with increased risk of preeclampsia.

Germain et al. reported poorer endothelium-dependent vasodilatory function in seemingly healthy postpartum women with a history of either preeclampsia or recurrent abortions, months after their complicated pregnancy, compared to women who had previously healthy pregnancies.[300] The study excluded subjects who had previous obesity or hypertension. In this study, both groups of women with a history of complicated pregnancies (the previously preeclamptic, or those with recurrent abortions) demonstrated impaired brachial reactivity and lower nitrite levels, suggestive of a defect in endothelial nitric oxide release. The authors concluded that the presence of the vascular abnormalities in women with recurrent miscarriage, who were not exposed to the vascular damage of an established maternal syndrome, supports the hypothesis that, in a subgroup of women, endothelial dysfunction constitutes a cause of placentation defects, which in women with preeclampsia is later aggravated by factors liberated by the ischemic placenta into the maternal circulation.[300]

The CHAMPS (Cardiovascular Health After Maternal Placental Syndromes) population-based retrospective cohort study is also consistent with the hypothesis that pre-existing MetS and endothelial dysfunction may contribute to both poor placentation and later cardiovascular disease.[301] This study found a two-fold increased risk of premature vascular disease in women who had had a pregnancy affected by maternal placental syndromes, the latter defined as having had either preeclampsia, gestational hypertension, placental abruption, or placental infarction. The primary endpoints were hospital admission or revascularization for coronary artery, cerebrovascular, or peripheral artery disease at least 90 days after the delivery discharge date. The risk (adjusted hazard ratio) of cardiovascular disease was profoundly elevated (11.7 [confidence interval 4.9–28.3]) in women who had a placental syndrome in combination with three or four of the following MetS-related features: prepregnancy obesity, prepregnancy or gestational dyslipidemia, prepregnancy or gestation diabetes mellitus, or prepregnancy hypertension.

A study of a Dutch population noted a significantly increased risk of International Diabetes Federation-defined metabolic syndrome (central obesity plus any two of four additional factors), not only in women with a history of preeclampsia compared to women with no history of hypertensive complications during pregnancy but also in mothers of women with a history of preeclampsia compared to mothers of women with no history of hypertensive complications during pregnancy.[302] These data further implicate the MetS as a constitutional trait in the risk of both preeclampsia and future cardiovascular disease.

The placental morphologic and vascular profile may cast light on the risk of future cardiovascular disease, for both mother and baby.[303,304] A lesion sometimes found in the uteroplacental spiral arteries in preeclampsia or IUGR (albeit less often than failed spiral artery transformation) has been termed "acute atherosis" because of the presence of foam cells and lipid inclusions resembling the atherosclerotic lesions of vascular disease.[6,305] Higher concentrations of total cholesterol, phospholipids, and lipid peroxides have been observed in placenta decidua basalis, the area that contains the spiral arteries, from women with preeclampsia.[306] The vascular lesions of atherosclerosis develop over decades, whereas the placental vascular lesions of preeclampsia accumulate over a few months of pregnancy – a temporal difference that must be reconciled for any theory proposing that MetS-related pathways are involved in spiral artery mal-transformation or acute atherosis. However, pregnancy renders the maternal endothelium uniquely sensitive to a spectrum of potentially injurious agents, including increased homocysteine,[307] low-dose endotoxin,[308] inhibition of endothelial NOS by L-NAME treatment,[309] and subnormal plasma vitamin C concentrations.[310] In a strain of vitamin C-dependent rats, for example, lowering plasma vitamin C reserves by 50% by dietary restriction had no evident pathogenic effects in non-pregnant rats but in the setting of pregnancy it significantly increased mesenteric artery myogenic contractility and reactive oxygen species production, and lowered fetal birth weights.[310] It would seem biologically plausible, therefore, that accelerated vascular damage by atherogenic factors could occur in the setting of pregnancy.

Harking back to the Introduction of this chapter, it has therefore been proposed by several research groups that, in at least a sub-group of women, chronic endothelial dysfunction and/or the underlying maternal dyslipidemic, insulin-resistant, and inflammatory milieu could actually contribute to abnormal cytotrophoblast invasion/differentiation and failure of the maternal spiral arteries to undergo their normal physiologic transformation into large conduits, in turn leading to reduced uteroplacental blood flow/perfusion.[39,99,297,300,301,311–314] The latent endothelial dysfunction that many of these women presumably enter pregnancy with might become aggravated by anti-angiogenic factors (soluble vascular endothelial growth factor receptor-1, syncytiotrophoblast microvillous membrane fragments, etc.) liberated by the ischemic placenta into the maternal circulation, creating a "feed-forward" cycle of inflammation and endothelial disrepair.

One potential "fly in the ointment" of MetS as a *primum movens* of poor placentation is that women with pregnancies complicated by IUGR but without preeclampsia, who have spiral artery lesions similar to preeclampsia (with or without IUGR), do not generally develop hypertriglyceridemia and, as a group, tend to have lower total cholesterol and LDL cholesterol concentrations than healthy pregnant controls.[315,316] Higher concentrations of triglyceride and inflammatory cytokines (leptin, IL-6 and vascular cell adhesion molecule (VCAM)-1) were observed in plasma from women with preeclampsia compared to normotensive controls matched for BMI, but this dyslipidemic/inflammatory phenotype was not seen in women with IUGR.[316] Women with IUGR had a four-unit smaller BMI on average compared to women with preeclampsia in the latter study, prompting the authors to speculate that molecules liberated from a hypoxic placenta require the additional insult of high maternal adiposity to express their systemic effects to the full.[316] However, mean maternal plasma concentrations of total free fatty acids are reportedly greater in pregnancies complicated by IUGR alone compared to uncomplicated pregnancies, with values similar in magnitude to IUGR plus preeclampsia.[317] As with preeclampsia, elevated markers of oxidative stress have been noted in the bloodstream of women with IUGR.[318]

SUMMARY AND PERSPECTIVES

Evidence points to preeclampsia and later-life cardiovascular disease having common antecedents in metabolic syndrome related factors – obesity, insulin resistance, dyslipidemia, and perhaps novel factors such as ADMA, all of which promote inflammation. In this scenario, chronic inflammation is heavily implicated. In cases of preeclampsia primarily driven by poor placentation and hypoxia, trophoblast debris and other antiangiogenic factors shed into the maternal circulation, and maternal neutrophils activated during transit through the placenta, might induce the widespread maternal systemic inflammatory response.[158–161] Obese/insulin-resistant women are more likely to enter pregnancy in a subclinical inflammatory state that becomes amplified by pregnancy.[8] A reduction in the bioavailability of nitric oxide, either by oxidative destruction or reduced synthesis, i.e., failure of nitric oxide to reach its intended targets, might be a convergence point of these candidate interacting placental and maternal factors.

The adverse effects of these factors are likely to be malleable; not just pre-programmed *in utero* but modulated by lifestyle, psychosocial stress, activity and diet. It is possible, therefore, that early interventions thought to be helpful in preventing endothelial dysfunction and atherosclerosis (glucose control, pregravid weight control, exercise and healthy diet) will also help to prevent preeclampsia and its cardiovascular sequelae (Fig. 7.9). The aggregate data imply that it should be possible for women who are overweight or obese to meaningfully lower their risk of preeclampsia by achievable reductions in body weight before pregnancy. For instance, assuming no unmeasured confounding, published risk curves predict that a lowering of BMI from 29 to 27.4 (10 pounds in a women 65 inches tall) will translate to nearly a 50% reduction in preeclampsia risk.[319] However, reduction of preeclampsia by prepregnancy weight loss is by no means assured. For example, one could envision that chronic deficiency or excess of a critical factor contributes to development of both pregravid obesity and preeclampsia; in this scenario weight loss alone (i.e., by caloric restriction) might

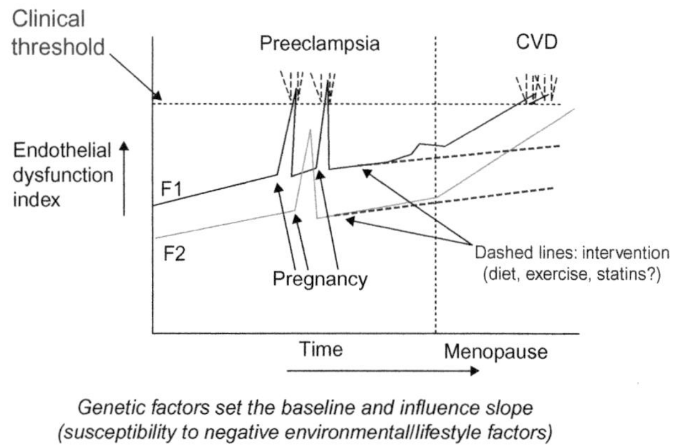

FIGURE 7.9 Intervention strategies against time-dependent progression of endothelial disease. F1, female #1 (high-risk); F2, female #2 (low-risk); CVD, cardiovascular disease. Genetic factors contribute to both the endothelial health baseline and the susceptibility to negative environmental factors (slope). Risk factors for endothelial disease may become identifiable during the "stress-test" of pregnancy. Clinically evident disease manifests once a threshold is crossed (as is the case for F1 during first pregnancy and post-menopause). Interventions such as exercise or statin drugs may significantly delay progression of endothelial disease. First presented by Hubel, C. A., at the 12th World Congress of the International Society for the Study of Hypertension in Pregnancy, Paris, 12 July 2000.

not have the expected degree of effect on the mechanisms driving preeclampsia risk. Furthermore, even if reducing body weight before conception lowers the risk of preeclampsia, weight loss during pregnancy is not recommended. For these reasons, it is imperative to better understand the role of metabolic syndrome-related factors in the pathogenesis of preeclampsia (what is insulin resistance mechanistically doing, for example). It is equally important to point out that, even with a two- to three-fold increased risk, the majority of obese women do not develop preeclampsia. It will thus be important to investigate what differentiates obese women who do and do not develop preeclampsia.

Adipose tissue is a metabolically active endocrine and paracrine organ, and hence more than just a repository for fat. Experimentation on this tissue would undoubtedly increase our understanding of the preeclampsia phenotype. Abdominal fat from women with normal pregnancy exhibits an increase in basal and hormone-stimulated rates of lipolysis in explant culture compared to abdominal fat from non-pregnant women.[320] To date, however, there has been little study of adipose tissue of women with normal, preeclamptic, or fetal growth-restricted pregnancy. The role of lipid-regulatory factors such as lipid droplets,[228] peroxisome proliferator activated receptors,[321] fatty acid binding proteins,[322] resistin, and adiponectin deserves further evaluation. Fascinating interactions between adipogenesis and angiogenesis have also been noted.[323] The study of such pathways before, during and after pregnancy may prove helpful.

References

1. Redman CWG, Sacks GP, Sargent IL. Preeclampsia: an excessive maternal inflammatory response to pregnancy. *Am J Obset Gynecol.* 1999;180(2 Part 1):499–506.
2. Roberts JM, Hubel CA. Is oxidative stress the link in the two-stage model of pre-eclampsia? *Lancet.* 1999;354:788–789.
3. Haig D. Genetics conflicts in human pregnancy. *Q Rev Biol.* 1993;68:495–532.
4. Kim YM, Bujold E, Chaiworapongsa T, et al. Failure of physiologic transformation of the spiral arteries in patients with preterm labor and intact membranes. *Am J Obset Gynecol.* 2003;189:1063–1069.
5. Arias F, Victoria A, Cho K, Kraus F. Placental histology and clinical characteristics of patients with preterm premature rupture of membranes. *Obstet Gynecol.* 1997;89:265–271.
6. Pijnenborg R, Vercruysse L, Hanssens M, Van Assche AF. Trophoblast invasion in pre-eclampsia and other pregnancy disorders. In: Lyall F, Belfort M, eds. *Pre-eclampsia Etiology and Clinical Practice.* Cambridge: Cambridge University Press; 2007:1–19.
7. Suhonen L, Teramo K. Hypertension and pre-eclampsia in women with gestational glucose intolerance. *Acta Obstet Gynecol Scand.* 1993;72:269–272.
8. King JC. Maternal obesity, metabolism, and pregnancy outcomes. *Annu Rev Nutr.* 2006;26:271–291.
9. Ness RB, Roberts JM. Heterogeneous causes constituting the single syndrome of preeclampsia: a hypothesis and its implications. *Am J Obset Gynecol.* 1996;175:1365–1370.
10. Kylin E. Studies of the hypertension-hyperglycemia hyperuricemia syndrome. *Zentral--blatt fuer Innere Medizin.* 1923;44:105–127.
11. Reaven GM. Banting lecture 1988. Role of insulin resistance in human disease. *Diabetes.* 1988;37:1595–1607.
12. Alberti KG, Zimmet PZ. Definition, diagnosis and classification of diabetes mellitus and its complications. Part 1: diagnosis and classification of diabetes mellitus provisional report of a WHO consultation. *Diabet Med.* 1998;15(7):539–553.
13. Grundy SM. Metabolic syndrome pandemic. *Arterioscler Thromb Vasc Biol.* 2008;28(4):629–636.
14. Grundy SM. Metabolic syndrome: connecting and reconciling cardiovascular and diabetes worlds. *J Am Coll Cardiol.* 2006;47(6):1093–1100.
15. Gustafson B, Hammarstedt A, Andersson CX, Smith U. Inflamed adipose tissue: a culprit underlying the metabolic syndrome and atherosclerosis. *Arterioscler Thromb Vasc Biol.* 2007;27(11):2276–2283.
16. Hauner H. Secretory factors from human adipose tissue and their functional role. *Proc Nutr Soc.* 2005;64(2):163–169.
17. Holvoet P. Relations between metabolic syndrome, oxidative stress and inflammation and cardiovascular disease. *Verh K Acad Geneeskd Belg.* 2008;70(3):193–219.
18. Chudek J, Wiecek A. Adipose tissue, inflammation and endothelial dysfunction. *Pharmacol Rep.* 2006;58(suppl):81–88.
19. Jonk AM, Houben AJ, de Jongh RT, Serne EH, Schaper NC, Stehouwer CD. Microvascular dysfunction in obesity: a potential mechanism in the pathogenesis of obesity-associated insulin resistance and hypertension. *Physiology (Bethesda).* 2007;22:252–260.
20. Frayn KN. Obesity and metabolic disease: is adipose tissue the culprit? *Proc Nutr Soc.* 2005;64(1):7–13.
21. Alderman M, Aiyer KJ. Uric acid: role in cardiovascular disease and effects of losartan. *Curr Med Res Opin.* 2004; 20(3):369–379.
22. Segura J, Ruilope LM. Obesity, essential hypertension and renin-angiotensin system. *Public Health Nutr.* 2007; 10(10A):1151–1155.
23. de Ferranti S, Mozaffarian D. The perfect storm: obesity, adipocyte dysfunction, and metabolic consequences. *Clin Chem.* 2008;54(6):945–955.
24. Gluckman PD, Hanson MA, Cooper C, Thornburg KL. Effect of in utero and early-life conditions on adult health and disease. *N Engl J Med.* 2008;359(1):61–73.
25. Hofman PL, Regan F, Jefferies CA, Cutfield WS. Prematurity and programming: are there later metabolic sequelae? *Metab Syndr Relat Disord.* 2006;4(2):101–112.
26. Myatt L. Placental adaptive responses and fetal programming. *J Physiol.* 2006;572(Pt 1):25–30.
27. Armitage JM, Poston L, Taylor PD. Developmental origins of obesity and the metabolic syndrome: the role of maternal obesity. In: Korbonits M, ed. *Obesity and Metabolism.* Basel: Karger; 2008:73–84.
28. Ridker PM, Buring JE, Cook NR, Rifai N. C-reactive protein, the metabolic syndrome, and risk of incident cardiovascular events: an 8-year follow-up of 14 719 initially healthy American women. *Circulation.* 2003;107(3):391–397.
29. Rutter MK, Meigs JB, Sullivan LM, D'Agostino Sr RB, Wilson PW. C-reactive protein, the metabolic syndrome, and prediction of cardiovascular events in the Framingham Offspring Study. *Circulation.* 2004;110(4):380–385.
30. Qiao Q, Gao W, Zhang L, Nyamdorj R, Tuomilehto J. Metabolic syndrome and cardiovascular disease. *Ann Clin Biochem.* 2007;44(Pt 3):232–263.
31. Kahn R. Metabolic syndrome – what is the clinical usefulness? *Lancet.* 2008;371(9628):1892–1893.
32. Sattar N, McConnachie A, Shaper G, et al. Can metabolic syndromes usefully predict cardiovascular disease and diabetes? Outcome data from two prospective studies. *Lancet.* 2008;371:1927–1935.
33. Kahn R, Buse J, Ferrannini E, Stern M. American Diabetes A, European Association for the Study of D. The metabolic syndrome: time for a critical appraisal: joint statement from the American Diabetes Association and the European Association for the Study of Diabetes. *Diabetes Care.* 2005;28(9):2289–2304.
34. Hossain P, Kawar B, El Nahas M. Obesity and diabetes in the developing world - a growing challenge. *N Engl J Med.* 2007;356:213–215.
35. Calle EE, Thun MJ, Petrelli JM, Rodriguez C, Heath Jr. CW. Body-mass index and mortality in a prospective cohort of U.S. adults.[see comment]. *N Engl J Med.* 1999;341(15):1097–1105.
36. Ogden CL, Carroll MD, Kit BK, Flegal KM. Prevalence of obesity among adults: United States, 2011–2012. *NCHS Data Brief.* 2013;131:1–8.
37. Galili O, Versari D, Sattler KJ, et al. Early experimental obesity is associated with coronary endothelial dysfunction and oxidative stress. *Am Physiol Soc.* 2007;292(2): H904–H911.
38. Mather KJ, Baron AD. Vascular dysfunction and obesity. In: Robinson MK, Thomas A, eds. *Obesity and Cardiovascular Disease.* London: Taylor & Francis; 2006.

39. Hubel CA. Dyslipidemia and pre-eclampsia. In: Belfort MA, Lydall F, eds. *Pre-eclampsia -Aetiology and Clinical Practice*. Cambridge: Cambridge University Press; 2006:164–182.
40. Wildman RP, Muntner P, Reynolds K, et al. The obese without cardiometabolic risk factor clustering and the normal weight with cardiometabolic risk factor clustering: prevalence and correlates of 2 phenotypes among the US population (NHANES 1999–2004). *Arch Intern Med.* 2008; 168(15):1617–1624.
41. Anim-Nyame N, Sooranna SR, Steer PJ, Johnson MR. Longitudinal analysis of maternal plasma leptin concentrations during normal pregnancy and pre-eclampsia. *Hum Reprod.* 2000;15(9):2033–2036.
42. Mazaki-Tovi S, Romero R, Vaisbuch E, et al. Maternal serum adiponectin multimers in preeclampsia. *J Perinat Med.* 2009; 37(4):349–363.
43. Kupferminc MJ, Peaceman AM, Wigton TR, Rehnberg KA, Socol ML. Tumor necrosis factor-alpha is elevated in plasma and amniotic fluid of patients with severe preeclampsia. *Am J Obstet Gynecol.* 1994;170(6):1752–1757. discussion 7–9.
44. Conrad KP, Benyo DF. Placental cytokines and the pathogenesis of preeclampsia. *Am J Reprod Immunol.* 1997; 37(3):240–249.
45. Hamai Y, Fujii T, Yamashita T, et al. Evidence for an elevation in serum interleukin-2 and tumor necrosis factor-alpha levels before the clinical manifestations of preeclampsia. *Am J Reprod Immunol.* 1997;38(2):89–93.
46. Nanda S, Poon LC, Muhaisen M, Acosta IC, Nicolaides KH. Maternal serum resistin at 11 to 13 weeks' gestation in normal and pathological pregnancies. *Metabolism.* 2012; 61(5):699–705.
47. Scifres CM, Catov JM, Simhan H. Maternal serum fatty acid binding protein 4 (FABP4) and the development of preeclampsia. *J Clin Endocrinol Metab.* 2012; 97(3):E349–E356.
48. Chappell LC, Seed PT, Briley A, et al. A longitudinal study of biochemical variables in women at risk of preeclampsia. *Am J Obset Gynecol.* 2002;187(1):127–136.
49. Dechend R, Muller DN, Wallukat G, et al. Activating auto-antibodies against the AT1 receptor in preeclampsia. *Autoimmun Rev.* 2005;4(1):61–65.
50. Berg AH, Scherer PE. Adipose tissue, inflammation, and cardiovascular disease. *Circ Res.* 2005;96(9):939–949.
51. Fain JN, Madan AK, Hiler ML, Cheema P, Bahouth SW. Comparison of the release of adipokines by adipose tissue, adipose tissue matrix, and adipocytes from visceral and subcutaneous abdominal adipose tissues of obese humans. *Endocrinology.* 2004;145(5):2273–2282.
52. Sattar N, Clark P, Holmes A, Lean ME, Walker I, Greer IA. Antenatal waist circumference and hypertension risk. *Obstet Gynecol.* 2001;97(2):268–271.
53. Yamamoto S, Douchi T, Yoshimitsu N, Nakae M, Nagata Y. Waist to hip circumference ratio as a significant predictor of preeclampsia, irrespective of overall adiposity. *J Obstet Gynaecol Res.* 2001;27(1):27–31.
54. Adiels M, Olofsson SO, Taskinen MR, Boren J. Overproduction of very low-density lipoproteins is the hallmark of the dyslipidemia in the metabolic syndrome. *Arterioscler Thromb Vasc Biol.* 2008;28(7):1225–1236.
55. Schonfeld P, Wojtczak L. Fatty acids as modulators of the cellular production of reactive oxygen species. *Free Radic Biol Med.* 2008;45(3):231–241.
56. Wojtczak L, Schonfeld P. Effect of fatty acids on energy coupling processes in mitochondria. *Biochim Biophys Acta.* 1993;1183(1):41–57.
57. Kim F, Tysseling KA, Rice J, et al. Free fatty acid impairment of nitric oxide production in endothelial cells is mediated by IKKbeta. *Arterioscler Thromb Vasc Biol.* 2005;25(5):989–994.
58. Roche M, Rondeau P, Singh NR, Tarnus E, Bourdon E. The antioxidant properties of serum albumin. *FEBS Lett.* 2008;582:1783–1787.
59. Vigne JL, Murai JT, Arbogast BW, Jia W, Fisher SJ, Taylor RN. Elevated nonesterified fatty acid concentrations in severe preeclampsia shift the isoelectric characteristics of plasma albumin. *J Clin Endocrinol Metab.* 1997;82:3786–3792.
60. Yuan H, Antholine WE, Subczynski WK, Green MA. Release of CuPTSM from human serum albumin after addition of fatty acids. *J Inorg Biochem.* 1996;61:251–259.
61. Kagan VE, Tyurin VA, Borisenko GG, et al. Mishandling of copper by albumin: role in redox-cycling and oxidative stress in preeclampsia plasma. *Hypertens Pregnancy.* 2001; 20:221–241.
62. Gryzunov YA, Arroyo A, Vigne JL, et al. Binding of fatty acids facilitates oxidation of cysteine-34 and converts copper-albumin complexes from antioxidants to prooxidants. *Arch Biochem Biophys.* 2003;413(1):53–66.
63. Picchi A, Gao X, Belmadani S, et al. Tumor necrosis factor-alpha induces endothelial dysfunction in the prediabetic metabolic syndrome. *Circ Res.* 2006;99(1):69–77.
64. Umemura S, Nyui N, Tamura K, et al. Plasma angiotensinogen concentrations in obese patients. *Am J Hypertens.* 1997;10(6):629–633.
65. Massiera F, Bloch-Faure M, Ceiler D, et al. Adipose angiotensinogen is involved in adipose tissue growth and blood pressure regulation. *FASEB J.* 2001;15(14):2727–2729.
66. Griendling KK, Minieri CA, Ollerenshaw JD, Alexander RW. Angiotensin II stimulates NADH and NADPH oxidase activity in cultured vascular smooth muscle cells. *Circ Res.* 1994;74:1141–1148.
67. Rajagopalan S, Kurz S, Munzel T, et al. Angiotensin II-mediated hypertension in the rat increases vascular superoxide production via membrane NADH/NADPH oxidase activation - Contribution to alterations of vasomotor tone. *J Clin Invest.* 1996;97(8):1916–1923.
68. Zhang H, Schmeisser A, Garlichs CD, et al. Angiotensin II-induced superoxide anion generation in human vascular endothelial cells: role of membrane-bound NADH-/NADPH-oxidases. *Cardiovasc Res.* 1999;44(1):215–222.
69. Warnholtz A, Nickenig G, Schulz E, et al. Increased NADH-oxidase-mediated superoxide production in the early stages of atherosclerosis - Evidence for involvement of the renin-angiotensin system. *Circulation.* 1999;99:2027–2033.
70. Tham DM, Martin-McNulty B, Wang YX, et al. Angiotensin II is associated with activation of NF-kappaB-mediated genes and downregulation of PPARs. *Physiol Genomics.* 2002;11(1):21–30.
71. Shah DM. Role of the renin-angiotensin system in the pathogenesis of preeclampsia. *Am J Physiol Renal Physiol.* 2005;288(4):F614–F625.

72. Yang X, Wang F, Chang H, et al. Autoantibody against AT1 receptor from preeclamptic patients induces vasoconstriction through angiotensin receptor activation. *J Hypertens.* 2008;26(8):1629–1635.
73. Dechend R, Viedt C, Muller DN, et al. AT1 receptor agonistic antibodies from preeclamptic patients stimulate NADPH oxidase. *Circulation.* 2003;107(12):1632–1639.
74. Dechend R, Gratze P, Wallukat G, et al. Agonistic autoantibodies to the AT1 receptor in a transgenic rat model of preeclampsia. *Hypertension.* 2005;45(4):742–746.
75. Hubel CA, Wallukat G, Wolf M, et al. Agonistic angiotensin II type 1 receptor autoantibodies in postpartum women with a history of preeclampsia. *Hypertension.* 2007;49(3):612–617.
76. Drolet R, Richard C, Sniderman AD, et al. Hypertrophy and hyperplasia of abdominal adipose tissues in women. *Int J Obes (Lond).* 2008;32(2):283–291.
77. Harman-Boehm I, Bluher M, Redel H, et al. Macrophage infiltration into omental versus subcutaneous fat across different populations: effect of regional adiposity and the comorbidities of obesity. *J Clin Endocrinol Metab.* 2007;92(6):2240–2247.
78. Redman CW. The endoplasmic reticulum stress of placental impoverishment. *Am J Pathol.* 2008;173(2):311–314.
79. Burton GJ, Yung HW, Cindrova-Davies T, Charnock-Jones DS. Placental endoplasmic reticulum stress and oxidative stress in the pathophysiology of unexplained intrauterine growth restriction and early onset preeclampsia. *Placenta.* 2009;30 (suppl A):S43–S48.
80. Poston L, Igosheva N, Mistry HD, et al. Role of oxidative stress and antioxidant supplementation in pregnancy disorders. *Am J Clin Nutr.* 2011;94(6 suppl):1980S–1985.
81. Catalano PM, Thomas AJ, Huston LP, Fung CM. Effect of maternal metabolism on fetal growth and body composition. *Diabetes Care.* 1998;21(suppl 2):B85–B90.
82. Catalano PM. Increasing maternal obesity and weight gain during pregnancy: the obstetric problems of plentitude [comment]. *Obstet Gynecol.* 2007;110(4):743–744.
83. Herrera E. Lipid metabolism in pregnancy and its consequences in the fetus and newborn. *Endocrine.* 2002;19:43–55.
84. Knopp RH, Bonet B, Lasuncion MA, Montelongo A, Herrera E. Lipoprotein metabolism in pregnancy. In: Herrera E, Knopp RH, eds. *Perinatal Biochemistry.* Boca Raton: CRC Press; 1992:19–51.
85. Catalano PM, Tyzbir ED, Roman NM, Amini SB, Sims EA. Longitudinal changes in insulin release and insulin resistance in non-obese pregnant women. *Am J Obset Gynecol.* 1991;165(6 Pt 1):1667–1672.
86. Murai JT, Muzykanskiy E, Taylor RN. Maternal and fetal modulators of lipid metabolism correlate with the development of preeclampsia. *Metabolism.* 1997;46(8):963–967.
87. Sivan E, Homko CJ, Whittaker PG, Reece EA, Chen X, Boden G. Free fatty acids and insulin resistance during pregnancy. *J Clin Endocrinol Metab.* 1998;83(7):2338–2342.
88. Kirwan JP, Hauguel-deMouzon S, Leqercq J, et al. TNFα is a predictor of insulin resistance in human pregnancy. *Diabetes.* 2002;51:2207–2213.
89. Ryan EA. Hormones and insulin resistance during pregnancy. *Lancet.* 2003;362:1777–1778.
90. Hubel CA, Roberts JM. Lipid metabolism and oxidative stress in preeclampsia. In: Lindheimer MD, Roberts JM, Cunningham FG, eds. *Chesley's Hypertensive Disorders in Pregnancy.* 2nd ed.. Stamford, CT: Appleton & Lange; 1999:453–486.
91. Potter JM, Nestel PJ. The hyperlipidemia of pregnancy in normal and complicated pregnancies. *Am J Obset Gynecol.* 1979;133(2):165–170.
92. Montelongo A, Lasuncion MA, Pallardo LF, Herrera E. Longitudinal study of plasma lipoproteins and hormones during pregnancy in normal and diabetic women. *Diabetes.* 1992;41:1651–1659.
93. Alvarez JJ, Montelongo A, Iglesias A, Lasuncion MA, Herrera E. Longitudinal study on lipoprotein profile, high density lipoprotein subclass, and postheparin lipases during gestation in women. *J Lipid Res.* 1996;37:299–308.
94. Coleman RA. The role of the placenta in lipid metabolism and transport. *Semin Perinatol.* 1989;13:180–191.
95. Herrera E, Lasuncion MA, Coronado DG, Aranda P, Luna PL, Maier I. Role of lipoprotein lipase activity on lipoprotein metabolism and the fate of circulating triglycerides in pregnancy. *Am J Obset Gynecol.* 1988;158:1575–1583.
96. Magnusson-Olsson AL, Lager S, Jacobsson B, Jansson T, Powell TL. Effect of maternal triglycerides and free fatty acids on placental LPL in cultured primary trophoblast cells and in a case of maternal LPL deficiency. *Am J Physiol Endocrinol Metab.* 2007;293(1):E24–E30.
97. Magnusson AL, Waterman IJ, Wennergren M, Jansson T, Powell TL. Triglyceride hydrolase activities and expression of fatty acid binding proteins in the human placenta in pregnancies complicated by intrauterine growth restriction and diabetes. *J Clin Endocrinol Metab.* 2004;89(9):4607–4614.
98. Haggarty P. Placental regulation of fatty acid delivery and its effect on fetal growth--a review. *Placenta.* 2002;23 (suppl A):S28–S38.
99. Lorentzen B, Henriksen T. Plasma lipids and vascular dysfunction in preeclampsia. *Semin Reprod Endocrinol.* 1998;16(1):33–39.
100. Leeners B, Rath W, Kuse S, Irawan C, Imthurn B, Neumaier-Wagner P. BMI: new aspects of a classical risk factor for hypertensive disorders of pregnancy. *Clin Sci.* 2006;111(1):81–86.
101. Cedergren MI. Maternal morbid obesity and the risk of adverse pregnancy outcome. *Obstet Gynecol.* 2004;103(2):219–224.
102. Raatikainen K, Heiskanen N, Heinonen S. Transition from overweight to obesity worsens pregnancy outcome in a BMI-dependent manner. *Obesity (Silver Spring).* 2006; 14(1):165–171.
103. Viswanathan M, Siega-Riz AM, Moos MK, et al. Outcomes of maternal weight gain. *Evid Report/technology Assess.* 2008;168:1–223.
104. Abenhaim HA, Kinch RA, Morin L, Benjamin A, Usher R. Effect of prepregnancy body mass index categories on obstetrical and neonatal outcomes. *Arch Gynecol Obstet.* 2007;275(1):39–43.
105. Hilson JA, Rasmussen KM, Kjolhede CL. Excessive weight gain during pregnancy is associated with earlier termination of breast-feeding among White women. *J Nutr.* 2006; 136(1):140–146.
106. Kugyelka JG, Rasmussen KM, Frongillo EA. Maternal obesity is negatively associated with breastfeeding success among Hispanic but not Black women. *J Nutr.* 2004; 134(7):1746–1753.

107. Doherty DA, Magann EF, Francis J, Morrison JC, Newnham JP. Pre-pregnancy body mass index and pregnancy outcomes. *Int J Gynaecol Obstet*. 2006;95(3):242–247.
108. Institute of Medicine and National Research Council. *Weight gain during pregnancy: reexamining the guidelines*. Washington DC: The National Academies Press; 2009.
109. Johnson J, Clifton RG, Roberts JM, et al. Pregnancy outcomes with weight gain above or below the 2009 Institute of Medicine guidelines. *Obstet Gynecol*. 2013;121(5):969–975.
110. de la Torre L, Flick AA, Istwan N, et al. The effect of new antepartum weight gain guidelines and prepregnancy body mass index on the development of pregnancy-related hypertension. *Am J Perinatol*. 2011;28(4):285–292.
111. Macdonald-Wallis C, Tilling K, Fraser A, Nelson SM, Lawlor DA. Gestational weight gain as a risk factor for hypertensive disorders of pregnancy. *Am J Obstet Gynecol*. 2013;209(4):327. e1–17.
112. Gibson KS, Waters TP, Catalano PM. Maternal weight gain in women who develop gestational diabetes mellitus. *Obstet Gynecol*. 2012;119(3):560–565.
113. Nehring I, Schmoll S, Beyerlein A, Hauner H, von Kries R. Gestational weight gain and long-term postpartum weight retention: a meta-analysis. *Am J Clin Nutr*. 2011; 94(5):1225–1231.
114. McClure CK, Catov JM, Ness R, Bodnar LM. Associations between gestational weight gain and BMI, abdominal adiposity, and traditional measures of cardiometabolic risk in mothers 8 y postpartum. *Am J Clin Nutr*. 2013;98(5):1218–1225.
115. Fraser A, Tilling K, Macdonald-Wallis C, et al. Associations of gestational weight gain with maternal body mass index, waist circumference, and blood pressure measured 16 y after pregnancy: the Avon Longitudinal Study of Parents and Children (ALSPAC). *Am J Clin Nutr*. 2011;93(6):1285–1292.
116. Wolf M, Kettyle E, Sandler L, Ecker JL, Roberts J, Thadhani R. Obesity and preeclampsia: the potential role of inflammation. *Obstet Gynecol*. 2001;98(5 Pt 1):757–762.
117. Sebire NJ, Jolly M, Harris JP, et al. Maternal obesity and pregnancy outcome: a study of 287,213 pregnancies in London. *Int J Obes Relat Metab Disord*. 2001;25(8):1175–1182.
118. O'Brien TE, Ray JG, Chan WS. Maternal body mass index and the risk of preeclampsia: a systematic overview [see comment]. *Epidemiology*. 2003;14(3):368–374.
119. Ness RB, Hubel CA. Risk for coronary artery disease and morbid preeclampsia: a commentary. *Ann Epidemiol*. 2005;15(9):726–733.
120. Eskenazi B, Fenster L, Sidney S. A multivariate analysis of risk factors for preeclampsia. *JAMA*. 1991;266(2):237–241.
121. Stone JL, Lockwood CJ, Berkowitz GS, Alvarez M, Lapinski R, Berkowitz RL. Risk factors for severe preeclampsia. *Obstet Gynecol*. 1994;83(3):357–361.
122. Wallis AB, Saftlas AF, Hsia J, Atrash HK. Secular trends in the rates of preeclampsia, eclampsia, and gestational hypertension, United States, 1987–2004. *Am J Hypertens*. 2008;21:521–526.
123. Catov JM, Ness RB, Kip KE, Olsen J. Risk of early or severe preeclampsia related to pre-existing conditions. *Int J Epidemiol*. 2007;36(2):412–419.
124. Bodnar LM, Catov JM, Klebanoff MA, Ness RB, Roberts JM. Prepregnancy body mass index and the occurrence of severe hypertensive disorders of pregnancy. *Epidemiology*. 2007;18(2):234–239.
125. Belogolovkin V, Eddleman KA, Malone FD, et al. The effect of low body mass index on the development of gestational hypertension and preeclampsia. *J Matern Fetal Neonatal Med*. 2007;20(7):509–513.
126. Villamor E, Cnattingius S. Interpregnancy weight change and risk of adverse pregnancy outcomes: a population-based study [see comment]. *Lancet*. 2006;368(9542):1164–1170.
127. Callaway LK, McIntyre HD, O'Callaghan M, Williams GM, Najman JM, Lawlor DA. The association of hypertensive disorders of pregnancy with weight gain over the subsequent 21 years: findings from a prospective cohort study. *Am J Epidemiol*. 2007;166(4):421–428.
128. Daly L. Confidence limits made easy: internal estimation using a substitution method. *Am J Epidemiol*. 1998; 147:783–790.
129. Kaaja R. Insulin resistance syndrome in preeclampsia. *Semin Reprod Endocrinol*. 1998;16(1):41–46.
130. Solomon CG, Graves SW, Greene MF, Seely EW. Glucose intolerance as a predictor of hypertension in pregnancy. *Hypertension*. 1994;23:717–721.
131. Solomon CG, Carroll JS, Okamura K, Graves SW, Seeley EW. Higher cholesterol and insulin levels in pregnancy are associated with increased risk for pregnancy-induced hypertension. *Am J Hypertens*. 1999;12:276–282.
132. Wolf M, Sandler L, Munoz K, Hsu K, Ecker JL, Thadhani R. First trimester insulin resistance and subsequent preeclampsia: a prospective study. *J Clin Endocrinol Metab*. 2002;87:1563–1568.
133. Laivuori H, Tikkanen MJ, Ylikorkala O. Hyperinsulinemia 17 years after preeclamptic first pregnancy. *J Clin Endocrinol Metab*. 1996;81:2908–2911.
134. Nisell H, Erikssen C, Persson B, Carlstrom K. Is carbohydrate metabolism altered among women who have undergone a preeclamptic pregnancy? *Gynecol Obstet Invest*. 1999;48:241–246.
135. Barden AE, Beilin LJ, Ritchie J, Walters BN, Michael C. Does a predisposition to the metabolic syndrome sensitize women to develop pre-eclampsia? *J Hypertens*. 1999; 17:1307–1315.
136. Wolf M, Hubel CA, Lam C, et al. Preeclampsia and future cardiovascular disease: potential role of altered angiogenesis and insulin resistance. *J Clin Endocrinol Metab*. 2004;89(12):6239–6243.
137. Hubel CA, Snaedal S, Ness RB, et al. Dyslipoproteinemia in postmenopausal women with a history of eclampsia. *Brit J Obstet Gynaecol*. 2000;107:776–784.
138. Meigs JB, Wilson PWF, Fox CS, et al. Body mass index, metabolic syndrome, and risk of type 2 diabetes or cardiovascular disease. *J Clin Endocrinol Metab*. 2006; 91(8):2906–2912.
139. Haggerty CL, Ferrell RE, Hubel CA, Markovic N, Harger G, Ness RB. Association between allelic variants in cytokine genes and preeclampsia. *Am J Obstet Gynecol*. 2005; 193(1):209–215.
140. Hubel CA. Dyslipidemia, iron, and oxidative stress in preeclampsia: assessment of maternal and feto-placental interactions. *Semin Reprod Endocrinol*. 1998;16(1):75–92.
141. Sattar N, Greer IA. Pregnancy complications and maternal cardiovascular risk: opportunities for intervention and screening? *Br Med J*. 2002;325:157–160.

142. Sattar N, Greer I. Insulin sensitivity in pre-eclampsia. *Brit J Obstet Gynaecol.* 1999;106(8):874–875.
143. Laraia BA, Bodnar LM, Siega-Riz AM. Pregravid body mass index is negatively associated with diet quality during pregnancy. *Public Health Nutr.* 2007;10:920–926.
144. Bodnar LM, Tang G, Ness RB, Harger G, Roberts JM. Periconceptional multivitamin use reduces the risk of preeclampsia. *Am J Epidemiol.* 2006;164(5):470–477.
145. Stewart FM, Freeman DJ, Ramsay JE, Greer IA, Caslake MJ, Ferrell WR. Longitudinal assessment of maternal endothelial function and markers of inflammation & placental function throughout pregnancy in lean and obese mothers. *J Clin Endocrinol Metab.* 2007;92(3):969–975.
146. Hotamisligil GS. Inflammation and metabolic disorders. *Nature.* 2006;444(7121):860–867.
147. Ghanim H, Aljada A, Hofmeyer D, Syed T, Mohanty P, Dandona P. Circulating mononuclear cells in the obese are in a proinflammatory state. *Circulation.* 2004;110(12):1564–1571.
148. Weisberg SP, McCann D, Desai M, Rosenbaum M, Leibel RL, Ferrante Jr. AW. Obesity is associated with macrophage accumulation in adipose tissue. *J Clin Invest.* 2003; 112(12):1796–1808.
149. Charriere G, Cousin B, Arnaud E, et al. Preadipocyte conversion to macrophage. Evidence of plasticity. *J Biol Chem.* 2003;278(11):9850–9855.
150. Cousin B, Munoz O, Andre M, et al. A role for preadipocytes as macrophage-like cells. *FASEB J.* 1999;13(2):305–312.
151. Challier JC, Basu S, Bintein T, et al. Obesity in pregnancy stimulates macrophage accumulation and inflammation in the placenta. *Placenta.* 2008;29:274–281.
152. Bodnar LM, Ness RB, Harger GF, Roberts JM. Inflammation and triglycerides partially mediate the effect of prepregnancy body mass index on the risk of preeclampsia. *Am J Epidemiol.* 2005;162(12):1198–1206.
153. Kaaja R, Laivuori H, Pulkki P, Tikkanen MJ, Hiilesmaa V, Ylikorkala O. Is there any link between insulin resistance and inflammation in established preeclampsia. *Metabolism.* 2004; 53:1433–1435.
154. Sacks GP, Studena K, Sargent II, Redman CWG. Normal pregnancy and preeclampsia both produce inflammatory changes in peripheral blood leukocytes akin to those of sepsis. *Am J Obset Gynecol.* 1998;179:80–86.
155. Tsukimori K, Maeda H, Ishida K, Nagata H, Koyanagi T, Nakano H. The superoxide generation of neutrophils in normal and preeclamptic pregnancies. *Obstet Gynecol.* 1993;81:536–540.
156. Tsukimori K, Fukushima K, Tsushima A, Nakano K. Generation of reactive oxygen species by neutrophils and endothelial cell injury in normal and preeclamptic pregnancies. *Hypertension.* 2005;46(4):696–700.
157. Greer IA, Leask R, Hodson BA, Dawes J, Kilpatrick DC, Liston WA. Endothelin, elastase, and endothelial dysfunction in preeclampsia. *Lancet.* 1991;337:558.
158. Mellembakken JR, Aukrust P, Olafsen MK, Ueland T, Hestdal K, Videm V. Activation of leukocytes during the uteroplacental passage in preeclampsia. *Hypertension.* 2002; 39(1):155–160.
159. Aly AS, Khandelwal M, Zhao J, Mehmet AH, Sammel MD, Parry S. Neutrophils are stimulated by syncytiotrophoblast microvillous membranes to generate superoxide radicals in women with preeclampsia. *Am J Obstet Gynecol.* 2004;190(1):252–258.
160. Luppi P, Deloia JA. Monocytes of preeclamptic women spontaneously synthesize pro-inflammatory cytokines. *Clin Immunol.* 2006;118(2–3):268–275.
161. Redman CW, Sargent IL. Latest advances in understanding preeclampsia. *Science.* 2005;308:1592–1594.
162. Walsh SW. Obesity: a risk factor for preeclampsia. *Trends Endocrinol Metab.* 2007;18(10):365–370.
163. Greenberg AS, Obin MS. Obesity and the role of adipose tissue in inflammation and metabolism. *Am J Clin Nutr.* 2006;83(2):461S–465.
164. Benyo DF, Smarason A, Redman CW, Sims C, Conrad KP. Expression of inflammatory cytokines in placentas from women with preeclampsia. *J Clin Endocrinol Metab.* 2001;86(6):2505–2512.
165. Founds SA, Powers RW, Patrick TE, et al. A comparison of circulating TNF-alpha in obese and lean women with and without preeclampsia. *Hypertens Pregnancy.* 2008;27:573–589.
166. Grimble RF. Inflammatory status and insulin resistance. *Curr Opin Clin Nutr Metab Care.* 2002;5(5):551–559.
167. Correia ML, Haynes WG. Leptin, obesity and cardiovascular disease. *Curr Opin Nephrol Hypertens.* 2004;13(2):215–223.
168. Matsuzawa Y. The metabolic syndrome and adipocytokines. *FEBS Lett.* 2006;580(12):2917–2921.
169. Henson MC, Castracane VD. Leptin in pregnancy. *Biol Reprod.* 2000;63(5):1219–1228.
170. Jakimiuk AJ, Skalba P, Huterski D, Tarkowski R, Haczynski J, Magoffin DA. Leptin messenger ribonucleic acid (mRNA) content in the human placenta at term: relationship to levels of leptin in cord blood and placental weight. *Gynecol Endocrinol.* 2003;17(4):311–316.
171. Laivuori H, Gallaher MJ, Collura L, et al. Relationships between maternal plasma leptin, placental leptin mRNA and protein in normal pregnancy, pre-eclampsia and intrauterine growth restriction without pre-eclampsia. *Mol Hum Reprod.* 2006;12(9):551–556.
172. de Mouzon SH, Lepercq J, Catalano PM. The known and unknown of leptin in pregnancy. *Am J Obset Gynecol.* 2006;194:1537–1545.
173. Tessier DR, Ferraro ZM, Gruslin A. Role of leptin in pregnancy: consequences of maternal obesity. *Placenta.* 2013;34(3):205–211.
174. Jansson N, Greenwood SL, Johansson BR, Powell TL, Jansson T. Leptin stimulates the activity of the system A amino acid transporter in human placental villous fragments. *J Clin Endocrinol Metab.* 2003;88(3):1205–1211.
175. von Versen-Hoynck F, Rajakumar A, Parrott MS, Powers RW. Leptin affects system A amino acid transport activity in the human placenta: evidence for STAT3 dependent mechanisms. *Placenta.* 2009;30(4):361–367.
176. Wolf HJ, Ebenbichler CF, Huter O, et al. Fetal leptin and insulin levels only correlate inlarge-for-gestational age infants. *Eur J Endocrinol.* 2000;142(6):623–629.
177. Catalano PM, Presley L, Minium J, Hauguel-de Mouzon S. Fetuses of obese mothers develop insulin resistance in utero. *Diabetes Care.* 2009;32(6):1076–1080.
178. McCarthy JF, Misra DN, Roberts JM. Maternal plasma leptin is increased in preeclampsia and positively correlates with fetal cord concentration. *Am J Obset Gynecol.* 1999;180:731–736.

179. Hendler I, Blackwell SC, Metha SH, et al. The levels of leptin, adiponectin, and resistin in normal weight, overweight, and obese pregnant women with and without preeclampsia. *Am J Obset Gynecol.* 2005;193:979–983.
180. Catalano PM, Hoegh M, Minium J, et al. Adiponectin in human pregnancy: implications for regulation of glucose and lipid metabolism. *Diabetologia.* 2006;49(7):1677–1685.
181. Mazaki-Tovi S, Kanety H, Pariente C, et al. Maternal serum adiponectin levels during human pregnancy. *J Perinatol.* 2007; 27(2):77–81.
182. Hedderson MM, Darbinian J, Havel PJ, et al. Low prepregnancy adiponectin concentrations are associated with a marked increase in risk for development of gestational diabetes mellitus. *Diabetes Care.* 2013;36:3930–3937.
183. Retnakaran R, Hanley AJ, Raif N, Connelly PW, Sermer M, Zinman B. Reduced adiponectin concentration in women with gestational diabetes: a potential factor in progression to type 2 diabetes. *Diabetes Care.* 2004;27(3):799–800.
184. Meyer BJ, Stewart FM, Brown EA, et al. Maternal obesity is associated with the formation of small dense LDL and hypoadiponectinemia in the third trimester. *J Clin Endocrinol Metab.* 2013;98(2):643–652.
185. Corbetta S, Bulfamante G, Cortelazzi D, et al. Adiponectin expression in human fetal tissues during mid- and late gestation. *J Clin Endocrinol Metab.* 2005;90(4):2397–2402.
186. Haugen F, Ranheim T, Harsem NK, Lips E, Staff AC, Drevon CA. Increased plasma levels of adipokines in preeclampsia: relationship to placenta and adipose tissue gene expression. *Am J Physiol Endocrinol Metab.* 2006;290(2):E326–E333.
187. Kajantie E, Kaaja R, Ylikorkala O, Andersson S, Laivuori H. Adiponectin concentrations in maternal serum: elevated in preeclampsia but unrelated to insulin sensitivity. *J Soc Gynecol Investig.* 2005;12(6):433–439.
188. Ramsay JE, Jamieson N, Greer IA, Sattar N. Paradoxical elevation in adiponectin concentrations in women with preeclampsia. *Hypertension.* 2003;42(5):891–894.
189. Fasshauer M, Waldeyer T, Seeger J, et al. Circulating high-molecular-weight adiponectin is upregulated in preeclampsia and is related to insulin sensitivity and renal function. *Eur J Endocrinol.* 2008;158(2):197–201.
190. Takemura Y, Osuga Y, Koga K, et al. Selective increase in high molecular weight adiponectin concentration in serum of women with preeclampsia. *J Reprod Immunol.* 2007; 73(1):60–65.
191. Smith SR, Bai F, Charbonneau C, Janderova L, Argyropoulos G. A promoter genotype and oxidative stress potentially link resistin to human insulin resistance. *Diabetes.* 2003;52(7):1611–1618.
192. Patel L, Buckels AC, Kinghorn IJ, et al. Resistin is expressed in human macrophages and directly regulated by PPAR gamma activators. *Biochem Biophys Res Commun.* 2003; 300(2):472–476.
193. Kim KH, Lee K, Moon YS, Sul HS. A cysteine-rich adipose tissue-specific secretory factor inhibits adipocyte differentiation. *J Biol Chem.* 2001;276(14):11252–11256.
194. Silswal N, Singh AK, Aruna B, Mukhopadhyay S, Ghosh S, Ehtesham NZ. Human resistin stimulates the pro-inflammatory cytokines TNF-alpha and IL-12 in macrophages by NF-kappaB-dependent pathway. *Biochem Biophys Res Commun.* 2005;334(4):1092–1101.
195. Degawa-Yamauchi M, Bovenkerk JE, Juliar BE, et al. Serum resistin (FIZZ3) protein is increased in obese humans. *J Clin Endocrinol Metab.* 2003;88(11):5452–5455.
196. Ntaios G, Gatselis NK, Makaritsis K, Dalekos GN. Adipokines as mediators of endothelial function and atherosclerosis. *Atherosclerosis.* 2013;227(2):216–221.
197. Choi SH, Hong ES, Lim S. Clinical implications of adipocytokines and newly emerging metabolic factors with relation to insulin resistance and cardiovascular health. *Front Endocrinol.* 2013;4:97.
198. Mazaki-Tovi S, Kanety H, Pariente C, et al. Insulin sensitivity in late gestation and early postpartum period: the role of circulating maternal adipokines. *Gynecol Endocrinol.* 2011; 27(9):725–731.
199. Nien JK, Mazaki-Tovi S, Romero R, et al. Resistin: a hormone which induces insulin resistance is increased in normal pregnancy. *J Perinat Med.* 2007;35(6):513–521.
200. Miehle K, Stepan H, Fasshauer M. Leptin adiponectin and other adipokines in gestational diabetes mellitus and preeclampsia. *Clin Endocrinol (Oxf).* 2012;76(1):2–11.
201. Furuhashi M, Hotamisligil GS. Fatty acid-binding proteins: role in metabolic diseases and potential as drug targets. *Nat Rev Drug Discov.* 2008;7(6):489–503.
202. Xu A, Wang Y, Xu JY, et al. Adipocyte fatty acid-binding protein is a plasma biomarker closely associated with obesity and metabolic syndrome. *Clin Chem.* 2006;52(3):405–413.
203. Xu A, Tso AW, Cheung BM, et al. Circulating adipocyte-fatty acid binding protein levels predict the development of the metabolic syndrome: a 5-year prospective study. *Circulation.* 2007;115(12):1537–1543.
204. Fasshauer M, Seeger J, Waldeyer T, et al. Serum levels of the adipokine adipocyte fatty acid-binding protein are increased in preeclampsia. *Am J Hypertens.* 2008;21(5):582–586.
205. Winn VD, Gormley M, Fisher SJ. The impact of Preeclampsia on Gene Expression at the Maternal-Fetal Interface. *Pregnancy Hypertens.* 2011;1(1):100–108.
206. Winn VD, Haimov-Kochman R, Paquet AC, et al. Gene expression profiling of the human maternal-fetal interface reveals dramatic changes between midgestation and term. *Endocrinology.* 2007;148(3):1059–1079.
207. Biron-Shental T, Schaiff WT, Ratajczak CK, Bildirici I, Nelson DM, Sadovsky Y. Hypoxia regulates the expression of fatty acid-binding proteins in primary term human trophoblasts. *Am J Obstet Gynecol.* 2007;197(5):516e1–516e6.
208. Murata M, Kodama H, Goto K, Hirano H, Tanaka T. Decreased very-low-density lipoprotein and low-density lipoprotein receptor messenger ribonucleic acid expression in placentas from preeclamptic pregnancies. *Am J Obset Gynecol.* 1996;175:1551–1556.
209. Scifres CM, Catov JM, Simhan HN. The impact of maternal obesity and gestational weight gain on early and mid-pregnancy lipid profiles. *Obesity (Silver Spring).* 2014;22:932–938.
210. Hubel CA, Lyall F, Weissfeld L, Gandley RE, Roberts JM. Small low-density lipoproteins and vascular cell adhesion molecule-1 are increased in association with hyperlipidemia in preeclampsia. *Metabolism.* 1998;47(10):1281–1288.
211. Lorentzen B, Drevon CA, Endressen MJ, Henriksen T. Fatty acid pattern of esterfied and free fatty acids in sera of women with normal and pre-eclamptic pregnancy. *Brit J Obstet Gynaecol.* 1995;102:530–537.

212. Hubel CA, McLaughlin MK, Evans RW, Hauth BA, Sims CJ, Roberts JM. Fasting serum triglycerides, free fatty acids, and malondialdehyde are increased in preeclampsia, are positively correlated, and decrease within 48 hours post partum. *Am J Obset Gynecol.* 1996;174:975–982.
213. Endresen MJ, Lorentzen B, Henriksen T. Increased lipolytic activity and high ratio of free fatty acids to albumin in sera from women with preeclampsia leads to triglyceride accumulation in cultured endothelial cells. *Am J Obstet Gynecol.* 1992;167(2):440–447.
214. Sattar N, Bedomir A, Berry C, Shepherd J, Greer IA, Packard CJ. Lipoprotein subfraction concentrations in preeclampsia: pathogenic parallels to atherosclerosis. *Obstet Gynecol.* 1997;89(3):403–408.
215. Kaaja R, Tikkanen MJ, Viinikka L, Ylikorkala O. Serum lipoproteins, insulin, and urinary prostanoid metabolites in normal and hypertensive pregnant women. *Obstet Gynecol.* 1995;85(3):353–356.
216. Arbogast BW, Leeper SC, Merrick RD, Olive KE, Taylor RN. Plasma factors that determine endothelial cell lipid toxicity in vitro correctly identify women with preeclampsia in early and late pregnancy. *Hypertens Pregnancy.* 1996;15(3):263–279.
217. Lorentzen B, Endresen MJ, Clausen T, Henriksen T. Fasting serum free fatty acids and triglycerides are increased before 20 weeks of gestation in women who later develop preeclampsia. *Hypertens Pregnancy.* 1994;13(1):103–109.
218. Gratacos E, Casals E, Sanllehy C, Cararach V, Alonso PL, Fortuny A. Variation in lipid levels during pregnancy in women with different types of hypertension. *Acta Obstet Gynecol Scand.* 1996;75:896–901.
219. Vrijkotte TG, Krukziener N, Hutten BA, Vollebregt KC, van Eijsden M, Twickler MB. Maternal lipid profile during early pregnancy and pregnancy complications and outcomes: the ABCD study. *J Clin Endocrinol Metab.* 2012; 97(11):3917–3925.
220. Clausen T, Djurovic S, Henriksen T. Dyslipidemia in early second trimester is mainly a feature of women with early onset pre-eclampsia. *Brit J Obstet Gynaecol.* 2001; 108(10):1081–1087.
221. Lamarche B, Rashid S, Lewis GF. HDL metabolism in hypertriglyceridemic states: an overview. *Clin Chim Acta.* 1999;286:145–161.
222. Hubel CA, Shakir Y, Gallaher MJ, McLaughlin MK, Roberts JM. Low-density lipoprotein particle size decreases during normal pregnancy in association with triglyceride increases. *J Soc Gynecol Investig.* 1998;5(5):244–250.
223. van den Elzen HJ, Waldimiroff JW, Cohen-Overbeek TE, de Bruijn AJ, Grobbee DE. Serum lipids in early pregnancy and risk of preeclampsia. *Brit J Obstet Gynaecol.* 1996;103:117–122.
224. Magnussen EB, Vatten LJ, Lund-Nilsen TI, Salvesen KA, Davey Smith G, Romundstad PR. Prepregnancy cardiovascular risk factors as predictors of pre-eclampsia: population based cohort study [see comment]. *BMJ.* 2007; 335(7627):978.
225. Endresen MJ, Tosti E, Heimli H, Lorentzen B, Henriksen T. Effects of free fatty acids found increased in women who develop pre-eclampsia on the ability of endothelial cells to produce prostacyclin, cGMP and inhibit platelet aggregation. *Scand J Clin Lab Invest.* 1994;54:549–557.
226. Bilodeau JF, Hubel CA. Current concepts in the use of antioxidants for the treatment of preeclampsia. *J Obstet Gynaecol Can.* 2003;25:742–750.
227. Beckman M. Cell biology. Great balls of fat. *Science.* 2006;311(5765):1232–1234.
228. Martin S, Parton RG. Lipid droplets: a unified view of a dynamic organelle. *Nat Rev Mol Cell Biol.* 2006;7(5):373–378.
229. Winkler KJ, Wetzka B, Hoffmann MM, et al. Triglyceride-rich lipoproteins are associated with hypertension in preeclampsia. *J Clin Endocrinol Metab.* 2003;88:1162–1166.
230. Lewis GF, Steiner G. Hypertriglyceridemia and its metabolic consequences as a risk factor for atherosclerotic cardiovascular disease in non-insulin-dependent diabetes mellitus. *Diabetes Metab Rev.* 1996;12:37–56.
231. Gardner CD, Fortmann SP, Krauss RM. Association of small low-density lipoprotein particles with the incidence of coronary artery disease in men and women. *JAMA.* 1996;276:875–881.
232. Lamarche B, Tchernof A, Moorjani S, et al. Small, dense low-density lipoprotein particles as a predictor of the risk of ischemic heart disease in men. *Circulation.* 1997;95:69–75.
233. Lamarche B, Tchernof A, Mauriege P, et al. Fasting insulin and apolipoprotein B levels and low-density lipoprotein particle size as risk factors for ischemic heart disease. *JAMA.* 1998;279:1955–1961.
234. Stampfer MJ, Krauss RM, Ma J, et al. A prospective study of triglyceride level, low-density lipoprotein particle diameter, and risk of myocardial infarction. *JAMA.* 1996;276:882–888.
235. Silliman K, Shore V, Forte TM. Hypertriglyceridemia during late pregnancy is associated with the formation of small dense low-density lipoproteins and the presence of large buoyant high-density lipoproteins. *Metabolism.* 1994;43:1035–1041.
236. Belo L, Caslake M, Gaffney D, et al. Changes in LDL size and HDL concentration in normal and preeclamptic pregnancies. *Atherosclerosis.* 2002;162(2):425–432.
237. Ogura K, Miyatake T, Fukui O, Nakamura T, Kameda T, Yoshino G. Low-density lipoprotein particle diameter in normal pregnancy and preeclampsia. *J Atheroscler Thromb.* 2002;9(1):42–47.
238. Wakatsuki A, Ikenoue N, Okatani Y, Shinohara K, Fukaya T. Lipoprotein particles in preeclampsia: susceptibility to oxidative modification. *Obstet Gynecol.* 2000;96:55–59.
239. Levine RJ, Maynard SE, Qian C, et al. Circulating angiogenic factors and the risk of preeclampsia. *New Engl J Med.* 2004;350(7):672–683.
240. Levine RJ, Lam C, Qian C, et al. Soluble endoglin and other circulating antiangiogenic factors in preeclampsia. *New Engl J Med.* 2006;355(10):992–1005.
241. Maynard SE, Min JY, Merchan J, et al. Excess placental soluble fms-like tyrosine kinase 1 (sFlt1) may contribute to endothelial dysfunction, hypertension, and proteinuria in preeclampsia. *J Clin Invest.* 2003;111(5):649–658.
242. Venkatesha S, Toporsian M, Lam C, et al. Soluble endoglin contributes to the pathogenesis of preeclampsia. *Nat Med.* 2006;12(6):642–649.
243. Miyazawa-Hoshimoto S, Takahashi K, Bujo H, Hashimoto N, Saito Y. Elevated serum vascular endothelial growth factor is associated with visceral fat accumulation in human obese subjects. *Diabetologia.* 2003;46(11):1483–1488.
244. Powers RW, Roberts JM, Plymire D, et al. High prepregnancy body mass index is associated with low PlGF in

Preeclampsia and Uncomplicated Pregnancies. *Reprod Sci.* 2010;17(3 suppl):128A.
245. Mijal RS, Holzman CB, Rana S, Karumanchi SA, Wang J, Sikorskii A. Midpregnancy levels of angiogenic markers in relation to maternal characteristics. *Am J Obstet Gynecol.* 2011;204(3):244 e1–12.
246. Faupel-Badger JM, Staff AC, Thadhani R, et al. Maternal angiogenic profile in pregnancies that remain normotensive. *Eur J Obstet Gynecol Reprod Biol.* 2011;158(2):189–193.
247. Johnson RJ, Rideout BA. Uric acid and diet--insights into the epidemic of cardiovascular disease [comment]. *New Engl J Med.* 2004;350(11):1071–1073.
248. Garbagnait E, Boschetti M. Uric acid homeostasis in lean and obese girls during pubertal development. *Metabolism.* 1994;43(7):819–821.
249. Lee J, Sparrow D, Vokonas PS, Landsberg L, Weiss ST. Uric acid and coronary heart disease risk: evidence for a role of uric acid in the obesity-insulin resistance syndrome. *Am J Epidemiol.* 1995;142:288–294.
250. Mazzali M, Hughes J, Kim YG, et al. Elevated uric acid increases blood pressure in the rat by a novel crystal-independent mechanism. *Hypertension.* 2001;38(5):1101–1106.
251. Mazzali M, Kanellis J, Han L, et al. Hyperuricemia induces a primary renal arteriolopathy in rats by a blood pressure-independent mechanism. *Am J Physiol Renal Physiol.* 2002;282(6):F991–F997.
252. Shi Y, Evans JE, Rock KL. Molecular identification of a danger signal that alerts the immune system to dying cells. *Nature.* 2003;425:516–521.
253. Patschan D, Patschan S, Gobe GG, Chintala S, Goligorsky MS. Uric acid heralds ischemic tissue injury to mobilize endothelial progenitor cells. *J Am Soc Nephrol.* 2007;18(5):1516–1524.
254. Roberts JM, Bodnar LM, Lain KY, et al. Uric acid is as important as proteinuria in identifying fetal risk in women with gestational hypertension [see comment]. *Hypertension.* 2005;46(6):1263–1269.
255. Powers RW, Bodnar LM, Ness RB, et al. Uric acid concentrations in early pregnancy among preeclamptic women with gestational hyperuricemia at delivery. *Am J Obstet Gynecol.* 2006;194(1):160.
256. Many A, Hubel CA, Roberts JM. Hyperuricemia and xanthine oxidase in preeclampsia, revisited. *Am J Obset Gynecol.* 1996;174:288–291.
257. Bainbridge SA, Roberts JM. Uric acid as a pathogenic factor in preeclampsia. *Placenta.* 2008;29(suppl A):S67–S72.
258. Eid HM, Arnesen H, Hjerkinn EM, Lyberg T, Seljeflot I. Relationship between obesity, smoking, and the endogenous nitric oxide synthase inhibitor, asymmetric dimethylarginine. *Metabolism.* 2004;53(12):1574–1579.
259. Boger RH. The emerging role of asymmetric dimethylarginine as a novel cardiovascular risk factor. *Cardiovasc Res.* 2003;59(4):824–833.
260. Fard A, Tuck CH, Donis JA, et al. Acute elevations of plasma asymmetric dimethylarginine and impaired endothelial function in response to a high-fat meal in patients with type 2 diabetes. *Arterioscler Thromb Vasc Biol.* 2000;20(9):2039–2044.
261. Palm F, Onozato ML, Luo Z, Wilcox CS. Dimethylanginine dimethylaminohydrolase (DDAH): expression, regulation, and function in the cardiovascular and renal systems. *Am J Physiol Heart Circ Physiol.* 2007;293:H3227–H3245.
262. Cooke JP. Asymmetrical dimethylarginine: the Uber marker? *Circulation.* 2004;109(15):1813–1818.
263. Boger RH. Association of asymmetric dimethylarginine and endothelial dysfunction. *Clin Chem Lab Med.* 2003;41(11):1467–1472.
264. Boger RH, Bode-Boger SM. Asymmetric dimethylarginine, derangements of the endothelial nitric oxide synthase pathway, and cardiovascular diseases. *Semin Thromb Hemost.* 2000;26(5):539–545.
265. Boger RH, Bode-Boger SM, Szuba A, et al. Asymmetric dimethylarginine (ADMA): a novel risk factor for endothelial dysfunction: its role in hypercholesterolemia. *Circulation.* 1998;98(18):1842–1847.
266. Thum T, Tsikas D, Stein S, et al. Suppression of endothelial progenitor cells in human coronary artery disease by the endogenous nitric oxide synthase inhibitor asymmetric dimethylarginine [see comment]. *J Am Coll Cardiol.* 2005; 46(9):1693–1701.
267. Thum T, Bauersachs J. ADMA, endothelial progenitor cells, and cardiovascular risk.[comment]. *Circ Res.* 2005; 97(8):e84.
268. Thum T, Bauersachs J. Spotlight on endothelial progenitor cell inhibitors: short review. *Vasc Med.* 2005;10(suppl 1): S59–S64.
269. Savvidou MD, Hingorani AD, Tsikas D, Frolich JC, Vallance P, Nicolaides KH. Endothelial dysfunction and raised plasma concentrations of asymmetric dimethylarginine in pregnant women who subsequently develop pre-eclampsia. *Lancet.* 2003;361(9368):1511–1517.
270. Speer PD, Powers RW, Frank MP, Harger G, Markovic N, Roberts JM. Elevated asymmetric dimethylarginine concentrations precede clinical preeclampsia, but not pregnancies with small-for-gestational-age infants. *Am J Obstet Gynecol.* 2008;198(1):112e1–112e7.
271. Roberts JM, Bodnar LM, Patrick TE, Powers RW. The role of obesity in preeclampsia. *Pregnancy Hypertens.* 2011;1(1):6–16.
272. Sibai BM, Ewell M, Levine RJ, et al. Risk factors associated with preeclampsia in healthy nulliparous women. The Calcium for Preeclampsia Prevention (CPEP) Study Group. *Am J Obset Gynecol.* 1997;177:1003–1010.
273. Caritis S, Sibai B, Hauth J, et al. Low-dose asprin to prevent preeclampsia in women at high risk. *New Engl J Med.* 1998;338(11):701–705.
274. Queiro R, Weruaga A, Riestra JL. C4 deficiency state in antiphospholipid antibody-related recurrent preeclampsia evolving into systemic lupus erythematosus. *Rheumatol Int.* 2002;22(3):126–128.
275. Boomsma CM, Eijkemans MJ, Hughes EG, Visser GHA, Fauser BC, Macklon NS. A meta-analysis of pregnancy outcomes in women with polycystic ovary syndrome. *Hum Reprod Update.* 2006;12(6):673–683.
276. Sorensen TK, Williams MA, Lee IM, Dashow EE, Thompson ML, Luthy DA. Recreational physical activity during pregnancy and risk of preeclampsia. *Hypertension.* 2003;41(6):1273–1280.
277. Ness RB, Markovic N, Bass D, Harger G, Roberts JM. Family history of hypertension, heart disease, and stroke

among women who develop hypertension in pregnancy. *Obstet Gynecol*. 2003;102:1366–1371.
278. Jonsdottir LS, Arngrimsson R, Geirsson RT, Sigvaldason H, Sigfusson N. Death rates from ischemic heart disease in women with a history of hypertensionin pregnancy. *Acta Obstet Gynecol Scand*. 1995;74:772–776.
279. Irgens HU, Reisaeter L, Irgens LM, Lie RT. Long term mortality of mothers and fathers after pre-eclampsia: population based cohort study. *BMJ*. 2001;323(7323):1213–1217.
280. Funai EF, Friedlander Y, Paltiel O, et al. Long-term mortality after preeclampsia. *Epidemiology*. 2005;16(2):206–215.
281. Bellamy L, Casas JP, Hingorani AD, Williams DJ. Preeclampsia and risk of cardiovascular disease and cancer in later life: systematic review and meta-analysis [see comment]. *BMJ*. 2007;335(7627):974.
282. Chesley LC. Hypertension in pregnancy: definitions, familial factor, and remote prognosis. *Kidney Int*. 1980;18: 234–240.
283. Fisher KA, Lluger A, Spargo BH, Lindheimer MD. Hypertension in pregnancy: clinical-pathological correlations and remote prognosis. *Medicine*. 1981;60:267–276.
284. Pouta A, Hartikainen A-L, Sovio U, et al. Manifestations of metabolic syndrome after hypertensive pregnancy. *Hypertension*. 2004;43:825–831.
285. Chambers JC, Fusi L, Malik IS, Haskard DO, De Swiet M, Kooner JS. Association of maternal endothelial dysfunction with preeclampsia. *JAMA*. 2001;285(12):1607–1612.
286. Agatisa PK, Ness RB, Roberts JM, Costantino JP, Kuller LH, McLaughlin MK. Impairment of endothelial function in women with a history of preeclampsia: an indicator of cardiovascular risk. *Am J Physiol*. 2004;286(4):1389–1393.
287. Ramsay JE, Stewart F, Greer IA, Sattar N. Microvascular dysfunction: a link between pre-eclampsia and maternal coronary heart disease. *BJOG*. 2003;110(11):1029–1031.
288. Hubel CA, Powers RW, Snaedal S, et al. C-reactive protein is elevated 30 years after eclamptic pregnancy. *Hypertension*. 2008;51(6):1499–1505.
289. He S, Silveira A, Hamsten A, Blomback M, Bremme K. Haemostatic, endothelial and lipoprotein parameters and blood pressure levels in women with a history of preeclampsia. *Thromb Haemost*. 1999;81:538–542.
290. Barden A. Pre-eclampsia: contribution of maternal constitutional factors and the consequences for cardiovascular health. *Clin Exp Pharmacol Physiol*. 2006;33:826–830.
291. Freeman DJ, McManus F, Brown EA, et al. Short- and long-term changes in plasma inflammatory markers associated with preeclampsia [see comment]. *Hypertension*. 2004;44(5):708–714.
292. Portelinha A, Belo L, Tejera E, Rebelo I. Adhesion molecules (VCAM-1 and ICAM-1) and C-reactive protein in women with a history of preeclampsia. *Acta Obstet Gynecol Scand*. 2008;87:969–971.
293. Vikse BE, Irgens LM, Leivestad T, Skjaerven R, Iversen BM. Preeclampsia and the risk of end-stage renal disease. *New Engl J Med*. 2008;359:800–809.
294. Thadhani R, Solomon CG. Preeclampsia - A glimpse into the future? *New Engl J Med*. 2008;359:858–860.
295. Mazar RM, Srinivas SK, Sammel MD, Andrela CM, Elovitz MA. Metabolic score as a novel approach to assessing preeclampsia risk. *Am J Obset Gynecol*. 2007;197:411.e1–411.e5.
296. Romundstad PR, Magnussen EB, Smith GD, Vatten LJ. Hypertension in pregnancy and later cardiovascular risk: common antecedents? *Circulation*. 2010;122(6):579–584.
297. Catalano PM, Ehrenberg HM. The short- and long-term implications of maternal obesity on the mother and her offspring. *BJOG*. 2006;113(10):1126–1133.
298. Pell JP, Smith GC, Walsh D. Pregnancy complications and subsequent maternal cerebrovascular events: a retrospective cohort study of 119,668 births. *Am J Epidemiol*. 2004;159(4):336–342.
299. Catov JM, Bodnar LM, Ness RB, Barron SJ, Roberts JM. Inflammation and dyslipidemia related to risk of spontaneous preterm birth. *Am J Epidemiol*. 2007;166(11):1312–1319.
300. Germain AM, Romanik MC, Guerra I, et al. Endothelial dysfunction. A link among preeclampsia, recurrent pregnancy loss, and future cardiovascular events? *Hypertension*. 2007;49:1–6.
301. Ray JG, Vermeulen MJ, Schull MJ, Redelmeier DA. Cardiovascular health after maternal placental syndromes (CHAMPS): population-based retrospective cohort study. *Lancet*. 2005;366(9499):1797–1803.
302. Berends AL, de Groot CJ, Sijbrands EJ, et al. Shared constitutional risks for maternal vascular-related pregnancy complications and future cardiovascular disease. *Hypertension*. 2008;51:1034–1041.
303. Staff AC, Dechend R, Pijnenborg R. Learning from the placenta: acute atherosis and vascular remodeling in preeclampsia-novel aspects for atherosclerosis and future cardiovascular health. *Hypertension*. 2010;56(6):1026–1034.
304. Barker DJ, Thornburg KL, Osmond C, Kajantie E, Eriksson JG. The surface area of the placenta and hypertension in the offspring in later life. *Int J Dev Biol*. 2010;54(2–3):525–530.
305. Sheppard BL, Bonnar J. An ultrastructural study of uteroplacental spiral arteries in hypertensive and normotensive pregnancy and fetal growth retardation. *Brit J Obstet Gynaecol*. 1981;88:695–705.
306. Staff AC, Ranheim T, Khoury J, Henriksen T. Increased contents of phospholipids, cholesterol, and lipid peroxides in decidua basalis in women with preeclampsia. *Am J Obset Gynecol*. 1999;180(3):587–592.
307. Powers RW, Gandley RE, Lykins DL, Roberts JM. Moderate hyperhomocysteinemia decreases endothelial-dependent vasorelaxation in pregnant but not nonpregnant mice. *Hypertension*. 2004;44(3):327–333.
308. Faas MM, Schuiling GA, Baller JF, Visscher CA, Bakker WW. A new animal model for human preeclampsia: ultra-low-dose endotoxin infusion in pregnant rats. *Am J Obstet Gynecol*. 1994;171(1):158–164.
309. Martinez-Orgado J, Gonzalez R, Tovar S, Marin J, Salaices M, Alonso MJ. Administration of N(omega)-L-arginine methyl ester (L-NAME) impairs endothelium-dependent relaxation in gravid but not nongravid rats. *J Soc Gynecol Investig*. 2003;10(2):74–81.
310. Ramirez RJ, Hubel CA, Novak J, DiCianno JR, Kagan VE, Gandley RE. Moderate ascorbate deficiency increases myogenic tone of arteries from pregnant but not virgin ascorbate-dependent rats. *Hypertension*. 2006;47(3):454–460.
311. Ray JG, Diamond P, Singh G, Bell CM. Brief overview of maternal triglycerides as a risk factor for pre-eclamsia. *BJOG*. 2006;113:379–386.

312. Ray JG. Dysmetabolic syndrome, placenta-mediated disease and future risk of cardiovascular disease. *Fetal Matern Med Rev*. 2004;15(3):231–246.
313. Gammill HS, Lin C, Hubel CA. Endothelial progenitor cells and preeclampsia. *Front Biosci*. 2007;12:2383–2394.
314. Roberts JM, Hubel CA. The two stage model of preeclampsia: variations on the theme. *Placenta*. 2009;30(suppl A):S32–S37.
315. Sattar N, Greer IA, Galloway PJ, et al. Lipid and lipoprotein concentrations in pregnancies complicated by intrauterine growth restriction. *J Clin Endocrinol Metab*. 1999; 84(1):128–130.
316. Ramsay JE, Ferrell WR, Crawford L, Wallace AM, Greer IA, Sattar A. Divergent metabolic and vascular phenotypes in pre-eclampsia and intrauterine growth restriction: relevance of adiposity. *J Hypertens*. 2004;22:2177–2183.
317. Alvino G, Cozzi V, Radaelli T, Ortega H, Cetin EH. Maternal and fetal fatty acid profile in normal and IUGR pregnancies with and without preeclampsia. *Pediatr Res*. 2008;64:615–620.
318. Karowicz-Bilinska A, Kedziora-Kornatowska K, Bartosz G. Indices of oxidative stress in pregnancy with fetal growth restriction. *Free Radic Res*. 2007;41(8):870–873.
319. Bodnar LM, Ness RB, Markovic N, Roberts JM. The risk of preeclampsia rises with increasing prepregnancy body mass index. *Ann Epidemiol*. 2005;15(7):475–482.
320. Williams C, Coltart TM. Adipose tissue metabolism in pregnancy: the lipolytic effect of human placental lactogen. *Brit J Obstet Gynaecol*. 1978;85:43–46.
321. Waite LL, Atwood AK, Taylor RN. Preeclampsia, an implantation disorder. *Rev Endocr Metab Disord*. 2002;3:151–158.
322. Biron-Shental T, Schaiff WT, Ratajczak CK, Bildirici I, Nelson M, Sadovsky Y. Hypoxia regulates the expression of fatty acid-binding proteins in primary term human trophoblasts. *Am J Obset Gynecol*. 2007;197:e1–e6.
323. Fukumura D, Ushiyama A, Duda DG, et al. Paracrine regulation of angiogenesis and adipocyte differentiation during in vivo adipogenesis. *Circ Res*. 2003;93:e88–e97.

CHAPTER **8**

Immunology of Normal Pregnancy and Preeclampsia

CHRISTOPHER W.G. REDMAN, IAN L. SARGENT AND ROBERT N. TAYLOR

Editors' comment: *Chesley in his single-authored first edition wrote little regarding immunology and preeclampsia, though he noted that as early as 1902 Viet had suggested that deported trophoblastic fragments are antigenic and could elicit antibodies that he named 'syncytiolysin (edition 1 p. 467). He further noted that Dienst had suggested incompatibility of maternal and fetal blood groups in 1905 (edition 1 p. 470). Despite such early beginnings the first edition's index had a little over 30 citations under the subheading "Immunologic Factors." Perhaps this represented immunology's state of the art at that time or, as the current chapter's authors imply, investigators then aware that preeclampsia rarely reoccurred in subsequent conceptions were dissuaded from exploring immunological factors in the genesis of preeclampsia. However, our knowledge of the intricacy and complexity of the immune system has expanded since Chesley's first edition, as the authors of this chapter show us. This report was, to our knowledge, the first to suggest a role for placental debris, or microvesicles, in preeclampsia. This initially made many an investigator smile, but microvesicles are now being appreciated everywhere, and are elegantly discussed here, as are other aspects of growing information regarding the immunological significance of preeclampsia.*

INTRODUCTION

In this chapter we describe the role of immune mechanisms in preeclampsia. The issue has always centered on the need to explain why women in their first pregnancies are most susceptible to the condition, which has been combined with the perception that the presence within the mother of a genetically foreign fetus must pose a challenge to the maternal immune system.[1] It has been postulated that immune accommodation to the fetus needs to be "learnt" or immunoregulated. In preeclampsia the adaptation may be relatively defective in a first pregnancy but less so in subsequent pregnancies. There is also the issue of partner specificity and primipaternity.[2,3]

MATERNAL ADAPTATION TO A FOREIGN FETUS

Maternal immune accommodation to paternal antigens expressed by their fetus might be acquired from a previous successful pregnancy or abortion, in the case of the same partner, or exposure to paternal seminal plasma. Partner specificity (primipaternity) was first suggested when a change of partner by parous women seemed to increase the risks of preeclampsia.[2] This study and others were anecdotal or uncontrolled but suggested that the issue was relevant. Subsequent reports seemed to confirm this (for example ref. 3) but a systematic review concluded that there are still substantial uncertainties.[4] This was confounded by the finding that longer inter-pregnancy intervals are associated with both a change of partner and preeclampsia[5,6] Which was the relevant variable: a new partner or delayed conception? One possible modifier is cigarette smoking, which is known to be negatively associated with preeclampsia but positively associated with a change of partner.[5] Statistical adjustment for these associated factors has not been necessarily helpful because of confusion about the causal relationships of associated variables.[7]

The concept that maternal immune adaptation to a partner's fetus might not be restricted to pregnancy but be learnt before conception by exposures to sperm or seminal fluid (reviewed by Saito et al. 2007[8]) was stimulated by a study from Guadeloupe, which showed that a short interval between first coitus and conception significantly increased the risk of preeclampsia[9] regardless of parity but specifically with a new partner (Table 8.1). A more recent investigation[12] confirms the issue of the short interval but only primiparous women were studied.

ISBN: 978-0-12-407866-6

DOI: http://dx.doi.org/10.1016/B978-0-12-407866-6.00008-0

TABLE 8.1 Prepregnancy Priming of Maternal Immune System

		Preeclampsia risk
First pregnancy	Short duration of coitus, pre-conception	↑[1]
Previous term pregnancy	New partner	↑[2]
Previous abortion	Short previous pregnancy	Probably↓[3]
Barrier contraception	Reduced exposure to sperm/seminal fluid:	Possibly↑[4]
Intracytoplasmic sperm injection	No prior immune exposure to sperm/seminal fluid	Possibly↑[5]
Donor insemination	No prior exposure to sperm/seminal fluid	Probably↑[6]
Donor oocyte	Fully allogeneic fetus	Probably↑[6]

Sources:[10,11]
[1]Compared to first pregnancy with longer exposure.
[2]Compared to second pregnancy with same partner.
[3]Compared to no previous abortion.
[4]Barrier contraception compared to no barrier contraception.
[5]Compared to no prior exposure to sperm/seminal fluid.
[6]Compared to matched control group.

In summary there is indirect evidence that tolerization to a foreign fetus can occur in prior pregnancies or even before conception, but if tolerance is not achieved, preeclampsia may ensue. Such tolerization would be expected to involve the adaptive immune system. However the more primitive innate immune system, which controls inflammatory responses, is also involved.

INNATE AND ADAPTIVE IMMUNITY

Innate and adaptive immunity both contribute to the pathogenesis of preeclampsia. Innate immunity is a more primitive, rapid-acting, early-response system for global and relatively nonspecific protection. It evolved to protect single and multicellular organisms from "danger."[13] Danger takes many forms: physical, for example high temperature, chemical, for example excessive oxidation, trauma, infection, or neoplasia. Of these, oxidative stress is a relevant proinflammatory trigger in both normal pregnancy and preeclampsia.

The innate immune system depends on a network of "danger" receptors, called pattern recognition receptors (PRRs), which are germ-line encoded and recognize many different danger signals. Their ligands may be endogenous, often in the form of modified self-molecules, or exogenous, typically derived from bacteria or viruses (Table 8.2). The same receptor frequently recognizes multiple antigens.[14] Most danger receptors are either soluble or bound to the cell surface but some function intracellularly, for example binding intracellular ligands that are derived from bacteria or viruses. When inflammatory cells are activated they release signals such as cytokines or chemokines that, in turn, attract and "instruct" adaptive immune cells (T or B lymphocytes) to generate antigen-specific responses by means of antibodies or cytotoxic cells. Hence the two systems, innate and adaptive, operate together and in sequence.

TABLE 8.2 Danger Signals that Activate the Innate (Inflammatory) Immune System

Stimulants of Inflammation	Corresponding Receptors
Bacterial products	Toll-like receptors
Products of oxidative stress	Scavenger receptors
Products of cell trauma	Toll-like and scavenger receptors
Thrombin	Protease-activated receptors
Heat shock proteins	Various
Soluble (viral) DNA	Toll-like receptor 9

Adaptive immunity develops slowly but delivers precise antigen-specific responses with immunological "memory." It evolved more recently than innate immunity and is present only in vertebrates. The innate and adaptive systems are asymmetrically interdependent; in effect the latter is a functional elaboration of the former. The innate system does not need the adaptive system to function whereas adaptive immunity has an absolute requirement for signals from the innate system. Adaptive immunity is strongly but not exclusively influenced by the so-called "transplantation antigens," the human leukocyte antigens (HLA), which determine transplant rejection. The distribution of fetal (foreign) HLA on trophoblast is therefore a key issue that has informed much of our insights into pregnancy immunology (see below). The main effectors of adaptive immunity are T cells and B cells. They are triggered by professional antigen-presenting cells expressing class II MHC antigens (HLA-D) that enable them to activate naïve T cells. They comprise dendritic cells, macrophages, and B cells.

In this chapter the inflammatory system is deemed to be the sum of events involving innate rather than adaptive immune changes. It is one component of a much larger system that responds to all cellular stresses, called the integrated stress response.[15]

NATURE'S TRANSPLANT

Within the fetoplacental unit, the placenta is the tissue that is most directly exposed to maternal immune cells, which have contact with trophoblast in the placental bed, intervillous space and amniochorion. Fetal cells may traffic across the placenta but in general are confined to the fetal compartment.

Maternal adaptive immunity appears to be most important during placentation, when the placenta and its uteroplacental circulation become established during the first stage

Tissue	Classical Class Ia			Non-Classical Class Ia	
	HLA-A	HLA-B	HLA-C	HLA-E	HLA-G
Adult and fetal cells	■	■	■	▩	
Extravillous, cytotrophoblast (Interface 1)			■	▩	
Syncytiotrophoblast (Interface 2)					

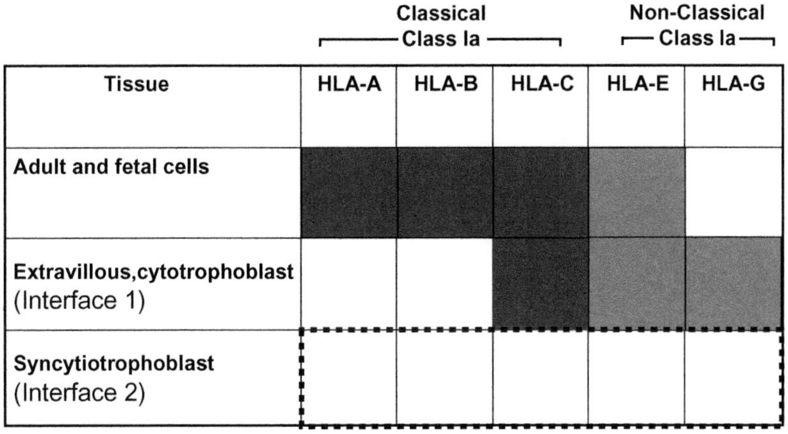

FIGURE 8.1 The unusual distribution of Class 1 HLA antigens on trophoblast. Human trophoblast does not express Class II (HLA-D) antigens. Syncytiotrophoblast expresses no HLA.

of preeclampsia. The major contribution to the maternal syndrome in the second stage of preeclampsia appears to derive from a systemic (or vascular) inflammatory response (innate immunity).

A key aspect of the maternal–placental immune interactions is the restricted expression of HLA by trophoblast. Trophoblast has a uniquely different pattern of HLA expression from other somatic cells. The decidual (extravillous) trophoblast in the placental bed does not express the strong polymorphic HLA-A, HLA-B (HLA-Class I) or HLA-D (HLA-Class II) antigens, the principal stimulators of T-cell-dependent graft-rejection responses. Instead, it expresses a unique combination of HLA-C and non-classical HLA-type 1b antigens: HLA-E and HLA-G (Fig. 8.1). Of these only HLA-C is polymorphic, meaning that it alone expresses paternal identity. HLA-E is a ubiquitous signal for self, which inhibits NK cells via specific receptors. Recognition of "missing-self" is immunostimulatory and contributes to immune recognition of malignant cells lacking HLA-E.[16] HLA-G is not expressed in normal tissues other than the placenta, where it is confined to extravillous trophoblast.[17] It is also expressed in some malignancies and may contribute to escape from immune surveillance. It is generally agreed that it confers a degree of immune privilege on trophoblast.[18] A soluble form of HLA-G is also released and is detectable in blood.

Maternal exposure to trophoblast varies with gestational age. The maternal-fetal immune interfaces[19] are not fully characterized but include two that are relevant to preeclampsia, which we have called Interfaces 1 and 2 (Fig. 8.2), and one that might be relevant but is largely unstudied, namely the chorionic component of the placental membranes.

Interface 1 is between maternal immune cells and invasive, extravillous trophoblast in the decidua of the placental bed. It dominates during the first half of pregnancy when placentation is established (see Chapter 5) and the placental bed is infiltrated with invasive cytotrophoblast.

Interface 2 is anatomically distinct and comprises syncytiotrophoblast, the surface layer of chorionic villi, which is in contact with maternal blood borne immune cells. The interface becomes active when the intervillous circulation is established (weeks 8–10) and expands with the growth of the placenta to become the dominant interface towards the end of pregnancy. Because the syncytium is in direct contact with the maternal circulation, if it promoted immune responses they would be systemic, not local. At Interface 2, the syncytiotrophoblast is HLA-negative and immunologically neutral.

The key point is that paternal HLA alloantigens are only expressed at Interface 1, which is most active in the first half of pregnancy. This constrains the space and time when abnormal placentation can cause the later development of preeclampsia and fetal growth restriction.

CLASSICAL TWO-STAGE MODEL OF PREECLAMPSIA

In 1991 we first proposed that preeclampsia evolves in two stages, pre-clinical and clinical[21] (Fig. 8.3). Each stage involves the maternal immune system in different ways. In the first stage there is an important element localized to the placental bed where placentation is typically inhibited. In the second stage, a diffuse vascular inflammation predominates. The two stages are associated with different patterns of maternal exposure to fetal tissues and potentially different consequences of immune maladaptation.

Immune Interface 1 needs to be considered in relation to Stage 1 preeclampsia and immune events that affect placentation could explain its apparent partner specificity. Immune Interface 2 is active through trimesters 2 and 3 and seems to be the dominant component of Stage 2 preeclampsia.

The two stage model has now been superseded by a more detailed multi-staged model as explained in a later section.

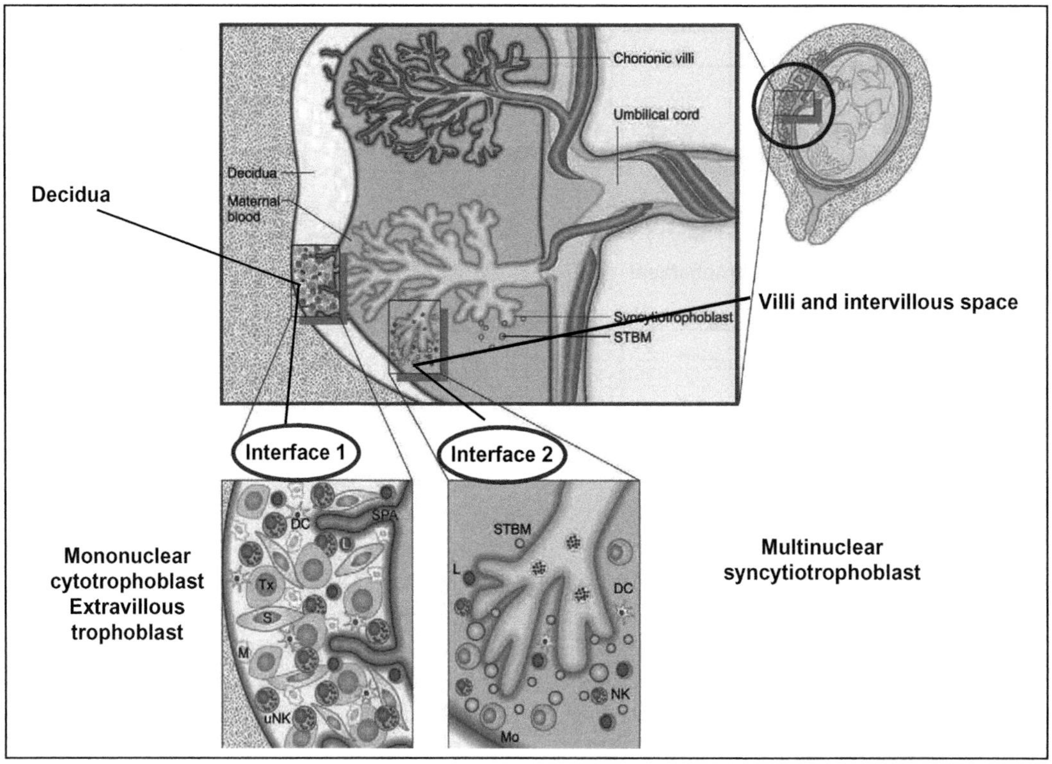

FIGURE 8.2 The two immune interfaces of human pregnancy. Adapted from[20] with permission. Immune events at Interface 1 drive the first stage of the two-stage model (see Fig. 8.3) of preeclampsia, whereas Interface 2 drives the second stage. Interface 1 comprises maternal immune cells, including uterine NK cells (uNK), T lymphocytes (L), macrophages (M) and dendritic cells (DC), and the invasive extravillous cytotrophoblast (Tx), between large stromal cells (S) and spiral arteries (SPA; not to scale). Interface 2 lies between circulating maternal immune cells, including T lymphocytes (L), NK cells (NK), monocytes (Mo) and dendritic cells (DC), and syncytiotrophoblast microparticles (STBM). (This figure is reproduced in color in the color plate section.)

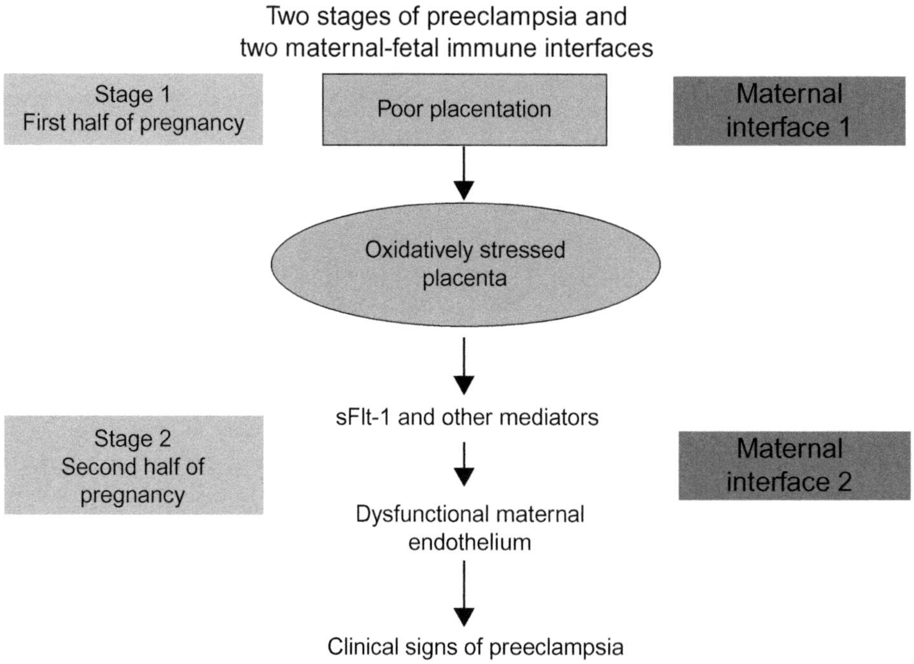

FIGURE 8.3 Two-stage model of development of preeclampsia. (This figure is reproduced in color in the color plate section.)

STAGE 1 PREECLAMPSIA, INTERFACE 1 AND MATERNAL IMMUNE RESPONSES TO TROPHOBLAST

The endometrium is an immune tissue. During the luteal phase it differentiates and begins to transform by infiltration of leukocytes. Once early pregnancy is established, decidua leukocytes are mainly (75%) natural killer cells. Macrophages comprise a smaller, but still abundant population. There are also rare dendritic cells and T cells. B cells are conspicuous by their absence. Natural killer cells are part of the innate immune system. Uterine NK (uNK) cells differ in phenotype from most circulating NK cells, are more activated, less cytotoxic and have an enhanced capacity to secrete cytokines and angiogenic factors. The latter promote infiltration of the spiral arteries by invasive trophoblast.[22] uNK cells bear receptors that interact with the unique repertoire of HLA expressed by invasive cytotrophoblast in the placental bed (HLA-C, -E and -G). HLA-C, the only polymorphic HLA expressed by trophoblast, apparently confined to invasive cytotrophoblast, is the ligand for killer immunoglobulin-like receptors (KIR) expressed by uNK cells. The main receptors for HLA-G are the inhibitory leukocyte immunoglobulin-like receptors (LIR -1 and LIR-2) which are expressed by monocytes, NK cells, T cells, and macrophages.[23] It is crucial to the issue of partner specificity of preeclampsia that HLA-C can signal fetal paternity, which can be recognized by the uNK cells.

The KIR receptors that recognize HLA-C on invasive trophoblast are themselves extremely polymorphic such that two individuals are unlikely to have the same genotype. The variability is not simply that of polymorphic genes, but of different patterns of inheritance of up to 17 different genes each with its own polymorphism, some that activate, others that inhibit. The number of KIR genes in different genotypes varies. The expression of the genotype is itself variable through differential expression of KIR genes, which becomes fixed by methylation, with the phenotype passed to daughter cells.[24] HLA-C has more than 1000 haplotypes.[25] In other words, maternal-fetal immune recognition at the site of placentation involves two polymorphic gene systems, maternal KIRs and fetal HLA-C molecules. Hence partner specificity will be high, if not unique (see Chapter 4).

KIR haplotypes can be divided into two groups, A and B, the latter being distinguished by additional activating (presumably beneficial) receptors. It is presumed that uNK cells need to be activated to produce cytokines and angiogenic factors to promote placentation. A maternal haplotype B would be predicted to protect from preeclampsia. In any pregnancy, the maternal KIR genotype could be AA (no activating KIR) or AB/BB (presence of one or more activating KIRs).

HLA-C haplotypes can also be grouped as C1 and C2 depending on an amino acid dimorphism at one position in the alpha-1 domain. It is considered that HLA-C2 interacts with KIRs more strongly than HLA-C1[26] so the combination of maternal HLA-C2 with fetal KIR B/B could be the best for promoting adequate placentation and avoiding preeclampsia. This is what is observed. Kir AA mothers confronted with HLA-C2 fetuses are the most susceptible to preeclampsia,[27,28] a similar pattern being shown with recurrent spontaneous miscarriages and normotensive fetal growth restriction.[29]

This form of maternal-fetal immune recognition can explain the partner specificity for preeclampsia but not the protection conferred by a previous pregnancy by the same partner, namely the immunological memory. As far as is known, immune memory is provided only by the adaptive immune system.

The issue with respect to trophoblast is one of T-cell recognition of HLA-C. HLA-C has unique features that distinguish if from HLA-A and B. It is less polymorphic for example, and in general its surface expression is lower. However, trophoblast uniquely proves the exception to this rule. Foreign antigens are recognized by T cells when they are bound to HLA-A and -B proteins on the cell surface of antigen-presenting cells – so-called HLA-restriction. HLA-C-restricted presentation of antigens has been less readily demonstrable except in specific viral infections and a few self-peptides. It is important that in the decidua, fetal (paternal) HLA-C, expressed by trophoblast, can bind to both NK and T cells (although not coincidentally). It is also relevant that the KIR receptors that dominate perceptions of interactions between uNK cells and invasive cytotrophoblast are also expressed by T cells where they appear to be able to modulate (inhibit or enhance) T cell responses.[30]

Maternal T cell responsivenes to foreign fetal HLA-C has been detected in the deciduas of normal pregnancies but is thought to be kept in check by T-regulatory cells.[31] Although this is a key issue, T cell reactivity to HLA-C on invasive cytotrophoblast in preeclampsia remains undefined.

Techniques in assisted reproduction create new immune challenges for pregnant women, with hitherto novel immune mating combinations from donated gametes, sperm, oocytes or embryos. Some methods of assisted reproduction increase the risk of preeclampsia. Such modes of conception or a short interval between first coitus and conception may hinder T-cell-dependent tolerance to paternal antigens, which facilitates the establishment of normal pregnancy at Interface 1. Otherwise, abnormal placentation and uteroplacental perfusion can lead to abnormalities at Interface 2 and Stage 2 disease.

STAGE 2 PREECLAMPSIA AND INTERFACE 2

Normal pregnancy and preeclampsia are both associated with a low-grade systemic (vascular) inflammatory response, which is more intense in preeclampsia.[32] It is believed this is secondary to syncytiotrophoblast stress[33] induced by hypoxia, oxidative stress or both. A generalized response to any form of cellular or tissue damage is inflammation,[13] a hypothesis which has been subsequently validated in many different contexts, not associated with pregnancy. In preeclampsia the maternal response to the oxidatively damaged placenta is what could have been predicted and what has been found. The systemic inflammatory response has massive and wide-ranging consequences, all of which are seen in the second stage of preeclampsia.[19] The nature of the vascular inflammation has unique attributes induced by release of factors from the placenta which are not present in nonpregnant individuals. These include the angiogenic placental growth factor (PlGF) and the antiangiogenic factors, soluble vascular endothelial growth factor receptor-1 (sVEGFR-1) and soluble endoglin (sEng) as described in Chapter 6.

ENDOTHELIAL CELLS ARE INFLAMMATORY CELLS

It has long been known that preeclampsia can be largely explained by diffuse maternal endothelial cell dysfunction[34] (see Chapter 9). In the circulation, endothelial cells are key players in systemic inflammatory responses as well as mediating local inflammation by upregulation of adhesion molecules that tether and then anchor marginated leukocytes. In the field of atherosclerosis this is well researched. Atherosclerosis is a focal large-vessel disease; whereas microvascular endothelium promotes diffuse systemic inflammation.[35] Both are examples of vascular inflammation. If preeclampsia is an endothelial disorder then the corollary is that it is also an inflammatory disorder. The latter is a more generalized concept that subsumes the former. Inflammation is one part of a more generalized and integrated stress response, hence other systems are inevitably involved and the stresses distributed between different cellular systems. Cellular oxidative stress will for example induce endoplasmic reticulum (ER) stress, the unfolded protein response and an inflammatory response (reviewed in[19]).

INFLAMMATION AND THE INTEGRATED STRESS RESPONSE

The integrated stress response (ISR) is evolutionarily ancient.[36] Central to the response is ER stress and the unfolded protein response, which enhances removal of unfolded proteins and re-programs protein translation in the ER. While most protein synthesis is reduced, that of specific transcription factors, promoting production of stress response proteins that restore homeostasis, is augmented via numerous sensors of cell damage or dysfunction. Among the stress response proteins are molecular chaperones, antioxidants, transcription factors and so on. Inflammation is part of the ISR; it activates and is activated by oxidative stress and ER stress.[36]

Protein folding consumes energy. Not surprisingly ER stress is precipitated by energy deficiency, for example from hypoxia or glucose deprivation. The ISR has profound effects on metabolism that help cells survive under stress and return to homeostasis. ER stress in the liver is particularly sensitive to inflammatory stimuli and underlies the acute-phase response.

WIDESPREAD IMPLICATIONS OF VASCULAR INFLAMMATION

It is self-evident that vascular inflammation involves endothelium as well as inflammatory leukocytes (granulocytes, macrophages, and natural killer lymphocytes). However, the coagulation system, liver and adipose tissue also directly contribute soluble factors to the inflammatory response (Table 8.3). The full extent of vascular inflammation on diverse systems may not be appreciated, nor the two-way interactions between its components. For example, blood coagulation is not only activated by inflammatory processes but thrombin, the final trigger to coagulation, also stimulates inflammation via specific receptors. Angiogenesis,[37] oxidative stress,[38] and obesity[39] are all tightly linked to inflammatory responses and all highly relevant to preeclampsia, as is described subsequently. The acute-phase response is a complex inflammatory stress response originating from the liver.

Nuclear factor-κB (NF-κB) is a transcription factor expressed by nearly all cells, which is central to many inflammatory responses. It is activated by numerous stressors, including inflammatory and oxidative stresses, and suppressed by hCG, estrogen, IL-4, and IL-10.[40] It is also activated by oxidative and endoplasmic reticulum (ER) stress.[41,42] These stresses also are prominent features of the syncytium in the preeclampsia placenta.[33] The interaction between NF-κB and hypoxia inducible factor (HIF)-1α, the transcription factor that controls cellular responses to low oxygen tension, is particularly close. HIF-1α activates transcription of over 400 genes involved in the regulation of immune and inflammatory responses and related functions. The acute-phase response is a complex inflammatory stress response originating from the liver.

TABLE 8.3 Components of the Inflammatory or Innate Immune System

Inflammatory leukocytes:
Granulocytes
Monocytes
Natural killer lymphocytes
Certain B cells producing "natural antibodies"
Endothelium
Platelets
Coagulation cascade
Complement system
Cytokines and chemokines
Adipocytes
Hepatocytes

Vascular inflammation brings in its trail the secretion of many factors that orchestrate its expression. Many of these are altered in preeclampsia.

The origins of preeclampsia lie in the placenta, specifically the syncytium, which demonstrates all the indicators of an ISR.[33] Three major stresses – ER, inflammatory and oxidative – are particularly relevant to preeclampsia and tightly interlinked. A number of trophoblast-derived mediators, discussed below and in Chapter 6 (sVEGR-1, PlGF), disseminate the placental problem by causing vascular inflammation, which provokes widespread dysfunction in many maternal systems. Ultimately, the dysfunction is more than simply vascular and has major metabolic effects.[43]

CYTOKINES, CHEMOKINES, GROWTH FACTORS, ADIPOKINES AND ANGIOGENIC FACTORS

A diverse group of secreted proteins and glycoproteins coordinate the inflammatory response. The term cytokine originally referred to peptides that are produced by and act on immune and hematopoietic cells. Although they are critical for both innate and adaptive immune function, they can also be secreted by non-immune cells and have non-immune functions that are relevant to the pathophysiology of preeclampsia. Some are chemokines, for example interleukin-8 (IL-8), that stimulate the migration of inflammatory cells. Adipokines include proteins that are secreted from (and synthesized by) adipocytes but exclude products of other cell types in adipose tissue, such as macrophages.[44] They include cytokines such as IL-6, classical adipokines (leptin, resistin, adiponectin) and even acute-phase proteins (PAI-1, angiotensinogen). The classical adipokines all have actions on immune cells, that is they have cytokine-like activity, as well as angiogenic activity, as reviewed by Ribatti et al.[45] Angiogenic factors (see Chapter 6) include VEGF, which is also a growth factor for endothelium. Other biologically active proteins or peptides can have cytokine-like activity, such as angiotensin II (Ang II); which is directly proinflammatory.[46] By its induction of VEGF, Ang II is also indirectly angiogenic.[47] Insulin on the other hand, while it is a mitogen, is antiinflammatory, as reviewed by Dandona et al.[48] Despite these cytokine-like actions, Ang II and insulin are not considered to be cytokines. In summary, the nomenclature surrounding these factors can be confusing and limit an understanding of their wider actions and their involvement with inflammatory responses in particular.

METABOLISM AND VASCULAR INFLAMMATION

Circulating proinflammatory factors such as endotoxin or tumor necrosis factor (TNF)-α cause insulin resistance and hence hyperlipidemia.[49] The hyperlipemia of sepsis has been known for many years (see review by Harris et al.).[50] Lipolysis releases free fatty acids, which contribute to the insulin resistance peripherally.

Obesity, which is a risk factor for preeclampsia (Chapter 7), is characterized by a vascular inflammatory response. It arises because adipose tissue is not simply an energy store but a source of proinflammatory cytokines and other metabolic mediators (adipokines) as mentioned in the preceding section and is the principal source of leptin. Leptin is secreted during acute inflammation and has an important action on immune cells, all of which express the leptin receptor. Leptin thereby causes or enhances proinflammatory responses, as reviewed by Matarese et al.[51] It can be classified as a cytokine as well as an adipokine. The net effect is that obesity is a state of chronic systemic inflammation.

The importance of obesity in generating the inflammatory response is demonstrated by its reversal after weight loss. Visceral rather than subcutaneous fat is more important in this context. Obesity is a key part of the metabolic syndrome,[52] which also includes insulin resistance with impaired glucose tolerance or overt diabetes, dyslipidemia, and hypertension. Obesity-associated IL-6 can induce the acute-phase response including production of circulating C-reactive protein (CRP). Elevated CRP is a typical feature of the metabolic syndrome but other plasma acute-phase reactants, such as fibrinogen, are similarly increased. Some acute-phase proteins are listed in Table 8.4. In the mouse liver the acute-phase response involves nearly 10% of the genome,[53] indicating the complexity and magnitude of the response.

Vascular inflammation is accompanied by an increase in triglyceride-rich lipoproteins, a reduction in high-density lipoprotein cholesterol, and impairment of cholesterol transport. These metabolic alterations, which promote atherosclerosis, may explain an epidemiological link between chronic inflammation and cardiovascular disease.[54]

The term metaflammation has been applied to the combination of low-grade chronic vascular inflammation and metabolic changes. It underlies chronic conditions such as

TABLE 8.4 Changes in Concentrations of Plasma Proteins in the Acute-Phase Response

Positive acute-phase reactants, increased in the circulation
- CRP (C-reactive protein)
- Angiotensinogen
- Fibrinogen, prothrombin, Factor VIII, plasminogen
- Complement components: (eg C3)
- Alpha-1-antitrypsin, alpha-1-antichymotrypsin
- Haptoglobin, hemopexin, ceruloplasmin
- Soluble phospholipase A2
- Sialic acid, α_1-acid glycoprotein

Negative acute-phase reactants, reduced in the circulation
- Albumin
- Transferrin
- Retinol-binding protein

obesity, atherosclerosis and type 2 diabetes,[55] and, as discussed above, shares many features with preeclampsia.

ACUTE-PHASE RESPONSE

The acute-phase response may also be a chronic response to local or systemic inflammation. It comprises variable changes in circulating plasma protein concentrations and other phenomena such as fever, anemia, leukocytosis and metabolic adaptations especially involving the liver and adipose tissue.[56] Proteins linked to this response, acute-phase proteins, are synthesized in the liver. They are classified as positive if they increase with systemic inflammation (e.g., C-reactive protein, CRP) or negative if they decrease (e.g., albumin) (Table 8.4).

The concentrations of CRP increase rapidly in response to inflammatory stimuli. Human CRP binds with high affinity to phosphocholine residues and other intrinsic and extrinsic ligands, including native and modified plasma lipoproteins, damaged cell membranes and various phospholipids and constituents of microorganisms and plant products. CRP is therefore a pattern-recognition receptor for a range of "danger" molecules.[57]

VASCULAR INFLAMMATION IN NORMAL PREGNANCY AND PREECLAMPSIA

A subtle systemic inflammatory response precedes conception in the luteal phase of the menstrual cycle[58] but becomes overt in the first trimester of pregnancy. Features of the response are wide-ranging. Some that are deemed to be physiological adaptations of pregnancy are in fact components of the acute-phase response such as reduced plasma albumin[59] and increased fibrinogen concentrations.[60] Many such examples have been summarized by Redman and Sargent.[61] CRP is modestly elevated in pregnancy, starting in the first trimester,[62] when the long-recognized leukocytosis of pregnancy, another sign of systemic maternal inflammation, is established.[63]

As pregnancy advances the vascular inflammation strengthens to peak during the third trimester. The concept was consolidated by flow cytometric analyses of circulating inflammatory leukocytes.[32,50,64] The changes are associated with increases in circulating inflammatory cytokines in the second half of pregnancy (several reports, cited in[65]; Table 8.5).

The inflammatory changes are associated with evidence of increasing systemic oxidative stress, in terms of several circulating markers particularly oxidized lipids.[71,72] In this regard, it is key that oxidative stress and chronic inflammation are related processes in ISR.[15] The inflammatory response generates oxidative stress (reviewed in[73]) and, in a converse fashion, oxidative stress can stimulate an inflammatory response.[74]

TABLE 8.5 The Vascular Inflammatory Network is Stimulated in Normal Pregnancy Relative to Non-Pregnancy

*Leukocytosis[66]	
*Increased leukocyte activation[64]	
*Complement activation[67]	
*Activation of the clotting system[68]	See Chapter 17
*Activation of platelets[69]	See Chapter 17
*Markers of endothelial activation[70]	See Chapters 9 and 17

*Significant change(s) relative to normal non-pregnant women.
Not all authors agree; see text. There are usually multiple references to justify each change.

TABLE 8.6 The Systemic Inflammatory Network is Stimulated in Preeclampsia Relative to Normal Pregnancy

*Leukocytosis[77]	
*Increased leukocyte activation[64]	
§*Complement activation[78]	
*Activation of the clotting system	See Chapter 17
*Activation of platelets	See Chapter 17
*Markers of endothelial activation	See Chapters 9 and 17
Increased circulating proinflammatory cytokines	
* Plasma tumor necrosis factor-α[65]	
§*Plasma interleukin-6[79]	
§*Plasma interleukin-8[80]	

*Significant change(s) relative to normal pregnant women.
§Not all authors agree; see text.
There are usually multiple references to justify each change

Measurements of circulating inflammatory cytokines during pregnancy can give complex results, owing in part to circulating binding proteins. Direct measurement of plasma concentrations of the proinflammatory cytokines interleukin-6 (IL-6) and tumor necrosis factor-α (TNF-α) show increased levels relative to non-pregnancy towards the end of gestation. Another method is to measure their production within peripheral blood mononuclear cells ex vivo, before or after stimulation. Intracellular cytokines can be measured flow cytometrically, but the intense stimulation required can introduce artifacts. Studies of peripheral blood cells exclude the possible contributions of endothelial cells which secrete IL-6, TNF-α and several other cytokines, as well as chemokines.[75]

Stage 2 preeclampsia, which appears to originate from the syncytial surface (Interface 2) of the placenta, is characterized by an exaggerated maternal vascular inflammatory response involving the inflammatory biomarkers of normal pregnancy,[76] which are more severely affected in preeclampsia (Table 8.6).

THE CONTINUUM BETWEEN NORMAL PREGNANCY AND PREECLAMPSIA

The systemic inflammatory response of preeclampsia is not a unique condition but represents the extreme part of a spectrum that is common to all pregnancies. Preeclampsia develops when vascular inflammation overcomes protective maternal compensation.[32] Because of the continuum, preeclampsia is clinically distinguished by arbitrary and artifactual thresholds. This explains why preeclampsia has evaded diagnostic categories or identification of clear pathological lesions distinct from normal pregnancy. This in turn has important implications for prediction, screening, and studies of genetic susceptibility or treatment.[32]

IMMUNOREGULATION

Inflammation stimulates adaptive immunity, which requires modulation to minimize potentially damaging transplantation reactions directed against the antigenically foreign fetus. During pregnancy, immunoregulation affects the gamut of T helper cell activity, namely Th1, Th2, Th17, memory T, and T regulatory cells (Tregs). There are other changes associated with preeclampsia described below.

T helper (Th) cells are ubiquitous lymphocytes that have no direct actions against pathogens, tumor cells or transplanted cells. Instead they direct other lymphocytes to deliver antigen-specific effector responses; via either cytotoxic (T cells) or antibody-producing B cells. T helper cells secrete cytokines and related proteins that interact with other leukocytes. Subsets of Th cells orchestrate different patterns of signals to achieve distinct activities (reviewed by Basu et al.[81]). Th1 cells promote cell-mediated immunity and secrete "Type 1" cytokines such as interleukin-2 (IL-2), and interferon-γ (IFN-γ); Th2 cells promote humoral immunity via "Type 2" cytokines such as IL-4, IL-5, IL-10, and IL-13. A third subset of IL-17-producing Th17 cells has been added to this list. Th17 cells share programs of differentiation with naïve T cells and T regulatory cells (Tregs) and are discussed below. Other subsets, Th3 and Th22, are less well defined and not considered here.

For many years normal pregnancy has been associated with the concept of a Th2 bias in immune function[82] characterized by reduced lymphocyte secretion of IFN-γ[83] and IL-2,[84] and increased production of IL-4 and IL-10.

The same shift is evident in decidual lymphocytes[85] and in circulating NK and NKT cells, but not detectable in unstimulated T cells,[86] leading to the suggestion that NK or NKT cells may be the first to impose this bias in pregnancy.[86] The Th2 cytokine balance is superimposed on the normal innate immune stimulation of pregnancy, restraining and redirecting lymphocyte activity toward a systemic humoral inflammatory response.

IFN-γ and IL-2 are produced by Th1 cells. Cytokines that are derived from monocyte/macrophages or other sources are more likely to be stimulated by the systemic inflammation of normal pregnancy. For example, pregnancy peripheral blood monocytes are primed to produce more of the type 1 cytokine, IL-12, than non-pregnancy monocytes.[87] IL-12 is a potent stimulator of other Type 1 responses (IFN-γ production), which implies that strong specific stimuli might rapidly convert the Type 2 bias of normal pregnancy to Type 1. Hence, the inflammatory system in pregnancy can be said to be meta-stable. It is restrained from producing Type 1 cytokines, but is primed to do so if it is challenged.

The Th1/Th2 paradigm can be extended to include Th17.[81] In general Th subsets inhibit each other: IFN-γ inhibits proliferation of Th2 cells; conversely IL-4 from Th2 cells inhibits expansion of Th1 cells. IL-17 from Th17 cells suppresses Th1 differentiation, whereas both Th1 and Th2 inhibit Th17.[88]

Th17 and Tregs are very closely related and therefore are considered together.

T REGULATORY CELLS, TH17 AND T-CELL MEMORY

Normal Pregnancy

Tregs are a heterogeneous class of suppressor T cells, which share the transcription factor Foxp3 but lack other specific markers. They are antigen specific and suppress activation of cytotoxic T cells but may stimulate B cells, for example to produce IgA. They may be thymically derived natural Tregs (nTregs) or induced iTregs. iTregs suppress T cell differentiation in both antigen-specific and nonspecific ways.[88]

Tregs suppress other T cells, NK cells, macrophages and dendritic cells.[89] They restrain autoimmunity and are the primary mediators of tolerance, for which reason they are highly relevant to the immunology of pregnancy, for example, essential for the success of murine pregnancy.[90] Tregs act in tissues by direct cell contact, hence, their concentrations in peripheral blood at best are only indirectly informative. Of interest is that their levels are stimulated by estrogens, with increases in both the follicular phase of the menstrual cycle[91] and early pregnancy.[92]

Tregs in the decidua are probably the most relevant, where they are enriched relative to concentrations in peripheral blood.[93] It is speculated that decidual Tregs downregulate maternal responses to HLA-C-incompatible extravillous cytotrophoblast.[94] This is a very important concept; if confirmed, it points to the involvement of T cells in partner-specific immunity, the involvement of T cell memory, and a potential mechanism for the second pregnancy phenomenon of preeclampsia. To date, in preeclampsia, only peripheral blood Tregs have been studied. The results are inconsistent: both reduced[95] and unchanged[96] numbers have been reported. However, the importance of Tregs, particularly those residing at Interface 1 in Stage 1 preeclampsia, has not been adequately assessed. Here, they would be predicted to play an important role in the control of placentation. Tregs have a reciprocal relationship with Th17 cells, the latter promoting autoimmunity.[97] Both T cell subtypes require TGFβ for differentiation from naïve T cells, and therefore share many early features in their transcriptional programs. Pathogenic Th17 cells, in addition, need IL-6. High concentrations of TGFβ favor Treg differentiation whereas lower concentrations combined with IL-6 promote Th17 cell production.[81] Both Th1 and Th2 cells inhibit their development. There is no change in the proportion of circulating Th17 lymphocytes during normal pregnancy.[98]

A key feature of Tregs and Th17 cells is their plasticity.[88] Th17 cells can transform into Th1 cells, while Tregs can transform into Th17, although the converse does not happen.[88] Thus the Th biases or states that are typical of normal pregnancy or preeclampsia are dynamic.

Preeclampsia has been associated with a Th1[83] and Th17[99] bias in peripheral blood lymphocytes with a relative deficiency in Tregs.[100,101] Since Th17 cells are promoted by lower availability of TGFβ it is possible that circulating sEng (a TGFβ-binding protein; see Chapter 6), interferes with T cell differentiation. TGFβ is proposed to promote differentiation of Th17 cells at the expense of Tregs, contributing to a proinflammatory environment with increased IL-17.[99]

In preeclampsia the specific Type 2 bias of normal pregnancy is lost with increased production of IFN-γ. and IL-2.[71,72] Circulating IL-17 concentrations appear to be unchanged[102] despite an increase in Th17 cell counts. There are increases in a range of other cytokines, chemokines, adipokines and antiangiogenic factors, which all reflect an inflammatory response. Circulating adipokines with cytokine activity, either proinflammatory (leptin[51]) or antiinflammatory (adiponectin[103]) activities are also increased in preeclampsia.

Finally, the central issue for this chapter is whether partner-specific memory develops and persists to enhance maternal immune tolerance in the next pregnancy by the same partner. Whereas such memory could be associated with NK cells, whose responses to specific stimuli can be "educated" or "licensed,"[104] memory is classically associated with T cells. In relation to pregnancy, what needs to be sought is partner-specific memory carried by Tregs. There is no evidence for such in humans but there is in mice. Mating-specific Tregs accumulate in the secondary lymphoid tissues of pregnant mice, persist after pregnancy, and give an accelerated response in a second pregnancy. If a similar mechanism applied to human pregnancy this would be the most likely explanation for the "preeclampsia second pregnancy effect."[105]

ANGIOTENSIN II (ANG II), THE IMMUNE SYSTEM AND PREECLAMPSIA

Ang II is a classical mediator of hypertension and of salt and fluid retention. It is also an important contributor to hypertensive end-organ damage. Its pathophysiology in preeclampsia is discussed extensively in Chapter 15. As already mentioned, in addition to its vasoactive properties, Ang II is potently proinflammatory. It activates NF-κB through the AT1R (angiotensin II type 1 receptor) to induce inflammatory cytokines, chemokines, and growth factors that mediate the tissue damage.[106] In fact, Ang II affects all aspects of immunity, innate and adaptive, cellular and humoral.[107] For example T cells and NK cells both have their own functional renin-angiotensin systems.[108] Furthermore at least some of the hypertension induced by Ang II depends on its direct action on T cells. An indirect effect, possibly involving Ang II stimulation of dendritic cells, may also contribute. The specific T cells involved are atypical, being negative for CD4 and CD8, and represent an inflammatory subtype that produces IL-17.[109]

In preeclampsia the renin-angiotensin system is typically suppressed relative to normal pregnancy.[110] But an alternative ligand of ATR1 is its agonistic autoantibody (AT-AA), which is highly associated with the disorder[111] and potentially contributes to its hypertension and other features (see Chapter 15). It represents a more general phenomenon of autoantibodies (agonistic or antagonistic) against G-protein-coupled receptors, for example TSH receptor antibodies in relation to Graves disease.[112]

As indicated above, a Th17 bias, such as occurs in preeclampsia, would be expected to stimulate autoimmunity. In addition experimental, Ang II-induced hypertension is not sustained in IL-17(−/−) mice,[113] indicating possible participation of IL-17 not only before but after the appearance of such antibodies. But the precise mechanisms of the generation of this antibody are not known. Its activity declines relatively quickly (within the first two weeks after pregnancy) but does not disappear[111] and persists to cause recurrent problems in further pregnancies. Hence AT-AA would be predicted to cause recurrent problems in later pregnancies, as do antiphospholipid autoantibodies. It is not known whether AT-AA are present before conception of eventually preeclamptic pregnancies, but their persistence would not be consistent with the first-pregnancy preponderance of preeclampsia.

SYSTEMIC IMMUNOREGULATION IN NORMAL PREGNANCY AND PREECLAMPSIA

Humoral factors also contribute to immunoregulation during normal pregnancy. Indoleamine 2,3-dioxygenase (IDO) is a widely distributed enzyme that catabolizes tryptophan. It is characterized best in monocytes and macrophages but can be produced by other cell types, such as endothelial cells. It is primarily induced by IFN-γ, but other proinflammatory stimulants are also effective. By reducing the availability of tryptophan, indoleamine 2,3-dioxygenase inhibits immune cell function (summarized in[95]). IDO is reduced in preeclampsia as judged by direct and indirect measures in the placenta or indirect measures of maternal plasma.[95] The effect would be to diminish the inhibitory action of tryptophan depletion, resulting in immune cell activation.

HLA-G is a non-classical HLA molecule expressed almost exclusively by non-villous trophoblast. Unlike classical HLA molecules it has limited polymorphism (reviewed by Apps et al.[18]). It is immunosuppressive rather than stimulatory (reviewed by Sargent[96]) by way of its interactions with receptors expressed by NK cells, antigen-presenting cells and some T cells.[102] Expression of HLA-G by extravillous cytotrophoblast is reduced in preeclampsia[114] and may contribute by undefined immune mechanisms to the associated restricted invasion of spiral arteries in the placental bed (see Chapter 5). Soluble HLA-G is released by protelolytic cleavage of the membrane-bound protein.[115] It is antiangiogenic[116] and appears to be immunosuppressive[116] and is reduced in preeclampsia,[117] consistent with the view that downregulation of the immune system is impaired in this condition.

ACUTE ATHEROSIS: A SECOND INFLAMMATORY LESION OF PREECLAMPSIA

Acute atherosis is a uteroplacental spiral artery lesion, characterized by subendothelial lipid-filled foam cells, fibrinoid necrosis and leukocyte infiltration.[118] The foam cells are predominantly CD68-positive macrophages. Acute atherosis most often occurs distally, downstream of inadequately remodeled spiral arteries in the myometrium, namely at the fully remodeled tips of their decidual segments. The lesion is focal, not necessarily affecting all spiral arteries or their entire length. The time course of acute atherosis is not well defined. It appears to regress after delivery.

Its incidence depends on patient selection and tissue sampling.[119] It affects 20–40% of preeclamptic women and is associated with more severe forms of the disorder.[120] It also occurs without preeclampsia – in normotensive fetal growth restriction or with underlying medical problems such as diabetes mellitus. It is an incidental finding in up to 15% of normal pregnancies.[121]

The arterial wall lesion can reduce spiral arterial caliber and blood flow, which would be expected to exacerbate placental dysfunction and oxidative stress especially as it is associated with spiral arterial thrombosis and placental infarction (reviewed in[121,122]).

Acute atherosis resembles atherosclerosis, which is a chronic inflammatory lesion of large and medium-sized arteries, characterized by lipid deposition and oxidative stress.[123] The trigger seems to be chronic, focal endothelial activation from the shear stress of turbulent blood flow[124] or other undefined factors. Enhanced focal endothelial adhesiveness attracts immune cells, including monocyte/macrophages, to the subendothelium, where increased endothelial permeability enables lipoproteins, with their cargo of lipids, to be scavenged by macrophages that transform into foam cells. Foam cells are not specific to atherosclerosis, but form in other inflammatory states, such as tuberculomas or graft rejection (see below).

Other than macrophages, dendritic, T and B cells, as well as NKT cells and possibly NK cells contribute to maturation of atherosclerosis. Th1 cells are proatherogenic, while subsets of T regulatory cells (Tregs) are protective (reviewed by[122]). This implies that specific antigens are important.

Atherosclerosis has a longer time course, with terminal plaque formation and rupture. Acute atherosis affects only small spiral arteries as does the atherosis of graft vascular disease (GVD), which complicates the integrity of solid allografts (such as of the kidney or heart), even when donors and recipients are HLA-compatible.[125] GVD depends on both innate and adaptive immune responses.[125] The similarities between acute atherosis and GVD have been highlighted previously and include fibrinoid necrosis (medial smooth muscle cells), intimal hyperplasia, and foam cell formation.[126] Intimal hyperplasia is more prominent in GVD[125] than in acute atherosis. Both lesions involve perivascular lymphocyte infiltration, complement and immunoglobulin deposits and occur at the boundaries between tissues from genetically different individuals. Both forms of atherosis are associated with conditions where there are circulating agonistic AT-AA.[127]

Whatever its cause it is reasonable to conclude that acute atherosis is a secondary consequence of the vascular inflammation that occurs in preeclampsia and even normal pregnancy and may reflect a true, low-grade maternal immune rejection of the fetus. Its development is likely to involve maternal specific constitutional factors. By its impact on uteroplacental perfusion it adds a further stage to the pathogenesis of preeclampsia, affecting a subset of all mothers (Fig. 8.4).

THE ROLE OF THE PLACENTA AND NON-PLACENTAL FACTORS

Since the placenta is ultimately the cause of preeclampsia (see Chapter 5) it would be predicted that one or more factors released from the syncytial surface of the placenta into the maternal circulation are the proinflammatory causes.

It is possible that syncytiotrophoblast could be a direct source of proinflammatory cytokines. Of most interest is secretion of VEGF, which is not only angiogenic, but strongly proinflammatory[128]; this is to be expected since

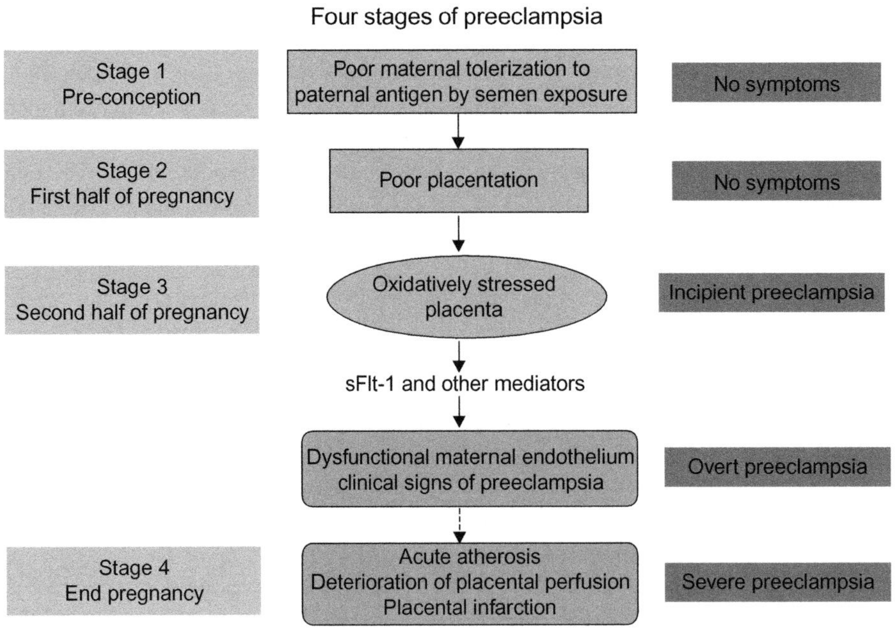

FIGURE 8.4 Two further stages are added to this adaptation of the two-stage model. Pre-conceptual maternal tolerization to partner's antigens is achieved by pre-conceptual exposure to paternal semen (new stage 1). Stages 2 and 3 are, respectively, stages 1 and 2 of the original two-stage model. Stage 4 comprises the development of acute atherosis in the spiral arteries which causes further adverse effects on uteroplacental perfusion, particularly from thrombosis. Acute atherosis probably affects less than half of preeclampsia cases. (This figure is reproduced in color in the color plate section.)

angiogenesis is driven by inflammatory mechanisms. VEGF is strongly expressed in syncytiotrophoblast and upregulated by hypoxia. Its circulating levels are measured as either higher or lower than normal in preeclampsia: total VEGF is increased, but free VEGF is reduced, because it induces, and is bound to, a circulating soluble receptor (sVEGFR-1 or sFlt-1). sVEGFR-1 is also produced by the placenta (see Chapter 6).

Whereas sVEGFR-1 is antiinflammatory in other medical contexts,[129] sVEGFR-1 release from trophoblast[130] and macrophages[129] is enhanced by inflammatory stimuli. While hypoxia has previously been given prominence as the trigger for release from the preeclampsia placenta (Chapter 6), inflammatory mechanisms may contribute or even predominate, especially as the activation of the hypoxia response requires the action of the NF-κB. The latter transcriptionally regulates stress responses, including those mediated by HIF-1α.[74]

VEGF-R1 is also the primary receptor for placental growth factor (PlGF), which is another cytokine and angiogenic factor produced by the placenta. Its circulating concentrations are reduced in preeclampsia (see Chapter 6). One relevant function of PlGF is, like VEGF, proinflammatory. It can activate monocytes and increase mRNA levels of proinflammatory cytokines (TNF-α, IL-1β) and chemokines.[131] However, in that PlGF production is diminished, it does not appear to be a placental factor contributing to the excessive inflammatory response of preeclampsia.

Agonistic AT-R1 autoantibodies would be expected, as does Ang II. to stimulate vascular NADPH oxidase and cause oxidative stress, reducing the bioavailability of nitric oxide by uncoupling its synthesis. In this context dysregulation of nitric oxide availability ascribable to VEGF antagonism is likely to be only one part of a bigger picture in which oxidative and inflammatory stresses are critical. The proximal cause of sVEGFR-1-induced hypertension is endothelin,[132] a classical vasoconstrictor and inflammatory mediator. Circulating endothelin-1 is increased in preeclampsia relative to normal pregnancy.[132,133] It is worth remembering that most endothelin is produced from the abluminal aspect of endothelial cells[134] where it reacts locally with vascular smooth muscle. So circulating levels are unlikely to be more than distant reflections of important constrictor mechanisms. Paradoxically, altlhough sVEGF-R1 might be expected to be antiinflammatory because it is antiangiogenic, in the context of oxidative stress it becomes proinflammatory because it amplifies vascular NO deficiency in endothelium and/or vascular smooth muscle.

There are several other circulating factors derived from syncytiotrophoblast that could affect systemic inflammation or angiogenesis or both. They are summarized in Table 8.7. Note that their circulating levels are increased in normal pregnancy and in, most instances, increased further in preeclampsia, consistent with the hypothesis that pregnancy is an intermediate state between non-pregnancy and preeclampsia. Note also that they include not only classical cytokines, but growth factors and adipokines. Their relative contributions to systemic inflammation in normal pregnancy have yet to be defined.

TABLE 8.7 Inflammatory Responses and Factors Secreted by Syncytiotrophoblast

Factor	Secreted by Trophoblast	*Pro (↑) or Anti (↓) Inflammatory	Normal Pregnancy Relative to Non-Pregnancy	Preeclampsia Relative to Normal Pregnancy
PlGF	✓Chapter 6	↑[130]	↑Chapter 6	↓Chapter 6
sVEGFR-1	✓Chapter 6	Variable	↑Chapter 6	↑Chapter 6
s-Endoglin	✓Chapter 6	Not defined	↑Chapter 6	↑Chapter 6
Activin-A	✓[135]	↑[136]	↑[137]	↑[138]
Inhibin-A	✓[135]	↓[136]	↑[137]	↑[138]
CRH	✓[139]	↑ circulating[38]	↑[140]	↑[140]
Leptin	✓[141]	↑[51]	↑[141]	↑[142]

TROPHOBLAST EXTRACELLULAR VESICLES

Cell membrane microvesicles are shed during apoptosis or cell activation. In healthy individuals, microvesicles from platelets, endothelium, neutrophils, and erythrocytes normally circulate, with platelet microvesicles being the most abundant. In inflammatory conditions they increase. Their concentrations are listed in Table 8.8. Microvesicles may be proinflammatory, antiinflammatory or procoagulant (reviewed by Redman and Sargent[144]). Procoagulant platelet-derived microvesicles (platelet dust) can disseminate and amplify localized clotting. The syncytial surface of the placenta is a transitory intrusion into the maternal circulation, where it sheds a wide range of cellular and subcellular debris, including syncytiotrophoblast microvesicles (reviewed in[145]). We have proposed that clearance of the debris provokes a systemic inflammatory stimulus in normal and preeclamptic pregnancies.

Syncytiotrophoblast microvesicles are shed in significantly increased amounts in preeclampsia, but not in normotensive intrauterine growth restriction.[146] They have a profound antiendothelial effect[147] and are also proinflammatory *in vitro*.[148]

Shedding of debris from the syncytial surface would be expected to increase in two situations. The first is with increased placental size. Preeclampsia is predominantly a disorder of the third trimester, when the placenta reaches its greatest size. The prevalence of preeclampsia also increases with multi-fetal pregnancies and with placental oxidative stress. Severe preeclampsia, typically of early onset, is associated with abnormally small placentas and intense fetal growth restriction. These are likely to be affected by an alteration in the quality of inflammatory stimuli, for example, increased content of peroxidized placental lipids.[149]

The maternal circulation also contains trophoblast-derived nanovesicles, more correctly termed exosomes. These are too small (~100 nm) to be reliably detected by conventional flow cytometry.[150] They are considered to be physiological intercellular signals and are notable for their cargo of microRNAs, which are small non-coding molecules that regulate gene expression, transcriptionally and post-transcriptionally. Little is known about differences in circulating exosomes in preeclamptic relative to normal pregnancies. miRNAs are emerging that are biomarkers for preeclampsia, for example, miR-210,[151] but none that have been linked so far to inflammatory or immune functions.

TABLE 8.8 Circulating Microvesicles in Healthy Non-Pregnant Individuals (n=15)

Platelets	237×10^6/L
Endothelial cells	64×10^6/L
Granulocytes	46×10^6/L
Erythrocytes	28×10^6/L

Berckmans et al. 2001.[143]
Constitute 5–50 µg protein/mL of plasma.

MATERNAL PREDISPOSING FACTORS

Some medical conditions are well known to predispose to preeclampsia, including obesity, diabetes, and chronic hypertension. The confluence of these three signs is referrred to as the metabolic syndrome and its relationship with preeclampsia is explored extensively in Chapter 7. The effect of these medical conditions is to elevate the prepregnancy baseline of systemic inflammation upon which the changes of pregnancy are superimposed (Fig. 8.5). We propose that, in pregnancy complicated by such conditions, the decompensation from excessive systemic inflammation will happen earlier, accounting for the predisposition of affected women to preeclampsia. It will also persist from pregnancy to pregnancy, leading to recurrent preeclampsia.

Low-grade systemic inflammation is also a feature when obesity,[152] diabetes,[153] and chronic hypertension[154] coexist in men or non-pregnant women. Thus, it is not surprising that the metabolic syndrome predisposes to preeclampsia.[154] Indeed the metabolic effects of pregnancy, with or without preeclampsia, have been likened to this syndrome.[155,156]

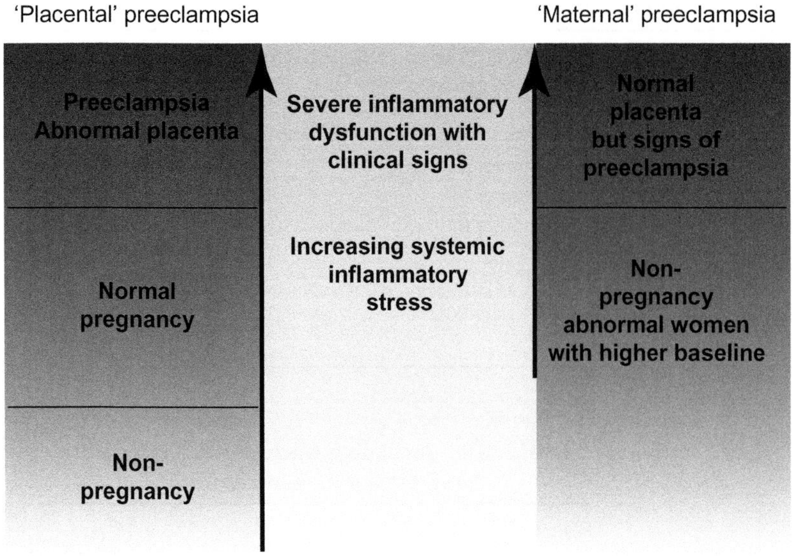

FIGURE 8.5 Systemic inflammation in pregnancy with and without conditions that cause a chronic systemic inflammatory response such as diabetes, obesity, chronic hypertension or chronic maternal infection. These conditions predispose to preeclampsia as the inflammation of pregnancy starts from a higher baseline. Because of this they also predispose to preeclampsia in subsequent pregnancies.

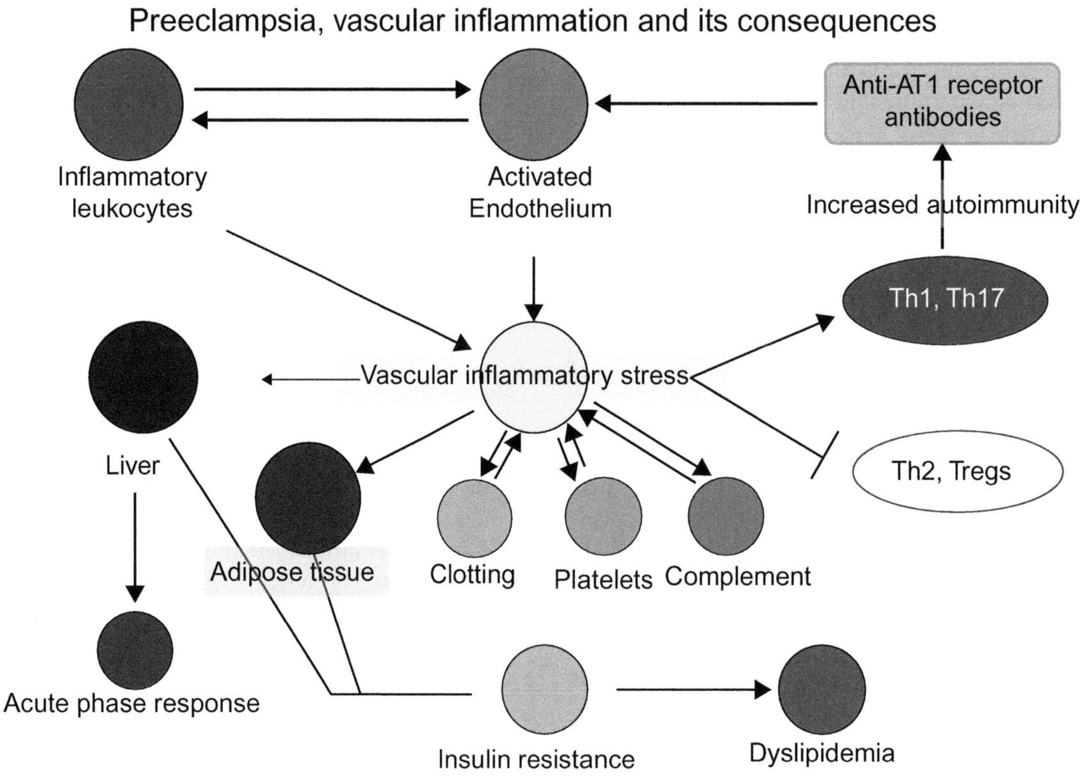

FIGURE 8.6 Vascular inflammation involves not only inflammatory leukocytes, but the endothelium, platelets, clotting and complement systems. Metabolic changes lead to insulin resistance and dyslipidemia. The inflammatory stress deviates away from the Th2 bias of normal pregnancy towards Th1/Th17. Th17 and loss of Tregs predispose to autoimmunity and may help stimulate formation of anti-AT1 receptor autoantibodies. These are agonistic and therefore proinflammatory. These processes contribute towards many recognized features of preeclampsia. (This figure is reproduced in color in the color plate section.)

CONCLUSIONS

Normal pregnancy involves two broad aspects of maternal immune function: adaptive immunity, with recognition and accommodation of the genetically foreign fetus (tolerization), and innate immunity, with vascular inflammation secondary to the stresses imposed by normal and abnormal placentas. There is evidence that both contribute to the pathogenesis of preeclampsia. Innate immunity has no memory and cannot therefore explain the first-pregnancy preponderance and possible partner specificity of preeclampsia. These features imply maternal tolerization to fetal (paternal) antigens and suggest the existence of fetus-specific maternal T regulatory cells, which persist after pregnancy as memory cells (adaptive immunity). These have been demonstrated in murine but not yet in human pregnancies.

To accommodate all immune aspects into a coherent model of preeclampsia, we have extended the classical two-stage model of preeclampsia to include four stages (Fig. 8.4). Before pregnancy there may be poor tolerization to paternal antigens, due to limited exposure to seminal antigens (Stage 1). At the end of pregnancy uteroplacental circulation may be impaired by acute atherosis (Stage 4). Immune recognition of the foreign fetus occurs at maternal-fetal Interface 1. Dysregulated recognition leads to poor placentation (new Stage 2) and abnormal uteroplacental perfusion.

Normal pregnancy imposes substantial systemic inflammatory stress on all pregnant women in the second half of pregnancy, which is associated with a Th2 immune bias. In preeclampsia the vascular inflammatory response is excessive with a Th1/Th17 bias. Many features that are considered to be physiological responses to pregnancy are best considered to be part of an acute-phase inflammatory response. Angiogenic imbalance, created by factors released from the oxidatively stressed placenta, adds to this burden. Th17 cells promote autoimmunity and may predispose to the generation of anti-ATR1 autoantibodies, which by their agonistic activity are strongly proinflammatory and exacerbate vascular oxidative stress.

In Stage 3, oxidative and inflammatory stresses coalesce in a vascular inflammatory response. The many consequences affect numerous circulating components of the inflammatory network and cause metabolic shifts including insulin resistance and dyslipidemia (Fig. 8.6). At the end of pregnancy the uteroplacental circulation may be impaired by acute atherosis (Stage 4), which may result from localized arterial inflammation (adaptive immunity) as in atherosclerosis or even graft vascular disease (innate immunity).

References

1. Billington WD. The immunological problem of pregnancy: 50 years with the hope of progress. A tribute to Peter Medawar. *J Reprod Immunol*. 2003;60:1–11.
2. Feeney JG, Scott JS. Pre-eclampsia and changed paternity. *Eur J Obstet Gynecol Reprod Biol*. 1980;11:35–38.
3. Robillard PY, Hulsey TC, Alexander GR, Keenan A, de Caunes F, Papiernik E. Paternity patterns and risk of pre-eclampsia in the last pregnancy in multiparae. *J Reprod Immunol*. 1993;24:1–12.
4. Zhang J, Patel G. Partner change and perinatal outcomes: a systematic review. *Paediatr Perinat Epidemiol*. 2007;21(Suppl 1): 46–57.
5. Basso O, Christensen K, Olsen J. Higher risk of pre-eclampsia after change of partner. An effect of longer interpregnancy intervals? *Epidemiology*. 2001;12:624–629.
6. Skjaerven R, Wilcox AJ, Lie RT. The interval between pregnancies and the risk of preeclampsia. *N Engl J Med*. 2002;346:33–38.
7. Zhang J. Partner change, birth interval and risk of pre-eclampsia: a paradoxical triangle. *Paediatr Perinat Epidemiol*. 2007;21(Suppl 1):31–35.
8. Saito S, Sakai M, Sasaki Y, Nakashima A, Shiozaki A. Inadequate tolerance induction may induce pre-eclampsia. *J Reprod Immunol*. 2007;76:30–39.
9. Robillard PY, Hulsey TC. Association of pregnancy-induced-hypertension, pre-eclampsia, and eclampsia with duration of sexual cohabitation before conception. *Lancet*. 1996;347 619–619.
10. Robillard PY, Dekker G, Chaouat G, Hulsey TC. Etiology of preeclampsia: maternal vascular predisposition and couple disease–mutual exclusion or complementarity? *J Reprod Immunol*. 2007;76:1–7.
11. Dekker G. The partner's role in the etiology of preeclampsia. *J Reprod Immunol*. 2002;57:203–215.
12. Kho EM, McCowan LM, North RA. Duration of sexual relationship and its effect on preeclampsia and small for gestational age perinatal outcome. *J Reprod Immunol*. 2009;82:66–73.
13. Matzinger P. The danger model: a renewed sense of self. *Science*. 2002;296:301–305.
14. Gordon S. Pattern recognition receptors: doubling up for the innate immune response. *Cell*. 2002;111:927–930.
15. Harding HP, Zhang Y, Zeng H, et al. An integrated stress response regulates amino acid metabolism and resistance to oxidative stress. *Mol Cell*. 2003;1:619–633.
16. Ljunggren HG, Karre K. In search of the missing self: MHC molecules and NK recognition. *Immunol Today*. 1990;11:237–244.
17. Hunt JS. Stranger in a strange land. *Immunol Rev*. 2006;213:36–47.
18. Apps R, Gardner L, Moffett A. A critical look at HLA-G. *Trends Immunol*. 2008;29:313–321.
19. Redman CW, Sargent IL. Placental stress and pre-eclampsia: a revised view. *Placenta*. 2009;30(Suppl A):S38–S42.
20. Sargent IL, Borzychowski AM, Redman CW. NK cells and human pregnancy – an inflammatory view. *Trends Immunol*. 2006;27:399–404.
21. Redman CW. Current topic: pre-eclampsia and the placenta. *Placenta*. 1991;12:301–308.
22. Hanna J, Goldman-Wohl D, Hamani Y, et al. Decidual NK cells regulate key developmental processes at the human fetal-maternal interface. *Nat Med*. 2006;2:1065–1074.
23. Clements CS, Kjer-Nielsen L, McCluskey J, Rossjohn J. Structural studies on HLA-G: implications for ligand and receptor binding. *Hum Immunol*. 2007;68:220–226.

24. Parham P. MHC class I molecules and KIRs in human history, health and survival. *Nat Rev Immunol*. 2005;5:201–214.
25. Chazara O, Xiong S, Moffett A. Maternal KIR and fetal HLA-C: a fine balance. *J Leukoc Biol*. 2011;90:703–716.
26. Rajagopalan S, Long EO. Understanding how combinations of HLA and KIR genes influence disease. *J Exp Med*. 2005;201:1025–1029.
27. Moffett A, Hiby SE. How Does the maternal immune system contribute to the development of pre-eclampsia? *Placenta*. 2007;28(Suppl A):S51–S56.
28. Wilczyński JR. Immunological analogy between allograft rejection, recurrent abortion and pre-eclampsia – the same basic mechanism? *Hum Immunol*. 2006;67:492–511.
29. Hiby SE, Apps R, Sharkey AM, et al. Maternal activating KIRs protect against human reproductive failure mediated by fetal HLA-C2. *J Clin Invest*. 2010;120:4102–4110.
30. van Bergen J, Koning F. The tortoise and the hare: slowly evolving T-cell responses take hastily evolving KIR. *Immunology*. 2010;131:301–309.
31. Tilburgs T, Roelen DL, van der Mast BJ, et al. Differential distribution of CD4(+)CD25(bright) and CD8(+)CD28(−) T-cells in decidua and maternal blood during human pregnancy. *Placenta*. 2006;27(Suppl A):S47–S53.
32. Redman CW, Sacks GP, Sargent IL. Preeclampsia: an excessive maternal inflammatory response to pregnancy. *Am J Obstet Gynecol*. 1999;180:499–506.
33. Redman CW, Sargent IL, Staff AC. Making sense of pre-eclampsia: two placental causes of preeclampsia? *Placenta*. 2014;35(Suppl):S20–S25.
34. Roberts JM, Taylor RN, Musci TJ, Rodgers GM, Hubel CA, McLaughlin MK. Preeclampsia: an endothelial cell disorder. *Am J Obstet Gynecol*. 1989;161:1200–1204.
35. Stokes KY, Granger DN. The microcirculation: a motor for the systemic inflammatory response and large vessel disease induced by hypercholesterolaemia? *J Physiol*. 2005;562:647–653.
36. Rath E, Haller D. Inflammation and cellular stress: a mechanistic link between immune-mediated and metabolically driven pathologies. *Eur J Nutr*. 2011;50(4):219–233.
37. Noonan DM, De Lerma Barbaro A, Vannini N, Mortara L, Albini A. Inflammation, inflammatory cells and angiogenesis: decisions and indecisions. *Cancer Metastasis Rev*. 2008;27:31–40.
38. Hensley K, Robinson KA, Gabbita SP, Salsman S, Floyd RA. Reactive oxygen species, cell signaling, and cell injury. *Free Radic Biol Med*. 2000;28:1456–1462.
39. Bastard JP, Maachi M, Lagathu C, et al. Recent advances in the relationship between obesity, inflammation, and insulin resistance. *Eur Cytokine Netw*. 2006;17:4–12.
40. Ahn KS, Aggarwal BB. Transcription factor NF-kappaB: a sensor for smoke and stress signals. *Ann N Y Acad Sci*. 2005;1056:218–233.
41. Dodson M, Darley-Usmar V, Zhang J. Cellular metabolic and autophagic pathways: traffic control by redox signalling. *Free Radic Biol Med*. 2013;63:207–221.
42. Gregor MF, Hotamisligil GS. Thematic review series: adipocyte biology adipocyte stress: the endoplasmic reticulum and metabolic disease. *J Lipid Res*. 2007;48:1905–1914.
43. Hotamisligil GS. Endoplasmic reticulum stress and the inflammatory basis of metabolic disease. *Cell*. 2010;140:900–917.
44. Trayhurn P, Wood IS. Adipokines: inflammation and the pleiotropic role of white adipose tissue. *Br J Nutr*. 2004;92:347–355.
45. Ribatti D, Conconi MT, Nussdorfer GG. Nonclassic endogenous novel regulators of angiogenesis. *Pharmacol Rev*. 2007;59:185–205.
46. Suzuki Y, Ruiz-Ortega M, Lorenzo O, Ruperez M, Esteban V, Egido J. Inflammation and angiotensin II. *Int J Biochem Cell Biol*. 2003;35:881–900.
47. Zhao Q, Ishibashi M, Hiasa K, Tan C, Takeshita A, Egashira K. Essential role of vascular endothelial growth factor in angiotensin II-induced vascular inflammation and remodelling. *Hypertension*. 2004;44:264–270.
48. Dandona P, Aljada A, Chaudhuri A, Mohanty P, Rajesh G. A novel view of metabolic syndrome. *Metab Syndr Relat Disord*. 2004;2:2–8.
49. Sethi JK, Hotamisligil GS. The role of TNF alpha in adipocyte metabolism. *Semin Cell Dev Biol*. 1999;10:19–29.
50. Harris HW, Gosnell JE, Kumwenda ZL. The lipemia of sepsis: triglyceride-rich lipoproteins as agents of innate immunity. *J Endotoxin Res*. 2001;6:421–430.
51. Matarese G, Moschos S, Mantzoros CS. Leptin in immunology. *J Immunol*. 2005;174 3137–3131.
52. Isomaa B. A major health hazard: the metabolic syndrome. *Life Sci*. 2003;73:2395–2411.
53. Yoo JY, Desiderio S. Innate and acquired immunity intersect in a global view of the acute-phase response. *Proc Natl Acad Sci U S A*. 2003;100:1157–1162.
54. Hansson GK, Libby P, Schönbeck U, Yan ZQ. Innate and adaptive immunity in the pathogenesis of atherosclerosis. *Circ Res*. 2002;91:281–291.
55. Hotamisligil GS. Inflammation and metabolic disorders. *Nature*. 2006;444:860–867.
56. Gabay C, Kushner I. Acute-phase proteins and other systemic responses to inflammation. *New Engl J Med*. 1999;340:448–454.
57. Pepys MB, Hirschfield GM. C-reactive protein: a critical update. *J Clin Invest*. 2003;111:1805–1812.
58. Willis C, Morris JM, Danis V, Gallery ED. Cytokine production by peripheral blood monocytes during the normal human ovulatory menstrual cycle. *Hum Reprod*. 2003;18:1173–1178.
59. Studd JW, Blainey JD, Bailey DE. Serum protein changes in the pre-eclampsia-eclampsia syndrome. *J Obstet Gynaecol Br Commonw*. 1970;77:796–801.
60. Gatti L, Tenconi PM, Guarneri D, et al. Hemostatic parameters and platelet activation by flow-cytometry in normal pregnancy: a longitudinal study. *Int J Clin Lab Res*. 1994;24:217–219.
61. Redman CW, Sargent IL. Pre-eclampsia and the systemic inflammatory response. In: Belfort M, Lyall F, eds. *Pre-eclampsia – Aetiology and Clinical Practice*. Cambridge: Cambridge University Press; 2004:103–120.
62. Sacks GP, Seyani L, Lavery S, Trew G. Maternal C-reactive protein levels are raised at 4 weeks gestation. *Hum Reprod*. 2004;19:1025–1030.
63. Smarason AK, Gunnarsson A, Alfredsson JH, Valdimarsson H. Monocytosis and monocytic infiltration of decidua in early pregnancy. *J Clin Lab Immunol*. 1986;21:1–5.
64. Sacks GP, Studena K, Sargent K, Redman CW. Normal pregnancy and preeclampsia both produce inflammatory changes in peripheral blood leukocytes akin to those of sepsis. *Am J Obstet Gynecol*. 1998;179:80–86.

65. Vince GS, Starkey PM, Austgulen R, Kwiatkowski D, Redman CW. Interleukin-6, tumour necrosis factor and soluble tumour necrosis factor receptors in women with preeclampsia. *Br J Obstet Gynaecol*. 1995;102:20–25.
66. Pitkin RM, Witte DL. Platelet and leukocyte counts in pregnancy. *JAMA*. 1979;242:2696–2698.
67. Richani K, Soto E, Romero R, et al. Normal pregnancy is characterized by systemic activation of the complement system. *J Matern Fetal Neonatal Med*. 2005;17:239–245.
68. Chabloz P, Reber G, Boehlen F, Hohlfeld P, de Moerloose P. TAFI antigen and D-dimer levels during normal pregnancy and at delivery. *Br J Haematol*. 2001;115:150–152.
69. Janes SL, Goodall AH. Flow cytometric detection of circulating activated platelets and platelet hyper-responsiveness in preeclampsia and pregnancy. *Clin Sci Colch*. 1994;86:731–739.
70. Sørensen JD, Secher NJ, Jespersen J. Perturbed (procoagulant) endothelium and deviations within the fibrinolytic system during the third trimester of normal pregnancy A possible link to placental function. *Acta Obstet Gynecol Scand*. 1995;74:257–261.
71. Belo L, Caslake M, Santos-Silva A, et al. LDL size, total antioxidant status and oxidised LDL in normal human pregnancy: a longitudinal study. *Atherosclerosis*. 2004;177:391–399.
72. Little RE, Gladen BC. Levels of lipid peroxides in uncomplicated pregnancy: a review of the literature. *Reprod Toxicol*. 1999;13:347–352.
73. Victor VM, Rocha M, Esplugues JV, De la Fuente M. Role of free radicals in sepsis: antioxidant therapy. *Curr Pharm Des*. 2005;11:3141–3158.
74. Rius J, Guma M, Schachtrup C, et al. NF-kappaB links innate immunity to the hypoxic response through transcriptional regulation of HIF-1alpha. *Nature*. 2008;453:807–811.
75. Peritt D, Robertson S, Gri G, Showe L, Aste-Amezaga M, Trinchieri G. Differentiation of human NK cells into NK1 and NK2 subsets. *J Immunol*. 1998;161:5821–5824.
76. Redman CW, Sargent IL. Immunology of pre-eclampsia. *Am J Reprod Immunol*. 2010;63:534–543.
77. Terrone DA, Rinehart BK, May WL, Moore A, Magann EF, Martin NJ. Leukocytosis is proportional to HELLP syndrome severity: evidence for an inflammatory form of preeclampsia. *South Med J*. 2000;93:768–771.
78. de Messias-Reason IJ, Aleixo V, de Freitas H, Nisihara RM, Mocelin V, Urbanetz A. Complement activation in Brazilian patients with preeclampsia. *J Investig Allergol Clin Immunol*. 2000;10:209–214.
79. Conrad KP, Miles TM, Benyo DF. Circulating levels of immunoreactive cytokines in women with preeclampsia. *Am J Reprod Immunol*. 1998;40:102–111.
80. Ellis J, Wennerholm UB, Bengtsson A, et al. Levels of dimethylarginines and cytokines in mild and severe preeclampsia. *Acta Obstet Gynecol Scand*. 2001;80:602–608.
81. Basu R, Hatton RD, Weaver CT. The Th17 family: flexibility follows function. *Immunol Rev*. 2013;252:89–103.
82. Wegmann TG, Lin H, Guilbert L, Mosmann TR. Bidirectional cytokine interactions in the materal-fetal relationship: is successful pregnancy a Th2 phenomenon? *Immunol Today*. 1993;14:353–356.
83. Saito S, Sakai M, Sasaki Y, Tanebe K, Tsuda H, Michimata T. Quantitative analysis of peripheral blood Th0, Th1, Th2 and the Th1:Th2 cell ratio during normal human pregnancy and preeclampsia. *Clin Exp Immunol*. 1999;117:550–555.
84. Marzi M, Vigano A, Trabattoni D, et al. Characterization of type 1 and type 2 cytokine production profile in physiologic and pathologic human pregnancy. *Clin Exp Immunol*. 1996;106:127–133.
85. Wilczynski JR, Tchorzewski H, Banasik M, et al. Lymphocyte subset distribution and cytokine secretion in third trimester decidua in normal pregnancy and preeclampsia. *Eur J Obstet Gynecol Reprod Biol*. 2003;109:8–15.
86. Borzychowski AM, Croy BA, Chan WL, Redman CW, Sargent IL. Changes in systemic type 1 and type 2 immunity in normal pregnancy and pre-eclampsia may be mediated by natural killer cells. *Eur J Immunol*. 2005;35:3054–3063.
87. Sacks GP, Redman CW, Sargent IL. Monocytes are primed to express the Th1 type cytokine IL-12 in normal human pregnancy: an intracellular flow cytometric analysis of peripheral blood mononuclear cells. *Clin Exp Immunol*. 2003;131:490–497.
88. Corthay A. How do regulatory T cells work? *Scand J Immunol*. 2009;70:326–336.
89. Sakaguchi S, Yamaguchi T, Nomura T, Ono M. Regulatory T cells and immune tolerance. *Cell*. 2008;133:775–787.
90. Aluvihare VR, Kallikourdis M, Betz AG. Regulatory T cells mediate maternal tolerance to the fetus. *Nat Immunol*. 2004;5:266–271.
91. Arruvito L, Sanz M, Banham AH, Fainboim L. Expansion of CD4 + CD25 + and FOXP3+ regulatory T cells during the follicular phase of the menstrual cycle: implications for human reproduction. *J Immunol*. 2007;178:2572–2578.
92. Somerset DA, Zheng Y, Kilby MD, Sansom DM, Drayson MT. Normal human pregnancy is associated with an elevation in the immune suppressive CD25 CD4 regulatory T-cell subset. *Immunology*. 2004;112:38–43.
93. Mjösberg J, Berg G, Jenmalm MC, Ernerudh J. FOXP3+ regulatory T cells and T helper 1, T helper 2, and T helper 17 cells in human early pregnancy decidua. *Biol Reprod*. 2010;82:698–705.
94. Tilburgs T, Scherjon SA, van der Mast BJ, et al. Fetal-maternal HLA-C mismatch is associated with decidual T cell activation and induction of functional T regulatory cells. *J Reprod Immunol*. 2009;82:148–157.
95. Kudo Y, Boyd CA, Sargent IL, Redman CW. Decreased tryptophan catabolism by placental indoleamine 2,3-dioxygenase in preeclampsia. *Am J Obstet Gynecol*. 2003;188:719–726.
96. Sargent IL. Does 'soluble' HLA-G really exist? Another twist to the tale. *Mol Hum Reprod*. 2005;11:695–698.
97. Bettelli E, Carrier Y, Gao W, et al. Reciprocal developmental pathways for the generation of pathogenic effector TH17 and regulatory T cells. *Nature*. 2006;441:235–238.
98. Nakashima A, Ito M, Yoneda S, Shiozaki A, Hidaka T, Saito S. Circulating and decidual Th17 cell levels in healthy pregnancy. *Am J Reprod Immunol*. 2010;63:104–109.
99. Santner-Nanan B, Peek MJ, Khanam R, et al. Systemic increase in the ratio between Foxp3+ and IL-17-producing CD4+ T cells in healthy pregnancy but not in preeclampsia. *J Immunol*. 2009;183:7023–7030.
100. Sasaki Y, Darmochwal-Kolarz D, Kludka-Sternik M, et al. The predominance of Th17 lymphocytes and decreased number and function of Treg cells in preeclampsia. *J Reprod Immunol*. 2012;93(2):75–81.
101. Toldi G, Svec P, Vásárhelyi B, et al. Decreased number of FoxP3+ regulatory T cells in preeclampsia. *Acta Obstet Gynecol Scand*. 2008;87:1229–1233.

102. Jonsson Y, Ruber M, Matthiesen L, et al. Cytokine mapping of sera from women with preeclampsia and normal pregnancies. *J Reprod Immunol.* 2006;70:83–89.
103. Park PH, Huang H, McMullen MR, Mandal P, Sun L, Nagy LE. Suppression of lipopolysaccharide-stimulated tumor necrosis factor-alpha production by adiponectin is mediated by transcriptional and post-transcriptional mechanisms. *J Biol Chem.* 2008;283:26850–26858.
104. Long EO, Kim HS, Liu D, Peterson ME, Rajagopalan S. Controlling natural killer cell responses: integration of signals for activation and inhibition. *Annu Rev Immunol.* 2013;31:227–258.
105. Rowe JH, Ertelt JM, Xin L, Way SS. Pregnancy imprints regulatory memory that sustains anergy to fetal antigen. *Nature.* 2012;490(7418):102–106.
106. Li XC, Zhuo JL. Nuclear factor-kappaB as a hormonal intracellular signaling molecule: focus on angiotensin II-induced cardiovascular and renal injury. *Curr Opin Nephrol Hypertens.* 2008;17:37–43.
107. Luft FC, Dechend R, Müller DN. Immune mechanisms in angiotensin II-induced target-organ damage. *Ann Med.* 2012;44(Suppl 1):S49–S54.
108. Hoch NE, Guzik TJ, Chen W, et al. Regulation of T-cell function by endogenously produced angiotensin II. *Am J Physiol Regul Integr Comp Physiol.* 2009;296:R208–R216.
109. Crispín JC, Tsokos GC. Human TCR-alpha beta+ CD4- CD8- T cells can derive from CD8+ T cells and display an inflammatory effector phenotype. *J Immunol.* 2009;183:4675–4681.
110. Merrill DC, Karoly M, Chen K, Ferrario CM, Brosnihan KB. Angiotensin-(1–7) in normal and preeclamptic pregnancy. *Endocrine.* 2002;18(3):239–245.
111. Wallukat G, Homuth V, Fischer T, et al. Patients with preeclampsia develop agonistic antibodies against the angiotensin AT1 receptor. *J Clin Invest.* 1999;1103:945–952.
112. Dragun D, Philippe A, Catar R, Hegner B. Autoimmune mediated G-protein receptor activation in cardiovascular and renal pathologies. *Thromb Haemost.* 2009;101(4):643–648.
113. Madhur MS, Lob HE, McCann LA, et al. Interleukin 17 promotes angiotensin II-induced hypertension and vascular dysfunction. *Hypertension.* 2010;55:500–507.
114. Le Bouteiller P, Pizzato N, Barakonyi A, Solier C. HLA-G, pre-eclampsia, immunity and vascular events. *J Reprod Immunol.* 2003;59:219–234.
115. Blaschitz A, Juch H, Volz A, et al. The soluble pool of HLA-G produced by human trophoblasts does not include detectable levels of the intron 4-containing HLA-G5 and HLA-G6 isoforms. *Mol Hum Reprod.* 2005;11:699–710.
116. Fons P, Chabot S, Cartwright JE, et al. Soluble HLA-G1 inhibits angiogenesis through an apoptotic pathway and by direct binding to CD160 receptor expressed by endothelial cells. *Blood.* 2006;108:2608–2615.
117. Hackmon R, Koifman A, Hyodo H, Glickman H, Sheiner E, Geraghty DE. Reduced third-trimester levels of soluble human leukocyte antigen G protein in severe preeclampsia. *Am J Obstet Gynecol.* 2007;197(255):e1–e5.
118. Hanssens M, Pijnenborg R, Keirse MJ, Vercruysse L, Verbist L, Van Assche FA. Renin-like immunoreactivity in uterus and placenta from normotensive and hypertensive pregnancies. *Eur J Obstet Gynecol Reprod Biol.* 1998;81:177–184.
119. Harsem NK, Roald B, Braekke K, Staff AC. Acute atherosis in decidual tissue: not associated with systemic oxidative stress in preeclampsia. *Placenta.* 2007;28:958–964.
120. Stevens DU, Al-Nasiry S, Bulten J, Spaanderman ME. Decidual vasculopathy and adverse perinatal outcome in preeclamptic pregnancy. *Placenta.* 2013;34:805–809.
121. Staff AC, Dechend R, Pijnenborg R. Learning from the placenta. Acute atherosis and vascular remodeling in preeclampsia – novel aspects for atherosclerosis and future cardiovascular health. *Hypertension.* 2010;56:1026–1034.
122. Staff AC, Johnsen GM, Dechend R, Redman CW. Preeclampsia and uteroplacental acute atherosis: immune and inflammatory factors. *J Reprod Immunol.* 2013;S0165-0378(13):00105–00108.
123. Libby P. Inflammation in atherosclerosis. *Arterioscler Thromb Vasc Biol.* 2012;32:2045–2051.
124. Tsou JK, Gower RM, Ting HJ, et al. Spatial regulation of inflammation by human aortic endothelial cells in a linear gradient of shear stress. *Microcirculation.* 2008;15:311–323.
125. Mitchell RN. Graft vascular disease: immune response meets the vessel wall. *Annu Rev Pathol.* 2009;4:19–47.
126. Nickeleit V, Vamvakas EC, Pascual M, Poletti BJ, Colvin RB. The prognostic significance of specific arterial lesions in acute renal allograft rejection. *J Am Soc Nephrol.* 1998;9(7):1301–1308.
127. Herse F, Dechend R, Harsem NK, et al. Dysregulation of the circulating and tissue-based renin-angiotensin system in preeclampsia. *Hypertension.* 2007;49:604–611.
128. Angelo LS, Kurzrock R. Vascular endothelial growth factor and its relationship to inflammatory mediators. *Clin Cancer Res.* 2007;13:2825–2830.
129. Tsao PN, Chan FT, Wei SC, et al. Soluble vascular endothelial growth factor receptor-1 protects mice in sepsis. *Crit Care Med.* 2007;35:1955–1960.
130. Ahmad S, Ahmed A. Elevated placental soluble vascular endothelial growth factor receptor-1 inhibits angiogenesis in preeclampsia. *Circ Res.* 2004;95:884–891.
131. Selvaraj SK, Giri RK, Perelman N, Johnson C, Malik P, Kalra VK. Mechanism of monocyte activation and expression of proinflammatory cytochemokines by placenta growth factor. *Blood.* 2003;102:1515–1524.
132. Taylor RN, Varma M, Teng NNH, Roberts JM. Women with preeclampsia have higher plasma endothelin levels than women with normal pregnancies. *J Clin Endocrinol Metab.* 1990;71:1675–1677.
133. Bernardi F, Constantino L, Machado R, Petronilho F, Dal-Pizzol F. Plasma nitric oxide, endothelin-1, arginase and superoxide dismutase in pre-eclamptic women. *J Obstet Gynaecol Res.* 2008;34:957–963.
134. Kleinz MJ, Davenport AP. Emerging roles of apelin in biology and medicine. *Pharmacol Ther.* 2005;107:198–211.
135. Manuelpillai U, Schneider-Kolsky M, Thirunavukarasu P, Dole A, Waldron K, Wallace EM. Effect of hypoxia on placental activin A, inhibin A and follistatin synthesis. *Placenta.* 2003;24:77–83.
136. Jones KL, Mansell A, Patella S, et al. Activin A is a critical component of the inflammatory response, and its binding

protein, follistatin, reduces mortality in endotoxemia. *Proc Natl Acad Sci U S A*. 2007;104:16239–16244.
137. Fowler PA, Evans LW, Groome NP, Templeton A, Knight PG. A longitudinal study of maternal serum inhibin-A, inhibin-B, activin-A, activin-AB, pro-alphaC and follistatin during pregnancy. *Hum Reprod*. 1998;13:3530–3536.
138. Muttukrishna S, Knight PG, Groome NP, Redman CW, Ledger WL. Activin A and inhibin A as possible endocrine markers for pre-eclampsia. *Lancet*. 1997;349:1285–1288.
139. Perkins AV, Linton EA. Identification and isolation of corticotrophin-releasing hormone-positive cells from the human placenta. *Placenta*. 1995;16:233–243.
140. Perkins AV, Linton EA, Eben F, Simpson J, Wolfe CD, Redman CW. Corticotrophin-releasing hormone and corticotrophin-releasing hormone binding protein in normal and pre-eclamptic human pregnancies. *Br J Obstet Gynaecol*. 1995;102:118–122.
141. Zavalza-Gómez AB, Anaya-Prado R, Rincón-Sánche AR, Mora-Martínez JM. Adipokines and insulin resistance during pregnancy. *Diabetes Res Clin Pract*. 2008;80:8–15.
142. Ramsay JE, Ferrell WR, Crawford L, Wallace AM, Greer IA, Sattar N. Divergent metabolic and vascular phenotypes in pre-eclampsia and intrauterine growth restriction: relevance of adiposity. *J Hypertens*. 2004;22:2177–2183.
143. Berckmans RJ, Neiuwland R, Böing AN, Romijn FP, Hack CE, Sturk A. Cell-derived microvesicles circulate in healthy humans and support low grade thrombin generation. *Thromb Haemost*. 2001;85:639–646.
144. Redman CW, Sargent IL. Circulating microvesicles in normal pregnancy and pre-eclampsia. *Placenta*. 2008;29(Suppl A):S73–S77.
145. Redman CW, Sargent IL. Placental debris, oxidative stress and pre-eclampsia. *Placenta*. 2000;21:597–602.
146. Goswami D, Tannetta DS, Magee LA, et al. Excess syncytiotrophoblast microvesicle shedding is a feature of early-onset pre-eclampsia, but not normotensive intrauterine growth restriction. *Placenta*. 2006;27:56–61.
147. Smarason AK, Sargent IL, Starkey PM, Redman CW. The effect of placental syncytiotrophoblast microvillous membranes from normal and pre-eclamptic women on the growth of endothelial cells in vitro. *Br J Obstet Gynaecol*. 1993;100:943–949.
148. Southcombe J, Tannetta D, Redman C, Sargent I. The immunomodulatory role of syncytiotrophoblast microvesicles. *PLoS One*. 2011;6:e20245.
149. Cester N, Staffolani R, Rabini RA, et al. Pregnancy induced hypertension: a role for peroxidation in microvillus plasma membranes. *MolCell Biochem*. 1994;131:151–155.
150. Simons M, Raposo G. Exosomes – vesicular carriers for intercellular communication. *Curr Opin Cell Biol*. 2009;21:575–581.
151. Anton L, Olarerin-George AO, Schwartz N, et al. miR-210 inhibits trophoblast invasion and is a serum biomarker for preeclampsia. *Am J Pathol*. 2013;18:1437–1445.
152. Visser M, Bouter LM, McQuillan GM, Wener MH, Harris TB. Elevated C-reactive protein levels in overweight and obese adults. *JAMA*. 1999;282:2131–2135.
153. Pickup JC, Chusney GD, Thomas SM, Burt D. Plasma interleukin-6, tumour necrosis factor alpha and blood cytokine production in type 2 diabetes. *Life Sci*. 2000;67:291–300.
154. Lacy F, O' Connor DT, Schmid-Schönbein GW. Plasma hydrogen peroxide production in hypertensives and normotensive subjects at genetic risk of hypertension. *J Hypertens*. 1998;16:291–303.
155. Mazar RM, Srinivas SK, Sammel MD, Andrela CM, Elovitz MA. Metabolic score as a novel approach to assessing preeclampsia risk. *Am J Obstet Gynecol*. 2007;197(411):e1–e5.
156. Karalis K, Sano H, Redwine J, Listwak S, Wilder RL, Chrousos GP. Autocrine or paracrine inflammatory actions of corticotropin-releasing hormone in vivo. *Science*. 1991;254:421–423.

CHAPTER 9

Endothelial Cell Dysfunction

SANDRA T. DAVIDGE, CHRISTIANNE J.M. DE GROOT AND ROBERT N. TAYLOR

Editors' comment: *It was not until the early 1980s that physicians and scientists gained an appreciation of the physiological importance of the endothelium, that simple unicellular layer lining the luminal surface of blood vessels. Indeed, in Chesley's first, single-authored edition of this text, the reference to this term was as it related to the "endotheliosis" lesion of the renal glomerulus. We now recognize that endothelial cells are critical sensors of the milieu interieur and potent regulators of vascular tone, organ perfusion, and ischemia. The "endothelial hypothesis" of preeclampsia etiology provides for a convergence of several factors thought to play fundamental roles in its pathogenesis: leukocytes, cytokines, fatty acids, oxygen free radicals, placental microvesicles, "antiangiogenic" factors, and autoantibodies are all considered. Moreover, a variety of therapeutic interventions has been conceived based on the principle of endothelial cell protection, and the clinical trials emanating from these concepts are reviewed in the chapter.*

INTRODUCTION

The clinical manifestations of preeclampsia–eclampsia have been recognized since antiquity, but the pathophysiology of this syndrome remained completely obscure for nearly two millennia. Beginning in the mid-19th century, abnormal renal handling of nitrogen and water, referred to at that time as "dropsy," was first reported among eclamptic women.[1] Recognition that these signs were a manifestation of endothelial cell injury took another 90 years. Renal glomerular capillary endotheliosis, swelling of endothelial cytoplasm and obliteration of endothelial fenestrae, was originally observed in 1924 by Mayer,[2] described further by Bell,[3] and later refined by Spargo et al.[4] The latter investigator was among the first to apply electron microscopy in this condition and is credited for introducing the terminology "glomerular capillary endotheliosis" as a characteristic feature of preeclampsia.

Investigators have documented morphological evidence of vascular injury in various organs of preeclamptic women. Sheehan and Lynch[5] described hepatic periportal vascular lesions, characterized by arterial media infiltration and capillary thrombosis, in fatal cases of preeclampsia and eclampsia. Shanklin and Sibai[6] observed ultrastructural defects in the mitochondria of myometrial venules in women with preeclampsia, further supporting a widespread endothelial cell disorder in this syndrome. As described in Chapters 6 and 10, several animal models of preeclampsia, including those established in nonhuman primates,[7] have manifestations of systemic endothelial cell activation. Moreover, Grundmann et al.[8] demonstrated that concentrations of freely circulating endothelial cells, markers of vascular injury, were more than 4-fold higher in women with preeclampsia compared to normal pregnant controls.

Abnormal trophoblast invasion and impaired placental perfusion appear to be requisite precursors to the development of preeclampsia. If we imagine the placental bed as the root of the disease process, this might be a good place to begin to look for evidence of endothelial cell dysfunction. Indeed, the endothelium lining the uterine spiral arterioles, which normally undergoes denudation and replacement by invading endovascular cytotrophoblasts, may manifest the first pathological changes associated with preeclampsia.[9,10] The failure of trophoblasts to undergo "pseudovasculogenesis" and assume an "endothelial" phenotype in preeclampsia is discussed in the accompanying Chapters 5 and 6. Moreover, other vascular manifestations have been noted at this site. The acute atherosis lesion, initially described by Zeek and Assali[11] and refined by Nadji and Sommers[12] and DeWolf et al.,[13] demonstrates endothelial cell vacuolization, myointimal proliferation and foam cell infiltration of the tunica media. These histological changes also are observed in the vascular pathology of atherosclerosis[14]; the similarities suggest that endothelial cell dysfunction may be a mechanism responsible for diffuse vascular disease that may manifest in patients with preeclampsia.[15]

Endothelial cells are strategically positioned at the interface between circulating blood and vascular smooth muscle or the extravascular space, where they occupy a surface area of more than $1000\,m^2$. These cells are able to secrete a variety of signaling molecules directly into the

Chesley's Hypertensive Disorders in Pregnancy.
ISBN: 978-0-12-407866-6

DOI: http://dx.doi.org/10.1016/B978-0-12-407866-6.00009-2

circulation, potentially reaching every cell in the body.[16] In turn, endothelial cells are themselves targets for cellular and soluble plasma constituents (e.g., cytokines, lipoproteins, platelets, leukocytes, placental microvesicles, antibodies, and other circulating factors). Endothelial cells modulate vascular tone, coagulation, permeability, and the targeting of immune cells. Under normal, physiological conditions the endothelium maintains homeostatic balance. Vascular tone is controlled by vasoconstrictors (e.g., endothelin and thromboxane A2) and vasodilators (e.g., nitric oxide (NO), prostacyclin (PGI2) and endothelial-derived hyperpolarizing factors (EDHF)). Hemostasis is maintained in equilibrium by procoagulant and anticoagulant influences and endothelial tight junctions control vascular permeability. Maternal endothelial cell dysfunction is believed to result in vasospasm, microthrombosis, and vascular permeability, which lead to the classical signs and symptoms observed in women with preeclampsia.

Findings relevant to endothelial cell biology in preeclampsia will be reviewed in this chapter in five parts. In Part I, we review the evidence favoring endothelial cell dysfunction and present a pivotal hypothesis for the pathophysiology of preeclampsia. Part II reviews the concept that circulating factors causally induce disturbances in maternal endothelial cell function. Part III proposes that prooxidant principles lead to a common convergence that creates an oxidative stress environment within maternal endothelia of preeclamptic women. Part IV discusses clinical trials that have been focused on improving vascular function. Finally, in Part V we conclude with a summary of the findings, speculations about future diagnostic and therapeutic approaches to preeclampsia, and some suggestions for the direction of new investigations into the mechanisms of this enigmatic condition.

PART I: ENDOTHELIAL CELL FUNCTION AND PREECLAMPSIA

Over the past decade, the attention of scholars of preeclampsia has focused on the triad of impaired trophoblast invasion, uteroplacental ischemia, oxidative stress and generalized maternal endothelial cell activation or injury as major mechanistic factors in its pathogenesis.[17,18] It is likely that several etiologies underlie the development of this syndrome, but these pathogenetic mechanisms are shared. As described in Chapter 5 (and detailed further in its new Appendix), failure of trophoblast differentiation, invasion, and vascular remodeling of the placental bed are now believed to be critical initiating factors in the development of preeclampsia and reduced angiogenic factor production and action (Chapter 6).

We will review recent evidence that supports the concept that one of the most important targets of circulating "toxins" in preeclampsia is the maternal vascular endothelium, which ultimately leads to vascular oxidative stress. Indeed, the historical moniker "toxemia" should be revisited as an appropriate appellation for this syndrome. When clinically manifested in later pregnancy, the vessels of essentially all maternal organs are involved in preeclamptic women. Cerebral ischemia, edema and convulsions, pleural effusion and ascites, and hepatic dysfunction, in addition to the classical proteinuria and peripheral edema reflecting renal and subcutaneous vasculatures, respectively, are evidence of the many maternal vascular beds at risk. Endothelial cells within these compromised vessels not only lose their normal constitutive, homeostatic functions but also acquire new pathological properties (e.g., vasoconstrictor and procoagulant production) that increase vascular resistance and exacerbate ischemia. In this chapter we will present the hypotheses and evidence that support a major role of endothelial cell dysfunction due to a variety of factors in the circulation creating endothelial oxidative stress as a common downstream converging mechanism in the clinical manifestations of preeclampsia. Studies suggest that circulating factors are elaborated in response to uteroplacental ischemia. These factors are postulated to alter both maternal and placental endothelial cell phenotypes.

Endothelial Cell Dysfunction in Preeclampsia

Endothelial cell dysfunction and/or "activation" is a term used to define an altered state of endothelial cell differentiation, typically induced as a result of cytokine stimulation.[19] It represents an inflammatory response to sublethal injury of these cells and is proposed to play a major role in the pathophysiology of atherosclerosis.[20] Endothelial cell activation in preeclampsia may result from a variety of circulating factors, including many discussed in the previous chapters such as angiogenic factors, metabolic factors, and inflammatory mediators.

Endothelial activation is manifest biochemically by the synthesis and secretion of a variety of endothelial cell products including prostanoids, endothelin-1 (ET-1), platelet-derived growth factor (PDGF), fibronectin, selectins, and other molecules that influence vessel tone and remodeling.[21,22] When these factors are elaborated in response to acute mechanical or biochemical endothelial cell damage they facilitate efficient wound healing. However, when activated by a chronic pathological process, such as preeclampsia, these responses can create a vicious circle of vasospasm, microthrombosis, and disruption of vascular integrity, creating serious physiologic disturbances, which persist until the inciting factor(s) is eliminated.

Some insights into the vascular pathophysiology of preeclampsia are derived from clinical experience with thrombotic microangiopathic disorders occurring during pregnancy. It has been noted that, while rare, pregnancy appears to predispose or exacerbate the development of these syndromes of microvascular thrombosis.[23] Thrombotic thrombocytopenic purpura (TTP) is characterized by thrombocytopenia, microangiopathic hemolytic anemia, renal involvement, neurological symptoms, and fever. Like preeclampsia, the signs of TTP during pregnancy most commonly manifest in the late second trimester.[24] Intravascular platelet agglutination, widespread hyaline thrombosis, and endothelial cell proliferation within capillaries and arterioles have been observed. PGI_2 normally suppresses platelet aggregation induced by cytokines and shear stress, but this vasodilator eicosanoid is reduced in TTP,[25,26] in part, from endothelial apoptosis and also from accelerated metabolism.

Circulating Markers of Endothelial Cell Activation

Procoagulant Proteins and Plasminogen Activators

Preeclampsia is characterized by a maternal hypercoagulable state, resulting in intravascular coagulation, microthromboses in several organs including the placenta, with further impairment of the uteroplacental circulation. The clinical significance of this problem is reviewed in Chapter 17, but hypercoagulability also underlies the principle of endothelial cell dysfunction in preeclampsia. Excessive fibrin deposition in the placenta was reported by Kitzmiller and Benirschke,[27] suggesting that a disordered balance of placental coagulation and fibrinolysis may play a role in the activation of hemostasis. Roberts et al. posited that the hypercoagulable state was due, in part, to diffuse endothelial cell activation.[15] The reduced expression of several relevant endothelial cell-associated anticoagulant proteins has been shown in preeclampsia, including antithrombin III, protein C, and protein S. The significance of these proteins as markers of maternal endothelial cell dysfunction is reviewed below.

Resistance to activated protein C is an inherited mutation of the coagulation factor V gene. The presence of the factor V Leiden mutation predisposes to thromboembolic events and its prevalence is increased in women with severe, early-onset preeclampsia compared to women with normal pregnancies.[28,29] In addition, in women with severe, early-onset preeclampsia, 25% had evidence of functional protein S deficiency, 18% demonstrated hyperhomocysteinemia, and 29% had detectable anticardiolipin IgG and IgM antibodies.[30,31]

Increased endothelial expression of other procoagulant proteins, including tissue factor, von Willebrand factor, platelet-activating factor, β-thromboglobulin, cellular fibronectin, and thrombomodulin, has also been reported. The latter two endothelial cell markers have been shown to differentiate preeclampsia from other forms of hypertension in pregnancy.[32,33] Similarly, inhibitors of fibrinolytic or antithrombotic proteins appear to play a role in the imbalance of the coagulation cascade. Plasminogen activator inhibitor type 1 (PAI-1), whose synthesis during pregnancy is predominantly placental in origin, was observed to be increased in the plasma of women with preeclamptic pregnancies.[34,35] Increased decidual and amniotic fluid concentrations of PAI-1 also have been reported in preeclampsia.[36] Circulating levels of PAI-1, thrombomodulin and fibronectin were found to correlate directly with severity of the syndrome.[37]

Endothelial Cell Adhesion Molecules

In addition to reduced anticoagulant synthesis, endothelial cells express extracellular matrix glycoproteins with procoagulant activities as a response to injury or activation. Two examples are fibronectin and von Willebrand factor.[38] Both of these proteins are predominantly localized to the abluminal extracellular matrix of human endothelium,[39] but as discussed below these can be actively secreted from endothelial cells under conditions of cellular activation. In addition, exposed fibronectin is stimulatory for neutrophil attachment and has important pathophysiological consequences.[40]

Elevated concentrations of fibronectin in women with preeclampsia have been recognized for decades[41] and have been confirmed in other clinical conditions associated with endothelial cell dysfunction.[42] A specific cellular fibronectin isoform (cFN), which is non-hepatic in origin and almost exclusively localized to the vascular endothelium, has been of particular interest. It can be distinguished at the protein level by two extra domains (ED-A and ED-B) generated by differential mRNA splicing in endothelial cells. While it is a major component of the endothelial extracellular matrix, cFN is normally only a minor component of circulating fibronectin and thus is an accurate marker of endothelial cell injury.[32]

Lockwood and Peters[43] showed that ED-A fibronectin levels were increased in a cross-sectional study of women who later developed preeclampsia. Taylor et al. used a monoclonal antibody that recognizes a conformational epitope near the ED-B region of cFN and showed in a prospective, longitudinal study that plasma cFN concentrations were increased as early as the second trimester in women destined to develop preeclampsia but the analyte was not elevated in women developing transient hypertension without proteinuria.[32]

The concentrations of other endothelial cell adhesion molecules, including VCAM-1, and P-selectin, are elevated in cases of preeclampsia.[44] ICAM-1 and VCAM-1 also appear to have some value as predictors of preeclampsia as early as the midtrimester of pregnancy.[45]

Mitogenic Activities and Growth Factors

Endothelial cells also respond to injury or activation with the release or secretion of mitogenic proteins or peptides. With acute vascular trauma, this response teleologically encourages the proliferation of vascular smooth muscle, allowing vessel remodeling and repair. However, in illnesses such as atherosclerosis or preeclampsia, mitogenic factors cause reduced blood flow by promoting vessel wall hypertrophy. Indeed, plasma from women with preeclampsia collected prior to delivery had more mitogenic activity than plasma from the same women obtained 48 hours post-partum.[46] By contrast, plasma from normal pregnant women had essentially the same mitogenic activity before and after delivery. These "mitogenic indices" were elevated from as early as the first trimester of pregnancy, compared to women who proceeded to have normal pregnancy outcomes. Other studies indicate that the mitogenic activity is protease-, heat- and acid-labile and has an apparent molecular mass of ~150,000.[47] To our knowledge the following hypothesis has not been tested directly, but activating Ang II type 1 receptor (AT1R) autoantibodies, which are present in women with preeclampsia as discussed in detail below, would be expected to have the identical biochemical characteristics since myocardial fibroblasts are induced to undergo mitogenesis through activation of the AT1R.[48]

Growth Factor Binding Proteins in Preeclampsia

Based on the high apparent molecular mass of the mitogenic activator(s) described above, the plasma activity was postulated to be attributable to IGF binding protein complexes. However, most studies failed to show differences in circulating concentrations of these complexes in preeclampsia.[49] In a longitudinal study comparing 20 primipara who developed preeclampsia with 20 matched, normal pregnant controls, de Groot et al. found that midtrimester maternal plasma IGFBP-1 concentrations were significantly reduced in the group of women who developed preeclampsia approximately 20 weeks later.[50] They interpreted their data as reflecting reduced trophoblastic invasion of the maternal decidual stroma and decreased vascular deportation of IGFBP-1 in pregnancies that result in preeclampsia. This observation was confirmed by Hietala et al.;[51] but it should be noted that some reports of third-trimester maternal serum IGFBP-1 show elevations in women with active, symptomatic severe preeclampsia.[52,53]

The growth factor binding protein that to date has had the greatest impact on the preeclampsia field is soluble VEGF receptor 1 (sFlt-1). The history and role of this glycoprotein are presented comprehensively in Chapter 6. Briefly, it is now broadly established that elevated sFlt-1 concentrations in women with preeclampsia are associated with decreased circulating levels of free VEGF and free PlGF. Soluble endoglin is a proteolytic cleavage product of the cell surface TGF-β co-receptor, released into the plasma by matrix metalloproteinases.[54] This extracellular-domain protein binds BMP9 and BMP10 with high affinity, scavenging these TGF-β ligands in the circulation, blocking their angiogenic signaling in the microvasculature.[55] Overall, many of the biomarkers of endothelial cell activation described above are also functional mediators of endothelial cell dysfunction. Their activities are described in the following section.

PART II: CIRCULATING FACTORS INDUCE ENDOTHELIAL CELL DYSFUNCTION

Efforts to identify the putative circulating toxic factors responsible for endothelial cell dysfunction in preeclampsia have been ongoing for the past decade with disappointing results. The effort continues, but while specific candidates have been identified, most data suggest that multiple molecular species account for endothelial activation. Because of their direct contact with the vascular endothelial cell monolayer, various plasma constituents are likely candidates for endothelial cell activation in preeclampsia. Formed blood elements (e.g., platelets and neutrophils), placental membrane microvesicles, soluble proteins (e.g., autoantibodies and cytokines), lipids, cytokines, and matrix metalloproteinases (MMPs) have all been identified in the plasma of women with preeclampsia. In this section, we systematically discuss some of these postulated mediators of endothelial cell dysfunction in preeclampsia.

Formed Elements in Blood as Activators of Endothelium

Platelets

Clinical manifestations of abnormal platelet function include hypercoagulability, as reviewed in Chapter 17. *In vitro* evidence of abnormal platelet aggregation in women with preeclampsia[56] suggests that this phenonmenon could lead to endothelial cell activation. Prostacyclin receptor concentrations (B_{max}) in platelet membranes were similar in women with normal pregnancy, preeclampsia, and transient hypertension of pregnancy. However, their binding affinity (K_a) was considerably reduced in preeclampsia compared to the other two groups,[57] resulting in enhanced aggregation. Retrospective cohort studies of women with renal disease in pregnancy indicate that the incidence of preeclampsia was significantly less common in women treated prophylactically with heparin or aspirin.[58]

The platelet-specific NO donor, *S*-nitrosoglutathione, was administered to 10 women with severe preeclampsia at 21–33 weeks gestation. Significant, dose-dependent reductions in mean arterial pressure and uterine artery resistance indices were observed. Platelet activation, as measured by P-selectin expression, also was found to be decreased by *S*-nitrosoglutathione treatment. Thus, activated platelets may be one of the circulating factors in preeclampsia that mediate maternal endothelial cell activation and platelet-specific NO donors may prove beneficial in the management of severe preeclampsia.[59]

Neutrophils

Innate immune cells bind to selectins on the surface of activated endothelial cells and modify the function of the vascular intima. Neutrophil activation was studied in 20 eclamptic and 10 preeclamptic patients and compared to 10 normotensive controls. Plasma levels of neutrophil elastase were increased significantly in eclamptic and preeclamptic women. Elastase values in cases of eclampsia were highly correlated with mean blood pressure, serum ET-1 levels, and endothelial cytotoxicity, measured by fura-2 release from human umbilical vein endothelial (HUVE) cell cultures.[60] Lactoferrin also has been used as an indicator of neutrophil activation in normal and preeclamptic pregnancy. A comparative study between 40 normal and 42 preeclamptic women in the third trimester of pregnancy demonstrated that predelivery ratios of lactoferrin per neutrophil were higher in preeclamptic than in normal women.[61]

These and other data suggest that neutrophil activation plays a role in this syndrome.[62] Using whole blood cytometric analysis, Studena et al.[63] observed an increase in leukocyte surface antigen expression in normal pregnancy relative to nonpregnant women and a further increase in activation antigen expression in cells from preeclamptic pregnancies. The findings support the hypothesis of Redman and Sargent[64] that vascular inflammation plays a primary role. This theory is discussed in detail in Chapter 8. By contrast, eosinophils do not appear to have a compelling effect in preeclampsia.[65]

Activated monocytes and neutrophils release the hemoprotein myeloperoxidase (MPO). Interestingly, MPO induces low-density lipoprotein oxidation, activates metalloproteinases, and oxidatively consumes endothelium-derived NO, which are all reported to be involved in the vascular pathophysiology of preeclampsia (see sections below). Gandley et al. have shown that MPO levels are significantly increased in the circulation and placenta of women with preeclampsia and they speculate that MPO contributes to oxidative stress in the endothelium and placenta of women with preeclampsia.[66]

Placental Membrane Microvesicles

Circulating syncytiotrophoblast microvesicles have been postulated as endothelial toxic factors derived from the placenta.[67] These are discussed in detail in Chapter 8. It was observed that dilutions of 20% sera from women with preeclampsia did not inhibit endothelial cell proliferation, whereas 20% plasma from the same preeclamptic women significantly suppressed endothelial cell growth compared with 20% normal pregnancy plasma. The suppression was even more pronounced in cases of severe preeclampsia. The authors discovered that syncytiotrophoblast microvesicles added to blood could not be recovered from serum, but only from plasma, and suggested that the plasma-borne endothelial cell suppressive factor in preeclampsia may be derived from particles released from the microvillous surface of preeclamptic placentas. Indeed, these microvesicles can be isolated by ultracentrifugation from the circulation of women with preeclampsia. These same particles have been shown to induce ultrastructural evidence of endothelial cell injury when incubated with isolated human arteries.[68,69]

Endothelial Progenitor Cells (EPCs)

Recent research highlights the potential role of endothelial progenitor cells (EPCs) in the pathology of preeclampsia. EPCs encompass two distinct types of cells, circulating angiogenic cells (CACs) and endothelial colony forming cells (ECFCs), both of which are involved in *de novo* vessel formation and repair. ECFCs are highly proliferative and differentiate into mature endothelial cells at the site of vessel formation, while CACs are hematopoietic cells which promote migration and proliferation of ECFCs via the release of paracrine factors (reviewed in[70]). A decline in circulating EPCs is associated with endothelial dysfunction and cardiovascular disease.[71–74] Compared to normal pregnancies, in which the level of circulating EPCs increases with gestational age,[75,76] women with preeclampsia have significantly reduced numbers of EPCs.[77–79] It has been suggested that limited bioavailability of NO, which is required for mobilization of EPCs, and an increase in antiangiogenic factors in preeclampsia may contribute to EPC-mediated endothelial dysfunction.[80] Interestingly, diminished levels of EPCs persist in the circulation of preeclamptic mothers postpartum, and are suggested to be associated with long-term cardiovascular risk.[81]

Immune Complexes: Antiphospholipid and Antiendothelial Cell Antibodies

As preeclampsia occurs more commonly in primigravidae[82] and in women with underlying collagen-vascular diseases,[83] an immunological component has long been suspected. Moreover, studies by Brosens et al.[84] and Kitzmiller and Benirschke[27] demonstrated histological changes in the placental beds of women with preeclampsia that resemble those of acute allograft rejection. Antipaternal epitope recognition of the placenta has been postulated as a stimulus to maternal immune complex deposition in preeclampsia.[85]

This hypothesis is consistent with several epidemiological observations including a higher prevalence of preeclampsia in first pregnancies[86] and when couples use condoms as their primary form of contraception.[87] Some studies suggest that pregnancies conceived after ovum donation, where both gametes are immunologically foreign to the host, have a high incidence of preeclampsia.[88] By contrast, previous exposure to foreign or paternal antigens appears to reduce the risk of preeclampsia. A history of prior blood transfusion,[89] the practice of oral sex[90] or an extended length of cohabitation prior to conception[91] all are associated with a lower than expected prevalence of preeclampsia. These data have been comprehensively reviewed.[92]

Elevated levels of IgG or IgM antibodies to cardiolipin and phosphatidylserine suggest that antiphospholipid antibodies may play a pathogenic role in some women with preeclampsia.[93] In case-control studies, antineutrophil autoantibodies were significantly increased in preeclamptic and eclamptic subjects compared to normal controls[94]; particularly those associated with β_2-glycoprotein I.[95]

Rodgers et al. initially postulated that immune complexes might be involved in the process of endothelial cell activation, but were unsuccessful in detecting specific endothelial cell antigens by Western blotting with sera from preeclamptic women.[96] However, using a more sensitive enzyme-linked immunosorbent assay (ELISA) method, Rappaport et al.[97] observed elevated serum concentrations of antiendothelial cell antibodies in women with severe preeclampsia. It has been proposed that these antibodies interfere with endothelial cell function via inhibition of prostacyclin production or stimulation of procoagulant synthesis.[98] However, prostacyclin production has been shown to be greater in cultured human endothelial cells incubated with plasma or sera from preeclamptic women than cells exposed to blood from normal pregnant controls.[99,100] Preeclampsia plasma appeared to stimulate biosynthesis of a variety of prostanoids in several different cell types,[101] including placental thromboxane synthesis, as suggested by Rote et al.[102] to result in vasospasm and thrombosis. Peaceman and Rehnberg[103] showed that the addition of 100 μM aspirin to IgG fractions from women with functional antiphospholipid antibodies inhibited thromboxane production by placental explants. Thus, autoantibodies may contribute functionally to immune cell activation and vascular and placental dysfunction associated with preeclampsia, and interference with these actions may be therapeutic.

Importantly, the identification of activating anti-AT1R autoantibodies (AT1-AA) adds a novel dimension to the potential role of immune complexes in preeclampsia.[104] These immunoglobulins are found to be present in the serum of preeclamptic women at much higher concentrations than sera from nonpregnant or normal pregnant women. The antibodies bind and activate the AT1R, inducing AP-1 signaling and NF-κB activation. Reactive oxygen species, sFlt-1, and plasminogen activator inhibitor-1 are produced as a result. Their production and consequences are presented in detail in Chapter 15

Cytokines

Activation of the maternal immune system plays an important role in the development of preeclampsia.[105,106] Excessive inflammation is central to this response and is believed to be a mediator of maternal endothelial dysfunction.[107]

Activated immune cells including neutrophils, monocytes, NK cells, and T cells are also significantly elevated in women with preeclampsia[106,108,109]; leading to increased production of inflammatory molecules which further modulate the immune response. Elevated levels of many cytokines and chemokines have been identified in the maternal circulation of women with preeclampsia, including TNF-α, IL-6, IL-2,[110,111] IL-8, IL-10, IP-10, MCP-1,[112] and IL-12.[113] Recent research shows that peripheral NK and T cells, although capable of producing VEGF, actually produce significantly less of this angiogenic factor under preeclamptic conditions.[114] While the mechanism is unknown, this change further contributes to the angiogenic imbalance, promoting endothelial dysfunction. Many cytokines, particularly TNF-α, are known mediators of endothelial activation and dysfunction.[19]

TNF-α has been postulated as an important mediator of endothelial cell activation in preeclampsia. Several studies have reported higher serum concentrations of TNF-α in preeclamptic women compared to normal pregnant women[115,116] that normalize after delivery.[117] Moreover, placentas from preeclamptic patients have increased mRNA and protein expression of TNF-α compared with placentas from normal pregnant women,[118] implicating the placenta as a production source for this cytokine. However, based on uterine to peripheral vein TNF-α ratios, Benyo et al.[119] suggested that the placenta is not the primary source of TNF-α in the maternal circulation and proposed that activated leukocytes and endothelium were responsible.

TNF-α alters the balance between endothelium-derived vasoconstrictors and vasodilators and impairs endothelium-dependent relaxation, in part, by activation of NAD(P)H oxidase, leading to production of superoxide anions that can scavenge NO.[120] Moreover, TNF-α can induce vasoconstrictors such as ET-1,[121] which are increased in preeclampsia patients.[122] TNF-α can also stimulate mitochondrial production of free radicals.[123] In addition, TNF-α increases the expression of endothelial adhesion molecules such as ICAM-1[124] and tissue factor,[125] inhibits the thrombomodulin/protein C anticoagulation pathway,[126] and blocks fibrin dissolution by stimulation of PAI-1.[127]

Interestingly, TNF-α has been shown to have direct vascular effects during pregnancy. Chronic infusion of TNF-α into rats starting day 14 of gestation results in hypertension and decrease in renal iNOS expression.[128] Furthermore, Vizi et al. reported that intraperitoneal injection of lipopolysaccharide, a potent stimulus for endogenous TNF-α production, induces greater TNF-α plasma levels and higher mortality in pregnant mice compared with non-pregnant animals.[129] The data suggest that pregnancy is associated with a higher susceptibility to the effects of this cytokine, which may be a potential target for therapeutic intervention.

TNF-α also has been shown to activate type II phospholipase A2 (PLA2) and prostanoid biosynthesis in a variety of cultured cells.[130,131] Thus, increased concentrations of

PLA2 in placental homogenates from women with preeclampsia[132] and elevated plasma levels of type II PLA2 observed in women with severe preeclampsia[133] may be responses to increased TNF-α action in these women. Complicating these observations are reports that concentrations of soluble p55 TNF-α receptor, a TNF-α binding protein and potential inhibitor of TNF-α action, also are increased in preeclampsia serum.[134,135]

Other cytokines, such as interleukin (IL)-2, a pleiotropic cytokine produced by activated T-lymphocytes, have potent autocrine and paracrine effects. Its ability to stimulate T cell, B cell and natural killer cell mitogenesis and activity have led investigators to believe that excessive IL-2 action may be involved in the pathogenesis of preeclampsia. It has been reported that normal maternal serum inhibits IL-2 production during pregnancy.[136] Serum concentrations of the IL-2 receptor also have been evaluated as an indicator of preeclampsia risk.[137]

Circulating plasma concentrations of IL-1 receptor antagonist, an inhibitor of IL-1 action, and IL-6 were noted to be increased in concentration by 77% and 24%, respectively, in women with preeclampsia compared to normal pregnant controls.[22] It was suggested that these increased cytokine concentrations might contribute to the endothelial damage associated with preeclampsia. Other cytokines, including GM-CSF, G-CSF, IL-1, and IL-8, in maternal blood and amniotic fluid have been associated with preeclampsia and IUGR.[138]

Circulating Lipids and Lipoproteins

Endothelial cell dysfunction may be caused by an imbalance between circulating VLDL particles[139] or other lipids[140] and a protective, basic isoform of plasma albumin (TxPA).[141] One family with two cases of severe preeclampsia/eclampsia was found to have very high levels of Lp(a) lipoprotein. As the serum concentration of Lp(a) lipoprotein is genetically determined and the Lp(a) apolipoprotein has a close homology to plasminogen, very high levels of Lp(a) lipoprotein might interfere with the fibrinolytic/thrombolytic processes and could represent a genetically determined risk factor for preeclampsia.[142]

Triglycerides are a major feature of metabolic syndrome, which is increasing due to the rising number of reproductive-aged women who are obese. In a systematic review by Ray et al.,[143] they determined that there was a consistent positive association between elevated maternal triglycerides and the risk of preeclampsia. Interestingly, normotensive pregnancies are characterized by progressive increases in serum LDL levels accompanied by parallel increases in serum triglyceride levels[144,145] followed by sharp fall post-partum.[144,146] Interestingly, increased triglyceride content of LDL results in smaller but denser LDL particles.[147] These small dense LDL particles are more atherogenic[148] and are more susceptible to oxidative modification.[148,149] Sattar et al. first reported that smaller, denser LDL particles are increased in preeclamptic plasma when compared to normotensive controls[150] which was subsequently confirmed by Hubel.[145] In contrast to the above studies, one recent study has shown that plasma levels of oxLDL are in fact decreased in preeclampsia compared to age-matched normotensive pregnancy.[151] The authors suggest that the decreased plasma oxLDL levels in preeclamptic women could be due to increased levels of autoantibody to oxLDL.[151] A few studies have reported increased serum levels of autoantibodies to oxLDL[152,153] while some studies have found no differences.[154,155] As with many of the circulating factors measured in women with preeclampsia, differences in the literature may be attributed to differences in gestational age at sampling.

Oxidized LDL binds to its lectin-like oxidized LDL receptor (LOX-1) receptor[156] and contributes from 50% to 70% of oxLDL uptake in endothelial cells,[157] resulting in generation of superoxide anion via activation of NAD(P)H oxidase.[158] While basal levels of LOX-1 expression are low, factors such as TNF-α,[159] transforming growth factor-β,[159] oxLDL,[158] Ang II,[154,160] ET-1,[161] C-reactive protein,[162] and 8-iso-prostane[163] have all been shown to upregulate LOX-1 expression. Since all of these factors have been reported to be elevated in plasma of women with preeclampsia enhanced LOX-1 expression and uptake of oxLDL mediated via LOX-1 could be important in endothelial cell activation and/or dysfunction. LOX-1 mRNA expression showed highest expression in human placenta when compared to other human tissues.[156] Halvorsen et al.[163] identified the presence of LOX-1 receptor in trophoblastic cells and demonstrated that 8-iso-PGF2α, a product of lipid peroxidation and a marker of oxidative stress, could enhance oxLDL uptake via LOX-1 using a choriocarcinoma cell line. The Davidge laboratory recently demonstrated increased LOX-1 expression in the maternal vasculature of women with preeclampsia.[164] They have further identified a decrease in endothelium-dependent relaxation in the vasculature of the reduced uteroplacental perfusion pressure (RUPP) model of preeclampsia that was accompanied by increased levels of LOX-1 and eNOS.[165] Moreover, plasma from women with preeclampsia impaired endothelial-dependent vascular function that was prevented by LOX-1 inhibition.[166] Taken together, these studies would further support the role of dyslipidemia in the pathophysiology of preeclampsia.

Non-Esterified Fatty Acids

Non-esterified fatty acids (NEFA) are organic acids containing an even number of carbon atoms. They are produced in the small intestine when fat is digested. Chains with only single bonds are "saturated" with hydrogen (e.g., palmitic and stearic acids) whereas chains with one or more double bonds are "unsaturated" (oleic (one double bond), linoleic (two double bonds), and linolenic (three double bonds)).

Several studies indicate that plasma NEFA are increased 2–3-fold in women with preeclampsia,[146] even early in gestation[167,168] and when controlled for race, BMI, and age.[169] Interestingly, pregnant women with transient hypertension had plasma NEFA levels significantly lower than preeclamptic patients with similar degrees of hypertension.[150] These observations reiterate that the pathophysiology of preeclampsia is more than pregnancy-induced hypertension.

NEFA levels in the plasma of pregnant women were found to be inversely proportional to a basic isoelectric isoform of albumin (p*I*=5.6) referred to as toxicity-preventing albumin (TxPA).[139,170] TxPA was so named because it could prevent lipid-induced endothelial cell injury *in vitro* and *in vivo*.[171] This relationship is consistent with the hypothesis that fatty acid binding to plasma albumin is responsible for its shift in p*I*. Vigne et al. directly quantified the amount of fatty acid bound to plasma albumin in normal pregnant and preeclamptic women, finding a 3-fold increase in the NEFA/albumin ratios (2.5 vs. 0.8) in severely preeclamptic vs. normal pregnant women.[169] These values are very similar to the 1.6 and 0.9 molar ratios previously estimated by Endresen and colleagues.[171] Albumin with <1.3 mol NEFA bound focuses at p*I* 5.6, whereas albumin with 2.5–10 mol NEFA focuses at p*I* 4.8, resulting in a conformational change that alters the redox activity and buffering capacity of plasma albumin. Recapitulating the elevated NEFA/albumin ratios observed in preeclamptic women[172] demonstrated lipid accumulation and apoptosis in human endothelial cell cultures, apparently as a result of alterations in mitochondrial membrane potential. Plasma albumin plays an important role in safeguarding against the prooxidant effects of copper in the circulation through its Cys-34 residue, which directly participates in free radical scavenging.[173] Modifications of Cys-34, resulting in the loss of its electron-donating SH-group, disrupt its antioxidant capacity. Kagan et al. have reported that the copper-binding capacity of albumin is impaired in women with preeclampsia.[174] The high concentration (~600 μM) of this protein in the circulation and perpetual catalytic nature of the redox-cycling processes make these reactions a powerful source of oxidative stress. These data indicate that reducing antioxidants, such as ascorbate, cannot effectively prevent or ameliorate oxidative stress induced by Cu/albumin, as the continuous redox-cycling process depletes any exogenously added ascorbate and does not remove the source of redox-cycling.[175] This observation may explain, in part, the failure of therapeutic trials of antioxidant vitamins to improve clinical outcomes in preeclampsia as discussed later in this chapter.

Among the essential fatty acids described above, linoleic acid (18:2 *n*−6) is necessary for the production of important long-chain polyunsaturated derivatives, particularly arachidonic acid (20:4 *n*−6). Not only is arachidonic acid a key structural element of cell membranes, it is the obligate precursor to a variety of vasoregulatory lipids. Cyclooxygenase and peroxidase catalyze the formation of prostaglandins, prostacyclin, and thromboxanes, whereas 5-lipoxygenase allows the production of leukotrienes. A third pathway is catalyzed by cytochrome P-450 epoxygenases, converting the substrate to epoxyeicosatrienoic acids (EETs). The latter are potent vasodilators and antiinflammatory compounds[176] that can be inactivated by soluble epoxide hydrolase (sEH) to corresponding dihydroxyeicosatrienoic acids (DHETs).[177] In a recent pilot study, urinary concentrations of total 14,15-DHET were found to be 2.6-fold higher in seven preeclamptic women at term, compared to nine uncomplicated, normotensive pregnant women ($P<0.02$).[178] The implication is that excessive catabolism of EETs in preeclampsia may contribute to reduced vasorelaxation and increased inflammation. These arachadonic acid metabolites also are known ligands for PPARs.[179]

Peroxisome Proliferator-Activated Receptors (PPARs)

PPARs are major regulators of lipid and glucose metabolism, inflammation, and angiogenesis, which allow adaptation of the mother to the nutritional and perfusion requirements of the fetus.[180] PPARs, members of the nuclear hormone receptor superfamily, are ligand-activated transcription factors. There are three PPAR isotypes, PPARα, PPARγ, and PPARβ/δ, which are structurally conserved across species.

The PPAR system is intimately involved in adipose and vascular tissues, and hence is highly relevant to preeclampsia.[181] All three PPAR isotypes are expressed in human vascular and placental trophoblast cells.[182,183] These receptors can be activated by natural ligands, e.g., prostaglandins (PGs), fatty acids and their derivatives, as well as by synthetic ligands. A number of naturally occurring PPAR ligands have been identified, including long-chain fatty acids (C16 and greater), eicosanoids such as 8(S)-HETE (PPARα) and 9- and 13-HODE (PPARγ), and PGs such as PGA1, which binds to PPARα and PPARβ/δ, and 15-deoxy-delta12, 14-prostaglandin J2 (15dPGJ2), which binds to PPARγ.[184]

As preeclampsia is marked by hyperlipidemia, and characterized by a state of oxidative stress (see below), it has been postulated that pathological changes may be regulated by the PPAR system.[185] The Taylor laboratory found that serum from women with severe preeclampsia had reduced levels of PPAR activating lipids compared with serum of parity and gestational age-matched, normal pregnant women.[186] The reduction of transcriptional activity observed in trophoblast cells by sera of women with preeclampsia was shown for PPARγ and PPARα, but not for PPARβ/δ or RXR, and was manifested weeks and sometimes months before the clinical diagnosis was made. These results are consistent with other clinical evidence of a "hyperinflammatory" state of preeclampsia.[187] It has been shown that the PPARγ agonist, rosiglitazone, ameliorated hypertension, improved vascular function and reduced the elevated microalbumin:creatinine ratio in pregnant rats that had undergone RUPP surgery. Some of these beneficial effects were abrogated in the presence of the heme oxygenase 1 inhibitor as a potential mechanism. Moreover, administration of a PPARγ antagonist induced features of preeclampsia (hypertension, proteinuria, and endothelial dysfunction) in healthy rats.

Genetic variations in the PPARγ gene may modify the risk to develop preeclampsia. The Pro467Leu mutation of PPARγ is a dominant negative mutant resulting from a C to T transition in exon 6. In a woman with this mutation, both of her two pregnancies were complicated by severe preeclampsia.[188] However, a preliminary study of PPARγ gene variations in a Finnish population showed no association with preeclampsia or its severity.[189]

Angiogenic Factors

That angiogenic factors play a critical role in the development of preeclampsia is now well accepted and this topic is reviewed comprehensively in the accompanying Chapter 6. Vascular endothelial growth factor (VEGF) belongs to a family of glycoproteins that express their primary biologic effects by binding to VEGF receptor-1 (fetal liver tyrosine-like; Flt-1)[190] and VEGF receptor-2 (kinase domain related receptor; KDR).[191] VEGF promotes angiogenesis, induces endothelial cell production of vasodilatory NO and prostacyclin,[192] and may be important in the regulation of vascular tone. Although VEGF has vasodilatory activity, some studies indicate detrimental effects following intravascular administration of this factor. VEGF injection increased ICAM-1 and VCAM-1 expression, leukocyte adhesion, vascular leakage, and inflammation.[194] It is tempting to speculate that such proinflammatory actions of VEGF could be involved in the pathogenesis of endothelial dysfunction in preeclampsia.

Brockelsby et al. showed that incubation of myometrial resistance vessels with VEGF resulted in a blunted endothelium-dependent relaxation, mimicking the endothelial dysfunction observed in preeclampsia.[192] By contrast, several studies suggest that low serum concentrations of VEGF-family angiogenic proteins are associated with preeclampsia.[195,196] In view of the latter hypothesis it is interesting to note that the dose-limiting toxicities of bevacizumab (Avastin®), humanized anti-VEGF monoclonal antibodies used to treat colon, breast and other carcinomas, include hypertension and proteinuria.[197]

Some of the conflicting reports regarding the role of VEGF in the pathophysiology of preeclampsia have been complicated by technical aspects of the available assays as well as differences in free vs. total VEGF measurements.[198] Most investigators have failed to detect free VEGF in the maternal circulation. In a careful study of women undergoing IVF therapy, free VEGF was undetectable in the circulation within a month following embryonic implantation.[199] Soluble Flt is a splice variant of the Flt-1 receptor that lacks transmembrane and cytoplasmic domains. sFlt-1 is produced by a number of organs including the placenta,[192,200] lung, liver, kidney, and uterus.[192] It acts as a potent VEGF and placental growth factor (PlGF) antagonist[201] with an important role in preeclampsia pathophysiology.[202] Maynard et al. performed gene expression profiling of placental tissue and demonstrated upregulated mRNA expression of sFlt-1 in preeclampsia and subsequently showed elevated serum levels of sFlt-1 levels in preeclamptic women,[203] which has been confirmed by other investigators.[196] There is a positive correlation between the severity of preeclampsia and serum sFlt-1 levels.[203] Also, sFlt-1 levels vary inversely with serum VEGF and PlGF levels,[196,203] an effect attributed to its adsorption of these factors. However, one study has shown decreased PlGF levels during the first trimester without an increase in sFlt-1 levels,[196] suggesting that decreased PlGF levels observed in PE could also arise as a result of inadequate placental production.[204] There is a strong association between decreased urinary PlGF and the subsequent development of preeclampsia.[205] Several research groups are currently developing algorithms to use circulating sFlt-1 and PlGF concentrations as predictors of preeclampsia risk.[196,206]

Abnormal expression of platelet-derived growth factor (PDGF), another vasculogenic factor primarily responsible for pericyte recruitment and new vessel maturation,[207] also has been implicated in preeclampsia.[208]

As noted above, endoglin is an extracellular component of the TGF-β receptor II complex. It is expressed in high concentrations by endothelial cells and syncytiotrophoblasts. Its primary function in endothelial cells appears to be in the regulation of TGF-β receptor II serine/threonine kinase activity. Endothelial cells that lack endoglin fail to migrate or proliferate.[209] Mutations in endoglin have been associated with hereditary hemorrhagic telangiectasia (Weber-Osler-Rendu syndrome) and elevated serum concentrations of soluble endoglin are associated with preeclampsia.[210]

Matrix Metalloproteinases

Matrix metalloproteinases (MMPs) are a large family of zinc-dependent endopeptidases expressed ubiquitously throughout the body. *In vivo*, MMP activity is regulated by the four tissue inhibitors of MMPs (TIMPs 1 through 4), which bind and inhibit the zinc-catlaytic domain of these enzymes (reviewed in[211]). An imbalance between MMP activity and TIMP inhibition has long been recognized to allow degradation of extracellular matrix, and in this manner contribute to vascular remodeling in both long-term physiological processes (e.g., embryogenesis[212]) and chronic pathological processes (e.g., heart failure, atherosclerosis[213]). More recently, however, MMPs have also been shown to cleave non-extracellular matrix proteins and contribute to acute processes such as vascular reactivity[214] and platelet aggregation.[215] Since the latter processes contribute to the pathology of preeclampsia, the potential role of MMPs in preeclampsia warrants attention.

MMP-2 may be of particular relevance in preeclampsia since its circulating levels have been found to be elevated in preeclamptic women compared to normal pregnant women.[216] Myers et al. demonstrated that the MMP-2 to TIMP-1 ratio is imbalanced in women who develop preeclampsia.[217] This imbalance preceded the onset of the condition and continued up to the time of delivery. Kolben et al. also found that circulating TIMP-1 is decreased in preeclampsia.[218] The circulating levels of other TIMPs in preeclampsia still remain to be determined.

In vitro studies also suggest an imbalance between MMPs and TIMPs in preeclampsia. Human umbilical vein endothelial cells from preeclamptic pregnancies release more MMP-2 than endothelial cells from normal pregnancies.[219] In this case, the increase in MMP-2 activity occurred with no change in TIMP-1 or -2 protein levels, predicting a net increase in proteolysis. These *in vitro* experiments also indirectly suggest that the source of increased circulating MMP-2 in preeclampsia may be the dysfunctional endothelium.

A number of studies have demonstrated that many of the circulating factors already discussed in this article increase MMP activity. For instance, VEGF was found to markedly increase MMP-2 release from endothelial cells in a time- and concentration-dependent manner.[216] Other experiments indicate that peroxynitrite, TNF-α, Ang II, and oxLDL also can activate MMPs *in vitro*.[220–222] Thus, it is possible that multiple molecular mediators of preeclampsia may converge to promote MMP production and release in preeclampsia.

The functional significance of elevated MMP-2 may lie in accumulating evidence suggesting that this enzyme can promote vasoconstriction. The Davidge laboratory demonstrated that MMP-2 can cleave big ET-1 to yield ET(1–32), a 32-amino-acid vasoconstrictor peptide.[214] Recently, using the RUPP model, vasoconstriction to big ET was greater in the RUPP animals compared to normal pregnancy and MMP inhibition normalized this enhanced vasoconstriction response.[223]

Endothelin

Endothelin is an endothelial cell marker of interest in preeclampsia. Of the three isoforms of endothelin, ET-1 is the predominant one produced in endothelial cells[224] and exerts its biological actions in a paracrine fashion on subjacent vascular smooth muscle cells. Elevated ET-1 concentrations have been observed in clinical conditions associated with decreased intravascular volume such as that seen in preeclampsia. Moreover, ET-1 is a potent vasoconstrictor of the human uterine[225] and renal[226] vascular beds, both of which are affected in this syndrome.

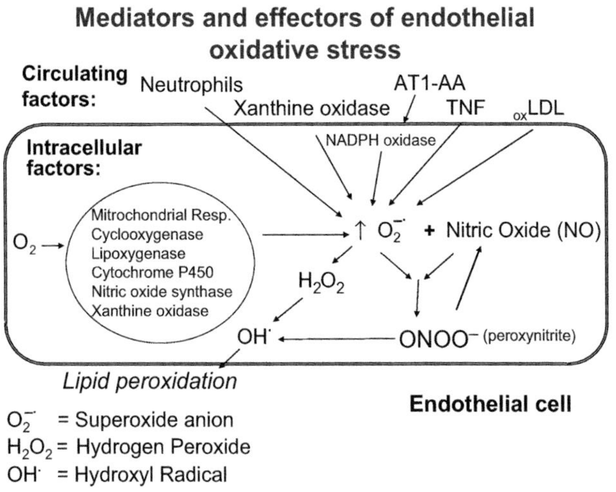

FIGURE 9.1 Oxidant sources on the endothelium. In this scheme, some of the potential sources of reactive oxygen species (ROS) derived from the circulation or within endothelial cells are shown. In the circulation, neutrophils, xanthine oxidase, antiangiotensin receptor autoantibodies (AT1-AA), and proinflammatory cytokines (TNF-α) can produce free radicals that react with endothelial cells. Further, these sources of oxidants can interact with each other to increase the prooxidant insult on the endothelial cells. Within the endothelial cell, mitochondrial respiration (resp.) and other metabolic pathways produce superoxide anions ($O_2\cdot^-$). Superoxide anions may be reduced (via superoxide dismutase) to hydrogen peroxide (H_2O_2). Hydrogen peroxide produces (via a metal-catalyzed reaction) hydroxyl radicals (OH·) that are potent oxidants that can be damaging to endothelial cells. Nitric oxide (NO) is produced by the nitric oxide synthase (NOS) enzymes that are present in the endothelium. Superoxide anions will react with NO to produce the oxidant peroxynitrite ($ONOO^-$). Peroxynitrite can alter endothelial cell function as well as decompose to the highly reactive hydroxyl radical.

There are two endothelin receptor isoforms: ETA and ETB. ETA has a selective affinity for ET-1, is found primarily on vascular smooth muscle cells, and is thought to mediate vasocontriction. ETB, which binds both ET-1 and ET-3 ligands with a similar affinity, is found on endothelial cells and may be involved in the autoregulation of NO and PGI2 release. In the human placenta, the ETB receptor isoform predominates, theoretically allowing vasodilatation of this important vascular bed.[227]

Circulating concentrations of immunoreactive ET-1 have been shown to be elevated in women with overt signs of preeclampsia[122]; however, no differences were observed in ET-1 levels during the second trimester of pregnancy (several months prior to the onset of clinical symptomatology), compared to a matched, normal pregnant group. Elevated concentrations of ET-1 in the blood of women with preeclampsia have been confirmed by other groups,[228,229] although Benigni et al.[230] failed to find significant differences between women with preeclampsia and normal pregnant women. Although the results do not support an early, predisposing role for ET-1 in preeclampsia etiology, as a potent vasoconstrictor it likely has a role in increased vascular resistance and reduced perfusion to organs. This is supported by studies in the RUPP model whereby ET has a central downstream role for vasoconstriction.[231] Thus, it has been suggested that manipulations of ET receptor may afford new strategies to control blood pressure and placental ischemia in preeclampsia.[232]

Relaxin

In recent years, a definite link has been established between relaxin and NO in mediating the vasodilatation of pregnancy.[233] Relaxin is typically considered a reproductive tract hormone active in mammalian parturition. However, new research indicates additional physiological roles for relaxin in the cardiovascular, renal and respiratory systems. Strong arguments to implicate the ETB receptor and NO pathway in relaxin action are presented in Chapter 16.

Angiotensin II

Angiotensin II (Ang II) is well known for its vasoconstrictive properties. Since preeclampsia is a state of enhanced vasoconstriction and hypertension, one would expect higher Ang II levels in the plasma of women with preeclampsia. However, circulating Ang II levels in preeclamptic women have been reported to be similar or even decreased relative to normal pregnant controls.[234] Thus, enhanced expression of the Ang II receptor (AT1R) has been invoked to explain the increased vascular tone in preeclampsia. Indeed, AT1R signaling in endothelial cells has been shown to result in cytokine production, increased extracellular matrix activity, and generation of reactive oxygen species through direct activation of endothelial NAD(P)H-oxidase.[235] Interestingly, AT1R activation induces the production of TNF-α from endothelial cells leading to increased MMP-2 and decreased TIMP-2 release.[236]

It has been reported that women with preeclampsia have increased expression of AT1R (specifically AT1R platelet binding sites), in comparison with normal pregnant women.[237,238] Abdalla et al. found that increased formation of heterodimers between the bradykinin receptor and AT1R in preeclampsia resulted in enhanced Ang II signaling.[239] These data support a role for AT1R levels in the alterations in Ang II sensitivity in preeclampsia that were observed in the seminal study by Gant et al.[240] Their study demonstrated that normal pregnancy was associated with decreased sensitivity to the pressor effects of Ang II, whereas women who later developed preeclampsia showed increased sensitivity to Ang II starting at 22 weeks of gestation.[240] Moreover, agonistic autoantibodies for AT1R (AT1-AA) have been measured in the circulation of women with preeclampsia that may account for enhanced AT1R activation in the absence of elevated Ang II levels in preeclampsia.[241–243] These are discussed in detail in the accompanying Chapter 15.

As mentioned briefly above, the discovery of circulating, agonistic autoantibodies directed against the AT1R represents a remarkable pathophysiological derangement in preeclampsia. The identification of these activating antibodies serves to align immunological theories with the rennin-angiotensin system.[244,245] Similar effects were noted in a rat model of preeclampsia.[246]

Toll-Like Receptors

Toll-like receptors (TLRs) modify the innate immune system via single membrane-spanning proteins that recognize ligands with structurally conserved molecular patterns. Pathogenic ligands of TLRs include bacterial cell lipopolysaccharides, viral double-stranded RNA, and fragmented host molecules such as fibrinogen and DNA with unmethylated CpG islands. These receptors are abundantly expressed in trophoblast and endothelial cells.[247] TLR9, for example, has been identified in HUVE cells, where it transduces inflammatory changes in response to free, particularly hypomethylated, DNA.[248] It has been demonstrated that concentrations of free fetal DNA in the maternal circulation may be more than 3-fold increased relative to the normal pregnant state,[249] presumably as a result of placental villus shedding and widespread placental apoptosis of anchoring cytotrophoblasts.[250] Furthermore, as placental DNA is globally hypomethylated,[251] this is likely to be a potent stimulus to maternal endothelial TLR9, activating the downstream NF-κB pathway and leading to adhesion molecule expression and vascular permeability.

PART III: OXIDATIVE STRESS: A POINT OF CONVERGENCE FOR ENDOTHELIAL CELL DYSFUNCTION

As noted, circulating factors such as TNF-α, oxLDL, and AT1-AA acting on the vasculature result in oxidative stress, inflammation, and vasoconstriction, features that have been well documented in preeclampsia. Moreover, these factors can act individually and in concert to disturb maternal endothelial function. The following section will provide evidence to support the hypothesis that intracellular oxidative stress promotes endothelial dysfunction in preeclampsia.

Oxidative Stress as a Mediator of Endothelial Cell Dysfunction

Oxidative stress is an imbalance between prooxidant and antioxidant forces resulting in an overall oxidant insult. In preeclampsia, oxidative stress has been postulated to lead to altered endothelial cell function. Indeed, the prooxidant environment of endothelial cells is extensive, as these cells are constantly exposed to extracellular factors in the circulation that are capable of inducing an oxidative insult and can produce their own oxidants as well. Prooxidants include free radicals such as superoxide anions ($O_2\cdot^-$), hydroxyl radicals (OH·), nitric oxide (NO^-) and other reactive oxygen species (ROS)[252] as well as reactive nitrogen species (RNS) such as peroxynitrite anion ($ONOO^-$). ROS and RNS are continuously produced *in vivo* by a number of cell types including endothelial cells. Oxidative stress is an example of a pathological process whereby multiple factors converge to cause endothelial cell dysfunction (see Fig. 9.1).

Reactive oxygen species contribute to normal metabolic processes in the body and have physiologic roles as second messengers in addition to their potentially pathogenic role. There are several metabolic pathways that can lead to the production of oxygen-derived free radicals. Mitochondria, endoplasmic reticulum, and nuclear membranes have been shown to produce superoxide anions as a consequence of autooxidation of electron transport chain components.[253] Oxygen-derived free radicals also are produced as the result of arachidonate metabolism by prostaglandin H (PGH) synthase, lipoxygenase, and cytochrome P450. NO synthase can generate superoxide anions and hydrogen peroxide, particularly if the intracellular concentrations of L-arginine (the precursor for the synthesis of NO) or its co-factor tetrahydrobiopterin are low.[254] Oxidation of hypoxanthine by xanthine oxidase also produces superoxide radicals within endothelial cells.[255] It is important to note that oxygen-derived free radicals can be produced at a number of subcellular compartments, such as the mitochondria, endoplasmic reticulum, peroxisomes, phagosomes, plasma membrane, nuclear membrane, and cytoplasm. Therefore, the ability of antioxidants to quench free radicals depends on the production, localization as well as the type of oxidative insult.

Several factors in the circulation (e.g., neutrophils, xanthine oxidase, and cytokines) can act on endothelial cells to produce oxygen free radicals. This raises the possibility of free radical-dependent vascular dysfunction at sites distant from the primary source or insult. This is important since the placenta is central for the development of preeclampsia. As a consequence of endothelial cell dysfunction, the perfusion of many organs, including the placenta, is reduced. This, in turn, could lead to a feed-forward progression for further endothelial cell dysfunction thereby accelerating the symptoms of preeclampsia until the placenta has been removed.

In women with preeclampsia, superoxide generation from circulating neutrophils is enhanced.[256] As previously noted, cytokines such as TNF-α can either directly or indirectly initiate oxidative stress. TNF-α, which is elevated in women with preeclampsia, has been shown to directly induce oxidative damage[257] as well as increase endothelial cell-induced oxidation of LDL.[258] TNF-α can also increase free radical production through the xanthine oxidase pathway[259] and concentrations of the latter enzyme are increased in women with preeclampsia[260] (see Fig. 9.1).

NAD(P)H oxidase appears to represent the most significant source of superoxide anion production in endothelial cells.[261] NADPH oxidase is a complex membrane-integrated b-type cytochrome, cytochrome b558, which is composed of 91 and 22 kDa subunits (gp91phox and p22phox, respectively), and at least three cytosolic proteins (p47phox, p67phox, and p21rac).[262] As previously noted, many of the circulating factors that are increased in women with preeclampsia can activate NADPH oxidase.

Free radicals thus generated have numerous deleterious downstream effects including lipid peroxidation of biological membranes and oxidative modification of lipoproteins. Some markers indicate that in women with preeclampsia, there is evidence for increased lipid peroxidation,[263,264] whereas Regan et al. disputed this claim.[265] However, their study only analyzed a single biochemical marker of oxidative stress, urinary 8,12-iso-iPF2alpha-VI, and hence their report should be interpreted with caution. Also, this marker has not been widely validated for oxidative stress. A subsequent study examined various biomarkers of oxidative stress in preeclampsia and concluded that preeclampsia is indeed a state of mild oxidative stress.[266] Decreased plasma vitamin C levels and significantly elevated lipid hydroperoxides, indicative of oxidative stress, as well as elevated serum levels of soluble vascular cell adhesion molecule-1 (VCAM-1), all are consistent with ROS-induced endothelial dysfunction.[263,266]

In preeclampsia, it has been suggested that certain antioxidant mechanisms are not adequate to compensate for an overwhelmed oxidant response. Re-establishing a balance between ROS and antioxidant protection in women with preeclampsia may protect the vascular endothelium. Unfortunately, this is a complex system and thus the results of antioxidant vitamin trials are equivocal, as discussed later in this chapter. Ultimately, antioxidant protection will depend on the compartment and type of oxidative insult imposed on a cell. Moreover, oxidative stress in preeclampsia is most likely the result of multiple prooxidant intracellular pathways. Understanding of the ultimate converging pathways leading to vascular oxidant stress and the downstream effects is needed so that rational interventions can be developed. Nonetheless, there is substantial evidence for elevated oxidants within the vascular endothelium of women with preeclampsia that can alter a number of vasoactive pathways, specifically involving endothelial-dependent relaxing pathways such as prostacyclin, NO, and endothelial-derived hyperpolarization.

Eicosanoid/Prostacylin Production

The production of eicosanoids, such as prostacyclin and thromboxane, is regulated by the availability of arachidonic acid and the activity of PGH synthase.[267] Liberation of arachidonate from membrane phospholipids is mediated through phospholipases. The primary pathway is through activation of PLA2, which releases arachidonate directly from membrane phospholipids. Elevated lipid peroxides are known to increase PLA2 activity[267] and play an important regulatory role in PGH synthase activity. The activity of PGH synthase requires low levels of lipid peroxides for activation and continued catalysis whereas higher levels of peroxides are inhibitory and can inhibit prostacyclin synthase.[268]

Prostacyclin (PGI2) is known to cause vasorelaxation and inhibit platelet aggregation by receptor-mediated mechanisms. While cyclic (c)AMP acts as a second messenger for platelet aggregation, vasorelaxation by hyperpolarization may provide an explanation, in addition to stimulation of cAMP, for the PGI2 mechanism of action on blood vessels. PGI2 released from the endothelium is capable of relaxing vascular smooth muscle. PGI2 has a role in determining vasodilatation in different vascular beds, especially in relation to sex steroid status and in pregnancy.

In a rat model of oxidative stress (produced by depletion of vitamin E), increased PGH synthase-dependent vasoconstrictor modification of vascular responses was observed.[268] Further, in rat models of hypertension (also believed to reflect a state of oxidative stress), a PGH synthase-dependent vasoconstrictor predominates to alter vascular function.[269] However, one research group has studied the effect of indomethacin (an inhibitor of PGH synthase activity) on the acetylcholine- and bradykinin-induced relaxation response of subcutaneous arteries from preeclamptic and normal pregnant women.[270,271] Indomethacin attenuated the relaxation response, but the degree of shift was not different between normotensive women and preeclamptic women.

Data suggest that there is an overall increase in the vasoconstrictor eicosanoids (as measured by their stable metabolites) in women with preeclampsia. For example, in the placentas of women with preeclampsia, more thromboxane is produced compared to prostacyclin.[272] Urinary and plasma levels of thromboxane are elevated while prostacyclin levels are reduced in women with preeclampsia compared to women with uncomplicated pregnancies.[273,274] However, the system is complex. Biphasic temporal patterns of PGI2 production in response to endothelial activators have been described. Wang et al.[275] noted an initial increase followed by a persistent diminution of PGI2 production when bovine endothelial cells were exposed to hyperlipidemic sera. Similar kinetics of PGI2 release by HUVE cells were reported by Baker et al.,[276] after exposure to preeclampsia plasma. During the first 48 hours the secretion of PGI2 was enhanced over baseline, whereas more chronic exposure (>72 h) appeared to deplete PGI2 production.

Nitric Oxide

One mechanism by which oxygen free radicals may alter endothelial cell function is by scavenging NO, thereby reducing bioavailability of this potent vasodilator. Normal pregnancy is maintained in a state of vasodilatation, mediated in part by enhanced NO production. Nitric oxide synthase (NOS) catalyzes the conversion of L-arginine to L-citrulline and NO, a powerful vasorelaxing molecule.[277] While a decrease in NO modulation of vascular tone could partly explain the pathogenesis of hypertension in preeclampsia, there are conflicting reports of increased,[278] decreased[279] or sometimes unchanged[280] metabolites of NO in the maternal circulation. Decreased plasma levels of nitrite and nitrate have been attributed to increased plasma levels of *S*-nitrosoalbumin in preeclampsia. Elevated concentrations of non-esterified fatty acids alter the conformation of serum albumin[139] and affect the redox potential of a critical cysteine residue in the protein. These changes impair the release of NO, ultimately resulting in reduced NO bioavailability.[175] This decreased release of NO could also be partly due to decreased vitamin C levels, since *S*-nitrosoalbumin-mediated relaxation of resistance sized-vessels is enhanced by ascorbate.[193]

Another pathway that may result in reduced NO levels is regulated by arginase. The two isoforms of this enzyme, arginase I and II, are found in cytosol and mitochondria, respectively. They catalyze the conversion of L-arginine to L-ornithine and urea.[281] Arginase reciprocally regulates NOS,[282] hence an increase in arginase activity could decrease L-arginine levels, the common substrate for arginase and NOS, and result in decreased NO production. Furthermore, a decrease in the substrate could result in uncoupling of endothelial NOS (eNOS) and generate superoxide anion.[283] Arginase mRNA expression is increased in villous tissue of preeclamptic women, which inversely correlates with fetal L-arginine levels, suggesting excess consumption of the substrate.[284] In preeclamptic women, arginase II expression, as assessed by immunohistochemistry, was found to be increased in syncytiotrophoblasts and endothelial cells of villous vessels of preeclamptic placenta.[284] In regard to the maternal vasculature, arginase II expression is increased in arteries from omental fat biopsies of women with preeclampsia compared to normotensive pregnant women. Furthermore, HUVE cells treated with 2% plasma from preeclamptic women show increased arginase II expression and activity. Preeclamptic plasma also increased superoxide and peroxynitrite levels; however, inhibition of arginase or NOS reduced superoxide levels. Thus, arginase may result in eNOS uncoupling, contributing to oxidative stress by reducing the production of NO and promoting the production of superoxide. In addition, enhanced circulating asymmetric dimethylarginine (ADMA), an endogenous inhibitor of eNOS, has been observed in women with preeclampsia[285–287] as yet another mechanism for reduced NO bioavailabilty in the maternal vasculature in preeclampsia.

Interestingly, endothelial cells acutely exposed to preeclamptic plasma showed an increase in NOS activity,[288] enhanced eNOS protein expression,[289] and stimulated nitrite/nitrate production.[290] Although NO is an important vasorelaxant, an elevation of NO in the face of oxidative stress may be deleterious. NO can react with superoxide anions yielding the powerful oxidant peroxynitrite, which may alter vascular function.[291] Myatt et al.[292] have demonstrated increased peroxynitrite in the placentas of women with preeclampsia, while the Davidge laboratory demonstrated enhanced peroxynitrite formation in the vessels of women with preeclampsia.[290] Moreover, circulating factors in women with preeclampsia can enhance peroxynitrite formation in cultured endothelial cells. These data imply that greater NO scavenging within the endothelium occurs in preeclampsia. It is interesting to note that decomposition of peroxynitrite at physiological pH also gives rise to nitrates and nitrites[293] and thus may be a source of the elevated NO metabolites that have been observed in endothelial cells exposed to preeclamptic plasma. Peroxynitrite is not only an indicator of NO scavenging, but it also affects endothelial function. Peroxynitrite increases inducible NOS (iNOS) protein expression in isolated endothelial cells via activation of NF-κB.[294] Thus a peroxynitrite-induced increase in NO could be deleterious in the face of increased oxidative stress, whereby there is a feed-forward mechanism to generate more peroxynitrite. Furthermore, peroxynitrite decreases prostacyclin synthase expression, which could alter the balance towards vasoconstriction.[294]

Interestingly, it seems that in normal pregnancy there is a compensatory pathway that causes relaxation when NO and prostacyclin are inhibited,[295] which may not occur in the vasculature of women with preeclampsia. This relaxation is mediated through endothelium-derived hyperpolarization.

Endothelium-Derived Hyperpolarization

Endothelium-dependent vasodilatation is primarily mediated by NO, prostaglandins, and endothelium-derived hyperpolarization factor (EDHF). However, the exact nature of EDHF is complex as there are multiple pathways or factors that result in hyperpolarization of the endothelial and/or smooth muscle cells thereby resulting in vasodilatation. Thus the current literature simply refers to this relaxation as EDHF with a number diverse factors/mechanisms involved, such as epoxyeicosatrienoic acid (EETs), cannabinoids, potassium ions, myoendothelial gap junctions, and hydrogen peroxide, which have all been identified as putative EDHF in various species and vascular beds.[296]

Hyperpolarization generated in endothelial cells spreads to adjacent vascular smooth muscle cells through myoendothelial gap junctions. Moreover, calcium-activated potassium channels, most probably the SK4 (IKCa) and SK3 (SKCa) expressed either on the endothelium or on vascular smooth muscle cells, are the end-cellular gateway mediating hyperpolarization and subsequent EDHF-mediated relaxation.[297] EDHF mechanisms have been reported to account for ~50% of endothelium-dependent vasodilatation in subcutaneous resistance arteries from healthy pregnant women and are dependent on Cx43 and gap junctions.[298] In women with preeclampsia, decreased vasodilator responsiveness in myometrial arteries has been attributed to a reduction in EDHF contribution,[299–301] potentially due to physical disruption of myoendothelial gap junctions that was also characterized by a diversification of EDHF mechanisms that includes EET and H_2O_2.[300,301] Notably, an increase in superoxide anions that has been demonstrated in compromised pregnancies[165] can produce H_2O_2 by the enzyme superoxide dismutase, thus providing a potential explanation for the increased involvement of this molecule. Further, PlGF, an activator of EDHF, has been consistently shown to be reduced in women with preeclampsia.[302] Thus EDHF via gap junctions and connexins are important mediators of pregnancy-induced vasodilatation that is impaired in women with preeclampsia.

In summary, numerous mechanisms act in concert to permit the vasodilatation featured in normal pregnancy, which is crucial for a successful outcome. Ultimately, there is reduced endothelium-dependent relaxation in the vasculature of women with preeclampsia. Enhanced generation of superoxide anion could scavenge NO, and this coupled with an absence of a compensatory pathway such as EDHF may lead to the symptoms observed in women with preeclampsia. Overall, it is likely that oxidative stress is a point of convergence for multiple factors affecting endothelial-mediated vasoactive pathways.

Other Vasodilators

Carbon monoxide and hydrogen sulfide are other vasodilators implicated in the pathophysiology of preeclampsia. Carbon monoxide is produced endogenously from heme-oxygenase-1 (HO-1), which is significantly decreased in placentas of women with preeclampsia.[303] Women with preeclampsia have significantly decreased CO concentration in exhaled breath compared to normal healthy women.[304] It is interesting to note that maternal cigarette smoking is associated with a lower risk of preeclampsia that may be due, in part, to CO (reviewed in[305]). In support of this hypothesis, Zhai et al. demonstrated an inverse association between CO concentration and preeclampsia risk.[306]

In addition to NO and CO as gaseous mediators of vascular function, hydrogen sulfide (H_2S) is another more recently identified gaseous mediator. It is produced endogenously from the activity of two enzymes, cystathionine-γ-lyase (CSE) and cystathionine-B-synthetase (CBS).[307] CSE is the primary H_2S-synthesizing enzyme in the vasculature. It was recently demonstrated that H_2S is a powerful vasodilator of the placental vasculature and that expression of CSE is reduced in placentas from women with preeclampsia displaying abnormal umbilical artery Doppler waveforms.[308] Moreover, plasma levels of H_2S are significantly decreased in women with preeclampsia.[309] Overall, it has been suggested that endogenous H_2S is important for a healthy placental vasculature and may be impaired in women with preeclampsia.

PART IV: CLINICAL TRIALS

Preeclampsia has been considered to be a disorder associated with abnormal placentation or perfusion resulting in an increased inflammatory response and endothelial dysfunction, which in turn lead to the characteristic maternal syndrome. As discussed in this chapter, oxidative stress is one of several mechanisms that have been proposed to cause manifestations of the disease. In this section we will discuss the clinical trials based on these principles that are focused on improving vascular function.

Aspirin

As described above from a pathophysiological view, antiplatelet agents such as aspirin (acetylsalicylic acid) are among the most promising candidates for prevention of preeclampsia. They have a positive effect on the balance between PGI2, a vasodilator, and thromboxane, a vasoconstrictor and stimulant of platelet aggregation. Initial trials of low-dose aspirin in high-risk subjects appeared promising in the prevention of preeclampsia. Unfortunately, with the exception of the study by Hauth et al.,[310] other trials of unselected pregnant women indicate that this antiplatelet drug has minimal effects on the incidence of preeclampsia.[311,312] A meta-analysis of the over 32,000 women included in aspirin trials indicates that there is some efficacy, although the effect was too modest to warrant routine use in all women. In addition, a Cochrane review of antiplatelet agents for prevention of preeclampsia describes a 17% reduction in the incidence when studies of different designs were combined. Although study results are inconsistent, many experts believe that, if started early in pregnancy, high-risk women are likely to benefit from aspirin therapy.[313,314]

As described, *antioxidants* maintain cellular function in pregnancy through inhibition of peroxidation, thus protecting enzymes, proteins, and cell integrity. Vitamins C and E are antioxidants: vitamin C scavenges free radicals and vitamin E acts to prevent lipid peroxidation and protect against oxidative stress.[315] These mechanisms led to the hypothesis that early supplementation with antioxidants might be effective in decreasing oxidative stress and improving endothelial integrity, thereby preventing preeclampsia. In a recent meta-analysis involving almost 20,000 women, no differences in the prevalence of preeclampsia were noted between women using vitamins C and E (in daily doses of 1000 mg and 400 Units) vs. placebo (9.6% vs. 9.6%).[316] In addition, vitamin C and E usage did not decrease the prevalence of preeclampsia in women either at low-moderate risk[317] or high risk of developing preeclampsia, including chronic hypertension, renal disease, pregestational diabetes, multiple pregnancy, or history of preeclampsia.[316] Although the expected results were not achieved, several options might explain why antioxidant therapy is not clinically effective, including timing of administration, doses used in the trials, and selection and characteristics of women participants in the trials.[317] In addition, the cellular and subcellular sites of oxidative stress, as well as the specific oxidants involved, must be considered when choosing the antioxidants.

L-Arginine

In line with the important role of oxidative stress in preeclampsia and evidence from experimental data in animals and humans, indicating that L-arginine could have a beneficial effect on hemodynamics, L-arginine was studied to prevent preeclampsia. As previously noted, the main site of production of NO is by NOS in endothelial cells, which uses circulating L-arginine as a substrate. Hence, the local availability of this amino acid is critical to the endothelial adaptive regulatory mechanisms opposing vasoconstrictors in preeclampsia. In a randomized trial including women at high risk of preeclampsia (women with a personal history of preeclampsia or in a first-degree relative) supplementation of L-arginine and antioxidant vitamins reduced the incidence of preeclampsia.[318] However, other studies showed no effects of L-arginine on the prevalence or clinical course of preeclampsia.[319] An explanation for the variability in outcome may be the diversity of women included and gestational age at intervention; women starting treatment beyond 24 weeks gestation showed no benefits.

Fish Oil

Another proposed preventive intervention is the supplementation of fish oil ($n-3$ long-chain PUFA). The rationale for supplementation of fish oil is that $n-3$ precursors are predicted to reverse abnormal plasma fatty acid profiles and endothelial injury, improving vascular function and reducing inflammatory response. Recently, Zhou et al. reported a large, double blind, randomized controlled trial that showed no reduction in the prevalence of preeclampsia between women using docosahexaenoic acid (DHA)-enriched fish oil (800 mg daily) vs. vegetable oil capsules as a control.[320] These data are consistent with recent systemic reviews of marine oil/$n-3$ PUFA supplementation in both high- and low-risk pregnancies.[321]

Calcium

Calcium supplementation has been suggested to have benefits in prevention of preeclampsia.[322] A recent meta-analysis described calcium supplementation as an effective measure to reduce the incidence of preeclampsia especially in women with low calcium intake.[323,324] However, a large, randomized trial failed to find any benefit from supplemental calcium (2 mg daily) given to women in early pregnancy in prevention of preeclampsia.[322,325] Heterogeneity in the pooled estimate can be explained by differences in effect size contributed by the increase of intervention of trails with high-risk women or low calcium intake populations.

Anticoagulants

Given that the pathophysiology of preeclampsia is complex, including placenta-mediated complications, especially placental thrombosis, it has been hypothesized that anticoagulants may prevent these complications. Several randomized controlled trials comparing the use of low molecular weight heparin (LMWH) in women who have had a complicated (placenta-mediated) pregnancy versus no prophylaxis found inconsistent results. In a multi-center, multi-national, randomized, controlled trial no differences in the incidence of preeclampsia were found comparing antepartum use of LMWH 5000 units per day and aspirin to aspirin alone in women with thrombophilia and prior-onset preeclampsia enrolled at less than 12 weeks gestation age. However, LMWH did reduce the incidence of recurrence of early-onset preeclampsia.[326] Therefore, it might be possible that prophylactic LMWH is beneficial in a sub-group of women with prior severe preeclampsia.[327]

Summary

Despite the variety of possible prophylactic interventions proposed, based on plausible scientific hypotheses, clinical studies have produced disappointing results. Primary prevention strategies applied to whole populations, such as antioxidant (vitamins C and E) and fish oil, as well as secondary prevention applied to high-risk populations, such as aspirin, L-arginine, calcium, and low molecular weight heparin, have not proven successful. Other pharmacologic approaches for reducing adverse platelet–endothelium interaction may show some promise. These include selective thromboxane-A2 synthase and thromboxane-A2 receptor antagonists, serotonin 2-receptor blockers, hirudin, endothelin antagonists, and ticlopidine.[328]

PART V: SPECULATIONS AND DIRECTIONS OF FUTURE INVESTIGATIONS

The evidence suggests that the pathophysiology of preeclampsia involves circulating plasma constituents that induce endothelial cell activation and/or dysfunction. Given

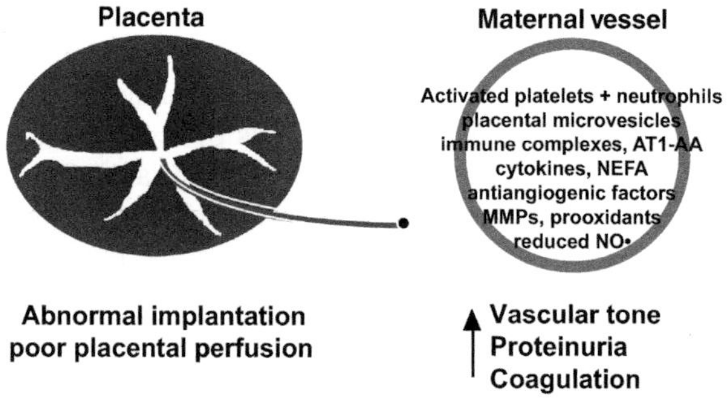

FIGURE 9.2 Endothelial dysfunction in the pathogenesis of preeclampsia. The hypothesis proposes that reduced placental perfusion results in the release into the circulation of a variety of factors, which may differ in different women, that alter endothelial function. Endothelial cell activation results in increased sensitivity to pressors, stimulation of the coagulation cascade, and loss of fluid from the intravascular compartment. These changes lead to reduced perfusion of maternal organs including the uterus, with a subsequent further reduction of placental perfusion. A feed-forward loop is generated whereby endothelial injury and dysfunction lead to the release of lipids (e.g., non-esterified fatty aids (NEFA)) and intrinsic vascular factors (e.g., cytokines, matrix metalloproteinases (MMPs), reactive oxygen species, etc.) that further compromise endothelial function.

our understanding of this syndrome, it is highly probable that the placenta contributes to the plasma factors responsible for endothelial cell dysfunction (see Fig. 9.2). In addition, considerable evidence suggests that maternal constitutional factors (e.g., genetic disposition of lipid disorders, antioxidant reserves, obesity) also are involved. It is possible that interactions between initiating events (e.g., relative placental ischemia, secretion of cytokines, and embolization of syncytiotrophoblast microvesicles) and maternal phenotypic factors (e.g., genetic polymorphisms, total body fat, and circulating lipoproteins) ultimately give rise to oxidative stress and endothelial cell dysfunction, culminating in the maternal syndrome of preeclampsia. All of this is augmented by the increased inflammatory activation of normal pregnancy that is accentuated in preeclampsia. It is evident that endothelial dysfunction in preeclampsia is auto-amplifying. Alteration of endothelial function by circulating products results in the production of factors (i.e., procoagulants, cytokines, reactive oxygen species, etc.) that themselves lead to further endothelial dysfunction. Moreover, the increased sensitivity to contractile agonists, activation of the coagulation cascade, and loss of intravascular volume induced by endothelial dysfunction lead to reduced perfusion of most if not all maternal organs. This includes the uterus, with subsequent further compromise of placental perfusion.

It is puzzling why such a disastrous biological scenario would have survived evolutionarily; but the maintenance of the disorder since early humanity suggests some benefit to mother or infant. There are controversial data indicating that the risk of perinatal intraventricular hemorrhage (IVH) and cerebral palsy are reduced in infants born of pregnancies with preeclampsia or eclampsia.[329–331] Perhaps, preeclampsia represents an inappropriate maternal response, predetermined in part by maternal sensitivities, to an appropriate fetal adaptation to reduced placental perfusion. While highly speculative, this concept should be kept in mind for the development of new mechanistic hypotheses and the design of future therapeutic trials.

When the concept of a disease-inducing "toxin" re-emerged in the 1980s, considerable effort was directed toward identifying the circulating factor(s) in preeclampsia that leads to endothelial dysfunction. However, over the past 20 years such a plethora of potential injurious agents has been identified that it is very unlikely that a single responsible toxin will be discovered. Nevertheless, it remains seductive to search for converging pathways and to focus on major pathophysiological events. For example, the large trials of antioxidant vitamins were stimulated by the concept, as we have advanced here, that multiple maternal and fetal factors converge to generate oxidative stress. Angiogenic activities and antagonists also have been posited as a similar cross-road. However, investigators have yet to identify a single pathophysiological marker that is abnormal in all women with preeclampsia (let alone in women destined to develop preeclampsia, where biomarkers could be utilized as a predictor of disease). The data reviewed in this chapter suggest that it might be useful in future studies to search for therapeutic targets that incorporate multiple convergence points, proximal to endothelial cell injury.

References

1. Lever JCW. Cases of puerperal convulsions, with remarks. *Guy's Hosp Rep*. 1843;1:495–517.
2. Mayer A. Changes in the endothelium during eclampsia and their significance (translated from German). *Klin Wochenzeitschrift*. 1924:H27.
3. Bell ET. Renal lesions in the toxemias of pregnancy. *Am J Pathol*. 1932;8:1–42.
4. Spargo BH, Lichtig C, Luger AM, Katz AI, Lindheimer MD. The renal lesion in preeclampsia: examination by light-, electron- and immunofluorescence-microscopy. In: Lindheimer MD, Katz AI, Zuspan FP, eds. *Hypertension in Pregnancy*. New York: Wiley; 1976:129–137.
5. Sheehan HL, Lynch JB. *Pathology of Toxaemia of Pregnancy*. London: Churchill; 1973.
6. Shanklin DR, Sibai BM. Ultrastructural aspects of preeclampsia. II. Mitochondrial changes. *Am J Obstet Gynecol*. 1990;163:943–953.
7. de Groot CJM, Merrill DC, Taylor RN, Kitzmiller JL, Goldsmith PC, Roberts JM. Increased von willebrand factor expression in an experimental model of preeclampsia produced by reduction of uteroplacental perfusion pressure in conscious rhesus monkeys. *Hypertens Pregnancy*. 1997;16:177–185.
8. Grundmann M, Woywodt A, Kirsch T, et al. Circulating endothelial cells: a marker of vascular damage in patients with preeclampsia. *Am J Obstet Gynecol*. 2008;198:317.e311–317.e315.
9. Khong TY, Sawyer IH, Heryet AR. An immunohistologic study of endothelialization of uteroplacental vessels in human pregnancy: evidence that endothelium is focally disrupted by trophoblast in preeclampsia. *Am J Obstet Gynecol*. 1992;167:751–756.
10. Zhou Y, Damsky CH, Fisher SJ. Preeclampsia is associated with failure of human cytotrophoblasts to mimic a vascular adhesion phenotype. One cause of defective endovascular invasion in this syndrome? *J Clin Invest*. 1997;99:2152–2164.
11. Zeek PM, Assali NS. Vascular changes in the decidua associated with eclamptogenic toxemia. *Am J Clin Pathol*. 1950;20:1099–1109.
12. Nadji P, Sommers SC. Lesions of toxemia in first trimester pregnancies. *Am J Clin Pathol*. 1973;59:344–348.
13. De Wolf F, Robertson WB, Brosens I. The ultrastructure of acute atherosis in hypertensive pregnancy. *Am J Obstet Gynecol*. 1975;123:164–174.
14. Raines EW, Ross R. Smooth muscle cells and the pathogenesis of the lesions of atherosclerosis. *Br Heart J*. 1993;69:S30–S37.
15. Roberts JM, Taylor RN, Musci TJ, Rodgers GM, Hubel CA, McLaughlin MK. Preeclampsia: an endothelial cell disorder. *Am J Obstet Gynecol*. 1989;161:1200–1204.
16. Jaffe EA. Cell biology of endothelial cells. *Hum Pathol*. 1987;18:234–239.
17. de Groot CJ, Taylor RN. New insights into the etiology of pre-eclampsia. *Ann Med*. 1993;25:243–249.
18. Roberts JM, Gammill HS. Preeclampsia: recent insights. *Hypertension*. 2005;46:1243–1249.
19. Pober JS, Cotran RS. Cytokines and endothelial cell biology. *Physiol Rev*. 1990;70:427–451.
20. Nachman RL, Silverstein R. Hypercoagulable states. *Ann Intern Med*. 1993;119:819–827.
21. Baxter RC, Martin JL. Radioimmunoassay of growth hormone-dependent insulin-like growth factor binding protein in human plasma. *J Clin Invest*. 1986;78:1504–1512.
22. Greer IA, Lyall F, Perera T, Boswell F, Macara LM. Increased concentration of cytokines interleukin-6 and imterleukin-1 receptor antagonist in plasma of women with preeclampsia: a mechanism for endothelial dysfunction? *Obstet Gynecol*. 1994;84:937–940.
23. Bukowski RM. Thrombotic thrombocytopenic purpura: a review. *Prog Hemostat Thromb*. 1982;6:287–337.
24. Weiner CP. Thrombotic microangiopathy in pregnancy and the postpartum period. *Semin Hematol*. 1987;24:119–129.
25. Chen YC, McLeod B, Hall ER, Wu KK. Accelerated prostacyclin degradation in thrombotic thrombocytopenic purpura. *Lancet*. 1981;2:267–269.
26. Mitra D, Jaffe EA, Weksler B, Hajjar KA, Soderland C, Laurence J. Thrombotic thrombocytopenic purpura and sporadic hemolytic-uremic syndrome plasmas induce apoptosis in restricted lineages of human microvascular endothelial cells. *Blood*. 1997;89:1224–1234.
27. Kitzmiller JL, Benirschke K. Immunofluorescent study of placental bed vessels in pre-eclampsia of pregnancy. *Am J Obstet Gynecol*. 1973;115(2):248–251.
28. Dizon-Townson DS, Nelson LM, Easton K, Ward K. The factor v leiden mutation may predispose women to severe preeclampsia. *Am J Obstet Gynecol*. 1996;175:902–905.

29. Lindoff C, Ingemarsson I, Martinsson G, Segelmark M, Thysell H, Astedt B. Preeclampsia is associated with a reduced response to activated protein C. *Am J Obstet Gynecol.* 1997;176:457–460.
30. Dekker GA, de Vries JI, Doelitzsch PM, et al. Underlying disorders associated with severe early-onset preeclampsia. *Am J Obstet Gynecol.* 1995;173:1042–1048.
31. Abou-Nassar K, Carrier M, Ramsay T, Rodger MA. The association between antiphospholipid antibodies and placenta mediated complications: a systematic review and meta-analysis. *Thromb Res.* 2011;128:77–85.
32. Taylor RN, Crombleholme WR, Friedman SA, Jones LA, Casal DC, Roberts JM. High plasma cellular fibronectin levels correlate with biochemical and clinical features of preeclampsia but cannot be attributed to hypertension alone. *Am J Obstet Gynecol.* 1991;165:895–901.
33. Hsu CD, Copel JA, Hong SF, Chan DW. Thrombomodulin levels in preeclampsia, gestational hypertension, and chronic hypertension. *Obstet Gynecol.* 1995;86:897–899.
34. Estelles A, Gilabert J, Espana F, Aznar J, Galbis M. Fibrinolytic parameters in normotensive pregnancy with intrauterine fetal growth retardation and in severe preeclampsia. *Am J Obstet Gynecol.* 1991;165:138–142.
35. Friedman SA, Schiff E, Emeis JJ, Dekker GA, Sibai BM. Biochemical corroboration of endothelial involvement in severe preeclampsia. *Am J Obstet Gynecol.* 1995;172:202–203.
36. Gao M, Nakabayashi M, Sakura M, Takeda Y. The imbalance of plasminogen activators and inhibitor in preeclampsia. *J Obstet Gynaecol Res.* 1996;22:9–16.
37. Shaarawy M, Didy HE. Thrombomodulin, plasminogen activator inhibitor type 1 (PAI-1) and fibronectin as biomarkers of endothelial damage in preeclampsia and eclampsia. *Int J Gynaecol Obstet.* 1996;55:135–139.
38. Hynes RO. *Cell Biology of the Extracellular Matrix.* New York: Plenum Press; 1981: 295–334.
39. Aznar-Salatti J, Bastida E, Buchanan MR, Castillo R, Ordinas A, Escolar G. Differential localization of von willebrand factor, fibronectin and 13-HODE in human endothelial cell cultures. *Histochemistry.* 1990;93:507–511.
40. Forsyth KD, Levinsky RJ. Fibronectin degradation; an in-vitro model of neutrophil mediated endothelial cell damage. *J Pathol.* 1990;161:313–319.
41. Lazarchick J, Stubbs TM, Romein L, Van Dorsten JP, Loadholt CB. Predictive value of fibronectin levels in normotensive gravid women destined to become preeclamptic. *Am J Obstet Gynecol.* 1986;154:1050–1052.
42. Ogawa S, Shreeniwas R, Butura C, Brett J, Stern DM. Modulation of endothelial function by hypoxia: perturbation of barrier and anticoagulant function, and induction of a novel factor X activator. *Adv Exp Med Biol.* 1990;281:303–312.
43. Lockwood CJ, Peters JH. Increased plasma levels of ED1+ cellular fibronectin precede the clinical signs of preeclampsia. *Am J Obstet Gynecol.* 1990;162:358–362.
44. Lyall F, Greer IA, Boswell F, Fleming R. Suppression of serum vascular endothelial growth factor immunoreactivity in normal pregnancy and in pre-eclampsia. *Br J Obstet Gynaecol.* 1997;104:223–228.
45. Krauss T, Kuhn W, Lakoma C, Augustin HG. Circulating endothelial cell adhesion molecules as diagnostic markers for the early identification of pregnant women at risk of developing preeclampsia. *Am J Obstet Gynecol.* 1997;177:443–449.
46. Taylor RN, Heilbron DC, Roberts JM. Growth factor activity in the blood of women in whom preeclampsia develops is elevated from early pregnancy. *Am J Obstet Gynecol.* 1990;163:1839–1844.
47. Taylor RN, Musci TJ, Kuhn RW, Roberts JM. Partial characterization of a novel growth factor from the blood of women with preeclampsia. *J Clin Endocrinol Metab.* 1990;70:1285–1291.
48. Booz GW, Baker KM. Molecular signalling mechanisms controlling growth and function of cardiac fibroblasts. *Cardiovasc Res.* 1995;30:537–543.
49. Varma M, de Groot CJM, Lanyi S, Taylor RN. Evaluation of plasma insulin-like growth factor-binding protein-3 as a potential predictor of preeclampsia. *Am J Obstet Gynecol.* 1993;169:995–999.
50. de Groot CJM, O'Brien TJ, Taylor RN. Biochemical edivence of impaired trophoblastic invasion of decidual stroma in omwne destined to have preeclampsia. *Am J Obstet Gynecol.* 1996;175:24–29.
51. Hietala R, Pohja-Nylander P, Rutanen EM, Laatikainen T. Serum insulin-like growth factor binding protein-1 at 16 weeks and subsequent preeclampsia. *Obstet Gynecol.* 2000;95:185–189.
52. Howell RJ, Economides D, Teisner B, Farkas AG, Chard T. Placental proteins 12 and 14 in pre-eclampsia. *Acta Obstet Gynecol Scand.* 1989;68:237–240.
53. Giudice LC, Martina NA, Crystal RA, Tazuke S, Druzin M. Insulin-like growth factor binding protein-1 at the maternal-fetal interface and insulin-like growth factor-I, insulin-like growth factor-II, and insulin-like growth factor binding protein-1 in the circulation of women with severe preeclampsia. *Am J Obstet Gynecol.* 1997;176:751–757.
54. Dijke P, Goumans M-J, Pardali E. Endoglin in angiogenesis and vascular diseases. *Angiogenesis.* 2008;11:79–89.
55. Castonguay R, Werner ED, Matthews RG, et al. Soluble endoglin specifically binds bone morphogenetic proteins 9 and 10 via its orphan domain, inhibits blood vessel formation, and suppresses tumor growth. *J Biol Chem.* 2011;286:30034–30046.
56. Ahlawat S, Pati HP, Bhatla N, Fatima L, Mittal S. Plasma platelet aggregating factor and platelet aggregation studies in pre-eclampsia. *Acta Obstet Gynecol Scand.* 1996;75:428–431.
57. Klockenbusch W, Hohlfeld T, Wilhelm M, Somville T, Schror K. Platelet PGI_2 receptor affinity is reduced in pre-eclampsia. *Br J Clin Pharmacol.* 1996;41:616–618.
58. North RA, Ferrier C, Gamble G, Fairley KF, Kincaid-Smith P. Prevention of preeclampsia with heparin and antiplatelet drugs in women with renal disease. *Aust N Z J Obstet Gynaecol.* 1995;35:357–362.
59. Lees C, Langford E, Brown AS, et al. The effects of S-nitrosoglutathione on platelet activation, hypertension, and uterine and fetal doppler in severe preeclampsia. *Obstet Gynecol.* 1996;88:14–19.
60. Halim A, Kanayama N, El Maradny E, Maehara K, Bhuiyan AB, Terao T. Correlated plasma elastase and sera cytotoxicity in eclampsia. A possible role of endothelin-1 induced neutrophil activation in preeclampsia-eclampsia. *Am J Hypertens.* 1996;9:33–38.

61. Rebelo I, Carvalho-Guerra F, Pereira-Leite L, Quintanilha A. Comparative study of lactoferrin and other blood markers of inflammatory stress between preeclamptic and normal pregnancies. *Eur J Obstet Gynec Reprod Biol.* 1996;64:167–173.
62. Barden A, Graham D, Beilin LJ, et al. Neutrophil CD11b expression and neutrophil activation in pre-eclampsia. *Clin Sci.* 1997;92:37–44.
63. Studena K, Sacks GP, Sargent IL, Redman CWG. Leucocyte phenotypic and functional changes in preeclampsia and normal pregnancy. Combined detection by whole blood flow cytometry. *J Soc Gynecol Invest.* 1997;4(suppl):664. [abst.].
64. Redman CW, Sargent IL. Latest advances in understanding preeclampsia. *Science.* 2005;308:1592–1594.
65. Salafia CM, Ghidini A, Minior VK. Uterine allergy: a cause of preterm birth? *Obstet Gynecol.* 1996;88:451–454.
66. Gandley RE, Rohland J, Zhou Y, et al. Increased myeloperoxidase in the placenta and circulation of women with preeclampsia. *Hypertension.* 2008;52:387–393.
67. Smarason AK, Sargent IL, Redman CWG. Endothelial cell proliferation is suppressed by plasma but not serum from women with preeclampsia. *Am J Obstet Gynecol.* 1996;174:787–793.
68. Cockell AP, Learmont JG, Smarason AK, Redman CW, Sargent IL, Poston L. Human placental syncytiotrophoblast microvillous membranes impair maternal vascular endothelial function. *Br J Obstet Gynaecol.* 1997;104:235–240.
69. Redman CW, Sargent IL. Circulating microparticles in normal pregnancy and pre-eclampsia. *Placenta.* 2008;29(suppl A):S73–S77.
70. Sipos PI, Crocker IP, Hubel CA, Baker PN. Endothelial progenitor cells: their potential in the placental vasculature and related complications. *Placenta.* 2010;31:1–10.
71. Loomans CJM, de Koning EJP, Staal FJT, et al. Endothelial progenitor cell dysfunction: a novel concept in the pathogenesis of vascular complications of type 1 diabetes. *Diabetes.* 2004;53:195–199.
72. Vasa M, Fichtlscherer S, Aicher A, et al. Number and migratory activity of circulating endothelial progenitor cells inversely correlate with risk factors for coronary artery disease. *Circ Res.* 2001;89:e1–e7.
73. Werner N, Kosiol S, Schiegl T, et al. Circulating endothelial progenitor cells and cardiovascular outcomes. *New Engl J Med.* 2005;353:999–1007.
74. Murphy C, Kanaganayagam GS, Jiang B, et al. Vascular dysfunction and reduced circulating endothelial progenitor cells in young healthy UK South Asian men. *Arterioscl Throm Vas.* 2007;27:936–942.
75. Buemi M, Allegra A, D'Anna R, et al. Concentration of circulating endothelial progenitor cells (EPC) in normal pregnancy and in pregnant women with diabetes and hypertension. *Am J Obstet Gynecol.* 2007;196:68.e61–68.e66.
76. Sugawara J, Mitsui-Saito M, Hoshiai T, Hayashi C, Kimura Y, Okamura K. Circulating endothelial progenitor cells during human pregnancy. *J Clin Endocrinol Metab.* 2005;90:1845–1848.
77. Luppi P, Powers RW, Verma V, Edmunds L, Plymire D, Hubel CA. Maternal circulating CD34+VEGFR-2+ and CD133 + VEGFR-2 + progenitor cells increase during normal pregnancy but are reduced in women with preeclampsia. *Reprod Sci.* 2010;17:643–652.
78. Sugawara J, Mitsui-Saito M, Hayashi C, et al. Decrease and senescence of endothelial progenitor cells in patients with preeclampsia. *J Clin Endocrinol Metab.* 2005;90:5329–5332.
79. Lin C, Rajakumar A, Plymire DA, Verma V, Markovic N, Hubel CA. Maternal endothelial progenitor colony-forming units with macrophage characteristics are reduced in preeclampsia. *Am J Hypertens.* 2009;22:1014–1019.
80. Hubel CA, Sipos PI, Crocker IP. Endothelial progenitor cells: their potential role in pregnancy and preeclampsia. *Pregnancy Hypertens.* 2011;1:48–58.
81. Murphy MSQ, Casselman RC, Smith GN. Postpartum alterations in circulating endothelial progenitor cells in women with a history of pre-eclampsia. *Pregnancy Hypertens.* 2013 [in press].
82. Sibai BM, Gordon T, Thom E, et al. Risk factors for preeclampsia in healthy nulliparous women: a prospective multicenter study. The National Institute of Child Health and Human Development network of maternal-fetal medicine units. *Am J Obstet Gynecol.* 1995;172:642–648.
83. Houser MT, Fish AJ, Tagatz GE, Williams PP, Michael AF. Pregnancy and systemic lupus erythematosus. *Am J Obstet Gynecol.* 1980;138:409–413.
84. Brosens IA, Robertson WB, Dixon HG. The role of the spiral arteries in the pathogenesis of preeclampsia. *Obstet Gynecol Annu.* 1972;1:177–191.
85. Robillard PY, Dekker G, Chaouat G, Hulsey TC. Etiology of preeclampsia: maternal vascular predisposition and couple disease – mutual exclusion or complementarity? *J Reprod Immunol.* 2007;76:1–7.
86. Robillard PY, Hulsey TC, Alexander GR, Keenan A, de Caunes F, Papiernik E. Paternity patterns and risk of preeclampsia in the last pregnancy in multiparae. *J Reprod Immunol.* 1993;24:1–12.
87. Klonoff-Cohen HS, Savitz DA, Cefalo RC, McCann MF. An epidemiologic study of contraception and preeclampsia. *JAMA.* 1989;262:3143–3147.
88. Soderstrom-Anttila V, Hovatta O. An oocyte donation program with goserelin down-regulation of voluntary donors. *Acta Obstet Gynecol Scand.* 1995;74:288–292.
89. Feeney JG, Tovey LA, Scott JS. Influence of previous blood-transfusion on incidence of pre-eclampsia. *Lancet.* 1977;1:874–875.
90. Koelman CA, Coumans AB, Nijman HW, Doxiadis II, Dekker GA, Claas FH. Correlation between oral sex and a low incidence of preeclampsia: a role for soluble HLA in seminal fluid? *J Reprod Immunol.* 2000;46:155–166.
91. Robillard PY, Hulsey TC, Perianin J, Janky E, Miri EH, Papiernik E. Association of pregnancy-induced hypertension with duration of sexual cohabitation before conception. *Lancet.* 1994;344:973–975.
92. Taylor RN. Review: immunobiology of preeclampsia. *Am J Reprod Immunol.* 1997;37:79–86.
93. Allen JY, Tapia-Santiago C, Kutteh WH. Antiphospholipid antibodies in patients with preeclampsia. *Am J Reprod Immunol.* 1996;36:81–85.
94. Shaarawy M, El-Mallah SY, El-Yamani AMA. The prevalence of serum antineutrophil cytoplasmic autoantibodies in preeclampsia and eclampsia. *J Soc Gynecol Invest.* 1997;4:34–39.

95. Katano K, Aoki A, Sasa H, Ogasawara M, Matsuura E, Yagami Y. Beta 2-glycoprotein I-dependent anticardiolipin antibodies as a predictor of adverse pregnancy outcomes in healthy pregnant women. *Hum Reprod.* 1996;11:509–512.
96. Rodgers GM, Taylor RN, Roberts JM. Preeclampsia is associated with a serum factor cytotoxic to human endothelial cells. *Am J Obstet Gynecol.* 1988;159:908–914.
97. Rappaport VJ, Hirata G, Yap HK, Jordan SC. Anti-vascular endothelial cell antibodies in severe preeclampsia. *Am J Obstet Gynecol.* 1990;162:138–146.
98. Tannenbaum SH, Finko R, Cines DB. Antibody and immune complexes induce tissue factor production by human endothelial cells. *J Immunol.* 1986;137:1532–1537.
99. de Groot CJ, Davidge ST, Friedman SA, McLaughlin MK, Roberts JM, Taylor RN. Plasma from preeclamptic women increases human endothelial cell prostacyclin production without changes in cellular enzyme activity or mass. *Am J Obstet Gynecol.* 1995;172:976–985.
100. Branch DW, Dudley DJ, LaMarche S, Mitchell MD. Sera from preeclamptic patients contain factor(s) that stimulate prostacyclin production by human endothelial cells. *Prostaglandins Leukot Essent Fatty Acids.* 1992;45:191–195.
101. de Groot CJM, Murai JT, Vigne J-L, Taylor RN. Eicosanoid secretion by human endothelial cells exposed to normal pregnancy and preeclampsia plasma *in vitro*. *Prostaglandins Leukot Essent Fatty Acids.* 1998;58:91–97.
102. Rote NS, Walter A, Lyden TW. Antiphospholipid antibodies – lobsters or red herrings? *Am J Reprod Immunol.* 1992;28:31–37.
103. Peaceman AM, Rehnberg KA. The effect of aspirin and indomethacin on prostacyclin and thromboxane production by placental tissue incubated with immunoglobulin g fractions from patients with lupus anticoagulant. *Am J Obstet Gynecol.* 1995;173:1391–1396.
104. Dechend R, Homuth V, Wallukat G, et al. Agonistic antibodies directed at the angiotensin ii, at1 receptor in preeclampsia. *J Soc Gynecol Invest.* 2006;13:79–86.
105. LaMarca B. The role of immune activation in contributing to vascular dysfunction and the pathophysiology of hypertension during preeclampsia. *Minerva Ginecol.* 2010;62:105–120.
106. Saito S, Shiozaki A, Nakashima A, Sakai M, Sasaki Y. The role of the immune system in preeclampsia. *Mol Aspects Med.* 2007;28:192–209.
107. Redman CWG, Sacks GP, Sargent IL. Preeclampsia: an excessive maternal inflammatory response to pregnancy. *Am J Obstet Gynecol.* 1999;180:499–506.
108. Luppi P, Tse H, Lain KY, Markovic N, Piganelli JD, DeLoia JA. Preeclampsia activates circulating immune cells with engagement of the NF-κb pathway. *Am J Reprod Immunol.* 2006;56:135–144.
109. Saito S, Umekage H, Sakamoto Y, et al. Increased T-helper-1-type immunity and decreased T-helper-2-type immunity in patients with preeclampsia. *Am J Reprod Immunol.* 1999;41:297–306.
110. Conrad KP, Miles TM, Benyo DF. Circulating levels of immunoreactive cytokines in women with preeclampsia. *Am J Reprod Immunol.* 1998;40:102–111.
111. Hamai Y, Fujii T, Yamashita T, et al. Evidence for an elevation in serum interleukin-2 and tumor necrosis factor-α levels before the clinical manifestations of preeclampsia. *Am J Reprod Immunol.* 1997;38:89–93.
112. Szarka A, Rigo J, Lazar L, Beko G, Molvarec A. Circulating cytokines, chemokines and adhesion molecules in normal pregnancy and preeclampsia determined by multiplex suspension array. *BMC Immunol.* 2010;11:59.
113. Daniel Y, Kupferminc MJ, Baram A, et al. Plasma interleukin-12 is elevated in patients with preeclampsia. *Am J Reprod Immunol.* 1998;39:376–380.
114. Molvarec A, Ito M, Shima T, et al. Decreased proportion of peripheral blood vascular endothelial growth factor–expressing T and natural killer cells in preeclampsia. *Am J Obstet Gynecol.* 2010;203:567.e561–567.e568.
115. Hamai Y, Fujii T, Yamashita T, et al. Evidence for an elevation in serum interleukin-2 and tumor necrosis factor-alpha levels before the clinical manifestations of preeclampsia. *Am J Reprod Immunol.* 1997;38:89–93.
116. Williams MA, Farrand A, Mittendorf R, et al. Maternal second trimester serum tumor necrosis factor-alpha-soluble receptor p55 (sTNFp55) and subsequent risk of preeclampsia. *Am J Epidemiol.* 1999;149:323–329.
117. Kupferminc MJ, Mullen TA, Russell TL, Silver RK. Serum from patients with severe preeclampsia is not cytotoxic to endothelial cells. *J Soc Gynecol Invest.* 1996;3:89–92.
118. Rinehart BK, Terrone DA, Lagoo-Deenadayalan S, et al. Expression of the placental cytokines tumor necrosis factor alpha, interleukin 1beta, and interleukin 10 is increased in preeclampsia. *Am J Obstet Gynecol.* 1999;181:915–920.
119. Benyo DF, Smarason A, Redman CW, Sims C, Conrad KP. Expression of inflammatory cytokines in placentas from women with preeclampsia. *J Clin Endocr Metab.* 2001;86:2505–2512.
120. Yoshizumi M, Perrella MA, Burnett Jr JC, Lee ME. Tumor necrosis factor downregulates an endothelial nitric oxide synthase mrna by shortening its half-life. *Circ Res.* 1993;73:205–209.
121. Marsden PA, Brenner BM. Transcriptional regulation of the endothelin-1 gene by tnf-alpha. *Am J Physiol.* 1992;262:C854–C861.
122. Taylor RN, Varma M, Teng NNH, Roberts JM. Women with preeclampsia have higher plasma endothelin levels than women with normal pregnancies. *J Clin Endocrinol Metab.* 1990;71:1675–1677.
123. Stark JM. Pre-eclampsia and cytokine induced oxidative stress. *Brit J Obstet Gynaecol.* 1993;100:105–109.
124. Pober JS, Gimbrone Jr MA, Lapierre LA, et al. Overlapping patterns of activation of human endothelial cells by interleukin 1, tumor necrosis factor, and immune interferon. *J Immunol.* 1986;137:1893–1896.
125. Bevilacqua MP, Pober JS, Majeau GR, Cotran RS, Gimbrone Jr. MA. Interleukin 1 (IL-1) induces biosynthesis and cell surface expression of procoagulant activity in human vascular endothelial cells. *J Exp Med.* 1984;160:618–623.
126. Nawroth PP, Stern DM. Modulation of endothelial cell hemostatic properties by tumor necrosis factor. *J Exp Med.* 1986;163:740–745.
127. van Hinsbergh VW, Kooistra T, van den Berg EA, Princen HM, Fiers W, Emeis JJ. Tumor necrosis factor increases the production of plasminogen activator inhibitor in

human endothelial cells in vitro and in rats in vivo. *Blood.* 1988;72:1467–1473.

128. Alexander BT, Cockrell KL, Massey MB, Bennett WA, Granger JP. Tumor necrosis factor-alpha-induced hypertension in pregnant rats results in decreased renal neuronal nitric oxide synthase expression. *Am J Hypertens.* 2002;15:170–175.
129. Vizi ES, Szelenyi J, Selmeczy ZS, Papp Z, Nemeth ZH, Hasko G. Enhanced tumor necrosis factor-alpha-specific and decreased interleukin-10-specific immune responses to lps during the third trimester of pregnancy in mice. *J Endocrinol.* 2001;171:355–361.
130. Jacobson PB, Schrier DJ. Regulation of cd11b/cd18 expression in human neutrophils by phospholipase a2. *J Immunol.* 1993;151:5639–5652.
131. Pfeilschifter J, Schalkwijk C, Briner VA, van den Bosch H. Cytokine-stimulated secretion of group ii phospholipase a2 by rat mesangial cells. Its contribution to arachidonic acid release and prostaglandin synthesis by cultured rat glomerular cells. *J Clin Invest.* 1993;92:2516–2523.
132. Jendryczko A, Drózdz M. Increased placental phospholipase A_2 activities in pre-eclampsia. *Zentralbl Gynakol.* 1990;112:889–891.
133. Lim KH, Rice GE, deGroot CJ, Taylor RN. Plasma type ii phospholipase A2 levels are elevated in severe preeclampsia. *Am J Obstet Gynecol.* 1995;172:998–1002.
134. Opsjon SL, Novick D, Wathen NC, Cope AP, Wallach D, Aderka D. Soluble tumor necrosis factor receptors and soluble interleukin-6 receptor in fetal and maternal sera, coelomic and amniotic fluids in normal and pre-eclamptic pregnancies. *J Reprod Immunol.* 1995;29:119–134.
135. Vince GS, Starkey PM, Austgulen R, Kwiatkowski D, Redman CW. Interleukin-6, tumour necrosis factor and soluble tumour necrosis factor receptors in women with pre-eclampsia. *Br J Obstet Gynaecol.* 1995;102:20–25.
136. Domingo CG, Domenech N, Aparicio P, Palomino P. Human pregnancy serum inhibits proliferation of T8-depleted cells and their interleukin-2 synthesis in mixed lymphocyte cultures. *J Reprod Immunol.* 1985;8:97–110.
137. Tjoa ML, Cindrova-Davies T, Spasic-Boskovic O, Bianchi DW, Burton GJ. Trophoblastic oxidative stress and the release of cell-free feto-placental DNA. *Am J Pathol.* 2006;169:400–404.
138. Stallmach T, Hebisch G, Joller H, Kolditz P, Engelmann M. Expression pattern of cytokines in the different compartments of the feto-maternal unit under various conditions. *Reprod Fertil Dev.* 1995;7:1573–1580.
139. Arbogast BW, Leeper SC, Merrick RD, Olive KE, Taylor RN. Which plasma factors bring about disturbance of endothelial function in pre-eclampsia? *Lancet.* 1994;343:340–341.
140. Endresen MJ, Tosti E, Heimli H, Lorentzen B, Henriksen T. Effects of free fatty acids found increased in women who develop pre-eclampsia on the ability of endothelial cells to produce prostacyclin, cgmp and inhibit platelet aggregation. *Scand J Clin Lab Invest.* 1994;54:549–557.
141. Dekker GA, van Geijn HP. Endothelial dysfunction in pre-eclampsia. Part II: reducing the adverse consequences of endothelial cell dysfunction in preeclampsia; therapeutic perspectives. *J Perinat Med.* 1996;24:119–139.
142. Husby H, Roald B, Schjetlein R, Nesheim BI, Berg K. High levels of Lp(a) lipoprotein in a family with cases of severe pre-eclampsia. *Clin Genet.* 1996;50:47–49.
143. Ray JG, Diamond P, Singh G, Bell CM. Brief overview of maternal triglycerides as a risk factor for pre-eclampsia. *BJOG.* 2006;113:379–386.
144. Potter JM, Nestel PJ. The hyperlipidemia of pregnancy in normal and complicated pregnancies. *Am J Obstet Gynecol.* 1979;133:165–170.
145. Hubel CA. Dyslipidemia, iron, and oxidative stress in pre-eclampsia: assessment of maternal and feto-placental interactions. *Semin Reprod Endocrinol.* 1998;16:75–92.
146. Hubel CA, McLaughlin MK, Evans RW, Hauth BA, Sims CJ, Roberts JM. Fasting serum triglycerides, free fatty acids, and malondialdehyde are increased in preeclampsia, are positively correlated, and decrease within 48 hours post partum. *Am J Obstet Gynecol.* 1996;174:975–982.
147. Hubel CA, Lyall F, Weissfeld L, Gandley RE, Roberts JM. Small low-density lipoproteins and vascular cell adhesion molecule-1 are increased in association with hyperlipidemia in preeclampsia. *Metabolism.* 1998;47:1281–1288.
148. Witztum JL. Susceptibility of low-density lipoprotein to oxidative modification. *Am J Med.* 1993;94:347–349.
149. Anber V, Griffin BA, McConnell M, Packard CJ, Shepherd J. Influence of plasma lipid and LDL-subfraction profile on the interaction between low density lipoprotein with human arterial wall proteoglycans. *Atherosclerosis.* 1996;124:261–271.
150. Sattar N, Bendomir A, Berry C, Shepherd J, Greer IA, Packard CJ. Lipoprotein subfraction concentrations in pre-eclampsia: pathogenic parallels to atherosclerosis. *Obstet Gynecol.* 1997;89:403–408.
151. Raijmakers MT, van Tits BJ, Hak-Lemmers HL, Roes EM, Steegers EA, Peters WH. Low plasma levels of oxidized low density lipoprotein in preeclampsia. *Acta Obstet Gynecol Scand.* 2004;83:1173–1177.
152. Branch DW, Mitchell MD, Miller E, Palinski W, Witztum JL. Pre-eclampsia and serum antibodies to oxidised low-density lipoprotein. *Lancet.* 1994;343:645–646.
153. Uotila J, Solakivi T, Jaakkola O, Tuimala R, Lehtimaki T. Antibodies against copper-oxidised and malondialdehyde-modified low density lipoproteins in pre-eclampsia pregnancies. *Br J Obstet Gynaecol.* 1998;105:1113–1117.
154. Gratacos E, Casals E, Deulofeu R, et al. Serum antibodies to oxidized low-density lipoprotein in pregnant women with preeclampsia and chronic hypertension: lack of correlation with lipid peroxides. *Hypertens Pregnancy.* 2001;20:177–183.
155. Armstrong VW, Wieland E, Diedrich F, et al. Serum antibodies to oxidised low-density lipoprotein in pre-eclampsia and coronary heart disease. *Lancet.* 1994;343:1570.
156. Sawamura T, Kume N, Aoyama T, et al. An endothelial receptor for oxidized low-density lipoprotein. *Nature.* 1997;386:73–77.
157. Yamada Y, Doi T, Hamakubo T, Kodama T. Scavenger receptor family proteins: roles for atherosclerosis, host defence and disorders of the central nervous system. *Cell Mol Life Sci.* 1998;54:628–640.
158. Cominacini L, Rigoni A, Pasini AF, et al. The binding of oxidized low density lipoprotein (ox-LDL) to ox-LDL receptor-1 reduces the intracellular concentration of nitric oxide in

endothelial cells through an increased production of superoxide. *J Biol Chem.* 2001;276:13750–13755.

159. Hofnagel O, Luechtenborg B, Stolle K, et al. Proinflammatory cytokines regulate LOX-1 expression in vascular smooth muscle cells. *Arterioscler Thromb Vasc Biol.* 2004;24:1789–1795.
160. Morawietz H, Rueckschloss U, Niemann B, et al. Angiotensin II induces LOX-1, the human endothelial receptor for oxidized low-density lipoprotein. *Circulation.* 1999;100:899–902.
161. Morawietz H, Duerrschmidt N, Niemann B, Galle J, Sawamura T, Holtz J. Augmented endothelial uptake of oxidized low-density lipoprotein in response to endothelin-1. *Clin Sci (Lond).* 2002;103(suppl 48):9S–12S.
162. Alexander BT, Llinas MT, Kruckeberg WC, Granger JP. L-arginine attenuates hypertension in pregnant rats with reduced uterine perfusion pressure. *Hypertension.* 2004;43:832–836.
163. Halvorsen B, Staff AC, Henriksen T, Sawamura T, Ranheim T. 8-iso-prostaglandin F(2alpha) increases expression of lox-1 in jar cells. *Hypertension.* 2001;37:1184–1190.
164. Sankaralingam S, Arenas IA, Lalu MM, Davidge ST. Preeclampsia: current understanding of the molecular basis of vascular dysfunction. *Exp Rev Mol Med.* 2006;8:1–20.
165. Morton JS, Abdalvand A, Jiang Y, Sawamura T, Uwiera RRE, Davidge ST. Lectin-like oxidized low-density lipoprotein 1 receptor in a reduced uteroplacental perfusion pressure rat model of preeclampsia. *Hypertension.* 2012;59:1014–1020.
166. English FA, McCarthy FP, McSweeney CL, et al. Inhibition of lectin-like oxidized low-density lipoprotein-1 receptor protects against plasma-mediated vascular dysfunction associated with pre-eclampsia. *Am J Hypertens.* 2013;26:279–286.
167. Lorentzen B, Drevon CA, Endresen MJ, Henriksen T. Fatty acid pattern of esterified and free fatty acids in sera of women with normal and pre-eclamptic pregnancy. *Br J Obstet Gynaecol.* 1995;102:530–537.
168. Arbogast B, Leeper S, Merrick R, Olive K, Taylor R. Plasma factors that determine endothelial cell lipid toxicity in vitro correctly identify women with preeclampsia in early and late pregnancy. *Hypertens Pregnancy.* 1996;15:263–279.
169. Vigne J-L, Murai JT, Arbogast BW, Jia W, Fisher SJ, Taylor RN. Elevated non-esterified fatty acid concentrations in severe preeclampsia shift the isoelectric characteristics of plasma albumin. *J Clin Endocrinol Metab.* 1997;83
170. Dekker GA, Sibai BM. Etiology and pathogenesis of preeclampsia: current concepts. *Am J Obstet Gynecol.* 1998;179:1359–1375.
171. Endresen MJ, Lorentzen B, Henriksen T. Increased lipolytic activity and high ratio of free fatty acids to albumin in sera from women with preeclampsia leads to triglyceride accumulation in cultured endothelial cells. *Am J Obstet Gynecol.* 1992;167:440–447.
172. Robinson NJ, Minchell LJ, Myers JE, Hubel CA, Crocker IP. A potential role for free fatty acids in the pathogenesis of preeclampsia. *J Hypertens.* 2009;27:1293–1302.
173. Gryzunov YA, Arroyo A, Vigne JL, et al. Binding of fatty acids facilitates oxidation of cysteine-34 and converts copper-albumin complexes from antioxidants to prooxidants. *Arch Biochem Biophys.* 2003;413:53–66.
174. Kagan VE, Tyurin VA, Borisenko GG, et al. Mishandling of copper by albumin: role in redox-cycling and oxidative stress in preeclampsia plasma. *Hypertens Pregnancy.* 2001;20:221–241.
175. Tyurin VA, Liu SX, Tyurina YY, et al. Elevated levels of S-nitrosoalbumin in preeclampsia plasma. *Circ Res.* 2001;88:1210–1215.
176. Spector AA, Fang X, Snyder GD, Weintraub NL. Epoxyeicosatrienoic acids (EETs): metabolism and biochemical function. *Prog Lipid Res.* 2004;43:55–90.
177. Roman RJ. P-450 metabolites of arachidonic acid in the control of cardiovascular function. *Physiol Rev.* 2002;82:131–185.
178. Joiakim A, Park J-A, Kaplan DJ, Putt D, Taylor RN, Kim H. Soluble epoxide hydrolase- and UDP-glucuronosyltransferase-dependent hypertension in pregnancy. *Annual Meeting of the American Society for Biochemistry and Molecular Biology.* 2013.
179. Ng VY, Huang Y, Reddy LM, Falck JR, Lin ET, Kroetz DL. Cytochrome P450 eicosanoids are activators of peroxisome proliferator-activated receptor α. *Drug Metab Dispos.* 2007;35:1126–1134.
180. Barak Y, Nelson MC, Ong ES, et al. Ppar gamma is required for placental, cardiac, and adipose tissue development. *Mol Cell.* 1999;4:585–595.
181. Kersten S, Desvergne B, Wahli W. Roles of ppars in health and disease. *Nature.* 2000;405:421–424.
182. Schaiff WT, Barak Y, Sadovsky Y. The pleiotropic function of ppar gamma in the placenta. *Mol Cell Endocrinol.* 2006;249:10–15.
183. Fournier T, Tsatsaris V, Handschuh K, Evain-Brion D. PPARs and the placenta. *Placenta.* 2007;28:65–76.
184. Wieser F, Waite L, Depoix C, Taylor RN. PPAR action in human placental development and pregnancy and its complications. *PPAR Res.* 2008;2008:527048.
185. Waite LL, Atwood AK, Taylor RN. Preeclampsia, an implantation disorder. *Rev Endocr Metab Disord.* 2002;3:151–158.
186. Waite LL, Louie RE, Taylor RN. Circulating activators of peroxisome proliferator-activated receptors are reduced in preeclamptic pregnancy. *J Clin Endocr Metab.* 2005;90:620–626.
187. Redman CW, Sargent IL. Pre-eclampsia, the placenta and the maternal systemic inflammatory response – a review. *Placenta.* 2003;24(suppl A):S21–S27.
188. Barroso I, Gurnell M, Crowley VE, et al. Dominant negative mutations in human ppargamma associated with severe insulin resistance, diabetes mellitus and hypertension. *Nature.* 1999;402:880–883.
189. Laasanen J, Heinonen S, Hiltunen M, Mannermaa A, Laakso M. Polymorphism in the peroxisome proliferator-activated receptor-gamma gene in women with preeclampsia. *Early Hum Dev.* 2002;69:77–82.
190. Vaisman N, Gospodarowicz D, Neufeld G. Characterization of the receptors for vascular endothelial growth factor. *J Biol Chem.* 1990;265:19461–19466.
191. Quinn TP, Peters KG, De Vries C, Ferrara N, Williams LT. Fetal liver kinase 1 is a receptor for vascular

endothelial growth factor and is selectively expressed in vascular endothelium. *Proc Natl Acad Sci U S A*. 1993;90: 7533–7537.

192. Brockelsby J, Hayman R, Ahmed A, Warren A, Johnson I, Baker P. VEGF via VEGF receptor-1 (Flt-1) mimics preeclamptic plasma in inhibiting uterine blood vessel relaxation in pregnancy: implications in the pathogenesis of preeclampsia. *Lab Invest*. 1999;79:1101–1111.
193. Gandley RE, Tyurin VA, Huang W, et al. S-nitrosoalbumin-mediated relaxation is enhanced by ascorbate and copper: effects in pregnancy and preeclampsia plasma. *Hypertension*. 2005;45:21–27.
194. Kim I, Moon SO, Kim SH, Kim HJ, Koh YS, Koh GY. Vascular endothelial growth factor expression of intercellular adhesion molecule 1 (ICAM-1), vascular cell adhesion molecule 1 (VCAM-1), and E-selectin through nuclear factor-kappa b activation in endothelial cells. *J Biol Chem*. 2001;276:7614–7620.
195. Taylor RN, Grimwood J, Taylor RS, McMaster MT, Fisher SJ, North RA. Longitudinal serum concentrations of placental growth factor: evidence for abnormal placental angiogenesis in pathologic pregnancies. *Am J Obstet Gynecol*. 2003;188:177–182.
196. Levine RJ, Maynard SE, Qian C, et al. Circulating angiogenic factors and the risk of preeclampsia. *New Engl J Med*. 2004;350:672–683.
197. Hurwitz H, Saini S. Bevacizumab in the treatment of metastatic colorectal cancer: safety profile and management of adverse events. *Semin Oncol*. 2006;33:S26–S34.
198. Jelkmann W. Pitfalls in the measurement of circulating vascular endothelial growth factor. *Clin Chem*. 2001;47:617–623.
199. Molskness TA, Stouffer RL, Burry KA, Gorrill MJ, Lee DM, Patton PE. Circulating levels of free and total vascular endothelial growth factor (VEGF)-a, soluble VEGF receptors-1 and -2, and angiogenin during ovarian stimulation in non-human primates and women. *Hum Reprod*. 2004;19:822–830.
200. Clark D, Smith S, Licence D, Evans A, Charnock-Jones D. Comparison of expression patterns for placenta growth factor, vascular endothelial growth factor (VEGF), VEGF-b and VEGF-c in the human placenta throughout gestation. *J Endocrinol*. 1998;159(3):459–467.
201. Shibuya M. Structure and function of VEGF/VEGF-receptor system involved in angiogenesis. *Cell Struct Funct*. 2001;26:25–35.
202. Vuorela P, Helske S, Hornig C, Alitalo K, Weich H, Halmesmaki E. Amniotic fluid-soluble vascular endothelial growth factor receptor-1 in preeclampsia. *Obstet Gynecol*. 2000;95:353–357.
203. Maynard SE, Min JY, Merchan J, et al. Excess placental soluble fms-like tyrosine kinase 1 (sflt1) may contribute to endothelial dysfunction, hypertension, and proteinuria in preeclampsia. *J Clin Invest*. 2003;111:649–658.
204. Gu Y, Lewis DF, Wang Y. Placental productions and expressions of soluble endoglin, soluble fms-like tyrosine kinase receptor-1, and placental growth factor in normal and preeclamptic pregnancies. *J Clin Endocr Metab*. 2008;93:260–266.
205. Levine RJ, Thadhani R, Qian C, et al. Urinary placental growth factor and risk of preeclampsia. *JAMA*. 2005;293:77–85.
206. De Vivo A, Baviera G, Giordano D, Todarello G, Corrado F, D'Anna R. Endoglin, PLGF and sFlt-1 as markers for predicting pre-eclampsia. *Acta Obstet Gynecol Scand*. 2008;87:837–842.
207. Hanahan D. Signaling vascular morphogenesis and maintenance. *Science*. 1997;277:48–50.
208. Taylor RN, Musci TJ, Rodgers GM, Roberts JM. Preeclamptic sera stimulate increased platelet-derived growth factor mrna and protein expression by cultured human endothelial cells. *Am J Reprod Immunol*. 1991;25:105–108.
209. Lebrin F, Deckers M, Bertolino P, Ten Dijke P. TGF-beta receptor function in the endothelium. *Cardiovasc Res*. 2005;65:599–608.
210. Staff AC, Braekke K, Johnsen GM, Karumanchi SA, Harsem NK. Circulating concentrations of soluble endoglin (cd105) in fetal and maternal serum and in amniotic fluid in preeclampsia. *Am J Obstet Gynecol*. 2007;197:176.e171–176.e176.
211. Visse R, Nagase H. Matrix metalloproteinases and tissue inhibitors of metalloproteinases: structure, function, and biochemistry. *Circ Res*. 2003;92:827–839.
212. Fernandez-Patron C, Stewart KG, Zhang Y, Koivunen E, Radomski MW, Davidge ST. Vascular matrix metalloproteinase-2-dependent cleavage of calcitonin gene-related peptide promotes vasoconstriction. *Circ Res*. 2000;87:670–676.
213. Galis ZS, Khatri JJ. Matrix metalloproteinases in vascular remodeling and atherogenesis: the good, the bad, and the ugly. *Circ Res*. 2002;90:251–262.
214. Fernandez-Patron C, Radomski MW, Davidge ST. Vascular matrix metalloproteinase-2 cleaves big endothelin-1 yielding a novel vasoconstrictor. *Circ Res*. 1999;85:906–911.
215. Sawicki G, Salas E, Murat J, Miszta-Lane H, Radomski MW. Release of gelatinase a during platelet activation mediates aggregation. *Nature*. 1997;386:616–619.
216. Narumiya H, Zhang Y, Fernandez-Patron C, Guilbert LJ, Davidge ST. Matrix metalloproteinase-2 is elevated in the plasma of women with preeclampsia. *Hypertens Pregnancy*. 2001;20:185–194.
217. Myers JE, Merchant SJ, Macleod M, Mires GJ, Baker PN, Davidge ST. Mmp-2 levels are elevated in the plasma of women who subsequently develop preeclampsia. *Hypertens Pregnancy*. 2005;24:103–115.
218. Kolben M, Lopens A, Blaser J, et al. Proteases and their inhibitors are indicative in gestational disease. *Eur J Obstet Gynecol Reprod Biol*. 1996;68:59–65.
219. Merchant SJ, Davidge ST. The role of matrix metalloproteinases in vascular function: implications for normal pregnancy and pre-eclampsia. *BJOG*. 2004;111:931–939.
220. Robbesyn F, Auge N, Vindis C, et al. High-density lipoproteins prevent the oxidized low-density lipoprotein-induced endothelial growth factor receptor activation and subsequent matrix metalloproteinase-2 upregulation. *Arterioscler Thromb Vasc Biol*. 2005.
221. Rajagopalan S, Kurz S, Munzel T, et al. Angiotensin II-mediated hypertension in the rat increases vascular

superoxide production via membrane nadh/nadph oxidase activation. Contribution to alterations of vasomotor tone. *J Clin Invest.* 1996;97:1916–1923.
222. Okamoto T, Akaike T, Sawa T, Miyamoto Y, van der Vliet A, Maeda H. Activation of matrix metalloproteinases by peroxynitrite-induced protein s-glutathiolation via disulfide S-oxide formation. *J Biol Chem.* 2001;276:29596–29602.
223. Abdalvand A, Morton JS, Bourque SL, Quon AL, Davidge ST. Matrix metalloproteinase enhances big-endothelin-1 constriction in mesenteric vessels of pregnant rats with reduced uterine blood flow. *Hypertension.* 2013;61:488–493.
224. Watanabe Y, Naruse M, Monzen C, et al. Is big endothelin converted to endothelin-1 in circulating blood? *J Cardiovasc Pharmacol.* 1991;17:S503–S505.
225. Bodelsson G, Sjoberg NO, Stjernquist M. Contractile effect of endothelin in the human uterine artery and autoradiographic localization of its binding sites. *Am J Obstet Gynecol.* 1992;167:745–750.
226. Remuzzi G, Benigni A. Endothelins in the control of cardiovascular and renal function. *Lancet.* 1993;342:589–593.
227. Kilpatrick SJ, Roberts JM, Lykins DL, Taylor RN. Characterization and ontogeny of endothelin receptors in human placenta. *Am J Physiol.* 1993;264:E367–E372.
228. Nova A, Sibai BM, Barton JR, Mercer BM, Mitchell MD. Maternal plasma level of endothelin is increased in preeclampsia. *Am J Obstet Gynecol.* 1991;165:724–727.
229. Mastrogiannis DS, O'Brien WF, Krammer J, Benoit R. Potential role of endothelin-1 in normal and hypertensive pregnancies. *Am J Obstet Gynecol.* 1991;165:1711–1716.
230. Benigni A, Orisio S, Gaspari F, Frusca T, Amuso G, Remuzzi G. Evidence against a pathogenetic role for endothelin in pre-eclampsia. *Br J Obstet Gynaecol.* 1992;99:798–802.
231. Tam Tam KB, George E, Cockrell K, et al. Endothelin type a receptor antagonist attenuates placental ischemia-induced hypertension and uterine vascular resistance. *Am J Obstet Gynecol.* 2011;204:330.e331–330.e334.
232. George EM, Palei AC, Granger JP. Endothelin as a final common pathway in the pathophysiology of preeclampsia: therapeutic implications. *Curr Opin Nephrol Hypertens.* 2012;21:157–162.
233. Jeyabalan A, Shroff SG, Novak J, Conrad KP. The vascular actions of relaxin. *Adv Exp Med Biol.* 2007;612:65–87.
234. Hanssens M, Keirse MJ, Spitz B, van Assche FA. Angiotensin II levels in hypertensive and normotensive pregnancies. *Brit J Obstet Gynaecol.* 1991;98:155–161.
235. Weiss D, Sorescu D, Taylor WR. Angiotensin II and atherosclerosis. *Am J Cardiol.* 2001;87:25C–32C.
236. Arenas IA, Xu Y, Lopez-Jaramillo P, Davidge ST. TNFalpha mediates angiotensin II-induced MMP-2 release from endothelial cells. *Am J Physiol.* 2004;286:779–784.
237. Pawlak MA, Macdonald GJ. Altered number of platelet angiotensin II receptors in relation to plasma agonist concentrations in normal and hypertensive pregnancy. *J Hypertens.* 1992;10:813–819.
238. AbdAlla S, Lother H, Quitterer U. At1-receptor heterodimers show enhanced g-protein activation and altered receptor sequestration. *Nature.* 2000;407:94–98.
239. Abdalla S, Lother H, el Massiery A, Quitterer U. Increased at(1) receptor heterodimers in preeclampsia mediate enhanced angiotensin ii responsiveness. *Nat Med.* 2001;7:1003–1009.
240. Gant NF, Daley GL, Chand S, Whalley PJ, MacDonald PC. A study of angiotensin II pressor response throughout primigravid pregnancy. *J Clin Invest.* 1973;52:2682–2689.
241. Bobst SM, Day MC, Gilstrap III LC, Xia Y, Kellems RE. Maternal autoantibodies from preeclamptic patients activate angiotensin receptors on human mesangial cells and induce interleukin-6 and plasminogen activator inhibitor-1 secretion. *Am J Hypertens.* 2005;18:330–336.
242. Thway TM, Shlykov SG, Day MC, et al. Antibodies from preeclamptic patients stimulate increased intracellular ca2+ mobilization through angiotensin receptor activation. *Circulation.* 2004;110:1612–1619.
243. Xia Y, Wen H, Bobst S, Day MC, Kellems RE. Maternal autoantibodies from preeclamptic patients activate angiotensin receptors on human trophoblast cells. *J Soc Gynecol Invest.* 2003;10:82–93.
244. Xia Y, Zhou CC, Ramin SM, Kellems RE. Angiotensin receptors, autoimmunity, and preeclampsia. *J Immunol.* 2007;179:3391–3395.
245. Herse F, Staff AC, Hering L, Muller DN, Luft FC, Dechend R. At1-receptor autoantibodies and uteroplacental ras in pregnancy and pre-eclampsia. *J Mol Med.* 2008;86:697–703.
246. Dechend R, Gratze P, Wallukat G, et al. Agonistic autoantibodies to the at1 receptor in a transgenic rat model of preeclampsia. *Hypertension.* 2005;45:742–746.
247. Koga K, Mor G. Expression and function of toll-like receptors at the maternal-fetal interface. *Reprod Sci.* 2008;15:231–242.
248. Jurk M, Schulte B, Kritzler A, et al. C-class CpG ODN: sequence requirements and characterization of immunostimulatory activities on mRNA level. *Immunobiology.* 2004;209:141–154.
249. Zhong XY, Holzgreve W, Hahn S. The levels of circulatory cell free fetal DNA in maternal plasma are elevated prior to the onset of preeclampsia. *Hypertens Pregnancy.* 2002;21:77–83.
250. DiFederico E, Genbacev O, Fisher SJ. Preeclampsia is associated with widespread apoptosis of placental cytotrophoblasts within the uterine wall. *Am J Pathol.* 1999;155:293–301.
251. Gama-Sosa MA, Wang RY, Kuo KC, Gehrke CW, Ehrlich M. The 5-methylcytosine content of highly repeated sequences in human DNA. *Nucleic Acids Res.* 1983;11:3087–3095.
252. Grisham M, McCord J. Chemistry and cytotoxicity of reactive oxygen metabolites. In: Taylor AE, Matalon S, Ward P, eds. *Physiology of Oxygen Radicals.* Bethesda, MD: American Physiological Society; 1986:1–19.
253. Rubanyi GM. Vascular effects of oxygen-derived free radicals. *Free Rad Biol Med.* 1988;4:107–120.
254. Pou S, Pou WS, Bredt DS, Snyder SH, Rosen GM. Generation of superoxide by purified brain nitric oxide synthase. *J Biol Chem.* 1992;267:24173–24176.
255. Parks DA, Granger DN. Xanthine oxidase: biochemistry, distribution and physiology. *Acta Physiol Scand Suppl.* 1986;548:87–99.
256. Tsukimori K, Maeda H, Ishida K, Nagata H, Koyanagi T, Nakano H. The superoxide generation of neutrophils in normal and preeclamptic pregnancies. *Obstet Gynecol.* 1993;81:536–540.

257. Zimmerman RJ, Chan A, Leadon SA. Oxidative damage in murine tumor cells treated in vitro by recombinant human tumor necrosis factor. *Cancer Res*. 1989;49:1644–1648.
258. Maziere C, Auclair M, Maziere J. Tumour necrosis factor enhances low density protein oxidative modification by monocytes and endothelial cells. *FEBS Lett*. 1994;338:43–46.
259. Tan S, Yokoyama Y, Dickens E, Cash TG, Freeman BA, Parks DA. Xanthine oxidase activity in the circulation of rats following hemorrhagic shock. *Free Rad Biol Med*. 1993;15:407–414.
260. Many A, Hubel CA, Roberts JM. Hyperuricemia and xanthine oxidase in preeclampsia, revisited. *Am J Obstet Gynecol*. 1996;174:288–291.
261. Heitzer T, Wenzel U, Hink U, et al. Increased NAD(P)H oxidase-mediated superoxide production in renovascular hypertension: evidence for an involvement of protein kinase C. *Kidney Int*. 1999;55:252–260.
262. Meyer JW, Schmitt ME. A central role for the endothelial NADPH oxidase in atherosclerosis. *FEBS Lett*. 2000;472:1–4.
263. Hubel CA, Roberts JM, Taylor RN, Musci TJ, Rogers GM, McLaughlin MK. Lipid peroxidation in pregnancy: new perspectives on preeclampsia. *Am J Obstet Gynecol*. 1989;161:1025–1034.
264. Davidge ST, Hubel CA, Brayden RD, Capeless EC, McLaughlin MK. Sera antioxidant activity in uncomplicated and preeclamptic pregnancies. *Obstet Gynecol*. 1992;79:879–901.
265. Regan CL, Levine RJ, Baird DD, et al. No evidence for lipid peroxidation in severe preeclampsia. *Am J Obstet Gynecol*. 2001;185:572–578.
266. Llurba E, Gratacos E, Martin-Gallan P, Cabero L, Dominguez C. A comprehensive study of oxidative stress and antioxidant status in preeclampsia and normal pregnancy. *Free Rad Biol Med*. 2004;37:557–570.
267. Smith WL, Marnett LJ. Prostaglandin endoperoxide synthase: structure and catalysis. *Biochim Biophys Acta*. 1991;1083:1–17.
268. Davidge ST, Hubel CA, McLaughlin MK. Cyclooxygenase-dependent vasoconstrictor alters vascular function in the vitamin E-deprived rat. *Circ Res*. 1993;73:79–88.
269. Davidge ST, Everson WV, Parisi VM, McLaughlin MK. Pregnancy and lipid peroxide-induced alterations of eicosanoid-metabolizing enzymes in the aorta of the rat. *Am J Obstet Gynecol*. 1993;169:1338–1344.
270. McCarthy AL, Woolfson RG, Raju SK, Poston L. Abnormal endothelial cell function of resistance arteries from women with preeclampsia. *Am J Obstet Gynecol*. 1993;168:1323–1330.
271. Knock GA, Poston L. Bradykinin-mediated relaxation of isolated maternal resistance arteries in normal pregnancy and preeclampsia. *Am J Obstet Gynecol*. 1996;175:1668–1674.
272. Walsh SW. Preeclampsia: an imbalance in placental prostacyclin and thromboxane production. *Am J Obstet Gynecol*. 1985;152:335–340.
273. Fitzgerald DJ, Entman SS, Mulloy K, FitzGerald GA. Decreased prostacyclin biosynthesis preceding the clinical manifestation of pregnancy-induced hypertension. *Circulation*. 1987;75:956–963.
274. Satoh K, Seki H, Sakamoto H. Role of prostaglandins in pregnancy-induced hypertension. *Am J Kidney Dis*. 1991;17:133–138.
275. Wang JA, Zhen EZ, Guo ZZ, Lu YC. Effect of hyperlipidemic serum on lipid peroxidation, synthesis of prostacyclin and thromboxane by cultured endothelial cells: protective effect of antioxidants. *Free Rad Biol Med*. 1989;7:243–249.
276. Baker PN, Davidge ST, Barankiewicz J, Roberts JM. Plasma of preeclamptic women stimulates and then inhibits endothelial prostacyclin. *Hypertension*. 1996;27:56–61.
277. Ignarro LJ. Nitric oxide. A novel signal transduction mechanism for transcellular communication. *Hypertension*. 1990;16:477–483.
278. Nobunaga T, Tokugawa Y, Hashimoto K, et al. Plasma nitric oxide levels in pregnant patients with preeclampsia and essential hypertension. *Gynecol Obstet Invest*. 1996;41:189–193.
279. Seligman SP, Buyon JP, Clancy RM, Young BK, Abramson SB. The role of nitric oxide in the pathogenesis of preeclampsia. *Am J Obstet Gynecol*. 1994;171:944–948.
280. Davidge ST, Stranko CP, Roberts JM. Urine but not plasma nitric oxide metabolites are decreased in women with preeclampsia. *Am J Obstet Gynecol*. 1996;174:1008–1013.
281. Li H, Meininger CJ, Hawker Jr JR, et al. Regulatory role of arginase I and II in nitric oxide, polyamine, and proline syntheses in endothelial cells. *Am J Physiol*. 2001;280:E75–E82.
282. Berkowitz DE, White R, Li D, et al. Arginase reciprocally regulates nitric oxide synthase activity and contributes to endothelial dysfunction in aging blood vessels. *Circulation*. 2003;108:2000–2006.
283. Heinzel B, John M, Klatt P, Bohme E, Mayer B. Ca^{2+}/calmodulin-dependent formation of hydrogen peroxide by brain nitric oxide synthase. *Biochem J*. 1992;281(Pt 3):627–630.
284. Noris M, Todeschini M, Cassis P, et al. L-arginine depletion in preeclampsia orients nitric oxide synthase toward oxidant species. *Hypertension*. 2004;43:614–622.
285. Savvidou MD, Hingorani AD, Tsikas D, Frölich JC, Vallance P, Nicolaides KH. Endothelial dysfunction and raised plasma concentrations of asymmetric dimethylarginine in pregnant women who subsequently develop pre-eclampsia. *Lancet*. 2003;361:1511–1517.
286. Rizos D, Eleftheriades M, Batakis E, et al. Levels of asymmetric dimethylarginine throughout normal pregnancy and in pregnancies complicated with preeclampsia or had a small for gestational age baby. *J Matern Fetal Neonat Med*. 2012;25:1311–1315.
287. Speer PD, Powers RW, Frank MP, Harger G, Markovic N, Roberts JM. Elevated asymmetric dimethylarginine concentrations precede clinical preeclampsia, but not pregnancies with small-for-gestational-age infants. *Am J Obstet Gynecol*. 2008;198:112.e11–112.e117.
288. Baker PN, Davidge ST, Roberts JM. Plasma from women with preeclampsia increases endothelial cell nitric oxide production. *Hypertension*. 1995;26:244–248.
289. Davidge ST, Baker PN, Roberts JM. Nos expression is increased in endothelial cells exposed to plasma from women with preeclampsia. *Am J Physiol*. 1995;269:1106–1112.

290. Roggensack AM, Zhang Y, Davidge ST. Evidence for peroxynitrite formation in the vasculature of women with preeclampsia. *Hypertension*. 1999;33:83–89.
291. Beckman JS, Koppenol WH. Nitric oxide, superoxide, and peroxynitrite: the good, the bad, and ugly. *Am J Physiol*. 1996;271:C1424–C1437.
292. Myatt L, Rosenfield RB, Eis AL, Brockman DE, Greer I, Lyall F. Nitrotyrosine residues in placenta. Evidence of peroxynitrite formation and action. *Hypertension*. 1996;28:488–493.
293. Pfeiffer S, Gorren ACF, Schmidt K, et al. Metabolic fate of peroxynitrite in aqueous solution. Reaction with nitric oxide and ph-dependent decomposition to nitrite and oxygen in a 2:1 stoichiometry. *J Biol Chem*. 1997;272:3465–3470.
294. Cooke CL, Davidge ST. Peroxynitrite increases inos through NF-kappab and decreases prostacyclin synthase in endothelial cells. *Am J Physiol Cell Physiol*. 2002;282:C395–C402.
295. Gillham JC, Kenny LC, Baker PN. An overview of endothelium-derived hyperpolarising factor (EDHF) in normal and compromised pregnancies. *Eur J Obstet Gynecol Reprod Biol*. 2003;109:2–7.
296. Busse R, Edwards G, Feletou M, Fleming I, Vanhoutte PM, Weston AH. EDHF: bringing the concepts together. *Trends Pharmacol Sci*. 2002;23:374–380.
297. Morton JS, Davidge ST. Arterial endothelium-derived hyperpolarization: potential role in pregnancy adaptations and complications. *J Cardiovasc Pharmacol*. 2013;61:197–203.
298. Lang NN, Luksha L, Newby DE, Kublickiene K. Connexin 43 mediates endothelium-derived hyperpolarizing factor-induced vasodilatation in subcutaneous resistance arteries from healthy pregnant women. *Am J Physiol Heart Circ Physiol*. 2007;292:H1026–H1032.
299. Kenny LC, Baker PN, Kendall DA, Randall MD, Dunn WR. Differential mechanisms of endothelium-dependent vasodilator responses in human myometrial small arteries in normal pregnancy and pre-eclampsia. *Clin Sci*. 2002;103:67–73.
300. Luksha L, Luksha N, Kublickas M, Nisell H, Kublickiene K. Diverse mechanisms of endothelium-derived hyperpolarizing factor-mediated dilatation in small myometrial arteries in normal human pregnancy and preeclampsia. *Biol Reprod*. 2010;83:728–735.
301. Luksha L, Nisell H, Luksha N, Kublickas M, Hultenby K, Kublickiene K. Endothelium-derived hyperpolarizing factor in preeclampsia: heterogeneous contribution, mechanisms, and morphological prerequisites. *Am J Physiol Reg I*. 2008;294:R510–R519.
302. Widmer M, Villar J, Benigni A, Conde-Agudelo A, Karumanchi SA, Lindheimer M. Mapping the theories of preeclampsia and the role of angiogenic factors: a systematic review. *Obstet Gynecol*. 2007;109:168–180.
303. Ehsanipoor RM, Fortson W, Fitzmaurice LE, et al. Nitric oxide and carbon monoxide production and metabolism in preeclampsia. *Reprod Sci*. 2013;20:542–548.
304. Baum M, Schiff E, Kreiser D, et al. End-tidal carbon monoxide measurements in women with pregnancy-induced hypertension and preeclampsia. *Am J Obstet Gynecol*. 2000;183:900–903.
305. Ahmed A. New insights into the etiology of preeclampsia: identification of key elusive factors for the vascular complications. *Thromb Res*. 2011;127(suppl 3):S72–S75.
306. Zhai D, Guo Y, Smith G, Krewski D, Walker M, Wen SW. Maternal exposure to moderate ambient carbon monoxide is associated with decreased risk of preeclampsia. *Am J Obstet Gynecol*. 2012;207:57.e51–57.e59.
307. Bhatia M. Hydrogen sulfide as a vasodilator. *IUBMB Life*. 2005;57:603–606.
308. Cindrova-Davies T, Herrera EA, Niu Y, Kingdom J, Giussani DA, Burton GJ. Reduced cystathionine y-lyase and increased mir-21 expression are associated with increased vascular resistance in growth-restricted pregnancies: hydrogen sulfide as a placental vasodilator. *Am J Pathol*. 2013;182:1448–1458.
309. Wang K, Ahmad S, Cai M, et al. Dysregulation of hydrogen sulfide producing enzyme cystathionine gamma-lyase contributes to maternal hypertension and placental abnormalities in preeclampsia. *Circulation*. 2013;127:2514–2522.
310. Hauth JC, Goldenberg RL, Parker CJ, et al. Low-dose aspirin therapy to prevent preeclampsia. *Am J Obstet Gynecol*. 1993;168:1083–1091.
311. Sibai BM, Caritis SN, Thom E, et al. Prevention of preeclampsia with low-dose aspirin in healthy, nulliparous pregnant women. The National Institute of Child Health and Human Development network of maternal-fetal medicine units. *New Engl J Med*. 1993;329:1213–1218.
312. CLASP A randomised trial of low-dose aspirin for the prevention and treatment of pre-eclampsia among 9364 pregnant women. CLASP (collaborative low-dose aspirin study in pregnancy) collaborative group. *Lancet*. 1994;343:619–629.
313. Askie LM, Duley L, Henderson-Smart DJ, Stewart LA. Antiplatelet agents for prevention of pre-eclampsia: a meta-analysis of individual patient data. *Lancet*. 2007;369:1791–1798.
314. Roberts JM, Catov JM. Aspirin for pre-eclampsia: compelling data on benefit and risk. *Lancet*. 2007;369:1765–1766.
315. Conde-Agudelo A, Romero R, Kusanovic JP, Hassan SS. Supplementation with vitamins C and E during pregnancy for the prevention of preeclampsia and other adverse maternal and perinatal outcomes: a systematic review and metaanalysis. *Am J Obstet Gynecol*. 2011;204:503.e501–503.e512.
316. Basaran A, Basaran M, Topatan B. Combined vitamin C and E supplementation for the prevention of preeclampsia: a systematic review and meta-analysis. *Obstet Gynecol Surv*. 2010;65:653–667.
317. Roberts JM, Myatt L, Spong CY. Vitamins c and e to prevent complications of pregnancy-associated hypertension. *New Engl J Med*. 2010;362:1282–1291.
318. Vadillo-Ortega F, Perichart-Perera O, Espino S, et al. Effect of supplementation during pregnancy with L-arginine and antioxidant vitamins in medical food on pre-eclampsia in high risk population: randomised controlled trial. *BMJ*. 2011;342.
319. Staff AC, Berge L, Haugen G, Lorentzen B, Mikkelsen B, Henriksen T. Dietary supplementation with L-arginine or placebo in women with pre-eclampsia. *Acta Obstet Gynecol Scand*. 2004;83:103–107.

320. Zhou SJ, Yelland L, McPhee AJ, Quinlivan J, Gibson RA, Makrides M. Fish-oil supplementation in pregnancy does not reduce the risk of gestational diabetes or preeclampsia. *Am J Clin Nutr*. 2012;95:1378–1384.
321. Makrides M, Duley L, Olsen SF. Marine oil, and other prostaglandin precursor, supplementation for pregnancy uncomplicated by pre-eclampsia or intrauterine growth restriction. *Cochrane Database Syst Rev*. 2006:CD003402.
322. Imdad A, Jabeen A, Bhutta ZA. Role of calcium supplementation during pregnancy in reducing risk of developing gestational hypertensive disorders: a meta-analysis of studies from developing countries. *BMC Public Health*. 2011;11(suppl 3):S18.
323. Patrelli TS, Dall'asta A, Gizzo S, et al. Calcium supplementation and prevention of preeclampsia: a meta-analysis. *J Matern Fetal Neonatal Med*. 2012;25:2570–2574.
324. Hofmeyr GJ, Lawrie TA, Atallah AN, Duley L. Calcium supplementation during pregnancy for preventing hypertensive disorders and related problems. *Cochrane Database Syst Rev*. 2010:CD001059.
325. Levine RJ, Hauth JC, Curet LB, et al. Trial of calcium to prevent preeclampsia. *New Engl J Med*. 1997;337:69–77.
326. De Vries JIP, Van Pampus MG, Hague WM, Bezemer PD, Joosten JH, On Behalf Of Fruit I Low-molecular-weight heparin added to aspirin in the prevention of recurrent early-onset pre-eclampsia in women with inheritable thrombophilia: the Fruit-RCT. *J Thromb Haemost*. 2012;10:64–72.
327. Rodger MA. An update on thrombophilia and placenta mediated pregnancy complications: what should we tell our patients? *Thromb Res*. 2013;131(suppl 1):S25–S27.
328. Dekker GA, van Geijn HP. Endothelial dysfunction in pre-eclampsia. Part I: primary prevention. Therapeutic perspectives. *J Perinat Med*. 1996;24:99–117.
329. Martin DGJ, Gardner MO, Izquierdo LA, Tobey K, Curet LB. Incidence of intraventricular hemorrhage in neonates under 32 weeks of gestation delivered to mothers with severe pre-eclampsia. *Prenat Neonat Med*. 1998;3:250–254.
330. Xiong X, Saunders LD, Wang FL, Davidge ST, Buekens P. Preeclampsia and cerebral palsy in low-birth-weight and preterm infants: implications for the current "ischemic model" of preeclampsia. *Hypertens Pregnancy*. 2001;20:1–13.
331. O'Shea TM, Klinepeter KL, Dillard RG. Prenatal events and the risk of cerebral palsy in very low birth weight infants. *Am J Epidemiol*. 1998;147:362–369.

CHAPTER 10

Animal Models for Investigating Pathophysiological Mechanisms of Preeclampsia

JOEY P. GRANGER, ERIC M. GEORGE AND JAMES M. ROBERTS

Editors' comment: *Animal models are not mentioned in Chesley's initial edition, and the word does not appear in his extensive appendix. At that time there were few models and little use of them to search for cause, pathophysiology, or treatment. Early strategies used uterine ischemia to produce hypertension and proteinuria; this approach was perfected in rodents by the new first author of this chapter. His group has used a model where uterine perfusion in pregnant rats is decreased by constricting the aorta (the RUPP model) that has produced a plethora of data pertinent to a better understanding of preeclampsia in humans. Starting with the second and expanding with the third edition other models have been described, most recently using the gene manipulation technique. Thus from "no press" in edition one and then steadily increasing, animal models have become an integral part of research both to understand and treat preeclampsia, and will be even more important as specific treatments are sought and tested.*

INTRODUCTION

Preeclampsia, estimated to affect 5–7% of all pregnancies worldwide, is normally characterized by hypertension, endothelial dysfunction, angiogenic imbalance and proteinuria.[1–7] Though preeclampsia is a leading cause of maternal death and a major contributor to maternal and perinatal morbidity, the paucity of knowledge of its causal mechanisms remains a major obstacle to developing specific prevention and/or treatment modalities. Since the spontaneous development of preeclampsia is essentially limited to the human species, the study of preeclampsia in humans is of critical importance to identify biomarkers and potential pathogenic factors that correlate with the progression of the syndrome. However, experimental studies in pregnant women have obvious limitations that prevent complete investigation of many pathophysiological mechanisms involved in this syndrome. Moreover, studies in humans often limit the ability to establish cause and effect relationships. In contrast, experimental studies in animal models, despite their limitations, allow investigators to directly test whether certain factors found in preeclamptic women can indeed lead to hypertension and other manifestations of preeclampsia.

Here we review some of the more widely studied and/or recently developed animal models used to investigate mechanisms that link placental pathology with maternal endothelial activation/dysfunction and vascular abnormalities of preeclampsia. A wide variety of animal models have been used to investigate pathophysiological mechanisms of preeclampsia. These mechanisms and resulting models include hypoxia (reduced uterine perfusion pressure models), impaired angiogenesis (sFlt-1 and sEng infusion models), excessive maternal immune activation (TNF-α and AT1-AA infusion) and genetic mouse models that target specific pathogenic pathways.

MODELS USED TO INVESTIGATE LINKS BETWEEN PLACENTAL ISCHEMIA AND ENDOTHELIAL AND CARDIOVASCULAR DYSFUNCTION

Preeclampsia develops during pregnancy and remits after delivery, implicating the placenta as a central culprit in the disease.[2–7] An initiating event in preeclampsia is postulated to be reduced placental perfusion that leads to widespread dysfunction of the maternal vascular endothelium by mechanisms that remain to be fully elucidated.[2–7] Experimental induction of chronic uteroplacental ischemia appears to be the most promising animal model to study potential mechanisms of preeclampsia since reductions in uteroplacental

Chesley's Hypertensive Disorders in Pregnancy.
ISBN: 978-0-12-407866-6

DOI: http://dx.doi.org/10.1016/B978-0-12-407866-6.00010-9

blood flow in a variety of animal models lead to a hypertensive state that closely resembles preeclampsia in women, including endothelial dysfunction, angiogenic imbalance, and proteinuria.[8,9] In early studies of the relationship between placental blood flow and preeclampsia, Ogden et al. observed that partially occluding the infrarenal aorta of pregnant but not non-pregnant dogs resulted in hypertension.[10] In subsequent studies, the bilateral ligation of the utero-ovarian arteries and placement of non-constrictive bands around the uterine arteries resulted in hypertension in dogs only after they achieved pregnancy.[11] The animals uniformly developed hypertension and proterinuria. This hypertensive state persisted until the postpartum period, at which time pressures returned to normal. These findings were confirmed by studies where precise constriction of the aorta below the dog's renal arteries resulted in hypertension and proteinuria.[12,13] The investigators claimed that the animals also displayed glomerular endothelial lesions similar to those classically described in the kidneys of preeclamptic women. Losonczy et al. constricted the aortas of rabbits and demonstrated elevated blood pressure in chronically catheterized pregnant animals.[14] These studies demonstrated elevated arterial pressures accompanied by increased peripheral resistance, a feature commonly observed in preeclamptic women.

The relationship between reduced uteroplacental perfusion and hypertension during pregnancy has also been demonstrated in sub-human primates. Cavanagh and co-workers studied baboons using interventions similar to those used by Hodari; ligation of the utero-ovarian arteries and placement of bands around the uterine arteries.[11,15,16] Females bred after the surgery exhibited higher blood pressures than non-pregnant or sham-operated control pregnant animals, and were also reported to manifest glomerular endothelial swelling. An aortic constriction model was also reported in rhesus monkeys to study the effects of reduced uteroplacental perfusion from early pregnancy through delivery.[17] In this latter model, the degree of aortic constriction was precisely controlled, and progressive hypertension, proteinuria, and glomerular endotheliosis developed. More recently, Makris et al. produced a uteroplacental ischemia model in radiotelemetered pregnant baboons by selective ligation of one uterine artery resulting in a 40% decrease in uteroplacental blood flow as determined by angiography.[18] Hypertension, proteinuria, and increased production of anti-angiogenic markers by the placenta and peripheral blood mononuclear cells were reported in the pregnant animals compared to control animas who underwent a sham procedure (Fig. 10.1). Endothelial histological changes consistent with those seen in human preeclampsia were also reported.

Eder and MacDonald reduced utero placental perfusion pressure in gravid rats, the intervention starting on the 14th day of a normally 21 day gestation.[19] More recently, Granger and colleagues modified and further characterized this rat model in order to examine potential pathophysiological mechanisms that mediate the hypertension during chronic reductions in uteroplacental perfusion pressure (RUPP).[9,20] They reduced uterine perfusion pressure in the gravid rat by approximately 40% by placing a silver clip around the aorta below the renal arteries.[9,20] Because this procedure causes an adaptive increase in uterine blood flow via the ovarian artery, they also placed a silver clip on both the right and left uterine arcade at the ovarian end just before the first segmental artery. They reported that reducing uteroplacental perfusion pressure results in significant and consistent elevations in arterial pressure of 20–30 mm Hg as compared to control pregnant rats at day 19 of gestation.[9,20] Reducing uteroplacental perfusion pressure in non-pregnant rats had no effect on blood pressure.[20] RUPP-induced hypertension was also associated with proteinuria, reductions in renal plasma flow and glomerular filtration rate, and a hypertensive shift in the pressure–natriuresis relationship.[20] Moreover, they found important evidence indicating that endothelial function is significantly impaired in the RUPP hypertensive rat.[21] They compared the relaxation responses to acetylcholine between aortic vessel strips of pregnant and RUPP hypertensive rats, observing that endothelial-dependent vasodilatation was significantly attenuated in the RUPP hypertensive rats. In addition, the production of nitric oxide is reduced in vascular tissue while the syntheses of thromboxane, endothelin, and 8-isoprostane, a marker of oxidative stress, are elevated in the RUPP hypertensive rat as compared to normal pregnant rats at day 19 of gestation.[22–26] Moreover, immune factors such as tumor necrosis factor alpha (TNF-α),[27–31] interleukin-6 (IL-6), and angiotensin II type I receptor autoantibodies (ATI-AA) are significantly elevated in the plasma of the RUPP hypertensive rat. Intrauterine growth restriction is also present in the RUPP hypertensive rats since the average pup size in this group is smaller than in normal pregnant rats. Thus, RUPP-induced hypertension in the pregnant rat has many of the features of preeclampsia in women (Table 10.1).

While uteroplacental ischemic models have many of the features of human preeclampsia they do have limitations. Although the rat RUPP model exhibits increases in protein excretion and consistent increases in blood pressure in response to reductions in uterine perfusion pressure, the proteinuric response is somewhat variable in the pregnant rat. The reason for the variability in the rat RUPP model is unknown but may be due to the short time frame (4–5 days) of exposure to placental ischemia. In contrast, proteinuria in the baboon uteroplacental ischemic model has a robust increase within the second and third week of uterine artery occlusion (Fig. 10.1). However, the primate model has a long gestation, very costly and labor-intensive, and has been used by a limited number of laboratories.

While animal models of placental ischemia do have some limitations, studies in a variety of animal species suggest the reduced uteroplacental perfusion models have

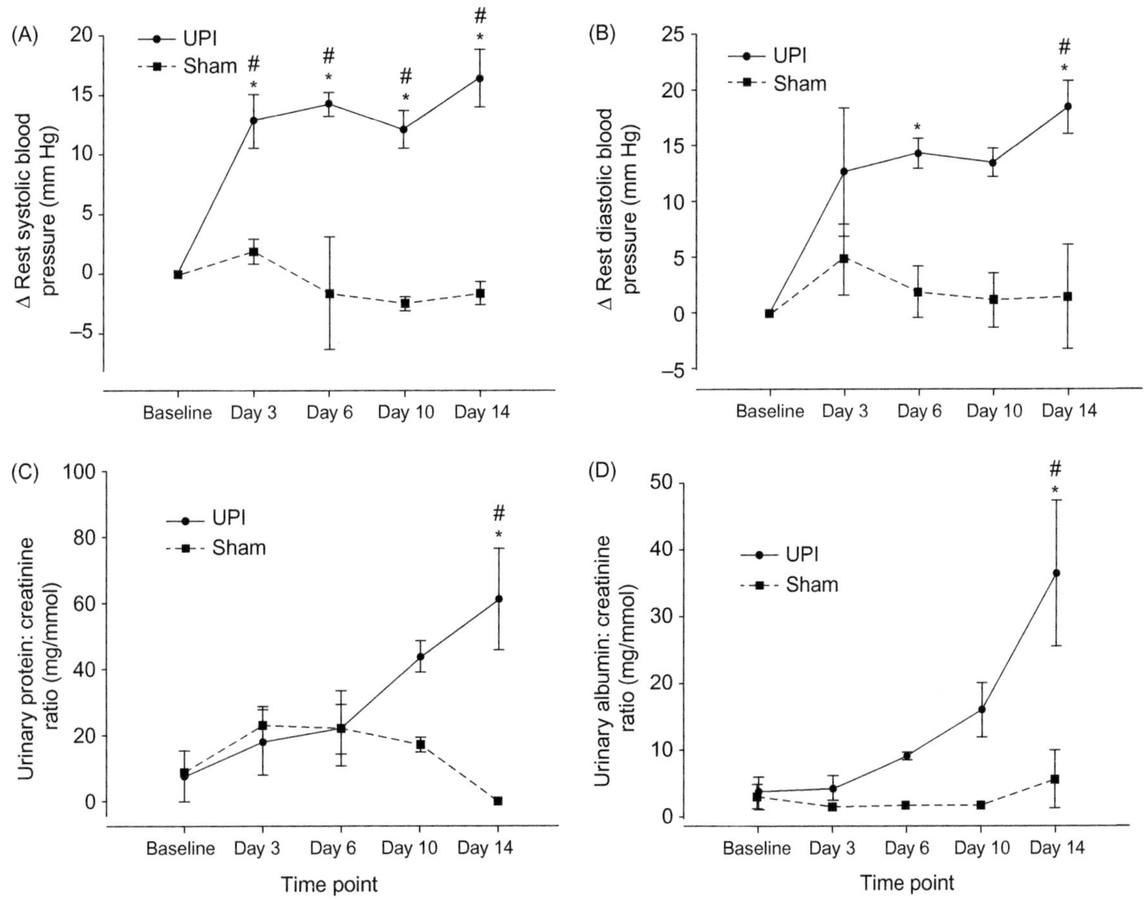

FIGURE 10.1 Pregnant baboons were subjected to unilateral uterine artery ligation (UPI) on gestational day 135 of a normal 182 day gestation, resulting in ~30–50% decrease in placental perfusion. Within 3 days, blood pressure in both sham and UPI animals showed an increase in both systolic and diastolic blood pressure both awake (A,B) and while asleep (C,D) as determined by implanted radiotelemetry. This effect was persistent through 2 weeks post-surgery. Reprinted with permission from[18].

TABLE 10.1 Comparison of RUPP Model with Features Seen in PE Women

Symptom	RUPP	Preeclampsia	Reference
Hypertension	+	+	Alexander et al.[20]
Proteinuria	+	+	Alexander et al.[20]
Decreased renal plasma Flow	+	+	Alexander et al.[20]
Decreased GFR	+	+	Alexander et al.[20]
Endothelial dysfunction	+	+	Crews et al.[21]
Angiogenic imbalance	+	+	Gilbert et al.[32]
Enhanced ET-1 expression	+	+	Alexander et al.[22]
Oxidative stress	+	+	Sedeek et al.[9]
Increased Inflammatory cytokines	+	+	LaMarca et al.,[27,28] Gadonski et al.[33]
Agonistic AT1-AA production	+	+	LaMarca et al.[30]
Increased CD4+ T-cells	+	+	Wallace et al.[34]
Decreased NO availability	+	+	Khalil et al.[21]
Increased total peripheral resistance	+	+	Sholook et al.[9]
Decreased cardiac output	+	+	Sholook et al.[9]
Intrauterine growth restriction	+	+	Alexander et al.[20]

many features of preeclampsia in women. These models, along with appropriate pharmacological tools, provide an opportunity for investigators to quantify the relative importance of complex pathophysiological factors in mediating cardiovascular and renal dysfunction in response to placental ischemia during pregnancy. The models also provide an important tool to test the efficacy of novel therapeutic approaches for the prevention and treatment of preeclampsia.

ANIMAL MODELS USED TO STUDY ROLE OF ANGIOGENIC FACTORS (SEE ALSO CHAPTER 6)

One of the most promising systems to receive scrutiny in recent preeclampsia research is the vascular endothelial growth factor (VEGF) signaling pathway.[35] VEGF, a powerful proangiogenic protein, is an essential factor in the maintenance of endothelial cell health. Additionally, VEGF is necessary for the maintenance of glomerular ultrastructure through the maintenance of its fenestrated endothelium.[36] Transgenic VEGF knockout in glomerular podocytes resulted in proteinuria and glomerular endotheliosis, two common findings in preeclampsia.[36,37] Similar findings are seen in cancer patients treated with VEGF monoclonal antibodies, as hypertension and proteinuria are common side effects. It is clear, then, that proper levels of VEGF are necessary for endothelial and vascular health.[37]

One factor shown to interfere with VEGF signaling is the soluble form of the VEGF receptor, sFlt-1. sFlt-1 is an alternately spliced variant of the full-length receptor in which the transmembrane and cytosolic domains have been excised, leaving only the extracellular recognition domain. This recognition domain acts as a VEGF antagonist by binding free VEGF, and making it unavailable for proper signaling. Of particular interest for preeclampsia, sFlt-1 is positively regulated by hypoxia, specifically through the actions of HIF-1α, and has been shown to be produced by both placental trophoblasts and human placental villous explants in response to low oxygen tension.[38,39] While sFlt-1 production is regulated by the hypoxia-inducible factor-1, studies using animal models (RUPP, TNF or AT1-AA infusion models) suggest that factors such as tumor necrosis factor and the agonistic autoantibody to the angiotensin II type I receptor also appear to be involved in the regulation of sFlt-1 production.[31]

Several experimental models have demonstrated a causative role for sFlt-1 in the pathology of preeclampsia. Viral ectopic expression of sFlt-1 in pregnant rats led to a preeclampsia-like state, with hypertension, glomerular endotheliosis, and proteinuria[40] (Fig. 10.2). Models of reduced uterine perfusion pressure in both nonhuman primates and rats show significant increases in circulating sFlt-1, and concurrent decreases in bioavailable circulating VEGF.[18,32] In agreement with the viral expression experiments, direct infusion of sFlt-1 into pregnant mice or rats induces preeclamptic-like symptoms, including hypertension and reduced fetal weight.[41–44] As with several other experimental forms of hypertension, endothelin is one of the major effector molecules in sFlt-1-induced hypertension, as administration of an ET_A receptor antagonist completely normalizes blood pressure.[42]

It does not appear that the pathological manifestations seen in the sFlt-1-induced preeclampsia-like syndrome in animals are a direct result of circulating sFlt-1 but rather are due to the reduced availability of free circulating VEGF. When excess VEGF is administered simultaneously with the dose of sFlt-1 known to cause a preeclampsia-like status in pregnant rats the hypertensive response is muted, and there is less evidence of abnormal renal function, than when the antiangiogenic protein was administered alone.[44,45] Additionally, VEGF administration in rats with placental ischemia restores normal blood pressure, renal function,

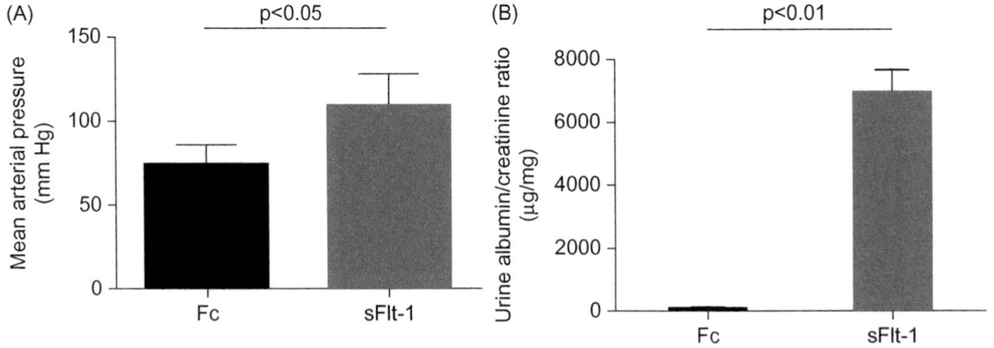

FIGURE 10.2 Adenoviral vectors encoding either Fc protein or sFlt-1 were injected into pregnant rats on gestation day 8 or 9 of a normal 21 day gestation. On gestational days 16 or 17, direct blood pressure measurements were performed on anesthetized animals. Animals receiving the Ad-sFlt-1 had significantly elevated blood pressure compared to the Ad-Fc controls (A). At the same time point, urine samples were obtained, and the urinary albumin/creatinine ratio was determined by ELISA. Dramatic increases in urinary albumin/creatinine were noted in the Ad-sFlt-1 group compared to the Ad-Fc controls (B), suggesting renal injury. Adapted from[40].

and vascular activity.[45] These and other data from animal models have led to the hypothesis that the ratio between sFlt-1 and VEGF is critical to maintain a healthy endothelium and normal vascular activity. The role of these proteins in the development of preeclampsia is a promising avenue in the search for new therapeutics.

Soluble endoglin (sEng), an antiangiogenic factor and soluble receptor for transforming growth factor-beta (TGF-β), is also implicated in the pathogenesis of PE. sEng is present during normal pregnancy, but is upregulated significantly in preeclampsia and correlates with disease severity.[46] Utilizing adenoviral delivery of sEng in pregnant rats, Venkatesha et al. reported the development of hypertension, intrauterine growth restriction, and focal endotheliosis.[47] These findings were more severe in rats which received the combination of both adenoviral sEng and sFlt-1 whereas animals also presented with hemolysis, elevated liver enzymes, and low platelets. The authors suggested that this model may reflect the syndrome of HELLP (hemolysis, elevated liver enzymes, and low platelets) seen in some PE women and could be utilized to investigate the pathophysiology of HELLP.

While compelling data derived from animal and human studies suggest an important role for angiogenic imbalance in the pathophysiology of preeclampsia, there are many unanswered questions and many opportunities for future research in animal models. For example, the molecular mechanisms involved in the regulation of sFlt-1 production have yet to be fully elucidated. Moreover, while sFlt-1 appears to play an important role in the pathogenesis of preeclampsia, research delineating the roles of specific inhibitors of sFlt-1 production had just started as of 2014. No doubt the finding of ways to inhibit sFlt-1 or stimulate greater production of VEGF and PlGF is of critical importance, with animal models needed to advance such research.

MODELS USED TO INVESTIGATE THE ROLE OF IMMUNE MECHANISMS IN PREECLAMPSIA

There is a growing interest in maternal immunity and inflammatory processes as mediators of the clinical manifestations of preeclampsia.[48,49] Preeclampsia is associated with a dysregulation of natural killer cells, activation of CD4 T lymphocytes, the release of proinflammatory factors such as TNF-α and IL-6 and IL-17, and the production of angiotensin II type-1 receptor autoantibody (AT1-AA).[49] However, the importance of these immune factors in the pathogenesis of preeclampsia in humans is unknown. In recent years, a number of animal models were developed in an attempt to dissect the relative importance of each of these immune factors.

Of interest is work suggesting that the inflammatory response is triggered by particles, ranging from large deported multinuclear fragments to sub-cellular components, shed from the syncytial surface of the human placenta.[50] Using an *in vitro* model (see cover photomicrograph and Chapter 6), Chamley and co-workers observed that when syncytial knots are shed by an apoptosis-like programmed cell death process, then phagocytosed by macrophages, the macrophages produce a tolerogenic response.[51] However, necrotic syncytial knots, when phagocytosed, appear to be immunostimulatory. While there is a strong correlation between trophoblast debris and immune activation in preeclampsia, it is unclear whether these particles could indeed elicit a pathogenic response *in vivo*. To address this issue, Chamley et al. recently reported chronically administered necrotic trophoblast debris to pregnant rats and observed increased blood pressure relative to pregnant controls in late gestation.[52] This report appears to be the first demonstration that necrotic trophoblast debris can alter blood pressure *in vivo* and, while the mechanism for this response remains to be determined, this observation is consistent with the growing body of *in vitro* evidence suggesting that necrotic trophoblast debris can contribute to the hypertension of preeclampsia.

In normal pregnancies, there is a heightened inflammatory state when compared to non-pregnant women, but in preeclampsia there is further elevation of inflammation markers, e.g., IL-6 and TNF-α, when compared to healthy pregnancies.[48,49,53] While inflammatory cytokines such as IL-6 and TNF-α are elevated in preeclamptic women, the importance of these cytokines in mediating the cardiovascular and renal dysfunction has yet to be fully elucidated. Serum levels of TNF-α and IL-6 are elevated in RUPP rats and chronic infusion of TNF-α or IL-6 into pregnant rats increases arterial pressure and decreases renal plasma flow and glomerular filtration rate.[27,28,33] Blockade of TNF-α with the soluble receptor Etanercept attenuated the associated hypertension and decreased tissue ET-1 expression in RUPP rats.[27] These findings indicate that TNF-α and IL-6 may play a role in mediating the hypertension and reduction in renal hemodynamics observed during RUPP in pregnant rats.

Similar findings were recently demonstrated in a primate model of cytokine-induced preeclampsia. Hennessey and co-workers reported that continuous administration of TNF-α to pregnant baboons for 2 weeks at mid-gestation caused increased blood pressure, proteinuria, and plasma sFlt-1.[54] The source of circulating sFlt-1 was determined to be at least partially placental, confirming its role as a link between placental dysfunction and the maternal syndrome. The placenta accrete seen in one baboon in the treatment group potentially suggests a fundamental disruption of placental structure and function as a result of TNF-α infusion.[54]

Numerous antiinflammatory cytokines are decreased in women with preeclampsia but the role of these cytokines in blood pressure regulation during pregnancy is unknown. Mitchell and co-workers recently addressed this issue by examining whether the lack of the potent antiinflammatory cytokine interleukin-4 (IL-4) would be sufficient to elicit a preeclampsia-like syndrome in mice, and when coupled with immune system activation would further augment these symptoms.[55] They reported that pregnant IL-4-deficient mice exhibited altered splenic immune cell subsets, increased levels of proinflammatory cytokines, placental inflammation, mild hypertension, endothelial dysfunction, and proteinuria compared to pregnant control mice. Compared to pregnant control mice treated with with the Toll-like receptor 3 agonist polyinosinic:polycytidylic (poly I:C).which exhibit preeclampsia-like symptoms, poly I:C-treated pregnant IL-4-deficient mice exhibited a further increase in proinflammatory cytokine levels, which was associated with augmented SBP and endothelial dysfunction. Collectively, these data show that the absence of IL-4 is sufficient to induce mild preeclampsia-like symptoms in mice due to excessive inflammation. Thus, the antiinflammatory effects of IL-4 may be important in preventing hypertension during pregnancy.

Another factor produced by the maternal immune system that has received a great deal of attention in recent years is the angiotensin-1 receptor autoantibody (AT1-AA), which has been identified in the circulation of preeclamptic women.[49,56–58] The AT1-AA has been purified and investigators have shown that AT1-AA signaling, via the AT1 receptor, results in a variety of physiological effects.[59] AT1-AA induces signaling in vascular cells and trophoblasts including transcription factor activation.[56,60] The signaling increases TNF-α and reactive oxygen species generation, both of which have been implicated in preeclampsia. Although these novel findings implicate AT1-AA in the pathogenesis of hypertension during preeclampsia, the specific mechanisms that lead to excess production of these autoantibodies and the mechanisms whereby AT1-AA increases blood pressure during pregnancy remain unclear.

Utilizing a bioassay for AT1-AA, LaMarca and colleagues determined that reductions in placental perfusion in pregnant rats had a profound effect to stimulate AT1-AA production. In contrast, the AT1-AA was not detected in normal pregnant rats.[30] In addition, they were able to suppress the production of AT1-AA in the RUPP model of placental ischemia by administration of CD20 blockade. CD20 blockade depletes B lymphocytes and suppresses secretion of antibody.[34] AT1-AA suppression via B cell depletion results in a blunted blood pressure response to placental ischemia. Collectively, these novel findings suggest that a reduction in placental perfusion may be an important stimulus for AT1-AA production. In addition to placental ischemia as a stimulus for the AT1-AA, chronic infusion of TNF-α in normal pregnant rats results in AT1-AA production.[31] The findings suggest the immune mechanisms stimulated by placental ischemia may play an important role in AT1-AA production.

In support of a role for AT1-AA in causing hypertension, the LaMarca lab reported that infusion of purified rat AT1-AA, isolated from serum collected from a pregnant transgenic rat overproducing components of the renin-angiotensin system, into pregnant rats from day 12 to day 19 of gestation, increased serum AT1-AA and blood pressure.[29] Zhou et al. also demonstrated that immunoglobulins isolated from preeclamptic women increase systolic pressure in pregnant mice.[61] This phenotype was ameliorated with co-injection of an AT1 receptor antagonist or the seven-amino-acid peptide that selectively blocks the actions of the AT1-AA.[61] While these collective findings suggest that AT1-AA has important hypertensive actions, the contribution of AT1-AA to the pathophysiology of hypertension in response to placental ischemia or preeclampsia remains an important area of investigation. Although animal studies suggest that increasing plasma AT1-AA concentrations in pregnant rats to levels observed in preeclamptic women or placental ischemic rats result in increased arterial pressure, the quantitative importance of AT1-AA in the pathophysiology of preeclampsia in humans has yet to be fully elucidated. Clinical studies utilizing specific antagonists of the AT1-AA or studies that block the formation of AT1-AA in preeclamptic women are the only approaches to truly determine the role of AT1-AA in human preeclampsia.

Utilizing adoptive transfer methods and pharmacological tools in the RUPP model, a role for CD4+ T cells in the hypertension in response to placental ischemia has been identified.[62] Adoptively transferred CD4+ T cells from RUPP rats increased blood pressure in normal pregnant rats. Moreover, the hypertension that developed in response to adoptive transfer of RUPP CD4+ T cells was associated with elevated TNF, sFlt-1, AT1-AA, and endothelin in normal pregnant recipient rats, none of which were elevated in normal pregnant control rats. To further establish a potential role of CD4+ T cells in the pathophysiology of preeclampsia, the LaMarca lab suppressed T cell activity by administration of abatacept (Orencia), which is a fusion molecule of CTLA-4. CTLA-4 is a marker on T cells used to stimulate an immune response to antigens.[63] Administration of Orencia on gestational day 13 decreased T cells and the blood pressure response to RUPP in pregnant rats, supporting the hypothesis that T cells are important in causing hypertension in response to placental ischemia.[63] (See also Chapters 8 and 15 regarding discussion of immunology and volume regulation in normal pregnancy and preeclampsia.)

GENETIC MODELS

BPH/5

Perhaps the best-characterized genetically linked model for the study of preeclampsia is the BPH/5 mouse. The BPH/5 is a substrain derivation of the BPH/2 "borderline hypertensive" mouse, and exhibits mildly elevated blood pressure throughout the adult lifespan of the animal. However, Davisson et al. demonstrated that the BPH/5 strain also demonstrates late gestational acute elevations in blood pressure (up to ~25 mm Hg), which resolve immediately following parturition, mimicking the effects on blood pressure seen in the typical preeclampsia patient. This was accompanied by proteinuria, intrauterine growth restriction, maternal endothelial dysfunction, and increased fetal mortality.[64] Work from the Davisson lab in subsequent years has uncovered a number of additional characteristics similar to human preeclampsia. For instance, the BPH/5 mouse has diminished cytotrophoblast invasion and placental abnormalities that coincide with increased uterine artery vascular resistance. Importantly, these effects are seen prior to onset of hypertension, as is postulated to occur in the human disorder.[65] Also, both angiogenic imbalance and oxidative stress have been implicated in the pathophysiology of BPH/5 gestational hypertension, as administration of the SOD mimetic Tempol or viral delivery of $VEGF_{121}$ were capable of attenuating the hypertension, proteinuria, and some associated effects of the model.[66,67] It should be noted, however, that while total circulating angiogenic potential is decreased in the BPH/5, this seems to be associated with a primary decrease in VEGF and PlGF rather than increased sFlt-1, an interesting difference from the human syndrome. Nevertheless, the BPH/5 remains one of the most intriguing models for the identification of new pathogenic factors in an animal model in which a preeclampsia-like syndrome occurs spontaneously.

Genetic Modification of the Renin-Angiotensin System

Though its role in long-term maintenance of blood pressure is well established, the exact pathophysiological role of the renin-angiotensin system (RAS) in the development of preeclampsia is less clear. During normal pregnancy, plasma renin concentration/activity as well as Ang II levels are all elevated, but vascular Ang II sensitivity appears to be decreased. In preeclampsia, however, there appears to be an increased Ang II sensitivity, possibly due in part to production of agonistic AT1-AA in the preeclampsia patient.[68,69] To investigate the role of the RAS in pregnancy-induced hypertension, several groups have utilized transgenic mouse and rodent strains in which females overexpressing human angiotensinogen (hAGN) are crossed with males overexpressing human renin hREN.[70–73] This model has been shown to exhibit late gestational hypertension and proteinuria.[70,71] Importantly, when the gender of the transgenic animals is swapped (i.e., male angiotensinogen and female renin), no phenotypic effect is noted, probably due to species specificity of renin acitivity and lack of secretion of AGN from the placenta itself. A related model has been utilized by the Lavoie laboratory which uses chronically hypertensive mice constitutively overexpressing both REN and AGN, as a model of superimposed preeclampsia.[73] Similarly to the transgenic hREN/hAGN cross model these animals develop late gestational hypertension and fetal growth restriction.[73] Both of these models have proved versatile tools for investigating the role of the RAS in the etiology of preeclampsia and are likely to be of great utility in future studies.

STOX1 Overexpression

Another new model that has recently been reported is transgenic overexpression of the transcription factor storkhead box 1 (STOX1) in a murine model. A large, although often contradictory, body of evidence has previously implicated STOX1 dysregulation in the etiology of preeclampsia. Increased risk of preeclampsia was originally linked to STOX1 mutations, particularly the Y153H missense mutation.[74] A number of ex vivo studies supported a role for STOX1's importance. For instance, overexpression of STOX1 in choriocarcinoma cells altered their transcriptional profile in a manner similar to that seen in preeclamptic placentas.[75] Further, overexpression of STOX1 inhibited trophoblast migration *in vitro*, while knockdown of the protein led to outgrowth of extravillous trophoblasts from placental explants.[76] It should be noted, however, that not all epidemiological studies have supported these findings, and it remains unclear whether STOX1 mutation is a universal marker, or limited to specific subgroups of patients.[77–80]

However, a recent report by Doridot et al. lends strong support for a role of STOX1 in the maternal symptoms of preeclampsia.[81] In this study, STOX1 was overexpressed in the placenta by crossing wild-type murine females with STOX1 homozygous knock-in males. Systolic blood pressure measured by tail cuff plethysmyography indicated an ~80 mm Hg increase in systolic blood pressure beginning from very early gestation, even prior to implantation of the embryos (Fig. 10.3). This was accompanied by increased proteinuria, renal capillary swelling, and fibrin deposition. Promisingly, levels of sFlt-1 and sEng were both shown to be increased in the maternal circulation, similar to the situation in the human patient, though at lower levels than seen in either the patient population or other experimental models. This suggests that they are unlikely to the major source of the rise in blood pressure. The increase in pressure seen prior to implantation also suggests that alterations in trophoblast invasion leading to hypoperfusion are also unlikely direct causes. It is more likely that as yet undiscovered pathogenic factors are being produced by the increased STOX1 expression in the placenta, and future studies examining these pathways should prove enlightening. While it is still unknown how fully this model recapitulates the human syndrome, further characterization of this model could be an interesting avenue to elucidating novel pathogenic mechanisms in the etiology of preeclampsia.

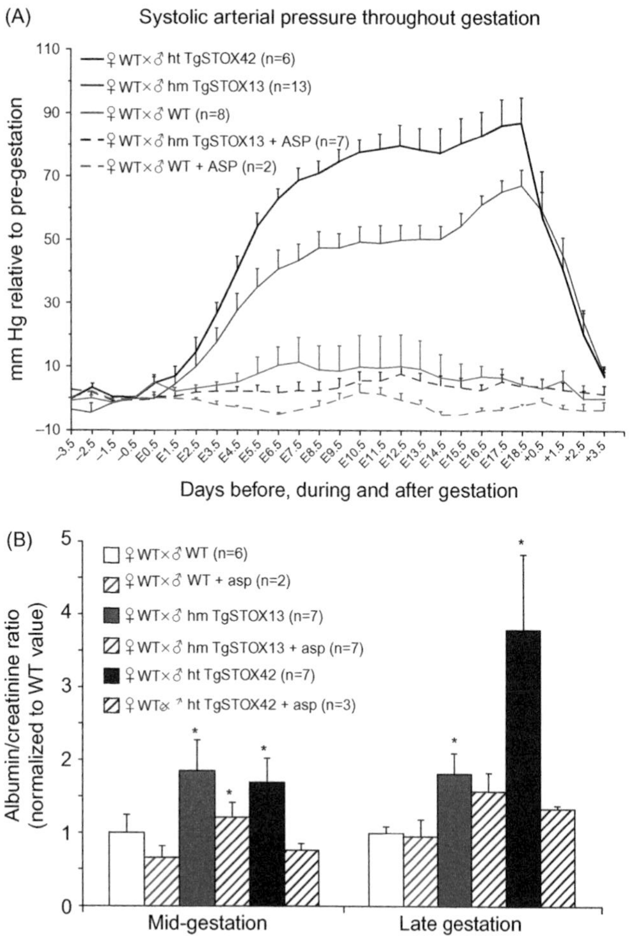

FIGURE 10.3 Fetal/placental STOX1 overexpression causes gestational hypertension in mice. Wild-type female mice were crossed with two STOX1 trangenic lines to overexpress fetally derived STOX1. In response the pregnant mice exhibited a dramatic increase in systolic blood pressure beginning from the time of conception, which remitted immediately after parturition. In contrast, no significant change in systolic pressure was noted when the mice were crossed with corresponding wild-type males (A). Interestingly, aspirin administration in a small cohort abolished this increase in blood pressure. These changes in blood pressure were accompanied by corresponding trends in albumin excretion, which was elevated in the transgenic crosses in both early and late gestation (B). Likewise, aspirin administration blocked this increase. Reprinted with permission from[81].

COMT Knockout

Several lines of evidence suggested a correlation between reduced 2-methoxyestradiol (2-ME), produced by placental catechol-*O*-methyltransferaces (COMT), and preeclampsia in the clinical population. Specifically, while 2-ME increases during normal pregnancy, possibly playing a role to protect the placenta from hypoxia-induced injury, circulating 2-ME levels are reduced in preeclampsia patients.[82–84] To investigate the direct effects of 2-ME deficiency, COMT$^{-/-}$ mice were utilized by multiple groups. The first reports from Kanasaki et al. demonstrated late hypertension, abuminuria, and growth restriction, among other features.[84] A more recent study of the COMT$^{-/-}$ from Stanley et al., while demonstrating fetal growth restriction and minor changes in vascular reactivity, showed only a transient increase in blood pressure in early pregnancy which normalized in late gestation. Further, they demonstrated that the differences could be reversed by administration of sildenafil citrate.[85] Whether this discrepancy is due to strain differences or as yet undetermined environmental differences remains unresolved.

Potential Models with Placental Abnormalities

Several animal models have been reported in the literature with abnormalities in placental development or trophoblast migration, which, with further study, may provide new models for studying various aspects of preeclampsia. Recently the Fisher laboratory demonstrated an important role for Notch2 signaling in human placental trophoblast migration. Conditional deletion of Notch2 in the mouse significantly reduced arterial invasion and placental perfusion, causing significant fetal death.[86] Whether these effects also cause a maternal syndrome similar to preeclampsia remains unknown, and is a fertile area of investigation. Likewise, the Soares lab has demonstrated that the brown Norway rat has highly deficient trophoblast invasion compared to the Dahl Salt Sensitive rat. This was associated with a significantly smaller placental junctional zone and a significant decrease in fetal viability.[87] A more recent study found that the brown Norway rat has alterations in genes involved in the RAS, and angiogenic pathways.[88] Given their effects on the placenta, both of these genetic models have the potential to be intriguing new animal models of preeclampsia. Further studies into both should prove enlightening.

Another potential, non-genetic, approach to the development of preeclampsia models is through dietary-dependent mechanisms. Early observations linking nutrient deprivation, largely in farm animals, with preeclampsia-like toxemia led to the use of several animal models which employed deprivation of one or more nutrients as a model of preeclampsia or toxemia, though these models have generally fallen out of favor.[89–91] One new interesting dietary linkage relies on the observation that obesity and Western diet are associated with an increased risk of preeclampsia and that placentas from obese women have reduced vascularity when compared to those of lean controls.[92,93] Indeed, recent experimental model pregnant animals fed high-fat diets exhibit some of the molecular and physiological characteristics of the preeclampsia patient. Frias et al. demonstrated that monkeys maintained for 4 years on a high-fat diet prior to gestation had significantly reduced uterine artery blood flow independent of weight gain, and those

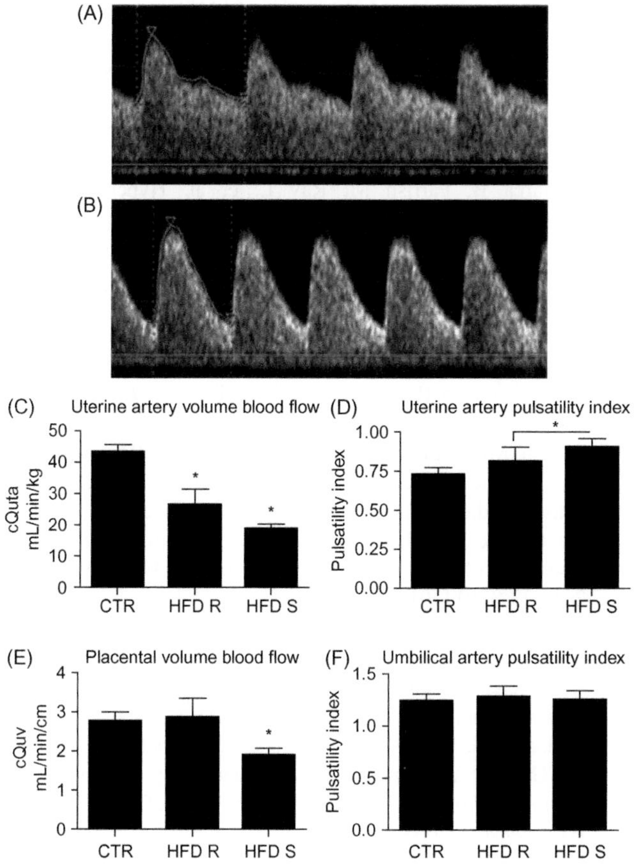

FIGURE 10.4 Japanese macaques maintained for 4 years on a high-fat diet (32% caloric fat) have significant placental alterations compared to those maintained on a control (CTR) (16% caloric fat) diet. Uterine artery Doppler waveforms from control (A) and high-fat diet (HFD) (B) are shown. These waveforms are consistent with decreased diastolic flow and increased impedance. This was concomitant with decreased uterine artery volume blood flow (C) and an increase in the uterine artery pulsatility index, particularly in animals with weight sensitivity (S) to the high-fat diet compared to weight resistant animals (D). In these animals, there was also a decrease in the placental volume blood flow (E). However, no change was seen in the umbilical artery pulsatility index in any of the groups (F). Reprinted with permission from[94].

which were susceptible to diet-induced weight gain had significantly reduced placental volume blood flow (Fig. 10.4). This was accompanied by placental infarction and calcification, although blood pressure in these animals was not determined.[94] Other studies of rats fed a long-term high-fat diet prior to conception also found evidence of placental hypoxia, altered placental vasculature, and elevated systolic blood pressure.[95] These early data suggest several potential diet-induced animal models which could shed light on the linkage between diet/obesity and the incidence of preeclampsia, as well as provide new models for the identification of novel pathogenic pathways in the etiology of preeclampsia.

SUMMARY

Although our understanding of the pathophysiology of preeclampsia has improved over the last decade, interventions to prevent hypertensive disorders in pregnancy have been disappointing. It is only through the continued interplay between basic research including animal models and clinical research in pregnant women that the discovery of new approaches for the prevention and treatment of preeclampsia can be realized. The aim of this chapter was to review some of the more widely studied and/or recently developed animal models used to investigate mechanisms that link placental pathology with maternal endothelial activation/dysfunction and vascular abnormalities of preeclampsia. We attempted to highlight the advantages and limitations of theses models. While studies in animal models have limitations, utilization of these models allows investigators to directly test whether certain factors found in preeclamptic women can indeed lead to hypertension and other manifestations of preeclampsia. Thus, animal models are useful in advancing our knowledge of preeclampsia and should undoubtedly be of greater use in future studies on mechanisms and the development of specific treatments.

References

1. August P, Lindheimer MD. Pathophysiology of preeclampsia. In: Laragh JL, Brenner BM, eds. *Hypertension.* 2nd ed.. New York: Raven Press; 1995:2407–2426.
2. Roberts JM, Taylor RN, Musci TJ, Rodgers GM, Hubel CA, McLaughlin MK. Preeclampsia: an endothelial cell disorder. *Am J Obstet Gynecol.* 1989;161:1200–1204.
3. Saftlas AF, Olson DR, Franks AL, Atrash HK, Pokras R. Epidemiology of preeclampsia and eclampsia in the United States, 1979–1986. *Am J Obstet Gynecol.* 1990;163:460–465.
4. Levine RJ, Maynard SE, Qian C, et al. Circulating angiogenic factors and the risk of preeclampsia. *New Engl J Med.* 2004;350:672–683.
5. Wang A, Rana S, Karumanchi SA. Preeclampsia: the role of angiogenic factors in its pathogenesis. *Physiology (Bethesda).* 2009;24:147–158.
6. Warrington JP, George EM, Palei AC, Spradley FT, Granger JP. Recent advances in the understanding of the pathophysiology of preeclampsia. *Hypertension.* 2013;62(4):666–673.
7. Palei AC, Spradley FT, Warrington JP, George EM, Granger JP. Pathophysiology of hypertension in pre-eclampsia: a lesson in integrative physiology. *Acta Physiol (Oxf).* 2013;208(3):224–233.
8. Conrad KP. Animal models of pre-eclampsia: do they exist? *Fetal Med Rev.* 1990;2:67–88.
9. Granger JP, LaMarca BBD, Cockrell K, et al. Reduced uterine perfusion pressure (RUPP) model for studying cardiovascular-renal dysfunction in response to placental ischemia. *Methods Mol Med.* 2006;122:383–392.
10. Ogden E, Hildebrand GJ, Page EW. Rise in blood pressure during ischemia of the gravid uterus. *Proc Soc Exp Biol Med.* 1940;43:49–51.

11. Hodari AA. Chronic uterine ischemia and reversible experimental "toxemia of pregnancy". *Am J Obstet Gynecol.* 1967;97:597–607.
12. Abitbol MM, Gallo GR, Pirani CL, Ober WB. Production of experimental toxemia in the pregnant rabbit. *Am J Obstet Gynecol.* 1976;124:460–470.
13. Abitbol MM, Pirani CL, Ober WB, Driscoll SG, Cohen MW. Production of experimental toxemia in the pregnant dog. *Obstet Gynecol.* 1976;48:537–548.
14. Losonczy G, Brown G, Venuto RC. Increased peripheral resistance during reduced uterine perfusion pressure hypertension in pregnant rabbits. *Am J Med Sci.* 1992;303:233–240.
15. Cavanagh D, Rao PS, Tung KS, Gaston L. Eclamptogenic toxemia: the development of an experimental model in the subhuman primate. *Am J Obstet Gynecol.* 1974;120:183–196.
16. Cavanagh D, Rao PS, Tsai CC, O'Connor TC. Experimental toxemia in the pregnant primate. *Am J Obstet Gynecol.* 1977;128:75–85.
17. Combs CA, Katz MA, Kitzmiller JL, Brescia RJ. Experimental preeclampsia produced by chronic constriction of the lower aorta: validation with longitudinal blood pressure measurements in conscious rhesus monkeys. *Am J Obstet Gynecol.* 1993;169:215–223.
18. Makris A, Thornton C, Thompson J, et al. Uteroplacental ischemia results in proteinuric hypertension and elevated sFlt-1. *Kidney Int.* 2007;71(10):977–984.
19. Eder DJ, McDonald MT. A role for brain angiotensin II in experimental pregnancy-induced hypertension in laboratory rats. *Clin Exp Hyper Preg.* 1987;B6:431–451.
20. Alexander BT, Kassab SE, Miller MT, et al. Reduced uterine perfusion pressure during pregnancy in the rat is associated with increases in arterial pressure and changes in renal nitric oxide. *Hypertension.* 2001;37:1191–1195.
21. Crews JK, Herrington JN, Granger JP, Khalil RA. Decreased endothelium-dependent vascular relaxation during reduction of uterine perfusion pressure in pregnant rats. *Hypertension.* 2000;35:367–372.
22. Alexander BT, Rinewalt AN, Cockrell KL, Bennett WA, Granger JP. Endothelin-A receptor blockade attenuates the hypertension in response to chronic reductions in uterine perfusion pressure. *Hypertension.* 2001;37:485–489.
23. Alexander BT, Cockrell KL, Sedeek M, Granger JP. Role of the renin-angiotensin system in meditating the hypertension produced by chronic reductions in uterine perfusion pressure in the pregnant rat. *Hypertension.* 2001;38:742–745.
24. Llinas MT, Alexander BT, Abram SR, Sedeek M, Granger JP. Enhanced production of thromboxane A2 in response to chronic reductions in uterine perfusion pressure in pregnant rats. *Am J Hypertens.* 2002;15:793–797.
25. Llinas MT, Alexander BT, Capparelli M, Carroll MA, Granger JP. Cytochrome P-450 inhibition attenuates hypertension induced by reductions in uterine perfusion pressure in pregnant rats. *Hypertension.* 2004;43:623–628.
26. Alexander BT, Llinas MT, Kruckeberg WC, Granger JP. L-arginine attenuates hypertension in pregnant rats with reduced uterine perfusion pressure. *Hypertension.* 2004;43:832–836.
27. LaMarca B, Bennett W, Alexander B, Cockrell K, Granger J. Hypertension produced by reductions in uterine perfusion in the pregnant rat: role of tumor necrosis factor-alpha. *Hypertension.* 2005;46:1022–1025.
28. Lamarca B, Wallukat G, Llinas M, Herse F, Dechend R, Granger J. Autoantibodies to the angiotensini type I receptor in response to placental ischemia and tumor necrosis factor alpha in pregnant rats. *Hypertension.* 2008;52:1168–1172.
29. LaMarca B, Parrish M, Ray L, et al. Hypertension in response to autoantibodies to the angiotensin II type I receptor (AT1-AA) in pregnant rats: a role of endothelin-1. *Hypertension.* 2009;54:905–909.
30. LaMarca B, Wallace K, Herse F, Wallukat G, Weimer A, Dechend R. Hypertension in response to placental ischemia during pregnancy: role of agonistic autoantibodies to the angiotensin II type I receptor (AT1-AA). *Hypertension.* 2010;56(5):e56.
31. Parrish MR, Murphy SR, Rutland S, et al. The effect of immune factors, tumor necrosis factor-alpha, and agonistic autoantibodies to the angiotensin II Type I receptor on soluble fms-like tyrosine-1 and soluble endoglin production in response to hypertension during pregnancy. *Am J Hypertens.* 2010;23(8):911–916.
32. Gilbert JS, Babcock SA, Granger JP. Hypertension produced by reduced uterine perfusion in pregnant rats is associated with increased soluble fms-like tyrosine kinase-1 expression. *Hypertension.* 2007;50:1142–1147.
33. Gadonski G, LaMarca BB, Sullivan E, Bennett W, Chandler D, Granger JP. Hypertension produced by reductions in uterine perfusion in the pregnant rat: role of interleukin 6. *Hypertension.* 2006;48:711–716.
34. Wallace K, Novotny S, Heath J, et al. Hypertension in response to CD4(+) T cells from reduced uterine perfusion pregnant rats is associated with activation of the endothelin-1 system. *Am J Physiol Regul Integr Comp Physiol.* 2012;303(2):R144–R149.
35. Hladunewich M, Karumanchi SA, Lafayette R. Pathophysiology of the clinical manifestations of preeclampsia. *Clin J Am Soc Nephrol.* 2007;2:543–549.
36. Ballermann BJ. Glomerular endothelial cell differentiation. *Kidney Int.* 2005;67:1668–1671.
37. Zhu X, Wu S, Dahut WL, Parikh CR. Risks of proteinuria and hypertension with bevacizumab, an antibody against vascular endothelial growth factor: systematic review and meta-analysis. *Am J Kidney Dis.* 2007;49:186–193.
38. Nagamatsu T, Fujii T, Kusumi M, et al. Cytotrophoblasts up-regulate soluble fms-like tyrosine kinase-1 expression under reduced oxygen: an implication for the placental vascular development and the pathophysiology of preeclampsia. *Endocrinology.* 2004;145:4838–4845.
39. Nevo O, Soleymanlou N, Wu Y, et al. Increased expression of sFlt-1 in in vivo and in vitro models of human placental hypoxia is mediated by HIF-1. *Am J Physiol Regul Integr Comp Physiol.* 2006;291:R1085–R1093.
40. Maynard SE, Min JY, Merchan J, et al. Excess placental soluble fms-like tyrosine kinase 1 (sFlt1) may contribute to endothelial dysfunction, hypertension, and proteinuria in preeclampsia. *J Clin Invest.* 2003;111:649–658.
41. Bridges JP, Gilbert JS, Colson D, et al. Oxidative stress contributes to soluble fms-like tyrosine kinase-1 induced vascular dysfunction in pregnant rats. *Am J Hypertens.* 2009;22:564–568.

42. Murphy SR, LaMarca BB, Cockrell K, Granger JP. Role of endothelin in mediating soluble fms-like tyrosine kinase 1-induced hypertension in pregnant rats. *Hypertension.* 2010;55:394–398.
43. Li Z, Zhang Y, Ying Ma, J, et al. Recombinant vascular endothelial growth factor 121 attenuates hypertension and improves kidney damage in a rat model of preeclampsia. *Hypertension.* 2007;50:686–692.
44. Bergmann A, Ahmad S, Cudmore M, et al. Reduction of circulating soluble Flt-1 alleviates preeclampsia-like symptoms in a mouse model. *J Cell Mol Med.* 2010;14:1857–1867.
45. Gilbert JS, Verzwyvelt J, Colson D, Arany M, Karumanchi SA, Granger JP. Recombinant vascular endothelial growth factor 121 infusion lowers blood pressure and improves renal function in rats with placentalischemia-induced hypertension. *Hypertension.* 2010;55:380–385.
46. Levine RJ, Lam C, Qian C, et al. Soluble endoglin and other circulating antiangiogenic factors in preeclampsia. *New Engl J Med.* 2006;355:992–1005.
47. Venkatesha S, Toporsian M, Lam C, et al. Soluble endoglin contributes to the pathogenesis of preeclampsia. *Nat Med.* 2006;12(6):642–649.
48. Redman CW, Sargent IL. Immunology of pre-eclampsia. *Am J Reprod Immunol.* 2010;63:534–543.
49. LaMarca B, Cornelius D, Wallace K. Elucidating immune mechanisms causing hypertension during pregnancy. *Physiology (Bethesda).* 2013;28(4):225–233.
50. Redman CW, Tannetta DS, Dragovic RA, et al. Review: Does size matter? Placental debris and the pathophysiology of preeclampsia. *Placenta.* 2012;33(Suppl):S48–54.
51. Chamley LW, Chen Q, Ding J, Stone PR, Abumaree M. Trophoblast deportation: just a waste disposal system or antigen sharing? *J Reprod Immunol.* 2011;88(2):99–105.
52. Lau SY, Barrett CJ, Guild SJ, Chamley LW. Necrotic trophoblast debris increases blood pressure during pregnancy. *J Reprod Immunol.* 2013;97(2):175–182.
53. Redman CW, Sargent IL. Preeclampsia and the systemic inflammatory response. *Semin Nephrol.* 2004;24:565–570.
54. Sunderland NS, Thomson SE, Heffernan SJ, et al. Tumor necrosis factor α induces a model of preeclampsia in pregnant baboons (Papio hamadryas). *Cytokine.* 2011;56(2):192–199.
55. Chatterjee P, Kopriva SE, Chiasson VL, et al. Interleukin-4 deficiency induces mild preeclampsia in mice. *J Hypertens.* 2013;31(7):1414–1423. [discussion 1423].
56. Xia Y, Ramin S, Kellems R. Potential roles of angiotensin receptor-activating autoantibody in the pathophysiology of preeclampsia. *Hypertension.* 2007;50:269–275.
57. LaMarca B. The role of immune activation in contributing to vascular dysfunction and the pathophysiology of hypertension during preeclampsia. *Minerva Ginecol.* 2010;62:105–120.
58. Wallukat G, Homuth V, Fischer T, et al. Patients with preeclampsia develop agonistic autoantibodies against the angiotensin AT1 receptor. *J Clin Invest.* 1999;103(7):945–952.
59. Wenzel K, Rajakumar A, Haase H, et al. Angiotensin II type 1 receptor antibodies and increased angiotensin II sensitivity in pregnant rats. *Hypertension.* 2011;58(1):77–84.
60. Xia Y, Wen H, Bobst S, Day M, Kellems R. Maternal autoantibodies from preeclamptic patients activate angiotensin receptors on human trophoblast cells. *Reprod Sci.* 2003;10:82–93.
61. Zhou CC, Zhang Y, Irani RA, et al. Angiotensin receptor agonistic autoantibodies induce pre-eclampsia in pregnant mice. *Nat Med.* 2008;14(8):855–862.
62. Novotny SR, Wallace K, Heath J, et al. Activating autoantibodies to the angiotensin II type I receptor play an important role in mediating hypertension in response to adoptive transfer of CD4+ T lymphocytes from placental ischemic rats. *Am J Physiol Regul Integr Comp Physiol.* 2012;302(10):R1197–R1201.
63. Novotny S, Wallace K, Herse F, et al. $CD4^+$ T cells play a critical role in mediating hypertension in response to placental ischemia. *J Hypertens.* 2013 [in press].
64. Davisson RL, Hoffmann DS, Butz GM, et al. Discovery of a spontaneous genetic mouse model of preeclampsia. *Hypertension.* 2002;39(2 Pt 2):337–342.
65. Dokras A, Hoffmann DS, Eastvold JS, et al. Severe feto-placental abnormalities precede the onset of hypertension and proteinuria in a mouse model of preeclampsia. *Biol Reprod.* 2006;75(6):899–907.
66. Hoffmann DS, Weydert CJ, Lazartigues E, et al. Chronic tempol prevents hypertension, proteinuria, and poor feto-placental outcomes in BPH/5 mouse model of preeclampsia. *Hypertension.* 2008;51(4):1058–1065.
67. Woods AK, Hoffmann DS, Weydert CJ, et al. Adenoviral delivery of VEGF121 early in pregnancy prevents spontaneous development of preeclampsia in BPH/5 mice. *Hypertension.* 2011;57(1):94–102.
68. Shah DM. Role of the renin-angiotensin system in the pathogenesis of preeclampsia. *Am J Physiol Renal Physiol.* 2005;288(4):F614–F625.
69. Wallukat G, Neichel D, Nissen E, Homuth V, Luft FC. Agonistic autoantibodies directed against the angiotensin II AT1 receptor in patients with preeclampsia. *Can J Physiol Pharmacol.* 2003;81(2):79–83.
70. Takimoto E, Ishida J, Sugiyama F, Horiguchi H, Murakami K, Fukamizu A. Hypertension induced in pregnant mice by placental renin and maternal angiotensinogen. *Science.* 1996;274(5289):995–998.
71. Bohlender J, Ganten D, Luft FC. Rats transgenic for human renin and human angiotensinogen as a model for gestational hypertension. *J Am Soc Nephrol.* 2000;11(11):2056–2061.
72. Verlohren S, Niehoff M, Hering L, et al. Uterine vascular function in a transgenic preeclampsia rat model. *Hypertension.* 2008;51(2):547–553.
73. Falcao S, Stoyanova E, Cloutier G, Maurice RL, Gutkowska J, Lavoie JL. Mice overexpressing both human angiotensinogen and human renin as a model of superimposed preeclampsia on chronic hypertension. *Hypertension.* 2009;54(6):1401–1407.
74. van Dijk M, Mulders J, Poutsma A, et al. Maternal segregation of the Dutch preeclampsia locus at 10q22 with a new member of the winged helix gene family. *Nat Genet.* 2005;37(5):514–519.
75. Rigourd V, Chauvet C, Chelbi ST, et al. STOX1 overexpression in choriocarcinoma cells mimics transcriptional alterations observed in preeclamptic placentas. *PLoS One.* 2008;3(12):e3905.
76. van Dijk M, van Bezu J, van Abel D, et al. The STOX1 genotype associated with pre-eclampsia leads to a reduction of trophoblast invasion by alpha-T-catenin upregulation. *Hum Mol Genet.* 2010;19(13):2658–2667.

77. Iglesias-Platas I, Monk D, Jebbink J, et al. STOX1 is not imprinted and is not likely to be involved in preeclampsia. *Nat Genet*. 2007;39(3):279–280. [author reply 280–271].
78. Berends AL, Bertoli-Avella AM, et al. STOX1 gene in pre-eclampsia and intrauterine growth restriction. *BJOG*. 2007;114(9):1163–1167.
79. Kivinen K, Peterson H, Hiltunen L, et al. Evaluation of STOX1 as a preeclampsia candidate gene in a population-wide sample. *Eur J Hum Genet*. 2007;15(4):494–497.
80. George EM, Bidwell GL. STOX1: a new player in preeclampsia? *Hypertension*. 2013;61(3):561–563.
81. Doridot L, Passet B, Mehats C, et al. Preeclampsia-like symptoms induced in mice by fetoplacental expression of STOX1 are reversed by aspirin treatment. *Hypertension*. 2013;61(3):662–668.
82. Berg D, Sonsalla R, Kuss E. Concentrations of 2-methoxyoestrogens in human serum measured by a heterologous immunoassay with an 125I-labelled ligand. *Acta Endocrinol*. 1983;103(2):282–288.
83. Barnea ER, MacLusky NJ, DeCherney AH, Naftolin F. Catechol-O-methyl transferase activity in the human term placenta. *Am J Perinatol*. 1988;5(2):121–127.
84. Kanasaki K, Palmsten K, Sugimoto H, et al. Deficiency in catechol-O-methyltransferase and 2-methoxyoestradiol is associated with pre-eclampsia. *Nature*. 2008;453(7198):1117–1121.
85. Stanley JL, Andersson IJ, Poudel R, et al. Sildenafil citrate rescues fetal growth in the catechol-O-methyl transferase knockout mouse model. *Hypertension*. 2012;59(5):1021–1028.
86. Hunkapiller NM, Gasperowicz M, Kapidzic M, et al. A role for Notch signaling in trophoblast endovascular invasion and in the pathogenesis of pre-eclampsia. *Development*. 2011;138(14):2987–2998.
87. Konno T, Rempel LA, Arroyo JA, Soares MJ. Pregnancy in the brown Norway rat: a model for investigating the genetics of placentation. *Biol Reprod*. 2007;76(4):709–718.
88. Goyal R, Yellon SM, Longo LD, Mata-Greenwood E. Placental gene expression in a rat 'model' of placental insufficiency. *Placenta*. 2010;31(7):568–575.
89. O'Brien PM, Broughton Pipkin F. The effects of deprivation of prostaglandin precursors on vascular sensitivity to angiotensin II and on the kidney in the pregnant rabbit. *Br J Pharmacol*. 1979;65(1):29–34.
90. Vanderlelie J, Venardos K, Perkins AV. Selenium deficiency as a model of experimental pre-eclampsia in rats. *Reproduction*. 2004;128(5):635–641.
91. Seidl DC, Hughes HC, Bertolet R, Lang CM. True pregnancy toxemia (preeclampsia) in the guinea pig (*Cavia porcellus*). *Lab Anim Sci*. 1979;29(4):472–478.
92. Mbah AK, Kornosky JL, Kristensen S, et al. Super-obesity and risk for early and late pre-eclampsia. *BJOG*. 2010;117(8):997–1004.
93. Dubova EA, Pavlov KA, Borovkova EI, Bayramova MA, Makarov IO, Shchegolev AI. Vascular endothelial growth factor and its receptors in the placenta of pregnant women with obesity. *Bull Exp Biol Med*. 2011;151(2):253–258.
94. Frias AE, Morgan TK, Evans AE, et al. Maternal high-fat diet disturbs uteroplacental hemodynamics and increases the frequency of stillbirth in a nonhuman primate model of excess nutrition. *Endocrinology*. 2011;152(6):2456–2464.
95. Hayes EK, Lechowicz A, Petrik JJ, et al. Adverse fetal and neonatal outcomes associated with a life-long high fat diet: role of altered development of the placental vasculature. *PLoS One*. 2012;7(3):e33370.

CHAPTER **11**

Tests to Predict Preeclampsia

AGUSTIN CONDE-AGUDELO, ROBERTO ROMERO AND JAMES M. ROBERTS

Editors' comment: *As noted by the authors in their Introduction, prediction tests had hardly been studied when Chesley published his single-authored edition. However, this concept was never far from his mind, and he was amongst the pioneers in the field suggesting urate clearance tests to predict the disease. As noted in the current chapter, this approach proved imprecise. One problem with the prediction literature is its reliance primarily on clinical criteria alone to establish a diagnosis of preeclampsia. As discussed further in Chapter 6 and elsewhere, diagnosis by purely clinical criteria can be erroneous in a considerable number of patients, especially in multiparous women, when histopathological confirmation is sought (e.g,. renal biopsy). Suggestions are made elsewhere in the text that specific biomarkers be sought and utilized in conjunction with classical clinical signs and symptoms; these are likely to aid in the accurate diagnosis of preeclampsia and should enhance the precision of predictive tests.*

INTRODUCTION

The prediction of preeclampsia is a worthwhile goal because identification of patients at risk could result in earlier diagnosis of the disease, monitoring the mother and fetus at risk, and testing/implementing preventive strategies. In addition, longitudinal studies of patients at risk are a means to gain insight into the pathogenesis of the disease and the development of mechanistic-based strategies for prevention and treatment.

In the single-authored first edition of this textbook, Chesley foresaw the importance of predictive tests; yet, the available tests were few. Chesley highlighted three possible methods: indices of exchangeable sodium, the flicker fusion test (the point at which a rapidly flickering light is perceived as steady), and a variety of pressor responses ranging from angiotensin infusion to the roll-over test. Much work has been conducted since the publication of the first edition of the textbook, and this chapter will review tests which have been proposed for the prediction of the syndrome of preeclampsia.

ASSESSING THE QUALITY OF TESTS TO PREDICT DISEASE

"Prediction" refers to the administration of a test to asymptomatic individuals with the goal of assessing the *likelihood* of developing a particular disease. The evaluation of predictive tests in preeclampsia is important and necessary for several reasons. Preeclampsia is associated with increased maternal and perinatal morbidity and mortality. Interventions to prevent preeclampsia (e.g., low-dose aspirin and calcium) can have small/moderate benefits that caregivers may wish to offer women at higher risk of developing the disease. Intensive maternal and fetal surveillance could be offered, and earlier interventions may decrease the severity of maternal morbidity.

The properties of a predictive test are traditionally assessed using sensitivity, specificity, and predictive values. Sensitivity is the probability of a test result being positive in a patient who will develop the disease, and specificity is the likelihood of a test result being negative in a patient who will not develop the disease. The performance of a test can also be measured in terms of positive and negative predictive values. Sensitivity and specificity cannot be used to estimate the probability of disease for an individual patient.

Predictive values (positive and negative) allow estimation of the probability of subsequently developing the disorder or not developing the disorder; yet they depend on the prevalence of the disorder in the population. Likelihood ratios are alternative indices for summarizing the performance of a test, and are independent of the prevalence of the disease. Likelihood ratios depend on the inherent value of the test to distinguish between patients who will and will not develop the disease, and are calculated by combining sensitivity and specificity.[1,2]

The likelihood ratio of a positive test is the ratio of the probability of a positive predictive test result in women who subsequently develop preeclampsia to the probability of a positive predictive test result in women who do not develop the disease (sensitivity/[1−specificity]). The likelihood ratio of a negative test is the ratio of the probability of a negative

Chesley's Hypertensive Disorders in Pregnancy.
ISBN: 978-0-12-407866-6

DOI: http://dx.doi.org/10.1016/B978-0-12-407866-6.00011-0

predictive test result in women who subsequently develop preeclampsia to the probability of the negative predictive test result in women who do not develop the disease ([1 − sensitivity]/specificity). The greater the positive likelihood ratio, the larger is the increase in the post-test probability of developing disease, while the smaller the negative likelihood ratio, the larger is the decrease in the probability of developing disease.

Tests with a high positive and a low negative likelihood ratio are considered to be clinically useful. Tests with positive likelihood ratios of 10 or greater and negative likelihood ratios of 0.1 or lower are most likely to be useful in clinical practice. Moderate prediction can be obtained with tests with a likelihood ratio value of 5–10 and 0.1–0.2, respectively, whereas those below 5 and above 0.2, respectively, yield only minimal prediction.[2]

The Bayes theorem permits the use of likelihood ratios in conjunction with pre-test probability of preeclampsia to estimate the post-test probability that an individual will develop this disorder once the result of a test is known. The use of odds rather than risk makes the calculation slightly complex; however, a nomogram can be used to avoid having to make conversions between odds and probabilities. The Fagan nomogram[3] is a useful and convenient graphical tool that permits estimation of the post-test probability of disease based on the pre-test or anterior probability of disease and the likelihood ratio.

Likelihood ratios indicate by how much a given test result will increase or decrease the probability of developing preeclampsia. To use the Fagan nomogram (as depicted in Fig. 11.1), a line is drawn from the estimated pre-test probability (left vertical line) through the likelihood ratio of the test result (center vertical line) and the intersection of the line with the right vertical line provides the post-test probability. For example, if a test has a likelihood ratio of a positive result of 15 and a negative likelihood ratio of 0.1, and if a pregnant woman has a pre-test probability of 5% and the test is positive, the post-test probability of disease would be 44% (red line). On the other hand, if the test is negative, the post-test probability would be approximately 0.6% (blue line).

When assessing a test during pregnancy, its accuracy is a function of the prevalence of the disease in the population under study. Since incidence rates of preeclampsia for all women in developing (1.3 to 6.7%) and developed countries (0.4 to 2.8%) are relatively low,[4] any predictive test for this disorder would require a very high positive likelihood ratio (>15) to increase the probability that preeclampsia will occur, and a very low negative likelihood ratio (<0.1) to confidently exclude the probability that the woman will develop the disorder. Criteria for the ideal predictive test for preeclampsia are shown in Table 11.1.

Table 11.2 depicts tests proposed for the prediction of preeclampsia. Such methods have been chosen on the basis of proposed mechanisms of disease implicated in the pathophysiology of preeclampsia.

We will summarize the accuracy of predictive tests for preeclampsia based on the best available evidence (published and updated systematic reviews and meta-analyses on the topic).[5–7] In accordance with recent recommendations, the terms "early preeclampsia" or "early-onset preeclampsia" will refer to preeclampsia that resulted in delivery before 34 weeks of gestation, whereas the terms "late preeclampsia" or "late-onset preeclampsia" will refer to preeclampsia requiring delivery at or after 34 weeks of gestation.[8]

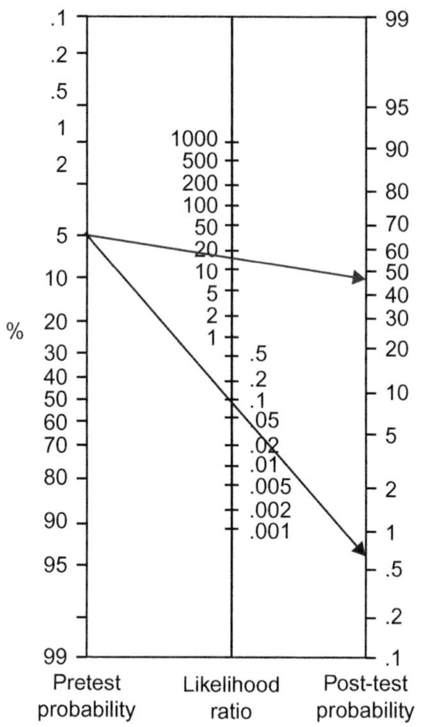

FIGURE 11.1 Use of Fagan's nomogram to calculate post-test probability of preeclampsia. (Adapted from Fagan[3]; This figure is reproduced in color in the color plate section.)

TABLE 11.1 Criteria for a Useful Clinical Predictive Test for Preeclampsia

1. Simple
2. Innocuous
3. Impose minimal discomfort on the women
4. Technology for its measurement must be widely available
5. Rapid
6. Inexpensive
7. Noninvasive
8. Easy to perform early in pregnancy
9. Results of the test must be valid, reliable, and reproducible
10. Very high likelihood ratio for a positive test result (>15)
11. Low likelihood ratio for a negative test result (<0.1)

TABLE 11.2 Predictive Tests for Preeclampsia

1. Placental perfusion/vascular resistance dysfunction-related tests
Roll-over test
Isometric handgrip exercise test
Cold pressor test
Flicker-fusion threshold test
Pressor response to aerobic exercise
Intravenous infusion of angiotensin II
Mean arterial blood pressure
Assessment of arterial stiffness
Platelet angiotensin II binding
Platelet calcium response to arginine vasopressin
Renin
24-hour ambulatory blood pressure monitoring
Transcranial Doppler velocimetry
Ophthalmic artery Doppler velocimetry
Ultrasonographic placental volume, location, and vascularization
3D power Doppler of the uteroplacental circulation space
Uterine artery Doppler velocimetry

2. Fetal and placental unit endocrinology dysfunction-related tests
Human chorionic gonadotropin
Alpha fetoprotein
Estriol
2-Methoxyestradiol
Inhibin A
Pregnancy-associated plasma protein A (PAPP-A)
Activin A
Placental protein 13 (PP-13)
Corticotrophin-releasing hormone
A disintegrin and metalloprotease 12 (ADAM-12)
Kisspeptin
Pregnancy-specific β_1-glycoprotein (SP1)

3. Renal dysfunction-related tests
Serum uric acid
Microalbuminuria
Urinary calcium excretion
Urinary kallikrein
Microtransferrinuria
N-Acetyl-β-glucosaminidase
Cystatin C
Podocyturia

4. Endothelial/oxidant stress dysfunction-related tests
Platelet count and volume
Fibronectin
Endothelin
Neurokinin B
Prostacyclin
Thromboxane
C-reactive protein
Cytokines
Matrix metalloproteinase-9 (MMP-9)
Homocysteine
Isoprostanes
Serum lipids
Ceruloplasmin
Insulin resistance
Adiponectin
Resistin
Inositol phosphoglycan P-type
Antiphospholipid antibodies
Plasminogen activator inhibitor
Leptin
Dimethylarginine
Pentraxin 3
Paraoxonase 1
Tryptophan
Endothelial cell adhesion molecules (P- and E-selectin, vascular cell adhesion molecule-1 (VCAM-1), and intercellular adhesion molecule-1 (ICAM-1))
Angiogenic factors (placental growth factor; vascular endothelial growth factor; fms-like tyrosine kinase receptor-1 [sFlt-1]; endoglin; angiopoietin)

5. Others
Hematocrit
Carboxyhemoglobin
Total proteins
Antithrombin III
Magnesium
Calcium
Ferritin
Transferrin
Haptoglobin
Atrial natriuretic peptide
β_2-Microglobulin
High temperature requirement A3 (HtrA3) protease
25-Hydroxyvitamin D
Thyroid function-related tests
Liver function-related tests
Histidine-rich glycoprotein
Lymphocyte micronucleus (maternal chromosomal damage)
Hydroxysteroid (17-β) dehydrogenase 1
Testosterone
Cell-free fetal DNA
Insulin-like growth factors/insulin-like growth factor binding protein-1
Proteomic, metabolomic and transcriptomic markers
Combination of tests and maternal characteristics

PLACENTAL PERFUSION AND VASCULAR RESISTANCE DYSFUNCTION-RELATED TESTS

Roll-Over Test, Isometric Exercise Test, and Angiotensin II Sensitivity Test

These tests were developed to detect the presence of abnormally increased vascular reactivity between 28 and 32 weeks of gestation and before the clinical onset of preeclampsia. In the roll-over test, maternal blood pressure is measured in the upper arm with the patient in the recumbent position, and when values stabilize the patient is placed in a supine position and the blood pressure is re-measured. The test is considered positive if the diastolic blood pressure increases ≥20 mm Hg. In the isometric hand-grip exercise test, the patient holds a ball in her dominant hand, exerting maximal compressive force for a 2-minute period. The test is considered positive if her diastolic blood pressure rises by ≥20 mm Hg. In the angiotensin II sensitivity test, a baseline blood pressure is established, after which angiotensin II is infused intravenously, the dose increased in a stepwise fashion until a diastolic blood pressure rise of 20 mm Hg is achieved, the angiotensin infusion rate at that time termed the "effective pressor dose."

In 2009, we performed three meta-analyses that included 11 studies (n = 1,126) evaluating the roll over test, three reports assessing the isometric handgrip test (n = 413), and five studies evaluating the angiotensin II sensitivity test (n = 898).[9] This analysis revealed that pooled sensitivities, specificities, and positive and negative likelihood ratios ranged between 54 and 71%, 84 and 85%, 3.7 and 4.5, and 0.3 and 0.5, respectively. We concluded that none of the three different "pressor" tests are clinically useful because they have poor predictive performance for preeclampsia. Additionally, the tests were typically performed late in gestation and, even if their predictive potential were to improve, they would have modest effect, if any, on the implementation of preventive strategies. The angiotensin II sensitivity test is not practical because it is time-consuming, relatively invasive, and expensive. Therefore, none of these tests have gained clinical support.

Blood Pressure

Measurements of maternal blood pressure in early pregnancy are easily performed and recorded. Thus, systolic and diastolic blood pressure, mean arterial pressure, and 24-hour ambulatory pressure monitoring have been proposed as predictors of preeclampsia. In 2008, Cnossen et al.[10] reported a comprehensive meta-analysis, including 34 studies (n = 60,599 women), to determine the value of using systolic and diastolic blood pressures, mean arterial pressure, and the increase over time in blood pressure to predict preeclampsia. A second-trimester mean arterial pressure of ≥90 mm Hg had a pooled sensitivity of 62% and a pooled specificity of 82% (positive and negative likelihood ratios of 3.5 and 0.5, respectively), whereas for a mean arterial pressure threshold of ≥85 mm Hg the pooled sensitivity and specificity were 52 and 84% (positive and negative likelihood ratios of 3.3 and 0.6, respectively). At a specificity of 90%, the sensitivities of diastolic and systolic blood pressures were only 35 and 24%, respectively. Poor predictive accuracy was also found for an increase of systolic or diastolic blood pressures. Subgroup analysis in which the test was evaluated in populations considered to be at high risk of preeclampsia, a diastolic blood pressure ≥75 mm Hg between 13 and 20 weeks of gestation, had a positive likelihood ratio of 2.8 and a negative likelihood ratio of 0.4.

A recent study evaluated mean arterial pressure measured by validated automated devices at 11–13 weeks in a large population of women with singleton gestations (n = 8,366).[11] At a fixed specificity of 90%, the sensitivities for early-onset (delivery at <34 weeks) and late-onset (delivery at ≥34 weeks) preeclampsia were 60 and 37%, respectively (positive and negative likelihood ratios of 6.0 and 0.4, respectively, for early preeclampsia, and 3.7 and 0.7, respectively, for late preeclampsia). Therefore, blood pressure measurement, as a single test during the first and second trimesters, has a modest predictive ability for preeclampsia.

Transcranial Doppler Velocimetry

Transcranial Doppler velocimetry, a noninvasive test, has been used to assess cerebral blood flow velocity in the middle cerebral artery. The procedure is performed by placing an ultrasound transducer over the temporal window and interrogating the brain circulation at a depth of 35–66 mm. Women with overt preeclampsia have a higher cerebral perfusion pressure and cerebrovascular resistance than non-affected women.[12] It has been suggested that some of these changes may be present prior to the development of overt disease, and, therefore, could have predictive value.

We reviewed five studies evaluating the ability of transcranial Doppler velocimetry to predict the development of preeclampsia.[13–17] Moutquin and Williams[13] measured cerebral blood flow velocities in a cohort of 1,400 primiparous women at 20–24 and 28–32 weeks of gestation in order to assess whether early changes in cerebral perfusion pressure or cerebrovascular resistance could predict preeclampsia. When women with a normal blood pressure were compared to those who developed preeclampsia, there were some differences; however, none achieved statistical significance. The sensitivity ranged from 10% to 15% for predicting preeclampsia, depending on the cut-off used. Riskin-Mashiah et al.,[14] using a case-control design, evaluated transcranial Doppler velocimetry findings in 166 women who were normotensive when first studied from 19 to 28 weeks of gestation. The middle cerebral artery pulsatility and resistance indices were lower in women destined to develop preeclampsia (n = 10) compared with those who remained normotensive (n = 20). Based on a receiver operator characteristic (ROC) curve, a middle cerebral artery resistance index value of <0.52 in the second trimester had a sensitivity of 80% and a specificity of 75% (positive and negative likelihood ratios of 3.2 and 0.3, respectively) for the prediction of preeclampsia.

In a prospective cohort study, 38 normotensive pregnant women were studied with Doppler velocimetry of the middle cerebral artery between 18 and 22 weeks of gestation. Using a resistance index <0.93 multiple of the median (MoM) as cut-off,[15] the reported sensitivity was

TABLE 11.3 Summary of Predictive Accuracy of Uterine Artery Doppler Velocimetry for Preeclampsia

Parameter	No. of Studies	No. of Women	Pooled Sensitivity (%)	Pooled Specificity (%)	Pooled Positive Likelihood Ratio	Pooled Negative Likelihood Ratio
Low-risk population						
RI >0.58 or >90–95th percentile	12	7171	56	83	3.4	0.5
Bilateral notch	19	48,571	34	93	5.1	0.7
Any notch	11	9801	76	83	4.5	0.3
[RI > 0.58 or > 90–95th percentile] or any notch	6	6992	71	84	4.3	0.4
RI ≥ 0.55 and bilateral notch or RI ≥ 0.65 and unilateral notch	4	2450	60	87	4.5	0.5
PI > |1.38–1.54|	4	21,974	43	92	5.0	0.6
[PI > |1.45–1.60| or > 90–95th percentile] and/or bilateral notch	3	21,709	46	89	4.1	0.6
High-risk population						
RI > 0.58 or > 90–95th percentile	18	2229	71	71	2.5	0.4
Bilateral notch	8	1202	40	88	3.3	0.7
Any notch	9	1019	61	76	2.6	0.5
[RI > 0.58 or > 90–95th percentile] and [any notch or bilateral notch]	6	1046	32	89	2.9	0.8

RI = resistance index; PI = pulsatility index.

100% and specificity 65% (positive and negative likelihood ratios of 2.9 and 0.0, respectively) for the prediction of preeclampsia ($n = 7$). In 2007, Belfort et al.[16] measured middle cerebral artery Doppler parameters in 181 low-risk women at 16–24 weeks of gestation, and found that a cerebral perfusion pressure >46 mm Hg had a sensitivity of 63% and specificity of 73% (positive and negative likelihood ratios of 2.3 and 0.5, respectively) for predicting preeclampsia at term. Recently, a study by the same group assessed the predictive value of decreased middle cerebral artery Doppler resistance in 405 low-risk women at a mean gestational age of 19.0 ± 1.3 weeks (range 12–26 weeks).[17] Either a resistance index of <0.54 or a pulsatility index of <0.81 had a sensitivity of 86% and a specificity of 93% for predicting preeclampsia (positive and negative likelihood ratios of 12.3 and 0.2, respectively). In summary, studies of the use of transcranial Doppler velocimetry for the prediction of preeclampsia have yielded conflicting results. Additional longitudinal studies are needed.

Uterine Artery Doppler Velocimetry

Failure of trophoblast invasion of the spiral arteries results in increased vascular resistance of the uterine artery and decreased perfusion of the placenta, which may subsequently result in fetal growth restriction and/or preeclampsia[18] (reviewed in Chapter 5). Uterine artery Doppler velocimetry, a noninvasive method to examine the impedance to flow in the uterine circulation, provides indirect evidence of whether failure of trophoblast invasion has occurred.[19] The uterine artery is identified with the aid of color Doppler velocimetry, and then pulsed-wave Doppler velocimetry is performed to obtain waveforms. Increased impedance is associated with an abnormal waveform pattern assessed by either an increased resistance index or pulsatility index, or by the persistence of a unilateral or bilateral diastolic "notch."

Table 11.3 depicts several meta-analyses (studies reviewed through 2009) that evaluated 71 studies (~90,000 women) assessing the accuracy of uterine artery Doppler velocimetry to predict preeclampsia.[9] In general, prediction accuracy in low-risk populations was moderate to minimal, while that in high-risk populations was minimal. In low-risk populations, the sensitivities and specificities of the uterine artery Doppler indices varied between 34 and 76%, and between 83 and 93%, respectively. The presence of a bilateral notch and an increased pulsatility index were the best predictors of preeclampsia (positive likelihood ratio of 5.1). These indices appeared to have high specificity, but at the expense of low sensitivity. In high-risk populations, all uterine artery Doppler indices had a poor predictive ability, with positive likelihood ratios ranging between 2.5 and 3.3 and negative likelihood ratios between 0.4 and 0.8. The combination of the presence of a notch and an increased pulsatility or resistance index did not improve the predictive accuracy of a single index. Subsequent studies confirmed that uterine artery Doppler velocimetry, in the first or second trimester, has low to moderate predictive ability for preeclampsia in low-risk populations[20–26] and low prediction accuracy in high-risk populations.[23,27–29] Recently, Akolekar et al.[30] reported on the predictive accuracy of the uterine artery pulsatility index at 11–13 weeks of gestation in 45,885 singleton gestations in the United Kingdom. At a fixed specificity of 90%, the sensitivity for the identification of the patient who would subsequently develop preeclampsia was 42% (positive and negative likelihood ratios of 4.2 and 0.6, respectively).

A systematic review and meta-analysis published in 2008 concluded that an increased pulsatility index with a notch in the second trimester is the best predictor of preeclampsia and strongly recommended the routine use of these measurement parameters for predicting preeclampsia.[31] Our review of this analysis, however, calls for caution – this recommendation was based on only two studies, one of which included 1,757 low-risk women and the other 351 high-risk women. There is no evidence that heterogeneity and publication bias were formally investigated.[32]

The predictive accuracy of uterine artery Doppler velocimetry for early-onset preeclampsia is presented in Table 11.4.[20–22,24,25,28–30,33–38] Overall, among women at low to moderate risk of preeclampsia, it appears that, irrespective of the index or combinations of indices used, uterine artery Doppler velocimetry is a moderate to good predictor of preeclampsia requiring delivery before 34 weeks of gestation. However, the presence of bilateral uterine artery notch, the test with the highest likelihood ratio for a positive test result (17.2), would only increase the pre-test probability of early-onset preeclampsia from 0.5% to 8%. Among high-risk women, the prediction accuracy was moderate to minimal.

Overall, although uterine artery Doppler velocimetry fulfills several of the criteria for an ideal predictive test for preeclampsia, current evidence does not support its routine clinical use for the prediction of this disorder. Nonetheless, this method could have value for the prediction of early-onset preeclampsia.

FETAL AND PLACENTAL UNIT ENDOCRINOLOGY DYSFUNCTION-RELATED TESTS

Various guidelines for routine prenatal care include measuring first- and/or second-trimester maternal serum levels of human chorionic gonadotropin (hCG), alpha fetoprotein (AFP), unconjugated estriol, inhibin A, pregnancy-associated plasma protein-A (PAPP-A), and activin A for screening of fetal aneuploidy, neural tube defects, and other fetal abnormalities. Since the results of these tests are available by early second trimester, it has been proposed that some of these analytes, alone or in combination, can be used to predict preeclampsia. Several groups have evaluated the utility of these fetoplacental proteins, but the results have shown limited accuracy to predict preeclampsia.

Human Chorionic Gonadotropin

Human chorionic gonadotropin, a glycoprotein composed of α- and β-subunits, is produced by the syncytiotrophoblast cells, whose primary function is the maintenance of the corpus luteum in early pregnancy.[39] The role of this hormone in the second and third trimesters is not as well understood. In 1993, Sorensen et al.[40] noted that women with elevated second-trimester hCG levels had a higher risk of preeclampsia. They suggested hypoperfusion of the villous tree would lead to an increase in hCG production. Several groups have investigated the value of hCG as a predictive test for preeclampsia. In 2009, we pooled results from 13 studies ($n = 56{,}067$) that used 2.0 multiples of the median (MoM) and five ($n = 49{,}572$) that used 2.3–2.5 MoM as cut-offs.[9] hCG concentrations above 2 MoM showed a pooled sensitivity of 21% and a pooled specificity of 91% (positive and negative likelihood ratios of 2.4 and 0.9, respectively). Similar pooled estimates were obtained for studies that used 2.3 to 2.5 MoM as cut-offs. Further studies have reported similar predictive values.[41–43]

Alpha Fetoprotein

Alpha fetoprotein is produced at first primarily by the yolk sac and later by the fetal liver.[44] In 1989, Milunsky et al.[45] reported that high levels of maternal serum AFP were associated with an increased risk of preeclampsia. Of the prediction studies we identified and assessed, most used 2.0 and 2.5 MoM as the cut-off level for defining abnormality. Irrespective of the cut-off used, the estimates of predictive accuracy for preeclampsia were very low. The pooled sensitivity, specificity, and positive and negative likelihood ratios from 11 studies ($n = 115{,}419$) that used 2.0 MoM as the cut-off were 13%, 96%, 3.3, and 0.9, respectively.[41,42,45–53]

Estriol

The placenta produces estriol by the desulfation of 16-hydroxydehydroepiandrosterone sulfate to androgens, which are subsequently aromatized to estriol. It has been hypothesized that impaired or reduced activity of placental sulfatase may lower the mid-trimester maternal levels of estriol.[54] In 1996, Santolaya-Forgas et al.[55] reported that women with unexplained low second-trimester maternal serum unconjugated estriol had an increased risk of adverse pregnancy outcomes, including preeclampsia, and suggested that measurement of this analyte could be a useful predictor. The accuracy of unconjugated estriol in predicting preeclampsia (using cut-offs of 0.50, 0.85, 0.86, and 0.90 MoM) was reported in five studies ($n = 56{,}513$).[47,48,56–58] Irrespective of the cut-off used, the predictive accuracy was very low. In fact, the pooled sensitivities ranged from 6 to 33%, specificities from 75 to 96%, and positive and negative likelihood ratios from 1.3 to 1.5 and 0.9 to 1.0, respectively.

TABLE 11.4 Uterine Artery Doppler Velocimetry for the Prediction of Early-Onset Preeclampsia

Author, Year	No of Women	Risk of Women	Gestational Week	Abnormal Test Result	Sensitivity (%)	Specificity (%)	Positive Likelihood Ratio	Negative Likelihood Ratio
Bower, 1993[33]	2058	Low to moderate	18–22	Any notch or RI > 95th percentile	79	84	5.0	0.3
Mires, 1998[34]	6579	Low to moderate	18–20	Bilateral notch	26	99	17.2	0.8
Coleman, 2000[35]	116	High	22–24	RI > 0.58	95	31	1.4	0.2
				Bilateral notch	62	89	5.6	0.4
Yu, 2005[36]	15,392	Low to moderate	22–24	PI > 1.6	78	95	15.6	0.2
Prefumo, 2004[37]	4149	Low to moderate	18–23	Any notch	77	85	5.0	0.3
				Bilateral notch	65	95	13.5	0.4
Pilalis, 2007[38]	1123	Low to moderate	11–14	PI > 95th percentile	33	95	6.8	0.7
Palma-Dias, 2008[20]	1057	Low to moderate	22–24	PI > 1.55	50	88	4.2	0.6
Llurba, 2009[21]	6035	Low to moderate	19–22	PI > 1.66 and/or bilateral notch	75	91	8.3	0.3
Espinoza, 2010[22]	3950	Low to moderate	23–25	Bilateral notch	53	93	7.6	0.5
				PI > 95th percentile	58	95	11.6	0.4
				Bilateral notch and/or PI > 95th percentile	74	91	8.2	0.3
Myatt, 2012[24]	2188	Low to moderate	13–20	Any notch or MoM of RI ≥ 75th percentile	78	66	2.3	0.3
Herraiz, 2012[28]	135	High	11–13	PI > 90th percentile	33	90	3.3	0.7
			19–22	PI > 90th percentile	67	90	6.7	0.4
Arcangeli, 2013[25]	382	Low to moderate	20–22	PI > 1.9	40	90	4.0	0.7
Demers, 2013[29]	1810	High	11–13	PI ≥ 1.0 MoM	95	53	2.0	0.1
Akolekar, 2013[30]	45,885	Low to moderate	11–13	PI (unreported cut-off)	59	95	11.8	0.4

RI = resistance index; PI = pulsatility index; MoM = multiple of the median.

Inhibin A

Inhibin A, a member of the transforming growth factor β family, is predominantly secreted by the placenta and localized within the syncytiotrophoblast.[59] Elevated concentrations of inhibin A have been observed in women with preeclampsia compared to normotensive controls.[60] In addition, Cuckle et al.[61] reported that pregnant women destined to develop preeclampsia had significantly higher inhibin A levels than those who remained normotensive, suggesting that this placental protein might be a useful predictive test for preeclampsia. We pooled results from seven studies[41,42,48,61–64] (n = 45,274) that evaluated inhibin A as a predictor of preeclampsia using a cut-off of 2 MoM. The pooled sensitivity, specificity, and positive and negative likelihood ratios were 21%, 95%, 4.2, and 0.8, respectively.

Pregnancy-Associated Plasma Protein A

PAPP-A, an insulin-like growth factor binding protein metalloproteinase complex predominantly produced by the placenta, has been implicated in the autocrine and paracrine control of trophoblast invasion of the decidua.[65] Maternal serum PAPP-A concentrations have been shown to be relatively low during the first trimester of pregnancies complicated by preeclampsia.[66] Based on this association, several groups of investigators have examined the value of low maternal serum levels of PAPP-A in the prediction of preeclampsia. Pooling results from seven studies[64,67–72] (n = 99,449) that evaluated the predictive accuracy of PAPP-A using a cut-off of about 0.4 MoM we noted a very low predictive accuracy, with pooled sensitivity, specificity, and positive and negative likelihood ratios of 10%, 95%, 2.0, and 0.9, respectively.

Activin A

Activin A, a homodimeric glycoprotein produced by the decidua, placenta, and fetal membranes, is involved in cellular differentiation, proliferation, remodeling, and morphogenesis.[73] In 1995, Petraglia et al.[74] reported that maternal serum activin A concentration was elevated in patients with preeclampsia. Some suggested that such elevations may be due to increased cytotrophoblast proliferation in response to ischemic damage in the placenta.[60] We analyzed 10 studies[75–84] (n = 1,985), using several cut-offs for defining abnormality, in which estimates of the accuracy of activin A to predict preeclampsia were available. Sensitivity ranged from 11 to 93%, specificity from 71 to 95%, and positive and negative likelihood ratios from 2.2 to 11.0 and 0.1 to 0.9, respectively.

Placental Protein 13

Placental protein 13 (PP-13), a member of the galectin family predominantly expressed by the syncytiotrophoblast, is involved in both implantation and maternal artery remodeling.[85] Preliminary data suggested that maternal serum PP-13 concentrations are significantly reduced in the first trimester in women who subsequently develop preeclampsia, particularly in those destined to present with early-onset disease.[86] The mechanisms responsible for this reduction have not been determined. We reviewed 10 studies (n = 4,468) in which maternal serum PP-13 measured during the first trimester was evaluated as a predictor of preeclampsia using several cut-offs for defining abnormality.[87–96] Seven studies reported on predictive accuracy for all cases of preeclampsia[88–92,95,96] and six for early-onset preeclampsia.[87,89,91–94]

For the prediction of preeclampsia as a whole, sensitivities ranged from 24 to 80% (median, 60%), specificities from 80 to 90% (median, 80%), and positive and negative likelihood ratios from 1.2 to 7.9 (median 4.0) and 0.2 to 1.0 (median 0.4), respectively. To predict early-onset preeclampsia, sensitivities ranged from 21 to 100% (median 60%), specificities from 80 to 95% (median 84%), and positive and negative likelihood ratios from 2.5 to 8.0 (median 4.9) and 0.0 to 0.8 (median 0.5), respectively. In addition, in the studies by Audibert et al.[97] (n = 893) and Myatt et al.[98] (n = 683) (which did not report predictive values for preeclampsia), the median levels of PP-13 in the first trimester did not differ significantly between women who later developed preeclampsia and normotensive controls. In conclusion, maternal serum first-trimester PP-13 has moderate to minimal predictive accuracy for both overall preeclampsia and early-onset preeclampsia.

A Disintegrin and Metalloprotease 12

A disintegrin and metalloprotease 12 (ADAM-12) is a placenta-derived glycoprotein produced by the trophoblast that could be involved in placental and fetal growth. In 2005, Laigaard et al.[99] reported that maternal serum levels of ADAM-12 were significantly lower during the first trimester in women who subsequently developed preeclampsia when compared with levels in women with normal pregnancies, and suggested that this glycoprotein might be a useful early marker for preeclampsia. We analyzed 4 studies[43,96–98] (n = 2,425), using several cut-offs for defining abnormality, in which estimates of the accuracy of ADAM-12 to predict preeclampsia were available. Irrespective of the cut-off used, the predictive accuracy was very low with sensitivities that ranged from 16 to 41%, specificities from 80 to 90%, and positive and negative likelihood ratios from 1.0 to 2.1, and 0.7 to 1.0, respectively. In addition, two studies reported that median ADAM-12 concentrations in the first[100] or early second trimester[101] were not significantly different between women who developed preeclampsia and normotensive controls. In summary, ADAM-12 is not a useful test to predict preeclampsia.

RENAL DYSFUNCTION-RELATED TESTS

Serum Uric Acid

Hyperuricemia was among the earliest biochemical abnormalities described in preeclampsia, and has been attributed to reduced uric acid clearance secondary to a fall in glomerular filtration, increased tubular reabsorption combined with decreased secretion of urate, tissue injury, and oxidative stress[102] (see Chapter 16). These observations, made after the diagnosis of disease, led to studies to determine whether measuring serum uric acid concentrations earlier in pregnancy could be used to predict preeclampsia.

In 2006, Cnossen et al.[103] published a systematic review of the accuracy of serum uric acid determination for the prediction of preeclampsia, identifying five primary articles (n = 572, of whom 44 developed preeclampsia). Sensitivities ranged from 0 to 56%, specificities from 77 to 95%, and positive and negative likelihood ratios from 0.0 to 9.8 and 0.5 to 1.2, respectively. A recent study evaluating the predictive ability of serum uric acid in the mid-second trimester (n = 1,000)[104] reported a sensitivity of 54% and a specificity of 65% (positive and negative likelihood ratios of 1.5 and 0.7, respectively). The low predictive values of serum uric acid render this test unsuitable for the prediction of preeclampsia.

Microalbuminuria

Ten studies[105–114] (n = 5,024 women), which used different cut-offs, have evaluated the potential value of microalbuminuria as a predictive test for preeclampsia. Sensitivities and specificities ranged between 7 and 92% and 29 and 97%, respectively, whereas the positive and negative likelihood ratios ranged between 0.8 and 3.1, and 0.1 and 1.1, respectively. Poon et al.[111] measured urine albumin concentration and albumin/creatinine ratios in random midstream urine collected between gestational weeks 11 to 14 from 2,679 unselected women. Using a urine albumin concentration ≥5 mg/L to define microalbuminuria, the sensitivity and specificity for predicting preeclampsia were 75 and 43%, respectively (positive and negative likelihood ratios of 1.3 and 0.6, respectively). It seems that evaluation of microalbuminuria has limited value in the prediction of preeclampsia.

Urinary Calcium Excretion

In 1987, Taufield et al.[115] reported that preeclampsia is associated with hypocalciuria. Since then, several studies have been performed to determine whether measurement of urinary calcium excretion might predict the subsequent development of preeclampsia. In six studies[112,116–120] (n = 965) that used cut-offs of approximately 200 mg/24 hours, the pooled sensitivity, specificity, and positive and negative likelihood ratios were 60%, 75%, 2.4, and 0.5, respectively. In 12 studies[105,117,118,121–129] (n = 3,298) in which investigators evaluated urinary calcium/creatinine ratio using several cut-offs, the sensitivity and specificity varied between 31 and 89%, and 55 and 95%, respectively. The positive and negative likelihood ratios ranged between 0.7 and 9.5, and 0.2 and 1.3, respectively. Overall, urinary calcium excretion, measured in 24-hour urine or as calcium/creatinine ratio, performed poorly in predicting preeclampsia.

Urinary Kallikrein

In patients with hypertensive disorders of pregnancy, and particularly in preeclampsia, urinary kallikrein excretion is lower than in normotensive pregnancy.[130] The kallikrein–kinin system induces vasodilatation, and is believed to play an important paracrine role in the regulation of local blood flow.[131] The potential value of measuring urinary kallikrein as a predictor of preeclampsia has been assessed by two groups of investigators. Millar et al.[132] measured inactive urinary kallikrein/creatinine ratios in spot urine samples from 307 healthy women at 16–20 weeks of gestation. They defined a positive test as an inactive urinary kallikrein/creatinine ratio of ≤200. The sensitivity and specificity were 83 and 99%, respectively (positive and negative likelihood ratios of 83.0 and 0.2, respectively) for "proteinuric preeclampsia."

Kyle et al.[133] determined the inactive urinary kallikrein/creatinine ratio at 28 weeks of gestation in 458 healthy nulliparas. They reported a sensitivity of 80% and a specificity of 71% (positive and negative likelihood ratios of 2.8 and 0.3, respectively) for "proteinuric preeclampsia." In that study, an inactive urinary kallikrein/creatinine ratio less than 170 was used as the cut-off value. The limited number of studies precludes firm conclusions about the predictive accuracy of the inactive urinary kallikrein/creatinine ratio as a test to predict preeclampsia.

Podocyturia (See Chapter 16)

Podocytes play a crucial role in maintaining the integrity of the glomerular slit diaphragm. Urinary shedding of podocytes (podocyturia) may indicate their loss from the glomerulus, which could lead to a disruption of the glomerular filtration barrier and consequent proteinuria in preeclampsia.[134]

In 2007, Garovic et al.[135] reported a small case-control study of 15 women with preeclampsia and 16 normotensive controls in which urinary podocyte excretion (as determined by podocin-positive cells) was present in all patients with preeclampsia at the time of delivery, and in none of the controls. A subsequent case-control study[136] reported that podocyturia was present in 11 out of 29 (38%) patients with preeclampsia. In contrast, none of the nine women with uncomplicated pregnancies had podocyturia. It has been suggested that detection of podocyturia could be a useful early test for the prediction of preeclampsia. To date, only one study evaluated the accuracy of podocyturia for the prediction of preeclampsia.[137] Podocyturia was detected in all 15 women (median gestational age 27 weeks) who developed preeclampsia compared with none of the 44 normotensive controls (both sensitivity and specificity of 100%). These findings should be confirmed in prospective cohort studies with a large sample size.

Podocyturia is discussed in detail in Chapter 16. The authors review the specificity (presence of podocyturia in other renal diseases that can mimic preeclampsia), debate about the methodology, and some evidence that podocyturia can be present in normal pregnant women.

ENDOTHELIAL DYSFUNCTION AND OXIDANT STRESS-RELATED TESTS

Fibronectin

Fibronectins, a class of high-molecular-weight glycoproteins, have been implicated in a variety of cellular functions, including adhesion, cytoskeletal organization, oncogenic transformation, cell migration, phagocytosis, hemostasis, and embryonic differentiation.[138] Cellular fibronectin represents only 5% of circulating fibronectins and is synthesized by a wide variety of cells, including endothelial cells, fibroblasts, and macrophages. It is released from endothelial cells and extracellular matrix after endothelial injury.[139] In 1984, Stubbs et al.[140] reported that plasma fibronectin concentrations were elevated in women with preeclampsia, and subsequently the role of fibronectin as an early predictor for preeclampsia has been evaluated by several investigators.

In 2007, Leeflang et al.[141] published a systematic review on the accuracy of fibronectin for the prediction of preeclampsia that included 12 studies, of which only five (n = 573) reported sufficient data to permit calculation of predictive accuracy estimates. Total plasma fibronectin was measured in three studies, cellular fibronectin in one, and both in another. The sensitivities vary widely, depending on the cut-off chosen. Requiring a sensitivity of at least 50%, the specificity achieved with total fibronectin ranged between 43 and 94%. For cellular fibronectin, specificities varied from 72 to 96%. The positive likelihood ratios ranged from 1.2 to 11.0 for total fibronectin and from 1.6 to 11.5 for cellular fibronectin. The negative likelihood ratios varied from 0.4 to 0.7 for total fibronectin and from 0.0 to 0.6 for cellular fibronectin. Such results suggest that plasma fibronectin concentrations are not clinically useful for the prediction of preeclampsia.

Homocysteine

Homocysteine, a metabolite of the essential amino acid methionine, has been postulated to lead to oxidative stress and endothelial cell dysfunction, alterations associated with preeclampsia.[142] A systematic review by Mignini et al.[143] concluded that homocysteine concentrations are increased in women who subsequently develop preeclampsia, the highest levels noted with overt disease. We pooled results from six studies (n = 2,449) which examined the predictive accuracy of serum homocysteine. Investigators used cut-off values that ranged from 5.5 to 6.3 μmol/L.[9] The pooled sensitivity, specificity, and positive and negative likelihood ratios were 36%, 88%, 3.1, and 0.7, respectively. Recently, Bergen et al.,[144] in a population-based cohort study of 5,805 women, found no association between the serum levels of homocysteine at 11–16 weeks and the risk of preeclampsia. There was no association between high homocysteine levels and preeclampsia. Thus, the determination of serum homocysteine levels does not appear to be useful in predicting preeclampsia.

Endothelial Cell Adhesion Molecules

Adhesion molecules play a central role in the adherence of leukocytes to endothelial cells and the subsequent migration of white blood cells into perivascular tissue.[145] Cellular forms of adhesion molecules mediate specific steps of leukocyte–endothelial cell interaction, and have been implicated in the pathophysiology of preeclampsia.[146] Leukocyte–endothelial cell interactions are mediated mainly by three families of cell adhesion molecules (CAMs): selectins (E-, P-, and L-selectin), integrins, and members of the immunoglobulin gene superfamily (i.e., vascular cell adhesion molecule-1 (VCAM-1), and intercellular adhesion molecule-1 (ICAM-1)).[145] Soluble forms of these molecules can be detected in plasma, and their concentrations are believed to reflect the degree of activation of a particular cell type. Elevations in soluble P-selectin reflect platelet activation; changes in soluble L-selectin suggest leukocyte activation; and an increase in soluble forms of E-selectin, VCAM-1, and ICAM-1 indicates endothelial cell activation/dysfunction.[145]

We identified five studies reporting estimates of predictive accuracy of cell adhesion molecules for preeclampsia,[77,147–150] of which four reported on VCAM-1[77,148–150] and two each on ICAM-1[148,150] and P-selectin.[147,149] For VCAM-1, sensitivities, specificities, and positive and negative likelihood ratios ranged between 27 and 91%, 52 and 100%, 1.1 and infinity, and 0.1 and 1.0, respectively. For ICAM-1 the sensitivity and specificity varied between 43 and 50%, and 33 and 98%, respectively. The positive and negative likelihood ratios ranged between 1.0 and 19.6, and 0.6 and 1.5, respectively. For P-selectin, the sensitivities and specificities varied between 76 and 80% and 83 and 96%, respectively, whereas the positive and negative likelihood ratios ranged between 4.5 and 19.0, and 0.2 and 0.3, respectively. More recently, Carty et al.[151] reported no significant differences in mean concentrations of VCAM-1,

ICAM-1, and P-selectin at 12–16 weeks of gestation between women who developed preeclampsia (n = 49) and normotensive controls (n = 74). In contrast, Akolekar et al.[152] reported that at 11–13 weeks of gestation, women who subsequently developed preeclampsia (n = 121) had higher serum levels of P-selectin than controls (n = 208). However, its predictive accuracy for preeclampsia was low.

In summary there is insufficient evidence to recommend the measurement of endothelial cell adhesion molecules as predictors of preeclampsia.

Circulating Angiogenic Factors

There is strong evidence that an imbalance between angiogenic factors, such as vascular endothelial growth factor (VEGF) or placental growth factor (PlGF), and anti-angiogenic factors, such as soluble fms-like tyrosine kinase 1 (sFlt-1, also referred to as soluble VEGF receptor-1) and the soluble form of endoglin (sEng), is related closely to the pathogenesis of preeclampsia.[153,154] The role of angiogenic factors in preeclampsia is the subject of Chapter 6, where it is noted that the plasma/serum concentration of angiogenic factors are lower, and antiangiogenic factors higher, in the maternal circulation both before the diagnosis and at the time of clinical recognition of the disease.[153–156] Therefore, plasma levels of angiogenic factors have been suggested as markers for the prediction of preeclampsia.

In 2012, Kleinrouweler et al.[157] published a comprehensive systematic review and meta-analysis on the accuracy of angiogenic factors for the prediction of preeclampsia. Eligible studies (through October 2010) were those that measured PlGF, VEGF, sFlt-1 or sEng in serum or plasma of pregnant women before 30 weeks of gestation. The authors used the diagnostic odds ratio (the ratio of the odds of positive test results in women who later developed preeclampsia relative to the odds of positive test results in those who did not develop preeclampsia) as the measure of predictive accuracy. Higher diagnostic odds ratios represent higher predictive accuracies. Thirty-four studies met the inclusion criteria. Twenty-seven studies reported on PlGF, three studies on VEGF, 19 studies on sFlt-1 and 10 studies on sEng. Overall, the concentrations of PlGF and VEGF were lower among women who subsequently developed preeclampsia whereas the levels of sFlt-1 and sEng were higher. The pooled diagnostic odds ratio of PlGF for predicting preeclampsia was 9.0 (95% confidence interval [CI] 5.6–14.5), which corresponds to both sensitivities and specificities of 75% (positive and negative likelihood ratios of 3.0 and 0.3, respectively), or a sensitivity of 50% and specificity of 90% (positive and negative likelihood ratios of 5.0 and 0.6, respectively), or a sensitivity of 32% and specificity of 95% (positive and negative likelihood ratios of 6.4 and 0.7, respectively). The pooled diagnostic odds ratio of sFlt-1 for predicting preeclampsia was 6.6 (95% CI 3.1–13.7), which corresponds to sensitivities and specificities of 72% (positive and negative likelihood ratios of 2.6 and 0.4, respectively), or a sensitivity of 42% and specificity of 90% (positive and negative likelihood ratios of 4.2 and 0.6, respectively), or a sensitivity of 26% and specificity of 95% (positive and negative likelihood ratios of 5.2 and 0.8, respectively). The pooled diagnostic odds ratio of sEng for predicting preeclampsia was 4.2 (95% CI 2.4–7.2), which corresponds to sensitivities and specificities of 67% (positive and negative likelihood ratios of 2.0 and 0.5, respectively), or a sensitivity of 32% and specificity of 90% (positive and negative likelihood ratios of 3.2 and 0.8, respectively), or a sensitivity of 18% and specificity of 95% (positive and negative likelihood ratios of 3.6 and 0.9, respectively). No studies reporting on the sensitivity and specificity of VEGF were identified. Subgroup analyses according to onset (early versus late) or severity (severe versus mild) of preeclampsia were not performed due to the small number of studies that reported results for these subgroups. The authors of this review concluded that PlGF, sFlt-1, and sEng had a poor predictive accuracy for preeclampsia.

After October 2010, one large prospective cohort study and two nested case-control studies of high methodologic quality assessing the predictive performance of angiogenic factors for preeclampsia have been published. In 2012, McElrath et al.[158] evaluated the predictive value of plasma PlGF and sFlt-1 concentrations at 10, 17, and 25 weeks of gestation in 2,246 healthy women with singleton gestations. The predictive ability of PlGF and sFlt-1 was poor across gestational ages. In fact, the sensitivities and specificities ranged between 47 and 62% and 40 and 62%, respectively, whereas the positive and negative likelihood ratios ranged between 1.0 and 1.6, and 0.6 and 1.1, respectively. In 2013, Myatt et al.[159] conducted a nested case-control study of 158 pregnancies complicated by preeclampsia and 468 normotensive nonproteinuric controls to determine whether the change in angiogenic factor levels between the first and second trimesters predicts preeclampsia in low-risk nulliparous women. At a fixed 80% specificity, changes in PlGF, sFlt-1, or sEng concentrations between the first and early or late second trimesters had a sensitivity that ranged between 42 and 51% for predicting preeclampsia (positive and negative likelihood ratios between 2.1 and 2.6, and between 0.6 and 0.7, respectively), 34 and 59% for predicting severe preeclampsia (positive and negative likelihood ratios between 1.7 and 3.0, and between 0.5 and 0.8, respectively), and 73 and 88% for predicting early onset preeclampsia (positive and negative likelihood ratios between 3.7 and 4.4 and 0.2 and 0.3, respectively).

Another recent nested case-control study assessed the ability of PlGF, sFlt-1, and free VEGF at 11–13 weeks of gestation to predict preeclampsia in 63 women who later developed preeclampsia and 252 unaffected controls.[160] At a fixed specificity of 90%, the sensitivities of these angiogenic factors varied from 26 to 45% (positive and negative likelihood ratios between 2.6 and 4.5 and 0.3 and 0.5, respectively). In addition, a secondary analysis of a trial of aspirin to prevent preeclampsia in high-risk pregnancies investigated the differences in concentrations of angiogenic factors at 9–26 weeks of gestation in patients at high risk who subsequently developed preeclampsia (n = 213) and similarly at-risk women who did not develop preeclampsia (n = 780).[161] Serum samples were available from 194 women with pre-existing diabetes, 313 with chronic hypertension, 234 with multiple gestations, and 252 with a history of preeclampsia in a previous pregnancy. Although there were significant differences in the levels of PlGF, sFlt-1, and sEng between women who later developed preeclampsia and those who did not, the authors concluded that the small differences in these angiogenic factors make it unlikely they are clinically useful in predicting preeclampsia in these high-risk populations. Overall, the predictive performance of angiogenic factors for preeclampsia reported in these recent high-quality studies is in accordance with that reported in the meta-analysis by Kleinrouweler et al.[157]

We analyzed 12 studies[43,94,159,162–170] (n = 10,583) that reported on the predictive accuracy of angiogenic factors for early-onset preeclampsia (Table 11.5). In summary, among women at a low to moderate risk of preeclampia, the predictive accuracy of angiogenic factors was moderate to high when measured during the second trimester (sensitivities ranging from 17 to 100%, specificities from 51 to 97%, and positive and negative likelihood ratios from 1.6 to 90.0 and 0.0 to 0.9, respectively) and moderate to minimal when measured during the first trimester (sensitivities ranging from 17 to 88%, specificities from 66 to 95%, and positive and negative likelihood ratios from 1.7 to 6.6 and 0.2 to 0.8, respectively). In high-risk populations, it appears that angiogenic factors have a moderate to good predictive ability, with sensitivities ranging from 50 to 100%, specificities from 51 to 95%, and positive and negative likelihood ratios from 1.4 to 16.6 and 0 to 0.6, respectively. However, the small total number of patients with early-onset preeclampsia (n = 23) who were included in the three studies that assessed women at high risk precludes conclusions about the predictive performance of angiogenic factors in these women.

Considering the above evidence, we currently consider that the determination of maternal circulating concentrations of angiogenic factors, as a single test, is not clinically useful to predict preeclampsia as a whole. Nevertheless, it appears that the ratio of angiogenic and antiangiogenic factors could be of value in the identification of women destined to develop early-onset preeclampsia.

OTHER TESTS

Cell-Free Fetal Deoxyribonucleic Acid

Cell-free fetal deoxyribonucleic acid (DNA), which comprises approximately 10–20% of the total cell-free maternal DNA, is thought to be placental in origin, resulting from apoptotic trophoblasts.[171] Analysis of cell-free fetal DNA in maternal blood for non-invasive prenatal testing has been shown to be highly accurate in the detection of common fetal autosomal trisomies.[171] In 1998, Holzgreve et al.[172] reported that fetal–maternal cell trafficking is significantly altered in pregnancies complicated by preeclampsia, with elevated numbers of fetal cells detected in the maternal circulation during these pregnancies. In a separate study, Lo et al.[173] showed that women diagnosed with preeclampsia had a statistically significant five-fold increase in circulating cell-free fetal DNA plasma levels compared with normotensive pregnant women. It was suggested that the increase in cell-free fetal DNA levels could precede the onset of clinical symptoms and signs of preeclampsia.

We reviewed nine studies that assessed the predictive accuracy of cell-free fetal DNA for preeclampsia, each using different quantification methods and cut-offs.[174–182] The three earliest case-control studies[174–176] included a total of 112 cases of preeclampsia and 239 controls, and reported a sensitivity, specificity, and positive and negative likelihood ratios that ranged between 33 and 67%, 82 and 95%, 3.7 and 6.6, and 0.4 and 0.7, respectively. Two subsequent case-control studies[177,178] with a total of 60 cases of preeclampsia and 248 controls found that there were no significant differences in median cell-free fetal DNA levels between preeclampsia and control groups at first or early second trimesters. These five studies were performed in women carrying male fetuses because cell-free fetal DNA was determined by quantifying Y-chromosome specific sequences. Two large prospective cohort studies assessed the predictive performance for preeclampsia of cell-free fetal DNA measured at the time of determining fetal RhD status in RhD-negative women with a singleton gestation and RhD-positive fetus at late second trimester.[179,180] One study (n = 611) found that median cell-free fetal DNA concentrations did not differ significantly between women with or without preeclampsia.[179] The second study, which included 1,020 women, reported that cell-free fetal DNA levels below the 5th or above the 90th percentile were associated with a significantly increased risk of preeclampsia.[180] However, cell-free fetal DNA levels >90th percentile had a sensitivity and specificity of 47% and 90%, respectively, to predict severe preeclampsia. Poon et al.[181] assessed the association between preeclampsia and plasma cell-free fetal DNA levels measured in 1,949 singleton gestations at 11–13 weeks of gestation by using digital analysis of selected regions assays. There was no significant difference in median plasma concentration of cell-free fetal DNA between women with or without preeclampsia. A recently published study by Papantoniou et al.[182] quantified the cell-free fetal DNA levels in maternal plasma by determining the *RASSF1A* levels at 11–13 weeks of gestation in 24 women who subsequently developed preeclampsia and 48 women with uncomplicated pregnancies. The promoter of the *RASSF1A*

TABLE 11.5 Angiogenic Factors for the Prediction of Early-Onset Preeclampsia

Author, Year	Design (Risk Population)	No. of Women [No. of Cases of Early-Onset Preeclampsia]	Gestational Week	Angiogenic Factor	Sensitivity (%)	Specificity (%)	Positive Likelihood Ratio	Negative Likelihood Ratio
Chaiworapongsa, 2005[162]	Case-control	10 cases 44 controls [10]	24–28	sFlt-1	17	97	6.4	0.9
Stepan, 2007[163]	Cohort (high)	63 [9]	19–24	PlGF	83	62	2.2	0.3
				sFlt-1	67	89	6.1	0.4
				sFlt-1/PlGF	67	51	1.4	0.6
				sFlt-1+PlGF	83	95	16.6	0.2
Espinoza, 2007[164]	Cohort (low to moderate)	3296 [25]	22–26	PlGF	80	51	1.6	0.4
Stepan, 2008[165]	Cohort (high)	77 [6]	19–24	sEng	50	95	10.9	0.5
				sEng + sFlt-1	100	93	14.3	0.0
Diab, 2008[166]	Cohort (high)	108 [8]	23	PlGF	100	76	4.2	0.0
				sFlt-1	100	87	7.7	0.0
				sFlt-1/PlGF	100	90	10.0	0.0
Akolekar, 2008[167]	Case-control	29 cases 609 controls [29]	11–13	PlGF	28	95	5.6	0.8
Kusanovic, 2009[168]	Cohort (low to moderate)	1622 [9]	6–15	PlGF	78	70	2.6	0.3
				PlGF/sEng	78	72	2.8	0.3
				PlGF/sEng + sFlt-1	78	73	2.9	0.3
			20–25	PlGF	100	96	25.0	0.0
				sFlt-1	67	93	9.6	0.4
				sEng	100	90	10.0	0.0
				PlGF/sEng	100	98	58.0	0.0
				Slope PlGF/sEng	100	99	90.0	0.0
Wortelboer, 2010[94]	Case-control	88 cases 480 controls [88]	8–13	PlGF	31	95	6.2	0.7
Abdelaziz, 2012[169]	Cohort (low to moderate)	1898 [16]	11–14	sEng	66	90	6.6	0.4
Ghosh, 2013[170]	Cohort (low to moderate)	1206 [19]	11–14	PlGF	58	66	1.7	0.6
			22–24		84	78	3.8	0.2
Kuc, 2013[43]	Case-control	68 cases 500 controls [68]	9–13	PlGF	17	95	3.4	0.9
Myatt, 2013[159]	Case-control	17 cases 468 controls [17]	11–14	Δ PlGF	77[a]	80[a]	3.9[a]	0.3[a]
					73[b]	80[b]	3.7[b]	0.3[b]
				Δ sFlt-1	77[a]	80[a]	3.9[a]	0.3[a]
					80[b]	80[b]	4.0[b]	0.3[b]
				Δ sEng	88[a]	80[a]	4.4[a]	0.2[a]

sFlt-1 = soluble fms-like tyrosine kinase 1; PlGF = placental growth factor; sEng = soluble endoglin.
[a]Changes in concentrations between first and early second trimesters.
[b]Changes in concentrations between first and late second trimesters.

gene is hypermethylated in the placenta and hypomethylated in maternal plasma. This methylation pattern allows the specific cutting of maternally derived hypomethylated sequences, leaving intact the hypermethylated fetal ones. The median cell-free fetal DNA levels were significantly increased in women with preeclampsia in comparison with normal controls (934.5 gEq/ml vs. 62 gEq/ml, $p = 0.000$). Using a cut-off value of 512 gEq/ml, cell-free fetal DNA plasma levels had a sensitivity and specificity of 100%.

In conclusion, there is no evidence supporting the quantification of cell-free fetal DNA for the prediction of preeclampsia. However, the quantification of cell-free fetal DNA by gender-independent approaches such as the promoter sequence of *RASSF1A* is promising, and large prospective cohort studies involving pregnant women at both low and high risk will be needed.

Proteomic, Metabolomic, and Transcriptomic Markers

In the last 10 years there has been a growing interest in the use of proteomic, metabolomic, and transcriptomic analysis of first-trimester maternal serum or urine for identifying biomarkers that could be used for the prediction of preeclampsia. Proteomics is the discipline which studies the global set of proteins and their expression, function, and structure.[183] Metabolomics is a discipline that aims to identify and quantify the global composition of metabolites of a biological fluid, tissue, or organism.[183] The metabolome refers to the comprehensive catalogue of metabolites in a specific organ or compartment under a set of conditions. The transcriptome is the full complement of mRNA in a cell or tissue at any given moment. A transcriptome forms the template for protein synthesis, resulting in a corresponding protein complement or proteome. The term transcriptomics has been used to describe the global mRNA expression of a particular tissue, yielding information about the transcriptional differences between two or more states.[183]

In 2006, Farina et al.[184] measured concentrations of a panel of circulating mRNAs in maternal blood from six women with preeclampsia and 30 controls. All mRNA markers except PAPP-A showed a statistically different distribution between cases and controls. Inhibin A, P-selectin, and VEGF receptor values were higher in preeclampsia, whereas human placental lactogen, KISS-1, and plasminogen activator inhibitor type 1 were lower, compared to normotensive controls. The investigators suggested that aberrant quantitative expression of these circulating placenta-specific mRNAs in serum from women with preeclampsia may be useful for the prediction of this disorder. We identified nine studies that assessed the predictive accuracy of proteomic, metabolomic or transcriptomic markers for preeclampsia.[185–193]

In 2009, Purwosunu et al.[185] reported on the predictive ability for preeclampsia of the analysis of cell-free mRNA in maternal plasma at 15–20 weeks. The mRNA expression levels for plasminogen activator inhibitor-1, tissue-type plasminogen activator, VEGF, VEGF receptor 1, endoglin, placenta-specific 1, and P-selectin were measured in the plasma of 62 patients with preeclampsia and 310 controls. Expression of all seven mRNAs was significantly increased in the preeclampsia group. At a fixed 95% specificity, the sensitivity of the individual markers varied from 18 to 58% to predict preeclampsia (positive and negative likelihood ratios from 3.6 to 11.6 and 0.2 to 0.6, respectively). A multivariate model that combined all markers increased the sensitivity to 84% (positive and negative likelihood ratios of 16.8 and 0.2, respectively). When stratified according to severity, the multimarker model had a sensitivity of 65%, 96%, and 100% for predicting mild preeclampsia ($n = 12$), severe preeclampsia ($n = 24$), and HELLP syndrome ($n = 12$), respectively.

Rasanen et al.[186] assessed the predictive accuracy for preeclampsia of maternal serum proteome profiles in serum obtained at 8–14 weeks of gestation from 70 women with preeclampsia (30 with mild preeclampsia and 40 with severe preeclampsia) and 79 normotensive controls. Maternal serum proteome analysis was performed by fluorescence 2D gel analysis, multidimensional liquid chromatography tandem mass spectrometry, and label-free quantification. Multiple serum proteins were significantly differentially expressed between women who subsequently developed preeclampsia and women who remained normotensive and included placental, vascular, and/or transport, and matrix and/or acute phase proteins. Maternal serum concentrations of complement factor D, VCAM-1, cystatin C, and β_2-microglobulin were higher and PAPP-A concentration lower in women who subsequently developed preeclampsia than in the control group. Serum markers such as endoglin, VEGF, sFlt-1, and fibronectin did not differ between the groups. Multiprotein logistic regression models that included known biomarkers for active disease showed low predictive accuracy for preeclampsia (unreported predictive values).

The Screening for Pregnancy Endpoints (SCOPE) study reported the results of a two-phase study that assessed metabolomic biomarkers in the prediction of preeclampsia.[187] In the discovery phase, a nested case-control study was conducted using samples obtained at 15 weeks of gestation from 60 women who subsequently developed preeclampsia and 60 controls with uncomplicated pregnancies. Plasma samples were analyzed using high-performance liquid chromatography–mass spectrometry. A multivariate predictive model combining 14 metabolites had a sensitivity of 77% at a fixed 90% specificity to predict preeclampsia (positive and negative likelihood ratios of 7.7 and 0.3, respectively). These findings were then validated using an independent case-control study that involved 39 women with preeclampsia and 40 controls. At 15 weeks of gestation, the same 14 metabolites had a sensitivity of 73% at a fixed 90% specificity to predict preeclampsia (positive and negative likelihood ratios of 7.3 and 0.3, respectively).

In 2013, the SCOPE study reported the results of a three-phase study that included 10 women with preeclampsia and 9 controls in the phase of selection of candidate markers, 100 women with preeclampsia and 200 controls in the verification phase, and 50 women with preeclampsia and 250 controls in the validation phase.[188] Logistic regression analyses identified eight multimarker validated models that yielded sensitivities ranging from 53 to 67% (median 60%) at a fixed specificity of 80% for predicting preeclampsia (positive and negative likelihood ratios between 3.0 and 3.4 and 0.4 and 0.6, respectively).

A case-control study evaluated the predictive performance of a panel of cellular mRNA markers in maternal blood at 10–14 weeks of gestation in 11 women who developed preeclampsia and 88 controls with uncomplicated pregnancies.[189] Women with preeclampsia had higher levels of mRNA for sFlt-1, endoglin, and transforming growth factor-β1, and lower levels of mRNA for PP-13 and PlGF than controls. A combined model of mRNAs for sFlt-1, endoglin, and transforming growth factor-β1 and parity had a sensitivity of 72% at a fixed 95% specificity for predicting preeclampsia (positive and negative likelihood ratios of 14.4 and 0.3, respectively).

Carty et al.[190] used a proteomic approach to identify biomarkers in maternal urine to predict preeclampsia. Samples from gestational weeks 12–16 (n = 45), 20 (n = 50), and 28 (n = 18) from women who subsequently developed preeclampsia were matched to controls (n = 86, n = 49, and n = 17, respectively). Candidate biomarkers were sequenced by liquid chromatography–tandem mass spectrometry. At 28 weeks of gestation, a model containing 50 biomarkers, mainly breakdown products of fibrinogen, collagen, and uromodulin, had a sensitivity and specificity of 100% to predict preeclampsia. However, when this panel of 50 biomarkers was assessed in a set of samples from 12–16 and 20 weeks of gestation, it did not reliably predict preeclampsia (predictive values were not reported).

In 2011, Odibo et al.[191] reported the findings of a discovery phase study of first-trimester prediction of preeclampsia using metabolomic biomarkers that involved 41 women with preeclampsia and 41 controls. Logistic regression modeling showed that four metabolites (hydroxyhexanoylcarnitine, alanine, phenylalanine, and glutamate) were significantly higher in women with preeclampsia than in controls. At a fixed specificity of 90%, the combination of these metabolites had a sensitivity of 50% for predicting all preeclampsia and early-onset preeclampsia (positive and negative likelihood ratios of 5.0 and 0.6, respectively).

Using a case-control design, Bahado-Singh et al. assessed the predictive ability of first-trimester maternal serum metabolomic markers for both early-onset[192] and late-onset[193] preeclampsia. The first study was carried out in singleton pregnancies that subsequently developed early preeclampsia (n = 30) and unaffected controls (n = 60).[192] The second study included 30 singleton pregnancies that subsequently developed late preeclampsia and 59 unaffected controls.[193] Nuclear magnetic resonance spectrometry was used to identify and quantify metabolites in serum samples from cases and controls of both studies. Significant differences between early-onset preeclampsia and control groups were found for 20 metabolites. A combination of four of these metabolites (citrate, glycerol, hydroxyisovalerate, and methionine) had a sensitivity of 76% at a fixed specificity of 95% for predicting early-onset preeclampsia (positive and negative likelihood ratios of 15.2 and 0.3, respectively). A total of 17 metabolites were present in significantly different concentrations in late-onset preeclampsia vs. control groups. Fourteen of these 17 metabolite concentrations were increased in the late-onset preeclampsia group while three were reduced compared to normotensive controls. A complex model consisting of multiple metabolites and maternal demographic characteristics had a sensitivity of 77% and a specificity of 100% for predicting late-onset preeclampsia (positive and negative likelihood ratios of ∞ and 0.2, respectively). A simplified model using fewer predictors had a sensitivity and specificity of 60 and 97% (positive and negative likelihood ratios of 20.0 and 0.4, respectively). By using a variable importance in project (VIP) analysis, which ranks the metabolites based on their importance in discriminating study from control groups, significant discrimination between early- and late-onset preeclampsia was achieved with the metabolites glycerol, acetate, trimethylamine, and succinate.

Overall, the use of proteomics, metabolomics, and transcriptomics analysis to identify biomarkers for the prediction of preeclampsia seems promising. Discovery approaches are considered unbiased. However, large prospective cohort studies are needed to validate the high predictive ability of novel biomarkers reported in the studies mentioned above. It would be very important to investigate whether differences in the first-trimester maternal serum concentrations of these markers could predict what some consider as different subtypes of preeclampsia. In addition, future studies could help to elucidate the molecular pathophysiology of this disorder.

THE USE OF COMBINED TESTS

This chapter has focused on the predictability of single tests. However, preeclampsia is a syndrome and not a single disorder, and can result from multiple mechanisms of disease. Thus, in order to improve the predictive ability of individual tests, several investigators have proposed combinations of different tests.[87,89,94,104,164,169,194–206] The general approach has been to combine biophysical tests (such as uterine artery Doppler velocimetry) with the results of biochemical parameters which provide information about the state of the syncytiotrophoblast (placental proteins) or endothelial cell function (angiogenic factors, homocysteine). Table 11.6 depicts the predictive values of several combinations of tests that have been proposed for the

TABLE 11.6 Combinations of Tests to Predict Preeclampsia

Author, Year	Country	Design (Risk Population)	No. of Women	Gestational Week	Tests Combined	Outcome	Sensitivity (%)	Specificity (%)	Positive LR	Negative LR
Aquilina, 2001[194]	United Kingdom	Cohort (low to moderate)	640	15–21	Uterine artery Doppler and inhibin A	All preeclampsia	71	93	10.1	0.3
						Preterm preeclampsia[a]	60	97	20.0	0.4
Audibert, 2005[195]	France	Cohort (low to moderate)	2,615	14–26	Uterine artery Doppler, β-hCG, and AFP	All preeclampsia	41	92	5.1	0.6
Spencer, 2005[196]	United Kingdom	Cohort (low to moderate)	4,390	11–24	Uterine artery Doppler PI and PAPP-A	All preeclampsia	62	95	12.4	0.4
Nicolaides, 2006[87]	United Kingdom	Case-control	10 cases 423 controls	11–13	Uterine artery Doppler PI and PP-13	Early preeclampsia[b]	91	90	9.1	0.1
Onalan, 2006[197]	Turkey	Cohort (low to moderate)	406	15–21	Uterine artery Doppler and homocysteine	All preeclampsia	61	98	31.0	0.4
Espinoza, 2007[164]	Chile	Cohort (low to moderate)	3348	22–26	Uterine artery Doppler and PlGF	All preeclampsia	27	96	6.8	0.8
						Early preeclampsia[b]	73	96	18.3	0.3
						Severe preeclampsia	64	96	16.0	0.4
Spencer, 2007[89]	United Kingdom	Case-control	44 cases 446 controls	11–24	Uterine artery Doppler PI, PP-13, and PAPP-A	All preeclampsia	74	80	3.7	0.3
						Early preeclampsia[c]	70	80	3.5	0.4
						Late preeclampsia[d]	73	80	3.7	0.3
Khalil, 2010[198]	United Kingdom	Case-control	42 cases 210 controls	11–13	Uterine artery Doppler PI, PP-13, and pulse wave analysis	All preeclampsia	86	90	8.6	0.2
						Early preeclampsia[b]	93	90	9.3	0.1
Kuromoto, 2010[199]	Japan	Case-control	25 cases 172 controls	9 ± 3	Urinary creatinine and blood pressure	All preeclampsia	75	95	15.0	0.3
Thilaganathan, 2010[200]	United Kingdom	Case-control	45 cases 125 controls	10–22	Uterine artery Doppler RI, and serum cystatin C and C-reactive protein	All preeclampsia	44	90	4.4	0.6

Wortelboer, 2010[94]	The Netherlands	Case-control	88 cases 480 controls	8–13	PlGF, PP-13 and PAPP-A	Early preeclampsia[b]	45	95	9.0	0.6
Anderson, 2011[201]	United Kingdom	Case-control	60 cases 36 controls	11–16	Fetal hemoglobin and α_1-microglobulin	All preeclampsia	73	90	7.3	0.3
Youssef, 2011[202]	Italy	Cohort (low to moderate)	528	11–13	PlGF, sFlt-1, and NGAL	Late preeclampsia[e]	77	90	7.7	0.3
Yu, 2011[203]	China	Case-control	31 cases 93 controls	12–16	Uterine artery Doppler PI, activin A, inhibin A, and PlGF	All preeclampsia	93	80	4.7	0.1
Abdelaziz, 2012[169]	Saudi Arabia	Case-control	76 cases 178 controls	11–14	Uterine artery Doppler PI, MAP, and sEng	Early preeclampsia[b] Late preeclampsia[e]	84 80	90 90	8.4 8.0	0.2 0.2
Bolin, 2012[204]	United Kingdom	Case-control	89 cases 86 controls	14	Uterine artery Doppler PI, histidine-rich glycoprotein	Preterm preeclampsia[a]	62	91	6.7	0.4
Ghosh, 2012[205]	India	Cohort (low to moderate)	1104	20–22	Uterine artery Doppler and PlGF	All preeclampsia	61	92	7.6	0.4
Wald, 2012[206]	United Kingdom	Case-control	88 cases 275 controls	15–18	AFP, β-hCG, uE_3, inhibin A, PlGF, and sEng	All preeclampsia Early preeclampsia[f] Late preeclampsia[g]	50 55 39	95 95 95	10.0 11.0 7.8	0.5 0.5 0.6
Zhou, 2012[104]	China	Cohort (low to moderate)	1000	20	Lipids and uric acid	All preeclampsia	92	50	1.8	0.2

LR=likelihood ratio; hCG=human chorionic gonadotropin; AFP = alpha fetoprotein; PI = pulsatility index; PAPP-A = pregnancy-associated plasma protein A; PP-13 = placental protein 13; PlGF = placental growth factor; RI = resistance index; sEng = soluble endoglin; sFlt-1 = soluble fms-like tyrosine kinase-1; NGAL = neutrophil gelatinase-associated lipocalin; MAP = mean arterial pressure; uE3 = unconjugated estriol.

[a]Delivery at <37 weeks.
[b]Delivery at <34 weeks.
[c]Delivery at <35 weeks.
[d]Delivery at ≥35 weeks.
[e]Delivery at ≥34 weeks.
[f]Delivery at <36 weeks.
[g]Delivery at ≥36 weeks.

prediction of preeclampsia. The most common combinations evaluated are those of uterine artery Doppler and fetoplacental proteins. Overall, it appears that the predictive accuracy of the combinations of tests is low to moderate for all preeclampsia, and moderate to high for early-onset preeclampsia.

We identified four systematic reviews that evaluated combinations of tests to predict preeclampsia.[207–210] In 2010, Giguère et al.[207] assessed the predictive performance of combinations of biochemical markers and uterine artery Doppler indices. Thirty-seven studies assessing 71 different combinations were identified. No given combination was evaluated more than once. In summary, combinations of tests generally led to increased sensitivity and/or specificity compared with single tests. In low-risk populations, combinations of PP-13, PAPP-A, ADAM-12, activin A, or inhibin A with uterine artery Doppler had sensitivities that varied between 60 and 80% with specificities >80%. The systematic review performed by Kuc et al.[208] analyzed the predictive ability of the combinations of uterine artery Doppler indices and seven first-trimester serum markers (β-hCG, inhibin A, activin A, PP-13, PAPP-A, ADAM-12, and PlGF). A total of 35 studies were included. The sensitivities of the single markers for the prediction of early preeclampsia, at a fixed specificity of 90%, ranged between 22 and 83%, whereas the sensitivities of the combinations of tests varied between 38 and 100%. Pedrosa et al.[209] evaluated the predictive value of the combination of uterine artery Doppler with other tests. A total of 33 studies were included in this review. Several combinations of tests provided moderate to convincing evidence for the prediction of early-onset preeclampsia. In contrast, the predictive ability for late-onset preeclampsia was poor. Moreover, the predictive ability of the combination of tests was consistently lower among populations at high risk of preeclampsia than among low-risk populations. In 2012, Hui et al.[210] assessed the accuracy of combined serum markers used in prenatal screening for aneuploidy and open neural-tube defects for predicting preeclampsia. Eight studies including ~115,000 pregnancies were included in the review. The pooled positive and negative likelihood ratios obtained from three relatively homogeneous studies of the combinations of hCG and AFP were 5.7 and 1.0, respectively.

Collectively, the evidence presented above indicates that a combination of placental proteins or angiogenic factors with uterine artery Doppler improves the predictive ability for preeclampsia – mainly for early-onset preeclampsia. Large and methodologically rigorous prospective cohort studies assessing a widely agreed combination of tests that report on the predictive accuracy for several subtypes of preeclampsia are necessary before combinations of tests are considered effective to predict preeclampsia.

MULTIVARIABLE PREDICTION MODELS DERIVED FROM COMBINATIONS OF MATERNAL CHARACTERISTICS AND TESTS

There has been a growing interest in multivariable prediction models for preeclampsia based on combinations of maternal demographic characteristics and medical and obstetrical history with biophysical and biochemical tests performed in the first or early second trimesters. The approach is similar to that used for combining maternal age with sonographic and maternal serum biomarkers in early screening for aneuplody and open neural-tube defects. Logistic regression analysis is used to determine which of the factors among the maternal characteristics and biophysical and biochemical tests have a significant contribution to predict preeclampsia. Then, a patient-specific risk of the development of preeclampsia is derived by algorithms based on combinations of the most important maternal characteristics and tests selected in the final model.

We reviewed 23 studies[30,36,43,71,97,98,167,211–226] that reported on the predictive accuracy for preeclampsia of multivariable models derived from combinations of maternal characteristics with biophysical and/or biochemical tests (Table 11.7). Most studies (n = 21) assessed combinations of maternal characteristics and tests in populations at a low to moderate risk of preeclampsia. Nineteen studies evaluated the combinations of tests performed during the first trimester. The biophysical and biochemical tests most frequently included in multivariable prediction models were uterine artery Doppler and mean arterial pressure, and PlGF and PAPP-A, respectively. Of the 21 studies performed in low/moderate-risk populations, nine provided data on the predictive accuracy of multivariable models for all preeclampsia,[30,36,71,97,98,211,219,220,226] 17 for early-onset preeclampsia,[30,36,43,97,167,211–214,216–222,224] and 13 for late-onset preeclampsia.[36,43,167,211–214,216,217,220–222,224] The median (range) sensitivity, specificity, and positive and negative likelihood ratios for predicting all preeclampsia were 38% (32–62%), 90% (80–95%), 6.2 (2.3–9.0), and 0.7 (0.4–0.7), respectively. The corresponding values for predicting early-onset preeclampsia were 82% (33–100%), 95% (90–95%), 13.6 (6.6–18.6), and 0.2 (0.2–0.7), respectively; and for predicting late-onset preeclampsia: 40% (20–57%), 95% (90–95%), 5.9 (4.0–9.4), and 0.6 (0.5–0.8), respectively. Only two studies reported predictive values for severe preeclampsia,[97,98] and both found a sensitivity of 55% with specificities of 80 and 90% (positive and negative likelihood ratios between 2.8 and 5.5 and 0.5 and 0.6). Two studies involving a total of 387 women reported predictive values of multivariable models for preeclampsia in high-risk populations.[215,225] The accuracy for predicting all preeclampsia, early-onset preeclampsia, late-onset preeclampsia, or severe preeclampsia was minimal (sensitivities between 23 and 43%, specificity of 90%, and positive

TABLE 11.7 Multivariable Prediction Models for Preeclampsia Derived from Combinations of Maternal Characteristics and Tests

Author, Year	Country	Design (Risk Population)	No. of Women	Gestational Week	Predictors Combined	Outcome	Sensitivity (%)	Specificity (%)	Positive LR	Negative LR
Yu, 2005[36]	United Kingdom	Cohort (low to moderate)	30,784	22–24	Maternal characteristics and uterine artery Doppler	All preeclampsia	45	95	9.0	0.6
						Early preeclampsia[a]	74	95	14.8	0.3
						Late preeclampsia[b]	35	95	7.0	0.7
Plasencia, 2007[211]	United Kingdom	Cohort (low to moderate)	6015	11–13	Maternal characteristics and uterine artery Doppler PI	All preeclampsia	62	90	6.2	0.4
						Early preeclampsia[a]	82	90	8.2	0.2
						Late preeclampsia[b]	52	90	5.2	0.5
Akolekar, 2008[167]	United Kingdom	Case-control	29 cases 609 controls	11–13	Maternal characteristics, uterine artery Doppler PI, PlGF, and PAPP-A	Early preeclampsia[a]	76	95	15.2	0.3
					Maternal characteristics, uterine artery Doppler PI, and PlGF	Late preeclampsia[b]	30	95	5.9	0.7
Onwudiwe 2008[212]	United Kingdom	Cohort (low to moderate)	3359	22–24	Maternal characteristics, uterine artery Doppler PI, and MAP	Early preeclampsia[a]	100	90	10.0	0.0
						Late preeclampsia[b]	56	90	5.6	0.5
Poon, 2009[213]	United Kingdom	Cohort (low to moderate)	8366	11–13	Maternal characteristics, uterine artery Doppler PI, and MAP	Early preeclampsia[a]	89	90	8.9	0.1
						Late preeclampsia[b]	57	90	5.7	0.5
Poon, 2009[214]	United Kingdom	Cohort (low to moderate)	7797	11–13	Maternal characteristics, uterine artery Doppler PI, MAP, PAPP-A, and PlGF	Early preeclampsia[a]	93	95	18.6	0.1
						Late preeclampsia[b]	45	95	9.0	0.6
Herraiz, 2009[215]	Spain	Cohort (high)	152	11–13	Maternal characteristics and uterine artery Doppler PI	Early preeclampsia[a]	43	90	4.3	0.6
						Late preeclampsia[b]	23	90	2.3	0.9
Poon, 2010[216]	United Kingdom	Case-control	116 cases 201 controls	11–13	Maternal characteristics, uterine artery Doppler PI, MAP, PlGF, activin A, P-selectin	Early preeclampsia[a]	89	95	17.8	0.1
						Late preeclampsia[b]	47	95	9.4	0.6
Poon, 2010[217]	United Kingdom	Cohort (low to moderate)	8366	11–13	Maternal characteristics, uterine artery Doppler PI, MAP, and PAPP-A	Early preeclampsia[a]	84	95	16.8	0.2
						Late preeclampsia[b]	42	95	8.4	0.6
Audibert, 2010[97]	Canada	Cohort (low to moderate)	893	11–13	Maternal characteristics, uterine artery Doppler PI, PAPP-A, inhibin A, and PlGF	All preeclampsia	40	90	4.0	0.7
						Early preeclampsia[c]	100	90	10.0	0.0
						Severe preeclampsia	55	90	5.5	0.5
Goetzinger, 2010[71]	United States	Cohort (low to moderate)	3716	11–13	Maternal characteristics and PAPP-A	All preeclampsia	36	87	2.8	0.7
Akolekar, 2011[218]	United Kingdom	Cohort (low to moderate)	33,602	11–13	Maternal characteristics and history, uterine artery Doppler PI, MAP, PAPP-A, PlGF, inhibin A, activin A, sEng, PP-13, pentraxin-3, and P-selectin	Early preeclampsia[a]	91	95	18.2	0.1
						Intermediate preeclampsia[d]	79	95	15.8	0.2
						Late preeclampsia[e]	61	95	12.2	0.4

(Continued)

TABLE 11.7 (Continued)

Author, Year	Country	Design (Risk Population)	No. of Women	Gestational Week	Predictors Combined	Outcome	Sensitivity (%)	Specificity (%)	Positive LR	Negative LR
Odibo, 2011[219]	United States	Case-control	42 cases 410 controls	11–14	Maternal characteristics, uterine artery Doppler, PAPP-A, and PP-13,	All preeclampsia Early preeclampsia	35 68	95 95	7.0 13.6	0.7 0.3
Di Lorenzo, 2012[220]	Italy	Cohort (low to moderate)	2118	11–13	Maternal characteristics and history, uterine artery Doppler, β-hCG PAPP-A, PP-13, and PlGF	All preeclampsia Early preeclampsia[a] Late preeclampsia[b]	32 67 23	95 95 95	6.4 13.4 4.6	0.7 0.4 0.8
Myatt, 2012[98]	United States	Cohort (low to moderate)	2394	9–13	Maternal characteristics, systolic blood pressure, PAPP-A, PlGF, and ADAM-12	All preeclampsia Severe preeclampsia	46 55	80 80	2.3 2.8	0.7 0.6
Akolekar, 2013[30]	United Kingdom	Cohort (low to moderate)	58,884	11–13	Maternal characteristics, uterine artery Doppler PI, MAP, PAPP-A, and PlGF	All preeclampsia Early preeclampsia[a]	38 93	95 95	7.6 18.6	0.7 0.1
Scazzocchio, 2013[221]	Spain	Cohort (low to moderate)	5170	8–13	Maternal characteristics, uterine artery Doppler PI, MAP, PAPP-A, and β-hCG	Early preeclampsia[a] Late preeclampsia[b]	69 29	95 95	13.8 5.8	0.3 0.8
Kuc, 2013[43]	The Netherlands	Case-control	167 cases 500 controls	9–13	Maternal characteristics, MAP, PAPP-A, PlGF, and ADAM-12	Early preeclampsia[a] Late preeclampsia[b]	56 40	95 95	11.2 8.0	0.5 0.6
Parra-Cordero, 2013[222]	Chile	Case-control	70 cases 289 controls	11–13	Maternal characteristics, uterine artery Doppler PI, and PlGF	Early preeclampsia[a] Late preeclampsia[b]	33 20	95 95	6.6 4.0	0.7 0.8
Myers, 2013[223]	New Zealand, Australia, United Kingdom, and Ireland	Case-control	47 cases 188 controls	14–21	Maternal characteristics, uterine artery Doppler RI, PlGF, and sEndoglin	Preterm preeclampsia[f]	52	95	10.4	0.5
Caradeux, 2013[224]	Chile	Cohort (low to moderate)	627	11–14	Maternal characteristics, uterine artery Doppler PI, and MAP	Early preeclampsia[a] Late preeclampsia[b]	63 32	95 95	12.6 6.4	0.4 0.7
Diguisto, 2013[225]	France	Cohort (high)	235	22 ± 1	Maternal characteristics, uterine artery Doppler, PlGF, sFlt-1, and lipid-related biomarkers	All preeclampsia Severe preeclampsia	40 40	90 90	4.0 4.0	0.7 0.7
Boucoiran, 2013[226]	Canada	Cohort (low to moderate)	893	11–13	Maternal characteristics and PlGF	All preeclampsia	35	90	3.5	0.7

LR = likelihood ratio; PI = pulsatility index; PlGF = placental growth factor; PAPP-A = pregnancy-associated plasma protein-A; MAP = mean arterial pressure; PP-13 = placental protein-13; RI = resistance index; sEng = soluble endoglin; sFlt-1 = soluble fms-like tyrosine kinase-1; NGAL = neutrophil gelatinase-associated lipocalin; hCG = human chorionic gonadotropin; ADAM-12 = a disintegrin and metalloproteinase-12; IPD = individual patient data.

[a]Delivery at <34 weeks.

[b]Delivery at ≥34 weeks.

[c]Diagnosed at <34 weeks.

[d]Delivery at 34–37 weeks.

[e]Delivery at >37 weeks.

[f]Delivery at <37 weeks.

and negative likelihood ratios between 2.3 and 4.3, and 0.6 and 0.9, respectively).

Research in prediction modeling has three major phases: (1) development and internal validation of a predictive model; (2) assessment of the model's predictive performance in new individuals, and if necessary, adjusting or updating the model to local circumstances or with new predictors (external validation); and (3) assessment of the prediction model's impact on clinical decision-making and patient outcomes.[227,228] Multivariable prediction models based on the combination of maternal characteristics and tests have a high predictive accuracy for early-onset preeclampsia. However, demonstration that a multivariable prediction model is of value requires replication of the initial results. Performance of developed and internally validated prediction models should be tested in new populations before they are implemented in daily clinical practice. Only one study has validated partially some of the multivariable models proposed for the prediction of preeclampsia. Farina et al.[229] assessed the predictive ability for late-onset preeclampsia of eight previously published multivariable models in 554 women at a low to moderate risk of preeclampsia. At a fixed specificity of 90%, the sensitivities of the models to predict late-onset preeclampsia in this new population ranged from 39 to 85% (positive and negative likelihood ratios between 3.9 and 8.5, and 0.7 and 0.2), respectively. The findings of this study were somewhat disappointing because only four of the eight multivariable models yielded a similar sensitivity to that reported previously. Two models yielded a lower sensitivity and two yielded a higher sensitivity. To date, none of the multivariable models proposed for the prediction of early-onset preeclampsia has been externally validated.

The assessment of multivariable prediction models on management behavior and individual health outcomes requires a comparative study.[228] A control group should be randomly assigned to the usual care or management without the use of prediction models, whereas in the intervention group the results of the prediction models are made available to individuals and/or healthcare professionals to guide their behavior and decision-making. However, an alternative approach would be that individuals with a positive result of a test derived from a multivariable prediction model are randomly assigned to receive an intervention or placebo/no intervention.

In conclusion, there is compelling evidence that multivariable prediction models derived from the combination of maternal characteristics with biophysical and/or biochemical tests have high predictive values for early-onset preeclampsia in populations with a low to moderate risk of developing this disorder. Further research should focus on the external validation of these models and in designing large prospective randomized controlled trials of prevention in women classified at high risk of developing preeclampsia by a multivariable prediction model. Additional studies are needed to assess the accuracy of multivariable models to predict late-onset preeclampsia in low/moderate-risk populations as well as for predicting preeclampsia in high-risk populations.

PERSPECTIVES AND CONCLUSIONS

The predictive accuracy of tests proposed for the prediction of preeclampsia is summarized in Table 11.8. At the present time, there is no clinically useful single test to predict preeclampsia. It has been suggested that preeclampsia is a heterogeneous disorder, and therefore no single test would be able to predict all cases of preeclampsia, as is the case with the other "great obstetrical syndromes."[230] Some believe that it is necessary to improve the phenotypic characterization of the disease in patients at term, with early- or late-onset preeclampsia.[231] This subject is discussed in detail in Chapter 6.

Multivariable models derived from combinations of maternal characteristics and tests show high predictive accuracy for early-onset preeclampsia in populations at a low to moderate risk of developing this disorder. Notwithstanding, prospective large-scale studies that externally validate these models are urgently needed. If a model is successfully validated, the next step will be to demonstrate that it is clinically useful through the design of randomized controlled trials of the prevention of preeclampsia in patients with a positive test result. Moreover, cost-effectiveness studies will also be required. Second-trimester circulating levels of angiogenic factors and uterine artery Doppler velocimetry appear to have a moderate to high predictive accuracy for early-onset preeclampsia.

Some caution is prudent. In many case-control studies on prediction of preeclampsia, the control group is women with completely normal pregnancy outcomes, excluding those with disorders such as preterm labor or small for gestational age fetuses – it has been argued that this approach is not ideal, and that the appropriate control group for a prediction study should include women who did not develop preeclampsia, regardless of whether they had other pregnancy complications.

In many studies on prediction of preeclampsia, there is little or no information about the results of the predictive test at the time of diagnosis of disease. It would be useful, as a part of the assessment of the predictive value of biochemical markers in longitudinal comparisons and for indicating whether there might be subtypes of preeclampsia, to know whether abnormal test results in early pregnancy are also present when the woman develops preeclampsia. For example, there is evidence that women who had low levels of PlGF in early pregnancy also had low levels of PlGF when they developed preeclampsia while the women who had normal concentrations of PlGF in early pregnancy also had normal concentrations of PlGF at the time of the diagnosis of preeclampsia.[232]

TABLE 11.8 Predictive Values of Tests Proposed for the Prediction of Preeclampsia

Test	No of Studies	No of Women	Cut-off	Outcome	Sensitivity (%)	Specificity (%)	Positive Likelihood Ratio	Negative Likelihood Ratio
Roll-over test[a]	11[9]	1126	20 mm Hg	All preeclampsia	54	85	3.7	0.5
Isometric handgrip test[a]	3[9]	413	20 mm Hg	All preeclampsia	71	84	4.5	0.3
Angiotensin II sensitivity test[a]	5[9]	898	10 ng/kg/min	All preeclampsia	57	85	3.7	0.5
Mean arterial pressure[a]	34[10]	60,599	90 mm Hg	All preeclampsia	62	82	3.5	0.5
			85 mm Hg		52	84	3.3	0.6
Transcranial Doppler velocimetry	5[13–17]	2190	Several	All preeclampsia	10–100	65–96	2.3–12.3	0.0–0.5
Human chorionic gonadotropin[a]	13[9]	56,067	2.0 MoM	All preeclampsia	21	91	2.4	0.9
	5[9]	49,572	[2.3–2.5] MoM		20	92	2.7	0.9
Alpha fetoprotein[a]	11[41,42,45–53]	115,419	2.0 MoM	All preeclampsia	13	96	3.3	0.9
	5[9]	117,711	2.5 MoM		9	97	2.5	1.0
Estriol[a]	2[48,56]	53,399	0.5 MoM	All preeclampsia	6	96	1.5	1.0
	3[47,57,58]	3114	0.85–0.90 MoM		33	75	1.3	0.9
Inhibin A[a]	7[41,42,48,61–64]	45,274	2.0 MoM	All preeclampsia	21	95	4.2	0.8
PAPP A[a]	7[64,67–72]	99,449	0.4 MoM	All preeclampsia	10	95	2.0	0.9
Activin A	10[75–84]	1985	Several	All preeclampsia	11–93	71–95	2.2–11.0	0.1–0.9
Placental protein 13	7[88–92,95,96]	3003	Several	All preeclampsia	24–80	80–90	1.2–7.9	0.2–1.0
	6[87,89,91–94]	2435		Early preeclampsia	21–100	80–95	2.5–8.0	0.0–0.8
ADAM-12	4[43,96–98]	2425	Several	All preeclampsia	16–41	80–90	1.0–2.1	0.7–1.0
Serum uric acid	6[103,104]	1572	Several	All preeclampsia	0–56	65–95	0.0–9.8	0.5–1.2
Microalbuminuria	10[105–114]	5024	Several	All preeclampsia	7–92	29–97	0.8–3.1	0.1–1.1
Urinary calcium excretion	6[a,b,112,116–120]	965	~200 mg/24 h	All preeclampsia	60	75	2.4	0.5
	12[c,105,117,118,121–129]	3298	Several		31–89	55–95	0.7–9.5	0.2–1.3

Urinary kallikrein/creatinine ratio[a]	2[132,133]	765	170–200	All preeclampsia	81	83	4.7	0.2
Fibronectin	4[d,141]	495	Several	All preeclampsia	50	43–94	1.2–11.0	0.4–0.7
	2[e,141]	135	Several		50	72–96	1.6–11.5	0.0–0.6
Homocysteine[a]	6[9]	2449	5.5–6.3 μmol/L	All preeclampsia	36	88	3.1	0.7
Endothelial cell adhesion molecules	4[f,77,148–150]	1720	Several	All preeclampsia	27–91	52–100	1.1–∞	0.1–1.0
	2[g,48,150]	1488	Several		43–50	33–98	1.0–19.6	0.6–1.5
	2[h,147,149]	244	Several		76–80	83–96	4.5–19.0	0.2–0.3
Placental growth factor[a]	27[157]	Unreported	Several	All preeclampsia	50	90	5.0	0.6
sFlt-1[a]	19[157]	Unreported	Several	All preeclampsia	42	90	4.2	0.6
Endoglin[a]	10[157]	Unreported	Several	All preeclampsia	32	90	3.2	0.8
Cell-free fetal DNA	5[174–176,180,182]	1441	Several	All preeclampsia	33–100	82–100	3.7–∞	0.0–0.7
Proteomics/metabolomics/transcriptomics	8[185,187–193]	1187	Several	All preeclampsia	50–100	80–100	3.0–∞	0.0–0.6
Multivariable prediction models	9[30,36,71,97,98,211,219,220,226]	NA[i]	NA	All preeclampsia	38 (32–62)[j]	90 (80–95)[j]	6.2 (2.3–9.0)[j]	0.7 (0.4–0.7)[j]
	17[30,36,43,97,167,211–214,216–222,224]	NA[i]	NA	Early preeclampsia	82 (33–100)[j]	95 (90–95)[j]	13.6 (6.6–18.6)[j]	0.2 (0.2–0.7)[j]
	13[36,43,167,211–214,216,217,220–222,224]	NA[i]	NA	Late preeclampsia	40 (20–57)[j]	95 (90–95)[j]	5.9 (4.0–9.4)[j]	0.6 (0.5–0.8)[j]

MoM = multiple of the median; PAPP A = pregnancy-associated plasma protein A; ADAM-12 = a disintegrin and metalloprotease 12; sFlt-1 = soluble fms-like tyrosine kinase-1; NA = not applicable.
[a]Pooled estimates calculated from meta-analytic techniques.
[b]24 h urinary calcium excretion.
[c]Calcium/creatinine ratio.
[d]Total fibronectin.
[e]Cellular fibronectin.
[f]Vascular cell adhesion molecule-1 (VCAM-1).
[g]Intercellular adhesion molecule-1 (ICAM-1).
[h]P-selectin.
[i]Several studies were derived from the same database.
[j]Median (range).

Finally, as combined testing becomes more complex with the use of sophisticated analyses and technology, attention must be given to their cost-effectiveness and applicability to general practice. This is especially true in developing nations where resources are limited, and morbidity and mortality associated with preeclampsia are high.

Acknowledgement

This work was supported (in part) by the Perinatology Research Branch, Division of Intramural Research, Eunice Kennedy Shriver National Institute of Child Health and Human Development, NIH, DHHS.

References

1. Simel DL, Samsa GP, Matchar DB. Likelihood ratios with confidence: sample size estimation for diagnostic test studies. *J Clin Epidemiol.* 1991;44:763–770.
2. Jaeschke R, Guyatt GH, Sackett DL. Users' guides to the medical literature. III. How to use an article about a diagnostic test. B. What are the results and will they help me in caring for my patients? *JAMA.* 1994;271:703–707.
3. Fagan TJ. Letter: Nomogram for Bayes theorem. *New Engl J Med.* 1975;293:257.
4. Villar J, Say L, Shennan A, et al. Methodological and technical issues related to the diagnosis, screening, prevention, and treatment of pre-eclampsia and eclampsia. *Int J Gynaecol Obstet.* 2004;85(suppl 1):S28–S41.
5. Conde-Agudelo A, Lede R, Belizán J. Evaluation of methods used in the prediction of hypertensive disorders of pregnancy. *Obstet Gynecol Surv.* 1994;49:210–222.
6. Conde-Agudelo A, Villar J, Lindheimer M. World Health Organization systematic review of screening tests for preeclampsia. *Obstet Gynecol.* 2004;104:1367–1391.
7. Meads CA, Cnossen JS, Meher S, et al. Methods of prediction and prevention of pre-eclampsia: systematic reviews of accuracy and effectiveness literature with economic modelling. *Health Technol Assess.* 2008;12:1–270.
8. Staff AC, Benton SJ, von Dadelszen P, et al. Redefining preeclampsia using placenta-derived biomarkers. *Hypertension.* 2013;61:932–942.
9. Conde-Agudelo A, Romero R, Lindheimer M. Test to predict preclampsia. In: Lindheimer MD, Roberts JM, Cunningham FG, eds. *Chesley's Hypertensive Disorders in Pregnancy.* Boston: Academic Press/Elsevier; 2009:189–211.
10. Cnossen JS, Vollebregt KC, de Vrieze N, et al. Accuracy of mean arterial pressure and blood pressure measurements in predicting pre-eclampsia: systematic review and meta-analysis. *BMJ.* 2008;336:1117–1120.
11. Poon LC, Kametas NA, Valencia C, Chelemen T, Nicolaides KH. Hypertensive disorders in pregnancy: screening by systolic diastolic and mean arterial pressure at 11–13 weeks. *Hypertens Pregnancy.* 2011;30:93–107.
12. Williams K, Galerneau F. Maternal transcranial Doppler in pre-eclampsia and eclampsia. *Ultrasound Obstet Gynecol.* 2003;21:507–513.
13. Moutquin JM, Williams KP. Can maternal middle cerebral velocity changes predict pre-eclampsia? A preliminary analysis. *Am J Obstet Gynecol.* 1999;180:S54.
14. Riskin-Mashiah S, Belfort MA, Saade GR, Herd JA. Transcranial doppler measurement of cerebral velocity indices as a predictor of preeclampsia. *Am J Obstet Gynecol.* 2002;187:1667–1672.
15. Belfort M, Riskin-Mashiah S, Lacoursiere Y, Varner M. Low second trimester maternal middle cerebral artery resistance index and subsequent preeclampsia. *Am J Obstet Gynecol.* 2004;191:S32.
16. Belfort M, Saade G, Clark S, Dildy G, Allred J, Ludlow S. Mid-trimester maternal middle cerebral artery (MCA) Doppler for prediction of preeclampsia in a low risk population. *Am J Obstet Gynecol.* 2007;197:S211.
17. Belfort M, Van Veen T, White GL, et al. Low maternal middle cerebral artery Doppler resistance indices can predict future development of pre-eclampsia. *Ultrasound Obstet Gynecol.* 2012;40:406–411.
18. Meekins JW, Pijnenborg R, Hanssens M, McFadyen IR, van Asshe A. A study of placental bed spiral arteries and trophoblast invasion in normal and severe pre-eclamptic pregnancies. *Br J Obstet Gynaecol.* 1994;101:669–674.
19. Olofsson P, Laurini RN, Marsál K. A high uterine artery pulsatility index reflects a defective development of placental bed spiral arteries in pregnancies complicated by hypertension and fetal growth retardation. *Eur J Obstet Gynecol Reprod Biol.* 1993;49:161–168.
20. Palma-Dias RS, Fonseca MM, Brietzke E, et al. Screening for placental insufficiency by transvaginal uterine artery Doppler at 22–24 weeks of gestation. *Fetal Diagn Ther.* 2008;24:462–469.
21. Llurba E, Carreras E, Gratacós E, et al. Maternal history and uterine artery Doppler in the assessment of risk for development of early- and late-onsetpreeclampsia and intrauterine growth restriction. *Obstet Gynecol Int.* 2009;2009:275613.
22. Espinoza J, Kusanovic JP, Bahado-Singh R, et al. Should bilateral uterine artery notching be used in the risk assessment for preeclampsia, small-for-gestational-age, and gestational hypertension? *J Ultrasound Med.* 2010;29:1103–1115.
23. Bhattacharyya SK, Kundu S, Kabiraj SP. Prediction of preeclampsia by midtrimester uterine artery Doppler velocimetry in high-risk and low-risk women. *J Obstet Gynaecol India.* 2012;62:297–300.
24. Myatt L, Clifton RG, Roberts JM, et al. The utility of uterine artery Doppler velocimetry in prediction of preeclampsia in a low-risk population. *Obstet Gynecol.* 2012;120:815–822.
25. Arcangeli T, Giorgetta F, Farina A, et al. Significance of uteroplacental Doppler at midtrimester in patients with favourable obstetric history. *J Matern Fetal Neonatal Med.* 2013;26:299–302.
26. Jamal A, Abbasalizadeh F, Vafaei H, Marsoosi V, Eslamian L. Multicenter screening for adverse pregnancy outcomes by uterine artery Doppler in the second and third trimester of pregnancy. *Med Ultrason.* 2013;15:95–100.
27. Pongrojpaw D, Chanthasenanont A, Nanthakomon T. Second trimester uterine artery Doppler screening in prediction of adverse pregnancy outcome in high risk women. *J Med Assoc Thai.* 2010;93(suppl 7):S127–S130.
28. Herraiz I, Escribano D, Gómez-Arriaga PI, Hernández-García JM, Herraiz MA, Galindo A. Predictive value of sequential models of uterine artery Doppler in pregnancies at high risk for pre-eclampsia. *Ultrasound Obstet Gynecol.* 2012;40:68–74.

29. Demers S, Bujold E, Arenas E, Castro A, Nicolaides KH. Prediction of recurrent preeclampsia using first-trimester uterine artery Doppler. *Am J Perinatol.* 2014;31:99–104.
30. Akolekar R, Syngelaki A, Poon L, Wright D, Nicolaides KH. Competing risks model in early screening for preeclampsia by biophysical and biochemical markers. *Fetal Diagn Ther.* 2013;33:8–15.
31. Cnossen JS, Morris RK, ter Riet G, et al. Use of uterine artery Doppler ultrasonography to predict pre-eclampsia and intrauterine growth restriction: a systematic review and bivariable meta-analysis. *CMAJ.* 2008;178:701–711.
32. Conde-Agudelo A, Lindheimer M. Use of Doppler ultrasonography to predict pre-eclampsia. *CMAJ.* 2008;179:53.
33. Bower S, Schuchter K, Campbell S. Doppler ultrasound screening as part of routine antenatal scanning: prediction of preeclampsia and intrauterine growth retardation. *Br J Obstet Gynaecol.* 1993;100:989–994.
34. Mires GJ, Williams FL, Leslie J, Howie PW. Assessment of uterine arterial notching as a screening test for adverse pregnancy outcome. *Am J Obstet Gynecol.* 1998;179:1317–1323.
35. Coleman MA, McCowan LM, North RA. Mid-trimester uterine artery Doppler screening as a predictor of adverse pregnancy outcome in high-risk women. *Ultrasound Obstet Gynecol.* 2000;15:7–12.
36. Yu CK, Smith GC, Papageorghiou AT, Cacho AM, Nicolaides KH, Fetal Medicine Foundation Second Trimester Screening Group An integrated model for the prediction of preeclampsia using maternal factors and uterine artery Doppler velocimetry in unselected low-risk women. *Am J Obstet Gynecol.* 2005;193:429–436.
37. Prefumo F, Bhide A, Sairam S, Penna L, Hollis B, Thilaganathan B. Effect of parity on second-trimester uterine artery Doppler flow velocity and waveforms. *Ultrasound Obstet Gynecol.* 2004;23:46–49.
38. Pilalis A, Souka AP, Antsaklis P, et al. Screening for preeclampsia and small for gestational age fetuses at the 11–14 weeks scan by uterine artery Dopplers. *Acta Obstet Gynecol Scand.* 2007;86:530–534.
39. Spencer K. Screening for Down's syndrome. The role of intact hCG and free subunit measurement. *Scand J Clin Lab Invest.* 1993;216:79–96.
40. Sorensen TK, Williams MA, Zingheim RW, Clement SJ, Hickok DE. Elevated second-trimester human chorionic gonadotropin and subsequent pregnancy-induced hypertension. *Am J Obstet Gynecol.* 1993;169:834–838.
41. Ree PH, Hahn WB, Chang SW, et al. Early detection of preeclampsia using inhibin A and other second-trimester serum markers. *Fetal Diagn Ther.* 2011;29:280–286.
42. Olsen RN, Woelkers D, Dunsmoor-Su R, Lacoursiere DY. Abnormal second-trimester serum analytes are more predictive of preterm preeclampsia. *Am J Obstet Gynecol.* 2012;207:228.e1–228.e7.
43. Kuc S, Koster MP, Franx A, Schielen PC, Visser GH. Maternal characteristics, mean arterial pressure and serum markers in early prediction of preeclampsia. *PLoS One.* 2013;8:e63546.
44. Jauniaux E, Gulbis B, Jurkovic D, Gavriil P, Campbell S. The origin of alpha-fetoprotein in first-trimester anembryonic pregnancies. *Am J Obstet Gynecol.* 1995;173:1749–1753.
45. Milunsky A, Jick SS, Bruell CL, et al. Predictive values, relative risks, and overall benefits of high and low maternal serum alpha-fetoprotein screening in singleton pregnancies: new epidemiologic data. *Am J Obstet Gynecol.* 1989;161:291–297.
46. Pouta AM, Hartikainen AL, Vuolteenaho OJ, Ruokonen AO, Laatikainen TJ. Midtrimester N-terminal proatrial natriuretic peptide, free beta hCG, and alpha-fetoprotein in predicting preeclampsia. *Obstet Gynecol.* 1998;91:940–944.
47. Stamilio DM, Sehdev HM, Morgan MA, Propert K, Macones GA. Can antenatal clinical and biochemical markers predict the development of severe preeclampsia? *Am J Obstet Gynecol.* 2000;182:589–594.
48. Dugoff L, Hobbins JC, Malone FD, FASTER Trial Research Consortium Quad screen as a predictor of adverse pregnancy outcome. *Obstet Gynecol.* 2005;106:260–267.
49. Capeless EL, Kelleher PC, Walters CP. Elevated maternal serum alpha-fetoprotein levels and maternal risk factors. Their association with pregnancy complications. *J Reprod Med.* 1992;37:257–260.
50. Simpson JL, Palomaki GE, Mercer B, et al. Associations between adverse perinatal outcome and serially obtained second- and third-trimester maternal serum α-fetoprotein measurements. *Am J Obstet Gynecol.* 1995;173:1742–1748.
51. Waller DK, Lustig LS, Cunningham GC, Feuchtbaum LB, Hook EB. The association between maternal serum alpha-fetoprotein and preterm birth, small for gestational age infants, preeclampsia, and placental complications. *Obstet Gynecol.* 1996;88:816–822.
52. Räty R, Koskinen P, Alanen A, Irjala K, Matinlauri I, Ekblad U. Prediction of pre-eclampsia with maternal mid-trimester total renin, inhibin A, AFP and free beta-hCG levels. *Prenat Diagn.* 1999;19:122–127.
53. Leung TN, Chung TK, Madsen G, et al. Analysis of mid-trimester corticotrophin-releasing hormone and alpha-fetoprotein concentrations for predicting preeclampsia. *Hum Reprod.* 2000;15:1813–1818.
54. Shenhav S, Gemer O, Sassoon E, Volodarsky M, Peled R, Segal S. Mid-trimester triple test levels in early and late onset severe pre-eclampsia. *Prenat Diagn.* 2002;22:579–582.
55. Santolaya-Forgas J, Jessup J, Burd LI, Prins GS, Burton BK. Pregnancy outcome in women with low midtrimester maternal serum unconjugated estriol. *J Reprod Med.* 1996;41:87–90.
56. Yaron Y, Cherry M, Kramer RL, et al. Second-trimester maternal serum marker screening: maternal serum alpha-fetoprotein, beta-human chorionic gonadotropin, estriol, and their various combinations as predictors of pregnancy outcome. *Am J Obstet Gynecol.* 1999;181:968–974.
57. Wald NJ, Morris JK, Ibison J, Wu T, George LM. Screening in early pregnancy for pre-eclampsia using Down syndrome quadruple test markers. *Prenat Diagn.* 2006;26:559–564.
58. Sayin NC, Canda MT, Ahmet N, Arda S, Süt N, Varol FG. The association of triple-marker test results with adverse pregnancy outcomes in low-risk pregnancies with healthy newborns. *Arch Gynecol Obstet.* 2008;277:47–53.
59. McCluggage WG, Ashe P, McBride H, Maxwell P, Sloan JM. Localization of the cellular expression of inhibin in trophoblastic tissue. *Histopathology.* 1998;32:252–256.
60. Muttukrishna S, Knight PG, Groome NP, Redman CW, Ledger WL. Activin A and inhibin A as possible endocrine markers for pre-eclampsia. *Lancet.* 1997;349:1285–1288.
61. Cuckle H, Sehmi I, Jones R. Maternal serum inhibin A can predict pre-eclampsia. *Br J Obstet Gynaecol.* 1998;105:1101–1103.

62. Aquilina J, Barnett A, Thompson O, Harrington K. Second-trimester maternal serum inhibin A concentration as an early marker for preeclampsia. *Am J Obstet Gynecol.* 1999;181:131–136.
63. D'Anna R, Baviera G, Corrado F, Leonardi I, Buemi M, Jasonni VM. Is mid-trimester maternal serum inhibin-A a marker of preeclampsia or intrauterine growth restriction? *Acta Obstet Gynecol Scand.* 2002;81:540–543.
64. Kang JH, Farina A, Park JH, et al. Down syndrome biochemical markers and screening for preeclampsia at first and second trimester: correlation with the week of onset and the severity. *Prenat Diagn.* 2008;28:704–709.
65. Irwin JC, Suen LF, Martina NA, Mark SP, Giudice LC. Role of the IGF system in trophoblast invasion and pre-eclampsia. *Hum Reprod.* 1999;14:90–96.
66. Ong CY, Liao AW, Spencer K, Munim S, Nicolaides KH. First trimester maternal serum free beta human chorionic gonadotrophin and pregnancy associated plasma protein A as predictors of pregnancy complications. *BJOG.* 2000;107:1265–1270.
67. Dugoff L, Hobbins JC, Malone FD, et al. First-trimester maternal serum PAPP-A and free-beta subunit human chorionic gonadotropin concentrations and nuchal translucency are associated with obstetric complications: a population-based screening study (the FASTER Trial). *Am J Obstet Gynecol.* 2004;191:1446–1451.
68. Pilalis A, Souka AP, Antsaklis P, et al. Screening for pre-eclampsia and fetal growth restriction by uterine artery Doppler and PAPP-A at 11–14 weeks' gestation. *Ultrasound Obstet Gynecol.* 2007;29:135–140.
69. Spencer K, Cowans NJ, Nicolaides KH. Low levels of maternal serum PAPP-A in the first trimester and the risk of pre-eclampsia. *Prenat Diagn.* 2008;28:7–10.
70. Poon LC, Maiz N, Valencia C, Plasencia W, Nicolaides KH. First-trimester maternal serum pregnancy-associated plasma protein-A and pre-eclampsia. *Ultrasound Obstet Gynecol.* 2009;33:23–33.
71. Goetzinger KR, Singla A, Gerkowicz S, Dicke JM, Gray DL, Odibo AO. Predicting the risk of pre-eclampsia between 11 and 13 weeks' gestation by combining maternal characteristics and serum analytes, PAPP-A and free β-hCG. *Prenat Diagn.* 2010;30:1138–1142.
72. Ranta JK, Raatikainen K, Romppanen J, Pulkki K, Heinonen S. Decreased PAPP-A is associated with preeclampsia, premature delivery and small for gestational age infants but not with placental abruption. *Eur J Obstet Gynecol Reprod Biol.* 2011;157:48–52.
73. Luisi S, Florio P, Reis FM, Petraglia F. Expression and secretion of activin A: possible physiological and clinical implications. *Eur J Endocrinol.* 2001;145:225–236.
74. Petraglia F, Aguzzoli L, Gallinelli A, et al. Hypertension in pregnancy: changes in activin A maternal serum concentration. *Placenta.* 1995;16:447–454.
75. Muttukrishna S, North RA, Morris J, et al. Serum inhibin A and activin A are elevated prior to the onset of pre-eclampsia. *Hum Reprod.* 2000;15:1640–1645.
76. Florio P, Reis FM, Pezzani I, Luisi S, Severi FM, Petraglia F. The addition of activin A and inhibin A measurement to uterine artery Doppler velocimetry to improve the early prediction of pre-eclampsia. *Ultrasound Obstet Gynecol.* 2003;21:165–169.
77. Hanisch CG, Pfeiffer KA, Schlebusch H, Schmolling J. Adhesion molecules, activin and inhibin – candidates for the biochemical prediction of hypertensive diseases in pregnancy? *Arch Gynecol Obstet.* 2004;270:110–115.
78. Ong CY, Liao AW, Munim S, Spencer K, Nicolaides KH. First-trimester maternal serum activin A in pre-eclampsia and fetal growth restriction. *J Matern Fetal Neonatal Med.* 2004;15:176–180.
79. Madazli R, Kuseyrioglu B, Uzun H, Uludag S, Ocak V. Prediction of preeclampsia with maternal mid-trimester placental growth factor, activin A, fibronectin and uterine artery Doppler velocimetry. *Int J Gynaecol Obstet.* 2005;89:251–257.
80. Ay E, Kavak ZN, Elter K, Gokaslan H, Pekin T. Screening for pre-eclampsia by using maternal serum inhibin A, activin A, human chorionic gonadotropin, unconjugated estriol, and alpha-fetoprotein levels and uterine artery Doppler in the second trimester of pregnancy. *Aust N Z J Obstet Gynaecol.* 2005;45:283–288.
81. Spencer K, Yu CK, Savvidou M, Papageorghiou AT, Nicolaides KH. Prediction of pre-eclampsia by uterine artery Doppler ultrasonography and maternal serum pregnancy-associated plasma protein-A, free beta-human chorionic gonadotropin, activin A and inhibin A at 22 + 0 to 24 + 6 weeks' gestation. *Ultrasound Obstet Gynecol.* 2006;27:658–663.
82. Diesch CH, Holzgreve W, Hahn S, Zhong XY. Comparison of activin A and cell-free fetal DNA levels in maternal plasma from patients at high risk for preeclampsia. *Prenat Diagn.* 2006;26:1267–1270.
83. Spencer K, Cowans NJ, Nicolaides KH. Maternal serum inhibin-A and activin-A levels in the first trimester of pregnancies developing pre-eclampsia. *Ultrasound Obstet Gynecol.* 2008;32:622–626.
84. Akolekar R, Etchegaray A, Zhou Y, Maiz N, Nicolaides KH. Maternal serum activin A at 11–13 weeks of gestation in hypertensive disorders of pregnancy. *Fetal Diagn Ther.* 2009;25:320–327.
85. Than NG, Pick E, Bellyei S, et al. Functional analyses of placental protein 13/galectin-13. *Eur J Biochem.* 2004;271:1065–1078.
86. Burger O, Pick E, Zwickel J, et al. Placental protein 13 (PP-13): effects on cultured trophoblasts, and its detection in human body fluids in normal and pathological pregnancies. *Placenta.* 2004;25:608–622.
87. Nicolaides KH, Bindra R, Turan OM, et al. A novel approach to first-trimester screening for early pre-eclampsia combining serum PP-13 and Doppler ultrasound. *Ultrasound Obstet Gynecol.* 2006;27:13–17.
88. Chafetz I, Kuhnreich I, Sammar M, et al. First-trimester placental protein 13 screening for preeclampsia and intrauterine growth restriction. *Am J Obstet Gynecol.* 2007;197:35.e1–35.e7.
89. Spencer K, Cowans NJ, Chefetz I, Tal J, Meiri H. First-trimester maternal serum PP-13, PAPP-A and second-trimester uterine artery Doppler pulsatility index as markers of pre-eclampsia. *Ultrasound Obstet Gynecol.* 2007;29:128–134.

90. Gonen R, Shahar R, Grimpel YI, et al. Placental protein 13 as an early marker for pre-eclampsia: a prospective longitudinal study. *BJOG*. 2008;115:1465–1472.
91. Romero R, Kusanovic JP, Than NG, et al. First-trimester maternal serum PP13 in the risk assessment for preeclampsia. *Am J Obstet Gynecol*. 2008;199:122.e1–122.e11.
92. Khalil A, Cowans NJ, Spencer K, Goichman S, Meiri H, Harrington K. First trimester maternal serum placental protein 13 for the prediction of pre-eclampsia in women with a priori high risk. *Prenat Diagn*. 2009;29:781–789.
93. Akolekar R, Syngelaki A, Beta J, Kocylowski R, Nicolaides KH. Maternal serum placental protein 13 at 11–13 weeks of gestation in preeclampsia. *Prenat Diagn*. 2009;29:1103–1108.
94. Wortelboer EJ, Koster MP, Cuckle HS, Stoutenbeek PH, Schielen PC, Visser GH. First-trimester placental protein 13 and placental growth factor: markers for identification of women destined to develop early-onset pre-eclampsia. *BJOG*. 2010;117:1384–1389.
95. Moslemi Zadeh N, Naghshvar F, Peyvandi S, Gheshlaghi P, Ehetshami S. PP13 and PAPP-A in the first and second trimesters: predictive factors for preeclampsia? *ISRN Obstet Gynecol*. 2012;2012:263871.
96. Deurloo KL, Linskens IH, Heymans MW, Heijboer AC, Blankenstein MA, van Vugt JM. ADAM12s and PP13 as first trimester screening markers for adverse pregnancy outcome. *Clin Chem Lab Med*. 2013;51:1279–1284.
97. Audibert F, Boucoiran I, An N, et al. Screening for preeclampsia using first-trimester serum markers and uterine artery Doppler in nulliparous women. *Am J Obstet Gynecol*. 2010;203:383.e1–383.e8.
98. Myatt L, Clifton RG, Roberts JM, et al. First-trimester prediction of preeclampsia in nulliparous women at low risk. *Obstet Gynecol*. 2012;119:1234–1242.
99. Laigaard J, Sørensen T, Placing S, et al. Reduction of the disintegrin and metalloprotease ADAM12 in preeclampsia. *Obstet Gynecol*. 2005;106:144–149.
100. Poon LC, Chelemen T, Granvillano O, Pandeva I, Nicolaides KH. First-trimester maternal serum a disintegrin and metalloprotease 12 (ADAM12) and adverse pregnancy outcome. *Obstet Gynecol*. 2008;112:1082–1090.
101. Bestwick JP, George LM, Wu T, Morris JK, Wald NJ. The value of early second trimester PAPP-A and ADAM12 in screening for pre-eclampsia. *J Med Screen*. 2012;19:51–54.
102. Bainbridge SA, Roberts JM. Uric acid as a pathogenic factor in preeclampsia. *Placenta*. 2008;suppl A:S67–S72.
103. Cnossen JS, de Ruyter-Hanhijärvi H, van der Post JA, Mol BW, Khan KS, ter Riet G. Accuracy of serum uric acid determination in predicting pre-eclampsia: a systematic review. *Acta Obstet Gynecol Scand*. 2006;85:519–525.
104. Zhou J, Zhao X, Wang Z, Hu Y. Combination of lipids and uric acid in mid-second trimester can be used to predict adverse pregnancy outcomes. *J Matern Fetal Neonatal Med*. 2012;25:2633–2638.
105. Soltan MH, Ismail ZA, Kafafi SM, Abdulla KA, Sammour MB. Values of certain clinical and biochemical tests for prediction of pre-eclampsia. *Ann Saudi Med*. 1996;16:280–284.
106. Rodriguez MH, Masaki DI, Mestman J, Kumar D, Rude R. Calcium/creatinine ratio and microalbuminuria in the prediction of preeclampsia. *Am J Obstet Gynecol*. 1988;159:1452–1455.
107. Konstantin-Hansen KF, Hesseldahl H, Pedersen SM. Microalbuminuria as a predictor of preeclampsia. *Acta Obstet Gynecol Scand*. 1992;71:343–346.
108. Massé J, Giguère Y, Kharfi A, Girouard J, Forest JC. Pathophysiology and maternal biologic markers of preeclampsia. *Endocrine*. 2002;19:113–125.
109. Lara González AL, Martínez Jaimes A, Romero Arauz JF. Microalbuminuria: early prognostic factor of preeclampsia? [in Spanish]. *Ginecol Obstet Mex*. 2003;71:82–86.
110. Salako BL, Olayemi O, Odukogbe AT, et al. Microalbuminuria in pregnancy as a predictor of preeclampsia and eclampsia. *West Afr J Med*. 2003;22:295–300.
111. Poon LC, Kametas N, Bonino S, Vercellotti E, Nicolaides KH. Urine albumin concentration and albumin-to-creatinine ratio at 11(+0) to 13(+6) weeks in the prediction of preeclampsia. *BJOG*. 2008;115:866–873.
112. Sirohiwal D, Dahiya K, Khaneja N. Use of 24-hour urinary protein and calcium for prediction of preeclampsia. *Taiwan J Obstet Gynecol*. 2009;48:113–115.
113. Baweja S, Kent A, Masterson R, Roberts S, McMahon LP. Prediction of pre-eclampsia in early pregnancy by estimating the spot urinary albumin:creatinine ratio using high-performance liquid chromatography. *BJOG*. 2011;118:1126–1132.
114. Singh R, Tandon I, Deo S, Natu SM. Does microalbuminuria at mid-pregnancy predict development of subsequent preeclampsia? *J Obstet Gynaecol Res*. 2013;39:478–483.
115. Taufield PA, Ales KL, Resnick LM, Druzin ML, Gertner JM, Laragh JH. Hypocalciuria in preeclampsia. *New Engl J Med*. 1987;316:715–718.
116. Sanchez-Ramos L, Jones DC, Cullen MT. Urinary calcium as an early marker for preeclampsia. *Obstet Gynecol*. 1991;77:685–688.
117. Baker PN, Hackett GA. The use of urinary albumin-creatinine ratios and calcium-creatinine ratios as screening tests for pregnancy-induced hypertension. *Obstet Gynecol*. 1994;83:745–749.
118. Suarez VR, Trelles JG, Miyahira JM. Urinary calcium in asymptomatic primigravidas who later developed preeclampsia. *Obstet Gynecol*. 1996;87:79–82.
119. Nisell H, Kublickas M, Lunell N-O, Pettersson E. Renal function in gravidas with chronic hypertension with and without superimposed preeclampsia. *Hypertens Pregnancy*. 1996;15:127–134.
120. Pal A, Roy D, Adhikary S, Roy A, Dasgupta M, Mandal AK. A prospective study for the prediction of preeclampsia with urinary calcium level. *J Obstet Gynaecol India*. 2012;62:312–316.
121. Conde-Agudelo A, Belizan JM, Lede R, Bergel E. Prediction of hypertensive disorders of pregnancy by calcium/creatinine ratio and other laboratory tests. *Int J Gynaecol Obstet*. 1994;47:285–286.
122. Raniolo E, Phillipou G. Prediction of pregnancy-induced hypertension by means of the urinary calcium:creatinine ratio. *Med J Aust*. 1993;158:98–100.
123. Phuapradit W, Manusook S, Lolekha P. Urinary calcium/creatinine ratio in the prediction of preeclampsia. *Aust N Z J Obstet Gynaecol*. 1993;33:280–281.

124. Rogers MS, Chung T, Baldwin S, Ho CS, Swaminathan R. A comparison of second trimester urinary electrolytes, microalbumin, and N-acetyl-beta-glucosaminidase for prediction of gestational hypertension and preeclampsia. *Hypertens Pregnancy*. 1994;13:179–192.
125. Ozcan T, Kaleli B, Ozeren M, Turan C, Zorlu G. Urinary calcium to creatinine ratio for predicting preeclampsia. *Am J Perinatol*. 1995;12:349–351.
126. Izumi A, Minakami H, Kuwata T, Sato I. Calcium-to-creatinine ratio in spot urine samples in early pregnancy and its relation to the development of preeclampsia. *Metabolism*. 1997;46:1107–1108.
127. Robinson CA, Amirthanayagan M, Goodman D, Roth N, Morgan MA. Urinary calcium-creatinine ratios fall from 16–20 to 28–32 weeks in healthy primigravid patients who subsequently develop preeclampsia. *Am J Obstet Gynecol*. 1997;176:S101.
128. Kazerooni T, Hamze-Nejadi S. Calcium to creatinine ratio in a spot sample of urine for early prediction of pre-eclampsia. *Int J Gynaecol Obstet*. 2003;80:279–283.
129. Vahdat M, Kashanian M, Sariri E, Mehdinia M. Evaluation of the value of calcium to creatinine ratio for predicting of pre-eclampsia. *J Matern Fetal Neonatal Med*. 2012;25:2793–2794.
130. Elebute OA, Mills IH. Urinary kallikrein in normal and hypertensive pregnancies. *Perspect Nephrol Hypertens*. 1976;5:329–338.
131. Bhoola KD, Figueroa CD, Worthy K. Bioregulation of kinins: kallikreins, kininogens, and kininases. *Pharmacol Rev*. 1992;44:1–80.
132. Millar JG, Campbell SK, Albano JD, Higgins BR, Clark AD. Early prediction of pre-eclampsia by measurement of kallikrein and creatinine on a random urine sample. *Br J Obstet Gynaecol*. 1996;103:421–426.
133. Kyle PM, Campbell S, Buckley D, et al. A comparison of the inactive urinary kallikrein:creatinine ratio and the angiotensin sensitivity test for the prediction of pre-eclampsia. *Br J Obstet Gynaecol*. 1996;103:981–987.
134. Wagner SJ, Craici IM, Grande JP, Garovic VD. From placenta to podocyte: vascular and podocyte pathophysiology in preeclampsia. *Clin Nephrol*. 2012;78:241–249.
135. Garovic VD, Wagner SJ, Turner ST, et al. Urinary podocyte excretion as a marker for preeclampsia. *Am J Obstet Gynecol*. 2007;196:320.e1–320.e7.
136. Jim B, Jean-Louis P, Qipo A, et al. Podocyturia as a diagnostic marker for preeclampsia amongst high-risk pregnant patients. *J Pregnancy*. 2012:984630.
137. Craici IM, Wagner SJ, Bailey KR, et al. Podocyturia predates proteinuria and clinical features of preeclampsia: longitudinal prospective study. *Hypertension*. 2013;61:1289–1296.
138. Ouaissi MA, Capron A. Fibronectins: structure and function [in French]. *Ann Inst Pasteur Immunol*. 1985;136C:169–185.
139. Peters JH, Maunder RJ, Woolf AD, Cochrane CG, Ginsberg MH. Elevated plasma levels of ED1+ ("cellular") fibronectin in patients with vascular injury. *J Lab Clin Med*. 1989;113:586–597.
140. Stubbs TM, Lazarchick J, Horger III EO. Plasma fibronectin levels in preeclampsia: a possible biochemical marker for vascular endothelial damage. *Am J Obstet Gynecol*. 1984;150:885–887.
141. Leeflang MM, Cnossen JS, van der Post JA, Mol BW, Khan KS, ter Riet G. Accuracy of fibronectin tests for the prediction of pre-eclampsia: a systematic review. *Eur J Obstet Gynecol Reprod Biol*. 2007;133:12–19.
142. Powers RW, Evans RW, Majors AK, et al. Plasma homocysteine concentration is increased in preeclampsia and is associated with evidence of endothelial activation. *Am J Obstet Gynecol*. 1998;179:1605–1611.
143. Mignini LE, Latthe PM, Villar J, Kilby MD, Carroli G, Khan KS. Mapping the theories of preeclampsia: the role of homocysteine. *Obstet Gynecol*. 2005;105:411–425.
144. Bergen NE, Jaddoe VW, Timmermans S, et al. Homocysteine and folate concentrations in early pregnancy and the risk of adverse pregnancy outcomes: the Generation R Study. *BJOG*. 2012;119:739–751.
145. Krieglstein CF, Granger DN. Adhesion molecules and their role in vascular disease. *Am J Hypertens*. 2001;14:44S–54S.
146. Chaiworapongsa T, Romero R, Yoshimatsu J, et al. Soluble adhesion molecule profile in normal pregnancy and pre-eclampsia. *J Matern Fetal Neonatal Med*. 2002;12:19–27.
147. Bosio PM, Cannon S, McKenna PJ, O'Herlihy C, Conroy R, Brady H. Plasma P-selectin is elevated in the first trimester in women who subsequently develop pre-eclampsia. *BJOG*. 2001;108:709–715.
148. Krauss T, Emons G, Kuhn W, Augustin HG. Predictive value of routine circulating soluble endothelial cell adhesion molecule measurements during pregnancy. *Clin Chem*. 2002;48:1418–1425.
149. Chavarría ME, Lara-González L, García-Paleta Y, Vital-Reyes VS, Reyes A. Adhesion molecules changes at 20 gestation weeks in pregnancies complicated by preeclampsia. *Eur J Obstet Gynecol Reprod Biol*. 2008;137:157–164.
150. Parra-Cordero M, Turan OM, Kaur A, Pearson JD, Nicolaides KH. Maternal serum soluble adhesion molecule levels at 11 + 0–13 + 6 weeks and subsequent development of pre-eclampsia. *J Matern Fetal Neonatal Med*. 2007;20:793–796.
151. Carty DM, Anderson LA, Freeman DJ, et al. Early pregnancy soluble E-selectin concentrations and risk of pre-eclampsia. *J Hypertens*. 2012;30:954–959.
152. Akolekar R, Veduta A, Minekawa R, Chelemen T, Nicolaides KH. Maternal plasma P-selectin at 11 to 13 weeks of gestation in hypertensive disorders of pregnancy. *Hypertens Pregnancy*. 2011;30:311–321.
153. Maynard SE, Karumanchi SA. Angiogenic factors and pre-eclampsia. *Semin Nephrol*. 2011;31:33–46.
154. Romero R, Nien JK, Espinoza J, et al. A longitudinal study of angiogenic (placental growth factor) and anti-angiogenic (soluble endoglin and soluble vascular endothelial growth factor receptor-1) factors in normal pregnancy and patients destined to develop preeclampsia and deliver a small for gestational age neonate. *J Matern Fetal Neonatal Med*. 2008;21:9–23.
155. Vaisbuch E, Whitty JE, Hassan SS, et al. Circulating angiogenic and antiangiogenic factors in women with eclampsia. *Am J Obstet Gynecol*. 2011;204:152.e1–152.e9.
156. Chaiworapongsa T, Romero R, Korzeniewski SJ, et al. Maternal plasma concentrations of angiogenic/antiangiogenic factors in the third trimester of pregnancy to identify the patient at risk for stillbirth at or near term and severe late pre-eclampsia. *Am J Obstet Gynecol*. 2013;208:287.e1–287.e15.

157. Kleinrouweler CE, Wiegerinck MM, Ris-Stalpers C, EBM CONNECT Collaboration Accuracy of circulating placental growth factor, vascular endothelial growth factor, soluble fms-like tyrosine kinase 1 and soluble endoglin in the prediction of pre-eclampsia: a systematic review and meta-analysis. *BJOG*. 2012;119:778–787.
158. McElrath TF, Lim KH, Pare E, et al. Longitudinal evaluation of predictive value for preeclampsia of circulating angiogenic factors through pregnancy. *Am J Obstet Gynecol*. 2012;207:407.e1–407.e7.
159. Myatt L, Clifton R, Roberts J, et al. Can changes in angiogenic biomarkers between the first and second trimesters of pregnancy predict development of pre-eclampsia in a low-risk nulliparous patient population? *BJOG*. 2013;120:1183–1191.
160. Odibo AO, Rada CC, Cahill AG, et al. First-trimester serum soluble fms-like tyrosine kinase-1, free vascular endothelial growth factor, placental growth factor and uterine artery Doppler in preeclampsia. *J Perinatol*. 2013;33:670–674.
161. Powers RW, Jeyabalan A, Clifton RG, et al. Soluble fms-Like tyrosine kinase 1 (sFlt1), endoglin and placental growth factor (PlGF) in preeclampsia among high risk pregnancies. *PLoS One*. 2010;5:e13263.
162. Chaiworapongsa T, Romero R, Kim YM, et al. Plasma soluble vascular endothelial growth factor receptor-1 concentration is elevated prior to the clinical diagnosis of pre-eclampsia. *J Matern Fetal Neonatal Med*. 2005;17:3–18.
163. Stepan H, Unversucht A, Wessel N, Faber R. Predictive value of maternal angiogenic factors in second trimester pregnancies with abnormal uterine perfusion. *Hypertension*. 2007;49:818–824.
164. Espinoza J, Romero R, Nien JK, et al. Identification of patients at risk for early onset and/or severe preeclampsia with the use of uterine artery Doppler velocimetry and placental growth factor. *Am J Obstet Gynecol*. 2007;196:326.e1–326.e13.
165. Stepan H, Geipel A, Schwarz F, Krämer T, Wessel N, Faber R. Circulatory soluble endoglin and its predictive value for preeclampsia in second-trimester pregnancies with abnormal uterine perfusion. *Am J Obstet Gynecol*. 2008;198:175.e1–175.e6.
166. Diab AE, El-Behery MM, Ebrahiem MA, Shehata AE. Angiogenic factors for the prediction of pre-eclampsia in women with abnormal midtrimester uterine artery Doppler velocimetry. *Int J Gynaecol Obstet*. 2008;102:146–151.
167. Akolekar R, Zaragoza E, Poon LC, Pepes S, Nicolaides KH. Maternal serum placental growth factor at 11 + 0 to 13 + 6 weeks of gestation in the prediction of pre-eclampsia. *Ultrasound Obstet Gynecol*. 2008;32:732–739.
168. Kusanovic JP, Romero R, Chaiworapongsa T, et al. A prospective cohort study of the value of maternal plasma concentrations of angiogenic and anti-angiogenic factors in early pregnancy and midtrimester in the identification of patients destined to develop preeclampsia. *J Matern Fetal Neonatal Med*. 2009;22:1021–1038.
169. Abdelaziz A, Maher MA, Sayyed TM, Bazeed MF, Mohamed NS. Early pregnancy screening for hypertensive disorders in women without a-priori high risk. *Ultrasound Obstet Gynecol*. 2012;40:398–405.
170. Ghosh SK, Raheja S, Tuli A, Raghunandan C, Agarwal S. Is serum placental growth factor more effective as a biomarker in predicting early onset preeclampsia in early second trimester than in first trimester of pregnancy? *Arch Gynecol Obstet*. 2013;287:865–873.
171. Benn P, Cuckle H, Pergament E. Non-invasive prenatal testing for aneuploidy: current status and future prospects. *Ultrasound Obstet Gynecol*. 2013;42:15–33.
172. Holzgreve W, Ghezzi F, Di Naro E, Gänshirt D, Maymon E, Hahn S. Disturbed feto-maternal cell traffic in preeclampsia. *Obstet Gynecol*. 1998;91:669–672.
173. Lo YM, Leung TN, Tein MS, et al. Quantitative abnormalities of fetal DNA in maternal serum in preeclampsia. *Clin Chem*. 1999;45:184–188.
174. Leung TN, Zhang J, Lau TK, Chan LY, Lo YM. Increased maternal plasma fetal DNA concentrations in women who eventually develop preeclampsia. *Clin Chem*. 2001;47:137–139.
175. Farina A, Sekizawa A, Sugito Y, et al. Fetal DNA in maternal plasma as a screening variable for preeclampsia. A preliminary nonparametric analysis of detection rate in low-risk nonsymptomatic patients. *Prenat Diagn*. 2004;24:83–86.
176. Cotter AM, Martin CM, O'Leary JJ, Daly SF. Increased fetal DNA in the maternal circulation in early pregnancy is associated with an increased risk of preeclampsia. *Am J Obstet Gynecol*. 2004;191:515–520.
177. Crowley A, Martin C, Fitzpatrick P, et al. Free fetal DNA is not increased before 20 weeks in intrauterine growth restriction or pre-eclampsia. *Prenat Diagn*. 2007;27:174–179.
178. Sifakis S, Zaravinos A, Maiz N, Spandidos DA, Nicolaides KH. First-trimester maternal plasma cell-free fetal DNA and preeclampsia. *Am J Obstet Gynecol*. 2009;201:472.e1–472.e7.
179. Stein W, Müller S, Gutensohn K, Emons G, Legler T. Cell-free fetal DNA and adverse outcome in low risk pregnancies. *Eur J Obstet Gynecol Reprod Biol*. 2013;166:10–13.
180. Jakobsen TR, Clausen FB, Rode L, Dziegiel MH, Tabor A. Identifying mild and severe preeclampsia in asymptomatic pregnant women by levels of cell-free fetal DNA. *Transfusion*. 2013;53:1956–1964.
181. Poon LC, Musci T, Song K, Syngelaki A, Nicolaides KH. Maternal plasma cell-free fetal and maternal DNA at 11–13 weeks' gestation: relation to fetal and maternal characteristics and pregnancy outcomes. *Fetal Diagn Ther*. 2013;33:215–223.
182. Papantoniou N, Bagiokos V, Agiannitopoulos K, et al. RASSF1A in maternal plasma as a molecular marker of preeclampsia. *Prenat Diagn*. 2013;33:682–687.
183. Romero R, Espinoza J, Gotsch F, et al. The use of high-dimensional biology (genomics, transcriptomics, proteomics, and metabolomics) to understand the preterm parturition syndrome. *BJOG*. 2006;113:118–135.
184. Farina A, Sekizawa A, Purwosunu Y, et al. Quantitative distribution of a panel of circulating mRNA in preeclampsia versus controls. *Prenat Diagn*. 2006;26:1115–1120.
185. Purwosunu Y, Sekizawa A, Okazaki S, et al. Prediction of preeclampsia by analysis of cell-free messenger RNA in maternal plasma. *Am J Obstet Gynecol*. 2009;200:386.e1–386.e7.
186. Rasanen J, Girsen A, Lu X, et al. Comprehensive maternal serum proteomic profiles of preclinical and clinical preeclampsia. *J Proteome Res*. 2010;9:4274–4281.

187. Kenny LC, Broadhurst DI, Dunn W, et al. Robust early pregnancy prediction of later preeclampsia using metabolomic biomarkers. *Hypertension*. 2010;56:741–749.
188. Myers JE, Tuytten R, Thomas G, et al. Integrated proteomics pipeline yields novel biomarkers for predicting preeclampsia. *Hypertension*. 2013;61:1281–1288.
189. Farina A, Zucchini C, Sekizawa A, et al. Performance of messenger RNAs circulating in maternal blood in the prediction of preeclampsia at 10–14 weeks. *Am J Obstet Gynecol*. 2010;203:575.e1–575.e7.
190. Carty DM, Siwy J, Brennand JE, et al. Urinary proteomics for prediction of preeclampsia. *Hypertension*. 2011;57:561–569.
191. Odibo AO, Goetzinger KR, Odibo L, et al. First-trimester prediction of preeclampsia using metabolomic biomarkers: a discovery phase study. *Prenat Diagn*. 2011;31:990–994.
192. Bahado-Singh RO, Akolekar R, Mandal R, et al. Metabolomics and first-trimester prediction of early-onset preeclampsia. *J Matern Fetal Neonatal Med*. 2012;25:1840–1847.
193. Bahado-Singh RO, Akolekar R, Mandal R, et al. First-trimester metabolomic detection of late-onset preeclampsia. *Am J Obstet Gynecol*. 2013;208:58.e1–58.e7.
194. Aquilina J, Thompson O, Thilaganathan B, Harrington K. Improved early prediction of pre-eclampsia by combining second-trimester maternal serum inhibin-A and uterine artery Doppler. *Ultrasound Obstet Gynecol*. 2001;17:477–484.
195. Audibert F, Benchimol Y, Benattar C, Champagne C, Frydman R. Prediction of preeclampsia or intrauterine growth restriction by second trimester serum screening and uterine Doppler velocimetry. *Fetal Diagn Ther*. 2005;20:48–53.
196. Spencer K, Yu CK, Cowans NJ, Otigbah C, Nicolaides KH. Prediction of pregnancy complications by first-trimester maternal serum PAPP-A and free beta-hCG and with second-trimester uterine artery Doppler. *Prenat Diagn*. 2005;25:949–953.
197. Onalan R, Onalan G, Gunenc Z, Karabulut E. Combining 2nd-trimester maternal serum homocysteine levels and uterine artery Doppler for prediction of preeclampsia and isolated intrauterine growth restriction. *Gynecol Obstet Invest*. 2006;61:142–148.
198. Khalil A, Cowans NJ, Spencer K, Goichman S, Meiri H, Harrington K. First-trimester markers for the prediction of pre-eclampsia in women with a-priori high risk. *Ultrasound Obstet Gynecol*. 2010;35:671–679.
199. Kuromoto K, Watanabe M, Adachi K, Ohashi K, Iwatani Y. Increases in urinary creatinine and blood pressure during early pregnancy in pre-eclampsia. *Ann Clin Biochem*. 2010;47:336–342.
200. Thilaganathan B, Wormald B, Zanardini C, Sheldon J, Ralph E, Papageorghiou AT. Early-pregnancy multiple serum markers and second-trimester uterine artery Doppler in predicting preeclampsia. *Obstet Gynecol*. 2010;115:1233–1238.
201. Anderson UD, Olsson MG, Rutardóttir S, et al. Fetal hemoglobin and α1-microglobulin as first- and early second-trimester predictive biomarkers for preeclampsia. *Am J Obstet Gynecol*. 2011;204:520.e1–520.e5.
202. Youssef A, Righetti F, Morano D, Rizzo N, Farina A. Uterine artery Doppler and biochemical markers (PAPP-A, PIGF, sFlt-1, P-selectin, NGAL) at 11 + 0 to 13 + 6 weeks in the prediction of late (>34 weeks) pre-eclampsia. *Prenat Diagn*. 2011;31:1141–1146.
203. Yu J, Shixia CZ, Wu Y, Duan T. Inhibin A, activin A, placental growth factor and uterine artery Doppler pulsatility index in the prediction of pre-eclampsia. *Ultrasound Obstet Gynecol*. 2011;37:528–533.
204. Bolin M, Wikström AK, Wiberg-Itzel E, et al. Prediction of preeclampsia by combining serum histidine-rich glycoprotein and uterine artery Doppler. *Am J Hypertens*. 2012;25:1305–1310.
205. Ghosh SK, Raheja S, Tuli A, Raghunandan C, Agarwal S. Combination of uterine artery Doppler velocimetry and maternal serum placental growth factor estimation in predicting occurrence of pre-eclampsia in early second trimester pregnancy: a prospective cohort study. *Eur J Obstet Gynecol Reprod Biol*. 2012;161:144–151.
206. Wald NJ, Bestwick JP, George LM, Wu T, Morris JK. Screening for pre-eclampsia using serum placental growth factor and endoglin with Down's syndrome Quadruple test markers. *J Med Screen*. 2012;19:60–67.
207. Giguère Y, Charland M, Bujold E, et al. Combining biochemical and ultrasonographic markers in predicting preeclampsia: a systematic review. *Clin Chem*. 2010;56:361–375.
208. Kuc S, Wortelboer EJ, van Rijn BB, Franx A, Visser GH, Schielen PC. Evaluation of 7 serum biomarkers and uterine artery Doppler ultrasound for first-trimester prediction of preeclampsia: a systematic review. *Obstet Gynecol Surv*. 2011;66:225–239.
209. Pedrosa AC, Matias A. Screening for pre-eclampsia: a systematic review of tests combining uterine artery Doppler with other markers. *J Perinat Med*. 2011;39:619–635.
210. Hui D, Okun N, Murphy K, Kingdom J, Uleryk E, Shah PS. Combinations of maternal serum markers to predict pre-eclampsia, small for gestational age, and stillbirth: a systematic review. *J Obstet Gynaecol Can*. 2012;34:142–153.
211. Plasencia W, Maiz N, Bonino S, Kaihura C, Nicolaides KH. Uterine artery Doppler at 11 + 0 to 13 + 6 weeks in the prediction of pre-eclampsia. *Ultrasound Obstet Gynecol*. 2007;30:742–749.
212. Onwudiwe N, Yu CK, Poon LC, Spiliopoulos I, Nicolaides KH. Prediction of pre-eclampsia by a combination of maternal history, uterine artery Doppler and mean arterial pressure. *Ultrasound Obstet Gynecol*. 2008;32:877–883.
213. Poon LC, Kametas NA, Maiz N, Akolekar R, Nicolaides KH. First-trimester prediction of hypertensive disorders in pregnancy. *Hypertension*. 2009;53:812–818.
214. Poon LC, Karagiannis G, Leal A, Romero XC, Nicolaides KH. Hypertensive disorders in pregnancy: screening by uterine artery Doppler imaging and blood pressure at 11–13 weeks. *Ultrasound Obstet Gynecol*. 2009;34:497–502.
215. Herraiz I, Arbués J, Camaño I, Gómez-Montes E, Grañeras A, Galindo A. Application of a first-trimester prediction model for pre-eclampsia based on uterine arteries and maternal history in high-risk pregnancies. *Prenat Diagn*. 2009;29:1123–1129.

216. Poon LC, Akolekar R, Lachmann R, Beta J, Nicolaides KH. Hypertensive disorders in pregnancy: screening by biophysical and biochemical markers at 11–13 weeks. *Ultrasound Obstet Gynecol.*. 2010;35:662–670.
217. Poon LC, Stratieva V, Piras S, Piri S, Nicolaides KH. Hypertensive disorders in pregnancy: combined screening by uterine artery Doppler, blood pressure and serum PAPP-A at 11–13 weeks. *Prenat Diagn.* 2010;30:216–223.
218. Akolekar R, Syngelaki A, Sarquis R, Zvanca M, Nicolaides KH. Prediction of early, intermediate and late pre-eclampsia from maternal factors, biophysical and biochemical markers at 11–13 weeks. *Prenat Diagn.* 2011;31:66–74.
219. Odibo AO, Zhong Y, Goetzinger KR, et al. First-trimester placental protein 13, PAPP-A, uterine artery Doppler and maternal characteristics in the prediction of pre-eclampsia. *Placenta.* 2011;32:598–602.
220. Di Lorenzo G, Ceccarello M, Cecotti V, et al. First trimester maternal serum PIGF, free β-hCG, PAPP-A, PP-13, uterine artery Doppler and maternal history for the prediction of pre-eclampsia. *Placenta.* 2012;33:495–501.
221. Scazzocchio E, Figueras F, Crispi F, et al. Performance of a first-trimester screening of preeclampsia in a routine care low-risk setting. *Am J Obstet Gynecol.* 2013;208:203.e1–203.e10.
222. Parra-Cordero M, Rodrigo R, Barja P, et al. Prediction of early and late pre-eclampsia from maternal characteristics, uterine artery Doppler and markers of vasculogenesis during first trimester of pregnancy. *Ultrasound Obstet Gynecol.* 2013;41:538–544.
223. Myers J, Kenny L, McCowan L, SCOPE consortium Angiogenic factors combined with clinical risk factors to predict preterm pre-eclampsia in nulliparous women: a predictive test accuracy study. *BJOG.* 2013;120:1215–1223.
224. Caradeux J, Serra R, Nien JK, et al. First trimester prediction of early onset preeclampsia using demographic, clinical, and sonographic data: a cohort study. *Prenat Diagn.* 2013;33:732–736.
225. Diguisto C, Le Gouge A, Piver E, Giraudeau B, Perrotin F. Second-trimester uterine artery Doppler, PlGF, sFlt-1, sEndoglin, and lipid-related markers for predicting preeclampsia in a high-risk population. *Prenat Diagn.* 2013;33:1070–1074.
226. Boucoiran I, Suarthana E, Rey E, Delvin E, Fraser WB, Audibert F. Repeated measures of placental growth factor, placental protein 13, and a disintegrin and metalloprotease 12 at first and second trimesters for preeclampsia screening. *Am J Perinatol.* 2013;30:681–688.
227. Moons KG, Kengne AP, Woodward M, et al. Risk prediction models: I. Development, internal validation, and assessing the incremental value of a new (bio)marker. *Heart.* 2012;98:683–690.
228. Moons KG, Kengne AP, Grobbee DE, et al. Risk prediction models: II. External validation, model updating, and impact assessment. *Heart.* 2012;98:691–698.
229. Farina A, Rapacchia G, Freni Sterrantino A, Pula G, Morano D, Rizzo N. Prospective evaluation of ultrasound and biochemical-based multivariable models for the prediction of late pre-eclampsia. *Prenat Diagn.* 2011;31:1147–1152.
230. Romero R, Kusanovic JP, Kim CJ. Placental bed disorders in the génesis of the great obstetrical síndromes. In: Pijnenbjorg R, Brosens I, Romero R, eds. *Placental Bed Disorders: Basic Science and its Translation to Obstetrics.* New York: Cambridge University Press; 2010.
231. Rana S, Karumanchi SA, Lindheimer MD. Angiogenic factors in diagnosis, management, and research in preeclampsia. *Hypertension.* 2014;63:198–202.
232. Powers RW, Roberts JM, Plymire DA, et al. Low placental growth factor across pregnancy identifies a subset of women with preterm preeclampsia: type 1 versus type 2 preeclampsia? *Hypertension.* 2012;60:239–246.

CHAPTER **12**

Prevention of Preeclampsia and Eclampsia

ANNE CATHRINE STAFF, BAHA M. SIBAI AND F. GARY CUNNINGHAM

Editors' comment: *The past two decades have seen many resources focused on the prevention of preeclampsia with minimal success (if prescribing low-dose aspirin to high-risk patients can be labeled a success), and even some are concerned that the small benefit attested to aspirin may be an artifact related to false positive significance associated with subset analysis, especially when not in the original study design. It is also fair to say that while the search for a better "preventive" goes on, some believe the dollar focus should be to better understand the causes, and only then think of prevention. The fourth edition introduces a new first author, Anne Cathrine Staff, who leads a premier preeclampsia research group in Oslo, Norway. She has suggested by her writings that redefining preeclampsia through placenta-derived biomarkers would hopefully be a better step towards eventual prevention.*

INTRODUCTION

Because of its sentinel position as a major cause of maternal and perinatal mortality and morbidity worldwide, and especially in developing countries, prevention of the hypertensive disorders of pregnancy has been an area of intense research interest.[1] As discussed throughout this text, however, the etiology of the preeclampsia syndrome remains elusive. The preeclampsia syndrome appears in many clinical forms and may have several pathophysiological pathways, and thus the search for its prevention is considerably hampered. Most studies designed on theoretical bases to reduce the incidence of preeclampsia in either low- or high-risk women have been disappointing.[2] Parallel to the quest for a preventive strategy, efforts have been made to develop a number of clinical, biophysical, or biochemical tests that would permit prediction or early detection of preeclampsia.[3,4] Most of these have low sensitivity and positive-predictive value, and thus are not suitable for routine clinical use (see Chapter 11). Thus, prevention studies have previously concentrated on women with demographic, familial, medical, and obstetrical factors that are associated with an increased risk of preeclampsia, as listed in Table 12.1.

The failure to find successful prevention for preeclampsia has not been from lack of effort. In the first edition of this book, Chesley[5] chronicled the abusive dietary restrictions used during the 20th century. And following the low-salt diet and thiazide diuretic fads of the 1960s and 1970s, randomized trials have described the use of various methods to prevent or reduce the incidence and/or severity of preeclampsia. The purpose of this chapter is to review some of these clinical trials to gain insight into the problems that have frustrated research for prevention measures. Shown in Table 12.2 is a list of regimens that have been studied by randomized controlled trials.

Prevention is traditionally classified as primary, avoiding disease occurrence, secondary, diagnosing and treating extant disease in early stages, and tertiary, reducing the negative impact of existent disease by restoring function and reducing disease-related complications. As will be discussed in further detail below when dissecting individual

TABLE 12.1 Some Risk Factors for Preeclampsia Syndrome

Nulliparity
Primipaternity
Advanced maternal age
IVF treatment
Multifetal gestation
Previous adverse obstetrical outcome
Gestational hypertension, especially preeclampsia
Fetal growth restriction
Abruptio placentae
Perinatal death
Abnormal midpregnancy uterine Doppler studies
Resistance index 0.58
Presence of diastolic notch
Medical conditions
Chronic hypertension
Renal disease
Pregestational diabetes
Thrombophilia syndromes
Connective-tissue disorders
Obesity and insulin resistance

Chesley's Hypertensive Disorders in Pregnancy.
ISBN: 978-0-12-407866-6

DOI: http://dx.doi.org/10.1016/B978-0-12-407866-6.00012-2

TABLE 12.2 Methods Evaluated in Randomized Trials to Prevent Preeclampsia

Dietary manipulations	
	Low-salt diet
	Fish oil supplementation
	Calcium supplementation
Cardiovascular drugs	
	Diuretics
	Antihypertensive drugs
Antioxidants	
	Ascorbic acid vitamin (C)
	α-Tocopherol vitamin (E)
Antithrombotic drugs	
	Low-dose aspirin
	Aspirin/dipyridamole
	Aspirin + heparin
	Aspirin + ketanserin
Physical activity	
NO donors	

studies to prevent preeclampsia, there have been many trials with many different approaches. We note here that the aims of such studies are not only to reduce the incidence of the disease but also to improve the clinical outcome for mother and offspring. One clinical primary prevention strategy used worldwide in 2014 is low-dose aspirin to women with high risk of preeclampsia, starting at the beginning of second trimester. The benefit of low-dose aspirin is modest at best, and some still question its efficacy, but its use has few risks. Also, magnesium sulfate is widely used to prevent eclampsia, and its efficacy is well documented, with substantial effects in reducing eclamptic convulsions. Most other intervention strategies are either not well documented or have proved to be without merit in randomized controlled trials (RCTs). As summarized from the recent UK NICE (National Institute for Health and Clinical Excellence) Guidelines on Hypertension in pregnancy (1), although several drugs, nitric oxide donors, progesterone, diuretics, and low-molecular weight heparin, and vitamin and nutrient supplements, vitamin C, vitamin E, folic acid, magnesium, fish oils, algae oils, and garlic, have been studied as preventive treatments for hypertensive disorders, these have not shown benefit and should not be given for this purpose. These are further discussed in the sections that follow, which included updated studies and meta-analyses done since the third edition of this book.

A challenge for any preeclampsia prevention and intervention study is how to identify the target population, to prevent preeclampsia in primiparous women, as well as recurrent preeclampsia in previously affected parous women. Preeclampsia is a heterogeneous syndrome, and clinical and laboratory prediction biomarkers in the general population are currently neither cost-effective nor sufficiently specific and sensitive, as reviewed in Chapter 11.

The only intervention that completely nullifies the risk of preeclampsia is to avoid pregnancy, which is not an option for most women.

DIETARY MANIPULATIONS

Low-Salt Diet

Because edema appeared to be an important component and even a forerunner to preeclampsia, it was natural that attempts would be made to prevent its progression. Mechanical methods, e.g., positional elevating of the feet, had been used, but a devoted following of salt restriction developed in the 1940s. This historical chapter was eclipsed by the introduction of chlorothiazide in 1957, when the use of diuretic agents for preeclampsia was advocated. Regarding sodium restriction, Steegers et al.[6] found no convincing evidence that it reduces the incidence of preeclampsia. Later, Knuist et al.[7] performed a randomized multicenter trial and found no benefits of prescribing a low-sodium intake in reducing the rate of preeclampsia. As of 2014, none of the major guidelines, including those of the World Health Organization, National Institute of Health and Clinical Excellence, and the American College of Obstetricians and Gynecologists, recommend salt restriction during pregnancy to prevent gestational hypertension or preeclampsia.

Fish Oil

The use of various fish oils to prevent or ameliorate a number of "diseases of civilization" has an interesting history that dates back to finding a link between certain types of fatty acids that mediate "inflammation." In this scheme, the atherogenic fatty acids are responsible for many of the ills that plague modern man in industrialized countries. As the diet drifted away from the hunter-gatherer to modern habits, so did the consumption of some heretofore protective fatty acids. The latter were found in oils of some fatty fishes in the diets of people from Scandinavian countries and in American Eskimos.

The fatty acid cascade begins with arachidonic acid, which follows a complex metabolic pathway of oxidation and decarboxylation. The variety of fatty acids produced is dependent on which precursors are incorporated, and many of their physiological properties stem from the position of the first double carbon bond in relation to the terminal CH group – this group is termed the "omega" carbon because it is furthest from the COOH – "alpha" – group. Thus, fatty acids with the first double bond located three carbon atoms away from the end are termed "omega minus 3" fatty acids – this is commonly written "omega-3." The average industrialized diet consists of a small percentage of omega-3 fatty acids, with the two most common being

TABLE 12.3 Randomized Trial of Fish Oil Versus Olive Oil Supplementation to Prevent Preeclampsia in High-Risk Women

Risk Factor and Study Group	Gestational Hypertension No. (%)	Preeclampsia No. (%)
Prior gestational hypertension		
Fish oil (n = 167)	55 (32.9)	11 (7.2)
Olive oil (n = 183)	61 (33.3)	17 (10.1)
Twin gestation		
Fish oil (n = 274)	38 (13.9)	14 (5.7)
Olive oil (n = 279)	29 (10.4)	6 (2.4)

All comparisons $p > 0.05$.
Data from Olsen et al.[12]

eicosapentaenoic acid (EPA), present in fish oil, and alpha-linolenic acid (ALA), derived from the omega-6 linoleic acid found in vegetable oil and some animal fats.

Proclamations soon followed that supplementation with omega-3 oils would prevent inflammation-mediated vascular disease, e.g., atherogenesis. Thus, it was not a quantum leap to posit that their ingestion might prevent preeclampsia as well as other adverse vascular disorders of pregnancy, such as fetal-growth restriction or even preterm delivery. Initial reports of beneficial effects of fish oil on lowering the incidence of preeclampsia arose from observational studies and a few small randomized trials. Beneficial effects of omega-3 fatty acids are typically ascribed to inhibition of platelet thromboxane A production with simultaneous production of limited amounts of physiologically inactive thromboxane A. Endothelial production of prostacyclin (PGI) is not inhibited, thus shifting the balance toward reduced platelet aggregation and increased vasodilatation. Another theoretical mechanism is through diminished vascular sensitivity to infused angiotensin II (Ang II). Adair et al.[8] evaluated the effects of omega-3 fatty acid supplementation on vascular sensitivity in 10 normotensive pregnant women between 24 and 34 weeks. Angiotensin infusions were performed before and 28 days after supplementation with 3.6 g per day of EPA. The effective pressor dose of angiotensin II was significantly increased after EPA supplementation – 35.8 ± 15.9 vs. 13.6 ± 6.3 ng/kg/min ($p = 0.001$).

There have been three randomized trials to study the effects of fish oil supplementation in women at high risk of preeclampsia.[9–11] In none was there a reduction in the incidence of preeclampsia. Olsen et al.[12] conducted a European multicenter trial to compare fish oil to olive oil in women with previous preeclampsia or multifetal gestation. As shown in Table 12.3, they found no significant differences between the groups in rates of either gestational hypertension or preeclampsia. Olafsdottir et al.[13] conducted a Cochrane review and reported no reduction in the incidence of gestational hypertension or preeclampsia with fish oil supplementation. Finally, Makrides et al.[14] reported results of a prospective study of the relationship between high consumption of marine fatty acids and hypertensive disorders during pregnancy. They found that consumption of large amounts of fish oil in early pregnancy actually *increases* the risk of hypertensive disorders.[14] Thus, fish oil supplementation is not recommended for the prevention of preeclampsia, which is reflected in recent NICE guidelines.[15]

Calcium Supplementation

The relationship between dietary calcium intake and pregnancy-associated hypertension has been of interest since the 1980s. In their review, Belizan et al.[16] reported an inverse association between calcium intake and maternal blood pressure, as well as the incidence of preeclampsia syndrome. Possible mechanisms by which calcium might prevent preeclampsia are unknown but these authors suggested that supplementation reduces maternal blood pressure by influencing parathyroid hormone release and intracellular calcium availability.

To test these hypotheses, a number of clinical studies have been designed to compare calcium supplementation during pregnancy with no treatment or a placebo. Protocols have varied considerably, and some studies compared women at low risk with those at high risk; there were varied types of randomization, placebo use, and blinding; gestational ages at enrollment ranged from 13 to 32 weeks; and calcium doses ranged from 156 to 2000 mg daily. In addition, these studies differed regarding the definition of hypertension and preeclampsia, and several did not differentiate between the two.[17–29]

A total of 10 randomized trials that studied the incidence of preeclampsia were placebo-controlled. Results from these are shown in Table 12.4. Of importance, the three trials with enrollment sizes exceeding 1000 healthy nulliparous women demonstrated no significant reduction in the incidence of preeclampsia with calcium supplementation. In two of these three, there was a beneficial effect with calcium supplementation to lower the incidence of maternal hypertension overall. Supplementation also had a salutary effect on the incidence of preeclampsia in a subgroup of women with *probable* calcium deficiency.[20,26] In the largest study,[26] secondary analysis with a composite morbidity index suggested that calcium supplementation had reduced the incidence of both severe disease and very adverse outcomes, although the number needed to treat to obtain an effect was high. By way of contrast, there were no effects on the incidence of maternal hypertension or preeclampsia in any of the subgroups analyzed by calcium intake in the NICHD Maternal Fetal Medicine Network trial of over 4500 women.[23]

These randomized trials have also been the subject of two systematic reviews. In a Cochrane review by Hofmeyr and Atallah[28] that included 15,206 women, for those assigned to treatment with calcium the relative risk of preeclampsia was 0.48 (95% CI 0.33–0.69) compared with

TABLE 12.4 Randomized Double-Blind Placebo Trials of Calcium Supplementation to Prevent Preeclampsia

Study	Risk Factors	Enrollment (wks)	Intervention (No.)		Preeclampsia (%)	
			Calcium	Placebo	Calcium	Placebo
Lopez-Jaramillo et al.[17]	Nulliparas	24	55	51	3.6	24*
Lopez-Jaramillo et al.[18]	Positive roll-over test	24–32	22	34	0	24*
Villar & Repke[19]	Nulliparas (85%)	24	90	88	0	3.4
Belizan et al.[20]	Nulliparas	20	579	588	2.6	3.9
Sanchez-Ramos et al.[21]	Positive roll-over test and A-II sensitivity	24–28	29	34	14	44*
Purvar et al.[22]	Low-calcium diet	<20	97	93	2.1	11.8
Levine et al.[23] [a]	Nulliparas	13–21	2295	2294	6.9	7.3
Lopez-Jaramillo et al.[24]	Nulliparas	20	125	135	3.2	16
Crowther et al.[25]	Nulliparas	<20	227	229	4.4	10
Villar et al.[26] [b]	Low-calcium diet	<20	4151	4161	4.1	4.5

[a]National Institute of Child Health Human Development Trial.
[b]World Health Organization Trial.
*$p < 0.05$.

nontreated women. These authors also concluded that calcium supplementation appeared to reduce the risk of severe hypertensive disorders, particularly in populations with poor calcium intake. In contrast, Trumbo and Elwood[29] reported an evidence-based review by the United States Food and Drug Administration. They concluded that "the relationship between calcium and risk of hypertension in pregnancy is inconsistent and inconclusive, and the relationship between calcium and the risk of pregnancy induced hypertension and preeclampsia is highly unlikely."

A Cochrane systematic review from 2010 concluded that calcium halved the risk of preeclampsia.[30] A later meta-analysis (2011) showed that calcium supplementation during pregnancy is associated with a reduced risk of gestational hypertension, preeclampsia, neonatal mortality and preterm birth in developing countries.[31] Another meta-analysis (2012) favors an additional intake of calcium during pregnancy to reduce the incidence of preeclampsia in populations at high risk of preeclampsia due to ethnicity, gender, age, or high BMI and in those with low baseline calcium intake.[32]

In summary, in the third edition of Chesley's book we reached a conclusion that remains valid currently in that preeclampsia prevention with calcium supplementation is unlikely to be achieved except perhaps in high-risk populations that are chronically calcium-deficient. Even in the latter groups the number needed to treat to obtain an effect was quite high. As the recent NICE Guidelines summarize: "The evidence for added calcium in the prevention of hypertensive disorders is conflicting and confusing, and more research is needed in this area."[33]

PHYSICAL ACTIVITY

Two Cochrane reviews from 2006 included only few trials and concluded that evidence is insufficient to recommend bed rest or reduced activity in preventing preeclampsia.[34] At the same time, it was concluded that there were no obvious beneficial effects of physical exercise.[35] A systematic review from 2012 indicates a trend towards a protective effect of physical activity in the prevention of preeclampsia.[36] However, the present documentation for recommending bed rest and/or exercise is weak, and the NICE Guidelines summarize that "advice on rest, exercise and work for women at risk of hypertensive disorders during pregnancy should be the same as for healthy pregnant women."[15] Likewise, the Task Force of the American College of Obstetricians and Gynecologists[37] does not recommend bed rest or increased physical activity for primary prevention of preeclampsia or its complications.

DIURETICS AND ANTIHYPERTENSIVE DRUGS

As discussed previously, there was a presumed, although never proved, efficacy of low-salt diets given to treat edema, hypertension, or preeclampsia. Thus, it was not surprising that diuretic therapy became popular with the introduction of chlorothiazide in the 1960s.[5,38,39] In a meta-analysis, Churchill et al.[40] summarized nine randomized trials that evaluated more than 7000 pregnant women. Those given diuretics had a decreased incidence of edema and hypertension, but not of preeclampsia.

According to most studies, women with preexisting chronic hypertension are at significantly high risk of preeclampsia compared with normotensive women.[41–43] There have been several randomized trials involving a total of 1055 participants – only a few were placebo-controlled – that evaluated the use of various antihypertensive drugs to reduce the incidence of superimposed preeclampsia in women with chronic hypertension. A critical analysis of

TABLE 12.5 Randomized Trials of Vitamin C and E Supplementation to Prevent Preeclampsia

Study	Risk Factors	Enrollment (wks)	Preeclampsia by Treatment Group	
			Vitamins No. (%)	Placebo No. (%)
Chappell et al.[46]	Abnormal UAD	16–22	141 (8)	142 (17)
Beazley et al.[47]	High-risk	14–21	52 (17.3)	48 (18.8)
Poston et al.[48]	High-risk	14–22	1196 (15)	1199 (16)
Rumbold et al.[49]	Nulliparas	14–22	935 (6)	942 (5)
Spinnato et al.[50]	High-risk	12–20	355 (13.8)	352 (15.6)
Villar et al.[51 a]	High-risk	<20	681 (14)	674 (23)

UAD = uterine artery Doppler.
[a]World Health Organization Study.

these trials failed to demonstrate any risk reduction.[41,42] We must caution that these trials did not have adequate power to show any potential benefits of therapy.

ANTIOXIDANT VITAMINS

There are inferential data that an imbalance between oxidant and antioxidant activity may have an important role in the pathogenesis of preeclampsia. For example, lipid peroxides and free radicals contribute to endothelial cell injury.[44,45] And there is evidence that elevated plasma concentrations of free radical oxidation products precede the development of preeclampsia. Two naturally occurring antioxidants – vitamins C and E – have been reported to decrease LDL oxidation and reduce both superoxide formation and cytokine production.[45] Moreover, women who developed preeclampsia had reduced plasma vitamin levels prior to its development. Thus, supplementation with these two vitamins was proposed to improve the oxidative capability of women at risk of preeclampsia.

In a pilot study, Chappell et al.[46] suggested a beneficial effect from pharmacological doses of vitamins C and E given to women identified as at risk of preeclampsia because of abnormal uterine artery Doppler flow velocimetry. Since then, a number of large randomized trials have been undertaken to evaluate vitamin C 1000 mg/d plus vitamin E 400 IU/d supplementation during pregnancy (Table 12.5). In the largest of these, conducted at 25 clinical sites in the United Kingdom, Poston et al.[48] enrolled 2410 women at risk of preeclampsia between 14 and 21 weeks. Heterogeneous risk factors included prior preeclampsia before 37 weeks, chronic hypertension, diabetes, antiphospholipid antibody syndrome, chronic renal disease, multifetal pregnancy, abnormal uterine artery Doppler findings, and nulliparas with body mass index (BMI) $>30\,kg/m^2$. As further shown in Table 12.5, the incidence of preeclampsia was similar in treatment and placebo groups – 15% compared with 16%. Of note, there were significantly more low-birthweight infants, fetal deaths, and low Apgar scores among babies born to mothers in the supplemented group, although small-for-gestational age infants did not differ between groups. Among participants who became preeclamptic, those in the antioxidant group developed it earlier. The studies by Rumbold et al.[49] and Spinnato et al.[50] were multicenter randomized placebo-controlled trials in women assigned to daily treatment with both vitamin C and E, or a placebo. As shown in Table 12.5, there were no differences in rates of preeclampsia between the study groups. In the study by Rumbold et al.,[49] hypertensive complications were significantly higher in the group given vitamins. In the study by Spinnato et al.,[50] however, there were no differences in mean gestational age at delivery or the incidence of perinatal mortality, abruptio placentae, preterm delivery, or growth-restricted or low-birth-weight infants. This trial failed to demonstrate a benefit of antioxidant supplementation in reducing the rate of preeclampsia among patients with chronic hypertension and/or prior preeclampsia.

The World Health Organization conducted a randomized trial of vitamin C and E supplementation[51] that included 1365 high-risk women who were randomly assigned before 20 weeks to either daily 1000 mg vitamin C plus 400 IU vitamin D or to placebo. The rates of preeclampsia were similar – 24% versus 23%. Subsequent meta-analyses and systematic reviews also concluded that there are no salutary protective effects of these antioxidant vitamins on the incidence of preeclampsia.[52,53]

ANTITHROMBOTIC AGENTS

There are several theoretical reasons to consider the use of antithrombotic agents to prevent preeclampsia. As discussed thoughout this fourth edition of Chesley, the preeclampsia syndrome is characterized by vasospasm, endothelial cell activation and dysfunction, and activation of the coagulation cascade. As discussed in Chapter 17, enhanced platelet activation and thromboxane production appear to play an important role in some of these pathophysiological abnormalities. It is likely that these abnormalities are caused at least partly by an imbalance in

TABLE 12.6 NICHD-Sponsored Trial of Low-Dose Aspirin in Women at High Risk of Preeclampsia

Risk Factors	Number	Preeclampsia (%)	
		Aspirin*	Placebo*
Normotensive, no proteinuria	1613	14.5	17.7
Proteinuria plus hypertension	119	31.7	22.0
Proteinuria only	48	25.0	33.3
Hypertension only	723	24.8	25.0
Insulin-dependent diabetes	462	18.3	21.6
Chronic hypertension	763	26.0	24.6
Multifetal gestation	678	11.5	15.9
Previous preeclampsia	600	16.7	19.0

NICHD = National Institute of Child Health and Human Development. *No statistical difference for any of the aspirin versus placebo groups.
Data from Caritis et al.[56]

prostaglandin production of those that are vasoconstrictors – thromboxane A – and others that are vasodilators – prostacyclin. Indeed, evidence suggests that thromboxane A production is markedly increased while that of prostacyclin is reduced even in women prior to the onset of clinical preeclampsia. Putative sequelae are placental infarction and spiral artery thrombosis, notably in pregnancies with severe fetal-growth restriction, fetal death, or both.[54]

Low-Dose Aspirin

The antithrombotic agents that have been most widely studied are low-dose aspirin regimens. In doses of 50–150 mg daily during pregnancy, aspirin effectively inhibits platelet thromboxane A biosynthesis with minimal effects on vascular prostacyclin reduction.[55] The first randomized placebo-controlled double-blind study to assess preeclampsia prevention with low-dose aspirin was reported in 1986 by Wallenburg et al.[55] This study included 46 nulliparas who had a positive angiotensin II sensitivity test at 28 weeks. In 21 women given low-dose aspirin, there was a significant reduction in the incidence of preeclampsia. These encouraging results led to numerous trials worldwide.

A National Institute of Child Health and Development (NICHD) randomized trial reported by Caritis et al.[56] evaluated the effects of low-dose aspirin in high-risk women with previous preeclampsia, multifetal gestation, chronic hypertension, or insulin-dependent diabetes. As shown in Table 12.6, in both groups the incidence of preeclampsia was particularly increased in those who had hypertension and proteinuria prior to 20 weeks. As also shown, however, low-dose aspirin failed to reduce the incidence of preeclampsia in any of these groups of extremely high-risk women.

The Paris Collaborative Group more recently performed a meta-analysis of the efficacy and safety of antiplatelet agents – predominantly aspirin – in preventing preeclampsia.[57] They included 31 randomized trials involving 32,217 women. For women assigned to antiplatelet agents, the relative risk of developing preeclampsia was slightly decreased – RR 0.90 (95% CI 0.84–0.96). For 6107 women with a previous history of hypertension or preeclampsia who were assigned to antiplatelet agents, the relative risk of developing preeclampsia was also lowered slightly – RR 0.86 (95% CI 0.77–0.97). There were marginal benefits of aspirin in reducing delivery before 34 weeks (RR 0.90; 95% CI 0.83–0.98) and reducing serious adverse outcomes (RR 0.90; 95% CI 0.85–0.96). Once again, the number needed to treat to obtain these results was very large. These modest benefits accrued without major adverse effects of low-dose aspirin.

Roberge et al.[58] performed an updated systematic review and meta-analysis and concluded that low-dose aspirin initiated at or before 16 weeks reduced the risk of severe preeclampsia, but not mild preeclampsia. But a small study by Villa et al.[59] did not show a significant effect on preeclampsia. When these latter data were included in the Roberge meta-analysis, however, they supported the concept of low-dose aspirin reducing the risk of preeclampsia in women at high risk.

One problem with all the approaches described above, and apparently neglected or not discussed in the recommendations of the various national and international working groups summarized below, is the problem inherent in subgroup analysis, especially when such analysis was not included in the original study design. These flaws have been discussed in greater detail by Klebenoff.[60]

Current recommendations restrict the use of low-dose aspirin prophylaxis for women at high risk of developing preeclampsia. Largely based on these marginal benefits with thus far no sustained deleterious effects, with these caveats, both NICE guidelines[15] and the American College of Obstetricians and Gynecologists[37] recommend the use of 60–80 mg aspirin beginning at the end of the first trimester for women at high risk. The latter includes those with hypertensive disease during a previous pregnancy, chronic kidney disease, autoimmune disease such as systemic lupus erythematosis or antiphospholipid syndrome, type 1 or type 2 diabetes, or chronic hypertension. Factors identifying moderate risk are: first pregnancy, age 40 years or older, pregnancy interval of more than 10 years, body mass index (BMI) of 35 kg/m^2 or more at first visit, a family history of preeclampsia, or multiple pregnancy.[61]

Low-Dose Aspirin Plus Heparin

Because of the high prevalence of placental thrombotic lesions associated with severe preeclampsia, combining heparin with aspirin prophylaxis seemed a logical strategy to prevent recurrent disease. There have been several observational trials to evaluate such treatment. Kupferminc et al.[62] studied 33 women with thrombophilia and a history of a previous pregnancy complicated by severe preeclampsia, placental abruption, fetal-growth restriction, or stillbirth. Those treated with low-molecular-weight heparin and low-dose aspirin had an 8% recurrence rate for preeclampsia and a 6% rate for recurrent fetal-growth restriction. In another observational study, Sergio et al.[63] assessed prophylaxis with low-molecular-weight heparin plus low-dose aspirin on pregnancy outcome in women with a history of severe preterm preeclampsia and low-birthweight infants. They compared pregnancy outcomes in 23 women given low-dose aspirin alone with those in 31 women given low-molecular-weight heparin plus low-dose aspirin. The incidence of recurrent preeclampsia in women given low-dose aspirin was 30% compared with only 3% in those given the combination ($p < 0.01$). Furthermore, the group given low-dose aspirin plus heparin had a greater mean gestational age at delivery ($p < 0.05$), higher birthweight ($p < 0.01$), and a higher birth percentile ($p < 0.01$) compared with the group given low-dose aspirin alone.

Despite these findings providing further evidence of benefit, current guidelines advise against the use of low-molecular heparin to prevent hypertensive disorders in pregnancy.[15,37]

Statins

At least one of these lipid-lowering agents – pravastatin – has been used in an attempt to decrease the placental production of antiangiogenic proteins in order to treat early-onset preeclampsia (StAmP trial: http://www.birmingham.ac.uk/research/activity/mds/trials/bctu/trials/womens/StAmP/index.aspx). The rationale is that statins stimulate hemoxygenase-1 expression and inhibit sFlt-1 release *in vivo* and *in vitro*, which may ameliorate the progression of early-onset preeclampsia.[64] If such studies are successful they will be the focus of future prevention trials. Importantly, statins as a class are traditionally considered to be contraindicated in pregnancy – class X FDA classification scheme – because of the theoretical risk of teratogenesis and of altered lipid accumulation during fetal brain development.

PREVENTION OF ECLAMPSIA

While the precise pathogenesis of convulsions with the preeclampsia syndrome remains enigmatic, the two prevailing hypotheses are vasospasm with ischemia and hypertensive encephalopathy.[65] These are discussed in detail in Chapter 13. Primary eclampsia prevention can be achieved by preventing worsening of preeclampsia; secondary prevention can be accomplished by using agents to prevent convulsions in women with established preeclampsia; and tertiary prevention targets recurrent convulsions in women with eclampsia.

The decision to use secondary prophylaxis is based on the likelihood of eclamptic convulsions in women with preeclampsia. This in turn is dependent on a number of factors, some of which are discussed in further detail in Chapter 20. The incidence of eclampsia ranges from 0.13% for women with gestational hypertension only to 1% for women with mild preeclampsia (Table 12.7).[66–68] The heterogeneity of women in these reports accounts for this range. Because seizure prophylaxis is recommended only during labor and for 12 to 24 hours postpartum, it can prevent only about 40% of eclampsia cases because 40% occur before labor, and 20% develop >24 hours postpartum.

Magnesium Sulfate for Mild Preeclampsia

The primary objective of magnesium sulfate prophylaxis in women with preeclampsia is to prevent or reduce the rate of eclampsia and associated complications. Other benefits

TABLE 12.7 Observational Studies of the Incidence of Eclampsia in Hypertensive Women Not Given Seizure Prophylaxis

Investigators	Inclusion Classification	Patients	Eclampsia	
			No.	(%)
Walker[66]	Hypertensive disorders	3885	7	0.18
Burrows & Burrows[67]	Gestational hypertension	745	1	0.13
Alexander et al.[68]	Gestational hypertension[a]	2496	27	1.0

[a]Includes women with mild preeclampsia.

TABLE 12.8 Randomized Trials of Magnesium Sulfate Versus Phenytoin or Placebo to Prevent Eclampsia in Women with Hypertensive Disorders of Pregnancy

Study	Eclampsia/Total Patients	
	$MgSO_4$	Phenytoin
Appleton et al.[69]	0/24	0/23
Friedman et al.[70]	0/60	0/43
Atkinson et al.[71]	0/28	0/26
Lucas et al.[72]	0/1049	10/1089*
Total	0/1228	10/1249 (0.8%)*

*$p < 0.001$ for $MgSO_4$ versus phenytoin comparison.

include reduced maternal and perinatal mortality and morbidity. In addition, in women with mild preeclampsia, a secondary benefit might include a decreased rate of progression to severe preeclampsia.

There have been four randomized trials that compared magnesium sulfate with phenytoin for women with various pregnancy hypertensive disorders (Table 12.8).[69–72] Only one of these was sufficiently powered to evaluate seizure prophylaxis in these women and the other three evaluated mostly side effects of magnesium therapy. In the trial reported by Lucas et al.[72] over 2000 women with gestational hypertension – only a small percentage had severe preeclampsia – were randomized to either intramuscular magnesium sulfate or an intravenous/oral phenytoin regimen. As shown in Table 12.8, there were no seizures among 1049 women assigned to magnesium sulfate, but there were 10 cases of eclampsia – 1% – among 1089 women assigned to phenytoin ($p = 0.004$). Four of the 10 women with seizures had severe preeclampsia. These findings suggest that the rate of seizures in women with mild hypertension or mild preeclampsia receiving phenytoin is 0.6% or 6 per 1000 treated women. Importantly, the trial indicates that magnesium sulfate is superior to phenytoin for seizure prophylaxis in such women.

There are no randomized trials with sufficient power to allow comparison of magnesium sulfate with placebo to prevent seizures in women with mild preeclampsia. The only two double-blind placebo-controlled trials that we could locate did, however, permit comparison for the percentage of women who progressed to severe preeclampsia, as shown in Table 12.9.[73,74] The differences in these studies were not significant. In fact there were higher rates of postpartum hemorrhage and two instances of magnesium toxicity in one of the trials.[73]

In order to determine the effectiveness or safety of magnesium sulfate with certainty, there is a need for a multicenter placebo-controlled trial of magnesium sulfate prophylaxis for mild preeclampsia. Based on a rate of eclampsia of 0.5%, and assuming 50% reduction by magnesium sulfate, to a 0.25% rate, with an α of 0.05 and a β of 0.2, at least 9383 women would need to be enrolled in each group to document a significant reduction in eclampsia.

Magnesium Sulfate for Severe Preeclampsia

There is now general agreement that magnesium sulfate should be given to prevent convulsion in women with severe preeclampsia. Studies that support this conclusion include two large prospective observational studies from the same medical center in South Africa that describe the rate of eclampsia in women with severe preeclampsia who did not receive prophylactic magnesium sulfate. Odendaal and Hall[75] reported 1001 such women, of whom 510 received magnesium sulfate prophylaxis based upon the clinical impression of impending eclampsia, while 491 women did not. Five patients (0.5%) developed eclampsia: two (0.4%) in the magnesium group – both before delivery – and three (0.6%) in the no magnesium group – all postpartum. Interestingly, two of the three postpartum seizures occurred beyond 48 hours and thus would not have been eligible for their standard magnesium sulfate regimen. In a subsequent report, Hall and associates[76] enrolled 318 women with preeclampsia – mostly severe and remote from term and not in labor – who were managed expectantly with antihypertensive drugs and without magnesium sulfate. Of these, 26 subsequently received magnesium sulfate during labor or because of antepartum seizures ($n = 4$). Five (1.5%) developed eclampsia: two (0.7%) within 24 hours of hospitalization, two (0.7%) at 8 and 14 days after hospitalization, and one at 4 days postpartum. Therefore, only two would have been prevented by their standard magnesium sulfate prophylaxis regimen. In addition, none of the five women had serious morbidity because of seizures. The authors of these studies therefore questioned the need for magnesium sulfate prophylaxis in patients with severe preeclampsia.[75,76] We add the caveat that selection bias was obvious regarding the decision to use magnesium sulfate in these nonrandomized studies.

Four large randomized controlled trials, shown in Table 12.10, compared magnesium sulfate with another treatment regimen to prevent convulsions in women with severe preeclampsia (see also Chapter 20).[77–80] There was considerable heterogeneity, best exemplified by the Magpie Trial,[79] which included 10,110 women with preeclampsia conducted in 175 hospitals in 33 countries. Thus, clinical characteristics, obstetrical care, and availability of maternal and neonatal intensive care units differed widely. Many aspects of clinical characteristics or management were poorly defined or controlled. Some of these women received study medications antepartum during expectant management for 24 hours only with no magnesium sulfate with subsequent labor or postpartum, some were discharged, some received drug during labor and delivery, and some were not randomized until postpartum. Half of the women received antihypertensive agents before randomization and 75% received antihypertensive medications afterwards. Finally, 9% received anticonvulsants before randomization to magnesium sulfate, 6% received magnesium sulfate or other anticonvulsants after randomization, and 17 women had eclampsia before randomization. Nonetheless, the Magpie

TABLE 12.9 Placebo-Controlled Trials of Magnesium Sulfate in Mild Eclampsia

Study	Progression to Severe No. (%)		Relative Risk (95% CI)
	Magnesium Sulfate	Placebo	
Witlin et al.[73]	8/67 (12)	6/68[a] (9)	1.35 (0.5–3.7)
Livingston et al.[74]	14/109 (12.8)	19/113 (16.8)	0.76 (0.4–2.4)
Total	22/176 (12.5)	25/181 (13.8)	0.90 (0.52–1.54)

[a]Magnesium toxicity in 2/67.

TABLE 12.10 Randomized Controlled Trials of Magnesium Sulfate in Severe Preeclampsia

Study	Site	Control Treatment	Magnesium No. (%)	Control No. (%)	RR (95% CI)[a]
Moodley & Moodley[77]	Single	None	1/112 (0.9)	0/116 (0)	N/A
Coetzee et al.[78]	Single	Placebo	1/345 (0.3)	11/340 (3.2)*	0.09 (0.01–0.69)
Magpie Trial Group[79]	Multi-	Placebo	40/5055 (0.8)	96/5055 (1.9)*	0.42 (0.26–0.60)
Belfort et al.[80]	Multi-	Nimodipine	7/831 (0.8)	21/819 (2.6)**	0.33 (0.14–0.77)
Total			49/6343 (0.6)	128/6330 (2.0)	0.39 (0.28–0.55)

N/A = not applicable; (%) = percent preeclampsia.
[a]Relative risk, 95% confidence interval.
*$p = 0.05$.
**$p = 0.001$.

TABLE 12.11 Effect of Magnesium Sulfate on the Rate of Abruptio Placentae in Women with Severe Preeclampsia

Study	Magnesium Sulfate No. (%)	Control No. (%)	RR (95% CI)
Moodley & Moodley[77]	0/112	0/116	N/A
Magpie Trial Group[79]	62/4387 (1.4)	113/4331 (2.6)	0.54 (0.40–0.74)
Belfort et al.[80]	8/831 (1.0)	6/819 (0.7)	1.30 (0.46–3.77)
Total	70/5330 (1.3)	119/5266 (2.3)	0.67 (0.38–1.19)

N/A = not applicable; RR = relative risk; CI = confidence interval.

Trial showed a significant reduction in the rate of eclampsia in women assigned to magnesium sulfate. This benefit was primarily found in women from developing countries and no significant eclampsia reduction was apparent in women from Western countries – RR = 0.67; 95% CI 0.19– 2.37.[79] That said, the number of women in these latter groups was small.

The trial by Belfort et al.[80] compared magnesium sulfate with nimodipine, a calcium-channel blocker with cerebrovascular vasodilating characteristics. These investigators enrolled women at 14 sites in eight countries who were given the study drugs during labor and for 24 hours postpartum. All had well-defined clinical characteristics prior to randomization.[80] There was a significant reduction in the rate of eclampsia in the magnesium sulfate group (Table 12.10). Most of this difference was due to the lower eclampsia rate postpartum among women assigned to magnesium sulfate. The overall results of these two studies, along with the other two, are listed in Table 12.10. In aggregate, they demonstrate that magnesium sulfate prophylaxis in severe preeclampsia is associated with a significantly lower rate of eclampsia compared with either nimodipine or no treatment.

Effects of Magnesium Sulfate on Maternal Mortality and Morbidity

The trials cited above provided information regarding maternal mortality and some morbidities. There were no maternal deaths reported in two of the trials,[77–80] and in another there was one death from pelvic sepsis among 340 women assigned to placebo.[78] In the Magpie Trial[79] there were 11 maternal deaths in the magnesium group and 20 deaths in the placebo group – RR 0.55, 95% CI 0.26–1.14. In the placebo group, three deaths were due to renal failure, three from pulmonary embolism, and two women died from sepsis. It seems likely that magnesium sulfate would not have prevented most of these. The rate of cesarean delivery was similar in the two groups: 50% magnesium compared with 48% placebo. While the numbers are too small to draw certain conclusions, the studies did not demonstrate a decided benefit of magnesium sulfate on maternal mortality.

Shown in Table 12.11 are data from randomized trials regarding the effects of magnesium sulfate on the rate of placental abruption. In aggregate, magnesium sulfate was not associated with a significant reduction – RR 0.67, 95% CI 0.38–1.19.

Summarized in Table 12.12 are randomized controlled trials that evaluated the effects of magnesium sulfate on the frequency of respiratory depression. Individually, the two smaller studies reported no significant differences. The two larger studies, and all studies in aggregate, however, showed that magnesium sulfate for severe preeclampsia is associated with a significant increase in the rate of respiratory depression – RR 2.06, 95% CI 1.33–3.18. Importantly, the frequency of ventilatory support was similar in the magnesium and control groups—59 versus 69%. Finally, the frequency of cerebrovascular accidents was similar in the magnesium and control groups – three versus six women.[77–80]

Effects of Magnesium Sulfate on Perinatal Mortality and Morbidity

The three trials that provided adequate information regarding perinatal mortality are shown in Table 12.13. Individually and in aggregate, they showed that magnesium sulfate for severe preeclampsia does not substantively alter the rate of perinatal deaths – RR 1.03, 95% CI 0.87–1.22. Two randomized trials provided information regarding neonatal morbidities.[79,80] Magnesium sulfate for severe preeclampsia did not significantly change the rates of Apgar scores at 5 minutes, a cord artery pH < 7.0, respiratory distress, need for intubation, hypotonia, or days in intensive care.

TREATMENT FOR ECLAMPSIA (SEE CHAPTER 20)

It is generally agreed that anticonvulsant treatment is necessary for the prevention of recurrent seizures in women with eclampsia. In the United States, magnesium sulfate has been used for this purpose for over 60 years. Until relatively recently, this was criticized as empirical and dogmatic because it had never been tested in randomized trials. The results of four large observational studies that address this are shown in Table 12.14.[81–84] The overall rate of recurrent seizures among these studies is 11%. This in part stimulated the design of randomized trials to compare magnesium sulfate with traditional anticonvulsants.

In apparently only one multicenter trial was there an adequate sample size to compute a significant difference between magnesium sulfate and the comparative drugs. The Collaborative Eclampsia Trial[85] was conducted in several centers in South Africa and South America. The trial included 1680 eclamptic women who were randomized to magnesium sulfate, phenytoin, or diazepam in two different randomization schemes. As shown in Table 12.15, magnesium sulfate was superior to both phenytoin and diazepam to prevent recurrent seizures in eclamptic women. In addition, the data by Bhalla et al.[88] indicate that magnesium sulfate is superior to lytic cocktail in this regard.

The protective effects of magnesium sulfate to decrease maternal mortality with eclampsia are shown in Table 12.16. Overall, magnesium sulfate therapy was associated with significantly lower maternal mortality – 3.0% versus 4.8% compared with other anticonvulsants ($p < 0.05$).

TABLE 12.12 Effects of Magnesium Sulfate on the Incidence of Respiratory Depression in Women with Severe Preeclampsia

Study	Magnesium Sulfate No. (%)	Control No. (%)	RR (95% CI)
Moodley & Moodley[77]	0/112	0/116	N/A
Coetzee et al.[78]	1/345 (0.3)	0/340	N/A
Magpie Trial Group[79]	51/4999 (1.0)*	26/4993 (0.5)*	1.96 (1.22–3.14)
Belfort et al.[80]	11/831 (1.3)	3/819 (0.4)	3.61 (1.01–12.90)
Total	63/6287 (1.0)	29/6268 (0.46)	2.06 (1.33–3.18)

N/A = not applicable; RR = relative risk; CI = confidence interval.
*30/51 in the magnesium group and 18/26 in the placebo group required ventilatory support.

TABLE 12.13 Effects of Magnesium Sulfate on the Perinatal Mortality Rate in Women with Severe Preeclampsia

Study	Magnesium Sulfate No. (%)	Control No. (%)	RR (95% CI)
Moodley & Moodley[77]	20/117 (17)	25/118 (21)	0.81 (0.47–1.37)
Coetzee et al.[78]	38/348 (11)	38/354 (8.0)	1.38 (0.87–2.20)
Magpie Trial Group[79]	576/4538 (13)	558/4486 (12)	1.02 (0.92–1.14)
Total	634/5003 (13)	601/4958 (13)	1.03 (0.87–1.22)

RR = Relative risk; CI = confidence interval.(%) = percent preeclampsia.

Side Effects and Toxicity

Magnesium sulfate has a high rate of minor side effects such as an intense warm feeling, flushing, nausea or vomiting, muscle weakness, dizziness, and irritation at the injection site. The reported rates of these effects in randomized trials ranged from 15 to 67%.[73,79,80,84] Side effects were the most common reason for discontinuation of treatment in the Magpie Trial.[79]

The most concerning major side effect of magnesium sulfate is respiratory depression.[79,80] While postpartum hemorrhage has been reported,[73] its incidence was not increased in the Magpie Trial.[79] Life-threatening magnesium toxicity is rare with proper dosing, administration, and monitoring during therapy. That said, maternal deaths or "near misses" from magnesium overdose have been reported from the United States and require vigilance to prevent.[83,90–92]

TABLE 12.14 Incidence of Recurrent Convulsions in Eclamptic Women Given Magnesium Sulfate

Study	Women with Eclampsia	Recurrent Convulsion(s) No. (%)
Pritchard et al.[81]	85	3 (3.5)
Gedekoh et al.[82]	52	1 (1.9)
Pritchard et al.[83]	83	10 (12)
Sibai & Ramanathan[84]	315	41 (13)
Total	535	55 (10.3)

Initiation, Dose, Duration, and Route of Administration

Randomized trials vary widely regarding the optimal time to initiate magnesium sulfate, loading and maintenance dosing, route of administration, and duration of therapy. In all trials except in some women enrolled in the Magpie Trial,[79] magnesium sulfate was started once the decision for delivery was made. In some trials, magnesium sulfate was given during labor and delivery, and for up to 24 hours postpartum.[73,74,77,80] In contrast, in two of the trials,[78,79] magnesium sulfate was only given for a maximum of 24 hours. In the Magpie Trial,[79] some women did not receive the drug during labor, delivery, or postpartum.[79] Among the trials utilizing the intravenous regimen, the loading dose ranged from 4 to 6 g, and the maintenance dose ranged from 1 to 2 g per hour. In most trials[73,74,78,80] magnesium sulfate was given by continuous infusion. In the trial by Moodley and Moodley,[77] the loading dose was given intravenously and the maintenance dose intramuscularly. In the Magpie Trial,[79] several of these combinations were used and side effects with the intramuscular regimen were more

TABLE 12.15 Randomized Trials Comparing Magnesium Sulfate with Other Anticonvulsants in Women with Eclampsia

Study	Recurrent Seizures			Relative Risk (95% CI)
	Antihypertensive Drugs	$MgSO_4$(%)	Other (%)	
Dommisse[86]	Dihydralazine	0/11 (0)	4/11 (37) Phenytoin	
Crowther[87]	Dihydralazine	5/24 (21)	7/27 (26) Diazepam	0.80 (0.29–2.2)
Bhalla et al.[88]	Nifedipine	1/45 (2.2)	11/45 (24) Lytic cocktail	0.09 (0.1–0.68)
Friedman et al.[89]	Nifedipine, labetalol	0.11 (0)	2/13 (15) Phenytoin	
Collaborative Eclampsia	Not reported	60/453 (13.2)	126/452 (28) Diazepam	0.48 (0.36–0.63)
Trial[85]		22/388 (5.7)	66/387 (17)	0.33 (0.21–0.53)
Total		88/922 (9.4)	216/935 (23)	0.41 (0.32–0.51)

TABLE 12.16 Maternal Deaths in Trials Comparing Magnesium Sulfate with Other Anticonvulsants in Women with Eclampsia

Study	Comparison Group	Maternal Deaths		Relative Risk (95% CI)
		$MgSO_4$(%)	Other (%)	
Dommisse[86]	Phenytoin	0/11	0/11	
Crowther[87]	Diazepam	1/24 (4.2)	0/27	
Bhalla et al.[88]	Lytic cocktail	0/45	2/45 (4.4)	
Friedman et al.[89]	Phenytoin	0/11	0/13	
Collaborative Eclampsia	Phenytoin	10/388 (2.5)	20/387 (5.2)	0.50 (0.24–1.00)
Trial[85]	Diazepam	17/453 (3.8)	23/452 (5.1)	0.74 (0.40–1.36)
Total		28/932 (3.0)	45/935 (4.8)	0.62 (0.39–0.99)

common – 28% versus 5%, respectively – and as a result more women in this group stopped the medication early (28% versus 5%). This variation in the route of administration and the total amount of magnesium sulfate used in the various trials possibly explains the differences in seizure rates and side effects among those assigned to magnesium sulfate.

Because of these protocol variations, investigators from the University of Mississippi Medical Center have suggested an individualized postpartum magnesium sulfate protocol based on clinical parameters in women with preeclampsia.[93,94] Their first study[93] included 103 women with mild and 55 with severe preeclampsia. Postpartum women with mild disease received a minimum of 6 hours of intravenous magnesium sulfate, and those with severe preeclampsia received a minimum of 12 hours infusion. This protocol was based on blood pressure levels, need for antihypertensive therapy, onset of diuresis, and presence of symptoms. Women with mild preeclampsia required an average duration of magnesium sulfate therapy of 9.5 ± 4.2 hours, whereas those with severe disease required an average infusion of 16 ± 5.9 hours. Those with HELLP syndrome required an average duration of therapy of 20 ± 6.7 hours. Although there were no cases of eclampsia, the sample size is inadequate to evaluate efficacy for convulsions.

In their second study, Isler et al.[94] evaluated an individualized protocol for postpartum magnesium sulfate therapy in 284 women with mild preeclampsia and 105 with severe preeclampsia. Like the first study, this protocol also was based on blood pressure levels, onset of use of antihypertensive drugs, diuresis, and symptoms. Magnesium sulfate was given for 2 to 72 hours in those with mild disease and up to 77 hours postpartum in those with severe disease. Magnesium sulfate therapy which had been discontinued was reinstituted based on clinical parameters in 6.3% of women with mild or severe disease and in 18% of those with superimposed preeclampsia. Again, there were no cases of eclampsia, but the number of women included in this study – most had mild disease – is inadequate to draw any conclusions regarding efficacy. Because such a protocol requires intensive postpartum monitoring, it is impractical compared with an empirical protocol and is not used in the United States.

Fontenot et al.[95] reported a randomized trial of magnesium sulfate given postpartum to 98 women with severe preeclampsia. One group of 50 were given 24 hours of therapy, whereas the other group of 48 were given therapy until the onset of diuresis. Women in the latter group had a shorter duration of therapy compared with those treated empirically for 24 hours – 507 ± 480 versus 1442 ± 158 minutes, respectively. There were no cases of eclampsia, and the postpartum hospital stays were not significantly different – 3.1 ± 1.1 versus 3.5 ± 1.1 days, respectively.

Ehrenberg and Mercer[96] performed a randomized trial comparing a 12-hour to a 24-hour course of postpartum magnesium sulfate for women with mild preeclampsia. In the 107 women assigned to the 12-hour regimen, magnesium sulfate therapy was extended in seven for progression to severe disease compared with only one in the 24-hour group ($p = 0.07$). There were no seizures, but women with chronic hypertension and insulin-requiring diabetes were at risk of progression to severe disease. Again, the small number of subjects in this study hampers the generaliziblity of these regimens.

Dayicioglu et al.[97] evaluated serum magnesium levels and efficacy of a standardized magnesium sulfate dose of 4.5 g loading over 15 minutes followed by 1.8 g/h in 183 women with preeclampsia. Serum magnesium levels were obtained within the first 2 hours, and every 6 hours in the subsequent 42 hours. In addition, serum creatinine levels and creatinine clearances were also studied to correlate with magnesium levels. They reported that most magnesium serum levels were <4.8 mg/dL in women whose BMI was ≥36. Nine women developed postpartum convulsions while still receiving magnesium sulfate, and four of these were women with a low BMI. They found no association between eclampsia treatment failures and BMI or with serum magnesium levels. They also found no association between serum magnesium levels and serum creatinine or creatinine clearance.

The effects of obesity on magnesium levels were further outlined in the study by Tudela et al.,[98] who reported that 40% of women whose BMI was above 30 kg/m^2 required a maintenance dose of 3 g/h of magnesium sulfate to achieve "therapeutic" levels.

Thus, a review of randomized trials indicates that magnesium sulfate is the best available agent to use as prophylaxis in women with severe preeclampsia and for treatment of eclamptic convulsions. A Cochrane review in 2010 concluded that magnesium sulfate therapy more than halved the the risk of an eclamptic convulsion, and appeared to reduce maternal death.[99] There is limited information regarding the efficacy of magnesium sulfate for prophylaxis in women with mild hypertension or preeclampsia, and there is a need for blinded placebo-controlled studies to address this. Questions remain regarding the optimal time to initiate magnesium sulfate as well as the dose and the duration of administration in the postpartum period. In sum, differences in approaches are used by practitioners regarding magnesium sulfate therapy, and this topic will be revisited in Chapter 20. At this time, magnesium prophylaxis for severe preeclampsia and treatment for eclampsia are recommended both by NICE guidelines[15] as well as by the American College of Obstetricians and Gynecologists.[37]

Prevention of Long-Term Maternal Health Risks Following Preeclampsia

Preeclampsia identifies a group of women with increased risk of cardiovascular disease later in life. This is discussed in detail in Chapter 3. At present there is no agreement on how best to follow up this group of women after preeclampsia in order to prevent or reduce the severity of long-term health complications. In addition, whether prevention strategies for preeclampsia also reduce the risk of future cardiovascular disease is unknown. The general recommendation of combined physical activity and weight control is applicable to the group of women having had preeclampsia, but its efficacy is not well documented. A major challenge again is identifying the optimal target group among the heterogeneous preeclampsia group for follow-up and potential cardiovascular disease prevention studies.[100]

References

1. Sibai BM. Diagnosis and management of gestational hypertension and preeclampsia. *Obstet Gynecol.* 2003;102:181–192.
2. Sibai BM, Dekker G, Kupferminc M. Preeclampsia. *Lancet.* 2005;365:785–799.
3. Conde-Agudelo A, Villar J, Lindheimer M. World Health Organization systemic review of screening tests for preeclampsia. *Obstet Gynecol.* 2004;104:1367–1391.
4. Widmer M, Villar J, Benigin A. Mapping the theory of preeclampsia and the role of angiogenic factors: a systemic review. *Obstet Gynecol.* 2007;109:168–180.
5. Chesley LC. *Hypertensive Disorders in Pregnancy.* New York: Appleton-Century-Croft; 1978.
6. Steegers EAP, Eskes TKAB, Jongsma HW, Hein PR. Dietary sodium restriction during pregnancy: a historical review. *Eur J Obstet Gynecol Reprod Biol.* 1991;40:83–90.
7. Knuist M, Bonsel GJ, Zondervan HA, Treffers PE. Low sodium diet and pregnancy-induced hypertension: a multi-centre randomized trial. *Br J Obstet Gynaecol.* 1998;105:430–434.
8. Adair CD, Sanchez-Ramos L, Briones DL, Ogburn P. The effect of high dietary N-3 fatty acid supplementation on angiotensin II pressor response in human pregnancy. *Am J Obstet Gynecol.* 1996;74:688–691.
9. Bulstra-Ramakers MTEW, Huisjes HJ, Visser GHA. The effects of 3 g eicosapentaenoic acid daily on recurrence of intrauterine growth retardation and pregnancy induced hypertension. *Br J Obstet Gynaecol.* 1994;102:123–126.
10. Onwude JL, Lilford RJ, Hjartardottier H. A randomized double blind placebo controlled trial of fish oil in high risk pregnancy. *Br J Obstet Gynaecol.* 1995;109:95–100.
11. Salvig JD, Olsen SF, Secher NJ. Effects of fish oil supplementation in late pregnancy on blood pressure: a randomized controlled trial. *Br J Obstet Gynaecol.* 1996;103:529–533.
12. Olsen SF, Secher NJ, Tabor A. Randomized clinical trials of fish oil supplementation in high risk pregnancies. *BJOG.* 2000;107:382–395.
13. Olafsdottir AS, Skuladottir GV, Thorsdottir I. Relationship between high consumption of marine fatty acids in early pregnancy and hypertensive disorders in pregnancy. *BJOG.* 2006;113:301–309.
14. Makrides M, Duley L, Olsen SF. Marine oil, and other prostaglandin precursor supplementation for pregnancy uncomplicated by pre-eclampsia or intrauterine growth restriction. *Cochrane Database Syst Rev.* 2006;3:CD003402.
15. National Institute for Health and Clinical Excellence. Hypertension in pregnancy: the management of hypertensive disorders during pregnancy (clinical guideline 107). 2010. 2013. Online source.
16. Belizan JM, Villar J, Repke J. The relationship between calcium intake and pregnancy induced hypertension: up-to-date evidence. *Am J Obstet Gynecol.* 1988;158:898–902.
17. Lopez-Jaramillo P, Narvaez M, Weigel RM, Yepez R. Calcium supplementation reduces the risk of pregnancy-induced hypertension in an Andes population. *Br J Obstet Gynaecol.* 1989;96:648–655.
18. Lopez-Jaramillo P, Narvaez M, Felix C, Lopez A. Dietary calcium supplementation and prevention of pregnancy hypertension (letter). *Lancet.* 1990;335:293.
19. Villar J, Repke J. Calcium supplementation during pregnancy may reduce preterm delivery in high-risk populations. *Am J Obstet Gynecol.* 1990;163:1124–1131.
20. Belizan JM, Villar J, Gonzalez L. Calcium supplementation to prevent hypertensive disorders of pregnancy. *New Engl J Med.* 1991;325:1399–1405.
21. Sanchez-Ramos L, Briones DK, Kaunitz AM. Prevention of pregnancy-induced hypertension by calcium supplementation in angiotensin II-sensitive patients. *Obstet Gynecol.* 1994;84:349–353.
22. Purwar M, Kulkarni H, Motghare V, Dhole S. Calcium supplementation and prevention of pregnancy induced hypertension. *J Obstet Gynaecol Res.* 1996;22:425–430.
23. Levine RJ, Hauth JC, Curet LB. Trial of calcium to prevent preeclampsia. *New Engl J Med.* 1997;337:69–76.
24. Lopez-Jaramillo P, Delgado F, Jacome P. Calcium supplementation and the risk of preeclampsia in Ecuadorian pregnant teenagers. *Obstet Gynecol.* 1997;90:162–167.
25. Crowther CA, Hiller JE, Pridmore B. Calcium supplementation in nulliparous women for the prevention of pregnancy-induced hypertension, preeclampsia and preterm birth: an Australian randomized trial. *Aust NZ J Obstet Gynaecol.* 1999;39:12–18.
26. Villar J, Abdel-Aleena H, Merialdi M. World Health Organization randomized trial of calcium supplementation among low calcium intake pregnant women. *Am J Obstet Gynecol.* 2006;194:639–649.
27. Herrera JA, Arevalo-Herrera M, Herrera S. Prevention of preeclampsia by linoleic acid and calcium supplementation: a randomized controlled trial. *Obstet Gynecol.* 1998;91:585–590.
28. Hofmeyr GJ, Atallah AN, Duley L. Calcium supplementation during pregnancy for preventing hypertensive disorders and related problems. *Cochrane Database Syst Rev.* 2006
29. Trumbo PR, Ellwood KC. Supplemental calcium and risk reduction of hypertension, pregnancy-induced hypertension, and preeclampsia: an evidence-based review by the US Food and Drug Administration. *Nutr Rev.* 2007;65:78–87.
30. Hofmeyr GJ, Lawrie TA, Atallah AN, Duley L. Calcium supplementation during pregnancy for preventing hypertensive disorders and related problems. *Cochrane Database Syst Rev.* 2010(8):CD001059.
31. Imdad A, Jabeen A, Bhutta ZA. Role of calcium supplementation during pregnancy in reducing risk of developing

gestational hypertensive disorders: a meta-analysis of studies from developing countries. *BMC Public Health.* 2011;11(suppl 3):S18.
32. Patrelli TS, Dall'asta A, Gizzo S, et al. Calcium supplementation and prevention of preeclampsia: a meta-analysis. *J Matern Fetal Neonatal Med.* 2012;25(12):2570–2574.
33. Visintin C, Mugglestone MA, Almerie MQ, Nherera LM, James D, Walkinshaw S. Management of hypertensive disorders during pregnancy: summary of NICE guidance. *BMJ.* 2010;341:c2207.
34. Meher S, Duley L. Rest during pregnancy for preventing pre-eclampsia and its complications in women with normal blood pressure. *Cochrane Database Syst Rev.* 2006(2):CD005939.
35. Meher S, Duley L. Exercise or other physical activity for preventing pre-eclampsia and its complications. *Cochrane Database Syst Rev.* 2006(2):CD005942.
36. Kasawara KT, do Nascimento SL, Costa ML, Surita FG, Silva JL. Exercise and physical activity in the prevention of pre-eclampsia: systematic review. *Acta Obstet Gynecol Scand.* 2012;91(10):1147–1157.
37. American College of Obstetricians and Gynecologists Hypertension in pregnancy. Report of the American College of Obstetricians and Gynecologists' Task Force on Hypertension in Pregnancy. *Obstet Gynecol.* 2013;122:1122–1131.
38. Fimerty FA, Bucholz JH, Tuckman J. Evaluation of chlorothiazide (Diuril) in the toxemias of pregnancy. *JAMA.* 1958;166:141–144.
39. Flowers CE, Grizzle JE, Easterling WE, Bonner OB. Chlorothiazide as a prophylaxis against toxemia of pregnancy. A double blind study. *Am J Obstet Gynecol.* 1962;84:919–927.
40. Churchill D, Beever GD, Meher S, Rhodes C. Diuretics for preventing preeclampsia. *Cochrane Database Syst Rev.* 2007:24.
41. Sibai BM. Chronic hypertension in pregnancy. *Obstet Gynecol.* 2002;100:369–377.
42. Abalos E, Duley L, Steyn DW, Henderson-Smart DJ. Antihypertensive drug therapy for mild to moderate hypertension during pregnancy. *Cochrane Database Syst Rev.* 2007:CD002252.
43. Powrie RO. A 30-year-old woman with chronic hypertension trying to conceive. *JAMA.* 2007;298:1548–1559.
44. Moretti M, Phillips M, Abouzeid A. Increased breath markers of oxidative stress in normal pregnancy and preeclampsia. *Am J Obstet Gynecol.* 2004;190:1184–1190.
45. Raijmakers MT, Dechend R, Poston L. Oxidative stress and preeclampsia: rationale for antioxidant clinical trials. *Hypertension.* 2004;44:374–380.
46. Chappell LC, Seed PT, Briley AL. Effect of antioxidants on the occurrence of preeclampsia in women at increased risk: a randomized trial. *Lancet.* 1999;354:810.
47. Beazley D, Ahokas R, Livingston J. Vitamin C and E supplementation in women at high risk for preeclampsia: a double-blind, placebo-controlled trial. *Am J Obstet Gynecol.* 2005;192:520–521.
48. Poston L, Briley AL, Seed PT. Vitamins in preeclampsia (VIP) trial consortium, Vitamin C and Vitamin E in pregnant women at risk for preeclampsia: randomized placebo-controlled trial. *Lancet.* 2006;367:1145–1154.
49. Rumbold AR, Crowther CA, Haslam RR. Supplementation with vitamins C and E and the risk of preeclampsia and perinatal complications. *New Engl J Med.* 2006;354:1796–1806.
50. Spinnato JA, Freire S, de Silva JLP. Antioxidant therapy to prevent preeclampsia. A randomised controlled trial. *Obstet Gynecol.* 2007;110:1311–1318.
51. Villar J, Purwar M, Merialdi M, Zavaleta N, Tien NN, Anthony J. World Health Organisation multicentre randomised trial of supplementation with vitamins C and E among pregnant women at high risk for pre-eclampsia in populations of low nutritional status from developing countries. *BJOG.* 2009;116(6):780–8.
52. Basaran A, Basaran M, Topatan B. Combined vitamin C and E supplementation for the prevention of preeclampsia: a systematic review and meta-analysis. *Obstet Gynecol Surv.* 2010;65(10):653–667.
53. Conde-Agudelo A, Romero R, Kusanovic JP, Hassan SS. Supplementation with vitamins C and E during pregnancy for the prevention of preeclampsia and other adverse maternal and perinatal outcomes: a systematic review and meta-analysis. *Am J Obstet Gynecol.* 2011;204(6):503–512.
54. Moldenhauer JS, Stanek J, Warshak C. The frequency and severity of placental findings in women with preeclampsia are gestational age dependent. *Am J Obstet Gynecol.* 2003;189:1173–1177.
55. Wallenburg HCS, Dekker A, Makovitz JW, Rotmans P. Low dose aspirin prevents pregnancy-induced hypertension and preeclampsia in angiotensin-sensitive primigravidae. *Lancet.* 1986;1:1–3.
56. Caritis SN, Sibai BM, Hauth J. Low-dose aspirin to prevent preeclampsia in women at high risk. *New Engl J Med.* 1998;338:701–705.
57. Askie LM, Henderson-Smart DJ, Stewart LA. Antiplatelet agents for the prevention of preeclampsia: a meta-analysis of individual data. *Lancet.* 2007;369:179–198.
58. Roberge S, Giguere Y, Villa P, et al. Early administration of low-dose aspirin for the prevention of severe and mild preeclampsia: a systematic review and meta-analysis. *Am J Perinatol.* 2012;29(7):551–556.
59. Villa PM, Kajantie E, Raikkonen K, et al. Aspirin in the prevention of pre-eclampsia in high-risk women: a randomised placebo-controlled PREDO Trial and a meta-analysis of randomised trials. *BJOG.* 2013;120(1):64–74.
60. Klebanoff MA. Subgroup analysis in obstetrics clinical trials. *Am J Obstet Gynecol.* 2007;197(2):119–122.
61. Coomarasamy A, Honest H, Papaioannou S. Aspirin for prevention of preeclampsia in women with historical risk factors: a systemic review. *Obstet Gynecol.* 2003;101:1319–1332.
62. Kupferminc MJ, Fait G, Many A. Low molecular weight heparin for the prevention of obstetric complications in women with thrombophilia. *Hypertens Pregnancy.* 2001;20:35–44.
63. Sergio F, Clara DM, Galbriella F. Prophylaxis of recurrent preeclampsia : low molecular weight heparin plus low-dose aspirin versus low-dose aspirin alone. *Hypertens Pregnancy.* 2006;25:115–127.
64. Ahmed A. New insights into the etiology of preeclampsia: identification of key elusive factors for the vascular complications. *Thromb Res.* 2011;127(suppl 3):S72–S75.
65. Sibai BM. Diagnosis, prevention, and management of eclampsia. *Obstet Gynecol.* 2005;105:402–410.

66. Walker JJ. Hypertensive drugs in pregnancy. *Hypertens Pregnancy*. 1991;18:845–872.
67. Burrows RF, Burrows EA. The feasibility of a control population for a randomized controlled trial of seizure prophylaxis in the hypertensive disorders of pregnancy. *Am J Obstet Gynecol*. 1995;173:929–935.
68. Alexander JM, McIntire DD, Leveno KJ, Cunningham FG. Selective magnesium sulfate prophylaxis for the prevention of eclampsia in women with gestational hypertension. *Obstet Gynecol*. 2006;108:826–832.
69. Appleton MP, Kuehl TJ, Raebel MA. Magnesium sulfate versus phenytoin for seizure prophylaxis in pregnancy-induced hypertension. *Am J Obstet Gynecol*. 1991;165:907–913.
70. Friedman SA, Lim KH, Baker CA, Repke JT. Phenytoin versus magnesium sulfate in preeclampsia: a pilot study. *Am J Perinatol*. 1993;10:233–238.
71. Atkinson MW, Guinn D, Owen J, Hauth JC. Does magnesium sulfate affect the length of labor induction in women with pregnancy-associated hypertension? *Am J Obstet Gynecol*. 1995;173:1219–1222.
72. Lucas MJ, Leveno KJ, Cunningham FG. A comparison of magnesium sulfate with phenytoin for the prevention of eclampsia. *New Engl J Med*. 1995;333:201–205.
73. Witlin AG, Friedman SA, Sibai BM. The effect of magnesium sulfate therapy on the duration of labor in women with mild preeclampsia at term: a randomized, double-blind, placebo-controlled trial. *Am J Obstet Gynecol*. 1997;176:623–627.
74. Livingston JC, Livingston LW, Ramsey R, Mabie BC, Sibai BM. Magnesium sulfate in women with mild preeclampsia: a randomized controlled trial. *Obstet Gynecol*. 2003;101:217–220.
75. Odendaal HJ, Hall DR. Is magnesium sulfate prophylaxis really necessary in patients with severe preeclampsia? *J Maternal Fetal Invest*. 1996;6:14–18.
76. Hall DR, Odendaal JH, Smith M. Is prophylactic administration of magnesium sulfate in women with pre-eclampsia indicated prior to labour? *Br J Obstet Gynaecol*. 2000; 107:903–908.
77. Moodley J, Moodley VV. Prophylactic anticonvulsant therapy in hypertensive crises of pregnancy – the need for a large randomized trial. *Hypertens Pregnancy*. 1994;13:245–252.
78. Coetzee EJ, Dommisse J, Anthony J. A randomized controlled trial of intravenous magnesium sulfate versus placebo in the management of women with severe pre-eclampsia. *Br J Obstet Gynaecol*. 1998;105:300–303.
79. Altman D, Carroli G, Duley L. Do women with preeclampsia, and their babies, benefit from magnesium sulfate? The Magpie Trial: a randomized placebo-controlled trial. *Lancet*. 2002;359:1877–1890.
80. Belfort M, Anthony J, Saade G. Interim report of the nimodipine vs. magnesium sulfate for seizure prophylaxis in severe preeclampsia study: an international, randomized, controlled trial. *Am J Obstet Gynecol*. 1998;178:1s.
81. Pritchard JA, Pritchard SA. Standardized treatment of 154 consecutive cases of eclampsia. *Am J Obstet Gynecol*. 1975;123:543–552.
82. Gedekoh RH, Hayashi TT, MacDonald HM. Eclampsia at Magee-Women's Hospital 1970 to 1980. *Am J Obstet Gynecol*. 1981;140:86–96.
83. Pritchard JA, Cunningham FG, Pritchard SA. The Parkland Memorial Hospital protocol for treatment of eclampsia: evaluation of 245 cases. *Am J Obstet Gynecol*. 1984;148:951–963.
84. Sibai BM, Ramanathan J. The case for magnesium sulfate in preeclampsia-eclampsia. *Int J Obstet Anesth*. 1992;1:167–175.
85. Which anticonvulsant for women with eclampsia: evidence from the Collaborative Eclampsia Trial. *Lancet*. 1995;345:1455–1463.
86. Dommisse J. Phenytoin sodium and magnesium sulfate in the management of eclampsia. *Br J Obstet Gynaecol*. 1990;97:104–109.
87. Crowther C. Magnesium sulfate versus diazepam in the management of eclampsia: a randomized controlled trial. *Br J Obstet Gynaecol*. 1990;97:110–117.
88. Bhalla AK, Dhall GI, Dhall K. A safer and more effective treatment regimen for eclampsia. *Aust NZ J Obstet Gynecol*. 1994;34:144–148.
89. Friedman SA, Schiff E, Kao L, Sibai BM. Phenytoin versus magnesium sulfate in patients with eclampsia: preliminary results from a randomized trial. Poster presented at 15th Annual Meeting of the Society of Perinatal Obstetricians, Atlanta; January 23–28, 1995 (abstr 452). *Am J Obstet Gynecol*. 1995;172 384, pt 2.
90. Richards A, Stather-Dunn L, Moodley J. Cardio-pulmonary arrest after the administration of magnesium sulfate. *S Afr Med J*. 1985;67:145.
91. McCubbin JH, Sibai BM, Abdella TN, Anderson GD. Cardio-pulmonary arrest due to maternal hypermagnesiumia. *Lancet*. 198;ii:1058.
92. Bohman VR, Cotton DB. Supralethal magnesemia with patient survival. *Obstet Gynecol*. 1990;76:984–986.
93. Ascarelli MG, Johnson V, May WL. Individually determined postpartum magnesium sulfate therapy with clinical parameters to safely and cost-effectively shorten treatment for preeclampsia. *Am J Obstet Gynecol*. 1998;179:952–956.
94. Isler CM, Barrilleaux PS, Rinehart BK. Postpartum seizure prophylaxis: using maternal clinical parameters to guide therapy. *Obstet Gynecol*. 2003;101:66–69.
95. Fontenot MF, Lewis DF, Frederick JB. A prospective randomized trial of magnesium sulfate in severe preeclampsia: use of diuresis as a clinical parameter to determine the duration of postpartum therapy. *Am J Obstet Gynecol*. 2005;192:1788–1793.
96. Ehrenberg HM, Mercer BM. Abbreviated postpartum magnesium sulfate therapy for women with mild preeclampsia. *Obstet Gynecol*. 2006;108:833–838.
97. Dayicioglu V, Sahinoglu Z, Kol E. The use of standard dose of magnesium sulfate in prophylaxis of eclamptic seizures: do body mass index alterations have any effect on success? *Hypertens Pregnancy*. 2003;22:257–265.
98. Tudela CM, McIntire DD, Alexander JM. Effect of maternal body mass index on serum magnesium levels given for seizure prophylaxis. *Obstet Gynecol*. 2013;121(2 Pt 1):314–320.
99. Duley L, Gulmezoglu AM, Henderson-Smart DJ, Chou D. Magnesium sulphate and other anticonvulsants for women with pre-eclampsia. *Cochrane Database Syst Rev*. 2010(11):CD000025.
100. Staff AC, Dechend R, Redman CW. Review: preeclampsia, acute atherosis of the spiral arteries and future cardiovascular disease: two new hypotheses. *Placenta*. 2013;34(suppl):S73–S78.

CHAPTER **13**

Cerebrovascular Pathophysiology in Preeclampsia and Eclampsia

MARILYN J. CIPOLLA, GERDA G. ZEEMAN AND F. GARY CUNNINGHAM

Editors' comment: *A single chapter focusing on the brain in normal pregnancy and preeclampsia appeared in edition 3 and this is its first revision. New additions include animal model observations that suggest explanations for changes in cerebral physiology during normal pregnancy and hypertensive complications. New findings in the human disease regarding volume status and the eclamptic convulsion are discussed and there is further clarification of that new but loosely thrown around term "posterior reversible encephalopathy syndrome" and in both areas our authors have made recent seminal contributions. One does not need an editors' comment to realize that cerebral involvement in preeclampsia, and especially eclampsia, is a purveyor of serious disease. In this regard discussions regarding prevention and management of cerebral symptoms, especially eclampsia, can be found in Chapters 12 and 20, the first reviewing trials with magnesium sulfate, the latter the treatment of impending or actual eclampsia.*

INTRODUCTION

The brain has a central role in the preeclampsia–eclampsia syndrome. For many centuries, convulsions in the pregnant woman were the most recognizable event of what we now know to be a generalized disorder that affects virtually every organ system. In his first edition of *Hypertensive Disorders in Pregnancy*, Chesley chronicled the historical evolution of theories concerning causes of convulsions in women with eclampsia. Since the times of Hippocrates and Galen, the two main theories were either *cerebral congestion – repletion*, or *cerebral anemia – depletion*. Beliefs concerning repletion led to the widespread practice of phlebotomy during the 1700s and 1800s. And throughout much of the last century, as therapeutic measures were aimed at either halting or preventing convulsions, evidence began to accrue leading to insights into the cerebrovascular pathophysiology of eclampsia. For example, during the renaissance of neuroanatomy and pathology, intracranial hemorrhages and generalized cerebral edema were prominently emphasized.

Neuroanatomical emphasis culminated in the seminal work of Sheehan and Lynch and their autopsy series of eclamptic women.[1] This work is unique as the brains studied were from autopsies performed within 2 hours after death, eliminating most of the postmortem changes that rapidly occur in brain tissue and might confound interpretation. By then, it was appreciated that women with fatal eclampsia frequently had brain abnormalities, but that these caused death in a minority of such cases. As deaths from eclampsia declined over the last half of the 20th century, interest in cerebral pathology waned also because there were few avenues from which to approach appropriate investigation. This gap was filled by the development of computed tomography (CT) in the 1970s, which allowed noninvasive brain imaging. Doppler studies of cerebral blood flow (CBF) velocity reawakened interests in cerebrovascular perturbations in the preeclampsia syndrome. More recently, the use of MRI has opened wider the vista to study neuroanatomical changes as well as to more accurately measure cerebral perfusion. These technologies, combined with reproducible animal models to better study cerebral blood flow and its alterations, have allowed a heretofore unknown look at cerebrovascular pathology provoked by the preeclampsia syndrome.

In this chapter we correlate the earlier neuropathological observations of eclampsia with the noninvasive imaging findings derived from CT scanning and MR imaging. We also correlate cerebrovascular abnormalities induced by preeclampsia–eclampsia and measured with direct and indirect noninvasive methods such as MRI technology and Doppler velocimetry. And finally, we review hypertensive effects on cerebral perfusion in both pregnant and nonpregnant animal models from which we draw a composite description of the effects of the preeclampsia syndrome on the brain. Finally this chapter is designed to specifically

Chesley's Hypertensive Disorders in Pregnancy.
ISBN: 978-0-12-407866-6

DOI: http://dx.doi.org/10.1016/B978-0-12-407866-6.00013-4

describe cerebral pathology and pathophysiology in preeclampsia/eclampsia. Clinical aspects of the preeclampsia syndrome are discussed in Chapters 2 (clinical spectrum), 12 (prevention), and 20 (management).

NEUROANATOMICAL FINDINGS WITH ECLAMPSIA

Most neuroanatomical descriptions of the brain in eclamptic women are taken from eras when mortality rates were quite high. One consistent finding was that brain pathology accounted for only about a third of fatal cases such as the one shown in Fig. 13.1.[2] In the majority of cases, however, death was from pulmonary edema, and the brain lesions were coincidental. Thus, while gross intracerebral hemorrhage was seen in up to 60% of eclamptic women, it was fatal in only half.[1,3,4] As shown in Fig. 13.2, other principal lesions found at autopsy consisted of cortical and subcortical petechial hemorrhages. Histologically, these are composed of numerous small hemorrhages, 0.3–1.0 mm in diameter, arranged in streaks of 2–4 cm running radially in the cortex. They may appear anywhere on the gyral surface and are most common in the occipital lobes and least common in the temporal lobes. Many occur in the border zones between cerebral arterial territories. Other frequently described major macroscopic lesions include subcortical edema, multiple nonhemorrhagic areas of "softening" throughout the brain, hemorrhagic areas in the white matter, and hemorrhage in the basal ganglia or pons, often with rupture into the ventricles. In some cases, numerous small cortical infarctions are described as well.[1,4] These infarcts vary from about 0.3 to 1.0 mm in diameter and are sometimes confluent. The classical microscopic vascular lesions consist of fibrinoid necrosis of the arterial wall and perivascular microinfarcts and hemorrhages.

NEUROIMAGING IN ECLAMPSIA

A number of neuroimaging techniques have been used to better understand the cerebrovascular mechanism(s) involved in the preeclampsia syndrome. These include angiography, CT, and MRI techniques. Specifically, the ever-increasing development of MRI techniques has especially been useful to provide information concerning the pathogenesis of cerebral manifestations of preeclampsia.

Computed Tomography (CT)

Localized hypodense lesions at the gray–white matter junction, primarily in the parieto-occipital lobes, are typically found in eclampsia (Fig. 13.3). Such lesions may also be seen in the frontal and inferior temporal lobes, as well as the basal ganglia and thalamus.[5–8] The spectrum of involvement is wide, and with increasing radiological involvement, either of the occipital lobes or with diffuse cerebral edema, symptoms such as lethargy, confusion, and blindness will develop.[9,10] In these cases, widespread edema shows as a marked compression or even obliteration of the cerebral ventricles (Fig. 13.4). Such women may develop signs of impending life-threatening transtentorial herniation.

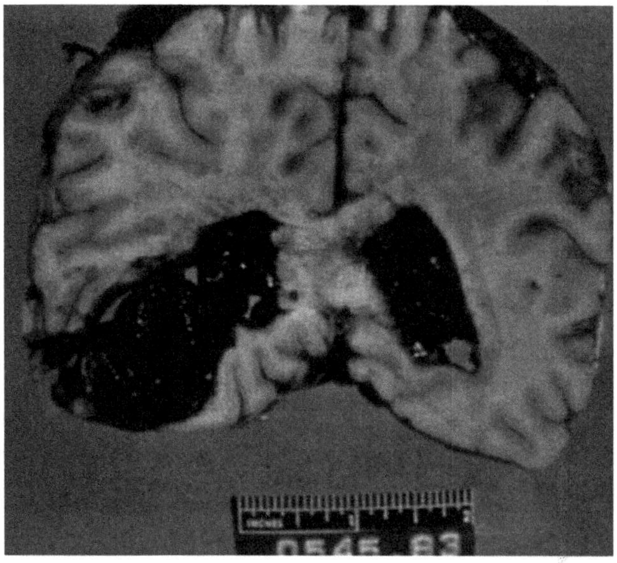

FIGURE 13.1 Hypertensive hemorrhage with eclampsia. (Reprinted, with permission, from Cunningham FG, et al.: *Williams Obstetrics*, 20th ed. Stamford, CT: Appleton & Lange; 1997.)

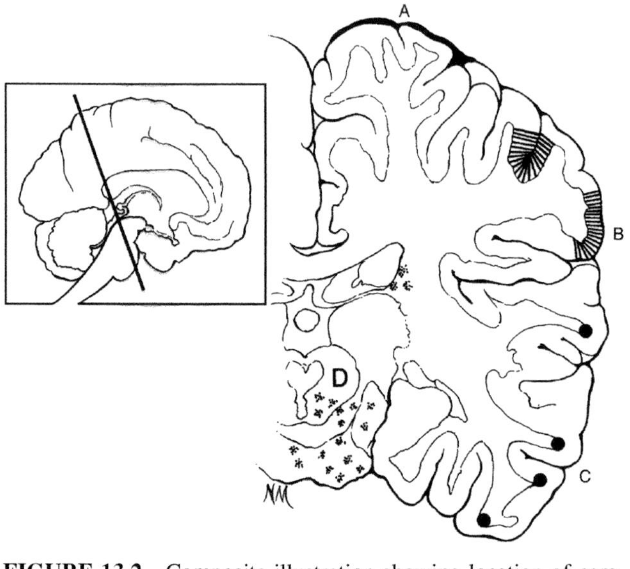

FIGURE 13.2 Composite illustration showing location of cerebral hemorrhages and petechiae in women with eclampsia: (A) pia-arachnoid hemorrhage; (B) cortical petechiae; (C) subcortical petechiae; and (D) focal softenings or petechiae in midbrain or white matter.[1]

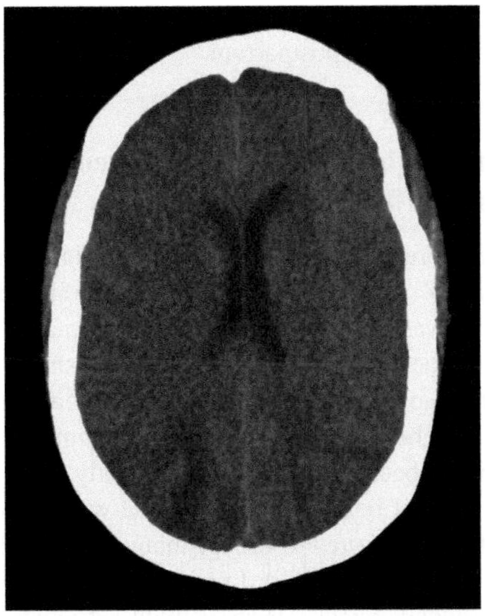

FIGURE 13.3 Cranial CT of a woman with eclampsia. Radiographic low-density areas (arrow) are seen in the right occipital lobe.

Most reports describe reversibility of cerebral edema.[11,12] In a few women with eclampsia, however, cerebral infarctions have been described.[13] And cerebral hemorrhagic transformation may also develop from areas of ischemic infarction.[14] These findings raise important issues regarding both the pathogenesis of preeclampsia-related intracranial hemorrhage as well as its prevention.

Magnetic Resonance Imaging (MRI)

There are a number of MRI acquisitions that are used to study cerebrovascular anatomy and function in eclamptic women. Common findings are hyperintense T2 lesions in the subcortical and cortical regions of the parietal and occipital lobes, with occasional involvement of basal ganglia and/or brainstem.[5,15,16] Some examples are shown in Figs. 13.5A and 13.6A. While these lesions of the *posterior reversible encephalopathy syndrome* are almost universal in women with eclampsia, their incidence in women with

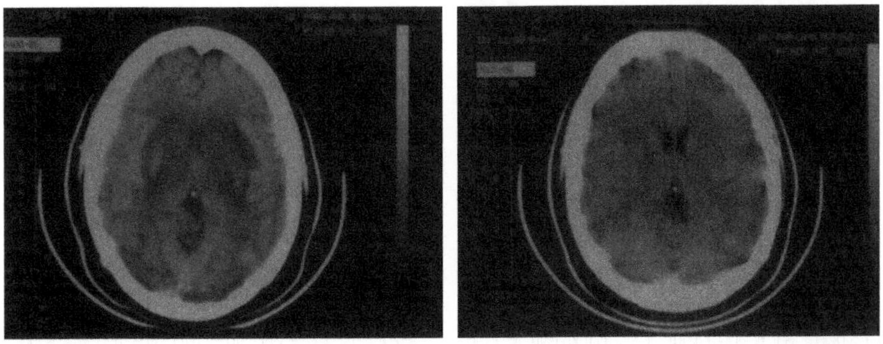

FIGURE 13.4 Computed tomographs in a woman with cerebral edema following acutely exacerbated severe hypertension. The radiograph on the left shows slit-like effaced ventricles as well as sharply demarcated gray–white interface, both indicating parenchymal swelling. The radiograph on the right taken 10 days later shows diminished edema manifest by larger ventricles and loss of gray–white interface demarcation.

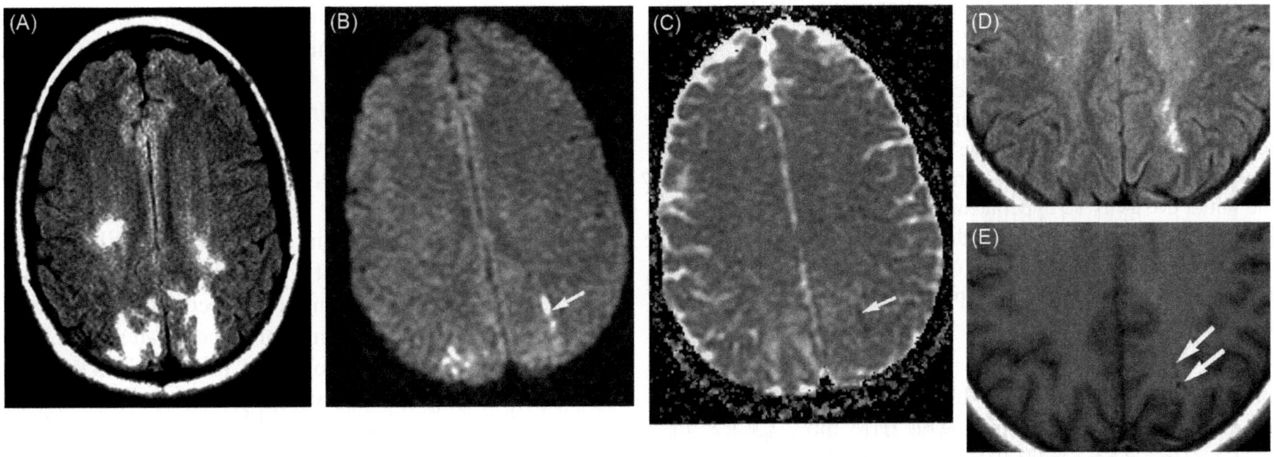

FIGURE 13.5 Classic MRI pattern of vasogenic edema in eclampsia with associated subcortical infarction. (A) T2 hyperintensity on FLAIR images indicates parieto-occipital distribution of vasogenic edema. (B) Within this volume is a smaller area of hyperintensity on DWI (arrow). (C) That this signal is due to restricted diffusion is confirmed by hypointensity on the ADC map (arrow). (D) These findings suggesting areas of subcortical infarction are supported by follow-up studies obtained 6 weeks later in which T2 hyperintensity on FLAIR image and (E) corresponding low signal intensity on T1-weighted image (arrows) indicated evolution to gliosis.[16]

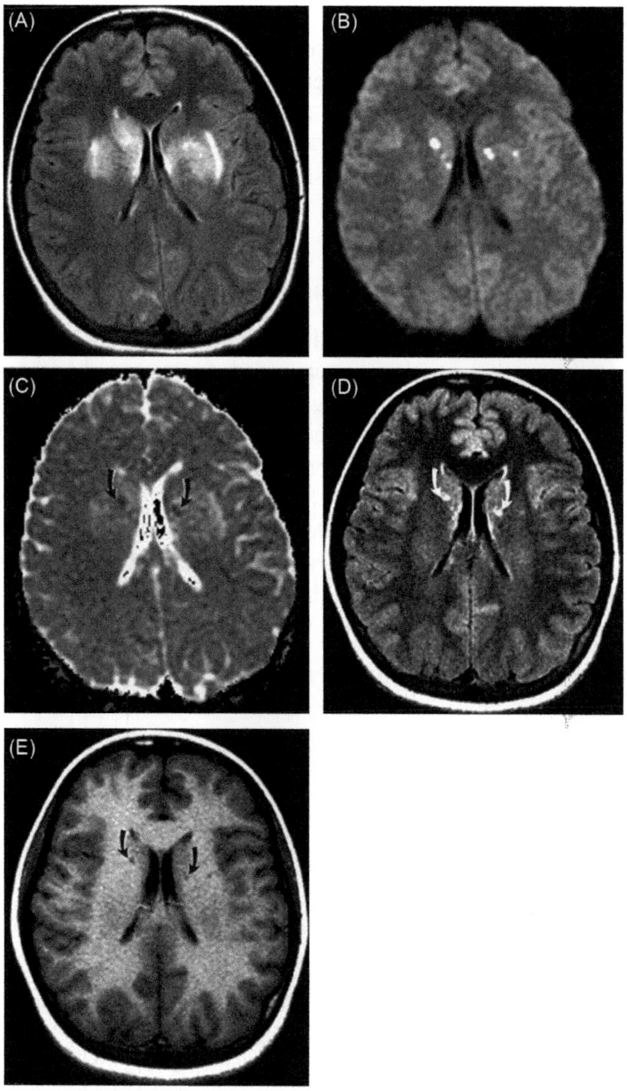

FIGURE 13.6 MRI evidence of hypertensive encephalopathy and lacunar infarctions in eclampsia. This case demonstrates the occasional atypical distribution of signal abnormalities away from the parieto-occipital region. (A) Note on T2-weighted FLAIR image that the predominant changes occur in the basal ganglia regions bilaterally. (B) Small foci of cytotoxic edema are indicated by the marked hyperintensity on DWIs. (C) Corresponding reduced ADC (arrows). (D) These small, presumed lacunar infarctions show a typical evolution on repeat examination 6 weeks after the initial events as T2 hyperintensity on FLAIR (arrows) and (E) low signal intensity on T1-weighted image (arrows) in regions corresponding to DWI evidence of infarct (see B and C) on the initial studies. Hypertensive encephalopathy is neither necessarily posterior nor reversible.[16]

preeclampsia is not known. Intuitively, however, they are more likely found in women who have more severe disease and who have neurological symptoms.[17–19] And they are although usually reversible, it is now known that some of these hyperintense lesions appear as infarctions and that there are persistent findings in up to a fourth of eclamptic women several weeks postpartum.[16,19,20]

Diffusion-Weighted MRI and Apparent Diffusion Coefficient

From the foregoing, it is apparent that there are two distinctly different types of cerebral edema in eclampsia – vasogenic and cytotoxic edema. Vasogenic edema is associated with increased hydrostatic pressure and ensuing capillary leak, while cytotoxic edema is associated with ischemia and cell death with infarction. This issue is critical because the former is usually reversible and the latter may not be.[21] It is not possible with conventional CT and MRI techniques to differentiate between these two forms of cerebral edema. To do so, a series of MRI acquisitions were developed. These include *diffusion-weighted imaging (DWI)* sequences and *apparent diffusion coefficient (ADC)* mapping. With these, it is possible to further characterize the hyperintense lesions seen on T2 imaging (Figs. 13.5B–E and 13.6B–E). Any predictions of the clinical course of eclampsia based on DWI and ADC findings remain currently speculative.[22]

DWI takes advantage of strong diffusion gradients that detect changes in water molecule distribution in tissue. Quantitative measurement of the diffusion property of a tissue is expressed as the ADC. Vasogenic edema is characterized by increased extracellular fluid with enhanced water diffusion and may be seen as a combination of normal DWI with hyperintense T2 signal lesions and increased apparent diffusion.[23] Conversely, in the presence of an ischemic event, cytotoxic edema is caused by sodium pump failure, cell swelling, and eventually cell death. This causes a reduction in proton diffusion, due to a shift of water from the extracellular to the intracellular space, and it elicits a hyperintense signal on DWI but with a decreased ADC.[24] It has been shown that ischemic brain regions can be identified within minutes to hours after the onset of neurological symptoms.[21,25] Furthermore, newer techniques such as *susceptibility-weighted imaging* and *diffusion tensor imaging* for assessing the structure of the white matter are being developed, but their benefit for eclampsia is yet unclear.[26]

In women with eclampsia, studies using DWI sequences showed that the origin of cerebral edema is primarily vasogenic, but that less commonly there are ischemic and cytotoxic changes, i.e., infarction. These latter cases have DWI hyperintense T2 lesions and decreased ADC superimposed on the pattern of vasogenic edema.[16,20] In related animal studies, Tamaki et al.[27] showed that blood–brain barrier (BBB) disruption due to marked local hydrostatic pressure precedes decreased tissue perfusion and ischemia.

PATHOGENESIS OF CEREBRAL MANIFESTATIONS IN (PRE)ECLAMPSIA

Pregnancy-induced changes of cerebrovascular hemodynamics have not been well studied, and thus the pathogenesis of cerebral manifestations of the preeclampsia syndrome is also unclear. Much of this is because of the various challenges associated with *in vivo* studies of cerebral blood flow in human pregnancy (see below). And while central nervous system histopathology is mainly based on autopsy data as discussed, most hemodynamic data are invariably from surviving women. Although this presents some difficulty in relating histopathological with hemodynamic findings, an accurate picture is emerging.

When taken clinically, data taken from the past several decades include pathological and neuroimaging findings that have led to two general theories to explain cerebral abnormalities associated with eclampsia. Importantly – and as emphasized throughout this edition – endothelial cell dysfunction that characterizes the preeclampsia syndrome may play a key role in both theories.

The first theory suggests that in response to acute severe hypertension cerebrovascular overregulation leads to vasospasm.[28] This presumption was based on the angiographic appearance of diffuse or multifocal segmental narrowings suggestive of vasospasm of the cerebral vasculature in women with severe preeclampsia and eclampsia.[29] In this scheme, diminished CBF is hypothesized to result in ischemia, cytotoxic edema, and eventually tissue infarction. Finally, the *reversible cerebral vasoconstriction syndrome* has been reported to be associated with preeclampsia,[30] but whether this causes eclampsia is not known.

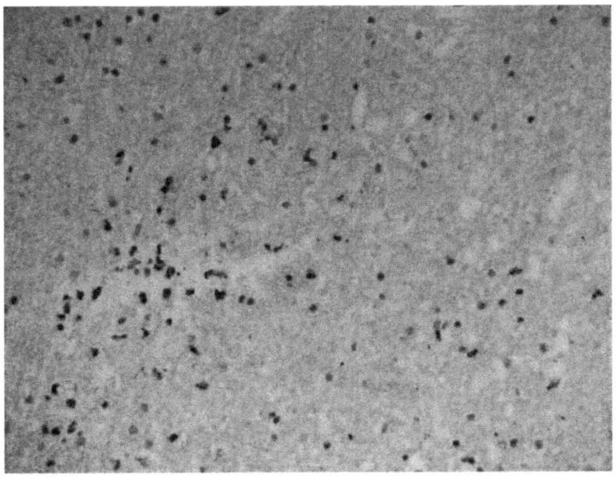

FIGURE 13.7 Cerebral biopsy of a 60-year-old man with hypertensive encephalopathy from blood pressure 220/120 mmHg. The white matter shows mild, diffuse vacuolization with minimal inflammatory reaction characterized by scattered macrophages. Abundant reactive astrocytes were evident. (From Schiff and Lopes. *Neurocrit Care.* 2005;2(3):303–305, with permission.)

The second theory is that eclampsia represents a form of hypertensive encephalopathy such that sudden elevations in systemic blood pressure exceed the normal cerebrovascular autoregulatory capacity.[8,31–33] The decrease in cerebrovascular resistance (CVR) causes disruption of end-capillary hydrostatic pressure, hyperperfusion, and extravasation of plasma and red cells through opening of the endothelial tight junctions with increased pinocytosis leading to the accumulation of vasogenic edema.[31–33] Regions of forced vasodilatation and vasoconstriction develop, especially in arterial boundary zones, as is prominent in hypertensive encephalopathy.[31] Figure 13.7 shows a brain biopsy of a patient with hypertensive encephalopathy, a condition that may relate to eclampsia. This mechanism has gained much attention over the last decade, especially since it was described as *reversible posterior leukoencephalopathy syndrome*.[34] More recently, it is usually referred to as the *posterior reversible encephalopathy syndrome* (PRES) to incorporate the posterior nature of the condition.[35] The normal structure of the neurovascular unit comprises capillary endothelial cells in close association with basal lamina, astrocytic endfeet, and pericytes (Fig. 13.8). It is largely

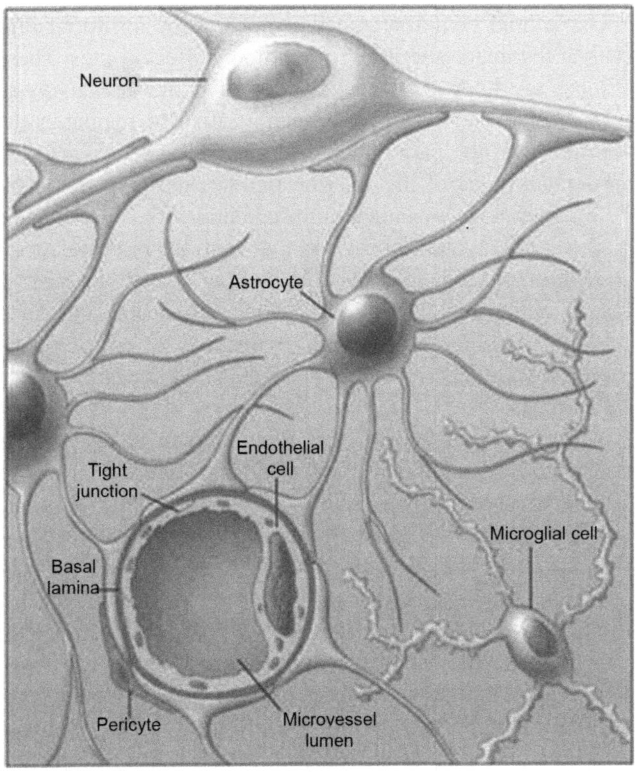

FIGURE 13.8 Schematic of the neurovascular unit. Increased hydrostatic (capillary lumen) pressure causes extravasation of plasma and red cells through endothelial tight junctions with perivascular edema. Reprinted with permission from del Zoppo GJ. *New Engl J Med.* 2006;354:553–555. (This figure is reproduced in color in the color plate section.)

unknown how these cells and structures are affected during preeclampsia that may contribute to brain pathologies including hemorrhage and edema.

ECLAMPSIA AS POSTERIOR REVERSIBLE ENCEPHALOPATHY SYNDROME (PRES)

It is currently thought that preeclampsia is one of the disorders that share a common etiopathogenesis and that comprise PRES. Its clinical, pathological, as well as neuroimaging features reflect the rapid and dynamic fluctuations in cerebral blood flow and water content.[34,36] In fact, nearly all patients with eclampsia had PRES as determined by MRI.[5,16] In nonpregnant and pregnant patients, PRES is usually an acute cerebral illness which may present with headache, nausea, altered mental function, visual disturbances, and seizures. [37] In preeclampsia–eclampsia related PRES, headaches are more frequent than altered mental status as the initial PRES-related symptom compared to nonpregnant PRES patients. The convulsions are commonly – but not exclusively – occipital in onset and correlate with the characteristic predominantly posterior imaging abnormalities seen with MRI in women with eclampsia. The arterial boundary zones, located at the territorial limits of the major arteries, are commonly affected sites. These zones are known as the *border zone* or *watershed areas*. In the human, the most frequently affected region in the cortex is at the parieto-occipital sulci, which represent the boundary zone of the anterior, middle, and posterior cerebral arteries. Involvement of the cerebral cortex may also be seen in PRES and lesions may extend to the brainstem, cerebellum, basal ganglia, and the more anterior brain regions such as the frontal lobes.[36,38,39] However, eclampsia-related PRES cases demonstrate less frequent involvement of the thalamus, midbrain, and pons compared to nonpregnant PRES cases.[37]

Some patients with PRES have only convulsions and do not manifest traditional prodromal signs and symptoms of hypertensive encephalopathy.[40] In addition, the dramatic blood pressure increases that typify hypertensive encephalopathy are not necessarily seen and there may be only mild to moderate blood pressure increases with PRES.[41] Importantly, PRES may develop with only mild hypertension,, which may involve endothelial damage. This has been described in the thrombotic microangiopathy syndromes – hemolytic uremic syndrome and thrombotic thrombocytopenic purpura – as well as with systemic lupus erythematosus, with immunosuppressive drug toxicity, or with the use of certain chemotherapeutic agents that include methotrexate and cisplatin.[37,42,43] Finally, it is important to recognize PRES because the neurological disorder is readily treatable by lowering any dangerously elevated blood pressures and correction of the underlying medical condition that caused endothelial cell injury.[44]

CEREBRAL BLOOD FLOW AUTOREGULATION

Autoregulation is the process by which cerebral blood flow (CBF) remains relatively constant in the face of alterations in cerebral perfusion pressure.[45] Put another way, when cerebral perfusion pressure declines, cerebrovascular resistance (CVR) decreases due to myogenic vasodilatation of pial arteries and arterioles and hypoxic vasodilatation, thus augmenting perfusion. Alternatively, if cerebral perfusion pressure increases, the autoregulatory response increases CVR by vasoconstriction, resulting in relatively constant CBF. Thus, autoregulation is a physiological protective mechanism that prevents brain ischemia during drops in pressure and prevents capillary damage and edema from hyperperfusion during pressure increases.[45] In normotensive adults, CBF is maintained at approximately 50 mL per 100 g of brain tissue per minute (mL/100 g/min), provided perfusion pressure is in the range ~60–160 mm Hg.[45,46] Above and below these limits, autoregulation is lost and CBF becomes dependent on mean arterial pressure in a linear fashion.[45,46]

Significant brain injury occurs when autoregulatory mechanisms are lost. For example, during acute hypertension at mean pressures above the autoregulatory limit – about 160 mm Hg in the otherwise healthy patient – the myogenic vasoconstriction of vascular smooth muscle is overcome by excessive intravascular pressure and forced dilatation of cerebral vessels occurs.[47,48] The loss of myogenic tone during forced dilatation decreases CVR and increases CBF, a result that produces hyperperfusion, BBB disruption, and acute edema formation.[47–49]

The relationship between autoregulation, CBF, and BBB disruption has been extensively studied. Numerous investigators have found a positive correlation between loss of autoregulation, increased perfusion, and BBB permeability that leads to cerebral edema.[45,46,49,50] Importantly, mechanisms that increase resistance, such as sympathetic nerve stimulation or inward remodeling of cerebral arterioles during chronic hypertension, attenuate increases in CBF during acute hypertension and are protective of the BBB.[50–52] In general, when CBF is compared in areas with and without albumin extravasation in the same brain, the regions of increased permeability have the highest blood flow, indicating loss of autoregulation and decreased CVR.[46,49,53] Together, these findings suggest that decreased resistance and hyperperfusion during acute hypertension cause BBB disruption, whereas increased resistance is protective of the microcirculation. In fact, decreased CVR that leads to hyperperfusion during acute hypertension is considered the primary cause of edema during hypertensive encephalopathy and eclampsia.[8,18,33,46,54–57]

When severe experimental hypertension is induced by intravenous infusion of vasogenic agents, arterioles develop a pattern of alternating constrictions and dilatations, giving rise to the so-called *sausage-string* appearance.[47–49,58] This vascular pattern has been demonstrated in small blood vessels in various vascular beds, including the brain. In the cerebral circulation, the development of the sausage-string pattern is linked to the development of vascular damage, specifically in the dilated regions of the vessel as they fail to maintain myogenic vasoconstriction, with resulting endothelial hyperpermeability and extravasation of macromolecules into the brain parenchyma.[47–49]

CEREBRAL BLOOD FLOW AUTOREGULATION AND HEMODYNAMICS IN PREGNANCY

The effect of pregnancy on cerebral hemodynamics and cerebral blood flow autoregulation is of significant interest mostly because impaired cerebral autoregulation is thought to be a major contributor to the development of eclampsia.[8,18,33,47,54,58,59] The adaptation of the cerebral circulation to pregnancy has recently been extensively reviewed.[59] From clinical observations, it is known that eclampsia can develop with only mild or even absent hypertension. While it is tempting to hypothesize that the upper limit of cerebral autoregulation is reduced with the preeclampsia syndrome, evidence for this is lacking. It seems much more likely that perivascular edema develops at a much lower capillary hydrostatic pressure, possibly as a function of endothelial activation known to accompany the preeclampsia syndrome. That said, failure of autoregulatory mechanisms may occur in response to either an acute and/or relatively large blood pressure increase, which seems more important than absolute blood pressure. Thus, it is possible that it is the acuteness of the blood pressure rise or relative change in pressure from baseline in the setting of endothelial dysfunction that disrupts the delicate balance between capillary and cerebral perfusion pressures in eclampsia. Understanding cerebral hemodynamic changes associated with pregnancy and preeclampsia is challenging, necessitating the use of animal models in some instances. Thus, the following sections will review both animal and clinical studies on changes in cerebral hemodynamics during pregnancy and preeclampsia.

Animal Studies

Many women who develop eclampsia do so at pressures that are considerably lower than those reported for posterior reversible encephalopathy syndrome or hypertensive encephalopathy.[58,60–63] These findings suggest that the cerebral blood flow autoregulatory curve is shifted to the lower range of pressures during pregnancy. Studies in anesthetized rats found that this was not the case. When the upper limit of CBF autoregulation was compared between nonpregnant and late-pregnant rats, there was no difference in the pressure of autoregulatory breakthrough[64] (Fig. 13.9A). In fact. more recent studies in rats found that both the upper and lower limits of CBF autoregulation were extended.[65,66] Importantly, however, only the pregnant animals developed significant edema formation in response to acute hypertension[64,65] and autoregulatory breakthrough, suggesting that the endothelium is more susceptible to hydrostatic edema during pregnancy (Fig. 13.9B).

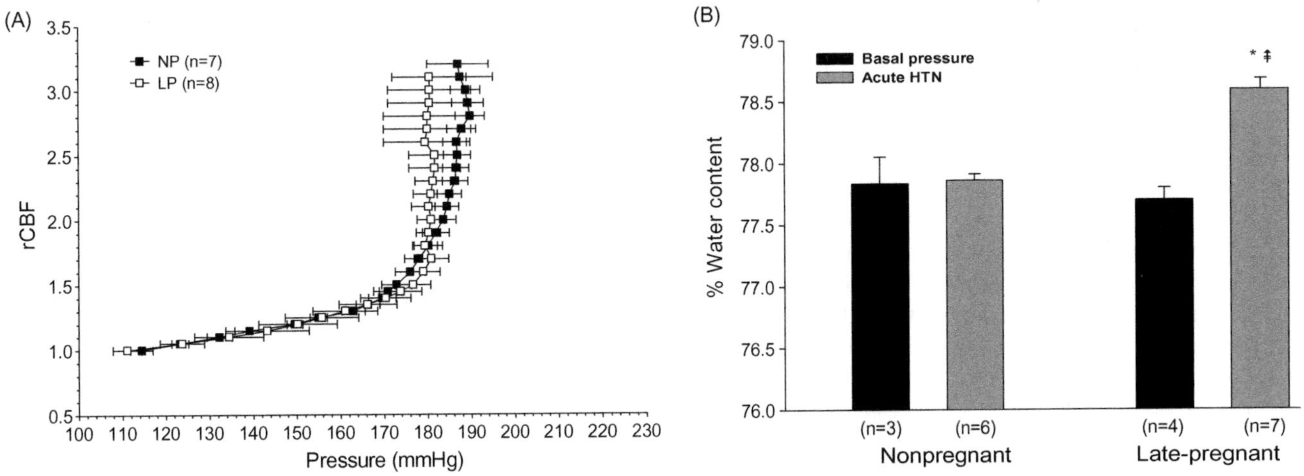

FIGURE 13.9 (A) Graph of cerebral blood flow autoregulatory curves in anesthetized nonpregnant (NP, closed squares) and late-pregnant (LP, open squares) rats. The curves were determined using laser Doppler to measure relative changes in cerebral blood flow during constant infusion of phenylephrine to raise mean arterial pressure. Notice there is no difference in autoregulation or the pressure at which breakthrough occurred. (B) Graph showing percent water content as a measure of cerebral edema formation in the same groups of animals as shown in (A). In nonpregnant animals, water content was similar at basal pressure (black bars) and after autoregulatory breakthrough (gray bars). In late-pregnant animals, however, acute hypertension that caused autoregulatory breakthrough caused a significant increase in water content (*$p < 0.05$ vs. basal; ‡$p < 0.05$ vs. NP). Thus, under these conditions, pregnancy alone predisposes the brain to edema formation.[64]

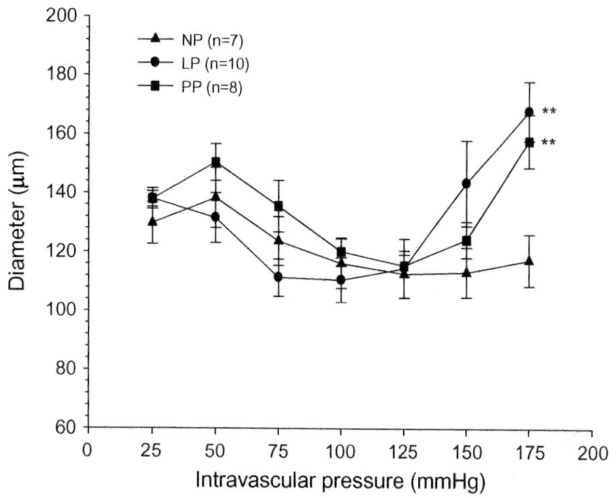

FIGURE 13.10 Graph showing pressure versus diameter curves of posterior cerebral arteries from nonpregnant (NP, closed triangles), late-pregnant (LP, closed circles), and postpartum (PP, closed squares) animals. Arteries from all groups of animals constricted in response to increased pressure within the pressure range 50–125 mm Hg, demonstrating myogenic reactivity. Arteries from LP and PP animals, however, underwent forced dilatation at significantly lower pressure compared with NP animals – this is noted by the large increase in diameter in response to increased pressure (**$p < 0.01$ vs. NP). The increase in diameter during forced dilatation is a primary event in the development of hydrostatic brain edema in which cerebrovascular resistance is decreased. Reprinted with permission.[71]

It is important to note that the autoregulatory curves in these studies were determined using laser Doppler methods to measure CBF and they thus only provide relative changes in blood flow. Other studies using microspheres to measure absolute CBF showed that acutely increased pressures in late pregnancy were associated with significantly decreased CVR and increased CBF when compared with nonpregnant animals.[67] Specifically, pregnancy was associated with a 40% decrease in CVR compared with that in nonpregnant animals with the same change in pressure. Because increased CVR in response to elevated cerebral perfusion pressure is a protective mechanism in the brain that prevents transmission of harmful hydrostatic pressure to the microcirculation, diminished CVR during pregnancy in response to acute hypertension could promote BBB disruption and vasogenic edema, similar to what is seen during eclampsia.

The mechanism by which pregnancy decreases CVR during acute hypertension is not clear, but may be related to structural changes that affect arterial and arteriolar diameter. Resistance and flow regulation are principally determined by vessel caliber because they are inversely related to the fourth power of vessel radius. The innate myogenic behavior of the cerebrovascular smooth muscle is crucial for establishment of an appropriate CVR, which serves to protect downstream arterioles and capillaries in the face of changing perfusion pressures and to maintain tissue perfusion when blood pressure falls.[68,69] The cerebral circulation is a unique vascular bed in that large extracranial and intracranial pial vessels contribute about 50% to total CVR.[70] Studies of isolated cerebral arteries from nonpregnant, late-pregnant, and postpartum animals suggest that forced dilatation occurs at lower pressures during pregnancy and postpartum (Fig. 13.10).[71] This may be a contributory mechanism by which pregnancy decreases CVR during acute increases in pressure.

Although these changes in larger pial arteries may contribute to decreased CVR and increased microvascular pressure noted in late-pregnant animals, the response of smaller parenchymal arterioles is a critical determinant of distal capillary pressure and a major determinant of BBB changes. In fact, previous studies have shown that differences in resistance of small vessels in the brain parenchyma can account for regional differences in BBB permeability during acute hypertension.[70,72] Because late-pregnant animals developed edema in response to autoregulatory breakthrough and elevated hydrostatic pressure, it seems likely that small-vessel resistance is reduced and contributes to edema. In studies to specifically examine brain parenchymal arterioles, vessels from late-pregnant animals were shown to have larger diameters compared with those from nonpregnant animals.[67] Thus, there is a gestation-induced effect to cause outward remodeling of parenchymal arterioles (Fig. 13.11A). Subsequent studies showed that selective enlargement of brain parenchymal arterioles was due to relaxin-induced activation of the transcription factor peroxisome proliferator-activated receptor-gamma (PPARγ).[73]Although such structural changes may not influence resting CBF, they would have a significant impact on local hemodynamics under conditions when vessels are markedly dilated such as during breakthrough of autoregulation – and the resulting forced dilatation. Importantly, outward remodeling of parenchymal arterioles in the brain appears to be at the expense of the vascular wall, which becomes significantly thinner during pregnancy (Fig. 13.11B). Therefore, the significance of outward remodeling of cerebral arteries and arterioles during pregnancy may not be limited to decreased CVR during acute hypertension, but may also predispose the brain to hemorrhage, another pathological finding of eclampsia, due to severely increased wall stress.

Human Studies

In 1949, McCall first described cerebral blood flow changes in women with eclampsia using an inhalation technique of a gaseous mixture containing nitrogen, nitric oxide, and oxygen.[74] Internal jugular arterial and venous blood was collected and the Fick principle applied by measuring serum concentrations. In eclamptic women, while the delivery of oxygen and CBF were normal, there was a 20% decrease in oxygen utilization. Other than these studies, there are none

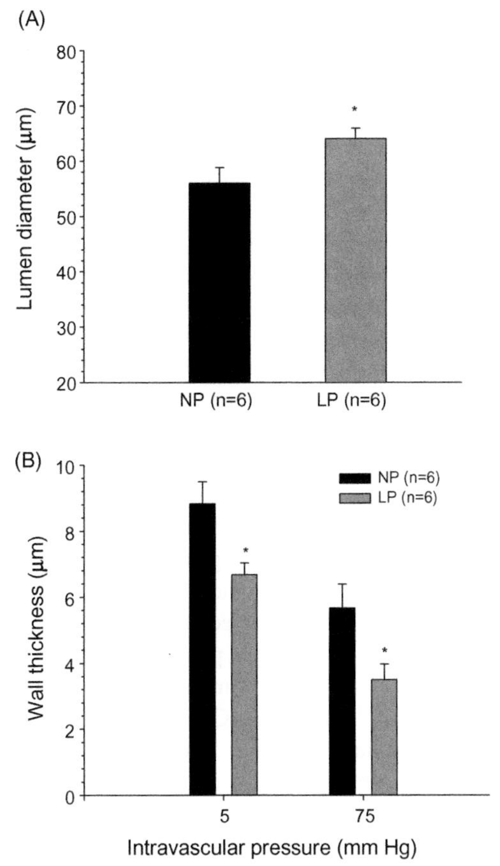

FIGURE 13.11 (A) Effect of pregnancy on the diameter of small penetrating arterioles in the brain. Measurements were taken using video microscopy, ex vivo and fully relaxed in papaverine solution to inhibit smooth muscle contraction at a pressure of 5 mm Hg. Measuring diameter at low pressure when changes in distensibility are minimal provides a more accurate assessment of arteriole remodeling and growth. Graph shows arteriolar lumen diameter from nonpregnant (NP, black bar) and late-pregnant (LP, gray bar) animals. Pregnancy causes outward remodeling of cerebral arterioles (*$p < 0.05$ vs. NP). This effect likely decreases small-vessel resistance under conditions when the vasculature is fully dilated, such as during acute hypertension and forced dilatation. (B) Effect of pregnancy on arteriolar wall thickness. Measurements were taken similarly to those shown in (A) at pressures of 5 and 75 mm Hg. Graph shows wall thickness of arterioles from nonpregnant (NP, black bar) and late-pregnant (LP, gray bar) animals. Notice that in addition to causing outward remodeling of arterioles that increases lumen diameter, pregnancy also causes arterioles to have significantly thinner walls (*$p < 0.05$ vs. NP). This effect on arteriolar wall thickness would be expected to cause a significant elevation in wall stress, especially under conditions of acute hypertension when both lumen diameter and arteriolar pressure are severely increased.[73]

reported that used invasive methods to assess pregnancy-related CBF in humans. In fact, there are obvious major challenges encountered when assessing CBF in the human, and accurate methods are either invasive or require radioactive substances.

Transcranial Doppler (TCD) Ultrasonography

Doppler ultrasonography is the most widely used noninvasive technique to assess the intracerebral circulation. It has been employed extensively in neurosurgical patients for the early detection of cerebral vasospasm following subarachnoid hemorrhage.[75] TCD studies in the middle cerebral artery provide information on changes in flow velocity of red blood cells and, when combined with blood pressure, an index of relative cerebral perfusion and cerebrovascular resistance is derived in the downstream arterioles.[76] It has been assumed that vascular constriction significantly increases the resistance met by blood inflowing from arteries supplying the microvasculature. Resistance to flow is inversely proportional to the fourth power of vessel radius provided that there is steady-state laminar flow. When extrapolating CBF using transcranial Doppler from velocity data, several assumptions are made, one of which is the caliber or cross-sectional area of the artery studied, which is likely to change dynamically.

Studies in normal pregnant women have demonstrated a decrease in mean velocity in the middle cerebral artery as pregnancy progresses, with return to nonpregnant values in the puerperium.[77–79] The decreased mean velocity is presumed secondary to decreased vascular resistance, which could imply the presence of more distal arteriolar vasodilatation and is in agreement with studies demonstrating outward remodeling of small vessels in the brain during pregnancy.[67] A recent cross-sectional study used dual-beam angle-independent digital ultrasound, which can measure changes in artery diameter and thus obtain absolute CBF measurements, to measure blood flow changes in the internal carotid artery over the course of pregnancy in healthy women.[80] This study found CVR decreased from a nonpregnant value of 0.141 to 0.112 mm Hg × mL/100 g/min in the third trimester, with CBF increasing 22% from 42.2 mL/100 g/min in nonpregnant women to 51.8 mL/100 g/min in the third trimester. This study is limited by a cross-sectional analysis and there were eight-fold more patients studied in the third trimester compared with nonpregnant women, and five-fold more than in the first trimester.[80] The discrepancies in these studies demonstrate the difficulties in measuring CBF during pregnancy especially because plasma volume and hematocrit change so dramatically, influencing velocity measurements.

A number of investigators have shown increased middle cerebral artery blood flow velocity in preeclampsia.[78,81,82] Moreover, symptomatic preeclamptic women with visual disturbances or headaches were found to have the highest velocities.[83] Increased velocity was also reported for women with eclamptic seizures.[84] Low maternal middle cerebral artery resistance indices in the second trimester may be predictive of the subsequent development of preeclampsia.[85] Along the same lines, a lower ophthalmic artery resistive index in preeclamptic women who presented with clinical characteristics of PRES is suggested as a clinically applicable biomarker for cerebral overperfusion.[86] Increasing velocity with preeclampsia is assumed secondary to high resistance in downstream arterioles. In chronically hypertensive women without preeclampsia, however, there was not substantively increased CBF velocity despite elevated mean arterial pressure.[87] Also, CBF velocity in preeclampsia was reduced by both antihypertensive therapy and magnesium sulfate.[88,89] Because of this observation, many favored the vasospasm model for the etiopathogenesis of eclampsia. Alternatively, these findings can be ascribed to relief of cerebral vasospasm.

Dynamic CBF autoregulation testing using TCD methods is based on the response of CBF velocity to small physiological changes in arterial blood pressure. Women with preeclampsia demonstrate impaired dynamic cerebral autoregulation capacity.[90] However, a recent study found that changes in dynamic CBF autoregulation in pregnant women were not predictive of preeclampsia.[91]

Velocity-Encoded Phase Contrast MRI and Perfusion-Weighted Imaging

In contrast to relative changes in CBF, velocity-encoded MRI can determine absolute blood flow and has been used to measure intracranial, renal, and cardiopulmonary circulations.[92] Velocity-encoded phase contrast MRI is based on the principle that hydrogen nuclei in blood moving through a magnetic field gradient accumulate a phase shift which is proportional to their velocity. Blood flow is then calculated by multiplying blood flow velocity and the cross-sectional area of the vessel under study. Due to the high spatial resolution for vessel localization and cross-sectional area, flow measurements are highly accurate.

Physiological normative data of CBF in the large cerebral vessels in both hemispheres longitudinally across normal pregnancy and postpartum have been described.[19,93] With this type of measurement, however, CBF is not corrected to mL/min/100 g cerebral tissue, so the numbers are not comparable and are larger than those shown for the animal models and other human studies that have measured CBF. As shown in Table 13.1, CBF was significantly reduced by the end of the first trimester. Flow then remained constant until 36–38 weeks, at which time there was another significant fall at term (Fig. 13.12). Taken together, there is a 20% decrease in CBF at term. Middle and posterior cerebral artery diameters remained unchanged throughout pregnancy and postpartum. These findings are in agreement with most TCD studies in which middle cerebral arterial flow velocities decreased and vessel wall tone diminished as pregnancy advanced.[77,79] A longitudinal TCD study demonstrated a decrease in cerebral perfusion pressure in mid pregnancy and after delivery in women with uncomplicated pregnancies.[94] The underlying cause of these changes that result in diminished CBF during pregnancy is not clear. It is hypothesized that downstream resistance arterioles become more dilated in order to maintain constant blood flow at the tissue level. However, changes in plasma volume during pregnancy that affect the hydrogen ion content of water may also have influenced the MR measurement. As newer MR sequences are developed, the physiological adaptation of pregnancy, especially changes in water, will need to be accounted for.

In women with severe preeclampsia, CBF determined at term with velocity-encoded phase contrast MRI is significantly increased when compared with normotensive pregnant controls.[95] Increased CBF is not related to vasodilatation of the large cerebral arteries because the diameter of the four main vessels remains unchanged. These observations also are similar to those obtained with TCD studies.[96] It remains speculative whether increased CBF is from changes in resistance of downstream resistance arterioles, increased cardiac output, increased mean arterial pressure, or central nervous system factors that control autoregulation.

MR perfusion-weighted imaging (PWI) has been used in preeclampsia and findings suggest that there is hyperperfusion-induced vasogenic edema without cerebrovascular spastic changes.[97] These conflicting observations with the

TABLE 13.1 Cerebral Blood Flow (mL/min) at Four Time Intervals

Artery	14–16 wks	28–32 wks	36–38 wks	Postpartum (6–8 wks)
MCA	135.2 ± 5.5	132.5 ± 4.6	118.2 ± 4.6	147.9 ± 5.0
PCA	52.4 ± 2.9	51.2 ± 2.4	44.2 ± 2.4	55.8 ± 2.7

From Zeeman et al., 2003,[93] with permission.
Values are expressed as the mean ± SE.
MCA = middle cerebral artery; PCA = posterior cerebral artery.

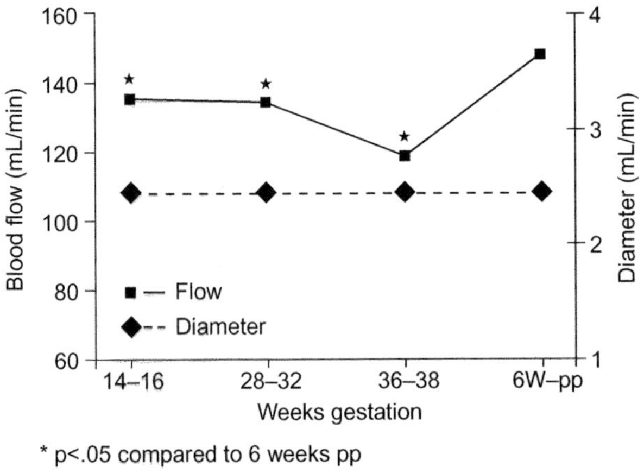

FIGURE 13.12 Middle cerebral arterial blood flow and vessel diameter determined longitudinally during pregnancy and compared with nonpregnant postpartum values in nine healthy women.[93]

different techniques probably indicate the difficulty of interpreting hemodynamic changes in preeclampsia.

Single Photon Emission Computed Tomography

The technique of single photon emission computed tomography (SPECT) involves intravenous injection of a radioisotope to determine alterations in regional CBF. The effect of early pregnancy was assessed in women between 7 and 19 weeks who planned termination. Regional flow in the cerebral frontal, temporal, and parietal lobes, as well as in the basal ganglia and cerebellum, was found to be increased about 10% during pregnancy compared with repeat studies done post abortion.[79] Case reports of SPECT used in women with preeclampsia described hyperemia in the posterior, temporal, lateral occipital, and inferior parietal cortex.[98] In a study of SPECT imaging in 63 eclamptic women, all demonstrated perfusion deficits in watershed areas.[99] And with a similar method, xenon CT, diffuse cerebral hyperperfusion and vasogenic edema without evidence of vasospasm were demonstrated in a woman with eclampsia.[100]

Proton MR Spectroscopy

The noninvasive method of proton MR spectroscopy (MRS) is used to investigate cerebral metabolism. Specifically, *intracellular metabolite diffusion* has been used to study women with preeclampsia. Increased choline was reported in edematous areas of the brain, and thought to reflect relative cerebral ischemia without infarction.[101,102] A lactate peak in eclampsia, even after complete reversibility of imaging abnormalities, suggested the presence of infarction. A marked decline in *N*-acetyl-aspartate (NAA) in a follow-up study of eclampsia correlated with the development of cerebral atrophy resulting from gross neuronal damage.

Near-Infrared Spectroscopy(NIRS)

This is an optical technique that allows real-time assessment of changes in tissue oxygenation and cerebral blood flow. Using this technique in women with several stages of hypertension, women with severe preeclampsia showed an increase in CBF with posture changes.[103]

MECHANISMS OF SEIZURE DURING PREGNANCY AND PREECLAMPSIA

Seizures consist of excessive release of excitatory neurotransmitters (especially glutamate), massive depolarization of network neurons, and bursts of action potentials.[104] Clinical and experimental evidence suggests that extended seizures can cause significant brain injury and later brain dysfunction.[104–106] In the brain, seizures can cause stroke, hemorrhage, edema, and brain herniation, and they also predispose to epilepsy and cognitive impairment later in life.[107–109] Acutely, seizures have significant effects on the cerebrovasculature, including BBB disruption and unregulated, excessive cerebral blood flow.[106,110] Glutamate receptor activation of vascular smooth muscle and endothelium can cause potent cerebral vasodilatation, diminished cerebrovascular resistance (CVR), and autoregulatory failure[106,110–113] In this regard, the effects of seizures are similar to those of hypertensive encephalopathy and PRES.

While the cause of the eclamptic seizure is not certain, it is thought to be stimulated by an acute rise in blood pressure superimposed on capillary interendothelial leakage from the preeclampsia syndrome, further discussed in Chapter 9. However, it is well established that *de novo* seizures can occur at relatively normal blood pressures and even in women with seemingly uncomplicated pregnancies, that is, without preeclampsia.[61,63,114–116] In fact, women who develop eclampsia exhibit a wide spectrum of signs and symptoms ranging from severe hypertension and proteinuria to mild or absent hypertension with no proteinuria.[57,61,63,115] For example, in one study of 53 pregnancies complicated by eclampsia, only seven women (13%) had severe preeclampsia prior to seizure,[61] and in another study 16% of women were considered normotensive.[63] It is therefore important to consider that the adaptation to normal pregnancy may predispose the brain to seizure.

An important aspect to consider is how pregnancy might change the excitability state of the brain. Pioneering work in the 1940s established that certain progesterone metabolites were potent sedatives and anesthetics that act too rapidly to be due to a genomic effect.[117,118] For example, some progesterone metabolites enhance the interaction of γ-aminobutyric acid (GABA) – the major inhibitory receptor in the brain – with its receptor, but have minimal action on glutamate receptors.[119] Fluctuations in neurosteroid levels during pregnancy result in selective changes in the expression and function of $GABA_A$ receptors that cause neuronal hyperexcitability.[120,121] Moreover, animal studies demonstrate late-pregnancy is associated with decreased expression of the δ subunit of the $GABA_A$ receptor and increased neuronal excitability of acute brain slices.[120,121] And a recent study of genome-wide methylation found that leukocytes obtained from women with preeclampsia had hypermethylation of *GABRA1*, the $\alpha 1$ subunit of the $GABA_A$ receptor.[122] Although implications of this are not clear, because the hypermethylation was genome-wide, it is possible that preeclampsia is a state of altered neuronal $GABA_A$ receptor function as well, further making the brain hyperexcitable. These studies suggest that pregnancy is a state of decreased seizure threshold that may partially explain "unheralded" seizure in seemingly normal pregnancy, whereas preeclampsia may have an even lowered seizure threshold due to altered $GABA_A$ function.

ROLE OF CIRCULATING FACTORS IN ECLAMPSIA

During pregnancy, large amounts of hormones, including growth factors, cytokines and chemokines, are produced, mostly by the fetal-placental unit, and secreted into the maternal circulation.[123,124] Eclamptic seizures occur almost exclusively in the last half of pregnancy, and often during late gestation when levels of these circulating hormones are highest.[61,63,115,116] In addition, there is a high propensity for seizure to occur in multi-fetal gestations when the levels of hormones are elevated even further, suggesting a role for these circulating proinflammatory mediators in eclampsia.[114,125] The effect of circulating factors on the excitability of hippocampal neurons was investigated by exposing brain slices in culture to serum from nonpregnant and late-pregnant rats.[126] After 48 hours of exposure, evoked network potentials were measured and quantified by a graded scale that represented a range from normal to seizure activity. Exposure to nonpregnant serum did not affect slice excitability relative to control conditions. But serum from late-pregnant animals caused hyperexcitability in the hippocampal slices.

When soluble TNF-α receptor type 1 (sTNF-R1) was added to the serum to inhibit TNF-α-dependent signaling, it prevented the increase in excitability. It was also determined that TNF-α levels were not different between pregnant vs. nonpregnant serum, but the late-pregnant serum caused microglial activation (Fig. 13.13). This study importantly demonstrates that when applied directly to hippocampal slices, serum from late-pregnant animals causes neuronal excitability mediated by TNF-α, likely from microglia. However, seizures do not normally occur in intact animals during pregnancy despite the presence of hyperexcitable factor(s) in serum. This finding points to a critical role of the BBB in protecting the brain from seizure-provoking circulating factors. How the BBB adapts to pregnancy or is altered by preeclampsia is just becoming understood. In addition, circulating factors have a significant effect on BBB permeability, which is discussed more below.

Blood–Brain Barrier

This unique structure is made up of the cerebral endothelium, which is characterized by high electrical resistance tight junctions and low hydraulic conductivity.[127,128] Together, these provide a barrier through which there is little ionic or solute flux, which minimizes the effect of

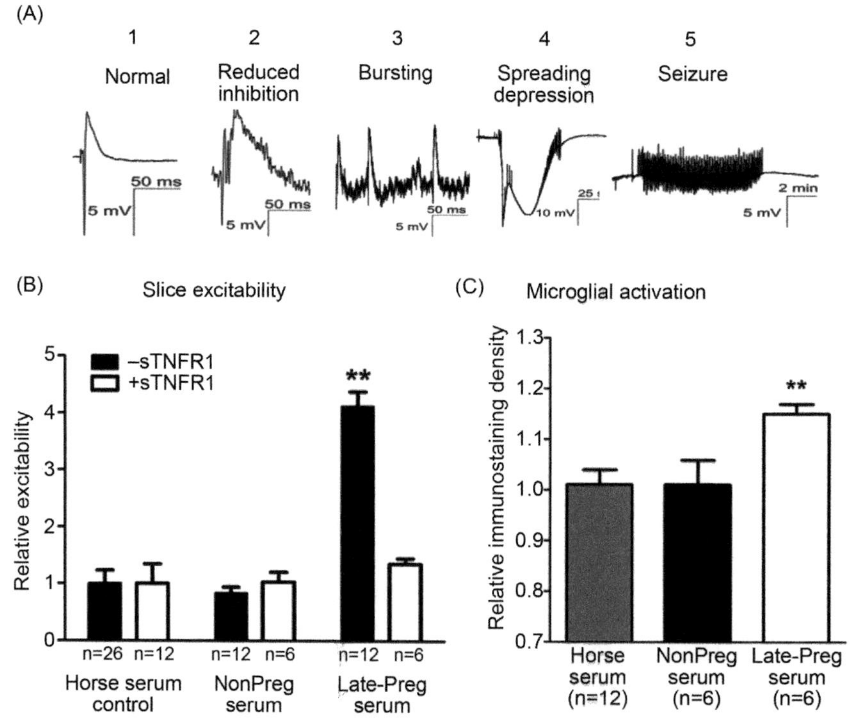

FIGURE 13.13 Relative neuronal excitability and microglial activation of hippocampal slices after exposure to nonpregnant or late-pregnant rat serum. (A) Graded scale for which neuronal network excitability was measured. (B) Graph showing relative excitability of neuronal networks after replacement of horse serum with nonpregnant or late-pregnant rat serum for 48 hours in the absence (– sTNFR1) and presence (+ sTNFR1) of soluble tumor necrosis factor receptor 1 to inhibit TNF-α signaling. Late-pregnant serum increased neuronal excitability, which was prevented by sTNFR1. (C) Microglial activation of slices in response to the different serums. Late-pregnant serum caused microglial activation. Reprinted with permission.[126]

hydrostatic pressure on capillary filtration and thus has a protective influence against vasogenic brain edema. An acute rise in blood pressure that causes cerebrovascular resistance to decrease, however, can severely increase hydrostatic pressure in the microcirculation, and this in turn disrupts the BBB, resulting in hydrostatic brain edema.[129] One mechanism by which cerebral edema forms with eclampsia is by increased BBB permeability due to pathologically increased vascular hydrostatic pressure – a form of vasogenic edema or hydrostatic brain edema.[129,130]

Blood–Brain Barrier in Pregnancy

It appears that cerebrovascular adaptation to normal pregnancy may actually predispose to development of hydrostatic brain edema when blood pressure is acutely elevated. Outward remodeling of parenchymal arterioles, discussed above, likely decreases small-vessel resistance in the brain. In addition, pregnancy appears to cause greater BBB permeability and/or hydraulic conductivity in response to *identical* hydrostatic pressures, which may play a major role in development of cerebral edema. A number of animal studies illustrate this principle. BBB permeability to Lucifer Yellow, a polar tracer that does not pass through tight junctions, was measured in pregnancy.[67] These studies were conducted on isolated and pressurized cerebral arteries from nonpregnant and late-pregnant rats such that permeability in response to the same change in hydrostatic pressure could be measured. The permeability of the BBB in response to pressure is shown in Fig. 13.14. It can be

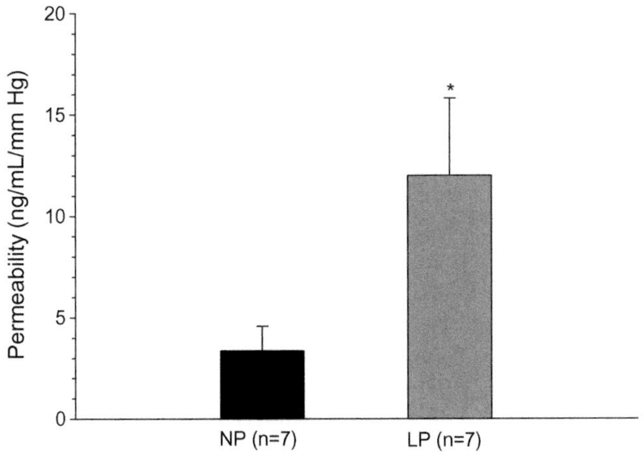

FIGURE 13.14 The effect of pregnancy on blood–brain barrier permeability in response to elevated hydrostatic pressure. Permeability to Lucifer Yellow was assessed in cerebral arteries in response to increases in pressure from 60 to 200 mm Hg. These data represent the average slope of the regression lines of the pressure versus permeability curves, or the rate of permeability in response to pressure, from nonpregnant (NP, black bar) and late-pregnant (LP, gray bar) animals. Notice that pregnancy causes a significant increase in blood–brain barrier permeability in response to the same change in hydrostatic pressure ($*p < 0.05$ vs. NP).[67]

seen that late-pregnant animals had greater permeability compared with nonpregnant animals. This suggests that at the same hydrostatic pressure, there is greater BBB permeability to Lucifer Yellow during pregnancy. These *in vitro* studies were followed up with *in vivo* measures of BBB permeability in nonpregnant and late-pregnant rats. It was found that pregnancy caused a significant increase in BBB permeability to both large and small tracers in response to acute hypertension.[67]

These observations are significant for several reasons. First, they suggest that, independent of changes in vascular resistance, pregnancy affects the cerebral endothelium, making it more permeable in response to increased pressure. Second, Lucifer Yellow is a polar compound commonly used to measure fluid-phase endocytosis. Therefore, increased permeability suggests that transcellular transport may be a major contributor to BBB permeability during acute elevations in pressure. In addition, while these experiments used cerebral arteries to measure BBB permeability, other studies have shown that it is the cerebral veins that disrupt first during acute hypertension.[131,132] Importantly, a recent study in rats found that pregnancy causes cerebral veins to enlarge with a thinner wall, similar to brain parenchymal arterioles.[133] This type of remodeling may promote rupture (due to the thinner wall) or stasis of blood (due to enlargement of the lumen) and contribute to brain pathologies during pregnancy.

Proinflammatory and Antiangiogenic Factors and the Blood–Brain Barrier

A major factor in the pathogenesis of preeclampsia is the presence of circulating proinflammatory and antiangiogenic factors that are thought to produce maternal vascular inflammation and cause the endothelial dysfunction that characterizes the preeclampsia syndrome.[134–136] Recent studies have investigated how circulating factors produced during pregnancy and preeclampsia affect the BBB. Plasma obtained from women with severe preeclampsia was found to increase BBB permeability, which was prevented by nonselective inhibition of vascular endothelial growth factor (VEGF) receptors.[137] A subsequent study found that plasma from women with early-onset preeclampsia, but not late-onset preeclampsia, had elevated levels of oxidized low-density lipoprotein (oxLDL) that were causative in increasing BBB permeability.[138] The increase in BBB permeability was prevented by inhibitory antibodies to lectin-like oxLDL (LOX-1) receptors. It was subsequently determined that the mechanism by which oxLDL activation of LOX-1 caused increased BBB permeability was through generation of the reactive oxygen and nitrogen species peroxynitrite.[139] This finding also explains why VEGF receptor inhibition also prevented the increase in permeability in response to preeclamptic plasma – VEGF inhibition would decrease nitric oxide, one of the two factors needed for peroxynitrite generation.

Aquaporins and Cerebral Edema During Pregnancy

The aquaporins are a family of channel-forming transmembrane proteins that facilitate the movement of water, glycerol, and other solutes across cell membranes.[140,141] Three aquaporins – AQP1, AQP4, and AQP9 – have been identified in the brain.[142–144] AQP4 is the predominant aquaporin in the brain, and it is mainly localized in the endfeet of astrocytes surrounding blood vessels, the glial limitans membranes, and ependyma.[145–147] Given its location, AQP4 is thought to facilitate the movement of water at the blood–brain interface and at the blood–cerebrospinal fluid barrier.[142,146,148]

All three aquaporins are expressed in the brain during pregnancy.[144,149] Importantly, AQP4 has expression levels that are significantly increased during pregnancy compared with the nonpregnant state when studied with quantitative PCR and Western blot.[144,149] These findings have led to the suggestion that pregnancy alone is a state of altered brain water homeostasis. Because AQP4 has not been found in cerebral endothelia, however, it is unlikely to affect hydraulic conductivity of the BBB. It is more likely that increased AQP4 expression during pregnancy is more related to edema resolution, as recently shown in brain injury models.[150] Genetic variations in AQP4 might serve to explain why some women with preeclampsia develop cerebral edema and eclampsia while others do not.

An increase in AQP4 in the brain during pregnancy has several implications related to eclampsia. AQP4 has been shown to affect K^+ homeostasis in the brain and modulate seizure activity. AQP4 knockout mice have a decreased seizure threshold when exposed to the chemoconvulsant pentylenetetrozol or to electrically stimulated seizure.[151,152] This is thought to be mediated by altered brain K^+ homeostasis.[153] Thus, increased AQP4 expression during pregnancy may be another mechanism by which the seizure threshold is lowered. Increased AQP4 may also be important for clearing potentially damaging and seizure-provoking serum factors that gain entry to the brain through a disrupted BBB during preeclampsia. AQP4 only transports water, not solute; however, a role for AQP4 in cerebrospinal fluid (CSF) and interstitial solute clearance has recently been demonstrated.[154] Because the brain lacks a lymphatic circulation, bulk fluid flow driven by AQP4 in astrocytic endfeet clears solute and regulates extracellular levels of protein in the brain. Thus, while normal pregnancy is associated with increased AQP4 that may serve to clear solute that is elevated in the brain during pregnancy, if AQP4 is decreased during preeclampsia there may be accumulation of damaging factors in the brain. The expression of AQP4 in the brain during preeclampsia is not known.

Effect of Magnesium Sulfate Treatment

As discussed in Chapters 12 and 20, magnesium sulfate is used for prevention and treatment of eclampsia. The mechanism by which magnesium acts to prevent convulsions is not clear. Some studies have suggested that magnesium acts on the cerebrovasculature to cause vasodilatation and relief of vasospasm.[155–157] Yet other studies have shown magnesium sulfate treatment has little effect on cerebral hemodynamics and CBF.[88,158] The mechanism of action of magnesium sulfate in prevention of eclampsia has been reviewed.[159] In a study using isolated and pressurized cerebral arteries, magnesium was shown to have a modest vasodilatory effect, but the sensitivity of the response decreased during pregnancy and postpartum.[160] Also, a clinical trial found that when compared with the calcium-channel blocker nimodipine, magnesium sulfate was more effective in preventing eclamptic convulsions.[161] These findings suggest that the primary action of magnesium sulfate to prevent eclampsia is not an effect on vasospasm.

There have been numerous reports that treatment with magnesium sulfate decreases BBB permeability and edema formation in a variety of brain injury conditions, including traumatic brain injury, septic encephalopathy, and hypoglycemia.[162–164] Euser et al.[165] demonstrated that magnesium sulfate treatment during pregnancy decreased BBB permeability to Evans blue in response to acute hypertension (Fig. 13.15). The effect was most pronounced in the posterior brain, a region that is most susceptible to edema formation during eclampsia.

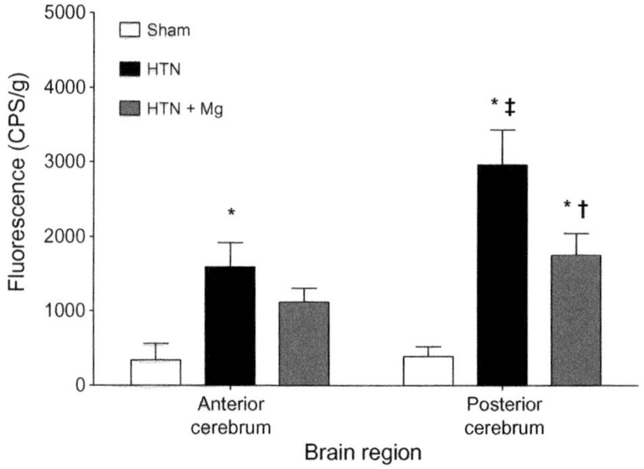

FIGURE 13.15 Graph showing blood–brain barrier permeability to Evans blue after acute hypertension and autoregulatory breakthrough in late-pregnant animals that were either sham controls (Sham, open bar), untreated (HTN, black bar), or treated with 270 mg/kg magnesium sulfate every 4 hours for 24 hours by intraperitoneal injection (HTN + Mag, gray bar). Permeability was measured by *in situ* perfusion in the anterior and posterior brain regions. Blood–brain barrier permeability was significantly increased with acute hypertension and autoregulatory breakthrough compared with sham controls ($*p < 0.05$). The increase in permeability was greater in the posterior versus anterior brain region ($‡p < 0.05$). Magnesium sulfate treatment significantly decreased blood–brain barrier treatment in posterior cerebrum in response to acute hypertension ($†p < 0.05$). Reprinted with permission.[165]

Cerebral Hemorrhage

As discussed previously, and shown in Fig. 13.2, nonlethal intracranial hemorrhage is also frequently found in women with eclampsia. In some women, sudden death occurs synchronously with the convulsion or follows shortly thereafter, and is the result of massive cerebral hemorrhage (see Fig. 13.1). Such hemorrhage is more common in older women with underlying chronic hypertension and is thought to be secondary to longstanding hypertension-induced lipohyalinosis, which damages small or medium-sized cerebral arteries in the striatocapsular area, thalamus, cerebellum, and brainstem.[166] These changes are termed Charcot–Bouchard aneuroforms or miliary aneuroforms.[167,168]

Hemorrhage may also develop in areas of cerebral ischemia or infarction that transform into a hemorrhagic infarction. These are probably more common in young women with HELLP syndrome and eclampsia.[10,14] Women with preeclampsia may also experience subarachnoid hemorrhage. In such cases a small amount of blood can be seen over the convexity of the frontal/parietal lobes extending into the sylvian fissure or interhemispheric tissue and is hypothesized to be the result of rupture of cortical petechiae over the surface of the brain or rupture of small pial veins.[167] Only rarely is intracerebral hemorrhage in women with preeclampsia due to a ruptured aneurysm or arteriovenous malformation.[169]

Cortical Blindness

Possibly due to the differential innervation of the posterior cerebral circulation as discussed, visual symptoms may manifest in 40% of preeclamptic women, and on rare occasions they may be the initial symptom.[170,171] The older term used to describe loss of vision was *amaurosis*, and in addition there are scotomata, blurred vision, diplopia, and chromatopsia. Retinal abnormalities, including edema, vascular changes such as arteriolar vasospasm, thrombosis of the central retinal artery, and retinal detachment can usually be ruled out. In addition, pupillary light reflexes and ocular movements remain intact, and there are normal ophthalmological findings. Focal occipital lobe edema, including bilateral edema of the lateral geniculate nuclei, can be seen on neuroimaging with this type of cortical blindness.[172] The majority of preeclamptic women with cortical blindness recover vision over a period varying from 2 hours to 21 days.[170] Clinical recovery typically precedes normalization of neuroimaging findings.

REMOTE CEREBROVASCULAR HEALTH FOLLOWING PREECLAMPSIA AND ECLAMPSIA

Until recently, it was commonly held that seizures with preeclampsia have no significant long-term sequelae. Findings have now accrued, however, that support the view that complete reversibility is not always the case.

Visual Functioning

Rarely, there may be permanent visual defects with severe preeclampsia or eclampsia. There are two possible causes of these. One is retinal artery ischemia and infarction – *Purtscher retinopathy* – and the other is associated with MR-imaging evidence of infarctions in the lateral geniculate nuclei.[172–174] In an examinination of visual fields in over 40 women who had suffered eclampsia several years previously none had evidence of visual field defects.[175] Vision-related quality of life impairment expressed by formerly eclamptic women may therefore be related to problems with higher-order visual functions.

Brain White Matter Lesions

In the strictest sense, *posterior reversible encephalopathy syndrome* (PRES) is inappropriate to describe the brain edema seen with MRI and CT in eclamptic women since it is not necessarily confined to the posterior cerebrum and there may be incomplete resolution and even ensuing infarction. Specifically, as discussed on p. 272, a quarter of eclamptic women have restricted diffusion at time of seizure. Persistent cerebral T2 hyperintensities can still be visualized 6–8 weeks postpartum with MRI.[16,20] Clinical parameters indicating eclampsia severity in women with or without such hyperintense lesions are shown in Table 13.2. Focal gliosis has been described in 40 % of eclamptic women on average 9 days following the eclamptic seizure.[176] The subsequent evolution of brain abnormalities in women who suffered eclampsia is unknown since longitudinal follow-up studies are lacking. For nonpregnant patients with PRES some information has become available describing neuroimaging follow-up. Imaging abnormalities frequently consist of nonspecific white matter lesions (WMLs), gliosis and infarction but may also include haemorrhage (unpublished data).

In a long-term follow-up study of women with eclampsia that averaged 7 years, there was a 41% prevalence of cerebral white matter lesions, which was significantly higher compared to the 17% found in age-matched women who had had a normotensive pregnancy[177] (Fig. 13.16). The total volume of such lesions was significantly larger in the previously eclamptic women. Women who suffered preeclampsia without eclamptic seizures, on average 5 years previously, also demonstrated such WMLs in 37 % of cases.[178] Current hypertension and a history of early-onset preeclampsia (<37 weeks) were independently associated with the presence of such lesions (Fig. 13.17). Women with term preeclampsia appeared to have an incidence of brain WMLs similar to control women who had normotensive pregnancies. These atypical WMLs were predominantly

TABLE 13.2 Comparison of Clinical Parameters that Indicate Eclampsia Severity in the Presence or Absence of Cerebral Infarctions

Factor	Infarctions Present (n=6)	Infarctions Absent (n=18)	p Value
Age (y)	23.0 ± 7.1	20.7 ± 6.4	.465
Gestational age (wk)	36.2 ± 2.4	39.0 ± 2.1	.012
MAP (mm Hg)	123 ± 10.8	112 ± 11.1	<.001
ΔMAP (mm Hg)	40.4 ± 15.1	30.1 ± 11.7	.096
Serum creatinine >0.9 mg/dL	4 (67%)	2 (11%)	.007
Proteinuria ≥3 + (dipstick)	4 (67%)	3 (17%)	.020
Platelet count (× 1000/μL)			
Mean	105 ± 53	203 ± 63	.002
<150	4 (67%)	4 (22%)	.046
<100	3 (50%)	1 (6%)	.012
HELLP syndrome	3 (50%)	1 (6%)	.012
Multiple seizures	5 (83%)	2 (11%)	<.001

From Zeeman et al. 2004,[16] with permission.
MAP = maximal mean arterial pressure in proximity to seizure(s); ΔMAP = change in MAP from most recent prenatal visit compared with MAP proximate to seizure; HELLP = hemolysis, elevated serum transaminase levels, low platelets.

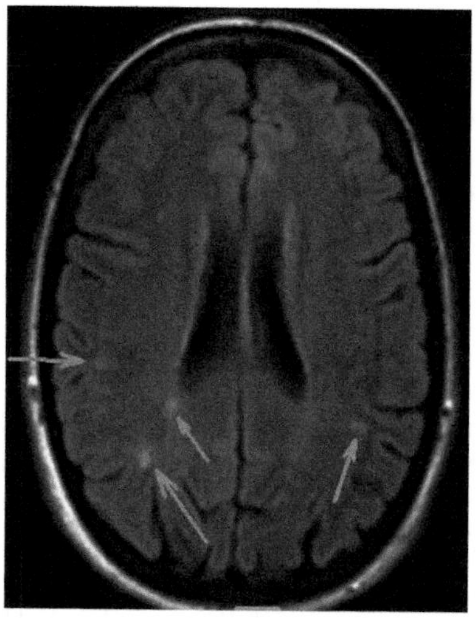

FIGURE 13.16 Magnetic resonance image (FLAIR) 7 years following eclampsia, demonstrating several white matter lesions (arrows).

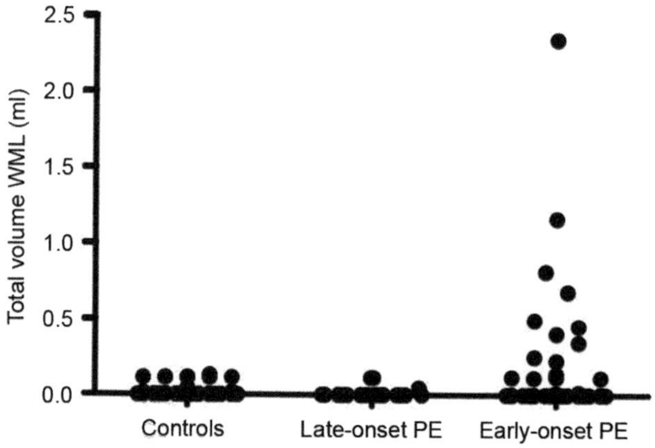

FIGURE 13.17 Total volume of white matter lesions (WMLs) in the control group, late-onset pre-eclamptic group, and early-onset pre-eclamptic group. Kruskal–Wallis test: $p < 0.01$. Post-hoc Mann–Whitney test: early-onset PE versus late-onset PE: $p = 0.01$. Post-hoc Mann–Whitney test: early-onset PE versus controls: $p < 0.01$. PE, pre-eclampsia. Reprinted with permission.[178]

found in the frontal lobes, followed by the parietal, insular, and temporal lobes, which are not the areas typically involved in PRES.[179]

The exact pathophysiology underlying these morphological brain changes in formerly preeclamptic and eclamptic women and their clinical relevance remain so far unknown. Cerebrovascular events in eclampsia appear to constitute a continuum characterized by an initial, reversible phase of vasogenic edema caused by hypertension along with endothelial dysfunction. In some cases, severe vasogenic edema may reduce cerebral perfusion to cause focal ischemia. Indeed, in a quarter of eclamptic women, MRI documents a transition from reversible vasogenic edema to cytotoxic edema, irreversible cerebral ischemia, and infarction.[16,20] In extreme cases, progressive global brain edema may lead to further neurological deterioration, including blindness, mental status changes, coma, transtentorial herniation, and death.[9,170]

The presence of atypical cerebral WMLs may indicate that preeclampsia might be a risk marker for early cerebrovascular damage. The predisposition of formerly preeclamptic women to later cardiovascular and cerebrovascular disease may be an important factor for the development of such lesions.[180] In particular, the presence of hypertension seems important for the development and progression of WMLs in elderly populations[181] but also in younger cohorts.[182,183] Whether a history of posterior reversible encephalopathy syndrome may be an additive risk factor for the development of these lesions remains unknown.

Neurocognitive Functioning

There is scarce literature concerning long-term clinical consequences of PRES in either obstetrical or non-obstetrical patients regarding neurocognitive status. Persistent brain lesions in formerly eclamptic women raise the possibility that they may develop evidence of subtle brain dysfunction. If so, this contradicts the predominant thinking that eclampsia is followed by full neurological recovery.

Recently, several studies were designed to assess this. Importantly, studies of neurocognitive functioning in these women were not available before eclampsia and longitudinal follow-up is lacking. Formerly preeclamptic and eclamptic women may report subjective cognitive difficulties in daily life, the interpretation of which is cumbersome, since they are affected by emotional factors such as anxiety and depression.[184–187] Whether these subjective reports could be interpreted as reflecting actual cognitive dysfunction has been assessed in some small pilot studies.[188,189] These only included preeclamptic women who were investigated shortly after the index pregnancy and demonstrate mixed results. We recently assessed cognitive functioning in a relatively large cohort of formerly preeclamptic and eclamptic women, as well as women who had normotensive pregnancies, several years following the index pregnancy. Formerly preeclamptic and eclamptic women demonstrated no clear cognitive impairment but for a slightly worse performance on visuomotor speed as compared to controls. Contrary to the well-structured test setting, they do report more cognitive failures, which are thought to reflect primarily neurocognitive dysfunctions in complex daily-life situations.[190]

Epilepsy

Some formerly eclamptic women have been described who developed chronic epilepsy. In this regard, hippocampal sclerosis has been reported in patients who developed temporal lobe epilepsy months to years after having PRES or eclampsia.[191–193] Several factors are involved in preeclampsia and PRES and may relate to epilepsy, including gliosis, immune cell activation, and persistent BBB disruption.

Stroke

It is not known whether strokes are more common in women who have experienced eclampsia but white matter hyperintensities define small-vessel disease and in general predict an increased risk of stroke, dementia, and death. Therefore, the presence of such lesions in formerly eclamptic and preeclamptic women may indicate an increased risk of cerebrovascular events.[194] Indeed, there are now epidemiological data that suggest long-term cerebrovascular consequences are more common in women who have preeclampsia compared with normotensive pregnant women. In addition, there is a three- to five-fold increase in fatal stroke in previously preeclamptic women.[169,195,196]

References

1. Sheehan JL, Lynch JB. *Pathology of Toxemia of Pregnancy.* New York: Churchill Livingstone; 1973.
2. Govan AD. The pathogenesis of eclamptic lesions. *Pathol Microbiol (Basel).* 1961;24:561–575.
3. Melrose EB. Maternal deaths at King Edward VIII Hospital, Durban. A review of 258 consecutive cases. *S Afr Med J.* 1984;65:161–165.
4. Richards A, Graham DI, Bullock MRR. Clinicopathological study of neurological complications due to hypertensive disorders of pregnancy. *J Neurol Neurosurg Psych.* 1988;51:416–421.
5. Brewer J, Owens MY, Wallace K, et al. Posterior reversible encephalopathy syndrome in 46 of 47 patients with eclampsia. *Am J Obstet Gynecol.* 2013;208(468):e1–e6.
6. Brown CE, Purdy P, Cunningham FG. Head computed tomographic scans in women with eclampsia. *Am J Obstet Gynecol.* 1988;159:915–920.
7. Kirby JC, Jaindl JJ. Cerebral CT findings in toxemia of pregnancy. *Radiology.* 1984;151:114.
8. Donaldson JO. Eclamptic hypertensive encephalopathy. *Semin Neurol.* 1988;8:230–233.
9. Cunningham FG, Twickler DM. Cerebral edema complicating eclampsia. *Am J Obstet Gynecol.* 2000;182:94–100.
10. Topuz S, Kalelioglu I, Iyibozkurt AC, et al. Cranial imaging spectrum in hypertensive disease of pregnancy. *Clin Exp Obstet Gynecol.* 2008;34:194–197.
11. Koyama M, Tsuchiya K, Hanaoka H, et al. Reversible intracranial changes in eclampsia demonstrated by MRI and MRA. *Eur J Radiol.* 1997;25:44–46.
12. Sarma GRK, Kumar A, Roy AK. Unusual radiological picture in eclamptic encephalopathy. *Neurol India.* 2003;5:127–128.
13. Moodley J, Bobat SM, Hoffman M, PLA Bill. Electroencephalogram and computerised cerebral tomography findings in eclampsia. *Brit J Obstet Gynaecol.* 1993;100:984–988.
14. Salerni A, Wald S, Flanagan M. Relationships among cortical ischemia, infarction and hemorrhage in eclampsia. *Neurosurgery.* 1988;22:408–410.
15. Bartynski WS. Posterior reversible encephalopathy syndrome, Part 1: fundamental imaging and clinical features. *AJNR.* 2008;29:1036–1042.
16. Zeeman GG, Fleckenstein JL, Twickler DM, Cunningham FG. Cerebral infarction in eclampsia. *Am J Obstet Gynecol.* 2004;190:714–720.
17. Matsuda H, Sakaguchi K, Shibasaki T, et al. Cerebral edema on MRI in severe preeclamptic women developing eclampsia. *J Perinat Med.* 2005;33:199–205.
18. Demirtas Ö, Gelal F, Vidinli BD, Demirtas LÖ, Uluç E, Baloglu A. Cranial MR imaging with clinical correlation in pre-eclampsia and eclampsia. *Diagn Intervent Radiol.* 2005;11:189–194.
19. Morriss MC, Twickler DM, Hatab MR, Clarke GD, Peshock RM, Cunningham FG. Cerebral blood flow and cranial magnetic resonance imaging in eclampsia and severe preeclampsia. *Obstet Gynecol.* 1997;89:561–568.

20. Loureiro R, Leite CC, Kahhale S, et al. Diffusion imaging may predict reversible brain lesions in eclampsia and severe preeclampsia: initial experience. *Am J Obstet Gynecol.* 2003;189:1350–1355.
21. Mangla R, Kolar B, Almast J, Ekholm SE. Border zone infarcts: pathophysiologic and imaging characteristics. *Radiographics.* 2011;31:1201–1214.
22. Ulrich K, Troscher-Weber R, Tomandl BF, Neundorfer B, Reinhardt F. Posterior reversible encephalopathy in eclampsia; diffusion-weighted imaging and apparent diffusion coefficient mapping as prognostic tools? *Eur J Neurol.* 2006;13:309–310.
23. Doelken M, Lanz S, Renenrt J, Alibek S, Richter G, Doerfler A. Differentiation of cytotoxic and vasogenic edema in a patient with reversible posterior leukoencephalopathy syndrome using diffusion-weighted MRI. *Diagn Interv Radiol.* 2007;13:125–128.
24. Edvinsson L, MacKenzie ET. General and comparative anatomy of the cerebral circulation. In: Edvinsson L, Krause DN, eds. *Cerebral Blood Flow and Metabolism.* 2nd ed. Philadelphia: Lippincott Williams & Wilkins; 2002:384–394. [Chapter 2].
25. Koch S, McClendon MS, Bhatia R. Imaging evolution of acute lacunar infarction: leukoariosis or lacune? *Neurology.* 2011;77:1091–1095.
26. Liu C, Murphy NE, Li W. Probing white-matter microstructure with higher-order diffusion tensors and susceptibility tensor MRI. *Front Integr Neurosci.* 2013;7:1–9.
27. Tamaki K, Sadoshima S, Baumbach GL, Iadecola C, Reis DJ, Heistad DD. Evidence that disruption of the blood-brain barrier precedes reduction in cerebral blood flow in hypertensive encephalopathy. *Hypertension.* 1984;6:I75–I81.
28. Trommer BL, Homer D, Mikhael MA. Cerebral vasospasm and eclampsia. *Stroke.* 1988;19:326–329.
29. Ito I, Sakai T, Inagawa S, Utsu M, Bun T. MR angiography of cerebral vasospasm in preeclampsia. *Am J Neuroradiol.* 1995;16:1344–1346.
30. Garcia-Reitboeck P, Al-Memar A. Reversible cerebral vasoconstriction after preeclampsia. *New Engl J Med.* 2013;369:3.
31. Hauser RA, Lacey DM, Knight MR. Hypertensive encephalopathy. *Arch Neurol.* 1988;45:1078–1083.
32. Paulson OB. Blood-brain barrier, brain metabolism and cerebral blood flow. *Eur Neuropsychopharmacol.* 2002;12:495–501.
33. Dinsdale HB, Mohr JP. Hypertensive encephalopathy. In: Henry JM, Barnett JP, Mohr JP, Stein BM, Yatsu FM, eds. *Stroke. Pathophysiology, Diagnosis and Management.* 3rd ed. New York: Churchill Livingston; 1998:869–874. [Chapter 34].
34. Hinchey J, Chaves C, Appignani B, et al. A reversible posterior leukoencephalopathy syndrome. *New Engl J Med.* 1996;334:494–500.
35. Narbone MC, Musolino R, Granata F, Mazzu I, Abbate M, Ferlazzo E. PRES: posterior or potentially reversible encephalopathy syndrome ? *Neurol Sci.* 2006;27:187–189.
36. Schwartz RB, Mulkern RV, Gudbjartsson H, Jolesz F. Diffusion-weighted MR imaging in hypertensive encephalopathy: clues to pathogenesis. *Am J Neurorad.* 1998;19: 859–862.
37. Liman TG, Bohner G, Heuschmann PU, Scheel M, Endres M, Siebert E. Clinical and radiological differences in posterior reversible encephalopathy syndrome between patients with preeclampsia-eclampsia and other predisposing diseases. *Eur J Neurol.* 2012;19:935–943.
38. Ahn KJ, You WJ, Jeong SL, et al. Atypical manifestations of reversible posterior leukoencephalopathy syndrome: findings on diffusion imaging and ADC mapping. *Neuroradiology.* 2004;46:978–983.
39. Hugonnet E, Da Ines D, Boby H, et al. Posterior reversible encephalopathy syndrome (PRES): features on CT and MR imaging. *Diagn Interv Imaging.* 2013;94:45–52.
40. Veltkamp R, Kupsch A, Polasek J, Yousry TA, Pfister HW. Late onset postpartum eclampsia without preeclamptic prodromi: clinical and neuroradiological presentation in two patients. *J Neurol Neurosurg Psych.* 2000;69:824–827.
41. Stevens CJ, Heran MK. The many faces of the posterior reversible encephalopathy syndrome. *Brit J Radiol.* 2012;85:1566–1575.
42. Schwartz RB, Bravo SM, Klufas RA, et al. Cyclosporine neurotoxicity and its relationship to hypertensive encephalopathy: CT and MR findings in 16 cases. *Am J Radiol.* 1995;165:627–631.
43. Feske SK. Posterior reversible encephalopathy syndrome; a review. *Semin Neurol.* 2011;31:202–215.
44. Legriel S, Schraub O, Azoulay E, et al. Critically ill posterior reversible encephalopathy syndrome study group (CYPRESS). Determinants of recovery from severe posterior reversible encephalopathy syndrome. *PLoS One.* 2012;7:e44534.
45. Paulson OB, Strandgaard S, Edvinsson L. Cerebral autoregulation. *Cerebrovasc Brain Metab Rev.* 1990;2:161–192.
46. Cipolla MJ. Brief review: cerebrovascular function during pregnancy and eclampsia. *Hypertension.* 2007;50:14–24.
47. Lassen NA, Agnoli A. Upper limit of autoregulation of cerebral blood flow: on the pathogenesis of hypertensive encephalopathy. *Scan J Clin Lab Invest.* 1972;30:113–115.
48. Johansson B, Li C-L, Olsson Y, Klatzo I. Effect of acute arterial hypertension on the blood-brain barrier to protein tracers. *Acta Neuropathol.* 1970;16:117–124.
49. Johansson B. The blood-brain barrier and cerebral blood flow in acute hypertension. *Acta Med Scand Suppl.* 1983;678:107–112.
50. Mayhan WG, Werber AH, Heistad DD. Protection of cerebral vessels by sympathetic nerves and vascular hypertrophy. *Circ Res.* 1987;75(suppl I):I-107–I-112.
51. Johansson B. Regional changes of cerebral blood flow in acute hypertension in cats. *Acta Neurol Scand.* 1974;50:366–372.
52. Sadoshima S, Busija DW, Heistad DD. Mechanisms of protection against stroke in stroke-prone spontaneously hypertensive rats. *Am J Physiol.* 1983;244:H406–H412.
53. Johansson BB. The blood brain barrier in acute and chronic hypertension. *Adv Exp Med Biol.* 1980;131:211–226.
54. Schafer PW, Buonnano FS, Gonzalez RG, Schwamm LH. Diffusion-weighted imaging discriminates between cytotoxic and vasogenic edema in patients with eclampsia. *Stroke.* 1997;29:1082–1085.
55. Jacobsen JC, Beierholm U, Mikkelsen R, Gustafsson F, Alstrom P, Holstein-Rathlou NH. "Sausage-string" appearance of arteries and arterioles can be caused by an instability of the blood vessel wall. *Am J Physiol Regul Integr Comp Physiol.* 2002;283:R1118–R1130.
56. Manfredi M, Beltramello A, Bongiovanni LG, Polo A, Pistoia L, Rizzuto N. Eclamptic encephalopathy: imaging and pathogenetic considerations. *Acta Neurol Scand.* 1997;96:277–282.

57. Mirza A. Posterior reversible encephalopathy syndrome: a variant of hypertensive encephalopathy. *J Clin Neurosci.* 2006;13:590–595.
58. Shah AK, Whitty JE. Brain MRI in periperum seizures: usefulness of combined T2 and diffusion-weighted MR imaging. *J Neurol Sci.* 1999;166:122–125.
59. Cipolla MJ. The adaptation of the cerebral circulation to pregnancy: mechanisms and consequences. *J Cereb Blood Flow Metab.* 2013;33:465–478.
60. Easton DJ. Severe preeclampsia/eclampsia hypertensive encephalopathy of pregnancy? *Cerebrovasc Dis.* 1998;8:53–58.
61. Sibai BM. Eclampsia. VI. Maternal-perinatal outcome in 254 consecutive cases. *Am J Obstet Gynecol.* 1990;163:1049–1055.
62. Mattar F, Sibai BM. Eclampsia. VIII. Risk factors for maternal mortality. *Am J Obstet Gynecol.* 2000;182:307–312.
63. Douglas KA, Redman CWG. Eclampsia in the United Kingdom. *BMJ.* 1994;309:1395–1400.
64. Euser AG, Cipolla MJ. Cerebral blood flow autoregulation and edema formation during pregnancy in anesthetized rats. *Hypertension.* 2007;49:334–340.
65. Cipolla MJ, Bishop N, Chan S-L. Effect of pregnancy on autoregulation of cerebral blood flow in anterior and posterior cerebrum. *Hypertension.* 2012;60:705–711.
66. Chapman AC, Chan S-L, Cipolla MJ. Effect of pregnancy on the lower limit of cerebral blood flow autoregulation and myogenic vasodilation of posterior cerebral arteries. *Reprod Sci.* 2012;20:1046–1054.
67. Cipolla MJ, Sweet JA, Chan S-L. Cerebral vascular adaptation to pregnancy and its role in the neurological complications of eclampsia. *J Appl Physiol.* 2011;110:329–339.
68. Phillips SJ, Whisnant JP. Hypertension and the brain. *Arch Intern Med.* 1992;152:938–945.
69. Johansson BB. Effect of an acute increase of the intravascular pressure on the blood-brain barrier. *Stroke.* 1978;9:588–590.
70. Faraci FM, Mayhan WG, Heistad DD. Segmental vascular responses to acute hypertension in cerebrum and brain stem. *Am J Physiol.* 1987;252:H738–H742.
71. Cipolla MJ, Vitullo L, McKinnon J. Cerebral artery reactivity changes during pregnancy and postpartum: a role in eclampsia? *Am J Physiol.* 2004;286:H2127–H2132.
72. Mayhan WG, Faraci FM, Heistad DD. Disruption of the blood-brain barrier in cerebrum and brainstem during acute hypertension. *Am J Physiol.* 1986;251:H1171–H1175.
73. Chan S-L, Cipolla MJ. Relaxin activates PPARγ and causes selective outward remodeling of brain penetrating arterioles. *FASEB J.* 2011;25:3229–3239.
74. McCall ML. Cerebral blood flow and metabolism in toxemias of pregnancy. *Surg Gynecol Obstet.* 1949;89:715–721.
75. Purkayastha F, Sorond F. Transcranial Dopper ultrasound: techniques and application. *Semin Neurol.* 2012;32:411–420.
76. Aaslid R. Cerebral hemodynamics. In: Newell DW, Aaslid R, eds. *Transcranial Doppler.* New York: Raven Press; 1992.
77. Belfort MA, Tooke-Miller C, Allen JC, et al. Changes in flow velocity, resistance indices, and cerebral perfusion pressure in the maternal middle cerebral artery distribution during normal pregnancy. *Acta Obstet Gynecol Scand.* 2001;80: 104–112.
78. Demarin V, Rundek T, Hodek B. Maternal cerebral circulation in normal and abnormal pregnancies. *Acta Obstet Gynecol Scand.* 1997;76:619–624.
79. Ikeda T, Ikenoue T, Moir N, et al. Effect of early pregnancy on maternal regional blood flow. *Am J Obstet Gynecol.* 1993;168:1303–1308.
80. Nevo O, Soustiel JF, Thaler I. Maternal cerebral blood flow during normal pregnancy: a cross-sectional study. *Am J Obstet Gynecol.* 2010;203:475.e1–475.e6.
81. Choi SJ, Lee JE, Oh SY, et al. Maternal cerebral blood flow and glucose metabolism in pregnancies complicated by severe preeclampsia. *Hypertens Preg.* 2012;31:177–188.
82. Zunker P, Happe S, Georgiadis AL, et al. Maternal cerebral hemodynamics in pregnancy-related hypertension. A prospective transcranial Doppler study. *Ultrasound Obstet Gynecol.* 2000;16:179–187.
83. Belfort MA, Saade GR, Grunewald C, et al. Association of cerebral perfusion pressure with headache in women with preeclampsia. *Brit J Obstet Gynaecol.* 1999;106:814–821.
84. Williams K, Galerneau F. Maternal transcranial Doppler in preeclampsia and eclampsia. *Ultrasound Obstet Gynecol.* 2003;21:507–513.
85. Belfort M, van Veen T, White GL, et al. Low maternal middle cerebral artery Doppler resistance indices can predict future development of preeclampsia. *Ultrasound Obstet Gynecol.* 2012;40:406–411.
86. Barbosa AS, Pereira AK, Reis ZS, Lage EM, Leite HV, Cabral AC. Ophthalmic artery resistive index and evidence of overperfusion-related encephalopathy in severe preeclampsia. *Hypertension.* 2010;55:189–193.
87. Riskin-Mashiah S, Belfort MA. Cerebrovascular hemodynamics in pregnant women with mild chronic hypertension. *Obstet Gynecol.* 2004;103:294–298.
88. Belfort MA, Saade GR, Yared M. Change in estimated cerebral perfusion pressure after treatment with nimodipine or magnesium sulfate in patients with preeclampsia. *Am J Obstet Gynecol.* 1999;181:402–407.
89. Serra-Serra V, Kyle PM, Chandran R, Redman CW. The effect of nifedipine and methyldopa on maternal cerebral circulation. *Brit J Obstet Gynaecol.* 1997;104:532–537.
90. van Veen TR, Panerai RB, Haeri S, Griffioen AC, Zeeman GG, Belfort MA. Cerebral autoregulation in normal pregnancy and preeclampsia. *Obstet Gynecol.* 2013;122:1064–1069.
91. Janzarik W, Ehlers E, Ehmann R, et al. Dynamic cerebral autoregulation in pregnancy and the risk of preeclampsia. *Hypertension.* 2014;63:161–166.
92. Markl M, Frydrychowicz A, Kozerke S, Hope M, Wieben O. 4D flow MRI. *J Mag Reson Imaging.* 2012;36:1015–1036.
93. Zeeman GG, Hatab MR, Twickler DM. Maternal cerebral blood flow changes in pregnancy. *Am J Obstet Gynecol.* 2003;189:967–972.
94. Lindqvist PG, Marsál K, Pirhonen JP. Maternal cerebral Doppler velocimetry before, during, and after a normal pregnancy: a longitudinal study. *Acta Obstet Gynecol Scand.* 2006;85:1299–1303.
95. Zeeman GG, Hatab MR, Twickler DM. Increased cerebral blood flow in preeclampsia with magnetic resonance imaging. *Am J Obstet Gynecol.* 2004;191:1425–1429.
96. Riskin-Mashiah S, Belfort MA. Preeclampsia is associated with global cerebral hemodynamic changes. *Soc Gynecol Invest.* 2005;12:253–256.
97. Takeuchi M, Matsuzaki K, Harada M, Nishitani H, Matsuda T. Cerebral hyperperfusion in a patient with eclampsia with

perfusion-weighted magnetic resonance imaging. *Radiat Med.* 2005;23:376–379.

98. Apollon KM, Robinson JN, Schwartz RB, Norwitz ER. Cortical blindness in severe preeclampsia: computed tomography, magnetic resonance imaging, and single-photon-emission computed tomography findings. *Obstet Gynecol.* 2000;95:1017–1019.
99. Naidu K, Moodley J, Corr P, Hoffman M. Single photon emission and cerebral computerized tomographic scan and transcranial Doppler sonographic findings in eclampsia. *Brit J Obstet Gynaecol.* 1997;104:1165–1172.
100. Ohno Y, Wakahara Y, Kawai M, Arii Y. Cerebral hyperperfusion in patient with eclampsia. *Acta Obstet Gynecol Scand.* 1999;78:555–556.
101. Rutherford JM, Moody A, Crawshaw S, Rubin PC. Magnetic resonance spectroscopy in preeclampsia: evidence of cerebral ischaemia. *BJOG.* 2003;110:416–423.
102. Sengar AR, Gupta RK, Dhanuka AK, Roy R, Das K. MR Imaging, MR angiography, and MR spectroscopy of the brain in eclampsia. *Am J Neuroradiol.* 1997;18:1485–1490.
103. Chipchase J, Peebles D, Rodeck C. Severe preeclampsia and cerebral blood volume response to postural change. *Obstet Gynecol.* 2003;101:86–92.
104. Meldrum BS. Concept of activity-induced cell death in epilepsy: historical and contemporary perspectives. *Prog Brain Res.* 2002;135:3–11.
105. Chapman AG. Cerebral energy metabolism and seizuresPedley T.A., Meldrum BS, editors. *Recent Advances in Epilepsy*, Vol 2. Edinburgh: Churchill Livingstone; 1985:19–63.
106. Cornford EM. Epilepsy and the blood brain barrier: endothelial cell responses to seizures. *Adv Neurol.* 1999;79:845–862.
107. Makhseed M, Musini VM. Eclampsia in Kuwait 1981–1993. *Aust N Z J Obstet Gynaecol.* 1996;36:258–263.
108. Vezzani A, Granata T. Brain inflammation in epilepsy: experimental and clinical evidence. *Epilepsia.* 2005;46:1724–1743.
109. Holmes GL, Lenck-Santini PP. Role of interictal epileptiform abnormalities in cognitive impairment. *Epilepsy Behav.* 2006;8:504–515.
110. Bolwig TG. Blood-brain barrier studies with special reference to epileptic seizures. *Acta Psychiatr Scand Suppl.* 1988;345:15–20.
111. Koenig H, Trout JJ, Goldstone AD, Lu CY. Capillary NMDA receptors regulate blood-brain barrier function and breakdown. *Brain Res.* 1992;588:297–303.
112. Parfenova H, Carratu P, Tcheranova D, Fedinec A, Pourcyrous M, Leffler CW. Epileptic seizures cause extended postictal cerebral vascular dysfunction that is prevented by HO-1 overexpression. *Am J Physiol.* 2005;288:H2843–H2850.
113. Faraci FM, Breese KR, Heistad DD. Nitric oxide contributes to dilatation of cerebralarterioles during seizures. *Am J Physiol.* 1993;265:H2209–H2212.
114. Sibai BM. Diagnosis, prevention, and management of eclampsia. *Obstet Gynecol.* 2005;105:402–410.
115. Katz VL, Farmer R, Kuller JA. Preeclampsia into eclampsia: toward a new paradigm. *Am J Obstet Gynecol.* 2000;182:1389–1396.
116. Sibai BM, Abdella TN, Spinnato JA, Anderson GD, Eclampsia V. The incidence of nonpreventable eclampsia. *Am J Obstet Gynecol.* 1986;154:581–586.
117. Selye H. Anesthetic effects of steroid hormones. *Proc Soc Exp Biol Med.* 1941;46:116–121.
118. McEwen BS, Coirini H, Schumacher M. Steroid effects on neuronal activity: when is the genome involved? *Ciba Found Symp.* 1990;153:3–12.
119. Callachan H, Cottrell GA, Hather NY, Lambert JJ, Nooney JM, Peters JA. Modulation of the GABAA receptor by progesterone metabolites. *Proc R Soc Lond B Biol Sci.* 1987;231:359–369.
120. Concas A, Mostallino MC, Porcu P, et al. Role of brain allopregnanolone in the plasticity of gamma-aminobutyric acid type A receptor in rat brain during pregnancy and after delivery. *Proc Natl Acad Sci USA.* 1998;95:13284–13289.
121. Maguire J, Ferando I, Simonsen C, Mody I. Excitability changes related to GABAA receptor plasticity during pregnancy. *J Neurosci.* 2009;29:9592–9601.
122. White WM, Brost B, Sun Z, et al. Genome-wide methylation profiling demonstrates hypermethylation in maternal leukocyte DNA in preeclamptic compared to normotensive pregnancies. *Hypertens Preg.* 2013;32:257–269.
123. Liu JH. Endocrinology of pregnancy. In: Creasy R, Resnik R, eds. *Maternal-Fetal Medicine, Principles and Practice.* Philadelphia, PA: Saunders; 2004:121–134.
124. Rusterholz C, Hahn S, Holzgreve W. Role of placentally produced inflammatory and regulatory cytokines in pregnancy and the etiology of preeclampsia. *Semin Immunopathol.* 2007;29:151–162.
125. Duley L. The global impact of pre-eclampsia and eclampsia. *Semin Perinatol.* 2009;33:130–137.
126. Cipolla MJ, Pusic AD, Grinberg YY, Chapman AC, Poynter ME, Kraig RP. Pregnant serum induces neuroinflammation and seizure activity via TNFα. *Exp Neurol.* 2012;234:398–404.
127. Rapoport SI. Brain edema and the blood-brain barrier. In: Welch KMA, Caplan LR, Reis DJ, Siesjo BK, Weir B, eds. *Primer on Cerebrovascular Disease.* San Diego, CA: Academic Press; 1997:25–28. [Chapter 39].
128. Kimelberg HK. Water homeostasis in the brain: basic concepts. *Neuroscience.* 2004;129:851–860.
129. Shima K. Hydrostatic brain edema: basic mechanisms and clinical aspect. *Acta Neurochir Suppl.* 2003;86:17–20.
130. Ishii S, Koike J, Hatashita Y. Brain edema in hypertensive intracranial hemorrhage. In: Mizukami M, ed. *Hypertensive Intracranial Hemorrhage.* New York: Raven Press; 1983.
131. Mayhan WG, Faraci FM, Heistad DD. Effects of vasodilatation and acidosis on the blood-brain barrier. *Microvasc Res.* 1988;35:179–192.
132. Mayhan WG, Heistad DD. Role of veins and cerebral venous pressure in disruption of the blood-brain barrier. *Circ Res.* 1986;59:216–220.
133. van der Wijk AE, Schreurs MP, Cipolla MJ. Pregnancy causes diminished myogenic tone and outward hypotrophic remodeling of the cerebral vein of Galen. *J Cereb Blood Flow Metab.* 2013;33:542–549.
134. Maynard SE, Min JY, Merchan J, et al. Excess placental soluble fms-like tyrosine kinase 1 (sFlt1) may contribute to endothelial dysfunction, hypertension, and proteinuria in preeclampsia. *J Clin Invest.* 2003;111:649–658.
135. Venkatesha S, Toporsian M, Lam C, et al. Soluble endoglin contributes to the pathogenesis of preeclampsia. *Nat Med.* 2006;12:642–649.

136. Roberts JM, Taylor RN, Musci TJ, Rodgers GM, Hubel CA, McLaughlin MK. Preeclampsia: an endothelial cell disorder. *Am J Obstet Gynecol*. 1989;161:1200–1204.
137. Amburgey O, Chapman AC, May V, Bernstein IM, Cipolla MJ. Plasma from preeclamptic women increases blood-brain barrier permeability: role of VEGF signaling. *Hypertension*. 2010;56:1003–1008.
138. Schreurs MP, Hubel CA, Bernstein IM, Jeyabalan A, Cipolla MJ. Increased oxidized low-density lipoprotein causes blood-brain barrier disruption in early-onset preeclampsia through LOX-1. *FASEB J*. 2013;27:1254–1263.
139. Schreurs MPH, Cipolla MJ. Cerebrovascular dysfunction and blood-brain barrier permeability induced by oxidized LDL are prevented by apocynin and magnesium sulfate. *J Cardiovasc Pharm*. 2014;63:33–39.
140. Ishibashi K, Kuwahara M, Sasaki S. Molecular biology of aquaporins. *Rev Physiol Biochem Pharmacol*. 2000;141:1–32.
141. Agre P, Bonhivers M, Borgnia MJ. The aquaporins, blueprints for cellular plumbing systems. *J Biol Chem*. 1998;273:14659–14662.
142. Amiry-Moghaddam M, Ottersen OP. The molecular basis of water transport in the brain. *Nat Neurosci*. 2003;4:991–1001.
143. Jung JS, Bhat RV, Preston GM, Guggino WB, Baraban JM, Agre P. Molecular characterization of an aquaporin cDNA from brain: candidate osmoreceptor and regulator of water balance. *Proc Natl Acad Sci USA*. 1994;91:13052–13056.
144. Wiegman MJ, Bullinger LV, Kholmeyer M, Hunter TC, Cipolla MJ. Regional expression of aquaporin -1, -4 and -9 in the brain during pregnancy. *Reprod Sci*. 2008;15:506–516.
145. Nielsen S, Smith BL, Christensen EL, Agre P. Distribution of the aquaporin CHIP in secretory and resorptive epithelia and capillary endothelia. *Proc Natl Acad Sci USA*. 1993;90:7275–7279.
146. Nielsen S, Nagelhus EA, Amiry-Moghaddam M, Bourque C, Agre P, Ottersen OP. Specialized membrane domains for water transport in glial cells: high-resolution immunogold cytochemistry of aquaporin-4 in rat brain. *J Neurosci*. 1997;17:171–180.
147. Amiry-Moghaddam M, Xue R, Haug FM, Neely JD, Bhardwaj A, Agre P. Alpha-syntrophin deletion removes the perivascular but not endothelial pool of aquaporin-4 at the blood-brain barrier and delays the development of brain edema in an experimental model of acute hyponatremia. *FASEB J*. 2004;18:542–544.
148. Amiry-Moghaddam M, Otuska T, Hurn PD, et al. An alpha-syntrophin-dependent pool of AQP4 in astroglial end-feet confers bidirectional water flow between blood and brain. *Proc Natl Acad Sci USA*. 2003;100:2106–2111.
149. Quick AM, Cipolla MJ. Pregnancy-induced upregulation of aquaporin-4 protein in brain and its role in eclampsia. *FASEB J*. 2005;19:170–175.
150. Papadopoulos MC, Manely GT, Krishna S, Verkman AS. Aquaporin-4 facilitates reabsorption of excess fluid in vasogenic brain edema. *FASEB J*. 2004;18:1291–1293.
151. Binder DK, Oshio K, Ma T, Verkman AS, Manley GT. Increased seizure duration in mice lacking aqauporin-4 water channels. *Acta Neurochir Suppl*. 2006;96:389–392.
152. Binder DK, Yao X, Verkman AS, Manley GT. Increased seizure threshold in mice lacking aquaporin-4 water channels. *Neuroreport*. 2004;15:259–262.
153. Binder DK, Yao X, Zador Z, Sick T, Verkman AS, Manley GT. Increased seizure duration and slowed potassium kinetics in mice lacking aquaporin-4 water channels. *Glia*. 2006;53:631–636.
154. Iliff JJ, Wang M, Liao Y, et al. A paravascular pathway facilitates CSF flow through the brain parenchyma and the clearance of interstitial solutes, including amyloid β. *Sci Transl Med*. 2012;4:147ra111.
155. Belfort MA, Moise KJ. Effect of magnesium sulfate on maternal brain blood flow in preeclampsia: a randomized, placebo-controlled study. *Am J Obstet Gynecol*. 1992;16:661–666.
156. Belfort M, Allred J, Dildy G. Magnesium sulfate decreases cerebral perfusion pressure in preeclampsia. *Hypertens Pregnancy*. 2008;27:315–327.
157. Naidu S, Payne AJ, Moodley J, Hoffmann M, Gouws E. Randomised study assessing the effect of phenytoin and magnesium sulphate on maternal cerebral circulation in eclampsia using transcranial Doppler ultrasound. *Brit J Obstet Gynaecol*. 1996;103:111–116.
158. Hatab MR, Zeeman GG, Twickler DM. The effect of magnesium sulfate on large cerebral artery blood flow in severe preeclampsia. *J Maternal-Fetal Neonat Med*. 2005;17:187–192.
159. Euser AG, Cipolla MJ. Magnesium sulfate for treatment of eclampsia: a brief review. *Stroke*. 2009;40:1169–1175.
160. Euser AG, Cipolla MJ. Resistance artery vasodilation to magnesium sulfate during pregnancy and the postpartum state. *Am J Physiol*. 2005;288:H1521–H1525.
161. Belfort MA, Anthony J, Saade GR, Allen Jr JC, for the Nimodipine Study Group A comparison of magnesium sulfate and nimodipine for the prevention of eclampsia. *New Engl J Med*. 2003;348:304–311.
162. Esen F, Erdem T, Aktan D, et al. Effects of magnesium administration on brain edema and blood-brain barrier breakdown after experimental traumatic brain injury in rats. *J Neurosurg Anesth*. 2003;15:119–125.
163. Esen F, Erdem T, Aktan D, et al. Effect of magnesium sulfate administration on blood-brain barrier in a rat model of intraperitoneal sepsis: a randomized controlled experimental study. *Crit Care*. 2005;9:R18–R23.
164. Kaya M, Kucuk M, Kalayci RB, et al. Magnesium sulfate attenuates increased blood-brain barrier permeability during insulin-induced hypoglycemia in rats. *Can J Physiol Pharmacol*. 2001;79:793–798.
165. Euser AG, Bullinger LV, Cipolla MJ. Magnesium sulfate decreases blood-brain barrier permeability during acute hypertension in pregnant rats. *Exp Physiol*. 2008;93:254–261.
166. Patel B, Lawrence AJ, Chung AW, et al. Cerebral microbleeds and cognition in patient with symptomatic small vessel disease. *Stroke*. 2013;44:356–361.
167. Kojima H, Eguchi H, Mizutani T, Tanaka K, Kikuchi Y, Fukudome N. Three-dimensional analysis of pathological characteristics of a microaneurysm. *Clin Neuropathol*. 2007;26:74–79.
168. Shah AK, Whitty JE. Brain MRI in peripartum seizures: usefulness of combined T2 and diffusion weighted MR imaging. *J Neurol Sci*. 1999;166:122–125.
169. Bushnell C, Chireau M. Preeclampsia and stroke: risks during and after pregnancy. *Stroke Res Treat*. 2011;585:1–9.

170. Cunningham FG, Fernandez CO, Hernandez C. Blindness associated with preeclampsia and eclampsia. *Am J Obstet Gynecol.* 1995;172:1291–1298.
171. Roos NM, Wiegman MJ, Jansonius NM, Zeeman GG. Visual disturbances in (pre)eclampsia. *Obstet Gynecol Surv.* 2012;67:242–250.
172. Moseman CP, Shelton S. Permanent blindness as a complication of pregnancy induced hypertension. *Obstet Gynecol.* 2002;100:943–945.
173. Stewart WM, Brazis PW, Guier CP, Thota SH, Wilson SD. Purtscher-like retinopathy in a patient with HELLP syndrome. *Am J Ophthalmol.* 2007;143:886–887.
174. Lara-Torre E, Lee MS, Wolf MA, Shah DM. Bilateral retinal occlusion progressing to long-lasting blindness in severe preeclampsia. *Obstet Gynecol.* 2002;100:940–942.
175. Wiegman MJ, de Groot JC, Jansonius NM, et al. Long-term visual functioning after eclampsia. *Obstet Gynecol.* 2012;119:959–966.
176. Liman TG, Bohner G, Heuschmann PU, Endres M, Siebert E. The clinical and radiological spectrum of posterior reversible encephalopathy syndrome: the retrospective berlin PRES study. *J Neurol.* 2012;259:155–164.
177. Aukes AM, de Groot JC, Aarnoudse JG, Zeeman ZZ. Brain lesions several years after eclampsia. *Am J Obstet Gynecol.* 2009;200(504):e1–e5.
178. Aukes AM, De Groot JC, Wiegman MJ, Aarnoudse JG, Sanwikarja G, Zeeman GG. Long-term cerebral imaging after pre-eclampsia. *BJOG.* 2012;119:1117–1122.
179. Wiegman MJ, Zeeman GG, Aukes AM, Faas MM, Aarnoudse JM, De Groot JC. Regional distribution of cerebral white matter lesions years after preeclampsia and eclampsia. *Obstet Gynecol.* 2014;123:790–795.
180. Bellamy L, Casas JP, Hingorani AD, Williams DJ. Pre-eclampsia and risk of cardiovascular disease and cancer in later life: systematic review and meta-analysis. *BMJ.* 2007;335:974–985.
181. De Leeuw FE, de Groot JC, Oudkerk M, et al. Hypertension and cerebral white matter lesions in a prospective cohort study. *Brain.* 2002;125:765–772.
182. Hopkins RO, Becke CJ, Burnett DL, Weaver LK, Victoroff J, Bigler ED. Prevalence of white matter hyperintensities in a young healthy population. *J Neuroimaging.* 2006;16:243–251.
183. Wen W, Sachdev PS, Li JJ, Chen X, Anstey KJ. White matter hyperintensities in the forties: their prevalence and topography in an epidemiological sample aged 44–48. *Hum Brain Mapp.* 2009;30:1155–1167.
184. Aukes AM, Wessel I, Dubois AM, Aarnoudse JG, Zeeman GG. Self-reported cognitive functioning in formerly eclamptic women. *Am J Obstet Gynecol.* 2007;197(365):e1–e6.
185. Postma IR, Groen H, Easterling TR, et al. The brain study: cognition, quality of life and social functioning following preeclampsia; an observational study. *Pregnancy Hypertens.* 2013;3:227–234.
186. Andersgaard A, Herbst A, Johansen M, Borgstrom A, Bille A, Øian P. Follow-up interviews after eclampsia. *Gynecol Obstet Invest.* 2009;67:49–52.
187. Porcel J, Feigal C, Poye L, et al. Hypertensive disorders of pregnancy and risk for screening positive for posttraumatic stress disorder: a cross sectional study. *Pregnancy Hypertens.* 2013;3:254–260.
188. Baecke M, Spaanderman MEA, van der Werf SP. Cognitive function after pre-eclampsia: an explorative study. *J Psychosom Obstet Gynaecol.* 2009;30:58–64.
189. Brussé I, Duvekot J, Jongerling J, Steegers E, de Koning I. Impaired maternal cognitive functioning after pregnancies complicated by severe pre-eclampsia: a pilot case-control study. *Acta Obstet Gynecol Scand.* 2008;87:408–412.
190. Postma IR, Bouma A, Ankersmit IF, Zeeman GG. Neurocognitive functioning following preeclampsia and eclampsia: a long-term follow-up study. *Am J Obstet Gynecol.* 2014 Feb 1. pii: S0002-9378(14)00113-6. doi: http://dx.doi.org/10.1016/j.ajog.2014.01.042. [Epub ahead of print].
191. Lawn N, Laich E, Ho S. Eclampsia, hippocampal sclerosis, and temporal lobe epilepsy: accident or association? *Neurology.* 2004;62:1352–1356.
192. Plazzi G, Tinuper P, Cerullo A, Provini F, Lugaresi E. Occipital lobe epilepsy: a chronic condition related to transient occipital lobe involvement in eclampsia. *Epilepsia.* 1994;35:644–647.
193. Solinas C, Briellmann RS, Harvey AS, Mitchell LA, Berkovic SF. Hypertensive encephalopathy: antecedent to hippocampal sclerosis and temporal lobe epilepsy? *Neurology.* 2003;60:1534–1536.
194. Debette S, Markus HS. The clinical importance of white matter hyperintensities on brain magnetic resonance imaging: systematic review and meta-analysis. *BMJ.* 2010;341:c3666.
195. Irgens HU, Reisæter L, Irgens LM, Lie RT. Long term mortality of mothers and fathers after pre-eclampsia: population based cohort study. *BMJ.* 2001;323:1213–1217.
196. Wilson BJ, Watson MS, Prescott GJ, et al. Hypertensive diseases of pregnancy and risk of hypertension and stroke in later life: results from cohort study. *BMJ.* 2003;326:845.

CHAPTER 14

Cardiovascular Alterations in Normal and Preeclamptic Pregnancy

JUDITH U. HIBBARD, SANJEEV G. SHROFF AND F. GARY CUNNINGHAM

Editors' comment: *The heart was a brief bystander in Chesley's original text. Evolving research on dissecting changes in cardiac function during gestation as well as a new school that claimed that exaggeration of the normal increase in cardiac output was a factor that led to preeclampsia emerged as the second edition went to press. These were refined and discussed further in the third edition, reflecting the fact that the technology by which we gauge the cardiac system had been markedly improved during the early years of this millennium. Progress has continued and the fourth edition contains interesting and exciting new material, some representing the research productivity of Drs. Hibbard and Shroff.*

INTRODUCTION

There are striking physiologic cardiovascular changes during pregnancy that ensure adequate uterine blood flow, as well as appropriate oxygenation and nutrient delivery to the fetus. These compensatory mechanisms allow the mother to function normally during this altered physiologic state. Both knowledge of and understanding the roles of these changes are particularly critical if strategies are to be developed to manage pregnant women with chronic or new hypertension and especially preeclampsia. Thus, how the preeclampsia syndrome impacts the cardiovascular system may be integral to appropriate therapy, but, as we shall see, studies designed to document effects of preeclampsia on the cardiovascular system have not always produced the same results. This chapter commences with a review of normal cardiac and hemodynamic function during pregnancy followed by a survey of knowledge regarding cardiac performance and vascular changes in preeclampsia, focusing on recent progress, much made possible by advances in noninvasive technology. Finally, we discuss the potential of using pregnancy-associated aberrant responses to predict cardiovascular disease risk later in life.

HEMODYNAMICS AND CARDIAC FUNCTION IN NORMAL PREGNANCY

Nearly 100 years ago, Lindhard,[1] reporting on normal cardiovascular adaptations in pregnancy, described a 50% increase in cardiac output as measured by a dye-dilution technique. Since then multiple methodologies have been employed to assess cardiovascular function in pregnancy and have resulted in a myriad of findings. The "gold standard" for such evaluation remains flow-directed pulmonary artery catheters using thermodilution methodology. Given the invasive nature of these techniques, only cross-sectional investigations are feasible. Fortunately, noninvasive M-mode echocardiography and continuous and pulsed-wave Doppler techniques, validated against an invasive technique, permit serial determinations throughout both normal and abnormal pregnancy.

One must be cautious in regard to several methodological issues that impact cardiovascular parameters in pregnancy. These include maternal posture during data collection, whether or not she received fluids or vasoactive medications prior to data acquisition, and whether there is active labor.[2] In addition, maternal body habitus can also affect cardiac measurements. Some, but not all, investigations have controlled for these potential confounding issues.

Systemic Arterial Hemodynamics in Normal Pregnancy

There are significant decreases in both systolic and diastolic blood pressure, noted as early as 5 weeks gestation[3] (Fig. 14.1). Interestingly, Chapman et al.[4] noted a significant decrease in blood pressure during the luteal compared with the follicular phase of the menstrual cycle – data suggesting a hormonal origin of the fall in blood pressure that persists and increases further after conception. The decrease in blood pressure during pregnancy is characterized as follows: Decrements in diastolic levels exceed those

Chesley's Hypertensive Disorders in Pregnancy.
ISBN: 978-0-12-407866-6

DOI: http://dx.doi.org/10.1016/B978-0-12-407866-6.00014-6

in systolic levels, the former averaging 10 mm Hg below baseline value. Mean blood pressure nadirs at 16–20 weeks, these changes persisting to the third trimester[3,5] (Figs. 14.1 and 14.2). In the mid-third trimester blood pressure rises gradually, often approaching prepregnancy values.[5–17] There are diurnal fluctuations in normal pregnancy, similar to patterns in the nonpregnant state, the nadir occurring at night.[18–20] Finally, all these observations have also been verified using 24-hour ambulatory blood pressure monitoring protocols[21] (Fig. 14.3A, B).

Cardiac output increases 35–50% during gestation, half or more of this increase being established by the eighth gestational week.[14] The earliest evidence of this change – in parallel with decreases in mean arterial pressure – can be detected in the luteal phase of the menstrual cycle.[3] If conception does not occur, then there is a significant reversal of all these changes, detectable in the next follicular phase (see Fig. 15.1). These data implicate the corpus luteum in the observed changes, and attest to a hormonal role in the early cardiovascular changes of pregnancy.

The significant increase in cardiac output is well established by gestational week 5, rising further to 50% above prepregnancy values by gestational weeks 16–20, then increasing slowly or plateauing until term[8–10,12,13,16,17,22,23]

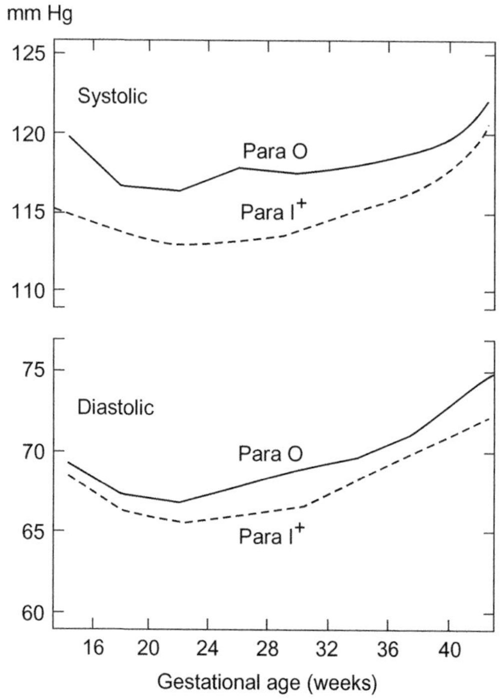

FIGURE 14.1 Mean blood pressure by gestational age in 6000 white women 25 to 34 years of age who delivered singleton term infants. (Reprinted with permission.[3])

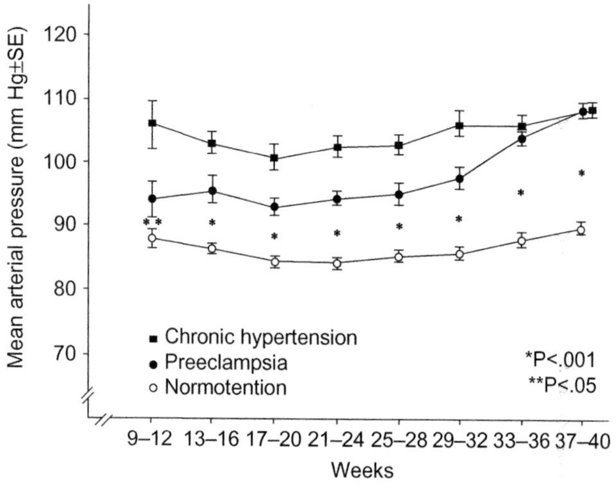

FIGURE 14.2 Averaged mean arterial blood pressures (±1 SE) in women who remained normotensive throughout pregnancy (O–O), in women who developed preeclampsia (●–●), and in women who developed hypertension (■–■) by periods of 4 weeks. (Reprinted with permission.[5])

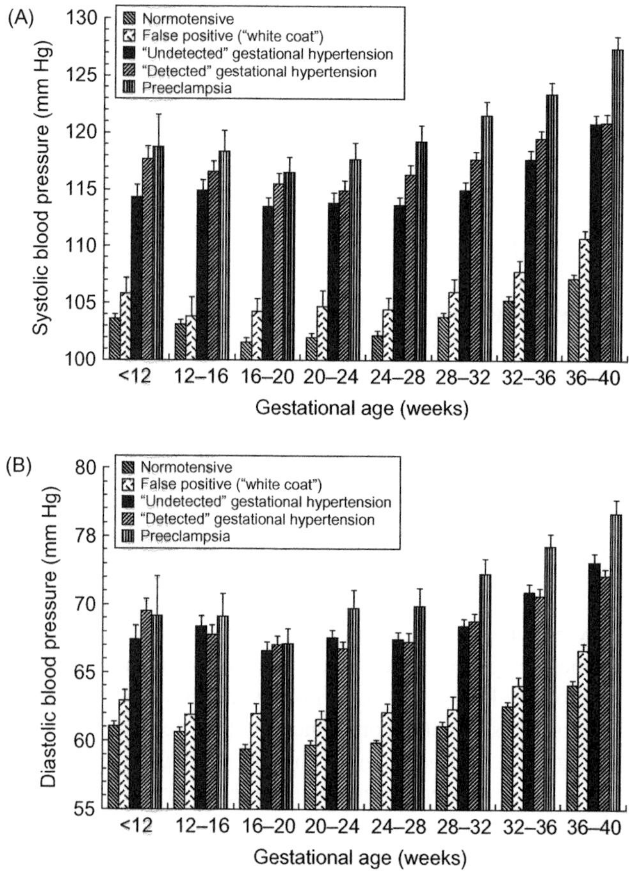

FIGURE 14.3 Differences in systolic (A) and diastolic (B) blood pressure between groups of pregnant women systematically measured by 48-hour ambulatory blood pressure monitoring throughout gestation. (Reprinted with permission.[21])

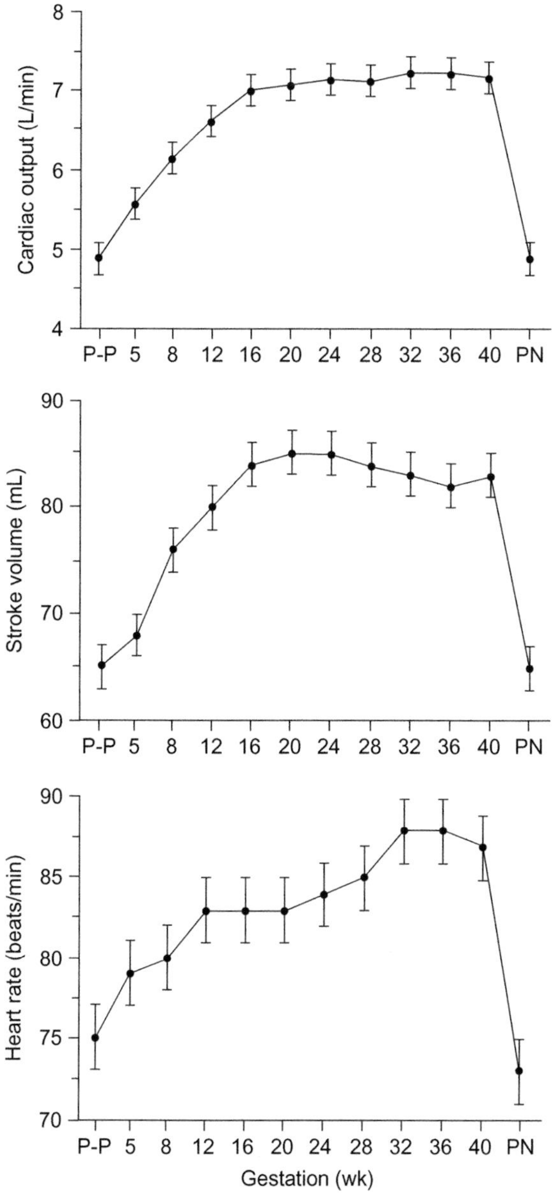

FIGURE 14.4 Longitudinal studies of cardiac output, stroke volume, and heart rate which began before conception and continued through to the postnatal (PN) period. (Reprinted with permission.[16])

(Fig. 14.4). Several investigators have noted different patterns, including a decrease in cardiac output from the peak pregnancy value when term approaches;[6,11,24,25] some of these discrepancies might relate to failing to control posture or to correct for body surface area.[6,11]

Stroke volume and heart rate, the two determinants of cardiac output, appear to rise sequentially, with the increases apparent by 5–8 weeks gestation.[6,9,14] Stroke volume continues to increase until gestational week 16, plateauing thereafter, while heart rate continues to increase slowly into the third trimester[9,10,14] (Fig. 14.4).

Capeless and Clapp[14] noted that 50% of the increase in cardiac output had occurred by gestational week 8, this early change primarily due to increased stroke volume (not to heart rate changes). These investigators also noted that multiparas had a greater rise and rate of change in stroke volume compared with nulliparas, as well as a greater drop in systemic vascular resistance, but there was no effect on heart rate.[8]

Postural changes can impact heart rate, blood pressure, and cardiac output. Both heart rate and blood pressure are significantly lower in lateral recumbency, while cardiac output is increased in this position. There is a reduction in cardiac output upon standing, noted in the first trimester, which becomes significantly attenuated in the second trimester and absent by the mid-third trimester.[26] Intravascular volume is progressively amplified up to 40%, perhaps contributing to the aforementioned changes (see Chapter 15).

To summarize, most evidence supports the following scenario of physiologic cardiovascular changes during pregnancy: there is an early decrease in systolic, diastolic, and mean blood pressures (the diastolic decrement exceeding the systolic decrement), an early increase in cardiac output that continues to rise or plateau into the third trimester, and increases in both stroke volume and heart rate contributing to this increase in cardiac output.

Venous System in Normal Pregnancy

Mean Circulatory Filling Pressure

Venous return to the right heart maintains filling pressure, permitting adaptation to changing cardiac output requirements. A prerequisite for such regulation is that the vascular bed with appropriate tone should be adequately filled with blood. The mean circulatory filling pressure (MCFP) characterizes this steady-state venocardiac filling.[27] The MCFP is the pressure recorded in the vascular tree at equilibrium and in the absence of any blood flow, which is the pressure in the circulation after the heart has been arrested and the system has come into equilibrium.[28] The MCFP thus provides an indication of the relationship between changes in blood volume compared with the size of the circulatory compartment and, as such, indicates to what extent the vascular compartment accommodates the large gestational increases in total blood volume.[29] Venous smooth muscle activation and changes in blood volume are mechanisms for changing the MCFP.[30]

Because measurement of MCFP requires stopping the heart, it follows that results must be derived from animal studies. Thus one must be circumspect about their relevance to human pregnancy, as not only may we be dealing with species differences, but also all of these experiments concerning MCFP were conducted in anesthetized animals.

The MCFP is slightly, but significantly, elevated in pregnant dogs, rabbits, sheep, and guinea pigs,[29,31–34] though one guinea pig study found no differences compared with the nonpregnant state.[35] Furthermore, pregnant dogs respond to epinephrine infusions with a rise in MCFP in a manner similar to nonpregnant controls, suggesting that they are able to increase their vascular tone normally. Also, the slope relating MCFP to changes in blood volume in most studies appears not to be altered by gestation,[31,32,34,35] rather the blood volume (BV)–MCFP relationship is merely shifted to the right, that is, there is increased unstressed volume. One group,[29] however, did suggest a decrease in the slope of the BV–MCFP relationship, interpreted as increased compliance. Both increased venous unstressed volume and compliance permit the large increases in intravascular volume, characteristic of gestation, to occur with very little rise in MCFP. As a whole, the increased MCFP with the aforementioned changes in venous unstressed volume and compliance can be interpreted as showing that the "stressed" – i.e., distending – component of the intravascular volume is elevated in pregnancy.

The higher MCFP has led some to suggest the increased cardiac output may be secondary to *relative overfilling of the circulatory system* during pregnancy[31,32] This is discussed in detail in Chapter 15 with considerations given to whether the gestational increase in blood volume should be considered "underfill," "overfill," or "normal fill." Because blood volume–MCFP relationships represent a measure of total circulatory compliance,[36] and for practical purposes total body venous compliance,[37] such data would suggest venous tone was unaltered in pregnancy, a finding that supports those who see pregnancy as vascular overfill. Concerning the meaning of MCFP in these discussions, one must recall the circumstances of how this index is measured, and the many interpretative problems with these anesthetized animal preparations.

In terms of cardiovascular homeostasis, the "active" regulation of capacitance remains a key question during pregnancy as this is one of the important mechanisms that influence venous return of blood to the heart, the systemic reflex capacitance in humans estimated at ~5 mL/kg.[27] The densely innervated mesenteric venous microcirculation appears to have a key function in controlling changes in vascular capacity. For example, most of the blood volume shift in the intestinal vascular bed during activation of the baroreceptor reflex occurs in the intestinal venules.[27]

Venous Tone Regulation

The venous system in pregnancy appears to be a neglected area of research. Hohmann et al.[38] studied adrenergic regulation of venous capacitance in pregnant and pseudopregnant rats, noting a progressive decline in the sensitivity to adrenergic nerve stimulation from cycling to late gestation. The reduced sympathetic nerve response was associated with marked increases in sensitivity to exogenously applied epinephrine during pregnancy, suggesting *denervation supersensitivity*.

It should be noted that venous pressure–volume or wall stress–strain relationships cannot accurately be characterized, either *in vivo* or *in vitro*, without defining the contractile state of the vascular smooth muscle.[27] Also, measurements made *in vivo* cannot distinguish between wall structural changes and those caused by differences in venous tone. In one study, where vascular smooth muscle was inactivated prior to assessing stress–strain relationships in pregnant rodents, the compliance of the mesenteric capacitance veins decreased by 40%.[39] The unstressed volume, however, doubled in comparison to the nonpregnant females.

In human studies, noninvasive measures suggest that venous distensibility increases with pregnancy in some investigations,[40–44] but not in others.[45–47] We could locate no longitudinal measures of human venous distensibility that included preconception values.

In summary, despite the importance of venous function to cardiovascular volume homeostasis, knowledge of either the normal or pathophysiologic status of this system is quite limited. It appears that venous distensibility (compliance) increases during gestation. Animal data, obtained under less than ideal experimental conditions, suggest that the increased vascular capacitance does not quite accommodate the increase in blood volume ("overfill"). Whether MCFP changes in human gestation is unknown. Information on sympathetic regulation of venous function in terms of regulating both venous return and fluid exchange at the capillary bed would be of interest, but again information is spotty. Thus, influences on the venous system by pregnancy are limited, and they are even more limited regarding changes in preeclampsia.

Systemic Arterial Properties in Normal Pregnancy

The gestational increase in cardiac output and decrement in blood pressure have traditionally been ascribed to the marked decrease in total systemic vascular resistance that is apparent early in gestation.[9,14,24,48] It should be recognized, however, that other changes may be involved. For example, both left ventricular and systemic arterial mechanical properties – ventricular afterload – have a potential to alter systemic hemodynamics. Afterload, or the arterial system load the heart experiences, is the mechanical opposition experienced by the blood ejected from the left ventricle. This opposition can be considered to have two components – one steady, the other pulsatile. The steady component, quantified in terms of total systemic vascular resistance, is determined by the properties of the small-caliber resistance vessels, for example, effective cross-sectional area, and blood rheological properties, for example, viscosity.

Due to the pulsatile nature of cardiac ejection, oscillations in pressure and flow exist throughout the arterial tree and thus the pulsatile component of the arterial load needs to be considered. Physically, the pulsatile arterial load is determined by the (visco)elastic properties of the arterial vessel wall, architectural features of the arterial circulation, that is, the network of branching tubes, and blood rheological properties.[49] Quantitative indices of pulsatile load include the global arterial compliance, aortic characteristic impedance, and measures of wave propagation and reflection.[49] *Global arterial compliance* is a measure of the reservoir properties of both the conduit and peripheral arterial tree. In contrast to arterial compliance, which is a global property (belonging to the entire circulation), characteristic impedance quantifies a local property (belonging to the site of pressure/flow measurement), being determined by local vascular wall stiffness and geometric properties. *Pulse wave velocity* and *global reflection coefficient* are indices often used to describe wave propagation and reflection within the arterial tree. Understanding the interplay of both steady and pulsatile components should lead to a better grasp of cardiovascular performance in pregnancy with its marked changes in both components.

To this end Poppas et al.[10] serially studied 14 normal, normotensive women throughout pregnancy and 8 weeks postpartum using noninvasive measures of instantaneous aortic pressure and flow velocities to assess both conduit and peripheral vessels. These investigators verified that systemic vascular resistance, the steady component of the arterial load, decreases very early in pregnancy, and continues to decrease significantly through the remainder of pregnancy, though less so in the latter weeks of gestation[10] (Fig. 14.5). Global arterial compliance increased by 30% during the first trimester and was maintained thereafter throughout pregnancy, temporally relating to the decreased systemic vascular resistance. By 8 weeks' postpartum, global compliance returned to normal levels[10] (Fig. 14.5). There was a tendency for aortic characteristic impedance to fall, and the magnitude of arterial wave reflections was reduced during late pregnancy, along with a delay in the timing of reflected waves. Similar results documenting increased compliance have been noted in pregnant animal models[29,35,50–53] and in other studies of pregnant humans,[44,54] as well as nonpregnant humans.[55] Mone et al.[24] also noted decreased characteristic impedance in a cohort of 33 normally pregnant women.

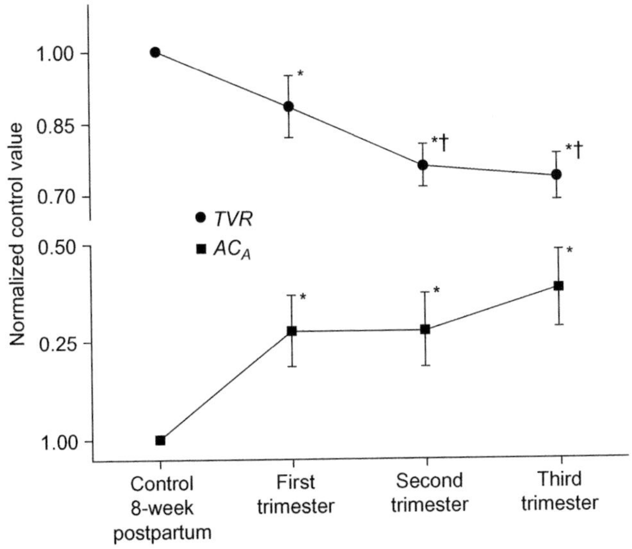

FIGURE 14.5 Temporal changes in global arterial compliance (AC) and total vascular resistance (TVR) during normal pregnancy. Data are normalized to 8-week postpartum control values (mean ± SEM; $*p < 0.05$, first, second, or third trimester vs. 8-week postpartum control; $^{\dagger}p < 0.05$, second or third vs. first trimester). (Reprinted with permission.[10])

It is thus clear that both steady and pulsatile arterial load decrease during normal pregnancy, indicating a state of peripheral vasodilatation and generalized vasorelaxation that involves both the peripheral (resistance) vessels and conduit vessels. The magnitudes of the fall in systemic vascular resistance and the rise in cardiac output seem to be equivalent, resulting in a very small change (fall) in mean arterial pressure. The decrement in pulsatile arterial load – that is, increased global compliance, decreased characteristic impedance, and decreased reflection coefficient – appears to be primarily due to a generalized increase in vascular distensibility, which, in turn, may be related to reduced smooth muscle tone and vascular remodeling.[10]

Left Ventricular Properties in Normal Pregnancy

Left ventricular mass increases in normal pregnancy. The increase has been described by some[6,10,23,24,56] as modest, averaging 10–20%, while others[9,17,22] have reported increments as great as 40%. An increase in ventricular mass should contribute to an increase in power as described below. Of note, in most studies the increase in mass does not meet criteria for ventricular hypertrophy – defined as >2 SD above the mean for normal population – as might occur, for example, in patients with chronic hypertension. Also, ventricular mass reverts to nonpregnant values postpartum.[2] There is a mild increase in left ventricular end-diastolic chamber diameter noted by many,[9,23,56,57] but not all[10,12,24,25,58] investigators.

Normal pregnancy is associated with an increase in the cross-sectional area of the left ventricular outflow tract, measured at the aortic annulus.[2,8–11,16,22,24,25,48,59] Thus, it is important to assess aortic diameter at the time of echocardiography in longitudinal studies.[2] These findings further highlight the risks the normal changes in pregnancy create for women with diseases known to be associated with a compromised aortic root, viz., Marfan or Turner syndromes. *In these women, pregnancy may precipitate aortic rupture or dissection.*[60,61]

Evaluation of left ventricular myocardial contractility in pregnancy has produced conflicting results. Use of traditional ejection-phase indices of left ventricular performance is problematic as these indices are unable to

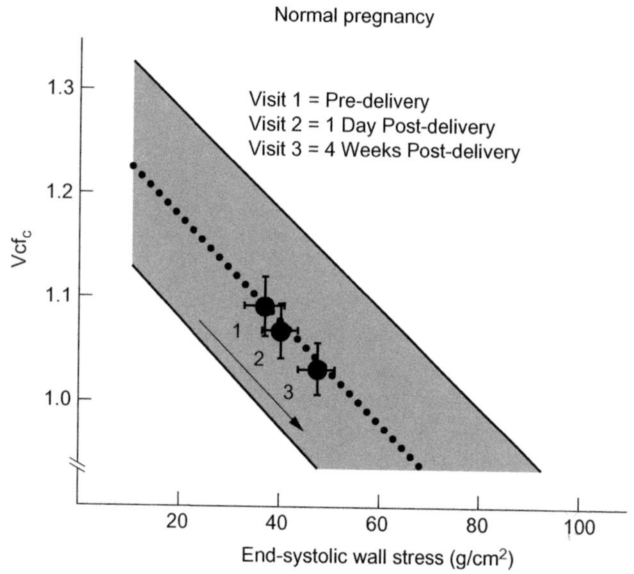

FIGURE 14.6 Average end-systolic wall stress (σ)-rate corrected velocity of fiber shortening (Vcf) obtained in normotensive pregnant control subjects before delivery and 1 day and 4 weeks after delivery. From visit 1 to visit 3, data points shifted rightward and downward (arrow) but still fell on the mean contractility line, indicating increase in afterload without changes in contractility. (Reprinted with permission.[2])

distinguish alterations in contractility from changes in ventricular load.[2,62,63] Thus, some of the variability in the results related to the assessment of left ventricular myocardial contractility may be attributable to the use of load-dependent indices.

Lang et al.[2] studied 10 normally pregnant women in early labor, and again at 1 day and 4 weeks postpartum. These investigators quantified left ventricular myocardial contractility using the measurements of end-systolic wall stress (σ) and rate-corrected velocity of fiber shortening (Vcf) (Fig. 14.6). Note, the σ –Vcf relationship yields a preload-independent and afterload-adjusted characterization of left ventricular myocardial contractility. The σ and Vcf data from individual subjects were compared to a nomogram, i.e., a σ–Vcf relationship constructed by studying a large group of normal, nonpregnant individuals, both in their basal state and after pharmacologic manipulation of afterload and preload. The σ–Vcf data points for the normal pregnant women were shifted rightward and downward, remaining superimposed on the nomogram, indicating a decrease in afterload without any changes in left ventricular contractility[2] (Fig. 14.6). These observations were verified by Poppas et al.,[10] who reported an invariant left ventricular myocardial contractility throughout pregnancy in a cohort of normal pregnant women. Finally, Simmons et al.[23] similarly demonstrated unchanged contractility in 44 pregnant women.

Some studies have claimed that left ventricular myocardial contractility changes during normal pregnancy. For example, Mone et al.[24] have reported a reversible fall in left ventricular myocardial contractility. This conclusion was based on the observation that Vcf progressively diminished during gestation – by 7% at term – even though σ was declining over the same time period – by 15% at term. However, a comparison with the nomogram indicated that the group-averaged values of σ–Vcf points during pregnancy were above the normal contractility line and were within the statistical bounds of the normal (nonpregnant) population. Similarly, Gilson et al.[12] have reported an enhanced left ventricular myocardial contractility. While there were no significant changes in Vcf, σ decreased by 12% over the observation period from early to late gestation. This observation, if anything, would imply decreased myocardial contractility. Their conclusion of enhanced contractility was based upon the observed decrease in σ / Vcf ratio, which they claim to be a load-independent index of myocardial contractility as proposed by Colan et al.[64] Interestingly, the latter investigators[64] never proposed the σ/Vcf ratio as an index of myocardial contractility; instead they used the position of the σ–Vcf point relative to the normal contractility line to quantify contractility in an individual subject. Furthermore, the σ/Vcf ratio is highly load-sensitive; it will change significantly as one moves along a given σ–Vcf relationship, that is, by definition, fixed myocardial contractility. Thus, the conclusion of enhanced myocardial contractility by Gilson et al.[12] appears to be erroneous.

To summarize, most of the evidence supports a conclusion that left ventricular myocardial contractility, as assessed by load-independent indices, is essentially unchanged in normal gestation. The data would be more secure, however, if the contractility nomogram used in future studies was derived exclusively from a female population of reproductive age.

In contrast to systolic function, left ventricular diastolic function has infrequently been the primary focus of studies in normal pregnancy, although this aspect is also important in evaluating cardiac function. Most common evaluation of left ventricular diastolic function is based on the measurement of transmitral inflow velocity by Doppler echocardiography. These transmitral inflow velocity-based indices, however, are load and heart rate dependent. Diastolic function of the left ventricle was assessed by Fok et al.[65] in a prospective longitudinal investigation of 29 pregnant women using tissue Doppler imaging (TDI). TDI-based indices are better suited for the evaluation of myocardial relaxation and are relatively independent of preload, although one pitfall of this technique is that it is angle dependent and may have high interobserver variability. Data obtained in each trimester and postpartum demonstrated that left ventricular diastolic function was preserved, with augmented late diastolic function accommodating the increased preload of normal gestation. Similar findings were noted in cross-sectional trials using the same methodology, but carried out to answer different questions, in normal pregnant control women.[66,67]

Coupling between Left Ventricle and Systemic Arterial Circulation in Normal Pregnancy

From the mean pressure-flow perspective, the coupled left ventricle–arterial circulation system produces significantly higher cardiac output during normal pregnancy, with little change in mean blood pressure, although a slight decrease is typically observed. This coupled equilibrium of mean pressure and flow is achieved by a significant peripheral vasodilatation (reduced systemic vascular resistance) and increases in heart rate, left ventricular preload (end-diastolic volume), and muscle mass, without any significant changes in left ventricular myocardial contractility.

From the perspective of pulsatile hemodynamics, mathematical simulation-based analysis indicates that if the fall in the steady arterial load – the decrease in systemic vascular resistance – was not accompanied by that in pulsatile arterial load, for example, increase in global compliance, then arterial pulse pressure would have increased significantly, with the decrement in diastolic pressure being significantly greater than that in systolic pressure.[10] Thus, the increase in global arterial compliance that accompanies the profound peripheral vasodilatation prevents the undesirable increase in pulse pressure and diastolic hypotension that can compromise myocardial perfusion.

Left ventricular hydraulic power, another functional index of the coupled left ventricle–arterial circulation system, has been evaluated during normal pregnancy. Two components constitute total power – steady and oscillatory power. The oscillatory component is considered to be wasted power as it does not result in net forward flow of the blood. Thus, the ratio of the oscillatory to total power is often used as an index of inefficiency of the coupled left ventricle–arterial circulation system. Although both total and oscillatory power increased significantly throughout pregnancy, peaking in the third trimester (Fig. 14.7), the ratio of the oscillatory to total power did not change significantly throughout gestation.[10] The aforementioned mathematical simulation-based analysis predicted that this ratio would have doubled, i.e., efficiency decreased by a factor of two, had the increase in global arterial compliance not accompanied the fall in systemic vascular resistance.

In summary, the various systemic arterial and left ventricular mechanical properties undergo coordinated changes in normal pregnancy that result in significantly increased cardiac output with little change in mean and pulse pressures and ventriculo-arterial coupling efficiency. The fall in the pulsatile arterial load (the increase in global arterial compliance) that accompanies the decrease in the steady load (systemic vascular resistance) in normal pregnancy is considered to be an adaptive response. There are at least three reasons for this: (1) markedly increased intravascular volume can be accommodated without a concomitant increase in mean arterial pressure, (2) increase in pulse pressure and diastolic hypotension are prevented, and (3) the efficiency of the mechanical energy transfer from the left ventricle to the arterial circulation is maintained.

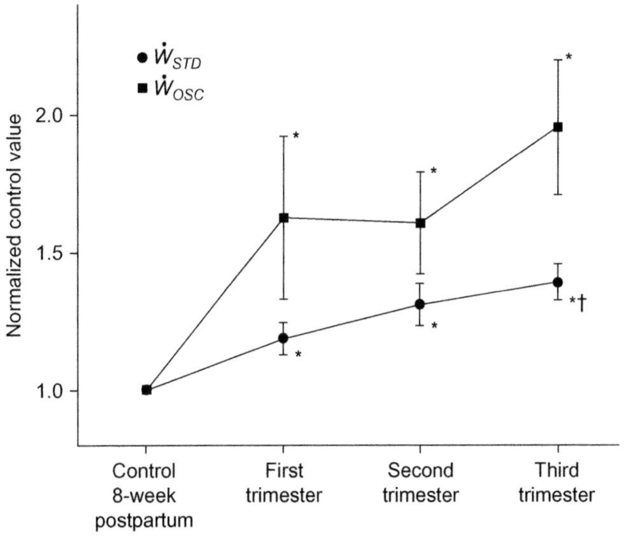

FIGURE 14.7 Temporal changes in steady (W_{STD}) and oscillatory (W_{OSC}) power during normal pregnancy. Data are normalized to 8-week postpartum control values (mean ± SEM; $*p < 0.05$, first, second or third trimester vs. 8-week postpartum control; $^{\dagger}p < 0.05$, second or third vs. first trimester). (Reprinted with permission.[10])

HEMODYNAMICS AND CARDIAC FUNCTION IN PREECLAMPSIA

As summarized above, assessment of the cardiovascular system in normal pregnancy is fraught with methodological and design pitfalls. Here we note emphatically that such problems and challenges are even greater when studying women with preeclampsia. For example, as discussed in Chapter 1, the true diagnosis of preeclampsia is never certain by clinical criteria alone.[68] There may be other concomitant pathology simultaneously influencing the cardiovascular system such as renal disease or diabetes. Finally, no matter how well the experiments are designed, certain confounders may be difficult to eliminate or circumvent, as for example treatment with magnesium sulfate or other antiseizure agents, antihypertensive medications, or parenteral fluid administration, as well as whether or not the preeclamptic woman is in active labor.

Systemic Arterial Hemodynamics in Preeclampsia

While preeclampsia is characterized by the development of hypertension, typically late in gestation, many, but not all, women destined to develop preeclampsia have been documented to have an elevated mean blood pressure when compared with normotensive women – an observation present as early as the ninth gestational week[5,15,45,69] (Fig. 14.2). These differences have also been shown using ambulatory blood pressure technology[21] (Fig. 14.3). Clearly, elevated blood pressure in early pregnancy is a risk marker for developing preeclampsia later; however, blood pressure values alone are a poor predictor for actually determining who will develop preeclampsia.[21,70–73]

A change in the normal diurnal pattern of blood pressure in women destined to develop preeclampsia has been noted, with either obliteration of the decrease in nocturnal pressure or a shift in the timing of the blood pressure nadir.[18,20,74] Recently alterations in cardiovascular regulatory behavior have been suggested to predict preeclampsia by assessing systolic as well as diastolic beat-to-beat pressures and heart rate by variability and coupling analysis. Discriminant function analysis of the parameters predicted preeclampsia with an 88% sensitivity and specificity, and if assessment of uterine artery resistance was added, a 70% positive-predictive value at 18–26 weeks gestation was realized.[75,76]

As discussed above, normal pregnancy is accompanied by increased intravascular volume, high cardiac output, and vasodilatation – a low-resistance systemic circulation. With the onset of overt preeclampsia there is a shift to a low-output, high-resistance state, and intravascular volume is significantly lower than in the normal pregnant state (Chapter 15). This traditional characterization of preeclampsia as a state of decreased intravascular volume, lower cardiac output, and vasoconstriction is not observed by all investigators. A myriad of problems may contribute to the variable observations, including manipulation of volume status prior to evaluation, that can alter the pathophysiological picture present before the treatment.

The dilemma alluded to above, for example, can be appreciated by a review of studies during the 1980s performed using "gold standard" invasive monitoring with Swan–Ganz pulmonary artery catheters that produced markedly contrasting results.[77–84] Some investigators noted decreased cardiac output, others increased or unchanged, and there was wide variation in the measured peripheral vascular resistance. That said, a landmark investigation by Visser and Wallenburg[85] appears to have pinpointed the reasons for these diverse findings. Two unique features distinguish this work. First, they compared a group of untreated or "virgin" preeclamptic women to a cohort of treated preeclamptics, and second, the number of subjects included – 87 untreated and 47 treated – appears to be the largest investigation using invasive technology and reported by 2014. It is also unlikely that this investigation will ever be repeated because pulmonary artery catheter monitoring is rarely used today. Instead, the wide availability of noninvasive techniques for assessment has replaced the invasive ones.

Visser and Wallenburg[85] carefully chose their subjects, using strict criteria and only selecting women with diastolic pressure of 100 mm Hg measured twice four hours apart, proteinuria ≥0.5 g/L, onset ≥20 weeks gestation, and complete recovery postpartum. Patients with medical disorders such as chronic hypertension, cardiac or renal disease were excluded. Patients receiving intravenous fluids, antihypertensive medication, antiseizure therapy or any other type of medication were considered "treated," while those women who had not yet received any of the aforementioned therapies were the "pure (or virgin) preeclamptic" group. These women were not randomized to treatment groups. Finally, a group of normotensive women studied in another protocol were used for further comparison.[81] (Of note, studying consenting normal pregnant volunteers with central monitoring was defended in the 1980s, the risk believed low, and the need to obtain normative data to improve intensive care monitoring and treatment of pregnant women was believed to be consistent with equipoise considerations. The status of noninvasive technology, we believe, would currently preclude such studies today.)

Table 14.1 summarizes the hemodynamic measurements of Visser and Wallenburg.[85] As expected, mean arterial pressure was highest in the untreated preeclamptic

TABLE 14.1 Hemodynamic Profile in Untreated and Treated Preeclamptic Patients and Normotensive Pregnant Women

	Preeclamptics, Untreated (n = 87)	p^a	Normotensive Controls (n = 10)	p^b	Preeclamptics, Treated (n = 47)
Mean intra-arterial pressure (mm Hg)	125 (92–156)	<0.001	83 (81–89)	<0.001	120 (80–154)[c]
Cardiac index (L min^{-1} m^{-2})	3.3 (2.0–5.3)	<0.001	4.2 (3.5–4.6)	NS	4.3 (2.4–7.6)[c]
Systemic vascular resistance index (dyne s cm^{-5} m^2)	3003 (1771–5225)	<0.001	1560 (1430–2019)	<0.005	2212 (1057–3688)[c]
Pulmonary capillary wedge pressure (mm Hg)	7 (1–20)	NS	5 (1–8)	<0.05	7 (0–25)

Values given are median (range).
(Reprinted with permission.[82])
NS, not significant.
[a]Differences between untreated preeclamptic patients and normotensive controls.
[b]Differences between pharmacologically treated preeclamptic patients and normotensive controls.
[c]$p < 0.05$ vs. untreated nulliparous patients.

group – 125 mm Hg – underscoring disease severity in relation to treated preeclamptics and normotensive pregnancies. Note the remarkably consistent hemodynamic parameters in the "virgin" preeclamptics, that is: a significantly decreased cardiac index, as well as the marked and significantly increased systemic vascular resistance. This contrasts with the results from treated preeclamptics. Note further that capillary wedge pressure remained normal. These findings strongly support the concept of preeclampsia being predominantly associated with low cardiac output, markedly increased systemic resistance, and an increased afterload, and suggest explanations for the variable findings of others. This signal work confirms the same group's preliminary findings in a smaller group studied previously.[81]

Lang et al.,[2] using noninvasive techniques to compare 10 severe preeclamptics with 10 normotensive pregnant women, each evaluated in labor, and again 1 day and 4 weeks postpartum, obtained similar results. They reported significantly decreased cardiac output and increased systemic vascular resistance during preeclampsia, and both groups had similar hemodynamic profiles at postpartum follow-up.

While the cardiovascular changes during overt preeclampsia seem clear, there are differences of opinions regarding the changes that precede clinical presentation of the disease. The general impression is that there is evidence of a vasoconstrictor state well before overt disease manifests. For instance increased sensitivity to pressor substances, increments in circulating antiangiogenic factors that affect the vasculature in a manner that opposes vasodilatation, and lower intravascular volume, precede the disease by many weeks.[86–90] These issues are explored further elsewhere in the text in Chapters 6 and 15.

Although this "vasoconstricted" state with preeclampsia is more widely accepted, there is an alternate view that women destined to develop preeclampsia have an exaggeration of the normal increase in cardiac output in early pregnancy. In this scenario, championed by Easterling[91] and Bosio[92] and their colleagues, preeclampsia is the end result of a pregnancy originally marked by an excessive increase in cardiac output with an exaggerated compensatory decrease in systemic peripheral resistance. In this regard, they liken the preclinical phase of preeclampsia to that preceding overt essential hypertension. In the latter, Messerli et al.[93] described this prehypertensive phase as one of increased cardiac output accompanied by a reflex decrease in systemic resistance, a protective mechanism against the appearance of elevated blood pressure. Eventually this autoregulation mechanism fails, afterload increases, cardiac output decreases, and hypertension becomes manifest. Thus in the combined views of Easterling[91] and Bosio[92] and their coworkers, preeclampsia would occur in pregnant women with exaggerated increases in cardiac output, and normal or slightly increased gestational falls in peripheral vascular resistance, but at the time the disease becomes overt there is a "crossover" to a high-resistance low-output state. If this theory were to prove correct, it would be logical to try to identify women with exaggerated increases in cardiac output early in pregnancy and treat them with drugs that lower output such as beta-blockers, and indeed such a study has been done.[94]

There are, however, a number of problems with the theories of Easterling[91] and Bosio.[92] First, Easterling et al.[91] studied nulliparous women serially using noninvasive techniques to measure mean arterial pressure and cardiac output. Of 179 women, 89 had normal pregnancies, 9 developed preeclampsia, and 81 had gestational hypertension. Before disease manifestations, the women destined to develop preeclampsia had higher mean arterial pressures and cardiac output, with lower systemic resistance, albeit the latter was not significant. In this study there were no significant changes detected in either the high cardiac output state or in the reduced systemic resistance when overt signs and symptoms of preeclampsia occurred. The higher initial blood pressure in the eventual preeclamptics is consistent with findings of other investigators, but the early elevated cardiac output with normal or reduced vascular resistance had not been described previously. Of concern, there was a high drop-out rate, and only nine women developed preeclampsia, mainly mild disease – they were delivered at a mean of 39.4 weeks. Also, neonatal birth weight was similar to the controls. Other concerns were the use of a 1 + qualitative determination to define proteinuria.

Second, Easterling et al.[91] used a continuous-wave Doppler ultrasound system with no range-gating or imaging capabilities, and measurements of the aortic annulus were obtained by A-mode ultrasound, measurements that are more prone to angulation errors.[58] The preeclamptic women had a much higher body mass than the control population, and there was no correction for this factor in calculating cardiac output. Specifically, the preeclamptics were on average 12 kg heavier than the nonpreeclamptic women – BMI 29.8 kg/m^2 vs. 24.9 kg/m^2 calculated from the data provided – and thus obesity, a factor known to increase cardiac output,[95] could explain the increments recorded. Finally, the authors did not provide results for the 89 women who developed gestational hypertension, a complication of pregnancy in many women who eventually manifest essential hypertension. Given this remarkably high incidence of women with gestational hypertension in the study, viz., nearly half, comparisons of cardiac output and vascular resistance between the two hypertensive groups would have been instructive.

There are similar problems with the study by Bosio et al.[92] It was much larger with 400 subjects, and it was a longitudinal survey that utilized similar noninvasive techniques. They too describe a high cardiac output state prior to evidence of clinical disease in the 20 women who developed preeclampsia, but here systemic vascular resistance in the latent phase was definitely normal.

After the "cross-over," however, and similar to Visser and Wallenberg[85] and Lang et al.,[2] but unlike Easterling et al.,[91] they noted markedly decreased cardiac output and increased systemic vascular resistance. Bosio et al.[92] did not correct cardiac output for maternal weight, but stated that even after adjusting for BMI statistical significance was maintained; however, these data are not presented. The group of 24 women who developed gestational hypertension remained with high cardiac output after disease manifestation[92] (Fig. 14.8). These investigators have suggested that the hemodynamic changes promote endothelial injury and trigger the low-output vasoconstricted state of clinical disease.

More recently, Dennis et al.[96] studied 40 "untreated" preeclamptic women and also reported increased cardiac output and increased cardiac contractility defined by fractional shortening, as well as reduced diastolic function. But the preeclamptic women had significantly greater BMIs. In addition, information regarding administration of intravenous fluids or magnesium sulfate was not given, and they did not correct for loading conditions in assessing contractility.

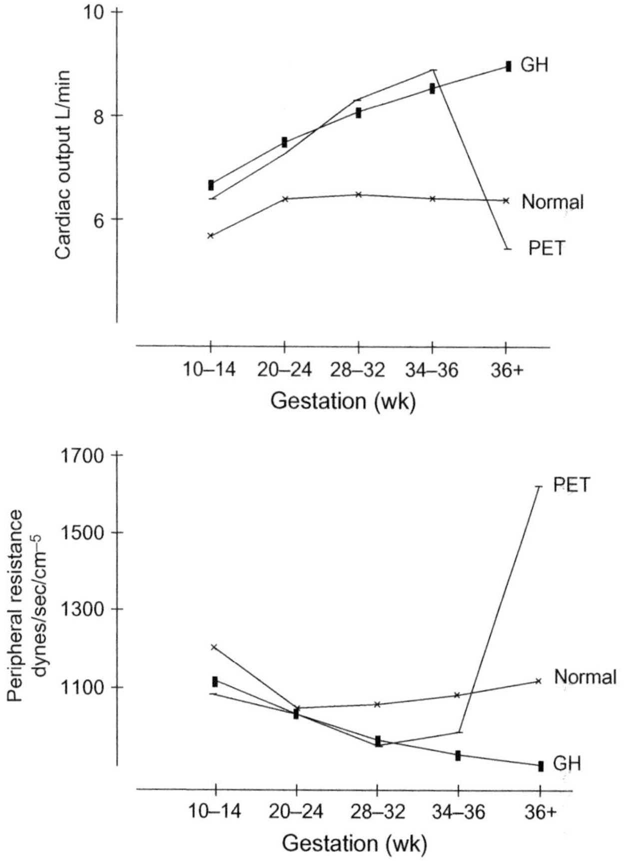

FIGURE 14.8 Median values for cardiac output and total peripheral resistance in the different patient populations plotted against gestation. Normal: normotensive cohort; PET: preeclamptic cohort; GH: gestational hypertensive cohort. (Reprinted with permission.[92])

Because of this "minority" theory, some have postulated that treatment with beta-receptor blocking drugs for women with exaggerated cardiac output in early pregnancy might prevent preeclampsia.[94] It appears that this suggestion might be premature. Carr et al.[97] reported that women at risk of preeclampsia with high cardiac output treated with atenolol have a blunted rise in sFlt levels compared with those at low risk. It is problematic, however, that this was a nonrandomized, secondary analysis that included a small number of women. More importantly, these investigators did not demonstrate prevention of preeclampsia. Thus, a prospective randomized trial is needed to further study this theorem. In addition, it is necessary to explain how this theory is compatible with overwhelming evidence that the preclinical state of preeclampsia is one marked by increased pressor responses, lower intravascular volume, and increased levels of circulating antiangiogenic proteins that impair vasodilatation.

Systemic Arterial Properties in Preeclampsia

Cross-Sectional Studies

The earlier studies of the effects of preeclampsia on systemic arterial load have been described mostly in terms of the steady component: peripheral vasoconstriction as indicated by an increase in systemic vascular resistance. In the previous edition, we introduced more recent approaches that included assessment of the pulsatile component of arterial load and the complex interactions of the components of the cardiovascular system in the face of a low-output, high-afterload disease state. Hibbard and colleagues[98] performed a cross-sectional study comparing preeclamptics, chronic hypertensives with superimposed preeclampsia, and normotensive women admitted in preterm labor. Because all women were given magnesium sulfate for either seizure prophylaxis or tocolysis, two additional control groups were included to eliminate confounding factors. One was that of normal laboring women with epidural analgesia and the second group included normotensive laboring women who were given neither epidural analgesia nor magnesium sulfate. This study confirmed that total vascular resistance, the steady component of the arterial load, was significantly elevated in both hypertensive groups, but more so in the chronic hypertensives with superimposed preeclampsia. Global arterial compliance was significantly lower in the pure preeclamptic group and with an even greater decrement in chronic hypertensive women with superimposed preeclampsia[99] (Fig. 14.9). The authors further showed that a substantial component of the decreased global compliance was not completely attributable to the rise in blood pressure, but independently related to preeclampsia per se[98] (Fig. 14.10). These findings indicated the overall reservoir properties of the systemic arterial circulation to be compromised. Lower compliance (or higher stiffness) was observed for large conduit arteries as well. For example,

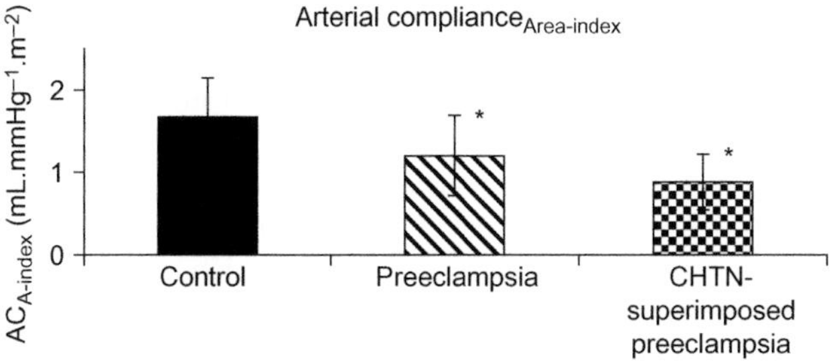

FIGURE 14.9 Global arterial compliance index, a measure of the elasticity component of the entire system, both conduit and peripheral vessels, is depicted for the three study groups: normal controls, solid bar; preeclamptic patients, striped bar; chronic hypertension with superimposed preeclampsia, checked bar. *$p < 0.05$ compared with normal control group. (Reprinted with permission.[99])

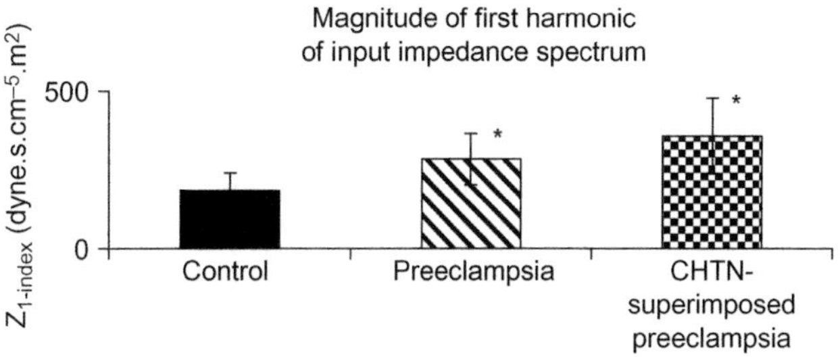

FIGURE 14.11 The magnitude of the first harmonic of the input impedance spectrum index, representing impedance properties of both large and small vessels, is depicted for the three study groups: normal controls, solid bar; preeclamptic patients, striped bar; chronic hypertension with superimposed preeclampsia, checked bar. *$p < 0.05$ compared with normal control group. (Reprinted with permission.[99])

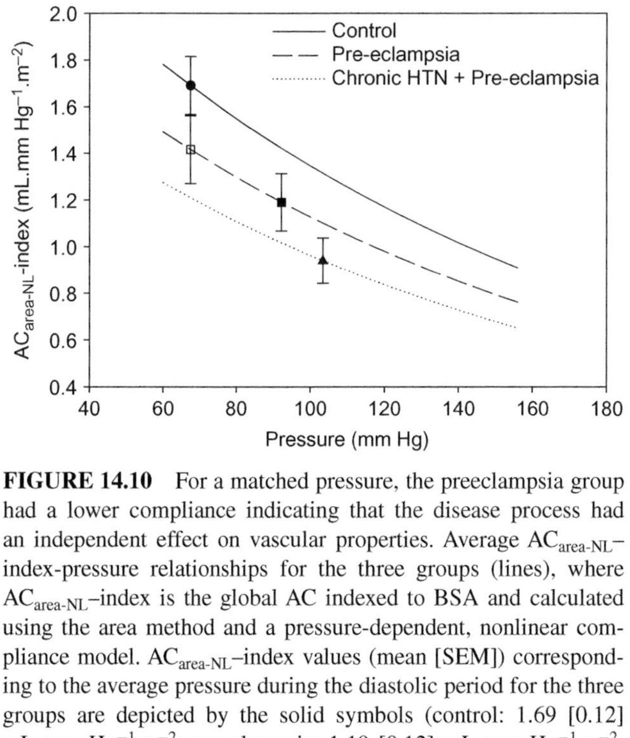

FIGURE 14.10 For a matched pressure, the preeclampsia group had a lower compliance indicating that the disease process had an independent effect on vascular properties. Average $AC_{area\text{-}NL}$–index-pressure relationships for the three groups (lines), where $AC_{area\text{-}NL}$–index is the global AC indexed to BSA and calculated using the area method and a pressure-dependent, nonlinear compliance model. $AC_{area\text{-}NL}$–index values (mean [SEM]) corresponding to the average pressure during the diastolic period for the three groups are depicted by the solid symbols (control: 1.69 [0.12] mL mm $Hg^{-1}m^{-2}$; preeclampsia: 1.19 [0.12] mL mm $Hg^{-1}m^{-2}$; chronic hypertension (HTN) + preeclampsia: 0.94 [0.10] mL mm $Hg^{-1}m^{-2}$). These three $AC_{area\text{-}NL}$–index values are almost identical to the values estimated using the linear (standard) model. The open symbol represents $AC_{area\text{-}NL}$–index for the preeclampsia group at a pressure equal to that of the control group. (Reprinted with permission.[98])

the magnitude of the first harmonic of aortic input impedance (*Z*) was significantly elevated in both hypertensive groups[99] (Fig. 14.11) and aortic characteristic impedance tended to be greater in both hypertensive groups, though not significantly so. Reflection index (RI), a measure of wave reflections within the arterial system, was increased in the preeclamptics and more so in those with chronic hypertension with superimposed preeclampsia. Thus, during the late stages of preeclampsia, both steady and pulsatile components of arterial load are elevated as compared to the normal pregnancy.

Other studies have also reported similar observations regarding lower arterial compliance and increased systemic vascular resistance during late gestation in preeclamptic subjects. For example, Elvan-Taspinar et al.[100] and Tihtonen et al.[101] have studied preeclamptic and nonpreeclamptic chronic hypertensive subjects during the third trimester. The carotid-to-femoral pulse wave velocity (PWV) was used to quantify aortic stiffness (inverse of compliance) in the Elvan-Taspinar study [100] and the ratio of stroke index-to-pulse pressure (SI/PP) was used to quantify arterial compliance in the Tihtonen study.[101] PWV was significantly higher and the SI/PP ratio was significantly lower in preeclamptic subjects compared with values for the control group i.e., normal pregnancy, indicating reduced arterial compliance in preeclampsia. Interestingly, stiffness (inverse of compliance) measures changed less in the chronic hypertension group (without superimposed preeclampsia) compared with the preeclamptic group. In both studies, systemic vascular resistance was greater in the preeclamptic and chronic hypertensive groups compared with the control group.

Global arterial compliance – or its inverse, arterial stiffness – has also been quantified by analyzing the central aortic pressure waveform in terms of augmentation pressure or augmentation index.[102,103] The central aortic pressure waveform is often calculated using noninvasively measured radial pressure waveform (applanation tonometry) and a generalized transfer function linking central aortic and radial pressure waveforms.[104] Using this technique, Khalil et al.[105] have shown that administration of α-methyldopa to preeclamptic women significantly reduced arterial stiffness (i.e., increased arterial compliance). Similarly, magnesium sulfate administration significantly increased arterial compliance in a cohort of 70 preeclamptic women.[106] These investigations suggest that decreased arterial compliance associated with preeclampsia can be improved in the short term. Cruz et al.[107] employed the augmentation index-based analysis to triage symptomatic women being evaluated for preeclampsia. They observed that the classification based on decreased arterial compliance (i.e., augmentation index above a certain threshold) had 80% sensitivity in correctly diagnosing preeclampsia, but the specificity was only 40%.

Longitudinal Studies

Serial measurements of systemic arterial properties have been reported. In nulliparous pregnant women studied longitudinally, those destined to develop preeclampsia were noted to have elevated pulse pressures early in pregnancy, indicating higher arterial stiffness or lower compliance.[108] Although pulse pressure is not as good a measure of pulsatile arterial load as global arterial compliance or aortic characteristic impedance, it is related to vascular compliance (stiffness) properties: high pulse pressure with similar stroke volume and heart rate corresponds to high stiffness or low compliance. Similarly, Oyama-Kato et al.[109] have reported increased brachial-to-ankle pulse wave velocity throughout pregnancy in subjects destined to develop preeclampsia compared with normotensive controls. Similar observations were made by Savvidou et al.[110] who reported that both carotid-to-femoral and carotid-to-radial pulse wave velocities (indices of arterial stiffness) measured at 22–24 weeks were increased in women who subsequently developed preeclampsia as compared with those who did not. In addition, Khalil et al.[111] have reported that the aortic augmentation index measured early in gestation – between 11 + 0 and 13 + 6 weeks – could predict women destined to develop preeclampsia with 79–88% sensitivity. The findings of these aforementioned studies suggest that the early increase in arterial compliance (or decrease in arterial stiffness) seen in normal pregnancy is absent or significantly diminished in preeclampsia. Furthermore, this aberrant compliance (stiffness) response is consistent with the views that alterations in vascular reactivity leading to a vasoconstricted state occur long before the development of preeclampsia.[86,112]

Differences in arterial properties noted in preeclamptic subjects during gestation may persist postpartum. Using venous occlusion plethysmography, studies have shown that endothelial dysfunction persists in prior preeclamptics for as long as up to 5–6 years postpartum.[113,114] Evans et al.[115] examined two groups of women at ~16 months postpartum. Group 1 were women who had preeclampsia during their first pregnancy and group 2 were women with an uncomplicated first pregnancy. Women in the prior-preeclampsia group had higher mean and diastolic arterial pressures, higher systemic vascular resistance, reduced endothelial function (lower stress-induced increase in forearm blood flow), and a tendency towards higher arterial stiffness (increased heart-to-brachial pulse wave velocity) and higher plasma glucose level. Furthermore, the logistic regression analysis indicated that a simultaneous evaluation of multiple derived indices better discriminated between the two groups. Melchiorre et al.[116] have reported higher systemic arterial systolic and diastolic blood pressures in preeclamptic women at 1 year postpartum.

There are a number of reports from the laboratory of Peeters and colleagues,[44,117,118] which suggest that formerly preeclamptic women have endothelial abnormalities, independent of whether or not they eventually develop chronic hypertension after an index gestation complicated by preeclampsia. Those who remained normotensive and had endothelial dysfunction were termed "latent" hypertensives.[118] Furthermore, when some of these women were restudied early in subsequent gestations at 5–7 weeks, none had the normal rise in femoral vascular compliance.[44] These results support observations by Hibbard et al.,[98,99] but also suggest that those with a history of preeclampsia have impaired vascular compliance long term. Haukkamaa et al.[119] presented pulse-wave velocity evidence for long-term vascular abnormalities in previously preeclamptic women, but not all investigators find support for this notion.[120]

One issue relevant to interpreting the data above, originally discussed by Chesley in his first edition, is the problem of comparing follow-up exams in preeclamptics to women with normotensive births. Chesley, and later Fisher et al.,[121] noted that women destined to have future cardiovascular disease, including unrecognized chronic hypertensives, masked by the early fall in blood pressure during pregnancy, segregate into the population that develop preeclampsia, as their incidence of superimposed preeclampsia is quite high. Chesley suggested that remote studies should be designed to compare preeclamptics with an age- and parity-matched general population, as they may have abnormalities that predate pregnancy. The studies reviewed above may underscore preeclampsia as a marker of such abnormalities, but they cannot be used to implicate preeclampsia in their etiology. The investigators would have to have studied the women prior to pregnancy to do so, but we could locate no such studies.

Left Ventricular Properties in Preeclampsia

In the study by Lang et al.,[2] left ventricular muscle mass and end-diastolic diameter during late gestation were similar between preeclamptic and control subjects, indicating that left ventricular structural changes in preeclampsia are similar to those observed in normal pregnancy. In contrast, Melchiorre et al.[122] have reported that left ventricular mass at term was significantly increased in preeclamptic women as compared with control subjects, particularly in those preeclamptic women who also had diastolic dysfunction.

Although a number of studies have addressed systolic function in preeclampsia, viz., cardiac output and left ventricular ejection phase indices such as fractional shortening,[77–80,123,124] few studies have examined ventricular myocardial contractility using techniques that eliminate the confounding effects of loading conditions.[2,62,63] Lang et al.[2] carefully selected 10 preeclamptic nulliparous women, strictly defining the disease,[87] without evidence of chronic hypertension, cardiac or renal disease. The subjects were studied at three time points: (1) prior to labor and prior to administering any antihypertensive medications, though magnesium sulfate therapy had been initiated, (2) one day postpartum when magnesium sulfate had been discontinued and subjects were still hypertensive, and (3) four weeks after delivery when the subjects were normotensive. Ten normotensive pregnant women who underwent identical protocols served as controls. As discussed before, end-systolic stress (σ)-rate corrected velocity of circumferential fiber shortening (Vcf) data were used to characterize left ventricular myocardial contractility. As depicted in Fig. 14.12, preeclamptic subjects had σ–Vcf values that were superimposed upon the normal contractility line (the nomogram) at each study point, results similar to those obtained in normotensive controls[2] (Fig. 14.6). Proceeding from visit 1 to 3 in preeclamptics, σ–Vcf points shifted leftward and upward along the nomogram, indicating decreased afterload over time without any change in left ventricular contractility. Put another way, there was decreased left ventricular performance with acute preeclampsia, but when afterload was eliminated as a confounding variable, it was obvious that contractility remained normal. In the face of increased afterload,

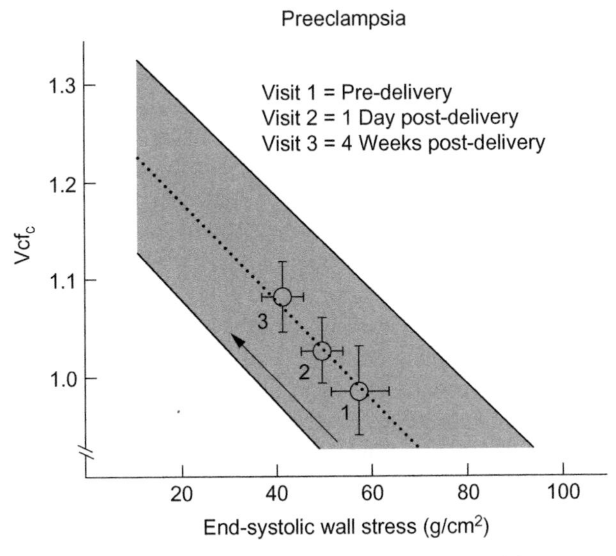

FIGURE 14.12 Average end-systolic wall stress (σ)–rate corrected velocity of fiber shortening (Vcf) data obtained in patients with preeclampsia before delivery, 1 day after delivery, and 4 weeks after delivery. From visit 1 to visit 3, data points shifted leftward and upward (arrow) but still fell on the mean contractility line, indicating decreased afterload without changes in contractility. (Reprinted with permission.[2])

this was the appropriate cardiovascular response and not a pathologic response.

In another investigation employing left ventricular σ–Vcf data to evaluate myocardial contractility in preeclampsia, Simmons et al.[23] studied 15 preeclamptic women and compared them with 44 normal controls. They reported similar myocardial contractility in the two groups, thereby confirming the findings of Lang et al.[2] that left ventricular contractility is unchanged in preeclampsia.

Melchiorre et al.[122] examined global and regional indices of left ventricular systolic function using echocardipgraphic tissue Doppler imaging (TDI). They reported that there were no differences in any of these systolic functional indices between women who develop preeclampsia at term and normotensive controls. In a subsequent study, however, they observed that left ventricular systolic dysfunction is evident in women who develop preterm preeclampsia before 37 weeks.[125] This is consistent with the notion that preterm preeclampsia is a more severe form of the disease.

Shahul et al.[126] have employed *speckle tracking*, a newer and presumably more sensitive technique, to assess global and regional ventricular function in women with preeclampsia. This group demonstrated that in 11 late-onset (32.7–37.4 weeks) preeclamptic women there was evidence for myocardial systolic dysfunction (based on myocardial strain data) in the presence of a normal ejection fraction. They conclude that myocardial strain imaging using echocardiographic speckle tracking may help detect subclinical left ventricular dysfunction in women with preeclampsia. We emphasize, however, that one has to be cautious about this conclusion because the sample size was small (11 preeclamptics) and myocardial strain is highly dependent on afterload (myocardial stress). In addition, information about potential confounders was not provided, viz., intravenous fluid administration or magnesium sulfate or antihypertensive drugs.

Left ventricular diastolic function in women with preeclampsia has been evaluated using TDI in cross-sectional study designs. Melchiorre et al. studied 50 women at term with preeclampsia compared with 50 matched normal pregnant women and found that 40% of women with preeclampsia had global diastolic dysfunction compared with 14% of the controls.[122] The same group subsequently reported that although diastolic dysfunction is evident late in gestation in both preterm and term preeclamptic women,[125] only preterm preeclamptic women have diastolic dysfunction at midgestation (20–23 weeks).[67] It should be noted that normative data from age-matched nonpregnant individuals were used to define normal versus abnormal diastolic function. As normal pregnancy is a volume-loaded state with sustained mild tachycardia and hypotension, the use of these normative data from nonpregnant individuals may be questionable. Although the TDI-based diastolic function indices used in these studies are relatively independent of preload, they do not take afterload or heart rate differences into account. Perhaps, this issue contributed to the relatively high incidence (14%) of diastolic dysfunction in the "normal" pregnancy cohort of young healthy women. These findings,[67,121,122] however, do support the notion that early (preterm) preeclampsia is a more severe or possibly different form of the disease that results in greater pathophysiological changes compared to late-onset preeclampsia.

Bamfo et al.[127] assessed maternal left ventricular diastolic function in normotensive and preeclamptic women who had fetal-growth restriction (FGR). Compared with the normotensive + FGR group, the ratio of E – peak blood velocity of early diastolic filling from Doppler-based transmitral flow – to E′ – peak tissue velocity of early diastolic filling from TDI of mitral annulus – was elevated in the preeclampsia + FGR group, indicating increased left ventricular filling pressure and impaired diastolic function in the preeclampsia group.[127] Melchiorre et al.[128] came to similar conclusions in women with FGR and early (preterm) preeclampsia.

FACTORS THAT MAY EXPLAIN VASCULAR CHANGES IN PREGNANCY

Normal Pregnancy

Autonomic Nervous System

Autonomic regulation of the cardiovascular system is a prime candidate for modification during gestation, but its study is complicated by baseline differences in heart rate, blood pressure, and blood volume that accompany pregnancy.[129] In the first edition of this text, Chesley reviewed striking observations made during the 1950s and 1960s, demonstrating the importance of the autonomic nervous system in maintaining blood pressure when changing posture. When a pregnant woman lies supine, autonomic blockade with tetramethyl ammonium or spinal analgesia resulted in marked hypotension, alleviated by assuming a lateral recumbent position, while similar treatment of nonpregnant subjects had but minimal effects. These studies suggested that the predominant effect of autonomic blockade was through a loss of vasomotor tone. As noted, a reduction in afferent outflow to the resistance vasculature could explain the reduction in systemic vascular resistance. That said, very few in-depth studies followed these observations.

Two studies in animal models are of interest. Pan et al.,[130] in an extensive study of blood pressure regulation in pregnant rats, observed similarly large decreases in blood pressure (40–50%) after either ganglionic or selective α-adrenoreceptor blockade in gravid and nonpregnant animals, suggesting that the adrenergic nervous system was of quantitatively similar importance in the pregnant and nonpregnant states of that species. Assessing nervous activity directly would be needed to confirm these differences. Such studies were performed by O'Hagen and Casey.[131] They evaluated renal sympathetic activity in chronically instrumented rabbits, eliminating anesthesia as a confounder, and observed no effect of pregnancy on renal sympathetic nerve activity in late gestation.

At first glance it would seem that postural differences should make comparisons between rodents and humans imprudent. However we have cited the above studies, in relation to the carefully conducted human studies by Schobel et al.[132] that are detailed further below and in Chapter 15. These investigators, utilizing peroneal nerve microneurology techniques, also noted no differences in basal sympathetic activity when age-matched pregnant and nonpregnant women were compared. Thus, their data also suggest that pregnancy does not affect basal regulation of vascular tone by the sympathetic nervous system.

There is a limited animal literature regarding autonomic nervous system function in pregnancy relating to pressure-mediated changes in heart rate and blood pressure, but baroreflex regulation of heart rate involves both vagal and sympathetic nerve effects.[129,131] More recent studies controlled for this by measuring the response to a variety of stimuli.[129,131] Also we could locate no studies relating to afferent inputs to the central nervous system in response to pressure stimuli.

The most meaningful studies, of course, are those in the conscious state. But published results in both animals and humans give discordant results, as the baroreflex-mediated heart rate response to increasing blood pressure has been reported as enhanced, unchanged, or depressed in pregnancy.[133–138] These discrepancies may relate to the period in pregnancy studied, as well as differences in control of heart rate between the pregnant and nonpregnant states. Of interest here is that in those experiments where the normal gestational difference in resting heart rate is present prior to imposed pressure steps, the reflex tachycardia is accentuated by gestational age,[129,138,139] suggesting that pregnancy augments sympathetic activity to the heart in response to elevations in blood pressure. Parenthetically, we note that pregnancy had little effect on heart rate in response to decreases in blood pressure.

The tachycardic response to hypotension appears unaffected by pregnancy, but this does not reflect the overall sympathetic response to hypotension. Pregnant rats show an attenuated ability to increase sympathetic nerve output in response to a hypotensive challenge.[135] Pregnancy does not seem to affect basal postganglionic sympathetic nerve activity in skeletal blood vessels, nor is there increased activity in response to a cold pressor test.[132]

Finally, while alterations in the autonomic nervous system during pregnancy appear more related to pressure stimuli than to changing basal tone maintenance per se, we note that vascular reactivity to neurotransmitters is reduced during normal gestation,[140–142] and call attention to a landmark study that addressed adrenergic vascular reactivity, performed in 1985 by Nisell and colleagues.[142] They combined measuring blood levels of administered agonists, calf blood flow, and cardiac output in both pregnant women and nonpregnant controls, and were able to simultaneously verify that both groups were receiving equivalent stimuli, and to determine the different components of the pressor response. The absolute pressor response to norepinephrine was similar between groups, but the rise in pregnant women was due solely to increased cardiac output, while the pressure rise in the nonpregnant state was due to vasoconstriction. This observation is an elegant example of why examining the change in blood pressure alone is not necessarily a measure of systemic vascular reactivity. Using the identical change in blood pressure alone, one might erroneously conclude there were no changes in pregnancy. But the simultaneous recording of cardiac output permitted the correct conclusion that normal pregnancy had blunted the systemic response to norepinephrine. Since as noted above basal sympathetic outflow does not decrease during human gestation, at least in the skeletal bed,[132] then the vascular response to vasoconstriction may be reduced. Thus, a given level of nervous activity may produce a different vascular response during pregnancy, compared with the nonpregnant state. There may also be local mechanisms within the arterial tree itself that account for the relaxation observed during pregnancy, perhaps due to humoral signals deriving from ovary, placenta, or possibly from the pituitary. In summary, we still have a lot to learn about the autonomic nervous system in pregnancy.

Vascular Wall Remodeling and Smooth Muscle Tone

Both increased vascular distensibility and the remarkably hypertrophied uteroplacental circulation could contribute to the observed increase in global arterial compliance during normal pregnancy. The second possibility is unlikely to be a major factor because compliance changes occur very rapidly – early in the first trimester – and several studies have shown that most of the pregnancy-associated hemodynamic changes can be reproduced simply by sex steroid[50,55] or other hormonal[143,144] administrations. Potential mechanisms for increased arterial distensibility can be divided into three categories: (1) passive changes in vessel wall properties secondary to reduced distending pressure, (2) vascular wall remodeling, and (3) reduced smooth muscle tone. The first factor is unlikely to play a major role because changes in distending pressure during normal pregnancy are very small. Vascular wall remodeling is composed of geometric (e.g., increase in vascular wall area) and compositional (e.g., relative amounts of wall constituents such as elastin and collagen) components. Osol and colleagues[145–147] have reported that uterine radial arteries from late pregnant rats were characterized by a significant increase in vascular wall cross-sectional area when compared with those from nonpregnant rats, and smooth muscle cell hypertrophy and hyperplasia were underlying factors in the observed increase in vascular wall area. Griendling et al.[148] have reported similar smooth muscle hypertrophy of uterine arteries during late pregnancy in sheep and this was accompanied by compositional remodeling, i.e., decreased collagen and unchanged elastin. Interestingly, carotid arteries did not show any geometric or compositional remodeling. Mackey et al.[53] observed reduced collagen and elastin concentrations in rat mesenteric resistance arteries during late pregnancy. Thus, geometric and/or compositional remodeling seems to be occurring during late pregnancy in an artery-specific manner. It is still not entirely clear whether these changes are present early in pregnancy and contribute to the observed alterations in the vascular mechanical properties. Finally, ample evidence exists that a reduction in smooth muscle tone contributes to the changes in arterial resistance and compliance properties in normal pregnancy. Factors responsible for this include hormonal signals, particularly estrogens[149–151] or relaxin,[143] or the release of endothelial relaxing factors such as nitric oxide or VEGF, PlGF, or other circulating angiogenic factors.[90]

Hormonal Signals

Estrogen decreases smooth muscle tone in animals and humans,[151–154] lowers blood pressure and carotid-femoral pulse wave velocity in postmenopausal women,[154,155] and also promotes endothelium-dependent vasodilatation in women with premature ovarian failure.[156] Activation of estrogen receptor-α modulates nitric oxide production and the response of smooth muscle relaxation in segments of isolated mouse aorta.[157]

The peptide relaxin, produced primarily by the corpus luteum, has been investigated as a candidate hormone promoting the vascular changes in normal pregnancy. As detailed in Chapter 16, relaxin may be the hormone most responsible for the physiologic gestational alterations in renal hemodynamics.[158,159] Similarly, chronic administration of recombinant human relaxin to conscious, female, nonpregnant rats (a) reduced the steady arterial load by decreasing systemic vascular resistance, (b) increased cardiac output, and (c) reduced the pulsatile arterial load as assessed by indices of systemic arterial compliance.[144] Interestingly, exogenous relaxin administration to male rats at doses that were efficacious in females increased both cardiac output and systemic arterial compliance, and reduced systemic arterial resistance.[160] Finally, neutralization of endogenous circulating relaxin through administration of specific rat relaxin antibodies during early gestation abolished the changes in cardiac output, systemic vascular resistance, and global arterial compliance observed in rats infused with an irrelevant antibody.[143] Thus, relaxin appears to play an important role in many of the systemic arterial (hemodynamics and mechanical properties) as well as renal changes during normal pregnancy,[161] and especially during early gestation. Mechanisms of how relaxin induces decreased smooth muscle tone are detailed in Chapter 16. Relaxin activates the endothelial ET-B receptor/nitric oxide vasodilatory pathway,[161] stimulates vascular endothelial growth factor (VEGF) *in vitro*,[162] and has angiogenic properties.[161]

Carbillon et al.[163] reviewed the evidence that nitric oxide plays a role in the regulation of vascular tone in pregnancy, contributing to the low systemic resistance. That a nitric oxide-mediated pathway may contribute to the cardiovascular changes described above was addressed by Williams et al.[164] in a pregnant dog model. These investigators, studying arterioles from the dogs, suggested that nitric oxide produced by endothelium enhanced release of nitrites, exerting greater control over shear stress-induced vasodilatation, promoting coupling of oxygen delivery and efficiency of the heart. Angiogenic factors such as VEGF and TGF-β1 maintain vascular homeostasis and endothelial health in normal pregnancy[90] and are discussed in Chapter 6. Their roles in gestational cardiovascular changes remain to be explored.

Preeclampsia

Imaging and other technologies used to probe the cardiovascular system were discussed above, and the data reviewed suggesting that all components of the arterial system are affected by preeclampsia. Factors responsible for the pathological cardiovascular changes are not completely understood.

An attempt to explain the vascular changes in preeclampsia would be mostly speculation, and we will focus on the meaning of a few experimental observations. As noted in the section devoted to the autonomic nervous system, Nisell et al.[142] infused norepinephrine in normal pregnant and nonpregnant women and measured cardiac output during the infusion. Blood pressure in the normal pregnant group increased secondary to an increase in cardiac output alone while in the nonpregnant state pressure increments were due to increased systemic vascular resistance and a small decrease in cardiac output. These investigators also studied preeclamptic women and in these subjects the enhanced pressure response was secondary to an exaggerated rise in systemic vascular resistance. These results are consistent with observations that vascular reactivity is attenuated in normal pregnancy, but augmented in preeclamptics.

In studies by Schobel,[132] described above and in Chapter 15, postganglionic sympathetic nerve activity was three-fold higher in preeclamptic women, the values normalizing postpartum, coinciding with the return of their blood pressures to normal nonpregnant levels. The observations of sympathetic overactivity correlate with the increments in total vascular resistance, decreased arterial compliance, and increased impedance that accompanied the increased blood pressure noted in our own studies.[98] We note though that these observations are inconsistent with old studies where pharmacological blockade of the autonomic system had no effect on the elevated blood pressure of preeclamptic women.[165,166] They are also at variance with data indicating lower heart rates in preeclampsia.

A final hint from human studies permits speculation that vascular remodeling may be involved in the cardiovascular findings of preeclampsia. Omental resistance arteries from preeclamptic women were analyzed, demonstrating irregular thickness of the elastic lamina, an incomplete basement membrane, and a changed location and arrangement of endothelial cells compared with normotensive controls as demonstrated by electron microscopy.[167] These findings, however, could be secondary to the disease process rather than the inciting event that stimulates cardiovascular changes.

The roles of genetic and circulating antiangiogenic factors in vascular responses in preeclampsia are discussed in Chapter 4 and Chapter 6, respectively.

PREGNANCY-ASSOCIATED RESPONSES AND THE ASSESSMENT OF CARDIOVACULAR DISEASE RISK LATER IN LIFE

Marked maternal cardiovascular as discussed above and metabolic (Chapter 7) changes are part of a complex adaptive response during normal pregnancy to ensure adequate uterine blood flow and nutrient delivery to the growing conceptus. For example, early pregnancy vasodilatation, viz., reduced total vascular resistance, is accompanied by an increase in global arterial compliance – reduced vascular stiffness. This accommodates the greater intravascular volume, stroke volume, and heart rate without increasing mean arterial pressure. In addition, transient excursions into insulin resistance and dyslipidemia, along with inflammation/innate immune system activation,[168] are observed during the latter half of normal pregnancy. As discussed above, the normal pregnancy-related (adaptive) cardiovascular changes are markedly attenuated both before and during preeclampsia. Similarly, pronounced insulin resistance, dyslipidemia and inflammation characteristically develop before and during preeclampsia, representing accentuations of normal pregnancy changes (Chapter 7). This cardio-metabolic profile of preeclampsia is strikingly similar to those associated with cardiovascular disease (CVD) generally.[169] In fact, many of the cardiovascular and insulin resistance-related alterations of preeclampsia, which in the nonpregnant state are predictors of CVD later in life,[170–172] continue or reemerge postpartum in women with prior preeclampsia.[115,116,169,173–179] Women with a history of preeclampsia are at two-to three-fold increased risk of developing hypertension, coronary artery disease or stroke in later life, and the association strengthens to three- to eight-fold with early onset, severe preeclampsia or recurrent preeclampsia.[180–182] Other pregnancy complications related to abnormal implantation or placental vascular disease *but without a hypertensive pregnancy syndrome* – placental abruption, recurrent spontaneous abortion, preterm birth, or normotensive fetal-growth restriction – are also associated with later-life CVD.[177,183] These data from women *who were not exposed to the vascular damage of an established maternal syndrome* suggest that the pregnancy syndromes unmask risk rather than cause CVD.[177,183–187] Many of the shared risk factors for preeclampsia and later-life CVD may be subclinical before the pregnancy.[188,189] This, however, does not exclude possible lasting effects of preeclampsia.[177,183,190]

Pregnancy can thus be considered as a cardio-metabolic "stress test" that unmasks underlying, subclinical cardiovascular and metabolic abnormalities that increase the risk of both pregnancy disorders and later-life CVD. The unique pregnancy window for early screening and risk management was noted in the *American Heart Association – 2011 Update*; it recommended a pregnancy history as part of cardiovascular health assessment, and summarized evidence that a "metabolic syndrome of pregnancy" may represent a "failed stress test" unmasking early or preexisting vascular endothelial dysfunction and vascular or metabolic disease.[191] The NICHD *Scientific Vision Workshop on Pregnancy and Pregnancy Outcomes*, White Paper, 2011, similarly noted that pregnancy *offers a unique opportunity to study adaptive and maladaptive responses to physiologic changes*, including *cardio-metabolic dysfunction*, that may lead to *interventions to improve pregnancy outcome* and *lifelong health for women.* One of us (SGG) and Dr. Hubel – coauthor of Chapter 7 – hypothesize that it is possible to construct a multivariable metric consisting of aberrant cardio-metabolic responses during pregnancy that can predict later-life CVD better than current prediction algorithms such as Framingham or Reynolds risk scores that are based on data collected in the nonpregnant state. Furthermore, we believe that this aberrant pregnancy response metric will be a continuum, with values for preeclamptic subjects – and perhaps women with other pregnancy-associated disorders mentioned above – being at the extreme. The concept that pregnancy metabolic responses are a continuum, with preeclampsia representing an extreme, was advanced by Redman and colleagues,[168] who showed that innate immune inflammatory changes in women with uncomplicated pregnancy can be nearly as profound as those during sepsis. As another example, there are indications that a supraphysiologic rise in maternal plasma lipids, even during uncomplicated pregnancy, serves as a marker of "prelipemia," that is it predicts later-life progression to overt lipemia in the same way that gestational diabetes is a marker for prediabetes.[192] Thus, we (Shroff and Hubel) propose that the amplifying nature of pregnancy with respect to latent cardio-metabolic flaws can be employed to identify subsets of women with a profile consistent with subclinical vascular disease postpartum which is known to be related to greater CVD risk later in life. We further propose that pregnancy-related aberrant response metrics can be used to predict later-life CVD even in the absence of any pregnancy-associated disorders. Our preliminary (unpublished) data support these hypotheses and we are currently conducting longitudinal studies aimed at developing and validating the multivariable metric. A recent review article underscores the need for the development and validation of such a tool to assess CVD risk in women and guide management.[193]

SUMMARY

A host of changes in the cardiovascular system occur during the course of a normal pregnancy: remarkable increases in cardiac output, systemic arterial compliance, and total blood volume while blood pressure and systemic vascular resistance fall, all in the face of normal myocardial contractility. Both steady and pulsatile arterial loads fail to decrease in preeclampsia, as occurs in normal pregnancy, involving changes in both conduit and small vessels. However, the ability of the cardiovascular system to adapt, and for the heart to continue functioning with normal myocardial contractility during preeclampsia, has been demonstrated. Although left ventricular systolic function in late-onset (term) preeclampsia is generally preserved throughout gestation, diastolic function in preeclampsia seems to be depressed during late gestation. In addition, this depression in diastolic function appears early and is more severe in the setting of early-onset preeclampsia. The attenuation or reversal of many of the cardiovascular responses occurs prior to and during preeclampsia, commencing with a marked reversal of the vasculature's resistance to pressor hormones and culminating in not only marked vascular sensitivity but hypertension characterized by a low cardiac output, high systemic vascular resistant state. There may be a hyperdynamic, low-resistance disease state preceding this but whether this represents a subset of women who have or will eventually develop chronic hypertension remains to be determined. Such a state, however, is hard to reconcile with the early change in vascular reactivity, the knowledge that volume decreases before the advent of hypertension, and the exaggerated rise in circulating antiangiogenic proteins that impede vasodilatation. Abnormal vascular responses during preeclampsia may be secondary to changes in vascular tone – autonomic nervous system, hormonal regulation – and/or vascular wall elements. Studies have convincingly shown that women with a history of preeclampsia and other pregnancy-associated disorders are at significantly higher risk of developing CVD in later life. Evidence is accumulating in support of the notion that pregnancy can be considered as a cardio-metabolic "stress test," capable of unmasking underlying subclinical cardiovascular and metabolic abnormalities that increase the risk of both pregnancy disorders such as preeclampsia and later-life CVD. Thus, a quantitative metric based on pregnancy-associated aberrant cardio-metabolic responses has a potential to predict later-life CVD risk in women, even in the absence of any pregnancy-associated disorders.

References

1. Lindhard J. Uber das Minutenvolume des Herzens bei Ruhe und bei Muskelarbeit. *Pflugers Arch.* 1915;161:223.
2. Lang RM, Pridjian G, Feldman T, Neumann A, Lindheimer M, Borow KM. Left ventricular mechanics in preeclampsia. *Am Heart J.* 1991;121:1768–1775.
3. Christianson RE. Studies on blood pressure during pregnancy. I. Influence of parity and age. *Am J Obstet Gynecol.* 1976;125:509–513.
4. Chapman AB, Zamudio S, Woodmansee W. Systemic and renal hemodynamic changes in the luteal phase of the menstrual cycle mimic early pregnancy. *Am J Physiol.* 1997;273:F777–F782.
5. Moutquin JM, Rainville C, Giroux L, et al. A prospective study of blood pressure in pregnancy: prediction of preeclampsia. *Am J Obstet Gynecol.* 1985;151:191–196.
6. Duvekot JJ, Cheriex EC, Pieters F, Menheere P, Peeters L. Early pregnancy changes in hemodynamics and volume homeostasis are consecutive adjustments triggered by a primary fall in systemic vascular tone. *Am J Obstet Gynecol.* 1993;169:1382–1392.
7. Duvekot JJ, Peeters LLH. Maternal cardiovascular hemodynamic adaptation to pregnancy. *Obstet Gynecol Surv.* 1994;49:S1–S14.
8. Clapp JF, Capeless E. Cardiovascular function before, during and after the first and subsequent pregnancies. *Am J Cardiol.* 1997;80:1469–1473.
9. Robson SC, Hunter S, Boys RJ, Dunlop W. Serial study of factors influencing changes in cardiac output during human pregnancy. *Am J Physiol.* 1989;256:H1060–H1065.
10. Poppas A, Shroff SG, Korcarz CE, et al. Serial assessment of the cardiovascular system in normal pregnancy: role of arterial compliance and pulsatile arterial load. *Circulation.* 1997;95:2407–2415.
11. van Oppen AC, Van der Tweel I, Alsbach GP, Heethaar RM, Bruinse HW. A longitudinal study of maternal hemodynamics during normal pregnancy. *Obstet Gynecol.* 1996;88:40–46.
12. Gilson GJ, Samaan S, Crawford MH, Qualls CR, Curet LB. Changes in hemodynamics, ventricular remodeling, and ventricular contractility during normal pregnancy: a longitudinal study. *Obstet Gynecol.* 1997;89:957–962.
13. Lees MM, Taylor SH, Scott DB, Kerr MG. A study of cardiac output at rest throughout pregnancy. *J Obstet Gynaecol Brit Commonw.* 1967;74:319–328.
14. Capeless EL, Clapp JF. Cardiovascular changes in early phase of pregnancy. *Am J Obstet Gynecol.* 1989;161:1449–1453.
15. Page EW, Christianson R. The impact of mean arterial pressure in the middle trimester upon the outcome of pregnancy. *Am J Obstet Gynecol.* 1976;125:740–746.
16. Hunter S, Robson SC. Adaptation of the maternal heart in pregnancy. *Brit Heart J.* 1992;68:540–543.
17. Desai DK, Moodley J, Naidoo DP. Echocardiographic assessment of cardiovascular hemodynamics in normal pregnancy. *Obstet Gynecol.* 2004;104:20–29.
18. Beilin LJ, Deacon J, Michael CA, et al. Circadian rhythms of blood pressure and pressor hormones in normal and hypertensive pregnancy. *Clin Exp Pharmacol Physiol.* 1982;9:321–326.
19. Seligman SA. Diurnal blood-pressure variation in pregnancy. *J Obstet Gynaecol Brit Commonw.* 1971;78:417–422.
20. Benedetto C, Zonca M, Marozio L, Dolci C, Carandente F, Massobrio M. Blood pressure patterns in normal pregnancy and in pregnancy-induced hypertension, preeclampsia, and chronic hypertension. *Obstet Gynecol.* 1996;88:503–510.
21. Hermida RC, Ayola DE. Prognostic value of office and ambulatory blood pressure measurements in pregnancy. *Hypertension.* 2002;40:298–303.
22. Mabie WC, DiSessa TG, Crocker LG, Sibai BM, Arheart KL. A longitudinal study of cardiac output in normal human pregnancy. *Am J Obstet Gynecol.* 1994;170:849–856.
23. Simmons LA, Gillin AG, Jeremy RW. Structural and functional changes in left ventricle during normotensive and preeclamptic pregnancy. *Am J Physiol Heart Circ Physiol.* 2002;283:H1627–H1633.
24. Mone SM, Sanders SP, Colan SD. Control mechanisms for physiological hypertrophy of pregnancy. *Circulation.* 1996;94:667–672.
25. Geva T, Mauer MB, Striker L, Kirshon B, Pivarnik JM. Effects of physiologic load of pregnancy on left ventricular contractility and remodelling. *Am Heart J.* 1997;133:53–59.
26. Del Bene R, Barletta G, Mello G. Cardiovascular function in pregnancy: effects of posture. *Brit J Obstet Gynaecol.* 2001;108:344–352.
27. Monos E, Berczi V, Nadasy G. Local control of veins: Biomechanical, metabolic, and humoral aspects. *Physiol Rev.* 1995;75:611–666.
28. Guyton AC, Polizo D, Armstrong GG. Mean circulatory filling pressure measured immediately after cessation of heart pumping. *Am J Physiol.* 1954;179:261–267.
29. Humphreys PW, Joels N. Effect of pregnancy on pressure-volume relationships in circulation of rabbits. *Am J Physiol.* 1994;267:R780–R785.
30. Rothe CF. Physiology of venous return. An unappreciated boost to the heart. *Arch Intern Med.* 1986;146:977–982.
31. Cha SC, Aberdeen GW, Nuwayhid BS, Quillen EW. Influence of pregnancy on mean systemic filling pressure and the cardiac function curve in guinea pigs. *Can J Physiol Pharmacol.* 1992;70:669–674.
32. Douglas BH, Harlan JC, Langford HG, Richardson TQ. Effect of hypervolemia and elevated arterial pressure on circulatory dynamics of pregnant animals. *Am J Obstet Gynecol.* 1967;98:889–894.
33. Goodlin RC, Niebauer MJ, Holmberg MJ, Zucker IM. Mean circulatory filling pressure in pregnant rabbits. *Am J Obstet Gynecol.* 1984;148:224–225.
34. Tabsh K, Monson R, Nuwayhid B. Regulation of cardiac output in ovine pregnancy. Society for Gynecologic Investigation 1995 32nd Meeting. 53p. [abstr].
35. Davis LE, Hohimer AR, Giraud GD, Paul MS, Morton MJ. Vascular pressure-volume relationships in pregnant and estrogen-treated guinea pigs. *Am J Physiol.* 1989;257:R1205–R1211.
36. Richardson TQ, Stallings JO, Guyton AC. Pressure-volume curves in live, intact dogs. *Am J Physiol.* 1961;201:471–474.

37. Yamamoto J, Trippodo NC, Ishise S, Frohlich ED. Total vascular pressure-volume relationship in the conscious rat. *Am J Physiol.* 1980;238:H823–H828.
38. Hohmann M, Keve TM, Osol G, McLaughlin MK. Norepinephrine sensitivity of mesenteric veins in pregnant rats. *Am J Physiol.* 1990;259:R753–R759.
39. Hohmann M, Mackey K, Davidge S, McLaughlin MK. Venous remodeling in the pregnant rat. *Clin Exp Hypertens B.* 1991;10:307–321.
40. Fawer R, Dettling A, Weihs D, Welti H, Schelling JL. Effect of the menstrual cycle, oral contraception and pregnancy on forearm blood flow, venous distensibility and clotting factors. *Eur J Clin Pharmacol.* 1978;13:251–257.
41. Goodrich SM, Wood JE. Peripheral venous distensibility and velocity of venous blood flow during pregnancy or during oral contraceptive therapy. *Am J Obstet Gynecol.* 1964;90:740–744.
42. Pickles CJ, Brinkman CR, Stainer K, Cowley AJ. Changes in peripheral venous tone before the onset of hypertension in women with gestational hypertension. *Am J Obstet Gynecol.* 1989;160:678–680.
43. Sakai K, Imaizumi T, Maeda H, et al. Venous distensibility during pregnancy: comparisons between normal pregnancy and preeclampsia. *Hypertension.* 1994;24:461–466.
44. Spaanderman MEA, Willekes C, Hoeks APG, Ekhart THA, Peeters LLH. The effect of pregnancy on the compliance of large arteries and veins in healthy parous control subjects and women with a history of preeclampsia. *Brit J Obstet Gynaecol.* 2000;183:1278–1286.
45. Smith AJ, Walters WA, Buckley NA, Gallagher L, Mason A, McPherson J. Hypertensive and normal pregnancy: a longitudinal study of blood pressure, distensiblility of dorsal hand veins and the ratio of the stable metabolites of thromboxane A2 and prostacyclin in plasma. *Brit J Obstet Gynaecol.* 1995;102:900–906.
46. Duncan SL, Bernard AG. Venous tone in pregnancy. *J Obstet Gynaecol Brit Commonw.* 1968;75:142–150.
47. Edouard DA, Pannier BM, London GM, Cuche JL, Safar ME. Venous and arterial behavior during normal pregnancy. *Am J Physiol.* 1998;274:H1605–H1612.
48. Robson SC, Dunlop W, Moore M, Hunter S. Combined Doppler and echocardiographic measurement of cardiac output: theory and application in pregnancy. *Brit J Obstet Gynaecol.* 1987;94:1014–1027.
49. Shroff SG. Pulsatile arterial load and cardiovascular function – facts, fiction and wishful thinking. *Therapeutic Res.* 1998;19:59–66.
50. Hart MV, Hosenpud JD, Hohimer AR, Morton MJ. Hemodynamics during pregnancy and sex steroid administration in guinea pigs. *Am J Physiol.* 1985;249:R179–R185.
51. Slangen BF, van Ingen Schenau DS, van Gorp AW, De Mey JG, Peeters LL. Aortic distensibility and compliance in conscious pregnant rats. *Am J Physiol.* 1997;272:H1260–H1265.
52. Danforth DN, Manalo-Estrella P, Buckingham JC. The effect of pregnancy and of enovid on the rabbit vasculature. *Am J Obstet Gynecol.* 1964;88:952–962.
53. Mackey K, Meyer MC, Stirewalt WS, Starcher BC, McLaughlin MK. Composition and mechanics of mesenteric resistance arteries from pregnant rats. *Am J Physiol.* 1992;263:R2–R8.
54. Hart MV, Morton MJ, Hosenpud JD, Metcalfe J. Aortic function during normal human pregnancy. *Am J Obstet Gynecol.* 1986;154:887–891.
55. Slater AJ, Gude N, Clarke IJ, Walters WAW. Haemodynamic changes and left ventricular performance during high-dose oestrogen administration to male transsexuals. *Brit J Obstet Gynaecol.* 1986;93:532–538.
56. Katz R, Karliner JS, Resnick R. Effects of a natural volume overload state (pregnancy) on left ventricular performance in normal human subjects. *Circulation.* 1978;58:434–441.
57. Rubler S, Damani PM, Pinto ER. Cardiac size and performance during pregnancy estimated with echocardiography. *Am J Cardiol.* 1977;40:534–540.
58. Lee W, Rokey R, Cotton DB. Noninvasive maternal stroke volume and cardiac output determinations by pulsed Doppler echocardiography. *Am J Obstet Gynecol.* 1988;158:505–510.
59. Easterling TR, Benedetti TJ, Schmucker BC, Carlson K, Millard SP. Maternal hemodynamics and aortic diameter in normal and hypertensive pregnancies. *Obstet Gynecol.* 1991;78:1073–1077.
60. Immer FF, Bansi AG, Immer-Bansi AS, et al. Aortic dissection in pregnancy: analysis of risk factors and outcome. *Ann Thorac Surg.* 2003;76:309–314.
61. Practice Committee of the American Society for Reproductive Medicine. Increased maternal cardiovascular mortality associated with pregnancy in women with Turner's syndrome. *Fertil Steril.* 2006;86:S127–S128.
62. Borghi C, Esposti DD, Immordio V, et al. Relationship of systemic hemodynamics, left ventricular structure and function, and plasma natriuretic peptide concentrations during pregnancy complicated by preeclampsia. *Am J Obstet Gynecol.* 2000;183:140–147.
63. Scardo J, Kiser R, Dillon A, Brost B, Newman R. Hemodynamic comparison of mild and severe preeclampsia: concept of stroke systemic vascular resistance index. *J Mater Fetal Med.* 1996;5:268–272.
64. Colan SD, Borow KM, Neumann A. The left ventricular end-systolic wall stress-velocity of fiber shortening relation: a load-independent index of myocardial contractility. *J Am Coll Cardiol.* 1984;4:715–724.
65. Fok WY, Chan LY, Wong JT, Yu CM, Lau TK. Left ventricular diastolic function during normal pregnancy: assessment by spectral tissue Doppler imaging. *Ultrasound Obstet Gynecol.* 2006;28:789–793.
66. Rafic HR, Larsson A, Pernow J, Bremme K, Eriksson MJ. Assessment of left ventricular structure and function in preeclampsia by echocardiography and cardiovascular biomarkers. *J Hypertens.* 2009;27:2257–2267.
67. Melchiorre K, Sutherland G, Sharma R, Nanni M, Thilaganathan B. Mid-gestational maternal cardiovascular profile in preterm and term pre-eclampsia: a prospective study. *Brit J Obstet Gynecol.* 2013;120:496–504.
68. Fisher SJ, Roberts JM. Defects in placentation and placental perfusion. In: Lindheimer MD, Roberts JM, Cunningham FG, eds. *Chesley's Hypertensive Disorders of Pregnancy.* Amsterdam: Elsevier; 1999:377–394. [Ch. 5].
69. Reiss RE, O'Shaughnessy RW, Quilligan TJ, Zuspan FP. Retrospective comparison of blood pressure course during preeclamptic and matched control pregnancies. *Am J Obstet Gynecol.* 1987;156:894–898.

70. Conde-Agudelo A, Belizan JM, Lede R, Bergel EF. What does an elevated mean arterial pressure in the second half of pregnancy predict-gestational hypertension or preeclampsia? *Am J Obstet Gynecol.* 1993;169:509–514.
71. Sibai BM, Gordon T, Thom E, Zuspan FP. Risk factors for preeclampsia in healthy nulliparous women: A prospective multicenter study. The NIH Maternal Fetal Medicine Units. *Am J Obstet Gynecol.* 1995;172:642–648.
72. Sibai BM, Ewell M, Levine RJ. Risk factors associated with preeclampsia in healthy nulliparous women. The Calcium for Preeclampsia Prevention (CPEP) Study Group. *Am J Obstet Gynecol.* 1997;177:1003–1010.
73. Conde-Agudelo A, Villar J, Lindheimer M. World Health Organization systematic review of screening tests for preeclampsia. *Obstet Gynecol.* 2004;104:1367–1391.
74. Redman CWG, Beilin LJ, Bonnar J. Reversed diurnal blood pressure rhythm in hypertensive pregnancies. *Clin Sci Mol Med Suppl.* 1976;51:687s–689s.
75. Malberg H, Bauernshcmitt R, Voss A, et al. Analysis of cardiovascular oscillations: a new approach to the early prediction of pre-eclampsia. *Chaos.* 2007;17:015113.
76. Walther T, Wessel N, Malberg H, Voss A, Stepan H, Faber R. A combined technique for predicting pre-eclampsia: concurrent measurement of uterine perfusion and analysis of heart rate and blood pressure variability. *J Hypertens.* 2006;24:747–750.
77. Benedetti TJ, Cotton DB, Read JC, Miller FC. Hemodynamic observations in severe pre-eclampsia with a flow-directed pulmonary artery catheter. *Am J Obstet Gynecol.* 1980;136: 465–470.
78. Strauss RG, Keefer JR, Burke T, Civetta JM. Hemodynamic monitoring of cardiogenic pulmonary edema complicating toxemia of pregnancy. *Obstet Gynecol.* 1980;55:170–174.
79. Rafferty TD, Berkowitz RL. Hemodynamics in patients with severe toxemia during labor and delivery. *Am J Obstet Gynecol.* 1980;138:263–270.
80. Phelan JP, Yurth DA. Severe preeclampsia: I. Peripartum hemodynamic observations. *Am J Obstet Gynecol.* 1982;144:17–22.
81. Groenendijk R, Trimbos JBMJ, Wallenburg HCS. Hemodynamic measurements in preeclampsia: preliminary observations. *Am J Obstet Gynecol.* 1984;150:232–236.
82. Clark SL, Greenspoon JSA, Aldahl D, Phelan JP. Severe preeclampsia with persistent oliguria: Management of hemodynamic subsets. *Am J Obstet Gynecol.* 1986;154:490–494.
83. Cotton DB, Lee W, Huhta JC, Dorman KF. Hemodynamic profile of severe preeclampsia-induced hypertension. *Am J Obstet Gynecol.* 1988;158:523–529.
84. Mabie WC, Ratts TE, Sibai BM. The central hemodynamics of severe preeclampsia. *Am J Obstet Gynecol.* 1989;161: 1443–1448.
85. Visser W, Wallenburg HCS. Central hemodynamic observations in untreated preeclamptic patients. *Hypertension.* 1991;17:1072–1077.
86. Gant NF, Daley GL, Chand S, Whalley PJ, MacDonald PC. A study of angiotensin II pressor response throughout primigravid pregnancy. *J Clin Invest.* 1973;52:2682–2689.
87. Lindheimer MD, Katz AI. Preeclampsia: pathophysiology, diagnosis, and management. *Annu Rev Med.* 1989;40:233–250.
88. Maynard SE, Min JY, Merchan J. Excess placental soluble fms-like tyrosine kinase 1 (sFlt1) may contribute to endothelial dysfunction, hypertension, and proteinuria in preeclampsia. *J Clin Invest.* 2003;111:649–658.
89. Levine RJ, Maynard SE, Qian C, et al. Circulating angiogenic factors and the risk of preeclampsia. *New Engl J Med.* 2004;350:672–683.
90. Maynard S, Epstein FH, Karumanchi SA. Preeclampsia and angiogenic imbalance. *Annu Rev Med.* 2008;59:61–78.
91. Easterling TR, Benedetti TJ, Schmucker BC, Millard SP. Maternal hemodynamics in normal and preeclamptic pregnancies: A longitudinal study. *Obstet Gynecol.* 1990;76:1061–1069.
92. Bosio PM, McKenna PJ, Conroy R, O'Herlihy C. Maternal central hemodynamics in hypertensive disorders of pregnancy. *Obstet Gynecol.* 1999;94:978–984.
93. Messerli FH, De Carvalho JGR, Christie B, Frohlich ED. Systemic and regional hemodynamics in low, normal and high cardiac output borderline hypertension. *Circulation.* 1978;58:441–448.
94. Easterling TR, Brateng D, Schmucker B, Brown Z, Millard SP. Prevention of preeclampsia: A randomized trial of atenolol in hyperdynamic patients before onset of hypertension. *Obstet Gynecol.* 1999;93:725–733.
95. Hall JE. Renal and cardiovascular mechanisms of hypertension in obesity. *Hypertension.* 1994;23:381–394.
96. Dennis AT, Castro J, Simmons S, Permezel M, Royse C. Haemodynamics in women with untreated pre-eclampsia. *Anaesthesia.* 2012;67:1105–1118.
97. Carr DB, Tran LT, Brateng DA, et al. Hemodynamically-directed atenolol therapy is associated with a blunted rise in maternal s-Flt-1 levels during pregnancy. *Hypertens Preg.* 2009;28:42–55.
98. Hibbard JU, Korcarz CE, Nendaz-Giardet G, Lindheimer MD, Lang RM, Shroff SG. Arterial system in preeclampsia and chronic hypertension with superimposed preeclampsia. *Brit J Obstet Gynecol.* 2005;112:897–903.
99. Hibbard JU, Shroff SG, Lang RM. Cardiovascular changes in preeclampsia. *Semin Nephrol.* 2004;24:580–587.
100. Elvan-Taspinar A, Franx A, Bots ML, Bruinse HW, Koomans HA. Central hemodynamics of hypertensive disorders in pregnancy. *Am J Hypertens.* 2004;17:941–946.
101. Tihtonen KMH, Koobi T, Uotila JT. Arterial stiffness in preeclamptic and chronic hypertensive pregnancies. *Eur J Obstet Gynecol Reprod Biol.* 2006;128:180–186.
102. Nichols WW, Singh BM. Augmentation index as a measure of peripheral vascular disease state. *Curr Opinion Cardiol.* 2002;17:543–551.
103. Nichols WW. Clinical measurement of arterial stiffness obtained from noninvasive pressure waveforms. *Am J Hypertens.* 2005;18:3S–10S.
104. O'Rourke MF, Gallagher DE. Pulse wave analysis. *J Hypertension.* 1996;14(Suppl 5):S147–S157.
105. Khalil A, Jauniaux E, Harrington K. Antihypertensive therapy and central hemodynamics in women with hypertensive disorders in pregnancy. *Obstet Gynecol.* 2009;113:646–654.
106. Rogers DT, Colon M, Gambala C, Wilkins I, Hibbard JU. Effects of magnesium on central arterial compliance in preeclampsia. *Am J Obstet Gynecol.* 2010;202:448–456.
107. Cruz M, Gambala C, Colon M, et al. Use of augmentation index in the diagnosis of preeclampsia. *Am J Obstet Gynecol.* 2013;208:S272–S273.

108. Thadhani R, Ecker JL, Kettyle E, Sandler L, Frigoletto FD. Pulse pressure and risk of preeclampsia: a prospective study. *Obstet Gynecol*. 2001;97:515–520.
109. Oyama-Kato M, Ohmichi M, Takahashi K, et al. Change in pulse wave velocity throughout normal pregnancy and its value in predicting pregnancy-induced hypertension: a longitudinal study. *Am J Obstet Gynecol*. 2006;195:464–469.
110. Savvidou MD, Kaihura C, Anderson JM, Nicolaides KH. Maternal arterial stiffness in women who subsequently develop pre-eclampsia. *PLoS One*. 2011;6:e18703.
111. Khalil AA, Cooper DJ, Harrington KF. Pulse wave analysis: a preliminary study of novel technique for the prediction of pre-eclampsia. *Brit J Obstet Gynecol*. 2009;116:268–276.
112. Roberts JM, Taylor RN, Musci TJ, Rodgers GM, Hubel CA, McLaughlin MK. Preeclampsia: an endothelial cell disorder. *Am J Obstet Gynecol*. 1989;161:1200–1204.
113. Agatisa PK, Ness RB, Roberts JM, Costantino JP, Kuller LH, McLaughlin MK. Impairment of endothelial function in women with a history of preeclampsia: an indicator of cardiovascular risk. *Am J Physiol Heart Circ Physiol*. 2004;286:H1389–H1393.
114. Lampinen KH, Ronnback M, Kaaja RJ, Groop PH. Impaired vascular dilation in women with a history of pre-eclampsia. *J Hypertens*. 2006;24:751–756.
115. Evans CS, Gooch L, Flott D, et al. Cardiovascular system during the postpartum state in women with history of pre-eclampsia. *Hypertension*. 2011;58:57–62.
116. Melchiorre K, Sutherland GR, Liberati M, Thilaganathan B. Preeclampsia is associated with persistent postpartum cardiovascular impairments. *Hypertension*. 2001;58:709–715.
117. Spaanderman MEA, Ekhart THA, van Eyck J, Cheriex EC, deLeeuw PW, Peeters LLH. Latent hemodynamic abnormalities in symptom-free women with a history of preeclampsia. *Am J Obstet Gynecol*. 2000;182:101–107.
118. Spaanderman ME, Schippers M, vander Graaf F, Thijssen HJ, Liem IH, Peeters LL. Subclinical signs of vascular damage relate to enhanced platelet responsiveness among nonpregnant formerly preeclamptics women. *Am J Obstet Gynecol*. 2006;194:855–860.
119. Haukkamaa L, Salminen M, Laivuori H, Leinonen H, Hiilesmaa V, Kaaja R. Risk for subsequent coronary artery disease after preeclampsia. *Am J Cardiol*. 2004;93:805–808.
120. Ronnback M, Lampinen K, Groop PH, Kaaja R. Pulse wave reflection in currently and previously preeclamptic women. *Hypertens Pregnancy*. 2005;24:171–180.
121. Fisher KA, Luger A, Spargo BH, Lindheimer MD. Hypertension in pregnancy: clinical pathological correlations and remote prognosis. *Medicine*. 1981;60:267–276.
122. Melchiorre K, Sutherland GR, Baltabaeva A, Liberati M, Thilaganathan B. Materanl cardiac dysfunction and remodeling in women with preeclampsia at term. *Hypertension*. 2011;57:85–93.
123. Assali NS, Holm LM, Parker HR. Systemic and regional hemodynamic alterations in toxemia. *Circulation*. 1964;30:53–57.
124. Wallenburg HCS. Hemodynamics in hypertensive pregnancy. In: Rubin PC, ed. *Handbook of Hypertension, vol. 10: Hypertension in Pregnancy*. Amsterdam: Elsevier; 1988: 66–101.
125. Melchiorre K, Sutherland GR, Watt-Coote I, Liberati M, Thilaganathan B. Severe myocardial impairment and chamber dysfunction in preterm preeclampsia. *Hypertens Pregnancy*. 2012;31:454–471.
126. Shahul S, Rhee J, Hacker MR, et al. Subclinical left ventricular dysfunction in preeclamptic women with preserved left ventricular ejection fraction, a 2D speckle-tracking imaging study. *Circ Cardiovasc Imaging*. 2012;5:734–739.
127. Bamfo JEAK, Kametas NA, Chambers JB, Nicolaides KH. Maternal cardiac function in normotensive and pre-eclamptic intrauterine growth restriction. *Ultrasound Obstet Gynecol*. 2008;32:682–686.
128. Melchiorre K, Sutherland GR, Liberati M, Thilaganathan B. Maternal cardiovascular impairment in pregnancies complicated by severe fetal growth restriction. *Hypertension*. 2012;60:437–443.
129. Heesch CM, Rogers RC. Effects of pregnancy and progesterone metabolites on regulation of sympathetic outflow. *Clin Exp Pharmacol Physiol*. 1995;22:136–142.
130. Pan ZR, Lindheimer MD, Bailin J, Barron WM. Regulation of blood pressure in pregnancy: pressor system blockade and stimulation. *Am J Physiol*. 1990;258:H1559–H1572.
131. O'Hagan KP, Casey SM. Arterial baroreflex during pregnancy and renal sympathetic nerve activity during parturition in rabbits. *Am J Physiol*. 1998;274:H1635–H1642.
132. Schobel HP, Fischer T, Heuszer K, Geiger H, Schmieder RE. Preeclampsia – a state of sympathetic overactivity. *New Engl J Med*. 1996;335:1480–1485.
133. Ekholm EM, Piha SJ, Antila KJ, Erkkola RU. Cardiovascular autonomic reflexes in mid-pregnancy. *Brit J Obstet Gynaecol*. 1993;100:177–182.
134. Leduc L, Wasserstrum N, Spillman T, Cotton DB. Baroreflex function in normal pregnancy. *Am J Obstet Gynecol*. 1991;165:886–890.
135. Masilamani S, Heesch CM. Effects of pregnancy and progesterone metabolites on arterial baroreflex in conscious rats. *Am J Physiol*. 1997;272:R924–R934.
136. Brooks VL, Keil LC. Changes in the baroreflex during pregnancy in conscious dogs: Heart rate and hormonal responses. *Endocrinology*. 1994;135:1894–1901.
137. Ekholm EMK, Erkkola RU. Autonomic cardiovascular control in pregnancy. *Eur J Obstet Gynecol Reprod Biol*. 1996;64:29–36.
138. Conrad KP, Russ RD. Augmentation of baroreflex-mediated bradycardia in conscious pregnant rats. *Am J Physiol*. 1992;262:R472–R477.
139. Gilson GJ, Mosher MD, Conrad KP. Systemic hemodynamics and oxygen transport during pregnancy in chronically instrumented, conscious rats. *Am J Physiol*. 1992;263:H1911–H1918.
140. Stock MK, Metcalfe J. Maternal physiology during gestation *The Physiology of Reproduction*. 2nd ed. New York: Raven Press; 1994: pp. 947–973.
141. Poston L, McCarthy AL, Ritter JM. Control of vascular resistance in the maternal and feto-placental arterial beds. *Pharmacol Ther*. 1995;65:215–239.
142. Nisell H, Hjemdahl P, Linde B. Cardiovascular responses to circulating catecholamines in normal pregnancy and in pregnancy induced hypertension. *Clin Physiol*. 1985;5:479–493.
143. Debrah DO, Novak JE, Matthews JE, Ramirez RJ, Shroff SG, Conrad KP. Relaxin is essential for systemic vasodilation and increased global compliance during early pregnancy in conscious rats. *Endocrinology*. 2006;147:5126–5131.

144. Conrad KP, Debrah DO, Novak J, Danielson LA, Shroff SG. Relaxin modifies systemic arterial resistance and compliance in conscious, nonpregnant rats. *Endocrinology.* 2004;145:3289–3296.
145. Osol G, Cipolla M. Pregnancy-induced changes in the three-dimensional mechanical properties of pressurized rat uteroplacental (radial) arteries. *Am J Obstet Gynecol.* 1993;168:268–274.
146. Cipolla M, Osol G. Hypertrophic and hyperplastic effects of pregnancy on the rat uterine arterial wall. *Am J Obstet Gynecol.* 1994;171:805–811.
147. Hammer ES, Cipolla MJ. Arterial wall hyperplasia is increased in placental compared with myoendometrial radial uterine arteries from late-pregnant rats. *Am J Obstet Gynecol.* 2005;192:302–308.
148. Griendling KK, Fuller EO, Cox RH. Pregnancy-induced changes in sheep uterine and carotid arteries. *Am J Physiol.* 1985;248:H658–H665.
149. Williams JK, Adams MR, Herrington DM, Clarkson TB. Short-term administration of estrogen and vascular responses of atherosclerotic coronary arteries. *J Am Coll Cardiol.* 1992;20:452–457.
150. Gilligan DM, Badar DM, Panza JA, Quyyumi AA, Cannon RO. Effects of estrogen replacement therapy on peripheral vasomotor function in postmenopausal women. *Am J Cardiol.* 1995;75:264–268.
151. Farhat MY, Lavigne MC, Ramwell PW. The vascular protective effects of estrogen. *FASEB J.* 1996;10:615–624.
152. Gilligan DM, Bader DM, Panza JA, Quyyumi AA, Cannon RO. Acute vascular effects of estrogen in postmenopausal women. *Circulation.* 1994;90:786–791.
153. Kublickiene K, Svedas E, Landgren BM, et al. Small artery endothelial dysfunction in postmenopausal women: in vitro function, morphology, and modification by estrogen and selective estrogen receptor modulators. *J Clin Endocrinol Metab.* 2005;90:6113–6122.
154. Masi CM, Hawkley LC, Berry JD, Cacioppo JT. Estrogen metabolites and systolic blood pressure in a population-based sample of postmenopausal women. *J Clin Endocrinol Metab.* 2006;91:1015–1020.
155. da Costa LS, de Oliveira MA, Rubim VS. Effects of hormone replacement therapy or raloxifene on ambulatory blood pressure and arterial stiffness in treated hypertensive postmenopausal women. *Am J Cardiol.* 2004;94: 1453–1456.
156. Kalantaridou SN, Naka KK, Papanikolaou E. Impaired endothelial function in young women with premature ovarian failure: normalization with hormone therapy. *J Clin Endocrinol Metab.* 2004;89:3907–3913.
157. Corbacho AM, Eiserich JP, Zuniga LA, Valacchi G, Villablanca AC. Compromised aortic vasoreactivity in male estrogen receptor-alpha-deficient mice during acute lipopolysaccharide-induced inflammation. *Endocrinology.* 2007;148:1403–1411.
158. Novak J, Danielson LA, Kerchner LJ, et al. Relaxin is essential for renal vasodilation during pregnancy in conscious rats. *J Clin Invest.* 2001;107:1469–1475.
159. Danielson LA, Sherwood OD, Conrad KP. Relaxin is a potent renal vasodilator in conscious rats. *J Clin Invest.* 1999;103:525–533.
160. Debrah DO, Conrad KP, Danielson LA, Shroff SG. Effects of relaxin on systemic arterial hemodynamics and mechanical properties in conscious rats: sex dependency and dose response. *J Appl Physiol.* 2004;98:1013–1020.
161. Conrad KP, Novak J. Emerging role of relaxin in renal and cardiovascular function. *Am J Physiol Regul Integr Comp Physiol.* 2004;287:R250–R261.
162. Unemori EN, Erikson ME, Rocco SE, et al. Relaxin stimulates expression of vascular endothelial growth factor in normal human endometrial cells in vitro and is associated with menometrorrhagia in women. *Hum Reprod.* 1999;14:800–806.
163. Carbillon L, Uzan M, Uzan S. Pregnancy, vascular tone, and maternal hemodynamics: a crucial adaptation. *Obstet Gynecol Surv.* 2000;55:574–581.
164. Williams JG, Rincon-Skinner T, Sun D, et al. Role of nitric oxide in the coupling of myocardial oxygen consumption and coronary vascular dynamics during pregnancy in the dog. *Am J Physiol Heart Circ Physiol.* 2007;293:H2479–H2486.
165. Brust AA, Assali NS, Ferris EB. Evaluation of neurogenic and humoral factors in blood pressure maintenance in normal and toxemic pregnancy using tetraethylammonium chloride. *J Lab Clin Med.* 1948;27:717–726.
166. Assali NS, Prystowsky H. Studies on autonomic blockade: I. Comparison between the effects of tetraethylammonium chloride (TEAC) and high selective spinal anesthesia on blood pressure of normal and toxemic pregnancy. *J Clin Invest.* 1950;29:1354–1366.
167. Suzuki Y, Yamamoto T, Mabuchi Y. Ultrastructural changes in omental resistance artery in women with preeclampsia. *Am J Obstet Gynecol.* 2003;189:216–221.
168. Sacks GP, Studena K, Sargent II, Redman CWG. Normal pregnancy and preeclampsia both produce inflammatory changes in peripheral blood leukocytes akin to those of sepsis. *Am J Obstet Gynecol.* 1998;179:80–86.
169. Ness RB, Hubel CA. Risk for coronary artery disease and morbid preeclampsia: a commentary. *Ann Epidemiol.* 2005;15:726–733.
170. Ross R. Atherosclerosis – an inflammatory disease. *New Engl J Med.* 1999;340:115–126.
171. Kahn BB, Flier JS. Obesity and insulin resistance. *J Clin Invest.* 2000;106:473–481.
172. Reaven GM. Syndrome X: 6 years later. *J Intern Med.* 1994;236:13–22.
173. Agatisa PK, Ness RB, Roberts JM, Costantino JP, Kuller LH, McLaughlin MK. Impairment of endothelial function in women with a history of preeclampsia: an indicator of cardiovascular risk. *Am J Physiol.* 2004;286:1389–1393.
174. Hubel CA, Powers RW, Snaedal S, et al. C-reactive protein is elevated 30 years after eclamptic pregnancy. *Hypertension.* 2008;51:1499–1505.
175. Hubel CA, Snaedal S, Ness RB, et al. Dyslipoproteinemia in postmenopausal women with a history of eclampsia. *Brit J Obstet Gynecol.* 2000;107:776–784.
176. Hubel CA, Wallukat G, Wolf M, et al. Agonistic angiotensin II type 1 receptor autoantibodies in postpartum women with a history of preeclampsia. *Hypertension.* 2007;49:612–617.
177. Roberts JM, Hubel CA. Pregnancy: a screening test for later life cardiovascular disease. *Womens Health Iss.* 2010;20:304–307.

178. Wolf M, Hubel CA, Lam C, et al. Preeclampsia and future cardiovascular disease: potential role of altered angiogenesis and insulin resistance. *J Clin Endocrinol Metab.* 2004;89:6239–6243.
179. Smith GN, Walker MC, Liu A, et al. A history of preeclampsia identifies women who have underlying cardiovascular risk factors. *Am J Obstet Gynecol.* 2009;200(58):e51–e58.
180. Bellamy L, Casas JP, Hingorani AD, Williams DJ. Pre-eclampsia and risk of cardiovascular disease and cancer in later life: systematic review and meta-analysis. *Brit Med J.* 2007;335:974–985.
181. McDonald SD, Malinowski A, Zhou Q, Yusuf S, Devereaux PJ. Cardiovascular sequelae of preeclampsia/eclampsia: a systematic review and meta-analyses. *Am Heart J.* 2008;156:918–930.
182. Valdiviezo C, Garovic VD, Ouyang P. Preeclampsia and hypertensive disease in pregnancy: their contributions to cardiovascular risk. *Clin Cardiol.* 2012;35:160–165.
183. Roberts JM, Catov JM. Pregnancy is a screening test for later life cardiovascular disease: now what? Research recommendations. *Womens Health Iss.* 2012;22:e123–e128.
184. Roberts JM, Hubel CA. The two stage model of preeclampsia: variations on the theme. *Placenta.* 2009;30(Suppl A):S32–S37.
185. Ray JG, Vermeulen MJ, Schull MJ, Redelmeier DA. Cardiovascular health after maternal placental syndromes (CHAMPS): population-based retrospective cohort study. *Lancet.* 2005;366:1797–1803.
186. Rodie VA, Freeman DJ, Sattar N, Greer IA. Pre-eclampsia and cardiovascular disease: metabolic syndrome of pregnancy? *Atherosclerosis.* 2004;175:189–202.
187. Romundstad PR, Magnussen EB, Smith GD, Vatten LJ. Hypertension in pregnancy and later cardiovascular risk: common antecedents? *Circulation.* 2010;122:579–584.
188. Magnussen EB, Vatten LJ, Lund-Nilsen TI, Salvesen KA, Davey Smith G, Romundstad PR. Prepregnancy cardiovascular risk factors as predictors of pre-eclampsia: population based cohort study. *Brit Med J.* 2007;335:978–986.
189. Maynard SE. Preeclampsia and subsequent cardiovascular disease: villain or innocent bystander? *Clin J Am Soc Nephrol.* 2013;8:1061–1063.
190. Karumanchi SA, Maynard SE, Stillman IE, Epstein FH, Sukhatme VP. Preeclampsia: a renal perspective. *Kidney Int.* 2005;67:2101–2113.
191. Mosca L, Benjamin EJ, Berra K, et al. Effectiveness-based guidelines for the prevention of cardiovascular disease in women – 2011 update: a guideline from the American Heart Association. *Circulation.* 2011;123:1243–1262.
192. Montes A, Walden CE, Knopp RH, Cheung M, Chapman MB, Albers JJ. Physiologic and supraphysiologic increases in lipoprotein lipids and apoproteins in late pregnancy and postpartum. Possible markers for the diagnosis of "prelipemia." *Arteriosclerosis.* 1984;4:407–417.
193. Wenger NK. Recognizing pregnancy-associated cardiovascular risk factors. *Am J Cardiol.* 2014;113(2):406–9.

CHAPTER **15**

The Renin-Angiotensin System, its Autoantibodies, and Body Fluid Volume in Preeclampsia

RALF DECHEND, BABBETTE LAMARCA AND ROBERT N. TAYLOR

Editors' comment: *The role of the renin-angiotensin-aldosterone system (RAAS) in preeclampsia has intrigued investigators for decades. Chesley had a chapter devoted to this topic in his first edition and he would have been delighted by some of the novel twists that have occurred in this important area. The discovery of agonistic autoantibodies (AT1-AA) is very much in line with his extreme interest on why preeclamptic women are hypersensitive to angiotensin II. He recognized that angiotensin II had a very short half-life in the circulation, and that peptide fragments (e.g., angiotensin III) also were vasoactive, but Chesley may have been surprised to learn that Ang 1-7 has potent vasodilatory activity. In the current edition we combine the subjects of the RAAS, AT1-AA, and body fluid volumes in preeclampsia as logical extensions of this intricate regulatory system.*

INTRODUCTION

The renin angiotensin-aldosterone system (RAAS) is one of the most evolutionarily conserved blood pressure and volume regulating systems in vertebrates. Importantly, during human pregnancy, there is a ~12kg weight gain, as well as a 30–50% increase in extracellular fluid, plasma, blood volume, and total body water. Also striking is a resetting of the thresholds for vasopressin secretion and thirst, plasma osmolality averaging 10mOsm below nonpregnant levels (see Chapter 16). There is a marked stimulation of the RAAS and other potent mineralocorticoids that accompanies these changes, while opposing salt-retaining influences are increases in GFR and a rise in the plasma concentration of several natriuretic hormones. Therefore, pregnancy is a sensitive state relying on a multifactoral autoregulation of blood pressure control mechanisms and body fluid volume homeostasis.

Alterations in the RAAS accompany the development of hypertension. Preeclamptic women have long been known to have increased vascular sensitivity to angiotensin II in the absence of elevated angiotensin II or plasma renin activity. Recently alterations in either the vasodilatory Ang 1-7 or activating autoantibodies to the angiotensin II type I receptor (AT1-AA) have been noted to occur during preeclampsia. AT1-AA were found to be present in the serum of preeclamptic women at much higher levels than sera from nonpregnant women or pregnant women who went on to have normal pregnancies. Therefore, in recent years much research has been performed to determine the contribution of AT1-AA to the pathophysiology associated with preeclampsia.[1–10] AT1-AA bind to and activate the AT1-receptor and induce signaling in vascular cells, including activating protein 1, calcineurin, and nuclear factor kappa-B activation, which can be blocked by an AT1 receptor antagonist.[3–8,11–14] This signaling results in increased reactive oxygen species, sFlt-1 production and plasminogen activator inhibitor-1, and endothelin-1, all of which have been implicated in preeclampsia.[3–8,11–16] More recent studies reveal an important role for AT1-AA in causing the increase in renal and vascular sensitivity to angiotensin II (Ang II).[15,16] In addition to being elevated during preeclampsia, AT1-AA have also been reported to be increased in postpartum women.[9] Hubel and colleagues demonstrated that AT1-AA correlated with insulin resistance and sFlt-1[17] and do not regress completely after delivery. Although these autoantibodies have been linked to poor placentation and abnormal renal function, their role in the hypertensive state of preeclampsia has yet to be fully elucidated.[18,19] Furthermore, the importance of AT1-AA after preeclampsia, especially in the context of increased cardiovascular risk, remains to be determined.

Resolving the ongoing debate concerning the cause of the increased plasma volume (does it represent "underfill," "overfill," or "normal fill"?) could have important implications for management of complications during preeclampsia. Plasma volume generally decreases in preeclampsia (most

Chesley's Hypertensive Disorders in Pregnancy.
ISBN: 978-0-12-407866-6

DOI: http://dx.doi.org/10.1016/B978-0-12-407866-6.00015-8

marked in eclampsia), while interstitial water (edema) may increase further or remain unchanged. Levels of all components of the RAAS are decreased compared to normal pregnancy while the increased incidence of AT1-AA is associated with increased severity of the disease.[14] Along with the controversial AT1-AA there is an evolving literature on the role of the dilating peptide Ang 1–7 in normal pregnancy and preeclampsia. Finally, evolving thoughts regarding treating preeclampsia with sodium loading or plasma volume expansion, a challenging view, were revisited in light of the recent discovery of genetic mutations leading to inefficient aldosterone production in preeclampsia. Contrarily, the possible presence of ouabain-like factors in animal models and women with preeclampsia is also noted. Furthermore, animal models of preeclampsia in which AT1-AA are suppressed are mentioned and are the subject of further investigation as we plunge forward to seek better, more innovative and safe therapeutics for managing edema and alterations in blood pressure during preeclampsia.

If preeclampsia is the "disease of theories," one of the most far-fetched of these is that agonistic autoantibodies may participate as direct mediators of increased vascular sensitivity and may induce alterations in volume homeostasis. The role of agonistic autoantibodies in Graves disease is well established and, increasingly, circulating immunoglobulins have been associated with hypertension. Their possible role in preeclampsia is explored in this chapter. Agonistic autoantibodies directed at the angiotensin II AT1 receptor were first detected in patients with malignant hypertension. When these were observed in a hypertensive woman with a history of preeclampsia, it prompted a cohort investigation for such antibodies in archived sera from preeclamptics. These antibodies induced AT1 receptor signaling and functioned in immuno blots as well as non-agonistic commercial antibodies; moreover, they were cleared rapidly after delivery. The epitope shows signaling events in vascular smooth muscle cells and trophoblasts that could contribute to the development and signs of preeclampsia. Moreover, AT1-AA have also been observed in a reduced uterine perfusion rat model of hypertension and in pregnant rat models of hypertension induced by elevated cytokines such as TNF-α, IL-6, and IL-17.[20–22] Once expressed in pregnancy, AT1-AA are capable of eliciting sFlt-1 production in human trophoblasts and in pregnant rodents.[7,8,13,23,24] Furthermore, hypertension resulting from AT1-AA during pregnancy is caused by activation of the endothelin-1 and placental oxidative stress pathways. Newer evidence suggests an interplay between the AT1-AA, Ang II and the AT1 receptor activating ET-1 and oxidative stress pathways and culminating in hypertension.[15,16] Knowledge gained from such research has validated the utility of rat and mouse disease models that are relevant for target identification and new therapies for preeclampsia.

The seemingly bizarre concept that circulating agonistic antibodies might mediate vasospasm in preeclampsia stemmed from a serendipitous observation in a single patient. When coupled to our knowledge of autoimmunity, immune tolerance, and the remarkable changes in the RAAS during pregnancy it led to a novel hypothesis. What has evolved in the past decade is an integrated framework of observations that fits into the concept of faulty angiogenesis as a precursor for phenotypic preeclampsia. The improbable agonistic autoantibody story warrants a brief review.

Autoantibodies that stimulate G protein-coupled receptors have long been accepted as causing diseases of the thyroid gland. Agonistic antithyrotropin receptor immunoglobulins (TSAb) that mimic the action of thyrotropin (TSH) mediate the endocrine manifestations of Graves disease and stimulate extraocular muscle fibroblast proliferation and differentiation in its associated ophthalmopathy. Nonetheless, the immunological mechanisms of TSAb production remain obscure. Recently, Kim-Saijo et al. produced a transgenic mouse using patient-derived TSAb[25] that may prove useful as a model for Graves disease. Autoantibodies directed against specific epitopes in the insulin receptor exist, but these only rarely cause recurrent hypoglycemia and a severe form of insulin resistance (type B insulin resistance).[26] In cancer chemotherapy, the generation of autoantibodies against the death receptors DR4 and DR5 of tumor necrosis factor-alpha-related apoptosis-inducing ligand (TRAIL) are interesting therapeutic targets, since agonistic antibodies against DR5 and DR2 induce apoptosis in cancer cells.[27] Furthermore, agonistic anti-CD40 antibodies profoundly suppress the immune response to infection with lymphocytic choriomeningitis virus.[28] Thus agonistic antibodies directed against cell-surface receptors appear to have an *in vivo* functional capacity to instill or suppress disease. A goal of this chapter will be to review the notion that such antibodies cause or amplify pathological cardiovascular responses in preeclampsia.

BODY FLUID VOLUMES

Pregnancy is a physiologic process whereby repeated adjustments in intracellular and extracellular volume occur to maintain the steady state. Each new steady-state value is then held within relatively narrow limits, that is, these changes are sensed as normal and "defended" in face of variations in fluid and sodium intake. There is a 30–50% increase in extracellular fluid (ECF), plasma, and blood volume, associated with 30–50% increases in cardiac output, glomerular filtration rate (GFR) and renal blood flow.

The cause and significance of such changes have been debated for decades by three schools of thought. One advocates that the alterations are secondary to primary arterial vasodilatation causing "underfill" or decreased effective volume. The second view, called "normal fill," implies that the gravida senses her volume as "normal" at every new steady state, and reacts appropriately to sodium restriction or surfeit. The final concept, "overfill," views pregnant women as overexpanded with intravascular volume. Each camp agrees, however, that primary renal sodium and water retention is presumably responsible for the volume changes associated with normal pregnancy.

Preeclampsia

The sources of peripheral edema, a sign of excessive interstitial ECF, are more complex in gravidas than in nonpregnant women. There are two forms of edema in pregnancy – "normal" edema reflects a physiological ECF volume increase, whereas "abnormal" or "pathological" edema in preeclampsia occurs when fluid has shifted from the vascular to the interstitial space (due, for example, to pathological "leakiness" of vessels (see Chapter 9)) – but the two etiologies are currently clinically indistinguishable. In women with "pathological" edema, total ECF is increased. In the second half of gestation, both pedal and pretibial edema can be detected in the majority of pregnant women, occurring more commonly as the day proceeds, often disappearing with recumbency.[29,30] In addition to this dependent edema, many women develop edema in the hands and face as pregnancy progresses; again, its frequency reflects the care with which edema is sought, an incidence of >60% being described in the classic 1941 monograph by Dexter and Weiss.[31] Even those women who fail to manifest overt edema have increases in lower limb volume (i.e., subclinical edema).[29,30]

Chesley, in the first edition of this text, suggested that the primary reason for increases in peripheral edema in normal pregnancy was the decrease in plasma oncotic pressure. Values for oncotic pressure decrease from an average level of 370 mm H_2O (27 mm Hg) in nonpregnant individuals to 345 mm H_2O (25 mm Hg) in early pregnancy and 300 mm H_2O (22 mm Hg) in late gestation. Figure 15.1 combines Chesley's summary during gestation with a study by Zinaman et al.[32] performed at delivery. Note that the lowest values are reached in the hours that follow delivery, the period of time when women with preeclampsia are most at risk of both pulmonary and cerebral edema, if fluid administration is not monitored with appropriate care.

Kogh et al.[33] have shown that reductions in plasma oncotic pressure on the order of that encountered in normal pregnancy can be associated with a significant increase in the rate of fluid extravasation from capillaries. The reduction in plasma oncotic pressure during pregnancy is mainly due to the ~1 g decrement in plasma albumin in normal gestation. The usual explanations of this decline are that it is a dilutional phenomenon related to the increase in plasma water, as well as in the decrease in plasma tonicity of ~10 mOsm. However, such explanations may be simplistic as little is known of the production and disposal of albumin in normal gestation, and while the levels of some

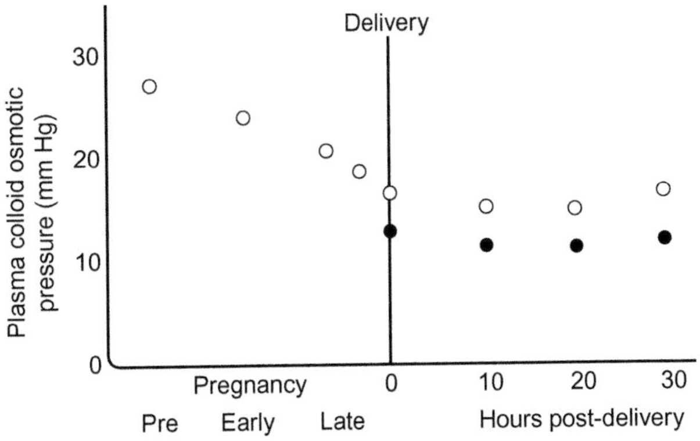

FIGURE 15.1 Mean plasma osmotic pressure measured before, during, and in the immediate puerperium. Data are from Chesley (the first edition of this text) combined with those from Zinaman et al.[32] o = normal pregnancy; • = preclamptics.

circulating solutes decline along with the decrease in tonicity, others do not. Finally, women with "normal" peripheral edema and otherwise normal pregnancies also have the greatest weight gain and plasma volume expansion as well as larger neonates with lower perinatal mortality rates than those with less fluid retention.[29,34]

Appreciation of changes in oncotic pressure during gestation is extremely relevant to clinical practice. Preeclamptic women with peripheral edema appear to be at risk of developing interstitial and frank pulmonary and cerebral edema.[34] This risk is increased by fluid challenges, particularly the intravenous infusion of crystalloids at the time of delivery, and the danger is enhanced in the immediate puerperium when fluid shifts from the expanded interstitial compartment of normal gestation back into the circulation. For example, Benedetti and Carlson[35] have suggested that the risk of pulmonary edema can be predicted by measuring colloid oncotic pressure, a suggestion supported by the observation that the lowest values are measured in the early postpartum period,[32,34] the period of greatest risk of this complication. Given that values in preeclamptic women are the lowest, they would naturally be the group at greatest risk of this life-threatening complication.

Finally, Hytten,[36] as far back as 1970, wrote that edema is so common in pregnancy that it is not a useful diagnostic criterion to use in the diagnosis of preeclampsia. Still, hypertension plus edema continued to be used to diagnose the disease (often compromising the value of research reports), only disappearing from the diagnostic criteria of all major classifications at the commencement of this millennium.

"Normal Fill" or Resetting of the "Volumestat"

This hypothesis is supported by animal experiments that evaluate relationships between intravascular volume depletion and vasopressin release during pregnancy compared to the nonpregnant state (described below and reviewed in[19]). Studies in pregnant animals and humans suggest that sodium and water reabsorption in the proximal nephron (determined by indices such as fractional lithium or solute-free water clearances) are unaltered, and these species dilute their urine normally when water-loaded during gestation.[37–39] Although osmotic thresholds are reset, experimental maneuvers aimed at abolishing "underfill" in both experimental animals and pregnant humans fail to alter the lower steady-state pressure, as well as the decreased osmotic threshold that occurs during gestation.[40,41]

The observations concerning normal dilution and the excretion of water loads are important because failure to dilute the urine maximally and a blunted excretory response to water loading are major pathophysiologic features of the "underfill" status in diseases accompanied by hyponatremia such as cirrhosis, cardiac failure, or nephrotic syndrome. These disorders are considered prototypes of diseases in which absolute extracellular volumes are increased and "effective arterial volume" is low.[42] Finally, most investigators describe similar sodium excretory responses to saline infusions in the pregnant and nonpregnant states[37] in animals and humans.[37,43,44]

Controversies about the various "fill" hypotheses are yet to be settled, and perhaps each is correct at particular stages during pregnancy. Hormone-induced vasodilatation creating temporary "underfill" in the early weeks, followed quickly by compensation to a "normal fill" state as pregnancy progresses, fits two of the three theories, whereas during the last trimester natriuretic factors predominate and restore Na balance, at least in some gravidas. The importance of settling this dispute, however, is not trivial, because a better understanding of how the pregnant woman "senses" her volume changes will impact management, particularly for gravidas with hypertensive complications and cardiac disorders.

Primary Arterial Vasodilatation ("Underfill")

The concept of "underfill" in pregnancy partially resembles views of how cirrhotic and heart failure patients "sense" their increased ECF volumes. The following observations suggest that primary arterial vasodilatation causing arterial underfilling with secondary sodium and water excretion occurs in early gestation. Supporters of "underfill" note that systolic and diastolic blood pressures decrease early in the first trimester of pregnancy despite an increase in blood volume.[37,42,45,46] The RAAS is activated early in pregnancy, an effect that would also occur with arterial underfilling due to peripheral arterial vasodilatation, while primary volume expansion would be expected to suppress these hormones.[37,45,47] The increase in GFR and renal blood flow may precede the expanded extracellular fluid volume in pregnancy, also suggesting primary vasodilatation.[45] Additional observations are the resetting of the osmostat and the volume depletion–vasopressin relationships (discussed below) in a direction also consistent with vascular underfilling due to systemic arterial vasodilatation.[48]

Nitric oxide (NO) is a prime candidate as the mediator of vasodilatation in pregnancy. The resistance to angiotensin (AII), norepinephrine, and vasopressin that characterizes normal pregnancy can also be reversed by blockade of nitric oxide synthase (NOS).[49]

Other evidence supporting "underfill" includes the observations that pregnant women manifest greater increases in aldosterone release than nonpregnant subjects in response to small quantities of AII[50] and display exaggerated decreases in blood pressure when treated with ACE inhibitors.[51]

Of particular interest is a study by Chapman et al.[45] They serially studied 13 women prior to and immediately following conception, and again during gestational weeks 6, 8, 10, 12, 24, and 36 (Fig. 15.2). Measured were blood pressure, cardiac output, and renal hemodynamics (inulin and para-amino hippurate clearances). Mean arterial blood pressure decreased by 6 weeks gestation, associated with an increase in cardiac output, decrease in systemic vascular resistance, and an increase in plasma volume (Fig. 15.2). Renal plasma flow and GFR increased by 6 weeks gestation (see Fig. 16.2, Chapter 16). Plasma renin activity (PRA) and aldosterone concentrations increased significantly by 6 weeks while norepinephrine levels did not change throughout pregnancy. Atrial natriuretic peptide (ANP) levels increased as well, first noted at 12 weeks. Plasma cyclic guanosine monophosphate (cGMP) levels decreased and cGMP clearance increased by 6 and 8 weeks, respectively. Chapman et al.[45] strongly suggest that peripheral vasodilatation occurs early in pregnancy in association with renal vasodilatation and activation of the renin–angiotensin–aldosterone system. Volume expansion occurs early, followed by later increases in ANP, suggesting that ANP increases in response to changes in intravascular volume. The authors also confirmed the decreases in Na, Cl, and HCO serum concentrations, as well as the lowered creatinine, blood urea nitrogen, and hematocrit concentrations throughout pregnancy.

Pronounced placental growth occurs at weeks 6–8 of gestation and is generally complete by week 12. The systemic and renal regulatory changes observed by Chapman et al.[45] occurred well before placentation. The authors speculated that maternal factors related to ovarian function are responsible for the peripheral vasodilatation, suggesting that NO might be largely responsible for the changes they observed. Renal production of NO or other natriuretic substances could in part explain these findings.

Excessive Expansion or "Overfill"

Supporters of the "overfill" theory note that there are absolute increments in ECF volumes, suggesting that this expansion may be an epiphenomenon to other changes, primarily a marked increase in mineralocorticoid hormone levels

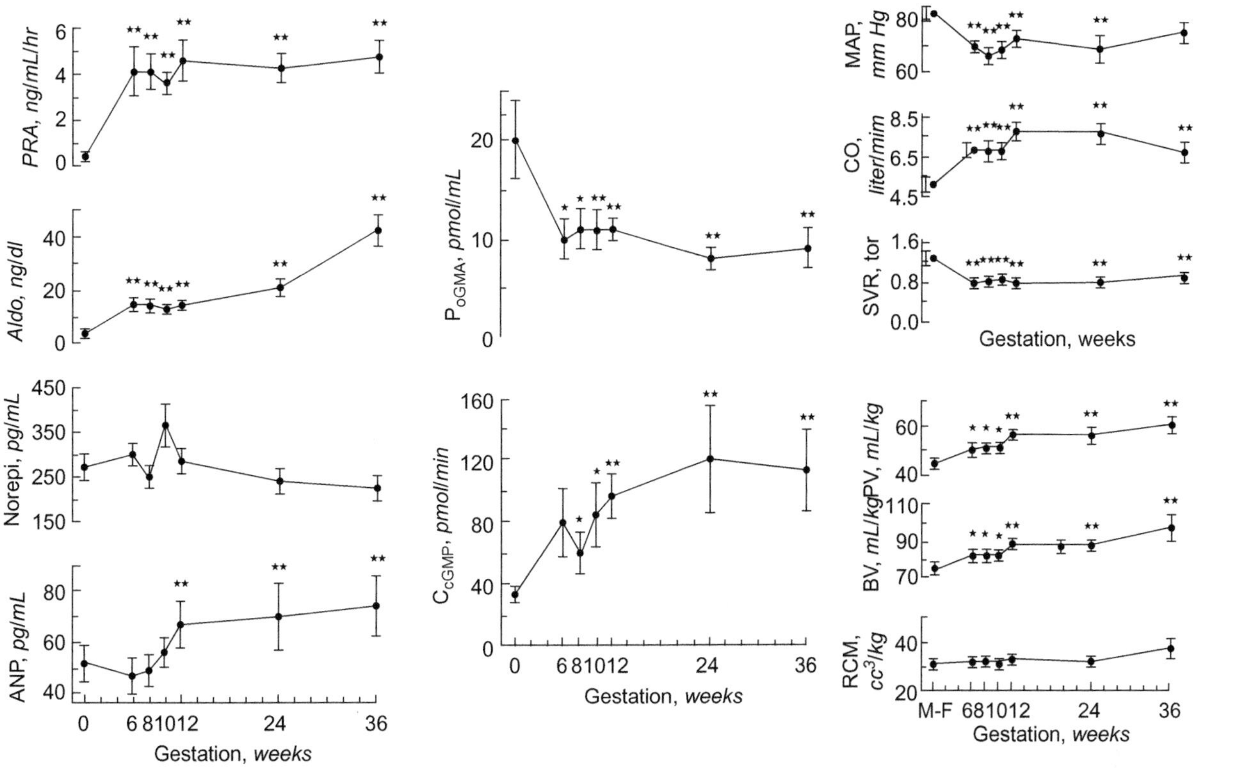

FIGURE 15.2 Serial study commenced over conception and repeated four times in the first, and once in the second and third trimesters, respectively, comparing the pregnancy course of many variables. They were mean arterial pressure (MAP), cardiac output (CO), systemic vascular resistance (SVR), plasma and blood volumes (PV & BV), red cell mass (RCM), plasma renin activity (PRA), aldosterone (Aldo), norepinephrine (Norepi), atrial natriuretic peptide (ANP), plasma levels and clearance of cGMP (P & C cGMP), the latter being the second messengers of nitric oxide and ANP. With the exception of RCM all changes occurred early, that is, shortly after conception and were near or at completion by gestational week 6. * represents a significant change from week 0, while M-F represents "mid-follicular," the phase of the cycle prior to conception when the study was initiated. (adapted from Figures 1–5 of[45]).

(see below). They also note the high levels of circulating natriuretic factors and various inhibitors of the membrane pump, which are expected responses to hypervolemia. There are also increases in renal hemodynamics, and two groups of investigators[52,53] have described an increased sodium excretory capacity in response to saline infusions. Although reports are anecdotal, women with cardiac or renal disorders appear more susceptible to volume overload complications, whereas healthy gravidas seem to tolerate blood loss better than nonpregnant women.

PLASMA VOLUME IN NORMAL PREGNANCY AND PREECLAMPSIA

Plasma volume is an especially clinically significant component of the ECF as it is a major determinant of organ perfusion. There are striking increments in intravascular volume during normal pregnancy, mainly the result of plasma water, but also due to a small increase in red blood cell mass.[34,37] Plasma volume has been measured in most studies with Evans blue dye,[37,45,54,55] and in the older literature values appeared to rise until gestational week 30 and then decline. Those studies were in error, because late in gestation the indicator cannot attain complete mixing during the 10 minutes the gravida is positioned in a supine or sitting position. When serial studies are performed with the pregnant woman positioned in lateral recumbency, plasma volume increases were observed to commence in the first trimester, accelerate in the second, peak near gestational week 32, and remain elevated until term.[54,56,57] The maximal increases are 1.1–1.3 L, and larger gains are observed when multiple fetuses are present.

Women with preeclampsia/eclampsia have significantly reduced plasma volumes (Table 15.1). The degree of contraction appears to be an index of severity, and in this respect the greatest decreases (approaching 50% of normal pregnant values) have been reported in nulliparas with eclampsia, the convulsive phase of the disease associated with extreme severity.[57]

The classic explanation of the decreased intravascular volume in hypertensive disorders is that the fluid is "chased" out by the increase in vasoconstrictor tone that results in increased blood pressure. The increase in

TABLE 15.1 Plasma Volume in Normal and Preeclamptic or Eclamptic Pregnancy

Author (Ref)	Normal Pregnancy		Preeclampsia/Eclampsia		
	Cases (n)	Mean (mL)[a]	Cases (n)	Mean (mL)[a]	% Change
Werko et al. [58]	4	3865	9	3145	−18
Freis & Kenny [59]	7	4287	5	3045	−29
Rottger[60,61]	20	3383	18	2890	−15
Cope[62]	29	3470	14	2820	−19
Freidberg & Lutz[63]	10	3104	17	3257	+5
Kolpakova[64]	20	3309	15	2918	−12
Honger[65]	20	3800	19	3300	−13
Haering et al.[66]	18	3721	21	3148	−15
Brody & Spetz[67]	46	4245	34	4010	−5
MacGillivray[68]	18	4040	35	3535	−12
Blekta et al.[69]	55	3133	14	2590	−17
Gallery et al.[55]	199	3878*	37	3383*	−13
MacGillivray[70]	55	3763	29	3524	−6
Brown et al.[71]	54	3912*	49	3260*	−17
Silver et al.[72]	20	4070	20	3416	−16
Zeeman et al.[57, b]	44	4505	29*	3215[c]	−29[d]

[a]Mean values except for those marked with *, which are median values. Values are listed as shown or were calculated from the publication listed.
[b]Total blood volume.
[c]All were eclamptic nulliparas.
[d]Fourteen eclamptics had a subsequent normotensive gestation in which their total blood volumes were 45% greater than that measured after the eclamptic seizure.

sympathetic tone that may occur during preeclampsia favors this notion.[73] Schobel et al.[73] concluded that preeclampsia is a state of sympathetic overactivity which reverts to normal after delivery. Their data suggest that the increases in peripheral vascular resistance and blood pressure that characterize this disorder are mediated, at least in part, by a substantial increase in sympathetic vasoconstrictor activity. Thus, sympathetic nervous system overactivity in preeclampsia may contribute to abrogated vasodilatation expected in normal pregnancy.

Contrary to the above, Gallery et al.[55] demonstrated that low plasma volume precedes the rise in blood pressure and other clinical manifestations of preeclampsia by several weeks (Fig. 15.3). Their serial study also demonstrated that the increment in plasma volume in preeclamptic women and those whose blood pressure remained normal was similar in the second trimester of pregnancy but plasma volume contraction occurred thereafter, still preceding the development of overt disease by weeks. These observations appear to counter explanations that the decreased volumes reflect the effect of vasoconstriction on efflux of fluid from vessel walls or via the kidney (so-called "pressure naturesis" associated with hypertension). Finally, once overt preeclampsia appears, further volume contraction occurs, and this appears proportional to the severity of the disorder (Fig. 15.3).

Intervention studies are rare in preeclampia; nevertheless one study compared low vs. high salt in 2077 pregnant women in 1958.[74] The interventions consisted of advice to either increase or reduce the salt intake with meals. Surprisingly, the authors observed a lower incidence of toxemia, edema, perinatal death, and bleeding during pregnancy in those told to consume more salt. Extra salt in the diet seems to be essential for the health of a pregnant woman, her fetus, and the placenta. In contrast to guidelines for nonpregnant women, salt restriction is not advised in pregnant women, even in those with a history of hypertension. In 2000 the Cochrane database investigated the effects of advice on salt consumption during pregnancy on prevention and treatment of preeclampsia and its complications and confirmed that advising reduced salt intake cannot be recommended.[75] Experimental *in vivo* data that salt reduction is not protective during pregnancy were obtained by Giradina et al.[76] A low-salt diet significantly increased arterial pressure and vascular reactivity in pregnant and hypertensive-pregnant rats (RUPP). The authors speculate that the observed phenotype is caused by an increase in calcium entry from the extracellular space with a low-salt diet.

Novel Salt Concept

Recently, novel findings in metabolism suggested that salt is implicated not only in blood pressure control and volume homeostasis, but also in immune regulation.

The traditional concept is that sodium excretion by the kidney is the critical pathway regulating fluid status, determining the level of intra-arterial pressure and blood presssure control. Machnik et al. proposed an alternative mechanism

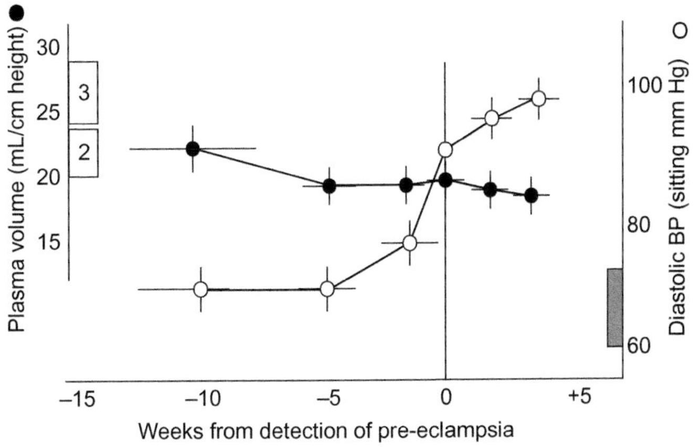

FIGURE 15.3 Time course depicting the decrease in plasma volume in pregnant women who developed preeclampsia (black circles). The right ordinate depicts diastolic blood pressure; the shaded rectangle is the 95% confidence interval for normal values. The left ordinate depicts plasma volume in the 2nd and 3rd trimester, with 95% confidence levels noted in open rectangles. (from ref[55]).

that implicates macrophages as mediators of sodium storage in the subdermis.[77] Sodium chloride can be stored without accumulation of water at hypertonic concentrations in interstitial proteoglycans. They investigated underlying mechanisms of salt-induced hypertension in rats and identified immune cells as principal sensors of salt load. Tissue macrophages express tonicity enhancer binding protein in response to the detected local hypertonicity and via activation of vascular endothelial growth factor-C (VEGF-C). As a result, they increase the density of the lymph-capillary network in the skin and enhance production of nitric oxide in the skin vessels, thus managing extracellular volume and blood pressure homeostasis. This finding does not overrule the renal regulatory function. It indicates rather that there also exists a local complementary extrarenal mechanism for electrolyte, volume, and blood pressure balance.

Human data on the role of VEGF-C-mediated salt homeostasis corroborate the findings in rats.[78] The researchers explored the role of the VEGF-C–macrophage–lymphangiogenesis pathway as an extrarenal homeostatic mechanism in proteinuric chronic kidney disease patients and in healthy controls under a high-salt diet. VEGF-C levels along with blood pressure were elevated both in patients and in healthy individuals who ingested a high-salt diet.

Another proof that salt activates the immune system has been recently found in a study looking into the triggering mechanisms of autoimmune diseases.[79] The authors showed that increased dietary salt intake is an environmental risk factor and generates pathogenic Th17 cells relevant for the development of autoimmune diseases. These data are based on the concept that the salt concentration in the interstitium and lymphoid tissue is considerable higher than in the plasma, approaching levels as high as 250 mM. However, the role of this concept in pregnancy and preeclampsia has yet to be determined.

Mineralocorticoids and the Renin-Angiotensin-Aldosterone Axis

Levels of several antinatriuretic hormones, primarily the mineralocorticoids aldosterone and desoxycorticosterone, increase markedly during gestation. Considerable research has also focused on the RAAS, because it is intimately involved in the renal control of salt and water balance.[1,55,80–88] Prorenin levels increase quite early, probably the reason for the increment in circulating total renin levels.[82,83] AII is also increased, an expected consequence of the increased renin and angiotensin production, and one reason for the high levels of aldosterone (an increased sensitivity of the adrenals to AII, noted previously,[50] is another reason). Figure 15.4 summarizes a detailed study by Wilson et al.[84] in which plasma renin substrate activity (PRA), plasma and urinary aldosterone, and urinary sodium and potassium excretion were measured throughout gestation. Sequential increases in substrate and activity, starting early in gestation, are shown more precisely in the study by Chapman and colleagues,[45] whose subjects were tested before conception and during the initial weeks of pregnancy (Fig. 15.2). Of interest in the study by Wilson et al.[84] is that salt excretion (reflecting intake) was similar throughout gestation compared to postpartum measurements, suggesting that inadequate sodium intake does not account for stimulation of the RAA axis.

Of further interest is that seemingly high levels of all components of the RAA axis, and other potent mineralocorticoids, are not constant but respond appropriately when volume status is manipulated. Thus circulating levels of renin, angiotensin, and aldosterone decrease after saline infusion or during a high-salt diet, or increase further following dietary sodium restriction or administration of diuretics.[37,50,52,80,81,85–87] Also, in one study, inhibition

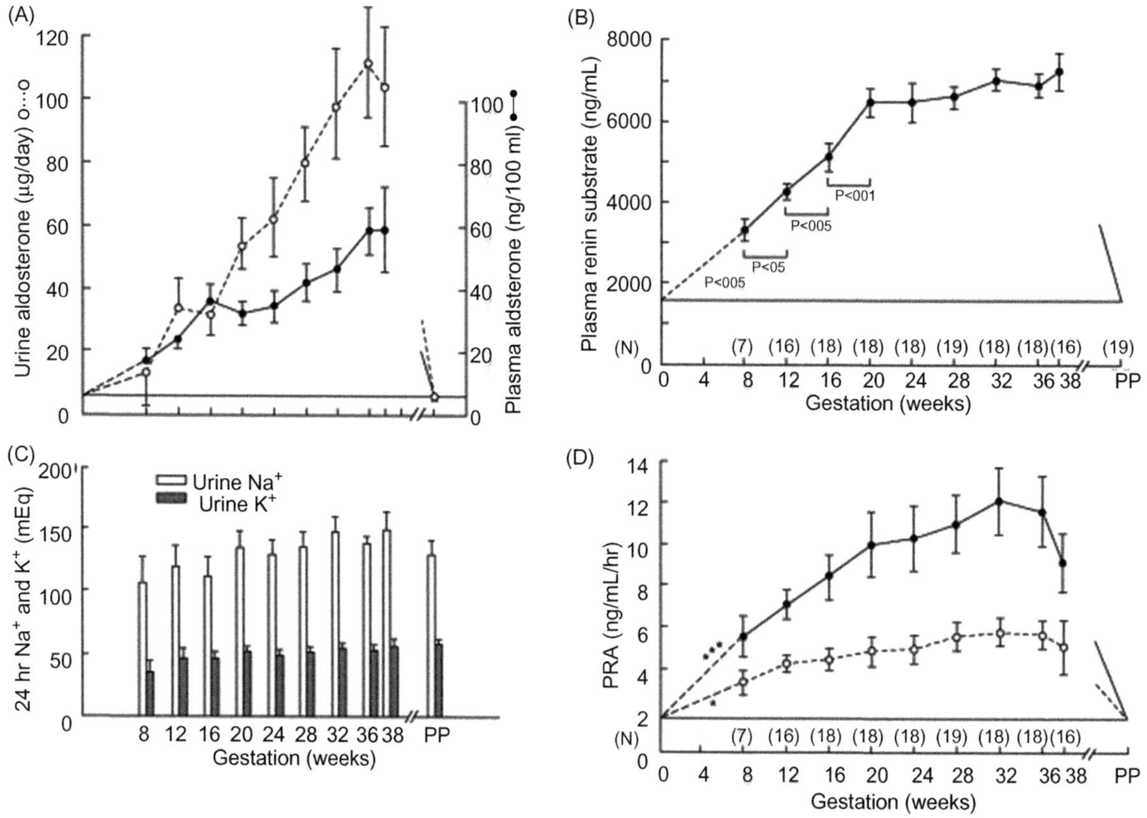

FIGURE 15.4 Sequential changes during pregnancy in (A) urinary aldosterone, (B) plasma renin substrate, (C) 24 h urinary Na and K excretion, and (D) plasma renin activity. Dashed lines in D are values normalized to postpartum substrate levels. (Data first published by Wilson et al. *Am J Med* 1980;68:97–104, the figure adapted from original graphs in August P, Lindheimer MD 1995, Pathology of preeclampsia. In Laragh JH, Brenner BM, eds, *Hypertension, Pathophysiology Diagnosis and Management*, 2nd edition. Raven Press, New York, 1995, pp. 2407–26.)

of aldosterone biosynthesis resulted in a diuresis and subtle signs of volume depletion, the salt loss already apparent when aldosterone excretion, though decreasing, was substantially greater than nonpregnant levels.[88] Thus, in pregnancy the renin–angiotensin system does not function autonomously as some have postulated, but around a new set point. In essence, the high levels of aldosterone, often exceeding those measured in nonpregnant patients with primary aldosteronism, are appropriate in pregnancy and respond to homeostatic demands. There is a corollary view, however, that this reset is a two-component system, nonsuppressible and unresponsive to physiologic stimuli – a view to our knowledge that remains untested.[89]

In addition to a marked rise in filtered sodium (possibly straining reabsorptive mechanisms), there are changes, too, in hormone levels and/or in autacoid function that theoretically enhance renal sodium excretion during pregnancy. These include increased circulating levels of oxytocin, vasodilating prostaglandins, melanocyte stimulating hormone, progesterone, and natriuretic peptides.[37,89,90] Progesterone levels increase markedly in pregnancy and because the affinity of this hormone for the mineralocorticoid receptor exceeds that of aldosterone, some have queried how the latter functions as a mineralocorticoid in the face of such high circulating progesterone levels.[91] The answer suggested is that progesterone actually becomes an antinatriuretic influence via metabolism to deoxycorticosterone (DOC) at extra-adrenal sites.[92–94] In fact, because renal steroid 21-hydroxylase activity (the enzyme that enhances conversion of progesterone to DOC) is particularly high in pregnancy, the considerable portion of maternal DOC produced in the vicinity of the renal receptors enhances sodium reabsorption.[95]

Studies by Shojaati et al.[80] suggest that women who develop preeclampsia have gene mutations leading to inefficient aldosterone production (primarily a decrease in aldosterone synthase, CYP11B2). These authors suggest this results in inefficient volume expansion and thus poor placental perfusion in early pregnancy, and preeclampsia later in gestation. This work reminds us of the 1958 paper by Robinson in the Lancet, where salt loading throughout gestation decreased the incidence of preeclampsia, and saline infusions temporarily improved blood pressure. Indeed, in a recent case report from the same group who suggested the aldosterone production may be inefficient in preeclampsia, salt loading throughout gestation was purported to have prevented preeclampsia.[79]

This concept – perturbations in steroid pathways – has been explored further.[96] Cortisol has the same affinity for the mineralocorticoid receptor as aldosterone. Cortisol availability is controlled by 11β-hydroxysteroid dehydrogenase type 2 (11β-HSD2), which inactivates cortisol into cortisone, which is unable to bind to the mineralocorticoid receptor. The 11β-HSD2 enzyme activity limits intracellular cortisol concentrations and, within the uteroplacental compartment, the transfer of cortisol into the fetal circulation. Mechanisms by which 11β-HSD2 activity is controlled include epigenetic regulation via methylation of genomic DNA, transcription, post-transcriptional modifications of 11β-HSD2 transcript half-life, and direct inhibition of enzymatic activity. Evidence exists that 11β-HSD2 expression and activity are reduced in preeclampsia and that enzyme activity correlates with factors associated with increased vasoconstriction, such as an increased Ang II receptor subtype 1 expression, and notably decreased fetal growth. Proinflammatory cytokines known to be present and/or elevated in preeclampsia regulate 11β-HSD2 activity. Shallow trophoblast invasion with the resulting hypoxemia seems to reduce 11β-HSD2 activity. A positive feedback loop exists as activated glucocorticoid receptors enhance 11β-HSD2 mRNA transcription and mRNA stability. These findings also implicate disturbed mineralocorticoid receptor signaling in preeclampsia.

Antinaturetic Peptides in Pregnancy

In normal pregnancy there is an increase in plasma atrial natriuretic peptide levels (ANP), though some investigators have failed to observe these increases.[90,97,98] Of interest is a study by Lowe et al.,[97] who prospectively controlled both sodium intake and posture and failed to find any increase during gestation. However, ANP's metabolic clearance rate increases, so that even were circulating levels not to rise, the hormone's production rate has increased.[99] ANP levels are even higher in women with preeclampsia, including those with the disease superimposed on chronic hypertension.[98–101] This situation seems paradoxical, as plasma volume decreases in preeclampsia. The answer may be that ANP production responds to stimuli such as contractility (e.g., brain natriuretic protein (BNP) is produced in the heart and increased in cardiac failure), and the ANP rise in preeclampsia is more related to cardiovascular events than to volume.

Of interest is a study by Irons and colleagues[102] noting that low doses of infused ANP produced a minimal natriuresis in normotensive third-trimester women but had no impact on sodium excretion when women were restudied 4 months postpartum. Such data suggest a heightened natriuretic system (more consistent with *over-* or *normal* rather than *underfill*).

On the other hand there are studies in pregnant rats suggesting loss of natriuretic responsiveness to administered ANP during normal pregnancy.[103–106] The acute natriuretic response to volume expansion is dependent on endogenous ANP release and this is also blunted in the pregnant rat.[106] Since both ANP and NO signal through cGMP, the observations in rodents suggested that the tubular response to cGMP might be blunted, and indeed Ni and colleagues noted increased cGMP breakdown in the inner medullary collecting duct of the pregnant rat kidney (a major site of natriuretic action of ANP and NO[105]). This effect is ascribed to selective, local increase in abundance/activity of phosphodiesterase 5 (PDE5).[106] These same collaborative laboratories have further reported that the natriuretic response to administered ANP in pregnant rats could be restored by local intrarenal PDE5 inhibition.[107] This natriuretic refractoriness in pregnant rats, therefore, is cGMP-specific, since dopamine-induced natriuresis (which signals via increased cAMP) remains unblunted.[108] As interesting as these recent findings appear, their relevance to human pregnancy needs to be established, and some differences already noted between the blunted natriuretic ANP observations in rodents[102,103] versus the enhancement of sodium excretion in pregnant humans[102] remain to be resolved. Nonetheless, a recent pilot study suggests salutary effects of sildenafil on intrauterine growth restriction.[109]

Preeclampsia

It appears paradoxical, given the decrease of intravascular volume associated with preeclampsia, that levels of all standard elements of the RAA axis decrease in this disorder.[37,47,55,82,110] Both aldosterone levels and AII concentrations are below those of normotensive gravidas. However, there may be relative increments in angiotensin II receptors in platelets, and in other tissues, a speculation used to explain the increased pressor sensitivity to infused AII, originally described by Gant and colleagues[111] in 1973. Of late it seems that the presence of the AT1-AA might explain the increased sensitivity observed by Gant.

Serial studies demonstrate that activity of the RAA axis is stimulated throughout pregnancy in pregnant women with chronic hypertension,[55] whereas in those who develop superimposed preeclampsia the RAAS functions normally early in gestation and then decreases, to levels below those in those women who do not develop preeclampsia.

The changes in the RAA axis described both in normal and in preeclamptic pregnancies are apparently not synchronized and thus the plasma aldosterone:renin ratio is increased during normal pregnancy though the slope across different salt intakes is similar[55,82] (Fig. 15.5). The highest ratio is observed in preeclamptic women, where the disorder seems to suppress renin and other system components more than aldosterone (Fig. 15.5).

Despite gross retention of the intravascular plasma volume with visible edema in patients with preeclampsia, plasma renin activity and aldosterone are paradoxically suppressed

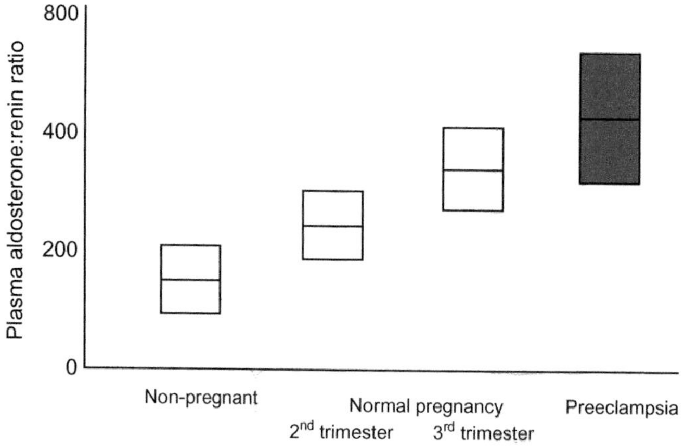

FIGURE 15.5 The plasma aldosterone:renin ratio in nonpregnant, pregnant second- and third-trimester subjects, and women developing preeclampsia (including mean and interquartile ranges). Lowest values occur in the nonpregnant state and rise as pregnancy progresses; the highest values were observed in preeclamptics. (Figure is a composite of data from[138–140]).

when compared to normotensive pregnant women. The compensatory activation of the RAAS in response to relative intravascular volume depletion does not happen. It is unclear, however, whether failure of RAAS activation and inappropriately low levels of aldosterone levels in pregnancy are a cause or simply a consequence of preeclampsia.

Relatively higher levels of aldosterone for a given level of renin are observed, indicating reduced aldosterone availability in established preeclampsia.[79,96,112–114] In normal pregnancy the ratio of aldosterone to renin increases substantially throughout gestation as most recently heavily investigated by the Mohaupt group.[79,96] They hypothesize that circulating aldosterone is the key regulator for the increase in maternal plasma volume during pregnancy that appears necessary for optimum fetal development. Reduced aldosterone production abrogates the pregnancy-associated expansion of circulating fluid volume in preeclampsia and contributes to poor placental perfusion and fetal substrate delivery.[112] The regulation of aldosterone in pregnancy and preeclampsia is poorly understood, but renin and Ang II levels insufficiently explain the aldosterone levels. Moreover, Mohaupt et al. have recently shown that VEGF is a strong inducer of adrenal aldosterone production.[113] VEGF stimulated aldosterone synthesis in H295R adrenal cells as assessed by the conversion of ^{3}H-deoxycorticosterone (DOC) to ^{3}H-aldosterone. They further showed that VEGF more specifically stimulated aldosterone production than angiotensin II. VEGF led primarily to an increased expression of the aldosterone synthesizing enzyme CYP11B2. The conclusion of these findings was that the unexpectedly low aldosterone availability in preeclampsia is caused by the sequestration of VEGF by sFlt-1, despite a reduction in plasma volume. This effect might be even more pronounced in individuals with an already reduced aldosterone synthesis caused by loss-of-function variants of the enzyme CYP11B2. Siddiqui et al. showed that AT1-AA caused sFlt-1-induced vascular impairment in adrenal glands using an adoptive transfer model of preeclampsia; this effect was attenuated by contemporaneous infusion of VEGF.[114]

In a further follow up, the Mohaupt group showed proliferation of trophoblasts was time- and dose-dependently stimulated by aldosterone and inhibited by glucocorticoids or spironolactone. These effects are independent of autocrine production of these steroid hormones.[115] Aldosterone-deficient mice suffer from severe placental insufficiency with necrotic placentas and massive lymphocyte infiltrations. Litter sizes were smaller and showed increased numbers of resorptions. In this mouse model, aldosterone deficiency did not cause preeclampsia, but high dietary salt intake improved placental function.

Agonistic Anti-AT1 Receptor Antibodies

Fu et al. produced antibodies against a synthetic peptide corresponding to amino acids 165–191 of the second extracellular loop of the AT1 receptor.[116] They observed that the antibodies did not interfere with ligand-binding properties of the receptor, rather, they exerted an agonist effect. When the investigators immunized rats with the AT1 receptor peptides over a 3-month period, the animals developed circulating anti-AT1 receptor antibodies that exerted a chronotropic effect in the neonatal rat cardiomyocyte bioassay.[117] However, no hypertension or target-organ damage was detected in the rats. Such results, however, may relate to not having utilized the most sophisticated methodology (e.g., tail-cuff measurements vs. the more sensitive radiotelemetric blood pressure measurements considered standard today). The animals may have also been subjected to immunization for a too short period. In this respect Jahns et al. observed that almost a year was required before immunization experiments caused a demonstrable effect.[118]

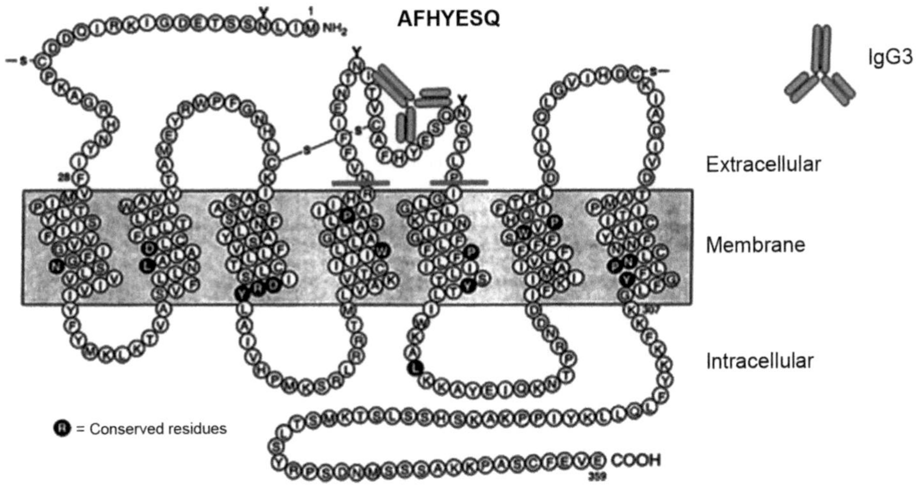

FIGURE 15.6 Schematic depiction of the AT1 receptor. The second AT1 receptor extracellular loop and the binding site (amino-acid sequence) of agonistic AT1 receptor antibodies are shown. The interaction can be blocked by a specific seven-amino-acid peptide (AFHYESQ). (This figure is reproduced in color in the color plate section.)

Agonistic Anti-AT1 Receptor Antibodies in Preeclampsia (AT1-AA)

Our interest in autoantibodies stems from our experience managing one 36-year-old woman with malignant hypertension who gave a history of having had preeclampsia 17 years earlier. In collaboration with Prof. G. Wallukat (Max Delbrück Center, Berlin-Buch, Germany) we discovered that her sera contained agonistic AT1 receptor autoantibodies. This led to our expanded studies of preeclamptic women, observing that the sera of these patients also contain agonistic AT1 receptor antibodies. These were retrospective studies, and in one single patient where samples were available for antibody testing prior to the development of overt disease, the autoantibody in her serum antedated preeclampsia. The AT1-AA were present during the active phase of the disease and regressed 4–6 weeks after pregnancy.

This initial report was received with considerable skepticism; however, we were also able to demonstrate that the IgG3 fraction from those patients functioned as well in Western blots as any available commercial antibody against the AT1 receptor.[1] We further verified specificity through co-immunoprecipitation experiments. These and subsequent studies relied on a bioassay for AT1-AA based on spontaneously beating neonatal rat cardiomyocytes exposed to serum from the pregnant women. In spite of the known problems with this technique, including subjective evaluation of the heart beat frequency and the fact that it is an expensive and time-consuming method, it has remained the method of choice to detect AT1-AA. With the help of this bioassay, the exact binding site of the AT1-AA to the AT1-receptor has been identified (the peptide AFHYESQ, corresponding to the second extracellular loop of the AT1 receptor) (Fig. 15.6).

Of interest, and shortly after our initial publication,[1] Fu et al. published an article stressing the presence of these agonistic autoantibodies in the sera of patients with malignant hypertension.[2] Such observations, of course, suggest options for potential therapies. The latter includes use of angiotensin converting enzyme (ACE) inhibitors, AT1 receptor blockers, or direct renin inhibitors, but most of these drugs are contraindicated for use in pregnancy. Another approach would be antibody removal by plasmapheresis, a complicated procedure for which considerably more experience is needed before studies would be attempted in human gestation.

Signal Transduction and Pathophysiological Role of AT1-AA

The studies described above were quickly followed by others demonstrating that the antibodies in patient sera exert a signaling effect. Since the initial description, AT1-AA have been shown to activate NADPH oxidase, and nuclear factor-kappa (NF-κB) and stimulate production of ROS, endothelin-1 (ET-1), and the antiangiogenic facors sFlt-1 and soluble endoglin (sEng) (Fig. 15.7).[3–8] The results of other studies focusing on human vascular smooth muscle cells (VSMC) and trophoblasts demonstrated that the antibodies activated the transcription factor AP-1 and caused vascular smooth muscle cells to produce tissue factor, supporting the hypothesis that these antibodies can elicit pathological effects, at least *in vitro*.[4] The activation of the NADPH oxidase by Ang II and AT1-AA is rapid and complex (Fig. 15.7). Various intracellular signal transduction pathways can lead to transactivation of the EGF receptor, which is necessary for sustained activation of NADPH oxidase.[119] We have shown that AT1-AA and Ang II have an additive effect on ET-1 release from human

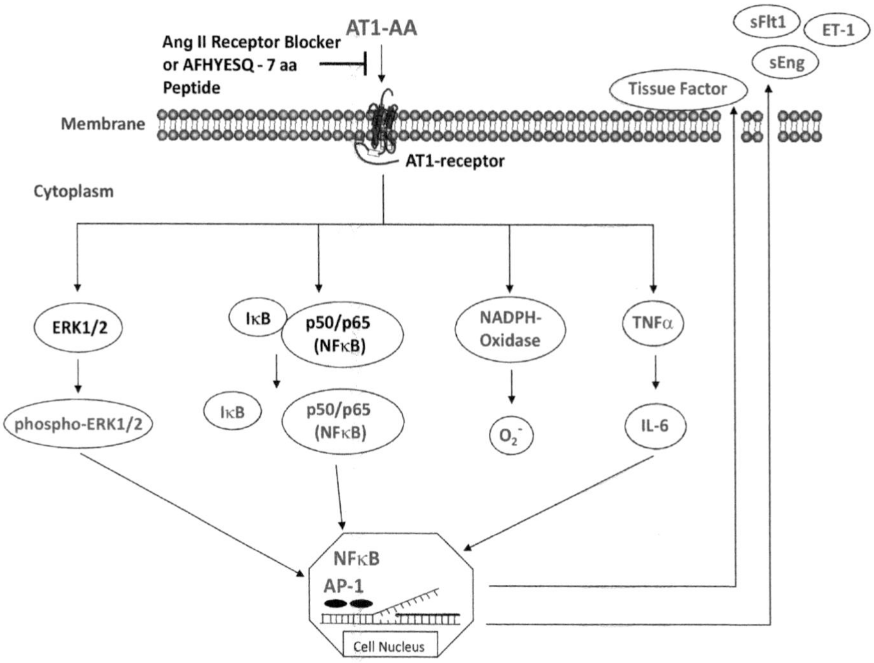

FIGURE 15.7 The agonistic AT1-AA induces signaling by the angiotensin II type 1 receptor (AT1-receptor), which can be inhibited by AT1-receptor blocker (ARB) or the seven-amino-acid peptide (AFHYESQ) mimicking the epitope of the AT1-AA in the second extracellular domain of the AT1-receptor. This activation then leads to NADPH oxidase, increased protein kinase C and calcineurin activity. Transcription factors such as activating protein-1, nuclear factor-κB (NF-κB), and nuclear factor activating T cell (NFAT) are activated. After translocating into the nucleus they lead to increased gene expression of target genes, i.e., interleukin-6, tissue factor, PAI-1, endothelin-1 (ET-1), soluble fms-like tyrosine kinase-1 (sFlt-1), soluble endoglin (sEng), and oxidative stress. (This figure is reproduced in color in the color plate section.)

endothelial cells in culture and isolated renal afferent arterioles. Effects in either cell culture system were blocked by a specific amino acid binding protein inhibiting the AT1-AA from binding to the AT1 receptor.[15] Furthermore, AT1-AA infused simultaneously with angiotensin II increased blood pressure but signicantly increased renal arteriole vasoconstriction, supporting the hypothesis that AT1-AA acts to enhance blood pressure and renal sensitivity to Ang II, observed by Gant many years earlier.

Other investigators have confirmed and extended our initial observations. Xia et al.[5] also detected agonistic AT1 receptor antibodies in preeclamptic women, further showing that the antibodies could activate AT1 receptors on a human trophoblast cell line to produce plasminogen activator inhibitor-1 (PAI-1) via the nuclear factor of activated T cells (NF-AT) pathway. This could account for increases in intracellular calcium that may occur in preeclampsia, and if it occurred in smooth muscle might explain the vasoconstrictive elements of the disease. In another report, Xia and colleagues showed that the antibodies can induce calcium signaling, and stimulate mesangial cells to produce interleukin-6 and PAI-1 (Fig. 15.8).[5,6] Intracellular calcium may also influence gene expression, and gene expression changes (i.e., hypoxia genes) have been speculated on when it comes to the hypotheses regarding preeclampsia. The authors speculated that this could be a mechanism for shallow implantation, which, if true, would place the effects of these autoantibodies at the doorstep of causality in relation to the preeclamptic syndrome. If so, one might predict that circulating antibodies would have to be present by the end of the first trimester in women destined to develop preeclampsia, but this is yet to be established.

AT1-AA in Animal Models of Hypertension During Pregnancy (See also Chapter 10)

We confirmed the studies by Takimoto et al., who had demonstrated that pregnancies conceived in rodents transgenic for the human renin and angiotensinogen genes display features of human preeclampsia.[120] When dams (mice or rats) harboring the human angiotensinogen gene are mated with sires carrying the human renin gene, the dams develop severe hypertension, proteinuria, and, in the last half of their 22-day pregnancy, target-organ damage that resembles that described in preeclampsia. We then demonstrated that rat dams in this model also develop agonistic antibodies directed at the AT1 receptor during the last half of pregnancy.[19] Other experiments revealed abnormal trophoblast invasion, altered vascular remodeling, endothelial dysfunction and the presence of arteriolosclerosis in this model, resembling several aspects of the pathology in the human disease (Fig. 15.8).[121,122]

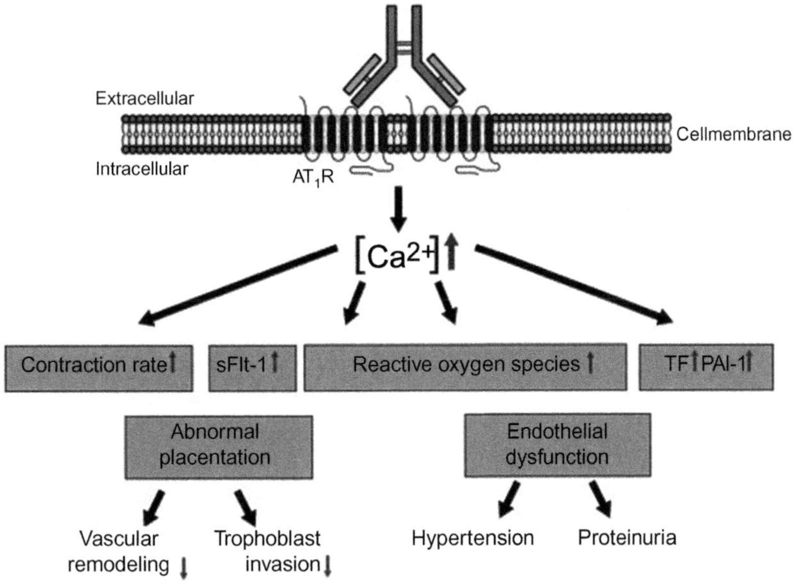

FIGURE 15.8 Molecular mechanisms of the interaction of AT1-AA with the AT1 receptor leading to the features of preeclampsia. (This figure is reproduced in color in the color plate section.)

Further AT1-AA have been detected in other experimental rat models of preeclampsia,[20] including the well-characterized reduced uteroplacental perfusion (RUPP) model. The RUPP model is created by placing silver surgical clips (0.203 mm ID) around the abdominal aorta above the iliac bifurcation and around both right and left ovarian arteries (silver clip, 0.100 mm) supplying the uterine horns. This procedure is performed on day 14 of gestation in the rat and hypertension, pup weight, and soluble and genetic factors are measured on day 19 of gestation.[20] The RUPP rat mimics numerous physiological features of preeclampsia in women. Some of these important pathophysiological characteristic include chronic immune activation, increased blood pressure, impaired renal function, and fetal growth reduction with decreased litter number and pup weight.

Most recently we have demonstrated a role for $CD4^+$ T cells in hypertension and in the production of the AT1-AA in response to placental ischemia induced by RUPP in pregnant rats.[122] We have shown that RUPP-induced CD4+ T cells increased blood pressure and decreased glomerular filtration rate when adoptively transferred into normal pregnant rats. Hypertension that developed in response to adoptive transfer of RUPP CD4+ T cells was associated with elevated TNF-α, sFlt-1, AT1-AA, and ET-1 in normal pregnant recipient rats, none of which were elevated in nonpregnant control rats.[122] Hypertension in response to RUPP CD4+ T cells was attenuated by AT1 receptor blockade and B cell depletion.[123]

We and others have shown that low-dose TNF-α administration to pregnant rats can produce proteinuria and hypertension but has no effect in nonpregnant controls.[20] Since uteroplacental and circulating TNF-α levels are also induced in preeclampsia and the RUPP model, we subsequently detected AT1-AA in pregnant rats treated with TNF-α but not in similarly treated nonpregnant control animals. Furthermore, studies indicate that inhibition of TNF-α decreased blood pressure, ET-1, sFlt-1, and AT1-AA in pregnant rodent models of preeclampsia. Studies from the Xia laboratory at the University of Texas at Houston demonstrated the AT1-AA to increase TNF-α in the circulation of AT1AA-injected pregnant mice, but not in nonpregnant mice.[125] Moreover, TNF-α blockade in AT1-AA-injected pregnant mice attenuated the key features of preeclampsia, such as hypertension, albuminuria, and circulating sFlt-1 and sEng. These data demonstrated an important role for TNF-α production subsequent to AT1-AA activation of the AT1 receptor to mediate hypertension during pregnancy.[125]

We have shown that infusion of IL-6 into pregnant rats increased blood pressure and plasma renin activity, decreased renal function and stimulated AT1-AA.[22] However, infusion of IL-6 into nonpregnant rats had no effect on blood pressure or AT1-AA production, again highlighting a very different effect of chronic inflammatory cytokines during pregnancy compared to the nonpregnant state. Similar findings from Zhou et al. confirm that IL-6 blockade decreased blood pressure and downstream ET-1 production in AT1-AA-induced hypertensive pregnant mice.[126] IL-6 is important in both antiinflammatory and proinflammatory processes and is a pivotal cytokine to influence activation of B cells as well as effector or regulatory T cells. IL-6 is elevated in preeclamptic women, the RUPP rat and in AT1-AA-induced hypertensive pregnant mice.

IL-17 is a cytokine that has mostly been associated with autoimmune diseases but has recently gained attention in preeclampsia research.[127] Recent studies show that circulating IL-17-secreting T_H 17 cells are increased in preeclamptic patients compared to those with normal pregnancies.[127] We recently showed that IL-17 increased blood pressure, placental reactive oxidative species (ROS), urinary isoprostanes, and AT1-AA in pregnant rats while having none of these effects in nonpregnant rats.[21] Furthermore, we found that administration of Tempol, an antioxidant, not only attenuated the placental oxidative stress but also decreased blood pressure and, surprisingly, AT1-AA produced in response to IL-17. These data indicate the importance of IL-17 and ROS as mediators of the pathology associated with preeclampsia.

AT1-AA Induced Hypertension

Investigators in the Xia laboratory provided additional evidence in murine models that support a direct role of AT1-AA in preeclampsia pathogenesis (Figs. 15.7 and 15.8).[13] Initially, the investigators injected total IgG from preeclamptic patients into pregnant mice and monitored the symptoms of preeclampsia. Following IgG injection, the pregnant mice showed hypertension, proteinuria, glomerular endotheliosis (the characteristic renal lesion of preeclampsia), as well as placental abnormalities coupled with increased secretion of antiangiogenic factors and small fetuses. These antibody-induced features of preeclampsia could be prevented by co-injection with an AT1 receptor antagonist, indicating that the antibody functioned via AT1 receptor activation. Subsequently, to test the direct role of AT1-receptor autoantibodies in the pathogenesis of preeclampsia, the investigators affinity-purified the antibodies and again injected them into pregnant mice. These mice displayed features of preeclampsia similar to those of mice injected with total IgG.[8,13] Zhou et al. then used an antibody-neutralizing seven-amino-acid epitope peptide corresponding to a site on the second extracellular loop of the AT1 receptor.[13] This peptide sequence prevented the preeclampsia features induced in the pregnant mice injected with total IgG or affinity-purified AT1-receptor autoantibodies. Their data provide direct evidence that features of preeclampsia seen in pregnant mice were induced by autoantibodies that bound to the second extracellular loop of the AT1receptor and activated this receptor. Parikh and Karumanchi point out that AT1-AAs may be an early event leading to the activation of a preeclampsia cascade, ending in the production of sFlt-1.[10] However, prospective clinical studies are warranted to evaluate whether AT1-AA are present before the onset of the clinical syndrome.

Interaction Between ATI-AA and sFlt-I

Investigators led by Drs Kellems and Xia at the University of Texas at Houston and several collaborators have produced a series of observations linking AT1-AA with the production of an antiangiogenic protein now believed to play a key roles in preeclampsia (see Chapter 6). This group demonstrated that Ang II can stimulate the production of soluble Fms-like tyrosine kinase receptor (sFlt-1) in a trophoblast cell line[6] and extended the findings to include evidence that AT1-AA and sFlt-1 exert unified pathogenic actions.[7] Zhou et al. observed that IgG from women with preeclampsia stimulates the synthesis and secretion of sFlt-1 via AT1 receptor activation in pregnant mice, human placental villous explants, and human trophoblast cells.[8] Of importance in the animal studies, infused antibody increased sFlt in pregnant but not nonpregnant mice, underscoring the contributory importance of the placenta. The investigators used FK506 or short-interfering RNA targeted to the calcineurin catalytic subunit mRNA and determined that NF-AT signaling downstream of the AT1 receptor was needed to induce sFlt-1 synthesis and secretion. AT1-AA-induced sFlt-1 secretion resulted in inhibition of endothelial cell migration and capillary tube formation *in vitro*. These effects of AT1-AA are summarized in Fig. 15.8.

Interaction Between ATI-AA and ROS

When evaluating the effects of AT1-AA on ROS in isolated vascular smooth muscle cells, AT1-AA were noted to stimulate ROS through NADPH oxidase.[3] When we infused AT1-AA into normal pregnant rats, achieving concentrations similar to those seen in preeclampsia, blood pressure increased approximately 20 mm Hg and a significant increase in placental production of ROS was noted.[23] Placental ROS production was increased further in the setting of NADPH stimulation, confirming previous findings indicating the NADPH increased AT1-AA in trophoblast cells. In AT1-AA-exposed rats treated with the SOD mimetic Tempol, we observed decreased basal and NADPH-stimulated placental production of ROS, compared to our Tempol-treated controls. Similarly, attenuation of hypertension in the AT1-AA + Tempol rats was noted, compared to normal pregnant control rats treated with Tempol. These observations and those from other laboratories suggest that AT1-AA produced in response to placental ischemia are an important mediator of placental oxidative stress and hypertension during pregnancy.

Interaction Between ATI-AA and ET-I

We have recently reported that infusion of purified rat AT1-AA into normal pregnant rats increased serum AT1-AA, blood pressure, and tissue levels of preproendothelin mRNA.[128] Preproendothelin is the first transcribed message of ET-1, and is the standard for measuring tissues levels of ET-1 transcripts utilizing real-time PCR. Preproendothelin mRNA increased 11-fold in the renal cortices and 4-fold in the placentas of AT1-AA-infused rats, compared to control pregnant rats. Furthermore, AT1-AA-induced hypertensive pregnant rats displayed renal vascular dysfunction.[24] Endothelial function was tested using isolated renal interlobar arteries in a pressure myograph system. Vasodilatory responses to the endothelial-dependent agonist acetylcholine were impaired in AT1-AA-infused rats compared to nonpregnant controls. To determine a role for endothelin in endothelial dysfunction of AT1-AA-treated animals, a selective ET_A receptor antagonist was administered. Normal pregnant rats treated with ET_A receptor antagonist alone served as controls. The ET_A receptor antagonist attenuated AT1-AA-induced hypertension and completely blocked the vascular dysfunction observed in hypertensive pregnant rats. These data support the ability of AT1-AA to cause hypertension by stimulating ET-1 and renal vascular endothelial dysfunction during pregnancy.

Inhibition of AT1-AA in the RUPP Rat Model

The effects of endogenous AT1-AA on hypertension in response to placental ischemia remained under-explored. To address this question, we recently utilized the technique of B cell depletion to suppress lymphocyte entry into the circulation and subsequent antibody secretion.[129] RUPP rats treated with rutuximab and having suppressed circulating B lymphocytes and AT1-AA exhibited less blood pressure increase in response to induced placental ischemia.[129] Furthermore, B-cell-depleted RUPP rats had lower tissue ET-1 transcript in renal cortices and placentas compared to RUPP control rats. Importantly, circulating B lymphocytes and AT1-AA decreased significantly in RUPP rats treated with rituximab.

To further examine the role of CD4+ T cells in the pathophysiology of preeclampsia, we suppressed T cells by administration of abatacept (Orencia), a fusion molecule of CTLA 4. CTLA-4 is an antigen-dependent co-stimulatory protein. Orencia was administered to pregnant rats on gestational day 13, prior to placental insult. RUPP was induced on day 14 and blood pressure and soluble factors were collected on day 19.[130] Administration of Orencia decreased T cells, blood pressure, and AT1-AA in response to RUPP in pregnant rats, confirming our hypothesis that T cells participate in hypertension and production of AT1-AA in response to placental ischemia.

AT1-AA, Preeclampsia and Lasting Effects (Update)

In our initial publication we observed that AT1-AA levels fall rapidly within 2 weeks after delivery, with levels hovering around the bioassay cut-off value (7.2 bpm). Hubel et al. compared the presence of sFlt-1 and agonistic AT1 receptor antibodies in the plasma of 29 previously preeclamptic women, 18 months postpartum.[9] Seventeen percent of these women still had circulating agonistic AT1 receptor antibodies. Women who did also had elevated circulating levels of sFlt-1.

Presence of AT1-AA in Other Diseases

Walther et al. evaluated whether agonistic AT1-AA were present during the second trimester in women presumably undergoing uneventful pregnancies.[18] They also monitored uterine perfusion using ultrasonographic Doppler flow and observed that antibodies were detectable prior to the development of overt preeclampsia. Antibodies were also present in some women who did not develop clinical preeclampsia. However, only women with abnormal Doppler examinations developed autoantibodies. The authors concluded that the antibodies were associated with impaired placental perfusion and perhaps fetal development. In a subsequent study, these investigators combined the detection of autoantibodies with simultaneously measured sFlt-1 levels, but failed to find any correlations among the two measurements, suggesting distinct pathogenic mechanisms.[19]

A vexing and persistent question is how these agonistic AT1 receptor antibodies arise. The epitope recognized by the antibodies, shown in Fig. 15.6, also occurs on parvovirus B19 capsid proteins. Stepan et al. tested and rejected the hypothesis that women developing the antibodies might also have a high prevalence of parvovirus B19 infection.[131] Thus, this specific example of epitope mimicry is not supported by current published evidence, though the possibility remains open and under investigation.

Dragun et al. made further serendipitous but elucidating observations.[132] Most humoral rejection after renal transplant involves antibodies against HLA moieties that recruit C4d complement deposition along the peritubular capillaries of the donor kidney. A woman with an unusual form of C4d-negative humoral rejection after kidney transplantion developed profound hypertension; she was noted to have had preeclampsia years earlier. The investigators evaluated other patients with C4d-negative humoral kidney rejection and identified they had AT1-AA involving two different epitopes. In a rat renal transplant model, passive transfer of human AT1-AA reproduced an increase in blood pressure. The data suggest that stimulating antibodies directed against the AT1 receptor contribute to malignant hypertension occurring in the setting of acute humoral rejection. We have preliminary clinical evidence that AT1 receptor blockers are helpful in patients with C4d-negative humoral rejection. However, controlled trials are needed to confirm this observation.

In related findings, Riemekasten et al. showed that AT1-AA were found in most patients with systemic sclerosis.[133] However, these AT1-AA can be detected by a commercially available ELISA (Celltrend), which does not detect AT1-AA in preeclamptic patients. Interestingly, autoantibodies against the endothelin receptor are found in the same patients. Higher levels of both autoantibodies were found in scerloderma patients who suffered from vascular complications. AT1-AA predicted scleroderma-related mortality. Interestingly, the obliterative vascular lesions observed in preeclampsia and acute kidney rejection are also found in systemic sclerosis.

The AT1 receptor autoantibody story has become more compelling as firm evidence links these autoantibodies to sFlt-1, ET-1, and ROS production *in vitro* and in several animal models that recapitulate features seen in the human disease. Also, as discussed above, a model of Ang II- and AT1-AA-induced sFlt-1 induction in normal pregnancy and in preeclampsia has been proposed by Kellems, Xia, and their collaborators. During preeclampsia AT1-AA can lead to increased sFlt-1 levels and Ang II induces sFlt-1 to inhibit ongoing angiogenesis late in pregnancy.

Nevertheless, many questions remain concerning these autoantibodies and others associated with different human diseases. A major question concerns the genesis of AT1-AA: are they causal, the result of the disease, secondarily produced but contributory to the pathogenesis or simply epiphenomena?

CONCLUDING PERSPECTIVES

There have been comprehensive reports that attest to the existence of agonistic antibodies directed against G protein-coupled and tyrosine-kinase receptors. Compelling evidence suggests that these antibodies are of pathogenic importance (Fig. 15.6); at a minimum they could play an ancillary accelerating role in relationship to preeclampsia.

In most instances, bioassay results have been required to demonstrate the relevance of agonistic antibodies. One reason includes their very low circulating concentrations, a fact that, by itself, does not detract from their potential pathogenic relevance. Technical problems and low titers of antibodies in these syndromes have precluded the establishment of a sufficiently sensitive ELISA to date.[134] Nikolaev et al. used a sensitive fluorescence resonance energy transfer (FRET) technique to detect functional β-adrenergic receptor antibodies in heart-failure patients.[135] The FRET assay detected antibody concentrations of 0.001–0.2 nmol/L.

Numerous questions remain to be explored. For example, while studies of the agonistic antibodies directed against the β-adrenergic receptor have largely fulfilled Koch's postulates in terms of establishing these antibodies as a cause of a disease,[118] evidence concerning agonistic antibodies directed against the AT1 receptor is compelling but not quite as convincing. Furthermore, recent evidence from other sources implicating sFlt-1 and sEng in preeclampsia development invokes alternative interactions of AT1-AA.[136] Antibodies to the AT1 receptor in post-transplant patients have greatly increased interest in C4d negative humoral rejection and the strong association between C4d positive rejection and post-transplant HLA antibodies.[132] However, much more work needs to be done in this area.

Will detection of agonistic antibodies in preeclampsia lead us to novel preventive and/or therapeutic options? Affinity columns, or less specifically plasmapheresis, are available to remove autoantibodies. These techniques clearly have potential for diseases such as scleroderma or kidney transplant rejection. For preeclampsia, antibody removal procedures have proved exceptionally difficult because of logistical problems and concerns for safety in pregnancy. Randomized controlled double-masked trials are unrealistic in this area. Evaluation of oral medications can be subjected to straightforward analysis, but trials of procedures and devices would require, for example, a closed plasmapheresis treatment room with arterial and venous lines coming through holes in the wall and operated by "blinded" technicians unable to observe the patients. Nonetheless, other possibilities exist. Our exploratory studies to develop a competitive fusion protein have encountered technical problems. However, aptamers and other small molecules of similar size might offer an alternative.[137] This chapter reviewed a new area that reflects the continued quest to understand what causes preeclampsia, and how to prevent or treat it. Knowledge about antibodies to the AT1 receptor in preeclampsia continues at a rapid and exciting pace, even as we review the galleys. We anticipate a much larger chapter in the next edition!

References

1. Wallukat G, Homuth V, Fischer T. Patients with preeclampsia develop agonistic autoantibodies against the angiotensin AT1 receptor. *J Clin Invest*. 1999;103:945–952.
2. Fu ML, Herlitz H, Schulze W. Autoantibodies against the angiotensin receptor (AT1) in patients with hypertension. *J Hypertens*. 2000;18:945–953.
3. Dechend R, Viedt C, Muller DN. AT1 receptor agonistic antibodies from preeclamptic patients stimulate NADPH oxidase. *Circulation*. 2003;107:1632–1639.
4. Dechend R, Homuth V, Wallukat G. AT(1) receptor agonistic antibodies from preeclamptic patients cause vascular cells to express tissue factor. *Circulation*. 2000;101:2382–2387.
5. Xia Y, Wen H, Bobst S, Day MC, Kellems RE. Maternal autoantibodies from preeclamptic patients activate angiotensin receptors on human trophoblast cell. *J Soc Gynecol Investig*. 2003;10:82–93.
6. Thway TM, Shlykov SG, Day MC. Antibodies from preeclamptic patients stimulate increased intracellular Ca2+ mobilization through angiotensin receptor activation. *Circulation*. 2004;110:1612–1619.
7. Zhou CC, Ahmad S, Mi T. Angiotensin II induces soluble fms-Like tyrosine kinase-1 release via calcineurin signaling pathway in pregnancy. *Circ Res*. 2007;100:88–95.
8. Zhou CC, Ahmad S, Mi T. Autoantibody from women with preeclampsia induces soluble Fms-like tyrosine kinase-1 production via angiotensin type 1 receptor and calcineurin/nuclear factor of activated T-cells signaling. *Hypertension*. 2008;51:1010–1019.
9. Hubel CA, Wallukat G, Wolf M. Agonistic angiotensin II type 1 receptor autoantibodies in postpartum women with a history of preeclampsia. *Hypertension*. 2007;49:612–617.
10. Parikh SM, Karumanchi A. Putting pressure on preeclampsia. *Nat Med*. 2008;14:2056–2061.
11. Xia Y, Ramin SM, Kellems RE. Potential roles of angiotensin receptor-activating autoantibody in the pathophysiology of preeclampsia. *Hypertension*. 2007;50:269–275.

12. Xia Y, Zhou CC, Ramin SM, Kellems RE. Angiotensin receptors, autoimmunity, and preeclampsia. *J Immunol.* 2007;179:3391–3395.
13. Zhou CC, Zhang Y, Irani RA. Angiotensin receptor agonistic autoantibodies induce pre-eclampsia in pregnant mice. *Nat Med.* 2008;14:855–862.
14. Herse F, Staff AC, Hering L, Muller DN, Luft FC, Dechend R. AT1-receptor autoantibodies and uteroplacental RAS in pregnancy and pre-eclampsia. *J Mol Med.* 2008;86:697–703.
15. Brewer J, Liu R, Lu Y, et al. B· Endothelin-1, oxidative stress and endogenous ANGII: mechanisms of AT1-AA Enhanced Renal and Blood Pressure Sensitivity during pregnancy. *Hypertension.* 2013;62:886–892.
16. Wenzel K, Rajakumar A, Haase H, et al. Angiotensin II type 1 receptor antibodies and increased angiotensin II sensitivity in pregnant rats. *Hypertension.* 2011;58:77–84.
17. Stepan H, Faber R, Wessel N, Wallukat G, Schultheiss HP, Walther T. Relation between circulating angiotensin II type 1 receptor agonistic autoantibodies and soluble fms-like tyrosine kinase 1 in the pathogenesis of preeclampsia. *J Clin Endocrinol Metab.* 2006;91:2424–2427.
18. Walther T, Wallukat G, Jank A. Angiotensin II type 1 receptor agonistic antibodies reflect fundamental alterations in the uteroplacental vasculature. *Hypertension.* 2005;46:1275–1279.
19. Verlohren S, Niehoff M, Hering L. Uterine vascular function in a transgenic preeclampsia rat model. *Hypertension.* 2008;51:547–553.
20. LaMarca B, Wallukat G, Llinas M, Herse F, Dechend R, Granger JP. Autoantibodies to the angiotensin type I receptor in response to placental ischemia and tumor necrosis factor alpha in pregnant rats. *Hypertension.* 2008;52:1168–1172.
21. Dhillon P, Wallace K, Herse F, et al. IL-17 mediated oxidative stress is an important stimulator of AT1-AA and hypertension during pregnancy. *Am J Physiol Regul Integr Comp Physiol.* 2012;303:R353–R358.
22. LaMarca BB, Speed J, Ray LF, et al. Hypertension in response to IL-6 during pregnancy: role of AT1-receptor activation. *Int J Interferon Cytokine Mediator Res.* 2011;3:65–70.
23. Parrish MR, Wallace K, Dechend R, et al. Hypertension in response to AT1-AA: role of reactive oxygen species (ROS) in preeclampsia. *Am J Hypertens.* 2011;24:835–840.
24. Parrish MR, Ryan MJ, Glover P, et al. Angiotensin II type 1 autoantibody induced hypertension during pregnancy is associated with renal endothelial dysfunction. *Gen Med.* 2011;8:184–188.
25. Kim-Saijo M, Akamizu T, Ikuta K. Generation of a transgenic animal model of hyperthyroid Graves' disease. *Eur J Immunol.* 2003;33:2531–2538.
26. Fareau GG, Maldonado M, Oral E, Balasubramanyam A. Regression of acanthosis nigricans correlates with disappearance of anti-insulin receptor autoantibodies and achievement of euglycemia in type B insulin resistance syndrome. *Metabolism.* 2007;56:670–675.
27. Takeda K, Stagg J, Yagita H, Okumura K, Smyth MJ. Targeting death-inducing receptors in cancer therapy. *Oncogene.* 2007;26:3745–3757.
28. Bartholdy C, Kauffmann SO, Christensen JP, Thomsen AR. Agonistic anti-CD40 antibody profoundly suppresses the immune response to infection with lymphocytic choriomeningitis virus. *J Immunol.* 2007;178:1662–1670.
29. Hytten FE. Weight gain in pregnancy. In: Hytten FE, Chamberlain G, eds. *Clinical Physiology in Obstetrics.* Oxford: Blackwell; 1991:173–203.
30. Robertson EG. The natural history of oedema during pregnancy. *J Obstet Gynaecol Br Commonw.* 1971;78:520–529.
31. Dexter L, Weiss S. *Preeclamptic and Eclamptic Toxemia of Pregnancy.* Boston: Little & Brown; 1941.
32. Zinaman M, Rubin J, Lindheimer MD. Serial oncotic pressure levels and echoencephalopathy during, and shortly after delivery in severe pre-eclampsia. *Lancet.* 1985:1245–1247.
33. Kogh A, Landis EM, Turner AH. The movement of fluid through the human capillary wall in relation to venous pressure and the colloid oncotic pressure of blood. *J Clin Invest.* 1932;11:63–95.
34. Thomson AM, Hytten FE, Billewicz WZ. The epidemiology of oedema during pregnancy. *J Obstet Gynaecol Br Commonw.* 1967;74:1–10.
35. Benedetti TJ, Carlson RW. Studies of of colloid osmotic pressure in pregnancy-induced hypertension. *Am J Obstet Gynecol.* 1979;135:308–313.
36. Hytten FE. Oedema in pregnancy. In: Rippman ET, ed. *Die Spatgestose.* Basel: Schwabe; 1970.
37. Lindheimer MD, Conrad K, Karumanchi SA. Renal physiology and disease in pregnancy. In: Halpern RJ, Hebert SC, eds. *Seldin and Giebisch's The Kidney: Physiology and Pathophysiology.* 4th ed. New York: Elsevier; 2008:2339–2398.
38. Lindheimer MD, Davison JM. Osmoregulation and the secretion of arginine vasopressin and its metabolism during pregnancy (rev). *Eur J Endocrinol.* 1995;132:133–143.
39. Durr JA, Miller NL, Alfrey AC. Lithium clearance derived from the natural trace blood and urine lithium levels. *Kidney Int.* 1990;28:S58–S62.
40. Davison JM, Sheills EA, Philips PR, Lindheimer MD. Influence of humoral and volume factors on altered osmoregulation of normal pregnancy. *Am J Physiol.* 1990;258:F900–F907.
41. Barron WM, Durr JA, Schrier RW, Lindheimer MD. Role of hemodynamic factors in the osmoregulatory alterations of rat pregnancy. *Am J Physiol.* 1989;257:R909–R916.
42. Schrier RW. Pathogenesis of sodium and water retention in high-output and low-output cardiac failure, nephrotic syndrome, cirrhosis, and pregnancy. *New Engl J Med.* 1988;319:1065–1072.
43. Chesley LC, Valenti LC, Rein H. Excretion of sodium loads by nonpregnant, and pregnant normal, hypertensive, and preeclamptic women. *Metabolism.* 1958;7:575–588.
44. Katz AI, Lindheimer MD. Renal handling of acute sodium loads in pregnancy. *Am J Physiol.* 1973;225:696–699.
45. Chapman AB, Abraham WT, Zamudio S. Temporal relationships between hormonal and hemodynamic changes in early human pregnancy. *Kidney Int.* 1998;54:2056–2063.
46. Duvekot JJ, Cheriex EC, Pieters FA, Menheere PP, Peeters LH. Early pregnancy changes in hemodynamics and volume homeostasis are consecutive adjustments triggered by a primary fall in systemic vascular tone. *Am J Obstet Gynecol.* 1993;169:1382–1413.

47. Weir RJ, Doig A, Fraser R, et al. Studies of the renin-angiotensin-aldosterone system, cortisol, DOC, and ADH, in normal and hypertensive pregnancy. In: Lindheimer MD, Katz AI, Zuspan FP, eds. *Hypertension in Pregnancy*. New York: John Wiley & Sons; 1976:251–260.
48. Robertson GL. Thirst and vasopressin function in normal and disordered states of water balance. *J Lab Clin Med*. 1983;101:351–371.
49. Molnar M, Hertelendy F. N-omega-nitro-L-arginine an inhibitor of nitric oxide synthesis, increases blood pressure in rats and reverses the pregnancy-induced refractoriness to vasopressor agents. *Am J Obstet Gynecol*. 1992;166:1560–1567.
50. Brown MA, Broughton Pipkin F, Symonds EM. The effects of intravenous angiotensin II upon blood pressure and sodium and urate excretion in human pregnancy. *J Hypertens*. 1988;6:457–464.
51. August P, Mueller FB, Sealey JE, Edersheim TG. Role of renin-angiotensin system in blood pressure regulation in pregnancy. *Lancet*. 1995;345:896–897.
52. Weinberger MH, Kramer NJ, Grim CE, Petersen LP. The effect of posture and saline loading on plasma renin activity and aldosterone concentration in pregnant, non-pregnant and estrogen-treated women. *J Clin Endocrinol Metab*. 1997;44:69–77.
53. Brown MA, Gallery ED, Ross MR, Eber RP. Sodium excretion in normal and hypertensive pregnancy; a prospective study. *Am J Obstet Gynecol*. 1988;159:297–307.
54. Zamudio S, Palmer SK, Dahms TE, Berman JC, McCullough RG, Moore LG. Blood volume expansion, preeclampsia, and infant weight at high altitude. *J Appl Physiol*. 1993;75:1566–1573.
55. Gallery ED, Hunyor SN, Gyory AZ. Plasma volume contraction: a significant factor in both pregnancy-associated hypertension (pre-eclampsia) and chronic hypertension in pregnancy. *Q J Med*. 1979;48:593–602.
56. Pirani BB, Campell DM, MacGillivray I. Plasma volume in normal first pregnancy. *J Obstet Gynaecol Br Commonw*. 1973;80:884–887.
57. Zeeman GG, Cunningham FG, Pritchard JA. The magnitude of the hemoconcentration in preeclampsia. *Hypertens Pregnancy*. 2009;28:127–137. [see Commentary by M Mahowold on ethical validation of this study.].
58. Werko L, Bucht H, Llagerlof H. Cirkulationen vid graviditet. *Nord Med*. 1948;14:1868–1869.
59. Freis ED, Kenny JF. Plasma volume, total circulating protein, and "available fluid abnormalities" in pre-eclampsia and eclampsia. *J Clin Invest*. 1948;27:282–289.
60. Rottger H. Uber den Wasserhaushalt in der physiologischen und toxischen Schwangershaft. I. Der Wasserhaushalt in der physiologishen Schwangerschaft. *Arch Gynaekol*. 1953;184:59–85.
61. Rottger H. Uber den Wasserhaushalt in der physiologischen und toxischen Schwangershaft. II. Der Wasserhaushalt bei Schwangerschaftsspatoxikosen. *Arch Gynaekol*. 1954;184:413–416.
62. Cope I. Plasma and blood volume changes in pregnancies complicated by preeclampsia. *J Obstet Gynaecol Br Commonw*. 1961;68:413–416.
63. Freidberg V, Lutz J. Untersuchungen uber die Cappillarpermeabilitat jn der Schwangerschaft (ein Beitra zur Urasche der Proteinure bei gestosen). *Arch Gynaekol*. 1963;199:96–106.
64. Kolpakova LL. Changes of the plasma volume and serum protein composition in late toxaemia of pregnancy. *Akush Ginekol*. 1965;1:130–133.
65. Honger PE. Intravascular mass of albumin in preeclampsia and normal pregnancy. *Scand J Lab Clin Invest*. 1967;19:283–287.
66. Haering M, Werners PH, Hemmerling J. Uber das Verhalten des Gesamthamoglobins bei Schwerin Praeklampsien. *Z Gebutsh Gynaekol*. 1967;166:271–279.
67. Brody S, Spetz S. Plasma, extracellular, and interstitial fluid volumes in pregnancy complicated by toxaemia. *Acta Obstet Gynecol Scand*. 1967;46:138–150.
68. MacGillivray I. The significance of blood pressure and body water changes in pregnancy. *Scot Med J*. 1967;12:237–245.
69. Blekta M, Hlavaty V, Trinkova M, Bedl J, Bendova L, Chytil M. Blood volume of whole blood and absolute amount of serum proteins in the early stage of late toxemia of pregnancy. *Am J Obstet Gynecol*. 1970;106:10–13.
70. MacGillivray I. *Pre-eclampsia. The Hypertensive Disease of Pregnancy*. London: WB Saunders; 1983.
71. Brown MA, Zammit VC, Mittar DM. Extracellular volumes in pregnancy induced hypertension. *J Hypertens*. 1992;10:61–68.
72. Silver HM, Seebeck MA, Carlson R. Comparison of total volume in normal, preeclamptic and nonproteinuric gestational hypertensive pregnancy by subcutaneous measurement of red blood cell and plasma volume. *Am J Obstet Gynecol*. 1998;179:87–93.
73. Schobel HP, Fischer T, Heusser K, Geiger H, Schmieder RE. Preeclampsia – a state of sympathetic overactivity. *New Engl J Med*. 1996;335:1480–1485.
74. Robinson M. Salt in pregnancy. *Lancet*. 1958;1:178–181.
75. Duley L, Henderson-Smart D. Reduced salt intake compared to normal dietary salt, or high intake, in pregnancy. *Cochrane Database Syst Rev*. 2000:CD001687.
76. Giardina JB, Cockrell KL, Granger JP, Khalil RA. Low-salt diet enhances vascular reactivity and Ca(2+) entry in pregnant rats with normal and reduced uterine perfusion pressure. *Hypertension*. 2002;39(2 Pt 2):368–374.
77. Machnik A, Neuhofer W, Jantsch J, et al. Macrophages regulate salt-dependent volume and blood pressure by a vascular endothelial growth factor-C-dependent buffering mechanism. *Nat Med*. 2009;15:545–552.
78. Slagman MC, Kwakernaak AJ, Yazdani S, et al. Vascular endothelial growth factor C levels are modulated by dietary salt intake in proteinuric chronic kidney disease patients and in healthy subjects. *Nephrol Dial Transplant*. 2012;27:978–982.
79. Kleinewietfeld M, Manzel A, Titze J, et al. Sodium chloride drives autoimmune disease by the induction of pathogenic TH17 cells. *Nature*. 2013;496(7446):518–522.
80. Shojaati K, Causevic M, Kadereit B, et al. Evidence for compromised aldosterone synthase enzyme activity in preeclampsia. *Kidney Int*. 2004;66:2322–2328.
81. Bay WH, Ferris TF. Factors controlling plasma renin and aldosterone during pregnancy. *Hypertension*. 1979;1:410–415.
82. Brown MA, Nicholson E, Gallery ED. Sodium-renin-aldosterone relations in normal and hypertensive pregnancy. *Br J Obstet Gynaecol*. 1988;95:1237–1246.

83. Sealy JE, von Lutterotti N, Rabattu S, et al. The greater renin system. Its prorenin-directed vasodilator limb. Relevance to diabetes mellitus, pregnancy, and hypertension. *Am J Hypertens*. 1991;4:972–977.
84. Wilson M, Morganti AA, Zervoudakis I. Blood pressure, the renin-aldosterone system and sex steroids throughout normal pregnancy. *Am J Med*. 1980;68:97–104.
85. Lindheimer MD, Del Greco F, Ehrlich EN. Postural effects on steroids and sodium excretion, and serum renin activity during pregnancy. *J Appl Physiol*. 1973;35:343–348.
86. Hseuh WA, Leutscher JA, Carlson EJ, Grislis G, Faze E, McHargue A. Changes in active and inactive renin throughout pregnancy. *J Clin Endocrinol Metab*. 1982;54:1010–1016.
87. Oliver WJ, Neel JV, Gerkin RJ, Cohen EL. Hormonal adaptation to the stresses imposed upon sodium balance by pregnancy and lactation in the Yanomama Indians, a culture without salt. *Circulation*. 1981;63:110–116.
88. Ehrlich EN. Heparinoid-induced inhibition of aldosterone secretion in pregnant women. The role of augmented aldosterone secretion in sodium conservation during normal pregnancy. *Am J Obstet Gynecol*. 1971;109:963–970.
89. Brown MA, Gallery ED. Volume homeostasis in normal pregnancy and pre-eclampsia: physiology and clinical implications. *Bailleres Clin Obstet Gynaecol*. 1994;8:287–310.
90. Castro LC, Hobel CJ, Gornbein J. Plasma levels of atrial natriuretic peptide in normal and hypertensive pregnancies: a meta-analysis. *Am J Obstet Gynecol*. 1994;171:1642–1651.
91. Quinkler M, Diederich S, Bahr V, Oelkers W. The role of progesterone metabolism and androgen synthesis in renal blood pressure regulation. *Horm Metab Res*. 2004;36:381–386.
92. Casey ML, MacDonald PC. Metabolism of deoxycorticosterone and deoxycorticosterone sulfate in men and women. *J Clin Invest*. 1982;70:312–319.
93. MacDonald PC, Cutrer S, MacDonald SC, Casey ML, Parker CR. Regulation of extraadrenal steroid 21-hydroxylase activity. Increased conversion of plasma progesterone to deoxycorticosterone during estrogen treatment of women pregnant with a dead fetus. *J Clin Invest*. 1982;69:469–478.
94. Winkel CA, Milewich L, Parker CR, Gant NF, Simpson ER, MacDonald PC. Conversion of plasma progesterone to deoxycorticosterone in men, nonpregnant and pregnant women, and adrenalectomized subjects. *J Clin Invest*. 1980;66:803–812.
95. Winkel CA, Simpson ER, Milewich L, MacDonald PC. Deoxycorticosterone biosynthesis in human kidney: potential for formation of a potent mineralocorticoid in its site of action. *Proc Natl Acad Sci USA*. 1980;77:7069–7073.
96. Escher G, Mohaupt M. Role of aldosterone availability in preeclampsia. *Mol Aspects Med*. 2007;28:245–254.
97. Lowe SA, MacDonald GJ, Brown MA. Acute and chronic regulation of atrial natriuretic factor in pregnancy. *J Hypertens*. 1992;10:821–829.
98. August P, Lenz T, Ales KL. Longitudinal study of the renin-angiotensin-aldosterone system in hypertensive pregnant women: deviations related to the development of superimposed preeclampsia. *Am J Obstet Gynecol*. 1990;163:1612–1621.
99. Irons DW, Baylis PH, Davison JM. The metabolic clearance of atrial natriuretic peptide during human pregnancy. *Am J Obstet Gynecol*. 1997;176:730–731.
100. Lowe SA, Zammit VC, Mitar D, Macdonald GJ, Brown MA. Atrial natriuretic peptide and plasma volume in pregnancy-induced hypertension. *Am J Hypertens*. 1991;4:897–903.
101. Malee MP, Malee KM, Azuma SD, Taylor RN, Roberts JM. Increases in plasma atrial natriuretic peptide concentration antedate clinical evidence of preeclampsia. *J Clin Endocrinol Metab*. 1992;74:1095–1100.
102. Irons D, Baylis PH, Butler TJ, Davison JM. Atrial natriuretic peptide in preeclampsia: metabolic clearance, sodium excretion and renal hemodynamics. *Am J Physiol*. 1997;273:F483–F487.
103. Masilamini S, Castro L, Baylis C. Pregnant rats are refractory to the natriuretic actions of atrial natriuretic factor. *Am J Physiol*. 1994;267:R1611–R1616.
104. Omer S, Mulay S, Cernacek P, Varma DR. Attenuation of renal effects of atrial natriuretic factor during rat pregnancy. *Am J Physiol*. 1995;268:F416–F422.
105. Mahaney J, Felton C, Taylor D, Fleming W, Kong JQ, Naylis C. Renal cortical NA^+K^+ATPase and abundance is decreased in normal pregnant rats. *Am J Physiol*. 1998;275:F812–F817.
106. Ni RishiR XN, Baylis C, Humphreys MH. Increased activity of cGMP-specific diesterase (PDE5) contributes to renal resistance to atrial natriuretic peptide in the pregnant rat. *JASN*. 2004;15:1254–1260.
107. Knight S, Snellen H, Humphreys M, Baylis C. Increased renal phosphodiesterase (PDE)-5 activity mediates the blunted natriuretic response to ANP in the pregnant rat. *Am J Physiol*. 2007;292:F655–F659.
108. Sasser JM, Baylis C. The natriuretic and diuretic response to dopamine is maintained during rat pregnancy. *Am J Physiol Renal Physiol*. 2008;294:F1342–F1344.
109. von Dadelszen P, Dwinnell S, Magee LA, et al. Sildenafil citrate therapy for severe early-onset intrauterine growth restriction. Research into Advanced Fetal Diagnosis and Therapy (RAFT) Group. *BJOG*. 2011;118:624–628.
110. Lindheimer MD, Katz AI, Zuspan FP, eds. *Hypertension in Pregnancy*. New York: John Wiley & Sons; 1976. 251–260.
111. Gant NF, Daley DL, Chand S, Whalley PJ, MacDonald PC. A study of angiotensin II pressor response throughout pregnancy. *J Clin Invest*. 1973;52:2682–2689.
112. Lindheimer MD, August P. Aldosterone, maternal volume status and healthy pregnancies: a cycle of differing views. *Nephrol Dial Transplant*. 2009;24:1712–1714.
113. Gennari-Moser C, Khankin EV, Escher G, et al. Vascular endothelial growth factor-A and aldosterone: relevance to normal pregnancy and preeclampsia. *Hypertension*. 2013;61:1111–1117.
114. Siddiqui AH, Irani RA, Zhang W, et al. Angiotensin receptor agonistic autoantibody-mediated soluble fms-like tyrosine kinase-1 induction contributes to impaired adrenal vasculature and decreased aldosterone production in preeclampsia. *Hypertension*. 2013;61:472–479.
115. Gennari-Moser C, Khankin EV, Schüller S, et al. Regulation of placental growth by aldosterone and cortisol. *Endocrinology*. 2011;152:263–271.
116. Fu ML, Schulze W, Wallukat G. Immunohistochemical localization of angiotensin II receptors (AT1) in the heart with anti-peptide antibodies showing a positive chronotropic effect. *Receptors Channels*. 1998;6:99–111.
117. Fu ML, Leung PS, Wallukat G. Agonist-like activity of antibodies to angiotensin II receptor subtype 1 (AT1) from

rats immunized with AT1 receptor peptide. *Blood Press.* 1999;8:317–324.

118. Jahns R, Boivin V, Hein L, et al. Direct evidence for a β1-adrenergic receptor-directed autoimmune attack as a cause of idiopathic dilated cardiomyopathy. *J Clin Invest.* 2004;113:1419–1429.
119. Griendling KK, Harrison DG. Out, damned dot: studies of the NADPH oxidase in atherosclerosis. *J Clin Invest.* 2001;108:1423–1424.
120. Takimoto E, Ishida J, Sugiyama F, Horiguchi H, Murakami K, Fukamizu A. Hypertension induced in pregnant mice by placental renin and maternal angiotensinogen. *Science.* 1996;274:995–998.
121. Geusens N, Verlohren S, Luyten C. Endovascular trophoblast invasion, spiral artery remodelling and uteroplacental haemodynamics in a transgenic rat model of pre-eclampsia. *Placenta.* 2008;29:614–623.
122. Wallace K, Richards S, Weimer A, Martin Jr JN, LaMarca BB. CD4+ T helper cells stimulated in response to placental ischemia mediate hypertension in pregnant rats. *Hypertension.* 2011;57:949–955.
123. Novotny SR, Wallace K, Heath J, et al. Activating autoantibodies to the angiotensin receptor are important in mediating hypertension in response to adoptive transfer of RUPP CD4+ T Lymphocytes. *Am J Physiol Regul Integr Comp Physiol.* 2012;302:R1197–R1201.
124. LaMarca BB, Speed J, Fournier L, et al. Hypertension in response to chronic reductions in uterine perfusion in pregnant rats: effect of TNF alpha blockade. *Hypertension.* 2008;52:1161–1167.
125. Irani RA, Zhang Y, Zhou CC, et al. Autoantibody-mediated angiotensin receptor activation contributes to preeclampsia through tumor necrosis factor-alpha signaling. *Hypertension.* 2010;55:1246–1253.
126. Zhou CC, Irani RA, Dai Y, et al. Autoantibody-mediated IL-6-dependent endothelin-1 elevation underlies pathogenesis in a mouse model of preeclampsia. *J Immunol.* 2011;186:6024–6034.
127. Santner-Nanan B, Peek MJ, Khanam R, et al. Systemic increase in the ratio between Foxp3+ and IL-17-producing CD4+ T cells in healthy pregnancy but not in preeclampsia. *J Immunol.* 2009;183:7023–7030.
128. LaMarca BB, Parrish M, Ray LF, et al. Hypertension in response to autoantibodies to the angiotensin II type I receptor (AT1-AA) in pregnant rats: role of endothelin-1. *Hypertension.* 2009;54:905–909.
129. LaMarca BB, Wallace K, Herse F, Wallukat G, Weimer A, Dechend R. Hypertension in response to placental ischemia during pregnancy: role of B lymphocytes. *Hypertension.* 2011;57:865–871.
130. Novotny S, Wallace K, Herse F, Moseley J, Darby M, Heath J, et al. CD4+ T cells play a critical role in mediating hypertension in response to placental ischemia. *J Hypertens.* 2013;2:116.
131. Stepan H, Wallukat G, Schultheiss HP, Faber R, Walther T. Is parvovirus B19 the cause for autoimmunity against the angiotensin II type receptor? *J Reprod Immunol.* 2007;73:130–134.
132. Dragun D, Muller DN, Brasen JH. Angiotensin II type 1-receptor activating antibodies in renal-allograft rejection. *New Engl J Med.* 2005;352:558–569.
133. Riemekasten G1, Philippe A, Näther M, Slowinski T, Müller DN, Heidecke H, Matucci-Cerinic M, Czirják L, Lukitsch I. Involvement of functional autoantibodies against vascular receptors in systemic sclerosis. *Ann Rheum Dis.* 2011;70:530–536. Available at: http://dx.doi.org/doi:10.1136/ard.2010.135772.
134. Fareau GG, Maldonado M, Oral E, Balasubramanyam A. Regression of acanthosis nigricans correlates with disappearance of anti-insulin receptor autoantibodies and achievement of euglycemia in type B insulin resistance syndrome. *Metabolism.* 2007;56:670–675.
135. Nikolaev VO, Boivin V, Stork S. A novel fluorescence method for the rapid detection of functional beta1-adrenergic receptor autoantibodies in heart failure. *J Am Coll Cardiol.* 2007;50:423–431.
136. Widmer M, Villar J, Benigni A, Conde-Agudelo A, Karumanchi SA, Lindheimer M. Mapping the theories of preeclampsia and the role of angiogenic factors: a systematic review. *Obstet Gynecol.* 2007;109:168–180.
137. Scornik JC, Guerra G, Schold JD, Srinivas TR, Dragun D, Meier-Kriesche HU. Value of posttransplant antibody tests in the evaluation of patients with renal graft dysfunction. *Am J Transplant.* 2007;7:1808–1814.
138. Gordon, et al. *Clin Sci Mol Med.* 1973;45:115–127.
139. Brown, et al. *Clin Exp Hypertens.* 1983:B5249–60.
140. Brown, et al. *Hypertens Preg.* 1993;12:37–51.

CHAPTER **16**

The Kidney in Normal Pregnancy and Preeclampsia

KIRK P. CONRAD, ISAAC E. STILLMAN AND MARSHALL D. LINDHEIMER

Editors' comment: *The revision of this chapter includes the welcoming of a new coauthor, Isaac Stillman, a pathologist who has recently added substantially to the literature regarding the pathological changes in the kidney during preeclampsia both in animal models and in humans.*[1,2] *It also contains new topics, gestational changes in osmoregulation, and urinary concentration and dilution. In earlier editions, this subject was relegated to the chapter on volume homeostasis, but as a true kidney function it has found its way home, chapter-wise!*

The chapter authors have also updated the section on renal biopsy in pregnancy. This procedure, already quite restricted indication-wise in pregnant women when discussed in the previous edition, appears even more so as of 2014, probably reflecting the ever-improving technology of non-invasive testing. The marked decline in performing renal biopsies in suspected preeclamptics following the large series that appeared during the last few decades of the 20th century is important to appreciate. That is, interpretive dilemma or misinformation is more likely to appear when the biopsy process focuses more and more on atypical or complicated preeclampsia. This point, already noted in our comments in the previous edition, underscores the importance of the older, larger and extensive series, as well as the signal work of Sheehan and Lynch.[3] *The latter is a unique and unrepeatable study whose first author, then at the Glasgow Royal Maternal Hospital, performed most of his autopsies within 2 hours after death (between 1935 and 1946), eliminating substantial postmortem tissue autolysis. The late Harrold Sheehan, then near his 80th year, published an article comparing the value of this autopsy material to renal biopsy descriptions in preeclamptic women, an article that bears reading.*[4]

Finally, this chapter contains the exciting progress made by the first author regarding relaxin's role in mediating the striking increases in renal hemodynamics during pregnancy, findings that have been extended to understanding the general vasodilated state of pregnancy, and possible new therapy for preeclampsia.

INTRODUCTION

Leon Chesley was among a special group of investigators who, between 1930 and 1960, pioneered the modern era of renal physiology. One of his earliest contributions was a formula for calculating urea clearance at low urine flow rates. Thus, it was only natural that Dr. Chesley's interest in normal and pathological pregnancies centered on the kidney. Indeed, his description of renal physiology and pathophysiology of pregnancy in the first edition of this book was encyclopedic in scope. Therefore, it is only fitting to Dr. Chesley's memory that we try to be as thorough as he was in the following discussion of the kidney in normal pregnancy and preeclampsia.

Familiarity with the changes in maternal renal and cardiovascular physiology as well as osmoregulation and volume homeostasis during normal pregnancy is prerequisite to complete understanding, proper diagnosis, and medical management of preeclampsia. General cardiovascular and volume alterations in normal pregnancy are reviewed elsewhere in this book (Chapters 14 and 15). Here, the alterations in renal hemodynamics, glomerular filtration, and osmoregulation during normal pregnancy will be considered first. Then, the dysregulation of renal hemodynamics and glomerular filtration occurring in preeclampsia will be addressed. Although filtration, reabsorption, and excretion of many solutes change in pregnancy, only the renal handling of uric acid and of proteins will be presented here because of their special clinical significance to preeclampsia. Last, we discuss the pathology of the kidney in preeclampsia with focus on its characteristic lesion, glomerular endotheliosis.

RENAL HEMODYNAMICS AND GLOMERULAR FILTRATION RATE DURING NORMAL PREGNANCY

Decreased vascular resistance of *nonreproductive* organs is one of the earliest adaptations to transpire in normal pregnancy, leading to a marked decrease in total systemic

Chesley's Hypertensive Disorders in Pregnancy.
ISBN: 978-0-12-407866-6

DOI: http://dx.doi.org/10.1016/B978-0-12-407866-6.00016-X

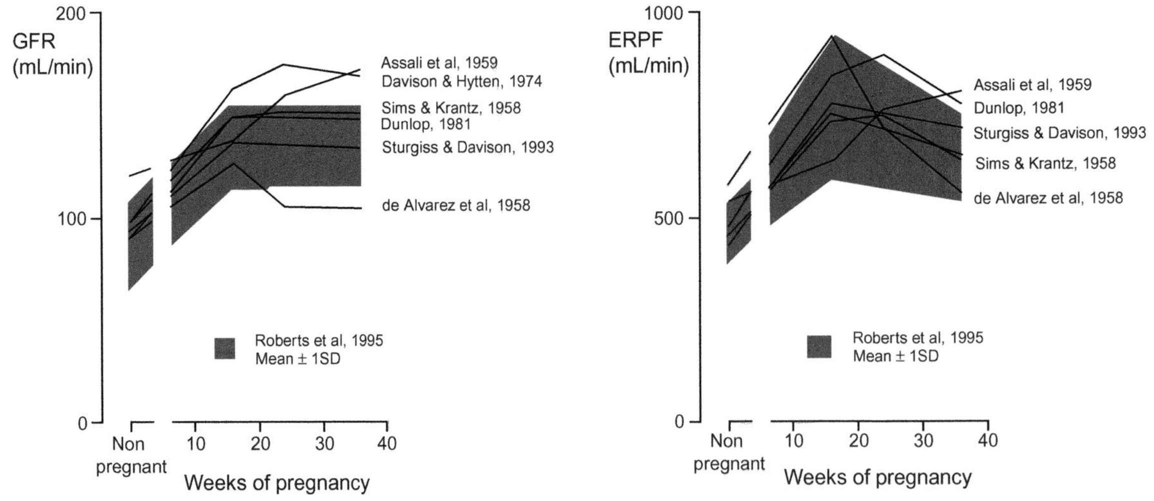

FIGURE 16.1 (A) Serial studies of glomerular filtration rate (GFR; C_{IN}) during pregnancy and in the postpartum period (nonpregnant values). (B) Serial studies of effective renal plasma flow (ERPF; C_{PAH}) or renal plasma flow[10] during pregnancy and in the postpartum period (nonpregnant values).[10,12,13]

vascular resistance. The kidneys contribute to this reduction in total systemic vascular resistance in a major way. Indeed, the nadir in renal and systemic vascular resistances and peak in renal blood flow, glomerular filtration and cardiac output coincide, and are attained by the end of the first or beginning of the second trimester. The early gestational rise in cardiac output occurs well before major increases in absolute uteroplacental blood flow, and oxygen and nutrient demands of the nascent fetus and placenta. This temporal dissociation is reflected by the oxygen content difference between arterial and mixed-venous blood, which narrows during early pregnancy in both humans and rats.[5] In summary, the reduction in vascular resistance of nonreproductive organs such as the kidney is one of the earliest and most fundamental maternal adaptations to occur in pregnancy, and insight into the hormonal signals and molecular mechanisms may be particularly critical because in preeclampsia, both renal and systemic vasodilatation are compromised.

Renal Clearances of para-Aminohippurate and Inulin

The most comprehensive investigations of renal hemodynamics and glomerular filtration rate (GFR) were published by Sims and Krantz,[6] de Alvarez,[7] Assali et al.,[8] Dunlop,[9] Roberts et al.,[10] and Chapman et al.[11] These studies stand out because of their superior experimental design and methodologies, i.e., (1) the same women were longitudinally studied during gestation and either preconception[11] or postpartum;[6–10] (2) the renal clearances of para-aminohippurate (C_{PAH}) and inulin (C_{IN}) that measure effective renal plasma flow (ERPF) and GFR, respectively, were determined by the constant infusion technique; and (3) the potential problem of urinary tract dead space, which can lead to inadequate collection of urine, thereby introducing error into the determination of C_{PAH} and C_{IN}, was avoided by instituting a water diuresis and/or by irrigating the bladder after each clearance period.* Taking precautions to avoid dead space error is especially important in pregnancy when the urinary tract is dilated and the bladder may fail to drain completely.[12,13]

To facilitate comparison of the studies, data of GFR and ERPF (or RPF ~ ERPF/0.9) from each investigation are illustrated in Fig. 16.1 with the exception of Chapman et al.[11] Because the work of Chapman et al. was particularly thorough, including the evaluation of renal function in women during the midfollicular phase before conception and then on six occasions throughout gestation, these data are presented separately in Fig. 16.2. On balance these studies reveal that both GFR and ERPF markedly increased during the first half of pregnancy. Peak levels were 40 to 65% and 50 to 85%, resepectively, above nonpregnant levels of GFR and ERPF. In general, the filtration fraction (FF; GFR/ERPF or GFR/RPF) fell during the first half of gestation. The pattern of change for GFR was comparable in all the studies except that of de Alvarez, in which GFR declined during the last half of pregnancy toward nonpregnant levels, whereas in the other investigations GFR remained elevated throughout gestation. The explanation for this discrepancy is unclear, but ERPF also fell earlier and more

* Although irrigation of the bladder with water and then air helps to improve urine collection, largely because of the increased risk of urinary tract infection, this procedure is currently considered to be inappropriate for research purposes. On the other hand, increasing urine flow rate by initiating a water diuresis is an acceptable and effective means to minimize urinary tract dead space error.

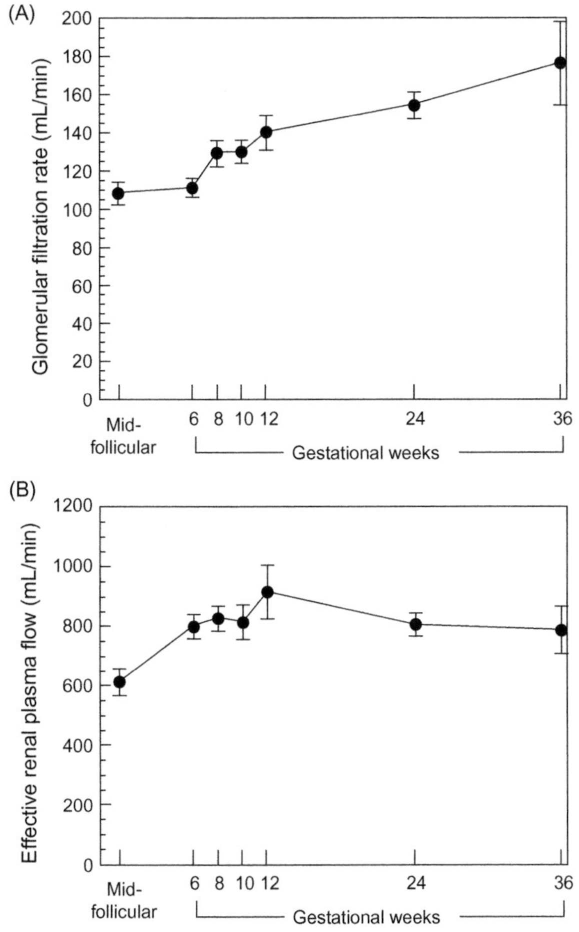

FIGURE 16.2 (A) Serial study of glomerular filtration rate before pregnancy in the midfollicular phase of the menstrual cycle and then throughout pregnancy in 10 women. (B) Serial study of effective renal plasma flow (ERPF) in the same women. All values during pregnancy are significantly different from those obtained during the midfollicular phase of the cycle before pregnancy.[11]

precipitously in the study by de Alvarez. ERPF declined modestly during the last stages of pregnancy in all studies except that of Assali et al. Thus, again with the exception of the investigation by Assali et al., FF gradually rose during the final stages of pregnancy, mostly because ERPF fell while GFR was relatively preserved. Certain body positions (e.g., supine, sitting, etc.) may compromise ERPF and GFR during the clearance experiments by mechanical effects of the enlarged gravid uterus, particularly in late gestation (see "Postural Infuences on Renal Function," below). Nevertheless, when Ezimokhai et al.[14] measured C_{IN} and C_{PAH} with subjects positioned in left lateral recumbency, a posture that helps prevent compression of major vessels by the gravid uterus, they observed that ERPF but not GFR significantly declined during the final stages of pregnancy (748 ± 20 to 677 ± 20 mL/min), suggesting the decline in ERPF is not solely an artifact of posture.

Elevated GFR during pregnancy is reflected by reciprocal changes in plasma concentration of creatinine,[6,10,15] which is decreased throughout gestation. The reason for this reciprocal relationship is that production of creatinine by skeletal muscle changes little during gestation and since glomerular filtration increases, plasma levels must fall. Although renal handling of urea is more complicated, it is freely filtered like creatinine; consequently, plasma levels are also lower during pregnancy, again primarily because renal clearance of urea is increased.[6,16] Average concentrations for plasma creatinine and urea nitrogen during gestation are 0.5 and 9.0 mg/dL, respectively, compared to nonpregnant values of 0.8 and 13.0 mg/ dL.[6,10,15,16] Of interest, GFR was noted to increase during pregnancy in renal allograft recipients and in women with a single kidney (albeit to a lesser degree), showing a pattern of change similar to that observed in normal pregnant women.[17,18] Thus, despite compensatory functional and anatomic hypertrophy, the renal allograft and single kidney adapt even further during pregnancy undergoing gestational hyperfiltration. Because the renal allograft is a denervated kidney, renal nerves are unlikely to be involved in the gestational increases in GFR. In the same vein, the kidneys of both pregnant women and rats demonstrated further elevation in GFR and ERPF in response to intravenous infusion of amino acids, and the percentage elevation was comparable to that observed in the nonpregnant condition.[19,20]

Creatinine Clearance

The 24-hour renal clearance of endogenous creatinine (C_{CR}) is not uncommonly used as a measure of GFR. However, this is due to a fortuitous chain of events. Creatinine undergoes glomerular filtration, and a small degree of proximal tubular secretion, but plasma levels are overestimated because of the presence of a chromagen, which is unavoidably detected in the assay along with creatinine. When GFR is normal, these two events cancel. However, as GFR falls, tubular secretion represents a greater proportion of urinary creatinine, and the influence of the chromagen on plasma levels diminishes. Under such circumstances, creatinine clearance overestimates GFR sometimes by as much as 25 to 50%.[21]

Measuring the 24-hour C_{CR}, Davison and Noble provided evidence that GFR rises 25% by the second week post-conception (or 4 weeks after the last menstrual period (Fig. 16.3).[22] This study is one of the first to show that the physiologic adaptations in the renal circulation during human pregnancy are among the earliest to occur. A report by Chapman et al. both supports and extends these findings, insofar as both C_{IN} (and C_{CR}) were significantly increased by 4 weeks post-conception (or 6 weeks after the last menstrual period), the earliest time-point investigated (Fig. 16.2).[11]

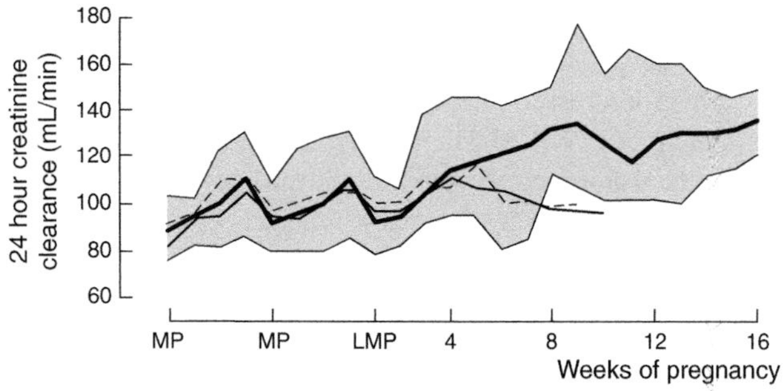

FIGURE 16.3 Alterations in the 24-hour renal creatinine clearance evaluated weekly before conception and throughout early pregnancy in 11 women. Solid line, mean; stippled area, range for nine women with normal pregnancy outcomes. Two women designated by the thin and dashed lines had uncomplicated spontaneous abortions.[22]

Several researchers measured the renal clearance of inulin throughout pregnancy and in the postpartum period using the constant infusion technique, and compared these values to the 24-hour endogenous C_{CR} in the same women.[6,7,23] The changes in GFR during pregnancy as measured by C_{IN} and the 24-hour C_{CR} were comparable except possibly in the last few weeks before delivery. Although 24-hour C_{CR} declined at 35 to 38 weeks of gestation in the study by Davison and Hytten, the short-term C_{CR}, as assessed by constant infusion of creatinine in the same study, did not decrease and was similar to C_{IN}.[23] This last finding suggested that the renal handling of creatinine did not change at this stage of pregnancy, and substantiated the C_{CR} as a valid measure of GFR in this setting.

In what appears to be the only comprehensive study of the last few weeks of pregnancy immediately before delivery that we could find, Davison et al. performed weekly, 24-hour C_{CR} measurements in 10 subjects, demonstrating that 24-hour C_{CR} decreased and plasma creatinine increased over this period to levels not significantly different from nonpregnant values.[24] Because creatinine is not only filtered but also secreted as discussed above, the authors chose not to conclude that this decline reflected a fall in GFR.[24] On the other hand, taken together with the study by Davison and Hytten,[23] (*vide supra*) it is likely that the fall in 24-hour C_{CR} which occurred prior to delivery did reflect a true decline in GFR. Because the 24-hour C_{CR} is performed while the subject goes about her normal daily activities, it may actually be a more realistic and physiologic measure of GFR. Extended periods of standing during the day or lying supine at night during late pregnancy may compromise renal perfusion and GFR, which is then reflected by a reduced 24-hour C_{CR} (also see next section). Such potentially protracted periods of reduced GFR would be missed by short-term measurements of C_{IN} or C_{CR} performed under the artificial conditions of a laboratory setting in the left lateral decubitus or sitting positions.

Postural Influences on Renal Function

Pritchard et al.[25] as well as Sims and Krantz[6] observed no compromise of C_{PAH} and C_{IN} when gravid subjects turned from the lateral recumbent to the supine position. In contrast, Chesley and Sloan studied 10 women between 34 and 43 weeks of gestation showing decreases of $19 \pm 3\%$ and $21 \pm 5\%$ in C_{PAH} and C_{IN}, respectively, upon assuming the supine position.[26] These decreases were accompanied by reciprocal increases in plasma para-aminohippurate and inulin concentrations, indicating a true reduction in renal function rather than an artifact of inadequate urine collection. The authors concluded that in late gestation when a subject assumes a supine position, the enlarged uterus compresses the great veins, which in turn impairs venous return, decreasing both cardiac output and renal perfusion. Similar findings and conclusions were made by Pippig.[27]

Dunlop studied 18 healthy women at approximately 36 weeks of gestation and again 8 weeks postpartum, measuring C_{PAH} and C_{IN} in three positions: supine, sitting, and lateral recumbency.[28] Although the subjects demonstrated the expected gestational elevations in ERPF and GFR, there was no significant influence of posture. In particular, these variables were not reduced by the supine position. Lindheimer and Weston, in a study designed to determine mechanisms of renal salt handling during pregnancy, noted decrements in GFR in 11 of 13 volume-expanded third-trimester women when they changed from a lateral recumbent to a supine position.[29] Assali et al. showed that ERPF and GFR declined markedly in response to quiet standing especially in the third trimester.[8] This decline in renal function persisted even after postural hypotension had subsided. *In summary*, change of position from lateral recumbency to supine or even standing has been frequently reported to compromise renal hemodynamics and GFR during late pregnancy.

Mechanisms for Alterations of Renal Hemodynamics and GFR

Our understanding of the mechanism(s) responsible for the increase of ERPF, and consequently of GFR, in pregnancy is improving. Ultimately, the fall in renal vascular resistance underlies the phenomena. An attractive and plausible theory is that the altered hormonal environment is causal. Unfortunately, so many hormones change during pregnancy that it has been difficult to know which ones to investigate first. Because of obvious ethical considerations as well as feasibility issues, many of the studies exploring the potential mechanisms of gestational changes in renal hemodynamics and GFR have employed animal models.

Renal Hyperfiltration During Pregnancy

The Munich-Wistar rat has been extensively investigated by renal physiologists because this strain has glomeruli belonging to superficial cortical nephrons at the kidney surface, which are accessible by micropuncture.[30] Thus, much of our current understanding of glomerular hemodynamics stems from studies of this rat strain. The single nephron GFR (SNGFR) is determined by Starling forces, both hydrostatic and oncotic pressures within the glomerular capillary and Bowman's space, as well as the ultrafiltration coefficient, K_f, which is the product of the glomerular capillary hydraulic permeability and surface area. Applying the renal micropuncture technique to Munich-Wistar rats during midgestation when whole kidney RPF and GFR are increased, Baylis showed that the gestational rise in SNGFR can be attributed to an increase in glomerular plasma flow and the transglomerular hydrostatic pressure difference remains unchanged.[31] Thus, the higher glomerular plasma flow effectively decreases the rate of rise of oncotic pressure along the glomerular capillary, leading to increased net pressure of ultrafiltration and SNGFR. In essence, comparable reductions in afferent and efferent arteriolar resistances account for both the unchanged glomerular hydrostatic pressure and increase in glomerular plasma flow and SNGFR during rat gestation. In this study, plasma oncotic pressure was not significantly different between nonpregnant and pregnant rats, and, because the animals were in filtration equilibrium, only a minimum value for K_f could be derived; nevertheless, these determinants of glomerular ultrafiltration most likely contributed little to the gestational rise in SNGFR in the gravid rat model. To summarize, SNGFR rises because glomerular plasma flow increases during pregnancy.[31]

Whether similar mechanisms occur in human gestation is difficult to determine, because glomerular dynamics cannot be directly evaluated as in the Munich-Wistar rat. Recent mathematical modeling[10,32] based on renal clearances and other measurements performed in pregnant women suggested that renal hyperfiltration of human pregnancy is mainly due to a rise in RPF. Although decrements in plasma oncotic pressure particularly during late pregnancy and increases in K_f especially during early pregnancy may contribute, there was no evidence for alterations in the transglomerular hydrostatic pressure. However, a note of caution is necessary here. The mathematical formulae used can skew the results when very small measurement errors occur. Such errors are most likely to occur when values entered into the equation are obtained indirectly, the case here. Finally, there is one group of investigators who cite changes of oncotic pressure as the sole cause of increased GFR.[33]

Conrad adapted the Gellai and Valtin method for chronic instrumentation of rats to the investigation of renal function in pregnancy.[34,35] Because renal, cardiovascular, and endocrine parameters are markedly perturbed by anesthesia and acute surgical stress, physiologic studies in chronically instrumented, conscious animals are critical for interrogation of underlying mechanisms. Thus, the same chronically instrumented, conscious rats were serially examined before, during, and after gestation.[35] Comparable to human pregnancy, the conscious rat demonstrates both renal vasodilatation and hyperfiltration throughout most of gestation. Thus, the gravid rat has been extensively investigated to determine the mechanisms underlying these remarkable changes in the renal circulation during pregnancy.

Plasma Volume Expansion

Pregnancy is accompanied by tremendous expansion of extracellular and plasma volume (see Chapter 15). *Acute* expansion of plasma volume by 10 to 15% failed to increase the GFR, SNGFR, or glomerular plasma flow in virgin female Munich-Wistar rats.[36] Volume expansion was previously shown to suppress tubuloglomerular feedback activity, which could conceivably permit the gestational increases in both glomerular plasma flow and SNGFR.[37] However, tubuloglomerular feedback activity was not suppressed in gravid Munich-Wistar rats, rather the mechanism was reset to the higher level of SNGFR manifested by the pregnant animals.[37] The authors concluded that the volume expansion of pregnancy may be perceived by the gravid rat as "normal." This contention logically follows from the concept that reductions in total peripheral vascular resistance (the "arteriolar underfilling" stimulus theory of normal pregnancy) and vascular refilling are tightly linked and temporally inseparable, although a dissociation has been discerned by some investigators. Thus in one study of humans and a second in baboons, increases in plasma volume and left atrial or left ventricular end diastolic dimensions indicative of plasma volume expansion lagged behind the decline in systemic vascular resistance.[38,39]

Whether *chronic* volume expansion comparable to that of pregnancy underlies the gestational changes in the renal circulation is difficult to test. Most instances of chronic volume expansion occurring in nature, other than pregnancy of course, result from pathology such as congestive heart failure or cirrhosis in which renal function is often reduced rather than elevated. However, in the rare cases of primary mineralocorticoid excess, which is associated with volume expansion, GFR rises but not by the same degree as observed in pregnancy.[40] Interestingly, prolonged administration of either arginine vasopressin or oxytocin to chronically instrumented rats allowed water *ad libitum* results in expansion of total body water, reduction in plasma osmolality, as well as increases in both ERPF and GFR comparable in magnitude to those observed in gestation.[41] Thus, chronic volume expansion may initiate, abet or maintain elevated ERPF and/or GFR during pregnancy.

Pseudopregnancy

Study of the renal circulation in rats that become pseudopregnant may provide insights into the mechanisms contributing to renal vasodilatation and hyperfiltration of pregnancy. By mating a female rat with a vasectomized male, pseudopregnancy – a condition that physiologically mimics the first half of gestation in rats, but lacks fetoplacental development – is produced. Pseudopregnancy mimics the increases in ERPF and GFR that are observed during early pregnancy in rats.[42,43] Thus, maternal factors alone may be sufficient to initiate the changes in the renal circulation during pregnancy.

Menstrual Cycle

Davison and Noble[22] demonstrated that the 24-hour endogenous C_{CR} increased by 20% in the luteal phase of the menstrual cycle (Fig. 16.3). This finding was corroborated by other investigators using C_{CR},[44,45] chromium 51-EDTA clearance,[44,46] or C_{IN}.[47,48] Moreover, ERPF measured either by C_{PAH} or renal clearance of iodine 125-hippuran was also reported to be increased during the luteal phase in two studies,[46,48] but not significantly so in another.[47] Therefore, the gestational increases in ERPF and GFR are observed, albeit to a lesser degree, in the luteal phase. This finding may shed light on the mechanisms underlying renal vasodilatation and hyperfiltration of pregnancy, because several hormones that increase in the luteal phase also rise during early gestation, e.g., the corpus luteal hormones, progesterone and relaxin.

Hormonal Regulation: Sex Steroids

Based on studies involving both acute and chronic administration of *estrogen* to humans and laboratory animals, this hormone appears to have little or no influence on ERPF or GFR, although it clearly increases blood flow to other nonreproductive and reproductive organs.[49–53] On the other hand, *progesterone* is a potential candidate. Chesley and Tepper administered 300 mg/d intramuscular (IM) progesterone to 10 nonpregnant women for 3.5 days, and found that the hormone produced a 15% increase in C_{IN} and C_{PAH}.[50] They speculated that more prolonged administration might elicit the magnitude of increase in GFR and ERPF observed in normal pregnancy. Similar findings were reported by Atallah et al.[54] In a 4-hour period following the IM administration of 200 mg of progesterone to nine nonpregnant women, plasma concentration rose on average from 7 to 30 ng/mL, and endogenous C_{CR} increased from 103 to 118 mL/min – a significant rise of 15%. By extrapolation, the authors suggested that the circulating levels of progesterone observed in pregnancy, which can be considerably higher than that attained in their study, might fully account for the 40 to 65% gestational increase of GFR. It should be noted, however, that these higher levels of circulating progesterone are not reached until well after the gestational peak in ERPF and GFR. Nevertheless, progesterone could help maintain elevated renal function later in pregnancy. In another report, 3 hours after IM administration of 310 μmol progesterone to male subjects, ERPF rose significantly by 15%, irrespective of the sodium content in the diet, although GFR was unaffected.[55] In the same report, IM administration of 155 μmol progesterone twice daily for 3 days produced comparable changes in ERPF, but, again, no change in the GFR. Finally, subcutaneous injection of 2 mg/kg/d progesterone for 3 days to intact female rats produced a 26% increase in GFR (ERPF was not measured in this study).[56] Based on the results from these reports, progesterone or its metabolites[57] may contribute in a small way to the early rise in ERPF and GFR of early pregnancy. They may also participate in maintaining elevated renal function in the second half of gestation when circulating levels of the steroid are considerably higher.

Hormonal Regulation: Peptide Hormones

It follows from the studies in pseudopregnant rats and women in the luteal phase (*vide supra*) that peptide hormones of maternal origin may participate in the early gestational increases of ERPF and GRF. *Prolactin* surges in both pseudopregnant and pregnant rats coincident with the gestational increases in ERPF and GFR. Unfortunately, whether prolactin can raise renal hemodynamics and GFR remains controversial[58] and, as such, requires further investigation. *Placental lactogen*, which activates the same receptor as prolactin, is another candidate peptide hormone at least in maintaining elevated ERPF and GFR later in pregnancy when the hormone circulates. However, based on the one relevant study that we could locate, if anything, human placental lactogen *decreases* RBF at least when administered acutely in the renal artery of anesthetized pigs and human pregnancies accompanied by placental lactogen gene deletion (Rygaard et al., Hum Genet, 1998) have been reported.[59]

Relaxin may contribute to the vasodilatory changes in renal and possibly other circulations during pregnancy. Circulating relaxin originates from the corpus luteum in both rats and humans.[60] In the latter, human chorionic gonadotrophin (hCG) is a major stimulus for relaxin secretion.[60] There were compelling, albeit circumstantial, reasons to consider relaxin as a potential mediator of renal vasodilatation and hyperfiltration during pregnancy.

First, plasma relaxin rapidly rises after conception in women,[60] corresponding with the large first-trimester increments in GFR and ERPF.[22,23] Second, relaxin also increases during the luteal phase[60–63] associated with a transient 10–20% increase in GFR and ERPF.[22,44–48] Third, the early gestational rise in plasma relaxin coincides with another early physiologic adaptation in human pregnancy; namely, changes in osmoregulation[64] (see below). Indeed, osmoregulatory changes were mimicked by administering hCG to women in the luteal phase and intact female rats, but not to men or ovariectomized rats, suggesting the intermediary role of an ovarian hormone.[65–67] In other studies, infusion of rat relaxin-neutralizing antibodies prevented the reduction in plasma osmolality at midterm pregnancy in the rat.[81] Furthermore, administration of synthetic human relaxin to ovariectomized rats for 7 days produced a significant decrease in plasma osmolality without a change in plasma arginine vasopressin similar to the osmoregulatory changes observed in normal pregnancy.[69] Similar findings were observed using recombinant human or porcine relaxin.[70] Finally, chronic administration of relaxin reduced blood pressure and vasoconstrictor responses in the mesenteric circulation of awake spontaneously hypertensive rats[71,72] while acute treatment increased coronary blood flow and reduced platelet aggregation via nitric oxide and guanosine 3′,5′-cyclic monophosphate.[73,74]

Although renal vasodilatation and hyperfiltration occur in rodent pregnancy before gestational day 8, when serum relaxin is undetectable, there is a marked jump in renal function between gestational days 8 and 12, at which time ovarian and plasma relaxin levels surge.[35,58,60] The increases in GFR and RPF that occur during rat gestation before gestational day 8 or during pseudopregnancy when circulating relaxin is undetectable are apparently mediated by other, as yet, undiscovered mechanisms.

Long-term administration of purified porcine or recombinant human relaxin (rhRLX) to chronically instrumented, conscious nonpregnant rats over a 2–5-day period increased both ERPF and GFR (and decreased plasma osmolality) to levels observed during midgestation when renal function peaks in this species,[35,70] and was also observed in ovariectomized female[70] and male rats.[75] Long-term relaxin administration also blunted the renal vasoconstrictor response to angiotensin II infusion,[70] thereby mimicking the diminished effect of the latter peptide during rat gestation.[76–78] Furthermore, myogenic constriction of small renal arteries isolated from relaxin-treated, nonpregnant rats was inhibited[68] and comparable to that observed in arteries harvested from midterm pregnant animals.[79] *Short-term* (1–4 hours) administration of rhRLX to chronically instrumented conscious rats also produced renal vasodilatation and hyperfiltration.[80] Administering relaxin-neutralizing antibodies or ovariectomy while maintaining pregnancy with exogenous sex steroids completely prevented gestational hyperfiltration, renal vasodilatation and inhibited myogenic constriction, and the osmoregulatory adaptations as well.[81] Thus, relaxin appears to be essential for the renal circulatory and osmoregulatory changes in midterm pregnant rats.

In normal human volunteers, *short-term* (~6 hours) intravenous infusion of rhRLX increased RPF by 60%, but, surprisingly, not GFR[82]. (See Discussion in ref. 82 for possible explanations of this apparent discrepancy.) Renal vasodilatation occurred both in men and in women, and as soon as 30 minutes after starting the infusion without any significant changes in blood pressure, heart rate or serum osmolality. After 26 weeks of rhRLX administration to patients with mild scleroderma, predicted creatinine clearance (calculated by the Cockcroft-Gault equation: GFR = (140 − age) × (Wt in kg) × (0.85 if female) / (72 × serum Cr) rose by 15–20%, and serum osmolality and blood pressure declined slightly but significantly throughout the study in a dose-dependent fashion.[83,84] Finally, in infertile women who conceived through donor eggs, IVF and embryo transfer, the gestational increase in GFR and decrease in serum osmolality were significantly attenuated.[85] Because these women lacked ovarian function and a corpus luteum, serum relaxin was undetectable. Thus, similar to gravid rats, circulating relaxin appears to have a role in establishing the renal and osmoregulatory responses to pregnancy in women. However, unlike gravid rats, partial responses may persist despite the absence of circulating relaxin.

Long-term administration of rhRLX for up to 10 days in either normotensive or hypertensive male and female rats mimicked the changes in systemic hemodynamic and global arterial compliance observed in pregnancy, and increased the passive compliance of isolated small renal arteries.[86–88] In contrast, *short-term* rhRLX administration over hours produced systemic vasodilatation only in the angiotensin II model of hypertension, and not in spontaneously hypertensive or normotensive rats.[87,88] Thus, by inference, relaxin apparently acted more rapidly in the renal circulation of normotensive rats (*vide supra*) compared to other organ circulations.[80,87,88] A critical role for relaxin in the alterations of systemic hemodynamics and global arterial compliance in midterm rats was also identified.[89] Specifically, relaxin-neutralizing antibodies prevented the gestational increases in cardiac output and global arterial compliance, as well as the reduction in systemic vascular resistance. Whether neutralization of circulating relaxin has similar inhibitory effects during late gestation is unknown. It is possible that hormones emanate from the placenta when it becomes sufficiently developed to conspire with relaxin in maintaining systemic vasodilatation and the osmoregulatory changes during late pregnancy.[90] (See Chapters 14 and 15 for further details.)

Cellular and Molecular Mechanisms of Renal Vasodilatation

Endothelium-derived relaxing factors have been postulated to mediate the gestational increases of ERPF and GFR including vasodilatory *prostaglandins* (PG) and *nitric oxide* (NO). A potential role for PG has been investigated in gravid animal models and in humans. The gestational increases in ERPF and/or GFR were unaffected by administration of PG synthesis inhibitors to chronically instrumented, conscious gravid rats and rabbits.[76,91,92] Further, vasodilatory PG synthesis *in vitro* by renal tissues from pregnant and nonpregnant animals was similar.[93,94] Intravenous infusion of prostacyclin to male volunteers failed to alter ERPF or GFR,[95] though parenteral administration may not mimic the actions of locally produced (autacoid functioning) prostanoids. Finally, the cyclooxygenase inhibitor indomethacin increased total peripheral vascular resistance by only 5% in pregnant women without significantly affecting either mean arterial pressure or cardiac output – a trivial increase compared to the marked decrease in total peripheral resistance characteristic of pregnancy.[96] Similarly, another cyclooxygenase inhibitor, meclofenamate, did not significantly augment peripheral vascular resistance in conscious, gravid guinea pigs.[97] These data support the conclusion that vasodilatory PG plays a minimal or no role in the rise of ERPF, GFR and cardiac output, as well as the reduction in renal and total peripheral vascular resistances during pregnancy.

Guanosine 3', 5-cyclic monophosphate (cGMP), a second messenger of nitric oxide, may contribute to the renal vasodilatation and hyperfiltration of pregnancy.[58,98] Because extracellular levels of cGMP generally reflect intracellular production, the plasma level, urinary excretion, and "metabolic production" of cGMP were investigated in conscious rats. Increases in all of these variables were observed throughout pregnancy and pseudopregnancy.[58,98,99] Similar elevations in urinary excretion and plasma concentration were described for human gestation.[100–102]

The 24-h urinary excretion of nitrite and nitrate (NOx), the stable metabolites of nitric oxide, also increased during pregnancy and pseudopregnancy in rats consuming a low-NOx diet, and they paralleled the rise in urinary cGMP excretion.[99] This gestational rise in urinary NOx was prevented by chronic administration of the NO synthase inhibitor nitro-L-arginine methyl ester (L-NAME), implicating nitric oxide as the source. Plasma NOx concentration was also increased during pregnancy, and NO-hemoglobin was detected in red blood cells from pregnant, but not from nonpregnant rats by electron paramagnetic resonance spectroscopy.[99] These data suggested that endogenous NO production is increased in gravid rats, and although the tissue source(s) was not identified, the authors suggested that the vasculature might contribute. Plasma level and urinary excretion of NOx were also reported to be increased in gravid ewes.[103] The status of NO biosynthesis during normal pregnancy in women (and in women with preeclampsia) is unclear.[100,104]

The renal circulation participates in the overall maternal vasodilatory response to pregnancy. In chronically instrumented conscious, midterm pregnant and virgin rats, the former having reached the peak of their gestational renal vasodilatory increase, *acutely* administered L-arginine analogs that inhibit NO synthase led to a convergence of renal hemodynamics and vascular resistance in the two animal groups.[77,104] That is, when compared to the virgin control animals, midterm pregnant rats responded more robustly to acute NO synthase inhibition, manifesting a larger decrease in GFR and ERPF, and a greater rise in effective renal vascular resistance. Consistent with these *in vivo* data was the myogenic constriction of small renal arteries isolated from midterm pregnant rats that was attenuated compared to virgin control animals, and inhibitors of NO synthase added to the bath or endothelial removal restored this attenuated myogenic constriction to robust virgin levels.[79]

A critical role for NO of endothelial origin in the renal circulation was also established for relaxin-treated nonpregnant rats.[68,70] Thus, NO plays an essential role in the renal circulatory changes in both midterm pregnant and in relaxin-treated nonpregnant rats. Although pregnancy and relaxin administration to nonpregnant rats both stimulate renal vasodilatation and hyperfiltration and inhibit myogenic constriction of small renal arteries, which depend upon NO (*vide supra*), the urinary excretion of cGMP and NO metabolites is only augmented during pregnancy.[70,98,99] Ironically, therefore, the increased production of cGMP and NO metabolites initially reported for rat gestation, and which stimulated further interrogation of this vasodilatory pathway in pregnancy, may not be of vascular origin or of hemodynamic consequence! Finally, the mechanism for increased NO in the renal circulation of midterm pregnant rats or of relaxin-infused nonpregnant rats is not a consequence of increased expression of endothelial nitric oxide synthase.[105–107]

Of additional interest in one study,[108] but not in another,[109] vasodilatory prostaglandins served a compensatory role by maintaining relative renal hyperfiltration and vasodilatation in gravid rats compared to virgin controls during *chronic* blockade of NO synthase with L-arginine analogs. That is, in the setting of chronic NO synthase inhibition, renal function converged in the virgin and pregnant rats, but only after acutely inhibiting prostaglandin synthesis with meclofenamate.[108] Prostaglandin blockade alone, however, did not affect renal function in conscious virgin or pregnant rats.[76,91,108]

Paradoxically, *endothelin* (ET) appears to play a critical role in the renal vasodilatation and hyperfiltration of midterm pregnant and relaxin-treated nonpregnant rats. ET, known primarily as a potent vasoconstrictor by interacting with vascular smooth muscle ET_A and ET_B receptor subtypes,[110] also increases cytosolic calcium in endothelial cells, thereby stimulating prostacyclin, NO and possibly other relaxing factors via an endothelial ET_B receptor subtype.[111–114] Blockade of the ET_B receptor subtype in chronically instrumented conscious male rats with the pharmacologic antagonist RES-701-1 elicited marked renal vasoconstriction.[115] This unexpected finding is consistent with a major contribution of endogenous ET towards maintaining the signature low vascular tone of the kidney by an RES-701-1-sensitive, endothelial ET_B receptor subtype through tonic stimulation of endothelial-derived relaxing factors and/or restraint of ET production.[111–116] RES-701-1 appeared to be relatively selective for the "vasodilator" ET_B receptor subtype on the endothelium. However, under pharmacological or pathophysiological conditions, ET-mediated vasoconstriction may predominate.[117]

Based on these findings in male rats, a logical question was whether the endothelial ET_B receptor-NO vasodilatory mechanism might be accentuated by pregnancy, thereby mediating the gestational changes in the renal circulation. [118] *Short-term* infusion of the endothelial ET_B receptor antagonist RES-701-1 to conscious virgin and gravid

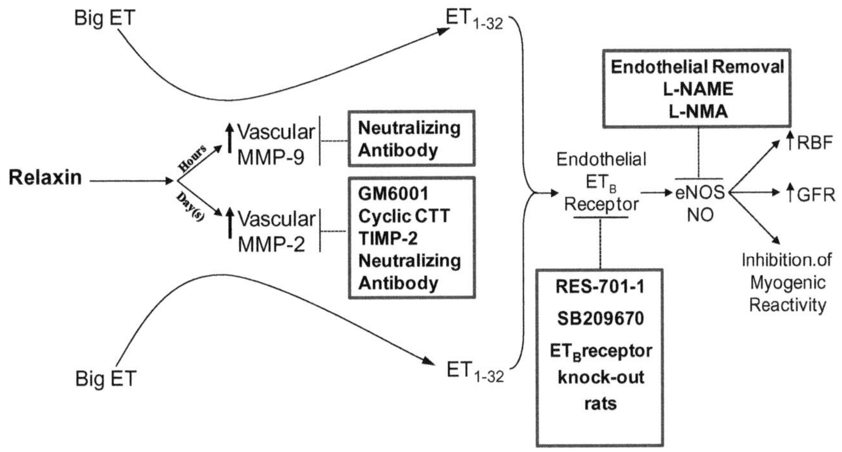

FIGURE 16.4 Working model for the sustained vasodilatory responses of relaxin. Inhibitors of relaxin vasodilatation are shown in the boxes. ET, endothelin; MMP, matrix metalloproteinase; RBF, renal blood flow; GFR, glomerular filtration rate; GM6001, a general MMP inhibitor; cyclic CTT, a specific peptide inhibitor of MMP-2; TIMP-2, tissue inhibitor of matrix metalloproteinase; RES-701-1, a specific ET_B receptor antagonist; SB209670, a mixed ET_A and ET_B receptor antagonist; L-NAME, nitro-L-arginine methyl ester; L-NMMA, N^G-monomethyl-L-arginine. Note that phosphoramidon (an inhibitor of the classic endothelin converting enzyme), STT (control peptide for cyclic CTT); heat inactivated TIMP-2, BQ-123 (a specific ET_A receptor antagonist), D-NAME and IgGs (control antibodies for MMP neutralizing antibodies) did not affect the slow vasodilatory responses of relaxin. Not depicted in this schema are the essential roles of the Lgr7 (RXFP1) receptor, and vascular endothelial and placental growth factors in mediating the vasodilatory actions of relaxin. See text for further details.

rats completely abolished gestational renal vasodilatation and hyperfiltration, producing convergence of GFR, ERPF, and effective renal vascular resistance in the two groups of animals.[118] The inhibited myogenic constriction of small renal arteries from midterm pregnant rats was also reversed by adding RES-701-1 or a mixed ET_B + ET_A antagonist (SB209670), but not a specific ET_A receptor antagonist (BQ123) to the bath.[79] These findings paralleled those observed using inhibitors of NO synthase (*vide supra*). Indeed, evidence was subsequently obtained supporting the role of the NO/cGMP pathway in mediating the vasodilatory action of endogenous ET in the renal circulation during rat pregnancy.[79,118] A critical role for the endothelial ET_B receptor subtype in mediating inhibited myogenic constriction of small renal arteries during rat pregnancy was also revealed by studies in ET_B receptor-deficient rats,[119] and a similar role for this receptor was also established for relaxin-treated nonpregnant rats.[68,75] Whether expression of the endothelial ET_B receptor increases, thereby representing a primary alteration that mediates renal vasodilatation, hyperfiltration and attenuated myogenic constriction during pregnancy or after chronic administration of rhRLX to nonpregnant rats, seems unlikely,[120] but is disputed.[121]

Jeyabalan and colleagues[122] proposed that relaxin enhances vascular gelatinase activity (*vascular matrix metalloproteinase-2* (MMP-2)) during pregnancy, thereby mediating renal vasodilatory changes through activation of the endothelial ET_B receptor-NO pathway (Fig. 16.4). This hypothesis was based on the confluence of several observations: (i) the essential role of relaxin, the endothelial ET_B receptor and NO in pregnancy-mediated renal vasodilatation as described above, (ii) the stimulation of MMP expression by relaxin at least in fibroblasts,[123,124] and (iii) the ability of vascular MMPs such as MMP-2 to hydrolyze big ET at a Gly-Leu bond to ET_{1-32} with subsequent activation of endothelin receptors.[125,126]

The ideal way to test the physiological role of MMP-2 is by blocking MMP-2 production or inhibiting its action. To this end, inhibition of gelatinase activity using a specific inhibitor of MMP-2/-9, a general inhibitor of MMPs, as well as TIMP-2 and specific neutralizing antibodies abrogated relaxin-mediated renal vasodilatation and hyperfiltration, and/or attenuated myogenic constriction of small renal arteries in relaxin-infused nonpregnant and/or pregnant rats.[122] In contrast, there was no effect of the traditional endothelin converting enzyme blocker, phosphoramidon, which inhibits the hydrolysis of big ET to ET_{1-21}. In small renal arteries harvested from relaxin-treated nonpregnant or midterm pregnant rats, vascular MMP-2 activity is increased by approximately 50%.[122,127] Thus, vascular gelatinase activity is not only a player in the endothelial ET_B receptor-NO vasodilatory pathway, but it is a major locus of regulation by relaxin and pregnancy, because neither endothelial NO synthase nor ET_B receptor abundance is increased by relaxin or pregnancy,[105–107,120] although not all agree.[121,128,129]

It is highly unlikely that vascular MMP-2 and endothelial ET_B receptor-NO are constituents of separate vasodilatory pathways working in parallel. If that was the case, then after inhibition of vascular MMP-2 or the endothelial ET_B–NO pathway, compensation of one for the other might be expected. However, not even partial compensation was observed. Each and every inhibitor of the ET_B receptor, nitric oxide synthase or MMP completely abolished the renal circulatory changes during pregnancy or in relaxin-treated nonpregnant rats (citations in[122]). Nevertheless, experimental confirmation of the link between vascular gelatinase activity and the endothelial ET_B receptor-NO vasodilatory pathway was sought. Small renal arteries isolated from relaxin-treated, ET_B receptor-deficient rats showed upregulation of vascular MMP-2 activity, but they failed to demonstrate the typical inhibition of myogenic constriction. This dissociation of the biochemical and functional consequences of relaxin in ET_B receptor-deficient rats, when taken in the context of the other results (*vide supra*), strongly suggests that vascular gelatinase is in series with, and upstream of, the endothelial ET_B receptor-NO signaling pathway.[122]

The mechanism for the increase in vascular MMP-2 induced by rhRLX or pregnancy is not completely understood. Both pro and active MMP-2 activities were increased to a similar extent on gelatin zymography, MMP-2 protein and mRNA were also elevated, and there was no apparent change in TIMP-1 or -2.[122,127] MMP-2 protein localized to both endothelium and vascular smooth muscle by immunohistochemistry, but further study is required to ascertain in which of these compartment(s) it increases in response to pregnancy or chronic rhRLX administration. (Another caveat is that immunohistochemical localization may be misleading because MMP-2 is a secreted protein.) More recently, arterial-derived vascular endothelial and placental growth factors were both identified to be essential players in the relaxin vasodilatory pathway as described above, although how they are involved remains unclear at the present time.[130] Finally, preliminary evidence implicates the major relaxin receptor, RXFP1, in mediating the inhibition of myogenic constriction by relaxin.[131]

An emerging concept is that the vasodilatory mechanisms of relaxin change according to the duration of exposure to the hormone. The mechanisms involved after *long-term* administration of hormone to nonpregnant rats or during pregnancy when endogenous relaxin circulates were detailed above. Jeyabalan and colleagues demonstrated that *matrix metalloproteinase-9* (MMP-9) rather than MMP-2 activity is increased in small renal and mesenteric arteries isolated from rats after more *short-term* administration of rhRLX for 4–6 hours[132] (Fig. 16.4). Small renal arteries manifested loss of myogenic constriction, but robust myogenic constriction was restored by incubation with a specific MMP-9, rather than specific MMP-2 antibody as observed following chronic relaxin administration (*vide supra*). Like MMP-2, MMP-9 can hydrolyze big ET at a Gly-Leu bond; the inhibition of myogenic constriction after short-term rhRLX administration was also mediated by the endothelial ET_B receptor-NO vasodilatory pathway. Finally, relaxin can *rapidly* relax within minutes some, but not all, pre-constricted arteries from humans and rats.[133,134] This rapid response ultimately involves NO stimulation via endothelial $G\alpha_{i/o}$ protein coupling to PI3 kinase, Akt, and eNOS.[134]

To summarize, relaxin accentuates the endothelial ET_B receptor-NO renal vasodilatory pathway during pregnancy by increasing arterial MMP-2 mRNA, protein and activity.[135] Interestingly, higher doses of the specific gelatinase or general MMP inhibitors administered *in vivo* to nonpregnant control rats also decreased GFR and ERPF and raised blood pressure, albeit to a lesser extent than in relaxin-treated rats, whereas phosphoramidon was without effect.[122] These results suggest that arterial gelatinase activity rather than the traditional endothelin converting enzyme might be the main mechanism for big ET processing, at least in the renal circulation of rats, leading to low vascular tone relative to other organ circulations, thus enabling 20–25% of the cardiac output to be distributed to the kidneys. Possibly, the colocalization of MMP-2 and associated proteins in the caveolae of endothelial cells with the ET_B receptor and eNOS facilitates this interaction ([122] and citations therein).

Other factors in addition to RLX are likely to contribute to gestational increases in GFR and RPF. PlGF may play a role particularly after the first trimester when serum concentrations begin to rise as the placenta grows and matures.[136] Another candidate is calcitonin gene related peptide (CGRP), which rises in the blood during early gestation in women.[137] CGRP is a potent vasodilator,[138] and therefore may contribute to systemic and renal vasodilatation of pregnancy, a possibility that could be tested by administration of CGRP antagonists in gravid animal models. Recent evidence supports an important role for the AT2 receptor in mediating the midterm decline in systolic blood pressure in mice,[139,140] and in attenuating constrictor responses to phenylephrine in aorta from gravid rats,[141] thus suggesting a potential role for AT2 receptor activation in the renal vasodilatation of pregnancy. Intriguingly, histidine decarboxylase and histamine, a potent vasodilator, were both reported to be increased in the superficial cortex of gravid mice.[142] Finally, renal production of epoxyeicosatrienoic acid may also contribute to renal vasodilatation and hyperfiltration of pregnancy.[143] To what degree, if any, these vasodilatory mechanisms may be activated by RLX in pregnancy is unknown.

In conclusion, evidence from both animal and human investigations suggests that renal hyperfiltration during pregnancy is mainly secondary to increased RPF, the latter being due to decrements in renal vascular resistance. However, increased K_f and reduced plasma oncotic pressure also contribute.[10,32,33,144] There has been considerable progress in identifying mechanisms responsible for this renal vasodilatation, and those where findings appear particularly promising were reviewed in detail above. Of importance, whatever the primary stimulus, it must be a powerful one, because the pregnancy-induced rise in GFR is not confined to women with two normally functioning kidneys, but occurs also in subjects with previously hypertrophied single kidneys (following uninephrectomy), and in transplant recipients.[17,18]

OSMOREGULATION IN NORMAL PREGNANCY

Plasma sodium levels decrease 4–5 mEq/L in normal gestation.[145] This decrement is often mistaken as a sign of changing sodium balance but in reality P_{Na} is but a surrogate for plasma osmolality (P_{osmol}), the latter decreasing ~10 mosm/kg during pregnancy.[67] The decrease commences during the luteal phase,[48] falling to a new steady state (~10 mosm/kg) very early in pregnancy and remaining at this new lower level until term.[11,64,65,67] Analysis of data from both human and rodent pregnancy leads to the following explanation of how this occurs.[146,147] The osmotic thresholds for thirst and arginine vasopressin (AVP) release decrease in parallel (Fig 16.5). Lowering the thirst threshold stimulates increased water intake and dilution of body fluids. Because inhibition of AVP release also occurs at a lower level of body tonicity, the hormone continues to circulate and the ingested water is retained. P_{osmol} then declines until it is below the new osmotic thirst threshold and a new steady state is established with only modest retention of water.

Figure 16.5 further reveals that the rate of rise in P_{AVP} as P_{osmol} increases decreases as pregnancy progresses. There are two possible interpertations for this finding. The first is that the sensitivity of the system may be decreasing (i.e., less secretion per unit rise in P_{osmol}), and the second that the disposal rate (the metabolic clearance, MCR) of AVP is increasing. It is the latter that appears to be the case, AVP's MCR rising four-fold between early and midgestation.[148] This rise in MCR also parallels the increase in trophoblast mass accompanied by a concomitant rise in circulating levels of cystine aminopeptidase (vasopressinase), the latter phenomenon being the most likely explanation of the striking increase in the MCR of AVP as gestation progresses. This hypothesis is strongly supported by the observation that the disposal rate of infused 1-desamino-8-D-arginine vasopressin (DDAVP: desmopressin), an AVP analog resistant to inactivation by vasopressinase, is virtually unaltered in pregnancy.[149]

Mechanisms responsible for altered osmogegulation are obscure though hCG,[65,67,150] constuitive NOS,[129] and relaxin[69,70] have all been implicated. Relaxin, though, being a hormone secreted by the corpus luteum, is the prime candidate as it explains the effects of hCG decreasing P_{osmol} and osmotic thresholds in premenopausal women, but not men,[65,149] the osmoregulatory changes during the menstrual cycle,[48] and the alterations in osmoregulation during rodent gestation described earlier in this chapter. Also the decrease in P_{osmol} is blunted in women with primary ovarian failure

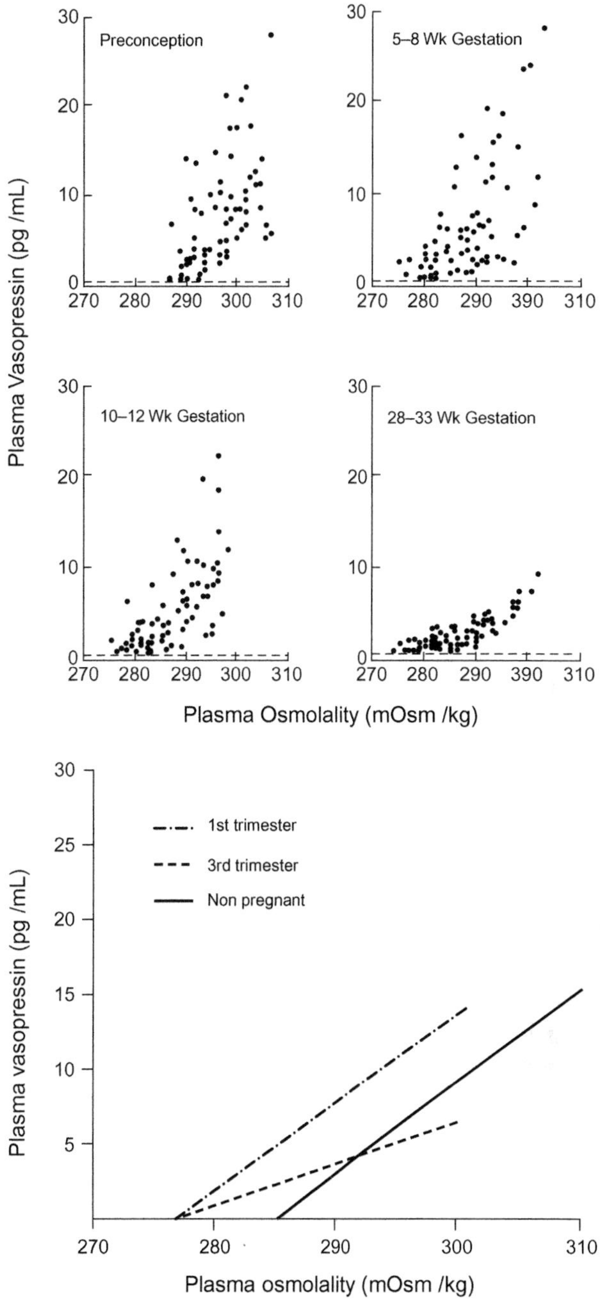

FIGURE 16.5 Relationship between plasma vasopressin (ordinate) and osmolality (abscissa) during serial hypertonic infusions loading through each trimester and postpartum (as a surrogate in the graph for "preconception"). Each dot represents a single plasma determination. The graph at the bottom contains the highly significant regression lines, the abscissal intercepts defining the marked and significant decrease in the osmotic threshold for vasopressin release. Of further interest is the decreased slope of the line (the gain in plasma vasopressin per unit rise of osmolality). This apparent decrease in the sensitivity of the system was not observed near term in rodents, a species that does not produce vasopressinase. This led to further human protocols that suggested the slope was related to the marked increases in plasma vasopressinase during the last half of pregnancy. (Modified from ref[65].)

who successfully carry donated ova,[85] and sheep, which lack relaxin due to a stop codon in the coding sequence, do not show reductions in P_{osm} during pregnancy.[151,152]

It has also been argued that the osmoregulatory changes reflect changes in how gravidas sense their altered volume, and that hypoosmolality relates to nonosmotic stimuli, the pregnant woman sensing her effective volume as "underfilled." This relative hypovolemia is then said to be the reason for AVP secretion at lower P_{osmol}. Volume regulation is detailed further in Chapter 15 but of interest here is a report that there are borderline or undetectable elevations in vasopressin levels in gravid rats that lead to upregulation of aquaporin-2 mRNA and its water channel protein in the apical membranes of collecting ducts.[153] These observations, though, appear inconsistent with patterns of renal water handling of both pregnant rodents and humans, who handle water loads similar to the they way they do in the nonpregnant state (unlike patients with cirrhosis and heart failure, who are prototypes for decreased "effective circulating volume" and where nonosmotic AVP secretion leads to decreased abilities to excrete water loads rapidly). That is, there should be a decreased response to water loading tests in human and rodent pregnancies if the density of the water channels was increased and this does not appear to be the case.

RENAL HEMODYNAMICS AND GLOMERULAR FILTRATION RATE IN PREECLAMPSIA

Chesley appears to have been the first to measure ERPF and GFR in women with normal gestations or with preeclampsia,[154,155] and his pioneering work was followed by numerous reports on the subject. Table 16.1 is a summary of those investigations where both women with preeclampsia and normal gravidae were studied as a control cohort. We located 23 publications that met these criteria, although in several instances the data from the two cohorts were published separately. In nine of 23 publications, a second control group of nonpregnant women was also studied. Inclusion in Table 16.1 required a reasonably clear presentation of the criteria for diagnosing preeclampsia. These criteria included onset of hypertension in late gestation and proteinuria, and, in many instances, evidence for the absence of hypertension before midpregnancy or normalization of blood pressure postpartum. Several investigators noted that most of their preeclamptic subjects were primiparous, which in retrospect may have improved their likelihood of correctly diagnosing the disease based on the clinical evidence alone.[181]

A word about the methodologies used. In the majority of cases, the subjects were hydrated by oral water intake to increase urine flow, thus decreasing dead space error and improving the accuracy of the urine collection, and consequently the renal clearance measurements. An indwelling bladder catheter was routinely used in most of the studies (a practice considered unacceptable today), which also permitted more accurate urine collections. In many instances, the renal clearances of inulin and para-aminohippurate were used to measure GFR and ERPF, respectively, but occasionally the clearance of thiosulfate, mannitol, or creatinine was employed, the latter being more "estimates" of GFR. Similarly, the renal clearance of diodrast was frequently used to estimate ERPF. Few investigators commented on how their subjects were positioned during the study, though for convenience one assumes most were probably studied in the supine position. However, lying supine, especially in late gestation, may depress renal function (see "Postural Influence on Renal Function," above). Finally, most investigators normalized GFR and ERPF for $1.73\,m^2$ body surface area as assessed at the time of measurement during pregnancy, a practice which has been abandoned in recent times because gestational changes in renal hemodynamics and GFR are believed to be functional in nature, and not related to renal hypertrophy.[182]

In Table 16.1, the means $\pm$ SEM have been recalculated from the individual measurements provided for all subjects, when they were presented in the original publications. In some instances, the measurements for each subject were not provided; rather, mean values and/or standard deviations were reported. These were incorporated directly into Table 16.1 after converting the standard deviations to standard errors. Perhaps the most meaningful way to assimilate all of the data from the 23 studies presented in Table 16.1 is to examine the percentage change in GFR, ERPF, and filtration fraction (FF) between preeclamptic and late-pregnant women, and between preeclamptic and nonpregnant women as depicted in Table 16.2. In all 23 reports, there was a reduction of GFR in preeclamptic subjects compared to late-pregnant women, on average by 32%. In all but one publication, there was a depression of ERPF, on average by 24%. In all but one of the nine studies that also included data for nonpregnant women, GFR and ERPF were reduced by comparison in the preeclamptic subjects both by 22%. Thus, both GFR and ERPF are impaired in preeclampsia compared to late pregnant and nonpregnant women – a conclusion that Chesley and Duffus reached based on their earlier survey of the literature in 1971.[180]

The study by Assali et al.[164] is noteworthy because of the succinct description of the diagnosis of preeclampsia, which was frequently lacking in other reports: "the presence of hypertension, edema, and proteinuria after the 24th week of gestation, together with the absence of a history of hypertension prior to pregnancy, and the return of the blood pressure to normal levels following delivery." In this study, GFR and ERPF were significantly reduced by 29.4% and 20.3%, respectively, compared to normal women in late pregnancy. Also deserving emphasis are the publications by McCartney et al.[176] and Sarles et al.,[178] in

TABLE 16.1 Renal Hemodynamics and Glomerular Filtration Rate in Preeclampsia[a]

1st Author & Reference No.	Nonpregnant Women				Normal Pregnancy – Last Trimester				Preeclampsia–eclampsia			
	No.	GFR	ERPF	FF (%)	No.	GFR	ERPF	FF (%)	No.	GFR	ERPF	FF (%)
Chesley[154,155]	8–9	57 ± 1	518 ± 15	11.4 ± 0.4	8	79 ± 6[b]	610 ± 38	12.2 ± 0.5	17	77 ± 4[b]	560 ± 39	14.8 ± 1.3
Wellen,[156] Welsh[157]	–	–	–	–	7–17[c]	121 ± 6	588 ± 40	19.9 ± 1.0	4–6[d]	94 ± 15	636 ± 70	14.7 ± 1.6
Dill[158]	–	–	–	–	7	113 ± 7	614 ± 45	18.6 ± 0.9	6–7[e]	83 ± 15	432 ± 58	17.7 ± 2.5
Kariher[159]	–	–	–	–	8[f]	150	754	20.9	17[f]	102	512	19.8
Schaffer[160]	–	–	–	–	9	117 ± 9	–	–	7[g]	86 ± 14	–	–
Chesley[161]	–	–	–	–	10	125 ± 9	–	–	8[h]	93 ± 7	–	–
Bucht[162,163]	23	122 ± 5	557 ± 30	22.7 ± 1.0	10	156 ± 10	571 ± 25	28.9 ± 1.2	8 (<33 wks)	102 ± 18	495 ± 72	22.0 ± 3.6
									18 (>33 wks)	98 ± 8	423 ± 36	24.6 ± 3.1
Assali[164]	–	–	–	–	7	109[i]	699	15.7	9	77 ± 6[l]	557 ± 47	14.7 ± 1.4
Brandstetter[165,166]	10	117 ± 4	584 ± 30	20.1 ± 0.7	11	128 ± 4	661 ± 27	19.4 ± 0.6	5 (preeclampsia)	77 ± 5	466 ± 27	16.5 ± 0.8
									3 (eclampsia)	59 ± 8	399 ± 36	14.5 ± 0.8
Friedberg[167]	–	–	–	–	6	112 ± 7	635 ± 58	18 ± 1	10	93 ± 5	571 ± 36	17 ± 1
Lanz[168]	12	115 ± 4[k]	585 ± 25	20 ± 1	5[j]	125[k]	700	18	8	91 ± 5[k]	529 ± 27	17 ± 1
Page[169]	–	–	–	–	12	181 ± 10	–	–	9	75 ± 6	–	–
Lovotti[170]	5	121 ± 7[k]	551 ± 54	22.3 ± 1.4	5	151 ± 9[k]	641 ± 34	23.6 ± 1.5	10	101 ± 5[k]	452 ± 40	23.6 ± 3.0
Hayashi[171]	–	–	–	–	14 (<32 wks)	129 ± 1	–	–	5 (<32 wks)	63 ± 9	–	–
					23 (>32 wks)	108 ± 6	–	–	21 (>32 wks)	99 ± 6	–	–
Chesley[172]	17	124 ± 6	–	–	11	145 ± 7	–	–	13	103 ± 8	–	–
Buttermann[173]	11	122 ± 5	624 ± 29	19.3 ± 0.9	26	132 ± 5	647 ± 28	20.9 ± 0.8	33	59 ± 5	341 ± 21	17.0 ± 0.6
Schlegel[174]	10	133 ± 8	659 ± 21	21 ± 1	12	132 ± 6	593 ± 28	23 ± 1	11	93 ± 11	454 ± 59	21.6 ± 1.7
Friedberg[175]	–	–	–	–	10	132 ± 5	586 ± 15	22.5 ± 0.8	14	105 ± 3	480 ± 16	22.2 ± 0.8
McCartney[176]	–	–	–	–	7	133 ± 8	–	–	6	90 ± 12	–	–
Bocci[177]	16	–	573 ± 36	–	14	–	560 ± 24	–	5	–	391 ± 43	–
Sarles[178]	–	–	–	–	5	175[l]	825[l]	21.2	9	115[l]	550[l]	20.9
Sismondi[179]	–	–	–	–	24	–	601 ± 6	–	12	–	406 ± 15	–
Chesley[180]	–	–	–	–	14	170 ± 9	755 ± 45	22.5	13	114 ± 9	606 ± 42	19.6 ± 1.3
Grand mean		114	581	19.5		133	649	20.4		90	487	18.7

[a] Mean ± SEM. All renal clearances are expressed as mL/min or mL/min × 1.73 m^2 body surface area. Glomerular filtration rate (GFR) was measured by the renal clearance of inulin, or estimated by the renal clearances of creatinine, mannitol, or thiosulfate. Effective renal plasma flow (ERPF) was measured by the renal clearances of para-aminohippurate or diodrast. Women with preeclampsia and eclampsia were evaluated separately or together. FF = filtration fraction (GFR/ERPF).

[b] GFR was measured by the renal clearance of creatinine.

[c] Four subjects who were in the first or second trimester were excluded from the analyses.

[d] Patients without proteinuria during pregnancy or persistent hypertension after pregnancy at a follow-up clinic were excluded from the analyses ($n = 8$).

[e] Two patients without proteinuria were excluded from the analyses; also, one subject with an average C_D of 1189 mL/min/1.73 m^2 was deleted as an outlier. Despite bladder catheterization and oral hydration, the authors suggested that low urine flow rate shown by some of the subjects may have compromised measurements of GFR and ERPF, Therefore, subjects with a urine flow of<1.0 mL/min were also excluded.

[f] Only mean values were provided. For preeclampsia, mild ($n = 7$), severe ($n = 7$), and eclamptic ($n = 3$) women were combined. However, the mean values for eclampsia provided by the authors were considerably lower than either mild or severe preeclampsia which were comparable.

[g] One subject without proteinuria was excluded from the analysis.

[h] Two subjects were excluded from the analysis due to oliguria.

[i] The renal clearance of mannitol was used as an estimate of GFR. Only mean values were provided for normal pregnancy.

[j] Only five subjects were apparently studied after 24 weeks of gestation, and the data points were plotted on a graph. Because mean values were not provided, they were estimated from the graph.

[k] GFR was estimated by the renal clearance of thiosulfate.

[l] Mean values were not provided, rather they were estimated from bar graphs.

TABLE 16.2 Percentage Change in GFR, ERPF, and Filtration Fraction between Preeclampsia/Eclampsia and Late-Pregnant or Nonpregnant Levels[a]

1st Author & Reference No.	% Change from Late Pregnant			% Change from Nonpregnant		
	GFR	ERPF	FF	GFR	ERPF	FF
Chesley[154,155]	−2.5	−8.2	+21.3	+35.1	+8.1	+29.8
Wellen,[156] Welsh[157]	−22.3	+8.2	−26.1	–	–	–
Dill[158]	−26.6	−29.6	−4.8	–	–	–
Kariher[159]	−32	−32.1	−5.3	–	–	–
Schaffer[160]	−26.5	–	–	–	–	–
Chesley[161]	−25.6	–	–	–	–	–
Bucht[162,163]						
<33 wks	−34.6	−13.3	−23.9	−16.4	−11.1	−3.1
>33 wks	−37.2	−25.9	−14.9	−19.7	−24.1	+8.4
Assali[164]	−29.4	−20.3	−6.4	–	–	–
Brandstetter[165,166]						
preeclampsia	−39.8	−29.5	−15.0	−34.2	−20.2	−17.9
eclampsia	−53.9	−39.6	−25.3	−49.6	−31.7	−27.9
Friedberg[167]	−17.0	−10.1	−5.6	–	–	–
Lanz[168]	−27.2	−24.4	−5.6	−20.9	−9.6	−15.0
Page[169]	−58.6	–	–	–	–	–
Lovotti[170]	−33.1	−29.5	0.0	−16.5	−18.0	+5.8
Hayashi[171]						
<32 wks	−51.2	–	–	–	–	–
>32 wks	−8.3	–	–	–	–	–
Chesley[172]	−29.0	–	–	−16.9	–	–
Buttermann[173]	−55.3	−47.3	−18.7	−51.6	−45.4	−11.9
Schlegel[174]	−29.6	−23.4	−6.1	−30.1	−31.1	+2.9
Friedberg[175]	−20.5	−18.1	−1.3	–	–	–
McCartney[176]	−32.3	–	–	–	–	–
Bocci[177]	–	−30.2	–	–	−31.8	–
Sarles[178]	−34.3	−33.3	−1.4	–	–	–
Sismondi[179]	–	−32.5	–	–	–	–
Chesley[180]	−32.9	−19.7	−12.9	–	–	–
Grand mean	−32%	−24%	−9%	−22%	−22%	−3%

[a]The percentage changes were calculated based on the mean values for GFR, ERPF, and filtration fraction shown in Table 17.1. See footnotes to Table 17.1 for an explanation of the abbreviations.

which the diagnosis of preeclampsia was based not only on clinical criteria, but also according to the presence of glomerular endotheliosis in renal biopsy obtained in the postpartum or intrapartum period, respectively. McCartney et al. also studied their subjects in the lateral recumbent position, which avoided the potential artifactual decrease in renal function due to compression of the great vessels by the gravid uterus.[176] Thus, using the renal clearance of inulin, McCartney et al. identified a 32.3% reduction of GFR in women with preeclampsia compared to normal women in late pregnancy (Fig. 16.6(A)). Similarly, Sarles et al.[178] observed a 34.3% and 33.3% reduction in the renal clearances of inulin and para-aminohippurate, respectively, in women with preeclampsia compared to normal women in late pregnancy. Last, but not least, is the well-controlled investigation by Chesley and Duffus in which women were studied in the lateral recumbent position, and GFR and ERPF were corrected for $1.73\,m^2$ based on *prepregnancy* body surface area. The investigators noted a 32.9% and 19.7% decline of GFR and ERPF, respectively, in the women with preeclampsia compared to normal women during late pregnancy.[180]

In more recent work, Irons et al.[183] measured ERPF and GFR by the renal clearances of para-aminohippurate and inulin, respectively, in women with normal pregnancy ($N = 10$) and women with preeclampsia ($N = 10$). The normotensive gravidae demonstrated an ERPF and GFR of 766 ± 52 and 153 ± 13 mL/min at 32 weeks of gestation, respectively, which decreased to 486 ± 17 and 87 ± 3 mL/min 4 months postpartum. The preeclamptic women, who were all primiparous and normotensive during gestational weeks 12–18, showed >2 g protein/24 h and blood pressure >140/90 mm Hg at 33.5 weeks of gestation. ERPF and GFR were 609 ± 24 and 97 ± 7 mL/min, respectively, which changed to 514 ± 22 and 109 ± 7 mL/min 4 months postpartum. Thus, consistent with previous investigations

(Table 16.1), both ERPF and GFR were compromised in preeclampsia, the latter to a greater extent.[183] Comparable findings were subsequently reported by the same group in another study.[184] However, concurrent analyses of the renal clearances of inulin, para-aminohippurate and neutral dextrans in this study also permitted calculation of K_f, which

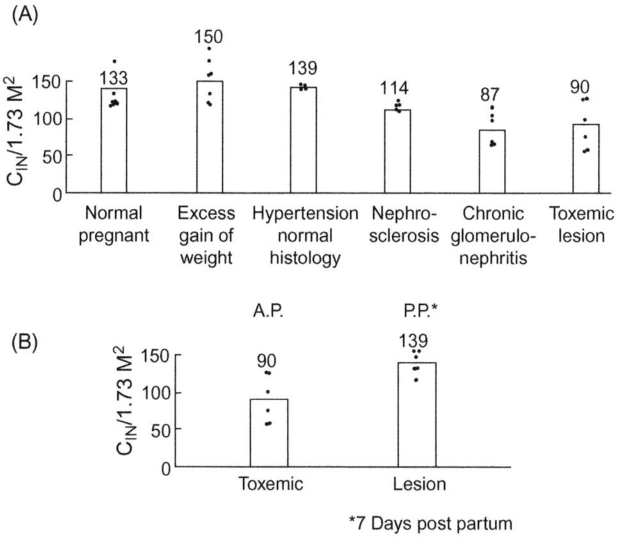

FIGURE 16.6 (A) Individual inulin clearance data and the mean values for six different patient groups. In particular, note the 32% reduction of glomerular filtration rate in pregnant women with preeclampsia (as documented by postpartum renal biopsy) compared to normal pregnant women. (B) Individual inulin clearance data and the mean values for pregnant women with preeclampsia documented by postpartum renal biopsy. AP = antepartum; PP = postpartum day 7. See text for further details.[176]

was approximately 50% lower in preeclamptic compared to normal pregnant women.[184]

The mechanism(s) ultimately responsible for the compromise of the renal circulation in preeclampsia remains unknown. It may stem, however, from vascular endothelial damage, first proposed by Stubbs et al.,[185] reflected by the finding of glomerular endothelial swelling first reported by Mayer in 1924.[186] Thus, "endothelial dysfunction" is widely believed to be a major mechanism in the pathogenesis of the disease leading to widespread vasospasm and organ hypoprofusion[187] (Chapter 9). Many candidate molecules causing endothelial damage, some arising from the placenta, have been identified in recent years (Chapters 6, 8, and 9). Endothelial mechanisms are also implicated in the renal vasodilatation and hyperfiltration of *normal* pregnancy as previously described (see "Renal Hemodynamics and Glomerular Filtration Rate during Normal Pregnancy"), and possibly these are compromised by the "endothelial dysfunction" which afflicts women with preeclampsia, thereby contributing to the impairment of GFR and ERPF. Irrespective of the inciting agent(s), renal vascular resistance is inappropriately high, which accounts for impaired renal blood flow. Based on indirect calculations of the renal afferent and efferent arteriolar, as well as venular resistances,[188] the pathologic increase in total renal vascular resistance is largely due to an increase in the afferent arteriolar resistance (Table 16.3). On the one hand, this finding is not entirely unexpected because the FF, which is used in the calculation of the renal segmental arteriolar resistances, shows only a 9% decline in preeclampsia (Table 16.2). That is, if GFR and ERPF are reduced in proportion such that the ratio or FF shows little change or is unchanged, then an increase only in the afferent arteriolar resistance is

TABLE 16.3 Estimated Segmental Renal Vascular Resistances in Preeclampsia[a]

1st Author & Reference No.	Diagnosis	Renal Vascular Resistance (Dynes × s × cm^{-5})[b]				
		No.	Total	Afferent Arteriolar	Efferent Arteriolar	Venular
Assali[164]	Normal pregnancy	7	5178	1883	1531	1751
	Preeclampsia	9	10,229	6666	1405	2228
Friedberg[167]	Normal pregnancy	6	6144	2085	1781	2333
	Preeclampsia	10	11,321	6887	2057	2470
Brandstetter[165]	Normal pregnancy	44	6640	2430	2080	2130
	Preeclampsia	5	17,643	12,896	1864	2884
	Eclampsia	3	21,522	16,611	1694	3417
Buttermann[173]	Normal pregnancy	26	7410	3055	2125	1960
	Preeclampsia	33	18,740	13,270	1788	3682
Bocci[177]	Normal pregnancy	14	7300	3167	1870	2275
	Preeclampsia	5	16,509	11,446	1867	3174
Grand mean	Normal pregnancy		6534	2524	1877	2090
	Preeclampsia/eclampsia		15,994	11,296	1779	2976

[a]Segmental renal vascular resistances were estimated according to the calculations of Gomez.[188]
[b]Mean values are shown.

TABLE 16.4 Antepartum and Postpartum Renal Function in Women with Preeclampsia[a]

1st Author & Reference No.	No.	Antepartum			Postpartum			Average Time After Delivery
		GFR	ERPF	FF(%)	GFR	ERPF	FF(%)	
Corcoran[192]	3–13	99 ± 7	659 ± 6	13.5 ± 1.2	109 ± 5	478 ± 11	22.3 ± 0.4	13 wks
Dill[158]	6–7	83 ± 15	432 ± 58	17.7 ± 2.5	129 ± 8	429 ± 38	28.5 ± 2.4	Not cited
Wellen[156]	4–6	94 ± 5	636 ± 70	14.7 ± 1.6	107 ± 5	540 ± 28	20.2 ± 0.9	5 wks
Kariher[159]	17	102	512	19.8	124	568	21.3	≥12 days
Schaffer[160]	3	93 ± 28	–	–	121 ± 15	–	–	8 days
Chesley[161]	8	93 ± 7	–	–	106 ± 6	–	–	6–9 days
Odell[193]	4	93 ± 17	–	–	182 ± 27	–	–	Not cited
Bucht[163]	18–26	98	423	24.6	119	426	27.9	2–7 wks
Lanz[168]	3	78 ± 1	598 ± 10	13 ± 0	108 ± 6	559 ± 50	19 ± 1	10 wks
Page[169]	9	75 ± 6	–	–	109 ± 6	–	–	Several weeks
Buttermann[173]	33	59 ± 5	341 ± 21	17.0 ± 0.6	98 ± 6	418 ± 17	24.0 ± 1.3	Immediately after delivery
Schlegel[174]	4	92 ± 24	504 ± 143	20.3 ± 2.9	121 ± 19	532 ± 56	23.3 ± 3.9	14–22 days
McCartney[176]	6	90 ± 12	–	–	139 ± 13	–	–	7 days
Grand mean		88	513	17.6	121	494	22.7	

[a]Mean ± SEM. All renal clearances are expressed as mL/min or mL/min × 1.73 m^2 body surface area. See footnotes to Table 16.1 for an explanation.

inferred.[189] On the other hand, although the decline in FF is small, it is a consistent finding in preeclampsia, because the impairment in GFR exceeds that of ERPF (Table 16.2). A simultaneous *reduction* in renal efferent arteriolar or venular resistance could theoretically account for this finding, but the calculated segmental resistances do not support this argument (Table 16.3). Rather, a reduction in the K_f is a more likely explanation stemming from perturbations in the glomerular capillary hydraulic conductivity, surface area available for filtration or both, a deduction recently supported by the study by Moran and co-workers (*vide supra*).[184] Indeed, neutralization of podocyte-derived vascular endothelial growth factor (VEGF) by pathological levels of circulating soluble vascular endothelial growth factor receptor-1 (VEGF-R1 or sFlt-1) emanating from the preeclamptic placenta leads to glomerular endotheliosis epitomized by loss of fenestrae, structures that are crucial for hydraulic conductivity.[2,190,191]

A final consideration is the recovery of renal function during the puerperium in women who suffered preeclampsia during pregnancy. Table 16.4 includes those investigations that reported antepartum and postpartum measurements of GFR and/or ERPF in women with preeclampsia. Also included is the average number of days or weeks after delivery when renal function was measured. Based on the studies by Schaffer et al.,[160] Chesley and Williams,[161] as well as McCartney et al.[176] (Fig. 16.6(B)), there was notable improvement of the GFR during the first 7 to 8 days postpartum. (Also compare with nonpregnant control values listed in Table 16.1.) Unfortunately, these investigators did not measure ERPF, but it appears from the other studies listed in Table 16.4 that ERPF may have remained depressed during the same postpartum period. Buttermann[173] studied the women immediately after delivery, and found GFR markedly and ERPF somewhat improved (Table 16.4). In the investigation by Moran et al., the decreased K_f observed in the preeclamptic women had recovered by at least 5 months postpartum, but earlier time points were not investigated.[144,184]

Summary

A review of 23 reports on renal function in preeclampsia shows that GFR and ERPF are decreased, on average, by 32% and 24%, respectively, from normal, late-pregnant values. GFR and ERPF are frequently decreased in preeclampsia compared to nonpregnant levels, but to a lesser degree. The work of Assali et al.[164] is exceptional because of the rigorous diagnostic criteria used. In this study, GFR and ERPF were reduced by 29% and 20%, respectively. In the studies by McCartney et al.[176] and Sarles et al.,[178] rigorous clinical diagnosis of preeclampsia was further corroborated by renal histology. In these reports, GFR and/or ERPF were also modestly reduced in women with preeclampsia compared to normal pregnant women. Several investigators calculated the renal segmental vascular resistances, and only the preglomerular arteriolar resistance was increased in preeclampsia. Because the reduction in GFR generally exceeded that of ERPF, yet only preglomerular arteriolar resistance was altered, when taken together these findings suggest a reduction in the K_f during preeclampsia. Interestingly, GFR appears to rapidly recover during the first week after delivery. On the other hand, ERPF may recover more slowly.

The mechanism(s) for compromised GFR and ERPF is unknown, but may relate to the generalized "endothelial dysfunction" believed to account for widespread vasospasm and organ hypoperfusion in preeclampsia (Chapter 9). Conceivably, the relaxin vasodilatory pathway that normally signals through the endothelium-derived relaxing factor, nitric oxide, thereby promoting renal vasodilatation and hyperfiltration during pregnancy (*vide supra*), is compromised by the endothelial dysfunction. Because local, arterial-derived vascular endothelial and placental growth factors (VEGF and PlGF) are likely to be proximal players in the relaxin vasodilatory pathway,[130] the increased circulating concentrations of sFlt-1 reported in preeclampsia (Chapter 6) may compromise relaxin-mediated renal vasodilatation by neutralizing these angiogenic growth factors in a *competitive* manner. Theoretically, upregulating the processes mediating vasodilatation in normal pregnancy could be one approach to ameliorating the vasoconstriction and organ hypoperfusion that occur in preeclampsia. An alternative, though not mutually exclusive, approach might be to neutralize the effect of the circulating factors injurious to the endothelium. Administration of relaxin to reach supraphysiologic serum levels is attractive as a potential therapeutic for preeclampsia, because it should target both of these endpoints. On the one hand, relaxin administration should increase vasodilatation and organ perfusion by accentuating its own vasodilatory pathway. On the other hand, relaxin may augment arterial VEGF and PlGF activity, which could partly or wholly counteract circulating sFlt-1 locally, thereby improving endothelial cell health.

RENAL HANDLING OF URIC ACID

Normal Pregnancy

Uric acid is the end product of purine metabolism in humans.[194] Purines are both dietary in origin and endogenously produced, the latter being the major source of uric acid. Most circulating uric acid is produced in the liver; in humans about 67% is excreted by the kidney and the remainder by the gastrointestinal tract. Five percent of circulating uric acid is bound to plasma proteins, thus almost all of this circulating solute is freely filtered by the glomeruli. Once filtered, it undergoes both reabsorption and secretion, mainly in the proximal tubule. In humans, net reabsorption occurs, and the bulk of filtered uric acid – 88 to 93% – is reabsorbed by the renal tubules back into the blood, and only 7 to 12% of the filtered load reaches the urine.[194]

Serum concentrations of uric acid are significantly decreased from nonpregnant values by approximately 25 to 35% throughout most of human pregnancy.[195–198] During late gestation, they rise toward nonpregnant values. Theoretically, serum concentrations of uric acid during pregnancy are determined by several factors, including dietary intake of purines, metabolic production of uric acid by mother and fetus, as well as renal and gastrointestinal excretion.[194,199,200] Alterations in any one or several of these factors could underlie the changes in serum uric acid that occur during normal pregnancy.

Of the factors discussed above that could determine serum uric acid concentration in pregnancy, renal handling of uric acid has been most investigated. Dunlop and Davison serially studied 24 healthy women at approximately 16, 26, and 36 weeks of gestation, as well as 8 weeks postpartum.[196] The patients were not fasted, but they were studied in the sitting position, thus minimizing postural decreases of renal function especially in late pregnancy. Following instigation of a water diuresis to minimize urinary tract dead space error, GFR was determined by the renal clearance of inulin using the constant infusion technique. Plasma and urine uric acid were measured by an enzymatic technique. In another investigation, Semple et al. serially studied 13 healthy women at approximately 14, 26, and 35 weeks of gestation, as well as 10 weeks postpartum.[198] Methodologies were similar to those used by Dunlop and Davison except that the patients were given a purine-restricted diet for 3 days before each experiment, and they were fasted overnight prior to each study.

The results are summarized in Table 16.5. The pattern of change in serum uric acid concentration in both investigations was as described above; namely, a decline of 25 to 35% throughout most of pregnancy, followed by a return toward nonpregnant levels during late gestation. The renal handling of uric acid, however, was not identical in the two studies. Dunlop and Davison reported increased renal clearance and urinary excretion of uric acid throughout gestation.[196] These variables were elevated, because the filtered load of uric acid was increased and fractional reabsorption decreased. A lower fractional reabsorption signified that the tubular transport of uric acid was reduced in normal pregnancy. Because urate undergoes bidirectional transport mainly in the proximal tubule (see above), either the absorptive component was decreased by pregnancy and/or the secretory component increased. In contrast, Semple et al. did not observe a significant rise in urinary excretion of uric acid, although enhanced renal clearance throughout pregnancy was noted.[198] The augmented renal clearance of uric acid was a consequence of reduced serum levels in the equation ($C_{urate} = U_{urate} \times V / P_{urate}$). The filtered load of uric acid was not increased by pregnancy in this study except at 30 to 40 weeks, and fractional reabsorption was not changed at any gestational period. These findings may be analogous to the situation of plasma creatinine in pregnancy: circulating levels decline secondary to increased glomerular filtration, which produces a transient rise in filtered load and urinary excretion. Then, a new steady state is reached in which circulating levels remain decreased

TABLE 16.5 Renal Handling of Uric Acid During Normal Human Pregnancy[a]

Variable	Calculation	Study	Weeks of Gestation 10–19	Weeks of Gestation 20–29	Weeks of Gestation 30–40	6–15 Weeks Postpartum
(1) Plasma or serum uric acid (μmol/L)[b]		Dunlop & Davison	168[c]	178[c]	202	219
		Semple et al.	180[c]	190[c]	230[c]	280
(2) Urinary excretion rate of uric acid (μmol/min)	$U_{urate} \times V$	Dunlop & Davison	3.5[c]	3.6[c]	3.5[c]	2.5
		Semple et al.	2.6	2.8	3.4	2.7
(3) Renal clearance of uric acid (mL/min)	(2)/(1)	Dunlop & Davison	22.0[c]	20.5[c]	17.6[c]	11.8
		Semple et al.	14.2[c]	15.3[c]	15.8[c]	9.8
(4) Renal clearance of inulin, GFR (mL/min)	$u_{IN} \times V/P_{IN}$	Dunlop & Davison	149[c]	153[c]	156[c]	98
		Semple et al.	135[c]	145[c]	145[c]	96
(5) Filtered load of uric acid (μmol/min)	(4) × (1)	Dunlop & Davison	24.9[c]	27.6[c]	31.0[c]	21.4
		Semple et al.[d]	24.3	27.6	33.4	26.9
(6) Fractional excretion of uric acid × 100(%)	(2)/(5) or (3)/(4)	Dunlop & Davison	14.8[c]	13.3[c]	11.5	12.0
		Semple et al.	10.5	10.6	10.9	10.2
(7) Fractional reabsorption of uric acid × 100(%)	100 − (6)	Dunlop & Davison	85.3[c]	86.8[c]	88.5	88.0
		Semple et al.[d]	89.5	89.4	89.1	89.8

[a]Mean values are depicted. Dunlop and Davison[196] serially studied 24 healthy women, and Semple et al.[198] investigated 13 healthy women throughout pregnancy and in the postpartum period. See text for details. Data obtained from Dunlop and Davison[196] and Semple et al.[198]
[b]To convert μmol/L to mg/100 mL multiply by 1.68×10^{-2}.
[c]Significantly different from postpartum values.
[d]Filtered load and fraction reabsorption of uric acid were not provided by Semple et al., but they were calculated from the data provided.

because the filtered load results in urinary creatinine excretion in amounts that virtually balance production.

Because the reports of Dunlop and Davison[196] and Semple et al.[198] are not in complete agreement, it is difficult to determine whether the handling of urate by the renal tubules changes during pregnancy. Nevertheless, pregnancy increased the renal clearance of uric acid in both studies by virtue of increased GFR, reduced tubular reabsorption or both, which accounted for the reduced circulating levels. A potential explanation for the elevated urinary excretion of uric acid during pregnancy in the study by Dunlop and Davison is that dietary intake was not controlled, and possibly during pregnancy women consumed more uric acid in the diet. The return of plasma uric acid concentrations toward nonpregnant values in late gestation can be explained by progressively increasing tubular reabsorption (and consequently, falling renal clearance), at least in the report by Dunlop and Davison.[196]

Preeclampsia

Slemons and Bogert reported in 1917, that serum concentration of uric acid in preeclampsia and eclampsia was elevated.[201] In primiparous women, four of whom had preeclampsia and two eclampsia, they observed a mean ± SEM of 7.6 ± 0.5 mg/dL compared to the values in uncomplicated pregnancies, which ranged from 2 to 5 mg/dL. In 1925, and again in 1934, Stander et al.[202,203] confirmed the discovery of Slemons and Bogert, and concluded that "blood uric acid is increased in eclampsia and preeclampsia ... and ... may be regarded as a fairly safe criterion of the severity of the disease." Subsequently, numerous investigators have substantiated hyperuricemia in preeclampsia and eclampsia,[161,171,199,200,204–228] and with one exception,[211] a relationship between the degree of hyperuricemia and disease severity was observed.[203,210,214,216–219,225]

Of particular note is the study by Pollak and Nettles.[214] They measured serum uric acid concentration in 30 healthy pregnant women in the third trimester, 10 pregnant women with hypertensive vascular disease, and 33 women with preeclampsia (Fig. 16.7). The diagnosis in the last two groups was corroborated by the absence and presence of histologic findings of "glomerular endotheliosis," respectively, on antepartum percutaneous renal biopsy. The mean values ± SEM of serum uric acid for the three groups were 3.6 ± 0.1, 3.7 ± 0.3, and 6.3 ± 0.3 mg/dL, respectively ($P < 0.001$ preeclampsia vs. other groups). Moreover, there was a significant correlation between the renal histologic severity of the preeclamptic lesion (degree of "glomerular swelling and ischemia") and the serum uric acid ($P < 0.014$ by rank correlation test). Although several investigators have suggested that elevated circulating uric acid concentration be included among the diagnostic criteria for preeclampsia,[210,216] one can see in Fig. 16.7 that several women with the renal histologic diagnosis of mild preeclampsia had serum uric acid concentrations in the normal range. As well, there was considerable overlap of values between preeclampsia and the other subject groups. Nevertheless, based mainly on this work of Pollak

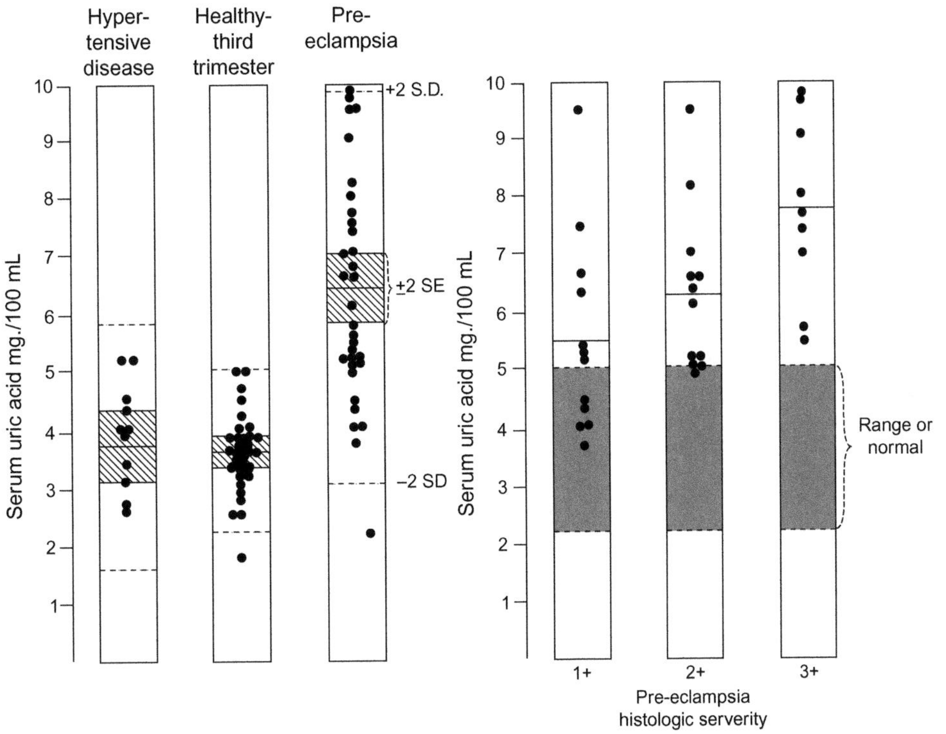

FIGURE 16.7 Left panel. Serum uric acid in normal pregnant women and pregnant women with hypertension or preeclampsia. The latter diagnosis was corroborated by the renal histological findings on antepartum percutaneous renal biopsy. Right panel. Serum uric acid according to the severity of the renal histological findings in women with preeclampsia. See text for details.[214]

and Nettles, inclusion of serum uric acid concentration as an additional criterion for the diagnosis of preeclampsia is reasonable, in addition to hypertension and proteinuria, particularly if one demands that the value should be one or two standard deviations above the mean for normal pregnant subjects adjusted for gestational age.[197] It is important to take into account the gestational age due to the changes in circulating uric acid observed in normal pregnancy (see "Renal Handling of Uric Acid – Normal Pregnancy," above).

Redman et al. reported that hyperuricemia is a relatively early event in preeclampsia.[223] Specifically, they observed at 25 ± 4 (±1 SD) weeks of gestation a 2.0 mg/dL rise in circulating uric acid above the first measurement made at 17 ± 3 weeks. In contrast, a 5.0 mg/dL rise in urea or the appearance of proteinuria did not occur until 28 ± 3 and 29 ± 3 weeks of gestation, respectively. Redman et al.[222] further noted that "perinatal mortality was markedly increased when maternal plasma urate concentrations were raised, generally in association with severe preeclampsia of early onset ... being ... a better indicator than blood pressure of prognosis for the fetus." Sagen et al.[226] corroborated this association between elevated circulating level of uric acid and poor fetal outcome, i.e., perinatal death, distress, or intrauterine growth restriction <5th percentile.

Theoretically, the potential mechanisms for increased circulating concentration of uric acid in preeclampsia may relate to increase in urate production or dietary purine intake, reduction in the metabolic or renal clearance of uric acid, or a combination thereof.

Production of Uric Acid

Both Slemons and Bogert[201] and Crawford[205] observed that the severity and length of labor were associated with increased circulating concentration of uric acid in normal pregnancy and in preeclampsia. Furthermore, the seizure activity of eclampsia was also reported to raise the serum uric acid level.[204,228] However, uric acid concentration is clearly increased during preeclampsia in the absence of convulsions and remote from labor, which suggests other etiologies.

In 1956, Seitchik evaluated the fate of injected[15] nitrogen-urate stable isotope in healthy pregnant women and women with preeclampsia, and concluded that the rate of production was increased in the latter.[199] However, two years later[200] Seitchik retracted this conclusion (citing inadequate control of protein intake and miscalculation of results in the 1956 study), and reported that "the rate of production of urate was not increased as previously reported ... and ... faulty renal function is the sole causative factor in urate accumulation in acute toxemia." In 1969, Fadel et al.[219] revisited the issue of increased uric acid production in preeclampsia, and postulated that production by the placenta may be increased in the disease due to cellular destruction caused by "multiple infarcts" leading to increased purines, the substrate for xanthine dehydrogenase/oxidase, and uric acid formation. However, the same investigators later showed that the uterine venous concentration of uric acid was actually less than (not greater than) that measured in simultaneously obtained peripheral venous plasma, providing no evidence for increased release of uric acid by the placenta in preeclampsia or a placental contribution to the raised systemic levels observed in the disease.[220] Similar negative data were reported by Hayashi et al., who found no significant difference in oxypurine levels between the peripheral blood and "intervillous space" blood of preeclamptic women.[212]

As a matter of fact, the human placenta was thought to be devoid of xanthine dehydrogenase/oxidase activity until recently,[229] and the concept that placental enzymatic activity is increased in preeclampsia was again considered.[230] Thus, increased production of uric acid from placental, fetal or maternal sources may contribute to the elevated circulating level of uric acid in the disease, although the evidence for a renal mechanism is strongly supported (see below).

Dietary Intake

Although dietary intake of purines can theoretically affect the circulating level of uric acid,[194] this mechanism seems unlikely to account for the elevation in preeclampsia. Many investigators around the world have reported increased circulating uric acid concentration in preeclampsia, and dietary intake has rarely been controlled, suggesting that it is not a critical factor. Even when preeclamptic and normal pregnant women were studied in a metabolic ward where their dietary intake of calories, protein, and purine was rigidly controlled, hyperuricemia associated with preeclampsia was still observed.[200]

Disposal of Uric Acid

Another possible mechanism for increased circulating uric acid in preeclampsia was considered, i.e., reduced metabolic clearance. Stander and Cadden[203,204] suggested that impairment in liver catabolism of uric acid occurred in preeclampsia. Later work, however, showed that this mechanism is untenable as the enzyme uricase, which degrades uric acid to allantoin, is absent in humans.[231]

Renal Excretion of Uric Acid in Preeclampsia

Virtually all instances of hyperuricemia in preeclampsia are accompanied by reduced renal clearance.[161,171,207–209,213,215,219,221,225,231] Table 16.6A,B present a tabulation of the results from 10 investigations in which the renal handling of uric acid was measured in the last trimester of normal pregnancy, as well as in women with preeclampsia and pregnant subjects with chronic hypertension. First, hyperuricemia was a characteristic feature of preeclampsia, and the degree of elevation generally correlated with disease severity ("mild" vs. "severe"), or as in the study by Hayashi[171] with the early onset of disease (before 32 weeks of gestation). Second, in all of the investigations, the renal clearance of uric acid was reduced in preeclamptic compared to healthy pregnant women or pregnant women with chronic hypertension. Roughly speaking, the reduction in renal clearance of uric acid by one-half observed in preeclampsia would be expected to produce the two-fold elevation in circulating level (see the grand means listed in Table 16.6A,B). Third, the decrease in uric acid clearance can be accounted for by two factors: increased net reabsorption of uric acid by the renal tubules and reduced glomerular filtration. The former is the more universal mechanism, because in several of the studies depicted in Table 16.6A,B, and upon close inspection of the data for individual subjects, there were instances where serum level was substantially increased despite little, if any, reduction in GFR, making the more likely explanation an increase in net renal reabsorption.

Studies by Hayashi[211] and Czaczkes et al.[213] using probenecid to inhibit tubular reabsorption of uric acid support the conclusion that abnormal renal handling of uric acid in preeclampsia contributes significantly to the elevated circulating level in the disease. The study by Czaczkes et al. is particularly thorough. Fifteen preeclamptic women with edema, proteinuria, and hypertension, as well as three healthy pregnant women were serially studied. Twenty-four-hour creatinine and uric acid clearances were assessed for 3–4 days (control), again for the next 5–6 days while the women received 0.5 g of probenecid t.i.d., and finally for several days after stopping the medicine. The renal clearance of creatinine in this study was not reduced in preeclampsia compared to normal pregnancy, nor was it affected by probenecid in either group of subjects. The plasma uric acid was modestly reduced by the drug during normal pregnancy from 3.7 to 3.3 mg/dL due to augmented renal clearance (8.6 to 12.5 mL/min) and reduced net renal reabsorption (91.8 to 87.0%). Probenecid had a more dramatic effect in preeclampsia; the renal clearance of uric acid was greatly augmented (3.7 to 18.8 mL/min) while net tubular reabsorption was reduced (96.3 to 81 %). Most important, the plasma uric acid concentration was completely restored to normal pregnancy levels by the drug, from 6.9 to 3.2 mg/dL. Thus, by blocking the pathologic renal reabsorption of uric acid in preeclampsia, the elevated circulating concentration of the organic anion was completely restored to the normal pregnancy range. Because probenecid inhibits tubular reabsorption of uric acid, these data suggest that abnormal renal retention of uric acid in preeclampsia is secondary to enhanced tubular reabsorption and not to reduced secretion. Finally, it is noteworthy to point out that Hayashi reported "no clinical change was observed in the women with preeclampsia receiving probenecid suggesting that hyperuricemia alone is not an important pathogenetic factor in pregnancy toxemia."[211]

TABLE 16.6A Renal Handling of Uric Acid in Preeclampsia[a]

1st Author & Reference No.	Normal Pregnancy – Last Trimester					Preeclampsia–eclampsia				
	No.	P_{UA} (mg%)	c_{UA} (mL/min)	GFR (mL/min)	FR_{UA} (%)	No.	P_{UA} (mg%)	C_{UA} (mL/min)	GFR (mL/min)	FR_{UA} (%)
Schaffer[207],[b]	8	3.7 ± 0.3	33.6 ± 2.7	117 ± 10	70.2 ± 3.0	6	6.0 ± 0.3	18.9 ± 4.0	85 ± 16	77.5 ± 2.0
Chesley[161]	10	3.8 ± 0.2	15.2 ± 1.1	125 ± 9	87.5 ± 0.8	8[c]	6.6 ± 0.5	7.5 ± 0.6	93 ± 7	91.8 ± 0.6
Bonsnes[208]	7[d]	–	12.2 ± 2.4	–	–	32[e]	–	6.4 ± 2.0	–	–
Chesley[231]	29[f]	–	16.7 ± 1.3	–	–	30	–	7.8 ± 0.5	–	–
Seitchik[209]	8	3.7 ± 0.3	16.3 ± 1.1	153 ± 8[g]	89.5 ± 0.5	12, mild	5.1 ± 0.4	8.6 ± 0.8	130 ± 9[g]	93.2 ± 0.4
						2, severe	9.0	7.2	102	93.3
Hayashi[171]	16 <32 wks	3.2 ± 0.2	18.4 ± 1.2	129 ± 11	85.7[h]	8 < 32 wks	7.5 ± 1.1	4.8 ± 0.6	63 ± 9	92.4[h]
	27 >32 wks	3.0 ± 0.3	13.0 ± 0.8	108 ± 6	88.0[h]	25 > 32 wks	6.0 ± 0.3	6.5 ± 0.5	99 ± 6	93.4[h]
Czaczkes[213],[i]	3	3.7 ± 0.2	8.6 ± 0.2	101 ± 4	91.8 ± 0.6	15	6.9 ± 0.4	3.7 ± 0.2	101 ± 1	96.3 ± 0.2
Handler[215]	10	4.3 ± 0.4	8.8 ± 1.1	109 ± 7	92.1 ± 0.8	11, preeclampsia	5.6 ± 0.6	8.1 ± 1.8	105 ± 13	92.9 ± 1.1
						3, eclampsia	9.0 ± 1.2	2.0 ± 0.7	66 ± 6	96.9 ± 1.0
Fadel[221],[j]	13	4.2	14.0	160	91.1	17	6.7	7.2	122	94.2
Dunlop[225]	13	258 ± 16[k]	10.5 ± 1.1	110 ± 7	89.4 ± 0.9	18, mild	329 ± 24[k]	7.7 ± 0.7	95 ± 7	91.9 ± 0.4
						10, severe	365 ± 25[k]	7.1 ± 1.0	90 ± 7	92.3 ± 0.7
Grand mean[l]		3.8	13.4	124	89.4		6.7	6.5	96	93.5

TABLE 16.6B Renal Handling of Uric Acid in Chronic Hypertension[a]

1st Author & Reference No.	Chronic Hypertension				
	No.	P_{UA} (mg%)	C_{UA} (mL/min)	GFR (mL/min)	FR_{UA} (%)
Hayashi[171]	22, <32 wks	3.7 ± 0.3	14.5 ± 1.2	114 ± 10	87.3[h]
	26, > 32 wks	4.1 ± 0.2	14.2 ± 1.3	128 ± 12	88.9[h]
Handler[215]	4	4.2 ± 0.3	9.0 ± 1.1	123 ± 5	92.7 ± 1.1
Fadel[221],[j]	22	5.5	9.9	135	92.4
Dunlop[225]	5	261 ± 25[k]	13.1 ± 2.6	132 ± 22	90.2 ± 0.9
Grand mean		4.4	12.6	126	90.3

[a] Mean ± SEM. All renal clearances are expressed as mL/min or mL/min × 1.73 m^2 body surface area. No. = number of subjects; P_{UA}, = plasma or serum uric acid; CUA = renal clearance of uric acid; GFR = glomerular filtration rate. Unless otherwise indicated, GFR was assessed by the renal clearance of inulin or creatinine. FRUA = fractional reabsorption of uric acid (1.0 – CUA/C_{IN}) × 100(%).

[b] Fractional reabsorption of uric acid was not provided by Schaffer. It was calculated from the CUA and C_{IN} presented for each subject. Only those subjects with simultaneous renal clearances for uric acid and inulin were tabulated, and one subject without proteinuria was omitted from the preeclamptic group. In this study, the generally high values for the renal clearance of uric acid may be attributable to infused diodrast (for measurement of ERPF), which interferes with renal uric acid reabsorption.[194]

[c] Two subjects were omitted from the analysis due to oliguria, which diminishes the accuracy of renal clearance measurements. One subject had antepartum eclampsia.

[d] An unspecified number of these seven subjects were pregnant. However, the authors stated that there was no difference in the renal clearances of uric acid between nonpregnant and pregnant women.

[e] Included are three subjects who had antepartum eclampsia and another two with postpartum eclampsia.

[f] Normal pregnant subjects ranged from 8 to 40 weeks gestation.

[g] The renal clearance of mannitol was used as an estimate of GFR.

[h] FRUA was not given. It was, therefore, calculated based on the average values for CUA and GFR.

[i] Renal clearance data were based on 24-h urine collections.

[j] Only mean values were reported.

[k] Expressed as μmol/L. To convert μmol/L to mg/100 mL, multiply by 1.68×10^{-2}.

[l] The CUA is high and the FRUA is low in the work by Schaffer et al.[207] because, in addition to inulin, diodrast was also infused to measure effective renal plasma flow. However, diodrast, a radiocontrast agent, interferes with the tubular reabsorption of uric acid.[194] Therefore, these results of Schaffer et al. were omitted from the grand means.

However, more recently, it has been suggested that uric acid may indeed play a positive feedback role in the pathogenesis of the disease.[232,233] See Chapter 7 for further details.

There are three studies in which the renal handling of uric acid was evaluated in the same preeclamptic women during the antepartum and postpartum period.[161,207,208] Similar to the reduction in GFR, the restoration to normal levels occurred quite rapidly in the postpartum period, generally within the first week. Of particular note is the work by Bonsnes and Stander, who studied women daily in the immediate postpartum period.[208] A large recovery in the plasma level and renal clearance of uric acid was evident between the 5th and 7th postpartum days.

The factors capable of influencing renal tubular handling of uric acid include various hormones such as angiotensin II and norepinephrine, which stimulate net tubular reabsorption.[194] In this regard, however, the circulating renin-angiotensin system is believed to be suppressed not augmented in preeclampsia.[234] It is possible that intrarenal generation of angiotensin II could be increased, or that circulating AT1 receptor agonist autoantibodies that have been identified in the disease could be involved.[235] On the other hand, sympathetic activity to the skeletal muscle has been reported to be increased in the disease.[236] But, whether *renal* sympathetic nerve activity is increased in preeclampsia is unknown. Finally, estrogens are believed to augment the renal clearance of uric acid by reducing the net renal reabsorption,[194] thereby accounting for the lower circulating levels observed in female compared to male subjects. A considerable amount of investigation over the years has not produced compelling evidence that estrogen levels are different in preeclampsia compared to normal pregnancy. Even if estrogen levels are somewhat reduced as reported by some investigators,[237] the overall capacity of estrogens to affect the renal handling and plasma levels of uric acid seems relatively modest compared to the large reductions in renal clearance and elevations in circulating levels observed in preeclampsia.

Handler et al. suggested that an increased plasma level of lactate, presumably derived from the ischemic placenta, accounted for increased circulating uric acid in preeclampsia by enhancing its net renal reabsorption.[215] Indeed, infusion of sodium lactate into normal pregnant women produced a marked decline in the renal clearance of uric acid secondary to increased net renal reabsorption.[215,221] However, whether circulating lactate is increased in preeclampsia is controversial.[215,221,238] Finally, lactate affects the renal handling of uric acid by inhibiting tubular secretion, not by enhancing reabsorption.[194] Thus, the lactate hypothesis conflicts with the reports by investigators who used probenecid and established that the abnormality of renal uric acid handling in preeclampsia was enhanced tubular reabsorption and not reduced secretion.

The most likely explanation for enhanced renal tubular reabsorption of uric acid in preeclampsia is the relative or absolute contraction of plasma volume. In general, volume contraction has been shown to be associated with increased renal tubular reabsorption of uric acid.[194] Because preeclampsia is associated with reduced plasma volume,[239,240] this mechanism for augmented renal tubular reabsorption of uric acid seems most plausible. To our knowledge, the mechanism linking altered volume status and renal tubular handling of uric acid is unknown, although it is most likely ultimately related to the coupling of sodium and urate transport in the proximal tubule.[194]

Summary

The serum concentration of uric acid is decreased throughout much of human pregnancy by 25 to 35% from nonpregnant values, and during late gestation returns toward nonpregnant values. Although several factors could theoretically contribute to the decline in circulating levels, altered renal handling likely plays a major role. That is, the renal clearance of uric acid is increased during gestation by virtue of raised GFR, reduced tubular reabsorption, or both. In one study, the restoration of serum uric acid toward nonpregnant levels near term was explained by progressively increasing renal tubular reabsorption.

Hyperuricemia is frequently associated with preeclampsia. In fact, some investigators include it as a diagnostic criterion because of the strong correlation between hyperuricemia and the histologic finding of "glomerular endotheliosis" on antepartum percutaneous renal biopsy, which is believed to be a characteristic pathologic lesion of preeclampsia. Moreover, a significant correlation between the renal histologic severity of the preeclamptic lesion and serum uric acid concentration has been noted. Hyperuricemia is a relatively early change in preeclampsia and correlates with poor fetal prognosis.

Altered renal handling is an important mechanism underlying hyperuricemia in preeclampsia. Specifically, the renal clearance of uric acid is decreased mainly because net tubular reabsorption is increased, although reduced GFR and filtered load also contribute in patients with decreased renal function. An interesting finding is that probenecid, which inhibits renal tubular reabsorption of uric acid, was reported to restore the renal clearance and plasma level of uric acid in preeclamptic women to normal, thus implicating enhanced tubular reabsorption rather than diminished secretion as the cause. More recently, it has been suggested that increased metabolic production of uric acid by the fetus, placenta or maternal tissues in preeclampsia may contribute to its elevated circulating level, which, in turn, plays a role in the pathogenesis of the disease (Chapter 7). Finally, the renal clearance and plasma level of uric acid were restored rapidly to normal values by postpartum day 5 in women who suffered preeclampsia.

The precise etiology of exaggerated renal tubular reabsorption of uric acid in preeclampsia is uncertain, but most likely relates to the plasma volume depletion that is also a typical feature of the disease. Presumably, the volume contraction leads to enhanced renal reabsorption of sodium, and therefore of uric acid by the proximal tubule, although the precise mechanism is unknown.

RENAL HANDLING OF PROTEINS

Normal Pregnancy

Circulating Proteins

Because GFR and plasma concentration determine filtered load, the circulating level of various proteins in normal pregnancy must be considered. In this regard, there are many reports,[241–244] but perhaps the most comprehensive series of investigations were carried out by Studd and Wood.[243] They serially measured serum protein concentrations in the same women at several stages during pregnancy and 3 months postpartum. Significant decreases in albumin, thyroxine binding prealbumin, IgG, and IgA were noted, whereas significant increases in α_1- antitrypsin, transferrin, β-lipoprotein, complement fraction β_1-A-C, IgD, and α_2-macroglobulin were observed. Hemopexin, haptoglobin, and IgM were unchanged. Many of these findings were corroborated in other investigations.[242,244]

Using iodine 131-albumin, Honger showed that the intravascular mass of albumin was comparable between late-pregnant and nonpregnant women, and he suggested an increase in plasma volume accounted for the reduction in serum albumin concentration.[241] The extravascular mass of albumin tended to be lower in pregnancy, suggesting that the capillary permeability was decreased or the lymphatic flow increased. Interestingly, the catabolic and synthetic rates were comparable, indicating suppression of synthesis in normal pregnancy, because low serum albumin concentration is normally countered by an increase in synthesis. Honger suggested that estradiol or progesterone may mediate the reduced synthesis of albumin, which was supported by later studies using derivatives of estradiol and progesterone.[243]

Urinary Excretion of Total Protein and Albumin

It is widely held that urinary protein excretion increases during normal pregnancy with an upper limit of approximately 300 mg/24 hours. Although some reports suggested that urinary albumin excretion decreases[242] or remains unchanged during normal pregnancy,[245–249] the majority support an increase.[10,250–262] In a cross-sectional study using spot urines, Cheung et a1.[255] demonstrated a significant rise in urinary excretion of albumin from a nonpregnant median value of 1.19 to 1.76, 1.90, and 1.91 g/mol of creatinine during the first, second, and third trimesters, respectively. In this study, the urinary excretion of albumin was approximately 30% of total protein excretion.[255] In another meticulously conducted study, Douma et al.[261] collected both 24-hour and separate 12-hour night- and daytime urines from nonpregnant women and women during the third trimester on an outpatient basis. Furthermore, they conducted a similar investigation in a metabolic ward with the subjects on strict bed rest. In both experimental settings, the 24-hour, as well as the nocturnal and diurnal urinary excretion of albumin was greater in pregnancy, and the difference between pregnant and nonpregnant subjects was greatest during the night. In the outpatient setting both the nonpregnant and pregnant women showed a significant day–night difference with *lower* excretion at night. Although this day–night difference was preserved for nonpregnant women on continuous bed rest in the metabolic ward, it was lost in pregnant subjects, which suggested a blunting of the intrinsic (i.e., independent of upright posture and normal daily activity) circadian variation in albumin excretion during pregnancy.[261] Taylor and Davison[262] studied five healthy women before pregnancy, then on 12 different occasions during gestation, and again 12 weeks postpartum. Urinary albumin excretion was significantly increased by 16 weeks gestation, showing overall a two- to three-fold increase by term, which returned to preconception levels by at least postpartum week 12. In contrast, the urinary excretion of total protein measured in the same samples was not significantly increased until the third trimester. Urinary albumin excretion was approximately 10% of the total protein excretion in this study. This work generally corroborated an earlier study from the same laboratory, except that in the latter the elevated urinary excretion of albumin persisted at least until 16 weeks postpartum.[10] This interesting finding of persisting albuminuria in the postpartum period was also reported by of McCance et al.,[256] Lopez-Espinoza et al.,[251] and Wright et al.[252] Thus, the alterations in renal function that affect urinary albumin excretion during normal pregnancy may not be completely restored to nonpregnant levels until some time after delivery.

Urinary Excretion of Low Molecular Weight (LMW) Proteins

Because β_2-microglobulin (β_2-m), retinol-binding protein (RBP), and Clara cell protein are LMW proteins of 11.8, 21.4, and 15.8 kDa, respectively, they are considered to be more or less freely filtered at the glomerulus. At the same time, they are virtually reabsorbed to completion in the proximal tubule. Thus, the concept has emerged that increased urinary excretion of these LMW proteins mainly reflects inadequate proximal tubular reabsorptive capacity.[263–265] Most[242,246,247,253,255,266] but not all[267,268] investigators observed increased urinary excretion of LMW proteins during normal pregnancy. Of particular note is the work of Bernard et al.,[247] who reported a progressive increase in the urinary excretion of all three of these LMW proteins throughout pregnancy, which significantly exceeded nonpregnant values by the second and third trimesters. Because the plasma concentrations of these LMW proteins did not increase during pregnancy, and the pattern of increase in GFR did not correspond with that of the LMW protein excretion, Bernard et al. suggested that their findings were consistent with a decrease in proximal tubular reabsorptive capacity, which recovered after delivery. The urinary excretion of α_1-microglobulin (26–33 kDa)[247] and of immunoglobulin light chains (25 kDa)[246] was also increased in normal pregnancy, although these proteins are of more intermediate molecular weight and are not so freely filtered. Interestingly, the absolute urinary excretion or fractional excretion of other substances which are primarily reabsorbed by the proximal tubule such as glucose, amino acids, calcium, and uric acid is also increased during gestation[12,269] (see "Renal Handling of Uric Acid," above). Taken together, these data indicate a general, physiologic compromise of proximal tubular reabsorption during normal pregnancy.

Urinary Excretion of Enzymes

Most,[235,254,255,270–272] but not all,[267] investigators have also reported increased urinary excretion of several large molecular weight enzymes during normal human gestation. These enzymes are presumably of proximal tubular origin being located in the brush border (γ-glutamyl transferase, alanine aminopeptidase, and tissue nonspecific alkaline phosphatase) or from lysosomal sources (β-glucoronidase, *N*-acetyl-β-D-glucosaminidase, α-galactosidase, β-galactosidase, and α-mannosidase). The urinary excretion of γ-glutamyl transferase was observed to be increased during pregnancy by Noble et al.,[270] Cheung et al.,[255] but not by Kelly et al.[267] Similar increases were reported for alanine aminopeptidase.[255,271] Urinary excretion of some of the lysosomal enzymes including *N*-acetyl-β-D-glucosaminidase and β-galactosidase was reported to be increased, too.[254,255,271,272] A raised filtered load could contribute to the increased urinary excretion of these enzymes during gestation: the plasma concentrations of *N*-acetyl-β-D-glucosaminidase and β-galactosidase are higher in pregnancy,[272] the GFR is also increased, and there may be subtle alterations in glomerular permeability or electrostatic charge allowing for greater filtration of these large molecular weight proteins. On the other hand, Jackson et al. observed that the molecular weights of the urinary lysosomal enzymes more closely matched those measured for the enzymes extracted from kidney homogenates than from the serum, suggesting a renal origin for the proteins.[272] On balance, these data further implicate physiologic alterations in the proximal tubule during normal gestation.

Glomerular Permselectivity and Proximal Tubular Reabsorption

The cause of increased urinary albumin excretion observed by most investigators during normal gestation is probably multifactorial. On the one hand, the filtered load is increased by virtue of elevated GFR, although this is offset somewhat by a decline in circulating albumin concentration. On the other hand, studies to date have failed to detect any change in the permselectivity of larger molecular weight proteins in human gestation, at least on the basis of size or molecular weight alone. Indeed, Roberts et al.[10] demonstrated *reduced* fractional clearance of smaller neutral dextran particles (30–39 Å, ~20 kDa) throughout normal human pregnancy, and no alteration in the fractional clearance of larger dextrans. Whether the anionic charge of the glomerular barrier is altered allowing for greater passage of plasma proteins has not been throughly investigated. In this regard, however, Cheung et al.[255] showed that the rise in urinary excretion of transferrin greatly exceeded the rise in urinary excretion of albumin during normal pregnancy despite similar molecular weights, possibly indicating an alteration in the charge of the glomerular membrane as transferrin is considerably less anionic than albumin. Finally, based on the data dealing with urinary excretion of LMW proteins (see above), inadequate proximal tubular reabsorption may contribute to increased urinary albumin excretion in normal pregnancy, too. This view is consistent with the argument that proximal tubular reabsorption and not glomerular permselectivity is the main factor regulating albuminuria in the normal kidney.[273] However, another possibility is that increased albumin excretion of normal pregnancy is also of exosomal origin.[274]

Podocyturia

Podocytes and proteins specific to their cells can be detected in the urine, their rate of excretion now being investigated as potentional biomarkers in renal disorders, especially in relation to preeclampsia (as will be discussed below).[275] As of 2014, however, it is fair to say that the methodologies used for the detection of podocyte specific markers (e.g., podocin, podocalyxin, synaptopodin, and nephrin [276]) and the development of feasible clinical testing remain technically challenging. In addition, norms for their rate of excretion in normotensive pregnant or nonpregnant women have yet to be determined (though as described below there has been considerable enthusiasm about their use to diagnose preeclampsia).[277]

Preeclampsia

Circulating Proteins

With one exception,[278] both serum albumin[244,279–281] and IgG[244,279,280,282] concentrations were further decreased in preeclampsia relative to normal pregnancy. Serum α_2-macroglobulin was further increased in preeclampsia relative to normal pregnancy.[244,279,280,283] Again, with one exception,[244] serum transferrin concentration fell during the disease.[279,280,282] In fact, Studd et al. reported that the pattern of change in serum proteins in preeclampsia was remarkably similar to the nephrotic syndrome, i.e., reduced serum concentrations of albumin, total protein, thyroxine-binding prealbumin, IgG, and transferrin, as well as increased levels of α_2-macroglobulin, β-lipoprotein and β_1-A-C complement fraction. They concluded that these alterations in serum proteins in part reflected heavy urinary loss of intermediate molecular weight proteins with relative retention of the larger species.[279,280]

Using iodine 131-albumin, Honger analyzed the metabolic fate of albumin in preeclampsia relative to normal pregnancy.[281] He reported that plasma volume and serum concentration of albumin as well as the intravascular and total exchangeable mass of albumin were all reduced in the disease. The fractional rate of albumin disappearance was increased, which he attributed approximately equally to urinary loss and hypercatabolism and/or increased gastrointestinal loss. Laakso and Paasio showed exaggerated loss of I 131-PVP (polyvinylpyrrolidone) into the feces of women who were preeclamptic 2 to 8 days postpartum, most likely due to increased capillary permeability.[284] Although Honger showed that the rate of albumin synthesis was increased in preeclampsia relative to normal pregnancy, evidently it was inadequate as serum albumin concentrations were further reduced in the disease.[281] Low serum albumin is consistent with the generalized state of "systemic inflammation" and activation of the "acute-phase response" which typifies preeclampsia, compromising hepatic albumin synthesis.[285] Interestingly, the distribution of albumin between the intravascular and extravascular compartments was not different,[281] which is counterintuitive since generalized capillary leakiness is believed to be a disease manifestation of preeclampsia.[286]

Urinary Excretion of Total Protein and Albumin

In 1843, Lever first reported protein in the urine of a woman with eclampsia.[287] Today, in addition to fulfilling blood pressure criteria, proteinuria in excess of 300 mg/24 h is part of the clinical diagnosis of preeclampsia.[288] In fact, preeclampsia is the most common cause of nephrotic-range proteinuria in pregnancy,[289] and the magnitude of protein excretion correlates significantly with the histologic severity of the renal lesion.[181] The proteinuria consists of a number of different proteins as identified by many investigators over the years.[282,290–296] These urinary proteins are mainly derived from the plasma, including albumin, α_1- and α_2-globulin, β-globulin, γ-globulin (IgG, IgA, and occasionally IgM), ceruloplasmin, pseudocholinesterase, and α_2-macroglobulin.

Glomerular Permselectivity

Normally, the glomerulus effectively retains proteins that are of the size of albumin or greater. The small amounts that are filtered are thought to be reabsorbed in the proximal tubule. Yet, there are a large number of membrane and high MW proteins in normal human urine[297] perhaps reflecting a tubular shedding process or secretion of exosomes.[274] For example, in a recent investigation of the urinary proteome in healthy humans, albumin constituted 25% of the protein and Tamm-Horsfall protein or uromodulin[298] 1.3%.[297] These proteins were also identified in another study of the exosome proteome in normal human urine.[274] In glomerular disease when urinary protein exceeds ~1 g/day, at which point reabsorptive capacity of the proximal tubule has been overwhelmed, one can compare the renal clearances of plasma proteins relative to their molecular weights, thereby evaluating the integrity of the glomerular barrier. In this setting, several investigators have reported the renal clearances of plasma proteins and of dextran or the inert polymer PVP in women with preeclampsia. Dextran and PVP are exogenous compounds, and therefore must be administered intravenously. They come in a variety of molecular sizes, and are not reabsorbed or secreted by the renal tubules, making their renal clearance solely determined by glomerular filtration and the integrity of the glomerular filter. In general, the relationship between the log of the renal clearance of plasma proteins ranging in size from 69 kDa (albumin) to 1000 kDa (IgM), and the log of the molecular weight or particle size is inverse and linear. The same holds true for dextran or PVP. By convention, the renal clearances of selected plasma proteins are factored by the renal clearance of one of the smaller plasma proteins such as transferrin, and plotted against their respective molecular weights. Thus, a steeper or flatter slope, respectively, represents a more or less "selective" or "unselective" proteinuria, signifying the relative retention or loss of larger molecular weight proteins by the glomerular filter (Fig. 16.8(A)). Using this technique, MacLean et al.[300] and Robson[299] noted that the protein selectivity was in an "intermediate range" for preeclampsia. The measurement of dextran selectivity in five of the same subjects by MacLean et al. yielded similar results. These corroborative findings using dextran suggested that the protein selectivity data were indeed reflecting a glomerular abnormality, and not tubular changes in protein processing during preeclampsia. As well, because comparable results were obtained with both charged (plasma proteins) and uncharged (neutral dextran) molecules, significant alteration in the glomerular electrostatic barrier seemed unlikely. Robson commented that, when viewed in the context of primary glomerular diseases, the intermediate values for selectivity of protein in preeclampsia seemed out of proportion relative to the minor anatomic changes in the glomerulus (Fig. 16.8(B)). Clearly, there is a paucity of data on the simultaneous measurements of the renal clearances of plasma proteins and of dextran or PVP in women with preeclampsia. In this regard, however, a recent investigation suggested *reduced* clearance of dextran molecules (33–49 Å) in preeclampsia compared to normal late-pregnancy values, indirectly supporting a loss

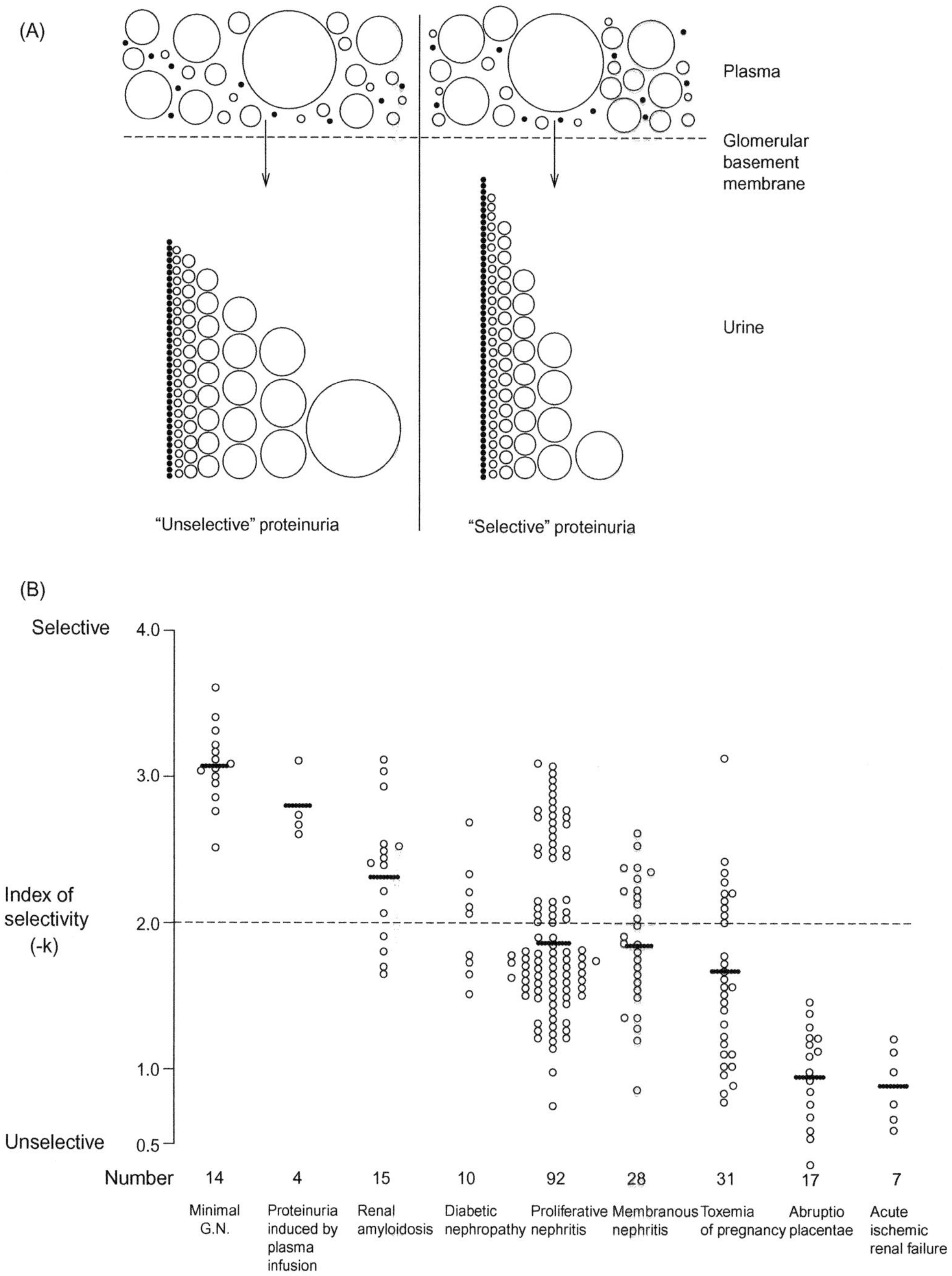

FIGURE 16.8 (A) Illustration of the concept of glomerular selectivity for proteinuria. The slope, $-k$ or the index of selectivity, is determined by the clearance of the proteins ranging in size from 60,000 to 2,000,000 daltons ($U_{protein}-V/P_{protein}$) expressed as a percentage of the clearance of the smaller plasma proteins such as albumin or transferrin. (B) Indices of protein selectivity for a variety of glomerular diseases, preeclampsia, and abruptio placentae. Again, "$-k$" relates the molecular size of selected proteins to their renal clearance, the latter being normalized to the renal clearance of relatively small molecular weight proteins such as albumin or transferrin.[299]

of charge-selectivity as the major cause of increased urinary albumin excretion in the disease (see below).[32]

Using similar methodology, Simanowitz et al.[295] substantiated the finding of intermediate-range protein selectivity in preeclampsia. By using an abbreviated technique of measuring the renal clearance of IgG relative to that of transferrin, Kelly and McEwan[282] as well as Simanowitz and MacGregor[301] again observed intermediate-range protein selectivity. Katz and Berlyne reported variable protein selectivity ranging from highly selective to unselective, but they did not actually calculate the renal clearances of the various plasma proteins.[296] It should be noted, however, that a "variable" pattern is not surprising given the microscopic picture of the disorder (see below). That is, while neither basement membrane nor foot process changes are noted in many renal biopsies, such changes can be seen with very severe disease.

Wood et al.[302] evaluated the glomerular filter by testing the renal clearance of PVP (10,000–100,000 MW). They reported a "vasoactive pattern" for the PVP clearance–molecular radius relationship in gestational hypertension and mild preeclampsia comparable to that observed with the infusion of angiotensin II, which increases glomerular capillary hydrostatic pressure, thereby increasing the filtration of intermediate-sized molecules, and is fully reversible. This finding is consistent with the intermittent nature of proteinuria in preeclampsia as first reported by Chesley.[303] In moderate to severe disease, a "membranous pattern" was noted, which Wood and colleagues postulated to be a consequence of intravascular coagulation and alterations in the glomerular basement membrane.[302]

An important component of the glomerular barrier is widely believed to be the negative charge associated with the polyanionic glycoproteins on the endothelial surface and podocytes, as well as possibly with the heparin sulfate-rich glycosaminoglycans in the lamina rara interna and externa of the basement membrane.[190,304] Disruption of this electrostatic barrier is not detected by the clearance of neutral dextrans and PVP. The work of MacLean discussed above, showing comparable results for both the plasma protein and dextran clearance techniques, is not supportive of a major change in the electrostatic properties of the glomerular barrier in preeclampsia.[300] In contrast, more recent publications by Naicker et al.[305,306] provided experimental evidence for fewer polyethyleneimine-labeled anionic sites in the glomerular basement membrane on renal biopsies procured 2 weeks after delivery from women with early-onset preeclampsia, at least when compared to nonpregnant subjects who underwent partial nephrectomy due to trauma.

Urinary Excretion of Low Molecular Weight Proteins and Enzymes

Weise et al. reported urine excretory rates for β_2-m of 0.11 and 0.5 mg/24 h for normal pregnant and preeclamptic women, respectively.[307] Although Pedersen et al. did not find a significant difference in renal β_2-m excretion between preeclamptic and normal pregnant women in the third trimester (both were increased relative to nonpregnant subjects; see "Normal Pregnancy" above), the increases persisted in the preeclamptic group 5 days postpartum with complete resolution not being observed until 3 months postpartum.[246] The same group of investigators noted substantial increases in immunoglobulin light-chain excretion in preeclampsia relative to third-trimester control subjects which also persisted 5 days postpartum and resolved by 3 months.[246] In contrast, Oian et al. reported reduced renal β_2-m excretion in preeclampsia when compared to normal third-trimester subjects.[266] Interestingly, Kreiger et al. observed elevated renal β_2-m excretion in subjects with pregnancy-induced hypertension, the majority of whom had proteinuria <0.3 g/24 h and normal serum uric acid.[268] Because the LMW proteins are freely filtered at the glomerulus, the additional increase in urinary excretion observed by many investigators in preeclampsia relative to normal pregnancy may reflect further impairment of proximal tubular function, above and beyond that of normal pregnancy (see "Renal Handling of Proteins – Normal Pregnancy," above).

The renal excretion of enzymes in preeclampsia has been reported by several groups of investigators. Goren et al. found a significant increase in both the renal excretion of *N*-acetyl-β-D-glucosaminidase and alanine aminopeptidase in preeclamptic women compared to normal third-trimester subjects.[271] Jackson et al. conducted an extensive investigation and observed that the urinary excretion and/or fractional excretion of five lysosomal enzymes were significantly increased in the disease.[308] Finally, Shaarawy et al. reported increased urinary excretion of nonspecific alkaline phosphatase, an enzyme localized to the proximal tubule, which correlated with the severity of preeclampsia.[309] The increase in renal excretion of enzymes relative to normal pregnancy reinforces the concept of pathologic proximal tubular dysfunction in the disease.

Podocyturia

Potential molecular underpinnings for the proteinuria of preeclampsia were noted in the previous edition, where then recent evidence that podocyte secretion of vascular endothelial growth factor (VEGF) maintained glomerular capillary integrity was cited. The current hypothesis is still that elevated circulating concentrations of soluble VEGF receptor 1 (sFlt-1) in preeclampsia, by neutralizing VEGF, disrupts the glomerular endothelial barrier, best illustrated by Ermina et al.,[310] who demonstrated that deletion of an allele of VEGF in podocytes of mice produces glomerular endotheliosis resembling the human pathology (Chapter 6). Subsequently, others using an animal model of preeclampsia produced by overexpression of sFlt-1 reversed this lesion by administering VEGF-121.[311] Podocyturia and podocyte proteins have been detected in urine of women who have or are destined to have preeclampsia by several groups, one claiming 100% sensitivity and specificity in predictive accuracy.[275,312,313] However, that claim is disputed[314,315] and further discussion of the controvery over podocyturia's predictive accuracy is detailed in Chapter 11. In this respect we note that podocyturia is not unique to preeclampsia as it occurs in other hypertensive proteinuric disorders.[275] Also the reliability of methodologies used either to detect or culture urinary podocytes, or to measure their specific proteins used in many of these studies remains uncertain. For a succinct discussion of podocyturia in normal and preeclamptic pregnancies see the review article by Polsani et al.[277]

Summary

The urinary excretion of total protein, albumin, LMW proteins, and of several renal tubular enzymes increases during normal human pregnancy. The increase in renal excretion of LMW proteins and of renal tubular enzymes is consistent with a physiologic impairment of proximal tubular function. This concept is also supported by the finding of elevated urinary excretion or fractional excretion of glucose, amino acids, uric acid, and calcium during normal human gestation[12,316] (see "Renal Handling of Uric Acid," above). In addition to impaired proximal tubular reabsorption, the elevated GFR of pregnancy and possibly an alteration in the electrostatic charge of the glomerular filter also contribute to gestational albuminuria. Available evidence does not indicate an increase in the glomerular permselectivity on the basis of size or molecular weight. Whether tubular shedding or secretion of exosomes containing membrane-associated and high molecular weight proteins like albumin increases during gestation has not been studied in detail.

In preeclampsia, the urinary excretion of total protein, albumin, LMW proteins, and of several renal tubular enzymes is exaggerated compared to normal pregnancy. The further increase in renal excretion of LMW protein and renal tubular enzymes suggests pathologic compromise of proximal tubular function. The gross albuminuria and excretion of other plasma proteins is likely to be secondary to alteration of both the molecular size constraints and of the electrostatic properties of the glomerular filter, and compromised proximal tubular reabsorptive capacity. Despite the calculated increase in afferent arteriolar resistance, an increase in glomerular hydrostatic pressure cannot be discounted, which may contribute to increased glomerular transit of albumin. Further investigation of both the permselective and electrostatic properties of the glomerular filter in preeclampsia is needed. An unresolved paradox is the reduced reabsorptive capacity for proteins in preeclampsia implicating proximal tubular dysfunction, and the enhanced reabsorptive capacity for uric acid by the proximal tubule in the disease (see "Renal Handling of Uric Acid," above).

Regarding potential molecular underpinnings for proteinuria in preeclampsia, recent evidence indicates that podocyte secretion of vascular endothelial growth factor (VEGF) maintains glomerular capillary integrity, and the elevated circulating concentrations of soluble VEGF receptor 1 or sFlt-1 in preeclampsia neutralizes VEGF, thereby disrupting the glomerular endothelial barrier[2] (see Chapter 6). Also, podocytoria has been noted in the urine of preeclamptic women, though its specificity is doubtful, but remains to be studied.[277]

RENAL MORPHOLOGY IN PREGNANCY AND PREECLAMPSIA

Introduction

Chapter 4 in Chesley's original single-authored edition, entitled "Structural Lesions," contained an extensive historical review of autopsies in preeclampsia. Although preeclampsia is a systemic vascular illness, its hallmark lesion involves the renal glomerulus, changes that were initially described by Löhlein in 1918.[317] Mayer, using autopsy material in 1924, may have been the first to mention glomerular endothelial swelling in eclampsia.[186] However, full characterization of the lesion required the application of electron microscopy, as described in 1959 by Farquhar[318] at Yale and Spargo[319] at the University of Chicago. The latter was responsible for the term that has become a pathological metonym for preeclampsia: "glomerular endotheliosis." The most extensive and modern series appeared in the 1973 publication of Sheenan and Lynch,[3] an autopsy study that focused on material obtained shortly following death (see Editors' note). Evaluation was limited to light microscopy.

The chapter written by Chesley in the singled-authored first edition of this book also stressed how our knowledge of the morphological changes in preeclampsia virtually exploded following the introduction of renal biopsy on pregnant women during the 1950s. The ability to evaluate "living tissue" not only eliminated postmortem changes, but also allowed for better correlation with physiologic parameters and outcomes. Of note, renal biopsies were used in both the study and management of hypertensive pregnancies from the 1950s through the 1970s. During this period, as investigators focused on describing lesions diagnostic of preeclampsia, postpartum biopsy was also heralded as useful in predicting the outcomes of future pregnancies. As clinicians became aware that such biopsy information was at best only marginally helpful, and could not be justified in terms of risk–benefit, the use of kidney biopsies either for management of hypertension in pregnancy or for prediction of future gestational outcomes was justly discontinued in most centers (as echoed in our recommendations in Chapter 17 of the previous edition). Thus publications from the 1990s onwards contain very small series, raising concerns regarding selection bias.

The renal lesions of preeclampsia (which are similar to those of eclampsia) fall within a spectrum of disease collectively referred to as "thrombotic microangiopathy" (TMA). These are a group of clinically diverse disorders (such as malignant hypertension and hemolytic uremic syndrome) defined by a common site of injury – the endothelium, and sharing a similar morphologic pattern of vascular injury (often accompanied by evidence of thrombosis). The lesions of preeclampsia share some similarities with other TMA, but there are some important and interesting differences that relate to its unique pathogenesis. Thrombosis, while a defining feature of most TMA, is decidedly unusual in preeclampsia. This likely reflects its specific pathogenesis – glomerular endothelial injury secondary to deprivation of proangiogenic factors (see below). Non-preeclamptic TMA, while incompletely understood and likely multifactorial, is not thought to reflect impaired vascular endothelial growth factor (VEGF) signaling. Furthermore, the fact that glomerular but not other renal vascular endothelium is targeted emphasizes the variable and site-specific phenotype of endothelium.

The glomerular pathology of preeclampsia is unusual – the changes develop in response to a relatively acute insult that is typically of short duration and resolves with delivery and restoration of angiogenic factor balance. Furthermore, the lesions are primarily intracellular, and do not involve the deposition of extracellular matrix or immune complexes. Consequently they can resolve more easily and normal anatomy can be restored relatively quickly. Failure to account for the particular point in the natural history of the disease at which the tissue was obtained for examination may account for some of the discrepancies between different pathologic series.

Renal pathology as a discipline has always relied on correlation with clinical and physiologic parameters for diagnostic, prognostic, and therapeutic purposes. With regard to the renal lesions of preeclampsia, we are on the cusp of an evolutionary advancement. As discussed elsewhere, cross-sectional and longitudinal studies have shown than high sFlt-1 and low PlGF are present during preeclampsia and prior to clinical disease.[320] The recent availability of automated platforms to measure them has allowed validation of these biomarkers in several cohort studies.[321] Furthermore, these tests could be used to distinguish other renal diseases that should be considered within the differential diagnosis of preeclampsia.[322] The introduction of these tests into routine clinical use has enormous potential. So too has this information, once available on individual patients for histopathological correlation, the potential to advance our understanding of pathologic findings and their significance. (See also Chapter 6.)

Gross Morphology

There is an obvious paucity of information regarding the gross appearance of kidneys in women undergoing normal pregnancy. In the series of Sheehan and Lynch[3] the combined kidney weights from normotensive pregnant women at autopsy were slightly above those published for nonpregnant populations. This may be due to the increased extracellular volume. In rats, where kidney weight also increases in pregnancy, this difference disappears when dry weights are examined.[323] The kidneys of women dying of preeclampsia/eclampsia usually had a pale broad cortex, and were variably enlarged compared to women dying from "normal" causes.

Light Microscopy

Glomeruli

There are few data concerning light or electron microscopic descriptions of the kidney in normal gestation that permit accurate comparisons to what occurs in preeclampsia. However, data from Sheehan and Lynch's autopsy series, and a more recent renal biopsy study performed on 12 normal third-trimester gravidas[324,325] (and see letter to the editor in BJOG 2004;111:191 re: ethical concerns regarding the study), suggest that glomerular size but not the number of cells is increased in pregnancy.

The primary focus of renal involvement in preeclampsia is the glomerulus, where tufts are enlarged, swollen, and solidified – "bloodless." Capillary lumens are partially or totally occluded by endothelial (and to a much lesser extent mesangial) swelling, without an overt increase in endocapillary cellularity – the definition of "endotheliosis"[326] (Fig. 16.9(A) and (B)) The change may be segmental but is often global and widespread, and, while characteristic, varies with the severity of the disease and most notably with the time course at which it is examined. While the swelling may create the appearance of capillary wall thickening, the basement membrane itself is not thickened. The increase in glomerular volume may be of such a magnitude that the tuft herniates through the urinary pole into the proximal tubule, termed "pouting"[3,326] (Fig. 16.10). In some particularly severe instances, the mesangial matrix is expanded and "mesangial interposition," i.e. cells extending into the peripheral capillary loops between the endothelium and basement membrane, may be seen. Subsequently, "double contours" ("tram tracks") may be seen, particularly in the resolving stage, as in other chronic TMA. While such lesions, suggesting endothelial injury of some duration, appear more frequently in a few reports,[327,328] they are unusual in larger series.[329] In addition deposition of fibrin-like material and foam cells are occasionally described.[326,330,331] Their frequency has varied greatly between different biopsy series. This at least in part reflects the evolution of disease over time and the relatively quick resolution of some of its morphologic expressions. True glomerular thrombosis is quite unusual, and when present suggests a superimposed non-preeclamptic TMA.

Visceral epithelial cells (podocytes) may show swelling and other reactive changes such as resorption droplets as a nonspecific response to the proteinuria. True crescents are seen in a small minority of cases, but mainly in association with atypical and severe disease.[332] The associated finding of segmental sclerosis is discussed below.

Tubulointerstitium and Vasculature

Proximal tubules may show nonspecific changes associated with heavy proteinuria, such as reabsorption droplets.[3,326] There may also be changes attributed to acute tubular injury or rarely to hemoglobinuria secondary to hemolysis.

(A)

(B)

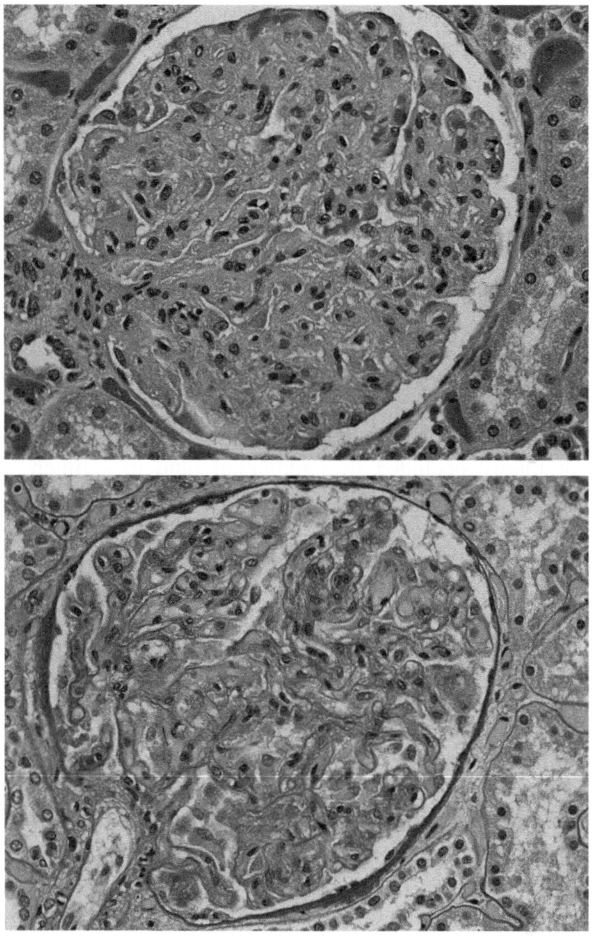

FIGURE 16.9 (A): Glomerular tuft with endotheliosis. Capillary lumens are occluded by swollen endothelial cells. Note that overall glomerular cellularity in not significantly increased. Vascular congestion, but not thrombosis, is present. (H&E, 400×). (B): Glomerular tuft with endotheliosis. PAS stain highlights the capillary basement membrane, which is within normal limits. In contrast, swollen endothelial cytoplasm only stains weakly. Endothelium at the vascular pole is characteristically uninvolved. (PAS, 400×). (This figure is reproduced in color in the color plate section.)

Interstitial fibrosis and tubular atrophy reflect chronic (usually preexisting) disease, as does vascular injury such as arteriosclerosis and arteriolosclerosis. Vascular thrombosis or other vascular changes of acute thrombotic microangiopathy are unusual. While they may be noted in severe cases or true eclampsia, their presence suggests the possibility of a superimposed process (such as HELLP syndrome or antiphospholipid antibodies). Hill et al., in a biopsy study, failed to demonstrate morphological changes in the juxtaglomerular apparatus that might have suggested activation of the renin-angiotensin system.[333]

Electron Microscopy

Preeclampsia may be the one exception to the insufficient attention that has been focused on the role of the healthy, uniquely fenestrated glomerular endothelium as an important component of filtration, and as a barrier to proteinuria.[334] While extensive evidence supports the key role of VEGF in inducing fenestrations, it bears recalling that acutely injured cells of all types swell and lose their specialized substructures (in this case fenestrations). Thus the possibility that there are other non-VEGF related pathways of endothelial injury in preeclampsia remains an open consideration.

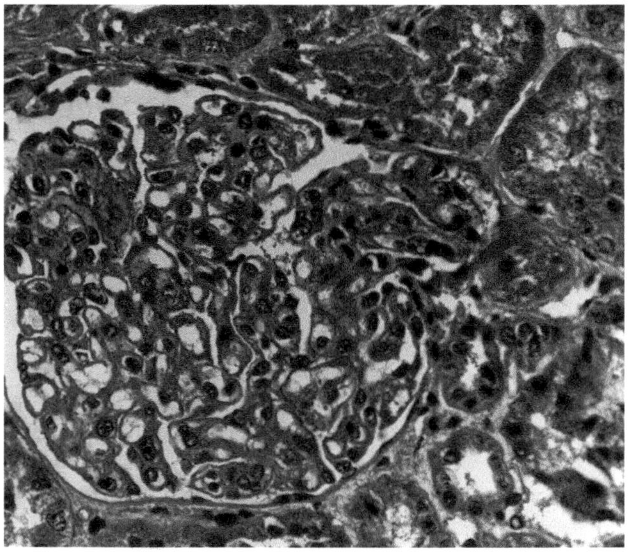

FIGURE 16.10 Light micrograph of a glomerulus from a patient with preeclampsia. The glomerulus is so swollen that it herniates into the proximal tubule, a phenomenon called "pouting" (×350).[1]

The characteristic lesion of preeclampsia, acute endothelial injury, or endotheliosis, is definitively appreciated by electron microscopy (Fig. 16.11). Indeed, mild cases may be difficult to appreciate by light microscopy alone. Endothelial cells, having lost their distinctive fenestrations (60–90 nm in diameter), are variably but usually markedly swollen, often occluding the capillary lumen and demonstrating other nonspecific ultrastructural signs of injury such as vacuolization and lipid accumulation.[191] This change is typically more pronounced than in other forms of TMA. However, it must be emphasized that poor tissue fixation can produce artifactual swelling of the endothelium that can be difficult to distinguish from the real thing. Foam cells (usually endothelial but occasionally mesangial) may be present, probably as a nonspecific response to proteinuria. Electron lucent ("fluff") expansion of the subendothelial space, a common finding in other acute TMA, may also be present but is not typically as prominent. The lamina densa itself shows normal thickness. Fibrin tactoids are rare unless the disease is severe. Granular electron-dense deposits are inconspicuous, in keeping with the lack of a role for immune complexes in this process. With increasing time and severity, mesangial interposition and neo-membrane formation can be seen.

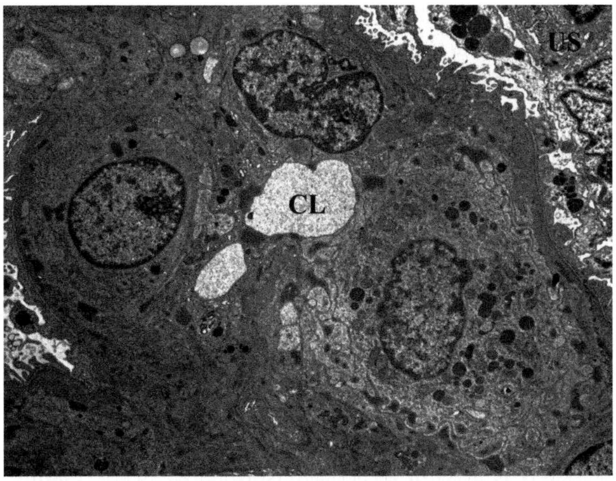

FIGURE 16.11 Endotheliosis by transmission Electron Microscopy. CL, residual glomerular capillary lumen, which has been markedly narrowed by swollen endothelial cells. US, urinary space, in which podocytes are noted to have vacuoles consistent with protein resorption. Note that their foot processes are largely intact. (Original magnification 3000×).

Interestingly, despite significant proteinuria, podocytes usually show relative preservation of their foot processes, particularly in the acute phase, a finding shared with other TMA.[335] Filtration slit frequency in preeclampsia is not significantly reduced below controls, an observation of significance in understanding mechanisms of proteinuria in general.[191] Despite this, some have suggested that podocyturia may be a marker for preeclampsia (see discussion in podocyturia section above).[275,312] Obviously the cell culture technique cannot serve as a diagnostic test but may be a useful research tool, particularly given the podocytes role in VEGF production.[336]

Immunohistology

The pathogenic role, if any, that fibrin and its related products might play remains incompletely understood. Vassali, Morris, and colleagues[337,338] underscored the presence of considerable glomerular fibrin deposition in preeclampsia, suggesting a causative role in the pathological changes. Subsequent reports were discrepant, our experience (summarized in Table 16.7)[181] being that fibrin is present in less than half the case material, and usually but 1+ in intensity. When seen, the glomerular tuft is diffusely positive; thrombi are unusual. The timing of the sample is critical as subendothelial fibrin is among the first lesions to regress after delivery, and our series reflects biopsies primarily performed in the immediate puerperium. It must be emphasized that segmental or even global/diffuse glomerular staining (? entrapment) for fibrin is not unusual in renal biopsies with acute injury, and by itself (i.e., without accompanying lesions on light microscopy) is not definitively diagnostic.

TABLE 16.7 Immunofluorescence Findings in Biopsies Depicting Glomerular Endotheliosis

	Frequency of Glomerular Deposition	
	All Cases	**Greater Than 1+**
Fibrin	20 of 45	8 of 45 (18)[a]
AHG[b]	5 of 10	1 of 10 (10)
IgG	10 of 43	3 of 43 (7)
IgM	23 of 45	16 of 45 (36)
IgA	6 of 43	2 of 43 (5)
IgE	9 of 36	2 of 36 (6)
C3	8 of 40	1 of 40 (3)
C4	15 of 36	10 of 36 (28)

Reprinted, with permission, from Fisher, K. A., Luger, A., Spargo, B. H., Lindheimer, M. D. (1981). Hypertension in pregnancy: Clinical-pathological correlations and late prognosis. *Medicine* 60, 267–276.
[a] Values in parentheses are percentages.
[b] AHG = antihemophilic globulin.

Variable deposition of other immunreactants (e.g., IgG and complement), usually in the mesangium, has also been observed.[339] That their presence is "insudative" (entrapment without immunological significance) is supported by the paucity of electron-dense deposits on ultrastructural study.

Morphometric Studies

Some authors have attempted to quantify glomerular swelling in preeclampsia. Using autopsy material Sheehan and Lynch[3] observed that glomerular diameters of women with severe preeclampsia (216 μm) were greater than those from normotensive women without parenchymal renal disease (193 μm), but failed to analyze their data statistically. Nochy et al.[340] demonstrated a significantly increased glomerular surface area (21,540 μm^2) but they used glomeruli from nonpregnant individuals biopsied for sporadic benign hematuria (17,950 μm^2) as controls. Finally, Strevens et al.[324] observed increased glomerular volume in their preeclamptic women compared to 12 normotensive gravidas biopsied in late pregnancy.

Of particular interest is a study by Lafayette and colleagues[191] that used scanning electron microscopy to analyze a small number of postpartum renal biopsies from preeclamptic women and made morphometric comparisons to biopsies from healthy female transplant donors. Their morphometric analysis was combined with a mathematical model used to estimate the glomerular ultrafiltration coefficient (K_f), which they concluded was decreased by ~40% in preeclampsia. They noted a reduction in the density and size of endothelial fenestrae, and subendothelial accumulation of fibrinoid deposits that they estimated had lowered glomerular hydraulic permeability in preeclampsia compared to controls. They also suggested that mesangial cell interposition had contributed to curtailed effective filtration surface area. The authors concluded that hypofiltration in preeclampsia had no hemodynamic basis as the reduction could be explained solely by the morphologic changes that decreased K_f and the decrease in oncotic pressure in preeclamptic women. However, the modeling methods used are subject to large errors especially when experimental conditions are aberrant. In their study, for instance, a decrease in renal plasma flow, typical for preeclampsia (*vide supra*), was not observed, perhaps due to volume loading or to the fluid shifts occurring in the immediate puerperium.

Glomerulosclerosis in Preeclampsia

The 1980s brought increasing reports of, and attention to, the association of focal segmental glomerular sclerosis (FSGS) with preeclampsia, with some authors implying FSGS was a consequence of the disorder.[341,342] This lesion, a nonspecific form of glomerular scarring that serves as a marker of chronic glomerular injury, may be seen in association with "essential" hypertensive injury and thus its significance in this setting is unclear. Gaber and Spargo reviewed this subject in detail, concluding that true FSGS when present had antedated pregnancy or might have resulted when very severe hypertension remained uncontrolled. (See Table 17.8 in edition 3 on the Elsevier web site.) Such views are supported by the careful studies by Gartner et al.[343] Some of the confusion in the literature may stem from the fact that some glomerular lesions seen in preeclamptics may appear "FSGS-like." Indeed, early lesions of this type (whatever the cause) can be difficult to identify and interobserver reproducibility for them is lower than for well-developed sclerotic lesions. For example, Sheehan and Lynch as well as others have described ballooning and capillary tip lesions, lesions that some might interpret as segmental sclerosis (cellular or tip variants). Insufficient attention may have been paid to the distinction between hyalinosis and true sclerosis. The former is easily reversible, and may help to account for the phenomenon of postpartum "disappearing FSGS."[344]

The simultaneous presence of chronic vascular injury might be taken to imply that the glomerulosclerosis is secondary to hypertension, and thus likely preceded the development of preeclampsia. The question of the diagnostic and prognostic significance of FSGS is further complicated by studies comparing the short-term outcome of preeclamptics with and without associated segmental sclerosis. Nochy et al. analyzed a series of biopsies obtained 8–10 days postpartum.[340] They found concurrent FSGS in 19 of 42 patients with endotheliosis. Note, however, how they defined FSGS – that these "'early lesions' of FSGS ... were always associated with the specific glomerular changes of preeclampsia ... all ... devoid of sclerosed glomeruli." In other words, these lesions were not those of overt and well-established sclerosis, and therefore reversibility of such lesions is less surprising. Indeed, while these women tended to have more severe clinical manifestations of disease, proteinuria resolved by 3 months in all of them, with only one showing persistent hypertension. Nishimoto et al., studying a Japanese population, showed similar results.[345] Clearly there are serious definitional issues on this topic, and despite insufficient clinico-pathological correlations, the natural history of these women is very different from primary FSGS. Population studies may shed some light, as most of the biopsy studies involved relatively short-term follow-up (a few years at most). Recent studies suggest a link between preeclampsia and subsequent ESRD.[346] Others, however, have questioned their validity.[347] A Taiwanese population study found the long-term risk of ESRD, while elevated for women with gestational hypertension alone, was markedly greater for women who had preeclampsia or eclampsia.[348] Our current understanding (such as it is) of the relationship between preeclampsia, FSGS (short-term) and chronic kidney disease (long-term) should undoubtedly be advanced through the clinical use of serum biomarkers such as sFlt-1.[349]

Reversibility of Preeclamptic Nephropathy

The consensus is that the lesions of preeclampsia resolve completely after delivery, regression seen as early as the first postpartum week.[350–352] Repeat biopsies within a month and up to 2 years after the initial diagnosis show complete resolution of the glomerular lesions, with the exception being a few reports of persistent glomerular adhesions or vascular pathology in the repeat biopsies. Unfortunately, preexisting pathology could not be excluded in those studies. Also, the restrictive indications for renal biopsy of gravid patients were bound to select women with atypical disease, a group more likely to have underlying disease. Studies where repeat biopsies were performed suggest that the lesion most likely to disappear during the first postpartum week is the subendothelial fibrin.[326,353] Subendothelial deposits within the capillary loops, substantial during pregnancy, and seeming to correlate with the impairment of glomerular permeability, are virtually absent in postpartum material.[326,353] In contrast, foam cells are most likely to be observed in biopsies performed after pregnancy and are believed related to the healing process.[326] Finally, the FSGS lesion, whatever its nature, does not appear to progress in the short term.

Persistent proteinuria (>3 months) merits further evaluation and consideration of biopsy, as the prevalence of nonpreeclamptic disease in that context is high – 71% in a recent biopsy series.[354] The question of renal biopsy during pregnancy is discussed below and a metaanalysis review was recently published.[355]

Specificity of Endotheliosis and Definition and Pathogenesis of Preeclampsia

Endotheliosis, when mild and/or focal, may be difficult to appreciate in paraffin sections, and of limited specificity for preeclampsia. Nevertheless, when present on biopsy in a diffuse and severe manner, *in the appropriate clinical context*, it is highly characteristic of preeclampsia. That said, the specificity of the lesion is in part a function of how the disease of preeclampsia is defined. This is underscored by the ability of anti-VEGF therapy to result in the development of endotheliosis and the clinical findings of preeclampsia (hypertension and proteinuria) in nonpregnant individuals.[356,357] Indeed, limited forms had been occasionally reported in association with other glomerular disorders and in gravidas with abruption.[299,358] The cause of glomerular endotheliosis appears to relate to the high levels of circulating sFlt-1 characteristic of preeclampsia. This was suggested by the elegant studies by Eremina et al. using recombinant gene technology to demonstrate that the lesion occurred in mice with deletion of one copy of VEGF-A from podocytes.[310] Li et al. in a rat model of preeclampsia induced by overexpression of sFlt-1 were able to reverse the lesion by administering VEGF 121.[311] These findings have been more recently confirmed.[359] The work by Strevens et al. (both papers published in 2003, and neither containing the word VEGF!) also deserves particular attention, its ethical issues (use of healthy pregnant women as controls, as noted above and in *BJOG* 2004;111:191 (letter) notwithstanding.[324,325] They evaluated the degree of endotheliosis using a semiquantitative scale: 0 (no endotheliosis), 1 (<20% of the lumina obliterated), 2 (20–80%), and 3 (>80%). Endotheliosis, while noted in five of 12 normotensive pregnant women, was graded as only 1+ in four and 2+ in the fifth. (Interestingly, women in such a category do not usually have significant podocyturia.)[275] In contrast, endotheliosis was found in all patients of the two hypertensive groups, both without (n = 8) and with proteinuria (n = 28) in increasing grades of severity. Most importantly, the degree of endotheliosis was significantly different between all three groups. (It is tempting to speculate that the mild endotheliosis was seen in some of the control samples that were taken at near term, when sFlt-1 levels are high in all gravidas and blood pressure is also rising, and represents an early (or preclinical) form of preeclampsia.) With regard to endotheliosis, the authors concluded "the transition between normal term pregnancy, gestational hypertension and preeclampsia appears to be a continuous process, perhaps of increasing adaptation to pregnancy. Preeclampsia may be the extreme of the adaptational process, rather than a separate abnormal condition." Through clinico-pathologic correlation we can now understand that this spectrum is the morphologic reflection of the pattern of circulating sFlt-1, which normally rises throughout pregnancy, especially after week 36 in normal pregnancy. While some overlap between values in normal and preeclamptic pregnancies exists, in most cases sFlt-1 levels are further greatly increased in women with established preeclampsia, even prior to the onset of clinical symptoms.[136] With the impending introduction of clinical testing of angiogenic parameters, similar pathologic correlations may only be beginning!

Indications for Renal Biopsy in Pregnancy

The use of renal biopsy in pregnancy or the immediate puerperium has greatly enhanced our understanding of the renal pathology associated with preeclampsia. However, as noted in the previous edition (plus in an editorial and several chapters written almost a decade before that edition), there are very few indications for renal biopsy during pregnancy, and rarely if at all in regard to diagnosing or managing the hypertensive disorders of gestation.

In a 1987 report Packham and and Fairely[360] noted the procedure had limited morbidity similar to that in nonpregnant populations and most women with antenatal undiagnosed hematuria and/or proteinuria should have antepartum biopsy, claiming this would certainly aid in diagnosis and management. In an accompanying editorial, Lindheimer and Davison[361] took issue with this and suggested only rare circumstances when biopsying during pregnancy should be considered.

Also in the editorial Lindheimer and Davison[361] suggested criteria that appear to have been endorsed by others ever since: (1) sudden deterioration of renal function without obvious cause, arguing that this would allow specific therapies for certain rapidly progressing glomerulonephritis; and (2) symptomatic nephrotic syndrome, remote from term, where the pathology could govern decisions regarding steroids or other therapy. On the other hand and especially when conditions were stable, the recommendation was for closer surveillance than routine prenatal care, deferring biopsy to the postpartum period. The reason given for this approach was the following: prognosis is determined primarily by the patient's renal functional status and the presence or absence of hypertension. The recommended approach to women with asymptomatic microhematuria and no pathological findings (e.g., stone or tumor) by ultrasonography was similar.

Also noted in the editorial was that renal biopsy should not be undertaken as term approaches, meaning after 34 weeks in earlier writing, and 32 weeks in the last and this edition. We underscored that at this late stage the decision to deliver is usually made rapidly and independent of biopsy results.

The above limitations on renal biopsies also relate to preeclampsia where, especially after weeks 30–32, decisions are based on clinical presentation, and unrelated to the biopsy. In addition, other laboratory tests and clinical features usually determine those instances where the correct diagnosis of a disease amenable to a specific therapy is the issue. Also, as stated above, the use of biopsy for prediction of the outcome of future pregnancy disappeared decades ago. Thus it was surprising to see examples in the current millennium in which investigators have biopsied both normal and prepartum women to determine whether the procedure had use in managing preeclampsia in pregnancy (see also discussions in BJOG [2004;111:191–195] regarding the ethical nature of such studies).

Finally, some speculation as this book goes to press. We now know that glomerular endothelosis in preeclampsia occurs when the very high levels of circulating sFlt-1 deplete the renal podocytes of active VEGF[310,362] (see Chapter 6). In fact a high sFlt-1/PIGF ratio or low free PIGF appears quite useful in the diagnosis of preeclampsia and in determining its severity.[362,363] Thus it would be of interest if sFlt-1 might also prove a surrogate for developing or already established glomeruloendotheliosis when evaluating patients for suspected preeclampsia.

References

1. Maynard SE, Min JY, Merchan J, et al. Excess placental soluble fms-like tyrosine kinase 1 (sFlt1) may contribute to endothelial dysfunction, hypertension, and proteinuria in preeclampsia. *J Clin Invest.* 2003;111:649–658.
2. Stillman IE, Karumanchi SA. The glomerular injury of preeclampsia. *J Am Soc Nephrol.* 2007;18:2281–2284.
3. Sheehan H, Lynch J. *Pathology of Toxaemia of Pregnancy.* Baltimore, MD: Williams & Wilkins; 1973.
4. Sheehan HL. Renal morphology in preeclampsia. *Kidney Int.* 1980;18:241–252.
5. Gilson GJ, Mosher MD, Conrad KP. Systemic hemodynamics and oxygen transport during pregnancy in chronically instrumented, conscious rats. *Am J Physiol.* 1992;263:H1911–H1918.
6. Sims EA, Krantz KE. Serial studies of renal function during pregnancy and the puerperium in normal women. *J Clin Invest.* 1958;37:1764–1774.
7. De Alvarez RR. Renal glomerulotubular mechanisms during normal pregnancy. I. Glomerular filtration rate, renal plasma flow, and creatinine clearance. *Am J Obstet Gynecol.* 1958;75:931–944.
8. Assali NS, Dignam WJ, Dasgupta K. Renal function in human pregnancy. II. Effects of venous pooling on renal hemodynamics and water, electrolyte, and aldosterone excretion during gestation. *J Lab Clin Med.* 1959;54:394–408.
9. Dunlop W. Serial changes in renal hemodynamics during normal human pregnancy. *Brit J Obstet Gynaecol.* 1981;88:1–9.
10. Roberts M, Lindheimer MD, Davison JM. Altered glomerular permselectivity to neutral dextrans and heteroporous membrane modeling in human pregnancy. *Am J Physiol.* 1996;270:F338–F343.
11. Chapman AB, Abraham WT, Zamudio S, et al. Temporal relationships between hormonal and hemodynamic changes in early human pregnancy. *Kidney Int.* 1998;54:2056–2063.
12. Conrad K. Renal changes in pregnancy. *Urol Ann.* 1992;6:313–340.
13. Davison J, Dunlop W. Changes in renal hemodynamics and tubular function induced by normal human pregnancy. *Semin Nephrol.* 1984;4:198–207.
14. Ezimokhai M, Davison JM, Philips PR, Dunlop W. Nonpostural serial changes in renal function during the third trimester of normal human pregnancy. *Brit J Obstet Gynaecol.* 1981;88:465–471.
15. Kuhlback B, Widholm O. Plasma creatinine in normal pregnancy. *Scand J Clin Lab Invest.* 1966;18:654–656.
16. Nice M. Kidney function during normal pregnancy I. The increased urea clearance of normal pregnancy. *J Clin Invest.* 1935;14:575–578.
17. Davison JM. Changes in renal function in early pregnancy in women with one kidney. *Yale J Biol Med.* 1978;51:347–349.
18. Davison JM. The effect of pregnancy on kidney function in renal allograft recipients. *Kidney Int.* 1985;27:74–79.
19. Baylis C. Effect of amino acid infusion as an index or renal vasodilatory capacity in pregnant rats. *Am J Physiol.* 1988;254:F650–F656.
20. Sturgiss SN, Wilkinson R, Davison JM. Renal reserve during human pregnancy. *Am J Physiol.* 1996;271:F16–F20.
21. Smith HW. *The Kidney.* New York: Oxford University Press; 1951.
22. Davison JM, Noble MC. Serial changes in 24 hour creatinine clearance during normal menstrual cycles and the first trimester of pregnancy. *Brit J Obstet Gynaecol.* 1981;88:10–17.
23. Davison JM, Hytten FE. Glomerular filtration during and after pregnancy. *J Obstet Gynaecol Brit Commonw.* 1974;81:588–595.
24. Davison JM, Dunlop W, Ezimokhai M. 24-hour creatinine clearance during the third trimester of normal pregnancy. *Brit J Obstet Gynaecol.* 1980;87:106–109.
25. Pritchard JA, Barnes AC, Bright RH. The effect of the supine position on renal function in the near-term pregnant woman. *J Clin Invest.* 1955;34:777–781.
26. Chesley LC, Sloan DM. The effect of posture on renal function in late pregnancy. *Am J Obstet Gynecol.* 1964;89:754–759.
27. Pippig L. Clinical aspects of renal disease during pregnancy. *Med Hygiene.* 1969:181.
28. Dunlop W. Investigations into the influence of posture on renal plasma flow and glomerular filtration rate during late pregnancy. *Brit J Obstet Gynaecol.* 1976;83:17–23.
29. Lindheimer M, Weston P. Effect of hypotonic expansion on sodium, water, and urea excretion in late pregnancy: the influence of posture on these results. *J Clin Invest.* 1969;48:947–956.
30. Brenner BM, Troy JL, Daugharty TM. The dynamics of glomerular ultrafiltration in the rat. *J Clin Invest.* 1971;50:1776–1780.
31. Baylis C. The mechanism of the increase in glomerular filtration rate in the twelve-day pregnant rat. *J Physiol.* 1980;305:405–414.
32. Milne JE, Lindheimer MD, Davison JM. Glomerular heteroporous membrane modeling in third trimester and postpartum before and during amino acid infusion. *Am J Physiol Renal Physiol.* 2002;282:F170–F175.
33. Lafayette RA, Malik T, Druzin M, Derby G, Myers BD. The dynamics of glomerular filtration after Caesarean section. *J Am Soc Nephrol.* 1999;10:1561–1565.
34. Gellai M, Valtin H. Chronic vascular constrictions and measurements of renal function in conscious rats. *Kidney Int.* 1979;15:419–426.
35. Conrad KP. Renal hemodynamics during pregnancy in chronically catheterized, conscious rats. *Kidney Int.* 1984;26:24–29.
36. Reckelhoff J, Samsell L, Baylis C. Failure of an acute 10–15% plasma volume expansion in the virgin female rat to mimic the increased glomerular filtration rate (GFR) and altered glomerular hemodynamics seen at midterm pregnancy. *Clin Exper Hypertens Pregnancy.* 1989;B8:533–549.

37. Baylis C, Blantz RC. Tubuloglomerular feedback activity in virgin and 12-day-pregnant rats. *Am J Physiol.* 1985;249:F169–F173.
38. Phippard A, Horvath J, Glynn E, et al. Circulatory adaptation to pregnancy–serial studies of haemodynamics, blood volume, renin and aldosterone in the baboon (*Papio hamadryas*). *J Hypertension.* 1986;4:773–779.
39. Robson SC, Hunter S, Boys RJ, Dunlop W. Serial study of factors influencing changes in cardiac output during human pregnancy. *Am J Physiol.* 1989;256:H1060–H1065.
40. Hall JE, Granger JP, Smith Jr MJ, Premen AJ. Role of renal hemodynamics and arterial pressure in aldosterone "escape". *Hypertension.* 1984;6:I183–I192.
41. Conrad KP, Gellai M, North WG, Valtin H. Influence of oxytocin on renal hemodynamics and sodium excretion. *Ann N Y Acad Sci.* 1993;689:346–362.
42. Atherton JC, Bu'lock D, Pirie SC. The effect of pseudopregnancy on glomerular filtration rate and salt and water reabsorption in the rat. *J Physiol.* 1982;324:11–20.
43. Baylis C. Glomerular ultrafiltration in the pseudopregnant rat. *Am J Physiol.* 1982;243:F300–F305.
44. Paaby P, Brochner-Mortensen J, Fjeldborg P, Raffn K, Larsen CE, Moller-Petersen J. Endogenous overnight creatinine clearance compared with 51Cr-EDTA clearance during the menstrual cycle. *Acta Med Scand.* 1987;222:281–284.
45. Paaby P. Endogenous overnight creatinine clearance, serum β-microglobulin and serum water during the menstrual cycle. *Acta Med Scand.* 1987;221:191–197.
46. Brochner-Mortensen J, Paaby P, Fjeldborg P, Raffn K, Larsen CE, Moller-Petersen J. Renal haemodynamics and extracellular homeostasis during the menstrual cycle. *Scand J Clin Lab Invest.* 1987;47:829–835.
47. van Beek E, Houben AJ, van Es PN, et al. Peripheral haemodynamics and renal function in relation to the menstrual cycle. *Clin Sci (Lond).* 1996;91:163–168.
48. Chapman AB, Zamudio S, Woodmansee W, et al. Systemic and renal hemodynamic changes in the luteal phase of the menstrual cycle mimic early pregnancy. *Am J Physiol.* 1997;273:F777–F782.
49. Christy N, Shaver J. Estrogens and the kidney. *Kidney Int.* 1974;6:366–376.
50. Chesley LC, Tepper IH. Effects of progesterone and estrogen on the sensitivity to angiotensin II. *J Clin Endocrinol Metab.* 1967;27:576–581.
51. Nuwayhid B, Brinkman C, Woods J, Martinek H, Assali N. Effects of estrogen on systemic and regional circulations in normal and renal hypertensive sheep. *Am J Obstet Gynecol.* 1975;123:495–504.
52. Rosenfeld C, Morriss F, Battaglia F, Makowski E, Meschia G. Effect of estradiol-17β on blood flow to reproductive and nonreproductive tissues in pregnant ewes. *Am J Obstet Gynecol.* 1976;124:618–629.
53. Magness RR, Parker Jr CR, Rosenfeld CR. Systemic and uterine responses to chronic infusion of estradiol-17 beta. *Am J Physiol.* 1993;265:E690–E698.
54. Atallah AN, Guimaraes JA, Gebara M, Sustovich DR, Martinez TR, Camano L. Progesterone increases glomerular filtration rate, urinary kallikrein excretion and uric acid clearance in normal women. *Braz J Med Biol Res.* 1988;21:71–74.
55. Oparil S, Ehrlich EN, Lindheimer MD. Effect of progesterone on renal sodium handling in man: relation to aldosterone excretion and plasma renin activity. *Clin Sci Mol Med.* 1975;49:139–147.
56. Omer S, Mulay S, Cernacek P, Varma DR. Attenuation of renal effects of atrial natriuretic factor during rat pregnancy. *Am J Physiol.* 1995;268:F416–F422.
57. Hagedorn KA, Cooke CL, Falck JR, Mitchell BF, Davidge ST. Regulation of vascular tone during pregnancy: a novel role for the pregnane X receptor. *Hypertension.* 2007;49:328–333.
58. Conrad KP. Possible mechanisms for changes in renal hemodynamics during pregnancy: studies from animal models. *Am J Kidney Dis.* 1987;9:253–259.
59. Grossini E, Molinari C, Battaglia A, et al. Human placental lactogen decreases regional blood flow in anesthetized pigs. *J Vasc Res.* 2006;43:205–213.
60. Sherwood O. In: Knobii E, Neill JD, eds. *The Phiosology of Reproduction.* New York: Raven Press; 1994:861–1009.
61. Stewart DR, Celniker AC, Taylor Jr CA, Cragun JR, Overstreet JW, Lasley BL. Relaxin in the peri-implantation period. *J Clin Endocrinol Metab.* 1990;70:1771–1773.
62. Johnson M, Carter G, Grint C, Lightman S. Relationship between ovarian steroids, gonadotrophins and relaxin during the menstrual cycle. *Acta Endocrinol.* 1993;129:121–125.
63. Wreje U, Kristiansson P, Aberg H, BystrÖm B, von Schoultz B. Serum levels of relaxin during the menstrual cycle and oral contraceptive use. *Gynecol Obstet Invest.* 1995;39:197–200.
64. Davison JM, Vallotton M, Lindheimer M. Plasma osmolality and urinary concentration and dilution during and after pregnancy: evidence that lateral recumbency inhibits maximal urinary concentrating ability. *Brit J Obstet Gynaecol.* 1981;88:472–479.
65. Davison JM, Shiells EA, Philips PR, Lindheimer MD. Serial evaluation of vasopressin release and thirst in human pregnancy. Role of human chorionic gonadotrophin in the osmoregulatory changes of gestation. *J Clin Invest.* 1988;81:798–806.
66. Lindheimer MD, Barron W, Davison J. Osmoregulation of thirst and vasopressin release in pregnancy. *Am J Physiol.* 1989;257:F159–F169.
67. Lindheimer MD, Davison JM. Osmoregulation, the secretion of arginine vasopressin and its metabolism during pregnancy. *Eur J Endocrinol.* 1995;132:133–143.
68. Novak J, Ramirez RJ, Gandley RE, Sherwood OD, Conrad KP. Myogenic reactivity is reduced in small renal arteries isolated from relaxin-treated rats. *Am J Physiol Regul Integr Comp Physiol.* 2002;283:R349–R355.
69. Weisinger RS, Burns P, Eddie LW, Wintour EM. Relaxin alters the plasma osmolality-arginine vasopressin relationship in the rat. *J Endocrinol.* 1993;137:505–510.
70. Danielson LA, Sherwood OD, Conrad KP. Relaxin is a potent renal vasodilator in conscious rats. *J Clin Invest.* 1999;103:525–533.
71. St-Louis J, Massicotte G. Chronic decrease of blood pressure by rat relaxin in spontaneously hypertensive rats. *Life Sci.* 1985;37:1351–1357.
72. Massicotte G, Parent A, St-Louis J. Blunted responses to vasoconstrictors in mesenteric vasculature but not in portal vein of spontaneously hypertensive rats treated with relaxin. *Proc Soc Exp Biol Med.* 1989;190:254–259.

73. Bani-Sacchi T, Bigazzi M, Bani D, Mannaioni PF, Masini E. Relaxin-induced increased coronary flow through stimulation of nitric oxide production. *Brit J Pharmacol.* 1995;116:1589–1594.
74. Bani D, Bigazzi M, Masini E, Bani G, Sacchi T. Relaxin depresses platelet aggregation: in vitro studies on isolated human and rabbit platelets. *Lab Invest.* 1995;73:709–716.
75. Danielson LA, Kercher LJ, Conrad KP. Impact of gender and endothelin on renal vasodilation and hyperfiltration induced by relaxin in conscious rats. *Am J Physiol Regul Integr Comp Physiol.* 2000;279:R1298–R1304.
76. Conrad KP, Colpoys M. Evidence against the hypothesis that prostaglandins are the vasodepressor agents of pregnancy. Serial studies in chronically instrumented, conscious rats. *J Clin Invest.* 1986;77:236–245.
77. Danielson LA, Conrad KP. Acute blockade of nitric oxide synthase inhibits renal vasodilation and hyperfiltration during pregnancy in chronically instrumented conscious rats. *J Clin Invest.* 1995;96:482–490.
78. Novak J, Reckelhoff J, Bumgarner L, Cockrell K, Kassab S, Granger JP. Reduced sensitivity of the renal circulation to angiotensin II in pregnant rats. *Hypertension.* 1997;30:580–584.
79. Gandley RE, Conrad KP, McLaughlin MK. Endothelin and nitric oxide mediate reduced myogenic reactivity of small renal arteries from pregnant rats. *Am J Physiol Regul Integr Comp Physiol.* 2001;280:R1–R7.
80. Danielson LA, Conrad KP. Time course and dose response of relaxin-mediated renal vasodilation, hyperfiltration, and changes in plasma osmolality in conscious rats. *J Appl Physiol.* 2003;95:1509–1514.
81. Novak J, Danielson LA, Kerchner LJ, et al. Relaxin is essential for renal vasodilation during pregnancy in conscious rats. *J Clin Invest.* 2001;107:1469–1475.
82. Smith MC, Danielson LA, Conrad KP, Davison JM. Influence of recombinant human relaxin on renal hemodynamics in healthy volunteers. *J Am Soc Nephrol.* 2006;17:3192–3197.
83. Erikson M, Unemori E. Relaxin clinical trials in systemic sclerosis. In: Tregear GW, Ivell R, Bathgate RA, Wade JD, eds. *Relaxin 2000.* : Springer; 2000:373–381.
84. Teichman SL, Unemori E, Dschietzig T, et al. Relaxin, a pleiotropic vasodilator for the treatment of heart failure. *Heart Fail Rev.* 2009;14:321–329.
85. Smith MC, Murdoch AP, Danielson LA, Conrad KP, Davison JM. Relaxin has a role in establishing a renal response in pregnancy. *Fertil Steril.* 2006;86:253–255.
86. Conrad KP, Debrah DO, Novak J, Danielson LA, Shroff SG. Relaxin modifies systemic arterial resistance and compliance in conscious, nonpregnant rats. *Endocrinology.* 2004;145:3289–3296.
87. Debrah DO, Conrad KP, Danielson LA, Shroff SG. Effects of relaxin on systemic arterial hemodynamics and mechanical properties in conscious rats: sex dependency and dose response. *J Appl Physiol.* 2005;98:1013–1020.
88. Debrah DO, Conrad KP, Jeyabalan A, Danielson LA, Shroff SG. Relaxin increases cardiac output and reduces systemic arterial load in hypertensive rats. *Hypertension.* 2005;46:745–750.
89. Debrah DO, Novak J, Matthews JE, Ramirez RJ, Shroff SG, Conrad KP. Relaxin is essential for systemic vasodilation and increased global arterial compliance during early pregnancy in conscious rats. *Endocrinology.* 2006;147:5126–5131.
90. Conrad KP, Baker VL. Corpus luteal contribution to maternal pregnancy physiology and outcomes in assisted reproductive technologies. *Am J Physiol Regul Integr Comp Physiol.* 2013;304:R69–R72.
91. Baylis C. Renal effects of cyclooxygenase inhibitaion in the pregnant rat. *Am J Physiol.* 1987;253:F158–F163.
92. Venuto RC, Donker AJ. Prostaglandin E2, plasma renin activity, and renal function throughout rabbit pregnancy. *J Lab Clin Med.* 1982;99:239–246.
93. Brown G, Venuto R. Eicosanoid production in rabbit vascular tissues and placentas. *Am J Physiol.* 1990;258:E418–E422.
94. Conrad KP, Dunn M. Renal synthesis and urinary excretion of eicosanoids during pregnancy in rats. *Am J Physiol.* 1987;253:F1197–F1205.
95. Gallery E, Ross M, Grigg R, Bean C. Are the renal functional changes of human pregnancy caused by prostacyclin? *Prostaglandins.* 1985;30:1019–1028.
96. Sorensen TK, Easterling TR, Carlson KL, Brateng DA, Benedetti TJ. The maternal hemodynamic effect of indomethacin in normal pregnancy. *Obstet Gynecol.* 1992;79:661–663.
97. Harrison GL, Moore LG. Blunted vasoreactivity in pregnant guinea pigs is not restored by meclofenamate. *Am J Obstet Gynecol.* 1989;160:258–264.
98. Conrad KP, Vernier KA. Plasma level, urinary excretion, and metabolic production of cGMP during gestation in rats. *Am J Physiol.* 1989;257:R847–R853.
99. Conrad KP, Joffe GM, Kruszyna H, et al. Identification of increased nitric oxide biosynthesis during pregnancy in rats. *FASEB J.* 1993;7:566–571.
100. Conrad KP, Kerchner LJ, Mosher MD. Plasma and 24-h NO(x) and cGMP during normal pregnancy and preeclampsia in women on a reduced NO(x) diet. *Am J Physiol.* 1999;277:F48–F57.
101. Kopp L, Paradiz G, Tucci J. Urinary excretion of cyclic 3',5'-adenosine monophosphate and cyclic 3',5'-guanosine monophosphate during and after pregnancy. *J Clin Endocrinol Metab.* 1977;44:590–594.
102. Sala C, Campise M, Ambroso G, Motta T, Zanchetti A, Morganti A. Atrial natriuretic peptide and hemodynamics changes during normal human pregnancy. *Hypertension Dallas.* 1995;25:631–636.
103. Yang D, Lang U, Greenberg SG, Myatt L, Clark KE. Elevation of nitrate levels in pregnant ewes and their fetuses. *Am J Obstet Gynecol.* 1996;174:573–577.
104. Sladek SM, Magness RR, Conrad KP. Nitric oxide and pregnancy. *Am J Physiol.* 1997;272:R441–R463.
105. Novak J, Rajakumar A, Miles TM, Conrad KP. Nitric oxide synthase isoforms in the rat kidney during pregnancy. *J Soc Gynecol Investig.* 2004;11:280–288.
106. Alexander BT, Miller MT, Kassab S, et al. Differential expression of renal nitric oxide synthase isoforms during pregnancy in rats. *Hypertension.* 1999;33:435–439.
107. Smith CA, Santymire B, Erdely A, Venkat V, Losonczy G, Baylis C. Renal nitric oxide production in rat pregnancy: role of constitutive nitric oxide synthases. *Am J Physiol Renal Physiol.* 2010;299:F830–F836.

108. Danielson LA, Conrad KP. Prostaglandins maintain renal vasodilation and hyperfiltration during chronic nitric oxide synthase blockade in conscious pregnant rats. *Circ Res.* 1996;79:1161–1166.
109. Baylis C. Cyclooxygenase products do not contribute to the gestational renal vasodilation in the nitric oxide synthase inhibited pregnant rat. *Hypertens Pregnancy.* 2002;21:109–114.
110. Haynes W. Endothelins as regulators of vascular tone in man. *Clin Sci.* 1995;88:509–517.
111. Hirata Y, Emori T, Eguchi S, et al. Endothelin receptor subtype B mediates synthesis of nitric oxide by cultured bovine endothelial cells. *J Clin Invest.* 1993;91:1367–1373.
112. Hirata Y, Hayakawa H, Suzuki E, et al. Direct measurements of endothelium-derived nitric oxide release by stimulation of endothelin receptors in rat kidney and its alteration in salt-induced hypertension. *Circulation.* 1995;91:1229–1235.
113. Tsukahara H, Ende H, Magazine HI, Bahou WF, Goligorsky MS. Molecular and functional characterization of the non-isopeptide-selective ETB receptor in endothelial cells. Receptor coupling to nitric oxide synthase. *J Biol Chem.* 1994;269:21778–21785.
114. Yokokawa K, Johnson J, Kohno M, et al. Phosphoinositide turnover signaling stimulated by ET-3 in endothelial cells from spontaneously hypertensive rats. *Am J Physiol.* 1994;267:R635–R644.
115. Gellai M, Fletcher T, Pullen M, Nambi P. Evidence for the existence of endothelin-B receptor subtypes and their physiological roles in the rat. *Am J Physiol.* 1996;271:R254–R261.
116. Kourembanas S, McQuillan LP, Leung GK, Faller DV. Nitric oxide regulates the expression of vasoconstrictors and growth factors by vascular endothelium under both normoxia and hypoxia. *J Clin Invest.* 1993;92:99–104.
117. Gellai M. Physiological role of endothelin in cardiovascular and renal hemodynamics: studies in animals. *Curr Opin Nephrol Hypertens.* 1997;6:64–68.
118. Conrad KP, Gandley RE, Ogawa T, Nakanishi S, Danielson LA. Endothelin mediates renal vasodilation and hyperfiltration during pregnancy in chronically instrumented conscious rats. *Am J Physiol.* 1999;276:F767–F776.
119. Novak J, Conrad K. Small renal arteries isolated from ETB receptor deficient rats fail to exhibit the normal maternal adaptation to pregnancy. *Faseb J.* 2004;18:205–232.
120. Kerchner LJ, Novak J, Hanley-Yanez K, Doty KD, Danielson LA, Conrad KP. Evidence against the hypothesis that endothelial endothelin B receptor expression is regulated by relaxin and pregnancy. *Endocrinology.* 2005;146:2791–2797.
121. Dschietzig T, Bartsch C, Richter C, Laule M, Baumann G, Stangl K. Relaxin, a pregnancy hormone, is a functional endothelin-1 antagonist: attenuation of endothelin-1-mediated vasoconstriction by stimulation of endothelin type-B receptor expression via ERK-1/2 and nuclear factor-kappaB. *Circ Res.* 2003;92:32–40.
122. Jeyabalan A, Novak J, Danielson LA, Kerchner LJ, Opett SL, Conrad KP. Essential role for vascular gelatinase activity in relaxin-induced renal vasodilation, hyperfiltration, and reduced myogenic reactivity of small arteries. *Circ Res.* 2003;93:1249–1257.
123. Palejwala S, Stein DE, Weiss G, Monia BP, Tortoriello D, Goldsmith LT. Relaxin positively regulates matrix metalloproteinase expression in human lower uterine segment fibroblasts using a tyrosine kinase signaling pathway. *Endocrinology.* 2001;142:3405–3413.
124. Unemori E, Pickford L, Salles A, et al. Relaxin induces an extracellular matrix-degrading phenotype in human lung fibroblasts in vitro and inhibits lung fibrosis in a murine model in vivo. *J Clin Invest.* 1985;2:190–192.
125. Fernandez-Patron C, Radomski MW, Davidge ST. Role of matrix metalloproteinase-2 in thrombin-induced vasorelaxation of rat mesenteric arteries. *Am J Physiol Heart Circ Physiol.* 2000;278:H1473–H1479.
126. Fernandez-Patron C, Radomski MW, Davidge ST. Vascular matrix metalloproteinase-2 cleaves big endothelin-1 yielding a novel vasoconstrictor. *Circ Res.* 1999;85:906–911.
127. Jeyabalan A, Kerchner LJ, Fisher MC, McGuane JT, Doty KD, Conrad KP. Matrix metalloproteinase-2 activity, protein, mRNA, and tissue inhibitors in small arteries from pregnant and relaxin-treated nonpregnant rats. *J Appl Physiol.* 2006;100:1955–1963.
128. Goetz RM, Morano I, Calovini T, Studer R, Holtz J. Increased expression of endothelial constitutive nitric oxide synthase in rat aorta during pregnancy. *Biochem Biophys Res Commun.* 1994;205:905–910.
129. Xu DL, Martin PY, St John J, et al. Upregulation of endothelial and neuronal constitutive nitric oxide synthase in pregnant rats. *Am J Physiol.* 1996;271:R1739–R1745.
130. McGuane JT, Danielson LA, Debrah JE, Rubin JP, Novak J, Conrad KP. Angiogenic growth factors are new and essential players in the sustained relaxin vasodilatory pathway in rodents and humans. *Hypertension.* 2011;57:1151–1160.
131. Debra J, Agoulnik A, Conrad K. Changes in arterial function by chronic relaxin infusion are mediated by the leucine rich repeat G coupled Lgr7 receptor. *Reprod Sci.* 2008;15:217A.
132. Jeyabalan A, Novak J, Doty KD, et al. Vascular matrix metalloproteinase-9 mediates the inhibition of myogenic reactivity in small arteries isolated from rats after short-term administration of relaxin. *Endocrinology.* 2007;148:189–197.
133. Fisher C, MacLean M, Morecroft I, et al. Is the pregnancy hormone relaxin also a vasodilator peptide secreted by the heart? *Circulation.* 2002;106:292–295.
134. McGuane JT, Debrah JE, Sautina L, et al. Relaxin induces rapid dilation of rodent small renal and human subcutaneous arteries via PI3 kinase and nitric oxide. *Endocrinology.* 2011;152:2786–2796.
135. Conrad KP. Unveiling the vasodilatory actions and mechanisms of relaxin. *Hypertension.* 2010;56:2–9.
136. Levine RJ, Maynard SE, Qian C, et al. Circulating angiogenic factors and the risk of preeclampsia. *New Engl J Med.* 2004;350:672–683.
137. Stevenson JC, Macdonald DW, Warren RC, Booker MW, Whitehead MI. Increased concentration of circulating calcitonin gene related peptide during normal human pregnancy. *Brit Med J (Clin Res Ed).* 1986;293:1329–1330.
138. Struthers AD, Brown MJ, Macdonald DW, et al. Human calcitonin gene related peptide: a potent endogenous vasodilator in man. *Clin Sci (Lond).* 1986;70:389–393.
139. Chen K, Merrill DC, Rose JC. The importance of angiotensin II subtype receptors for blood pressure control during mouse pregnancy. *Reprod Sci.* 2007;14:694–704.

140. Carey LC, Rose JC. The midgestational maternal blood pressure decline is absent in mice lacking expression of the angiotensin II AT2 receptor. *J Renin Angiotensin Aldosterone Syst.* 2011;12:29–35.
141. Stennett AK, Qiao X, Falone AE, Koledova VV, Khalil RA. Increased vascular angiotensin type 2 receptor expression and NOS-mediated mechanisms of vascular relaxation in pregnant rats. *Am J Physiol Heart Circ Physiol.* 2009;296:H745–H755.
142. Morgan TK, Montgomery K, Mason V, et al. Upregulation of histidine decarboxylase expression in superficial cortical nephrons during pregnancy in mice and women. *Kidney Int.* 2006;70:306–314.
143. Huang H, Chang HH, Xu Y, et al. Epoxyeicosatrienoic acid inhibition alters renal hemodynamics during pregnancy. *Exp Biol Med (Maywood).* 2006;231:1744–1752.
144. Moran P, Baylis PH, Lindheimer MD, Davison JM. Glomerular ultrafiltration in normal and preeclamptic pregnancy. *J Am Soc Nephrol.* 2003;14:648–652.
145. Newman RL. Serum electrolytes in pregnancy, parturition, and puerperium. *Obstet Gynecol.* 1957;10:51–55.
146. Durr JA, Stamoutsos B, Lindheimer MD. Osmoregulation during pregnancy in the rat. Evidence for resetting of the threshold for vasopressin secretion during gestation. *J Clin Invest.* 1981;68:337–346.
147. Barron WM, Durr J, Stamoutsos BA, Lindheimer MD. Osmoregulation and vasopressin secretion during pregnancy in Brattleboro rats. *Am J Physiol.* 1985;248:R29–R37.
148. Davison JM, Sheills EA, Barron WM, Robinson AG, Lindheimer MD. Changes in the metabolic clearance of vasopressin and in plasma vasopressinase throughout human pregnancy. *J Clin Invest.* 1989;83:1313–1318.
149. Davison JM, Sheills EA, Philips PR, Barron WM, Lindheimer MD. Metabolic clearance of vasopressin and an analogue resistant to vasopressinase in human pregnancy. *Am J Physiol.* 1993;264:F348–F353.
150. Davison JM, Shiells EA, Philips PR, Lindheimer MD. Influence of humoral and volume factors on altered osmoregulation of normal human pregnancy. *Am J Physiol.* 1990;258:F900–F907.
151. Roche PJ, Crawford RJ, Tregear GW. A single-copy relaxin-like gene sequence is present in sheep. *Mol Cell Endocrinol.* 1993;91:21–28.
152. Bell RJ, Laurence BM, Meehan PJ, Congiu M, Scoggins BA, Wintour EM. Regulation and function of arginine vasopressin in pregnant sheep. *Am J Physiol.* 1986;250:F777–F780.
153. Ohara M, Martin PY, Xu DL, et al. Upregulation of aquaporin 2 water channel expression in pregnant rats. *J Clin Invest.* 1998;101:1076–1083.
154. Chesley L, Chesley E. The diodrast clearance and renal blood flow in normal pregnant and nonpregnant women. *Am J Physiol.* 1939;127:731–739.
155. Chesley L, Connell E, Chesley E. The diodrast clearance and renal blood flow in toxemias of pregnancy. *J Clin Invest.* 1940;19:219–224.
156. Wellen I, Welsh C, Taylor H. The filtration rate, effective renal blood flow, tubular excretory mass and phenol red clearance in specific toxemia of pregnancy. *J Clin Invest.* 1942:21.
157. Welsh C, Wellen I, Taylor H. The filtration rate, effective renal blood flow, tubular excretory mass and phenol red clearance in normal pregnancy. *J Clin Invest.* 1942;21:51–61.
158. Dill L, Isenhour C, Cadden J, Schaffer N. Glomerular filtration and renal blood flow in the toxemias of pregnancy. *Am J Obstet Gynecol.* 1942;43:32–42.
159. Kariher D, George R. Toxemias of pregnancy and the inulin-diodrast clearance tests. *Proc Soc Exp Biol Med.* 1943;52: 245–247.
160. Schaffer NK, Dill LV, Cadden JF. Uric acid clearance in normal pregnancy and pre-eclampsia. *J Clin Invest.* 1943;22:201–206.
161. Chesley L, Williams L. Renal glomerular and tubular function in relation to the hyperuricemia of pre-eclampsia and eclampsia. *Am J Obstet Gynecol.* 1945;50:367–375.
162. Bucht H. Studies on renal function in man. *Scand J Clin Lab Invest.* 1951;3:5–64.
163. Bucht H, Werko L. Glomerular filtration rate and renal blood flow in hypertensive toxaemia of pregnancy. *J Obstet Gynaecol Brit Emp.* 1953;60:157–164.
164. Assali NS, Kaplan SA, Fomon SJ, Douglass Jr. RA. Renal function studies in toxemia of pregnancy; excretion of solutes and renal hemodynamics during osmotic diuresis in hydropenia. *J Clin Invest.* 1953;32:44–51.
165. Brandstetter F, Schuller E. Nierenclearance in der normalen schwangerschaft. *Zbl Gynakol.* 1954;76:181–190.
166. Brandstetter V, Schüller E. Die clearanceuntersuchung in der gravidität. Fortschritte der geburtshilfe und gynäkologie. *Bibliotheca Gynaecologica.* 1956;4:1–99.
167. Friedberg V. Über die clearancemethode als nierenfunktionsprüfung in der schwangerschaft. *Zbl Gynakol.* 1954;76:2135–2147.
168. Lanz V, Hochuli E. Uber die nierenclearance in der normalen schwangerschaft und bei hypertensiven spattoxikosen, ihre beeinflussung durch hypotensive medikamente. *Schweiz Med Wochenschr.* 1955;85:395–400.
169. Page EW, Glendening MB, Dignam W, Harper HA. The reasons for decreased histidine excretion in pre-eclampsia. *Am J Obstet Gynecol.* 1955;70:766–773.
170. Lovotti A. La filtrazione glomerulare ed il flusso plasmatico renale nella tossicosi gravidica. *Rev Obstet Ginecol Prat.* 1956;38:323–332.
171. Hayashi T. Uric acid and endogenous creatinine clearance studies in normal pregnancy and toxemias of pregnancy. *Am J Obstet Gynecol.* 1956;71:859–870.
172. Chesley LC, Valenti C, Rein H. Excretion of sodium loads by nonpregnant and pregnant normal, hypertensive and pre-eclamptic women. *Metabolism.* 1958;7:575–588.
173. Butterman K. Clearance-untersuchungen in der normalen und pathologischen schwangerschaft. *Arch Gynäkol.* 1958;190:448–492.
174. Schlegel V. Ergebnisse und prognosticsche bedeutung der nierenclearance bei spätschwangerschaftstoxikosen. *Zbl Gynakol.* 1959;81:869–893.
175. Friedberg V. Die veränderungen des wasser-und elektrolythaushaltes in der schwangerschaft. *Anaesthetist.* 1961;10: 334–339.
176. McCartney CP, Spargo B, Lorincz AB, Lefebvre Y, Newton RE. Renal structure and function in Pregnant Patients with Acute Hypertension; Osmolar Concentration. *Am J Obstet Gynecol.* 1964;90:579–592.

177. Bocci A, Bartoli E, Refelli E, et al. L'emodinamica renale nella gravidanza normale e nella sindrome getosica. *Minerva Ginelol.* 1966;18:203–207.
178. Sarles HE, Hil SS, LeBlanc AL, Smith GH, Canales CO, Remmers Jr. AR. Sodium excretion patterns during and following intravenous sodium chloride loads in normal and hypertensive pregnancies. *Am J Obstet Gynecol.* 1968;102:1–7.
179. Sismondi P, Massobrio M, Coppo F. Stuido delle correlazioni intercorrenti tra flusso plasmatico renale e flusso ematico miometriale nella gravidanza normale e nella sindrome gestosica. *Minerva Ginecol.* 1969;21:96–99.
180. Chesley LC, Duffus GM. Preeclampsia, posture and renal function. *Obstet Gynecol.* 1971;38:1–5.
181. Fisher K, Luger A, Spargo B, Lindheimer MD. Hypertension in pregnancy. Clinical-pathological correlations and late prognosis. *Medicine.* 1981;60:267–276.
182. Hytten F, Leitch I. *The Physiology of Human Pregnancy.* London: Blackwell Scientific; 1971. 193.
183. Irons DW, Baylis PH, Butler TJ, Davison JM. Atrial natriuretic peptide in preeclampsia: metabolic clearance, sodium excretion and renal hemodynamics. *Am J Physiol.* 1997;273:F483–F487.
184. Moran P, Lindheimer MD, Davison JM. The renal response to preeclampsia. *Semin Nephrol.* 2004;24:588–595.
185. Stubbs TM, Lazarchick J, Horger 3rd EO. Plasma fibronectin levels in preeclampsia: a possible biochemical marker for vascular endothelial damage. *Am J Obstet Gynecol.* 1984;150:885–887.
186. Mayer A. Changes of the endothelium during eclampsia and their significance. *Klinische Wochenzeitschrift.* 1924:H27.
187. Roberts JM, Taylor RN, Musci TJ, Rodgers GM, Hubel CA, McLaughlin MK. Preeclampsia: an endothelial cell disorder. *Am J Obstet Gynecol.* 1989;161:1200–1204.
188. Gomez DM. Evaluation of renal resistances, with special reference to changes in essential hypertension. *J Clin Invest.* 1951;30:1143–1155.
189. Valtin H, Schafer J. *Renal Function.* 3rd ed. Boston: Little, Brown; 1995: 98–101.
190. Obeidat M, Obeidat M, Ballermann BJ. Glomerular endothelium: a porous sieve and formidable barrier. *Exp Cell Res.* 2012;318:964–972.
191. Lafayette RA, Druzin M, Sibley R, et al. Nature of glomerular dysfunction in pre-eclampsia. *Kidney Int.* 1998;54:1240–1249.
192. Corcoran A, Page I. Renal function in late toxemia of pregnancy. *Am J Med Sci.* 1941;201:385–396.
193. Odell L. Renal filtration rates in pregnancy toxemia. *Am J Med Sci.* 1947;213:709–714.
194. Sica D, Schoolwerth A. Renal handling of organic anions and cations and renal excretion of uric acid. In: Brenner BM, ed. *The Kidney.* 5th ed. Philadelphia: W.B. Saunders; 1996:607–626.
195. Boyle JA, Campbell S, Duncan AM, Greig WR, Buchanan WW. Serum uric acid levels in normal pregnancy with observations on the renal excretion of urate in pregnancy. *J Clin Pathol.* 1966;19:501–503.
196. Dunlop W, Davison JM. The effect of normal pregnancy upon the renal handling of uric acid. *Brit J Obstet Gynaecol.* 1977;84:13–21.
197. Lind T, Godfrey KA, Otun H, Philips PR. Changes in serum uric acid concentrations during normal pregnancy. *Brit J Obstet Gynaecol.* 1984;91:128–132.
198. Semple PF, Carswell W, Boyle JA. Serial studies of the renal clearance of urate and inulin during pregnancy and after the puerperium in normal women. *Clin Sci Mol Med.* 1974;47:559–565.
199. Seitchik J. The metabolism of urate in pre-eclampsia. *Am J Obstet Gynecol.* 1956;72:40–47.
200. Seitchik J, Szutka A, Alper C. Further studies on the metabolism of N^{15} -labeled uric acid in normal and toxemic pregnant women. *Am J Obstet Gynecol.* 1958;76:1151–1155.
201. Slemons J, Bogert L. The uric acid content of maternal and fetal blood. *J Biol Chem.* 1971;32:63–69.
202. Stander H, Duncan E, Sisson W. Chemical studies on the toxemias of pregnancy. *Bull Johns Hopkins Hosp.* 1925;36:411–427.
203. Stander H, Cadden J. Blood chemistry in preeclampsia and eclampsia. *Am J Obstet Gynecol.* 1934;28:856–871.
204. Cadden J, Stander H. Uric acid metabolism in eclampsia. *Am J Obstet Gynecol.* 1939;37:37–47.
205. Crawford M. The effect of labour on plasma uric acid and urea. *J Obstet Gynaecol Brit Emp.* 1939;46:540–553.
206. Nayar A. Eclampsia. A clinical and biochemical study. *J Obstet Gynaecol Brit Emp.* 1940;47:404–436.
207. Schaffer N, Dill L, Cadden J. Uric acid clearance in normal pregnancy and preeclampsia. *J Clin Invest.* 1943;22:201–206.
208. Bonsnes RW, Stander HJ. A Survey of the Twenty-Four-Hour Uric Acid and Urea Clearances in Eclampsia and Severe Preeclampsia. *J Clin Invest.* 1946;25:378–385.
209. Seitchik J. Observations on the renal tubular reabsorption of uric acid. I. Normal pregnancy and abnormal pregnancy with and without pre-eclampsia. *Am J Obstet Gynecol.* 1953;65:981–985.
210. Lancet M, Fisher IL. The value of blood uric acid levels in toxaemia of pregnancy. *J Obstet Gynaecol Brit Emp.* 1956;63:116–119.
211. Hayashi TT. The effect of benemid on uric acid excretion in normal pregnancy and in pre-eclampsia. *Am J Obstet Gynecol.* 1957;73:17–22.
212. Hayashi TT, Gillo D, Robbins H, Sabbagha RE. Simultaneous measurement of plasma and erythrocyte oxypurines. I. Normal and toxemic pregnancy. *Gynecol Invest.* 1972;3:221–236.
213. Czaczkes WJ, Ullmann TD, Sadowsky E. Plasma uric acid levels, uric acid excretion, and response to probenecid in toxemia of pregnancy. *J Lab Clin Med.* 1958;51:224–229.
214. Pollak VE, Nettles JB. The kidney in toxemia of pregnancy: a clinical and pathologic study based on renal biopsies. *Medicine (Baltimore).* 1960;39:469–526.
215. Handler JS. The role of lactic acid in the reduced excretion of uric acid in toxemia of pregnancy. *J Clin Invest.* 1960;39:1526–1532.
216. McFarlane C. An evaluation of the serum uric acid level in pregnancy. *J Obstet Gynaecol Brit Emp.* 1963;70:63–68.
217. Widholm O, Kuhlback B. The prognosis of the fetus in relation to the serum uric acid in toxaemia of pregnancy. *Acta Obst Gynec Scand.* 1964;43:137–139.
218. Connon AF, Wadsworth RJ. An evaluation of serum uric acid estimations in toxaemia of pregnancy. *Aust N Z J Obstet Gynaecol.* 1968;8:197–201.
219. Fadel HE, Sabour MS, Mahran M, Seif el-Din D, el-Mahallawi MN. Serum uric acid in pre-eclampsia and eclampsia. *J Egypt Med Assoc.* 1969;52:12–23.

220. Fadel H, Osman L. Uterine-vein uric acid in EPH-gestosis and normal pregnancy. *Schweiz Z Gynak Geburtsh.* 1970;1:395–398.
221. Fadel H, Northrop G, Misenhimer H. Hyperuricemia in pre-eclampsia. *Am J Obstet Gynecol.* 1976;125:640–647.
222. Redman CW, Beilin LJ, Bonnar J, Wilkinson RH. Plasma-urate measurements in predicting fetal death in hypertensive pregnancy. *Lancet.* 1976;1:1370–1373.
223. Redman CW, Beilin LJ, Bonnar J. Renal function in pre-eclampsia. *J Clin Pathol Suppl (R Coll Pathol).* 1976;10:91–94.
224. Redman CW, Bonnar J, Beilin L. Early platelet consumption in pre-eclampsia. *Brit Med J.* 1978;1:467–469.
225. Dunlop W, Hill LM, Landon MJ, Oxley A, Jones P. Clinical relevance of coagulation and renal changes in pre-eclampsia. *Lancet.* 1978;2:346–349.
226. Sagen N, Haram K, Nilsen ST. Serum urate as a predictor of fetal outcome in severe pre-eclampsia. *Acta Obstet Gynecol Scand.* 1984;63:71–75.
227. Fischer RL, Bianculli KW, Hediger ML, Scholl TO. Maternal serum uric acid levels in twin gestations. *Obstet Gynecol.* 1995;85:60–64.
228. Crawford M. Plasma uric acid and urea findings in eclampsia. *J Obstet Gynaecol Brit Emp.* 1941;48:60–72.
229. Many A, Westerhausen-Larson A, Kanbour-Shakir A, Roberts JM. Xanthine oxidase/dehydrogenase is present in human placenta. *Placenta.* 1996;17:361–365.
230. Many A, Hubel CA, Roberts JM. Hyperuricemia and xanthine oxidase in preeclampsia, revisited. *Am J Obstet Gynecol.* 1996;174:288–291.
231. Chesley L. Simultaneous renal clearances of urea and uric acid in the differential diagnosis of the late toxemias. *Am J Obstet Gynecol.* 1950:59.
232. Lam C, Lim KH, Kang DH, Karumanchi SA. Uric acid and preeclampsia. *Semin Nephrol.* 2005;25:56–60.
233. Bainbridge SA, Roberts JM. Uric acid as a pathogenic factor in preeclampsia. *Placenta.* 2008;29(suppl A):S67–S72.
234. Brown MA, Wang J, Whitworth JA. The renin-angiotensin-aldosterone system in pre-eclampsia. *Clin Exp Hypertens.* 1997;19:713–726.
235. Wallukat G, Homuth V, Fischer T, et al. Patients with pre-eclampsia develop agonistic autoantibodies against the angiotensin AT1 receptor. *J Clin Invest.* 1999;103:945–952.
236. Schobel HP, Fischer T, Heuszer K, Geiger H, Schmieder RE. Preeclampsia – a state of sympathetic overactivity. *New Engl J Med.* 1996;335:1480–1485.
237. Zamudio S, Leslie KK, White M, Hagerman DD, Moore LG. Low serum estradiol and high serum progesterone concentrations characterize hypertensive pregnancies at high altitude. *J Soc Gynecol Investig.* 1994;1:197–205.
238. Schaffer N, Barker S, Summerson W, Stander H. Relation of blood lactic acid and acetone bodies to uric acid in pre-eclampsia and eclampsia. *Proc Soc Exp Biol Med.* 1941.
239. Redman CW. Maternal plasma volume and disorders of pregnancy. *Brit Med J (Clin Res Ed).* 1984;288:955–956.
240. Brown MA, Zammit VC, Mitar DM. Extracellular fluid volumes in pregnancy-induced hypertension. *J Hypertens.* 1992;10:61–68.
241. Honger PE. Albumin metabolism in normal pregnancy. *Scand J Clin Lab Invest.* 1968;21:3–9.
242. Beetham R, Dawnay A, Menabawy M, Silver A. Urinary excretion of albumin and retinol-binding protein during normal pregnancy. *J Clin Pathol.* 1988;41:1089–1092.
243. Studd JW, Wood S. Serum and urinary proteins in pregnancy. *Obstet Gynecol Annu.* 1976;5:103–123.
244. Horne CH, Howie PW, Goudie RB. Serum-alpha2-macroglobulin, transferrin, albumin, and IgG levels in pre-eclampsia. *J Clin Pathol.* 1970;23:514–516.
245. Misiani R, Marchesi D, Tiraboschi G, et al. Urinary albumin excretion in normal pregnancy and pregnancy-induced hypertension. *Nephron.* 1991;59:416–422.
246. Pederson E, Rasmussen A, Johannesen P, et al. Urinary excretion of albumin, beta-2-microglobulin and light chains in pre-eclampsia, essential hypertension in pregnancy and normotensive pregnant and non-pregnant control subjects. *Scand J Clin Lab Invest.* 1981;41:777–784.
247. Bernard A, Thielemans N, Lauwerys R, van Lierde M. Selective increase in the urinary excretion of protein 1 (Clara cell protein) and other low molecular weight proteins during normal pregnancy. *Scand J Clin Lab Invest.* 1992;52:871–878.
248. Brown MA, Wang MX, Buddle ML, et al. Albumin excretory rate in normal and hypertensive pregnancy. *Clin Sci (Lond).* 1994;86:251–255.
249. MacRury SM, Pinion S, Quin JD, et al. Blood rheology and albumin excretion in diabetic pregnancy. *Diabet Med.* 1995;12:51–55.
250. Irgens-Moller L, Hemmingsen L, Holm J. Diagnostic value of microalbuminuria in pre-eclampsia. *Clin Chim Acta.* 1986;157:295–298.
251. Lopez-Espinoza I, Dhar H, Humphreys S, Redman CW. Urinary albumin excretion in pregnancy. *Brit J Obstet Gynaecol.* 1986;93:176–181.
252. Wright A, Steele P, Bennett JR, Watts G, Polak A. The urinary excretion of albumin in normal pregnancy. *Brit J Obstet Gynaecol.* 1987;94:408–412.
253. Gero G, Anthony F, Davis M, Richardson M, Dennis K, Row D. Retinol-binding protein, albumin and total protein excretion patterns during normal pregnancy. *J Obstet Gynaecol.* 1987;8:104–108.
254. Skrha J, Perusicova J, Sperl M, Bendl J, Stolba P. N-acetyl-beta-glucosaminidase and albuminuria in normal and diabetic pregnancies. *Clin Chim Acta.* 1989;182:281–287.
255. Cheung CK, Lao T, Swaminathan R. Urinary excretion of some proteins and enzymes during normal pregnancy. *Clin Chem.* 1989;35:1978–1980.
256. McCance DR, Traub AI, Harley JM, Hadden DR, Kennedy L. Urinary albumin excretion in diabetic pregnancy. *Diabetologia.* 1989;32:236–239.
257. Helkjaer PE, Holm J, Hemmingsen L. Intra-individual changes in concentrations of urinary albumin, serum albumin, creatinine, and uric acid during normal pregnancy. *Clin Chem.* 1992;38:2143–2144.
258. Erman A, Neri A, Sharoni R, et al. Enhanced urinary albumin excretion after 35 weeks of gestation and during labour in normal pregnancy. *Scand J Clin Lab Invest.* 1992;52:409–413.
259. Konstantin-Hansen KF, Hesseldahl H, Pedersen SM. Microalbuminuria as a predictor of preeclampsia. *Acta Obstet Gynecol Scand.* 1992;71:343–346.

260. Higby K, Suiter CR, Phelps JY, Siler-Khodr T, Langer O. Normal values of urinary albumin and total protein excretion during pregnancy. *Am J Obstet Gynecol.* 1994;171:984–989.
261. Douma CE, van der Post JA, van Acker BA, Boer K, Koopman MG. Circadian variation of urinary albumin excretion in pregnancy. *Brit J Obstet Gynaecol.* 1995;102:107–110.
262. Taylor AA, Davison JM. Albumin excretion in normal pregnancy. *Am J Obstet Gynecol.* 1997;177:1559–1560.
263. Peterson PA, Evrin PE, Berggard I. Differentiation of glomerular, tubular, and normal proteinuria: determinations of urinary excretion of beta-2-macroglobulin, albumin, and total protein. *J Clin Invest.* 1969;48:1189–1198.
264. Strober W, Waldmann TA. The role of the kidney in the metabolism of plasma proteins. *Nephron.* 1974;13:35–66.
265. Maack T, Johnson V, Kau ST, Figueiredo J, Sigulem D. Renal filtration, transport, and metabolism of low-molecular-weight proteins: a review. *Kidney Int.* 1979;16:251–270.
266. Oian P, Monrad-Hansen HP, Maltau JM. Serum uric acid correlates with beta 2-microglobulin in pre-eclampsia. *Acta Obstet Gynecol Scand.* 1986;65:103–106.
267. Kelly AM, McNay MB, McEwan HP. Renal tubular function in normal pregnancy. *Brit J Obstet Gynaecol.* 1978;85:190–196.
268. Krieger MS, Moodley J, Norman RJ, Jialal I. Reversible tubular lesion in pregnancy-induced hypertension detected by urinary beta 2-microglobulin. *Obstet Gynecol.* 1984;63:533–536.
269. Jeyabalan A, Conrad KP. Renal physiology and pathophysiology in pregnancy. In: Schrier RW, ed. *Renal and Electrolyte Disorders.* Philadelphia: Lippincott Williams & Wilkins; 2010:462–518.
270. Noble MC, Landon MJ, Davison JM. The excretion of gamma-glutamyl transferase in pregnancy. *Brit J Obstet Gynaecol.* 1977;84:522–527.
271. Goren MP, Sibai BM, el-Nazar A. Increased tubular enzyme excretion in preeclampsia. *Am J Obstet Gynecol.* 1987;157:906–908.
272. Jackson DW, Carder EA, Voss CM, Fry DE, Glew RH. Altered urinary excretion of lysosomal hydrolases in pregnancy. *Am J Kidney Dis.* 1993;22:649–655.
273. Russo LM, Sandoval RM, McKee M, et al. The normal kidney filters nephrotic levels of albumin retrieved by proximal tubule cells: retrieval is disrupted in nephrotic states. *Kidney Int.* 2007;71:504–513.
274. Pisitkun T, Shen RF, Knepper MA. Identification and proteomic profiling of exosomes in human urine. *Proc Natl Acad Sci USA.* 2004;101:13368–13373.
275. Garovic VD, Wagner SJ, Turner ST, et al. Urinary podocyte excretion as a marker for preeclampsia. *Am J Obstet Gynecol.* 2007;196(320):e1–e7.
276. Garovic VD, Wagner SJ, Petrovic LM, et al. Glomerular expression of nephrin and synaptopodin, but not podocin, is decreased in kidney sections from women with preeclampsia. *Nephrol Dial Transplant.* 2007;22:1136–1143.
277. Polsani S, Phipps E, Jim B. Emerging new biomarkers of preeclampsia. *Adv Chronic Kidney Dis.* 2013;20:271–279.
278. McCartney CP, Schumacher GF, Spargo BH. Serum proteins in patients with toxemic glomerular lesion. *Am J Obstet Gynecol.* 1971;111:580–590.
279. Studd JW. Immunoglobulins in normal pregnancy, preeclampsia and pregnancy complicated by the nephrotic syndrome. *J Obstet Gynaecol Brit Commonw.* 1971;78:786–790.
280. Studd JW, Shaw RW, Bailey DE. Maternal and fetal serum protein concentration in normal pregnancy and pregnancy complicated by proteinuric pre-eclampsia. *Am J Obstet Gynecol.* 1972;114:582–588.
281. Honger PE. Albumin metabolism in preeclampsia. *Scand J Clin Lab Invest.* 1968;22:177–184.
282. Kelly AM, McEwan HP. Proteinuria in pre-eclamptic toxaemia of pregnancy. *J Obstet Gynaecol Brit Commonw.* 1973;80:520–524.
283. Horne CH, Briggs JD, Howie PW, Kennedy AC. Serum -macroglobulins in renal disease and preeclampsia. *J Clin Pathol.* 1972;25:590–593.
284. Laakso L, Paasio J. Gastrointestinal protein loss in toxaemic patients. *Acta Obstet Gynecol Scand.* 1969;48:357–361.
285. Conrad KP, Benyo DF. Placental cytokines and the pathogenesis of preeclampsia. *Am J Reprod Immunol.* 1997;37:240–249.
286. Brown MA. The physiology of pre-eclampsia. *Clin Exp Pharmacol Physiol.* 1995;22:781–791.
287. Lever C. Case of puerperal convulsions. *Guys Hosp Rep.* 1843;1:995.
288. National High Blood Pressure Education Program (NHBPEP). Working group report on High Blood Pressure in Pregnancy. *Am J Obstet Gynecol.* 1990;163:1689–1712.
289. Fisher KA, Ahuja S, Luger A, Spargo BH, Lindheimer MD. Nephrotic proteinuria with pre-eclampsia. *Am J Obstet Gynecol.* 1977;129:643–646.
290. Parviainen S, Soiva K, Ehrnrooth CA. Electrophoretic study of proteinuria in toxemia of late pregnancy. *Scand J Clin Lab Invest.* 1951;3:282–287.
291. Lorincz AB, McCartney CP, Pottinger RE, Li KH. Protein excretion patterns in pregnancy. *Am J Obstet Gynecol.* 1961;82:252–259.
292. Buzanowski Z, Chojnowska I, Myszkowski L, Sadowski J. The electrophoretic pattern of proteinuria in cases of normal labor and in the course of toxemia of pregnancy. *Pol Med J.* 1966;5:217–221.
293. McEwan HP. Investigation of proteinuria in pregnancy by immuno-electrophoresis. *J Obstet Gynaecol Brit Commonw.* 1968;75:289–294.
294. McEwan HP. Investigation of proteinuria associated with hypertension in pregnancy. *J Obstet Gynaecol Brit Commonw.* 1969;76:809–812.
295. Simanowitz MD, MacGregor WG, Hobbs JR. Proteinuria in pre-eclampsia. *J Obstet Gynaecol Brit Commonw.* 1973;80:103–108.
296. Katz M, Berlyne GM. Differential renal protein clearance in toxaemia of pregnancy. *Nephron.* 1974;13:212–220.
297. Nagaraj N, Mann M. Quantitative analysis of the intra- and inter-individual variability of the normal urinary proteome. *J Proteome Res.* 2011;10:637–645.
298. Kumar S, Muchmore A. Tamm-Horsfall protein–uromodulin (1950–1990). *Kidney Int.* 1990;37:1395–1401.
299. Robson JS. Proteinuria and the renal lesion in preeclampsia and abruptio placentae. *Perspect Nephrol Hypertens.* 1976;5:61–73.
300. MacLean PR, Paterson WG, Smart GE, Petrie JJ, Robson JS, Thomson D. Proteinuria in toxaemia and abruptio placentae. *J Obstet Gynaecol Brit Commonw.* 1972;79:321–326.

301. Simanowitz MD, MacGregor WG. A critical evaluation of renal protein selectivity in pregnancy. *J Obstet Gynaecol Brit Commonw*. 1974;81:196–200.
302. Wood SM, Burnett D, Studd J. Selectivity of proteinuria during pregnancy assessed by different methods. *Perspect Nephrol Hypertens*. 1976;5:75–83.
303. Chesley LC. The Variability of Proteinuria in the Hypertensive Complications of Pregnancy. *J Clin Invest*. 1939;18:617–620.
304. Anderson S, Kennefick T, Brenner B. Renal and systemic manifestations of glomerular disease. In: Brenner BM, ed. *The Kidney*. 5th ed. Philadelphia: W.B. Saunders; 1996:1981–2010.
305. Naicker T, Randeree I, Moodley J. Glomerular basement membrane changes in African women with early-onset preeclampsia. *Hyptens Pregn*. 1995;14:371–378.
306. Naicker T, Randeree I, Moodley J, Khedun S, Ramasroop R, Seedat Y. Correlation between histological changes and loss of anionic charge of the glomerular basement membrane in early-onset pre-eclampsia. *Nehpron*. 1997;75:201–207.
307. Weise M, Prufer D, Neubuser D. beta2-microglobulin and other proteins in serum and urine during preeclampsia. *Klin Wochenschr*. 1978;56:333–336.
308. Jackson DW, Sciscione A, Hartley TL, et al. Lysosomal enzymuria in preeclampsia. *Am J Kidney Dis*. 1996;27:826–833.
309. Shaarawy M, el Mallah SY, el-Yamani AA. Clinical significance of urinary human tissue non-specific alkaline phosphatase (hTNAP) in pre-eclampsia and eclampsia. *Ann Clin Biochem*. 1997;34(Pt 4):405–411.
310. Eremina V, Sood M, Haigh J, et al. Glomerular-specific alterations of VEGF-A expression lead to distinct congenital and acquired renal diseases. *J Clin Invest*. 2003;111:707–716.
311. Li Z, Zhang Y, Ying MaJ, et al. Recombinant vascular endothelial growth factor 121 attenuates hypertension and improves kidney damage in a rat model of preeclampsia. *Hypertension*. 2007;50:686–692.
312. Aita K, Etoh M, Hamada H, et al. Acute and transient podocyte loss and proteinuria in preeclampsia. *Nephron Clin Pract*. 2009;112:c65–c70.
313. Craici IM, Wagner SJ, Bailey KR, et al. Podocyturia predates proteinuria and clinical features of preeclampsia: longitudinal prospective study. *Hypertension*. 2013;61:1289–1296.
314. Jin B, Jean-Louis P, Qipo A, et al. Podocyturia as a Diagnostic marker for preeclampsia amongst high-risk pregnant patients. *J Pregnancy*. 2012
315. Jim BB, Metha S, et al. *Podocyturia is not Predictive of the Development of Preeclampsia*. Atlanta, GA: American Society of Nephrology; 2013.
316. Lindheimer M, Katz A. The normal and diseased kidney in pregnancy. In: Schrier RW, Gottshalk CW, eds. *Diseases of the Kidney*. 6th ed. Boston: Little Brown; 1997:2063–2097.
317. Lohlein Zur pathogense der Nierenkrankheiten: Nephritisund Nephrosemit besonderer Berucksichtigung der Nephropathiagravadarum. *Deutsche Medziniche Wochenschrift*. 1918;44:1187–1189.
318. Farquhar M. Review of normal and pathologic glomerular ultrastructure *Proceedings of the 10th Annual Conference of Nephritic Syndrome*. New York: National Kidney Foundation; 1959.
319. Spargo B, McCartney CP, Winemiller R. Glomerular capillary endotheliosis in toxemia of pregnancy. *Arch Pathol*. 1959;68:593–599.
320. Hagmann H, Thadhani R, Benzing T, Karumanchi SA, Stepan H. The promise of angiogenic markers for the early diagnosis and prediction of preeclampsia. *Clin Chem*. 2012;58:837–845.
321. Sunderji S, Gaziano E, Wothe D, et al. Automated assays for sVEGF R1 and PIGF as an aid in the diagnosis of preterm preeclampsia: a prospective clinical study. *Am J Obstet Gynecol*. 2010;202(40):e1–e7.
322. Verdonk K, Visser W, Russcher H, Danser AH, Steegers EA, van den Meiracker AH. Differential diagnosis of preeclampsia: remember the soluble fms-like tyrosine kinase 1/placental growth factor ratio. *Hypertension*. 2012;60:884–890.
323. Davison JM, Lindheimer MD. Changes in renal haemodynamics and kidney weight during pregnancy in the unanaesthetized rat. *J Physiol*. 1980;301:129–136.
324. Strevens H, Wide-Swensson D, Hansen A, et al. Glomerular endotheliosis in normal pregnancy and pre-eclampsia. *BJOG*. 2003;110:831–836. (See discussion *BJOG* 2004;111:191 regarding ethical considerations of the study.).
325. Strevens H, Wide-Swensson D, Grubb A, et al. Serum cystatin C reflects glomerular endotheliosis in normal, hypertensive and pre-eclamptic pregnancies. *BJOG*. 2003;110:825–830. (See discussion *BJOG* 2004;111:191 regarding ethical considerations of the study.).
326. Gaber LW, Spargo BH, Lindheimer MD. Renal pathology in pre-eclampsia. *Baillieres Clin Obstet Gynaecol*. 1994;8:443–468.
327. Seymour AE, Petrucco OM, Clarkson AR, et al. Morphological and immunological evidence of coagulopathy in renal complications of pregnancy. *Perspect Nephrol Hypertens*. 1976;5:139–153.
328. Tribe CR, Smart GE, Davies DR, Mackenzie JC. A renal biopsy study in toxaemia of pregnancy. *J Clin Pathol*. 1979;32:681–692.
329. Fisher KA, Luger A, Spargo BH, Lindheimer MD. Hypertension in pregnancy: clinical-pathological correlations and remote prognosis. *Medicine (Baltimore)*. 1981;60:267–276.
330. Altchek A, Albright NL, Sommers SC. The renal pathology of toxemia of pregnancy. *Obstet Gynecol*. 1968;31:595–607.
331. Pirani CL, Pollak VE, Lannigan R, Folli G. The renal glomerular lesions of pre-eclampsia: electron microscopic studies. *Am J Obstet Gynecol*. 1963;87:1047–1070.
332. Gaber L, Spargo B, Lindheimer M. The nephrology of pre-eclampsia-eclampsia. In: Tisher CC, Brenner BM, eds. *Renal Pathology*. 2nd ed. : Lippincott Williams & Wilkins; 1994.
333. Hill PA, Fairley KF, Kincaid-Smith P, Zimmerman M, Ryan GB. Morphologic changes in the renal glomerulus and the juxtaglomerular apparatus in human preeclampsia. *J Pathol*. 1988;156:291–303.
334. Satchell SC, Braet F. Glomerular endothelial cell fenestrations: an integral component of the glomerular filtration barrier. *Am J Physiol Renal Physiol*. 2009;296:F947–F956.
335. Mautner W, Churg J, Grishman E, Dachs S. Preeclamptic nephropathy. An electron microscopic study. *Lab Invest*. 1962;11:518–530.

336. Henao DE, Saleem MA, Cadavid AP. Glomerular disturbances in preeclampsia: disruption between glomerular endothelium and podocyte symbiosis. *Hypertens Pregnancy.* 2010;29:10–20.
337. Vassali P, Morris R, McCluskey R. The pathogenic role of fibrin deposition in the glomerular lesions of toxemia. *J Exp Med.* 1963;118:467–479.
338. Morris R, Vassalli P, Beller F, McCluskey R. Immunofluorescent studies of renal biopsies in the diagnosis of tosemia of pregnancy. *Obstet Gynecol.* 1964;24:32–46.
339. Petrucco O, Thompson N, Laurence J, Weldon M. Immunofluorescent studies in renal biopsies in preeclampsia. *Brit Med J.* 1974:473–476.
340. Nochy D, Heudes D, Glotz D, et al. Preeclampsia associated focal and segmental glomerulosclerosis and glomerular hypertrophy: a morphometric analysis. *Clin Nephrol.* 1994;42:9–17.
341. Kida H, Takeda S, Yokoyama H, Tomosugi N, Abe T, Hattori N. Focal glomerular sclerosis in pre-eclampsia. *Clin Nephrol.* 1985;24:221–227.
342. Nochy D, Gaudry C, Hinglais N, Rouchen M, Bariety J. Can focal segmental glomerulosclerosis appear in preeclampsia? *Adv Nephrol.* 1986; 15:71–85.
343. Gartner H, Sammoun A, Wehrmann M, Grossmann T, Junghans R, Weihing C. Preeclamptic nephropathy–and endothelial lesion. A morphological study with a review of the literature. *Eur J Obstet Gynecol Reprod Biol.* 1998;77:11–27.
344. Kincaid-Smith P. The renal lesion of preeclampsia revisited. *Am J Kidney Dis.* 1991;17:144–148.
345. Nishimoto K, Shiiki H, Nishino T, et al. Glomerular hypertrophy in preeclamptic patients with focal segmental glomerulosclerosis. A morphometric analysis. *Clin Nephrol.* 1999;51:209–219.
346. Vikse BE, Irgens LM, Leivestad T, Skjaerven R, Iversen BM. Preeclampsia and the risk of end-stage renal disease. *New Engl J Med.* 2008;359:800–809.
347. Davison JM, Lindheimer MD. Pregnancy and chronic kidney disease. *Semin Nephrol.* 2011;31:86–99.
348. Wang IK, Muo CH, Chang YC, et al. Association between hypertensive disorders during pregnancy and end-stage renal disease: a population-based study. *CMAJ.* 2013;185:207–213.
349. Rolfo A, Attini R, Nuzzo AM, et al. Chronic kidney disease may be differentially diagnosed from preeclampsia by serum biomarkers. *Kidney Int.* 2013;83:177–181.
350. Fadel HE, Sabour MS, Mahran M, Seif el-Din D, el-Mahallawi MN. Reversibility of the renal lesion and functional impairment in preeclampsia diagnosed by renal biopsy. *Obstet Gynecol.* 1969;33:528–534.
351. Oe PL, Ooms EC, Uttendorfsky OT, Stolte LA, van Delden L, Graaff P. Postpartum resolution of glomerular changes in edema-proteinuria-hypertension gestosis. *Ren Physiol.* 1980;3:375–379.
352. Pollak VE, Pirani CL, Kark RM, Muehrcke RC, Freda VC, Nettles JB. Reversible glomerular lesions in toxaemia of pregnancy. *Lancet.* 1956;271:59–62.
353. Packham DK, Mathews DC, Fairley KF, Whitworth JA, Kincaid-Smith PS. Morphometric analysis of pre-eclampsia in women biopsied in pregnancy and post-partum. *Kidney Int.* 1988;34:704–711.
354. Unverdi S, Ceri M, Unverdi H, Yilmaz R, Akcay A, Duranay M. Postpartum persistent proteinuria after preeclampsia: a single-center experience. *Wien Klin Wochenschr.* 2013;125:91–95.
355. Piccoli GB, Daidola G, Attini R, et al. Kidney biopsy in pregnancy: evidence for counselling? A systematic narrative review. *BJOG.* 2013;120:412–427.
356. Eremina V, Jefferson JA, Kowalewska J, et al. VEGF inhibition and renal thrombotic microangiopathy. *New Engl J Med.* 2008;358:1129–1136.
357. Robinson ES, Matulonis UA, Ivy P, et al. Rapid development of hypertension and proteinuria with cediranib, an oral vascular endothelial growth factor receptor inhibitor. *Clin J Am Soc Nephrol.* 2010;5:477–483.
358. Subramanya A, Houghton D, Watnick S. Steroid-responsive idiopathic glomerular capillary endotheliosis: case report and literature review. *Am J Kidney Dis.* 2005;45:1090–1095.
359. Mateus J, Bytautiene E, Lu F, et al. Endothelial growth factor therapy improves preeclampsia-like manifestations in a murine model induced by overexpression of sVEGFR-1. *Am J Physiol Heart Circ Physiol.* 2011;301:H1781–H1787.
360. Packham D, Fairley KF. Renal biopsy: indications and complications in pregnancy. *Brit J Obstet Gynaecol.* 1987;94:935–939.
361. Lindheimer MD, Davison JM. Renal biopsy during pregnancy: 'to b ... or not to b ...?'. *Brit J Obstet Gynaecol.* 1987;94:932–934.
362. Rana S, Karumanchi SA, Lindheimer MD. Angiogenic factors in diagnosis, management, and research in Preeclampsia. *Hypertension.* 2014;63:198–202.
363. Chappell LC, Duckworth S, Seed PT, et al. Diagnostic accuracy of placental growth factor in women with suspected preeclampsia: a prospective multicenter study. *Circulation.* 2013;128:2121–2131.

CHAPTER 17

Platelets, Coagulation, and the Liver

LOUISE C. KENNY, KEITH R. McCRAE AND F. GARY CUNNINGHAM

Editors' comment: *This chapter of the fourth edition of Chesley's textbook has been updated. In order to do so, the editors have enlisted the aid of Dr. Keith McCrae who has special expertise and clinical and research interests in platelets and their function in pregnancy and preeclampsia. As with the description of brain and liver pathology with the preeclampsia syndrome, we chose to discuss alterations in liver pathology and function in their clinical context rather than anatomically. Combined together, microangiopathic hemolysis, hepatocellular disruption, and thrombocytopenia make up the HELLP syndrome, which is discussed in this chapter as well as in Chapters 2 and 20.*

INTRODUCTION

In his first edition,[1] Chesley included a chapter entitled *Disseminated Intravascular Coagulation* and began by stating the contemporaneous thinking that this was a fundamental feature of preeclampsia–eclampsia. In his usual thorough fashion, he reviewed data that had accrued up to that time, and he concluded that there was evidence for slightly increased coagulation and fibrinolysis during normal pregnancy. He went on to say, however, that many women with severe preeclampsia and eclampsia show no detectable evidence of increased coagulation and fibrinolysis. He concluded that disseminated intravascular coagulation did not appear to be a fundamental feature of the disease.

Many of Chesley's predecessors and contemporaries who espoused activation of intravascular coagulation with preeclampsia syndrome drew their conclusions from autopsy findings that undoubtedly led to some of these erroneous findings. Although observed as early as 1924, it had been proven by that time that platelet concentrations were decreased in some women with preeclampsia syndrome – especially severe cases that included those with eclampsia. In the first large study, Pritchard et al.[2] reported a mean platelet count of 206,000/μL in 91 consecutive women with eclampsia. In a fourth of these, the platelet count was <150,000/μL, in 15% it was <100,000 μL, and in 3% it was <50,000/μL. From these and other studies, Chesley concluded that thrombocytopenia is a feature of the preeclampsia syndrome, but that it was not caused by consumptive coagulopathy.

Somewhat parallel to the coagulation story, it had been long known that severe preeclampsia and eclampsia were associated with gross and microscopical changes in the liver.[3] A search for a serum analyte for hepatocellular necrosis came with documentation of elevated serum glutamic oxaloacetic transaminase (SGOT) levels. In his first edition[1] chapter entitled *The Liver*, Chesley summarized 11 studies of SGOT measurements and he cited abnormal values in 84% of women with eclampsia, half of those with severe preeclampsia, and a fourth who had mild preeclampsia. And while hepatocellular damage is a known cause of coagulopathy, Chesley concluded that damage to the liver was generally not severe enough to cause significant liver dysfunction. But the link between thrombocytopenia and liver involvement characterized by elevated serum transaminase levels did evolve as a marker for the severity of preeclampsia. To call attention to this, Weinstein[4,5] coined the term *HELLP syndrome* – *H*emolysis, *E*levated *L*iver enzymes, and *L*ow *P*latelets.

Thus coagulation, thrombocytopenia, and hepatic changes of the preeclampsia syndrome became accepted as interrelated. As with any review concerning preeclampsia, a major difficulty is the use of variable or imprecise criteria for its diagnosis, as discussed in Chapter 1. This caveat must be considered even when comparing studies cited in this chapter.

PLATELETS

Platelets are the smallest of the formed blood elements, with a diameter of 1.5 to 3 μm, a volume approximating 7 fL and a lifespan *in vivo* of 9 to 10 days. Platelets are extremely complex morphologically and biochemically, and have myriad functions. These small discoid elements consist of a

Chesley's Hypertensive Disorders in Pregnancy.
ISBN: 978-0-12-407866-6

DOI: http://dx.doi.org/10.1016/B978-0-12-407866-6.00017-1

TABLE 17.1 Some Platelet Granule and Cytoplasmic Contents

Dense granules – ADP, ATP, GDP, Ca^{2+}, Mg^{2+}, serotonin, pyrophosphate
α-Granules – platelet-specific proteins (β-thromboglobulin family, platelet factor 4, multimerin)
Adhesive glycoproteins – fibrinogen, vWF, fibronectin, thrombospondin-1, vitronectin
Coagulation factors – factors V and X1, protein S
Mitogenic factors – PDGF, TGF-3, ECGF, EGF, IGF-1
Angiogenic factors – VEGF, PF4 inhibitor (α -PI), plasminogen-activator inhibitor-1 (PAI-1) Fibrinolytic inhibitors
Albumin
Immunoglobulin
Granule membrane-specific proteins – P-selectin (CD62P), CD63, GMP33 (thrombospondin fragment)
Others – proteases, interleukins, chemokines, inhibitors

plasma membrane, cytoskeletal elements, and several organelles, some of which communicate with the surface via an open canalicular system. Numerous membrane receptors serve to discharge platelet functions, the primary one being their adaptation to adhere to damaged blood vessels, with one another, and to stimulate thrombin generation, all of which generate a hemostatic plug – the clot.

Critical to platelet function are platelet surface receptors that bind adhesive glycoproteins. These receptors include the GPIb/IX/V complex that binds von Willebrand factor, the integrin GPIIb/IIIa (αIIbβ3) receptor which binds fibrinogen and von Willebrand factor, and GPIa/IIa, which binds collagen. Others receptors bind additional matrix glycoproteins, while the P-selectin receptor mediates interactions with leukocytes to incite a proinflammatory response.

Ligand binding by platelet cell surface receptors may induce platelet activation, through "outside-in" signaling. An idea of the complexity of this process comes from consideration of the multiple agonists shown in Table 17.1. Various stimuli, including collagen, thrombin, serotonin, epinephrine, thromboxane A_2, platelet-activating factor, and ADP, can stimulate platelet activation. When endothelium is disrupted, platelets adhere to exposed subendothelial collagen. This process requires von Willebrand factor and results in platelet shape change from a discoid to a tiny sphere with numerous fine filopodia or pseudopodia. Most platelets that accumulate at a site of injury do not adhere directly to subendothelial surfaces, but rather aggregate with each other, a process mediated primarily through the effects of platelet GPIIbIIIa, which adopts an active conformation capable of binding fibrinogen as a consequence of platelet activation. Inherent to the platelet activation process is secretion of the contents of dense granules and alpha granules that contain a variety of substances, such as some of those shown in Table 17.1.

Platelets in Normal Pregnancies and with Preeclampsia

There are a number of normal pregnancy-induced changes that relate to platelets and their various functions. Some of these myriad changes in platelet numbers, morphology, and function are shown in Table 17.2. Several large, population-based studies have demonstrated that in uncomplicated pregnancies, the platelet count decreases by about 10% by term.[6] In one study, the mean platelet count in 6770 pregnant women near term was 213,000/μL, compared with 248,000/μL in nonpregnant control women. The 2.5th percentile in the pregnant group, used to define the lower limit of normal platelet concentrations, was 116,000/μL. The genesis of this decrease is not known for certain, but is likely to be related to the larger total blood volume as well as the expanded splenic volume, which may be as much as 35%.[7]

Most studies have also observed that the platelet count decreases in preeclampsia/eclampsia.[6,8–12] The variation of frequency and intensity of thrombocytopenia between studies may reflect differences between methods and/or equipment used for automated blood cell analysis. Up to half of women with preeclampsia develop thrombocytopenia, the extent of which is generally proportional to the severity of disease.[13] The pathogenesis of preeclampsia-associated thrombocytopenia is likely multifactorial. The elevated levels of thromboxane A_2 metabolites in the urine of preeclamptic patients, as well as the increased plasma levels of the platelet α-granule proteins β-thromboglobulin and platelet factor 4, supports the argument that platelet activation contributes to accelerated platelet clearance in this disorder.[14]

Although only the most severe cases of preeclampsia are associated with a coagulation profile suggestive of disseminated intravascular coagulation (DIC), the plasma of most patients with preeclampsia contains increased levels of thrombin-antithrombin complexes, and approximately 40% of these plasmas contain increased levels of fibrin D-dimers.[13,15] As discussed on page 386, this suggests that there is at least subclinical activation of the coagulation system. Thus, increased generation of thrombin may be one mechanism that promotes platelet activation. Platelets may also be stimulated through contact with dysfunctional endothelium and/or exposed subendothelium underlying the injured placental vasculature. Platelet adhesion may be promoted by reduced levels of ADAMTS-13 as well as elevated levels and larger, more active multimers of von Willebrand factor (VWF) and other adhesive proteins such as cellular fibronectin.[16–19]

Platelet volume increases normally across pregnancy. As shown in Table 17.2, preeclampsia is associated with a further increase in mean platelet volume.[1,20,21] This increase reflects a population of larger platelets thought to be the result of increased platelet consumption or destruction with a concomitantly increased proportion of young platelets.[22] Other factors include complex changes in the pattern of

TABLE 17.2 Platelet Changes Associated with Preeclampsia Compared with Normally Pregnant Women

Factor	Preeclampsia vs. Normal Pregnancy	Comments
Circulating platelets		
Concentration	Decreased	Dependent on severity and duration
Volume	Increased	Younger, larger platelets
Lifespan	Decreased	
Platelet activation *in vivo*		
Beta-thromboglobulin	Increased (serum)	Associated with degranulation
Immune stimulation	Increased serum platelet-associated IgG	
Cell adherence molecule expression	Increased	Increased expression anti-P-selectin, CD63, CD40+, CD60L
Thromboxane A_2	Urinary metabolites increased	
Platelets *in vivo*		
Aggregation	Decreased compared with increase of normal pregnancy	Reduced in response to ADP, arachidonic acid, vasopressin, and epinephrine
Release	Decreased	Reduced release of 5-hydroxytryptamine in response to epinephrine
Membrane microfluidity	Decreased	
Nitric oxide synthase	Decreased iNOS and peroxynitrite [NO(x)] – increased	
Platelet second messengers		
Intracellular free Ca^{2+}	Increased over normal pregnancy increase	Causes platelet activation
cAMP	Reduced cAMP platelet response to prostacyclin	
Mg^{2+} increases cAMP levels via prostacyclin		
Platelet-binding sites		
Angiotensin II	Normal levels compared with decreased levels in normotensive pregnancy	Angiotensin II enhances platelet aggregation with ADP and epinephrine

platelet production and release by megakaryocytes. Optimal methods to investigate platelet lifespan require radiolabeling, which is prohibited in pregnancy. Using the method of platelet malondialdehyde production, disparate findings have been reported.[23,24] Specifically, in their longitudinal study, Pekonen et al.[23] did not find a significant reduction in platelet lifespan in preeclampsia. Conversely, Rakoczi et al.[24] reported a significant decrease. The demonstration of a shorter platelet production time is consistent with a shorter platelet half-life.[25] Importantly, the degree of thrombocytopenia is related to severity, and significant thrombocytopenia is usually associated with severe preeclampsia and eclampsia. These changes are found only with maternal platelets because, even with marked maternal thrombocytopenia with severe preeclampsia, neither cord blood nor fetal platelet counts are affected.[26]

Redman et al.[27] demonstrated that the platelet count fell at an early stage in the evolution of preeclampsia. Despite this, the absolute platelet count is of limited predictive or prognostic value.[28] Of clinical significance are the findings of Leduc et al.,[29] who showed that in the absence of thrombocytopenia, women with severe preeclampsia do not have significant clotting abnormalities. And Barron et al.[30] showed that a combination of a normal platelet count plus a normal serum lactate dehydrogenase level had a negative-predictive value of 100% for clinically significant clotting abnormalities in women with preeclampsia. Thus, studies to assess coagulation, viz. prothrombin and activated partial thromboplastin times and plasma fibrinogen concentration, can be reserved for women with platelet counts <100,000/μL.

HELLP Syndrome

This acronym defines a presumed preeclampsia variant manifest by hemolysis, elevated liver transaminases, and low platelets. Thrombocytopenia with HELLP syndrome is generally more severe than that encountered with uncomplicated preeclampsia syndrome. It has been reported that the rate of fall of the platelet count is a predictor of the eventual severity of HELLP, with women whose platelet counts decrease by >50,000/μL per day having a higher probability of developing moderate to severe thrombocytopenia with a platelet count <100,000/μL.[31] There appears to be a correlation between the extent of thrombocytopenia and the degree of liver dysfunction in women with HELLP syndrome, as discussed on page 389. The platelet nadir is usually reached approximately 24 hours postpartum, with normalization occurring within 6–11 days.[32,33] As with preeclampsia, the thrombocytopenia of the HELLP syndrome probably reflects a multifactorial pathogenesis. Thus, likely events operative include platelet activation by contact with damaged endothelium, platelet consumption secondary to thrombin generation, and microangiopathic hemolysis.

Platelet Activation *In Vivo*

Most studies have provided evidence that there is increased platelet activation across pregnancy. Moreover, this activation is increased further in women with preeclampsia (Table 17.2). Circulating levels of factors stored within platelets reflect platelet activation – specifically, platelet aggregation and release of granule contents. Plasma levels of β-thromboglobulin, a platelet α-granule protein, are increased in normal pregnant women.[23,25,34,35] Several studies have demonstrated higher plasma levels of β-thromboglobulin in preeclamptic women compared with normal pregnant controls.[34–38] Janes and Goodall[37] reported that these increased β-thromboglobulin concentrations were associated with degranulation, as evidenced by elevated levels of the lysosomal-granule membrane antigen, CD63. Socol et al.[38] found that this measure of platelet α-granule release correlated with increasing levels of proteinuria and serum creatinine, suggesting a link between platelet activation and renal microvascular changes.

There are increased expressions of other platelet membrane antigens that signify activation. Platelets taken from preeclamptic women express increased levels of CD40L and its circulating soluble component, sCD40L. There are also increased levels of the antigens CD41- and CD62P+ and their respective circulating microparticles along with platelet-monocyte aggregates.[39,40]

Elevation of β-thromboglobulin levels precedes the clinical development of preeclampsia by at least 4 weeks.[24] In contrast with normal pregnancy, levels of β-thromboglobulin did not correlate with increased fibrinopeptide A – a marker of thrombin generation – reported in preeclampsia.[25] These findings suggest that mechanisms other than thrombin-mediated platelet stimulation are responsible for platelet activation. An immune mechanism may be a contributory factor. Burrows et al.[41] reported increased serum levels of platelet-associated immunoglobulin G which correlated with disease severity. In a prospective study, Samuels et al.[42] measured platelet-bound and circulating platelet-bindable immunoglobulin. They reported a higher frequency of abnormal platelet antiglobulin found in preeclamptic women compared with normotensive pregnant women. Alterations in platelet-bound immunoglobulins might be from the deposition of autoreactive antibodies or immune complexes caused by placental tissue antigens. Alternatively, platelet activation at sites of microvascular injury could lead to the externalization of IgG and other proteins in platelet α-granules.

Serum levels of platelet factor 4 (PF4), another α-granule protein, are not significantly elevated in preeclampsia.[23] Because PF4 is cleared by binding to the endothelium rather than by renal excretion, a contribution of impaired renal function to the increased β-thromboglobulin levels cannot be excluded. Serotonin is also released when platelets aggregate, and Middelkoop et al.[43] found decreased serotonin concentrations in platelet-poor plasma from preeclamptic women compared with levels in plasma from normal pregnant women. These results are consistent with platelet aggregation and consumption.

Janes and Goodall[37] used whole blood flow cytometry to detect circulating activated platelets identified by bound fibrinogen or by CD63 antigen expression. They reported that activated platelets were detected prior to the development of preeclampsia. These findings were confirmed by Konijnenberg et al.,[44] who also demonstrated enhanced expression of anti-P-selectin, another marker of α-granule secretion.

The most reliable method of assessing *in vivo* thromboxane production is by measurement of urinary metabolites of thromboxane A. Urinary excretion of 2,3-dinor-thromboxane B and 11-dehydro-thromboxane B is increased in normal pregnancy.[45] These urinary metabolites are further increased in women with preeclampsia compared with normotensive pregnant women, and may increase before clinical signs develop.[46,47] These observations provide further evidence of increased platelet activation in preeclampsia and, moreover, indicate that thromboxane generation activates platelets.

Garzetti et al.[48] studied platelet membranes and found increased fluidity and cholesterol concentration with preeclampsia. These changes are consistent with increased unsaturated fatty acid content. These latter compounds are both a substrate for lipid oxidation and participate in thromboxane formation. Increased thromboxane production may thus reflect altered platelet membranes in preeclampsia.

There may also be an activated-platelet interaction with the ADAMTS-13 metalloproteinase system that cleaves von Willebrand factor. Platelets interact with factor VIII to promote normal ADAMTS-13 action; however, disruption of this impairs vWF proteolysis both *in vivo* and *in vitro*.[49] The effects of this on the normally deficient ADAMTS-13 state for pregnancy and the genesis of preeclampsia is speculative at this time.

Summary of Platelet Activation

Taken together, there is ample evidence for platelet activation in preeclampsia. There is reduced platelet concentration, increased size, reduced lifespan, increased α-granule release, enhanced expression of cell adhesion molecules, and increased thromboxane production. This increased activation, which occurs early in the disease, may either result from an extrinsic factor such as endothelial damage with platelet activation, or it might be intrinsic and antedate pregnancy. Evidence that intrinsic platelet alterations are at least partly responsible comes from findings of platelet binding-site alterations. One example of this is increased platelet angiotensin II binding sites which are apparent in early gestation in some women who subsequently develop preeclampsia.[35,36] Maki et al.[37] proposed an alternative mechanism caused by diminished circulating platelet-activating factor acetylhydrolase activity. This would lead to decreased platelet-activating factor in normally pregnant compared with nonpregnant women, but not in those with preeclampsia.

Platelet Behavior *In Vitro*

While the evidence discussed previously points to increased platelet activation *in vivo* in normal pregnancy, and even more so in preeclamptic women, the data are in contrast to findings from numerous *in vitro* studies. Specifically, these studies demonstrate increased platelet activation in normal pregnancy compared with nonpregnant women; however, *reduced* activation is consistently reported *in vitro* with preeclampsia. In a recent investigation, Burke et al.[50] studied platelet aggregation across normal pregnancy in response to exposure to collagen and arachidonic acid. Platelet aggregation decreased somewhat in early pregnancy, but thereafter it was increased throughout the latter two trimesters.

As indicated, however, others have reported decreased platelet aggregation in women with eclampsia. Horn et al.[51] studied radiolabeled serotonin release *in vitro* in platelet-rich plasma from preeclamptic patients in response to arachidonic acid. This response was diminished compared with nonhypertensive control subjects. When whole blood is studied *in vitro*, platelets from women with preeclampsia release less serotonin in response to epinephrine compared with platelets from normally pregnant women. This mirrors the *in vitro* reduction in platelet aggregation in response to this agonist in preeclampsia. Taken together, these findings suggest that the platelet content of serotonin is reduced in preeclampsia, suggesting that platelets have become activated and released their granule contents prior to *in vitro* studies.[52,53]

Platelets, like endothelial cells, contain a constitutive form of nitric oxide synthase, the activity of which was found to be significantly lower in platelets from women with preeclampsia compared with those from normally pregnant women.[54] At the same time, however, platelet nitric oxide and peroxynitrite levels are increased in preeclamptic women compared with normally pregnant controls.[55] Although platelet nitric oxide synthesis may contribute to the vasodilatation of normal pregnancy, it is unclear whether lower activity in preeclampsia results from diminished production, or if it even contributes to its pathogenesis.

Thus, most *in vitro* studies demonstrate reduced platelet reactivity in preeclampsia compared with normal pregnancy. One possible explanation of the disparate *in vivo* and *in vitro* observations is that preeclampsia leads to *in vivo* activation which results in circulating "exhausted" platelets which are hyporeactive when tested *in vitro*.

There have been other *in vitro* studies designed to elucidate mechanisms responsible for increased *in vivo* platelet activation in preeclampsia. For example, there is evidence that the inhibitory mechanisms that switch off platelet activation responses may be less effective in preeclampsia. Vascular endothelial cell production of prostacyclin, which acts via cyclic adenosine monophosphate (cAMP) to inhibit platelet aggregation, is diminished in preeclampsia compared with normal pregnancy.[56] The resulting tendency to vasoconstriction and platelet aggregation is accentuated by a diminution in platelet sensitivity to prostacyclin. Horn et al.[51] found that there was no alteration in sensitivity to prostacyclin – or other manipulators of cAMP such as thromboxane synthase inhibitors – when platelets from women with gestational hypertension were compared with platelets from normally pregnant women. In reports focusing on preeclampsia, however, pregnancy-induced diminished susceptibility to prostacyclin inhibition was significantly more marked – up to 50% – in preeclampsia.[57,58] The finding of increased numbers of platelet thromboxane A receptors in women with preeclampsia is also consistent with increased *in vivo* platelet activation.[59] Arachidonic acid, ADP, and epinephrine all have a thromboxane-dependent component in their mechanism of action.[60] Finally, increased thromboxane receptor density should lead to increases *in vivo* reactivity.

Expression of Platelet Receptors in Pregnancy and Preeclampsia

The platelet surface is decorated by a plethora of glycoprotein receptors that function to mediate signals from the extracellular milieu to the platelet signaling machinery, often using specific G-protein-coupled receptors as an intermediary.[61] Some receptors, such as GPIIbIIIa, the platelet fibrinogen receptor, undergo transition to an active conformation only upon platelet activation – "inside-out signaling" – enabling the binding of fibrinogen that bridges platelets during platelet activation. In addition to receptors that bind adhesive ligands, platelets display numerous receptors for platelet agonists, such as protease activated receptors, particularly PAR-1, which bind thrombin, P2Y(12) receptors, which bind ADP, and receptors for PGI2, TXA2 and other mediators. The integrated function of these receptors regulates the platelet activation response, and antiplatelet therapies, either approved or in development, target many of these receptors. However, there are very few data available on the expression of these receptors in normal pregnancy or hypertensive pregnancy disorders. A recent study demonstrated that pregnancy-specific glycoproteins, secreted by syncytiotrophoblast, bind GPIIbIIIa, and inhibit fibrinogen binding; however the role of this interaction in regulation of platelet aggregation during pregnancy is uncertain.[62] Polymorphisms in glycoprotein platelet receptors may regulate platelet responsiveness *in vitro*, and potentially impact thrombosis risk, but their roles in pregnancy outcomes have not been thoroughly studied; a polymorphism in platelet GP6, which mediated collagen signaling, has been recently associated with platelet hyperaggregability and an increased risk of fetal loss in a retrospective analysis, though an increased risk of pregnancy-associated hypertension was not reported.[63] Likewise, a polymorphism of platelet GPIaIIa (integrin $\alpha2\beta1$) was also retrospectively associated with early-onset fetal loss, but not hypertensive disease.[64]

Changes in the expression of P2Y(12), PGI2 or TXA2 receptors involved in platelet activation and signaling have not been reported in pregnancy or pregnancy-induced hypertensive disease, and are a potentially fruitful area of study. One report[65] described decreased affinity of PGI2 receptors in preeclampsia, but this finding has not been reproduced.

Platelet Second Messengers

In an attempt to further elucidate mechanisms underlying *in vivo* and *in vitro* changes in platelet behavior, studies have been done to investigate platelet second messenger systems in normal pregnancy and in preeclampsia. The majority of studies have been directed at intracellular free calcium ([Ca^{2+}]) and cyclic AMP. In one study of platelet [Ca^{2+}], Barr et al.[66] found no differences in basal or ADP-stimulated platelet [Ca^{2+}] in either normal pregnancy or preeclampsia. They used the calcium-sensitive indicator *quin-2*, which is known to quench increases in platelet [Ca^{2+}] resulting from platelet stimulation.[67] Although they used imprecise methods, they did demonstrate reduced serotonin-stimulated platelet [Ca^{2+}] in preeclampsia compared with normal pregnancy. It had been previously shown that 5-hydroxytryptamine responses were easily suppressed as a result of prior platelet activation.[68]

The calcium-sensitive fluorophore *fura-2* has the advantage of weaker calcium-chelating properties than *quin-2*. Using this indicator, Kilby et al.[69] found that basal platelet [Ca^{2+}] levels were increased in normal pregnancy and further increased in preeclampsia but not in nonproteinuric gestational hypertension. Whether this increase in platelet [Ca^{2+}] reflects a population of partially activated platelets in preeclampsia, or is a cause of altered platelet reactivity, is unclear. There is some evidence that alteration in stimulated platelet [Ca^{2+}] precedes clinical signs of preeclampsia. In a prospective study of nulliparous women, Zemel et al.[70] found that African-American women who subsequently developed preeclampsia had higher levels of platelet [Ca^{2+}] following stimulation with arginine vasopressin. But while racial differences in platelet [Ca^{2+}] have been reported,[71] Kyle et al.[72] did not find increased vasopressin stimulation of platelet [Ca^{2+}] levels with established preeclampsia in Caucasian women.

Basal cAMP levels do not appear to differ from the nonpregnant state in either normal pregnancy or preeclampsia.[73–75] Horn et al.[51] demonstrated reduced platelet production of cAMP in pregnancy in response to a range of adenylate cyclase stimulators. They used a sensitive assay based on prelabeling of the metabolic adenine nucleotide pool in platelets with hydrogen 3-adenine. These findings are consistent with the reduction in sensitivity of platelets during pregnancy to inhibition by prostaglandins and other agents which act by increasing levels of cAMP, the inhibitory second messenger. In a cross-sectional study, no differences were found between normal and a heterogeneous group of women including those with nonproteinuric gestational hypertension and preeclampsia.[75] These same investigators did a longitudinal comparative study in a small number of low-risk pregnant women and a group at high risk of developing preeclampsia. They reported that platelets from at-risk women accumulated less cAMP in response to adenylate cyclase stimulators.

Cyclic guanosine monophosphate (cGMP) is another second messenger that inhibits platelet activation.[76] It is synthesized from GTP by the cytosolic-soluble guanylate cyclase enzyme, and in platelets guanylate cyclase is stimulated by nitric oxide, a potent inhibitor of platelet activation. Hardy et al.[77] tested the hypothesis that platelet activation in preeclampsia resulted from underactivity of the inhibitory cGMP system. They found that both platelet cGMP responses to nitric oxide donors and the inhibitory effect of donors on platelet release were increased in platelets from women with preeclampsia compared with normotensive pregnant as well as nonpregnant women. They speculated that upregulation of platelet guanylate cyclase activity may be a compensatory response to impaired nitric oxide production in preeclampsia.

Platelet Angiotensin II-Binding Sites

Specific angiotensin II-binding sites with the characteristics of receptors have been demonstrated on the surface of platelets.[78] Their role is unclear, although it has been suggested that angiotensin II enhances the platelet aggregation response to ADP and epinephrine.[79,80] Moreover, *in vitro* studies suggest that angiotensin II increases platelet [Ca^{2+}] levels and this increase is greater in platelets from women with preeclampsia compared with normal pregnancy.[81] Platelets have many of the structural and biochemical characteristics of smooth muscle cells, and similarities between catecholamine-induced changes in both platelet behavior and vascular tone have been described.[82] Measurement of platelet angiotensin II-binding sites has been suggested as an alternative to the labor-intensive and invasive angiotensin II sensitivity test to predict preeclampsia and gestational hypertension.[49] Platelets from normotensive pregnant women exhibit reduced angiotensin II-binding sites,[83,84] whereas binding site concentrations revert toward nonpregnant levels in preeclampsia.[85] These results mirror the pressor sensitivity of vascular smooth muscle to intravenously infused angiotensin II. This is in contrast to normal pregnancy, in which there is a reduced pressor response to infused angiotensin II, but there is an increased pressor effect prior to the onset of clinical preeclampsia.[86]

COAGULATION

The process of coagulation involves a series of enzymatic reactions, which ultimately lead to the conversion of soluble plasma fibrinogen to fibrin clot. Classically, this enzyme sequence is divided into the intrinsic and extrinsic pathways, which both converge in a final common pathway. The major distinction between the two pathways is that the intrinsic pathway is activated from within the bloodstream while the extrinsic pathway begins in the blood vessel walls. In both pathways, clotting factors, numbered I to XIII, are responsible for different interactions and in turn activate the next clotting factor, resulting in a cascading reaction as illustrated in Fig. 17.1.

The extrinsic pathway is activated at the time of blood vessel injury. Thromboplastin released from damaged cells, together with factor VII and calcium ions, activates factor X, which is the beginning of the *common pathway*. The intrinsic pathway is initiated by activation of factor XII by collagen. This in turn leads to a series of reactions culminating with the activation of factor X. Activated factor XII also converts prekallikrein to kallikrein, which leads to further activation of XII and activation of the fibrinolytic pathway. Activated factor IX together with factor VIII, calcium ions and platelet factor 3 activate factor X. Activated factor X eventually leads to conversion of prothrombin to thrombin. Thrombin hydrolyzes the peptide bonds of fibrinogen molecules to form fibrin. At the same time thrombin, in the presence of calcium ions, activates factor XIII, which stabilizes the fibrin clot by cross-linking adjacent fibrin molecules. This process is regulated by the fibrinolytic system shown in Fig. 17.1, composed primarily of plasminogen and regulatory proteins, including antithrombin III, protein C, and protein S.

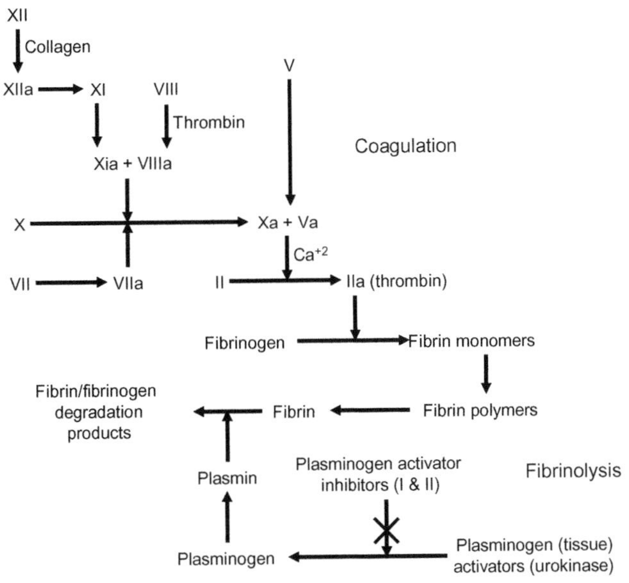

FIGURE 17.1 The coagulation and fibrinolysis cascades.

Coagulation Cascade Factors

In normal pregnancy, the coagulation cascade appears to be in an activated state. Evidence of activation includes increased concentrations of all the clotting factors except factors XI and XIII, with increased levels of high molecular weight fibrinogen complexes.[87–89] Considering the substantive physiological increase in plasma volume in normal pregnancy such increased concentrations represent a marked increase in production of these procoagulants.

There is considerable evidence that preeclampsia is accompanied by a number of coagulopathic changes when compared with normal pregnant women. Despite this, except for the rare case of preeclampsia complicated by overt clinical disseminated intravascular coagulation, routine coagulation tests are usually normal.[29,30] This is because clinical tests commonly used are relatively insensitive to minor changes in the coagulation system. There have been reports, however, of covert activation of both the intrinsic and extrinsic coagulation pathways in preeclampsia.[90,91] Vaziri et al.[91] found changes in plasma coagulation activity of all intrinsic pathway factors in preeclampsia and suggested that such activation was fundamental to the pathogenesis of preeclampsia. The mechanism for such activation is unclear, although endothelial injury and exposure of subendothelial tissue would facilitate activation of factor XII.

One of the most definitive tests of coagulation activity is factor VIII consumption. The test depends on simultaneous measurement of factor VIII clotting activity and factor VIII-related antigen. When the clotting system is activated, circulating levels of both factors increase rapidly as a secondary response, but because factor VIII clotting activity is destroyed by thrombin, its final level is lower than that of the related antigen and the difference between the two is a reflection of factor VIII consumption. During normal pregnancy the levels of factor VIII coagulation activity and factor VIII-related antigen show a proportional rise, thus their ratio remains constant.[92] In preeclampsia, there is an early rise in the factor VIII-related antigen:coagulation activity ratio, which correlates with severity of the disease and the degree of hyperuricemia.[93–95] While this increased ratio was initially thought to be due to factor VIII consumption, Scholtes et al.[95] found it to be almost entirely due to increased factor VIII-related antigen. This was most marked in cases of preeclampsia associated with fetal growth restriction. Because factor VIII-related antigen is synthesized by endothelial cells and megakaryocytes and is released by aggregating platelets, it is possible that increased levels result from endothelial damage and platelet aggregation rather than increased thrombin action.

During normal pregnancy, plasma fibrinogen concentration substantively increases. Although fibrinogen levels are the same or only slightly increased in women with preeclampsia compared with normal pregnant women,[96] the turnover of radiolabeled fibrinogen is increased in preeclamptic women.[97] Because increased fibrinogen turnover returned to normal with low-dose heparin therapy, it was concluded that it was thrombin-mediated.

The action of thrombin on fibrinogen is a crucial step in the coagulation cascade. Thrombin cleaves two pairs of peptides – fibrinopeptide A and B – from fibrinogen to produce soluble fibrin monomer, which rapidly polymerizes to fibrin. Determination of free fibrinopeptides in blood can be used to measure thrombin activity and fibrinopeptide concentrations are considered to be the best markers of accelerated thrombosis or coagulopathy. Levels of fibrinopeptide A are either elevated or they are unchanged in normal pregnancy; however, most investigators describe increased fibrinopeptide levels in women with preeclampsia compared with normal pregnant women.[60,88,98,99] Moreover, Borok et al.[99] found that the increased serial total fibrinopeptide measurements in preeclampsia correlated with the clinical manifestations of the disease and persisted for 3 to 7 days postpartum.

Regulatory Proteins and Thrombophilia

A number of plasma proteins regulate and maintain the coagulation cascade and its intricate balance between the liquid and solid phases (Fig. 17.1). It seems indisputable that many of the mutations of these factors are associated with a substantively increased risk of venous thromboembolism. Any association between these and adverse pregnancy outcomes is, however, controversial, as recently reviewed by the American College of Obstetricians and Gynecologists.[100]

Antithrombin III

This glycoprotein is manufactured by the liver. It is an important physiologic inhibitor of coagulation and forms irreversible complexes with all activated factors except VIIIa.[101] There is increased antithrombin III synthesis associated with coagulation activation. Decreased antithrombin III activity indicates increased thrombin binding secondary to increased thrombin generation and is found after thromboembolic events, in disseminated intravascular coagulopathy, and after major surgery.

Antithrombin III activity levels appear to be unchanged in normal pregnancy, although marginal decreases have been described.[89,102] Decreased antithrombin III activity has been demonstrated in most women with preeclampsia, but not in pregnant women with chronic hypertension.[88,103–106] Exacerbations and remissions of the disease were reflected in fluctuations of antithrombin III levels, and low antithrombin III concentrations were associated with placental infarctions as well as perinatal and maternal morbidity and mortality.[88,103–107] Changes in antithrombin III levels did not, however, correlate with clinical improvement during the puerperium.[105] Antithrombin III activity was reported to begin to decline as much as 13 weeks prior to the development of clinical manifestations in three women who were studied longitudinally and who developed preeclampsia.[103] Diminished antithrombin III activity in preeclampsia is not a consequence of liver dysfunction, but is caused by increased consumption. This led Paternoster et al.[108] to suggest that administration of antithrombin III to women with preeclampsia to normalize the chronic coagulopathy may improve fetal outcome. This was supported by studies in which administration of antithrombin III prevented renal dysfunction and hypertension induced by enhanced intravascular coagulation in pregnant rats.[109]. In a subsequent case-control prospective trial, antithrombin III administration to women with moderate to severe preeclampsia resulted in a significant prolongation of pregnancy, improved neonatal outcomes, and less maternal hemorrhage.[110] Finally, antithrombin III deficiency has been associated with preeclampsia in some but not all.[111,112] Because of its rarity, it is not possible to ascertain the individual impact of antithrombin deficiency on pregnancy outcome.

Protein S

This naturally occurring anticoagulant protein serves as a cofactor for activated protein C in the degradation of the activated factors V and VIII by binding to lipid and platelet surfaces.[108] Levels during pregnancy may decrease to those levels found in patients with congenital protein S deficiency.[113] Levels are further diminished in women with preeclampsia compared with normal pregnant women.[108] In one large prospective study of women with severe early-onset preeclampsia, 35% had protein S deficiency.[114] Recent systematic reviews have reported conflicting results regarding the strength of the association between protein S deficiency and preeclampsia.[103,115].

Protein C

This is a serine protease synthesized by the liver and it is activated by contact with thrombin and thrombomodulin on the surface of endothelial cells. It is a potent inhibitor of activated factor V and VIII, and is an activator of fibrinolysis.[116,117] Protein C is highly sensitive to consumption and reduced levels are found after surgery, thromboembolic events, and in disseminated intravascular coagulation. Protein C levels appear to be unchanged in normal pregnancy compared with the nonpregnant state.[118,119] Protein C levels have been found to be substantively reduced in preeclampsia compared with normal pregnancy.[103,115]

Activated Protein C Resistance

This mutation results in a limitation in anticoagulant response to activated protein C which predisposes to venous thromboembolism.[120] The cause was identified as a point mutation in the factor V gene at nucleotide position 1691, resulting in an arginine-to-glutamine substitution. This reduces the sensitivity of the factor V protein to inactivation by activated protein C – hence, *activated protein C resistance* – resulting in a procoagulant state and an increased risk of thrombosis.[121] The resultant mutation is termed factor V Leiden because the research was carried out in this Dutch town. The trait is inherited in an autosomal dominant manner with the risk of thrombosis increased seven-fold in heterozygotes and 80-fold in homozygotes.[122] Studies have shown that the distribution of the factor V Leiden mutation varies in different populations – it is found in about 5% of Caucasians, Europeans, Jews, Arabs, and Indians while it is virtually absent in Africans and Asians.[123]

Since the discovery and characterization of activated protein C resistance, a myriad of largely retrospective studies have examined the association between factor V Leiden mutation and a range of adverse pregnancy outcomes including preeclampsia. One metaanalysis of many of these studies suggested that factor V Leiden is associated with a 2.04-fold (95% CI 1.23–3.36) increased risk of severe preeclampsia.[124] Factor V Leiden accounts for approximately 70% of women with activated protein C resistance; the remainder are factor V Leiden negative ($APCR^{FVL-}$) and they are considered to have acquired activated protein C resistance.[125] The effect of $APCR^{FVL-}$ on pregnancy outcomes is not well characterized. Lindqvist et al. perfomed a prospective study of 2480 unselected women in early pregnancy.[126] The $APCR^{FVL-}$ cohort of 106 women did not have an increased risk of preeclampsia. Similar findings from a recent Australian study have also been reported.[127] According to the American College of Obstetricians and Gynecologists,[100] data are inconclusive that factor V Leiden mutations increase the risk of any adverse pregnancy outcome except for venous thromboembolism.

Prothrombin 20210

This procoagulant protein is the inactive precursor of thrombin. It results from a G-to-A mutation at nucleotide 20210 in the prothrombin gene that causes excessive production and increased circulating prothrombin levels. Like the factor V Leiden variant, this mutation shows significant ethnic variation, being most prevalent in southern European populations, with an overall population prevalence of 2–3%. Controversy remains as to whether it is associated with preeclampsia. In a number of case-control studies there does not appear to be a significant association.[127–131] In some small studies, however, the prothrombin mutation does appear to be associated with fetal-growth restriction and placental abruption.[128,130,132]

Antiphospholipid Syndrome

These autoimmune antibodies include lupus anticoagulant, anticardiolipin antibodies, and anti-β_2-glycoprotein. Amongst other adverse effects, these autoantibodies are associated with recurrent pregnancy loss, preeclampsia, and fetal-growth restriction.[82] They are also associated with a four-fold increase of thrombosis or thrombocytopenia.[133] The presence of antiphospholipid antibodies – anticardiolipin antibodies or lupus anticoagulant or both – significantly increases the risk of developing preeclampsia (RR 9.72, 94% CI 4.34–21.75).[134,135] There is evidence that treatment with aspirin and low-molecular-weight heparin (LMWH) can ameliorate these effects.[136] While large interventional trials have studied the effect of aspirin on the recurrence rate of preeclampsia,[137] only limited data are available regarding LMWH. In a small study from the Netherlands, the recurrence rate of preeclampsia was similar in 26 women with thrombophilia following treatment with LMWH and aspirin and 32 women receiving aspirin alone.[138] Several studies since, however, have demonstrated improved outcomes with a combination of LMWH and aspirin when compared with aspirin alone in women with a history of preeclampsia in a previous pregnancy.[139,140] Larger and appropriately powered randomized trials are needed to clarify this issue.

Fibrinolytic System

The end-product of the coagulation cascade is fibrin formation. The main function of the fibrinolytic system is to remove excess fibrin deposited as a consequence of thrombin activity (see Fig. 17.1). Plasmin, which is the main protease enzyme in this system, originates from plasminogen secreted by the liver. The activation of plasminogen into plasmin is through plasminogen activators which are serine proteases. These include tissue- and urokinase-plasminogen activators – t-PA and u-PA – and they are secreted by endothelial cells, monocytes, and macrophages. Both act upon plasminogen to convert it to plasmin and in so doing trigger a proteolysis cascade that causes thrombolysis. Plasmin cleaves and converts t-PA and u-PA into two-chain proteases, which exhibit higher proteolytic activity, implying a positive feedback for the fibrinolytic cascade.[141]

A negative feedback is therefore essential for the fibrinolytic pathway and thus the activity of plasminogen activators is balanced by plasminogen activator inhibitors. Plasminogen activator inhibitor type 1 (PAI-1) is a serine protease inhibitor that functions as the principal inhibitor of t-PA and u-PA. The other PAI, plasminogen activator inhibitor-2 (PAI-2), is secreted by the placenta and is only present in significant amounts during pregnancy. In addition, plasminogen activator inhibitor-3 (the protease nexin) acts as an inhibitor of t-PA and urokinase.

There are other fibrinolytic inhibitors. A2-antiplasmin (α2-AP) is a single-chain glycoprotein that reacts with plasmin to form a plasmin–α2-AP complex which is incapable of breaking down fibrin. Thrombin activatable fibrinolytic inhibitor – TAFI – is a glycoprotein synthesized by the liver and also found in platelet granules and it acts as an enzyme that may modulate fibrinolytic activity. Finally, α_2-macroglobulin – α2-M – is synthesized mainly by the liver and is a general inhibitor of both coagulation and fibrinolysis, acting as a scavenger. In the fibrinolytic system, α2-M inhibits the action of plasmin and kallikrein, while in coagulation it inhibits thrombin.

Fibrinolysis in Normal Pregnancy

Studies of the fibrinolytic system in pregnancy have produced conflicting results. The majority of early studies suggested that fibrinolytic activity is reduced in normal pregnancy, reaching a nadir in the third trimester.[142–144]

Conversely, investigators have reported that plasma plasminogen concentrations are increased in normal pregnancy[145] and more recent studies have shown that t-PA and u-PA levels increase in pregnancy, suggesting an activation of the fibrinolytic system.[146–148] To counter such activation, there is a several-fold increase in PAI-1 levels and placental production of PAI-2.[146,149] A progressive increase in serum fibrin degradation products and D-dimers has also been observed throughout pregnancy.[147,150,151] Since D-dimers reflect both fibrin polymerization and breakdown, fibrinolysis has been considered active during pregnancy.[152]

Fibrinolysis in Preeclampsia

The fibrinolytic pathway in preeclampsia has been the focus of intense investigation. Overall, increased levels of circulating t-PA levels have been described in preeclamptic compared with normal pregnancies.[106,153–156] Most, but not all, studies report increased levels of PAI-1 in preeclamptic pregnancies.[106,146,154–158] By way of contrast, however, other studies have reported a significant reduction or no difference[159,160] in PAI-1 plasma levels in preeclampsia.[159–162] Because both t-PA and PAI-1 are synthesized by endothelial cells, their increased levels may reflect endothelial dysfunction. In contrast, PAI-2 is produced by the placenta and has been reported as being significantly decreased in severe preeclampsia, possibly reflecting placental insufficiency.[160] In support of this, reduced levels of PAI-2 have also been reported in pregnancies complicated by fetal-growth restriction, with or without concomitant preeclampsia.[163]

The discrepant results for PAI-1 might in part be explained by the fact that preeclampsia may be a heterogeneous disorder with two principal phenotypes often characterized by the time of presentation (see Chapter 6). Early-onset disease presents between 24 and 34 weeks gestation and late-onset disease presents closer to term. Early-onset disease is thought to be predominantly driven by uteroplacental insufficiency, which is less obvious in women with term preeclampsia.[164] It is possible that increased t-PA and PAI-1 antigen levels found in preeclampsia could be regarded as markers for a phenotype which presents predominantly with endothelial dysfunction,[147] while reduced PAI-2 could reflect a uteroplacental insufficiency-driven phenotype. In support of this, Wikstrom et al.[165] demonstrated decreased PAI-2 levels, increased placental oxidative stress, and increased PAI-1:PAI-2 ratio in early-onset, but not in late-onset preeclampsia.

Studies of fibrin–fibrinogen degradation products in preeclamptic women have been conflicting. These have included reports of D-dimer levels, a well-established clinical laboratory marker of fibrin polymerization and breakdown *in vivo* (see Fig. 17.1). Most, but not all, studies have shown increased dimer levels in women with preeclampsia compared with those in normal pregnancy.[160,162–166] A recent metaanalysis[166] confirmed this association with third-trimester preeclampsia, but the authors highlighted the need for longitudinal studies throughout pregnancy in order to fully elucidate any prognostic value.

At this juncture, it is fair to conclude that assessment of the fibrinolytic system in preeclampsia is difficult to interpret. There have been reports of reduced fibrinogen levels, increases in t-PA but not u-PA, increases in PAI-1 but not PAI-2, and equivocal evidence of increased fibrin–fibrinogen degradation products. It is difficult to establish whether any of these changes contribute to or merely reflect changes induced by preeclampsia. Altered endothelial cell function in preeclampsia may result in increased release of both tPA and PAI-1 and reduced levels of PAI-2 might reflect either impaired placental function or a disorder in the fibrinolytic mechanism.

A polymorphism in the plasminogen activator inhibitor-1 gene has been described and has been examined in association with obstetrical complications.[167] Homozygosity for the 4G insertion in the promoter region results in higher circulating levels of PAI-1 and is associated with a higher risk of thrombosis. Glueck et al.[167] found homozygosity for the 4G allele to be twice as common in women with obstetrical complications including preeclampsia compared with normal pregnant control women. It was also seen more frequently in association with other recognized thrombophilias in women with complications. Two recent metaanalyses have confirmed that this polymorphism is a likely susceptibility variant for preeclampsia.[168,169]

THE LIVER IN PREECLAMPSIA

Hepatic involvement is common with the preeclampsia syndrome and this is especially evident when the disease is severe. Anatomical changes in the liver of women dying from eclampsia were reported in 1856 by Virchow, who described regions of periportal hemorrhage in the periphery.

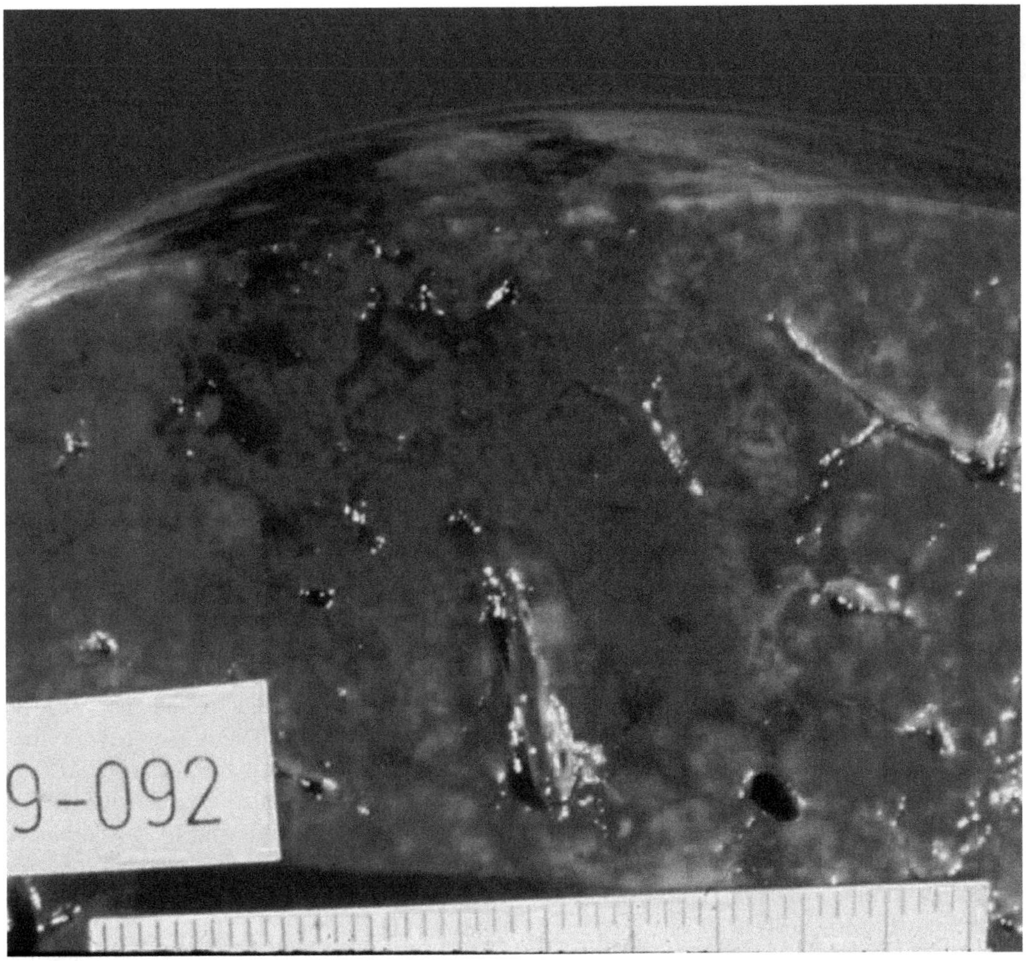

FIGURE 17.2 Cut edge of liver from a 36-week pregnant woman with severe preeclampsia. There are numerous areas of hemorrhagic infarction from the parenchyma and the surface of the liver.

In their extensive autopsy studies, Sheehan and Lynch[3] described "toxaemic lesions" that were of two basic types. The first are periportal lesions which begin as localized hemorrhages and which are later replaced by fibrin. The second are various grades of ischemic parenchymal lesions that range in size from microscopic to very large infarcts such as those shown in Fig. 17.2.

In these early studies, it was apparent that the extent of liver involvement was related to severity of preeclampsia–eclampsia. Specifically, Sheehan and Lynch[3] studied 75 women who died with eclampsia, and found that two-thirds had hepatic lesions – periportal lesions were seen in about 25%, periportal hemorrhages along with ischemic lesions in another 40%, but in no cases were ischemic lesions seen alone. By contrast they saw these lesions in only three of 38 women with "mild toxaemia," and neither type of lesion was seen in the livers of 97 normally pregnant women who died from unrelated causes. They also reviewed reports of 30 needle liver biopsies in eclamptic women and found periportal lesions in five and ischemic lesions in two. Thus, it is not possible to use these findings to estimate a contemporaneous prevalence of hepatic hemorrhage and necrosis.

Clinical Aspects of Liver Involvement

The clinical significance of hepatic involvement with the preeclampsia syndrome has at least four aspects:

1. Symptomatic involvement, typically manifest as moderate to severe right upper, midepigastric or substernal pain and tenderness. As discussed in Chapter 1, the presence of these symptoms defines severe preeclampsia. Many of these women are found to have abnormally elevated serum levels of hepatic transaminases – aspartate transaminase (AST) and alanine transferase (ALT). In some cases, the extent of hepatic involvement can be surprisingly extensive, yet be clinically insignificant.
2. Asymptomatic elevation of serum hepatic transaminase levels – AST and ALT – which are also considered defining markers for severe preeclampsia. These values

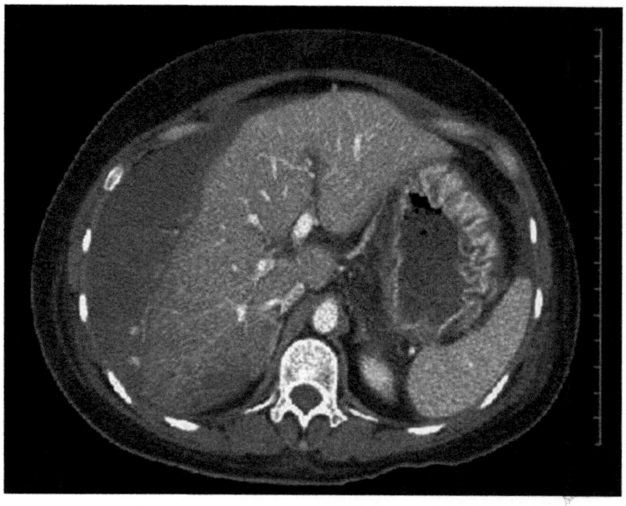

FIGURE 17.3 Computed-tomographic scan of upper abdomen shows large subcapsular hematoma of the right lobe of the liver. Flame-shaped hemorrhages are seen at the hepatic-hematoma interface. Another area of ischemic necrosis and intrahepatic hemorrhage is seen in the posterior region of the right lobe.

seldom exceed 500 U/L, but occasionally may be as high as 2000 U/L. In general, serum levels inversely follow platelet levels, and both usually normalize within 3 to 5 days following delivery.

3. Hepatic hemorrhage or infarction may extend to form a hepatic subcapsular hematoma under the Glisson capsule that may rupture into the peritoneal cavity. These can be diagnosed with CT- or MR-imaging as shown in Fig. 17.3. Intrahepatic or unruptured subcapsular hematomas are probably more common than realized or clinically suspected, and they are more likely to be found with HELLP syndrome and increase its related mortality risks.[133,170] Management depends on the clinical circumstances. In most cases, observation and conservative treatment are usually elected, unless hemorrhage is ongoing. That said, in some cases prompt surgical intervention may be lifesaving. There are reports of the use of an argon beam coagulator in one woman and use of recombinant VIIa in another to control hepatic hemorrhage. Rinehart et al.[171] reviewed 121 cases of spontaneous hepatic rupture associated with preeclampsia and reported the maternal mortality rate to be 30%. Vigil De Gracia et al.[170] reviewed 180 cases and 90% were complicated by HELLP syndrome and in 90% the capsule had ruptured. Maternal mortality was 22% and perinatal mortality was 31%. Hunter et al.[172] and Wicke et al.[173] each described similar women in whom liver transplant was considered life-saving.
4. Acute fatty liver of pregnancy is sometimes confused with the preeclampsia and the HELLP syndrome.[174–176] It too has its onset in late pregnancy, there is often hypertension, and there are elevated serum hepatic transaminase and creatinine levels as well as thrombocytopenia.

The Liver in HELLP Syndrome

The logical follow-up of these observations of liver involvement with preeclampsia was a search for a serum analyte that would reflect hepatic damage. In his first edition,[1] Chesley carefully reviewed methods that had been tried and abandoned – including, for example, thymol turbidity and cephalin flocculation tests. By that time, however, there were a number of studies reporting values for normally pregnant women of serum glutamic oxaloacetic transferase (SGOT) and serum glutamic pyruvic transaminase (SGPT) levels. Since that time, these have been renamed aspartate aminotransferase (AST) and alanine aminotransferase (ALT), respectively. Chesley found that about a fourth of mildly preeclamptic women had abnormally elevated transaminase levels and most women with eclampsia had such elevations. As this association continued to develop over the next 20 years, it became a marker for severity of preeclampsia. To call attention to its relationship to severe preeclampsia, Weinstein[4,5] coined the term *HELLP syndrome*, which is now used widely: *H*emolysis, *E*levated *L*iver enzymes, and *L*ow *P*latelets (see p. 379).

Currently, although some use the Mississippi or Tennessee classification of severity of the syndrome or other various monikers for *partial HELLP syndrome*, there is not a widely accepted classification. Because of this, the reported incidence of the HELLP syndrome varies. While it usually coexists with other markers for severe preeclampsia, at times these laboratory abnormalities are found alone. In one large study by Sibai et al.,[177] the syndrome was reported in 20% of women with severe preeclampsia or eclampsia. It seems to be more common with early-onset preeclampsia, and indeed some are of the view that it represents only one of several manifestations of the preeclampsia syndrome. There are a few studies that describe preterm pregnancies complicated by HELLP syndrome. In their review of severe preeclampsia between 24 and 36 weeks gestation, Sibai and Barton[141] reported a range of incidences from 4 to 23%. In the large study from Amsterdam, 25% of such women had HELLP syndrome.[178]

Both maternal and perinatal morbidity and mortality are increased in women with preeclampsia complicated by HELLP syndrome.[179,180] In a multicenter study, Haddad et al.[181] described 183 women with HELLP syndrome – almost 40% had adverse outcomes and two women died. Other complications included eclampsia (6%), placental abruption (10%), acute renal failure (5%), pulmonary edema (10%), and subcapsular liver hematoma (1.6%). In a study from Cali, Columbia,[182] of 132 women with HELLP syndrome, there were four maternal deaths – three of the women who died had liver failure. In their review of 693 women with HELLP syndrome, Keiser et al.[183] cited a 10% incidence of concurrent eclampsia.

There is some evidence that preeclampsia with HELLP syndrome is a different entity than preeclampsia without HELLP. Sep et al.[184] reported that preeclampsia

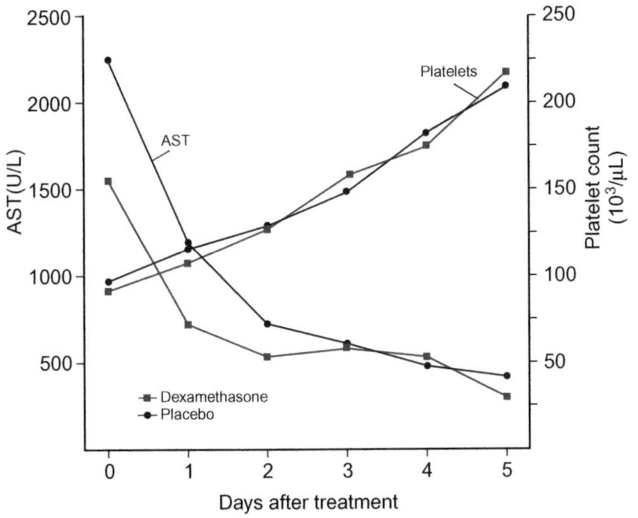

FIGURE 17.4 The recovery times for platelets and serum aspartate aminotransferase (AST) are shown for women with HELLP syndrome who were randomly assigned to receive dexamethasone or placebo. (Data from Katz et al.[189]) (This figure is reproduced in color in the color plate section.)

complicated by HELLP syndrome had worse morbidity than "isolated preeclampsia" and they postulated that the two disorders are distinctly different. Reimer et al.[185] found a difference in the ratio of antiangiogenic and inflammatory acute-phase proteins in these two conditions and reached similar conclusions. Sibai and Stella[186] have discussed some of these aspects under the rubric of "atypical preeclampsia-eclampsia."

Corticosteroid Therapy for HELLP Syndrome

Use of glucocorticoids to treat HELLP syndrome by ameliorating its severity has been advocated by some. Martin et al.[187] reviewed outcomes of almost 500 such women from the University of Mississippi. From 1994 to 2000, they gave glucocorticoid treatment to 90% of women with HELLP syndrome, and their perinatal outcomes were considered more favorable than the cohort from 1985 to 1991, of whom only 16% were treated. They conducted a randomized trial to compare betamethasone with dexamethasone; however, the study did not have a nontreated control group.[188] There have been two randomized studies that have addressed this question. In one, Fonseca and associates[182] randomized 132 women with HELLP syndrome to blinded treatment with either dexamethasone or placebo. Outcomes assessed included duration of hospitalization, time to recovery for abnormal laboratory tests, recovery of clinical parameters, and complications. None of these was statistically different between the two groups. The other study was a blinded trial from Recife, Brazil, which included 105 postpartum women with HELLP syndrome who were randomized to treatment with dexamethasone versus placebo.[189] Outcomes were analyzed similar to the Cali study and there were no advantages to dexamethasone to hasten recovery. The responses of platelets and AST are shown in Fig. 17.4. In a review of the English literature from 1990 through 2006, Vidaeff and Yeomans[190] cited nine studies, including seven randomized controlled trials. They concluded that glucocorticoid therapy for HELLP syndrome should remain investigational. Thus, it does appear that, at this time, corticosteroid therapy should be considered investigational.

References

1. Chesley L. *Hypertensive Disorders in Pregnancy*. New York: Appleton-Century-Crofts; 1978.
2. Pritchard RA, Cunningham FG, Mason RA. Coagulation changes in eclampsia: their frequency and pathogenesis. *Am J Obstet Gynecol*. 1976;124:855–864.
3. Sheehan HL, Lynch JB. *Cerebral Lesions. Pathology of Toxaemia of Pregnancy*. Baltimore: Williams & Wilkins; 1973.
4. Weinstein L. Syndrome of hemolysis, elevated liver enzymes and low platelet count: a severe consequence of hypertension in pregnancy. *Am J Obstet Gynecol*. 1982;142:159–160.
5. Weinstein L. Preeclampsia-eclampsia with hemolysis, elevated liver enzymes, and thrombocytopenia. *Obstet Gynecol*. 1985;66:657–660.
6. Boehlen F, Hohlfeld H, Extermann P, Perneger TV, de Moerloose P. Platelet count at term pregnancy: a reappraisal of the threshold. *Obstet Gynecol*. 2000;95:29–33.
7. Maymon RI, Zimerman AL, Strauss S, Gayer G. Maternal spleen size throughout normal pregnancy. *Semin Ultrasound CT MRI*. 2007;28:64–66.
8. McCrae KR. Thrombocytopenia in pregnancy. *Hematology Am Soc Hematol Educ Program*. 2010;2010:397–402.
9. Burrows RF, Andrew M. Neonatal thrombocytopenia in the hypertensive disorders of pregnancy. *Obstet Gynecol*. 1990;76:234–238.
10. Valera MC, Parant O, Vayssiere C, Arnal JF, Payrastre B. Physiologic and pathologic changes of platelets in pregnancy. *Platelets*. 2010;21:8587–8595.
11. Sainio S, Kekomäki R, Riikonon S, Teramo K. Maternal thrombocytopenia at term: a population-based study. *Acta Obstet Gynecol Scand*. 2000;79:744–749.
12. Burrows RF, Kelton JG. Fetal thrombocytopenia and its relation to maternal thrombocytopenia. *New Engl J Med*. 1993;329:1463–1466.
13. McCrae KR. Thrombocytopenia in pregnancy:differential diagnosis, pathogenesis and management. *Blood Rev*. 2003;17:7–14.
14. Pridjian G, Puschett JB. Preeclampsia. Part 1: clinical and pathophysiological considerations. *Obstet Gynecol Surv*. 2002;57:598–618.
15. McCrae KR, Cines DB. Thrombotic microangiopathy during pregnancy. *Semin Hematol*. 1997;34:148–158.
16. Stepanian A, Cohen-Moatti M, Sanglier T, et al. Von Willebrand factor and ADAMTS13: a candidate couple for preeclampsia pathophysiology. *Arteriosclerosis Thromb Vasc Biol*. 2011;31:71703–71709.

17. Brenner B, Zwang E, Bronshtein M, Seligsohn U. Von Willebrand factor multimer patterns in pregnancy-induced hypertension. *Thromb Haemost.* 1989;62:715–717.
18. Thorp JM, White GC, Moake JL, Bowes WA. Von Willebrand factor multimeric levels and patterns in patients with severe preeclampsia. *Obstet Gynecol.* 1990;75:163–167.
19. Lockwood CJ, Peters JH. Increased plasma levels of $ED1^+$ cellular fibronectin precede the clinical signs of preeclampsia. *Am J Obstet Gynecol.* 1990;162:358–362.
20. Stubbs TM, Lazarchick J, Van Dorsten JP, Cox J, Loadholt CB. Evidence of accelerated platelet production and consumption in nonthrombocytopenic preeclampsia. *Am J Obstet Gynecol.* 1986;155:263–265.
21. Ahmed Y, van Iddekinge B, Paul C, Sullivan HF, Elder MG. Retrospective analysis of platelet numbers and volumes in normal pregnancy and in pre-eclampsia. *BJOG.* 1993;100:216–220.
22. Paulus JM. Platelet size in man. *Blood.* 1975;46:321–336.
23. Pekonen F, Rasi V, Ammala M, Viinikka L, Ylikorkala O. Platelet function and coagulation in normal and preeclamptic pregnancy. *Thromb Res.* 1986;43:553–560.
24. Rakoczi I, Tallian F, Bagdany S, Gati I. Platelet life-span in normal pregnancy and pre-eclampsia as determined by a non-radioisotope technique. *Thromb Res.* 1979;15:553–556.
25. Inglis TC, Stuart J, George AJ, Davies AJ. Haemostatic and rheological changes in normal pregnancy and pre-eclampsia. *Brit J Haematol.* 1982;50:461–465.
26. Pritchard JA, Cunningham FG, Pritchard SA, Mason RA. How often does maternal preeclampsia-eclampsia incite thrombocytopenia in the fetus? *Obstet Gynecol.* 1987;69:292–295.
27. Redman CW, Bonnar J, Beilin L. Early platelet consumption in pre-eclampsia. *Br Med J.* 1978;1:467–469.
28. Brown MA, Buddle ML. Hypertension in pregnancy: maternal and fetal outcomes according to laboratory and clinical features. *Med J Aust.* 1996;165:360–365.
29. Leduc L, Wheeler JM, Kirshon B, Mitchell P, Cotton DB. Coagulation profile in severe preeclampsia. *Obstet Gynecol.* 1992;79:14–18.
30. Barron WM, Heckerling P, Hibbard JU, Fisher S. Reducing unnecessary coagulation testing in hypertensive disorders of pregnancy. *Obstet Gynecol.* 1999;94:364–370.
31. Rinehart BK, Terrone DA, May WL, et al. Change in platelet count predicts eventual maternal outcome with syndrome of hemolysis, elevated liver enzymes and low platelet count. *J Maternal-Fetal Med.* 2001;10:2.
32. Rath W, Faridi A, Dudenhausen JW. HELLP syndrome. *J Perinatal Med.* 2000;28:249–260.
33. Martin JN, Blake PG, Perry KG, et al. The natural history of HELLP syndrome: patterns of disease progression and regression. *Am J Obstet Gynecol.* 1991;164:1500–1513.
34. Douglas JT, Shah M, Lowe GD, Belch JJ, Forbes CD, Prentice CR. Plasma fibrinopeptide A and beta-thromboglobulin in pre-eclampsia and pregnancy hypertension. *Thromb Haemost.* 1982;47:54–55.
35. Hayashi M, Inoue T, Hoshimoto K, Negishi H, Ohkura T, Inaba N. Characterization of five marker levels of the hemostatic system and endothelial status in normotensive pregnancy and pre-eclampsia. *Eur J Haematol.* 2002;69:297–302.
36. Ballegeer VC, Spitz B, De Baene LA, Van Assche AF, Hidajat M, Criel AM. Platelet activation and vascular damage in gestational hypertension. *Am J Obstet Gynecol.* 1992;166:629–633.
37. Janes SL, Goodall AH. Flow cytometric detection of circulating activated platelets and platelet hyper-responsiveness in pre-eclampsia and pregnancy. *Clin Sci (Lond).* 1994;86:731–739.
38. Socol ML, Weiner CP, Louis G, Rehnberg K, Rossi EC. Platelet activation in preeclampsia. *Am J Obstet Gynecol.* 1985;151:494–497.
39. Alijolas-Reig J, Palacio-Garcia C, Farran-Codina I, Ruiz-Romance M, Llurba E, Vilardell-Tarres M. Circulating cell-derived microparticles in severe preeclampsia and in fetal growth restriction. *Am J Reprod Immunol.* 2012;67:140–151.
40. Macey MG, Bevan S, Alam S, et al. Platelet activation and endogenous thrombin potential in pre-eclampsia. *Thromb Res.* 2010;125:e76–e81.
41. Burrows RF, Hunter DJ, Andrew M, Kelton JG. A prospective study investigating the mechanism of thrombocytopenia in preeclampsia. *Obstet Gynecol.* 1987;70:334–338.
42. Samuels P, Main EK, Tomaski A, Mennuti MT, Gabbe SG, Cines DB. Abnormalities in platelet antiglobulin tests in preeclamptic mothers and their neonates. *Am J Obstet Gynecol.* 1987;157:109–113.
43. Middelkoop CM, Dekker GA, Kraayenbrink AA, Popp-Snijders C. Platelet-poor plasma serotonin in normal and preeclamptic pregnancy. *Clin Chem.* 1993;39:1675–1678.
44. Konijnenberg A, Stokkers EW, van der Post JA, et al. Extensive platelet activation in preeclampsia compared with normal pregnancy: enhanced expression of cell adhesion molecules. *Am J Obstet Gynecol.* 1997;176:461–469.
45. Fitzgerald DJ, Mayo G, Catella F, Entman SS, FitzGerald GA. Increased thromboxane biosynthesis in normal pregnancy is mainly derived from platelets. *Am J Obstet Gynecol.* 1987;157(2):325–330.
46. Fitzgerald DJ, Rocki W, Murray R, Mayo G, FitzGerald GA. Thromboxane A2 synthesis in pregnancy-induced hypertension. *Lancet.* 1990;335:751–754.
47. Van Geet C, Arnout J, Eggermont E, Vermylen J. Urinary thromboxane B2 and 2,3-dinor-thromboxane B2 in the neonate born at full-term age. *Eicosanoids.* 1990;3:39–43.
48. Garzetti GG, Tranquilli AL, Cugini AM, Mazzanti L, Cester N, Romanini C. Altered lipid composition, increased lipid peroxidation, and altered fluidity of the membrane as evidence of platelet damage in preeclampsia. *Obstet Gynecol.* 1993;81:337–340.
49. Zheng XL. Structure-function and regulation of ADAMTS-13 protease. *J Thromb Haemost.* 2013;11(suppl 1):11–23.
50. Burke SD, Karumanchi SA. Spiral artery remodeling in preeclampsia revisited. *Hypertension.* 2013;62:1013–1014.
51. Horn EH, Cooper J, Hardy E, Heptinstall S, Rubin PC. A cross-sectional study of platelet cyclic AMP in healthy and hypertensive pregnant women. *Clin Sci (Lond).* 1991;80:549–558.
52. Howie PW. The haemostatic mechanisms in pre-eclampsia. *Clin Obstet Gynaecol.* 1997;4:595–611.
53. Ahmed Y, Sullivan MH, Elder MG. Detection of platelet desensitization in pregnancy-induced hypertension is dependent on the agonist used. *Thromb Haemost.* 1991;65:474–477.
54. Delacretaz E, de Quay N, Waeber B. Differential nitric oxide synthase activity in human platelets during normal pregnancy and pre-eclampsia. *Clin Sci (Lond).* 1995;88:607–610.
55. Mazzanti L, Raffaelli F, Vignini A, et al. Nitric oxide and peroxynitrite platelet levels in gestational hypertension and preeclampsia. *Platelets.* 2012;23:26–35.

56. Bussolino F, Benedetto C, Massobrio M, Camussi G. Maternal vascular prostacyclin activity in pre-eclampsia. *Lancet*. 1980;2:702.
57. Briel RC, Kieback DG, Lippert TH. Platelet sensitivity to a prostacyclin analogue in normal and pathological pregnancy. *Prostaglandins Leukot Med*. 1984;13:335–340.
58. Dadak C, Kefalides A, Sinzinger H. Prostacyclin-synthesis stimulating plasma factor and platelet sensitivity in pre-eclampsia. *Biol Res Pregnancy Perinatol*. 1985;6:65–69.
59. Liel N, Nathan I, Yermiyahu T, et al. Increased platelet thromboxane A2/prostaglandin H2 receptors in patients with pregnancy induced hypertension. *Thromb Res*. 1993;70:205–210.
60. Charo IF, Feinman RD, Detwiler TC. Interrelations of platelet aggregation and secretion. *J Clin Invest*. 1989;60:866–873.
61. Stalker TJ, Newman DK, Ma P, Wannemacher KM, Brass LF. Platelet signaling. *Handb Exp Pharmacol*. 2012;210:59–85.
62. Shanley DK, Kiely PA, Golla K, et al. Pregnancy-specific glycoproteins bind integrin αIIbβ3 and inhibit the platelet-fibrinogen interaction. *PLoS One*. 2013;8(2):e57491.
63. Sokol J1, Biringer K, Skerenova M. Platelet aggregation abnormalities in patients with fetal losses: the GP6 gene polymorphism. *Fertil Steril*. 2012;98(5):1170–1174.
64. Gerhardt A, Goecke TW, Beckmann MW, et al. The G20210A prothrombin-gene mutation and the plasminogen activator inhibitor (PAI-1) 5G/5G genotype are associated with early onset of severe preeclampsia. *J Thromb Haemost*. 2005;3:686–691.
65. Klockenbusch W, Hohlfeld T, Wilhelm M, Somville T, Schrör K. Platelet PGI2 receptor affinity is reduced in pre-eclampsia. *Br J Clin Pharmacol*. 1996;41:616–618.
66. Barr SM, Lees KR, Butters L, O'Donnell A, Rubin PC. Platelet intracellular free calcium concentration in normotensive and hypertensive pregnancies in the human. *Clin Sci (Lond)*. 1989;76:67–71.
67. Rao GH, Peller JD, White JG. Measurement of ionized calcium in blood platelets with a new generation calcium indicator. *Biochem Biophys Res Commun*. 1985;132:652–657.
68. Erne P, Mittelholzer E, Burgisser E, Fluckiger R, Buhler FR. Measurement of receptor induced changes in intracellular free calcium in human platelets. *J Recept Res*. 1984;4:587–604.
69. Kilby MD, Broughton Pipkin F, Cockbill S, Heptinstall S, Symonds EM. A cross-sectional study of basal platelet intracellular free calcium concentration in normotensive and hypertensive primigravid pregnancies. *Clin Sci (Lond)*. 1990;78:75–80.
70. Zemel MB, Zemel PC, Berry S. Altered platelet calcium metabolism as an early predictor of increased peripheral vascular resistance and preeclampsia in urban black women. *New Engl J Med*. 1990;323:434–438.
71. Cho JH, Nash F, Fekete Z, Kimura M, Reeves JP, Aviv A. Increased calcium stores in platelets from African Americans. *Hypertension*. 1995;25:377–383.
72. Kyle PM, Jackson MC, Buckley DC, de Swiet M, Redman CW. Platelet intracellular free calcium response to arginine vasopressin is similar in preeclampsia and normal pregnancy. *Am J Obstet Gynecol*. 1995;172:654–660.
73. Roberts JM, Lewis V, Mize N, Tsuchiya A, Starr J. Human platelet alpha-adrenergic receptors and responses during pregnancy: no change except that with differing hematocrit. *Am J Obstet Gynecol*. 1986;154:206–210.
74. Louden KA, Broughton Pipkin F, Heptinstall S, Fox SC, Mitchell JR, Symonds EM. Platelet reactivity and serum thromboxane B2 production in whole blood in gestation hypertension and pre-eclampsia. *Br J Obstet Gynaecol*. 1991;98:1239–1244.
75. Horn EH, Cooper JA, Hardy E, Heptinstall S, Rubin PC. Longitudinal studies of platelet cyclic AMP during healthy pregnancy and pregnancies at risk of pre-eclampsia. *Clin Sci (Lond)*. 1995;89:91–99.
76. Mellion BT, Ignarro LJ, Ohlstein EH, Pontecorvo EG, Hyman AL, Kadowitz PJ. Evidence for the inhibitory role of guanosine 3′,5′-monophosphate in ADP-induced human platelet aggregation in the presence of nitric oxide and related vasodilators. *Blood*. 1981;57:946–955.
77. Hardy E, Rubin PC, Horn EH. Effects of nitric oxide donors in vitro on the arachidonic acid-induced platelet release reaction and platelet cyclic GMP concentration in pre-eclampsia. *Clin Sci (Lond)*. 1994;86:195–202.
78. Moore TJ, Williams GH. Angiotensin II receptors on human platelets. *Circ Res*. 1982;51:314–320.
79. Ding YA, MacIntyre DE, Kenyon CJ, Semple PF. Angiotensin II effects on platelet function. *J Hypertens Suppl*. 1982;3:S251–S253.
80. Poplawski A. The effect of angiotensin II on the platelet aggregation induced by adenosine diphosphate, epinephrine and thrombin. *Experientia*. 1970;26:86.
81. Haller H, Oeney T, Hauck U, Distler A, Philipp T. Increased intracellular free calcium and sensitivity to angiotensin II in platelets of preeclamptic women. *Am J Hypertens*. 1989;2:238–243.
82. American College of Obstetricians and Gynecologists: Antiphospholipid syndrome. Practice Bulletin No. 132, December 2012.
83. Poulio L, Forest JC, Moutquin JM, Coulombe N, Masse J. Platelet angiotensin II binding sites and early detection of preeclampsia. *Obstet Gynecol*. 1998;91:591–595.
84. Baker PN, Broughton Pipkin F, Symonds EM. Platelet angiotensin II binding and plasma renin concentration, plasma renin substrate and plasma angiotensin II in human pregnancy. *Clin Sci (Lond)*. 1990;79:403–408.
85. Baker PN, Broughton Pipkin F, Symonds EM. Platelet angiotensin II binding sites in normotensive and hypertensive women. *Br J Obstet Gynaecol*. 1991;98:436–440.
86. Gant NF, Daley GL, Chand S, Whalley PJ, MacDonald PC. A study of angiotensin II pressor response throughout primigravid pregnancy. *J Clin Invest*. 1973;52:2682–2689.
87. Fletcher AP, Alkjaersig NK, Burstein R. The influence of pregnancy upon blood coagulation and plasma fibrinolytic enzyme function. *Am J Obstet Gynecol*. 1979;134:743–751.
88. Maki M. Coagulation, fibrinolysis, platelet and kinin-forming systems during toxemia of pregnancy. *Biol Res Pregnancy Perinatol*. 1983;4:152–154.
89. Stirling Y, Woolf L, North WR, Seghatchian MJ, Meade TW. Haemostasis in normal pregnancy. *Thromb Haemost*. 1984;52:176–182.
90. Lox CD, Dorsett MM, Hampton RM. Observations on clotting activity during pre-eclampsia. *Clin Exp Hypertens B*. 1983;2:179–190.
91. Vaziri ND, Toohey J, Powers D, et al. Activation of intrinsic coagulation pathway in pre-eclampsia. *Am J Med*. 1986;80:103–107.

92. Fournie A, Monrozies M, Pontonnier G, Boneu B, Bierme R. Factor VIII complex in normal pregnancy, pre-eclampsia and fetal growth retardation. *Br J Obstet Gynaecol.* 1981;88:250–254.
93. Thornton CA, Bonnar J. Factor VIII-related antigen and factor VIII coagulant activity in normal and pre-eclamptic pregnancy. *Br J Obstet Gynaecol.* 1977;84:919–923.
94. Redman CW, Denson KW, Beilin LJ, Bolton FG, Stirrat GM. Factor-VIII consumption in pre-eclampsia. *Lancet.* 1977;2:1249–1252.
95. Scholtes MC, Gerretsen G, Haak HL. The factor VIII ratio in normal and pathological pregnancies. *Eur J Obstet Gynecol Reprod Biol.* 1983;16:89–95.
96. Howie PW, Prentice CR, McNicol GP. Coagulation, fibrinolysis and platelet function in pre-eclampsia, essential hypertension and placental insufficiency. *J Obstet Gynaecol Br Commonw.* 1971;78:992–1003.
97. Wallenburg HCS. Changes in the coagulation system and platelets in pregnancy-induced hypertension and pre-eclampsia. In: Sharp F, Symonds EM, eds. *Hypertension in Pregnancy Procedings of the 16th RCOG Study Group.* New York: Perinatology Press; 1986:227–248.
98. Douglas JT, Shah M, Lowe GD, et al. Plasma fibrinopeptide A and beta-thromboglobulin in pre-eclampsia and pregnancy hypertension. *Thromb Haemost.* 1982;47:54–55.
99. Borok Z, Weitz J, Owen J, Auerbach M, Nossel HL. Fibrinogen proteolysis and platelet alpha-granule release in preeclampsia/eclampsia. *Blood.* 1984;63:525–531.
100. American College of Obsetricians and Gynecologists: Inherited thrombophilias in pregnancy. Practice Bulletin No. 138, September 2013.
101. Seegers WH. Antithrombin III. Theory and clinical applications. H. P. Smith Memorial Lecture. *Am J Clin Pathol.* 1978;69:299–359.
102. Weenink GH, Treffers PE, Kahle LH, ten Cate JW. Antithrombin III in normal pregnancy. *Thromb Res.* 1982;26: 281–287.
103. Weiner CP, Brandt J. Plasma antithrombin III activity: an aid in the diagnosis of preeclampsia-eclampsia. *Am J Obstet Gynecol.* 1982;142:275–281.
104. Weenink GH, Borm JJ, Ten Cate JW, Treffers PE. Antithrombin III levels in normotensive and hypertensive pregnancy. *Gynecol Obstet Invest.* 1983;16:230–242.
105. Saleh AA, Bottoms SF, Welch RA, et al. Preeclampsia, delivery, and the hemostatic system. *Am J Obstet Gynecol.* 1987;157:331–336.
106. Yin KH, Koh SC, Malcus P, et al. Preeclampsia: haemostatic status and the short-term effects of methyldopa and isradipine therapy. *J Obstet Gynaecol Res.* 1998;24:231–238.
107. Schjetlein R, Haugen G, Wisloff F. Markers of intravascular coagulation and fibrinolysis in preeclampsia: association with intrauterine growth retardation. *Acta Obstet Gynecol Scand.* 1997;76:541–546.
108. Paternoster D, Stella A, Simioni P, et al. Clotting inhibitors and fibronectin as potential markers in preeclampsia. *Int J Gynaecol Obstet.* 1994;47:215–221.
109. Shinyama H, Akira T, Uchida T, et al. Antithrombin III prevents renal dysfunction and hypertension induced by enhanced intravascular coagulation in pregnant rats: pharmacological confirmation of the benefits of treatment with antithrombin III in preeclampsia. *J Cardiovasc Pharmacol.* 1996;27:702–711.
110. Paternoster DM, Fantinato S, Manganelli F, et al. Efficacy of AT in pre-eclampsia: a case-control prospective trial. *Thromb Haemost.* 2004;91:283–289.
111. Larciprete G, Gioia S, Angelucci PA, et al. Single inherited thrombophilias and adverse pregnancy outcomes. *J Obstet Gynaecol Res.* 2007;33:423–430.
112. Mello G, Parretti E, Marozio L, et al. Thrombophilia is significantly associated with severe preeclampsia: results of a large-scale, case-controlled study. *Hypertension.* 2005;46:1270–1274.
113. Fernandez JA, Estelles A, Gilabert J, Espana F, Aznar J. Functional and immunologic protein S in normal pregnant women and in full-term newborns. *Thromb Haemost.* 1989;61:474–478.
114. Dekker GA, de Vries JI, Doelitzsch PM, et al. Underlying disorders associated with severe early-onset preeclampsia. *Am J Obstet Gynecol.* 1995;173:1042–1048.
115. de Boer K, ten Cate JW, Sturk A, Borm JJ, Treffers PE. Enhanced thrombin generation in normal and hypertensive pregnancy. *Am J Obstet Gynecol.* 1989;160:95–100.
116. Brandt JT. Current concepts of coagulation. *Clin Obstet Gynecol.* 1985;28:3–14.
117. Comp PC, Jacocks RM, Ferrell GL, Esmon CT. Activation of protein C in vivo. *J Clin Invest.* 1982;70:127–134.
118. Gonzalez R, Alberca I, Vicente V. Protein C levels in late pregnancy, postpartum and in women on oral contraceptives. *Thromb Res.* 1985;39:637–640.
119. Weiner CP, Brandt J. Plasma antithrombin III activity in normal pregnancy. *Obstet Gynecol.* 1980;56:601–603.
120. Dahlback B, Carlsson M, Svensson PJ. Familial thrombophilia due to a previously unrecognized mechanism characterized by poor anticoagulant response to activated protein C: prediction of a cofactor to activated protein C. *Proc Natl Acad Sci U S A.* 1993;90:1004–1008.
121. Bertina RM, Koeleman BP, Koster T, et al. Mutation in blood coagulation factor V associated with resistance to activated protein C. *Nature.* 1994;369:64–67.
122. Spina V, Aleandri V, Morini F. The impact of the factor V Leiden mutation on pregnancy. *Hum Reprod Update.* 2000;6:301–306.
123. De Stefano V, Chiusolo P, Paciaroni K, Leone G. Epidemiology of factor V Leiden: clinical implications. *Semin Thromb Hemost.* 1998;24:367–379.
124. Robertson L, Wu O, Langhorne P, et al. Thrombophilia in pregnancy: a systematic review. *Br J Haematol.* 2006;132: 171–196.
125. Rai R, Shlebak A, Cohen H, et al. Factor V Leiden and acquired activated protein C resistance among 1000 women with recurrent miscarriage. *Hum Reprod.* 2001;16:961–965.
126. Lindqvist PG, Svensson P, Dahlback B. Activated protein C resistance—in the absence of factor V Leiden—and pregnancy. *J Thromb Haemost.* 2006;4:361–366.
127. Said JM, Higgins JR, Moses EK, et al. Inherited thrombophilias and adverse pregnancy outcomes: a case-control study in an Australian population. *Acta Obstet Gynecol Scand.* 2012;91:250–255.
128. Alfirevic Z, Roberts D, Martlew V. How strong is the association between maternal thrombophilia and adverse

pregnancy outcome? A systematic review. *Eur J Obstet Gynecol Reprod Biol.* 2002;101:6–14.
129. Higgins JR, Kaiser T, Moses EK, North R, Brennecke SP. Prothrombin G20210A mutation: is it associated with pre-eclampsia? *Gynecol Obstet Invest.* 2000;50:254–257.
130. Kupferminc MJ, Peri H, Zwang E, et al. High prevalence of the prothrombin gene mutation in women with intrauterine growth retardation, abruptio placentae and second trimester loss. *Acta Obstet Gynecol Scand.* 2000;79:963–967.
131. Livingston JC, Park V, Barton JR, et al. Lack of association of severe preeclampsia with maternal and fetal mutant alleles for tumor necrosis factor alpha and lymphotoxin alpha genes and plasma tumor necrosis factor alpha levels. *Am J Obstet Gynecol.* 2001;184:1273–1277.
132. Kupferminc MJ, Eldor A, Steinman N, et al. Increased frequency of genetic thrombophilia in women with complications of pregnancy. *New Engl J Med.* 1999;340:9–13.
133. McNeil HP, Chesterman CN, Krilis SA. Immunology and clinical importance of antiphospholipid antibodies. *Adv Immunol.* 1991;49:193–280.
134. Pattison NS, Chamley LW, McKay EJ, Liggins GC, Butler WS. Antiphospholipid antibodies in pregnancy: prevalence and clinical associations. *Br J Obstet Gynaecol.* 1993;100:909–913.
135. Yasuda M, Takakuwa K, Tokunaga A, Tanaka K. Prospective studies of the association between anticardiolipin antibody and outcome of pregnancy. *Obstet Gynecol.* 1995;86:555–559.
136. Rai R, Cohen H, Dave M, Regan L. Randomised controlled trial of aspirin and aspirin plus heparin in pregnant women with recurrent miscarriage associated with phospholipid antibodies (or antiphospholipid antibodies). *BMJ.* 1997;314:253–257.
137. CLASP: a randomised trial of low-dose aspirin for the prevention and treatment of pre-eclampsia among 9364 pregnant women. CLASP (Collaborative Low-dose Aspirin Study in Pregnancy) Collaborative Group. *Lancet.* 1994;343:619–629.
138. Kalk JJ, Huisjes AJ, de Groot CJ, et al. Recurrence rate of pre-eclampsia in women with thrombophilia influenced by low-molecular-weight heparin treatment? *Neth J Med.* 2004;62:83–87.
139. Mello G, Parretti E, Fatini C, et al. Low-molecular-weight heparin lowers the recurrence rate of preeclampsia and restores the physiological vascular changes in angiotensin-converting enzyme DD women. *Hypertension.* 2005;45:86–91.
140. Sergio F, Maria Clara D, Gabriella F, et al. Prophylaxis of recurrent preeclampsia: low-molecular-weight heparin plus low-dose aspirin versus low-dose aspirin alone. *Hypertens Pregnancy.* 2006;25:115–127.
141. Syrovets T, Lunov O, Simmet T. Plasmin as a proinflammatory cell activator. *J Leukoc Biol.* 2012;92:509–519.
142. Biezenski JJ, Moore HC. Fibrinolysis in normal pregnancy. *J Clin Pathol.* 1958;11:306–310.
143. Shaper AG, Macintosh DM, Evans CM, Kyobe J. Fibrinolysis and plasminogen levels in pregnancy and the puerperium. *Lancet.* 1965;2:706–708.
144. Menon IS. Fibrinolytic activity in pregnancy. *Br Med J.* 1970;2:239.
145. Bonnar J, Daly L, Sheppard BL. Changes in the fibrinolytic system during pregnancy. *Semin Thromb Hemost.* 1990;16:221–229.
146. Halligan A, Bonnar J, Sheppard B, Darling M, Walshe J. Haemostatic, fibrinolytic and endothelial variables in normal pregnancies and pre-eclampsia. *Br J Obstet Gynaecol.* 1994;101:488–492.
147. Belo L, Santos-Silva A, Rumley A, et al. Elevated tissue plasminogen activator as a potential marker of endothelial dysfunction in pre-eclampsia: correlation with proteinuria. *BJOG.* 2002;109:1250–1255.
148. Koh CL, Viegas OA, Yuen R, et al. Plasminogen activators and inhibitors in normal late pregnancy, postpartum and in the postnatal period. *Int J Gynaecol Obstet.* 1992;38:9–18.
149. Cadroy Y. Evaluation of six markers of haemostatic system in normal pregnancy and pregnancy complicatied by hpertension or pre-eclampsia. *BJOG.* 1993;100:416–420.
150. Sattar N, Greer IA, Rumley A, et al. A longitudinal study of the relationships between haemostatic, lipid, and oestradiol changes during normal human pregnancy. *Thromb Haemost.* 1999;81:71–75.
151. Ghirardini G, Battioni M, Bertellini C, et al. D-dimer after delivery in uncomplicated pregnancies. *Clin Exp Obstet Gynecol.* 1999;26:211–212.
152. Lee AJ, Fowkes GR, Lowe GD, Rumley A. Determinants of fibrin D-dimer in the Edinburgh Artery Study. *Arterioscler Thromb Vasc Biol.* 1995;15:1094–1097.
153. Bachmann F. The enigma PAI-2. Gene expression, evolutionary and functional aspects. *Thromb Haemost.* 1995;74:172–179.
154. Hellgren M. Hemostasis during pregnancy and puerperium. *Haemostasis.* 1996;26(suppl 4):244–247.
155. Schjetlein R, Abdelnoor M, Haugen G, et al. Hemostatic variables as independent predictors for fetal growth retardation in preeclampsia. *Acta Obstet Gynecol Scand.* 1999;78:191–197.
156. Tanjung MT, Siddik HD, Hariman H, Koh SC. Coagulation and fibrinolysis in preeclampsia and neonates. *Clin Appl Thromb Hemost.* 2005;11:467–473.
157. Estelles A, Gilabert J, Keeton M, et al. Altered expression of plasminogen activator inhibitor type 1 in placentas from pregnant women with preeclampsia and/or intrauterine fetal growth retardation. *Blood.* 1994;84:143–150.
158. Chappell LC, Seed PT, Briley A, et al. A longitudinal study of biochemical variables in women at risk of preeclampsia. *Am J Obstet Gynecol.* 2002;187:127–136.
159. Wikstrom AK, Nash P, Eriksson UJ, Olovsson MH. Evidence of increased oxidative stress and a change in the plasminogen activator inhibitor (PAI)-1 to PAI-2 ratio in early-onset but not late-onset preeclampsia. *Am J Obstet Gynecol.* 2009;201:597.e591–597.e598.
160. Roes EM, Sweep CG, Thomas CM, et al. Levels of plasminogen activators and their inhibitors in maternal and umbilical cord plasma in severe preeclampsia. *Am J Obstet Gynecol.* 2002;187:1019–1025.
161. Higgins JR, Walshe JJ, Darling MR, Norris L, Bonnar J. Hemostasis in the uteroplacental and peripheral circulations in normotensive and pre-eclamptic pregnancies. *Am J Obstet Gynecol.* 1998;179:520–526.
162. Estelles A, Gilabert J, Grancha S, et al. Abnormal expression of type 1 plasminogen activator inhibitor and tissue factor in severe preeclampsia. *Thromb Haemost.* 1998;79:500–508.
163. Estelles A, Gilabert J, Espana F, Aznar J, Galbis M. Fibrinolytic parameters in normotensive pregnancy with

intrauterine fetal growth retardation and in severe preeclampsia. *Am J Obstet Gynecol.* 1991;165:138–142.

164. Egbor M, Ansari T, Morris N, Green CJ, Sibbons PD. Morphometric placental villous and vascular abnormalities in early- and late-onset pre-eclampsia with and without fetal growth restriction. *BJOG.* 2006;113:580–589.
165. Wikstrom AK, Larsson A, Akerud H, Olovsson M. Increased circulating levels of the antiangiogenic factor endostatin in early-onset but not late-onset preeclampsia. *Reprod Sci.* 2009;16:995–1000.
166. Pinheiro Mde B, Junqueira DR, Coelho FF, et al. D-dimer in preeclampsia: systematic review and meta-analysis. *Clin Chim Acta.* 2012;414:166–170.
167. Glueck CJ, Kupferminc MJ, Fontaine RN, et al. Genetic hypofibrinolysis in complicated pregnancies. *Obstet Gynecol.* 2001;97:44–48.
168. Zhao L, Bracken MB, Dewan AT, Chen S. Association between the SERPINE1 (PAI-1) 4G/5G insertion/deletion promoter polymorphism (rs1799889) and pre-eclampsia: a systematic review and meta-analysis. *Mol Hum Reprod.* 2013;19:136–143.
169. Morgan JA, Bombell S, McGuire W. Association of plasminogen activator inhibitor-type 1 (-675 4G/5G) polymorphism with pre-eclampsia: systematic review. *PLoS One.* 2013;8:e56907.
170. Vigil de Gracia P, Ortega-Paz L. Pre-eclampsia/eclampsia and hepatic rupture. *Int J Gynaecol Obstet.* 2012;118:186–189.
171. Rinehart BK, Terrone DA, Magann EF, Martin RW, May WL, Martin JN. Preeclampsia-associated hepatic hemorrhage and rupture: mode of management related to maternal and perinatal outcome. *Obstet Gynecol Surv.* 1999;54:196–202.
172. Hunter SK, Martin M, Benda JA, Zlatnik FJ. Liver transplant after massive spontaneous hepatic rupture in pregnancy complicated by preeclampsia. *Obstet Gynecol.* 1995;85:819–822.
173. Wicke C, Pereira PL, Neeser E, Flesch I, Rodegerdts EA, Becker HD. Subcapsular liver hematoma in HELLP syndrome: evaluation of diagnostic and therapeutic options – a unicenter study. *Am J Obstet Gynecol.* 2004;190:106–112.
174. Nelson DB, Yost NP, Cunningham FG. Acute fatty liver of pregnancy: clinical outcomes and expected duration of recovery. *Am J Obstet Gynecol.* 2013;209:456.e1–456.e7.
175. Manas KJ, Welsh JD, Rankin RA, Miller DD. Hepatic hemorrhage without rupture in preeclampsia. *New Engl J Med.* 1985;312:424–426.
176. Sibai BM. Imitators of severe preeclampsia. *Obstet Gynecol.* 2007;109:956–966.
177. Sibai BM, Ramadan MK, Usta I, Salama M, Mercer BM, Friedman SA. Maternal morbidity and mortality in 442 pregnancies with hemolysis, elevated liver enzymes, and low platelets (HELLP syndrome). *Am J Obstet Gynecol.* 1993;169:1000–1006.
178. Sibai BM, Barton JR. Expectant management of severe preeclampsia remote from term: patient selection, treatment, and delivery indications. *Am J Obstet Gynecol.* 2007;196.
179. Kozic JR, Benton SJ, Hutcheon JA, Payne BA, Magee LA, von Dadelszen P. Abnormal liver function tests as predictors of adverse maternal outcomes in women with preeclampsia. Preeclampsia Integrated Estimate of Risk Study Group. *J Obstet Gynaecol Can.* 2011;33:995–1004.
180. Martin Jr JN, Brewer JM, Wallace K, et al. Hellp syndrome and composite major maternal morbidity: importance of Mississippi classification system. *J Matern Fetal Neonatal Med.* 2013;26:1201–1206.
181. Haddad B, Barton JR, Livingston JC, Chahine R, Sibai BM. Risk factors for adverse maternal outcomes among women with HELLP (hemolysis, elevated liver enzymes, and low platelet count) syndrome. *Am J Obstet Gynecol.* 2000;183:444–448.
182. Fonseca JE, Méndez F, Cataño C, Arias F. Dexamethasone treatment does not improve the outcome of women with HELLP syndrome: double blind placebo-controlled, randomized clinical trial. *Am J Obstet Gynecol.* 2005;193:1591–1598.
183. Keiser SD, Owens MY, Parrish MR, et al. HELLP syndrome with and without eclampsia. *Am J Perinatol.* 2011;28:187–194.
184. Sep S, Verbeek J, Koek G, Smits L, Spaanderman M, Peeters L. Clinical differences between early-onset HELLP syndrome and early-onset preeclampsia during pregnancy and at least 6 months postpartum. *Am J Obstet Gynecol.* 2010;202:271.e1.
185. Reimer T, Rohrmann H, Stubert J, et al. Angiogenic factors and acute-phase proteins in serum samples of preeclampsia and HELLP patients: a matched-pair analysis. *J Matern Fetal Neonatal Med.* 2013;26:263–269.
186. Sibai BM, Stella CL. Diagnosis and management of atypical preeclampsia-eclampsia. *Am J Obstet Gynecol.* 2009;200:481.e1–481.e7.
187. Martin JN, Thigpen BD, Rose CH. Maternal benefit of high-dose intravenous corticosteroid therapy for HELLP syndrome. *Am J Obstet Gynecol.* 2003;189:830.
188. Isler CM, Barrilleaux PS, Magann EF, Bass JD, Martin JN. A prospective, randomized trial comparing the efficacy of dexamethasone and betamethasone for the treatment of antepartum HELLP (hemolysis, elevated liver enzymes, and low platelet count) syndrome. *Am J Obstet Gynecol.* 2001;184:1332–1337.
189. Katz L, DeAmorim MMR, Figueiroa JN, de Silva JLP. Postpartum dexamethasone for women with hemolysis, elevated liver enzymes, and low platelets (HELLP) syndrome: a double-blind, placebo-controlled, randomized clinical trial. *Am J Obstet Gynecol.* 2008;198.
190. Vidaeff AC, Yeomans ER. Corticosteroids for the syndrome of hemolysis, elevated liver enzymes, and low platelets (HELLP): what evidence? *Minerva Ginecol.* 2007;59:183–190.
191. Baker PN, Broughton Pipkin F, Symonds EM. Comparative study of platelet angiotensin II binding and the angiotensin II sensitivity test as predictors of pregnancy-induced hypertension. *Clin Sci (Lond).* 1992;83:89–95.
192. Maki N, Magness RR, Miyaura S, Gant NF, Johnston JM. Platelet-activating factor-acetylhydrolase activity in normotensive and hypertensive pregnancies. *Am J Obstet Gynecol.* 1993;168:50–54.
193. Wallmo L, Karlsson K, Teger-Nilsson AC. Fibrinopeptide A and intravascular coagulation in normotensive and hypertensive pregnancy and parturition. *Acta Obstet Gynecol Scand.* 1984;63:637–640.
194. Ganzevoort W, Rep A, Bonsel GJ, et al. Randomised controlled trial comparing two temporising management strategies, one with and one without plasma volume expansion, for severe and early onset preeclampsia. *BJOG.* 2005;112:1358–1368.

CHAPTER **18**

Chronic Hypertension and Pregnancy

PHYLLIS AUGUST, ARUN JEYABALAN AND JAMES M. ROBERTS

Editors' comments: *The fourth edition adds two new authors to the chapter on chronic hypertension and pregnancy. All three authors were members of the recent ACOG task force that elucidated the College's guidelines published in 2013; the first and second authors particularly act on determining recommendations for managing chronic hypertension in pregnancy, and the diagnosing of superimposed preeclampsia, the editor author (JMR) having chaired the whole report (see Chapter 1).*

INTRODUCTION

Chronic hypertension in pregnancy received but modest attention in the first edition of this text. It was discussed with renal diseases in the final chapter, where Chesley noted briefly that most patients had the "essential variety," 85% of whom did well, and that many of these women had an exaggerated decrease in blood pressure in early gestation. He cited personal experience to support his view that "accelerated (malignant)" hypertension was an unusual pregnancy complication. He also discussed the difficulty in diagnosing "superimposed preeclampsia" in these women and, by applying his own stringent criteria, suggested an incidence lower than most studies. There were also short discussions of renovascular hypertension, aldosteronism, and pheochromocytoma. Chesley's summary reflected the paucity of contemporary data, not only in relation to chronic hypertension in pregnancy, but knowledge of the disease in non-pregnant populations as well. Research in both areas has progressed substantially in the last 35 years.

In this chapter, the natural history, diagnosis, and treatment of chronic hypertension complicating pregnancy are reviewed, followed by an overview of specific diagnostic categories including essential, renal, renovascular, and adrenal hypertension. Evaluation and treatment of hypertension postpartum are also discussed.

BACKGROUND

The prevalence of hypertension in premenopausal women may be as high as 25% in whites and 30% in blacks.[1] Poorly controlled hypertension is a leading cause of cardiovascular disease and global mortality. Elevated blood pressure also has significant implications for pregnancy outcome. Based on current estimates of the prevalence of hypertensive disorders during gestation, chronic hypertension likely complicates about 0.5 to 5% of pregnancies.[2,3] The wide variation is due to different populations surveyed and the use of different criteria to make the diagnosis. Rates are higher in African-Americans, older women, and those with increased body mass indices (BMI). The global obesity epidemic and the trend towards postponing childbearing, particularly in urban, industrialized regions, further contribute to the increasing prevalence of hypertension during pregnancy. The prevalence may also be higher, because some multiparous women with unrecognized chronic hypertension may be misdiagnosed with preeclampsia.[4,5]

Definition

Chronic hypertension complicating pregnancy is defined as high blood pressure known to predate conception, and thus not caused by pregnancy.[4] The US National High Blood Pressure Education Program Guidelines, summarized in the seventh Joint National Committee on Prevention, Detection, Evaluation and Treatment of High Blood Pressure (JNC 7), define normal blood pressure for nonpregnant populations as 120/80 mm Hg, a level also considered "optimal" by virtually every other hypertension society.[1] There is uniform agreement that levels of blood pressure higher than 120/80 mm Hg are associated with increased cardiovascular risk, but the various societies differ in their labeling of elevated blood pressure levels. The American (JNC) guidelines label those with blood pressures 120–139 mm Hg systolic or 80–89 mm Hg diastolic as having prehypertension

Chesley's Hypertensive Disorders in Pregnancy.
ISBN: 978-0-12-407866-6

DOI: http://dx.doi.org/10.1016/B978-0-12-407866-6.00018-3

TABLE 18.1 Classification of Blood Pressure for Adults Age 18 and Older

Category	Systolic (mm Hg)		Diastolic (mm Hg)
Normal	<120	and	<80
Prehypertension	120–139		80–89
Stage 1 hypertension	140–159	or	90–99
Stage 2 hypertension	≥160	or	≥100

whereas the European Guidelines, the British Hypertension Society, the World Health Organization, and the International Society of Hypertension Society consider <130/85 mm Hg as normal and 130–139 or 85–89 as "high normal." Stage 1 hypertension (previously designated "mild") is defined as a blood pressure of 140 to 159/90 to 99 mm Hg (Table 18.1).[1] The JNC 7 recognizes two stages of hypertension, whereas the European and British societies as well as the WHO distinguish grades 1, 2, and 3, as well as isolated systolic hypertension. Most women with chronic hypertension will have essential (also called primary) hypertension, but as many as 10% may have underlying renal or endocrine disorders, i.e., secondary hypertension. The next JNC report (JNC 8), in press when this book was published, is expected to focus on developing evidence-based systematic reviews focusing on nonpregnant populations only and will probably not be relevant to treatment of hypertension during pregnancy.

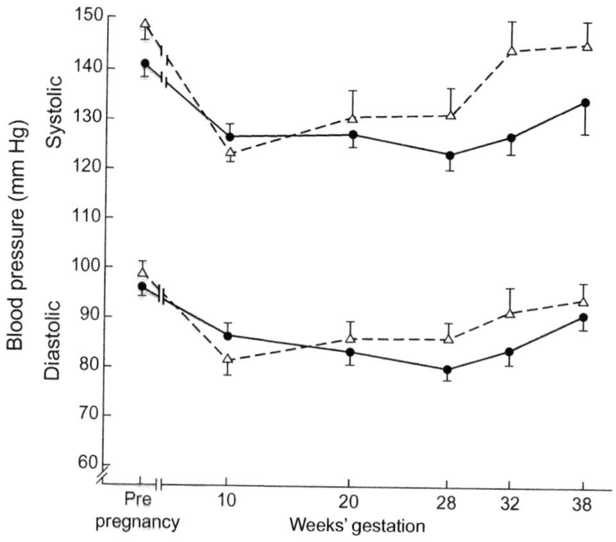

FIGURE 18.1 Sequential changes in systolic and diastolic blood pressure throughout pregnancy in women with uncomplicated chronic hypertension (n = 17, ●–●), and women with chronic hypertension with superimposed preeclampsia (n = 13, Δ–Δ) (mean ± SEM). In cases of uncomplicated chronic hypertension blood pressure decreases in the first and second trimesters, rising to early pregnancy or pregnancy levels in the third trimester. In contrast, blood pressure did not decrease after 10 weeks in women who developed superimposed preeclampsia, but remained constant or increased until the third trimester when preeclampsia was diagnosed. Reprinted with permission.[10]

Diagnosis

When hypertension has been clearly documented prior to conception, the diagnosis of chronic hypertension in pregnancy is straightforward. It is also the most likely diagnosis when hypertension is present prior to 20 weeks gestation, although isolated, rare cases of preeclampsia before this time have been reported, particularly in the presence of hydatidiform mole.[6] The clinician should also be aware that if blood pressure was clearly normal at the first antepartum visit (e.g. ~8–12 weeks), then a blood pressure of 140–150/90–100 at 15–20 weeks could represent a pregnancy-related exacerbation including early preeclampsia.

Difficulties may also arise when pregnant women with prepregnancy, undiagnosed stage 1 hypertension present initially in the second trimester with normal blood pressure after having experienced the pregnancy-associated "physiologic" decrease in blood pressure. These women will have been presumed to be normotensive, and if blood pressure rises in the third trimester they will be erroneously diagnosed with either gestational hypertension or, if proteinuria is present, preeclampsia.

Because 15% to 25% of women with chronic hypertension develop *superimposed* preeclampsia, it may be impossible to diagnose chronic hypertension in this setting until well after delivery. In other instances, women with well-documented hypertension prior to conception will demonstrate normal blood pressures throughout their entire pregnancy, only to return to prepregnancy hypertensive levels postpartum. Thus, an understanding of the normal pregnancy-induced physiologic changes, as well as an appreciation of the clinical and laboratory features of preeclampsia, is essential for correct diagnosis and management of women with chronic hypertension. These principles, reviewed here briefly, are more extensively discussed in Chapter 14.

Cardiac and Hemodynamic Alterations

Normal pregnancy is characterized by generalized vasodilatation, so marked that despite increases in cardiac output and blood volume of 40 to 50%, mean arterial pressure decreases by approximately 10 mm Hg (see Chapter 14). The decrement is apparent in the first trimester and reaches a nadir by midpregnancy. Blood pressure then increases gradually, approaching prepregnancy values at term, but may transiently increase to values above the woman's nonpregnant level during the puerperium.[7–9] Women with preexisting or chronic hypertension may also develop significantly lower BP during pregnancy with decrements as great as 15 to 20 mm Hg (Fig. 18.1).[10,11] Also, in light of our previous discussions that focus on the growing awareness that 120–139 mm Hg systolic or 80–89 mm Hg diastolic in nonpregnant populations is labeled "prehypertension," it may be reasonable to reconsider the designation of all pregnant women with values below 140/90 mm Hg as "normotensive." Increased clinical vigilance is prudent for all gravidas with blood pressure levels exceeding 120/80 mm Hg. Failure in this respect may result in a delay in diagnosing preeclampsia.

Blood pressure control in normal pregnancy, including the influence of pressor systems (autonomic nerves and catecholamines, renin–angiotensin–aldosterone, and vasopressin), baroreceptor function, endothelial cell function, and volume-mediated changes (detailed elsewhere in this text), has not been well evaluated in pregnant women. Animal studies are frequently contradictory, in part because of species differences. Thus, there continues to be a critical need to better explain normal vascular physiology before we completely understand the blood pressure alterations in chronically hypertensive pregnant women.

There are some data suggesting that the autonomic nervous system is more active in the control of blood pressure during pregnancy, although conflicting results have also been reported.[12] There is little dispute that the renin–angiotensin system is markedly stimulated, and normotensive pregnant women have an exaggerated hypotensive response to angiotensin-converting enzyme inhibition, compared with nonpregnant women.[13] These data suggest that this system is stimulated to help maintain normal blood pressure.

Some evidence exists, mainly from animal models, that basal and stimulated nitric oxide production increases in pregnancy, and may account for the marked vasodilatation and lower blood pressure.[14,15] There are conflicting data on endothelial cell function in human pregnancy as discussed in Chapter 9. The link between markedly increased placental hormones, particularly estrogen and progesterone, and alterations in vascular endothelial function is another area of active investigation. Finally, studies of the role of additional vasoactive peptides (vascular endothelial growth factor, placental growth factor, calcitonin gene-related peptide, relaxin) in normal blood pressure regulation or dysregulation in hypertension during pregnancy further suggest that the vasculature in pregnant women is modulated by multiple factors.[16,17]

Other pregnancy-associated cardiovascular alterations that impact on blood pressure are increased cardiac output and blood volume. The relationship between these changes in women with chronic hypertension has been studied using traditional echocardiography but results are conflicting and confounded by small numbers, varying gestational ages at the time of study, and concomitant use of antihypertensive medications.[18,19] A limited study of hemodynamic changes prospectively assessed by impedance cardiography in 20 women with chronic hypertension during pregnancy demonstrated increased peripheral vascular resistance and slightly lower stroke volume index compared to normotensive pregnant women, changes that would be expected in hypertension.[20] Larger, prospective, controlled studies are needed to better understand the physiologic alterations in this group of women and potentially to optimize management of individual patients.

The profound alterations in renal hemodynamics in normal pregnancy which include increases in glomerular filtration rate and renal blood flow of ~50% are also detailed elsewhere[21,22] (see Chapter 16). The impact of these changes on blood pressure control and on responsiveness to antihypertensive agents in women with chronic hypertension is not known. It does seem probable, though, that these hemodynamic alterations contribute to the greater ease of blood pressure control during pregnancy, that is, in the absence of superimposed preeclampsia. Finally, there are emerging data indicating that excessive gestational weight gain is associated with higher blood pressure, hypertensive disorders of pregnancy, and later-life cardiometabolic risk.[23,24] The extent to which excessive weight gain during gestation compromises optimum blood pressure control in pregnant women with chronic hypertension is poorly understood since women with chronic hypertension are usually underrepresented in studies of hypertension in pregnancy. Some dietary habits, such as increased consumption of dairy products high in sodium, may also affect blood pressure control in the chronically hypertensive pregnant woman. Credible investigations of dietary modification have not been performed.

Effect of Chronic Hypertension on the Mother

Chronic hypertension in pregnancy is associated with higher rates of maternal and perinatal morbidity and mortality. This has been confirmed in a number of recent large, population-based studies.[25,26] However, these and other reports have not consistently distinguished between women with preeclampsia superimposed on chronic hypertension and pregnant women with uncomplicated chronic hypertension alone.[27–30] While there is little doubt that women with superimposed preeclampsia have higher rates of adverse maternal and fetal/neonatal outcomes, the independent risks associated with uncomplicated chronic hypertension are less clear.

Vanek and colleagues reported outcomes of 1,807 deliveries after 36 weeks in women with chronic hypertension and observed that, controlling for superimposed preeclampsia, uncomplicated chronic hypertension was associated with a greater risk of cesarean delivery (OR 2.7, 95% CI 2.4–3.0) and postpartum hemorrhage (OR 2.2, 95% CI 1.4–3.7).[3] Tuuli and colleagues reported on pregnancy outcomes of 1,521 women with chronic hypertension.[31] Women with chronic hypertension without superimposed preeclampsia had a higher risk of cesarean delivery (OR 2.1, 95% CI 1.8–2.4) and prolonged maternal hospital stay (OR 1.6, 95% CI 1.3–1.9) compared to normotensive control subjects. In a prospective cohort of pregnant women with chronic hypertension, Chappell and colleagues also reported higher rates of cesarean delivery and longer maternal hospital stay compared to background rates in the general population.[32] The increased cesarean delivery rate may reflect the higher likelihood of iatrogenic intervention in response to elevated blood pressure rather than the disease process itself.[31]

Chronic hypertension is also associated with increased risk of gestational diabetes (OR 1.8, 95% CI 1.4–2.0). This may reflect common risk factors for both conditions such as obesity as well as similar pathogenic mechanisms, e.g., insulin resistance.

Placental abruption, which is associated with life-threatening maternal hemorrhage, is estimated to be three-fold higher in women with chronic hypertension, although most of this risk is associated with superimposed preeclampsia.[27,33,34] Other studies have not demonstrated an increased risk of abruption in women with chronic hypertension without superimposed preeclampsia.[28,31,35] Differences in sample size and study population may account for the varying results.

Other adverse maternal outcomes include accelerated hypertension during pregnancy with resultant target organ damage, e.g., kidneys, heart, and brain, although in the absence of superimposed preeclampsia this is extremely uncommon. One exception may be women with severe hypertension prior to conception, many of whom have underlying renal disease or secondary hypertension. Some women with secondary forms of hypertension, such as chronic renal disease and collagen disorders, may suffer from irreversible deterioration in renal function during and after pregnancy. In the case of systemic lupus erythematosus, there may be multi-organ morbidity, regardless of the development of superimposed preeclampsia.

Finally, although the expectation is that pregnancies in women with uncomplicated chronic hypertension will be successful, these women are at higher risk of hospitalization due to worsening blood pressures.[35]

Effects of Chronic Hypertension on Fetal/Neonatal Outcome

Perinatal mortality is also higher in pregnancies complicated by chronic hypertension, with most of this increased risk attributable to superimposed preeclampsia and fetal growth restriction.[27,29,35] As previously noted, many studies do not precisely distinguish between pregnancies with uncomplicated chronic hypertension alone and those with superimposed preeclampsia. Furthermore, many of these studies lack a normotensive comparison group. Tuuli and colleagues compared women with chronic hypertension only (without superimposed preeclampsia) to normotensive control subjects and reported increased risk of small-for-gestational-age infants (OR 1.8, 95% CI 1.5–2.2), preterm delivery less than 34 weeks (OR 1.93, 95% CI 1.5–2.8), stillbirth (1.9, 95% CI 1.2–2.9), and admission to neonatal intensive care unit (1.4, 95% CI 1.1–1.8).[31] Rey and colleagues also reported a higher relative risk of perinatal death (RR 2.3, 95% CI 1.2–4.6) and SGA (1.7, 95% CI 1.2–2.6) in women with chronic hypertension only compared to the normotensive control group.[35] These findings have also been confirmed by others.[3,36]

An additional risk to fetal well-being is the *in utero* exposure to antihypertensive medication. Overall, this risk has been poorly quantified, particularly in regard to long-term follow-up into childhood and beyond. Clinical trials have confirmed the safety of methyldopa with follow-up to 7 years of age.[37] (See Chapter 19 for detailed discussion.)

Chronic Hypertension with Superimposed Preeclampsia

As discussed, pregestational hypertension is a recognized risk factor for preeclampsia. The incidence of superimposed preeclampsia ranges from 13 to 40% among women with chronic hypertension, depending on the diagnostic criteria, etiology (essential versus secondary), duration, and the severity of hypertension.[27,38,39] Superimposed preeclampsia is more common in women with secondary causes of hypertension than with essential hypertension (discussed later in this chapter).

A major reason for this wide range in incidence is that the definition of superimposed preeclampsia is used liberally in some studies. The 2013 American College of Obstetricians and Gynecologists (ACOG) Hypertension Task Force described diagnostic criteria[39] similar to those proposed by the National Institutes of Health Working Group Report in 2000[4] (see below), the latter forming the basis of most studies reviewed here. The criteria included a significant increase in blood pressure, new onset of proteinuria ($\geq$300 mg over 24 hours) or features of HELLP syndrome (hemolysis, elevated liver enzymes, low platelet count). Using these criteria, several well-conducted large surveys report a risk between 15 and 25%.[27,29,32,35,40,41] This risk increases with the severity and duration of hypertension as well as the presence of underlying renal disease. Of note, the diagnosis of preeclampsia is often challenging due to the usual gestational increase in blood pressure during the third trimester. The diagnosis is especially problematic in women with preexisting proteinuria. We and others have observed that women with mild (stage 1) hypertension have a risk of superimposed preeclampsia of about 15%, whereas those with stage 2 hypertension have a risk of 25–30%. Analysis of a large database in New York State suggests that these rates may be declining.[2]

An intriguing question is why women with preexisting hypertension are at greater risk of the development of superimposed preeclampsia. There are few studies addressing this, but it has been suggested that women at risk of preeclampsia have genetic, biochemical, and metabolic abnormalities similar to women with essential hypertension.[42] This list includes a higher incidence of polymorphisms in the angiotensinogen gene, obesity, hypertriglyceridemia, and insulin resistance. Such observations raise the possibility that the genesis of "superimposed" preeclampsia in hypertensive pregnant women may be related to the underlying genetic and metabolic disturbances that led to hypertension, rather than the elevated blood pressure itself. This is clearly an exciting area worthy of further investigation.

Recent paradigms of the pathogenesis of preeclampsia emphasize that there are necessary fetal as well as maternal susceptibility factors. Elevated blood pressure, considered to be a "maternal susceptibility factor," clearly increases risk; however, most would agree that coincident fetal/placental pathologic abnormalities are necessary for the full expression of the disease. (Further aspects of this intriguing subject are discussed in several other chapters, including Chapters 4,5, 6, and 10.)

In 2013, the American College of Obstetricians and Gynecologists' Task Force on Hypertension in Pregnancy released updated and evidence-based definitions and guidelines for the hypertensive disorders in pregnancy including superimposed preeclampsia.[39] The Task Force, consisting of obstetricians, maternal-fetal medicine specialists, nephrologists, an anesthesiologist, a physiologist, and a patient representative, utilized the GRADE approach to evaluate the quality of the evidence and convey the strength of its recommendations.[43] Given the higher risk of adverse pregnancy outcomes with superimposed preeclampsia, overdiagnosis was deemed to be preferable, with the goal of increasing vigilance and preventing catastrophic maternal and fetal outcomes. However, it was recognized that a more specific and stratified approach along with predictors of adverse outcomes would be useful in guiding clinical management and avoiding unnecessary preterm births. Therefore, superimposed preeclampsia was stratified into two groups (superimposed preeclampsia and superimposed preeclampsia with severe features) to guide clinical management (Table 18.2).

Importantly, it should be recognized that there is often ambiguity in the diagnosis and that clinical vigilance should be high when superimposed preeclampsia is suspected. Furthermore, women with preeclampsia can progress and develop end-organ involvement and adverse outcomes. Superimposed preeclampsia without severe features mandates increased maternal and fetal surveillance given the progressive nature of preeclampsia. The presence of severe features directs further management and timing of delivery. Delivery decisions must carefully balance maternal and fetal well-being.

TABLE 18.2 2013 ACOG Hypertension in Pregnancy Working Group Definition of Superimposed Preeclampsia[39]

Superimposed preeclampsia is likely when any of the following are present:	• A sudden increase in blood pressure that was previously well controlled or escalation of antihypertensive medications to control blood pressure • New onset of proteinuria or sudden increase in proteinuria in a woman with known proteinuria before or early in pregnancy
The diagnosis of **superimposed preeclampsia with severe features** is confirmed when any of the following are present:	• Severe-range blood pressures despite escalation of antihypertensive therapy • Thrombocytopenia (platelet count less than 100,000/μL) • Elevated liver transaminases (two times the upper limit of normal concentration for a particular laboratory) • New onset and worsening of renal insufficiency • Pulmonary edema • Persistent cerebral or visual disturbances

Management considerations are detailed in the Task Force recommendations.

Effects of Superimposed Preeclampsia on the Mother and Fetus/Neonate

As noted previously, maternal and fetal/neonatal morbidity and mortality are clearly higher with superimposed preeclampsia compared to chronic hypertension alone. C-section rates, length of hospital stay, placental abruption with hemorrhage, and HELLP syndrome are all increased.[27,28,31,32,35,40] In addition, seizures, strokes, pulmonary edema, and renal dysfunction are all associated with superimposed preeclampsia.

Rey and colleagues reported a relative risk of perinatal death of 3.6 in women with superimposed preeclampsia compared to women with uncomplicated chronic hypertension.[35] Fetal growth restriction is also higher with superimposed preeclampsia. Chappell and colleagues reported fetal growth restriction less than the 5th percentile in 42% of women with superimposed preeclampsia compared to 14% in women with chronic hypertension alone.[32] Preterm delivery, with the associated neonatal complications of prematurity, is also higher with superimposed preeclampsia. Preterm delivery is usually indicated by worsening maternal status or fetal growth restriction. While the details vary, the pattern of these findings is fairly consistent across a number of studies and varying populations and is the basis for recommending increased maternal and fetal surveillance.[39]

SPECIFIC HYPERTENSIVE DISORDERS

Essential Hypertension

One in four adults has hypertension worldwide, which equates to approximately 1 billion individuals. Based on data from the US National Health and Nutrition Examination Survey (NHANES), which included 26,349 adults aged 30 years and older, and from the Behavioral Risk Factor Surveillance System from 1997 to 2009, including 1,283,722 adults aged 30 years and older, the prevalence of hypertension was estimated at 37.6% in men and 40.1% in women.[44] Hypertension is more prevalent in non-Hispanic blacks and older persons. The prevalence of hypertension in women is estimated to be 3.8% between ages 20 and 34, 14.2% between 35 and 44, and 31.2% between 44 and 54 years of age based on recent US Health statistics.[45] A recent study reported the prevalence of chronic hypertension in US reproductive-age women to be approximately 8%.[46] Overall, chronic hypertension is estimated to affect 0.5 to 5% of all pregnancies.[2,3] The wide variation is likely to be due to the population studied and the criteria used to establish the diagnosis. Using a database that included 56 million hospitalizations for delivery in the United States, approximately 1.8% of all pregnancies were complicated by chronic hypertension in 2007–2008, with a progressive increase since 1995.[25] The increasing rates of obesity as well as the trend towards postponing child-bearing, particularly in urban, industrialized regions, contribute to the increasing rates of hypertension during pregnancy. The rates may even be higher, because some multiparous women with unrecognized chronic hypertension may be misdiagnosed with preeclampsia.[4,5]

Before age 50 most people with hypertension have elevated diastolic pressure. After age 50, systolic pressure continues to rise and diastolic pressure tends to fall, and isolated systolic hypertension is prevalent. Poorly controlled hypertension is the leading cause of death globally.[47] Reduction of blood pressure reduces cardiovascular disease and death by amounts predicted from data associating elevations in blood pressure with adverse events.

Essential hypertension is a heterogeneous, polygenic disorder resulting from dysregulation of hormones, proteins, and neurogenic factors involved in blood pressure regulation interacting with diet, level of activity, and other environmental influences. About 30–60% of blood pressure variability is inherited; however, genetic studies of essential hypertension have not yet uncovered common genes with large effects on blood pressure. Rather, polymorphisms in alleles at many different loci, interacting with behavioral and environmental factors, contribute to the final disease trait.[48]

Using the candidate gene approach, studies where authors report polymorphisms in genes regulating the renin–angiotensin system, the adrenergic system, G protein signaling, nitric oxide synthase, and kallikreins have been inconclusive. Rather than focusing on simple single nucleotide polymorphism (SNP) based association studies of candidate genes, ongoing research is utilizing more complex haplotype-based association studies, genome-wide linkage analysis and genome-wide SNP analyses of multiple SNPs. Rare monogenic forms of hypertension have been described. The genes identified so far confirm the importance of renal sodium excretion in the pathogenesis of hypertension, as almost all cause alterations in renal sodium handling. These studies have also identified potential mechanisms that can be studied in essential hypertension, in particular, alterations in WNK kinases, aldosterone synthase, and the epithelial sodium channel (ENaC).[48]

Renal Basis for Essential Hypertension

Despite the complexity of blood pressure regulation, the central role of impaired renal sodium excretion in the genesis of hypertension is undisputed and continues to be supported by experimental and clinical studies. Subclinical renal abnormalities observed in some individuals with essential hypertension include focal renal ischemia leading to chronic non-suppressible renin secretion, renal sodium retention, reduced renal mass, decreased glomerular filtration rate, and a compromised sodium excretory capacity.[49]

Therapeutically, the role for renal sodium retention in the pathogenesis of hypertension is supported by the efficacy in many hypertensive individuals of diuretic therapy. Similarly, a role for non-suppressible renin secretion is supported by the efficacy of agents that interrupt the renin–angiotensin system.[50] The relationship between subtle abnormalities in renal function and pregnancy outcome has not been investigated in detail. Given the significant impact of reduced renal function on the risk of preeclampsia, this is an area of investigation of potential importance.

Hormonal Basis for Essential Hypertension

Alterations in the renin–angiotensin system are important aspects of the pathophysiology of primary aldosteronism and renovascular hypertension. Conversely, in essential hypertension there is considerable heterogeneity in circulating renin and aldosterone levels.[50] Some patients have normal or high plasma renin activity, whereas about one-third of hypertensive individuals have low or suppressed plasma renin activity. The latter values are those expected if volume overload is present, and, indeed, low-renin hypertension is often associated with increased sensitivity to salt restriction or diuretic therapy. Because pregnancy is characterized by activation of the renin–angiotensin system, adverse perinatal outcomes in women with essential hypertension who have alterations in the renin system is a real possibility.

Sympathetic Nervous System and Essential Hypertension

The role of the sympathetic nervous system in the pathogenesis of primary hypertension is supported by a large number of indirect experimental and clinical observations.

These include increased heart rate and plasma catecholamine levels in response to a variety of stimuli in hypertensive individuals and in animals.[51] While sympathetic nervous system activity and function are difficult to measure accurately in humans, there are a few studies that link preeclampsia to its activation.[12]

Vascular Structure and Function and Hypertension

It is beyond the scope of this chapter to review the expanding field of vascular biology and the role of abnormalities of vascular structure and function that are important in the genesis and the maintenance of elevated blood pressure. However, several key areas of investigation are worth mentioning, particularly those with relevance to the pathophysiology of preeclampsia. Mechanisms for vascular hypertrophy and blood vessel remodeling, as well as alterations in ion transport and signal transduction in vascular smooth muscle and endothelial cells are the focus of active investigation. Current evidence supports a role for dysregulation of these processes and suggests that there is significant overlap between processes involved in generalized atherosclerosis and the pathogenesis of preeclampsia.[52] (See Chapter 9.) The role of these processes in the increased tendency of women with preexisting hypertension to develop preeclampsia deserves study.

Metabolic Disturbances and Hypertension

The common occurrence of obesity, type 2 diabetes, and hypertension, all features of the metabolic syndrome, as well as the observation that a significant number of nonobese hypertensives will have insulin resistance and hyperinsulinemia have led to the concept of insulin resistance in the genesis of primary hypertension.[53] This has relevance to hypertension in pregnancy because insulin resistance appears more prevalent in preeclamptic women, and women with overt and gestational diabetes are reported to be at greater risk of preeclampsia.[27,54–56] (See Chapter 7 for additional discussion).

Dietary Factors

The role of sodium in the pathogenesis of some, if not most, cases of primary hypertension is established.[49] Other dietary components possibly involved are calcium and potassium.[57–59] Inverse relationships between dietary calcium intake and blood pressure have been documented, while calcium supplementation lowers blood pressure in experimental models and in clinical trials.[58] Calcium requirements are increased in normal pregnancy, and there are data suggesting an association between inadequate calcium intake and hypertension during pregnancy.[60] Supplemental calcium has been shown to prevent preeclampsia and adverse pregnancy outcomes in women ingesting a low-calcium diet.[61] However, a small study of calcium supplementation in pregnant women with essential hypertension but eating a calcium-replete diet failed to demonstrate a reduction in superimposed preeclampsia in the calcium-supplemented group.[62]

Environment and Behavior

Obesity and excess alcohol intake contribute to hypertension. Conversely, weight loss, increased physical activity, and decreased alcohol intake have been demonstrated to be effective strategies for lowering blood pressure.[1] As noted in Chapter 7, obesity is also an independent risk factor for preeclampsia.[63] Excessive gestational weight gain is also associated with adverse clinical outcomes including hypertensive disorders.[23,64,65] In 2009 the Institute of Medicine released updated guidelines for weight gain in pregnancy. These guidelines reduced the recommended weight gain in obese women during pregnancy to improve pregnancy and maternal outcomes.[66]

Physiology and Pathophysiology of Essential Hypertension During Pregnancy

The cardiovascular, renal, and hemodynamic alterations in pregnancy pertinent to blood pressure regulation were summarized above. Surprisingly, few detailed investigations of the physiology of essential hypertension in pregnancy have been performed. Women with essential hypertension are an intriguing group to investigate because they have a high incidence of superimposed preeclampsia. Thus, longitudinal studies of chronic hypertensive women may be helpful in elucidating early pregnancy phenomena important in the pathophysiology of preeclampsia.

Blood Pressure Patterns and Hemodynamic Measurements

Women with essential hypertension normally demonstrate the expected decrease in blood pressure in early and mid-pregnancy, although compared to normotensive women they have increased peripheral vascular resistance. In women who develop superimposed preeclampsia, blood pressure may start to increase in the late second trimester, whereas women who have uncomplicated pregnancies will frequently demonstrate even lower blood pressures at this time of gestation.[10] It has been suggested that absence of a second-trimester decrease in blood pressure is associated with more gestational complications and predictive of development of preeclampsia. We reported that systolic blood pressure greater than 140 mm Hg at 20 weeks gestation and increased uric acid and suppressed plasma renin activity predict superimposed preeclampsia with a high sensitivity and specificity.[62] Ambulatory blood pressure monitoring techniques have not been extensively studied in pregnant women with essential hypertension, although preliminary studies suggest that they will have limited utility in predicting superimposed preeclampsia.[67]

Cardiac function has not been investigated extensively during pregnancy in women with essential hypertension. Increased left ventricular mass has been reported in third-trimester chronic hypertensives.[68] In a small, longitudinal study of 20 pregnant women with chronic hypertension the authors reported increased peripheral vascular resistance and lower stroke volume index compared to normotensive women. Atrial natriuretic peptide levels were lower in the women with hypertension, suggesting decreased plasma volume expansion.[20] Larger longitudinal studies are needed. Similarly, there is a paucity of information concerning renal function and hemodynamics during pregnancies complicated by chronic hypertension (see Chapter 16). In one study there was a normal increase in creatinine clearance and urinary calcium excretion in most women with uncomplicated essential hypertension during pregnancy, but, as expected, when superimposed preeclampsia developed, renal function decreased modestly while marked hypocalciuria supervened.[69] We have observed increased mid-trimester hyperuricemia in women with chronic hypertension prior to development of superimposed preeclampsia, which raises the question of whether subtle alterations in renal hemodynamics precede the development of preeclampsia.[62] A preliminary study of Doppler analysis of the renal artery in hypertensive pregnant women noted an increase in renal blood flow with chronic hypertension, consistent with abnormal autoregulation.[70] These intriguing observations are not surprising given the alterations that have been reported in nonpregnant essential hypertensives.

Hormonal and Biochemical Alterations

Most investigations of pregnant women with chronic essential hypertension demonstrate that until superimposed preeclampsia develops, levels of hormones and other circulating substances associated with blood pressure regulation are similar to values in normotensive pregnancy. The stimulation of the renin–angiotensin systems in these two groups are also similar,[10] as is platelet angiotensin II binding.[71] Of further interest, pregnant women with essential hypertension destined to develop superimposed preeclampsia also manifest reductions in plasma renin activity in midpregnancy, again in a fashion similar to that described in nulliparas who subsequently become preeclamptic.[10,62] The decrease in plasma renin activity is accompanied by a decrease in urinary aldosterone excretion, and is an expected response to increased vasoconstriction. Furthermore, some women who developed superimposed preeclampsia and had low plasma renin activity also had a marked decrease in plasma estradiol.[10] Others have reported that plasma estradiol levels are decreased in women with uncomplicated chronic hypertension in pregnancy.[72] Finally, we have noted similar plasma progesterone levels in women with chronic hypertension, with and without preeclampsia, and in normotensive pregnant women.[10]

Several studies of nitric oxide production in normal and preeclamptic pregnant women have been conducted with conflicting results. As of the middle of 2013, we were aware of only one that focuses on essential hypertension in pregnancy. In this study, the authors reported significantly decreased levels of nitric acid metabolites compared to normotensive gestation.[73] This finding, consistent with observations in nonpregnant essential hypertensives, is surprising in view of the frequently observed pregnancy-induced vasodilatation and decrease in blood pressure in pregnant women.

Platelet and lymphocyte intracellular calcium concentrations are increased in nonpregnant essential hypertensives, as well as in preeclamptics.[74–76] The rationale for such studies is the belief that platelets are surrogates for vascular smooth muscle, an increase in their intracellular calcium implying similar increments in the vascular cells suggesting increased tone or vasoconstriction. Kilby et al.[75] reported increased platelet cytosolic calcium levels in five pregnant chronic hypertensives, while Hojo et al.[76] noted no differences in the lymphocyte intracellular free calcium levels of pregnant women with essential hypertension compared to normotensives, although levels were higher in preeclampsia.

Pathophysiology of Superimposed Preeclampsia

Women with essential hypertension are at greater risk of developing superimposed preeclampsia, but the reason for this is not known. Dysregulation of angiogenic factors is a feature of preeclampsia developing in previously normotensive women (see Chapter 6). Elevations of soluble fms-like tyrosine kinase-1 (sFlt-1) and endoglin, have been reported prior to and at the time of clinical disease.[77,78] Soluble Flt-1 is a circulating antagonist of vascular endothelial growth factor (VEGF) released from the placenta, and is hypothesized to contribute to maternal vascular endothelial dysfunction and glomerular endotheliosis.[79] Placental growth factor, an angiogenic factor similar to VEGF, is decreased in women with preeclampsia. We and others have recently reported that maternal serum levels of angiogenic factors are altered in women with chronic hypertension and superimposed preeclampsia, similar to women with preeclampsia without preexisting hypertension.[80–82] In particular, we found that the ratio of sFlt-1/PlGF was highest in midpregnancy in women with chronic hypertension who developed preterm preeclampsia, compared with those who developed preeclampsia close to term, or not at all. In a secondary analysis of the Maternal-Fetal Medicine Units Network trial of aspirin to prevent preeclampsia in high-risk pregnancies, we reported on 313 women with chronic hypertension.[83] Soluble Flt-1 and endoglin were significantly elevated between 26 and 30weeksgestation in women who later developed superimposed preeclampsia compared to women with chronic hypertension who did not develop superimposed preeclampsia. However, there were no significant differences at other time points in gestation, including at the time of diagnosis. The variability in these studies may be related to sample size, differing criteria for the diagnosis of superimposed preeclampsia, and varying populations. Overall, the aggregate of observations suggest possible similarities in pathogenesis between preeclampsia in previously normotensive women and those with superimposed preeclampsia. Further study is needed to determine whether angiogenic factors play a causal role in the pathogenesis of superimposed preeclampsia and the contribution of other etiologies in this subgroup of women.

The concept of "shared risk factors" for both essential hypertension and preeclampsia, e.g., insulin resistance, has already been mentioned. It is also worth considering whether some forms of essential hypertension, for example those characterized as high or low renin, may be particularly predisposed to preeclampsia. This appears to be the case with secondary hypertension: for instance, women with renovascular hypertension, in whom the renin–angiotensin system is activated, appear to have an unusually high incidence of preeclampsia.[84] Women with hyperaldosteronism, a low-renin form of hypertension, may have a lower incidence of preeclampsia compared to those with renovascular hypertension. Currently, there is little information regarding the association of baseline renin levels and development of superimposed preeclampsia in women with essential hypertension although we have reported that suppressed plasma renin activity and elevated serum uric acid levels at mid pregnancy (20 weeks) are associated with an increased risk of later development of superimposed preeclampsia.[62]

Secondary Hypertension

Secondary forms of hypertension are uncommon compared with essential hypertension, and comprise only 2 to 5% of hypertensives diagnosed and treated at specialized centers. In routine primary care settings, however, their numbers are even lower. The most common etiologies are renal disease, aldosteronism, renovascular hypertension, Cushing syndrome, and pheochromocytoma. The prevalence of secondary hypertension in women of childbearing age has not been precisely determined, but is estimated to be 10% of women with prepregnancy hypertension. Of importance, prognosis is best when a diagnosis of secondary hypertension is made prior to conception, because most forms of secondary hypertension are associated with increased maternal and fetal morbidity and mortality.

Renal Disease

A detailed discussion of renal disease and pregnancy has not been undertaken in this text, except those discussions relating to superimposed preeclampsia. We suggest ref. 22 for our review of this topic.[22] There are, however, points to emphasize regarding management of hypertension associated with intrinsic renal disease. Kidney disease is the most common cause of secondary hypertension in nonpregnant populations, in which high blood pressure is often found in patients with anatomic or congenital abnormalities, glomerulonephritis, diabetes, systemic lupus erythematosus, or interstitial nephritis.

All young women with newly diagnosed hypertension should be screened for intrinsic renal disease including measurements of renal function, and urinalysis for detection of proteinuria or red blood cells. Those with a strong family history of renal disease should be screened with renal ultrasound for polycystic kidney disease. This autosomal dominant disorder often presents with hypertension in the third and fourth decades. When renal disease is detected, regardless of its cause, these women should be counseled about the increased maternal and fetal risks associated with impaired renal function (preconception serum creatinine level $\geq$1.4 mg/dL) or poorly controlled hypertension. Differentiating between worsening renal disease and superimposed preeclampsia is often challenging in this group of women.

Women with renal disease are best managed by a multidisciplinary team including high-risk obstetricians and nephrologists. This is extremely important in the case of renal transplant recipients, in whom concerns regarding immunosuppression and risk of infection and rejection require coordinated specialty care. Therapy of hypertension is similar to that in gravidas with essential hypertension, although maintaining normal levels of blood pressure in women with renal disease may be more important in order to protect the kidneys from further deterioration.

Renovascular Hypertension

This entity refers to hypertension caused by stenotic lesions of the renal arteries. The narrowing of the arterial lumen leads to diminished blood flow to one or both kidneys, with resulting renal ischemia, stimulation of the renin–angiotensin system, and ensuing hypertension. It is important to emphasize that not all stenotic lesions of the renal artery result in hypertension, because they may not be severe enough to compromise renal blood flow. Thus, demonstration (by angiography) of a lesion does not necessarily establish the diagnosis of renovascular hypertension. Cure of hypertension after renal artery revascularization with either bypass surgery or angioplasty remains the most reliable proof of renovascular hypertension.[85]

The causes of renovascular hypertension are many. *Atherosclerotic renovascular disease* is primarily observed in postmenopausal women, especially those with a history of tobacco use and diffuse vascular disease, whereas *fibromuscular dysplasia* is more likely to be present in young women, making it the form most likely to be encountered in pregnancy.[86] Fibromuscular dysplasia is three times more common in women than men (but infrequent in black or Asian populations), and is a nonatherosclerotic, noninflammatory vascular occlusive disease, most often presenting as *medial fibroplasia*.[86] Fibromuscular dysplasia has a prevalence of approximately 1% in hypertensive populations. Although the renal arteries are most commonly involved, other vessels including carotid, coronary, abdominal aorta, and peripheral arteries may also be affected.

Clinical features suggestive of renovascular hypertension are severe hypertension, which may be resistant to medical therapy, and which first appears in the second, third, or fourth decades. Abdominal bruit and a high peripheral venous plasma renin activity are among the findings traditionally associated with the disorder. Several noninvasive screening tests are also useful. These include captopril renography, Doppler ultrasonography of the renal arteries, and magnetic resonance angiography with gadolinium contrast enhancement.[87] The gold-standard test remains conventional angiography.[87] These tests are justified for use in pregnant women because the hypertension is potentially curable with either angioplasty or surgery, and without adequate treatment these women are at increased risk of doing poorly in pregnancy. In addition the fetal radiation dose during angiography is small and in the range permissible for pregnancy. We prefer angiography to noninvasive testing, as the latter may not detect the smaller branch lesions which are more frequent when fibromuscular dysplasia is present. Angiography may reveal either a single area of stenosis, or multiple stenotic lesions with intervening aneurysmal outpouchings, the "string of beads" image.

Renal angioplasty is a highly effective method in treating non-pregnant patients with fibromuscular dysplasia. There is a high rate of technical success with cure, or significant improvement in blood pressure exceeding 80%.[88]

Knowledge of renovascular hypertension in pregnancy is based on a handful of case reports and a few limited series of patients totaling approximately 30 cases. Many of the patients presented with early and severe preeclampsia and poor pregnancy outcomes.[89–91] One retrospective comparison of pregnancy outcomes in four patients with known renovascular hypertension matched to 20 women with essential hypertension demonstrated that those with renovascular hypertension were younger (age 25 versus 36) and had higher blood pressure levels during pregnancy.[84] Of interest, the four women with renovascular hypertension all developed superimposed preeclampsia, in contrast to 30% of those with essential hypertension. Plasma renin activity in one woman with renovascular hypertension was, as expected, quite elevated when measured early in pregnancy, but decreased when preeclampsia developed. There have been a number of case reports of successful angioplasty in the setting of renal artery stenosis during pregnancy, with good pregnancy outcome.[90–95] We and others have also documented the experiences of women with untreated renovascular hypertension whose pregnancies were complicated by severe preeclampsia.[84,96] Both had successful angioplasty postpartum and their subsequent pregnancies were normal. In view of the dramatic clinical improvement which usually follows revascularization, as well as the anecdotal experience described above, it seems justified and preferable to rule out renovascular hypertension before conception in young women with suggestive clinical features. Furthermore, based on the observations that hypertension mediated by renal ischemia and stimulation of the renin–angiotensin system increases the risk of preeclampsia, the role of these factors in the pathogenesis of preeclampsia warrants further investigation.

In summary, pregnant women with renovascular hypertension who have not undergone renal artery revascularization are at considerable risk, especially for superimposed preeclampsia and fetal complications. Temporizing therapy with angiotensin-converting enzyme inhibitors or angiotensin II receptor blocking agents is not an option in pregnancy because these drugs are associated with major malformations and significant fetal toxicity (see Chapter 19), but treatment of hypertension with other drugs that suppress renin secretion is possible. Those most likely to be effective are methyldopa and beta-adrenergic receptor blockers.

Primary Aldosteronism

Primary aldosteronism consists of a heterogeneous group of disorders characterized by mineralocorticoid hypertension coexisting with anatomical abnormalities of the adrenal glands. Surgically curable forms of primary aldosteronism include unilateral aldosterone-producing adenoma (APA, or Conn syndrome), unilateral adrenal hyperplasia, and, rarely, aldosterone-producing carcinoma. Surgical cure (with adrenalectomy) is associated with younger age and shorter duration of hypertension, observations which emphasize the importance of early identification. Those with bilateral adrenal abnormalities – either adenomas or hyperplasia – are not surgically curable. A rare form of non-primary aldosteronism, glucocorticoid remediable aldosteronism (GRA), is an autosomal dominant familial condition that is characterized by excess aldosterone production which is suppressible by exogenous steroids. This disorder is not surgically curable.

Primary aldosteronism may be the most common curable endocrine form of hypertension, present in as many as 5 to 15% of patients with elevated blood pressure.[97] The classic clinical features are hypertension, hypokalemia, suppressed plasma renin activity, excessive urinary potassium excretion, hypernatremia, and metabolic alkalosis. Manifestation of each of these features is variable, and even hypokalemia, once thought of as the sine qua non for diagnosing aldosterone excess, may be initially absent in as many as 25–50% of patients.[97,98] Diagnosis is made by demonstrating biochemical and hormonal abnormalities, followed by computerized tomography (CT) or magnetic resonance imaging (MRI) of the adrenal glands (helpful in differentiating an adenoma from hyperplasia). If an adenoma is detected, radiologic imaging may be followed by adrenal vein sampling to document unilateral aldosterone secretion prior to surgery. Surgery is indicated for unilateral disease, with a cure rate of about 65%, and an improvement in an additional 33%. Medical therapy with aldosterone blockade (in nonpregnant individuals) is usually effective for patients with hyperplasia, although high doses may be necessary. Calcium-channel blockers have also been reported to be effective.[99]

There are a number of case reports describing primary aldosteronism in pregnancy.[100–111] Some of these have been complicated by considerable morbidity, including severe hypertension, hypokalemia, preeclampsia, and poor fetal outcome, but there are also instances where hypertension and hypokalemia have been ameliorated during gestation.[71] It is hypothesized that such improvement, when it occurs, is a consequence of the high levels of progesterone which antagonize the actions of aldosterone.[105,110]

Primary aldosteronism may be difficult to diagnose in pregnant women because of the marked alterations in the renin–angiotensin–aldosterone system that occur in normal pregnancy. Both renin and aldosterone production are markedly increased, with a five-fold increase in urine aldosterone excretion compared to that observed in nonpregnant patients with primary aldosteronism.[112] Moreover, mild hypokalemia is not unusual in the course of a normal pregnancy. Greater degrees of hypokalemia (≤3 mEq/L), however, are unusual and should be investigated. In one case involving a pregnant women with an aldosteronoma, reported by us[112] (Fig. 18.2), plasma renin activity was not totally suppressed, but became undetectable postpartum. Her plasma renin levels, however, were approximately one-third to one-half of those routinely observed in other pregnant women with essential hypertension.[10] Her urinary aldosterone excretion was well above the increases recorded in most women with uncomplicated gestations.

Treatment of aldosteronism diagnosed during pregnancy is controversial. If blood pressure improves spontaneously or is easily controlled with antihypertensive drugs, then it is reasonable to postpone surgical intervention until postpartum. Spironolactone has antiandrogenic effects and is reported to cause feminization in rodent male fetuses exposed *in utero*; therefore, it should be avoided in pregnancy.[113] We found one case report of eplerenone, a more selective mineralocorticoid antagonist, used in pregnancy starting at 27 weeks with no adverse effects.[114] Given the limited safety information, we do not recommend the routine use of this medication during pregnancy. Amiloride, a specfic blocker of the epithelial sodium channel (ENac), is also effective in patients with mineralocorticoid hypertension; however, its safe use in pregnancy is not well documented. The use of antihypertensive medications in pregnancy is detailed in Chapter 19, and in this respect there are reports of the beneficial effects of calcium-channel blockers in nonpregnant patients with aldosteronism.[99] Thus, we would prescribe these latter agents as our first choice, especially if methyldopa proved ineffective. Other agents may be necessary if blood pressure becomes difficult to control. Medical management of primary hyperaldosteronism should also include potassium replacement. When faced with severe hypertension resistant to therapy, and marked hypokalemia requiring very large replacement doses, it may be prudent to consider surgery. During pregnancy, magnetic resonance imaging (MRI) is preferable to CT scan, as the former does not use ionizing radiation. Adrenal vein sampling does not seem to have been reported in pregnancy, and is unadvisable because of the concern about excessive radiation exposure. Thus, documentation of a unilateral adenoma during pregnancy may be suboptimal, which is one reason why surgery may be indicated for treatment failure alone. In this respect, there are several reports of surgical removal of adenomas during the second trimester followed by favorable maternal and fetal outcomes.[110,115]

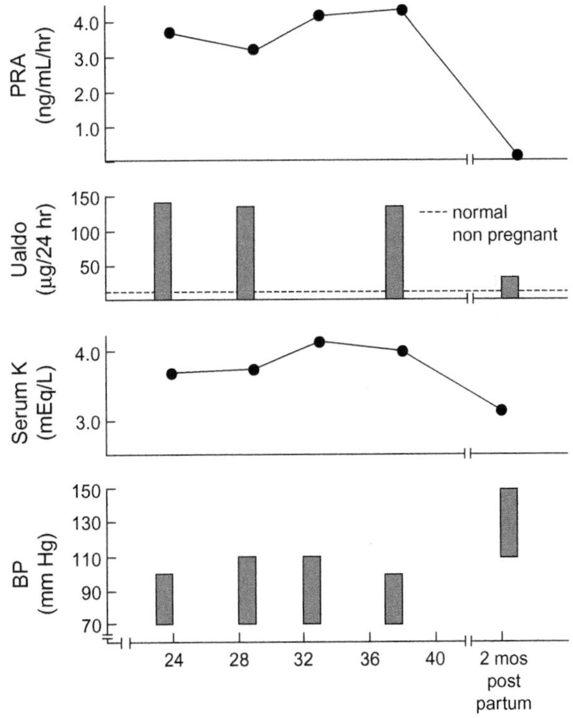

FIGURE 18.2 Plasma renin activity (PRA), urine aldosterone excretion, serum potassium, and blood pressure during pregnancy and 2 months postpartum in a woman with primary aldosteronism. Hypertension and hypokalemia resolved during pregnancy therapy and recurred after delivery. PRA was higher during pregnancy compared with postpartum values, although it was relatively low compared with women with essential hypertension. Urine aldosterone excretion was higher than normal for pregnancy. Reprinted with permission.[112]

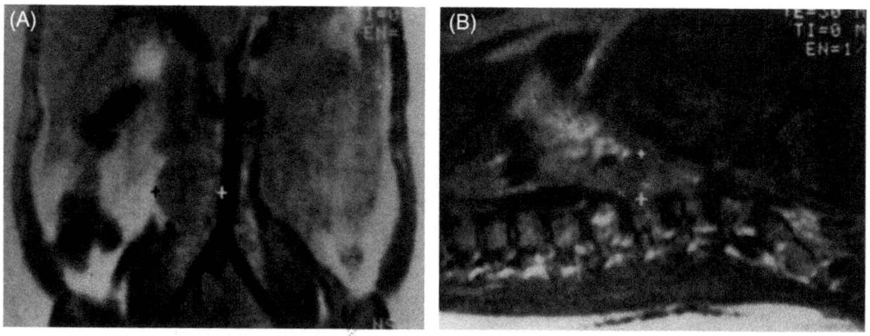

FIGURE 18.3 Magnetic resonance images of an extra-adrenal pheochromocytoma in a 27-week pregnant patient who presented during gestational week 19 with severe hypertension alternating with hypotension even at bed rest. Plasma norepinephrine levels were ≥8000 pg/mL in this patient, who was managed medically with phenoxybenzamine and propranolol through her 36th gestational week, at which time a 2,900-g healthy boy was delivered by cesarean section followed by successful extirpation of the tumor. Coronal section, A, demonstrates that the tumor, 4–5 cm in diameter (between the +s), is located above the bifurcation of the aorta, and just behind the enlarged uterus shown in the sagittal section (again between the +s) and virtually on the vena cava. Despite its proximity to the great vessels, the pheochromocytoma presented few problems at surgery. Modified, with permission, from Greenberg M, Moawad AH, Wieties BM, et al: Extra-adrenal pheochromocytoma: detection during pregnancy using MR imaging. *Radiology*. 1986;161:475–476.

Pheochromocytoma

Pheochromocytoma is rare, but when unrecognized may have fatal consequences.[116–130] On the other hand, there are many apparently benign pheochromocytomas revealed only as an incidental finding at autopsy. Pheochromocytomas arise from chromaffin cells, a tissue which has differentiated from neural crest stem cells and which synthesizes and stores catecholamines. Although the majority of these tumors are in the adrenal glands, as many as 10 to 15% are extra-adrenal in association with sympathetic nerves, mainly in the abdomen (Fig. 18.3) and pelvis, but very rarely in the thorax or in the neck.

The clinical manifestations of pheochromocytoma may be dramatic. They result from catecholamine excess or complications of severe hypertension, and include headache, sweating, palpitations, and anxiety in association with paroxysmal or sustained hypertension. Any patient presenting with such symptoms should be screened for pheochromocytoma, preferably prior to conception, thus avoiding concerns regarding fetal X-ray exposure, and making therapeutic strategies easier. Additional features that may be suggestive of pheochromocytoma are hyperglycemia, orthostatic hypotension, and weight loss. The two most reliable screening procedures are the measurement of plasma fractionated metanephrines or 24-hour urinary metanephrines and/or catecholamines. If biochemical evidence of pheochromocytoma is present, then CT, or preferably MRI, of the abdomen (Fig. 18.3) should be performed. Specialized nuclear medicine tests utilizing iodine 131-metaiodobenzylguanidine may be helpful in identifying extra-adrenal pheochromocytomas. Appropriate treatment in nonpregnant individuals is preoperative alpha-blockade, followed by surgical removal of the tumor.

Over 100 cases of pheochromocytoma presenting during pregnancy or in the immediate puerperium have been reported, and both maternal and fetal morbidity and mortality are extremely high when the presence of the tumor is unknown prior to delivery.[116–130] A systematic review of articles published from 2000 to 2011 reported average maternal and fetal mortality rates of 8% and 17%, respectively, with improved survival rates if the diagnosis can be established antenatally.[130] Cases of unsuspected pheochromocytoma presenting as myocardial infarction in pregnant women have been reported.[122,123] Other serious complications include cardiac arrhythmias, shock, pulmonary edema, cerebral hemorrhage, and hemorrhage into the tumor. In several instances the presenting signs and symptoms, late pregnancy-accelerated hypertension, proteinuria, and seizures, were indistinguishable from preeclampsia and/or eclampsia. In fact, there is a suggestion that the clinical manifestations may be more dramatic as pregnancy progresses because the enlarged uterus is more apt to compress the tumor.

Surgical removal is the therapy of choice when a pheochromocytoma is diagnosed in the initial two trimesters, although successful medical management throughout pregnancy has been reported.[119,120,127,130] Preoperative management includes alpha-blockade with either phenoxybenzamine,[121] or combinations of alpha- and beta-blockers. While phenoxybenzamine is generally safe, placental transfer has been reported and may lead to perinatal depression and hypotension in newborns.[126] Labetalol, an alpha- and beta-blocker, has also been used successfully during gestation.[119]

The approach to treatment in the third trimester is more variable and includes combined cesarean delivery and tumor resection, tumor resection followed by delivery, and delivery followed by tumor resection at a later date. Multidisciplinary decision-making including anesthesiology expertise is advised in these situations.[131] Once the predicted fetal survival is high, we prefer the combined procedure of cesarean delivery and immediate tumor resection.[132]

Cushing Syndrome

Cushing syndrome is also quite rare in pregnant hypertensives, possibly because patients with this disorder are oligo-ovulatory or anovulatory.[110,133–138] Also, the syndrome may be difficult to diagnose because the hormonal alterations of normal pregnancy mimic those of the disease.[110] Most of the pregnancy-associated cases have been due to either adrenal adenomas or pituitary-dependent adrenal hyperplasia. In several instances the disease has appeared during pregnancy, and remitted or improved postpartum, leading to suggestions that a placental ACTH-like compound may have caused the disease. Of importance, Cushing syndrome is associated with excessive maternal morbidity, with hypertension, superimposed preeclampsia, diabetes, and congestive heart failure being the most common complications. There is also a high incidence of preterm delivery, growth restriction, and fetal demise. Management includes surgical resection during the first or early second trimester or surgery after delivery in the third trimester. Therapeutic approaches in the late second and third trimester are more complex, as the risks of surgery must be weighed against the risks of medication to treat hypercortisolism.[139,140]

Postpartum Hypertension

Hypertension associated with preeclampsia may resolve in the first postpartum week, although with more severe cases blood pressure may continue to rise after delivery and the hypertension may persist for 2–4 weeks postpartum.[141,142] Return of blood pressure to prepregnancy levels has not been studied in women with chronic hypertension and superimposed preeclampsia. Ferrazzani et a1. stress that the time required for the blood pressure to improve postpartum correlates both with laboratory markers indicative of renal impairment and with early delivery.[142] We reported a retrospective cohort study of postpartum hypertension and found that women in whom blood pressure failed to normalize by 3 months postpartum were more likely to remain persistently hypertensive, and thus were likely undiagnosed chronic hypertensives.[143] Interestingly, these observations are consistent with the long-held view that chronic hypertension is defined as hypertension in pregnancy diagnosed prior to 20 weeks, or that fails to normalize by 3 months postpartum. Blood pressure in normotensive women often increases between the third and fifth day postpartum,[144] which may be due to the shifting of fluid from the interstitium to the intravascular space which occurs in the immediate puerperium. Importantly, this increase is likely to occur after a woman is discharged from the hospital and who will only be reexamined on her routine postpartum follow-up visit in 6 weeks. We are unaware of large clinical treatment trials that focus uniquely on hypertension in the immediate puerperium. This is unfortunate because severe hypertension in the setting of resolving vasospasm may potentially interfere with cerebral autoregulation, resulting in seizures or cerebrovascular accidents. Thus, in the absence of more definitive data we recommend antihypertensive therapy to maintain systolic blood pressure levels below 150 mm Hg and diastolic levels below 100 mm Hg in the postpartum period (see below).[39] Use of nonsteroidal antiinflammatory agents (NSAIDs) should be avoided in women with postpartum hypertension,[145] particularly those with chronic hypertension with superimposed preeclampsia. NSAIDs have been shown to increase blood pressure and sodium retention in nonpregnant populations.[146] Although bromocriptine is no longer routinely used, it should be particularly avoided in women with postpartum hypertension because of anecdotal reports that such therapy paradoxically exacerbates the hypertensive syndrome and may cause cerebrovascular accidents.[147]

New-Onset Hypertension in the Puerperium

Hypertension may first be diagnosed in the postpartum period. When this occurs, the differential diagnosis includes pregnancy-related hypertensive disorders that were not appreciated antepartum including unrecognized chronic hypertension that was masked by pregnancy, true postpartum preeclampsia or eclampsia, and, rarely, microangiopathic syndromes such as hemolytic uremic syndrome or thrombotic thrombocytopenic purpura.

The prevalence of *de novo* postpartum hypertension or preeclampsia is estimated to be between 0.3 and 27.5%,[148–151] albeit limited by the quality of research studies, which are often single-center and limited to readmissions and/or the immediate postpartum period (2–6 days). The occurrence of postpartum preeclampsia/eclampsia is estimated at 5.7% and associated with substantial maternal morbidity including eclampsia, cerebrovascular accidents, pulmonary edema, cardiomyopathy, HELLP syndrome, thromboembolism, and maternal death. In some cases, careful scrutiny of antepartum records reveals evidence of preeclampsia prior to delivery. However, there are instances of well-documented normotensive deliveries where hypertension in association with laboratory features of preeclampsia has appeared late in the first postpartum week or afterward. Late postpartum *eclampsia* has also been observed, the seizures occurring after the first postpartum week.[150,152–154] The pathogenesis of these phenomena is poorly understood, and certainly is at odds with the traditional concept of preeclampsia–eclampsia as disorders observed only during pregnancy or the immediate puerperium caused by abnormalities in placental development. Blaauw and colleagues have reported persistent endothelial dysfunction up to 11 months post-delivery in women with early-onset preeclampsia and some have hypothesized that persistent vascular dysfunction may contribute to the pathogenesis of postpartum preeclampsia/eclampsia.[155] We are not aware of any randomized trials that evaluate therapies to prevent postpartum hypertension/preeclampsia or that evaluated the use of magnesium sulfate therapy in women with postpartum preeclampsia. In the absence of evidence, the ACOG Hypertension in Pregnancy Task Force recommends treatment to keep blood pressures below 150 mm Hg systolic and 100 mm Hg diastolic.[39] Parenteral magnesium sulfate is recommended for women who present with hypertension or preeclampsia during the postpartum period that is associated with persistent neurologic symptoms including severe headache, visual changes or altered mental status, epigastic pain, or shortness of breath. Duration of therapy is generally for at least 24 hours after diagnosis.[39,151] Some have suggested the use of phenytoin due to concern about previously subclinical epilepsy, although the data to support this approach are lacking. Electroencephalogram (EEG) may be helpful in the decision-making. In most instances the disease resolves within 48 to 72 hours. Neuroimaging should be considered particularly if focal neurologic signs or symptoms are present. The most common finding with preeclampsia–eclampsia is posterior reversible encephalopathy syndrome (PRES) (see Chapter 13). However, intracranial hemorrhage, cerebrovascular accidents, central venous thrombosis, and infarction have also been reported in the postpartum period and treatment for these conditions should be instituted in a timely manner to prevent permanent injury.[151] Periodic suggestions that corticosteroids and/or uterine curettage be used are also offered, especially when severe systemic disease is present.[154] We do not recommend these approaches.

Of concern in regard to late postpartum hypertension is a review that notes that many of these women were not given adequate discharge information, ignored premonitory signs, and/or presented to a primary care setting or an emergency room, often, after convulsing, where the treating physician was unfamiliar with the entity of postpartum preeclampsia.[156] Thus, adequate discharge education is strongly recommended and should prevent some of the cases of postpartum eclampsia. Efforts should also be directed at educating emergency physicians, primary care providers, and obstetricians to facilitate early recognition and treatment. Again, although under-studied in pregnant women, the routine use of nonsteroidal antiinflammatory agents for peripartum analgesia contributes, in our opinion, to some cases of significant hypertension in the postpartum period. These drugs are well known to elevate blood pressure in subgroups of nonpregnant patients at risk of hypertension.[146]

Hypertension appearing *de novo* in the puerperium without laboratory abnormalities suggestive of preeclampsia is most likely to be due to undiagnosed chronic hypertension. In some cases the blood pressure had normalized during gestation as a response to the physiologic vasodilatation that accompanies pregnancy. In most instances, the underlying disorder is essential hypertension; however, secondary causes should be considered when there are unusual features such as severe hypertension, hypokalemia, or symptoms suggestive of pheochromocytoma (see above). Primary aldosteronism in particular may first become apparent in the postpartum period, since antepartum blood pressure is often ameliorated by the antimineralocorticoid effects of progesterone during pregnancy.

MANAGEMENT PRINCIPLES

The purpose of treating chronic hypertension in pregnancy is to ensure a successful term delivery of a healthy infant without jeopardizing maternal well-being. This is in contrast to the treatment philosophy for nonpregnant subjects where the primary concern is prevention of long-term cardiovascular morbidity and mortality. In the latter population blood pressure control is essential, the report of the Joint National Committee on Prevention, Detection, Evaluation, and Treatment of High Blood Pressure[4] recommending levels be maintained at 120 to 135 mm Hg systolic and 75 to 85 mm Hg diastolic, respectively. Management also includes aggressive attention to modifying other cardiovascular risk factors such as blood lipid and glucose levels, body weight, and smoking. Some of these concerns are relevant during pregnancy (e.g., smoking, blood glucose), but others are not. For example, the level of blood pressure control tolerated during pregnancy may be higher, because the risk of exposure of the fetus to additional antihypertensive agents may outweigh the small benefits to the mother of tight control of her blood pressure during a 9-month period. Weight loss is not recommended during pregnancy, nor is vigorous cardiovascular exercise, which may reduce uteroplacental perfusion.

Preconception Counseling

Management ideally begins prior to conception, and includes ruling out and treating, if detected, secondary causes of hypertension. Women in whom hypertension is known to have been present for 5 years or more require careful evaluation for evidence of target organ damage, i.e., left ventricular hypertrophy, retinopathy, and chronic kidney disease. Pregnant women 35 years of age or older, particularly those with chronic illnesses, should be screened for occult coronary disease, and this is particularly important in women with type 1 diabetes with evidence of vascular complications. Ideally, adjustment of medications should also precede conception, discontinuing, of course, drugs with known deleterious fetal effects (especially angiotensin-converting enzyme inhibitors and angiotensin II receptor antagonists). Risks posed by pregnancy are best discussed, and less emotionally evaluated, prior to conception. For example, women with stage 1 hypertension should be informed of the high likelihood of a favorable outcome, but should still be apprised of the risks of superimposed preeclampsia and the fetal complications associated with this disorder. This is also the best time to emphasize the importance of compliance and that frequent visits increase the likelihood of detecting preeclampsia and other complications well before they become life-threatening to mother or fetus. Patient education, by increasing compliance, should a priori improve outcome. Women with small children, and those in the workforce, should be informed of the possibility that lifestyle adjustments will be necessary especially if complications develop. This will allow them to plan ahead for increased support both at home and at work. Finally, early planning, including the assembling of a multidisciplinary team consisting of obstetrician, maternal-fetal medicine specialists, and internists, optimizes the chances of a successful outcome in hypertensive women with other medical complications (e.g., renal transplant recipients, diabetic nephropathy, systemic lupus erythematosus).

Non-Pharmacologic Management

Pregnant hypertensives, in contradistinction to their nonpregnant counterparts, are not advised to exercise vigorously, although careful studies of the effects of aerobic exercise on pregnancy outcome have not been performed. However, moderate exercise, such as a daily walking program, is acceptable. There is some evidence that exercise is associated with a reduced risk of developing preeclampsia in previously normotensive women.[157–160] The major concern regarding vigorous exertion is that women with chronic hypertension are at risk of preeclampsia, a condition characterized by decreased uteroplacental blood flow, and such exercise may compromise blood flow even further.

Excessive weight gain, of course, is not advisable, but again, in contradistinction to therapy in nonpregnant populations, obese women should not be advised to lose weight during pregnancy. Dietary adjustments in pregnant women with chronic hypertension have not been extensively investigated. Salt restriction, an important component of management in nonpregnant populations, is less so in gestation, where extremely low sodium intakes ($\leq$2 g NaCl) may even jeopardize the physiologic plasma volume expansion which normally occurs. However, in women with "salt-sensitive" hypertension, successfully managed with a low-sodium diet prior to conception, it is reasonable to continue such diets during pregnancy, limiting restriction to between 60 and 80 mEq/d. Adequate dietary calcium intake (i.e., 1200 mg/d) may be beneficial in preventing complications of hypertension during pregnancy. A large World Health Organization sponsored multicenter trial conducted in the developing world suggested improved pregnancy outcomes associated with calcium supplementation only in women with low dietary calcium.[61] We performed a placebo-controlled trial of calcium supplementation in 125 women with chronic hypertension and did not find any benefits of calcium on either maternal or fetal outcomes.[62] Other dietary approaches, such as supplementation with magnesium or fish oil, have been investigated in normotensive pregnant women with negative results, but have not been studied in women with chronic hypertension.

Pharmacologic Management (see Chapter 19)

Guidelines for antihypertensive therapy during gestation are less clear than those for nonpregnant hypertensives. For the latter, there are compelling data from large population studies to document the benefits of lowering blood pressure with medication, even in women with only mild hypertension. During pregnancy, however, though maternal safety remains the primary concern, there is also a desire to minimize exposure of the fetus to drugs, given their unknown long-term effects on growth and development. Therefore, permissible maternal blood pressure levels are analyzed in terms of preventing complications during the relatively short duration of the pregnancy, rather than the long-term cardiovascular risk. Another debatable issue is whether lowering blood pressure will prevent superimposed preeclampsia, but at present there is little or no convincing evidence to support this contention.[39,161] Thus, stage 1 hypertension (140–160/90–99 mm Hg) is considered acceptable during pregnancy. In this respect, many caregivers will not treat with antihypertensive medications unless blood pressure exceeds these levels. This approach, however, has not been rigorously evaluated with randomized controlled trials. There is even less evidence when considering the subgroup of women with chronic hypertension. The evidence-based approach recommended by the ACOG Hypertension in Pregnancy Task Force suggests that antihypertensive drug treatment should be commenced when maternal blood pressure reaches systolic levels of 160 mm Hg or diastolic levels of 105 mm Hg.[39] Some experts suggest a lower threshold and initiate therapy if blood pressures are consistently above 150 mm Hg systolic or 95 mm Hg diastolic. There are exceptions, however, including parenchymal renal disease, and evidence of target organ damage (e.g., retinopathy and cardiac hypertrophy), in which case therapy is recommended once diastolic levels are *equal to or greater than* 90 mm Hg.

Whether or not antihypertensive treatment of mild to moderate hypertension during pregnancy results in better fetal outcomes is not clear.[39,161] For instance, in one frequently cited trial, treatment with methyldopa was associated with a reduction in perinatal mortality, primarily midtrimester,[162] while a similar benefit was not evident in another large trial.[163] Two meta-regression analyses have reported an association between lowering blood pressure and decrease in infant birth weight [164,165]; however, a recent Cochrane review did not find evidence of fetal harm.[166] Concerns regarding the impact of aggressive blood pressure lowering on uteroplacental perfusion have tempered the use of antihypertensive medication during pregnancy. We encourage carefully conducted clinical trials to resolve these issues.

In summary, the unknown but potential hazards of antihypertensive treatment during pregnancy are sufficient reasons for withholding drug treatment when mild hypertension is present, particularly during the initial trimester. As noted, many of these patients experience a physiologic decrease in blood pressure which on occasion reaches normotensive levels. Patients whose levels are equal to or greater than 160/100–105 mm Hg, however, should be treated, while evidence of renal disease or end-organ damage requires initiation of treatment at lower levels (≥90 mm Hg).

Specific Antihypertensives

The pharmacology, safety, and efficacy of antihypertensive drugs in pregnancy are addressed in detail in Chapter 19. For most agents the evaluations are sporadic, and there are almost no follow-up data regarding the children exposed to the drug *in utero*. In fact, the only antihypertensive for which credible, prospective follow-up exists is methyldopa,[167] where no adverse effects were documented, an important reason why this agent is considered one of the safest drugs for use during pregnancy. Methyldopa is usually prescribed alone, but on occasion the direct-acting vasodilator hydralazine has been added to the regimen, and this drug also has a long history of use in gestation and appears safe. The reader is referred to Chapter 19, which discusses antihypertensive drug therapy in detail. Briefly, methyldopa remains a preferred drug, because it is the only drug with extended observations (7.5 years) in neonates, thus it is still considered the safest antihypertensive drug to prescribe to pregnant women. However, methyldopa may be less effective in preventing severe hypertension based on a Cochrane analysis of a subset of studies comparing it to beta-blockers and calcium-channel blocker classes.[166] Therefore, many practitioners use labetalol as their first-line drug, but the potential effects of drugs with beta-blocking potentials remain unresolved. There is some concern regarding the use of calcium-channel blocking agents given the high incidence of superimposed preeclampsia in chronically hypertensive women, and the possibility that magnesium sulfate therapy will be required. Here the concern is that magnesium sulfate combined with calcium-channel blocking drugs may lead to precipitous decreases in blood pressure and even neuromuscular blockade; however, this effect is quite rare and has not been substantiated by retrospective review.[168–170] Calcium-channel blockers are used widely during pregnancy, particularly long-acting nifedipine. Angiotensin-converting enzyme inhibitors and angiotensin receptor blockers are associated with major malformations, as well as fetopathy, and are considered by us as contraindicated in pregnancy.[171,172]

The following chapter also reviews an even more limited literature on effects of antihypertensive agents on breastfeeding infants. In general, drugs that are bound to plasma proteins are not transferred to breast milk.[173–175] Lipid-soluble drugs may achieve higher breast concentrations compared with water-soluble drugs. Methyldopa is considered safe, and preliminary data suggest that the levels in breast milk are low. Several beta-blockers are concentrated in breast milk, with atenolol and metoprolol resulting in high levels, and propranolol and labetalol resulting in very low levels. Although captopril levels in breast milk have been reported as low, in view of the adverse effects of ACE inhibitors on neonatal renal function we do not routinely recommend these agents to lactating women. There are only limited reports of calcium-channel blockers and their transfer into breast milk; however, no adverse effects have been reported. Finally, although the concentration of diuretics in breast milk is usually low, these agents may reduce the quantity of milk production and interfere with the ability to successfully breast feed.[173]

References

1. Chobanian AV, Bakris GL, Black HR, et al. The seventh report of the Joint National Committee on Prevention, Detection, Evaluation, and Treatment of High Blood Pressure. *JAMA*. 2003;289:2560–2572.
2. Lawler J, Osman M, Shelton JA, Yeh J. Population-based analysis of hypertensive disorders in pregnancy. *Hypertens Pregnancy*. 2007;26(1):67–76.
3. Vanek M, Sheiner E, Levy A, Mazor M. Chronic hypertension and the risk for adverse pregnancy outcome after superimposed pre-eclampsia. *Int J Gynaecol Obstet*. 2004;86(1):7–11.
4. Gifford RW, August PA, Cunningham FG, et al. Report of the National High Blood Pressure working group on research on Hypertension in Pregnancy. *Am Jf Obstet Gynecol*. 2000;183:S1–S22.
5. Fisher KA, Luger A, Spargo BH, Lindheimer MD. Hypertension in pregnancy: clinical-pathological correlations and remote prognosis. *Medicine (Baltimore)*. 1981;60(4): 267–276.
6. Berkowitz RS, Goldstein DP. Chorionic Tumors. *N Engl J Med*. 1996;335(23):1740–1748.
7. Bader ME, Bader RA. Cardiovascular hemodynamics in pregnancy and labor. *Clin Obstet Gynecol*. 1968;11(4):924–939.
8. Wilson M, Morganti AA, Zervoudakis I, et al. Blood pressure, the renin-aldosterone system and sex steroids throughout normal pregnancy. *Am J Med*. 1980;68(1):97–104.
9. Poppas A, Shroff SG, Korcarz CE, et al. Serial assessment of the cardiovascular system in normal pregnancy. Role of arterial compliance and pulsatile arterial load. *Circulation*. 1997;95(10):2407–2415.

10. August P, Lenz T, Ales KL, et al. Longitudinal study of the renin-angiotensin-aldosterone system in hypertensive pregnant women: deviations related to the development of superimposed preeclampsia. *Am J Obstet Gynecol.* 1990;163(5 Pt 1): 1612–1621.
11. Sibai BM, Abdella TN, Anderson G. Pregnancy outcome in 211 patients with mild chronic hypertension. *Obstet Gynecol.* 1983;61(5):571–576.
12. Schobel HP, Fischer T, Heuszer K, Geiger H, Schmieder RE. Preeclampsia — a state of sympathetic overactivity. *N Engl J Med.* 1996;335(20):1480–1485.
13. August P, Mueller FB, Sealey JE, Edersheim TG. Role of renin-angiotensin system in blood pressure regulation in pregnancy. *Lancet.* 1995;345(8954):896–897.
14. Baylis C, Beinder E, Suto T, August P. Recent insights into the roles of nitric oxide and renin-angiotensin in the pathophysiology of preeclamptic pregnancy. *Semin Nephrol.* 1998;18(2):208–230.
15. Lowe DT. Nitric oxide dysfunction in the pathophysiology of preeclampsia. *Nitric Oxide.* 2000;4(4):441–458.
16. Gangula PR, Supowit SC, Wimalawansa SJ, et al. Calcitonin gene-related peptide is a depressor in NG-nitro-L-arginine methyl ester-induced hypertension during pregnancy. *Hypertension.* 1997;29(1 Pt 2):248–253.
17. Jeyabalan A, Shroff SG, Novak J, Conrad KP. The vascular actions of relaxin. *Adv Exp Med Biol.* 2007;612:65–87.
18. de Mattia NC, Barbin RL, Borges VT, Peracoli JC, Matsubara BB. Doppler echocardiographic assessment of pregnant women with chronic arterial hypertension. *Arq Brasil Cardiol.* 2002;79(6):579–584. 3–8.
19. Borghi C, Cicero AF, Degli Esposti D, et al. Hemodynamic and neurohumoral profile in patients with different types of hypertension in pregnancy. *Intern Emerg Med.* 2011;6(3):227–234.
20. Tihtonen K, Koobi T, Huhtala H, Uotila J. Hemodynamic adaptation during pregnancy in chronic hypertension. *Hypertens Pregnancy.* 2007;26(3):315–328.
21. Sturgiss SN, Dunlop W, Davison JM. Renal haemodynamics and tubular function in human pregnancy. *Baillieres Clin Obstet Gynaecol.* 1994;8(2):209–234.
22. Jeyabalan A, Conrad KP. Renal physiology and pathophysiology in pregnancy. In: Schrier RW, ed. *Renal and Electrolyte Disorders.* 7th ed.. Philadelphia: Lippincott Williams & Wilkins; 2010:462–518.
23. Macdonald-Wallis C, Tilling K, Fraser A, Nelson SM, Lawlor DA. Gestational weight gain as a risk factor for hypertensive disorders of pregnancy. *Am J Obstet Gynecol.* 2013;209(4):327 e1–e17.
24. McClure CK, Catov JM, Ness R, Bodnar LM. Associations between gestational weight gain and BMI, abdominal adiposity, and traditional measures of cardiometabolic risk in mothers 8 y postpartum. *Am J Clin Nutr.* 2013;98(5):1218–1225.
25. Bateman BT, Bansil P, Hernandez-Diaz S, Mhyre JM, Callaghan WM, Kuklina EV. Prevalence, trends, and outcomes of chronic hypertension: a nationwide sample of delivery admissions. *Am J Obstet Gynecol.* 2012;206(2):134e1–134e8.
26. Gilbert WM, Young AL, Danielsen B. Pregnancy outcomes in women with chronic hypertension: a population-based study. *J Reprod Med.* 2007;52(11):1046–1051.
27. Sibai BM, Lindheimer M, Hauth J, et al. Risk factors for preeclampsia, abruptio placentae, and adverse neonatal outcomes among women with chronic hypertension. National Institute of Child Health and Human Development Network of Maternal-Fetal Medicine Units. *N Engl J Med.* 1998;339(10):667–671.
28. Giannubilo SR, Dell'Uomo B, Tranquilli AL. Perinatal outcomes, blood pressure patterns and risk assessment of superimposed preeclampsia in mild chronic hypertensive pregnancy. *Eur J Obstet Gynecol Reprod Biol.* 2006;126(1):63–67.
29. Ray JG, Burrows RF, Burrows EA, Vermeulen MJ. MOS HIP: McMaster outcome study of hypertension in pregnancy. *Early Hum Dev.* 2001;64(2):129–143.
30. Zetterstrom K, Lindeberg SN, Haglund B, Hanson U. Maternal complications in women with chronic hypertension: a population-based cohort study. *Acta Obstet Gynecol Scand.* 2005;84(5):419–424.
31. Tuuli MG, Rampersad R, Stamilio D, Macones G, Odibo AO. Perinatal outcomes in women with preeclampsia and superimposed preeclampsia: do they differ? *Am J Obstet Gynecol.* 2011;204(6):508e1–508e7.
32. Chappell LC, Enye S, Seed P, Briley AL, Poston L, Shennan AH. Adverse Perinatal Outcomes and risk factors for Preeclampsia in Women with Chronic Hypertension: a prospective study. *Hypertension.* 2008;51(4, Part 2 suppl):1002–1009.
33. Williams MA, Mittendorf R, Monson RR. Chronic hypertension, cigarette smoking, and abruptio placentae. *Epidemiology.* 1991;2(6):450–453.
34. Ananth CV, Smulian JC, Vintzileos AM. Incidence of placental abruption in relation to cigarette smoking and hypertensive disorders during pregnancy: a meta-analysis of observational studies. *Obstet Gynecol.* 1999;93(4):622–628.
35. Rey E, Couturier A. The prognosis of pregnancy in women with chronic hypertension. *Am J Obstet Gynecol.* 1994;171(2):410–416.
36. Sibai BM, Anderson GD. Pregnancy outcome of intensive therapy in severe hypertension in first trimester. *Obstet Gynecol.* 1986;67(4):517–522.
37. Cockburn J, Moar VA, Ounsted M, Redman CW. Final report of study on hypertension during pregnancy: the effects of specific treatment on the growth and development of the children. *Lancet.* 1982;1(8273):647–649.
38. Ferrer RL, Sibai BM, Mulrow CD, Chiquette E, Stevens KR, Cornell J. Management of mild chronic hypertension during pregnancy: a review. *Obstet Gynecol.* 2000;96(5 Pt 2): 849–860.
39. Executive summary: hypertension in pregnancy. American College of Obstetricians and Gynecologists. *Obstet Gynecol.* 2013;122(5):1122–1131.
40. McCowan LM, Buist RG, North RA, Gamble G. Perinatal morbidity in chronic hypertension. *Brit J Obstet Gynaecol.* 1996;103(2):123–129.
41. Sibai BM, Koch MA, Freire S, et al. The impact of prior preeclampsia on the risk of superimposed preeclampsia and other adverse pregnancy outcomes in patients with chronic hypertension. *Am J Obstet Gynecol.* 2011;204(4):345e1–345e6.
42. Ness RB, Roberts JM. Heterogeneous causes constituting the single syndrome of preeclampsia: a hypothesis and its implications. *Am J Obstet Gynecol.* 1996;175(5):1365–1370.
43. Brożek JL, Akl EA, Alonso-Coello P, et al. Grading quality of evidence and strength of recommendations in clinical practice guidelines. *Allergy.* 2009;64(5):669–677.

44. Olives C, Myerson R, Mokdad AH, Murray CJ, Lim SS. Prevalence, awareness, treatment, and control of hypertension in United States counties, 2001–2009. *PLoS One.* 2013;8(4):e60308.
45. NCfHS. Health, United States, 2012: With Special Feature on Emergency Care. Hyattsville, MD, 2013.
46. Bateman BT, Shaw KM, Kuklina EV, Callaghan WM, Seely EW, Hernandez-Diaz S. Hypertension in women of reproductive age in the United States: NHANES 1999–2008. *PLoS One.* 2012;7(4):e36171.
47. Perkovic V, Huxley R, Wu Y, Prabhakaran D, MacMahon S. The burden of blood pressure-related disease: a neglected priority for global health. *Hypertension.* 2007;50(6):991–997.
48. Mein CA, Caulfield MJ, Dobson RJ, Munroe PB. Genetics of essential hypertension. *Hum Mol Genet.* 2004;13 Spec No 1:R169–R175.
49. Johnson RJ, Feig DI, Nakagawa T, Sanchez-Lozada LG, Rodriguez-Iturbe B. Pathogenesis of essential hypertension: historical paradigms and modern insights. *J Hypertens.* 2008;26(3):381–391.
50. Sealey JE, Blumenfeld JD, Bell GM, Pecker MS, Sommers SC, Laragh JH. On the renal basis for essential hypertension: nephron heterogeneity with discordant renin secretion and sodium excretion causing a hypertensive vasoconstriction-volume relationship. *J Hypertens.* 1988;6(10):763–777.
51. Goldstein DS, Kopin IJ. The autonomic nervous system and catecholamines in normal blood pressure control and in hypertension. In: Laragh JH, Brenner BM, eds. *Hypertension: Pathophysiology, Diagnosis, and Management*; 1990: 711–747.
52. Powers RW, Catov JM, Bodnar LM, Gallaher MJ, Lain KY, Roberts JM. Evidence of endothelial dysfunction in preeclampsia and risk of adverse pregnancy outcome. *Reprod Sci.* 2008;15(4):374–381.
53. DeFronzo RA, Ferrannini E. Insulin resistance. A multifaceted syndrome responsible for NIDDM, obesity, hypertension, dyslipidemia, and atherosclerotic cardiovascular disease. *Diabetes Care.* 1991;14(3):173–194.
54. Kaaja R. Insulin resistance syndrome in preeclampsia. *Semin Reprod Endocrinol.* 1998;16(1):41–46.
55. Caritis S, Sibai B, Hauth J, et al. Low-dose aspirin to prevent preeclampsia in women at high risk. *N Engl J Med.* 1998;338(11):701–705.
56. Solomon CG, Graves SW, Greene MF, Seely EW. Glucose intolerance as a predictor of hypertension in pregnancy. *Hypertension.* 1994;23(6 Pt 1):717–721.
57. McCarron DA, Morris CD, Henry HJ, Stanton JL. Blood pressure and nutrient intake in the United States. *Science.* 1984;224(4656):1392–1398.
58. Grobbee DE, Hofman A. Effect of calcium supplementation on diastolic blood pressure in young people with mild hypertension. *Lancet.* 1986;2(8509):703–707.
59. Appel LJ, Moore TJ, Obarzanek E, et al. A clinical trial of the effects of dietary patterns on blood pressure. DASH Collaborative research group. *N Engl J Med.* 1997;336(16):1117–1124.
60. Ortega RM, Martinez RM, Lopez-Sobaler AM, Andres P, Quintas ME. Influence of calcium intake on gestational hypertension. *Ann Nutr Metab.* 1999;43(1):37–46.
61. Villar J, Abdel-Aleem H, Merialdi M, et al. World Health Organization randomized trial of calcium supplementation among low calcium intake pregnant women. *Am J Obstet Gynecol.* 2006;194(3):639–649.
62. August P, Helseth G, Cook EF, Sison C. A prediction model for superimposed preeclampsia in women with chronic hypertension during pregnancy. *Am J Obstet Gynecol.* 2004;191(5):1666–1672.
63. Sibai BM, Gordon T, Thom E, et al. Risk factors for preeclampsia in healthy nulliparous women: a prospective multicenter study. The National Institute of Child Health and Human Development Network of Maternal-Fetal Medicine Units. *Am J Obstet Gynecol.* 1995;172(2 Pt 1):642–648.
64. Johnson J, Clifton RG, Roberts JM, et al. Pregnancy outcomes with weight gain above or below the 2009 Institute of Medicine guidelines. *Obstet Gynecol.* 2013;121(5):969–975.
65. de la Torre L, Flick AA, Istwan N, et al. The effect of new antepartum weight gain guidelines and prepregnancy body mass index on the development of pregnancy-related hypertension. *Am J Perinatol.* 2011;28(4):285–292.
66. Institute of Medicine and National Research Council *Weight Gain during Pregnancy: Reexamining the Guidelines.* Washington DC: The National Academies Press; 2009.
67. Benedetto C, Zonca M, Marozio L, Dolci C, Carandente F, Massobrio M. Blood pressure patterns in normal pregnancy and in pregnancy-induced hypertension, preeclampsia, and chronic hypertension. *Obstet Gynecol.* 1996;88(4 Pt 1): 503–510.
68. Thompson JA, Hays PM, Sagar KB, Cruikshank DP. Echocardiographic left ventricular mass to differentiate chronic hypertension from preeclampsia during pregnancy. *Am J Obstet Gynecol.* 1986;155(5):994–999.
69. Taufield PA, Ales KL, Resnick LM, Druzin ML, Gertner JM, Laragh JH. Hypocalciuria in preeclampsia. *N Engl J Med.* 1987;316(12):715–718.
70. Kublickas M, Lunell NO, Nisell H, Westgren M. Maternal renal artery blood flow velocimetry in normal and hypertensive pregnancies. *Acta Obstet Gynecol Scand.* 1996;75(8):715–719.
71. Baker PN, Broughton Pipkin F. Platelet angiotensin II binding in pregnant women with chronic hypertension. *Am J Obstet Gynecol.* 1994;170(5 Pt 1):1301–1302.
72. Warren WB, Gurewitsch ED, Goland RS. Corticotropin-releasing hormone and pituitary-adrenal hormones in pregnancies complicated by chronic hypertension. *Am J Obstet Gynecol.* 1995;172(2 Pt 1):661–666.
73. Nobunaga T, Tokugawa Y, Hashimoto K, et al. Plasma nitric oxide levels in pregnant patients with preeclampsia and essential hypertension. *Gynecol Obstet Invest.* 1996;41(3):189–193.
74. Haller H, Oeney T, Hauck U, Distler A, Philipp T. Increased intracellular free calcium and sensitivity to angiotensin II in platelets of preeclamptic women. *Am J Hypertens.* 1989;2(4):238–243.
75. Kilby MD, Broughton Pipkin F, Symonds EM. Platelet cytosolic calcium in human pregnancy complicated by essential hypertension. *Am J Obstet Gynecol.* 1993;169(1):141–143.
76. Hojo M, Suthanthiran M, Helseth G, August P. Lymphocyte intracellular free calcium concentration is increased in preeclampsia. *Am J Obstet Gynecol.* 1999;180(5):1209–1214.
77. Levine RJ, Maynard SE, Qian C, et al. Circulating angiogenic factors and the risk of preeclampsia. *N Engl J Med.* 2004;350(7):672–683.
78. Levine RJ, Lam C, Qian C, et al. Soluble endoglin and other circulating antiangiogenic factors in preeclampsia. *N Engl J Med.* 2006;355(10):992–1005.

79. Maynard SE, Min JY, Merchan J, et al. Excess placental soluble fms-like tyrosine kinase 1 (sFlt1) may contribute to endothelial dysfunction, hypertension, and proteinuria in preeclampsia. *J Clin Invest*. 2003;111(5):649–658.
80. Perni U, Sison C, Sharma V, et al. Angiogenic factors in superimposed preeclampsia: a longitudinal study of women with chronic hypertension during pregnancy. *Hypertension*. 2012;59(3):740–746.
81. Sibai BM, Koch MA, Freire S, et al. Serum inhibin A and angiogenic factor levels in pregnancies with previous preeclampsia and/or chronic hypertension: are they useful markers for prediction of subsequent preeclampsia? *Am J Obstet Gynecol*. 2008;199(3):268e1–268e9.
82. Maynard SE, Crawford SL, Bathgate S, et al. Gestational angiogenic biomarker patterns in high risk preeclampsia groups. *Am J Obstet Gynecol*. 2013;209(1):53e1–53e9.
83. Powers RW, Jeyabalan A, Clifton RG, et al. Soluble fms-Like tyrosine kinase 1 (sFlt1), endoglin and placental growth factor (PlGF) in preeclampsia among high risk pregnancies. *PLoS One*. 2010;5(10):e13263.
84. Hennessey A, Helseth G, August P. Renovascular hypertension in pregnancy: increased incidence of severe preeclamspia. *J Am Soc Nephrol*. 1997;8(316A).
85. Wilkinson R. Epidemiology and clinical manifestations. In: Novick AC, Scoble J, Hamilton G, eds. *Renal Vascular Disease*. London: WB Saunders; 1996:171–184.
86. Stanley JC. Arterial fibrodysplasia. In: Novick AC, Scoble J, Hamilton G, eds. *Renal Vascular Disease*. London: WB Saunders; 1996:21–35.
87. Levin A, Linas S, Luft FC, Chapman AB, Textor S. Group AHA. Controversies in renal artery stenosis: a review by the American Society of Nephrology Advisory Group on Hypertension. *Am J Nephrol*. 2007;27(2):212–220.
88. Birrer M, Do DD, Mahler F, Triller J, Baumgartner I. Treatment of renal artery fibromuscular dysplasia with balloon angioplasty: a prospective follow-up study. *Eur J Vasc Endovasc Surg*. 2002;23(2):146–152.
89. Hotchkiss RL, Nettles JB, Wells DE. Renovascular hypertension in pregnancy. *South Med J*. 1971;64(10):1256–1258.
90. McCarron DA, Keller FS, Lundquist G, Kirk PE. Transluminal angioplasty for renovascular hypertension complicated by pregnancy. *Arch Intern Med*. 1982;142(9):1737–1739.
91. Easterling TR, Brateng D, Goldman ML, Strandness DE, Zaccardi MJ. Renal vascular hypertension during pregnancy. *Obstet Gynecol*. 1991;78(5 Pt 2):921–925.
92. Pollock CA, Gallery ED, Gyory AZ. Hypertension due to renal artery stenosis in pregnancy–the use of angioplasty. *Aust N Z J Obstet Gynaecol*. 1990;30(3):265–268.
93. Heyborne KD, Schultz MF, Goodlin RC, Durham JD. Renal artery stenosis during pregnancy: a review. *Obstet Gynecol Surv*. 1991;46(8):509–514.
94. Le TT, Haskal ZJ, Holland GA, Townsend R. Endovascular stent placement and magnetic resonance angiography for management of hypertension and renal artery occlusion during pregnancy. *Obstet Gynecol*. 1995;85(5 Pt 2):822–825.
95. Cohen DL, Townsend RR, Clark TW. Renal artery stenosis due to fibromuscular dysplasia in an 18-week pregnant woman. *Obstet Gynecol*. 2005;105(5 Pt 2):1232–1235.
96. Thorsteinsdottir B, Kane GC, Hogan MJ, Watson WJ, Grande JP, Garovic VD. Adverse outcomes of renovascular hypertension during pregnancy. *Nat Clin Pract Nephrol*. 2006;2(11):651–656.
97. Fogari R, Preti P, Zoppi A, Rinaldi A, Fogari E, Mugellini A. Prevalence of primary aldosteronism among unselected hypertensive patients: a prospective study based on the use of an aldosterone/renin ratio above 25 as a screening test. *Hypertens Res*. 2007;30(2):111–117.
98. Mulatero P, Stowasser M, Loh KC, et al. Increased diagnosis of primary aldosteronism, including surgically correctable forms, in centers from five continents. *J Clin Endocrinol Metab*. 2004;89(3):1045–1050.
99. Nadler JL, Hsueh W, Horton R. Therapeutic effect of calcium channel blockade in primary aldosteronism. *J Clin Endocrinol Metab*. 1985;60(5):896–899.
100. Crane MG, Andes JP, Harris JJ, Slate WG. Primary aldosteronism in pregnancy. *Obstet Gynecol*. 1964;23:200–208.
101. Neerhof MG, Shlossman PA, Poll DS, Ludomirsky A, Weiner S. Idiopathic aldosteronism in pregnancy. *Obstet Gynecol*. 1991;78(3 Pt 2):489–491.
102. Gordon RD, Fishman LM, Liddle GW. Plasma renin activity and aldosterone secretion in a pregnant woman with primary aldosteronism. *J Clin Endocrinol Metab*. 1967;27(3):385–388.
103. Lotgering FK, Derkx FM, Wallenburg HC. Primary hyperaldosteronism in pregnancy. *Am J Obstet Gynecol*. 1986;155(5):986–988.
104. Colton R, Perez GO, Fishman LM. Primary aldosteronism in pregnancy. *Am J Obstet Gynecol*. 1984;150(7):892–893.
105. Biglieri EG, Slaton Jr. PE. Pregnancy and primary aldosteronism. *J Clin Endocrinol Metab*. 1967;27(11):1628–1632.
106. Merrill RH, Dombroski RA, MacKenna JM. Primary hyperaldosteronism during pregnancy. *Am J Obstet Gynecol*. 1984;150(6):786–787.
107. Solomon CG, Thiet M, Moore Jr. F, Seely EW. Primary hyperaldosteronism in pregnancy. A case report. *J Reprod Med*. 1996;41(4):255–258.
108. Aboud E, De Swiet M, Gordon H. Primary aldosteronism in pregnancy--should it be treated surgically? *Ir J Med Sci*. 1995;164(4):279–280.
109. Webb JC, Bayliss P. Pregnancy complicated by primary aldosteronism. *South Med J*. 1997;90(2):243–245.
110. Abdelmannan D, Aron DC. Adrenal disorders in pregnancy. *Endocrinol Metab Clin*. 2011;40(4):779–794.
111. Baron F, Sprauve ME, Huddleston JF, Fisher AJ. Diagnosis and surgical treatment of primary aldosteronism in pregnancy: a case report. *Obstet Gynecol*. 1995;86(4 Pt 2):644–645.
112. August P, Sealey JE. The renin-angiotensin system in normal and hypertensive pregnancy and in ovarian function. In: Laragh JH, Brenner BM, eds. *Hypertension: Pathophysiology, Diagnosis and Management*. New York: Raven Press; 1990:1761–1778.
113. Hecker A, Hasan SH, Neumann F. Disturbances in sexual differentiation of rat foetuses following spironolactone treatment. *Acta Endocrinol*. 1980;95(4):540–545.
114. Cabassi A, Rocco R, Berretta R, Regolisti G, Bacchi-Modena A. Eplerenone use in primary aldosteronism during pregnancy. *Hypertension*. 2012;59(2):e18–e19.
115. Kosaka K, Onoda N, Ishikawa T, et al. Laparoscopic adrenalectomy on a patient with primary aldosteronism during pregnancy. *Endocr J*. 2006;53(4):461–466.

116. Lie JT, Olney BA, Spittell JA. Perioperative hypertensive crisis and hemorrhagic diathesis: fatal complication of clinically unsuspected pheochromocytoma. *Am Heart J*. 1980;100(5):716–722.
117. Leak D, Carroll JJ, Robinson DC, Ashworth EJ. Management of pheochromocytoma during pregnancy. *Can Med Assoc J*. 1977;116(4):371–375.
118. Schenker JG, Chowers I. Pheochromocytoma and pregnancy. Review of 89 cases. *Obstet Gynecol Surv*. 1971;26(11):739–747.
119. Lyons CW, Colmorgen GH. Medical management of pheochromocytoma in pregnancy. *Obstet Gynecol*. 1988;72(3 Pt 2): 450–451.
120. Burgess 3rd GE. Alpha blockade and surgical intervention of pheochromocytoma in pregnancy. *Obstet Gynecol*. 1979;53(2):266–270.
121. Stenstrom G, Swolin K. Pheochromocytoma in pregnancy. Experience of treatment with phenoxybenzamine in three patients. *Acta Obstet Gynecol Scand*. 1985;64(4):357–361.
122. Jessurun CR, Adam K, Moise Jr. KJ, Wilansky S. Pheochromocytoma-induced myocardial infarction in pregnancy. A case report and literature review. *Tex Heart I J*. 1993;20(2):120–122.
123. Hamada S, Hinokio K, Naka O, Higuchi K, Takahashi H, Sumitani H. Myocardial infarction as a complication of pheochromocytoma in a pregnant woman. *Eur J Obstet Gynecol Reprod Biol*. 1996;70(2):197–200.
124. Easterling TR, Carlson K, Benedetti TJ, Mancuso JJ. Hemodynamics associated with the diagnosis and treatment of pheochromocytoma in pregnancy. *Am J Perinatol*. 1992;9(5–6):464–466.
125. Freier DT, Thompson NW. Pheochromocytoma and pregnancy: the epitome of high risk. *Surgery*. 1993;114(6):1148–1152.
126. Santeiro ML, Stromquist C, Wyble L. Phenoxybenzamine placental transfer during the third trimester. *Ann Pharmacother*. 1996;30(11):1249–1251.
127. Kennelly MM, Ball SG, Robson V, Blott MJ. Difficult alpha-adrenergic blockade of a phaeochromocytoma in a twin pregnancy. *J Obstet Gynaecol*. 2007;27(7):729–730.
128. Kisters K, Franitza P, Hausberg M. A case of pheochromocytoma symptomatic after delivery. *J Hypertens*. 2007;25(9):1977.
129. Grodski S, Jung C, Kertes P, Davies M, Banting S. Phaeochromocytoma in pregnancy. *Intern Med J*. 2006; 36(9):604–606.
130. Biggar MA, Lennard TW. Systematic review of phaeochromocytoma in pregnancy. *Brit J Surg*. 2013;100(2):182–190.
131. Johnson RL, Arendt KW, Rose CH, Kinney MA. Refractory hypotension during spinal anesthesia for Cesarean delivery due to undiagnosed pheochromocytoma. *J Clin Anesth*. 2013;25(8):672–674.
132. Song Y, Liu J, Li H, Zeng Z, Bian X, Wang S. Outcomes of concurrent Caesarean delivery and pheochromocytoma resection in late pregnancy. *Intern Med J*. 2013;43(5):588–591.
133. Buescher MA, McClamrock HD, Adashi EY. Cushing syndrome in pregnancy. *Obstet Gynecol*. 1992;79(1):130–137.
134. Kreines K, Perin E, Salzer R. Pregnancy in cushing's syndrome. *J Clin Endocrinol Metab*. 1964;24:75–79.
135. Reschini E, Giustina G, Crosignani PG, D'Alberton A. Spontaneous remission of Cushing syndrome after termination of pregnancy. *Obstet Gynecol*. 1978;51(5):598–602.
136. van der Spuy ZM, Jacobs HS. Management of endocrine disorders in pregnancy. Part II. Pituitary, ovarian and adrenal disease. *Postgrad Med J*. 1984;60(703):312–320.
137. Yawar A, Zuberi LM, Haque N. Cushing's disease and pregnancy: case report and literature review. *Endocr Pract*. 2007;13(3):296–299.
138. Lindsay JR, Nieman LK. The hypothalamic-pituitary-adrenal axis in pregnancy: challenges in disease detection and treatment. *Endocr Rev*. 2005;26(6):775–799.
139. Sammour RN, Saiegh L, Matter I, et al. Adrenalectomy for adrenocortical adenoma causing Cushing's syndrome in pregnancy: a case report and review of literature. *Eur J Obstet Gynecol Reprod Biol*. 2012;165(1):1–7.
140. Lim WH, Torpy DJ, Jeffries WS. The medical management of Cushing's syndrome during pregnancy. *Eur J Obstet Gynecol Reprod Biol*. 2013;168(1):1–6.
141. Walters BN, Walters T. Hypertension in the puerperium. *Lancet*. 1987;2(8554):330.
142. Ferrazzani S, De Carolis S, Pomini F, Testa AC, Mastromarino C, Caruso A. The duration of hypertension in the puerperium of preeclamptic women: relationship with renal impairment and week of delivery. *Am J Obstet Gynecol*. 1994;171(2):506–512.
143. Podymow T, August P. Postpartum course of gestational hypertension and preeclampsia. *Hypertens Pregnancy*. 2010;29(3):294–300.
144. Walters BN, Thompson ME, Lee A, de Swiet M. Blood pressure in the puerperium. *Clin Sci (Lond)*. 1986;71(5):589–594.
145. Makris A, Thornton C, Hennessy A. Postpartum hypertension and nonsteroidal analgesia. *Am J Obstet Gynecol*. 2004;190(2):577–578.
146. White WB. Cardiovascular effects of the cyclooxygenase inhibitors. *Hypertension*. 2007;49(3):408–418.
147. Makdassi R, de Cagny B, Lobjoie E, Andrejak M, Fournier A. Convulsions, hypertension crisis and acute renal failure in postpartum: role of bromocriptine? *Nephron*. 1996;72(4):732–733.
148. Clark SL, Belfort MA, Dildy GA, et al. Emergency department use during the postpartum period: implications for current management of the puerperium. *Am J Obstet Gynecol*. 2010;203(1):38e1–38e6.
149. Matthys LA, Coppage KH, Lambers DS, Barton JR, Sibai BM. Delayed postpartum preeclampsia: an experience of 151 cases. *Am J Obstet Gynecol*. 2004;190(5):1464–1466.
150. Al-Safi Z, Imudia AN, Filetti LC, Hobson DT, Bahado-Singh RO, Awonuga AO. Delayed postpartum preeclampsia and eclampsia: demographics, clinical course, and complications. *Obstet Gynecol*. 2011;118(5):1102–1107.
151. Sibai BM. Etiology and management of postpartum hypertension-preeclampsia. *Am J Obstet Gynecol*. 2012;206(6):470–475.
152. Lubarsky SL, Barton JR, Friedman SA, Nasreddine S, Ramadan MK, Sibai BM. Late postpartum eclampsia revisited. *Obstet Gynecol*. 1994;83(4):502–505.
153. Brady WJ, DeBehnke DJ, Carter CT. Postpartum toxemia: hypertension, edema, proteinuria and unresponsiveness in an unknown female. *J Emerg Med*. 1995;13(5):643–648.
154. Magann EF, Martin Jr. JN. Complicated postpartum preeclampsia-eclampsia. *Obstet Gynecol Clin North Am*. 1995;22(2):337–356.

155. Blaauw J, Graaff R, van Pampus MG, et al. Abnormal endothelium-dependent microvascular reactivity in recently preeclamptic women. *Obstet Gynecol*. 2005;105(3):626–632.
156. Hirshfeld-Cytron J, Lam C, Karumanchi SA, Lindheimer M. Late postpartum eclampsia: examples and review. *Obstet Gynecol Surv*. 2006;61(7):471–480.
157. Saftlas AF, Logsden-Sackett N, Wang W, Woolson R, Bracken MB. Work, leisure-time physical activity, and risk of preeclampsia and gestational hypertension. *Am J Epidemiol*. 2004;160(8):758–765.
158. Rudra CB, Williams MA, Lee IM, Miller RS, Sorensen TK. Perceived exertion during prepregnancy physical activity and preeclampsia risk. *Med Sci Sports Exerc*. 2005;37(11):1836–1841.
159. Rudra CB, Sorensen TK, Luthy DA, Williams MA. A prospective analysis of recreational physical activity and preeclampsia risk. *Med Sci Sports Exerc*. 2008;40(9):1581–1588.
160. Sorensen TK, Williams MA, Lee IM, Dashow EE, Thompson ML, Luthy DA. Recreational physical activity during pregnancy and risk of preeclampsia. *Hypertension*. 2003;41(6):1273–1280.
161. Podymow T, August P. Update on the use of antihypertensive drugs in pregnancy. *Hypertension*. 2008;51(4):960–969.
162. Redman CW, Beilin LJ, Bonnar J. Treatment of hypertension in pregnancy with methyldopa: blood pressure control and side effects. *Brit J Obstet Gynaecol*. 1977;84(6):419–426.
163. Gallery ED, Ross MR, Gyory AZ. Antihypertensive treatment in pregnancy: analysis of different responses to oxprenolol and methyldopa. *Brit Med J (Clin Res Ed)*. 1985;291(6495):563–566.
164. von Dadelszen P, Ornstein MP, Bull SB, Logan AG, Koren G, Magee LA. Fall in mean arterial pressure and fetal growth restriction in pregnancy hypertension: a meta-analysis. *Lancet*. 2000;355(9198):87–92.
165. von Dadelszen P, Magee LA. Fall in mean arterial pressure and fetal growth restriction in pregnancy hypertension: an updated metaregression analysis. *J Obstet Gynaecol Can*. 2002;24(12):941–945.
166. Abalos E, Duley L, Steyn DW, Henderson-Smart DJ. Antihypertensive drug therapy for mild to moderate hypertension during pregnancy. *Cochrane Database Syst Rev*. 2007(1):CD002252.
167. Ounsted M, Cockburn J, Moar VA, Redman CW. Maternal hypertension with superimposed pre-eclampsia: effects on child development at 71/2 years. *Brit J Obstet Gynaecol*. 1983;90(7):644–649.
168. Waisman GD, Mayorga LM, Camera MI, Vignolo CA, Martinotti A. Magnesium plus nifedipine: potentiation of hypotensive effect in preeclampsia? *Am J Obstet Gynecol*. 1988;159(2):308–309.
169. Magee LA, Schick B, Donnenfeld AE, et al. The safety of calcium channel blockers in human pregnancy: a prospective, multicenter cohort study. *Am J Obstet Gynecol*. 1996;174(3):823–828.
170. Weber-Schoendorfer C, Hannemann D, Meister R, et al. The safety of calcium channel blockers during pregnancy: a prospective, multicenter, observational study. *Reprod Toxicol*. 2008;26(1):24–30.
171. Cooper WO, Hernandez-Diaz S, Arbogast PG, et al. Major congenital malformations after first-trimester exposure to ACE inhibitors. *N Engl J Med*. 2006;354(23):2443–2451.
172. Laube GF, Kemper MJ, Schubiger G, Neuhaus TJ. Angiotensin-converting enzyme inhibitor fetopathy: long-term outcome. *Arch Dis Child Fetal Neonatal Ed*. 2007;92(5):F402–F403.
173. Ghanem FA, Movahed A. Use of antihypertensive drugs during pregnancy and lactation. *Cardiovasc Ther*. 2008;26(1):38–49.
174. White WB. Management of hypertension during lactation. *Hypertension*. 1984;6(3):297–300.
175. Committee on Drugs. The transfer of drugs and other chemicals into human milk. *Pediatr*. 1994;93:137–150.

CHAPTER **19**

Antihypertensive Treatment

JASON G. UMANS, EDGARDO J. ABALOS AND F. GARY CUNNINGHAM

Editors' comment: *The authors in updating this chapter continue to emphasize the need for better study designs regarding the ideal antihypertensive drugs for pregnant women. A plethora of national guidelines offer contrasting views regarding the hypertensive level at which drug therapy should be given during pregnancy and the authors in the initial sections of the chapter succinctly chronicle the reasons for this confusion. Additionally, systematic reviews of trials are discussed that further explain dilemmas in interpretation and provide guidelines for better-designed trials.*

INTRODUCTION

In the initial edition of this text, discussion of antihypertensive therapy in pregnancy was mainly historical; the drugs noted included veratrum alkaloids, opium and its derivatives, a host of sedatives, and even spinal analgesia – the older literature was often unclear whether such treatment was prescribed for hypertension *per se* or for an eclamptic convulsion. More emphasis, however, was devoted to diuretics – parenthetically vehemently opposed by Chesley – and to the recurring theme that hypertension in preeclampsia might, paradoxically, be treated by volume expansion. Space prohibits republishing these historical vignettes, which bear rereading. Of interest though, use of veratrum viride was incorporated into the treatment of eclampsia even before physicians were aware that a rise of blood pressure accompanied the "puerperal convulsion." Also, for many years, use of veratrum was called the "Brooklyn treatment," noted here because Chesley spent most of his career on the faculty of the State University of New York in Brooklyn.

There are several reasons for the paucity of information on antihypertensive treatment in the original Chesley monograph. Effective and tolerable drugs to lower blood pressure only emerged during the last four decades. Even then, many considered hypertension as "protective," and that the increased pressure was needed to perfuse vital organs such as the kidney and brain in the setting of arteriosclerosis. This view persisted longer in the pregnancy literature, where it was feared that lowering blood pressure would decrease placental perfusion, and thus fetal nutrition and growth. This topic, still the subject of some controversy, will be addressed subsequently. More important was the absence of multicenter randomized trials of a scale large enough to assess the safety and efficacy of specific antihypertensive medications during pregnancy, the risks of congenital anomalies, and infant outcomes. This, too, will be revisited.

GOALS OF ANTIHYPERTENSIVE DRUG THERAPY

Prior to the availability of safe and effective antihypertensive drugs, hypertension could only be defined by evaluation of the normal distribution of blood pressures within the population or by associations with morbid sequelae at different levels of pressure; the former remains the basis for definitions of hypertension in children, while the latter approach is taken in nonpregnant adults.[1] Now, by contrast, hypertension could better be defined as a level of arterial pressure whose pharmacologic control would improve outcome for the population at risk.

In nonpregnant adults, blood pressure control can decrease the remote incidence of stroke, coronary heart disease, congestive heart failure, and cardiovascular mortality. Each of these outcomes – as well as others, for example, nephrosclerosis – is due only in part to hypertension per se, adding to or interacting with other risks such as smoking, dyslipidemia, diabetes mellitus, race, age, sex, and familial predisposition. In addition, the ability to detect improved outcome due to blood pressure control also depends on the severity of hypertension and the presence of target organ damage,[2] the duration of follow-up observation, the completeness of blood pressure control,[3] perhaps the pathophysiology leading to hypertension, and the drugs used to treat it. For example, treatment trials have been most easily designed to show decreases in stroke incidence in patients with severe hypertension; however, much larger, longer

Chesley's Hypertensive Disorders in Pregnancy.
ISBN: 978-0-12-407866-6

DOI: http://dx.doi.org/10.1016/B978-0-12-407866-6.00019-5

duration, or more restrictively designed trials were required to show benefits for coronary disease or in patients with only mild or moderate hypertension.

Hypertension in pregnant women is different. Here, the major goals of treatment are to safeguard the mother from *acute* dangers or irreversible insults during or immediately after the pregnancy while delivering a healthy infant. In this respect, one balances *short-term* maternal outcome against possible *long-term* consequences of intrauterine drug exposure or hemodynamic insult on fetal and childhood growth and development. While comorbidities such as diabetes mellitus, target organ dysfunction at baseline, and (uncommon) secondary causes of hypertension may certainly interact with maternal hypertension and alter strategies for its control, these concerns will not apply to most hypertensive pregnant women. The majority of women with blood pressures high enough to warrant treatment during pregnancy will either have chronic essential hypertension or preeclampsia leading to threatening levels of pressure which occur when there are compelling reasons to extend the length of the pregnancy.

It should be noted that the balance of risks and benefits for antihypertensive therapy will differ for women with chronic hypertension present from early in gestation, whose fetuses may have greater drug exposure during early stages of development, compared with women who develop hypertension, viz., either gestational hypertension or preeclampsia, closer to term. Well-designed and adequately powered trials, of which there are *shamefully few and too many poorly designed* – despite our pleas in previous editions – are still needed to account for this clinical heterogeneity.

Maternal risks which may justify pharmacotherapy include that of superimposed preeclampsia, which, with its morbid outcomes, appears to account for most complications ascribed to chronic hypertension.[4] Additional risks are those of placental abruption, accelerated hypertension leading to hospitalization or to target organ damage, and cerebrovascular catastrophes. Risks to the fetus include death, growth restriction, and early delivery, the latter occurring in many cases due to concerns regarding maternal safety. As will be discussed later, there are no convincing data to currently suggest that antihypertensive treatment of women with mild to moderate chronic hypertension prevents any of these outcomes, except perhaps for worsened hypertension or the need for additional antihypertensive therapy.

GENERAL PRINCIPLES IN THE CHOICE OF ANTIHYPERTENSIVE AGENTS

Blood pressure, in its simplest conceptualization, is determined as the product of cardiac output and systemic vascular resistance. The latter is sensitive to the structure of small arterial and arteriolar resistance vessels, activity of local vasodilator and vasoconstrictor systems, humoral influences such as the renin–angiotensin system (see Chapter 15), and the activity of the autonomic nervous system. The former is sensitive to changes in volume status and autonomic tone; as other influences on intrinsic myocardial contractility are usually minor in healthy women. These physiologic targets, sometimes obscured in the chronic state by vascular autoregulation or our limited ability to measure relevant volumes or pressures with precision,[5] provide the rationale for each of the available pharmacologic strategies for control of hypertension. Further, due to the homeostatic nature of blood pressure control, even when pathologically elevated, this simple physiologic construct suggests likely mechanisms of apparent resistance to antihypertensive drugs, especially when used as single agents. In nonpregnant hypertensive patients, choice of a specific antihypertensive drug is usually rationalized by the severity of hypertension and immediate risk of end-organ damage, the desired time–action characteristics of the drug, specific comorbidities, spectrum of possible adverse drug effects, cost, and known secondary causes of the hypertension. Therapy may also be based on outcomes of well-conducted, large clinical trials, on the systematic review of many smaller randomized trials, or on broad, population-based assumptions regarding the likely physiologic mechanisms leading to hypertension in a given patient. A brief review of the above considerations follows below, but it is tempered by the knowledge that, in mild to moderate hypertension, most available agents appear effective in a similar proportion of patients when used as monotherapy, and most hypertension can be adequately controlled by a combination of two rationally paired drugs.

In nonpregnant adults with hypertensive emergencies defined by systolic BP $\geq$180 mm Hg or diastolic BP $\geq$ 120 mm Hg with evolving target organ damage, blood pressure is usually controlled acutely, albeit only to a target of ~160 mm Hg systolic so as not to compromise organ (especially cerebral) perfusion, by use of rapidly and short-acting, easily titrated parenteral agents.[1] These include, most commonly, sodium nitroprusside (a nitric oxide donor), enalaprilat (an ACE inhibitor), hydralazine (a direct vasodilator), labetalol (a combined α- and β-adrenergic antagonist), nicardipine (a dihydropyridine calcium entry blocker), and fenoldopam (a specific dopamine-receptor antagonist). Short-acting (i.e., immediate-release) oral or sublingual nifedipine, used in the past for acute blood pressure control, is avoided due to unpredictable hypotensive responses, excessive autonomic activation, and precipitation of acute myocardial ischemia.

In the chronic setting, results of large randomized controlled trials provide compelling evidence that angiotensin-converting enzyme (ACE) inhibitors or angiotensin II (type 1; AT-1) receptor blockers (ARBs) should be the cornerstone of antihypertensive therapy in *nonpregnant* patients with established diabetic nephropathy because they significantly slow the progression of renal failure.[6,7] Extrapolation from these findings and the results of other studies support the use of these drugs in diabetic patients with less evident nephropathy or in patients with proteinuric nondiabetic renal insufficiency.[8] Similarly strong evidence supports the use of β-blockers (lacking intrinsic sympathomimetic activity), ACE inhibitors, ARBs, and aldosterone antagonists to prevent death in patients who have suffered a myocardial infarction[9–11] and to improve both survival and functional status in patients with congestive heart failure, whether or not they are hypertensive.[12–14] Finally, diuretics appear especially efficacious in avoiding a variety of morbid complications, especially stroke, in patients with isolated systolic hypertension, particularly in the elderly.[15] Indeed, based on the results of very large trials such as ALLHAT – Antihypertensive and Lipid Lowering Heart Attack Trial[16] – diuretics are considered first-line agents in most nonpregnant patients with essential hypertension who lack a "compelling indication" for a different drug.[1] Another evolving literature recognizes that the large population of patients with more severe hypertension – defined as >20 mm Hg systolic or >10 mm Hg diastolic above treatment targets in the Joint National Committee (JNC) 7 guidelines – will require two or more agents for blood pressure control, advocating initial therapy with rationally chosen drug combinations rather than focusing on the drugs most likely to achieve control when used as monotherapy. This change in thinking has further focused research on the comparative effectiveness of such drug combinations.[17]

In the absence of compelling data from controlled trials, drug choice may be influenced by reasoned extrapolation of the impact of known pharmacology of specific antihypertensive agents on medical comorbidities in individual patients. For example, β-blockers, even those with relative β selectivity, are routinely avoided in asthmatic patients due to their ability to provoke bronchospasm. They are also avoided in some patients due to their potential to acutely exacerbate systolic heart failure, their ability to mask autonomic symptoms of hypoglycemia in diabetics, and their capacity to worsen atrial–ventricular conduction defects. Conversely, β-blockers are rational agents in hypertensive patients also suffering from angina, supraventricular tachyarrhythmias, benign essential tremor, or migraine, as they are often useful in these latter conditions even when hypertension is absent. Thiazide diuretics are reasonably avoided in some patients due to their capacity to exacerbate hyperuricemia and gout, to impair glucose tolerance, or to worsen hypercalcemia in patients with primary hyperparathyroidism. Potassium-sparing diuretics, ACE inhibitors, or ARBs are all reasonably avoided in patients with a severe potassium excretory defect, most commonly due to hyperkalemic distal renal tubular acidosis and associated with diabetes mellitus, sickle-cell nephropathy, or obstructive uropathy.

Chronic hypertension in young *men* is more often associated with increased cardiac output and is often responsive to monotherapy with β-blockers.[18] Likewise, and in spite of a striking lack of conformity in the clinical assessment of salt sensitivity,[19] African-Americans and elderly women are more often apt to manifest low-renin hypertension, relative defects in the renal excretion of a salt load, and/or enhanced blood pressure increments with volume expansion. It is not surprising, therefore, that these groups demonstrate a greater tendency to effective blood pressure control with diuretic or calcium entry blocker monotherapy, with lesser rates of response to β-blockers or ACE inhibitors.[20] In spite of these population differences, careful titration of even haphazardly selected antihypertensive drugs usually leads to acceptable blood pressure control in individual patients, casting some doubt on the importance of such "physiologically based" drug choice strategies in the absence of outcome data from well-designed prospective treatment trials.

All of these insights should lead to some humility as we choose drugs for use in hypertensive pregnant women, in whom both pathophysiologic and clinical trial data are even more limited.

FETAL SAFETY AND DRUG USE IN PREGNANT WOMEN

In the past, both regulations and ethical norms led industry to scrupulously avoid testing drugs in pregnant women. While regulatory changes made by the United States Food and Drug Administration (FDA) in 1993 reversed this policy by encouraging testing of some drugs in pregnant women,[21] such clinical information remains unavailable for most currently prescribed agents. Indeed, rigorous evaluation of pharmacokinetics, biotransformation, maternal efficacy, fetal exposure, and long-term fetal effects of drugs used during pregnancy is generally lacking. Available information, save for assessment of teratogenicity in laboratory animals, is limited and selective. Over 5 years ago, the FDA proposed major changes to the pregnancy and lactation sections of drug labels, which would abandon use of its outmoded classification scheme to categorize potential fetal risks – shown in Table 19.1 – instead including more data and narrative rather than summary (letter) categories.[22] Despite much enthusiasm, final guidelines have not been implemented as of this time, leaving us in limbo, so we include the current risk categories in this chapter, stressing, however, the need to take all clinical information into account when making a prescribing decision in a particular patient.

Because data from human and animal studies are so limited, most drugs are listed in FDA category C, with the caveat that they should be used only if potential benefit justifies the potential risk to the fetus. This category is so broad as to be useless; it includes those antihypertensive drugs with the greatest history of safe use in pregnant women but, for many years, also included the ACE inhibitors, which are contraindicated in later pregnancy (Class D). Even when

TABLE 19.1 Current FDA Categories Describing Risks of Drug Use in Pregnancy

Category	Description
A	Adequate, well-controlled studies in pregnant women have not shown an increased risk of fetal abnormalities.
B	Animal studies have revealed no evidence of harm to the fetus; however, there are no adequate and well-controlled studies in pregnant women. *or* Animal studies have shown an adverse effect, but adequate and well-controlled studies in pregnant women have failed to demonstrate a risk to the fetus.
C	Animal studies have shown an adverse effect and there are no adequate and well-controlled studies in pregnant women. *or* No animal studies have been conducted and there are no adequate and well-controlled studies in pregnant women.
D	Studies, adequate well-controlled or observational, in pregnant women have demonstrated a risk to the fetus. However, the benefits of therapy may outweigh the potential risk.
X	Studies, adequate well-controlled or observational, in animals or pregnant women have demonstrated positive evidence of fetal abnormalities. The use of the product is contraindicated in women who are or may become pregnant.

drugs are placed in category B, their presumed safety may be a function of the insensitivity of animal tests to predict subtle clinical effects, such as fetal ability to withstand hypoxic stress, changes in functional physiologic development, and altered postnatal neurocognitive development. Limited evidence of clinical safety is often extended injudiciously, such that drugs that lack significant teratogenic potential in early pregnancy may exert devastating effects on fetal organ function nearer to term; conversely, drugs that are safe in the third trimester may have irreversible effects on fetal growth or development when used earlier. By contrast, one can easily err against drug use in pregnancy, for example, when discontinuation of category D drugs might jeopardize both mother and fetus. Examples of this latter scenario would include discontinuing immunosuppressives in a pregnant renal transplant recipient or of antiepileptic agents in a pregnant woman with a seizure disorder.

CHOICE OF AN ANTIHYPERTENSIVE DRUG FOR USE IN PREGNANCY

Following discussion of the rationale for choosing an antihypertensive drug, we turn to consideration of each class of agents available in common practice. *Each of these drug classes, their apparent mechanisms of action, and, in terms of this chapter's primary goals, their suitability for use during pregnancy are described separately.* Not considered are appropriate targets for blood pressure control during pregnancy, as these have not yet been established. Whether antihypertensive therapy in pregnancy should be targeted to specific hemodynamic endpoints other than systolic and diastolic arterial pressure, such as maternal cardiac output or measures derived from maternal pulse wave analysis,[23–25] is also not discussed.

Sympathetic Nervous System Inhibition

Introduction of agents to decrease peripheral activity of the sympathetic nervous system marks, perhaps, the beginning of the modern era of effective antihypertensive therapy. Strategies for sympathoinhibition have included ganglionic blockade (e.g., guanethidine), depletion of norepinephrine from sympathetic nerve terminals, such as reserpine which is probably the first antihypertensive to have proven benefit in a prospective clinical trial,[26] α-adrenergic agonists to decrease sympathetic outflow from the central nervous system (e.g., α-methyldopa and clonidine), and specific antagonists of α- or β-adrenergic receptors – including α-antagonists such as prazosin, terazosin, and doxazosin and β-blockers such as propranolol, atenolol, and metoprolol; and combined α/β antagonists including labetalol and carvedilol. Figure 19.1 shows a physiologic scheme for sympathetic nervous system influences on arterial pressure, along with sites of action for drugs such as α_2-adrenergic agonists and peripheral adrenergic antagonists, which remain widely used in pregnancy.

Centrally Acting α_2-Adrenergic Agonists

Methyldopa is the prototypical agent of this class. It is a prodrug metabolized to α-methylnorepinephrine, which then replaces norepinephrine in the neurosecretory vesicles of adrenergic nerve terminals. Because its efficacy is equivalent to that of norepinephrine at peripheral α_1 receptors, vasoconstriction is unimpaired. Centrally, however, it is resistant to degradation by monoamine oxidase, resulting in enhanced effect at the α_2 sites that regulate sympathetic outflow. Decreased sympathetic tone reduces systemic vascular resistance, accompanied by only minor decrements in cardiac output, at least in young, otherwise healthy hypertensive patients. Blood pressure control is gradual, over 6 to 8 hours, due to the indirect mechanism of action. There do not appear to be significant decreases in renin and, while the hypotensive effect is greater in the upright than supine posture, orthostatic hypotension is usually minor. Clonidine, a selective α_2 agonist, acts similarly.

Adverse effects are mostly predictable consequences of central α_2-agonism or decreased peripheral sympathetic tone. These drugs act at sites in the brainstem to decrease mental alertness, impair sleep, lead to a sense of fatigue or

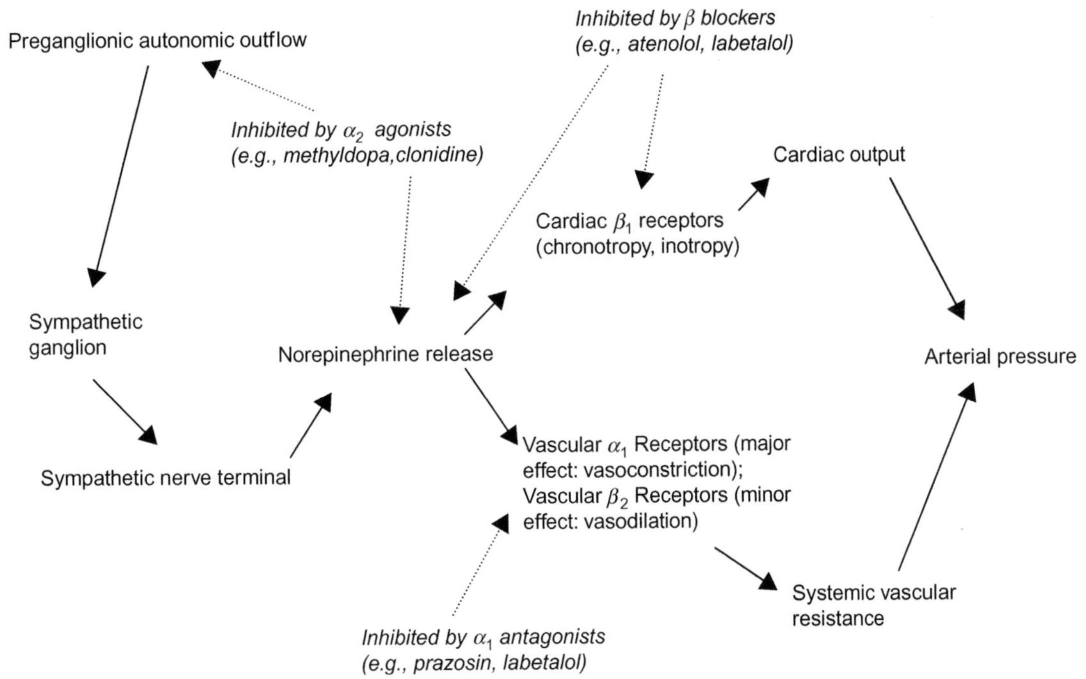

FIGURE 19.1 Scheme showing major influences of sympathetic nervous system on physiologic determinants of arterial pressure. Bold arrows denote endogenous mechanisms leading from central nervous system to maintenance of blood pressure. Dotted arrows show sites at which sympatholytic antihypertensive drugs used in pregnancy might exert their effects.

depression in some patients, and decrease salivation to cause xerostomia. Peripheral sympathoinhibition may impair cardiac conduction in susceptible patients. Methyldopa can induce hyperprolactinemia and Parkinsonian signs in some patients. In addition, it can cause some potentially serious dose-independent adverse effects. Approximately 5% of patients receiving methyldopa will have elevated liver enzymes, with some manifesting frank hepatitis and, rarely, hepatic necrosis. Likewise, many patients will develop a positive Coombs test with chronic use, a small fraction of these progressing to hemolytic anemia.

Pregnancy

Methyldopa has been compared with placebo or no treatment in eight trials focused on pregnant women,[27–29] or with alternative hypotensive agents in 20 trials.[28,30–33] The drug is also unique in that careful, albeit underpowered, studies have also assessed remote development of children exposed to this drug *in utero*[34] (see "Systematic Analysis" below). Methyldopa is one of our preferred agents for nonemergency blood pressure control during pregnancy since no "modern" antihypertensive has proven more efficacious, better tolerated, or possesses a superior history of clinical safety. Indeed, observations of increased sympathetic nerve activity to the skeletal muscle vasculature in preeclamptic women, reverting to normal along with blood pressure after delivery,[35] lends a compelling physiologic rationale to control of preeclamptic hypertension with agents that decrease sympathetic outflow.

Treatment with methyldopa decreases the subsequent incidence of severe hypertension,[36] but not preeclampsia, and it is well tolerated by the mother without any apparent adverse effects on uteroplacental or fetal hemodynamics[37] or on fetal well being.[27,28] One placebo-controlled trial that included over 200 women with diastolic BP $\geq$ 90 mm Hg at entry reported fewer midpregnancy losses in women randomized to methyldopa,[27] an observation that was not confirmed in a later trial of similar size.[28] It is important to note that birth weight, neonatal complications, and development during the first year were similar in children exposed to methyldopa compared with those in the placebo group.[38,39] While Cockburn et al.[34] noted somewhat smaller head circumference at 7 years of age in the subset of male offspring exposed to methyldopa at 16 to 20 weeks gestation, these children exhibited intelligence and neurocognitive development similar to controls.

Studies of clonidine have been more limited. One third-trimester comparative trial compared clonidine with methyldopa and showed similar efficacy and tolerability,[32] while a small, controlled follow-up study of 22 neonates reported an excess of sleep disturbance in clonidine-exposed infants.[40] Clonidine should be avoided in early pregnancy due to suspected embryopathy; later effects on fetal growth varied with drug-induced changes in maternal hemodynamics.[41] There appears to be little justification for its use in preference to methyldopa given their similar mechanisms of action and the proven safety of the latter agent.

Peripherally Acting Adrenergic-Receptor Antagonists

Cardiac β_1 receptors mediate the chronotropic and inotropic effects of sympathetic stimulation, while receptors in the kidney modulate renin synthesis in response to renal sympathetic input. Activation of β_2 receptors leads to relaxation of airway smooth muscle and to peripheral vasodilatation. Acutely, nonselective β-blockade decreases cardiac output; but with limited change in arterial pressure, due to increased systemic vascular resistance. Over time, vascular resistance falls to predrug levels, resulting in a persisting hypotensive response that parallels the decrease in cardiac output. Moderate decrements in renin, and thus in angiotensin II and aldosterone, may contribute to the chronic antihypertensive efficacy of β-blockers in some patients, likely accounting for their greater efficacy in patient groups not believed to have salt-sensitive hypertension.[18,20] Some agents, like pindolol or oxprenolol, are partial β-receptor agonists, viz. they possess some limited degree of "intrinsic sympathomimetic activity." These drugs lead to lesser decrements in cardiac output and β_2 stimulation may even result in significant decrements of vascular resistance. Individual drugs may also possess selective potency at $\beta_1 > \beta_2$ receptors (e.g., atenolol and metoprolol), the additional capacity to block vascular α_1 receptors (e.g., labetalol), and may differ in their lipid solubility, such that hydrophilic agents gain less access to the central nervous system.

β-receptor antagonists are currently considered second-line agents for hypertension in nonpregnant adults (after diuretics) except when specific comorbidities suggest their initial use.[1] They are preferred agents in patients who have experienced recent myocardial infarction,[1,9] but appear to provide less benefit than diuretics in elderly patients with isolated systolic hypertension, in the prevention of stroke, or when compared with angiotensin receptor blockers in patients with multiple cardiovascular risk factors.[42] Unlike with other agents, long-term use of β-blockers is not associated with (beneficial) remodeling of small resistance arteries which have been hypertrophied due to hypertension,[43] the clinical consequences of this observation being unknown.

Adverse effects are predictable results of β-receptor blockade. They include fatigue and lethargy, exercise intolerance due mostly to (nonselective) β effects in skeletal muscle vasculature, peripheral vasoconstriction secondary to decreased cardiac output, sleep disturbance with use of more lipid-soluble drugs, and bronchoconstriction. While negative inotropy could worsen acute decompensated congestive heart failure or lead to heart block in susceptible patients, β-blockers exert paradoxical benefit in all stages of chronic heart failure with decreased systolic function.[1]

Pregnancy

β-blocking drugs have been used extensively in pregnancy and subjected to several randomized trials in pregnancy versus no treatment[16] or placebo and versus alternative agents,[17] and these are discussed further in the section "Systematic Analysis."[30,31,44–50] Long ago, animal studies and anecdotal clinical observations led to concerns that these agents could cause fetal-growth restriction, impair uteroplacental blood flow, and exert detrimental cardiovascular and metabolic effects on the fetus. However, most prospective studies which focused on drug administration in the third trimester and included a mix of hypertensive disorders have demonstrated effective control of maternal hypertension without significant adverse effects to the fetus. Further reassurance is derived from a 1-year follow-up study, which showed normal development of infants exposed to atenolol *in utero*.[51] By contrast, when atenolol therapy for chronic hypertension was started between 12 and 24 weeks gestation, it resulted in clinically significant growth restriction, along with decreased placental weight;[47] this observation was supported by subsequent retrospective series and reviews comparing atenolol with alternative therapies.[52–54] Similarly, several more intensively monitored studies have demonstrated fetal and neonatal bradycardia, adverse influences on uteroplacental and fetal circulations, or evidence of other fetal insults following nonselective β-blockade; these effects may be mitigated with the partial agonist pindolol.[50,55] By contrast, a more recent and reassuring systematic review showed no clear evidence of fetal or neonatal bradycardia due to maternally administered β-blockers.[56]

Labetalol, which is also an antagonist at vascular α_1 receptors, appears to be the β-blocking drug most widely preferred and prescribed in pregnancy. The drug is administered parenterally to treat severe hypertension, especially the rapid rise that frequently occurs with the appearance or superimposition of preeclampsia. In such cases, its efficacy and tolerability appear equivalent to parenteral hydralazine.[57–61] When administered orally in chronic hypertension, it appeared safe[28,33,62–65] and equivalent to methyldopa, though high doses might result in neonatal hypoglycemia.[66] Of further concern, its use was associated with fetal-growth restriction or neonatal difficulties in one placebo-controlled study.[65]

The above discussion of β-blockers focuses on control of systolic and diastolic blood pressure, without any attempt to further dissect maternal hemodynamics. While of uncertain significance, we draw brief attention to a separate literature whose authors have used noninvasive – Doppler echocardiography or impedence cardiography – technology to longitudinally assess cardiac output and systemic vascular resistance in hypertensive pregnant women. Such studies have noted supra-normal increments in cardiac output in many women with hypertensive pregnancy and in those destined to develop preeclampsia.[67,68] These investigators have proposed that use of β-blockers – principally atenolol – to lower cardiac output towards gestational age-adjusted normal values would be beneficial and also that blood pressure control which did not excessively lower cardiac output would avoid uteroplacental hypoperfusion.[24,69,70] Despite suggestive small studies reporting decreased incidence of preeclampsia and blunted increases in sFlt-1 following atenolol therapy targeted to hemodynamic endpoints,[69,71] these interesting hypotheses remain largely untested.

Peripherally acting α_1-adrenergic antagonists are third-line antihypertensive drugs in nonpregnant adults which have their clearest indication during pregnancy in the management of hypertension due to suspected pheochromocytoma (see Chapter 18). Both prazosin and phenoxybenzamine have been used, along with β-blockers as adjunctive agents, in the management of these life-threatening disorders.[72] Because there is but limited and primarily anecdotal additional experience with these agents in pregnancy, their more routine use cannot be advocated.

Diuretics

These are probably the most commonly used antihypertensive agents worldwide. This is because they are among the drugs most clearly associated with beneficial outcome in randomized, controlled trials relating to hypertension in nonpregnant adults.[1] They lower blood pressure by promoting natriuresis and subtle decrements of intravascular volume. Evidence for this mechanism includes observations that the acute hypotensive effect of thiazide diuretics parallels net sodium loss (usually 100–300 mEq, equivalent to 1–2 L of extracellular fluid) over the first few days of therapy, that the hypotensive effect is absent in animals or patients with renal failure, that a high-salt diet blocks the fall in blood pressure, that salt restriction perpetuates the hypotensive response even following diuretic withdrawal, and that infusion of salt or colloid solutions reverses the hypotensive effect.[73–75] This early phase of volume depletion decreases cardiac preload and thus cardiac output. Decrements in blood pressure reflect the failure of counter-regulatory mechanisms, including activation of the renin–angiotensin and sympathetic nervous systems, to raise peripheral resistance enough to defend elevated arterial pressure.

Curiously, the chronic antihypertensive effect of diuretics is maintained despite partial restoration of plasma volume and normalization of cardiac output, due to persisting *decrements* in systemic vascular resistance. Numerous studies have failed to reveal neurohumoral mechanisms, mediators, or direct actions on the vasculature which lead to this prolonged secondary vasodilator response; at present it is best ascribed to the phenomenon of total body autoregulation.[5] Those patients most apt to have maintained responses to diuretics are those who most effectively achieve this secondary vasodilated state without persisting decrements in cardiac output leading to vasoconstrictor responses (i.e., salt-sensitive, low-renin hypertensives). When a patient appears diuretic-refractory, the addition of a β-blocker or ACE inhibitor is frequently effective; indeed, use of these pharmacologically rational combinations is encouraged in current guidelines.[1]

The adverse effects of diuretics are mainly due to their precipitation of fluid, electrolyte, and solute disturbances, complications that could be predicted from their known effects on renal salt handling. By blocking renal tubular sodium reabsorption, thiazides enhance sodium delivery to distal sites in the nephron. These are sites at which sodium may be exchanged for potassium or protons; modest volume contraction resulting from diuretic use may thus be accompanied by hypokalemia and metabolic alkalosis. Likewise, volume depletion may evoke both thirst and non-osmotic secretion of vasopressin, favoring hyponatremia in some patients. Similarly, decreased renal perfusion may lead to azotemia if severe, but this is usually preceded by increased proximal reclamation of salt, fluid, and solutes, which commonly causes hyperuricemia.

Pregnancy

There have been many prospective trials of diuretics or dietary salt restriction in pregnancy, primarily focused on prevention of preeclampsia rather than on treatment of hypertension. A meta-analysis published in 1985[76] suggested that diuretics did indeed prevent preeclampsia, noting, however, no decrease in the incidence of proteinuric hypertension; thus the claim of efficacy could be ascribed to use of an improper definition of the clinical outcome being studied. In addition, only limited data support the value of salt restriction, which some consider deleterious. Thus, neither of the above interventions appears successful, though diuretics may prevent much of the physiologic volume expansion which normally accompanies pregnancy.[77] While volume contraction might be expected to limit fetal growth,[78] outcome data have not supported these concerns.[76] However, diuretic-induced hyperuricemia, as described above, may complicate the already difficult clinical diagnosis of superimposed preeclampsia. Also, observations of volume depletion and primary systemic vasoconstriction in preeclampsia[79] make diuretics physiologically irrational agents in this disorder.

By contrast, diuretics are commonly prescribed in essential hypertension prior to conception and, given their apparent safety, there is general agreement, articulated first by the National High Blood Pressure Education Program (NHBPEP) Working Group on High Blood Pressure in Pregnancy, and retained in the recommendations of the Task Force convened by the American College of Obstetricians and Gynecologists and that included specialists from several disciplines. The Task Force, further described in Chapter 2 (p. 25), was specifically charged with updating the NHBPEP report, noted that diuretics may be continued through gestation or used in combination with other agents, especially for women deemed likely to have salt-sensitive hypertension.[80,81] In this sense, the greatest utility for diuretics may be as second or third agents which might act synergistically with other antihypertensives in drug-resistant patients, particularly those with co-morbid chronic kidney disease or heart failure. By contrast, we note that gestational vasodilatation itself often normalizes the modestly elevated blood pressures most often controlled by diuretics, leading many to suggest that these agents be discontinued in most hypertensive patients lacking another indication for their continued use. Indeed, one of several small trials of diuretics in chronically hypertensive gravidas found no greater need for addition of antihypertensive medications in patients withdrawn from thiazides than in those who continued to receive diuretics through pregnancy.[77]

Finally, there is an old literature, albeit anecdotal, reviewed in the first edition of this text, which called attention to cases of severe hyponatremia, hypokalemia, volume depletion, or thrombocytopenia in pregnant women treated with diuretics. Given that most of these are predictable consequences of diuretic pharmacology in susceptible patients, caution is still advised.

Calcium-Channel Antagonists

Each of the agents available in this drug class inhibits influx of Ca^{2+} via voltage-dependent, slow L-type calcium channels. Principal cardiovascular sites of action include smooth muscle cells of arterial resistance vessels, cardiac myocytes, and cells of the cardiac conduction system. These drugs appear to have little effect on the venous circulation, do not act via the endothelium and, despite the presence of target channels in the central nervous system, appear to exert their hemodynamic effects peripherally. Because contraction of vascular myocytes is a direct function of free cytosolic Ca^{2+}, the latter depending, in part, on influx via voltage-gated channels, these drugs act as direct vasodilators, antagonizing vasoconstriction, regardless of the original neural or humoral stimulus. Of the prototypical agents, verapamil is most selectively a negative chronotrope and negative inotrope, with significant but lesser effects as a direct vasodilator. Dihydropyridines, of which nifedipine is the prototype, are by contrast vasoselective agents with lesser effects on the heart; diltiazem is intermediate in its tissue selectivity.

Following calcium antagonist administration, blood pressure falls acutely due to decreased peripheral resistance; the response is blunted by reflex activation of the sympathetic nervous and renin–angiotensin systems. While these drugs often trigger dependent edema, probably due to local microvascular effects, they are only rarely the cause of compensatory renal volume retention. This observation suggests that the sustained antihypertensive effect of these agents includes substantial contributions by drug-induced natriuresis, the latter mediated primarily by effects on intrarenal hemodynamics including prominent afferent arteriolar vasodilatation, and perhaps at tubular sites as well. Indeed, increments in sodium excretion have paralleled decrements in blood pressure following administration of isradipine[82] and other dihydropyridines, an effect opposite that observed with other classes of direct vasodilators. This secondary natriuretic effect likely explains results from population studies which demonstrate that calcium antagonist monotherapy is most likely to prove efficacious in those groups that respond preferentially to diuretics rather than to β-blockers, viz., salt-sensitive hypertensive patients.[20] This is also consistent with some observations of limited benefit to combined therapy with calcium antagonists and diuretics; by contrast, either diuretics or calcium-channel blockers are effectively combined with ACE inhibitors.[1,17,83]

A controversial report noted an apparent excess incidence of myocardial infarction and death in hypertensive patients with concomitant coronary artery disease who received higher doses of short-acting dihydropyridine calcium antagonists.[84] Here the possibility that a precipitous fall in blood pressure, coupled with excessive sympathetic nervous system response, could lead to myocardial ischemia in patients with underlying coronary disease appears reasonable. Similarly convincing data fail to suggest such a risk for long-acting or sustained-release preparations of dihydropyridine calcium antagonists. While individual trials might be interpreted by some to suggest such risks, others appear to support the safety of these agents, even in high-risk populations.[85]

Pregnancy

Calcium-channel blockers have been used to treat both chronic hypertension and women with preeclampsia to prolong gestation, as well as women presenting late in gestation with accelerating hypertension due to preeclampsia. We could identify no adequate studies of calcium-channel blockers early in pregnancy, save for two multicenter retrospective studies whose results argue against significant teratogenic effect.[86,87] Most investigators have focused on use of nifedipine,[88,89] with sporadic reports on nicardipine,[49,90] isradipine,[91] nimodipine,[92] and verapamil.[93] Some have advocated oral or sublingual (immediate-release) nifedipine as a preferred agent in severely hypertensive preeclamptics;[94] however, there is no difference in nifedipine pharmacokinetics or time–effect curves by these two routes of administration.[95] Indeed, the results of small comparative studies suggest that this drug controls blood pressure in such patients with efficacy similar to that of parenteral hydralazine, with a similar spectrum of maternal adverse effects (mostly ascribed to vasodilatation).[60,61,96–99] A more recent report focused on use of intravenous nicardipine (3–9 mg/h) in this setting with good result.[90]

When treating severe hypertension, the data appear to conflict regarding the influence of calcium-channel blockers on uteroplacental blood flow and fetal well-being, especially in comparison with other agents. In two studies,[96,100] one of which involved maternal hemodynamic monitoring via a pulmonary artery catheter, the authors claimed less fetal distress with nifedipine than with hydralazine. However, in one of these very limited trials, this observation may have been due to a greater hypotensive effect of the hydralazine doses used; several other studies have failed to discern such differences, finding no significant fetal benefit of nifedipine or alternative agents. Of interest too is an animal study where chronically instrumented pregnant sheep demonstrated fetal hypoxia and acidosis following high-dose maternal nifedipine infusion, unexplained by changes in maternal or uteroplacental hemodynamics;[101] we are unaware of any corroborative clinical data for this worrisome observation.

Calcium-channel antagonists also relax uterine smooth muscle and have been used as tocolytics in preterm labor, but there appear to be no data to suggest that their use as antihypertensives compromises the progression of labor or leads to ineffective hemostasis following delivery. There have been concerns regarding use of calcium antagonists for urgent blood pressure control in preeclampsia because of the need to simultaneously infuse magnesium sulfate to prevent eclamptic seizures. Magnesium itself can interfere with calcium-dependent contractile signaling in excitable tissue and in muscle; its combined use with calcium antagonists could conceivably lead to increased risk of neuromuscular blockade or circulatory collapse. Indeed, there are isolated reports of such complications,[102,103] while others argue against such adverse outcomes with routine therapy, this more reassuring conclusion echoed in a retrospective review comparing 162 women who received both nifedipine and magnesium with 215 who received magnesium without calcium-channel blocking drugs.[104] In the Magpie Trial, in which magnesium was used to prevent preeclampsia (n = 10,110), 30% of women received nifedipine after trial entry and no associated adverse events were reported.[105] For the few women who did have hypotension there was no association with the combination of nifedipine and magnesium sulfate. Finally, in accord with observations in nonpregnant patients with proteinuric renal disease, dihydropyridine calcium channel antagonists may increase proteinuria. This would favor the clinical diagnosis of preeclampsia and possibly create misdiagnosis; indeed, this is supported by results in a recent Cochrane review (see below).

In summary, despite apparently widespread use, these agents are under-studied other than in late pregnancy. There are numerous, albeit size-limited, studies suggesting that they are relatively safe and effective antihypertensive agents in chronically hypertensive or preeclamptic pregnant women and only more mechanistically defined diagnosis would allow us to determine whether they increase preeclampsia. In spite of these reassuring reports, the limited literature leads us to consider them preferred second-line agents.

Direct Vasodilators

Commonly prescribed direct vasodilators exert their antihypertensive effects by one of three pharmacologic mechanisms. Diazoxide, like the active metabolite of minoxidil, opens ATP-sensitive K^+ channels, hyperpolarizing and relaxing arteriolar smooth muscle, with little effect on capacitance vessels. Sodium nitroprusside is a direct nitric oxide (NO) donor; the spontaneously released NO nonselectively relaxes both arteriolar and venular vascular smooth muscle, principally due to activation of soluble guanylyl cyclase and cyclic guanosine 3′,5′-monophosphate (cGMP) accumulation. Organic nitrates act similarly, though with greater effect on capacitance vessels than arterioles. By contrast, hydralazine and related phthalazine vasodilators selectively relax arteriolar smooth muscle by an as yet uncertain mechanism. These agents all have their greatest utility in the rapid control of severe hypertension, or as third-line agents for multidrug control of refractory hypertension.

Hydralazine-induced vasodilatation leads to striking reflex activation of the sympathetic nervous system, increments in plasma renin, and compensatory fluid retention. The sympathetic activation, combined with hydralazine's lack of effect on epicardial coronary arteries, can precipitate myocardial ischemia or infarction in hypertensive patients with coronary atherosclerosis. These same compensatory mechanisms rapidly attenuate the hypotensive effect, requiring combination therapy with sympatholytic agents and diuretics for long-term blood pressure control. Hydralazine is effective orally or intramuscularly; parenteral administration is used for rapid control of severe hypertension. Adverse effects are mainly those due to excessive vasodilatation or sympathetic activation, such as headache, nausea, flushing, or palpitations. Chronic use can lead to a pyridoxine-responsive polyneuropathy or to a variety of immunologic reactions including a drug-induced lupus syndrome.

Nitroprusside, administered only by continuous intravenous infusion, is easily titrated because it has a nearly immediate onset of action, whose duration of effect is only about 3 minutes. Cardiac output tends to fall during nitroprusside administration in patients with normal myocardial function due to decreased preload, while it increases in those with systolic heart failure due to afterload reduction. Reflex sympathetic activation is the rule. Nitroprusside metabolism releases cyanide, which can reach toxic levels with high infusion rates; cyanide is metabolized to thiocyanate, whose own toxicity usually occurs after 24 to 48 hours of nitroprusside infusion, unless its excretion is delayed due to renal insufficiency.

Use of diazoxide is limited to urgent parenteral control of severe hypertension. Minoxidil is the alternative agent for prolonged use due to intolerable adverse effects when diazoxide was used chronically. Since diazoxide leads to profound activation of the sympathetic and renin–angiotensin systems, its hypotensive effect is enhanced by use of sympatholytic agents and diuretics. An intravenous bolus of diazoxide lowers blood pressure within 30 seconds, with maximum effect in 5 minutes, allowing easy titration by repeated administration. Most side effects are due to excessive vasodilatation, as with hydralazine, though stimulation of ATP-sensitive K^+ channels in the pancreas can inhibit insulin secretion, leading to striking hyperglycemia.

PREGNANCY

Hydralazine is the direct vasodilator most often used in pregnant women, either as a second agent for hypertension uncontrolled following methyldopa (or a β-blocker) or, more commonly, as a parenteral agent for control of severe hypertension. Its use is justified not by pharmacologic selectivity but by long clinical experience with tolerable side effects. Several studies of hydralazine or related compounds in preeclamptic women monitored with pulmonary artery catheters have highlighted concerns regarding its safety, including the occurrence of precipitous decreases in cardiac output and blood pressure with oliguria; these complications are predictable with knowledge of its pharmacology in the setting of the primary vasoconstriction and relative volume contraction which is characteristic in these patients.[60,100,106] Effects on uteroplacental blood flow are unclear, likely due to variation in the degree of reflex sympathetic activation, though fetal distress may result with precipitous control of maternal pressure.[57,96,107] Neonatal thrombocytopenia has been reported following intrauterine hydralazine exposure.[108] Many investigators have suggested that urgent blood pressure control might be better achieved with less fetal risk by use of other agents such as labetalol or nifedipine; as will be discussed in greater detail below, objective outcome data currently fail to support the choice of one agent over another in this setting.[60,61]

Diazoxide, even when dosed carefully, can lead to excessive hypotension, arrested labor, and hyperglycemia[109] and interpretation of some animal studies suggests that this may compromise uterine blood flow.[110] It is possible, however, that slow infusion or small dose increments may avoid these complications. Indeed, in a well-conducted recent randomized trial comparing 15 mg "miniboluses" of diazoxide with hydralazine, there was quicker blood pressure control observed with diazoxide with no excess of maternal, fetal, or neonatal distress.[111] Finally, there are concerns from animal studies that repeated administration could result in pancreatic islet cell degeneration,[112] an important observation that does not seem to have been addressed by follow-up studies in humans. Currently, this drug seems to have fallen into relative disuse in both pregnant and nonpregnant populations.

Nitroprusside has only been used sporadically in pregnancy, usually in life-threatening refractory hypertension.[113] Adverse effects include those due to excessive vasodilatation, apparently including cardioneurogenic, i.e., paradoxically bradycardic, syncope in volume-depleted preeclamptic women.[114] The risk of fetal cyanide intoxication remains unknown. Given the long experience with hydralazine and alternative utility of calcium-channel blockers or parenteral labetalol, both nitroprusside and diazoxide must be considered agents of last resort.

Modulators of the Renin–Angiotensin–Aldosterone Axis

Angiotensin-converting enzyme inhibitors block ACE, which is a kininase II. This enzyme is responsible for conversion of angiotensin I – cleaved from angiotensinogen by renin – to angiotensin II (AII). Inhibition of ACE, localized to the endothelium and most abundant in the pulmonary microvasculature, predictably leads to decrements in AII and also in levels of aldosterone, whose synthesis in the zona glomerulosa of the adrenal cortex is AII-dependent. These agents lower blood pressure primarily by blocking AII-induced vasoconstriction; their efficacy in patients with low circulating renin and AII serves to reveal contributions of a parallel "tissue" renin–angiotensin axis distributed in the arterial wall.[115] In some hypertensive patients, accumulation of bradykinin, an endothelium-dependent vasodilator, and also an endogenous substrate for ACE, may contribute to the antihypertensive effect. Indeed, adverse effects, including the common ACE inhibitor-induced dry cough, appear to be due, at least in part, to bradykinin.[116] Long-term blood pressure control may be favored by an apparent natriuretic effect of these drugs, partially due to decrements in aldosterone synthesis, but mainly to a resetting of the normal relationship between salt excretion and renal perfusion pressure.[5] It is unclear what long-term benefits may be derived from nonhemodynamic effects of ACE inhibition, as AII is a potent growth factor for a variety of cardiovascular cells.

Antagonists acting at the AT-l subtype of AII receptors (ARBs) exhibit antihypertensive pharmacology virtually identical to that of ACE inhibitors, save for hypersensitivity reactions and those effects due to bradykinin accumulation. Indeed, the fidelity with which these agents recapitulate the beneficial effects of ACE inhibition casts doubt on significant contributions of either bradykinin accumulation or AT-2 receptor activation to the blood pressure-lowering effect of ACE inhibitors in most patients. An emerging literature with relevance to pregnancy has noted an alternative pathway, depending on the ability of ACE-2 to cleave angiotensin I to the Ang 1–7 fragment, which apparently exerts vasodilator activity via the *mas* receptor; its levels appear to be increased in normal pregnancy and decreased in preeclampsia.[117,118] Effective drugs modulating this system are not yet available in clinical practice. Aliskiren, a direct renin inhibitor in use since 2007, has been the subject of much study focused on achieving more complete blockade of the renin–angiotensin system with recent disappointments due to unanticipated harm when used in combination with other agents in high-risk patient populations;[119] given existing concerns regarding ACE inhibitors, we do not expect these drugs to be used (intentionally) during pregnancy.

As predicted, ACE inhibitors and ARBs are most efficacious in patients with renin-dependent hypertension, such as those with unilateral renal artery stenosis. Their benefit in diabetic nephropathy, their ability to decrease proteinuria in patients with glomerular diseases, and their apparent "renal protective" effect in other (principally proteinuric) diseases leading to progressive renal insufficiency derive from their effects in the renal microvasculature. Since AII acts selectively to constrict the efferent arteriole (its effect in the afferent arteriole being antagonized by locally synthesized NO),[120] its absence, or antagonism, lowers intraglomerular pressure out of proportion to any effect on systemic arterial pressure. It is important to note that in those cases where preservation of glomerular filtration rate in the face of decreased renal perfusion depends on selective efferent arteriolar vasoconstriction by AII – examples include renal artery stenosis, decreased effective arterial volume, and congestive heart failure – ACE inhibition may lead not only to exaggerated hypotension, but also to acute renal failure. In addition to having "compelling indications" for the treatment of hypertension in patients with diabetes and proteinuric renal disease, ACE inhibitors and ARBs are similarly indicated following myocardial infarction and in all stages of congestive heart failure.[1]

Finally, there has been a recent resurgence in the use of aldosterone antagonists such as spironolactone and eplerenone, with clinical trial data demonstrating their specific benefit in the postmyocardial infarction, congestive heart failure, and refractory hypertension populations.[11,14,121] The aldosterone antagonists are generally avoided in pregnancy due to concerns regarding possible antiandrogenic effects in the fetus.

PREGNANCY

All circulating elements of the renin–angiotensin system are increased in normal human pregnancy.[122] Preeclampsia, where AII levels are lower than in normal gestation, is characterized by the appearance of autoantibodies which activate the AT-1 receptor, which, itself, appears to be upregulated.[123] Thus, both ARBs and ACE inhibitors might have seemed attractive antihypertensive agents for treating hypertensive pregnant women, indeed, a conclusion apparently supported by one small case series.[124] Concerns regarding use of these drugs might have been anticipated, however, from excess fetal wastage in animal studies. Indeed, use of drugs from these classes in human pregnancy has been associated with frequent reports of a specific fetopathy – renal dysgenesis and calvarial hypoplasia; oligohydramnios – likely a result of fetal oliguria; fetal-growth restriction, and neonatal anuric renal failure leading to death.[125,126] For these reasons, these drugs are contraindicated following midpregnancy.

Then, in 2006, a study which linked (Tennessee) Medicaid prescription records with maternal and infant medical and vital records identified a 2.7-fold increase in congenital malformations, due entirely to increases in cardiovascular and central nervous system malformations[127] following ACE-inhibitor first-trimester exposures. These observations have led to an avoidance of their use in pregnancy as well as in women of childbearing potential. Subsequently, several much larger studies have failed to confirm concerns regarding first-trimester exposure,[128–131] some suggesting that any increased risks of malformation may be related to maternal hypertension, rather than its treatment. Likewise, secondary analysis of 208 pregnancies which occurred during a randomized clinical trial of candesartan in type 1 diabetics suggested no excess risk due to first-trimester ARB exposure.[132] So, while it seems prudent to discontinue their use when pregnancy is first confirmed, and many may argue against their use in women with prepregnancy essential hypertension, due to the high rates of unintended pregnancy and of delayed or limited antenatal care, these risks must be weighed against the significant potential benefits of their use in women of childbearing potential with compelling indications, such as those with underlying diabetes or proteinuric renal disease, and in patients with difficult to control hypertension, which might recur following drug withdrawal.

DRUG USE WHILE BREASTFEEDING

No well-designed studies are available that assess neonatal effects of maternally administered antihypertensive drugs delivered via breast milk. The pharmacokinetic principles which govern drug distribution to milk and subsequently infant exposure are well established.[133,134] Milk, secreted by alveolar cells, is a suspension of fat globules in a protein-containing aqueous solution whose pH is lower than that of maternal plasma. Factors that favor drug passage into milk are a small maternal volume of distribution, low plasma protein binding, high lipid solubility, and lack of charge at physiologic pH; several drugs are also substrates for specific transporters as well. Even when drugs are ingested by nursing infants, exposure depends on volume ingested, intervals between drug administration and nursing, oral bioavailability, and the capacity of the infant to clear the drug.

Transfer of antihypertensive drugs into milk has been the subject of a systematic review.[135] Neonatal exposure to methyldopa via nursing is likely low and it is generally considered safe. Atenolol and metoprolol are concentrated in breast milk, possibly to levels that could affect the infant; by contrast, exposure to labetalol and propranolol appears low. While milk concentrations of diuretics are limited, these agents may decrease milk production significantly. There are brief reports of calcium-channel blocker transfer into breast milk, apparently without adverse effects. While ACE inhibitors and ARBs are only of potential concern in very preterm neonates, there are reassuring data that captopril levels in milk appear extraordinarily low, even when studied at steady state.[136] Single-dose studies of quinapril and enalapril are also reassuring.[137,138]

EVIDENCE FROM RANDOMIZED TRIALS

So far, this chapter has approached the treatment of hypertension in pregnancy in terms of disease pathophysiology that requires reversal, focusing on discussions of the pharmacology of the agents administered, their potential to benefit pregnant hypertensive women, and their administration. However, another approach, in a sense the final analysis in terms of success and failure, is appropriately designed clinical trials. This final section reviews clinical trials to treat hypertension in pregnancy with special focus on their quality and homogeneity, and we note, up front, "We have a long way to go!" First we present a review of the appropriate approach to grouping and analyzing randomized trials.

Randomized controlled trials (RCTs) are the most valid means of evaluating medical or surgical treatments, screening or preventive maneuvers, as well as health, nutritional, social, and educational interventions.[139–141] Evaluation of different forms of care outside the context of a proper randomization strategy is prone to major biases that can provide an incorrect assessment of effectiveness. Randomization is probably the most certain way of eliminating selection bias, and control for any known or unknown factors that may influence outcomes other than the treatments.

The exponential expansion in the literature, however, makes it difficult, and occasionally virtually impossible, to review all trials addressing a clinical question. This is especially the case when a large literature exists concerning a single question, a dilemma encountered more and more in regard to antihypertensive drug trials. A powerful way to assess these data is by evidence-based decisions through *systematic reviews*, among the least biased approaches as of 2014. These reviews, when performed properly, should exhaustively review the literature on a clearly defined question. They should also use systematic and explicit methodology that identifies, selects, and critically evaluates all relevant studies, analyzing the data emerging from them. Statistical methods such as meta-analysis may or may not be used.[142] Pooled estimates, because of larger numbers, will have more statistical power than could be obtained using the individual trials. This is the context in which we now review the conclusions of those who have performed systematic reviews to determine the effectiveness of various antihypertensive treatment approaches to the management of chronic hypertension and preeclampsia. We focus first on treatment of mild to moderate (usually chronic) hypertension, first on the benefits of treatment per se, then on comparison of specific agents. Afterwards, we focus similarly on treatment of more severe hypertension, usually closer to term and due principally to preeclampsia.

Evidence for Antihypertensive Treatment in Mild to Moderate Hypertension

Several reviews of randomized trials have been conducted to assess the potential benefits and hazards of antihypertensive drugs for the treatment of mild to moderate hypertension during pregnancy and the puerperium.[143–148] A Cochrane review, updated in 2013, reports outcomes on 4723 women recruited in 49 small trials published largely in the 1980s and 1990s, with only seven publications located after 2000.

The antihypertensive drugs compared were α-agonists, β-blockers, calcium-channel blockers, vasodilators, the serotonin S receptor antagonist ketanserin, and glyceryl trinitrate. Obvious, at first glance, was the considerable heterogeneity especially in regard to criteria for treatment and the regimens utilized. For example there was a wide variation in gestational age at trial entry, the definition of mild to moderate hypertension, proteinuria at recruitment, and criteria for blood pressure measurement, viz., Korotkoff phase IV or phase V sound, across trials. Small-for-gestational-age was also defined in a variety of ways in the 29 trials reporting this outcome.

Drugs were compared with placebo or no antihypertensive drug in 29 trials, authors concluding that the main benefit of therapy in these pregnant women with mild to moderate hypertension was to decrease the rate of progression to severe hypertension, the outcome most expected from drugs known to lower blood pressure (Table 19.2). Other anticipated findings were a decreased need for additional therapy – including a second agent – and more side effects when treated women were compared with their controls. Similar findings were reported in another Cochrane review that focused on β-blockers.

There were no apparent reductions in outcomes such as admission to hospital, though this outcome was only reported in four trials (455 women; RR 0.87, 95% CI 0.73–1.03), placental abruption (11 trials, 1433 women; RR 1.41, 95% CI 0.65–3.07), or stroke (no data). Nor was there evidence to support a major impact on the risk of progression to preeclampsia by better blood pressure control per se (23 trials, 2851 women; RR 0.93, 95% CI 0.80–1.08). A small but clinically important reduction in the risk of fetal or neonatal death was observed but the effect did not attain statistical significance.

Clinicians worry that antihypertensive drugs may increase the risk of small-for-gestational-age neonates, mediated perhaps by a reduction in placental perfusion

TABLE 19.2 Treatment of Mild–Moderate Hypertension During Pregnancy

	Any Antihypertensive Drug Versus None		
Outcome	**No. of Studies**	**No. of Subjects**	**RR [95% CI]**
Severe hypertension	20	2558	0.49 [0.40–0.60]
Proteinuria/preeclampsia	23	2851	0.93 [0.80–1.08]
Eclampsia	5	578	0.34 [0.01–8.15]
Small for gestational age*	9	904	1.38 [0.99–1.92]
Fetal or neonatal death	27	3230	0.71 [0.49–1.02]
Preterm birth (<37 weeks)	15	2141	0.96 [0.85–1.10]

*Only β-blockers trials.
Reference: Abalos et al.[143]

following antihypertensive treatment, or by a direct action of the drug on fetal growth. This construct has been most strongly supported by a recently updated meta-regression analysis relating decreased birth weight to blood pressure lowering.[149,150] This hypothesis served as a key basis for the Control of Hypertension in Pregnancy Study (CHIPS) pilot trial,[151] which compared tight with less tight blood pressure control, though it apparently sacrificed the ability to detect drug-specific adverse effects.

When analyzing data from all trials together, there is no overall effect of using antihypertensive agents on the relative risk of a small-for-gestational-age baby (20 trials, 2586 babies; RR 0.97, 95% CI 0.80–1.17). When the analysis focuses on trials of β-blockers alone, an increased risk appears likely (nine trials, 904 women; RR 1.38, 95% CI 0.99–1.92). One could argue therefore that the observed association with fetal-growth restriction is related to β-blockers, rather than to reducing blood pressure, per se. This β-blocker subgroup, however, was itself heterogeneous, including six trials (633 women) which evaluated labetalol versus no treatment/placebo, two trials of atenolol versus placebo (153 women), and another two comparing propranolol or pindolol with no treatment (151 women). This effect, on the other hand, is not seen when β-blockers are compared with methyldopa (six trials, 577 women; RR 0.85, 95% CI 0.55–1.32). Concerns about β-blockers and fetal-growth restriction are partly due to one small trial,[42] which may have exaggerated the real effect. Although this trial was of poor quality, other potential explanations include a specific effect of atenolol, its hemodynamic effects[70] or, as noted previously, the fact that drug treatment was commenced quite early in gestation.

As a combined α/β-blocker, labetalol lowers blood pressure without reducing cardiac output, perhaps leading to less uteroplacental hypoperfusion and restricted fetal growth than other β-blockers. Also, labetalol may reduce cerebral perfusion pressure without reducing cerebral blood flow, an observation which led to the suggestion it may reduce the risk of eclampsia[152] (see Chapter 12, p. 263).

As discussed above, we and others consider α-methyldopa a preferred drug for treating mild to moderate hypertension, while acknowledging that appropriate trial evidence is limited.[80,81] Our recommendation is based on a long history of use supporting drug safety and tolerability, but of primary importance was one unique study that included a follow-up of 190 children of the original 230 born in the trial who were examined at 18 months and at 7½ years of life.[34] While underpowered, this single study, coupled with subsequent experience, makes methyldopa the reference standard against which all other agents should be judged. We note, however, the key caveat that it appears to have been evaluated in only eight randomized controlled trials, including a total of but 936 women. One study compared the drug with a placebo, and in seven to no drug treatment. Twenty studies compared methyldopa to a second antihypertensive medication, which in 16 of the trials was β-blockers (1507 women). In the remaining four, methyldopa was compared to nifedipine in two (49 women), to nimodipine in one (111 women) and to ketanserin in the other (20 women).

Methyldopa, like all other antihypertensive agents, proved effective in avoiding episodes of severe hypertension (two trials, 310 women, RR 0.32; 95% CI 0.17–0.58) when compared with placebo or no treatment. However, β-blockers appeared superior to methyldopa when both were used in individual trials (nine trials, 592 women, RR 0.74; 95% CI 0.59–0.93; RD–0.08 (–0.13–0.02); NNT 13 (8–50). Also patients appeared to tolerate β-blockers better than methyldopa (RR 0.69; 95% CI 0.52–0.91). We note that "tolerability" was reported in but half the trials, thus there is a distinct potential for bias – for example, drowsiness, a common side effect of methyldopa, might not be called a major disadvantage if the woman was hospitalized, but would be cited as problematic if she was at home.

Calcium-channel blockers were compared with no drug treatment or placebo in five trials (isradipine one, verapamil one, and nifedipine three) totaling 900 women. Of interest, rates of progression to preeclampsia were marginally, but significantly, higher with calcium-channel blockers than with placebo or no treatment (four trials, 725 women, RR 1.40; 95% CI 1.06–1.86, $p = 0.02$) perhaps due to prolongation of pregnancy until signs of superimposed preeclampsia became evident or to the known ability of calcium-channel blockers to dilate renal afferent arterioles, thus increasing the detection of proteinuria. In this respect there was no difference in the rates of preterm births (four trials, 742 babies, RR 1.03; 95% CI 0.88–1.21).

Calcium-channel blockers were compared with methyldopa in three small trials, two of nifedipine (49 women) and one of nimodipine (111 women), with ketanserin in one (20 women), and with glyceryl trinitrate in one (36 women). In one study (100 women), nicardipine was compared with metoprolol. There were no statistical differences for any of the outcomes reported in these three groups.

As noted briefly above, the most important message in all these reviews is the relatively poor quality of most studies, underscoring a critical need for large prospective trials specifically focused on the maternal and fetal effects of differing levels of targeted (and achieved) blood pressure control to decide whether mild hypertension during pregnancy requires treatment. Only four eligible trials (recruiting 511 women) appear to have been published since a Cochrane review published in 2001 and in which one of us (EA) participated.[141] Importantly, 25 of 49 trials included were published before 1990, i.e., more than 25 years ago! Evidence from these trials accounts for half of the total number of women recruited. It goes without saying that since these studies were done maternal and perinatal survival, availability of resources, technologies for diagnosis and treatment of complications, as well as indications for termination of pregnancy for maternal or perinatal conditions have changed substantially.

Evidence for Antihypertensive Treatment in Severe Hypertension

A Cochrane review to assess the ideal drug for hypertensive emergencies during gestation – most related to severe or rapidly accelerating hypertension in the puerperium – was updated in 2013 and included an analysis of 35 randomized studies (3573 women).[61] Here we note that the definition of hypertensive emergency in pregnancy differs from that in nonpregnant women and there is general agreement that blood pressure be controlled to <170/110 mm Hg (and in most cases to <160/100 mm Hg) in order to avoid cerebrovascular catastrophe. Most studies were limited in scope, with 22 consisting of data from 50 women or fewer. Additionally, one large study (1750 women) was actually designed to compare nimodipine with magnesium sulfate to prevent eclamptic convulsions in women with severe preeclampsia, i.e., prevention of eclampsia was its primary outcome.[92] In other trials, hydralazine, labetalol, atenolol, methyldopa, diazoxide, prostacyclin, ketanserin, urapidil, prazosin, isosorbide and other calcium-channel blockers (nifedipine, nimodipine, nicardipine, and isradipine) were also evaluated. Hydralazine was the most commonly designated reference drug, being compared with labetalol, calcium-channel blockers, prostacyclin, ketanserin, or urapidil. Most drugs were administered intravenously or intramuscularly, except nifedipine, nimodipine, isosorbide and prazosin, which were administered orally. Dosage varied considerably, in both amount and duration of therapy.

It is noteworthy that 13 of 15 trials assessing oral or sublingual preparations were conducted in low- and middle-income countries, for example, the multinational Nimodipine SG trial. Nifedipine was compared with intravenous hydralazine in five trials (273 women), with labetalol in two (70 women), with intravenous chlorpromazine in one (60 women), and with oral prazosin in another (150 women). In four (263 women) the sublingual route was used and in the remaining three (220 women) the drug was administered orally, but as noted above there are no discernible pharmacokinetic or pharmacodynamic differences between these two routes of administration for nifedipine.

Intravenous magnesium sulfate was compared with sublingual isosorbide spray in one trial (36 women) and with oral nimodipine in the multicenter Nimodipine SG trial (1750 women). More and better evaluations of the efficacy and safety of oral or sublingual antihypertensive medications could have important implications in low-resource settings, especially in developing parts of the world where hospital-based treatment may not be readily available.

The trials reviewed here all demonstrated that antihypertensive drugs reduce blood pressure, but there was insufficient evidence regarding other key outcomes to conclude that one medication is preferable to another. Diazoxide at high doses (though not, perhaps with minibolus regimens[111]) was associated with a greater risk of hypotension and cesarean section than labetalol and the serotonin receptor antagonist ketanserin, and led to more persistent hypertension than hydralazine. In one analysis, the authors suggested avoiding the use of intramuscular or intravenous hydralazine,[60] but this review included quasi-random studies and pooled all those comparing hydralazine with various drugs, regardless of whether they were likely to be better or worse. As trials were small, the confidence intervals for outcomes such as perinatal mortality are all very wide, so the differences between drugs were not statistically significant (Table 19.3). Few trials reported side effects.

CONCLUSION

Antihypertensive agents are used in pregnancy either for the urgent control of severe hypertension (Table 19.4) or for control of chronic hypertension (Table 19.5), realizing that this latter indication may include women with a variety of hypertensive disorders that will include early preeclampsia. Tables 19.4 and 19.5 summarize clinical data on the use of drugs discussed in this chapter, including FDA risk classification, usual doses, and special concerns.

TABLE 19.3 Treatment of Severe Hypertension During Pregnancy

Comparison	Eclampsia			Fetal or Neonatal Death		
	No. of Studies	No. of Subjects	RR (95% CI)	No. of Studies	No. of Subjects	RR [95% CI]
Labetalol vs. hydralazine	2	220	Not estimable	4	274	0.75 [0.17–3.21]
Calcium antagonists vs. hydralazine	–	–	–	4	161	1.36 [0.42–4.41]
Ketanserin vs. hydralazine	2	64	0.60 [0.08–4.24]	2	116	0.27 [0.05–1.64]
Prostacyclin vs. hydralazine	–	–	–	1	47	1.14 [0.08–17.11]
Urapidil vs. hydralazine	1	26	Not estimable	3	101	0.54 [0.10–3.03]
Urapidil vs. Nicardipine	–	–	–	–	–	–
Labetalol vs. calcium antagonists	2	70	0.72[0.05–10.26]	–	–	–
Labetalol vs. methyldopa	–	–	–	1	72	4.49 [0.22–90.30]
Labetalol vs. diazoxide	–	–	–	1	90	0.14 [0.01–2.69]
Nitrates vs. magnesium sulfate	1	36	Not estimable	–	–	–
Nimodipine vs. magnesium sulfate	2	1683	1.03 [0.07–16.03]	–	–	–
Nifedipine vs. chlorpromazine	1	55	2.52 [0.11–59.18]	–	–	–
Nifedipine vs. prazosin	1	145	Not estimable	1	149	0.46 [0.18–1.13]
Hydralazine vs. diazoxide	–	–	–	1	101	7.42 [0.39–140.06]
Methyldopa vs. atenolol	–	–	–	1	60	1.00 [0.07–15.26]

Reference: Duley et. al.[61]

TABLE 19.4 Antihypertensives Used Commonly for Urgent BP Control

Drug (FDA risk[a])	Dose and Route	Concerns or Comments[b]
Hydralazine (C)	5 mg, iv or im, then 5–10 mg every 20–40 min; or constant infusion of 0.5–10 mg/h	Preferred by NHBPEP working group. Higher doses or more frequent administration often precipitate maternal or fetal distress, which appears more common than with other agents.
Labetalol (C)	20 mg iv, then 20–80 mg every 20–30 min, up to maximum of 300 mg; or constant infusion of 1–2 mg/min	Probably less risk of tachycardia and arrhythmia than with other vasodilators, likely less BP control than hydralazine. Increasingly preferred as first-line agent.
Nifedipine (C)	5–10 mg po, repeat in 30 min if needed, then 10–20 mg every 2–6 h	Parenteral calcium-channel blockers, e.g., nicardipine at 3–9 mg/h, appear a reasonable alternative, but fewer data are available.

[a]See Table 19.1 for FDA risk categories.
[b]Adverse effects for all agents, except as noted, may include headache, flushing, nausea, and tachycardia (primarily due to precipitous hypotension and reflex sympathetic activation).

Currently, there is little evidence to support the notion that blood pressure control in pregnant women with chronic hypertension will prevent the subsequent occurrence of preeclampsia itself, the cause of most adverse outcomes. Indeed, given the pathophysiological hypotheses which ascribe this disorder to events in early pregnancy, it would seem unreasonable to expect such a benefit. As well, there are no data to suggest that antihypertensive therapy will lessen the incidence of placental abruption. There have been few studies that have rigorously assessed the prevention of severe or accelerated hypertension, focusing on avoidance of the perceived need for hospitalization or urgent early delivery. What is clear is that effective agents can control hypertension during pregnancy with acceptable risks to mother and fetus. Desperately needed, however, are adequately powered prospective clinical trials that distinguish between women with essential hypertension, gestational hypertension, or preeclampsia, and stratify treatment both by severity of hypertension and by gestational age. Such trials should include systematic assessment of maternal and uteroplacental hemodynamics, fetal well-being, as well as long-term functional and childhood developmental assessment. Only through such research will we ever gain the confidence to rationally select from a broader armamentarium of antihypertensive agents to benefit pregnant women.

TABLE 19.5 Oral Antihypertensives Used Commonly in Pregnancy

Drug (FDA risk[a])	Dose	Concerns or Comments
Most commonly used first-line agents		
Methyldopa (B)	0.5–3.0 g/d in 2 or 3 divided doses	Preferred agent based on follow-up of newborn. Maternal side effects sometimes limit use.
Labetalol (C) or other β-receptor antagonists	200–2400 mg/d in 2 or 3 divided doses	Preferred alternative to methyldopa. β-blockers with intrinsic sympathomimetic activity. Actually preferred by many practitioners and some National guidelines as of 2013. May cause fetal growth restriction when started early in pregnancy.
Nifedipine (C)	30–120 mg/d of a slow-release preparation	Less experience with other calcium entry blockers, though all appear effective and well-tolerated.
Adjunctive agents		
Hydralazine (C)	50–300 mg/d in 2–4 divided doses	Few controlled trials, long experience; use only in combination with sympatholytic agent (e.g., methyldopa or a β-blocker) to prevent reflex tachycardia.
Thiazide diuretics (C)	Depends on specific agent	Large experience demonstrates safety, mostly from studies in normotensive gravidas. Often key to BP control, usually in combination therapy, in women with CKD or in allograft recipients.
Contraindicated		
ACE inhibitors and AT1 receptor antagonists (D)		Use after first trimester can lead to fetopathy, oligohydramnios, growth retardation, and neonatal anuric renal failure.

[a]See Table 19.1 for description of FDA risk categories; no antihypertensive drug has been proven safe for use during the first trimester (i.e., FDA Category A).

References

1. Chobanian AV, Bakris GL, Black HR, et al. National high blood pressure education program coordinating committee. The seventh report of the Joint National Committee on Prevention, Detection, Evaluation, and Treatment of High Blood Pressure: the JNC 7 report. *JAMA*. 2003;289:2560–2572 [also reprinted as NIH publication 04-5230 and available at http://www.nhlbi.nih.gov/guidelines/hypertension/jnc7full.htm].
2. Black HR, Yi JY. A new classification scheme for hypertension based on relative and absolute risk with implications for treatment and reimbursement. *Hypertension*. 1996;28:719–724.
3. Hansson L, Zanchetti A, Carruthers SG. Effects of intensive blood-pressure lowering and low-dose aspirin in patients with hypertension: principal results of the Hypertension Optimal Treatment (HOT) randomised trial. *Lancet*. 1998;351:1755–1762.
4. Sibai BM, Abdella TN, Anderson GD. Pregnancy outcome in 211 patients with mild chronic hypertension. *Obstet Gynecol*. 1983;61:571–576.
5. Guyton AC. Blood pressure control – special role of the kidneys and body fluids. *Science*. 1991;252:1813–1816.
6. Lewis EJ, Hunsicker LG, Bain RP, Rohde RD. The effect of angiotensin converting enzyme inhibition on diabetic nephropathy. *New Engl J Med*. 1993;329:1456–1462.
7. Brenner BM, Cooper ME, de Zeeuw D. Effects of losartan on renal and cardiovascular outcomes in patients with type 2 diabetes and nephropathy. *New Engl J Med*. 2001;345:861–869.
8. Giatras I, Lau J, Levey AS. Effect of angiotensin converting enzyme inhibition on the progression of nondiabetic renal disease; a meta-analysis of randomized trials. *Ann Intern Med*. 1997;127:337–345.
9. Yusuf S, Peto R, Lewis J. Beta blockade during and after myocardial infarction: an overview of the randomized trials. *Prog Cardiovasc Dis*. 1985;27:335–371.
10. Pfeffer MA, Braunwald E, Moye LA. Effect of captopril on mortality and morbidity in patients with left ventricular dysfunction after myocardial infarction: results of the survival and ventricular enlargement trials. *New Engl J Med*. 1992;327:669–677.
11. Pitt B, Remme W, Zannad F. Eplerenone, a selective aldosterone blocker, in patients with left ventricular dysfunction after myocardial infarction. *New Engl J Med*. 2003;348:1309–1321.
12. Garg R, Yusuf S. Overview of randomized trials of angiotensin converting enzyme inhibitors on mortality and morbidity in patients with heart failure. *JAMA*. 1995;273:1450–1456.
13. Pitt B, Segal R, Martinez FA. Randomised trial of losartan versus captopril in patients over 65 with heart failure (Evaluation of Losartan in the Elderly Study, ELITE). *Lancet*. 1997;349:375–380.
14. Pitt B, Zannad F, Remme WJ. The effect of spironolactone on morbidity and mortality in patients with severe heart failure. Randomized Aldactone Evaluation Study Investigators. *New Engl J Med*. 1999;341:709–717.
15. Prevention of stroke by antihypertensive drug treatment in older persons with isolated systolic hypertension: final results of the Systolic Hypertension in the Elderly Program (SHEP). *JAMA*. 1991;265:3255–3264.
16. The Antihypertensive and Lipid-Lowering Treatment to Prevent Heart Attack Trial. Major outcomes in high-risk hypertensive patients randomized to angiotensin-converting enzyme inhibitor or calcium channel blocker vs diuretic: the Antihypertensive and Lipid-Lowering Treatment to Prevent Heart Attack Trial (ALLHAT). *JAMA*. 2002;288:2981–2997.
17. Jamerson K, Weber MA, Bakris GL, et al. ACCOMPLISH Trial Investigators. Benazepril plus amlodipine or hydrochlorothiazide for hypertension in high-risk patients. *New Engl J Med*. 2008;359:2417–2428.

18. Materson BJ, Reda DJ, Cushman WC. Single drug therapy for hypertension in men: a comparison of six antihypertensive agents with placebo. *New Engl J Med*. 1993;328:914–921.
19. Weinberger MH, Stegner JE, Fineberg NS. A comparison of two tests for the assessment of blood pressure responses to sodium. *Am J Hypertens*. 1993;6:179–184.
20. Kiowski W, Buhler FR, Fadayomi MO. Age, race, blood pressure and renin; predictors for antihypertensive treatment with calcium antagonists. *Am J Cardiol*. 1985;56:81H–85H.
21. Merkatz RB, Temple R, Subel S. Women in clinical trials of new drugs. A change in Food and Drug Administration policy. *New Engl J Med*. 1993;329:292–296.
22. http://www.fda.gov/Drugs/DevelopmentApprovalProcess/DevelopmentResources/Labeling/ucm093307.htm.
23. Easterling TR, Benedetti TJ, Schmucker BC, Carlson KL. Antihypertensive therapy in pregnancy directed by noninvasive hemodynamic monitoring. *Am J Perinatol*. 1989;6:86–89.
24. Chaffin DG, Webb DG. Outcomes of pregnancies at risk for hypertensive complications managed using impedance cardiography. *Am J Perinatol*. 2009;26:717–721.
25. Elvan-Taşpinar A, Franx A, Bots ML, Bruinse HW, Koomans HA. Central hemodynamics of hypertensive disorders in pregnancy. *Am J Hypertens*. 2004;17:941–946.
26. Effects of treatment on morbidity in hypertension: results in patients with diastolic blood pressures averaging 115 through 129 mm Hg. *JAMA*. 1967;202:1028–1034.
27. Redman CWG, Beilin LJ, Bonnar J, Ounsted MK. Fetal outcome in trial of antihypertensive treatment in pregnancy. *Lancet*. 1976;2:753–756.
28. Sibai BM, Mabie WC, Shamsa F. A comparison of no medication versus methyldopa or labetalol in chronic hypertension during pregnancy. *Am J Obstet Gynecol*. 1990;162:960–967.
29. Leather HM, Humphreys DM, Paker PB, Chadd MA. A controlled trial of hypotensive agents in hypertension in pregnancy. *Lancet*. 1968;2:488–490.
30. Fidler I, Smith V, Fayers P, de Swiet M. Randomized controlled comparative study of methyldopa and oxprenolol in treatment of hypertension in pregnancy. *Brit Med J*. 1983;286:1927–1930.
31. Gallery EDM, Ross MR, Gyory AZ. Antihypertensive treatment in pregnancy: analysis of different responses to oxprenolol and methyldopa. *Brit Med J*. 1985;291:563–566.
32. Horvath JS, Phippard A, Korda A. Clonidine hydrochloride – a safe and effective antihypertensive agent in pregnancy. *Obstet Gynecol*. 1985;66:634–638.
33. Plouin PF, Breart G, Maillard F. Comparison of antihypertensive efficacy and perinatal safety of labetalol and methyldopa in the treatment of hypertension in pregnancy: a randomized controlled trial. *Brit J Obstet Gynaecol*. 1988;95:868–876.
34. Cockburn J, Moar VA, Ounsted M, Redman CW. Final report of study on hypertension during pregnancy: the effects of specific treatment on the growth and development of the children. *Lancet*. 1982;1:647–649.
35. Schobel HF, Fischer T, Heuszer K. Preeclampsia-a state of sympathetic overactivity. *New Engl J Med*. 1996;335:1480–1485.
36. Redman CW, Beilin LJ, Bonnar J. Treatment of hypertension in pregnancy with methyldopa: blood pressure control and side effects. *Brit J Obstet Gynaecol*. 1977;84:419–426.
37. Montan S, Anandakumar C, Arulkumaran S. Effects of methyldopa on uteroplacental and fetal hemodynamics in pregnancy-induced hypertension. *Am J Obstet Gynecol*. 1993;168:152–156.
38. Mutch LM, Moar VA, Ounsted MK, Redman CW. Hypertension during pregnancy, with and without specific hypotensive treatment. 1. Perinatal factors and neonatal morbidity. *Early Hum Dev*. 1977;1:47–57.
39. Mutch LM, Moar VA, Ounsted MK, Redman CW. Hypertension during pregnancy, with and without specific hypotensive treatment. II. The growth and development of the infant in the first year of life. *Early Hum Dev*. 1977;1:59–67.
40. Huisjes HI, Hadders-Algra M, Touwen BC. Is clonidine a behavioural teratogen in the human? *Early Hum Dev*. 1986;14:43–48.
41. Rothberger S, Carr D, Brateng D, Hebert M, Easterling TR. Pharmacodynamics of clonidine therapy in pregnancy: a heterogeneous maternal response impacts fetal growth. *Am J Hypertens*. 2010;23:1234–1240.
42. Dahlöf B, Devereux RB, Kjeldsen SE. Cardiovascular morbidity and mortality in the Losartan Intervention for Endpoint reduction in hypertension study (LIFE): a randomised trial against atenolol. *Lancet*. 2002;359:995–1003.
43. Schiffrin EL. Vascular remodeling and endothelial function in hypertensive patients: effects of antihypertensive therapy. *Scand Cardiovasc J*. 1998;47:15–21.
44. Rubin PC, Butters L, Clark DM. Placebo-controlled trial of atenolol in treatment of pregnancy-associated hypertension. *Lancet*. 1983;1:431–434.
45. Wichman K, Ryden G, Karlberg BE. A placebo controlled trial of metoprolol in the treatment of hypertension in pregnancy. *Scand J Clin Lab Med Invest*. 1984;44:90–94.
46. Hogstedt S, Lindeberg S, Axelsson O. A prospective controlled trial of metoprololhydralazine treatment in hypertension during pregnancy. *Acta Obstet Gynecol Scand*. 1985;64:505–510.
47. Butters L, Kennedy S, Rubin PC. Atenolol in essential hypertension during pregnancy. *Brit Med J*. 1990;301:587–589.
48. Plouin PF, Breart G, Llado J. A randomized comparison of early with conservative use of antihypertensive drugs in the management of pregnancy-induced hypertension. *Brit J Obstet Gynaecol*. 1990;97:134–141.
49. Jannet D, Carbonne B, Sebban E, Milliez J. Nicardipine versus metoprolol in the treatment of hypertension during pregnancy: a randomized comparative trial. *Obstet Gynecol*. 1994;84:354–359.
50. Paran E, Holzberg G, Mazor M. Beta-adrenergic blocking agents in the treatment of pregnancy-induced hypertension. *Int J Clin Pharmacol Ther*. 1995;33:119–123.
51. Reynolds B, Butters L, Evans J. First year of life after the use of atenolol in pregnancy associated hypertension. *Arch Dis Child*. 1984;59:1061–1063.
52. Lip GY, Beevers M, Churchill D. Effect of atenolol on birthweight. *Am J Cardiol*. 1997;79:1436–1438.
53. Lydakis C, Lip GYH, Beevers M, Beevers DG. Atenolol and fetal growth in pregnancies complicated by hypertension. *Am J Hypertens*. 1999;12:541–547.
54. Bayliss H, Churchill D, Beevers M, Beevers DG. Antihypertensive drugs in pregnancy and fetal growth: evidence for "pharmacological programming" in the first trimester? *Hypertens Pregnancy*. 2002;21:161–174.
55. Montan S, Ingemarsson I, Marsal K, Sjoberg NO. Randomised controlled trial of atenolol and pindolol in

human pregnancy: effects on fetal haemodynamics. *Brit Med J*. 1992;304:946–949.

56. Waterman EJ, Magee LA, Lim KI, Skoll A, Rurak D, von Dadelszen P. Do commonly used oral antihypertensives alter fetal or neonatal heart rate characteristics? A systematic review. *Hypertens Pregnancy*. 2004;23:155–169.
57. Mabie WC, Gonzalez AR, Sibai BM, Amon E. A comparative trial of labetalol and hydralazine in the acute management of severe hypertension complicating pregnancy. *Obstet Gynecol*. 1987;70:328–333.
58. Michael CA. Intravenous labetalol and intravenous diazoxide in severe hypertension complicating pregnancy. *Aust N Z J Obstet Gynaecol*. 1986;26:26–29.
59. Ashe RG, Moodley J, Richards AM, Philpott RH. Comparison of labetalol and dihydralazine in hypertensive emergencies in pregnancy. *S Afr Med J*. 1987;71:354–356.
60. Magee LA, Cham C, Waterman EJ, Ohlsson A, von Dadelszen P. Hydralazine for treatment of severe hypertension in pregnancy: meta-analysis. *BMJ*. 2003;327:955.
61. Duley L, Meher S, Jones L. Drugs for treatment of very high blood pressure during pregnancy. *Cochrane Database Syst Rev*. 2013;7:CD001449. pub3.
62. Pickles CJ, Symonds EM, Pipkin FB. The fetal outcome in a randomized double-blind controlled trial of labetalol versus placebo in pregnancy-induced hypertension. *Brit J Obstet Gynaecol*. 1989;96:38–43.
63. Redman CWG. A controlled trial of the treatment of hypertension in pregnancy: Labetalol compared with methyldopa. In: Riley A, Symonds EM, eds. *The Investigation and Management of Hypertension in Pregnancy*. Amsterdam: Excerpta Medica; 1982:111–122.
64. el-Qarmalawi AM, Morsy AH, al-Fadly A. Labetalol vs. methyldopa in the treatment of pregnancy-induced hypertension. *Int J Gynaecol Obstet*. 1995;49:125–130.
65. Sibai BM, Gonzalez AR, Mabie WC, Moretti M. A comparison of labetalol plus hospitalization versus hospitalization alone in the management of preeclampsia remote from term. *Obstet Gynecol*. 1987;70:323–327.
66. Munshi UK, Deorari AK, Paul VK, Singh M. Effects of maternal labetalol on the newborn infant. *Indian Pediatr*. 1992;29:1507–1512.
67. Easterling TR, Benedetti TJ, Schmucker BC, Millard SP. Maternal hemodynamics in normal and preeclamptic pregnancies: a longitudinal study. *Obstet Gynecol*. 1990;76:1061–1069.
68. Bosio PM, McKenna PJ, Conroy R, O'Herlihy C. Maternal central hemodynamics in hypertensive disorders of pregnancy. *Obstet Gynecol*. 1999;94:978–984.
69. Easterling TR, Brateng D, Schmucker B, Brown Z, Millard SP. Prevention of preeclampsia: a randomized trial of atenolol in hyperdynamic patients before onset of hypertension. *Obstet Gynecol*. 1999;93:725–733.
70. Easterling TR, Carr DB, Brateng D, Diederichs C, Schmucker B. Treatment of hypertension in pregnancy: effect of atenolol on maternal disease, preterm delivery, and fetal growth. *Obstet Gynecol*. 2001;98:427–433.
71. Carr DB, Tran LT, Brateng DA, et al. Hemodynamically-directed atenolol therapy is associated with a blunted rise in maternal sFLT-1 levels during pregnancy. *Hypertens Pregnancy*. 2009;28:42–55.
72. Freier DT, Thompson NW. Pheochromocytoma and pregnancy: the epitome of high risk. *Surgery*. 1993;114:1148–1152.
73. Wilson IM, Freis IM. Relationship between plasma and extracellular fluid volume depletion and the antihypertensive effect of chlorothiazide. *Circulation*. 1959;20:1028–1036.
74. Roos JC, Boer P, Koomaans HA. Haemodynamic and hormonal changes during acute and chronic diuretic treatment in essential hypertension. *Eur J Clin Pharmacol*. 1981;19:107–112.
75. Bennett WM, McDonald WJ, Kuehnel E. Do diuretics have antihypertensive properties independent of natriuresis. *Clin Pharmacol Ther*. 1977;22:499–504.
76. Collins R, Yusuf S, Peto R. Overview of randomised trials of diuretics in pregnancy. *Brit Med J*. 1985;290:17–23.
77. Sibai BM, Grossman RA, Grossman HG. Effects of diuretics on plasma volume in pregnancies with long-term hypertension. *Am J Obstet Gynecol*. 1984;150:831–835.
78. Gallery EDM, Hunyor SN, Gyory AZ. Plasma volume contraction: a significant factor in both pregnancy-associated hypertension (preeclampsia) and chronic hypertension in pregnancy. *Q J Med*. 1979;192:593–602.
79. Visser W, Wallenburg HCS. Central hemodynamic observations in untreated preeclamptic patients. *Hypertension*. 1991;17:1072–1077.
80. National High Blood Pressure Education Program Working Group report on high blood pressure in pregnancy. *Am J Obstet Gynecol*. 2000;183:S1–S22. [Also reprinted as NIH Publication No. 00-3029 and available at: http://www.nhlbi.nih.gov/guidelines/archives/hbp_preg/index.htm].
81. American college of obstetricians and gynecologists, task force on Hypertension in Pregnancy. Hypertension in Pregnancy. Available at: http://www.acog.org/Resources%20And%20Publications/Task%20Force%20and%20Work%20Group%20Reports/Hypertension%20in%20Pregnancy.aspx.
82. Krusell LR, Jespersen LT, Schmitz A. Repetitive natriuresis and blood pressure. Longterm calcium entry blockade with isradipine. *Hypertension*. 1987;10:577–581.
83. Brouwer RM, Bolli P, Erne P. Antihypertensive treatment using calcium antagonists in combination with captopril rather than diuretics. *Cardiovasc Pharmacol*. 1985;7:S88–S91.
84. Furberg CD, Psaty BM, Meyer JV. *Nifedipine:* Dose-related increase in mortality in patients with coronary heart disease. *Circulation*. 1995;92:1326–1331.
85. Chobanian AV. Calcium channel blockers. Lessons learned from MIDAS and other clinical trials. *JAMA*. 1996;276:829–830.
86. Magee LA, Schick B, Donnenfeld AE. The safety of calcium channel blockers in human pregnancy: a prospective, multicenter cohort study. *Am J Obstet Gynecol*. 1996;174:823–828.
87. Weber-Schoendorfer C, Hannemann D, Meister R. The safety of calcium channel blockers during pregnancy: a prospective, multicenter, observational study. *Reprod Toxicol*. 2008;26:24–30.
88. Sibai BM, Barton JR, Akl S. A randomized prospective comparison of nifedipine and bed rest versus bed rest alone in the management of preeclampsia remote from term. *Am J Obstet Gynecol*. 1992;167:879–884.
89. Ismail AA, Medhat I, Tawfic TA, Kholeif A. Evaluation of calcium-antagonist (Nifedipine) in the treatment of preeclampsia. *Int J Gynaecol Obstet*. 1993;40:39–43.

90. Hanff LM, Vulto AG, Bartels PA, et al. Intravenous use of the calcium-channel blocker nicardipine as second-line treatment in severe, early-onset pre-eclamptic patients. *J Hypertens*. 2005;23:2319–2326.
91. Wide-Swensson DH, Ingemarsson I, Lunell NO. Calcium channel blockade (isradipine) in treatment of hypertension in pregnancy: a randomized placebo-controlled study. *Am J Obstet Gynecol*. 1995;173:872–878.
92. Belfort MA, Anthony J, Saade GR, Allen JC. A comparison of magnesium sulfate and nimodipine for the prevention of eclampsia. *New Engl J Med*. 2003;348:304–311.
93. Belfort MA, Anthony J, Buccimazza A, Davey DA. Hemodynamic changes associated with intravenous infusion of the calcium antagonist verapamil in the treatment of severe gestational proteinuric hypertension. *Obstet Gynecol*. 1990;75:970–974.
94. Gallery ED, Gyory AZ. Sublingual nifedipine in human pregnancy. *Aust N Z J Med*. 1997;27:538–542.
95. van Harten J, Burggraaf K, Danhof M. Negligible sublingual absorption of nifedipine. *Lancet*. 1987;2:1363–1365.
96. Fenakel K, Fenakel G, Appelman Z. Nifedipine in the treatment of severe preeclampsia. *Obstet Gynecol*. 1991;77:331–337.
97. Seabe SJ, Moodley J, Becker P. Nifedipine in acute hypertensive emergencies in pregnancy. *S Afr Med J*. 1989;76: 248–250.
98. Martins-Costa S, Ramos JG, Barros E. Randomized, controlled trial of hydralazine versus nifedipine in preeclamptic women with acute hypertension. *Clin Exp Hypertens*. 1992;11:25–44.
99. Jegasothy R, Paranthaman S. Sublingual nifedipine compared with intravenous hydralazine in the acute treatment of severe hypertension in pregnancy: potential for use in rural practice. *J Obstet Gynaecol Res*. 1996;22:21–24.
100. Visser W, Wallenburg HC. A comparison between the hemodynamic effects of oral nifedipine and intravenous dihydralazine in patients with severe preeclampsia. *J Hypertens*. 1995;13:791–795.
101. Blea CW, Barnard JM, Magness RR. Effect of nifedipine on fetal and maternal hemodynamics and blood gases in the pregnant ewe. *Am J Obstet Gynecol*. 1997;176:922–930.
102. Waisman GD, Mayorga LM, Camera MI. Magnesium plus nifedipine: Potentiation of hypotensive effect in preeclampsia? *Am J Obstet Gynecol*. 1988;159:308–309.
103. BenAmi M, Giladi Y, Shalev E. The combination of magnesium sulphate and nifedipine: a cause of neuromuscular blockade. *Brit J Obstet Gynaecol*. 1994;101:262–263.
104. Magee L, Miremadi S, Li J. Therapy with both magnesium sulphate and nifedipine does not increase the risk of serious magnesium-related maternal side effects in women with preeclampsia. *Am J Obstet Gynecol*. 2005;193:153–163.
105. Altman D, Carroli G, Duley L, et al. Do women with preeclampsia, and their babies, benefit from magnesium sulphate? The Magpie Trial: a randomised placebo-controlled trial. *Lancet*. 2002;359:1877–1890.
106. Wallenburg HCS. Hemodynamics in hypertensive pregnancy. In: Rubin PC, ed. *Hypertension in Pregnancy*. New York: Elsevier; 1988:66–101.
107. Vink GJ, Moodley J, Philpott RH. Effect of dihydralazine on the fetus in the treatment of maternal hypertension. *Obstet Gynecol*. 1980;55:519–522.
108. Widerlov E, Karlman I, Storsater J. Hydralazine-induced neonatal thrombocytopenia. *New Engl J Med*. 1980;303: 1235.
109. Dudley DKL. Minibolus diazoxide in the management of severe hypertension in pregnancy. *Am J Obstet Gynecol*. 1985;151:196–200.
110. Nuwayhid B, Brinkman CR, Katchen B. Maternal and fetal hemodynamic effects of diazoxide. *Obstet Gynecol*. 1975;46:197–203.
111. Hennessy A, Thornton CE, Makris A, et al. A randomised comparison of hydralazine and mini-bolus diazoxide for hypertensive emergencies in pregnancy: the PIVOT trial. *Aust N Z J Obstet Gynaecol*. 2007;47:279–285.
112. Boulos BM, Davis LE, Almond CH, Jackson RL. Placental transfer of diazoxide and its hazardous effect on the newborn. *J Clin Pharmacol J New Drugs*. 1971;11: 206–210.
113. Shoemaker CT, Meyers M. Sodium nitroprusside for control of severe hypertensive disease of pregnancy: a case report and discussion of potential toxicity. *Am J Obstet Gynecol*. 1984;149:171–173.
114. Wasserstrum N. Nitroprusside in preeclampsia. Circulatory distress and paradoxical bradycardia. *Hypertension*. 1991;18: 79–84.
115. Rosenthal J. Role of renal and extrarenal renin-angiotensin system in the mechanism of arterial hypertension and its sequelae. *Steroids*. 1993;58:566–572.
116. Takahama K, Araki T, Fuchikami J. Studies on the magnitude and the mechanism of cough potentiation by angiotensin-converting enzyme inhibitors in guinea pigs: involvement of bradykinin in the potentiation. *J Pharm Pharmacol*. 1996;48:1027–1033.
117. Raizada MK, Ferreira AJ. ACE2: a new target for cardiovascular disease therapeutics. *J Cardiovasc Pharmacol*. 2007;50:112–119.
118. Merrill DC, Karoly M, Chen K, Ferrario CM, Brosnihan KB. Angiotensin-(1-7) in normal and preeclamptic pregnancy. *Endocrine*. 2002;18:239–245.
119. Rajagopalan S, Bakris GL, Abraham WT, Pitt B, Brook RD. Complete renin-angiotensin-aldosterone system (RAAS) blockade in high-risk patients: recent insights from renin blockade studies. *Hypertension*. 2013;62:444–449.
120. Ito S, Arima S, Ren YL. Endothelium-derived relaxing factor/nitric oxide modulates angiotensin II action in the isolated microperfused rabbit afferent but not efferent arteriole. *J Clin Invest*. 1993;91:2012–2019.
121. Chapman N, Dobson J, Wilson S, et al. Effect of spironolactone on blood pressure in subjects with resistant hypertension. *Hypertension*. 2007;49:839–845.
122. August P, Mueller FB, Sealey JE, Edersheim TG. Role of renin-angiotensin system in blood pressure regulation in pregnancy. *Lancet*. 1995;345:896–897.
123. Baker PN, Pipkin FB, Symonds EM. Comparative study of platelet angiotensin II binding and the angiotensin II sensitivity test as predictors of pregnancy-induced hypertension. *Clin Sci (Colch)*. 1992;83:89–95.
124. Easterling TR, Carr DB, Davis C, Diederichs C, Brateng DA, Schmucker B. Low-dose, short-acting, angiotensin-converting enzyme inhibitors as rescue therapy in pregnancy. *Obstet Gynecol*. 2000;96:956–961.

125. Pryde PG, Sedman AB, Nugent CE, Barr M. Angiotensin-converting enzyme inhibitor fetopathy. *J Am Soc Nephrol.* 1993;3:1575–1582.
126. Serreau R, Luton D, Macher MA, Delezoide AL, Garel C, Jacqz-Aigrain E. Developmental toxicity of the angiotensin II type 1 receptor antagonists during human pregnancy: a report of 10 cases. *BJOG.* 2005;112:710–712.
127. Cooper WO, Hernandez-Diaz S, Arbogast PG. Major congenital malformations after first-trimester exposure to ACE inhibitors. *New Engl J Med.* 2006;354:2443–2451.
128. Caton AR, Bell EM, Druschel CM, et al. National Birth Defects Prevention Study. Antihypertensive medication use during pregnancy and the risk of cardiovascular malformations. *Hypertension.* 2009;54:63–70.
129. Lennestål R, Olausson PO, Källén B. Maternal use of antihypertensive drugs in early pregnancy and delivery outcome, notably the presence of congenital heart defects in the infants. *Eur J Clin Pharmacol.* 2009;65:615–625.
130. Nakhai-Pour HR, Rey E, Bérard A. Antihypertensive medication use during pregnancy and the risk of major congenital malformations or small-for-gestational-age newborns. *Birth Defects Res B Dev Reprod Toxicol.* 2010;89:147–154.
131. Li DK, Yang C, Andrade S, Tavares V, Ferber JR. Maternal exposure to angiotensin converting enzyme inhibitors in the first trimester and risk of malformations in offspring: a retrospective cohort study. *BMJ.* 2011;343:d5931.
132. Porta M, Hainer JW, Jansson SO, et al. DIRECT Study Group. Exposure to candesartan during the first trimester of pregnancy in type 1 diabetes: experience from the placebo-controlled DIabetic REtinopathy Candesartan Trials. *Diabetologia.* 2011;54:1298–1303.
133. Atkinson HC, Begg EJ, Darlow BA. Drugs in human milk: clinical pharmacokinetic considerations. *Clin Pharmacokinet.* 1988;14:217–240.
134. Breitzka RL, Sandritter TL, Hatzopoulos FK. Principles of drug transfer into breast milk and drug disposition in the nursing infant. *J Hum Lact.* 1997;13:155–158.
135. Beardmore KS, Morris JM, Gallery ED. Excretion of antihypertensive medication into human breast milk: a systematic review. *Hypertens Pregnancy.* 2002;21:85–95.
136. Devlin RG, Fleiss PM. Captopril in human blood and breast milk. *J Clin Pharmacol.* 1981;21:110–113.
137. Redman CW, Kelly JG, Cooper WD. The excretion of enalapril and enalaprilat in human breast milk. *Eur J Clin Pharmacol.* 1990;38:99.
138. Begg EJ, Robson RA, Gardiner SJ, et al. Quinapril and its metabolite quinaprilat in human milk. *Brit J Clin Pharmacol.* 2001;51:478–481.
139. Collins R, MacMahon S. Reliable assessment of the effects of treatment on mortality and major morbidity. I: clinical trials. *Lancet.* 2001;357:373–380.
140. Stepheson J, Imrie J. Why do we need randomized controlled trials to assess behavioural interventions. *BMJ.* 1998;316:611–613.
141. Villar J, Carroli G. Methodological issues of randomized controlled trials for the evaluation of reproductive health interventions. *Prev Med.* 1996;25:365–375.
142. Villar J, Mackey ME, Carroli G, Donner A. Meta-analyses in systematic reviews of randomized controlled trials in perinatal medicine: comparison of fixed and random effect model. *Statist Med.* 2001;20:3635–3647.
143. Abalos E, Duley L, Steyn DW. Antihypertensive drug therapy for mild to moderate hypertension during pregnancy. *Cochrane Database Syst Rev.* 2014;1:CD002252.
144. Magee LA, Duley L. Oral beta-blockers for mild to moderate hypertension during pregnancy. *Cochrane Database Syst Rev.* 2003;3:CD002863.
145. Umans JG, Lindheimer MD. Antihypertensive therapy in pregnancy. *Curr Hypertens Rep.* 2001;3:392–399.
146. Ferrer RL, Sibai BM, Mulrow CD. Management of mild chronic hypertension during pregnancy: a review. *Obstet Gynecol.* 2000;96:849–860.
147. Magee L, von Dadelszen P. Prevention and treatment of postpartum hypertension. *Cochrane Database Syst Rev.* 2013;4:CD004351.
148. Nabhan AF, Elsedawy MM. Tight control of mild-moderate pre-existing or non-proteinuric gestational hypertension. *Cochrane Database Syst Rev.* 2008;1:CD006907.
149. Von Dadelszen P, Ornstein M, Bull S, Logan A, Koren G, Magee L. Fall in mean arterial pressure and fetal growth restriction in pregnancy hypertension: a meta-analysis. *Lancet.* 2000;355:87–92.
150. von Dadelszen P, Magee LA. Fall in mean arterial pressure and fetal growth restriction in pregnancy hypertension: an updated metaregression analysis. *J Obstet Gynaecol Can.* 2002;24:941–945.
151. Magee LA, von Dadelszen P, Chan S, et al. The control of hypertension in pregnancy study pilot trial. *BJOG.* 2007;114(770):e13–e20.
152. Belfort MA, Tooke-Miller C, Allen JC, Dizon-Townson D, Varner MA. Labetalol decreases cerebral perfusion pressure without negatively affecting cerebral blood flow in hypertensive gravidas. *Hypertens Pregnancy.* 2002;21:185–197.

CHAPTER **20**

Clinical Management

JAMES M. ALEXANDER AND F. GARY CUNNINGHAM

Editors' comment: *Although not an obstetrician or a clinician, Leon Chesley, PhD, was associated through most of his career with academic clinical departments and hospitals. Still, in the single-authored first edition of this text he devoted two entire chapters to management of preeclampsia and eclampsia. His chapters bear rereading, for they display Chesley's critical assessment of the literature with an acumen that surpassed that of most obstetricians and hypertension experts of his time. The introductory chapter of this edition contains a historical compilation of treatment approaches through the age, and this Management chapter contains Chesley's pioneering studies concerning infused magnesium sulfate in preeclampsia, as well as prescient analyses of published treatment regimens used in the 1960s and 1970s.*

The goal of this chapter differs from most that precede it, which were more specifically aimed at describing and critiquing progress in the field, focusing primarily on possible etiopathogenesis of the preeclampsia syndrome. This chapter is designed to help physicians in the day-to-day management of women with preeclampsia, and to do so we describe some of the management schemes used at Parkland Hospital in Dallas. This tertiary-care center, in which one of the editors has been involved clinically for over 40 years, has been responsible for the obstetrical care of over 450,000 indigent women who had more than 50,000 pregnancies complicated by hypertension. Thus the approach of the chapters authors will stress the practical aspects of treatment, emphasizing those references that have had the most influence on our practice.

INTRODUCTION

Management of preeclampsia depends upon its severity as well as the gestational age at which it becomes clinically apparent. While in most cases diagnosis is made by the appearance of new-onset gestational hypertension accompanied by proteinuria, observations over the last two decades – which are discussed in detail in other chapters – have emphasized the importance of endothelial cell injury and multiorgan dysfunction as integral parts of the *preeclampsia syndrome*.

As emphasized in Chapter 2, there are instances where it is not possible to make a definitive diagnosis of preeclampsia. For this reason, and given the explosive nature of the disorder, the American College of Obstetricians and Gynecologists[1] ACOG Task Force[2] and the National High Blood Pressure Education Program (NHBPEP) Working Group[3] recommend close surveillance even if preeclampsia is only "suspected." *Increases in systolic and diastolic blood pressure can either be normal physiological changes or signs of developing pathology.* With increased surveillance, temporal changes in blood pressure and laboratory values, as well as the development of signs and symptoms are monitored. Thus, increased surveillance permits rapid recognition of ominous changes in blood pressure, critical laboratory findings, and development of clinical signs and symptoms.[1]

The basic management objectives for any pregnancy complicated by preeclampsia are:

1. Termination of pregnancy with the least possible trauma to mother and fetus.
2. Birth of an infant who subsequently thrives.
3. Complete restoration of health to the mother.

In certain women with preeclampsia, especially those at or near term, all three objectives are served equally well by induction of labor. *One of the most important clinical questions for successful management is precise knowledge of the age of the fetus.*

PREECLAMPSIA

Early Diagnosis and Evaluation

Traditionally, the frequency of prenatal visits is increased during the third trimester and this facilitates early detection of preeclampsia. Women with overt new-onset hypertension – 140/90 mm Hg or greater – are frequently admitted to the hospital for 2 to 3 days primarily to determine whether the rise in pressure is due to preeclampsia, and if so to evaluate its severity. Women with persistent severe disease are observed closely, and many are delivered. Conversely,

Chesley's Hypertensive Disorders in Pregnancy.
ISBN: 978-0-12-407866-6

DOI: http://dx.doi.org/10.1016/B978-0-12-407866-6.00020-1

women with apparently mild disease especially if remote from term may be carefully managed as outpatients.

Hospitalization is considered at least initially for women with new-onset hypertension, especially if there is persistent or worsening hypertension or development of proteinuria. A systematic evaluation is instituted to include the following:

1. Detailed examination followed by daily scrutiny for clinical findings such as headache, visual disturbances, epigastric pain, and rapid weight gain.
2. Weight on admittance and daily thereafter.
3. Analysis for proteinuria on admittance and at least every 2 days thereafter.
4. Blood pressure readings in the sitting position with an appropriate-size cuff every 4 hours, except during normal sleeping hours.
5. Measurements of plasma or serum creatinine and liver enzymes, and hemogram to include platelet quantification. The frequency of testing is determined by the severity of hypertension. Some recommend measurement of serum uric acid and lactic acid dehydrogenase levels as well as coagulation studies, but some investigations have called into question the value of these tests.[4,5]
6. Frequent evaluation of fetal size and well-being and amniotic fluid volume either clinically or using sonography.

Goals of such management include early identification of worsening preeclampsia and the development of a management plan for obstetrical care which includes a plan for timely delivery. If any of these observations leads to a diagnosis of severe preeclampsia, further management is the same as described subsequently for eclampsia.

Hospitalization Versus Outpatient Management

For women with non-severe stable hypertension – whether or not preeclampsia has been confirmed – continued surveillance either in hospital, or at home for reliable patients, or through a day-care unit is carried out. At least intuitively, reduced physical activity throughout much of the day seems beneficial, but absolute bed rest is unnecessary and it may predispose to venous thromboembolism.[6] A number of observational studies as well as randomized trials have addressed the benefits of inpatient care as well as outpatient management to include day-care unit observation and these are subsequently discussed. Somewhat related, Abenhaim et al.[7] conducted a retrospective cohort study of 677 women hospitalized for bed rest because of threatened preterm delivery. When outcomes of these women were compared with their general obstetrical population, bed rest was associated with a significantly reduced risk of developing preeclampsia – RR 0.27 (0.16–0.48). In a review of two small randomized trials totaling 106 women, prophylactic bed rest 4–6 hours daily at home in women at high risk of preeclampsia was successful in lowering the incidence of preeclampsia, but not gestational hypertension.[8] These and other observations support the claim that restricted activity alters the underlying pathophysiology of the preeclampsia syndrome.

Many cases of new-onset hypertension, either with or without proteinuria, prove to be sufficiently mild and have an onset near enough to term that they can be managed conservatively until labor commences spontaneously or until the cervix becomes favorable for labor induction. Complete abatement of all signs and symptoms, however, is uncommon until after delivery. *Almost certainly, the underlying disease persists until after delivery.*

High-Risk Pregnancy Unit

An inpatient antepartum unit was established in 1973 by Dr Peggy Whalley at Parkland Hospital in large part to provide care for women with hypertensive disorders. Initial results from this unit were reported by Hauth et al.[9] and Gilstrap et al.[10] The majority of women hospitalized have a beneficial response characterized by disappearance or improvement of hypertension. *These women are not "cured," because nearly 90% have recurrent hypertension before or during labor.* By the end of 2013, more than 10,000 nulliparous women with mild to moderate early-onset hypertension during pregnancy had been managed successfully in this unit. Provider costs (*not* charges) for this relatively simple physical facility, modest nursing care, no drugs other than iron and folate supplements, and the very few laboratory tests that are essential are minimal when compared with the cost of neonatal intensive care for a preterm infant.

Home Health Care

Many clinicians believe that further hospitalization is not warranted if hypertension abates within a few days, and this has unfortunately legitimized third-party payers to refuse hospital reimbursement. Consequently, most women with mild to moderate hypertension are managed at home. Outpatient management may continue as long as the disease does not worsen and if fetal jeopardy is not suspected. Sedentary activity throughout the greater part of the day is recommended. These women are instructed in detail about reporting symptoms. Home blood pressure and urine protein monitoring or frequent evaluations by a visiting nurse may be necessary. Lo et al.[11] and more recently Ostchega et al.[12] cautioned about the use of certain automated home blood pressure monitors that may fail to detect severe hypertension.

In an observational study by Barton et al.[13] 1182 women with mild "gestational hypertension" – 20% had proteinuria – were managed with home health care. These nulliparous women were a mean of 32–33 weeks pregnant at enrollment and 36–37 weeks at delivery. Pregnancy outcomes are shown in Table 20.1. Severe preeclampsia developed in about 20%, and although some developed HELLP syndrome, eclampsia, or abruptio placentae, perinatal outcomes were generally acceptable.

TABLE 20.1 Pregnancy Outcomes for 1182 Women Hospitalized for Mild Gestational Hypertension or Preeclampsia

Group	No.	Admission EGA (wks)	Admission Prot (%)	Delivery EGA (wks)	Delivery Prot (%)	Delivery CD (%)	Delivery HELLP (%)	Delivery SP (%)	Delivery ECL (%)	Perinatal Outcomes Mean BW (g)	Perinatal Outcomes SGA (%)	Perinatal Outcomes PMR (%)
Hispanic	92	33.2	22	36.8	65	38	4.3	26	1.1	2710	30	0
African-American	476	33.2	23	36.5	53	48	2.1	21	0	2735	22	1.1
White	614	33.5	19	36.9	54	45	2.9	16	0.2	2845	18	0

EGA = estimated gestational age; Prot = proteinuria ≤1+; CD = cesarean delivery; SP = severe preeclampsia; ECL = eclampsia; BW = birth weight; SGA = small-for-gestational-age; PMR = perinatal mortality rate.
Data from Barton and colleagues.[13]

Several prospective studies have been designed to compare continued hospitalization with either home health care or a day-care unit. In a pilot study from Parkland Hospital, Horsager et al.[14] randomly assigned 72 nulliparas with new-onset hypertension from 27 to 37 weeks to continued hospitalization versus outpatient care. In all of these women, proteinuria had receded to <500 mg per day when randomized. Outpatient management included daily blood pressure monitoring by the patient or her family, and weight and spot urine protein were determined three times weekly. A home health nurse visited twice weekly and the women were seen weekly in the obstetrical complications clinic. Perinatal outcomes were similar in each group. The only significant difference was that women in the home-care group developed severe preeclampsia more frequently than hospitalized women – 42% versus 25% ($p < 0.05$).

A larger randomized trial reported by Crowther et al.[15] included 218 women with mild gestational (nonproteinuric) hypertension, half of whom remained hospitalized after evaluation and the remainder were managed as outpatients. As shown in Table 20.2, the mean duration of hospitalization was 22.1 days for women with inpatient management compared with only 6.5 days in the home-care group, otherwise all other outcomes were similar.

Day-Care Unit

Another approach, now quite common in European countries, is day care.[18] This approach has been evaluated by several investigators. In the study designed by Tufnell et al.,[17] 54 women developing hypertension after 26 weeks were assigned to either day care or routine management (Table 20.2). Hospitalizations, progression to overt preeclampsia, and labor inductions were significantly increased in the routine management group. In another study performed in Australia, Turnbull et al.[16] enrolled 395 women who were randomly assigned to either day care or inpatient management (Table 20.2). Almost 95% had mild to moderate hypertension – 288 without proteinuria and 86 with ≥1+ proteinuria at baseline. Fetal outcomes overall were good and none of the women developed eclampsia or HELLP syndrome, and there were no neonatal deaths. Routes of delivery and neonatal complications were similar. Surprisingly, the costs of the two schemes were not significantly different. Perhaps not surprisingly, general satisfaction favored day care.

Antepartum Hospitalization Versus Outpatient Care

From the above, it appears that either inpatient or close outpatient management is appropriate for the woman with mild *de novo* hypertension, with or without non-severe preeclampsia. The key to success is close follow-up and a conscientious patient.

Antihypertensive Therapy for Mild to Moderate Hypertension

The use of antihypertensive drugs in attempts to prolong pregnancy or modify perinatal outcomes in pregnancies complicated by various types and severities of hypertensive disorders has been of considerable interest. Treatment for women with chronic hypertension complicating pregnancy is discussed in detail in Chapter 18 while the pharmacology and use during pregnancy of specific antihypertensive drugs are discussed in Chapter 19.

Drug treatment for early mild preeclampsia has been disappointing, as shown in representative studies listed in Table 20.3. Sibai et al.[19] performed a randomized study to evaluate the effectiveness of labetalol and hospitalization compared with hospitalization alone. They evaluated 200 nulliparous women with preeclampsia diagnosed between 26 and 35 weeks. Although women given labetalol had significantly lower mean blood pressures, there were no differences between the groups in terms of mean pregnancy prolongation, gestational age at delivery, or birth weight. The cesarean delivery rates were similar, as were the numbers of infants admitted to special-care nurseries. *It was problematic that growth-restricted infants were twice as*

TABLE 20.2 Randomized Clinical Trials Comparing Hospitalization Versus Routine Care for Women with Mild Gestational Hypertension or Preeclampsia

Study Groups	No.	Maternal Characteristics – Admission				Maternal Outcomes – Delivery								Perinatal Outcomes		
		Para0 (%)	Chronic Htn (%)	EGA (wks)	Prot. (%)	EGA (wks)	<37 wks (%)	<34 wks (%)	CD (%)	Mean Hosp (d)	HELLP (%)	SP (%)	ECL (%)	Mean BW (g)	SGA (%)	PMR (%)
Crowther et al.[15]	218[a]				–											
Hospitalization	110	13	14	35.3	0	38.3	12	1.8	21	22.2	–	–	–	3080	14	0
Outpatient	108	13	17	34.6	0	38.2	22	3.7	15	6.5	–	–	–	3060	14	0
Turnbull et al.[16]	374[b]															
Hospitalization	125	63	0	35.9	22	39	–	–	–	8.5	0	20.8	0.8	3330	3.8	0
Day unit	249	62	0	36.2	22	39.7	–	–	–	7.2	0	14.9	0	3300	2.3	0
Tuffnell et al.[17]	54[a]															
Day unit	24	57	23	36	0	39.8	–	–	–	1.1	–	–	–	3320	—	0
Usual care	30	54	21	36.5	21	39	–	–	–	5.1	–	–	–	3340	—	0

Para0 = never delivered a viable pregnancy before; Htn = Hypertension; EGA = estimated gestational age; Prot = proteinuria ≤1 +; CD = cesarean delivery; SP = severe preeclampsia; ECL = eclampsia; BW = birth weight; SGA = small-for-gestational-age; PMR = perinatal mortality rate.

[a]Excluded women with proteinuria at entry.

[b]Included women with ≤1 + proteinuria.

TABLE 20.3 Randomized Placebo-Controlled Trials of Antihypertensive Therapy for Early Mild Hypertension Due to Pregnancy

Study	Study Drug (No.)	Prolongation Pregnancy (Days)	Severe Hyper-tension[a] (%)	Cesarean Delivery (%)	Abruptio Placentae (%)	Mean Birth Weight (g)	Growth Restriction (%)	Neonatal Deaths
Sibai et al.[19]	Labetalol (100)	21.3	5	36	2	2205	19	1
200 inpatients	Placebo (100)	20.1	15[b]	32	0	2260	9[b]	0
Sibai et al.[20]	Nifedipine (100)	22.3	9	43	3	2405	8	0
200 outpatients	Placebo (100)	22.5	18[b]	35	2	2510	4	0
Pickles et al.[21]	Labetalol (70)	26.6	9	24	NS[c]	NS	NS	NS
144 outpatients	Placebo (74)	23.1	10	26	NS	NS	NS	NS
Wide-Swensson et al.[22]	Isradipine (54)	23.1	22	26	NS	NS	NS	0
111 outpatients	Placebo (57)	29.8	29	19	NS	NS	NS	0

[a] Includes postpartum hypertension.
[b] Significant ($p < 0.05$) when study drug compared with placebo.
[c] NS = not stated.

TABLE 20.4 Summary of Randomized Trials Comparing Antihypertensive Therapy Versus no Drug or Placebo for Women with Mild to Moderate Gestational Hypertension

Factor	Trials (No.)	Antihypertensive Versus None	
		RR	95% CI
Developed severe hypertension	19	0.50	(0.41–0.61)*
Developed preeclampsia	22	0.97	(0.83–1.13)
Stillborn infant	26	0.73	(0.50–1.08)
Preterm birth	14	1.02	(0.89–1.16)
Fetal growth restriction	19	1.04	(0.84–1.27)

*$p < 0.05$, all other comparisons nonsignificant.
Data from Abalos et al.[25]

frequent in women given labetalol compared with those treated by hospitalization alone – 19 versus 9%.

The three other studies listed in Table 20.3 were performed to compare the beta-blocking agent, labetalol, or calcium-channel blockers, nifedipine and isradipine, with placebo. In none of these studies were any benefits of antihypertensive treatment shown. Von Dadelszen et al.[23] performed a meta-analysis that included the aforementioned trials for the purpose of determining the relation between fetal growth and antihypertensive therapy, and this is further analyzed in the systematic review undertaken in Chapter 19. Von Dadelszen et al.[24] concluded that treatment-induced decreases in maternal blood pressure may adversely affect fetal growth.

Abalos et al.[25] performed a literature search for randomized trials of active antihypertensive therapy compared with either no treatment or placebo given to women with mild to moderate gestational hypertension. They included a total of 46 trials (4282 women) in their analysis, which is summarized in Table 20.4. As seen, except for a halving of the risk of developing severe hypertension, active antihypertensive therapy had no beneficial effects. Importantly, fetal-growth restriction was not increased in the treated women. These issues are further discussed in Chapter 19.

Indications for Delivery

Delivery is the only current cure known for preeclampsia. Headache, visual disturbances, or epigastric pain are indicative that convulsions may be imminent, and oliguria is another ominous sign. Most clinicians manage severe preeclampsia with prophylactic anticonvulsant magnesium sulfate infusion. Antihypertensive therapy is given to lower dangerously elevated blood pressure. Treatment is identical to that described subsequently for eclampsia. The prime objectives are to forestall convulsions, to prevent intracranial hemorrhage and serious damage to other vital organs, and to deliver a healthy infant.

At term, delivery is typically indicated as opposed to the preterm gestation where temporizing measures are often employed. Exactly when to temporize versus facilitate delivery is not agreed upon, however, and this has not been well studied. The HYPITAT trial demonstrated improved maternal outcome with induction of women with mild preeclampsia beyond 37 weeks in a randomized controlled trial.[26] Tavik et al.[27] later published a secondary analysis from the same trial showing that this improvement in maternal outcome occurred with induction even if the cervix was unfavorable. For women with a late preterm gestation, the decision to deliver is not clear. Barton et al. reported excessive neonatal morbidity in women delivered before 38 weeks despite having stable, mild nonproteinuric hypertension.[28] The Netherlands study of 4316 infants delivered between $34^{0/7}$ and $36^{6/7}$ weeks also described substantive neonatal morbidity.[29] When the decision has been made to delay delivery because of prematurity, the hope is that a few more weeks *in utero* will reduce the risk of neonatal death or serious morbidity. Such a policy certainly is justified with non-severe hypertension. Assessments of fetal well-being and placental function are carried out, especially when there is hesitation to deliver the fetus preterm. Most investigators recommend frequent performance of various tests currently used to assess fetal well-being as described by the American College of Obstetricians and Gynecologists.[30] These include the *nonstress test* or the *biophysical profile*. Measurement of the lecithin:sphingomyelin ratio in amniotic fluid will provide evidence of lung maturity in cases with unsure gestational dating.

With moderate or severe preeclampsia that does not improve after hospitalization, delivery is usually advisable for the welfare of both mother and fetus. Often labor is induced with intravenous oxytocin, usually along with cervical ripening techniques such as a prostaglandin or osmotic dilator. Whenever it appears that labor induction almost certainly will not succeed, or attempts at induction have failed, cesarean delivery is indicated. There is no evidence that failed efforts at induction followed by cesarean delivery are harmful to low-birthweight infants.[26,31]

Delayed Delivery with Early-Onset Preeclampsia

In the past, all women with severe preeclampsia were usually delivered without delay. Over the past 25 years, a different approach for women with severe preeclampsia remote from term has been advocated. This calls for conservative or "expectant" management in a selected group of women with the aim of improving neonatal outcome without compromising the safety of the mother. Aspects of such conservative management always include careful daily – and usually more frequent – monitoring of the pregnancy in the hospital, with or without use of drugs to control hypertension.

Theoretically, antihypertensive therapy has potential application when preeclampsia severe enough to warrant termination of pregnancy develops before neonatal survival is likely. The caveat is that such management is controversial, and it may be catastrophic. In an early study, the Memphis group[32] attempted to prolong pregnancy because of fetal immaturity in 60 women with severe preeclampsia diagnosed between 18 and 27 weeks. *The total perinatal mortality rate was 87%, and although no mother died, 13 suffered placental abruption, 10 had eclampsia, three developed renal failure, two had hypertensive encephalopathy, and one each had an intracerebral hemorrhage and a ruptured hepatic hematoma.*

Later, the Memphis group[33] redefined their study criteria and performed a randomized trial of expectant versus aggressive management of 95 women who had severe preeclampsia but with more advanced gestations of 28 to 32 weeks. *Women with HELLP syndrome were excluded from this trial.* Aggressive management included glucocorticoid administration for fetal lung maturation followed by delivery in 48 hours. Expectantly managed women were observed at bed rest and given either labetalol or nifedipine orally to control severe hypertension. Pregnancy was prolonged for a mean of 15.4 days in the expectant management group with an overall improvement in neonatal outcomes.

Following these experiences, expectant management became more commonly practiced but women with HELLP syndrome or growth-restricted fetuses were usually excluded. In a follow-up observational study, the Memphis group[34] compared outcomes for 133 women with HELLP syndrome and infants delivered between 24 and 36 weeks with 136 women with severe preeclampsia but no HELLP criteria. Women with HELLP syndrome were subdivided into three study groups. The first group included those with hemolysis, elevated liver enzymes, *and* low platelets. The second included those with *partial HELLP syndrome* – this was defined as either one or two, but not three, of these laboratory findings. The third group included women who had severe preeclampsia without any of the accoutrements of the HELLP syndrome. Perinatal outcomes were similar in each group and this led the investigators to conclude that women with partial HELLP syndrome, as well as those with severe preeclampsia alone, could be managed expectantly. They also concluded that infant outcomes were related to gestational age rather than the hypertensive disorder per se.

Women in the Abramovici study[35] were indeed severely hypertensive and had mean diastolic blood pressures of 110 mm Hg. The distinguishing feature between those with complete and those with partial HELLP syndrome appears to be the platelet count – the mean value was 52,000/μL in women with complete HELLP syndrome compared with 113,000/μL in those with partial HELLP syndrome. Gestational ages at delivery were about 2 weeks more advanced in women with severe preeclampsia alone compared with those with some degree of HELLP syndrome. Accordingly, neonatal outcomes, in terms of need for mechanical ventilation and neonatal death, were better in women with severe preeclampsia alone. Fetal-growth restriction was not related to the severity of maternal disease and was prevalent in all three groups. Maternal morbidity was not described. A most important observation was that the median elapsed time from admission to delivery was 0, 1, and 2 days for women with HELLP syndrome,

TABLE 20.5 Maternal and Perinatal Outcomes Reported Since 2005 with Expectant Management of Severe Preeclampsia from 24 to 34 Weeks

Study	EGA at Enrollment wks	No.	Days Gained	Maternal Outcomes (Percent)					Perinatal Outcomes (Percent)	
				Placental Abruption	HELLP Syndrome	Pulmonary Edema	ARF	Eclampsia	FGR	PMR
Oettle (2005)[38]	24–34	131[a]	11.6	23	4.6	0.8	2.3	2.3	NS	13.8
Shear (2005)[39]	24–34	155	5.3	5.8	27	3.9	NS	1.9	62	3.9
Ganzevoort (2005a,b)[40]	24–34	216	11	1.8	18	3.6	NS	1.8	94	18
Sarsam (2008)[41]	24–34	35	9.2	5.7	11	2.9	2.9	18	31	2.8
Bombrys (2009)[42]	27–34	66	5	11	8	9	3	0	27	1.5
Abdel-Hady (2010)[43]	24–34	211	12	3.3	7.6	0.9	6.6	0.9	NS	48
Range		814	5–12	1.8–23	4.6–27	0.9–3.9	2.3–6.6	0.9–18	27–94	1.5–48

ARF = acute renal failure; EGA = estimated gestational age; FGR = fetal-growth restriction; HELLP = hemolysis, elevated liver enzymes, low platelet count; NS = not stated; PMR = perinatal mortality rate.
[a]Includes one maternal death.

partial HELLP syndrome, or severe preeclampsia, respectively. Later reports describing this cohort indicated that growth restriction adversely affected survival in infants from that institution.[33,36]

Sibai and Barton[37] reviewed most reports since the early 1990s of expectant management of women with severe preeclampsia with early onset from 24 to 34 weeks gestation. Some of these studies, as well as those published since 2005, are shown in Table 20.5. While the average time gained ranged from 5 to 10 days, maternal morbidity was formidable and included placental abruption, HELLP syndrome, pulmonary edema, renal failure, and eclampsia. Moreover, perinatal mortality averaged from 15 to 480 per 1000. Fetal-growth restriction was common, and in the study from the Netherlands[44] it was an astounding 94%. Perinatal mortality is disproportionately high in these growth-restricted infants, but maternal outcomes are not appreciably different from those in women without growth-restricted fetuses.[36,39] Barber et al.[45] provided a 10-year review of 3408 women with severe preeclampsia from 24 to 28 weeks who had been entered into the California vital statistics database. They found correlation between increased lengths of stay and increased rates of both maternal and neonatal morbidity.

Expectant Management of Midtrimester Severe Preeclampsia

A number of small studies have focused on expectant management of the severe preeclampsia syndrome before 28 weeks. In their review, Bombrys et al.[42] found eight such studies, which included a total of nearly 200 women with severe preeclampsia from less than 24 to 26 completed weeks. Maternal complications were common and there were no infant survivors in those less than 23 weeks, and the authors recommend pregnancy termination for these. For those at 23 weeks, perinatal survival was 18% but morbidity was unknown. For those at 24–26 weeks, perinatal survival approaches 60%, and it averaged almost 90% for those at 26 weeks. Results from five studies published since 2005 are shown in Table 20.6. Again, there are extraordinarily high maternal and perinatal morbidity and mortality rates in these extremely preterm pregnancies. At this time, there are inadequate contemporaneous comparative studies attesting to the perinatal benefits of such expectant treatment versus early delivery in the face of serious maternal complications that approach 50%. Thus, the caveat for detailed maternal counseling is repeated here. The 2013 Task Force supports these findings and recommended that a fetus that is <23 weeks gestation be considered previable and that women with severe preeclampsia be delivered.

Glucocorticoids

In attempts to enhance fetal lung maturation, glucocorticoids have been administered to women with severe hypertension who are remote from term. Treatment does not seem to worsen maternal hypertension and a decrease in the incidence of respiratory distress and improved fetal survival have been cited. That said, we are aware of only one randomized trial of corticosteroids given to hypertensive women for fetal lung maturation.[50] This trial included 218 women with severe preeclampsia between 26 and 34 weeks who were randomly assigned to be given betamethasone or placebo. Neonatal complications, including

TABLE 20.6 Maternal and Perinatal Outcomes with Expectant Management in Women with Midtrimester Severe Preeclampsia

Study	No.	Maternal Complications (%)	Perinatal Mortality (%)
Hall et al.[46]	8	36	88
Gaugler-Senden et al.[47]	26	65*	82
Budden et al. [48]	31	71	71
Bombrys et al. [42]	46	38–64	43
Belghiti et al. [49]	51	43	58
Weighted average	152	55	61

*One maternal death.

TABLE 20.7 Randomized Clinical Trials of Intravenous Dexamethasone Versus Placebo for Women with Severe Preeclampsia with HELLP Syndrome

Study	No.	Before Treatment[a]			Maternal Outcomes (%)[a]			
		Platelets ($10^3/\mu L$)	AST (U/L)	LDH (U/L)	ARF	Pulmonary Edema	Eclampsia	Death
Fonseca et al.[57]	132[b]							
Dexamethasone	66	61 ± 19	573 ± 621	2124 ± 1849	10	4.6	14	4.6
Placebo	66	58 ± 21	492 ± 579	2242 ± 1671	13	1.5	15	1.5
Katz et al.[58]	105[c]							
Dexamethasone	56	91 ± 45	155 ± 241	1103 ± 1500	16.1	3.6	14	3.6
Placebo	49	95 ± 57	240 ± 371	1020 ± 1282	24.5	10.2	25	4.1

HELLP = Hemolysis, Elevated Liver enzymes, Low Platelets; AST = aspartate transferase; LDH = lactate dehydrogenase; ARF = acute renal failure.
[a]All comparisons $p > 0.05$.
[b]Includes 60 antepartum, 72 postpartum.
[c]All postpartum.

respiratory distress, intraventricular hemorrhage, and death, were decreased significantly when betamethasone was given compared with placebo. But there were two maternal deaths and 18 stillbirths. We add these findings to buttress our unenthusiastic acceptance of attempts to prolong gestation in many of these women.[51]

Corticosteroids to Ameliorate HELLP Syndrome

Over 20 years ago, Thiagarajah et al.[52] suggested that glucocorticoids might also play a role in treatment of the laboratory abnormalities associated with the HELLP syndrome. Subsequent investigators,[53,54] however, reported less than salutary effects. Investigators at the University of Mississippi have been the staunchest advocates of corticosteroids to treat HELLP syndrome. Martin et al.[55] reviewed outcomes of almost 500 such women. From 1994 to 2000, 90% were treated, and their outcomes considered more favorable than the cohort from 1985 to 1991, during which time only 16% were treated with steroids. Their randomized trial, while comparing two corticosteroid compounds, did not include a nontreated group.[56] There have since been at least two prospective randomized studies designed to address this and they are summarized in Table 20.7. Fonseca et al.[57] from Cali, Columbia, randomized 132 women with HELLP syndrome to blinded treatment with either dexamethasone or placebo. Outcomes assessed included duration of hospitalization, time to recovery for abnormal laboratory tests, recovery of clinical parameters, and complications. None of these was statistically different between the two groups. In a similar blinded study Katz et al.[58] randomized 105 postpartum women with HELLP syndrome to dexamethasone versus placebo. They analyzed outcomes similar to the Cali study and found no advantage to dexamethasone to hasten recovery (Table 20.7). Shown in Fig. 20.1 are recovery times for platelet counts and serum AST and LDH levels in the Katz study.[58] The responses are almost identical in women receiving the corticosteroid compared with those receiving placebo. The 2013 Task Force[2] does not recommend use of corticosteroids to treat women with HELLP syndrome.

Risks Versus Benefits – Recommendations

Taken *in toto*, these studies do not show overwhelming evidence of favorable risk-versus-benefits of expectant management of severe preeclampsia from 24 to 32 weeks. The Society of Maternal Fetal Medicine[37] as well as the 2013 Task Force[2] have determined that such management is a reasonable alternative in selected women with severe preeclampsia before 34 weeks. As shown in Table 20.8, this type of management calls for in-hospital maternal and fetal

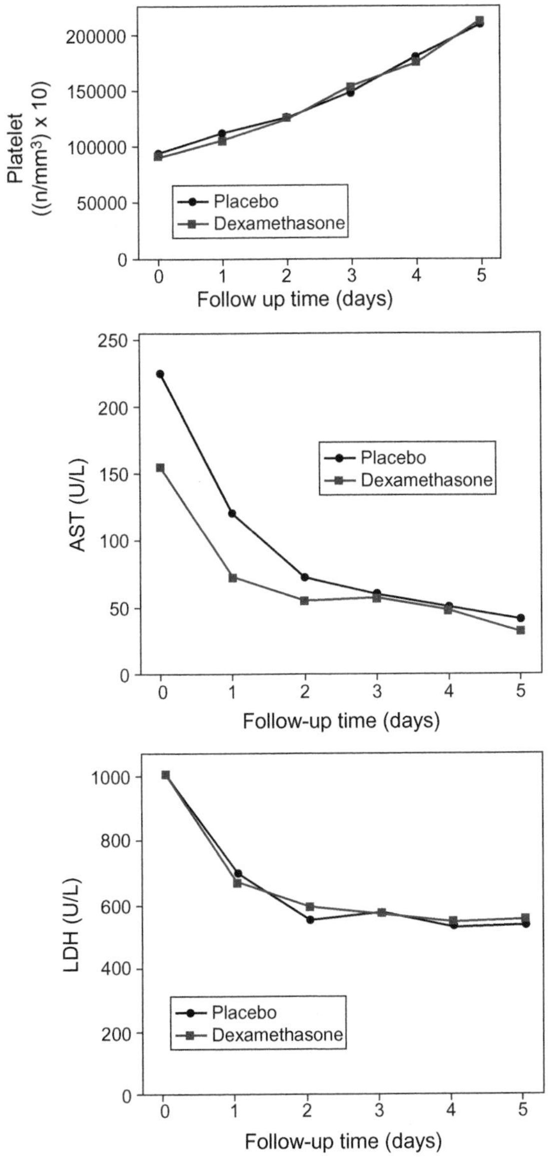

FIGURE 20.1 Serial laboratory values from initiation of therapy – dexamethasone versus placebo – in women with HELLP syndrome. (From Katz et al.[58], with permission.)

surveillance with delivery prompted by evidence of worsening severe preeclampsia or maternal or fetal compromise. While attempts are made for vaginal delivery in most cases, the likelihood of cesarean delivery increases with decreasing gestational age.

Undoubtedly, the overriding reason to terminate pregnancies with severe preeclampsia is maternal safety.[59] There are no data to suggest that expectant management is beneficial for the mother. Indeed, it seems obvious that a delay to prolong gestation in women with severe preeclampsia may have serious maternal consequences such as those shown in Table 20.8. Notably, placental abruption develops in up to 20% and pulmonary edema in up to 4%. Moreover, there are substantive risks for eclampsia,

TABLE 20.8 Indications for Delivery in Women with Severe Preeclampsia*

Maternal
Persistent severe headache or visual changes; eclampsia
Shortness of breath; chest tightness with rales and/or SaO_2 <94% breathing room air; pulmonary edema
Uncontrolled severe hypertension despite treatment
Oliguria <500 mL/24 hr or serum creatinine ≥1.5 mg/dL
Persistent platelet counts <100,000/μL
Suspected abruption, progressive labor, and/or ruptured membranes
Fetal
Severe growth restriction – <5th percentile for EGA
Persistent severe oligohydramnios – AFI <5 cm
Biophysical profile ≤4 done 6 hours apart
Reversed end-diastolic umbilical artery flow
Fetal death

From Sibai and Barton.[37]
AFI = amniotic fluid index.
*Indications represent recommendations of the article's authors and are not society guidelines.

cerebrovascular hemorrhage, and maternal death. These observations are especially pertinent when considered along with the absence of convincing evidence that perinatal outcomes are markedly improved by the average prolongation of pregnancy of about 1 week. If expectant management is undertaken, the caveats from the Society for Maternal Fetal Medicine and those shown in Table 20.8 should be strictly heeded. Thus, indications for delivery include worsening symptoms or hypertension; eclampsia, abruption, or pulmonary edema; renal dysfunction or oliguria; HELLP syndrome; fetal growth restriction, oligohydramnios, or other evidence of fetal compromise; and a gestational age of 34 weeks.

ECLAMPSIA

Preeclampsia complicated by generalized tonic-clonic convulsions increases appreciably the risk to both mother and fetus. Mattar and Sibai[60] described complications in 399 consecutive women with eclampsia from 1977 through 1998. Major adverse events included placental abruption (10%), neurological deficits (7%), aspiration pneumonia (7%), pulmonary edema (5%), cardiopulmonary arrest (4%), and acute renal failure (4%), and 1% of mothers died. These experiences were from a center that serves as a major referral hospital for obstetrical transfers from several neighboring states. In a report from Scandinavia encompassing a 2-year period through mid-2000, Andersgaard et al.[61] described 232 women with eclampsia. While there was but a single maternal death, a third of the women experienced major complications, including HELLP syndrome

(16%), renal failure (4.8%), pulmonary edema and pulmonary embolism (each 1.9%), and cerebrovascular accident (1.4%). In England over the past 20 years, maternal mortality has improved as evidenced by the 2005 audit by the United Kingdom Obstetric Surveillance System (UKOSS).[6] Of 214 eclamptic women in 229 hospitals, there were no maternal deaths, although five women experienced cerebrovascular hemorrhage. Akkawi et al. from Dublin, however, reported that there were four maternal deaths from a total of 247 eclamptic women.[62] Data from Australia are similar.[63]

Almost without exception, preeclampsia precedes the onset of eclamptic convulsions. Depending on whether convulsions appear before, during, or after labor, eclampsia is designated as antepartum, intrapartum, or postpartum. Eclampsia is most common in the last trimester and becomes increasingly more frequent as term approaches. In more recent years, there has been an increasing shift in the incidence of eclampsia toward the postpartum period. This is presumably related to improved access to prenatal care, earlier detection of preeclampsia, and prophylactic use of magnesium sulfate.[64] Importantly, other diagnoses should be considered in women with the onset of convulsions more than 48 hours postpartum or in women with focal neurological deficits, prolonged coma, or atypical eclampsia.[65] For example, epilepsy, encephalitis, meningitis, cerebral tumor, cysticercosis, and rupture of a cerebral aneurysm during late pregnancy and the puerperium may simulate eclampsia. *Until other such causes are excluded, however, all pregnant women with convulsions should be considered to have eclampsia.*

Immediate Management of Seizure

Eclamptic seizures may be quite violent, and during the convulsion the woman must be protected, especially the airway. So forceful are the muscular movements that the woman may throw herself out of her bed, and if not protected her tongue is bitten by the violent action of the jaws (Fig. 20.2). This phase, in which the muscles alternately contract and relax, may last about a minute. Gradually, the muscular movements become smaller and less frequent, and finally the woman lies motionless. After a seizure, a coma of variable duration ensues. When the convulsions are infrequent, the woman usually recovers some degree of consciousness after each attack. As the woman rouses, a semiconscious combative state may ensue. In very severe cases, the coma persists from one convulsion to another, and death may result before she awakens. In rare instances, a single convulsion may be followed by coma from which the woman may never emerge, although, as a rule, death does not occur until after frequent convulsions.

In antepartum eclampsia, labor may begin spontaneously shortly after convulsions ensue and progress rapidly. If the convulsion occurs during labor, contractions may increase in frequency and intensity, and the duration of labor may be shortened. Because of maternal hypoxemia

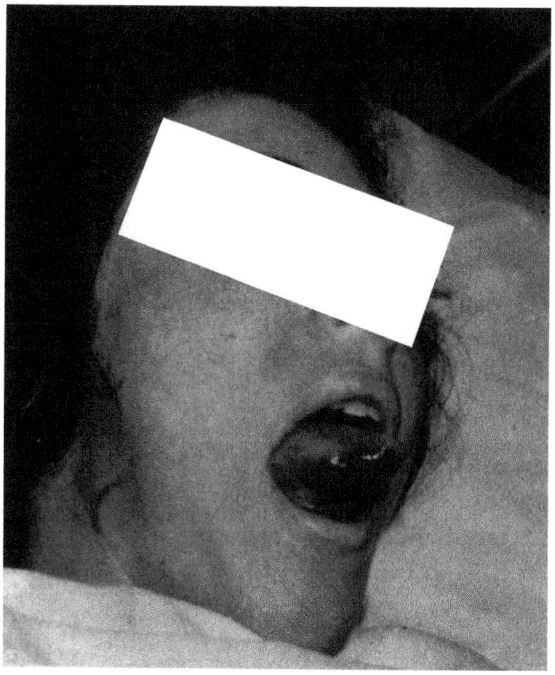

FIGURE 20.2 Hematoma of tongue from laceration during an eclamptic convulsion. Thrombocytopenia may have contributed to the bleeding.

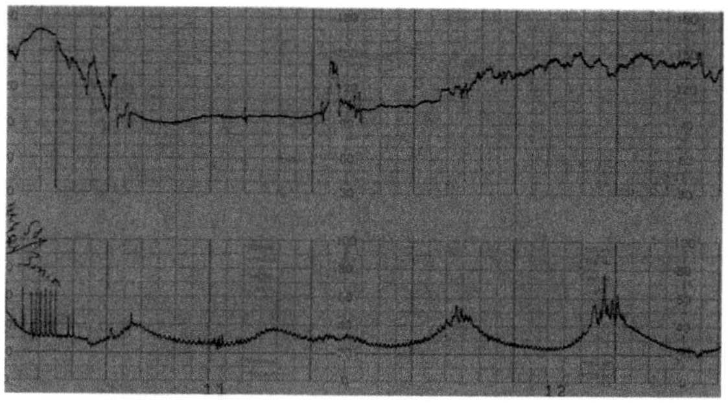

FIGURE 20.3 Fetal bradycardia following an intrapartum eclamptic convulsion. Bradycardia resolved and beat-to-beat variability returned about 5 minutes following the seizure.

and lactic acidemia caused by convulsions, it is not unusual for fetal bradycardia to follow a seizure (Fig. 20.3). The fetal heart rate tracing usually recovers back to baseline rate within 3 to 5 minutes, and if it persists more than about 10 minutes another cause, such as placental abruption or imminent delivery, must be considered. Baseline variability does not return to normal for an hour or so, provided there are no further convulsions.

Medical Treatment of Eclampsia

In his first edition, Chesley[66] recounts the history of magnesium sulfate to treat eclampsia. He wrote: "Magnesium sulfate is highly effective in preventing convulsions in women with preeclampsia and in stopping them in eclamptic patients." And later: "To repeat, magnesium sulfate prevents convulsions, it almost always stops them in eclampsia and it has minimal adverse effects on mother and fetus." He cited observational data by Pritchard et al.[67,68] from Parkland Hospital, as well as his own institution, Kings County Hospital in Brooklyn. At that time, most eclampsia regimens used in the United States adhered to a similar philosophy, the tenets of which include:

1. Control of convulsions using an intravenously administered loading dose of magnesium sulfate. This is followed either by a continuous infusion of magnesium sulfate or by an intramuscular loading dose and periodic intramuscular injections.
2. Intermittent intravenous or oral administration of an antihypertensive medication to lower blood pressure whenever the diastolic pressure is considered dangerously high. Some clinicians treat at 100 mm Hg, some at 105 mm Hg, and some at 110 mm Hg.
3. Avoidance of diuretics and limitation of intravenous fluid administration unless fluid loss is excessive. Hyperosmotic agents are avoided.
4. Delivery.

Magnesium Sulfate to Control Convulsions

In more severe cases of preeclampsia, as well as eclampsia, magnesium sulfate administered parenterally is an effective anticonvulsant without producing central nervous system depression in either the mother or the infant. It may be given intravenously by continuous infusion or intramuscularly by intermittent injection (Table 20.9). The dosage schedule for severe preeclampsia is the same as for eclampsia. Because labor and delivery is a more likely time for convulsions to develop, women with preeclampsia–eclampsia are usually given magnesium sulfate during labor and for 24 hours postpartum.

Magnesium sulfate is almost universally administered intravenously, and in most units the intramuscular route has been abandoned. Of concern, magnesium sulfate solutions, although inexpensive to prepare, are not readily available in all parts of the developing world. And even when the solutions are available, the technology to infuse them may not be. Therefore, it should not be forgotten that the drug can be administered intramuscularly and that this route is as effective as intravenous administration. In a recent report from India, Chowdhury and colleagues[70] showed that the two regimens were equivalent in preventing recurrent convulsions and maternal deaths in 630 women with eclampsia.

Magnesium sulfate is not given to treat hypertension. Based on a number of studies cited subsequently, as well as extensive clinical observations, magnesium most likely exerts a specific anticonvulsant action on the cerebral cortex. Typically, the mother stops convulsing after the initial administration of magnesium sulfate and, within an hour or two, regains consciousness sufficiently to be oriented as to place and time.

TABLE 20.9 Magnesium Sulfate Dosage Schedule for Severe Preeclampsia and Eclampsia

Continuous intravenous infusion

1. Give 4–6 g loading dose of magnesium sulfate diluted in 100 mL of IV fluid administered over 15–20 min
2. Begin 2 g/h in 100 mL of IV maintenance infusion. Some recommend 1 g/h (RCOG[69])
3. Monitor for magnesium toxicity:
   a. Assess deep tendon reflexes periodically
   b. Measure serum magnesium level at 4–6 h and adjust infusion to maintain levels between 4 and 7 mEq/L (4.8–8.4 mg/dL)
   c. Measure serum magnesium levels if serum creatinine ≥1.0 mg/mL
4. Magnesium sulfate is discontinued 24 h after delivery

Intermittent intramuscular injections

1. Give 4 g of magnesium sulfate ($MgSO_4 \cdot 7H_2O$ USP) as a 20% solution intravenously at a rate not to exceed 1 g/min
2. Follow promptly with 10 g of 50% magnesium sulfate solution, one-half (5 g) injected deeply in the upper outer quadrant of both buttocks through a 3-inch-long 20-gauge needle. (Addition of 1.0 mL of 2% lidocaine minimizes discomfort.) If convulsions persist after 15 min, give up to 2 g more intravenously as a 20% solution at a rate not to exceed 1 g/min. If the woman is large, up to 4 g may be given slowly
3. Every 4 h thereafter give 5 g of a 50% solution of magnesium sulfate injected deeply in the upper outer quadrant of alternate buttocks, but only after ensuring that:
   a. the patellar reflex is present
   b. respirations are not depressed
   c. urine output the previous 4 h exceeded 100 mL
4. Magnesium sulfate is discontinued 24 h after delivery

The magnesium sulfate dosage schedules presented in Table 20.9 usually result in plasma magnesium levels illustrated in Fig. 20.4. When magnesium sulfate is given to arrest and prevent recurrent eclamptic seizures, about 10–15% of women have a subsequent convulsion. An additional 2 g dose of magnesium sulfate in a 20% solution is administered slowly intravenously. In a small woman, an additional 2 g dose may be used once – and the same dose may be used twice if needed in a larger woman. In only 5 of 245 women with eclampsia at Parkland Hospital was it necessary to use supplementary medication to control convulsions.[71] An intravenous barbiturate such as amobarbital or thiopental is given slowly. Small doses of midazolam or lorazepam are also effective. These medications and diazepam are given in small single doses because prolonged use is associated with higher mortality because of aspiration from sedation.[69]

Maintenance magnesium sulfate therapy for eclampsia is continued for 24 hours after delivery. For eclampsia that develops postpartum, magnesium sulfate is administered for 24 hours after the onset of convulsions. Ehrenberg and Mercer[72] studied abbreviated postpartum magnesium administration in 200 women with *mild* preeclampsia. Of

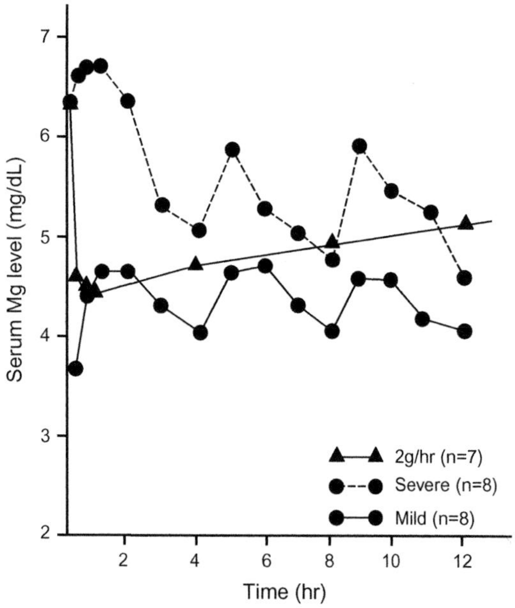

FIGURE 20.4 Comparison of serum magnesium levels following treatment for: (1) mild preeclampsia – 10 g intramuscular loading dose of magnesium sulfate and a 5 g maintenance dose every 4 hours (●- - ●); (2) severe preeclampsia – 4 g intravenous loading dose followed by intramuscular regimen (●—●); and (3) 4 g intravenous loading dose followed by a continuous infusion of 2 g/hr (▲—▲). Adapted from[74].

101 women randomized to 12-hour treatment, seven had worsening of preeclampsia and treatment was extended to 24 hours. None of these women and none of the 95 who received 24-hour magnesium infusion developed eclampsia. This abbreviated regimen needs further study before it is routinely administered for women with severe preeclampsia or eclampsia.

Pharmacology and Toxicology

Magnesium sulfate USP is $MgSO_4{\cdot}7H_2O$ and not $MgSO_4$. Parenterally administered magnesium is cleared almost totally by renal excretion, and magnesium intoxication is unusual when the glomerular filtration rate is maintained or only slightly decreased. Obstetricians strive to keep urine output adequate, and this usually correlates with preserved glomerular filtration rates. Importantly, however, magnesium excretion is not urine flow dependent, and urinary volume per unit time is not renal function. Thus, serum creatinine levels must be measured to detect signs of declining glomerular filtration rate. At the same time, in the absence of hemorrhage, vigorous intravenous crystalloid infusion to treat declining urine output with an unchanging serum creatinine level can lead to pulmonary edema. Other parameters should be monitored, including ensuring that a patellar or biceps reflex is present, and there is no respiratory depression.

Eclamptic convulsions are almost always prevented or arrested by plasma magnesium levels maintained at 4 to 7 mEq/L, or 4.8 to 8.4 mg/dL, or 2.0 to 3.5 mmol/L. Although laboratories typically report *total* magnesium levels, free or *ionized* magnesium is the active moiety for suppressing neuronal excitability. Taber et al.[73] found that there is a poor correlation between total and ionized magnesium levels. Further studies are necessary to determine whether measurement of ionized, rather than total, magnesium would provide a superior method for surveillance.

Sibai et al.[74] performed a prospective study in which they compared intermittent intramuscular and continuous intravenous magnesium sulfate. There was no significant difference between mean magnesium levels observed after intramuscular magnesium sulfate and those observed following a maintenance intravenous infusion of 2 g per hour (see Fig. 20.4). We routinely measure serum magnesium levels, and have recently reported that a number of women require an infusion rate of 3 g/h to maintain therapeutic magnesium levels.[75] Subtherapeutic levels are most often encountered in women whose BMI is >30. It is important to note that most guidelines do not recommend routine determination of serum levels.[1,65,69]

Patellar reflexes disappear when the plasma magnesium level reaches 10 mEq/L – about 12 mg/dL – presumably because of the curariform action of magnesium. This sign serves to warn of impending magnesium toxicity, because a further increase leads to respiratory depression. When plasma levels rise >10 mEq/L, respiratory depression develops, and at 12 mEq/L or more respiratory paralysis and arrest follow. Somjen et al.[76] induced in themselves, by intravenous infusion, marked hypermagnesemia, achieving plasma levels up to 15 mEq/L. Predictably, at such high plasma levels, respiratory depression developed that necessitated mechanical ventilation, but depression of the sensorium was not dramatic as long as hypoxia was prevented.

Treatment with calcium gluconate or chloride, 1 g intravenously, along with withholding further magnesium sulfate usually reverses mild to moderate respiratory depression. Unfortunately, the effects of intravenously administered calcium may be short-lived. For severe respiratory depression and arrest, prompt tracheal intubation and mechanical ventilation are life-saving. Direct toxic effects on the myocardium from high levels of magnesium are uncommon. It appears that the cardiac dysfunction associated with magnesium is due to respiratory arrest and hypoxia. With appropriate ventilation, cardiac action is satisfactory even when plasma levels are exceedingly high.[77] Magnesium safety and toxicity were recently reviewed by Smith et al.[78] In more than 9500 treated women, the overall rate of patellar tendon absence was 1.6%, respiratory depression 1.3%, and calcium gluconate administration 0.2%. They reported only one maternal death due to magnesium toxicity. Our anecdotal experiences are similar – in the estimated 50 years of its use in over 30,000 women, there has been only one maternal death with overdosage.[71]

Because magnesium is cleared almost exclusively by renal excretion, the doses described will become excessive if glomerular filtration is decreased substantively. The initial standard dose of magnesium sulfate can be safely administered without knowledge of renal function. We stress here the importance of using this standard dose as many clinicians erroneously reduce the loading dose under the mistaken conception that diminished renal function demands such action. This loading dose is meant to raise plasma levels to or near to the therapeutic level desired. It is only the maintenance infusion rate that should be lowered when glomerular filtration falls. *Thus, only the maintenance infusion rate is altered with diminished glomerular filtration rate.* Renal function is estimated by measuring plasma creatinine concentration, and whenever it is >1.0 mg/mL, serum magnesium levels are used to adjust the infusion rate.

Acute cardiovascular effects of parenteral magnesium ion in women with severe preeclampsia have been studied using data obtained by pulmonary and radial artery catheterization.[79] After a 4 g intravenous dose administered over a 15-minute period, mean arterial blood pressure fell slightly, accompanied by a 13% increase in cardiac index. Thus, magnesium decreased systemic vascular resistance and mean arterial pressure, and at the same time increased cardiac output, without evidence of myocardial depression. These findings were coincidental with transient nausea and flushing, and the cardiovascular effects persisted for only 15 minutes despite continued infusion of magnesium sulfate at 1.5 g per hour.

Thurnau et al.[80] showed that there was a small but highly significant increase in total magnesium concentration in the cerebrospinal fluid after magnesium therapy for preeclampsia. The magnitude of the increase was directly proportional to the corresponding serum concentration. This increase should not be ascribed to the disease itself, because magnesium levels in the cerebrospinal fluid appear unchanged in untreated women with severe preeclampsia when compared with normotensive controls.[81]

Magnesium is anticonvulsant and neuroprotective in a number of animal models. Some proposed mechanisms of action include reduced presynaptic release of the neurotransmitter glutamate, blocking glutamatergic *N*-methyl-D-aspartate (NMDA) receptors, potentiation of adenosine action, improved mitochondrial calcium buffering, and blockage of calcium entry via voltage-gated channels.[82] Because hippocampal seizures can be blocked by magnesium, it is believed that this implicates the NMDA receptor in eclamptic convulsions.[83] Importantly, results such as these suggest that magnesium has a central nervous system effect in blocking seizures.[84]

Uterine Effects

Magnesium ions in relatively high concentration depress myometrial contractility both *in vivo* and *in vitro*. With the regimen described and the plasma levels that have resulted, no evidence of myometrial depression has been observed beyond a transient decrease in activity during and immediately after the initial intravenous loading dose. Indeed, Leveno et al.[85] compared labor and delivery outcomes in 480 nulliparous women given phenytoin for preeclampsia with outcomes in 425 similar women given magnesium sulfate. Magnesium sulfate did not significantly alter oxytocin stimulation of labor, admission-to-delivery intervals, or route of delivery. Similar results have been reported by others.[86–88] Inhibition of uterine contractility is dose-dependent, because serum magnesium levels of 8 to 10 mEq/L are necessary to inhibit uterine contractions.[89] This likely explains why there is no uterine effect clinically when magnesium sulfate is given for treatment or prophylaxis of eclampsia. Nor is it an effective tocolytic agent to forestall preterm labor.[90]

Fetal Effects

Magnesium administered parenterally to the mother promptly crosses the placenta to achieve equilibrium in fetal serum and less so in amniotic fluid.[91] Neonatal depression occurs only if there is *severe* hypermagnesemia at delivery. That said, neonatal compromise after maternal therapy with magnesium sulfate has not been reported.[92–94] Current evidence supports the view that magnesium sulfate has small and significant effects on the fetal heart rate pattern – specifically beat-to-beat variability. Hallak et al.[93] compared an infusion of magnesium sulfate with saline and reported that magnesium was associated with a small and clinically insignificant decrease in variability. Similarly, in a retrospective study, Duffy and associates [94] reported a lower heart rate baseline that was within the normal range; decreased variability; and fewer prolonged decelerations, but with no evidence of adverse outcomes.

The results of two observational studies published during the mid-1990s suggested a protective effect of magnesium against the development of cerebral palsy in very-low-birthweight infants.[95,96] This led to at least three randomized trials designed to assess any protective fetal effects. In the trial from Australia and New Zealand, 1062 women ≤30 weeks for whom birth was planned or expected within 24 hours were randomly allocated to receive magnesium sulfate or placebo. [97] While mortality and cerebral palsy were less frequent for magnesium-exposed infants, the differences were not significant. Substantive gross motor dysfunction, however, was significantly less frequent in magnesium-treated fetuses – 3.4% versus 6.6%. From France, Marret et al.[98] published similar observations. The third trial was carried out by the National Institute of Child Health and Development Maternal-Fetal Medicine Units Network and its results were reported by Rouse et al.[99] A total of 2241 women with preterm pregnancy were randomized to receive magnesium sulfate or placebo. While the composite outcomes – perinatal deaths or moderate or severe cerebral palsy at 2 years of age – were similar, the most important finding was that fetuses exposed to magnesium had significantly decreased rates of cerebral palsy – 1.9% versus 3.5%.

Clinical Efficacy of Magnesium Sulfate Therapy

The multinational Eclampsia Trial Collaborative Group study[100] was funded in part by the World Health Organization and coordinated by the National Perinatal Epidemiology Unit in Oxford, England. This study involved 1687 women with eclampsia who were randomly allocated to different anticonvulsant regimens. In one cohort, 453 women were randomly assigned to be given magnesium sulfate and compared with 452 given diazepam. In a second cohort, another 388 eclamptic women were randomly assigned to be given magnesium sulfate and compared with 387 women given phenytoin. The results of these and other comparative studies are discussed in detail in Chapter 12 and results summarized in Table 20.10. In aggregate, magnesium sulfate therapy was associated with a significantly decreased rate of recurrent seizures – 9% versus 23% – and a lower maternal death rate – 3% versus 4.8% – when compared with either phenytoin or diazepam.

MANAGEMENT OF SEVERE HYPERTENSION

In addition to causing an eclamptic convulsion, dangerous hypertension can cause cerebrovascular hemorrhage in women with preeclampsia. Severe hypertension may also lead to ventricular dysfunction with congestive heart failure as well as placental abruption. The link between eclampsia and hemorrhagic strokes has been known since the early 1800s. As methods became available to estimate blood pressure, these morbid events then became linked to hypertension. As discussed in Chapter 13, Sheehan and Lynch[105] reported that half of women dying within 24 hours of eclamptic seizures had large hemorrhages in the basal ganglia and pons. These lesions were much more common in women with chronic hypertension, and their anatomical appearance was identical to hypertensive hemorrhage in nonpregnant individuals.

When effective blood pressure lowering drugs became available, the well-recognized association between cerebral hemorrhage and preeclampsia appropriately stimulated interest for treatment of severe hypertension accompanying the disorder. For example, in 1961[106] in the 12th and subsequent editions of *Williams Obstetrics*, one indicator of "severe preeclampsia" was blood pressure exceeding 160/110 mm Hg. By the 14th edition in 1971,[107] hydralazine treatment for unspecified levels of hypertension was recommended. In the next seven editions through the 24th edition in 2014,[108] hydralazine is recommended specifically to lower diastolic pressure to <110 mm Hg. Although still considered "severe," treatment of systolic pressures exceeding 160 mm Hg was only implied. And while precise data are lacking, in 2000, the Working Group[3] specifically recommended that treatment include lowering systolic pressures to ≤160 mm Hg. This too is the recommendation of the recent 2013 Task Force,[2] which recommended acute treatment of systolic pressures ≥160 mm Hg or diastolic pressures ≥110 mm Hg.

These recommendations are buttressed by the provocative observations of Martin et al.[109] that highlight the importance of treating systolic hypertension. These investigators described 28 selected women with severe preeclampsia who suffered an associated stroke. Most (93%) of these were hemorrhagic strokes, and all women had systolic pressures >160 mm Hg before the stroke. By contrast, only 20% of these same women had diastolic pressures >110 mm Hg. It seems likely that at least half of serious hemorrhagic strokes associated with preeclampsia are in women with chronic hypertension.[110] As discussed in Chapter 13, chronic hypertension results in development of *Charcot–Bouchard aneurysms* in the deeply penetrating branches of the lenticulostriate branch of the middle cerebral artery. These arterial branches supply the basal ganglia, putamen, thalamus, and adjacent deep white matter, as well as the pons and deep cerebellum. Aneurysmal weakening at the bifurcation of these small arteries predisposes to rupture when there is sudden hypertension.

Commonly Used Antihypertensive Agents

There are several drugs available for immediate lowering of dangerously elevated blood pressure in women with the gestational hypertensive disorders. These are discussed further in Chapter 19. The three most commonly employed in North

TABLE 20.10 Randomized Comparative Trials of Magnesium Sulfate with Another Anticonvulsant to Prevent Recurrent Eclamptic Convulsions

Study	Comparison Drug	Recurrent Seizures			Maternal Deaths		
		$MgSO_4$	Other	RR (95% CI)	$MgSO_4$	Other	RR (95% CI)
Dommisse [101]	Phenytoin	0/11	4/11		0/11	0/11	
Crowther[102]	Diazepam	5/24	7/27	0.80 (0.29–2.2)	1/24	0/27	
Bhalla et al.[103]	Lytic cocktail	1/45	11/45	0.09 (0.1–0.68)	0/45	2/45	
Friedman et al.[104]	Phenytoin	0/11	2/13 0/11	0/13			
Collaborative	Phenytoin	60/453	126/452	0.48 (0.36–0.63)	10/388	20/387	0.5 (0.24–1.00)
Eclampsia Trial[100]	Diazepam	22/388	66/387	0.33 (0.21–0.53)	17/453	24/452	0.74 (0.40–1.36)
Totals		88/932	216/935	0.41 (0.32–0.51)	28/932	45/935	0.62 (0.39–0.99)

America and Europe are hydralazine, labetalol, and nifedipine. For years, parenteral hydralazine was the only one of the three available and it was almost exclusively used in the United States. When parenteral labetalol was introduced, it was considered to be equally effective for obstetrical use. Orally administered nifedipine then gained popularity as first-line treatment for severe gestational hypertension.

Hydralazine

This is still probably the most commonly used antihypertensive agent in the United States for treatment of women with severe gestational hypertension. Hydralazine is administered intravenously in 5–10 mg doses at 15–20-minute intervals until a satisfactory response is achieved.[1] Some limit the total dose to 30 mg per treatment cycle.[111] A satisfactory response antepartum or intrapartum is defined as a decrease in diastolic blood pressure to 90 to 100 mm Hg, but not lower lest placental perfusion be compromised. Hydralazine so administered has proven remarkably effective in the prevention of cerebral hemorrhage. We note that the instruction to repeat administration of hydralazine every 15–20 minutes means that a repeat dose is given before the previous one has reached its peak effect. While theoretically such a regimen may lead to excess administration and undesirable hypotension, this has not been our experience. There is also a paucity of data regarding the pharmacokinetics of antihypertensive drugs in pregnant women, and thus one must be circumspect when using data derived from nonpregnant subjects.

At Parkland Hospital, approximately 8% of all women with hypertensive disorders are given hydralazine as described, and we estimate that at least 4000 women have been treated with this regimen. We do not limit the total dose per treatment cycle, and seldom has another antihypertensive agent been needed. Hydralazine is used less frequently in European centers.[69] From their meta-analysis, von Dadelszen et al.[24] concluded that the use of hydralazine as a first-line agent could not be supported. The Vancouver group after a systematic review[112] agreed with this conclusion; however, as discussed in Chapter 19, objective outcome data do not support the use of one antihypertensive agent versus another. The important caveat is that the drug administered should be one most often employed at that institution and with which there is greatest experience.

As with any antihypertensive agent, the tendency to give a larger initial dose of hydralazine when the blood pressure is extremely elevated must be avoided. The response to even 5–10 mg doses cannot be predicted by the level of hypertension, thus we always administer 5 mg as the initial dose. An example of very severe hypertension in a woman with chronic hypertension complicated by superimposed eclampsia that responded to repeated intravenous injections of hydralazine is shown in Fig. 20.5. Hydralazine was injected more frequently than recommended in the protocol, and blood pressure decreased in less than 1 hour from 240–270/130–150 mm Hg to 110/80 mm Hg. Fetal heart rate decelerations characteristic of uteroplacental insufficiency were evident when the pressure fell to 110/80 mm Hg, and persisted until maternal blood pressure was restored using intravenous crystalloid solutions. There is danger in that in some cases this fetal response to diminished uterine perfusion may be confused with placental abruption and result in emergency cesarean delivery with concerns subsequently discussed.

Labetalol

This is the other commonly used antihypertensive agent for gestational hypertension in the United States and Europe. Labetalol is an α- and nonselective β-blocker which is very effective in acutely lowering blood pressure and many prefer its use over hydralazine because of fewer side effects.[90] At Parkland Hospital we give 10 mg intravenously initially. If the blood pressure has not decreased to the desirable level in 10 minutes, then 20 mg is given. The next 10-minute incremental dose is 40 mg followed by another 40 mg, and then 80 mg if a salutary response is not yet achieved. Sibai[90]

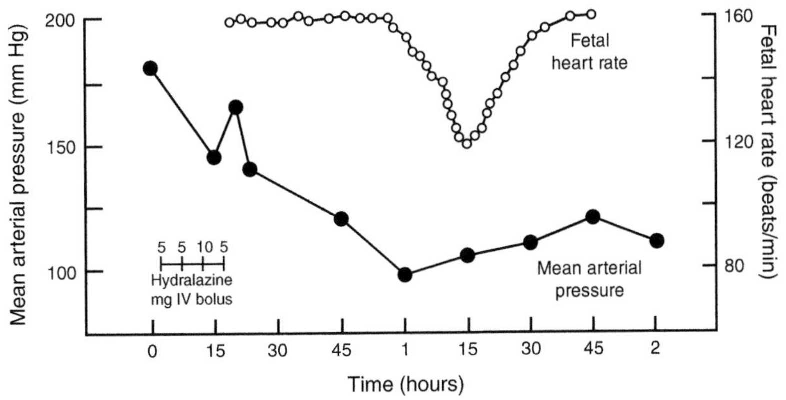

FIGURE 20.5 Effects of acute blood pressure decrease on fetal status. Hydralazine was given at 5-minute intervals instead of 15-minute intervals, and mean arterial pressure decreased from 180 to 90 mm Hg within 1 hour; this change was associated with fetal bradycardia.

recommends 20–40 mg every 10 to 15 minutes as needed for a maximum dose of 220 mg per treatment cycle. The NHBPEP Working Group[3] and the American College of Obstetricians and Gynecologists[1] recommend starting with a 20-mg intravenous bolus. If not effective within 10 minutes, this is followed by 40 mg, then 80 mg every 10 minutes but not to exceed a 220-mg total dose per episode treated.

Hydralazine Versus Labetalol

Vigil-De Gracia et al.[111] randomly assigned 200 severely hypertensive intrapartum women to either (1) intravenous 5-mg hydralazine which could be given every 20 minutes and repeated to a maximum of five doses, or (2) intravenous 20-mg labetalol initially, followed by 40 mg in 20 minutes, and then 80 mg every 20 minutes if needed up to a maximum 300 mg dose. Principal maternal and neonatal outcomes were similar. Hydralazine caused significantly more tachycardia and palpitations whereas labetalol more frequently caused hypotension and bradycardia.

Mabie et al.[113] compared intravenous hydralazine with labetalol for blood pressure control in 60 peripartum women. Labetalol lowered blood pressure more rapidly, and associated tachycardia was minimal, but hydralazine lowered mean arterial pressure to safe levels more effectively. Both drugs have been associated with a reduced frequency of fetal heart rate accelerations.[114]

Nifedipine

Obstetrical use of this calcium-channel blocking agent has also become popular because of its efficacy in control of acute gestational hypertension.[24] The Working Group[2] and the Royal College of Obstetricians and Gynaecologists[69] also recommend that it usually be given in a 10-mg oral dose to be repeated in 30 minutes if necessary. Nifedipine should not be given sublingually and the capsule should not be punctured. Randomized trials that compared nifedipine with labetalol found neither to be definitively superior to the other.[115,116]

Other Antihypertensive Agents

A few other available antihypertensive agents have been tested in clinical trials but are not widely used. Belfort and associates[117] administered *verapamil* – a calcium antagonist – by intravenous infusion at 5–10 mg per hour and reported this to lower mean arterial pressure by 20%. Belfort and co-workers[118,119] later reported that *nimodipine*, when given by continuous infusion or orally, was effective to lower blood pressure in women with severe preeclampsia. Bolte et al.[120,121] reported good results with intravenous *ketanserine*, a selective serotonin-receptor blocker, in preeclamptic women. In a small preliminary comparative trial, Wacker et al.[122] compared intravenous *dihydralazine* with *urapidil*. *Nitroprusside* or *nitroglycerine* is recommended by some if there is not optimal response to the first-line agents.[3,65] With either agent, fetal cyanide toxicity may develop after 4 hours. We have not had the need for either using first-line treatment with hydralazine or labetalol.

There are experimental antihypertensive drugs that some have suggested may become useful for preeclampsia treatment. Two of these are *calcitonin gene-related peptide (CGRP)*, a 37-amino-acid potent vasodilator, and *endogenous digitalis-like factors*, also called *cardiotonic steroids*.[123–125]

Diuretics

When given antepartum, potent diuretics can further compromise placental perfusion because their immediate effects include depletion of intravascular volume, which most often is already reduced compared with that of normal pregnancy. Therefore, before delivery, diuretics are not used to lower blood pressure lest they enhance the intensity of the maternal hemoconcentration and its adverse effects on the mother and the fetus.[126] Antepartum use of furosemide or similar drugs typically is limited to treatment of impending or frank pulmonary edema or congestive heart failure.

Fluid Therapy

Maintenance fluid therapy is provided with lactated Ringer solution administered routinely at the rate of 60 mL to no more than 125 mL per hour. This rate is increased if there is unusual fluid loss from vomiting, diarrhea, or fever, or, more likely, there is excessive blood loss with delivery. In women with oliguria, which is common with severe preeclampsia syndrome, and coupled with the knowledge that maternal blood volume is very likely constricted compared with that of normal pregnancy; it is tempting to administer intravenous fluids more vigorously. The rationale for controlled, conservative fluid administration is that the typical woman with eclampsia already has excessive extracellular fluid that is inappropriately distributed between the intravascular and extravascular spaces. Infusion of large fluid volumes could and does enhance the maldistribution of extravascular fluid and thereby appreciably increases the risk of pulmonary and cerebral edema.[127–129] These concepts are discussed further in Chapter 2 and are illustrated in Figs 2.1 and 2.2 (p. 27).

Pulmonary Edema

Women with severe preeclampsia–eclampsia who develop pulmonary edema most often do so postpartum.[128,130] Aspiration of gastric contents, the result of convulsions, anesthesia, or over-sedation, should be excluded. There are three common causes of pulmonary edema in women with severe preeclampsia syndrome:

1. Pulmonary capillary permeability edema
2. Cardiogenic edema
3. Combinations of cardiogenic and permeability edema.

Most women with severe preeclampsia will have mild pulmonary congestion from permeability edema. Normal pregnancy changes, magnified by the preeclampsia syndrome, predispose to pulmonary edema as discussed in Chapter 2. Importantly, plasma oncotic pressure decreases appreciably in normal term pregnancy because of decreased serum albumin concentration, and oncotic pressure falls even more with preeclampsia.[127,128] Moreover, both increased extravascular fluid oncotic pressure and increased capillary permeability have been described in women with preeclampsia.[131,132] Finally a correlation between plasma colloid osmotic pressure and fibronectin concentration[133] suggests that protein loss results from increased vascular permeability caused by endothelial injury.

Invasive Hemodynamic Monitoring

Much of what has been learned within the past decade about cardiovascular and hemodynamic pathophysiological alterations associated with severe preeclampsia–eclampsia has been made possible by invasive hemodynamic monitoring using a flow-directed pulmonary artery catheter. The need for clinical implementation of such technology for women with preeclampsia–eclampsia, however, has not been established. Gilbert et al.[134] retrospectively reviewed the records of 17 women with eclampsia who underwent pulmonary artery catheterization. Although they concluded that the procedure was subjectively "helpful" in clinical management, all of these women had undergone "multiple interventions," including volume expansion, prior to catheterization. Data obtained in this setting are less likely to be reproducible.

A number of reviews have addressed use of pulmonary artery catheterization in obstetrics.[135–138] Two conditions frequently cited as indications for such monitoring are preeclampsia associated with oliguria and preeclampsia associated with pulmonary edema. Perhaps somewhat paradoxically, it is usually vigorous treatment of the oliguria that results in most cases of pulmonary edema. The American College of Obstetricians and Gynecologists[1] recommends that such monitoring be reserved for preeclamptic women with accompanying severe cardiac disease, renal disease, or both, or in cases of refractory hypertension, oliguria, and pulmonary edema. This was also the recommendation of the 2013 Task Force.[2]

Plasma Volume Expansion

Severe hemoconcentration, along with reduced central venous and pulmonary capillary wedge pressures in women with severe preeclampsia, has tempted some investigators to infuse various fluids, starch polymers, albumin concentrates, or combinations thereof in attempts to expand blood volume. It was theorized that this would somehow relieve vasospasm and reverse organ deterioration. But there are observational studies that describe serious complications – especially pulmonary edema – with volume expansion. For example, López-Llera[139] reported that vigorous volume expansion in more than 700 eclamptic women was associated with a high incidence of pulmonary edema. Benedetti et al.[140] described pulmonary edema in 7 of 10 women with severe preeclampsia who were given colloid therapy. And Sibai et al.[129] cited excessive colloid and crystalloid infusions as causing most of their 37 cases of pulmonary edema associated with severe preeclampsia–eclampsia.

Because preeclampsia syndrome is associated with hemoconcentration directly proportional to severity, attempts to perform volume expansion seemed, at least intuitively, to be reasonable.[141] Over the years, various plasma expanders employed have included crystalloid solutions, albumin, dextran-40, starch polymers, and plasma colloid substitute. In general, these studies were not controlled or even comparative.[142] The Amsterdam randomized study reported by Ganzevoort et al.[44,143] is the largest reported investigation designed to evaluate the benefits and risks of plasma volume expansion. These investigators enrolled 216 women with severe preeclampsia who were between 24 and 34 weeks gestation. The study included women with HELLP syndrome, eclampsia, or fetal growth restriction. Both groups were given magnesium sulfate to prevent eclampsia, betamethasone to promote pulmonary maturity, and ketanserine to control dangerous hypertension. Normal saline infusions were restricted to deliver medications in each group. The group randomly assigned to volume expansion was given 250 mL of 6% hydroxyethyl starch infused over 4 hours twice daily. Maternal demographics and pregnancy outcomes are shown in Table 20.11 and there are no significant differences between the two groups. Perinatal outcomes are shown in Table 20.12 and these too are not significantly different between the two cohorts.

NEUROPROPHYLAXIS – PREVENTION OF ECLAMPSIA

There have been a number of large randomized trials designed to test the efficacy of seizure prophylaxis for women with gestational hypertension, with or without proteinuria. These studies are heterogeneous in that severity of preeclampsia was variable. In most of these studies, magnesium sulfate was compared with another active agent or a placebo. *In all studies, magnesium sulfate was reported to be superior to the comparator agent.* These studies are also discussed in detail in Chapter 12 and four of them are summarized in Table 20.13. Lucas et al.[144] reported that magnesium sulfate therapy was superior to phenytoin in *preventing* eclamptic seizures. *The MAGnesium Sulfate for Prevention of Eclampsia – the MAGPIE Trial*[146] –included more than 10,000 women with preeclampsia from 33 countries who were randomly allocated to be given magnesium sulfate or placebo. Women allocated to magnesium

TABLE 20.11 Maternal Outcomes: Randomized Controlled Trial of Plasma Volume Expansion Versus Saline Infusion in 216 Women with Severe Preeclampsia Between 24 and 34 Weeks

Maternal Factors	Control Group* ($n = 105$)	Treatment Group* ($n = 111$)
Demographics		
Age (mean)	30.8 years	29.0 years
Chronic hypertension	30%	33%
Nulliparous	67%	73%
Gestational age (mean)	30.0 weeks	30.0 weeks
Severe preeclampsia (percent):		
Alone	42	47
HELLP syndrome	26	24
Growth restriction	55	61
Eclampsia	3	2
Pregnancy Outcome		
Maternal death	0	0
Eclampsia (after enrollment)	1.9%	1.8%
HELLP (after enrollment)	19%	17%
Pulmonary edema	2.9%	4.5%
Placental abruption	3.8%	1%
Hepatic hematoma	1.0%	0
Encephalopathy	2.9%	1.8%

*All differences comparison, $p > 0.05$.
Data from Ganzevoort et al.[44,143]

TABLE 20.12 Perinatal Outcomes: Randomized Controlled Trial of Plasma Volume Expansion Versus Saline Infusion in 216 Women with Severe Preeclampsia Between 24 and 34 Weeks

Perinatal Outcomes	Control Group* ($n = 105$)	Treatment Group* ($n = 111$)
Fetal deaths	7%	12%
Prolongation pregnancy (mean)	11.6 days	6.7 days
Estimated gestational age at death (mean)	26.7 weeks	26.3 weeks
Birth weight (mean)	625 g	640 g
Live births	93%	88%
Prolongation pregnancy (mean)	10.5 days	7.4 days
Estimated gestational age at delivery (mean)	31.6 weeks	31.4 weeks
Assisted ventilation	40%	45%
Respiratory distress syndrome	30%	35%
Chronic lung disease	7.6%	8.1%
Neonatal death	7.6%	8.1%
Perinatal mortality rate	142/1000	207/1000

*All differences comparison, $p > 0.05$.
Data from Ganzevoort et al.[44,143]

had a 58% lower risk of eclampsia than those given placebo. In a follow-up study, Smyth et al.[147] reported that exposed children had normal behavior. Finally, Belfort et al.[119] compared magnesium sulfate with nimodipine – a calcium-channel blocker with specific cerebral vasodilator activity – for the prevention of eclampsia. In this unblinded randomized trial involving 1650 women with severe preeclampsia, the rate of eclampsia was more than three-fold higher for women allocated to the nimodipine group – 2.6% versus 0.8%.

Who Should Be Given Magnesium Sulfate?

Because the efficacy of seizure prevention with magnesium is related to severity of the preeclampsia syndrome, the drug will prevent more seizures in women with correspondingly worse disease. And because this is difficult to quantify, it is not always easy to decide which particular woman might benefit from prophylaxis. At least in the United States, the consensus is that women judged to have *severe preeclampsia* should be given magnesium sulfate prophylaxis.[1] Criteria that establish "severity" are not totally uniform and most investigators in the United States for years used the criteria promulgated by the NHBPEP Working Group.[3] As discussed in Chapter 1, these were superseded by criteria of the 2013 Task Force.[2] Our criteria currently in use at Parkland Hospital are shown in Table 20.14.

For women with "non-severe" disease – again variably defined – there are even less clear-cut guidelines. In some countries, there continues to be debate concerning whether any prophylactic magnesium sulfate should be given, and instead therapy is reserved for women who have an eclamptic seizure. We are of the opinion that eclamptic seizures are dangerous and cite maternal mortality rates of up to 5%, even in recent studies.[61,100,149–151] And while there were no maternal deaths from eclampsia in the 2005 audit by the UK Obstetric Surveillance System (UKOSS),[6] the perinatal mortality of 59 per 1000 was essentially unchanged compared with 54 per 1000 reported in the 1992 eclampsia audit. Others have reported similarly increased perinatal mortality rates in both industrialized countries as well as underdeveloped ones.[17,70,146,150,152,153] Finally, the possibility of adverse long-term neuropsychological and visual sequelae of eclampsia,[154,155]which are discussed in Chapter 13, has raised additional concerns that eclamptic seizures may not be so "benign."

Parkland Hospital Study of Selective Versus Universal Magnesium Sulfate Prophylaxis

As discussed, at least in the United States, debate currently centers on which women with preeclampsia should be given prophylaxis. At Parkland Hospital, for women with "non-severe" preeclampsia not given magnesium prophylaxis, Lucas et al.[144] calculated the risk of eclampsia to be about 1 in 100. An opportunity to address these questions was afforded by our change in treatment policy.

TABLE 20.13 Randomized Comparative Trials of Magnesium Sulfate Prophylaxis with Placebo or Active Drug in Women with Gestational Hypertension

Study	Seizures/Total (%)		Comparison
	Magnesium treatment	Control	
Lucas et al.[144] All hypertensives	0/1049 (0)	Phenytoin 10/1089 (0.9)	$p < 0.001$
Coetzee et al.[145] Severe preeclampsia	1/345 (0.3)	Placebo 11/340 (3.2)	RR = 0.09 (0.1–0.69)
Magpie Trial[146] Severe preeclampsia	40/5055 (0.8)	Placebo 96/5055 (1.9)	RR = 0.42 (0.26–0.60)
Belfort et al.[105] Severe preeclampsia	7/831 (0.8)	Nimodipine 21/819 (2.6)	RR = 0.33 (0.14–0.77)

TABLE 20.14 Selective Versus Universal Magnesium Sulfate Prophylaxis: Criteria to Define Severity

- Blood pressure ≥140/90 mm Hg after 20 weeks gestation in a woman not known to be chronically hypertensive
- Proteinuria ≥2+ as measured by dipstick in a catheterized urine specimen
- Serum creatinine >1.2 mg/dL
- Platelet count <100,000/μL
- Aspartate transaminase (AST) elevated two times above upper limit of normal range
- Persistent headache or scotomata
- Persistent midepigastric or right-upper quadrant pain

Criteria used at Parkland Hospital and cited by Alexander et al.[148]

In 2000, we instituted a standardized protocol for the intravenous administration regimen of magnesium sulfate for seizure prophylaxis.[148] Before this, all women with any form of gestational hypertension were given intramuscular magnesium sulfate maintenance treatment or prophylaxis as first described by Pritchard in 1955.[67] Coincidental with adoption of the intravenous magnesium regimen, we changed our practice of universal seizure prophylaxis for all women with gestational hypertension to one of selective prophylaxis given only to women who met our criteria for severe hypertension. These criteria, shown in Table 20.14, called for magnesium prophylaxis for women whose blood pressure was at least 140/90 mm Hg and who had criteria for severe disease as defined in Practice Bulletin 33 of the American College of Obstetricians and Gynecologists,[1] or in whom findings "increased the certainty" of the diagnosis of preeclampsia as set forth by the Working Group.[3] Thus, we included women with ≥2+ proteinuria measured by dipstick in a catheterized urine specimen, although this criterion was not considered a marker for severe disease by ACOG[1] or the Working Group.[3] While using these liberal criteria, about 60% of 6431 women with gestational hypertension over a 4½-year period were given magnesium sulfate prophylaxis. Of the 40% with non-severe hypertension not treated, 27 developed eclamptic seizures – 1 in 92 women. This compared with a seizure rate of 1 in 358 in 3935 women with criteria for severe disease who were given magnesium sulfate. To assess morbidity, outcomes in 87 eclamptic women were compared with all 6431 noneclamptic hypertensive women (Table 20.15). Although maternal outcomes were similar, almost a fourth of women with eclampsia underwent emergency cesarean delivery requiring general anesthesia. This is of great concern because eclamptic women have laryngotracheal edema and are at a higher risk of failed intubation, gastric acid aspiration, and death from the acute respiratory distress syndrome.[156] Neonatal outcomes likewise were of concern as the composite morbidity was increased ten-fold.

Thus, about 1 of 100 women with non-severe gestational hypertension not given magnesium sulfate prophylaxis can be expected to have an eclamptic seizure. If so, adverse perinatal outcomes can be expected to increase as much as ten-fold. A fourth of these women with seizures will require emergency cesarean delivery with attendant maternal morbidity and mortality from general anesthesia.[157]

DELIVERY

To avoid maternal risks from cesarean delivery, steps to effect vaginal delivery are used initially in women with eclampsia. After an eclamptic seizure, labor often ensues spontaneously or can be induced successfully even in women remote from term. An immediate cure does not immediately follow delivery by any route, but serious morbidity is less common during the puerperium in women delivered vaginally.

Blood Loss at Delivery

Hemoconcentration, or lack of normal pregnancy-induced hypervolemia, is an almost predictable feature of severe preeclampsia–eclampsia as quantified by Zeeman et al.[158] and discussed in Chapter 2 (see Figure 2.3). These women, who consequently lack normal pregnancy hypervolemia, are much less tolerant of even normal blood loss than are normotensive pregnant women. *It is of great importance to recognize that an appreciable fall in blood pressure very soon after delivery most often means excessive blood loss and not sudden resolution of vasospasm and endothelial damage.* When oliguria follows delivery, the hematocrit should be evaluated frequently to help detect excessive blood loss, which, if identified, should be treated appropriately by careful crystalloid and blood infusions.

TABLE 20.15 Pregnancy Outcomes in 6518 Women with Gestational Hypertension with and Without Eclampsia

Pregnancy Outcomes	Eclampsia No. (%)	Gestational Hypertension[a] No. (%)	*p* Value
Number	87	6431	
Maternal			
Cesarean delivery	32 (37)	2423 (38)	0.86
Placental abruption	1 (1)	72 (1)	0.98
General anesthesia[b]	20 (23)	270 (4)	<0.01
Neonatal			
Composite morbidity[c]	10 (12)	240 (1)	0.04

[a]Includes preeclampsia.
[b]Emergency cesarean delivery.
[c]One or more: cord artery pH <7.0; 5-minute Apgar score <4; perinatal death; or unanticipated admission of term infant to Intensive Care Nursery.
Data from Alexander et al.[148]

Analgesia and Anesthesia

Over the past 25 years, the use of conduction analgesia for women with preeclampsia syndrome has proven ideal. Initial problems were addressed regarding hypotension and diminished uterine perfusion caused by sympathetic blockade in a woman with attenuated pregnancy-induced hypovolemia from severe preeclampsia. For example, slow induction of epidural analgesia with dilute solutions of anesthetic agents mitigated the need for rapid infusion of large volumes of crystalloid or colloid. These had been given with rapid epidural induction to counteract the associated maternal hypotension and in some women caused pulmonary edema.[129,159,160] Some were even of the opinion that epidural analgesia for women with severe preeclampsia would ameliorate vasospasm and lower blood pressure.[161] Epidural blockade was also ideal because, with general anesthesia, stimulation caused by tracheal intubation could cause sudden severe hypertension, which in turn could cause pulmonary edema, cerebral edema, or intracranial hemorrhage.[162] Finally, as discussed previously, tracheal intubation may be particularly hazardous in women with airway edema due to preeclampsia.[1,106,163,164]

At least two randomized studies have been performed to evaluate these methods of analgesia and anesthesia. Wallace et al.[160] studied 80 women with severe preeclampsia at a mean of 34.8 weeks who were to undergo cesarean delivery. They were randomized to receive general anesthesia or epidural analgesia or combined spinal–epidural analgesia. Their mean preoperative systolic and diastolic blood pressures were approximately 170/110 mm Hg, and all had proteinuria. Anesthetic and obstetrical management included antihypertensive drug therapy and limited intravenous fluids and other drug therapy. Perinatal outcomes of the infants in each group were similar. Maternal hypotension resulting from regional analgesia was managed without excessive intravenous fluid administration. Similarly, maternal blood pressure was managed without severe hypertensive effects

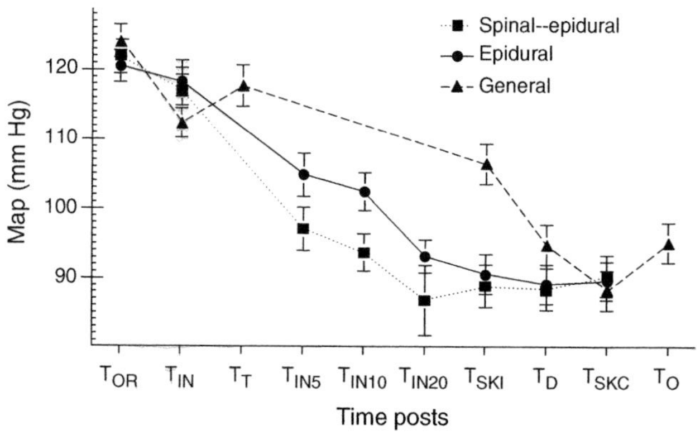

FIGURE 20.6 Blood pressure effects of general anesthesia versus epidural or spinal–epidural analgesia for cesarean delivery in 80 women with severe preeclampsia. MAP = mean arterial pressure. Time posts (T): OR = operating room, IN = anesthesia induction, T = intubation, SKI = skin incision, D = delivery, SKO = skin closure, O = extubation.[149]

in women undergoing general anesthesia (Fig. 20.6). There were no serious maternal or fetal complications attributable to any of the three anesthetic methods. It was concluded that all three were acceptable for use in pregnancies complicated by severe preeclampsia if steps are taken to ensure a careful approach to either method. A similar conclusion was reached by Dyer et al.[164] following their randomized study of spinal analgesia versus general anesthesia in 70 women with preeclampsia and a non-reassuring fetal heart rate tracing. Dyer et al.[165] later showed that decreased vascular resistance and mean arterial blood pressure was effectively counteracted by phenylephrine infusion given in such amounts as to keep the cardiac output unchanged.

Head and colleagues[166] from the University of Alabama at Birmingham randomly assigned 116 women with severe preeclampsia to receive either epidural or patient-controlled intravenous meperidine analgesia during labor. Again, a standardized protocol was used that limited intravenous

TABLE 20.16 Comparison of Cardiovascular Effects of Epidural Versus Patient-Controlled Meperidine Analgesia During Labor in Women with Gestational Hypertension

Hemodynamic Change	Labor Analgesia		*p* Value
	Epidural (*n* = 372)	Meperidine (*n* = 366)	
Mean arterial pressure change (mean)	−25 mm Hg	−15 mm Hg	<.001
Ephedrine for hypotension	11%	0	<.001
Severe hypertension after analgesia (BP ≥160/110 mm Hg)	<1%	1%	NS

NS = not significant.
Data from Lucas et al.[167]

fluids to 100 mL/hour. More women (9%) of the group assigned to epidural analgesia required ephedrine for hypotension. As expected, pain relief was superior in the epidural group compared with the group given intravenous analgesia. Otherwise, other maternal and neonatal complications were similar between groups. One woman in each group developed pulmonary edema.

Other studies have been performed to address the viewpoint that epidural analgesia is an important factor in the intrapartum *treatment* of women with preeclampsia. Lucas et al.[167] studied 738 laboring women from Parkland Hospital who were 36 weeks or more and who had gestational hypertension of varying severity. They were randomly assigned to receive epidural analgesia or patient-controlled intravenous meperidine analgesia. Maternal and infant outcomes were similar in the two study groups. As shown in Table 20.16, although epidural analgesia resulted in a lower mean maternal blood pressure compared with meperidine, this provided no significant benefit in terms of preventing severe hypertension later in labor. It was concluded that labor epidural analgesia should not be misconstrued to be a therapy for hypertension.

The caveat remains that judicious fluid administration in severely preeclamptic women is important in women who receive regional analgesia. Newsome et al.[168] showed that epidural blockade in women with severe preeclampsia caused elevation of pulmonary capillary wedge pressures. It is clear that aggressive volume replacement in these women increases their risk of pulmonary edema, especially in the first 72 hours postpartum.[169,170] When pulmonary edema develops there is also concern for development of cerebral edema. Finally, Heller et al.[171] demonstrated that the majority of cases of pharyngolaryngeal edema were related to aggressive volume therapy.

PERSISTENT SEVERE POSTPARTUM HYPERTENSION

The potential problem of antihypertensive agents causing serious compromise of placental perfusion and fetal well-being is obviated by delivery. If there is a problem after delivery in controlling severe hypertension, and intravenous hydralazine or labetalol is being used repeatedly early in the puerperium to control persistent severe hypertension, then other regimens can be used. If hypertension of appreciable intensity persists or recurs in these postpartum women, then oral labetalol or another beta-blocker, or nifedipine or another calcium-channel blocker, along with a thiazide diuretic has been used. The persistence or refractoriness of hypertension is likely to be due to either mobilization of edema fluid with redistribution into the intravenous compartment or underlying chronic hypertension, or both.[172] In women with chronic hypertension and left ventricular hypertrophy, severe postpartum hypertension can cause pulmonary edema from cardiac failure.[68,173]

We have used a simple method to estimate excessive extracellular/interstitial fluid. The postpartum weight is compared with the last recorded recent prenatal weight, either from the last clinic visit or upon admission for delivery. On average, soon after delivery, maternal weight should be reduced by at least 10 to 15 pounds depending on birth weight, amniotic fluid volume, placental weight, and blood loss. Because of various interventions, especially intravenous crystalloid infusions given during operative vaginal or cesarean delivery, women with severe preeclampsia often have a postpartum weight in excess of their last prenatal weight. If this weight increase is associated with severe persistent postpartum hypertension, diuresis with intravenous furosemide is usually helpful.

Furosemide

Because persistence of severe hypertension corresponds to the onset and length of extracellular fluid mobilization and diuresis, it seems logical that furosemide-augmented diuresis would improve blood pressure control. To study this, the Mississippi group[174] designed a randomized trial that included 264 postpartum preeclamptic women. After onset of spontaneous diuresis, they were randomly assigned to 20 mg furosemide orally versus no therapy. Compared with women with mild disease, those with severe preeclampsia had lower systolic blood pressure by 2 days – 142 ± 13 versus 153 ± 19 mm Hg – and less often required antihypertensive therapy – 14% versus 26%, respectively.

References

1. American College of Obstetricians and Gynecologists. Diagnosis and management of preeclampsia and eclampsia. *Practice Bulletin No. 33*, January 2002, reaffirmed 2012.
2. American College of Obstetricians and Gynecologists Task Force on Hypertension in Pregnancy. Washington DC, November 2013.
3. National High Blood Pressure Education Program: Working Group Report on High Blood Pressure in Pregnancy. *Am J Obstet Gynecol.* 2000;183:S1–S22.
4. Macdonald-Wallis C, Lawlor DA, Fraser A, et al. Blood pressure change in normotensive, gestational hypertensive, preeclamptic, and essential hypertensive pregnancies. *Hypertension.* 2012;59(6):1241.
5. Cnossen JS, de Ruyter-Hanhijarvi H, van der Post JA, et al. Accuracy of serum uric acid determination in predicting preeclampsia: a systematic review. *Acta Obstet Gynecol Scand.* 2006;85(5):519–525.
6. Knight M. Eclampsia in the United Kingdom2005. *BJOG.* 2007;114(9):1072–1078.
7. Abenhaim HA, Bujold E, Benjamin A, et al. Evaluating the role of bedrest on the prevention of hypertensive disease of pregnancy and growth restriction. *Hypertens Pregnancy.* 2008;27(2):197–205.
8. Meher S, Duley L. Rest during pregnancy for preventing preeclampsia and its complications in women with normal blood pressure. *Cochrane Database Syst Rev.* 2006;19:CD005939.
9. Hauth JC, Cunningham FG, Whalley PJ. Management of pregnancy-induced hypertension in the nullipara. *Obstet Gynecol.* 1976;48(3):253–259.
10. Gilstrap III LC, Cunningham FG, Whalley PJ. Management of pregnancy-induced hypertension in the nulliparous patient remote from term. *Semin Perinatol.* 1978;2(1):73–81.
11. Lo C, Taylor RS, Gamble G, et al. Use of automated home blood pressure monitoring in pregnancy: is it safe? *Am J Obstet Gynecol.* 2002;187(5):1321–1328.
12. Ostchega Y, Zhang G, Sorlie P, et al. Blood pressure randomized methodology study comparing automatic oscillometric and mercury sphygmomanometer devices: National Health and Nutrition Examination Survey, 2009-2010. *Natl Health Stat Report.* 2012;59:1.
13. Barton CB, Barton JR, O'Brien JM, et al. Mild gestational hypertension: differences in ethnicity are associated with altered outcomes in women who undergo outpatient treatment. *Am J Obstet Gynecol.* 2002;186(5):896–898.
14. Horsager R, Adams M, Richey S. Outpatient management of mild pregnancy induced hypertension. *Am J Obstet Gynecol.* 1995;172(1 Part 2):383.
15. Crowther CA, Bouwmeester AM, Ashurst HM. Does admission to hospital for bed rest prevent disease progression or improve fetal outcome in pregnancy complicated by non-proteinuric hypertension? *Br J Obstet Gynaecol.* 1992;99(1):13–17.
16. Turnbull DA, Wilkinson C, Gerard K. Clinical, psychosocial, and economic effects of antenatal day care for three medical complications of pregnancy: a randomized controlled trial of 395 women. *Lancet.* 2004;363:1104–1109.
17. Tuffnell DJ, Lilford RJ, Buchan PC, et al. Randomized controlled trial of day care for hypertension in pregnancy. *Lancet.* 1992;339:224–227.
18. Miline F, Redman C, Walker J, et al. Assessing the onset of preeclampsia I the hospital day unit: summary of the preeclampsia guideline (PRECOG II). *BMJ.* 2009;339:b3129.
19. Sibai BM, Gonzalez AR, Mabie WC. A comparison of labetalol plus hospitalization versus hospitalization alone in the management of preeclampsia remote from term. *Obstet Gynecol.* 1987;70:323.
20. Sibai BM, Barton JR, Akl S. A randomized prospective comparison of nifedipine and bed rest alone in the management of preeclampsia remote from term. *Am J Obstet Gynecol.* 1992;167:879–885.
21. Pickles CJ, Broughton F, Pipkin EM, et al. A randomised placebo controlled trial of labetalol in the treatment of mild to moderate pregnancy induced hypertension. *Br J Obstet Gynaecol.* 1992;99:964–969.
22. Wide-Swensson DH, Ingemarsson I, Lunnell NO. Calcium channel blockade (isradipine) in treatment of hypertension in pregnancy: a randomized placebo-controlled study. *Am J Obstet Gynecol.* 1995;173:872–878.
23. von Dadelszen P, Ornstein MP, Bull SB. Fall in mean arterial pressure and fetal growth restriction in pregnancy hypertension: a meta-analysis. *Lancet.* 2000;355:87–96.
24. von Dadelszen P, Menzies J, Gilgoff S, et al. Evidence-based management for preeclampsia. *Front Biosci.* 2007;12:2876–2889.
25. Abalos E, Duley L, Steyn DW, et al. Antihypertensive drug therapy for mild to moderate hypertension during pregnancy. *Cochrane Database Syst Rev.* 2007;24:CD002252.
26. Nassar AH, Adra AM, Chakhtoura N. Severe preeclampsia remote from term: labor induction or elective cesarean delivery? *Am J Obstet Gynecol.* 1998;179:1210–1218.
27. Tajik P, van der Tuuk K, Koopmans CM, et al. Should cervical favourability play a role in the decision for labour induction in gestational hypertension or mild preeclampsia at term? An exploratory analysis of the HYPITAT trial. *BJOG.* 2012;119(9):1123.
28. Barton J, Barton L, Istwan N, et al. The frequency of elective delivery at 34.0-36.9 weeks' gestation and its impact on neonatal outcomes in women with stable mild gestational hypertension (MGHTN). Society for Maternal-Fetal Medicine. *Am J Obstet Gynecol.* 2008;199(6):S82. [Abstract No. 252].
29. Langenveld J, Ravelli ACJ, van Kaam AH, et al. Neonatal outcome of pregnancies complicated by hypertensive disorders between 34 and 37 weeks of gestation: a 7 year retrospective analysis of national registry. *Am J Obstet Gynecol.* 2011;205(6):540.e1.
30. American College of Obstetricians and Gynecologists. Antepartum fetal surveillance. *Practice Bulletin No. 9*, 1999.
31. Alexander JM, Bloom SL, McIntire DD. Severe preeclampsia and the very low-birthweight infant: is induction of labor harmful? *Obstet Gynecol.* 1999;93:485–495.
32. Sibai BM, Taslimi M, Abdella TN. Maternal and perinatal outcome of conservative management of severe preeclampsia in midtrimester. *Am J Obstet Gynecol.* 1995;152:32–37.
33. Sibai BM, Mercer BM, Schiff E. Aggressive versus expectant management of severe preeclampsia at 28 to 32 weeks' gestation: a randomized controlled trial. *Am J Obstet Gynecol.* 1994;171:818–828.
34. Witlin AG, Saade GR, Mattar F. Predictors of neonatal outcome in women with severe preeclampsia or eclampsia

between 24 and 33 weeks' gestation. *Am J Obstet Gynecol.* 2000;182:607–612.

35. Abramovici D, Friedman SA, Mercer BM. Neonatal outcome in severe preeclampsia at 24 to 36 weeks' gestation: does the HELLP (hemolysis, elevated liver enzyme, and low platelet count) syndrome matter? *Am J Obstet Gynecol.* 1999;80:221–228.
36. Haddad B, Kayem G, Deis S, et al. Are perinatal and maternal outcomes different during expectant management of severe preeclampsia in the presence of intrauterine growth restriction? *Am J Obstet Gynecol.* 2007;196:237.e1–237.e5.
37. Sibai BM, Barton JR. Expectant management of severe preeclampsia remote from term: patient selection, management, and delivery indications. *Am J Obstet Gynecol.* 2007;196:514.e1–514.e9.
38. Oettle C, Hall D, Roux A, et al. Early onset severe pre-eclampsia: expectant management at a secondary hospital in close association with a tertiary institution. *BJOG.* 2005;112:84–88.
39. Shear RM, Rinfret D, Leduc L. Should we offer expectant management in cases of severe preterm preeclampsia with fetal growth restriction? *Am J Obstet Gynecol.* 2005;192:1119–1125.
40. Ganzevoort W, Rep A, de Vries JI, et al. Prediction of maternal complications and adverse infant outcome at admission for temporizing management of early-onset severe hypertensive disorders in pregnancy. *Am J Obstet Gynecol.* 2006;195:495–503.
41. Sarsam DS, Shamden M, Al Wazan R. Expectant versus aggressive management in severe preeclampsia remote from term. *Singapore Med J.* 2008;49(9):698–703.
42. Bombrys AE, Barton JR, Nowacki EA, et al. Expectant management of severe preeclampsia at less than 27 weeks' gestation: maternal and perinatal outcomes according to gestational age by weeks at onset of expectant management. *Am J Obstet Gynecol.* 2008;199:247.
43. Abdel-Hady el-S, Fawzy M, El-Negeri M, et al. Is expectant management of early-onset severe preeclampsia worthwhile in low-resource settings? *Arch Gynecol Obstet.* 2010;282(1):23–27.
44. Ganzevoort W, Rep A, Bonsel GJ, et al. A randomized controlled trial comparing two temporizing management strategies, one with and one without plasma volume expansion, for severe and early onset pre-eclampsia. *BJOG.* 2005;112:1337–1338.
45. Barbar D, Xing G, Towner D. Expectant management of severe eclampsia between 24–32.weeks gestation: a ten year review. Society for Maternal-Fetal Medicine. *Am J Obstet Gynecol.* 2009;199(6):S211. [Abstract No. 742].
46. Hall DR, Odendaal HJ, Steyn DW. Expectant management of severe pre-eclampsia in the mid-trimester. *Eur J Obstet Gynecol Reprod Biol.* 2001;96:168–172.
47. Gaugler-Senden IP, Huijssoon AG, Visser W, et al. Maternal and perinatal outcome of preeclampsia with an onset before 24 weeks' gestation: audit in a tertiary referral center. *Eur J Obstet Gynecol Reprod Biol.* 2006;128:216–221.
48. Budden A, Wilkinson L, Buksh MJ, et al. Pregnancy outcome in women presenting with pre-eclampsia at less than 25 weeks gestation. *Aust N Z J Obstet Gynaecol.* 2006;46:407–412.
49. Belghiti J, Kayem G, Tsatsaris V, et al. Benefits and risks of expectant management of severe preeclampsia at less than 26 weeks gestation: the impact of gestational age and severe fetal growth restriction. *Am J Obstet Gynecol.* 2011;205(5):465.e1–465.e6.
50. Amorim MMR, Santos LC, Faúndes A. Corticosteroid therapy for prevention of respiratory distress syndrome in severe preeclampsia. *Am J Obstet Gynecol.* 1999;180:1283.
51. Bloom SL, Leveno KJ. Corticosteroid use in special circumstances: preterm ruptured membranes, hypertension, fetal growth restriction, multiple fetuses. *Clin Obstet Gynecol.* 2003;46:150–158.
52. Thiagarajah S, Bourgeois FJ, Harbert GM. Thrombocytopenia in preeclampsia: associated abnormalities and management principles. *Am J Obstet Gynecol.* 1984;150:1–8.
53. O'Brien JM, Shumate SA, Satchwell SL. Maternal benefit of corticosteroid therapy in patients with HELLP (hemolysis, elevated liver enzymes, and low platelet count) syndrome: impact on the rate of regional anesthesia. *Am J Obstet Gynecol.* 2002;186:475–483.
54. Tompkins MJ, Thiagarajah S. HELLP (hemolysis, elevated liver enzymes, and low platelet count) syndrome: the benefit of corticosteroids. *Am J Obstet Gynecol.* 1999;181:304–309.
55. Martin JN, Thigpen BD, Rose CH. Maternal benefit of high-dose intravenous corticosteroid therapy for HELLP syndrome. *Am J Obstet Gynecol.* 2003;189:830–839.
56. Isler CM, Barrilleaux PS, Magann EF. A prospective, randomized trial comparing the efficacy of dexamethasone and betamethasone for the treatment of antepartum HELLP (hemolysis, elevated liver enzymes, and low platelet count) syndrome. *Am J Obstet Gynecol.* 2001;184:1332–1340.
57. Fonseca JE, Méndez F, Cataño C, et al. Dexamethasone treatment does not improve the outcome of women with HELLP syndrome: a double-blind, placebo-controlled, randomized clinical trial. *Am J Obstet Gynecol.* 2005;193:1591–1598.
58. Katz L, Ramos de Amorim MM, Natal Figueiroa J, et al. Postpartum dexamethasone for women with hemolysis, elevated liver enzymes, and low platelets (HELLP) syndrome: a double-blind, placebo-controlled, randomized clinical trial. *Am J Obstet Gynecol.* 2008;198:283.
59. Cunningham GC, Pritchard JA. How should hypertension during pregnancy be managed? Experience at Parkland Memorial Hospital. *Med Clin North Am.* 1984;68:505–526.
60. Mattar F, Sibai BM. Eclampsia: VIII. Risk factors for maternal morbidity. *Am J Obstet Gynecol.* 2000;182:307.
61. Andersgaard AB, Herbst A, Johansen M. Eclampsia in Scandinavia: incidence, substandard care, and potentially preventable cases. *Acta Obstet Gynecol.* 2006;85:929–936.
62. Akkawi C, Kent E, Geary M, et al. The incidence of eclampsia in a single defined population with a selective use of magnesium sulfate. Society for Maternal-Fetal Medicine. *Am J Obstet Gynecol.* 2009;199(6):S225. [Abstract No. 798].
63. Thornton C, Dahlen H, Korda A, et al. The incidence of preeclampsia and eclampsia and associated maternal mortality in Australia from population-linked datasets: 2000–2008. *Am J Obstet Gynecol.* 2013;280(6):476.e1.
64. Chames MC, Livingston JC, Ivester TS. Late postpartum eclampsia: a preventable disease? *Am J Obstet Gynecol.* 2002;186:1174–1179.
65. Sibai BM. Diagnosis, prevention, and management of eclampsia. *Obstet Gynecol.* 2005;105:402–410.

66. Chesley LC. *Chesley's Hypertensive Disorders in Pregnancy*. 1st ed. New York: Appleton; 1978.
67. Pritchard JA. The use of magnesium ion in the management of eclamptogenic toxemias. *Surg Gynecol Obstet*. 1955;100:131–140.
68. Pritchard JA, Pritchard SA. Standardized treatment of 154 consecutive cases of eclampsia. *Am J Obstet Gynecol*. 1975;123:543–552.
69. Royal College of Obstetricians and Gynaecologists: the management of severe pre-eclampsia/eclampsia. Guideline No. 10 (A). *London RCOG Press*. 2006;1–11.
70. Chowdhury JR, Chaudhuri S, Bhattacharyya N, et al. Comparison of intramuscular magnesium sulfate with low does intravenous magnesium sulfate regimen for treatment of eclampsia. *J Obstet Gynaecol Res*. 2009;35:119.
71. Pritchard JA, Cunningham FG, Pritchard SA. The Parkland Memorial Hospital protocol for treatment of eclampsia: evaluation of 245 cases. *Am J Obstet Gynecol*. 1983;148:951–961.
72. Ehrenberg HM, Mercer BM. Abbreviated postpartum magnesium sulphate therapy for women with mild preeclampsia: a randomized controlled trial. *Obstet Gynecol*. 2006;108:833–838.
73. Taber EB, Tan L, Chao CR, et al. Pharmacokinetics of ionized versus total magnesium in subjects with preterm labor and preeclampsia. *Am J Obstet Gynecol*. 2002;186:1017–1021.
74. Sibai BM, Graham JM, McCubbin JH. A comparison of intravenous and intramuscular magnesium sulfate regimens in preeclampsia. *Am J Obstet Gynecol*. 1984;150:728–733.
75. Tudela CM, McIntire DD, Alexander JM. Effect of maternal body mass index onserum magnesium levels given for seizure prophylaxis. *Obstet Gynecol*. 2013;121(2 Part 1): 314–320.
76. Somjen G, Hilmy M, Stephen CR. Failure to anesthetize human subjects by intravenous administration of magnesium sulfate. *J Pharmacol Exp Ther*. 1966;154:652–659.
77. McCubbin JH, Sibai BM, Abdella TN. Cardiopulmonary arrest due to acute maternal hypermagnesemia. *Lancet*. 1981;1:1058–1061.
78. Smith JM, Lowe RF, Fullerton J, et al. An integrative review of the side effects related to the use of magnesium sulfate for preeclampsia and eclampsia management. *BMC Pregnancy Childbirth*. 2013;13:34.
79. Cotton DB, Longmire S, Jones MM. Cardiovascular alterations in severe pregnancy-induced hypertension: effects of intravenous nitroglycerin coupled with blood volume expansion. *Am J Obstet Gynecol*. 1986;154:1053–1062.
80. Thurnau GR, Kemp DB, Jarvis A. Cerebrospinal fluid levels of magnesium in patients with preeclampsia after treatment with intravenous magnesium sulfate: a preliminary report. *Am J Obstet Gynecol*. 1987;157:1435–1442.
81. Fong J, Gurewitsch ED, Vipe L. Baseline serum and cerebrospinal fluid magnesium levels in normal pregnancy and preeclampsia. *Obstet Gynecol*. 1995;85:444–452.
82. Arango MF, Mejia-Mantilla JH. Magnesium for acute traumatic brain injury. *Cochrane Database Syst Rev*. 2006;18:CD005400.
83. Hallak M, Hotca JW, Evans JB. Magnesium sulfate affects the N-methyl-D-aspartate receptor binding in maternal rat brain. *Am J Obstet Gynecol*. 1998;178:S112.
84. Wang LC, Huang CY, Want HK, et al. Magnesium sulfate and nimesulide have synergistic effects on rescuing brain damage after transient focal ischemia. *J Neurotrauma*. 2012;29(7):1518.
85. Leveno KJ, Alexander JM, McIntire DD. Does magnesium sulfate given for prevention of eclampsia affect the outcome of labor? *Am J Obstet Gynecol*. 1998;178:707–714.
86. Atkinson MW, Guinn D, Owen J. Does magnesium sulfate affect the length of labor induction in women with pregnancy-associated hypertension. *Am J Obstet Gynecol*. 1995;173:1219.
87. Szal SE, Croughan-Minibane MS, Kilpatrick SJ. Effect of magnesium prophylaxis and preeclampsia on the duration of labor. *Am J Obstet Gynecol*. 1999;180:1475.
88. Witlin AG, Friedman SA, Sibai BA. The effect of magnesium sulfate therapy on the duration of labor in women with mild preeclampsia at term: a randomized, double-blind, placebo-controlled trial. *Am J Obstet Gynecol*. 1997;176:623–631.
89. Watt-Morse ML, Caritis SN, Kridgen PL. Magnesium sulfate is a poor inhibitor of oxytocin-induced contractility in pregnant sheep. *J Matern Fetal Med*. 1995;4:139–148.
90. Grimes DA, Nanda K. Magnesium sulfate tocolysis: time to quit. *Obstet Gynecol*. 2006;108:986–989.
91. Hallak M, Berry SM, Madincea F. Fetal serum and amniotic fluid magnesium concentrations with maternal treatment. *Obstet Gynecol*. 1993;81:185–192.
92. Cunningham FG, Pritchard JA. How should hypertension during pregnancy be managed? Experience at Parkland Memorial Hospital. *Med Clin North Am*. 1984;68:505–514.
93. Hallak M, Martinez-Poyer J, Kruger ML. The effect of magnesium sulfate on fetal heart rate parameters: a randomized, placebo-controlled trial. *Am J Obstet Gynecol*. 1999;181:1122.
94. Duffy CR, Odibo AO, Roehl KA, et al. Effect of magnesium sulfate on fetal heart rate patterns in the second stage of labor. *Obstet Gynecol*. 2012;119(6):1129.
95. Nelson KB, Grether JK. Can magnesium sulfate reduce the risk of cerebral palsy in very low birthweight infants? *Pediatrics*. 1995;95:263–272.
96. Schendel DE, Berg CJ, Yeargin-Allsopp M. Prenatal magnesium sulfate exposure and the risk for cerebral palsy or mental retardation among very low birthweight children aged 3 to 5 years. *JAMA*. 1996;276:1805–1814.
97. Crowther CA, Hilles JE, Doyle LW. Effect of magnesium sulfate given for neuroprotection before birth: a randomized controlled trial. *JAMA*. 2003;290:2669–2679.
98. Marret S, Marpeau L, Follet-Bouhamed C. Effect of magnesium sulphate on mortality and neurologic morbidity of the very-preterm newborn (of less than 33 weeks) with two-year neurological outcome: results of the prospective PREMAG trial. *Gynecol Obstet Fertil*. 2008;36:278–288.
99. Rouse DJ, Hirtz DG, Thom E. A randomized, controlled trial of magnesium sulfate for the prevention of cerebral palsy. *New Engl J Med*. 2008;359:895–905.
100. Which anticonvulsant for women with eclampsia? Evidence from the collaborative eclampsia trial. *Lancet*. 1995;345: 1455–1463.
101. Dommisse J. Phenytoin sodium and magnesium sulphate in the management of eclampsia. *Br J Obstet Gynaecol*. 1990;97:104–109.

102. Crowther C. Magnesium sulphate versus diazepam in the management of eclampsia: a randomized controlled trial. *Br J Obstet Gynaecol.* 1990;97:110–117.
103. Bhalla AK, Dhall GI, Dhall K. A safer and more effective treatment regimen for eclampsia. *Aust N Z J Obstet Gynaecol.* 1995;34:144–148.
104. Friedman SA, Lim KH, Baker CA, et al. Phenytoin versus magnesium sulfate in preeclampsia: a pilot study. *Am J Perinatol.* 1993;10:233–238.
105. Cerebral lesionsSheehan HL, Lynch JB, eds. *Pathology of Toxaemia of Pregnancy*. Baltimore: Williams & Wilkins; 1973.
106. *Williams Obstetrics*. 12th ed. New York: Appleton; 1961.
107. Hellman LM, Pritchard JA, Eastman NJ, et al. *Williams Obstetrics*; 1971.
108. Cunningham FG, Leveno KJ, Bloom SL, et al. *Williams Obstetrics*. 24th ed. New York: McGraw-Hill; 2014.
109. Martin JN, Thigpen BD, Moore RC, et al. Stroke and severe preeclampsia and eclampsia: a paradigm shift focusing on systolic blood pressure. *Obstet Gynecol.* 2005;105:246–254.
110. Cunningham FG. Severe preeclampsia and eclampsia: systolic hypertension is also important. *Obstet Gynecol.* 2005;105:237–238.
111. Vigil-De Gracia P, Ruiz E, López JC, et al. Management of severe hypertension in the postpartum period with intravenous hydralazine or labetalol: a randomized clinical trial. *Hypertens Pregnancy*. 2007;26:163–171.
112. Magee LA, Yong PJ, Espinosa V, et al. Expectant management of severe preeclampsia remote from term: a structured systematic review. *Hypertens Pregnancy*. 2009;25:1.
113. Mabie WC, Gonzalez AR, Sibai BM. A comparative trial of labetalol and hydralazine in the acute management of severe hypertension complicating pregnancy. *Obstet Gynecol.* 1987;70:328–337.
114. Cahill A, Odibo A, Roehl K, et al. Impact of intrapartum antihypertensives on electronic fetal heart rate (EFM) patterns in labor. Abstract No. 615. *Am J Obstet Gynecol.* 2013;208(1 suppl):S262.
115. Vermillion ST, Scardo JA, Newman RB. A randomized, double-blind trial of oral nifedipine and intravenous labetalol in hypertensive emergencies of pregnancy. *Am J Obstet Gynecol.* 1999;181:858–868.
116. Scardo JA, Vermillion ST, Newman RB. A randomized, double-blind, hemodynamic evaluation of nifedipine and labetalol in preeclamptic hypertensive emergencies. *Am J Obstet Gynecol.* 1999;181:862–869.
117. Belfort MA, Anthony J, Buccimazza A. Hemodynamic changes associated with intravenous infusion of the calcium antagonist verapamil in the treatment of severe gestational proteinuric hypertension. *Obstet Gynecol.* 1990;75:970–977.
118. Belfort MA, Taskin O, Buhur A. Intravenous nimodipine in the management of severe preeclampsia: double blind, randomized, controlled clinical trial. *Am J Obstet Gynecol.* 1996;174:451–462.
119. Belfort M, Anthony J, Saade G. A comparison of magnesium sulfate and nimodipine for the prevention of eclampsia. *New Engl J Med.* 2003;348:304–311.
120. Bolte AC, Gafar S, van Eyck J. Ketanserin, a better option in the treatment of preeclampsia? *Am J Obstet Gynecol.* 1998;178:S118.
121. Bolte AC, van Eyck J, Gaffar SF. Ketanserin for the treatment of preeclampsia. *J Perinat Med.* 2001;29:14.
122. Wacker JR, Wagner BK, Briese V, et al. Antihypertensive therapy in patients with preeclampsia: a randomized multicentre study comparing dihydralazine with urapidil. *Eur J Obstet Gynecol Reprod Biol.* 2006;127:160–165.
123. Márquez-Rodas I, Longo F, Rothlin RP, et al. Pathophysiology and therapeutic possibilities of calcitonin gene-related peptide in hypertension. *J Physiol Biochem.* 2006;62:45–56.
124. Bagrov AY, Shapiro JI. Endogenous digitalis: pathophysiologic roles and therapeutic applications. *Nat Clin Pract Nephrol.* 2008;4:378–392.
125. Lam GK, Hopoate-Sitake M, Adair CD, et al. Digoxin antibody fragment, antigen binding (Fab), treatment of preeclampsia in women with endogenous digitalis-like factor: a secondary analysis of the DEEP trial. *Am J Obstet Gynecol.* 2013;209(2):119.e1–119.e6.
126. Zondervan HA, Oosting J, Smorenberg-Schoorl ME. Maternal whole blood viscosity in pregnancy hypertension. *Gynecol Obstet Invest.* 1988;25:83–92.
127. Sciscione AC, Ivester T, Largoza M. Acute pulmonary edema in pregnancy. *Obstet Gynecol.* 2003;101:511–518.
128. Zinaman M, Rubin J, Lindheimer MD. Serial plasma oncotic pressure levels and echoencephalography during and after delivery in severe preeclampsia. *Lancet.* 1985;1:1245–1252.
129. Sibai BM, Mabie BC, Harvey CJ. Pulmonary edema in severe preeclampsia-eclampsia: analysis of thirty-seven consecutive cases. *Am J Obstet Gynecol.* 1987;156:1174–1185.
130. Cunningham FG, Pritchard JA, Hankins GDV. Peripartum heart failure: Idiopathic cardiomyopathy or compounding cardiovascular events? *Obstet Gynecol.* 1986;67:157–168.
131. Brown MA, Zammit VC, Lowe SA. Capillary permeability and extracellular fluid volumes in pregnancy-induced hypertension. *Clin Sci.* 1989;77:599–608.
132. Øian P, Maltau JM, Noddleland H. Transcapillary fluid balance in preeclampsia. *Br J Obstet Gynaecol.* 1986;93:235–242.
133. Bhatia RK, Bottoms SF, Saleh AA. Mechanisms for reduced colloid osmotic pressure in preeclampsia. *Am J Obstet Gynecol.* 1987;157:106–112.
134. Gilbert WM, Towner DR, Field NT. The safety and utility of pulmonary artery catheterization in severe preeclampsia and eclampsia. *Am J Obstet Gynecol.* 2000;182:1397.
135. Clark SL, Cotton DB. Clinical indications for pulmonary artery catheterization in the patient with severe preeclampsia. *Am J Obstet Gynecol.* 1988;158:453–461.
136. Hankins GDV, Cunningham FG. Severe preeclampsia and eclampsia: controversies in management. 18th ed. *Williams Obstetrics*, suppl 12 Norwalk, CT: Appleton & Lange; 1991: 1–12.
137. Nolan TE, Wakefield ML, Devoe LD. Invasive hemodynamic monitoring in obstetrics. A critical review of its indications, benefits, complications, and alternatives. *Chest.* 1992;101:1429–1438.
138. Young P, Johanson R. Haemodynamic, invasive and echocardiographic monitoring in the hypertensive parturient. *Best Pract Res Clin Obstet Gynaecol.* 2001;15:605–615.
139. López-Llera M. Complicated eclampsia: fifteen years' experience in a referral medical center. *Am J Obstet Gynecol.* 1982;142:28–39.

140. Benedetti TJ, Kates R, Williams V. Hemodynamic observations in severe preeclampsia complicated by pulmonary edema. *Am J Obstet Gynecol.* 1985;152:330.
141. Ganzevoort W, Rep A, Bonsel GJ, et al. Plasma volume and blood pressure regulation in hypertensive pregnancy. *J Hypertens.* 2004;22:1235–1242.
142. Habek D, Bobic MV, Habek JC. Oncotic therapy in management of preeclampsia. *Arch Med Res.* 2006;37:619–623.
143. Ganzevoort W, Rep A, Bonsel GJ, et al. A randomized trial of plasma volume expansion in hypertensive disorders of pregnancy: influence on the pulsatile indices of the fetal umbilical artery and middle cerebral artery. *Am J Obstet Gynecol.* 2005;192:233–239.
144. Lucas MJ, Leveno KJ, Cunningham FG. A comparison of magnesium sulfate with phenytoin for the prevention of eclampsia. *New Engl J Med.* 1995;333:201–205.
145. Coetzee EJ, Dommisse J, Anthony J. A randomized controlled trial of intravenous magnesium sulfate versus placebo in the management of women with severe preeclampsia. *Br J Obstet Gynaecol.* 1998;105:300–312.
146. Do women with pre-eclampsia, and their babies, benefit from magnesium sulphate? The Magpie Trial: a randomized placebo-controlled trial. *Lancet.* 2005;359:1877–1890.
147. Smyth RM, Spark P, Armstrong N, Duley L. Magpie Trial in the UK: methods and additional data for women and children at 2 years following pregnancy complicated by pre-eclampsia. *Pregnancy Childbirth.* 2009;9:15.
148. Alexander JM, McIntire DD, Leveno KJ, et al. Selective magnesium sulfate prophylaxis for the prevention of eclampsia in women with gestational hypertension. *Obstet Gynecol.* 2006;108:826–832.
149. Miguli M, Chekairi A. Eclampsia, study of 342 cases. *Hypertens Pregnancy.* 2008;27:103–111.
150. Schutte JM, Schuitemaker NW, van Roosmalen J, et al. Substandard care in maternal mortality due to hypertensive disease in pregnancy in the Netherlands. *BJOG.* 2008;115:732–736.
151. Basso O, Rasmussen S, Weinberg CR, et al. Trends in fetal and infant survival following preeclampsia. *JAMA.* 2006;296:1357–1362.
152. Ndaboine EM, Kihunrwa A, Rumanyika R, et al. Maternal and perinatal outcomes among eclamptic patients admitted to Bugando Medical Centre, Mwanza, Tanzania. *Afr J Reprod Health.* 2012;16(1):35.
153. Moodley J. Maternal deaths associated with eclampsia in South Africa: lessons to learn from the confidential enquiries into maternal deaths, 2005–2007. *S Afr Med J.* 2010;100(11):717.
154. Aukes AM, Wessel I, Dubois AM, et al. Self-reported cognitive functioning in formerly eclamptic women. *Am J Obstet Gynecol.* 2007;197:365.
155. Wiegman MJ, de Groot JC, Jansonius NM, et al. Long-term visual functioning after eclampsia. *Obstet Gynecol.* 2012;119(5):959–966.
156. Aukes AM, de Groot JC, Aarnoudse JG, et al. Brain lesions several years following eclampsia. *Am J Obstet Gynecol.* 2009;200:504.
157. American College of Obstetricians and Gynecologists. Obstetric analgesia and anesthesia. *Practice Bulletin No. 36*, July 2002.
158. Zeeman GG, Cunningham FG, Pritchard JA. The magnitude of hemoconcentration with eclampsia. *Hypertens Pregnancy.* 2009;28:127–137.
159. Hogg B, Hauth JC, Caritis SN. Safety of labor epidural anesthesia for women with severe hypertensive disease. *Am J Obstet Gynecol.* 1999;181:1096.
160. Wallace DH, Leveno KJ, Cunningham FG. Randomized comparison of general and regional anesthesia for cesarean delivery in pregnancies complicated by severe preeclampsia. *Obstet Gynecol.* 1995;86:193.
161. Gutsche BB, Cheek TG, Shnider SM, et al. Anesthesia considerations in preeclampsia-eclampsia. *Anesth Obstet.* 1993:321.
162. Lavies NG, Meiklejohn BH, May AE. Hypertensive and catecholamine response to tracheal intubation in patients with pregnancy-induced hypertension. *Br J Anaesth.* 1989;63:429–438.
163. Chadwick HS, Easterling T. Anesthetic concerns in the patient with preeclampsia. *Semin Perinatol.* 1991;15:397–411.
164. Dyer RA, Els I, Farbas J. Prospective, randomized trial comparing general with spinal anesthesia for cesarean delivery in preeclamptic patients with a nonreassuring fetal heart trace. *Anesthesiology.* 2003;99:561.
165. Dyer RA, Piercy JL, Reed AR, et al. Hemodynamic changes associated with spinal anesthesia for cesarean delivery in severe preeclampsia. *Anesthesiology.* 2008;108:802–811.
166. Head BB, Owen J, Vincent RD. A randomized trial of intrapartum analgesia in women with severe preeclampsia. *Obstet Gynecol.* 2002;99:452.
167. Lucas MJ, Sharma S, McIntire DD. A randomized trial of the effects of epidural analgesia on pregnancy-induced hypertension. *Am J Obstet Gynecol.* 2001;185:970.
168. Newsome LR, Bramwell RS, Curling PE. Severe preeclampsia: hemodynamic effects of lumbar epidural anesthesia. *Anesth Analg.* 1996;65:31.
169. Clark SL, Divon MY, Phelan JP. Preeclampsia/eclampsia: hemodynamic and neurologic correlations. *Obstet Gynecol.* 1985;66:337–343.
170. Cotton DB, Jones MM, Longmire S. Role of intravenous nitroglycerine in the treatment of severe pregnancy-induced hypertension complicated by pulmonary edema. *Am J Obstet Gynecol.* 1986;154:91–99.
171. Heller PJ, Scheider EP, Marx GF. Pharyngo-laryngeal edema as a presenting symptom in preeclampsia. *Obstet Gynecol.* 1983;62:523–532.
172. Tan LK, deSwiet M. The management of postpartum hypertension. *BJOG.* 2002;109:733–736.
173. Dennis AT, Solnordal CB. Acute pulmonary oedema in pregnant women. *Anaesthesia.* 2012;67(6):646–659.
174. Ascarelli MH, Johnson V, McCreary H, et al. Postpartum preeclampsia management with furosemide: a randomized clinical trial. *Obstet Gynecol.* 2005;105:29–33.

Index

Note: The page numbers followed by "*f*" and "*t*" refer to figures and tables, respectively.

N

O

P

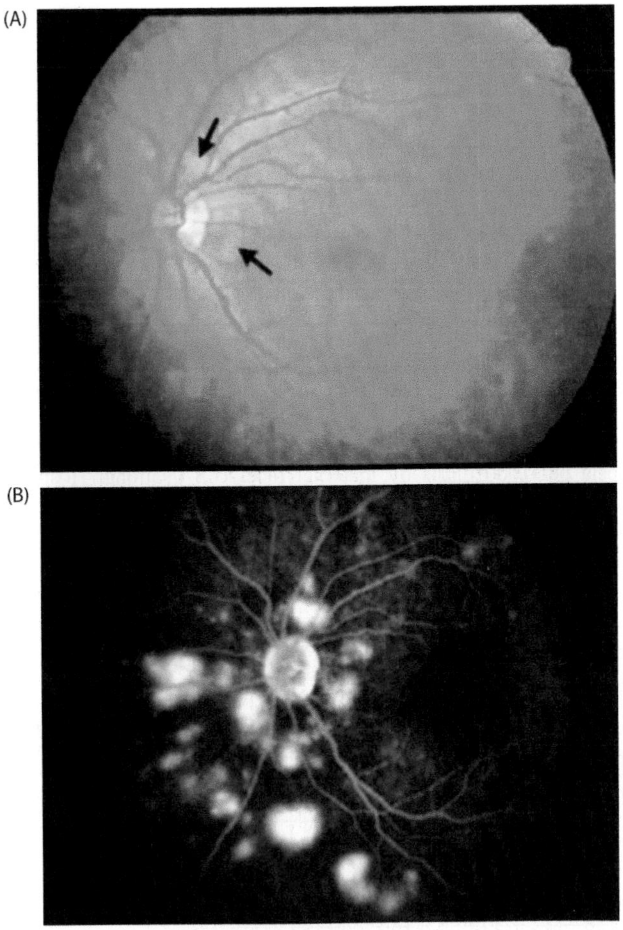

FIGURE 2.6 Purtscher retinopathy caused by choroidal ischemia in preeclampsia syndrome. (A) Ophthalmoscopy shows scattered yellowish, opaque lesions of the retina (arrows). (B) The late phase of fluorescein angiography shows areas of intense hyperfluorescence representing pooling of extravasated dye.[70]

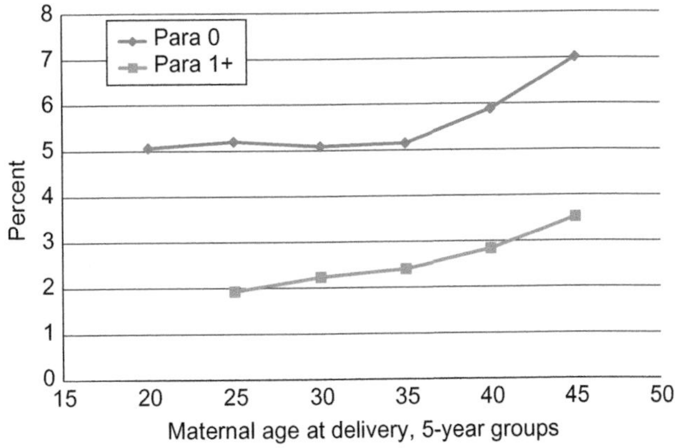

FIGURE 3.1 Prevalence of preeclampsia by age and parity, Norway, 1999–2012. (K. Klunsoyr, personal communication.)

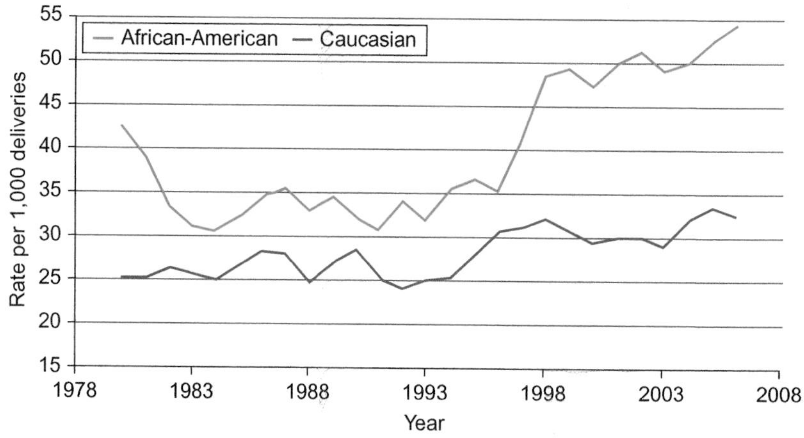

FIGURE 3.2 Cases of preeclampsia per 1000 births by maternal race, U.S. National Hospital Discharge Survey, 1979 to 2006.[82] Reprinted from Breathett K, Muhlestein D, Foraker R, Gulati M. The incidence of pre-eclampsia remains higher in African-American women compared to Caucasian women: trends from the National Hospital Discharge Survey 1979–2006. *Circulation*. 2013;127:AP192, with permission.

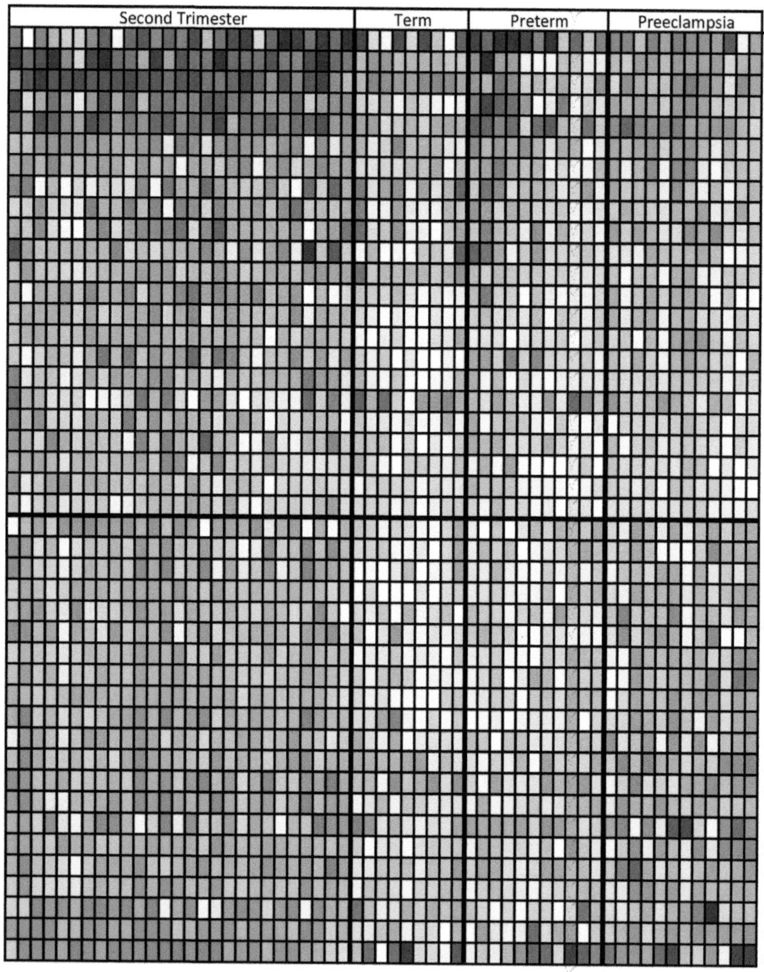

Symbol	Title	Δ
LEP	leptin	8.2
CRH	corticotropin releasing hormone	5.1
FABP4	fatty acid binding protein 4, adipocyte	4.9
INHBA	inhibin, beta A	4.0
LPL	lipoprotein lipase	3.6
FLT1	fms-related tyrosine kinase 1 (VEGF receptor)	3.1
SIGLEC6	sialic acid binding Ig-like lectin 6	2.9
INHA	inhibin, alpha	2.7
BCL6	B-cell CLL/lymphoma 6	2.6
BHLHE40	basic helix-loop-helix family, member e40	2.4
PSG11	pregnancy specific beta-1-glycoprotein 11	2.3
SPAG4	sperm associated antigen 4	2.2
LIMCH1	LIM and calponin homology domains 1	2.2
PPL	periplakin	2.2
LTF	lactotransferrin	2.1
PLIN2	perilipin 2	2.1
PAPPA2	pappalysin 2	2.1
FSTL3	follistatin-like 3 (secreted glycoprotein)	2.0
HTRA1	HtrA serine peptidase 1	1.9
EPS8L1	EPS8-like 1	1.9
DUSP1	dual specificity phosphatase 1	1.8
EFHD1	EF-hand domain family, member D1	1.6
SYDE1	synapse defective 1, Rho GTPase, homolog 1 (C. elegans)	1.6
COL1A1	collagen, type I, alpha 1	-2.2
PLAGL1	pleiomorphic adenoma gene-like 1	-2.2
COL14A1	collagen, type XIV, alpha 1	-2.3
COL3A1	collagen, type III, alpha 1	-2.3
AGPAT5	1-acylglycerol-3-phosphate O-acyltransferase 5	-2.3
COL21A1	collagen, type XXI, alpha 1	-2.3
COL6A1	collagen, type VI, alpha 1	-2.3
COL6A3	collagen, type VI, alpha 3	-2.4
COL1A2	collagen, type I, alpha 2	-2.4
C3orf58	chromosome 3 open reading frame 58	-2.4
ASB2	ankyrin repeat and SOCS box-containing 2	-2.4
ABCB1	ATP-binding cassette, sub-family B1	-2.4
ATP1B1	ATPase, Na+/K+ transporting, beta 1 polypeptide	-2.5
HAPLN1	hyaluronan and proteoglycan link protein 1	-2.5
MMP12	matrix metallopeptidase 12 (macrophage elastase)	-2.5
SPP1	secreted phosphoprotein 1	-2.5
SLC26A2	solute carrier family 26 (sulfate transporter), member 2	-2.6
LAMA2	laminin, alpha 2	-2.6
COL6A2	collagen, type VI, alpha 2	-2.7
CXCL14	chemokine (C-X-C motif) ligand 14	-3.3
ANGPT2	angiopoietin 2	-4.0

FIGURE 5.9 Heat map of the most highly upregulated and downregulated differentially expressed genes in basal plates of PE placentas as compared to the second-trimester, term and preterm labor samples. The normalized log intensity values for the differentially expressed probe sets were centered to the median value of each probe set and colored on a range of −2.5 to +2.5. Red denotes upregulated and blue denotes downregulated expression levels as compared with the median value. Columns contain data from a single basal plate specimen, and rows correspond to a single probe set. Samples within each category are arranged from left to right, ordered by increasing gestational age. Rows are ranked by fold change. Reprinted with permission from *Pregnancy Hypertension*.[51]

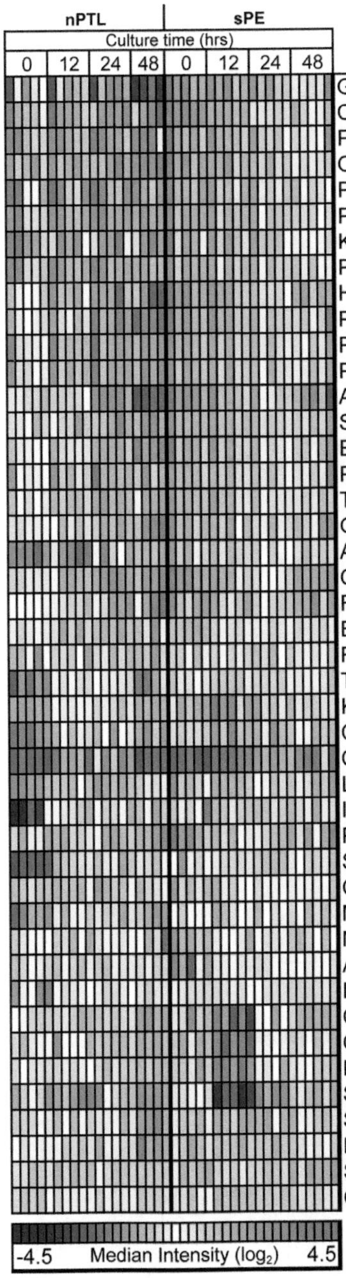

Symbol	Title	sPE v. nPTL Fold Change 0	12	24	48	
GH2	growth hormone 2	9	8	8	6	t
CRH	corticotropin releasing hormone	10	6	3	ns	t
PSG11	pregnancy specific beta-1-glycoprotein 11	6	6	4	ns	t
OLAH	oleoyl-ACP hydrolase	7	5	4	ns	t
PSG7	pregnancy specific beta-1-glycoprotein 7	6	5	4	ns	t
PSG1	pregnancy specific beta-1-glycoprotein 1	5	5	4	ns	t
KISS1	KiSS-1 metastasis-suppressor	6	4	4	ns	t
PSG4	pregnancy specific beta-1-glycoprotein 4	5	5	4	ns	t
HOPX	HOP homeobox	5	5	3	ns	t
PLAC1	placenta-specific 1	5	4	4	ns	t
PSG6	pregnancy specific beta-1-glycoprotein 6	4	4	4	ns	t
PSG9	pregnancy specific beta-1-glycoprotein 9	4	4	4	ns	t
ADAM12	ADAM metallopeptidase domain 12	4	4	4	ns	t
SEMA3B	semaphorin 3B	4	3	3	ns	t
ENDOU	endonuclease, polyU-specific	4	3	3	ns	t
PSG5	pregnancy specific beta-1-glycoprotein 5	3	3	3	ns	t
TPPP3	tubulin polymerization-promoting protein family member 3	3	3	3	ns	t
GPC3	glypican 3	2	3	3	ns	t
ALPP	alkaline phosphatase, placental (Regan isozyme)	7	4	ns	ns	t
CYP19A1	cytochrome P450,19A1	4	4	ns	ns	t
FRZB	frizzled-related protein	3	3	ns	ns	t
EFHA2	EF-hand domain family, member A2	2	2	ns	ns	t
RORA	RAR-related orphan receptor A	2	ns	3	ns	t
TFPI2	tissue factor pathway inhibitor 2	4	ns	ns	3	t
KCTD4	potassium channel tetramerisation domain containing 4	ns	4	3	ns	
CA10	carbonic anhydrase X	ns	3	3	ns	t
CYP1A1	cytochrome P450,1A1	ns	ns	5	ns	t
LOC284561	hypothetical protein LOC284561	5	ns	ns	ns	
INHBA	inhibin, beta A	5	ns	ns	ns	t
PLAC4	placenta-specific 4	5	ns	ns	ns	
STEAP4	STEAP family member 4	4	ns	ns	ns	t
C9orf27	chromosome 9 open reading frame 27	3	ns	ns	ns	
MUC1	mucin 1, cell surface associated	3	ns	ns	ns	t
MAN1C1	mannosidase, alpha 1C1	3	ns	ns	ns	
ASS1	argininosuccinate synthase 1	-3	ns	ns	ns	
DEFA4	defensin, alpha 4, corticostatin	-3	ns	ns	ns	
C19orf59	chromosome 19 open reading frame 59	ns	-3	ns	ns	t
CD14	CD14 molecule	ns	-3	ns	ns	
DOK3	docking protein 3	ns	-3	ns	ns	
S100A8	S100 calcium binding protein A8	ns	-5	ns	ns	t
SIGLEC10	sialic acid binding Ig-like lectin 10	ns	ns	ns	-4	
RETN	resistin	ns	ns	ns	-4	
S100A14	S100 calcium binding protein A14	ns	-2	ns	-3	
CHURC1	churchill domain containing 1	-2	-2	ns	-2	t

FIGURE 5.10 Severe preeclampsia-associated aberrations in cytotrophoblast gene expression returned to control values after 48 h of culture. RNA was analyzed immediately after the cells were isolated (0 h) and after 12, 24 and 48 h in culture. The relative gene expression levels for cytotrophoblasts (CTBs) isolated from placentas of patients who delivered due to preterm labor with no sign of infection (nPTL; $n = 5$) or a severe form of preeclampsia (sPE; $n = 5$) are shown as a heat map, ranging from high (red) to low (blue). The sPE CTBs were from the following cases (tiled from left to right): (1) hemolysis, elevated liver enzymes and low platelets (HELLP) syndrome and intrauterine growth restriction (IUGR); (2) sPE; (3) sPE and IUGR; (4) superimposed sPE; and (5) HELLP syndrome. One sample of nPTL CTBs collected at 48 h was omitted for technical reasons. The fold changes for each time point (sPE vs. nPTL) are shown on the right. ns, no significant difference (LIMMA); t, no significant difference in expression (sPE vs. nPTL) by 48 h (maSigPro). Reprinted with permission from *The Journal of Clinical Investigation.*[2]

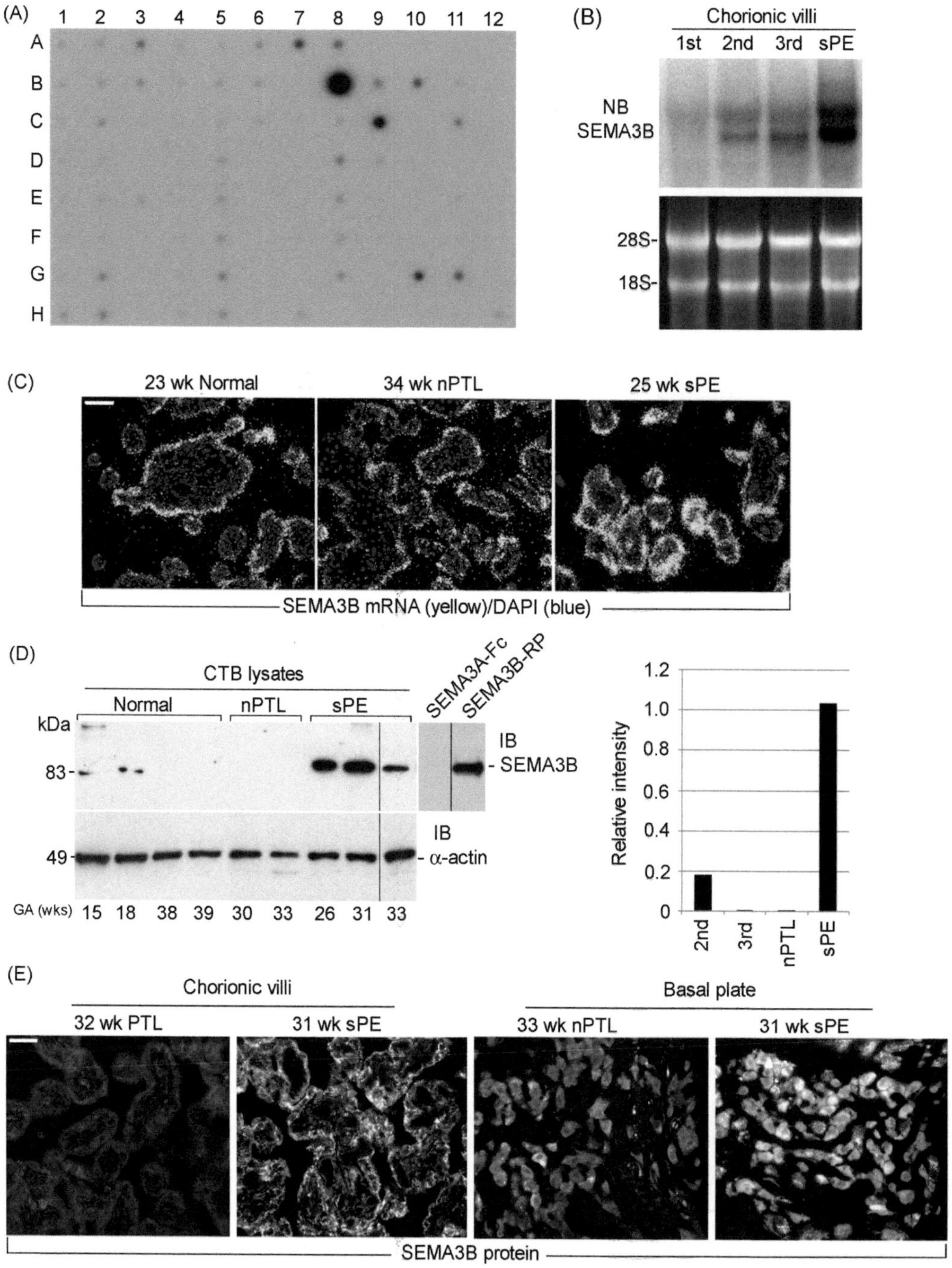

FIGURE 5.11 SEMA3B expression was high in the placenta and upregulated in severe preeclampsia. (A) Binding of a ^{32}P-SEMA3B probe to a multiple tissue expression array revealed high placental expression (coordinate B8). (B) Northern hybridization of polyA^{+} RNA extracted from chorionic villi and pooled from three placentas showed that SEMA3B expression increased over gestation and was highest in sPE (n = 3 replicates). (**C**) *In situ* hybridization (three placentas/group) confirmed enhanced SEMA3B mRNA expression in the syncytiotrophoblast layer of the chorionic villi in sPE (25 wk) as compared to normal pregnancy (23 wk) and nPTL (34 wk). (D, left panel) Immunoblotting of CTB lysates (15 μg/lane) showed that SEMA3B protein expression was low to undetectable in control cells from normal placentas (15–39 wks). In all cases, expression was higher in sPE (26–33 wks) as compared to nPTL (30, 33 wks). A protein of the expected M_r was detected in COS-1 cells transfected with SEMA3B, but not in the SEMA3A-Fc lane. Vertical lines denote noncontiguous lanes from the same gel. (D, right panel) The relative intensity of the bands was quantified by densitometry. The values for each sample type were averaged and expressed relative to the α-actin loading controls. The entire experiment was repeated twice. (E) Staining tissue sections with anti-SEMA3B showed a sPE-associated upregulation of immunoreactivity associated with the trophoblast components of chorionic villi and among extravillous CTBs within the basal plate (n = 5/group). Trophoblasts were identified by staining adjacent tissue sections with anti-cytokeratin-8/18 (data not shown). Scale bars (C, E): 100 μm. NB, northern blot; GA, gestational age; IB, immunoblot; RP, recombinant protein. Reprinted with permission from *The Journal of Clinical Investigation*.[2]

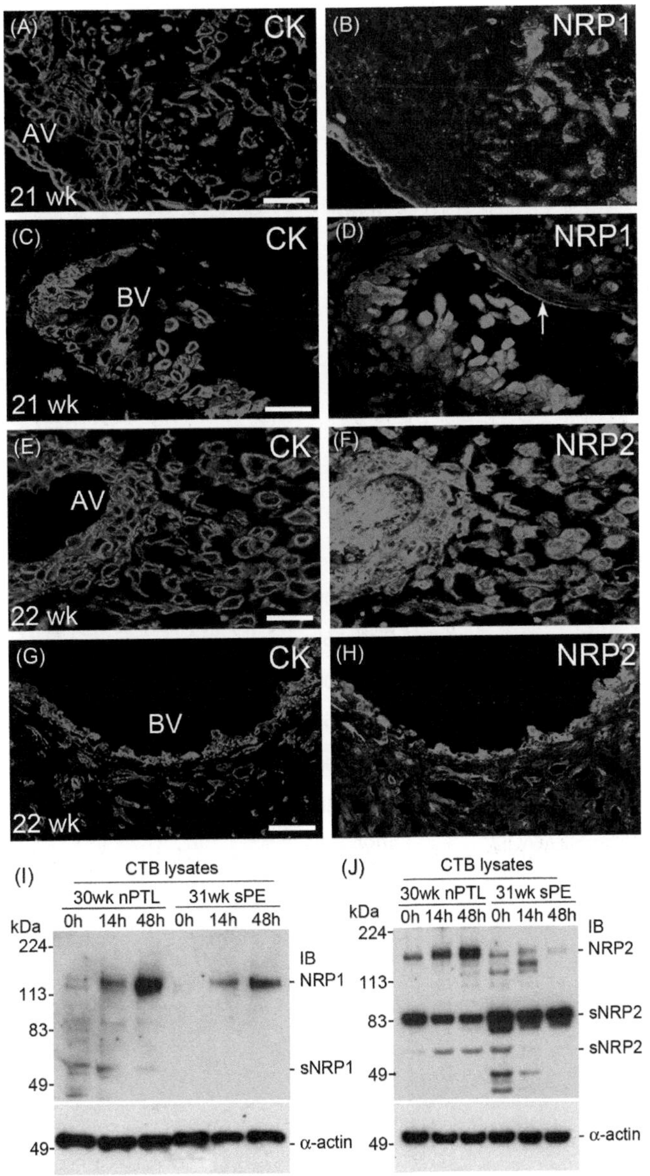

FIGURE 5.12 NRP-1 and -2 (protein) expression at the maternal–fetal interface in normal pregnancy and in sPE. Tissue sections were double-stained with anti-cytokeratin (CK)-7, which reacts with all trophoblast (TB) subpopulations, and anti-NRP-1 or -2. (A, B) NRP-1 expression was detected in association with villous TBs. Within the uterine wall, immunoreactivity associated with invasive CTBs was upregulated as the cells moved from the surface to the deeper regions. (C, D) Endovascular cytotrophoblasts (CTBs) that lined a maternal blood vessel (BV) also stained. (E-H) Anti-NRP-2 reacted with TB and non-TB cells in anchoring villi (AV) as well as interstitial and endovascular CTBs. Essentially the same staining patterns, but with weaker intensity, were observed in sPE (data not shown). CTBs were isolated from the placentas of control nPTL cases and from the placentas of women who experienced sPE. (I, J) Over 48 h in culture, NRP-1 expression was upregulated in both instances, but to a lesser degree in sPE. (K, L) Control nPTL CTBs also upregulated NRP2. Expression of this receptor was reduced in sPE and the soluble form was more abundant. (A–L) The data shown are representative of the analysis of a minimum of three samples from different placentas. Scale bars, 100 μm. IB, immunoblot. Reprinted with permission from *The Journal of Clinical Investigation*.[2]

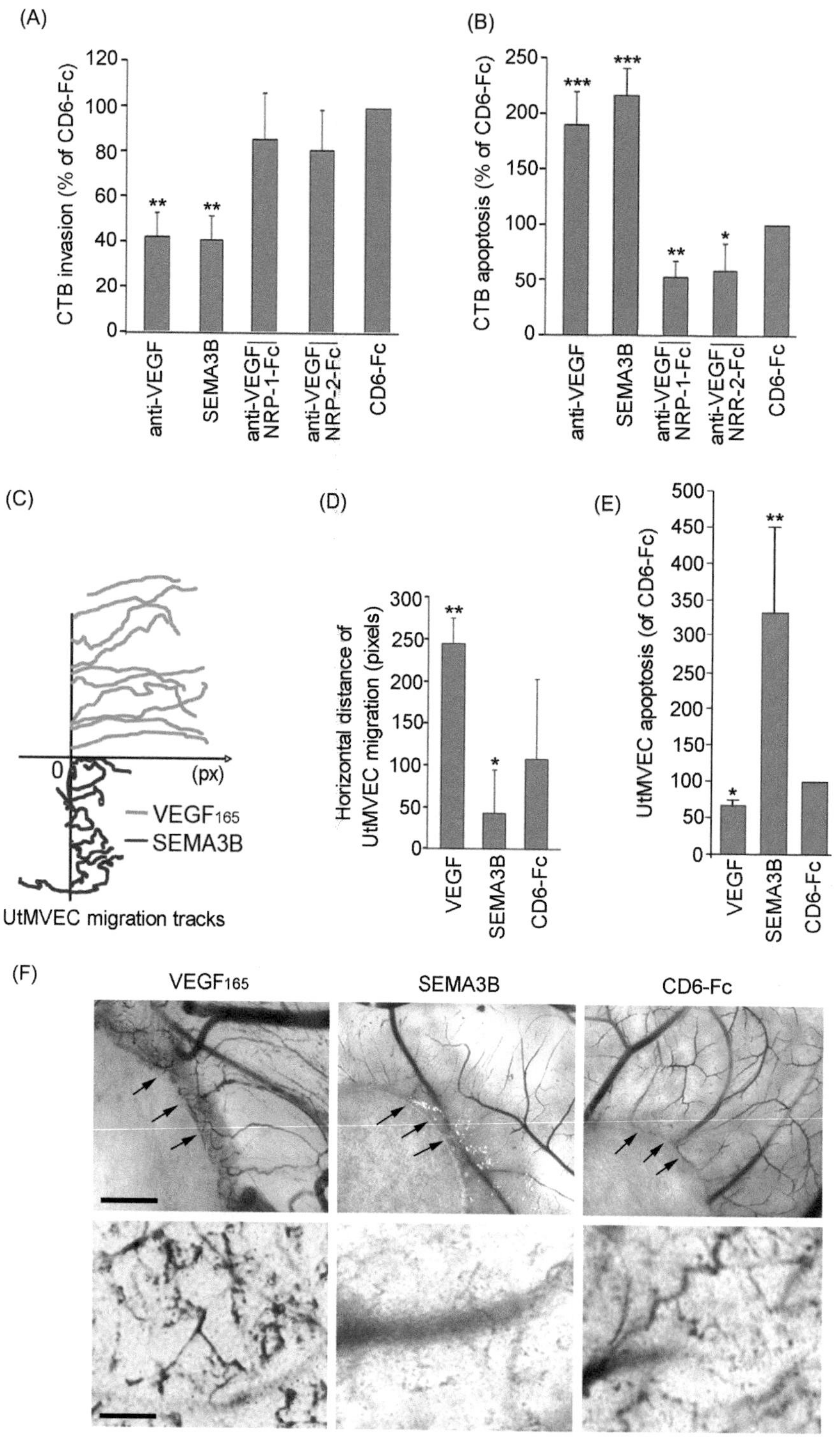

FIGURE 5.13 Exogenous SEMA3B mimicked the effects of sPE on CTBs and endothelial cells, and inhibited angiogenesis. (A) The addition of anti-VEGF or SEMA3B protein significantly inhibited cytotrophoblast (CTB) invasion as compared to the addition of a control protein, CD6-Fc. The removal of both ligands (anti-VEGF/NRP1-Fc, anti-VEGF/NRP2-Fc) restored invasion to control levels. (B) The variables tested in panel A had the opposite effects on CTB apoptosis, suggesting that increased programmed cell death contributed to decreased invasion. (C) Exogenous VEGF stimulated the migration of uterine microvascular endothelial cells (UtMVECs), which was inhibited by SEMA3B. (D) The results in panel C were quantified relative to the addition of CD6-Fc. (E) In UtMVECs, VEGF promoted survival and SEMA3B increased apoptosis relative to control levels. (F) In the chick chorioallantoic membrane (CAM) angiogenesis assay, VEGF promoted angiogenesis by ~3-fold and SEMA3B inhibited this process ~5-fold relative to the effects of CD6-Fc. Top row: arrows mark the edge of the filter paper used to apply the protein. Scale bar: 200 μm. Bottom row: the area of the CAM beneath the filter paper. Scale bar: 100 μm. A–D, $n = 6$ replicates; E, F, $n = 3$ replicates. A,B,D,E, mean ± SEM, two-tailed Student's t-test. $^{*}p < 0.05$, $^{**}p < 0.01$, $^{***}p < 0.001$. Reprinted with permission from *The Journal of Clinical Investigation.*[2]

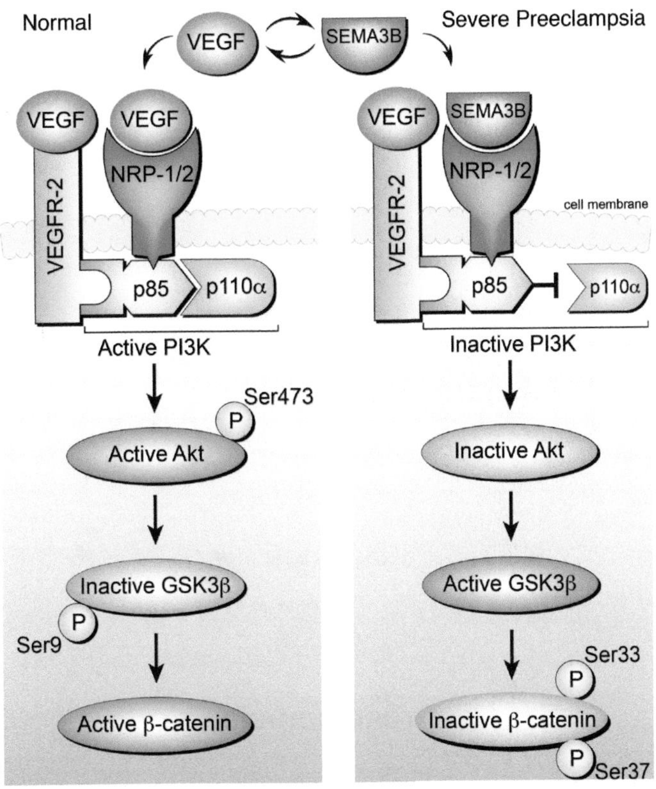

FIGURE 5.15 Model of SEMA3B effects on CTBs in sPE vs. normal pregnancy. Reprinted with permission from *The Journal of Clinical Investigation*.[2]

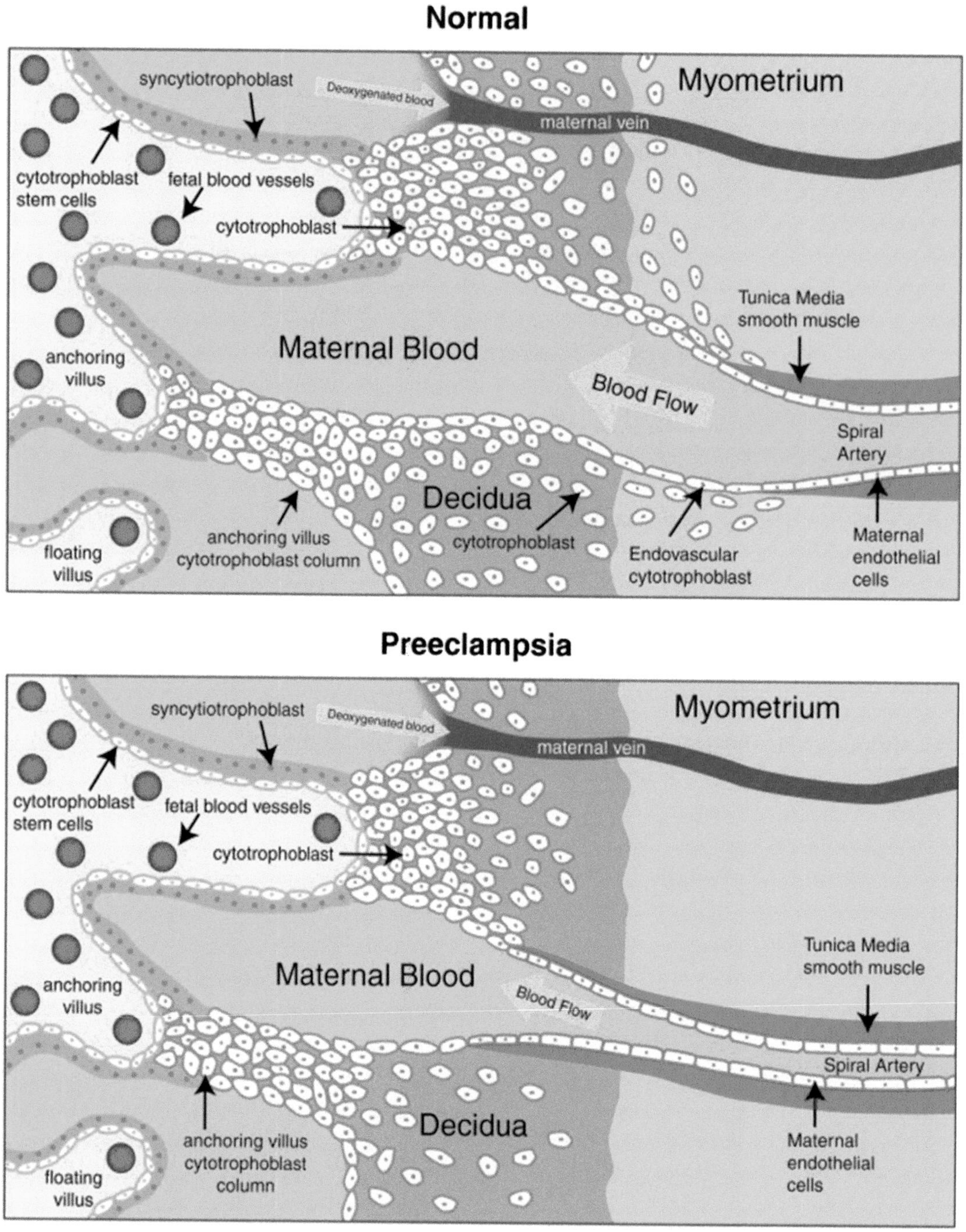

FIGURE 6.1 A schematic of placental vascular remodeling in health (upper panel) and in disease – preeclampsia (lower panel). Exchange of oxygen, nutrients, and waste products between the fetus and the mother depends on adequate placental perfusion by maternal spiral arteries. Blood from the intervillous space is returned to the mother's circulation via spiral maternal veins noted above. In normal placental development, cytotrophoblasts of fetal origin invade the maternal spiral arteries, transforming them from small-caliber resistance vessels to high-caliber capacitance vessels capable of providing adequate placental perfusion to sustain the growing fetus. During the process of vascular invasion, the cytotrophoblasts undergo a transformation from an epithelial to an endothelial phenotype, a process referred to as "pseudovasculogenesis" (upper panel). In preeclampsia, cytotrophoblasts fail to adopt an invasive endothelial phenotype. Instead, invasion of the spiral arteries is shallow and they remain small-caliber, resistance vessels (lower panel). This is thought to lead to placental ischemia and secretion of antiangiogenic factors. Figure reproduced with permission from Lam et al.[12]

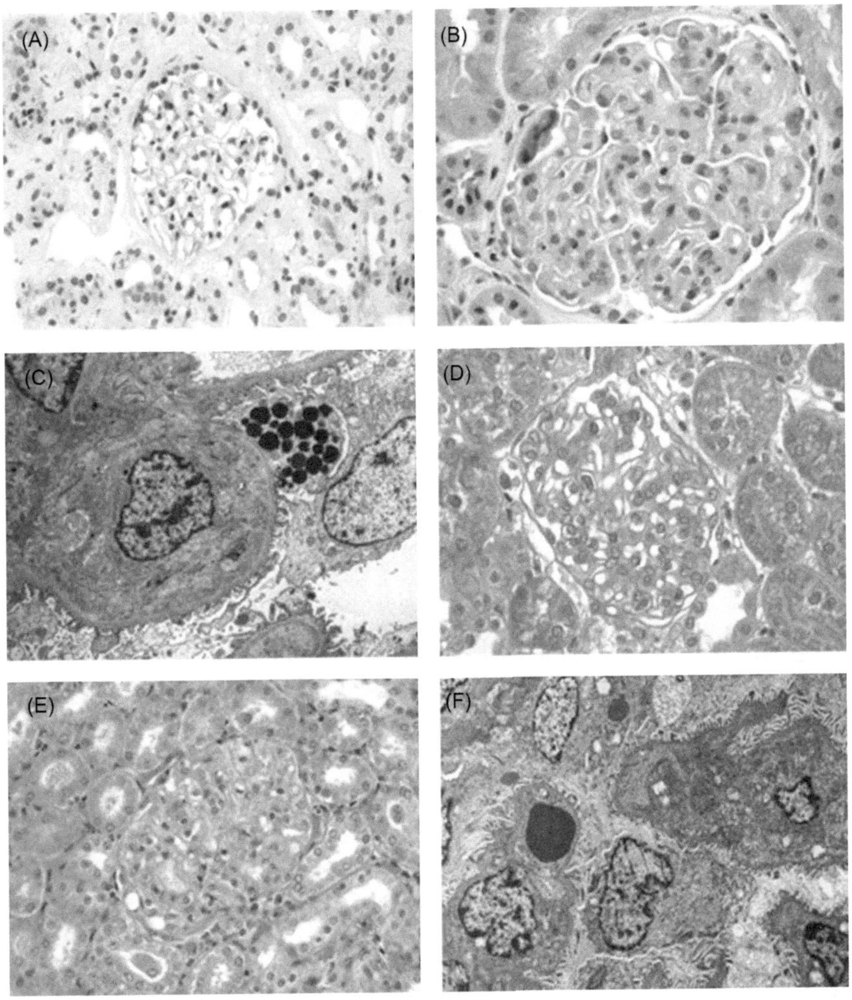

FIGURE 6.2 Glomerular endotheliosis. (A) Normal human glomerulus. (B) Human preeclamptic glomerulus of 33-yr-old woman with a twin gestation and severe preeclampsia at 26 weeks gestation. The urine protein/creatinine ratio was 26 at the time of biopsy. (C) Electron microscopy of a glomerulus from the same patient. Note the occlusion of the capillary lumens by the swollen cytoplasm of endocapillary cells. Podocyte cytoplasm shows protein resorption droplets but relatively intact foot processes. Original magnification 1500×. (D) Control rat glomerulus: note normal cellularity and open capillary loops. (E) sFlt-1 treated rat: note similar occlusion of the capillary lumens by swollen endothelial cells with minimal increase in cellularity. (F) Electron microscopy of a sFlt-1 treated rat: note similar occlusion of capillary loops by swollen endocapillary cell cytoplasm accompanied by the relative preservation of podocyte foot processes. Original magnification 2500×. All light photomicrographs are of H&E sections taken at the identical original magnification of 40×. Figure reproduced with permission from Karumanchi et al.[114]

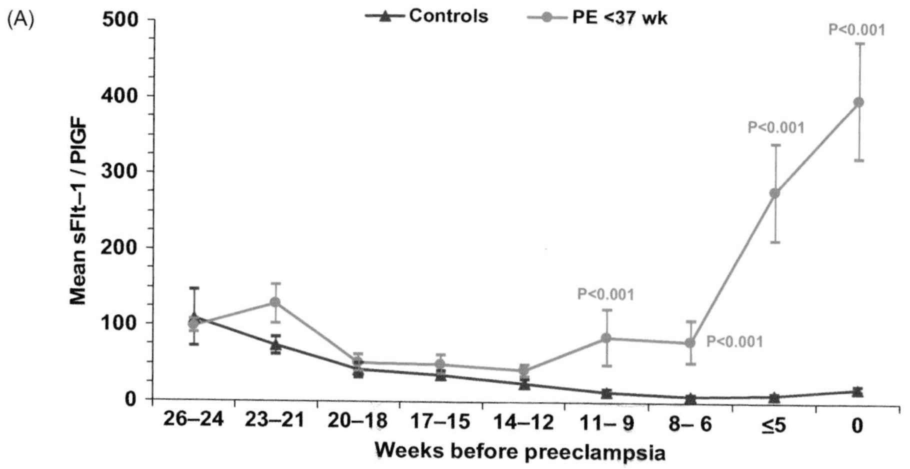

Specimen pairs	4	16	19	16	9	25	25	21	40
Specimen gestational age (day)									
Controls	86	98	118	123	122	171	196	198	241
PE < 37 wk	82	96	118	123	122	170	196	198	240

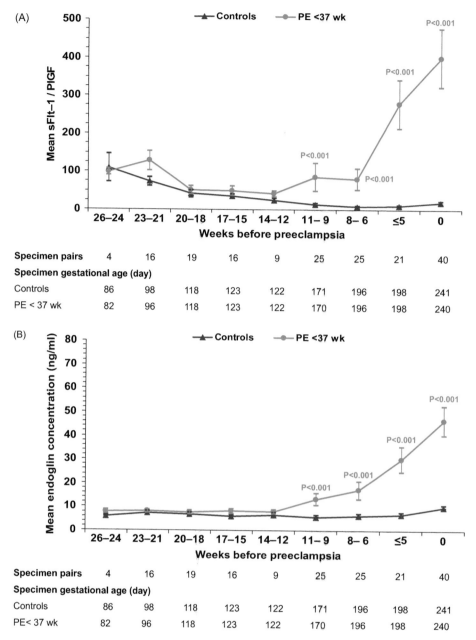

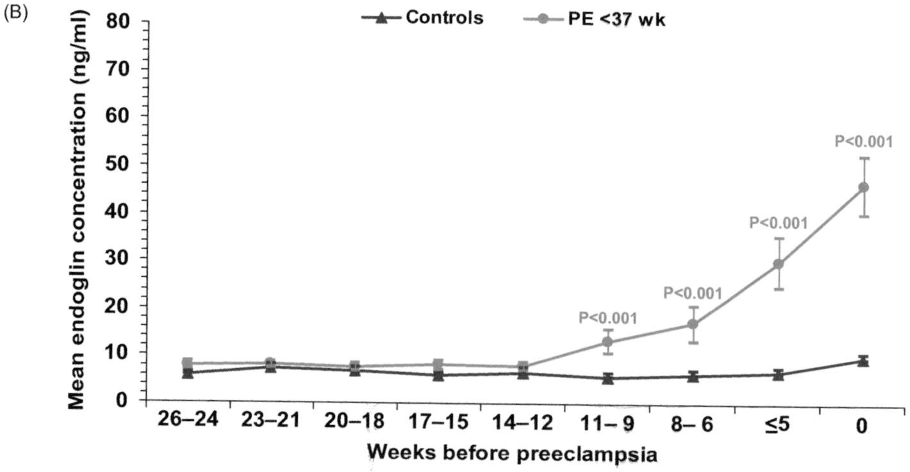

Specimen pairs	4	16	19	16	9	25	25	21	40
Specimen gestational age (day)									
Controls	86	98	118	123	122	171	196	198	241
PE< 37 wk	82	96	118	123	122	170	196	198	240

FIGURE 6.3 Mean levels of sFlt-1/PlGF and soluble endoglin (sEng) and by weeks before the onset of preeclampsia. (A) This panel shows the mean concentrations of sFlt-1/PlGF according to the number of weeks before the onset of preterm preeclampsia (PE < 37 weeks) and the mean concentrations in normotensive controls with appropriate- or large-for-gestational-age infants. Control specimens were matched within 1 week of gestational age to specimens from women who later developed preterm preeclampsia. (B) This panel shows the mean levels of sEng in case and control specimens shown in panel A according to the number of weeks before the onset of preterm preeclampsia (PE < 37 weeks). Figures reproduced with permission from Levine et al. and Hagmann et al.[125,139]

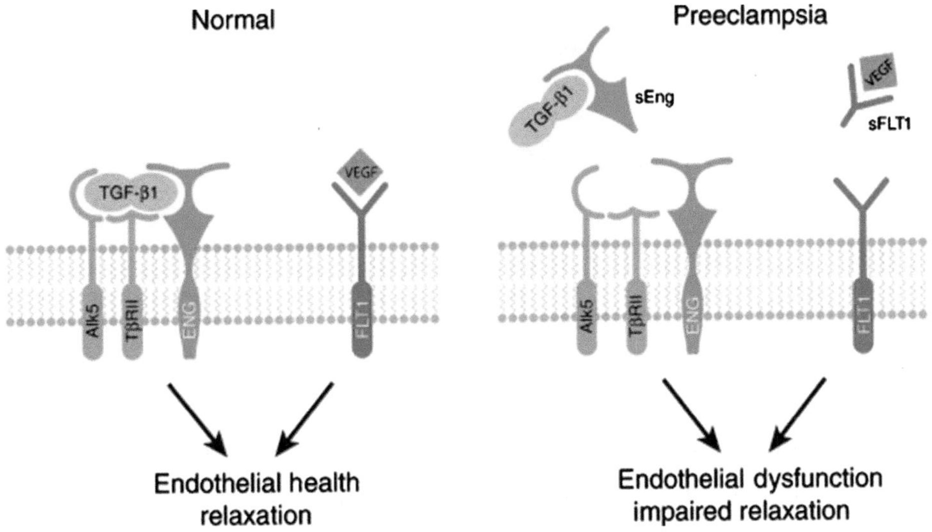

FIGURE 6.5 sFlt-1 and sEng causes endothelial dysfunction by antagonizing VEGF and TGF-β signaling. There is mounting evidence that VEGF and TGF-β are required to maintain endothelial health in several tissues including the kidney and perhaps the placenta. During normal pregnancy, vascular homeostasis is maintained by physiological levels of VEGF and TGF-β signaling in the vasculature. In preeclampsia, excess placental secretion of sFlt-1 and sEng (two endogenous circulating antiangiogenic proteins) inhibits VEGF and TGF-β1 signaling respectively in the vasculature. This results in endothelial cell dysfunction, including decreased prostacyclin, nitric oxide production and release of procoagulant proteins. Figure reproduced with permission from Karumanchi and Epstein.[145]

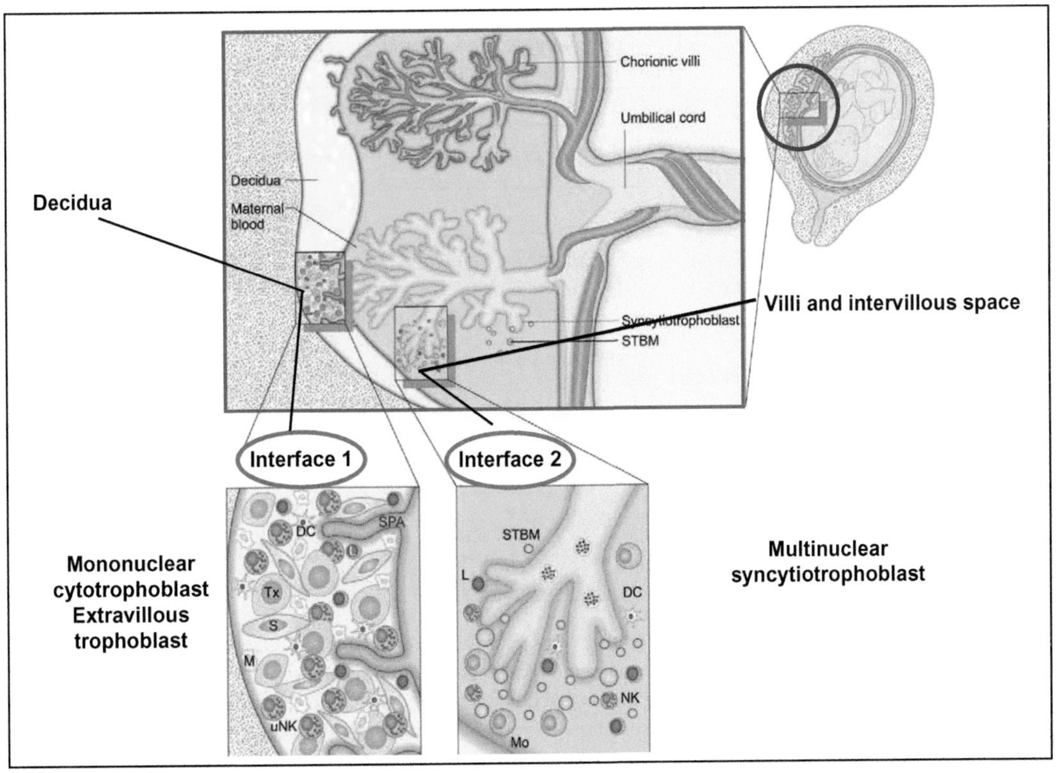

FIGURE 8.2 The two immune interfaces of human pregnancy. Adapted from[20] with permission. Immune events at Interface 1 drive the first stage of the two-stage model (see Fig. 8.3) of preeclampsia, whereas Interface 2 drives the second stage. Interface 1 comprises maternal immune cells, including uterine NK cells (uNK), T lymphocytes (L), macrophages (M) and dendritic cells (DC), and the invasive extravillous cytotrophoblast (Tx), between large stromal cells (S) and spiral arteries (SPA; not to scale). Interface 2 lies between circulating maternal immune cells, including T lymphocytes (L), NK cells (NK), monocytes (Mo) and dendritic cells (DC), and syncytiotrophoblast microparticles (STBM).

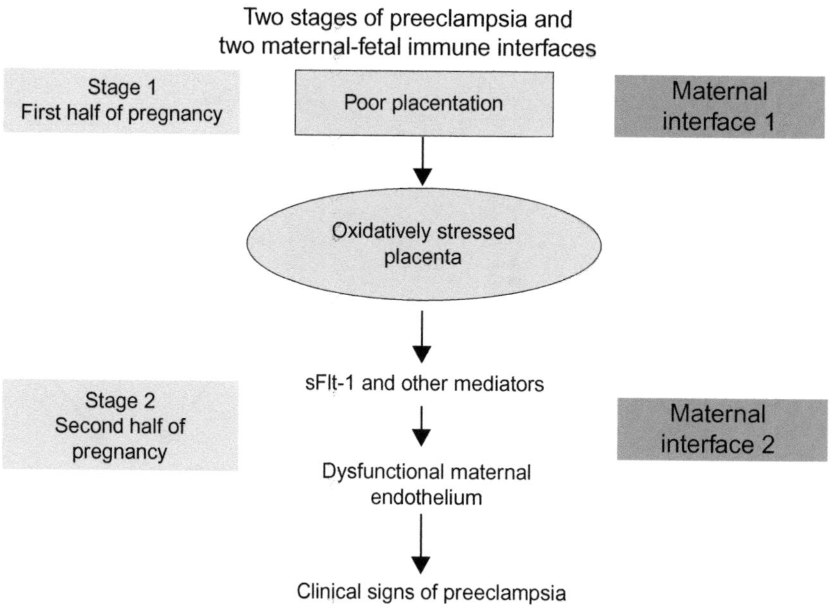

FIGURE 8.3 Two-stage model of development of preeclampsia.

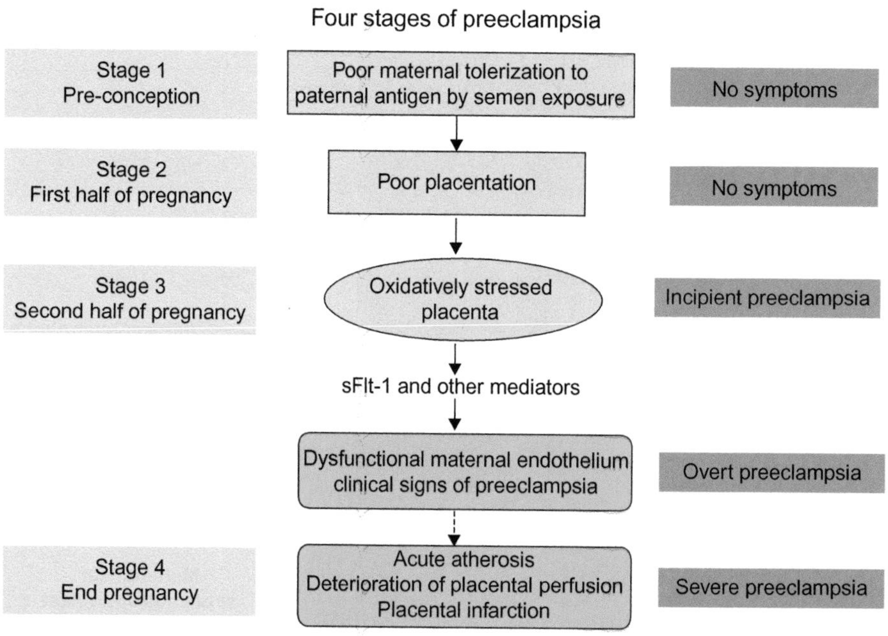

FIGURE 8.4 Two further stages are added to this adaptation of the two-stage model. Pre-conceptual maternal tolerization to partner's antigens is achieved by pre-conceptual exposure to paternal semen (new stage 1). Stages 2 and 3 are, respectively, stages 1 and 2 of the original two-stage model. Stage 4 comprises the development of acute atherosis in the spiral arteries which causes further adverse effects on uteroplacental perfusion, particularly from thrombosis. Acute atherosis probably affects less than half of preeclampsia cases.

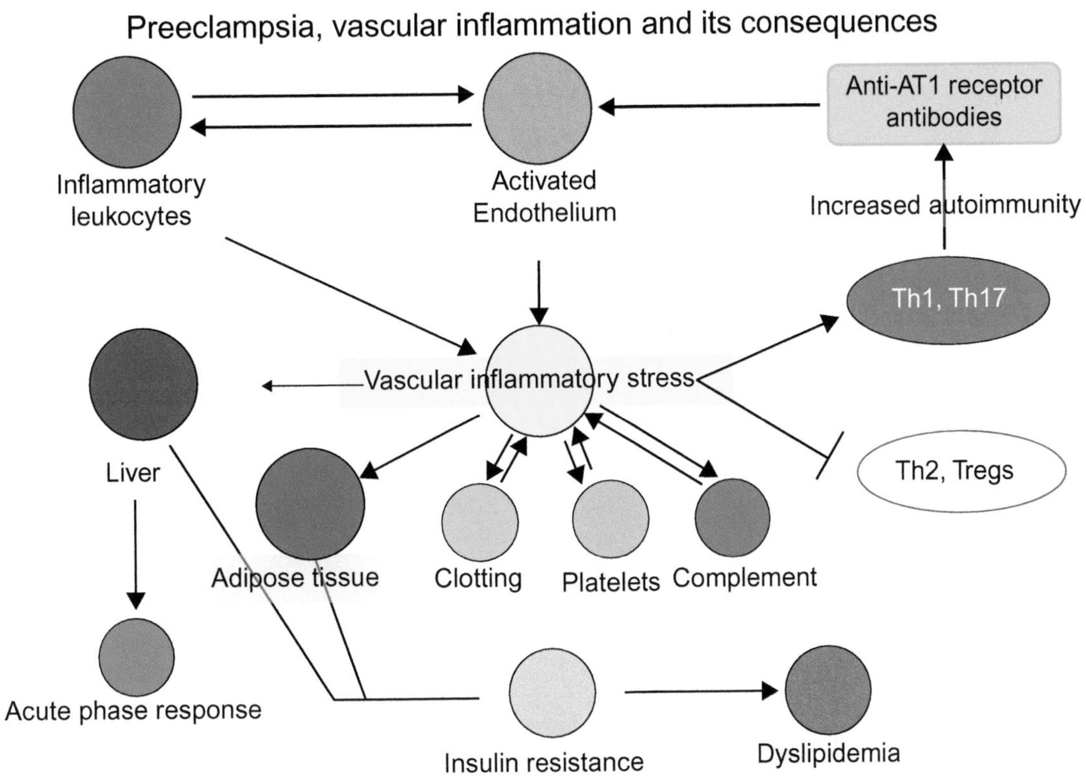

FIGURE 8.6 Vascular inflammation involves not only inflammatory leukocytes, but the endothelium, platelets, clotting and complement systems. Metabolic changes lead to insulin resistance and dyslipidemia. The inflammatory stress deviates away from the Th2 bias of normal pregnancy towards Th1/Th17. Th17 and loss of Tregs predispose to autoimmunity and may help stimulate formation of anti-AT1 receptor autoantibodies. These are agonistic and therefore proinflammatory. These processes contribute towards many recognized features of preeclampsia.

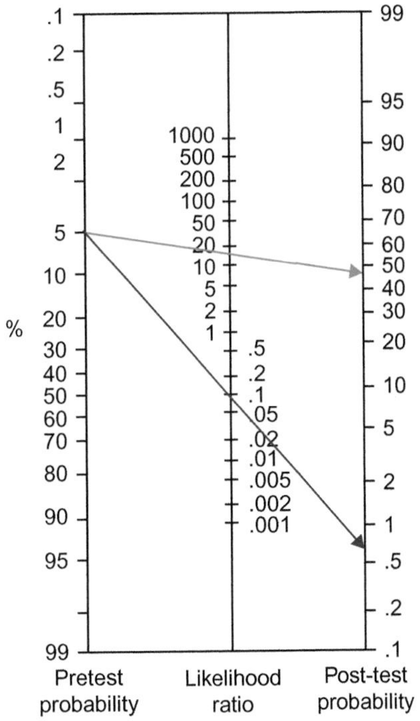

FIGURE 11.1 Use of Fagan's nomogram to calculate post-test probability of preeclampsia. (Adapted from Fagan[3]).

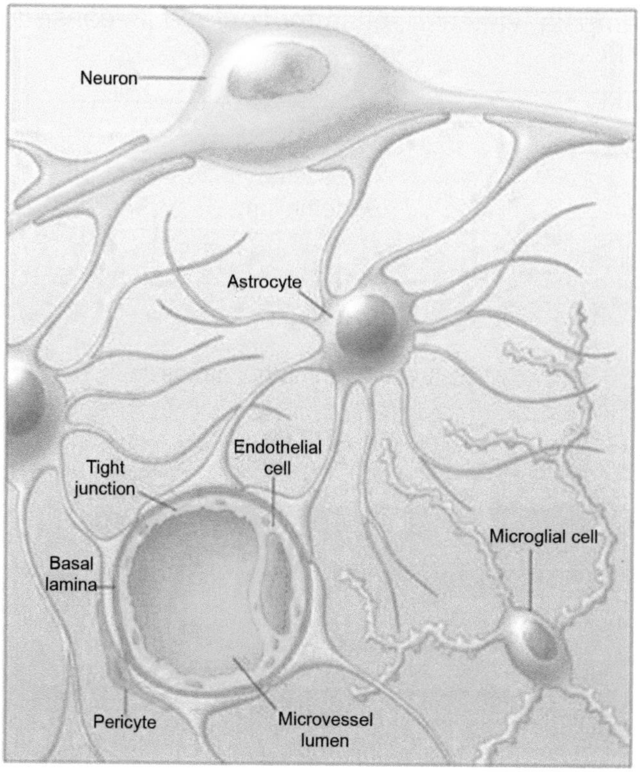

FIGURE 13.8 Schematic of the neurovascular unit. Increased hydrostatic (capillary lumen) pressure causes extravasation of plasma and red cells through endothelial tight junctions with perivascular edema. Reprinted with permission from del Zoppo GJ. *New Engl J Med.* 2006;354:553–555.

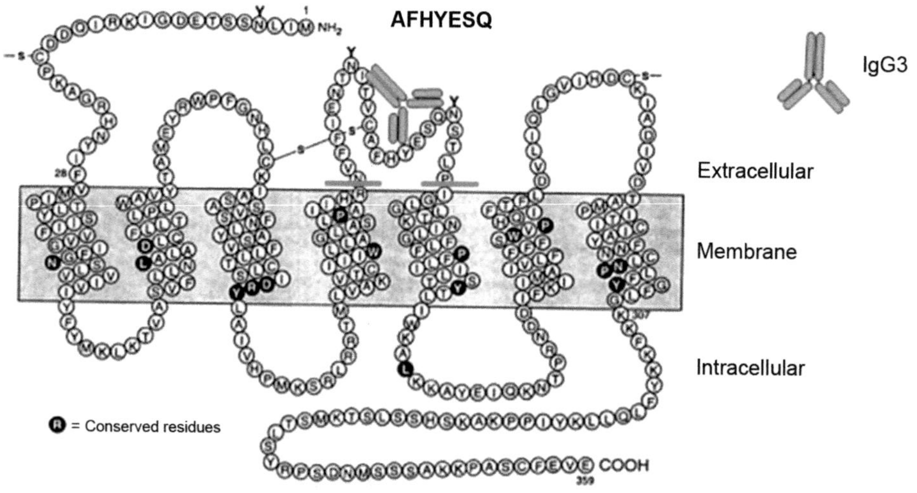

FIGURE 15.6 Schematic depiction of the AT1 receptor. The second AT1 receptor extracellular loop and the binding site (amino-acid sequence) of agonistic AT1 receptor antibodies are shown. The interaction can be blocked by a specific seven-amino-acid peptide (AFHYESQ).

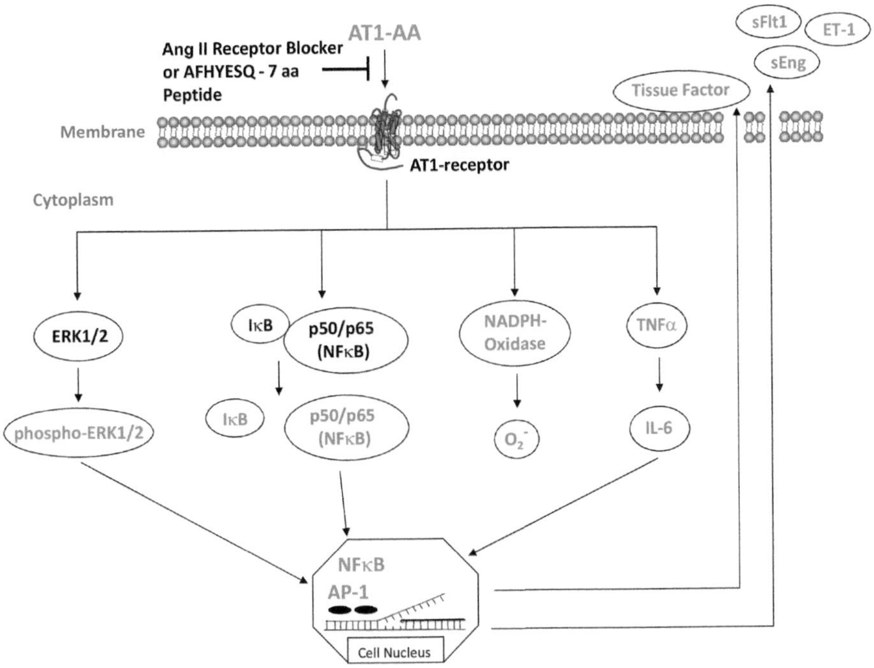

FIGURE 15.7 The agonistic AT1-AA induces signaling by the angiotensin II type 1 receptor (AT1-receptor), which can be inhibited by AT1-receptor blocker (ARB) or the seven-amino-acid peptide (AFHYESQ) mimicking the epitope of the AT1-AA in the second extracellular domain of the AT1-receptor. This activation then leads to NADPH oxidase, increased protein kinase C and calcineurin activity. Transcription factors such as activating protein-1, nuclear factor-κB (NF-κB), and nuclear factor activating T cell (NFAT) are activated. After translocating into the nucleus they lead to increased gene expression of target genes, i.e., interleukin-6, tissue factor, PAI-1, endothelin-1 (ET-1), soluble fms-like tyrosine kinase-1 (sFlt-1), soluble endoglin (sEng), and oxidative stress.

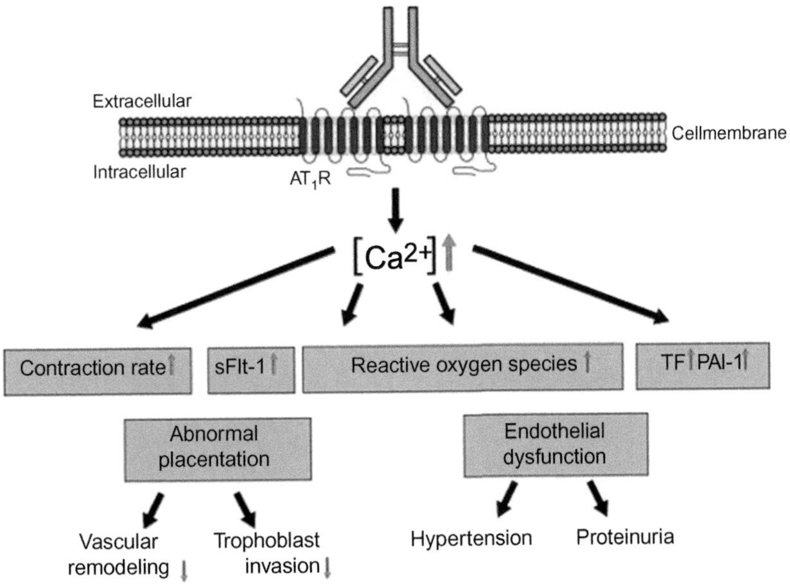

FIGURE 15.8 Molecular mechanisms of the interaction of AT1-AA with the AT1 receptor leading to the features of preeclampsia.

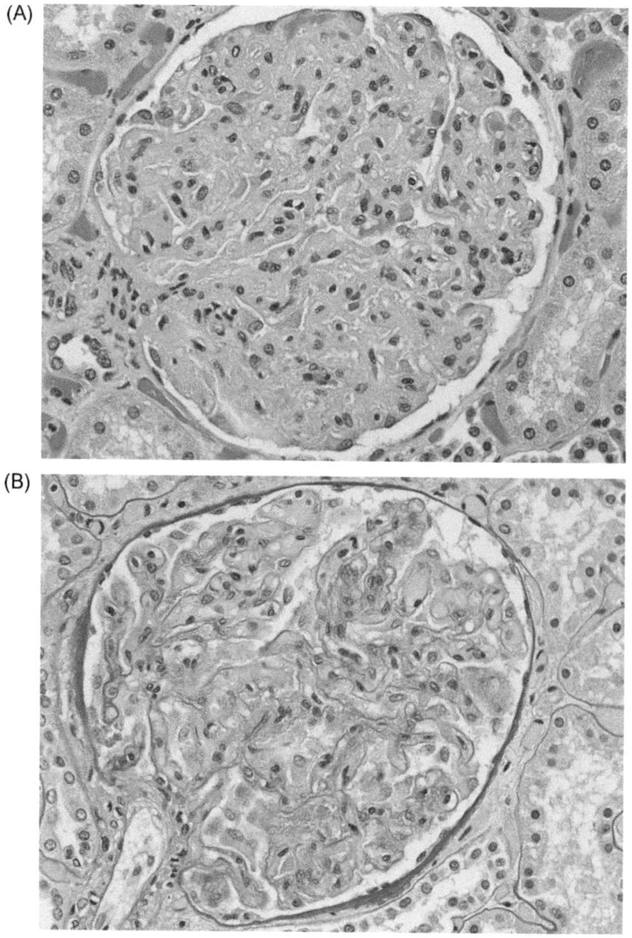

FIGURE 16.9 (A): Glomerular tuft with endotheliosis. Capillary lumens are occluded by swollen endothelial cells. Note that overall glomerular cellularity in not significantly increased. Vascular congestion, but not thrombosis, is present. (H&E, 400×). (B): Glomerular tuft with endotheliosis. PAS stain highlights the capillary basement membrane, which is within normal limits. In contrast, swollen endothelial cytoplasm only stains weakly. Endothelium at the vascular pole is characteristically uninvolved. (PAS, 400×).

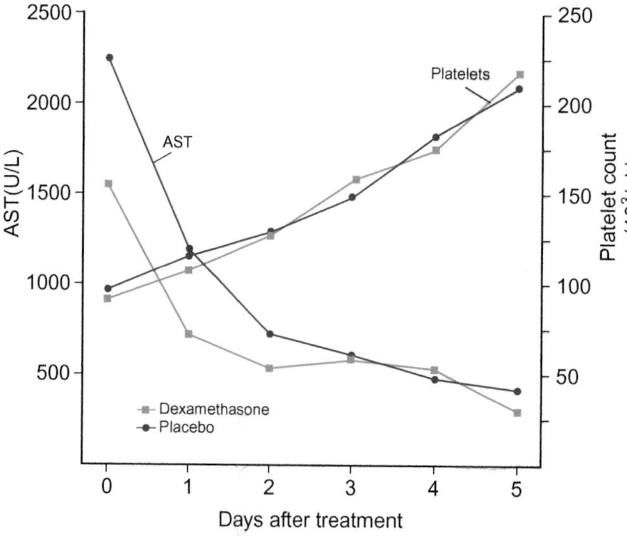

FIGURE 17.4 The recovery times for platelets and serum aspartate aminotransferase (AST) are shown for women with HELLP syndrome who were randomly assigned to receive dexamethasone or placebo. (Data from Katz et al.[189])

9780124078666